Carnitine Deficiency and Insufficiency in Medicine

Basillica di Santa Maria degli Angeli, Rome, July 2010, after licensing meeting with Sigma-tau

Natsuki Nakayama

Carnitine Deficiency and Insufficiency in Medicine

Springer

Natsuki Nakayama
Representative
Carnitine Science
Matsushige, Tokushima, Japan

ISBN 978-981-95-7796-5 ISBN 978-981-95-7797-2 (eBook)
https://doi.org/10.1007/978-981-95-7797-2

This Springer imprint is published by the registered company Springer Nature Singapore Pte Ltd.
The registered company address is: 152 Beach Road, #21-01/04 Gateway East, Singapore 189721, Singapore

Preface

Carnitine is a nutrient that is abundant in meat and dairy products, and it is also biosynthesized in the body. Carnitine is an essential molecule for the transport of long-chain fatty acids into mitochondria and is extremely important for energy production. In addition, carnitine excretes harmful acyl groups, which are accumulated in mitochondria, to outside of mitochondria in various pathological conditions. So that carnitine has buffering effects to maintain normal mitochondrial functions. Furthermore, recent studies have revealed that carnitine has new effects that are not related to fatty acid metabolism, such as anti-inflammatory, antioxidant, and antiapoptotic effects.

Carnitine is not usually deficient, but carnitine deficiency can occur in various medical situations, such as in various inherited metabolic diseases, kidney diseases, liver diseases, heart diseases, and digestive disorders. In addition, carnitine deficiency is likely to occur in children with severe physical and mental disabilities and patients after surgery, who are nutritionally managed for a long time by tube feeding or total parenteral nutrition that does not contain carnitine. In addition, many drug-induced carnitine deficiency have been reported. For example, sodium valproate, which is commonly used as antiepileptic drug and as a mood stabilizer in psychiatry, sometimes cause hyperammonemia and hepatotoxicity. Also, antibiotics containing pivalic acid moiety sometimes cause hypoglycemic attacks. Now it is revealed that these symptoms are due to drug-induced carnitine deficiency.

Since carnitine has important biochemical and physiological effects related to energy metabolism, carnitine deficiency may cause symptoms such as muscle weakness, muscle cramps, fatigue, and stunted growth, and in severe cases, life-threatening conditions such as hypoglycemic attacks, hyperammonemia, cardiac hypertrophy, and heart failure.

In Japan, L-carnitine as a pharmaceutical drug was approved in 1990 only for carnitine deficiency caused by methylmalonic acidemia and propionic acidemia. Physicians could not prescribe carnitine to carnitine deficiency due to inherited metabolic diseases other than methylmalonic and propionic acidemia, dialysis, and drugs-induced carnitine deficiency, because it was not indicated. In Japan, there are so-called drug-lag issues, which means many standard drugs in the world could not be prescribed because they are not indicated or approved in Japan. Carnitine was a typical lagged drug especially in pediatric fields. To solve these drug-lag issues, "The Evaluation Committee on Unapproved of Off-labeled Drug with High Medical Needs," under the Ministry of Health, Labour and Welfare (MHLW), was established in 2010 and reviewed many drugs that were requested from various academic societies. The Japanese Society for Inherited Metabolic Diseases requested L-carnitine and other drugs. The Committee evaluated the medical necessity and requested Otsuka Pharmaceutical Co., Ltd. to develop L-carnitine for general carnitine deficiency. In response to this request, Otsuka applied to change the indication of carnitine to "carnitine deficiency," and the application was approved in March 2011. As a result, carnitine can be used for a wide range of carnitine deficiency disorders in Japan. Subsequently, intravenous preparations that are necessary for acute-phase patients and dialysis patients, as well as oral liquids that are easy for children to intake, have been developed and introduced in Japan.

For 10 years since 2007, I engaged and played a central role in these carnitine projects in Otsuka, namely, the expansion of indications to "carnitine deficiency," the development of new dosage formulations such as intravenous and oral liquids, and the development of diagnostic agents for carnitine deficiency and their insurance coverage. In addition, I have held more than 500 of lectures, briefings, and academic conferences for medical professionals in various fields to promote the understanding of concept of carnitine deficiency, because "carnitine deficiency" was not widely known in Japan at that time. During these activities, I strongly felt that except for a few specialists in inherited metabolic diseases, many clinicians did not have enough understanding of carnitine and carnitine deficiency, and there were many misunderstandings such as "Is it like a vitamin?" or "Why do we need carnitine preparation as a pharmaceutical drug when there are many carnitine products as supplements?". Most medical professionals such as physicians, nurses, clinical engineers, nutritionists, clinical laboratory technicians, pharmacists, and physical therapists, who are deeply involved in the treatment of patients, complained to me that it is not easy to understand what carnitine or carnitine deficiency/insufficiency is. Since carnitine is deeply involved in energy metabolism, which is the basis of life, it is extremely important to have a deep understanding of the function of carnitine and the concept of carnitine deficiency/insufficiency. However, even if medical professionals in various fields try to learn deeply about carnitine deficiency/insufficiency, there are only two old books about carnitine published from Academic Press (Frenkel RA and McGarry JD eds. *Carnitine Biosynthesis, Metabolism and Functions*, 1980, ISBN-10: 0122670604, ISBN-13: 978-012670602, and Ferrari R, DiMauro S, Sherwood G eds. *L-Carnitine and Its Role in Medicine -From Function to Therapy*, 1992, ISBN-10: 0122539400, ISBN-13: 978-0122539404). The former was a proceeding of the conference meeting on carnitine held in 1979 at the University of Texas at Dallas. The latter book explained the basic role of carnitine, and described clinical aspects of carnitine deficiency and carnitine supplementation therapy in pediatrics, cardiology, nephrology, etc. They were excellent books to grasp the understanding and knowledge for the experts who are studying carnitine and carnitine deficiency/insufficiency, but these book are also too old considering the great progress of molecular understanding about molecules in carnitine systems and accumulated clinical data since then. Also, these books were for the experts, so it is difficult to understand for the general medical professionals who want to learn first-time about carnitine deficiency/insufficiency, also it is very difficult to obtain these old books now. During my efforts on carnitine projects mentioned above, I have learned a lot of knowledge from literatures about carnitine deficiency, from many physicians who are active in various academic societies such as the Japanese Society for Inherited Metabolic Diseases, the Japanese Society on Severe Motor and Intellectual Disabilities, the Japanese Society of Child Neurology, and so on. Discussing with these physicians in various fields, I learned about the diagnosis and treatment, about how to administrate carnitine in actual clinical practice. In addition, I had many opportunities to interact with clinicians and researchers outside of Japan when I attended several international conferences on carnitine deficiency sponsored by Sigma Tau, Italy, from which Otsuka Pharmaceutical Co., Ltd. is introducing its active pharmaceutical ingredient (API). Although I am not a physician and have no clinical experience to treat carnitine deficiency/insufficiency, I thought it was necessary to organize the scientific, medical, and pharmaceutical knowledge about carnitine deficiency that I have learned from a huge amount of medical and scientific reference papers during my activities so far, as well as my valuable experience in the development process, and summarize it as a book in an easy-to-understand manner. It took more than 5 years but finally I published in 2023 "All about Carnitine—To Prevent Carnitine Deficiency" (Igaku Tosho Shuppan, Tokyo, ISBN 978-4-86517-533-2) in Japanese. The same situation as Japan, where medical professionals do not well understand about carnitine deficiency/insufficiency, might spread in all over the world. Even in the EU and the United States where carnitine was approved as the indication for carnitine deficiency around the 1990s, many medical professionals sometimes despise or ignore the importance of carnitine. In addition, when physicians or medical professionals want to learn more about carnitine deficiency/insufficiency, there is no special textbook about carnitine deficiency/insufficiency in an easy-to-understand manner with recent basic and clinical findings. This is the reason why I want to publish the English version of my book. Through this book, I hope that many medical professionals, including clinicians in the world, will gain a deeper understanding of the function of carnitine and what carnitine deficiency/insufficiency is, and take care to prevent their own patients from developing carnitine deficiency. The history of carnitine research is also very interesting. Its history is full of curiosity, enthusiasm, and passion of scientists and clinicians working on this nutrient, which is deeply involved in energy metabolism. Through this book, I hope that you will also feel the deep excitement of research on a simple nutrient, "carnitine."

In compiling this book, I would like to express my appreciation to the many physicians who have guided and taught me so far.

Matsushige, Tokushima, Japan Natsuki Nakayama
July 1, 2026

Words of Recommendation

Fumio Endo

Natsuki Nakayama's *Carnitine Deficiency and Insufficiency in Medicine* provides a comprehensive and detailed review of carnitine and human health. Beginning with the discovery of carnitine, the final chapter describes the history and meaning of carnitine in the treatment of disease. There has never been a publication that covers all aspects of carnitine. Even if we look overseas, we can say that there are no books like it. It can be said that it was made possible for the first time by Dr. Natsuki Nakayama, a leading scientist on carnitine.

There are many good things about this book, but the first thing I would like to point out is that it is well structured. Biochemistry, multifaceted metabolism in living body, nutritional importance, significance of carnitine in the medical field, congenital carnitine deficiency and relationship with inherited metabolic diseases, and carnitine as a drug are described in detail. The information covered is truly worthy of the name "All about Carnitine" (the title of the Japanese original version).

If there is one feature of the content, it is the abundance of figures, tables, and literature. Important figures and tables are presented appropriately and in an easy-to-understand manner to help you better understand each topic. The bibliography is also neatly organized and presented at the end of each chapter. Through such descriptions and presentations, it is easy to understand the various effects of carnitine on the body, various effects on human health, and the status of administration to humans. This information is extremely useful for a wide range of people, from those who are knowledgeable about carnitine to those who are newly interested in carnitine.

L-carnitine is an important drug for the treatment of inherited intractable diseases in children, and it is covered by insurance in the United States and Europe. Despite this, it was not approved as a drug for congenital intractable diseases in Japan. As the president of the Japanese Society for Inherited Metabolic Diseases and as the head of the research group on inherited metabolic diseases at the Ministry of Health, Labour and Welfare, I have had the opportunity to work with Dr. Nakayama for a long time, including research and development of carnitine, such as clinical application and expansion of the indications for several diseases. I believe that Dr. Nakayama's contribution to the indication for the treatment of intractable diseases is very significant in Japan. I am very pleased that Dr. Nakayama has published this book. It's a great book. I would like to recommend it from the bottom of my heart.

Former President of Japanese Society for Inherited Metabolic Diseases
Professor Emeritus, Department of Pediatrics, School of Medicine, Kumamoto University
CEO, Kumamoto Ezuko Medical Center for the Severely Disabled

Makiko Osawa

Carnitine Deficiency and Insufficiency in Medicine is a book that falls into the reader's mind and seeps into the body!

Carnitine was first described in 1905 by Gulewitsch and Krimberg of the Institute of Medical Chemistry of Moscow State University. Carnitine is abundant in meat and dairy products, and it is also biosynthesized in the body. It is an essential molecule for the transport of long-chain fatty acids into the mitochondria and is extremely important for energy production. At the same time, harmful acyl groups accumulated in the mitochondria are excreted from the mitochondria as form of acylcarnitines, keeping and maintaining the mitochondrial functions normal. Recent studies have revealed new effects of carnitine that are not related to fatty acid metabolism, such as anti-inflammatory, antioxidant, and antiapoptotic effects.

"Guidelines for the Diagnosis and Treatment of Carnitine Deficiency 2018" was published by the Japan Pediatric Society, and there is a summary of the function of carnitine and the concept of carnitine deficiency, but this book covers the function of carnitine and the mechanism of carnitine deficiency so that it can be deeply understood. In 1990, L-carnitine was approved as a drug only for carnitine deficiency due to methylmalonic acidemia and propionic acidemia. It was used off-label for other inherited metabolic diseases and carnitine deficiency due to various causes such as dialysis and drugs. In order to solve this situation, Otsuka received a request from the "Review Council on Unapproved Drugs and Off-Label Drugs with High Medical Needs" and filed a public knowledge application. In March 2011, the indication was changed to "carnitine deficiency" and it can be used for a wide range of carnitine deficiency disorders. Carnitine deficiency is not usually present, but kidney disease, liver disease, heart disease, and gastrointestinal disorders can cause carnitine deficiency. Deficiency causes serious symptoms such as ▼ impaired consciousness ▼ convulsions ▼ rhabdomyolysis ▼ encephalopathy ▼ hypoglycemia ▼ abnormal liver function ▼ frequent vomiting ▼ psychomotor retardation ▼ cardiomegaly/cardiomyopathy/decreased cardiac function ▼ sudden death. The causes of carnitine deficiency are ▼ inherited metabolic diseases ▼ cirrhosis and liver failure ▼ anorexia ▼ severely mentally and physically handicapped children (persons) and patients after surgery are nutritionally managed for a long time with nutritional supplements that do not contain carnitine ▼ undernutrition such as the elderly ▼ patients with renal failure ▼ nutritional management with some milk allergen removal formulas ▼ administration of drugs that cause carnitine deficiency (valproic acid, antibacterial drugs containing a pivoxil group, anticancer drugs). Since it is important to measure the blood carnitine concentration in patients at risk of carnitine deficiency, the diagnostic tests for free carnitine and total carnitine have been covered by insurance since

February 2018. It is an easy-to-understand book that is worth reading based on the scientific results of numerous animal experiments.

Professor Emeritus, Department of Pediatrics, Tokyo Women's Medical University

Former President of Japan Epilepsy Society

Former President of The Japanese Society of Child Neurology

Shinobu Ida

Congratulations on the completion and publication of *Carnitine Deficiency and Insufficiency in Medicine*. This is a project that has been warmed up for a long time, and it is the culmination of Natsuki Nakayama's love for carnitine. Carnitine is a small substance biosynthesized from lysine and methionine, which was discovered in 1905. While it plays a major role in supporting the life of living organisms such as energy metabolism and detoxification, it is also a "behind-the-scenes" presence. Dr. Nakayama was fascinated by the fact that carnitine works amazingly and is a behind-the-scenes presence, and I believe that he himself has played a behind-the-scenes role in clinical settings involving carnitine in Japan. He has generously described and presented all his knowledge about carnitine since its discovery in 1905, to physicians and researchers in clinical practice.

Dr. Nakayama and I were involved in the creation of the "Guidelines for the Diagnosis and Treatment of Carnitine Deficiency 2016," and the acquisition of insurance coverage for the carnitine measurement method as a necessary clinical diagnosis agent. Dr. Takashi Igarashi, who was the president of the Japan Pediatric Society at the time, told that the Nutrition Committee of the Pediatric Society should create a medical guideline for diagnosis and treatment of carnitine deficiency. As the chairman of the Committee, I had long been well aware that carnitine is an essential nutrient in intravenous nutrition and in neonatal infants, so that I formed a working group and created the "Guidelines for the Diagnosis and Treatment of Carnitine Deficiency 2016." Thereafter, this Guideline was revised to "Guidelines for the Diagnosis and Treatment of Carnitine Deficiency 2018." These guidelines have become a kind of strong driving-force for approval of insurance coverage of carnitine measurement using easy-to-use diagnostic kits, in the clinical settings of diagnosis and treatment of carnitine deficiency thereafter. It is no exaggeration to say that these series of activities were driven by Dr. Nakayama's comprehensive knowledge of carnitine and his love for carnitine.

Carnitine Deficiency and Insufficiency in Medicine consists of Chaps. 1, 2, 3, 4, 5, 6, 7, 8, 9, 10, 11, 12, 13, 14, 15, 16, 17, 18, and 19, and you can understand everything about carnitine in this one book, from basic to clinical. Through the charm of the existence of carnitine, you might feel how life is supported delicately and interestingly. The description are light and come into your head quickly. I think that because Dr. Nakayama understands everything about carnitine, he is able to explain it in plain language. If you pick up this book and read it, you will be able to enjoy the big world of small carnitine.

Director of Department of Clinical Laboratory Department, Osaka Women's and Children's Hospital

Yoshihiko Kanno

Although carnitine has been the subject of research as a biological substance related to muscle and fatty acid metabolism, there are problems with the measurement method and the drug in terms of insurance treatment, and there are not many places where clinical discussions are held. However, in 2011, a drug indication was approved for carnitine deficiency in Japan, and in 2018, blood carnitine measurement was covered by insurance, and it attracted a great deal of clinical attention. Under such circumstances, "Guidelines for the Diagnosis and Treatment of Carnitine Deficiency 2018" (Japan Pediatric Society) has pointed out that carnitine deficiency also occurs in renal disease, liver disease, and nutritional disorders in various diseases in adults. This book is based on the knowledge about carnitine that Dr. Natsuki Nakayama, who has been witnessing these changes in the treatment of carnitine in Japan for a long time, and who can be said to be the father of the child's use of carnitine today, learned through his life. It is a content that is worthy of the title of the book, "All about Carnitine" (the title of original Japanese version). I recommend this book to all medical professionals in the hope that I can share with you the happiness of being able to read this book in Japanese, and English which covers everything from 1957, before I was born, to the latest literature.

Professor of Department of Nephrology, Tokyo Medical University
Former President of The Japanese Society of Clinical Nutrition

Professor Masahito Shimizu

In recent years, it has been reported that carnitine deficiency is involved in the onset and progression of various diseases, and that carnitine replacement therapy is useful for these pathologies. In the area of liver disease, which is my specialized area, it has been found that patients with liver cirrhosis are associated with a high frequency of (relative) carnitine deficiency, and that carnitine deficiency is associated with complications such as protein and energy undernutrition, hepatic encephalopathy, sarcopenia, and painful muscle spasms, and that it worsens the patient's QOL and prognosis. In the "Cirrhosis Clinical Treatment Guidelines 2020" (edited by the Japanese Society of Gastroenterology and the Japan Society of Hepatology), carnitine replacement therapy was proposed for the first time for hepatic encephalopathy with carnitine deficiency and muscle spasms complicated by cirrhosis. As can be seen from this, carnitine preparations are currently attracting a great deal of attention as one of the key drugs in the treatment of cirrhosis. However, there have been almost no specialized books that systematically explain the mechanism by which carnitine improves various pathologies, or even "what carnitine is" in the first place.

This book by the author, Dr. Natsuki Nakayama, is the "world's first" specialized book on carnitine, which comprehensively explains the discovery of carnitine, its biochemical characteristics, bioactivity, homeostasis, and the significance of carnitine deficiency in various diseases. In particular, this book, which transcends the framework of diseases and organs, and provides a bird's-eye view of the physiological changes caused by carnitine deficiency from newborns to the elderly, can be said to play a major role in understanding carnitine and promoting its research and clinical application. Another excellent point of this book is that it extensively touches on the tips of carnitine research, measurement, and replacement therapy, as well as its clinical pharmacological issues.

Through his work and academic activities, the author has clarified the significance of carnitine deficiency in clinical practice and the necessity of carnitine preparations as pharmaceuticals. Therefore, it goes without saying that the basis of this book is the author's deep knowledge of carnitine. In addition, the author's attitude of being attracted to the "fun" of carnitine and stepping into the world of carnitine with intellectual inquiry, curiosity, passion, and "love" makes this book even more attractive. It is hoped that the publication of this book will lead to a better understanding of carnitine deficiency and carnitine replacement therapy, that carnitine research will take the next step, and that many patients will be able to enjoy the benefits of carnitine treatment.

Gifu University Graduate School of Medicine
Department of Gastroenterology

Hiroshi Ozawa

My first encounter with carnitine goes back more than 20 years. Infusion through a nasogastric tube causes delayed gastric emptying which might results in malabsorption, so I used jejunal tube feeding to a severely mentally and physically handicapped child who was spending time at home. When a jejunal tube was inserted, the procedure took a long time, so suddenly metabolic acidosis occurred, and the general condition was deteriorated. The cause was carnitine deficiency. At that time, carnitine could be measured only in the laboratory, and I asked for it to over the phone. When the child was given carnitine, he no longer developed metabolic acidosis and was able to feed through a nasogastric tube. At that time enteral feeding products did not contain carnitine.

Since then, I was keenly aware of the importance of carnitine, and I was conducting clinical research on carnitine. On July 1, 2011, I met Dr. Natsuki Nakayama. He and I were excited, and we couldn't stop talking about carnitine with each other.

Currently, carnitine is included in most of the tube feeding products, and it is now possible to test for diagnosis through insurance, and carnitine deficiency due to insufficient intake has drastically decreased in Japan. This is also the result of Dr. Nakayama's efforts.

This book is a historical novel about carnitine, which is "*All* about carnitine" (the title of original Japanese version). In addition to clinical aspects, it is full of fascination, such as the discovery of the physiological effects of carnitine as an insect nutrient, the drama of the research in each component in carnitine cycle, homeostasis, and its relationship with various diseases.

This book should be read by all medical professionals, including physicians, nurses, nutritionists, pharmacists, and rehabilitation staffs. You can't talk about carnitine without reading this book. It is a great pleasure to come across this book.

Thank you very much, Dr. Natsuki Nakayama.

Director, Shimada Ryoiku Medical Center for Challenged Children, Hachioji

Contents

About the Author

Natsuki Nakayama obtained a PhD in 1980 (Department of Biochemistry, Faculty of Agriculture, Nagoya University), and worked for a while as an Assistant Professor of Forensic Science, Faculty of Medicine, Nagoya City University (1981–1982).

In 1982, he joined Tokushima Research Institute, Otsuka Pharmaceutical Co., Ltd. Soon, he was dispatched to the USA as a Visiting Assistant Professor in the Department of Pharmacology and Cell Biophysics, University of Cincinnati (1984–1986, Professor Arnold Schwarz). He returned to Tokushima Research Institute and engaged in research on drugs for cardiovascular diseases and diabetes (1986–1995). Then he engaged in the clinical development of cardiovascular drugs and in research project management in the institute (1995–2007).

In 2007, he became a Director of Development Department, Diagnostic Division, in Otsuka and engaged in the expansion of the indication of L-Cartin® (Otsuka's L-carnitine products) until 2011. Then he worked as Senior Science Manager and Scientific Advisor for L-Cartin®. During this period, he engaged in the development of internal liquid solutions, intravenous formulations, development and insurance coverage of in vitro diagnostic drugs for carnitine deficiency, etc.

After retirement in January 2018, he became a representative of Carnitine Science and engaged in activities to popularize "carnitine deficiency/insufficiency" through lectures and writing. In June 2023, he published a book titled *All About Carnitine—To Prevent Carnitine Deficiency* in Japanese (the original book of this English version).

E-mail: carnitinescience@db3.so-net.ne.jp

Abbreviations

A

AC	Acylcarnitine
AC/FC	Acylcarnitine/free carnitine
ACE	Acylcarnitine esterase
ACS	Acyl-CoA synthetase
ADHD	Attention-deficit/hyperactivity disorder
ADI	Acceptable daily intake
ADME	Absorption, distribution, metabolism, and excretion
ALC or ALCAR	Acetyl-L-carnitine
ALS	Amyotrophic lateral sclerosis
AMA	Arm muscle area
Ang II	Angiotensin II
ANT	Adenine nucleotide translocase
AOPP	Advanced oxidative protein products
ARDS	Acute respiratory distress syndrome
ARE	Antioxidant responsive element
ASD	Autism spectrum disorder
ASO	Arteriosclerosis obliterans or obstructive arteriosclerosis
ASS	Argininosuccinate synthetase or arginosuccinate synthetase
ATP	Adenosine triphosphate
AUC	Area under the blood concentration-time curve

B

BA	Biological availability
baPWV	Brachial-ankle pulse wave velocity
BBB	Blood-brain barrier
BCAA	Branched chain amino acid
BI	Barthel Index
BNP	Brain natriuretic peptide

C

CACT	Carnitine acylcarnitine translocase
CAPD	Continuous ambulatory peritoneal dialysis
CD	Crohn's disease
CDH	Carnitine dehydrogenase
CFS	Chronic fatigue syndrome
CGH	Comparative genomic hybridization
CMC	Critical micelle concentration
CNV	Copy number variant
CoA	Coenzyme A
CONUT score	Controlling nutritional status score
CPS1	Carbamyl (or carbamoyl) phosphate synthetase-1
CPT	Carnitine palmitoyltransferase
CPT1	Carnitine palmitoyltransferase 1
CPT2	Carnitine palmitoyltransferase 2
CRP	C-reactive protein
CRRT	Continuous renal replacement therapy

D

DCD	Dialysis-related carnitine disorder or deficiency

E

EBM	Evidence-based medicine
ecNOS	Endothelial constitutive nitric oxide synthase
ERI	Erythropoietin resistance index
ESA	Erythropoiesis stimulating agents

F

FADS	Fatty acid desaturase
FC	Free carnitine
FMO3	Flavin-containing monooxygenase form 3

G

GNRI	Geriatric Nutritional Risk Index
GSDH	Glutamate-5-semialdehyde dehydrogenase

H

hCETP Human cholesteryl ester transfer protein
HFpEF Heart failure with preserved ejection fraction
HFrEF Heart failure with reduced ejection fraction
HO-1 Heme oxygenase-1
HPLC High-performance liquid chromatography
hsCRP High sensitive CRP
HSP70 Heat shock protein 70

I

IBD Inflammatory bowel disease
IC Intermittent claudication
ICAM-1 Intercellular adhesion molecule 1
ICH International Council for Harmonization (of Technical requirements for Pharmaceuticals for Human Use)
IPD Intermittent peritoneal dialysis

J

JVS mouse Juvenile visceral steatosis mouse

L

LAT Lysophospholipid acyltransferase
LAVI Left atrial volume index
LBCCA Ligation of the bilateral common carotid arteries
LBM Lean body mass
LES Late evening snack
Lp(a) Lipoprotein (a)
LPL Lysophospholipid
LVAD Left ventricular assist device
LVEF Left ventricular ejection fraction
LVMI Left ventricular mass index

M

MACE Major adverse cardiac events
MCAD Medium-chain acyl-CoA dehydrogenase
MCD Methionine-choline deficient diet
MCD Muscle carnitine deficiency
MCT Middle-chain triglyceride
MDA Malondialdehyde

MELAS Mitochondrial myopathy, encephalopathy, lactic acidosis, and stroke-like episodes
MIA Malnutrition inflammation atherosclerosis
MPO Myeloperoxidase
MPT Mitochondrial permeability transition or mitochondrial membrane permeability transition
MS Multiple sclerosis
MS/MS Tandem mass spectrometry
MSUD Maple syrup urine disease

N

NAFLD Non-alcoholic fatty liver disease
NAG N-acetylglutamate
NAGS N-acetylglutamate synthase
NASH Non-alcoholic steatohepatitis
NCT Number connection test
NGAL Neutrophil gelatinase-associated lipocalin
NIRS Near-infrared spectroscopy
NOAEL No observed adverse effect level
NOEL No observed effect level
NPC/N Non-protein calorie/nitrogen
npRQ Non-protein respiratory quotient
NP test Neuro-psychological test
NQO1 NAD(P)H-quinone oxidoreductase 1
NT-proBNP N-terminal pro-brain natriuretic peptide

O

OCTN2 Organic cation/carnitine transporter 2
OTC Ornithine transcarbamylase

P

PAD Peripheral arterial disease
PAG Phosphate-activated glutaminase
PAS Peripheral arterial stiffness
PBC Primary biliary cholangitis
PCD Primary carnitine deficiency
PDH Pyruvate dehydrogenase
PEM Protein-energy malnutrition
PEW Protein-energy wasting
PGC1α Peroxisome proliferator-activated receptor gamma coactivator 1α
PK Pharmacokinetics
PKU Phenylketonuria

PLC	Propionyl-L-carnitine
PMI	Psoas muscle mass index
PS	Phosphatidyl serine
PS shunt	Portosystemic shunt

R

rHuEPO	Recombinant human erythropoietin

S

SAA	Serum amyloid A
SCD	Systemic carnitine deficiency
SMI	Skeletal muscle index
SOD	Superoxide dismutase
SVR	Sustained virological response

T

TACE	Transcatheter arterial chemoembolization
TAI	Transcatheter arterial infusion
T-AOC	Total antioxidant capacity
TMA	Trimethylamine
TMAO	Trimethylamine-N-oxide
TMLHE	Trimethyllysine hydroxylase epsilon
TNFα	Tumor necrosis factor α
TPN	Total parenteral nutrition

U

UC	Ulcerative colitis

V

VAS	Visual Analog Scale
VCAM-1	Vascular cell adhesion molecule 1
VDAC	Voltage-dependent anion channel
VHE	Valproate-induced hyperammonaemic encephalopathy
VHT	Valproate-induced hepatotoxicity
VLCAD	Very long-chain acyl-CoA dehydrogenase
VPA	Valproate, valproic acid

1 Discovery of Carnitine

1.1 Discovery of Carnitine from Muscle Extract

Carnitine was first described in 1905 by Gulewitsch and Krimberg at the Moscow University Institute of Medical Chemistry [1]. In 1900, they discovered carnosine, an imidazole dipeptide that has recently attracted attention for its anti-fatigue effects. In the course of their research, they focused on the filtrate that remained as a silver base after the separation of carnosine and isolated a new base from the filtrate using a precipitation method with potassium bismuth iodide solution, which was not widely used in the field of animal chemistry at that time. They proposed the name "carnitine" from the Latin word "carnus" meaning "meat" for this base and reported it [1, 2]. In their report, Gulewitsch and Krimberg reported that carnitine is a chloride salt of an unknown base with the chemical formula $C_{14}H_{32}N_2O_6C_{16}Pt$, and that the experimental formula of carnitine is $C_7H_{15}NO_3$, assuming that the quaternary ammonium base is not present. In addition, it is described that the free base exhibits a strong alkaline reaction, that this substance and its hydrochloric acid and nitrate are highly soluble in water, that the nitrate of this substance is obtained from its aqueous solution and alcohol solution in the form of radial crystals of needle-like small crystals, and that nitrate exhibits left rotation and its optical rotation is about 22 degrees [1].

At the same time, Kutscher isolated a new nitrogen compound from meat extract and named it Novain [3]. Kutscher proposed the experimental formula $C_7H_{18}NO_2AuCl_4$ from the analysis of the auric chloride. However, subsequent research by Krimberg showed that the experimental formula proposed by Kutscher was wrong, but that the carnitine isolated by Gulewitsch and Krimberg and the nitrogen compound called Novain isolated by Kutscher were identical substance [4, 5]. This is why Gulewitsch and Krimberg are now considered to be the discoverers of carnitine.

1.2 Determination of the Chemical Structure of Carnitine

After the discovery of carnitine in 1905, Gulewitsch and Krimberg began to study the chemical structure of carnitine, and until around 1910, many studies on the chemical structure of carnitine, including those of other research groups, were reported. However, they gradually lost interest in the study without much success, and the research was subsequently discontinued. Although the chemical formula of carnitine ($C_7H_{15}NO_3$) proposed by Gulewitsch and Krimberg was itself correct, the chemical structure of carnitine was not clarified despite the subsequent efforts of many researchers to study the chemical structure of car-

N. Nakayama, *Carnitine Deficiency and Insufficiency in Medicine*,
https://doi.org/10.1007/978-981-95-7797-2_1

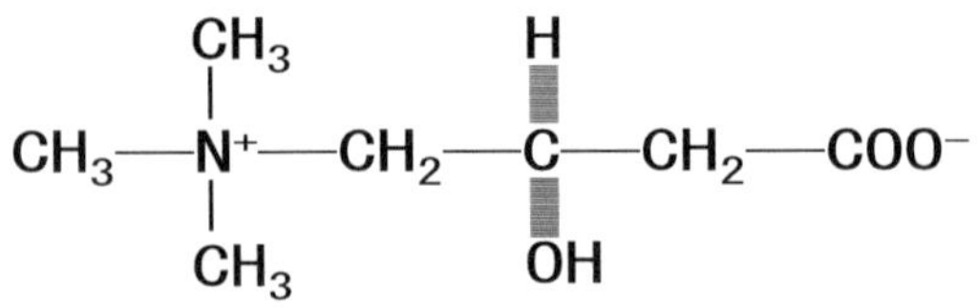

Fig. 1.1 Chemical structure of L-carnitine

nitine. The synthesis of carnitine (γ-trimethylamino-β-hydroxybutyrate) itself was carried out by other researchers in 1910, but it was not recognized that γ-trimethylamino-β-hydroxybutyrate itself is the carnitine discovered by Gulewitsch and Krimberg. However, in 1927, Tomita and Sendju in the Institute of Physiological Chemistry in Nagasaki (at that time) chemically synthesized DL-γ-trimethylamino-β-hydroxybutyrate and by optical resolution using brucine found that the levo-rotatory substance has the same rotation and melting point as carnitine in nature; this L-γ-trimethylamino-β-hydroxybutyrate proved to be the same substance as carnitine isolated by Gulewitsch and Krimberg [6]. It took 22 years from its discovery to determine the correct chemical structure of carnitine. This study established the chemical structure of carnitine (Fig. 1.1). Japanese chemists made a significant contribution to the determination of the chemical structure of carnitine. The details of the history of the study of the chemical structure of carnitine are omitted because they are beyond the scope of this paper. If you are interested, please refer to the review article on the chemistry of carnitine by Fraenkel et al. [7]

1.3 Subsequent Studies of the Physiological Actions of Carnitine

Once the chemical structure of carnitine was determined, interest was lost again, and research on carnitine was rarely reported. In the early days of physiology and pharmacology, Henry Hallett Dale discovered acetylcholine in 1914. Furthermore, in 1921, Otto Loewi showed the existence of a vagus nerve substance in an experiment using isolated heart preparation from two frogs, and he named it "Vagusstoff" (meaning substance which stimulates vagal nerve). Dale later proved that this vagal neurotransmitter was acetylcholine. For these achievements, Dale and Loewi were awarded the Nobel Prize in Physiology or Medicine in 1936 for their discovery of acetylcholine as a neurotransmitter. Therefore, at that time, the physiological functions of acetylcholine and other quaternary amines were of great interest and vigorously studied. The physiological effects of carnitine, which is also a quaternary amine very similar to acetylcholine, were also studied again, and in the 1920s and 1930s, research on the physiological and pharmacological actions of carnitine were actively studied. For example, experiments on whether carnitine and acetylcarnitine have neurotransmitter actions have been actively attempted and reported. In this series of studies, it was found that carnitine promotes the secretion of gastric and intestinal juices through parasympathetic nerve stimulation, which later led to the clinical application of carnitine's appetite-stimulating effects (see "Chapter 6: Appetite Enhancement and Digestive Enzyme Secretion Promotion Effects of Carnitine"). Other than that, however, no significant progress was made, and eventually carnitine was forgotten once again.

1.4 Discovery of the Physiological Actions of Carnitine in Insect Nutrition

1.4.1 What Is Insect Nutrition?

Clues to the physiological effects of carnitine were surprisingly found in studies on insects. Insects feed on a very wide variety of foods, but about half of them feed mainly on the plant leaves. Also, a specific insect feed only on the leaves of a specific plants. In addition, there are specific insects that feed on each specific food, for example, some insects feed only on grains such as wheat flour. Therefore, in the field of entomology at that time, the research works were carried out to clarify whether there were common and basic essential nutrients for the growth of insects, or whether the nutrient requirements were different for each insect species. In other words, it has been a major research topic in insect nutrition to study (1) what determines the feeding of different insects on different foods, and (2) what are the requirements of micronutrients and components such as vitamins and polyphenols for insects (these were called accessory food factors). These research topics are important in terms of reducing damage caused by insects in agriculture and protecting against pests that attach to stored grains, and it is also a basic issue of interest in the nutrition of mammals, including humans.

1.4.2 Research on Vitamin B_T by Dr. Fraenkel et al.

Gottfried Fraenkel and colleagues at the Imperial College of Science and Technology in London (later at the University of Illinois in the United States) raised insect larvae that naturally feed on only flour using a basic synthetic diet (refined casein, starch or glucose, cholesterol, salt, water) with the addition of yeast and B group of vitamins. Gottfried Fraenkel continued to study the nutritional aspects of the larvae using indicators such as the pupation rate of the larvae metamorphosed into pupae, the increase in body weight, the number of days they lived, and the number of surviving individuals. First, he studied the amino acid composition and vitamins necessary for the growth of six species of cereal pests, such as Mediterranean flour moss (*Ephestia kuehniella* Zeller) and confused flour beetle (*Tribolium confusum*). [8] Through a series of studies, it was found that one insect species examined required sterols [9] and eight kind of B group of vitamins [10]. Subsequently, as a nutrient for *Ephestia kuehniella* Zeller, *Tribolium confusum*, and *Tenebrio molitor*, the components adsorbed by activated charcoal in the yeast extract are necessary for growth. It was reported that the substance is folic acid (vitamin B_9), which is a B group of vitamins (Fig. 1.2) [11, 12].

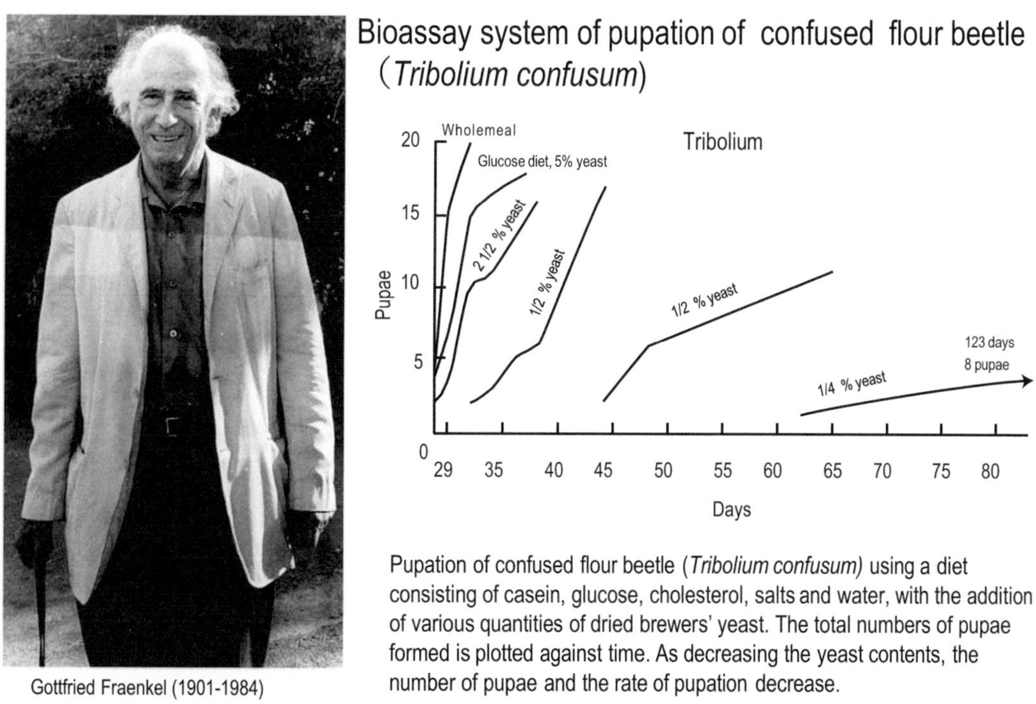

Fig. 1.2 Gottfried S. Fraenkel (1901–1984), who discovered the role of carnitine in insect nutrition (left), and their bioassay system of vitamin B_T. Growth of *Tribolium confusum* on a diet consisting of casein, glucose, cholesterol, salts, and water, with the addition of graded quantities of dried brewers' yeast. (From Ref. [24])

Furthermore, in 1948, Fraenkel et al. reported that a water-soluble fraction of yeast extract was necessary for the growth and pupation of the yellow mealworm beetle (*Tenebrio molitor*), and that this nutrient was an essential nutrient factor for a new B group of vitamins different from folic acid, and tentatively named the nutrient vitamin B_T [13]. At that time, only the yellow mealworm beetle showed vitamin B_T requirements, so the name B_T was taken from the T in *Tenebrio*. Subsequently, Fraenkel et al. found that vitamin B_T was essential for the pupation and growth of seven species of cereal pests in the same family as *Tenebrio molitor* (Tenebrionidae), such as confused flour beetle (*Tribolium confusum*) and red flour beetle (*Palorus ratzeburgi* Wissmann). [14, 15] Subsequent studies by Fraenkel et al. have shown that vitamin B_T is abundant in yeast, milk, whey, liver, etc., and is widely distributed in the animal kingdom [14, 16–18]. In 1952, Carter et al. in the same group purified vitamin B_T from whey and liver extracts, while carnitine was purified from muscle extracts, and he compared their chemical and physical properties. It was found that the optical rotation and the melting points of vitamin B_T and its various salts coincide with those of carnitine, the experimental formula of vitamin B_T is the same $C_7H_{15}NO_3$ as carnitine, and when vitamin B_T was alkaline decomposed, trimethylamine is produced as in the case of carnitine, while crotonobetaine is produced by dehydration by sulfuric acid. From these experimental results, they concluded that vitamin B_T was the same substance as carnitine, which was discovered by Gulewitsch and Krimberg in 1905. [19, 20]

McFarlane in Fraenkel's group divided the normally bred larvae of yellow mealworm beetle into two groups, one group fed on a synthetic diet without vitamin B_T and the other group fed no feed at all, and compared the carbohydrate, protein, and fat content of the two groups just before death. It was found that the fat content was significantly reduced in the latter starvation group, whereas the fat content was retained in the synthetic diet group without vitamin B_T (carnitine) and died. [21] This suggests that in the absence of carnitine, the fat stored in the larvae's body cannot be used, namely, carnitine is somehow involved in fatty acid oxidation (degradation of fatty acids). In addition, Fraenkel et al. found and reported the presence of an enzyme in pigeon and sheep liver extracts that reversibly transfers acetyl group from acetylcarnitine to coenzyme A (CoA) in the presence of ATP and acetic acid. [22] In other words, Fraenkel et al. had almost grasped the fact that the physiological function of carnitine seemed to be related to fatty acid metabolism and acyl group transfer, but perhaps because they were entomologists, they did not pursue the relationship between carnitine and fatty acid metabolism further. Presumably, due to the rapid increase in energy demand when the larvae metamorphose into pupae, it is necessary to change drastically the energy metabolism and energy production based on fatty acid metabolism to accommodate this critical situation.

In this way, Fraenkel experimentally demonstrated and reported carnitine deficiency in a very limited number of insect species. However, when he switched the purified casein used in the experiment to a different lot of casein around 1953, the insect metamorphosed into a pupae and could grow into an adult insect without supplementation with vitamin B_T, and the carnitine deficiency could not be reproduced. Researchers from Belgium and Germany, who were rivals in the study, were also unable to reproduce the carnitine deficiency in insects that had been confirmed so far. Therefore, in cooperation with a rival Belgian researcher, Fraenkel thoroughly investigated the cause of this problem, and found that the new lot of purified casein contained a small amount of carnitine, and also that trace mineral components such as zinc and potassium in addition to carnitine are essential for the transformation of the larvae into pupae. [23] By using highly purified casein that does not contain carnitine and adding sufficient concentrations of zinc and potassium, it became possible to reproduce carnitine deficiency in the yellow mealworm beetle. In other words, when the larvae of the yellow mealworm beetle become pupae, not only carnitine but also folic acid, zinc, potassium, etc. are essential, and even if one of them is deficient, it will not be enough to become pupae. It was also found that even if they become adults, they die early (Fig. 1.3). In the carnitine deficiency of insects studied by Fraenkel et al., it is necessary to consider not only carnitine but also deficiencies in other nutrients. In plant nutrition, there is a law called Liebig's minimum rule (or Dobeneck's tub; see Chap. 14, Fig. 14.65). "Plants need multiple nutrients for growth, but they are affected by the fewest nutrients." Carnitine deficiency is also often deficient in other nutrients, and the clinical symptoms may not be improved by carnitine supplementation alone. For example, vitamin D_3 and zinc deficiency along with carnitine in dialysis patients cause erythropoietin-resistant anemia, and carnitine deficiency is often found in severely mentally and physically handicapped people who are nutritionally controlled for a long time with carnitine-free tube feeding. Even in the nutritional science of insects with short lifespans, it is not easy to determine the significance of a certain nutrient in a pathology in which multiple nutrient deficiencies occur. The pathology of deficiency of multiple nutrients is an issue that should be carefully noted when considering nutrient deficiencies in human clinical nutrition.

If you are interested in the history of Fraenkel's research on carnitine in insect nutrition, which is introduced in this chapter, please refer to an excellent review article written by Wolf. [24] In addition, a review article on vitamin B_T by Fukamachi is also instructive, although it is quite old and written in Japanese. [25]

Fraenkel studied the yellow mealworm beetle as a pest of stored cereals, but now the larvae are sold as a "mealworm" for birds, reptiles, small mammals such as mice and hedgehogs, and as bait for fishing. It is also attracting attention as a source of insect-derived protein. You might find out many Internet sites dealing these topics.

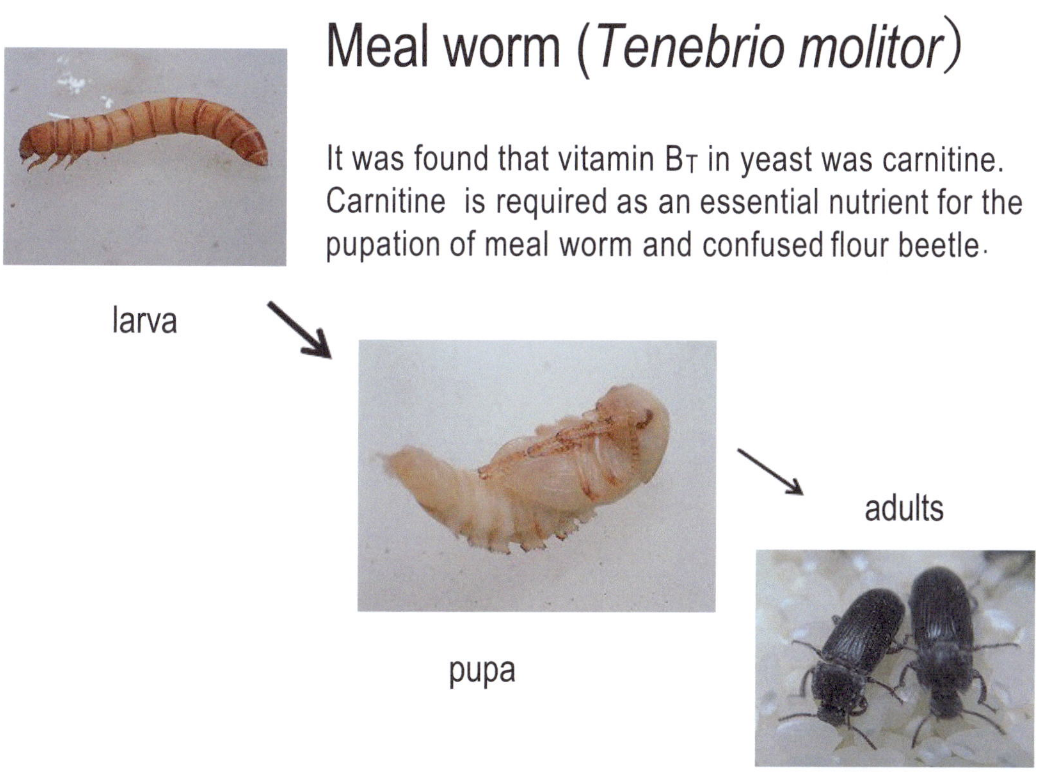

Fig. 1.3 The process of transformation of meal worm (*Tenebrio molitor*) from larva → pupa → adult. Vitamin B_T (L-carnitine) was found to be an essential nutrient for transformation from larva to pupa. (From Sankyo Remake Co., Ltd. Home page, with permission, http://www.sankyoremake.com/syokug_control_s.html)

As described above, Fraenkel's achievements include showing that carnitine is an essential component for metamorphosis and growth, although only in a limited number of insect species, and demonstrating for the first time a pathology called "carnitine deficiency." In addition, his research suggested that carnitine is involved in fatty acid metabolism, which led to subsequent studies on the role of carnitine in fatty acid metabolism by Fritz, Bremer, and others (see "Chapter 2: Establishment of the Carnitine Cycle").

References

1. Gulevich WI, Krimberg R. Zur kenntnis der extraktivstoffe der muskeln. II. Mitteilung. Über das carnitine. Hoppe Seylers Z Physiol Chem. 1905;45:326–30 (Article in German).
2. Alesci S, Manoli I, Costello R, et al. Preface: carnitine: lessons from one hundred years of research. Ann N Y Acad Sci. 2004;1033:ix.
3. Kutscher F: Uber Liebig's fleischextract. Mitteiung I-Z Untersuch Nähr u Genussm. 1905;10: 528–37 (Article in German).
4. Krimberg R: Zur kenntnis der extraktivstoffe der muskeln. IV. Mitteilung. Über das vorkommen des carnosins, carnitins und methylguanidins im fleisch. Z Physiol Chem. 1906;48:412–8 (Article in German).
5. Krimberg R: Zur kenntnis der extraktivstoffe der muskeln. X. Mitteilung. Über die identität des novains mit dem carnitin. Z Physiol Chem. 1908;55:466–80 (Article in German).
6. Tomita M, Sendju Y: Über die oxyaminoverbindungen, welche die biuret-reaktion zeigen. III. Spaltung der γ-amino-β-oxy-buttersäure in die optisch-activen komponenten. Hoppe Seylers Z Physiol Chem. 1927;169:263–77 (Article in German).
7. Fraenkel G, Friedman S. Carnitine. Vitam Horm. 1957;15:73–118. https://doi.org/10.1016/s0083-6729(08)60508-7.
8. Fraenkel G, Blewett M. The basic food requirements of several insects. J Exp Biol. 1943;20:28–34.
9. Fraenkel G, Blewett M. The sterol requirements of several insects. Biochem J. 1943;37(6):692–5. https://doi.org/10.1042/bj0370692.
10. Fraenkel G, Blewett M. The vitamin B-complex requirements of several insects. Biochem J. 1943;37(6):686–92. https://doi.org/10.1042/bj0370686.
11. Fraenkel G, Blewett M. Folic acid in the nutrition of certain insects. Nature. 1946;157:697. https://doi.org/10.1038/157697b0.
12. Fraenkel G, Blewett M. The importance of folic acid and unidentified members of the vitamin B complex in the nutrition of certain insects. Biochem J. 1947;41(3):469–75. https://doi.org/10.1042/bj0410469.
13. Fraenkel G, Blewett M, Coles M. BT, a new vitamin of the B-group and its relation to the folic acid group, and other anti-anaemia factors. Nature. 1948;161:981–3. https://doi.org/10.1038/161981a0.
14. French EW, Fraenkel G. Carnitine (vitamin B_T) as a nutritional requirement for the confused flour beetle. Nature. 1954;173:173. https://doi.org/10.1038/173173a0.
15. Friedman S, Fraenkel GS. Carnitine. In: Sebrell Jr WH, Harris RS, editors. The vitamins, vol. 5. New York: Academic Press; 1972. p. 329–55.
16. Fraenkel G. Effect and distribution of vitamin B_T. Arch Biochem Biophys. 1951;34(2):457–67. https://doi.org/10.1016/0003-9861(51)90026-4.
17. Fraenkel G. Isolation procedures and certain properties of vitamin B_T. Arch Biochem Biophys. 1951;34(2):468–77. https://doi.org/10.1016/0003-9861(51)90027-6.
18. Fraenkel G. The distribution of vitamin B_T(carnitine) throughout the animal kingdom. Arch Biochem Biophys. 1954;50(2):486–95. https://doi.org/10.1016/0003-9861(54)90064-8.
19. Carter HE, Bhattacharyya PK, Weidman KR, et al. Chemical studies on vitamin B_T. Isolation and characterization as carnitine. Arch Biochem Biophys. 1952;38(1):405–16. https://doi.org/10.1016/0003-9861(52)90047-7.
20. Carter HE, Bhattacharyya PK, Weidman KR, et al. The identity of vitamin B_T with carnitine. Arch Biochem Biophys. 1952;35(1):241–2. https://doi.org/10.1016/s0003-9861(52)80069-4.
21. McFarlane JE. Studies on the physiology of carnitine. University of Illinois the Graduate College; 1955.
22. Friedman S, Fraenkel G. Reversible enzymatic acetylation of carnitine. Arch Biochem Biophys. 1955;59(2):491–501. https://doi.org/10.1016/0003-9861(55)90515-4.
23. Fraenkel GS. The effect of zinc and potassium in the nutrition of Tenebrio molitor, with observations on the expression of a carnitine deficiency. J Nutr. 1958;65(3):361–95. https://doi.org/10.1093/jn/65.3.361.
24. Wolf G. The discovery of a vitamin role for carnitine: the first 50 years. J Nutr. 2006;136(8):2131–4. https://doi.org/10.1093/jn/136.8.2131.
25. Fukamachi C. On vitamin B_T. Vitamin. 1954;7:317–22. (Article in Japanese).

2 Establishing the Carnitine Cycle

2.1 What Is the Carnitine Cycle?

The most important biochemical and physiological functions of carnitine are to transport long-chain fatty acids into the mitochondria to promote energy production through β-oxidation of fatty acids, and, conversely, to reduce the toxicity of acyl groups by transporting acyl groups accumulated in the mitochondria for some reason to the outside of the mitochondria. In other words, the most important physiological action is to carry the acyl group of long-chain fatty acids into the mitochondria and contribute to energy production, and to carry the acyl group of harmful acyl-CoA accumulated in the mitochondria out of the mitochondria due to inborn metabolic disorders or other pathological conditions, by functioning as an inward and outward acyl group carrier. As will be described later, CoA, which is very important in energy metabolism, is a coenzyme with a relatively large molecular weight that cannot be synthesized quickly. When acyl groups accumulate in mitochondria, they bind to CoA to form acyl-CoA, resulting in reduction of free CoA contents, which is important for various metabolism. In addition, acyl-CoA is known to inhibit various enzymes and inhibit the metabolic function of mitochondria (details of acyl group and CoA are explained in Chap. 3). If free carnitine is abundant in the mitochondria, these acyl groups can be transferred to carnitine and excreted as acylcarnitine outside of the mitochondria, reducing the toxicity of harmful acyl groups.

By the time the role of the acyl group carrier of carnitine was established, carnitine palmitoyltransferase (CPT) 1 and 2, carnitine transporter (OCTN2), carnitine acylcarnitine translocase (CACT), and other molecules were hypothesized to exist, and their existence was demonstrated by many biochemical studies; thus the current concept of the carnitine cycle (also called carnitine shuttle) has been established (Fig. 2.1).

Initially, only one type of CPT was assumed, but later it was found that there are two types, CPT1 and CPT2. In addition, facts that could not be explained without the existence of CACT came out and its existence was demonstrated. It was also found that the transport of acylcarnitine via CACT was in a 1:1 stoichiometry as an exchange reaction with the transport of free carnitine in the opposite direction to each other. On the other hand, the existence of carnitine transporters has been assumed for a long time, but its existence was actually proven only in the last of the four molecules of the carnitine cycle. The history of the research on CPT 1 and 2, CACT, and OCTN2, which are responsible for these carni-

N. Nakayama, *Carnitine Deficiency and Insufficiency in Medicine*,
https://doi.org/10.1007/978-981-95-7797-2_2

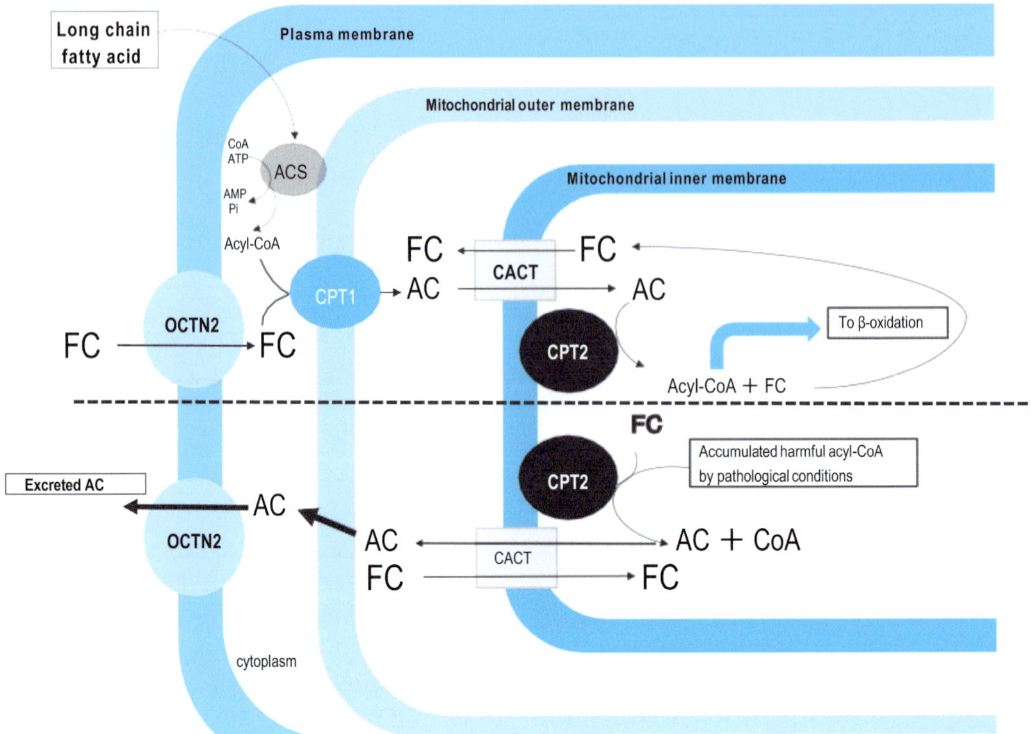

Fig. 2.1 Carnitine cycle. Carnitine cycle is also called carnitine shuttle. *Above the dotted line*: Uptake of long-chain fatty acid into mitochondria. The acyl-group of long-chain fatty acid is converted to acyl-CoA by ACS, and acyl group is transferred to carnitine to form acylcarnitine by CPT1 located in mitochondrial outer membrane, then transferred into mitochondria by CACT located in mitochondrial inner membrane. Acylcarnitine is again backed to acyl-CoA and free carnitine by CPT2 located at inner side of mitochondrial inner membrane, and this acyl-CoA is transferred to β-oxidation and produces energy, and free carnitine back to outside of inner membrane through exchange reaction of CACT with acylcarnitine. *Under the dotted line*: Accumulated harmful acyl-CoA by metabolic abnormality or various pathological conditions is converted to acyl-CoA by CPT2 and transferred to outside of mitochondrial inner membrane by CACT, and finally excreted outside of cells, and the toxicity of acyl-CoA is removed, and released free CoA can be used for various mitochondrial metabolism such as energy metabolism. In this situation, free carnitine and acylcarnitine are exchanged through CACT reaction. *AC* acylcarnitine, *ACS* acyl-CoA synthetase, *CACT* carnitine acylcarnitine translocase, *CoA* coenzyme A, *CPT1* carnitine palmitoyltransferase 1, *CPT2* carnitine palmitoyltransferase 2, *FC* free carnitine, *OCTN2* carnitine transporter

tine cycles, was established by researchers who steadily unraveled the mechanism of these components like a mystery-solving puzzle game, and finally established that carnitine not only carried acyl groups to the inner membrane of the mitochondria, but also carried acyl groups outside the mitochondria. To understand the history is also very useful for understanding the biochemical and physiological actions of carnitine, so I would like to introduce the outline of the research history.

2.2 The Discovery of Carnitine Palmitoyltransferase (CPT)

2.2.1 Research by Fraenkel

Fraenkel, who discovered the nutritional effects of carnitine in insects, reported the presence of transfer enzyme of acetyl group to carnitine (acetyltransferase) in 1955, which performs the following reversible reaction in pigeon and sheep liver extracts [1].

Acetylcarnitine + CoA ⇄ Acetyl CoA + Carnitine

McFarlane in Fraenkel's group divided larvae of the normally bred yellow mealworm beetle into two groups, one group fed on a synthetic diet without vitamin B_T, and the other group fed without any feed (starvation group), and measured and compared the carbohydrate, protein, and fat content of the two groups just before death. He found that the fat content was significantly reduced in the starvation group, whereas the fat content was retained in the synthetic diet group without vitamin B_T (carnitine) [2]. This means that without carnitine, the fat stored in the larvae could not be utilized. This finding was very important because it suggested that carnitine is not only a nutrient for certain insects, but also participates in fatty acid metabolism. However, perhaps because Fraenkel was a scientist specialized to entomology, he did not change his stance to focus on insects as a research subject, and he continued to study the effects of carnitine derivatives such as acetylcarnitine in insects, and he did not further pursue the role of carnitine in fatty acid metabolism in higher animals. Subsequently, many research groups participated in the study of the role of carnitine in fatty acid metabolism. Among them, the research by Fritz et al. and by Bremer et al. contributed greatly to understand the role of carnitine in fatty acid metabolism.

2.2.2 Studies by Fritz et al.

Around the same time as the discovery of physiological function of carnitine in insects by Fraenkel et al., Lundsgaard in the University of Copenhagen found that there was a factor in fresh blood that increased oxygen consumption in feline liver perfused preparation [3]. Subsequent attempts by Lundsgaard to identify this factor were unsuccessful, but he also found that there was a factor that maximized oxygen consumption in liver perfusion preparation, not only in fresh blood but also in the fluid that perfused the hind limb muscles. Fritz (later the University of Michigan and the University of Toronto), who happened to be studying in the Lundsgaard laboratory, worked on investigating the substance that caused this phenomenon after returning to the United States. In order to simplify the experimental system, Fritz added muscle extracts to rat liver slices and rat liver homogenates instead of the liver perfusion preparation, and he investigated the effects of the muscle extracts on oxygen consumption, oxidation of ^{14}C-palmitic acid, and ketone production [4]. The addition of muscle extract increased the oxygen consumption in liver slices and homogenates, and also promoted the oxidation of ^{14}C-palmitic acid and ketone production. Since the activity of this muscle extract remained even after the protein was precipitated and removed with ferric sulfate and barium carbonate, the active substance was considered to be a low-molecular substance rather than a high-molecular substance such as a protein. First, he investigated the effects of 20 low-molecular substances such as methionine, alanine, carnosine, anserine, betaine, taurine, creatine, and urea, which were known to exist in muscles at the time, on palmitic acid oxidation, but no increase in oxygen consumption or ketone production was observed even if the concentration and combination were changed. Then, he decided to investigate carnitine, which has been known to be present in high concentrations in muscles for a long time. The addition of racemic DL-carnitine or synthetic L-carnitine along with ^{14}C-palmitic acid to rat liver slices and rat liver homogenates promoted ketone formation from palmitic acid. Through this study, it was discovered that L-carnitine has the effect of promoting the oxidation of long-chain fatty acids [4]. Subsequently, Fritz investigated the effect of carnitine and acetylcarnitine on long-chain fatty acid oxidation in skeletal muscles and myocardium, which do not have a ketone synthesis system unlike the liver. Palmitic acid oxidation in the mitochondrial fraction of skeletal muscles was suppressed by increasing the ratio of albumin to palmitic acid, but when carnitine was added to it, palmitic acid was oxidized even if albumin was present at high concentrations [5]. When fatty acid oxidation in rat myocardial mitochondria was examined, it was found that by increasing the concentration of carnitine ($10^{-6} \sim 10^{-2}$ M), the oxidation of palmitic acid reached its peak at 10^{-3} M, and when the length of the carbon chain of fatty acids was changed from C_2 to C_{18}, the fatty acid oxidation promotion effect of carnitine was remarkable in long-chain fatty acids of C_{12} or more [6]. The reason why carnitine does not promote the oxidation of short-chain fatty acids is thought to be that short-chain fatty acids are transported into the mitochondria bypassing the fatty acid transport system (carnitine cycle) that requires carnitine. It has been found that the hydroxyl group in carnitine is important for the effect of promoting palmitic acid oxidation in rat myocardium [7], and that carnitine itself is not degraded, so it acts as a catalyst in fatty acid oxidation [7]. The fatty acid oxidation promotion effect of carnitine was lost when the mitochondrial fraction was solubilized with deoxycholic acid, etc [6].

In summary, Fritz found that [8]:

1. Carnitine palmitoyltransferase, which transfers a palmitoyl group to carnitine, is present in mitochondria and in cytoplasm, which are outside of mitochondria.
2. Palmitoyl carnitine is oxidized more efficiently than palmitoyl-CoA (this indicates that carnitine acts promotively in the fatty acid oxidation system in mitochondria before the formation of palmitoyl-CoA).
3. Carnitine increases the utilization of acyl-CoA without the need for ATP.
4. The long-chain fatty acid oxidation promoting effect of carnitine is lost when the complex mitochondrial system is dissolved.

Although the whole picture has not yet been drawn, it is thought that carnitine, which is contained in high concentrations of 1–2 mg/g dry weight in muscles, may have a role in transporting fatty acids into mitochondria.

Fritz then studied the properties of carnitine acetyltransferase (CAT) and carnitine palmitoyltransferase (CPT) in detail. First, CAT was partially purified from porcine myocardium and the substrate specificity was investigated in detail. CAT reacted well with acetyl-CoA, propionyl-CoA, and butyryl-CoA, but the reaction rate was slow with acyl-CoA with a long-chain length such as decanoyl-CoA, and it hardly reacted with palmitoyl-CoA. In other words, CAT is an enzyme that reacts with short-chain to medium-chain acyl-CoA, and CPT which reacts well with long-chain acyl-CoA as a substrate is a different enzyme [9]. Next, he examined CPT in detail [10]. In addition to substrates and enzymes, ATP and CoA are required for the reaction of palmitic acid + carnitine → palmitoyl carnitine in rat myocardium. On the other hand, ATP and CoA are not required for the reaction of palmitoyl-CoA + carnitine → palmitoyl carnitine. The addition of palmitoyl carnitine to myocardial mitochondria promoted respiration, and the effect was much greater than that of palmitoyl-CoA. These results suggest that palmitoyl carnitine is more suitable for fatty acid oxidation in mitochondria in terms of the use of acyl groups. Based on these studies, Fritz proposed the hypothesis that carnitine binds to long-chain fatty acids to form acylcarnitine, and that it has the role of a carrier that brings the acyl group to the fatty acid oxidation site and passes the acyl group to free CoA, as shown in Fig. 2.2. At this time, the existence of carnitine acylcarnitine translocase (CACT) was not yet known, but in 1963, the role of carnitine in long-chain fatty acid transport and the basic image of the carnitine cycle had already been proposed.

2.2.3 Studies by Bremer

Bremer (University of Oslo, Norway) was doing similar research around the same time as Fritz. He showed that mitochondria isolated from various tissues of rats had the activity of acetylation of carnitine or deacetylation of acetylcarnitine and suggested that carnitine may act as a carrier of active acetic acid [11]. Since all organisms have carnitine in their cells, he hypothesized that carnitine may play a role in metabolism as a carrier of acyl groups as well as acetyl groups. Therefore, he synthesized acylcarnitines with different chain lengths, such as palmitoyl carnitine, octanoylcarnitine, and butyrylcarnitine, and examined them. Incubation

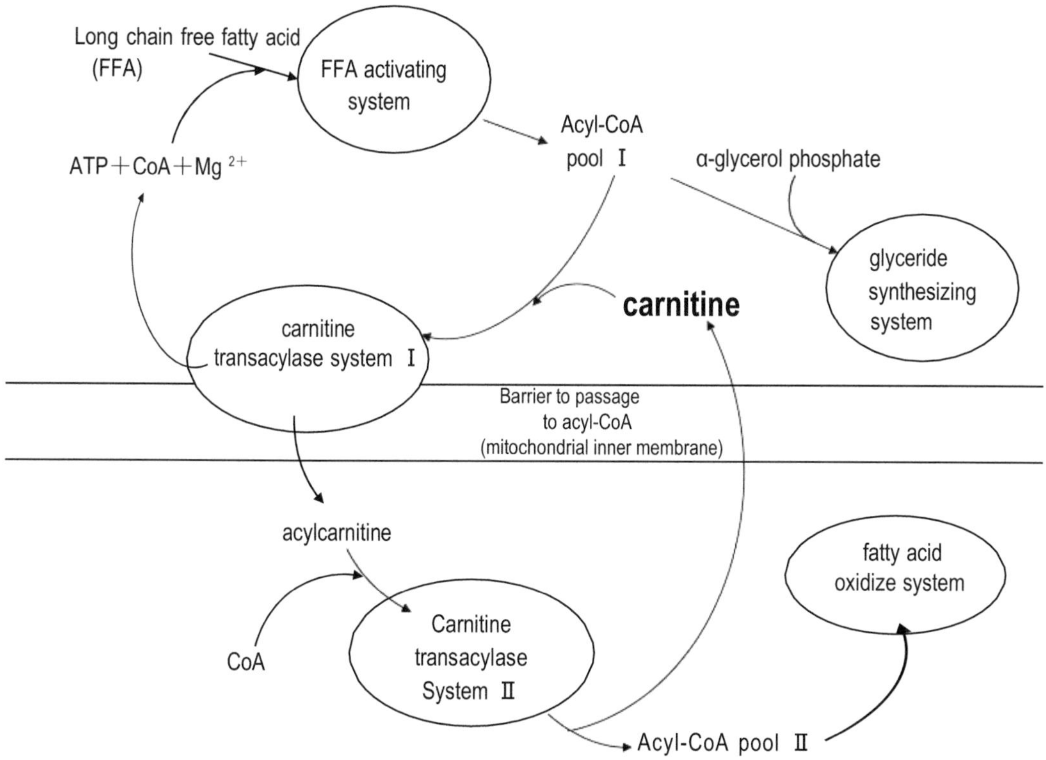

Fig. 2.2 Role of carnitine in long-chain fatty acid oxidation proposed by Fritz. Hypothetical site of action of carnitine on fatty acid metabolism. This scheme shows functional compartment of acyl-CoA pools postulated to account for carnitine stimulation of fatty acid oxidation. (From Ref. [10])

of acylcarnitine with mitochondria increased the formation of acetoacetate, but incubation in the presence of trace amounts of succinate decreased acetoacetate formation and increased oxygen uptake. These reactions were also observed to occur in the L-body of carnitine [11, 12]. When mitochondria and palmitoyl carnitine or palmitic acid + L-carnitine are incubated, oxygen uptake is significantly increased, and in the case of synthesized DL-palmitoyl carnitine, only 40–45% of the expected value of oxygen uptake is observed. It was thought that palmitoyl carnitine is an intermediate when L-carnitine receives a palmitoyl group and becomes palmitoyl carnitine, and then it is transported into mitochondria to cause fatty acid oxidation. He proposed the role of carnitine as an acyl carrier a little earlier than Fritz [13].

2.2.4 Presence of Two Types of CPTs

According to these studies by Fritz, Bremer, and others, it was suggested that there may be two types of CPTs: CPTo ("o" for outer), CPT which is easily solubilized and perhaps located in the outer mitochondrial membrane or in the cytoplasmic side, and CPTi ("I" for inner), CPT which is strongly bound to the matrix side of inner mitochondrial membrane [14]. Then the question of whether these are the same enzyme or different enzymes arose, and numerous studies were reported [14]. There was a temporary confusion because various studies reported various CPT activity, stability, and strength of binding to mitochondrial membranes, but finally it became clear that these discrepancies were due to differences in experimental condition and contamination with peroxisomal CPT. Comparison of the properties of purified CPT, studies using antibodies against them [15–17], and fractionation using the surfactant digitonin strongly suggested that there are two types of CPT with different localization and roles [18–21]. McGarry et al. showed that cytoplasmic CPT1 is loosely bound to the outer membrane of the mitochondria and is strongly inhibited by malonyl-CoA, but CPT2 in the mitochondria is not inhibited by malonyl-CoA. It has been established that there are two types of CPTs: CPT1, which is bound to the outer mitochondrial membrane, and CPT2, which is bound to the inner mitochondrial membrane [14, 22, 23]. For more information on the history of research about the existence of these two CPTs, please refer to the review article by Bieber [14].

2.3 Discovery of Carnitine-Acylcarnitine Translocase (CACT)

Mainly according to the research of Fritz and Bremer, it is thought that carnitine plays a role in transporting the acyl group of long-chain fatty acids into mitochondria, and the acyl group of the long-chain fatty acid is activated as acyl-CoA, and it is transported into the mitochondria as acylcarnitine by CPT1 outside of the inner mitochondrial membrane. It was assumed that CPT2 inside the inner mitochondrial membrane would again return the acylcarnitine to acyl-CoA for β-oxidation.

However, it has become clear that these assumptions alone do not fit the experimental facts. At that time, many researchers were studying the material permeability of the inner mitochondrial membrane. When the permeability of carnitine and acylcarnitine through the inner membrane was examined using ^{3}H-water, which can freely pass through into the mitochondrial matrix, sucrose, which can pass through the outer membrane of the mitochondria but not the inner membrane, and dextran and blue dextran, which cannot enter the mitochondria, the data from all researchers showed that carnitine and acylcarnitine could not pass through the mitochondrial inner membrane as well as CoA and acyl CoA [19, 24, 25].

2.3.1 Studies by Pande et al.

Pande et al. examined pyruvate-dependent respiration in the presence of malonic acid in isolated rat myocardial mitochondria and found that the addition of L-carnitine enhanced oxygen consumption and produced almost the equal amount of acetylcarnitine, while the addition of decanoyl-CoA and palmitoyl-CoA inhibited respiration. When incubated with carnitine or acylcarnitine, ^{14}C-carnitine comes out from mitochondria (efflux), and the efflux is saturated with low concentrations of carnitine and acylcarnitine, and the effect of acylcarnitine is higher in the L-form. Since it is temperature-dependent, they thought that there is a carnitine acylcarnitine translocase (CACT) that transports carnitine and acylcarnitine in the opposite direction through the inner mitochondrial membrane as exchange reaction, and that it is responsible for the transport of acylcarnitine and carnitine (Fig. 2.3). In other words, L-carnitine and acylcarnitine cannot pass through the inner mitochondrial membrane as it is, same as CoA and acyl-CoA, but equimolar carnitine and acylcarnitine pass through via CACT in an exchange reaction. As a result, it was thought that acylcarnitine produced by CPT outside the inner membrane would be transported to the CPT present inside the inner membrane, and the β-oxidation of the acyl group of the long-chain fatty acid would occur [26].

2.3.2 Studies by Ramsay et al.

At almost the same time, Ramsay et al. used bovine cardiac mitochondria to study how carnitine is involved in the transport of acyl groups despite the inner mitochondrial membrane being a barrier for carnitine [27]. Since the mitochondrial inner membrane is a barrier that neither carnitine nor CoA can penetrate, they hypothesized that the CPT outside the inner membrane utilizes CoA added to the mitochondrial suspension,

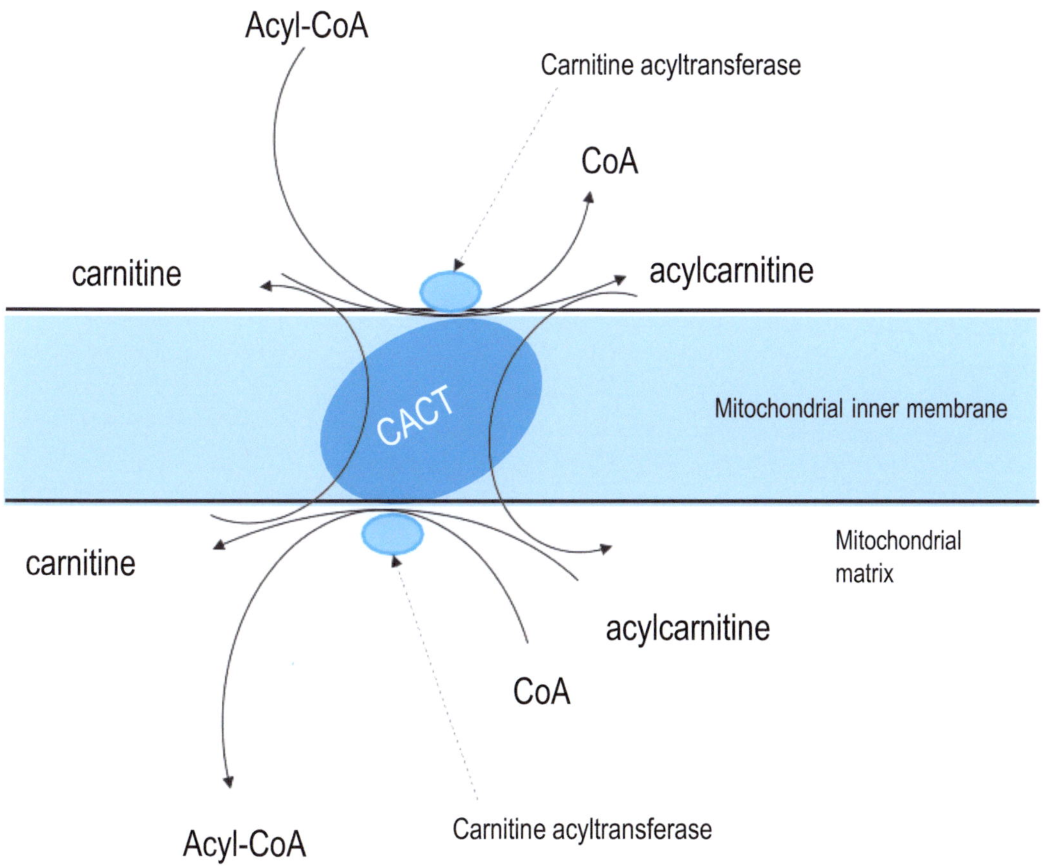

Fig. 2.3 Role of carnitine and transport of acyl-group through CACT proposed by Pande. A scheme for the L-carnitine-dependent transport of fatty acyl groups across mitochondrial inner membrane. Although only the inward transport is depicted above, the reactions involved, being readily reversible, would facilitate outward transport as well. *CACT* carnitine acylcarnitine translocase. (From Ref. [26])

while the CPT inside the inner membrane utilizes CoA from within the mitochondria. They thought that this inner CPT may be responsible for the transport of acyl groups through the inner membrane between the outer carnitine and inner matrix CoA. There are 2–4 nmol/mg protein of total carnitine in the bovine myocardial mitochondria. Most of it was acylcarnitine, but when the mitochondrial suspension was incubated at 0°C, the ratio of free carnitine increased over time. Next, ^{14}C-carnitine was added to the mitochondrial suspension, and the migration of carnitine inside and outside the mitochondria was measured, and it was found that the ratio of ^{14}C-carnitine export to import was 1:1. Furthermore, when the inner carnitine of mitochondria was labeled with ^{3}H and the outer carnitine was labeled with ^{14}C, the exchange reaction was confirmed to occur in a 1:1 ratio. Based on these experiments, Ramsay et al. also assumed that there is a carrier in the inner mitochondrial membrane that exchanges carnitine and acylcarnitine in a 1:1 ratio, as shown in Fig. 2.4 [27].

After these findings, Pande et al. also reported several things on CACT as below [28, 29]:

- A mercury compound mersalyl is an inhibitor of CACT activity.
- CACT reactions do not require energy.
- CACT is used to incorporate carnitine into the mitochondria, or to modulate the rapid exchange of carnitine and acylcarnitine when

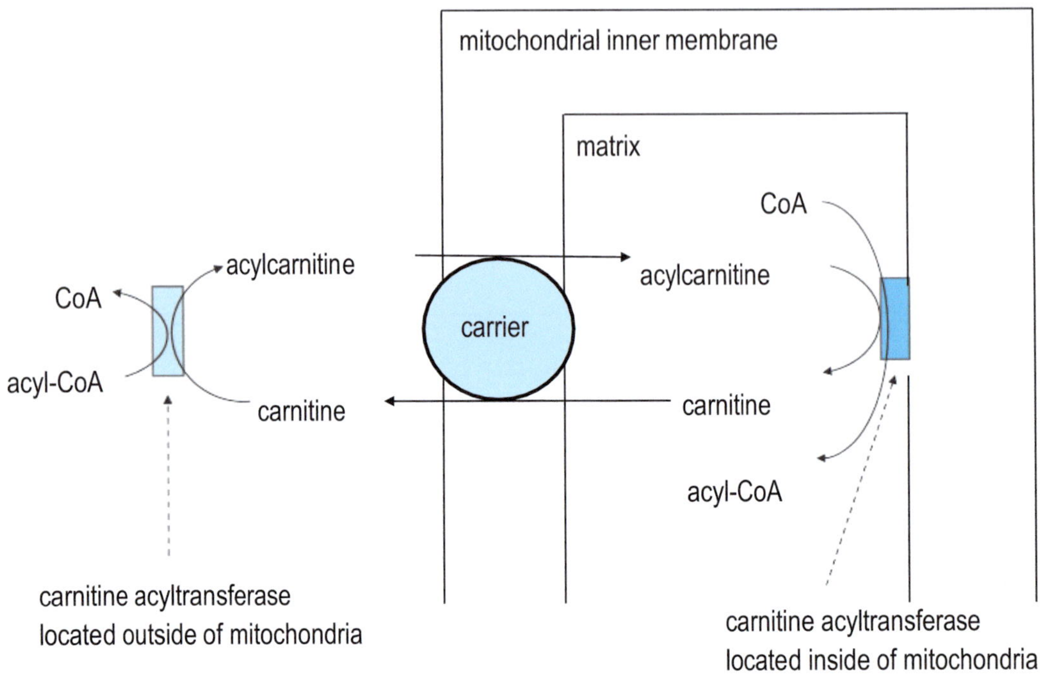

Fig. 2.4 Postulated acylcarnitine-carnitine exchange carrier in the mitochondrial inner membrane proposed by Ramsay. (From Ref. [27])

there is a change in tissue carnitine concentrations, although the reaction rate is very slow.

2.4 Discovery of Carnitine Transporter

As shown so far, the role of carnitine is transporting acyl groups of long-chain fatty acids into mitochondria to promote β-oxidation, and it was established the responsible molecules that form the carnitine cycle were CPT1, CPT2, and CACT. It can be said that the outline of carnitine cycle was elucidated until the first half of the 1970s, except for the carnitine transporter on the cell membrane, which is described below. Although the carnitine transporter molecule was not identified until the 1990s, the existence of the carnitine transporter was assumed from earlier than that based on the following facts.

2.4.1 Assumption of Carnitine Transporter

Since carnitine is present in skeletal and cardiac muscles at concentrations tens to hundreds of times higher than blood levels, and in liver, kidney, and other tissues at concentrations several times higher than blood levels, it was early thought that the cell membranes of these tissues had a mechanism for actively taking up carnitine [30]. In addition, studies on tissue distribution and renal excretion of carnitine by injecting ^{14}C- and ^{3}H-labeled carnitine into rats and dogs also reported that carnitine was distributed at high concentrations in tissues such as skeletal and cardiac muscles [31–34]. Also, studies on carnitine biosynthesis by Broquist et al. have ruled out the possibility that carnitine is biosynthesized from precursors to carnitine in skeletal muscles and is present in high concentrations, since carnitine biosynthesis in skeletal muscles is limited to the intermediate, and the enzyme in the last step of the biosynthesis is present only in the liver and kidney, thus denying the possibility of high concentrations of carnitine biosynthesized from precursors in skeletal muscles [35]. From these, it was assumed that carnitine transporters were present in the cell membranes of each tissue. Therefore, carnitine uptake has been actively studied in tissues and cells in myocardium [36–38], skeletal muscles [39, 40], liver [41], kidney [42], and small intestine [43].

2.4.2 Muscle Carnitine Deficiency (MCD)

In addition, reports of clinical cases of carnitine deficiency have also triggered these carnitine transporter studies. In 1973, Engel et al. first reported the existence of patients with extremely low levels of carnitine in skeletal muscles and with fat-accumulating myopathy symptoms [44]. The patient's muscles contained only 20% carnitine of the control, and the fatty acid oxidation capacity of palmitic acid and oleic acid in the muscle homogenate was extremely low compared to the control, and the supplementation of carnitine restored it to normal levels. By this study, a new type of disease called muscle carnitine deficiency (MCD) was proposed for the first time. Later, many muscle carnitine deficiencies have been reported since this Engel's report [45–51].

2.4.3 Systemic Carnitine Deficiency (SCD)

Subsequently, systemic carnitine deficiency (SCD), in which carnitine is extremely deficient not only in muscles but also in the blood and liver, has also been reported. Chapoy et al. examined a 3.5-year-old boy with frequent somnolence and hypoketotic hypoglycemic attacks, as well as hepatomegaly and cardiac hypertrophy, and found that blood carnitine concentrations were extremely low at 2–5 μmol/L (control: 28–44 μmol/L). Carnitine in muscles and liver were also less than 5% of control. After 6 months of oral carnitine supplementation (initially 4 g/day of DL-carnitine divided into three times a day, and later 2 g/day of L-carnitine orally three times a day, when the L-carnitine became available), hepatomegaly and cardiac hypertrophy were improved, and Reye-like syndrome, which had recurred repeatedly, was treatable [52]. McGarry and colleagues, who had originally been studying carnitine, took up the report of Chapoy and colleagues and discussed the role of carnitine in vivo and the reasons for the intermittent onset of encephalopathy, hypoglycemic attacks, and other conditions [53]. According to his discussion, when carnitine is deficient the bypass mechanism of energy production such as gluconeogenesis and ketone production does not work, and it is thought that these defects in energy production become apparent especially during starvation and fasting. Glasgow et al. also reported on a girl at 11 months who developed Reye-like syndrome after an upper respiratory tract infection and rhinorrhea [54]. The affected child continued to have Reye-like syndrome, and muscle weakness was observed at the age of 5 years; and when examined in detail, the serum carnitine concentration was 3.3 nmol/mL and the carnitine content in the liver and muscles was extremely low compared to normal values. Tripp et al. also reported familial SCD in five cases of one male and four females from the same family [55]. The first child died of respiratory arrest at the age of 15 months. The second child also died of respiratory infection at the age of 22 months. The third child was 11 months old and had cardiac hypertrophy, cardiac contractile failure, anemia, etc., but no hepatomegaly. The fourth child was a normal child. The fifth child, at the age of 26 months, developed an upper respiratory tract infection and died of vomiting and hyperventilation. The plasma concentrations of carnitine in the third and fifth children were in the single digits, and the carnitine content in skeletal muscle, myocardium, and liver was also extremely low compared to normal controls. When L-carnitine supplementation therapy was started in the third child (initially 0.66 g/day, later increased to 1.0 g/day), the concentration of carnitine in the tissues did not increase much, but the heart failure score improved from New York Heart Association (NYHA) II or III to IIa. Until then, cardiomyopathy was thought to be an incurable disease, but it was reported as "treatable cardiomyopathy." Similarly, Waber et al. reported a case of "treatable cardiomyopathy" in which carnitine supplementation therapy was given to boys with SCD aged 3.5 years who had cardiomyopathy, cardiac hypertrophy and congestive heart failure from 5.5 to 6.5 years of age (174 mg/kg/day), and cardiac symptoms improved and myasthenia normalized [56]. Subsequently, Treem et al. investigated the ability to uptake carnitine in the kidneys, muscles, and fibroblasts of 3.5-year-old girls with primary SCD. It was shown that it was extremely low compared to the control [57]. Stanley et al. in the same group reported 15 cases of SCD together [58]. The clinical features include repeated fasting hypoglycemic attacks by about 2 years of age, progressive cardiomyopathy by about 3 years of age, and sometimes with myasthenia or sometimes without myasthenia. The ability of patients to uptake carnitine in skin fibroblasts and leukocytes was extremely impaired, and that of the parents was intermediate between the patient and normal controls, suggesting that the inheritance form of this disease is autosomal latent (recessive) inheritance. Tein et al. reported four similar cases of SCD [59].

Engel et al., who reported MCD in 1973, also reported four cases of primary SCD in 1981 [60] and postulated that the cause of MCD was (1) decreased activity of carnitine transporters in muscle cell membranes and (2) decreased reabsorption of carnitine in the kidney. However, Rebouche et al. in the same group examined the uptake capacity of carnitine using cultured muscle cells and fibrocytes of the patients, and surprisingly they found that it was only slightly reduced and almost the same as the control [61]. Rebouche and colleagues further hypothesized that the in vitro uptake activity of carnitine was not different from that of controls, but in vivo uptake might be different. Therefore, Rebouche et al. further injected intravenously ^{3}H-labeled carnitine into the patients with MCD and their SCD patients, analyzed using a three-compartment model of extracellular fluid, muscle, and other tissues, and compared them with normal controls [62]. They concluded, however, that the transport of carnitine to each tissue in SCD was decreased, but that the decrease was dependent on a decrease in blood concentration, a result that in the case of MCD could be due to a defect in the carnitine transport process. Karpati et al. in the same group also reported on an 11-year-old boy with SCD with extremely low concentrations of carnitine in plasma, muscle, and liver [63]. The child grew up to age 11 years without cardiomyopathy, suggesting that it may have been a different disease type from the SCD in Chapoy's report. Treem et al. also stated, in the introduction of the paper, that the cases identified as primary SCD by Engel et al. and Rebouche et al. may have been secondary carnitine deficiency due to medium-chain acyl-CoA dehydrogenase deficiency (MCADD) [57].

Although there was a period of some confusion, a clinical disease type called SCD or primary carnitine deficiency (PCD), which is thought to be a genetic abnormality of carnitine transporters, was established. Clinical characteristics include the development of cardiomyopathy with cardiac hypertrophy by the age of 3 years, and although the concentration of carnitine in serum, muscle, and liver is extremely low, the cardiomyopathy responds well to high-dose carnitine supplementation therapy and can be treated, and the ability of fibroblasts to uptake carnitine in the skin of these patients is extremely low.

2.4.4 Discovery of Juvenile Visceral Steatosis (JVS) Mice and Subsequent Research

Another important discovery and series of studies from Japan led to the discovery of the carnitine transporter. In 1988, Koizumi et al. at the Animal Laboratory of Kanazawa University School of Medicine discovered and reported that C3H-H-2°

strain mice had visceral fat deposits in the liver and abdominal cavity, with symptoms such as growth retardation, abdominal distention, hyperammonemia, and hypoglycemia similar to Reye's syndrome [64]. These mice were later named juvenile visceral steatosis mice (JVS mice). Kuwajima et al. thought that these mice resembled human SCD and measured carnitine levels in blood, liver, and skeletal muscles, and found that the carnitine content was extremely low compared to that of control mice [65]. Subsequently, carnitine reuptake experiments using kidney slices from homozygous JVS mice showed that it was extremely reduced compared to that of control mice, while that of heterozygous mice was intermediate between homozygous and control values [66], and that the carnitine uptake capacity of fibroblasts was only 18% of that of control mice [67]. Decreased carnitine transport activity has also been reported in liver cells [68]. In addition, it was found that mitochondrial abnormalities were observed due to the deposition of lipid droplets not only in the liver and abdominal cavity, but also in the myocardium and skeletal muscles [69], and cardiomyopathy with cardiac hypertrophy was observed [70], and carnitine supplementation increased the short-chain acylcarnitine in the heart and reduced cardiac hypertrophy. From these results, it was considered that JVS mice is a good animal model for human SCD [71].

Hashimoto et al. generated B6-jvs mice that retained the autosomal recessive (latent) jvs gene by backcrossing the original JVS mice to C57BL/6 J mice [72]. Fibroblasts were isolated and cultured from homozygotes, heterozygotes, and control mouse embryos of B6-jvs mice, and it was found that the carnitine uptake activity was significantly reduced in homozygotes compared to control mice, and the values were intermediate between the two in the heterozygous. This study revealed that the high-affinity carnitine transporter responsible for carnitine uptake is Na^+ and temperature dependent and has a higher affinity for the L-form than for the D-form.

It has also been found that the locus of the jvs gene (probably the carnitine transporter gene) that causes SCD in JVS mice is located within the genetic distance no greater than 1.6 cM of chromosome 11 [73, 74].

2.4.5 Discovery of Carnitine Transporter (OCTN2)

Tamai et al. found and reported the characteristics of an organic cation/carnitine transporter (OCTN2) with a nucleotide sequence approximately 76% homologous to OCTN1, an organic cation transporter that had already been cloned from a cDNA library in a human kidney [75]. This cDNA encodes 557 amino acids and was highly expressed in the kidneys, skeletal muscles, heart, and placenta of human adults in northern blot. In addition, when OCTN2 were expressed in HEK293 cells and L-^{3}H-carnitine uptake was examined, Na^+-dependent high-affinity uptake activity was observed. Prior to the cloning of OCTN2, Tamai et al. cloned an organic cationic transporter such as tetraethylammonium (TEA) and named it organic cation transporter with nucleotide binding site/domain 1 (OCTN1). This is because there was an ATP-binding sequence motif in the amino acid sequence. After that, the ATP dependence could not be confirmed, and the "N" in OCTN became meaningless, but it is still called OCTN. However, some clinicians have since interpreted N as "novel" and described it as "novel organic cation transporter," as the full spelling of OCTN. Currently, three genes are known as the carnitine/organic cation transporter (OCTN) family—OCTN1, OCTN2, and OCTN3—and these genes are classified as SLC22A4 (human), SLC22A5 (human), and Slc22a21 (mouse) [76]. OCTN1 can transport D- and L-carnitine, but its affinity is low, and OCTN2 is thought to function for carnitine uptake into skeletal and cardiac muscles and for reabsorption of carnitine in renal tubules. Although the transport of carnitine, an amphiphilic compound, is Na^+-dependent, it is also known to have the effect of transporting various organic cations such as the antihistamine pyrilamine and the calcium antagonist verapamil in an Na^+-independent manner in addition to carnitine [77].

2.4.6 Mutation in Human and Mice OCTN2

Furthermore, Nezu et al., who were in the same group, examined the slc22a5 gene in JVS mice based on human OCTN2 information and found that there was a missense mutation [78]. In addition, they also analyzed three families of human SCD found in the Department of Pediatrics at Akita University, Kurume University, and Tohoku University, and found four mutations. In a homozygous child with SCD in one family, 113 bases including the initiation codon were deleted. The offspring of another family was a complex heterozygote of two mutations, a flame shift and an immature stop codon. The third family was a homozygote that caused an immature stop codon due to a splice site mutation. This study proved that OCTN2 is the causative gene for SCD in humans and mice.

Koizumi et al. at Akita University (later Kyoto University) conducted a further genetic epidemiological study of OCTN2 mutations [79]. First, as a screening, free carnitine concentration in the blood of 973 patients in Akita Prefecture was measured, and it was found that 14 patients were below the 5th percentile. Then they examined and analyzed the nucleotide sequences of these 14 patients who were below the 5th percentile, of 22 patients who were below the 5th percentile in the first screening but normal in the second screening, and of 69 patients who were normal in both the first and second screenings. As a result, gene mutations were found in 9 out of 14 patients. When these mutated OCTN2 genes were expressed in HEK293 cells and the carnitine uptake activity was examined, the carnitine uptake activity was significantly reduced in the two mutations of Ser467Cys and Trp283Cys compared to normal OCTN2. This study revealed that the prevalence of heterozygotes was 1.01%, and the prevalence of homozygotes, primary systemic carnitine deficiency (MIM 212140), was estimated at the rate of 1 per 40,000 live births, a fairly high incidence. Primary systemic carnitine deficiency responds well to high-dose carnitine supplementation therapy if detected early after birth and the patient can live a normal life. However, if detected late, it causes hypoglycemic attacks, recurrent Reye-like syndrome, cardiomyopathy, and sudden death by about 3 years of age. In addition, since the frequency of cardiac hypertrophy increases in carriers (heterozygotes) when they become adults and the elderly, it is expected that research on OCTN2 mutations in patients with heart failure and decreased cardiac function will be an important issue in the future.

In his review article on genetic epidemiology (modified from the 11th Japan Society of Hygiene Award Lecture, written in Japanese), Koizumi reflected on why these series of studies were successful [80]. According to this review, in general, autosomal latent (recessive) genetic disorders require many family lineages because carriers cannot be distinguished from wild types, but sometimes the traits can be detected even in hetero carriers (called "carrier state"). In the case of carnitine transporters (OCTN2), blood carnitine concentrations in hetero carriers of SCD ranged from 6 to 6 in normal subjects. Although the value was as low as 70%, it was found that a large amount of it was released in the urine [81]. In other words, in carriers (heterozygotes), the carnitine reabsorption threshold in the kidneys was reduced by about half, and it was thought that a large amount of carnitine was outflowed into the urine. Since the uptake activity of carriers who are considered to be heterozygotes in HEK293 cells is intermediate between normal and homozygotes, it was possible to predict that they were heterozygotes even from the analysis of a single family lineage. The existence of such a gene dosage effect was a major clue.

In a review article commemorating the Japan Pharmaceutical Society of Japan Award in 2002, Prof. Akira Tsuji in Kanazawa University, who had been researching drug transporters for many years, described the circumstances that led to the discovery of OCTN2 in collaboration with researchers of human SCD and JVS mice as following [77]:

> The existence of the carnitine transporter had been speculated for a long time, but its actual existence was unknown, and it was said that it is a "phantom transporter". It could be said that the discovery of JVS mice at the same experimental animal facility

at Kanazawa University, the fact that the JVS mice exhibited clinical symptoms similar to SCD, and the joint research with pediatricians and genetic epidemiologists who were conducting clinical research on human SCD led to major results.

Thus, an interdisciplinary collaboration between transporter researchers in the field of pharmacy, pediatricians specializing in inherited metabolic diseases, and genetic epidemiologists in Japan proved that OCTN2 is the causative gene of SCD in humans and JVS mice. Japanese research groups contributed greatly to the discovery of the carnitine transporter (OCTN2) and its significance in pathological conditions.

In the 2016 report, 138 types of genetic mutations were reported for OCTN2, including 97 missense mutations, 29 mutations causing premature insertion of a stop codon as a result of direct mutation events (14) or result of a frameshift (15), 7 mutations that predicted to cause abnormal splicing, 2 in-frame deletions of single amino acids, and 3 large gene deletions one of which affected the whole *SLC22A5* gene, and it was thought that the clinical phenotype is determined by the effect of these mutations on transporter activity [82]. However, it has also been reported that two homozygous patients with the same mutation (R399Q) from the same family have completely different clinical phenotypes [83]. In this way, even if the same gene mutation occurs, the phenotype may be different, such as undiagnosed cases or severe cases, and the actual clinical picture is not simple, and there are many parts where the pathophysiology is not elucidated. Subsequently, Rose et al. from the same Longo group examined 14 patients with symptoms of cardiomyopathy (7 cases), hypoglycemic attacks (4 cases), and Reye-like syndrome (3 cases) with OCTN2 gene mutations and 14 asymptomatic women with OCTN2 gene mutations between the ages of 5 months and 6 years. The mutated OCTN2 gene was expressed in Chinese hamster ovary cells and carnitine uptake activity was compared [84] (Fig. 2.5). As a result, the carnitine uptake activity of symptomatic and asymptomatic women with these missense mutations was significantly reduced compared to normal OCTN2, but there was no significant difference between symptomatic patients and asymptomatic women. In other words, although the cause of SCD is a decrease in carnitine uptake activity due to mutations in the OCTN2 gene, it is thought that environmental factors and other factors affect the onset of SCD, rather than determining the presence or absence of symptoms only by the uptake activity.

In any case, carnitine actively taken up by OCTN2 at the plasma membrane thus receives acyl groups of long-chain fatty acids by the action of CPT1 to become acylcarnitine, enters the mitochondrial inner membrane by exchange transport of free carnitine and acylcarnitine through CACT, and is reassociated by CPT2 with acyl-CoA to be supplied for β-oxidation. The barrier to long-chain acyl groups, the mitochondrial inner membrane, is that these molecular devices form a carnitine cycle (carnitine shuttle) that allows long-chain acyl groups to pass inward or outward across the mitochondrial inner membrane.

If there is a congenital genetic abnormality in the four molecules that make up the carnitine cycle, the carnitine cycle cannot run smoothly, resulting in serious conditions. These diseases are grouped into the category of carnitine cycle abnormalities among inherited metabolic diseases as “carnitine cycle disorders” (“designated intractable diseases No. 316” in Japan, Intractable Disease Information Center, https://www.nanbyou.or.jp/entry/5429).

Thus, transporting long-chain acyl groups of long-chain fatty acids into the mitochondria and using them for β-oxidation was immediately accepted. On the other hand, when a short-chain acyl group such as a propionyl group accumulates in the mitochondria due to organic acidemia, or when a long-chain acyl group accumulates in the mitochondria in various pathologies (carnitine insufficiency), carnitine and the carnitine cycle also play a role in excreting these from the mitochondria. The function of carnitine cycle is not only to carry the acyl groups of long-chain fatty acids into the mitochondria and contribute to energy production, but also to excrete acyl groups outside the mitochondria as a reverse

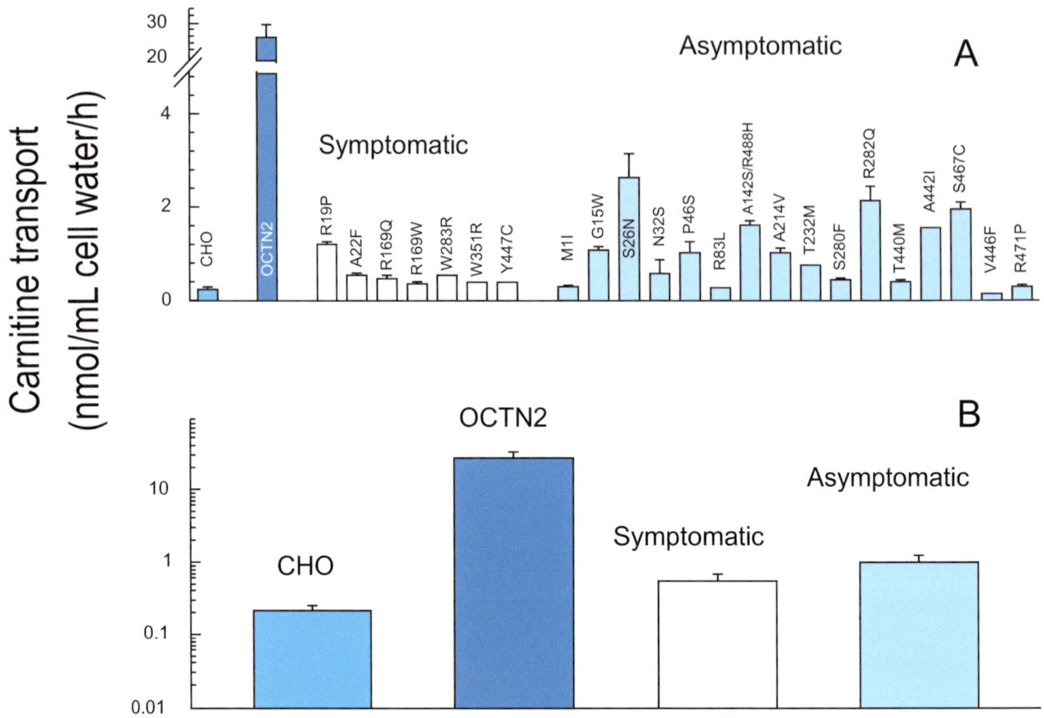

Fig. 2.5 Carnitine transport by Chinese hamster ovary (CHO) cells expressing normal and mutant carnitine transporters identified from symptomatic patients and asymptomatic mothers. Panel **a**: Data for individual missense mutations of symptomatic patients and asymptomatic mothers. Points are averages ± SE of six samples. Panel **b**: The average for mutations identified in symptomatic patients and asymptomatic mothers. Note the logarithmic scale in panel **b**. In both groups, carnitine transport was significantly ($P < 0.01$ using ANOVA) reduced as compared to wild-type OCTN2, but there was no significant difference between the groups of missense mutations identified in symptomatic patients versus asymptomatic mothers using ANOVA. *ANOVA* analysis of variance, *SE* standard error. (From Ref. [84])

reaction and finally outside the cell, when the harmful acyl groups accumulate in mitochondria and disturb mitochondrial function. However, this role is not easy to understand and has not yet been fully accepted. In recent years, it has been found that metabolic abnormalities occur due to the accumulation of acylcarnitine in mitochondria not only in organic acidemia but also in various pathologies in common diseases such as liver cirrhosis and heart failure, and the significance of carnitine supplementation therapy for such diseases is being recognized. The author hopes that this book will deepen your understanding of these issues.

References

1. Friedman S, Fraenkel G. Reversible enzymatic acetylation of carnitine. Arch Biochem Biophys. 1955;59(2):491–501. https://doi.org/10.1016/0003-9861(55)90515-4.
2. McFarlane JE. Studies on the physiology of carnitine. University of Illinois the Graduate College; 1955.
3. Lundsgaard E. Observations on a factor determining the metabolic rate of the liver. Biochim Biophys Acta. 1950;4(1–3):322–9. https://doi.org/10.1016/0006-3002(50)90038-2.
4. Fritz IB. The effects of muscle extracts on the oxidation of palmitic acid by liver slices and homogenates. Acta Physiol Scand. 1955;34(4):367–85. https://doi.org/10.1111/j.1748-1716.1955.tb01256.x.

5. Fritz IB, McEwen B. Effects of carnitine on fatty-acid oxidation by muscle. Science. 1959;129(3345):334–5. https://doi.org/10.1126/science.129.3345.334.
6. Fritz IB. Facts and speculations about the function of carnitine. Med Bull (Ann Arbor). 1959;25:382–9.
7. Fritz IB, Kaplan E, Yue KTN. Specificity of carnitine action on fatty acid oxidation by heart muscle. Am J Phys. 1962;202:117–21. https://doi.org/10.1152/ajplegacy.1962.202.1.117.
8. Fritz IB. Carnitine and its role in fatty acid metabolism. Adv Lipid Res. 1963;1:285–334.
9. Fritz IB, Schultz SK, Srere PA. Properties of partially purified carnitine acetyltransferase. J Biol Chem. 1963;238:2509–17.
10. Fritz IB, Yue KTN. Long-chain carnitine acyltransferase and the role of acylcarnitine derivatives in the catalytic increase of fatty acid oxidation induced by carnitine. J Lipid Res. 1963;4:279–88.
11. Bremer J. Carnitine in intermediary metabolism. Reversible acetylation of carnitine by mitochondria. J Biol Chem. 1962;237:2228–31.
12. Bremer J. Carnitine in the intermediary metabolism. The metabolism of fatty acid esters of carnitine by mitochondria. J Biol Chem. 1962;237:3628–32.
13. Bremer J. Carnitine as a fatty acid carrier in intermediary metabolism. Nature. 1962;196:993–4. https://doi.org/10.1038/196993a0.
14. Bieber LL. Carnitine. Annu Rev Biochem. 1988;57:261–83. https://doi.org/10.1146/annurev.bi.57.070188.001401.
15. Kopec B, Fritz IB. Properties of a purified carnitine palmitoyltransferase, and evidence for the existence of other carnitine acyltransferases. Can J Biochem. 1971;49(8):941–8. https://doi.org/10.1139/o71-136.
16. Kopec B, Fritz IB. Comparison of properties of carnitine palmitoyltransferase I with those of carnitine palmitoyltransferase II, and preparation of antibodies to carnitine palmitoyltransferases. J Biol Chem. 1973;248(11):4069–74.
17. Brosnan JT, Kopec B, Fritz IB. The localization of carnitine palmitoyltransferase on the inner membrane of bovine liver mitochondria. J Biol Chem. 1973;248(11):4075–82.
18. Norum KR, Farstad M, Bremer J. The submitochondrial distribution of acid: CoA ligase (AMP) and palmityl-CoA: carnitine palmitoyltransferase in rat liver mitochondria. Biochem Biophys Res Commun. 1966;24(5):797–804. https://doi.org/10.1016/0006-291x(66)90397-4.
19. Yates DW, Garland PB. The partial latency and intramitochondrial distribution of carnitine-palmitoyltransferase (e.c.2.3.1.-), and the CoASH and carnitine permeable space of rat liver mitochondria. Biochem Biophys Res Commun. 1966;23(4):460–5. https://doi.org/10.1016/0006-291x(66)90750-9.
20. West DW, Chase JF, Tubbs PK. The separation and properties of two forms of carnitine palmitoyltransferase from ox liver mitochondria. Biochem Biophys Res Commun. 1971;42(5):912–8. https://doi.org/10.1016/0006-291x(71)90517-1.
21. Hoppel CL, Tomec RJ. Carnitine palmityltransferase. Location of two enzymatic activities in rat liver mitochondria. J Biol Chem. 1972;247(3):832–41.
22. McGarry JD, Leatherman GF, Foster DW. Carnitine palmitoyltransferase I. The site of inhibition of hepatic fatty acid oxidation by malonyl-CoA. J Biol Chem. 1978;253(12):4128–36.
23. McGarry JD, Foster DW. In support of the roles of malonyl-CoA and carnitine acyltransferase I in the regulation of hepatic fatty acid oxidation and ketogenesis. J Biol Chem. 1979;254(17):8163–8.
24. Haddock BA, Yates DW, Garland PB. The localization of some coenzyme A-dependent enzymes in rat liver mitochondria. Biochem J. 1970;119(3):565–73. https://doi.org/10.1042/bj1190565.
25. Brosnan JT, Fritz IB. The permeability of mitochondria to carnitine and acylcarnitine. Biochem J. 1971;125(4):94P–5P. https://doi.org/10.1042/bj1250094pb.
26. Pande SV. A mitochondrial carnitine acylcarnitine translocase system. Proc Natl Acad Sci USA. 1975;72(3):883–7. https://doi.org/10.1073/pnas.72.3.883.
27. Pande SV, Parvin R. Characterization of carnitine acylcarnitine translocase system of heart mitochondria. J Biol Chem. 1976;251(21):6683–91.
28. Pande SV, Parvin R. Carnitine-acylcarnitine translocase catalyzes an equilibrating unidirectional transport as well. J Biol Chem. 1980;255(7):2994–3001.
29. Ramsay RR, Tubbs PK. The mechanism of fatty acid uptake by heart mitochondria: an acylcarnitine-carnitine exchange. FEBS Lett. 1975;54(1):21–5. https://doi.org/10.1016/0014-5793(75)81059-3.
30. Bremer J. Carnitinemetabolism and functions. Physiol Rev. 1983;63(4):1420–80. https://doi.org/10.1152/physrev.1983.63.4.1420.
31. Brooks DE, McIntosh JEA. Turnover of carnitine by rat tissues. Biochem J. 1975;148(3):439–45. https://doi.org/10.1042/bj1480439.
32. Yue KT. Fate of tritium-labeled carnitine administered to dogs and rats. Am J Phys. 1962;202:122–8. https://doi.org/10.1152/ajplegacy.1962.202.1.122.
33. Cederblad G. Metabolism of labeled carnitine in the rat. Arch Biochem Biophys. 1976;175(1):173–80. https://doi.org/10.1016/0003-9861(76)90496-3.
34. Rebouche CJ, Engel AG. Kinetic compartmental analysis of carnitine metabolism in the dog. Arch Biochem Biophys. 1983;220(1):60–70. https://doi.org/10.1016/0003-9861(83)90387-9.
35. Broquist HP, Borum PR. Carnitine biosynthesis: nutritional implications. Adv Nutr Res. 1982;4:181–204. https://doi.org/10.1007/978-1-4613-9934-6_7.
36. Böhmer T, Eiklid K, Jonsen J. Carnitine uptake into human heart cells in culture. Biochim Biophys Acta. 1977;465(3):627–33. https://doi.org/10.1016/0005-2736(77)90278-4.
37. Mølstad P, Böhmer T, Eiklid K. Specificity and characteristics of the carnitine transport in human heart cells (CCL 27)in culture. Biochim

Biophys Acta. 1977;471(2):296–304. https://doi.org/10.1016/0005-2736(77)90257-7.
38. Bahl J, Navin T, Manian AA, et al. Carnitine transport in isolated adult rat heart myocytes and the effect of 7, 8-diOH chlorpromazine. Circ Res. 1981;48(3):378–85. https://doi.org/10.1161/01.res.48.3.378.
39. Willner JH, Ginsburg S, Dimauro S. Active transport of carnitine into skeletal muscle. Neurology. 1978;28(7):721–4. https://doi.org/10.1212/wnl.28.7.721.
40. Rebouche CJ. Carnitine movement across muscle cell membranes. Studies in isolated rat muscle. Biochim Biophys Acta. 1977;471(1):145–55. https://doi.org/10.1016/0005-2736(77)90402-3.
41. Christiansen RZ, Bremer J. Active transport of butyrobetaine and carnitine into isolated liver cells. Biochim Biophys Acta. 1976;448(4):562–77. https://doi.org/10.1016/0005-2736(76)90110-3.
42. Huth PJ, Shug AL. Properties of carnitine transport in rat kidney cortex slices. Biochim Biophys Acta. 1980;602(3):621–34. https://doi.org/10.1016/0005-2736(80)90340-5.
43. Hamilton JW, Li BU, Shug AL, et al. Carnitine transport in human intestinal biopsy specimens. Gastroenterology. 1986;91(1):10–6. https://doi.org/10.1016/0016-5085(86)90432-4.
44. Engel AG, Angelini C. Carnitine deficiency of human skeletal muscle with associated lipid storage myopathy: a new syndrome. Science. 1973;179(4076):899–902. https://doi.org/10.1126/science.179.4076.899.
45. Markesbery WR, McQuillen MP, Procopis PG, et al. Muscle carnitine deficiency. Association with lipid myopathy, vacuolar neuropathy, and vacuolated leukocytes. Arch Neurol. 1974;31(5):320–4. https://doi.org/10.1001/archneur.1974.00490410068007.
46. VanDyke DH, Griggs RC, Markesbery WR, et al. Hereditary carnitine deficiency of muscle. Neurology. 1975;25(2):154–9. https://doi.org/10.1212/wnl.25.2.154.
47. Smyth DPL, Lake BD, MacDermot J, et al. Inborn error of carnitine metabolism ("carnitine deficiency") in man. Lancet. 1975;1(7917):1198–9. https://doi.org/10.1016/s0140-6736(75)93191-8.
48. Angelini C, Lücke S, Cantarutti F. Carnitine deficiency of skeletal muscle: report of a treated case. Neurology. 1976;26(7):633–7. https://doi.org/10.1212/wnl.26.7.633.
49. Boudin G, Mikol J, Guillard A, et al. Fatal systemic carnitine deficiency with lipid storage in skeletal muscle, heart, liver and kidney. J Neurol Sci. 1976;30(2–3):313–25. https://doi.org/10.1016/0022-510x(76)90137-4.
50. Ware AJ, Burton WC, McGarry JD, et al. Systemic carnitine deficiency. Report of a fatal case with multisystemic manifestations. J Pediatr. 1978;93(6):959–64. https://doi.org/10.1016/s0022-3476(78)81219-0.
51. Levitan MD, Murphy JT, Sherwood WG, et al. Adult onset systemic carnitine deficiency: favorable response to L-carnitine supplementation. Can J Neurol Sci. 1987;14(1):50–4. https://doi.org/10.1017/s0317167100026172.
52. Chapoy PR, Angelini C, Brown WJ, et al. Systemic carnitine deficiency – a treatable inherited lipid-storage disease presenting as Reye's syndrome. N Engl J Med. 1980;303(24):1389–94. https://doi.org/10.1056/NEJM198012113032403.
53. McGarry JD, Foster JD. Systemic carnitine deficiency. N Engl J Med. 1980;303(24):1413–4. https://doi.org/10.1056/NEJM198012113032410.
54. Glasgow AM, Eng G, Engel AG. Systemic carnitine deficiency simulating recurrent Reye syndrome. J Pediatr. 1980;96(5):889–91. https://doi.org/10.1016/s0022-3476(80)80571-3.
55. Tripp ME, Katcher ML, Peters HA, et al. Systemic carnitine deficiency presenting familial endocardial fibroelastosis. A treatable cardiomyopathy. N Engl J Med. 1981;305(7):385–90. https://doi.org/10.1056/NEJM198108133050707.
56. Waber LJ, Valle D, Neill C, et al. Carnitine deficiency presenting as familial cardiomyopathy: a treatable defect in carnitine transport. J Pediatr. 1982;101(5):700–5. https://doi.org/10.1016/s0022-3476(82)80294-1.
57. Treem WR, Stanley CA, Finegold DN, et al. Primary carnitine deficiency due to a failure of carnitine transport in kidney, muscle, and fibroblasts. N Engl J Med. 1988;319(20):1331–6. https://doi.org/10.1056/NEJM198811173192006.
58. Stanley CA, Deleeuw S, Coates PM, et al. Chronic cardiomyopathy and weakness or acute coma in children with a defect in carnitine uptake. Ann Neurol. 1991;30(5):709–16. https://doi.org/10.1002/ana.410300512.
59. Tein I, De Vivo DC, Bierman F, et al. Impaired skin fibroblast carnitine uptake in primary systemic carnitine deficiency manifested by childhood carnitine-responsive cardiomyopathy. Pediatr Res. 1990;28(3):247–55. https://doi.org/10.1203/00006450-199009000-00020.
60. Engel A, Rebouche CJ, Wilson DM, et al. Primary systemic carnitine deficiency II. Renal handling of carnitine. Neurology. 1981;31:819–25. https://doi.org/10.1212/wnl.31.7.819.
61. Rebouche CJ, Engel AG. Carnitine transport in cultured muscle cells and skin fibroblasts from patients with primary systemic carnitine deficiency. In Vitro. 1982;18(5):495–500. https://doi.org/10.1007/BF02796479.
62. Rebouche CJ, Engel AG. Kinetic compartmental analysis of carnitine metabolism in the human carnitine deficiency syndromes evidence for alterations in tissue carnitine transport. J Clin Invest. 1984;73(3):857–67. https://doi.org/10.1172/JCI111281.
63. Karpati G, Carpenter S, Engel AG, et al. The syndrome of systemic carnitine deficiency. Clinical,

morphologic, biochemic, and pathophysiologic features. Neurology. 1975;25(1):16–24. https://doi.org/10.1212/wnl.25.1.16.

64. Koizumi T, Nikaido H, Hayakawa J, et al. Infantile disease with microvesicular fatty infiltration of viscera spontaneously occurring in the C3H-H-2° strain of mouse with similarities to Reye's syndrome. Lab Anim. 1988;22(1):83–7. https://doi.org/10.1258/002367788780746511.
65. Kuwajima M, Kono N, Horiuchi M, et al. Animal model of systemic carnitine deficiency: analysis in C3H-H-2° strain of mouse associated with juvenile visceral steatosis. Biochem Biophys Res Commun. 1991;174(3):1090–4. https://doi.org/10.1016/0006-291x(91)91532-h.
66. Horiuchi M, Kobayashi K, Yamaguchi S, et al. Primary defect of juvenile visceral steatosis (*jvs*) mouse with systemic carnitine deficiency is probably in renal carnitine transport system. Biochim Biophys Acta. 1994;1226(1):25–30. https://doi.org/10.1016/0925-4439(94)90054-x.
67. Kuwajima M, Lu KM, Harashima H, et al. Carnitine transport defect in fibroblast of juvenile visceral steatosis (JVS) mouse. Biochem Biophys Res Commun. 1996;223(2):283–7. https://doi.org/10.1006/bbrc.1996.0885.
68. Yokogawa K, Yonekawa M, Tamai I, et al. Loss of wild-type carrier-mediated L-carnitine transport activity in hepatocytes of juvenile visceral steatosis mice. Hepatology. 1999;30(9):997–1001. https://doi.org/10.1002/hep.510300423.
69. Miyagawa J, Kuwajima M, Hanafusa T, et al. Mitochondrial abnormalities of muscle tissue in mice with juvenile visceral steatosis associated with systemic carnitine deficiency. Virchows Arch. 1995;426(3):271–9. https://doi.org/10.1007/BF00191365.
70. Horiuchi M, Yoshida H, Kobayashi K, et al. Cardiac hypertrophy in juvenile visceral steatosis (jvs) mice with systemic carnitine deficiency. FEBS Lett. 1993;326(1–3):267–71. https://doi.org/10.1016/0014-5793(93)81805-a.
71. Kuwajima M, Horiuchi M, Harashima H, et al. Cardiomegaly in the juvenile visceral steatosis (JVS) mouse is reduced with acute elevation of heart short-chain acyl-carnitine level after L-carnitine injection. FEBS Lett. 1999;443(3):261–6. https://doi.org/10.1016/s0014-5793(98)01732-3.
72. Hashimoto N, Suzuki F, Tamai I, et al. Gene-dose effect on carnitine transport activity in embryonic fibroblasts of JVS mice as a model of human carnitine transporter deficiency. Biochem Pharmacol. 1998;55(10):1729–32. https://doi.org/10.1016/s0006-2952(97)00670-9.
73. Nikaido H, Horiuchi M, Hashimoto N, et al. Mapping of jvs (juvenile visceral steatosis) gene, which causes systemic carnitine deficiency in mice, on chromosome 11. Mamm Genome. 1995;6(5):369–70. https://doi.org/10.1007/BF00364804.
74. Okita K, Tokino T, Nishimori H, et al. Definition of the locus responsible for systemic carnitine deficiency within a 1.6-cM region of mouse chromosome 11 by detailed linkage analysis. Genomics. 1996;33(2):289–91. https://doi.org/10.1006/geno.1996.0194.
75. Tamai I, Ohashi R, Nezu J, et al. Molecular and functional identification of sodium ion-dependent, high affinity human carnitine transporter OCTN2. J Biol Chem. 1998;273(32):20378–82. https://doi.org/10.1074/jbc.273.32.20378.
76. Tamai I. Pharmacological and pathophysiological roles of carnitine/organic cation transporters (OCTNs: SLC22A4, SLC22A5 and Slc22a21). Biopharm Drug Dispos. 2013;34(1):29–44. https://doi.org/10.1002/bdd.1816.
77. Tsuji A. Biopharmaceutical studies on molecular mechanisms of membrane transport. J Pharm Soc Jpn. 2002;122(12):1037–58. Article in Japanese.
78. Nezu J, Tamai I, Oku A, et al. Primary systemic carnitine deficiency is caused by mutations in a gene encoding sodium ion-dependent carnitine transporter. Nat Genet. 1999;21(1):91–4. https://doi.org/10.1038/5030.
79. Koizumi A, Nozaki J, Ohura T, et al. Genetic epidemiology of the carnitine transporter OCTN2 gene in a Japanese population and phenotypic characterization in Japanese pedigrees with primary systemic carnitine deficiency. Hum Mol Genet. 1999;8(12):2247–54. https://doi.org/10.1093/hmg/8.12.2247.
80. Koizumi A. Toward a more rational field-genetic epidemiology. Jpn J Hyg. 2010;65(1):37–47. Article in Japanese.
81. Shoji Y, Koizumi A, Kayo T, et al. Evidence for linkage of human primary systemic carnitine deficiency with D5S436: a novel gene locus on chromosome 5q. Am J Hum Genet. 1998;63(1):101–8. https://doi.org/10.1086/301911.
82. Longo N, Frigeni M, Pasquali M. Carnitine transport and fatty acid oxidation. Biochim Biophys Acta. 2016;1863(10):2422–35. https://doi.org/10.1016/j.bbamcr.2016.01.023.
83. Wang Y, Korman SH, Ye J, et al. Phenotype and genotype variation in primary carnitine deficiency. Genet Med. 2001;3(6):387–92. https://doi.org/10.1097/00125817-200111000-00002.
84. Rose EC, di San A, Filippo C, Ndukwe Erlingsson UC, et al. Genotype-phenotype correlation in primary carnitine deficiency. Hum Mutat. 2012;33(1):118–23. https://doi.org/10.1002/humu.21607.

3 What Is Carnitine?

3.1 Introduction

So far, the history of the discovery of carnitine and the carnitine cycle have been described. In this chapter, I will explain once again in detail what carnitine is, as well as acyl groups and coenzyme A (CoA), which appear frequently in this book but are very important for understanding the function of carnitine, even though they are often used in biochemistry textbooks. I will also discuss common misconceptions about carnitine, because carnitine is considered to be a vitamin-like substance. The biochemical and physiological actions of carnitine known to date are summarized.

3.2 Chemical Structure of Carnitine

As described in Chap. 1, carnitine (4-trimethylamino-3-hydroxybutyric acid, 4-trimethyl-3-hydroxybutyrobetaine) is a low molecular weight (molecular weight 161.20) biological substance discovered in muscles in 1905. It is distributed in a wide range of organisms and chemically has two optical isomers, L- and D-forms, but only the L-form is biologically active in nature, while the D-form has antagonistic effects. In this book, unless otherwise noted, the term "carnitine" refers to L-carnitine. The "levo" in "levocarnitine" means L-carnitine. Figure 3.1 shows the chemical structure of L-carnitine.

Generic name: Levocarnitine
Chemical name: (*R*)-3-Hydroxy-4-trimethylammoniobutanoate
Chemical structure:

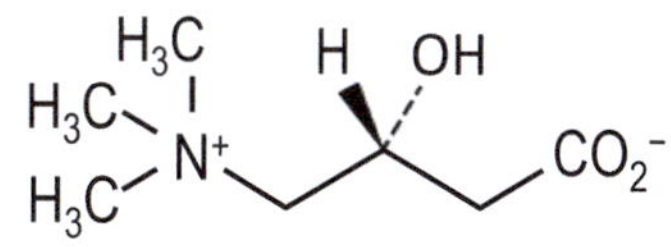

Molecular formula: $C_7H_{15}NO_3$
Molecular weight: 161.20

Fig. 3.1 Chemical structure of L-carnitine

3.3 Salts of Carnitine Preparations

The carnitine preparation named "L-Cartin® Tablets" launched by Otsuka Pharmaceutical Co., Ltd. in Japan in 1990 was a levocarnitine chloride, which is a hydrochloride salt of L-carnitine, due to stability and other considerations. Later, when the indication for carnitine preparations was expanded to "carnitine deficiency" by "Kouchi-Shinsei" in Japanese (meaning public knowledge-based application), Japan Society for Inherited Metabolic Diseases and others requested development of liquid formulations for oral use and the intravenous injections, and to respond promptly to these requests, these liquid formulations were introduced from Sigma Tau (currently Alpha Sigma) of Italy, which manufactures and sells carnitine preparations worldwide. The active pharmaceutical ingredient (API)

N. Nakayama, *Carnitine Deficiency and Insufficiency in Medicine*,
https://doi.org/10.1007/978-981-95-7797-2_3

of these liquid formulations was a free form of L-carnitine, not hydrochloride salt. In February 2013, free form of L-Cartin® FF oral solution 10% (10 mL glass vial preparation) and a free form of L-Cartin FF intravenous injection 1000 mg (glass ampoule preparation) were launched. Therefore, for a while, there was a period when the tablet was levocarnitine chloride (molecular weight 197.66) and the oral solution and intravenous agent were free form levocarnitine (molecular weight 161.2). Thus, there was a difference in active pharmaceutical ingredient (API) between drug formulations. In December 2014, Otsuka Pharmaceutical Co., Ltd. launched free L-Cartin® FF tablets 100 mg and L-Cartin® FF tablets 250 mg. Since then, all of Otsuka Pharmaceutical Co.'s products have been free levocarnitine (L-carnitine). According to Sigma Tau S.p.A. (now Alpha sigma S.p.A.), the manufacturer of the API, this free levocarnitine is stabilized by forming an intramolecular salt (inner salt). Since free L-carnitine is not chloride salt, it does not cause an overload of chlorine when administered intravenously, unlike hydrochloride salt of L-carnitine. Metabolic acidosis is widely observed in the acute phase of inherited metabolic diseases for which intravenous levocarnitine is often used. Further chloral overload may exacerbate the acidosis due to hyperchloremic acidosis. For this reason, free levocarnitine is considered suitable for intravenous preparations. In addition, since excessive salt intake is always a concern in the dietary management of dialysis patients with reduced urinary output due to decreased renal function, and some dialysis physicians are also concerned about excessive intake of chloroform, hence all carnitine formulations of Otsuka Pharmaceutical Co., Ltd. use free levocarnitine as API.

Subsequently, generic levocarnitine chloride tablets (both 100 mg and 300 mg tablets) were launched by Nichi-Iko, Fuso Pharmaceutical Co., Ltd., and Koaisei Co., Ltd. in December 2015, and Koa Biotech Bay (Yoshindo) in June 2016. On the other hand, in December 2014, Otsuka Pharmaceutical Co., Ltd.'s L-Cartin® tablets were switched to free form of levocarnitine (L-Cartin® FF tablets 100 mg and 250 mg tablets) as mentioned above, and levocarnitine chloride tablet products have been off the market since March 2016. Therefore, these generic levocarnitine chloride tablets were in a state of "generic drugs without original brand-name drug." In June 2022, Towa Pharmaceutical launched generic free levocarnitine tablets (100 mg and 250 mg). In addition, a generic drug (1000 mg prefilled syringe preparation) of an intravenous levocarnitine-free form has been released by Towa Yakuhin, Nipro, and Fuso Pharmaceutical Industries. The active pharmaceutical ingredient of generic levocarnitine chloride and levocarnitine-free form is manufactured in the People's Republic of China.

3.4 Physical Properties of Carnitine

The physical properties of L-carnitine are as follows (see the L-Cartin® Interview Form: lfo_if.pdf [otsuka-elibrary.jp]) [1]:

1. Appearance and properties: white crystalline powder
2. Solubility: extremely soluble in water and slightly soluble in ethanol; the pH of an aqueous solution (1 → 20) is 6.5–8.5
3. Hygroscopicity: The product is hygroscopic, with approximate 10% increase of moisture content under storage conditions of 25 °C/80% room humidity (RH) for 28 days
4. Melting point (decomposition point): about 200 °C (decomposition)
5. Acid-base dissociation constant: pKa = 3.72
6. Optical rotation: $[\alpha]_D^{20}$: −30.0 ∽ −32.0°

Its physical properties are characterized by its extremely high hygroscopicity. It is sometimes described as "deliquescent." Therefore, Otsuka Pharmaceutical's carnitine tablets are film-coated and individually packaged in press through package (PTP) packaging. The fumarate and tartrate salts of carnitine, which are commonly used in supplements, are said to have improved hygroscopicity, but the details of the degree of hygroscopicity is unknown. Therefore, it is necessary to pay attention to the hygroscopicity of bottled products that are not individually packaged after opening the package. On the other hand, before the introduction of liquid oral solutions, film-coated tablets were sometimes dissolved and used for tube feedings, causing problems such as clogged tubes because the film did not dissolve completely.

3.5 Misconceptions About Carnitine: An Amino Acid? B Vitamin?

Carnitine is sometimes called an "amino acid-like substance" because of its structure. This is because it has a trimethylamine group, which are almost homologous in nature to an amino group ($-NH_2$) and a carboxylic acid (-COOH) in its chemical structural formula. So it fits the definition of amino acids in a broad sense. Another reason carnitine is called an amino acid-like substance is that it is biosynthesized from the essential amino acids, lysine and methionine. However, the term "amino acid" usually refers to the 22 amino acids such as lysine, glutamic acid, leucine, and phenylalanine that make up proteins, but carnitine is not one of those amino acids that make up proteins.

Carnitine is also sometimes referred to as a "vitamin-like substance." The reason for this is that, as described in Chap. 1, early physiological studies by Fraenkel et al. in the 1940s and 1950s showed that carnitine was an essential nutrient for the larval metamorphosis of a mealworm beetle (*Tenebrio molitor*) into a pupa, and it was thought to be a new vitamin in the B vitamin group. The name "vitamin B_T" was given to this new vitamin, which was thought to belong to the vitamin B group. However, it was later found to be biosynthesized in vivo in mammals and is no longer considered a vitamin. Carnitine is not a vitamin for mammals, but it is nutritionally a very important substance, and a deficiency of carnitine can cause severe carnitine deficiency diseases; so it is called a "conditionally essential nutrient" because a deficiency is a rare occurrence, but when it does occur, it can lead to serious, even life-threatening conditions. However, because it is a "vitamin-like substance," it is often thought of in the same way as vitamin deficiency diseases. However, it is important to note that although it is true that carnitine deficiency can occur, it is a misconception to think of carnitine deficiency in the same way as vitamin deficiency. The mechanism of action of carnitine is completely different from that of many vitamins, and if this is not the case, carnitine deficiency will not be fully understood. Many people say, "I do not understand carnitine deficiency," mostly because they view carnitine deficiency in the same way as vitamin deficiencies (especially B vitamins). Many vitamins, especially the B vitamins, act as cofactors for various enzymes (Table 3.1).

For example, Wernicke's encephalopathy, which develops in alcoholics and other patients, is a typical vitamin B_1 deficiency, and vitamin B_1 acts as a cofactor for several metabolic enzymes, including pyruvate dehydrogenase (PDH) complex, and enzyme activity is exerted only when vitamin B_1 is bound to the enzyme protein itself (Fig. 3.2).

In addition to vitamin deficiencies, there are often nutritionally problematic trace metal deficiencies, such as zinc deficiency and selenium deficiency, which, like coenzymes, bind to the enzyme body as cofactors to stabilize the active center and maintain its structure, thereby enhancing enzyme activity. A complete enzyme with such coenzymes or cofactors bound is called a holoenzyme, while the protein portion without coenzymes or cofactors is called an apoenzyme. Thus, in vitamin B_1 deficiency and zinc deficiency, various symptoms develop because the coenzyme vitamin B_1 and the cofactor zinc are deficient and thus cannot exhibit their original enzymatic activity. However, it is a mistake to view carnitine deficiency in this way. Carnitine has no function as a coenzyme or cofactor. So, what is the role of carnitine? Carnitine is a carrier.

Table 3.1 Main B group vitamins and enzymes to which these vitamins bound

Vitamins	Compounds	Cofactors	Enzymes bound	Another name
Vitamin B_1	Thiamine	TPP	Pyruvate dehydrogenase; transketolase; α-ketoglutarate dehydrogenase	Oryzanin
Vitamin B_2	Riboflavin	FMN, FAD	Succinate dehydrogenase; acyl-CoA dehydrogenase; glutathione reductase; etc.	
Niacin	Nicotinic acid; nicotinic acid amide	NAD, NADP	Lactate dehydrogenase; glycerol-3-phosphate dehydrogenase; etc. (nearly 500 of oxidoreductases)	Vitamin B_3
Pantothenic acid	Pantothenic acid	Coenzyme A (CoA)	A component of Coenzyme A	Vitamin B_5
Vitamin B_6	Pyridoxine; pyridoxal; pyridoxamine	PNP, PLP, PMP	Many enzymes in amino acid, sugar, and lipid metabolism such as glycogen phosphorylase or transaminases	
Biotin	Biotin	Biotin	Pyruvate carboxylase; acetyl-CoA carboxylase; propionyl-CoA carboxylase; 3-methyl crotonyl-CoA carboxylase	Vitamin B_7; vitamin H
Folic acid	Folic acid	THF	Methyl or methylene group transfer enzymes in the process of amino acid or nucleic acid synthesis	Vitamin B_9; vitamin M
Vitamin B_{12}	Cyanocobalamin	Adenosylcobalamin; methylcobalamin	Methylmalonyl-CoA mutase; 5-methyltetrahydro folic acid; homocysteine methyltransferase; etc.	

These B group vitamins act as a cofactor or coenzyme of many important metabolic enzymes. Also, pantothenic acid is a component of Coenzyme A (CoA)

CoA coenzyme A, *FAD* flavin adenine dinucleotide, *FMN* flavin mononucleotide, *NAD* nicotinic acid amide adenine dinucleotide, *NADP* nicotinic acid amide adenine dinucleotide phosphate, *PLP* pyridoxal phosphate, *PMP* pyridoxamine phosphate, *PNP* pirydoxine-5′-phosphate, *THF* tetrahydrofolate, *TPP* thiamine diphosphate

- **vitamin B_1 deficiency** (beriberi, Wernicke encephalopathy etc.)
  ⇒vitamin B_1 act as a cofactor of enzymes such as PDH, etc.
- **zinc deficiency** (taste disorder, bedsores, etc.)
  ⇒zinc act as a cofactor of enzymes such as SOD,ADH, etc.
- **selenium deficiency** (cardiomyopathy, myalgia, susceptibility to infection, abnormality of nail and hair)
  ⇒selenium act as a cofactor of enzymes such as GPx, iodothyronine deiodinase, etc.

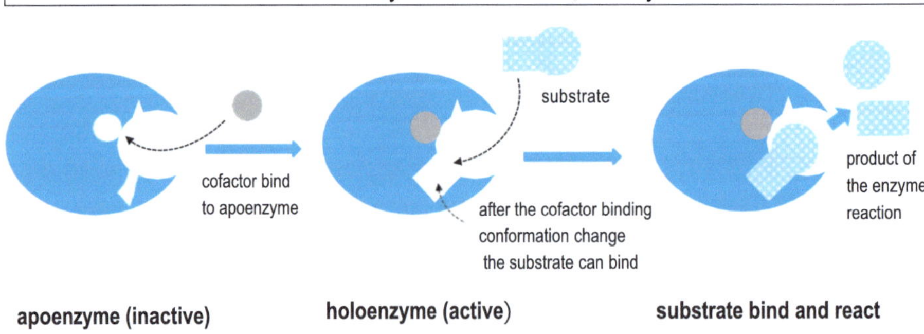

Fig. 3.2 Role of coenzymes and cofactors. Most vitamins or trace metals, whose deficiencies are important in nutrition, act as a prosthetic factor of various metabolic enzymes. These coenzymes or cofactors bind near the active center of enzyme and transform from the inactive form to the active form and progress the enzyme reactions. Carnitine does not have these roles as coenzymes or cofactors

3.6 What Is the Acyl Group Carried by Carnitine?

In a word, the most important function of carnitine is to be a carrier of acyl groups. First, the acyl group carried by the carnitine will be described. An acyl group is the general term for the portion of an organic acid or fatty acid bonded to a carboxylic acid (-COOH) without the -OH moiety (R-CO- portion) (Fig. 3.3). Officially, it is called an alkanoyl group. For example, the shortest organic acid is acetic acid (CH_3COOH), whose acyl group is acetyl group (CH_3CO-); in the case of caprylic acid International Union of Pure and Applied Chemistry (IUPAC) system name: octanoic acid), a fatty acid with eight carbons, its acyl group is octanoyl group ($CH_3(CH_2)_6CO$. In the case of palmitic acid (IUPAC system name is hexadecanoic acid), which is a fatty acid that is a major component of animal fats and vegetable oils, it has 16 carbons and its acyl group is palmitoyl group ($CH_3(CH_2)_{14}CO$-), while stearic acid has 18 carbons and its acyl group is stearoyl group ($CH_3(CH_2)_{16}CO$ -), and so on. Valproic acid, discussed in detail in Chap. 12, is also a branched-chain fatty acid, and its acyl group is called valproyl group (Fig. 3.3).

Table 3.2 lists the structural formulas, systematic names, common names, and acyl group names of the major organic acids and fatty acids that appear frequently in this book.

Acyl groups of long-chain fatty acids with 14 or more carbons are transferred to coenzyme A (CoA) by acyl CoA synthetase (ACS) on the mitochondrial outer membrane to form acyl CoA, which is activated. Carnitine is also taken up into the cell by the carnitine transporter (OCTN2) on the cell membrane. The acyl group of acyl-CoA is transferred to carnitine by the action of carnitine palmitoyltransferase 1 (CPT1) in the outer mitochondrial membrane to form acylcarnitine. It is then transported into the inner mitochondrial membrane by carnitine acylcarnitine translocase (CACT) in the inner mitochondrial membrane, and extremely efficient energy production (ATP production) is performed by β-oxidation of fatty

- acyl group: a generic name of **R-CO-** (the dotted frame area) in the organic acid or fatty acid (**R-CO-**OH)
- acyl group bound with CoA by the action of acyl-CoA synthetase forming **acyl-CoA**
- **carnitine carries acyl group** in acyl-CoA into mitochondria by the action of CPT1

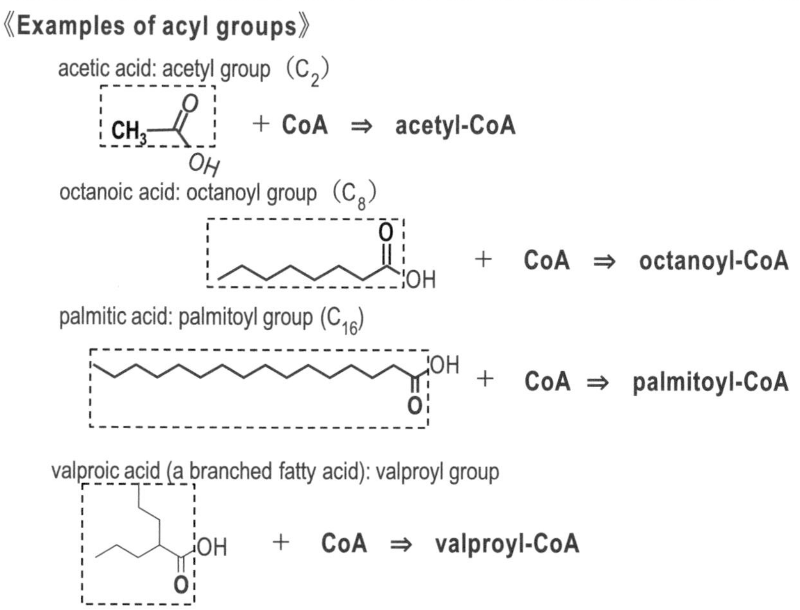

Fig. 3.3 Acyl group

Table 3.2 Structure, systematic name, common name, and acyl group of main organic acids and fatty acids

Carbon number	Structure	Systematic name	Common name	Acyl group
Saturated fatty acids				
2	CH_3COOH	Ethanoic acid	Acetic acid	Acetyl
3	CH_3CH_2COOH	Propanoic acid	Propionic acid	Propionyl
4	$CH_3(CH_2)_2COOH$	Butanoic acid	Butyric acid	Butyryl
5	$CH_3(CH_2)_3COOH$	Pentanoic acid	Valeric acid	Valeryl
6	$CH_3(CH_2)_4COOH$	Hexanoic acid	Caproic acid	Caproyl
7	$CH_3(CH_2)_5COOH$	Heptanoic acid	Enantoic acid	Heptanoyl
8	$CH_3(CH_2)_6COOH$	Octanoic acid	Caprylic acid	Octanoyl
9	$CH_3(CH_2)_7COOH$	Nonanoic acid	Pelargonic acid	Nonanoyl
10	$CH_3(CH_2)_8COOH$	Decanoic acid	Capric acid	Decanoyl
12	$CH_3(CH_2)_{10}COOH$	Dodecanoic acid	Lauric acid	Dodecanoyl
14	$CH_3(CH_2)_{12}COOH$	Tetradecanoic acid	Myristic acid	Myristoyl (tetradecanoyl)
16	$CH_3(CH_2)_{14}COOH$	Hexadecanoic acid	Palmitic acid	Palmitoyl
18	$CH_3(CH_2)_{16}COOH$	Octadecanoic acid	Stearic acid	Stearoyl
20	$CH_3(CH_2)_{18}COOH$	Eicosanoic acid	Arachidic acid	Eicosanoyl
22	$CH_3(CH_2)_{20}COOH$	Docosanoic acid	Behenic acid	Docosanoyl
Monounsaturated fatty acid				
18	$CH_3(CH_2)_7CH{=}CH(CH_2)_7COOH$	Cis-9-octadecenoic acid	Oleic acid	Oleoyl

acids. In other words, carnitine transports acyl groups of long-chain fatty acids, which are an important energy source for cells, into the inner mitochondrial matrix where β-oxidation takes place. Fatty acids with 8–12 carbons are called medium-chain fatty acids. Triglycerides composed of medium-chain fatty acids are called medium-chain triglycerides (MCT), and carnitine is not required for the transport of acyl groups of medium-chain fatty acids into the mitochondria.

Acyl groups of organic acids such as propionyl and methylmalonyl groups and acyl groups of medium-chain fatty acids accumulated in mitochondria due to inherited metabolic diseases such as organic acidemia and fatty acid metabolism disorders increase as acyl CoA, resulting in a relative deficiency in free CoA, which is important for energy metabolism, etc. In addition, acyl CoA inhibits various enzymes related to energy metabolism, resulting in mitochondrial dysfunction. As described below, carnitine also removes the accumulated acyl groups from the mitochondria as acylcarnitine.

3.7 What Is Coenzyme A (CoA)?

Coenzyme A (CoA) is also extremely important in fatty acid metabolism and energy metabolism, and is closely related to carnitine, and is often mentioned in this book, so I will explain it in detail here. CoA is a coenzyme discovered by Fritz Albert Lipmann in 1945 and plays an extremely important role in energy metabolism and fatty acid metabolism. For this discovery, Lipmann was awarded the Nobel Prize in Physiology or Medicine in 1953, together with Hans Krebs, who discovered the tricarboxylic acid (TCA) cycle. The "A" in coenzyme A was because it is a coenzyme involved in acetylation, and then Lynen (Feodor Lynen, awarded the Nobel Prize in Physiology or Medicine in 1964 for research on fatty acid and cholesterol metabolism) found that active acetic acid = acetyl-CoA. The fact that the discoverer of CoA was jointly awarded the Nobel Prize with the discoverer of the TCA cycle shows that CoA is extremely important for various metabolisms necessary to maintain life, such as fatty acid oxidation, biosynthesis of fatty acids, cholesterol, amino acid, and energy metabolism such as the TCA cycle.

CoA is a very important substance present in all living organisms (the biological kingdom is classified into three categories: archaea, bacteria, and eukaryotes) and is an essential substance involved in their metabolism to sustain life. CoA is a relatively large molecule composed of cysteine, pantothenic acid, and ADP (formed from ATP) with molecular weight of 767.5; its chemical structure is shown in Fig. 3.4.

CoA has a highly reactive thiol group (-SH) and a nucleotide moiety, which exert various chemical reactions and regulatory effects. For example, organic acids and fatty acids having carbonyl groups such as acetic acid (acetyl group) and long-chain fatty acids (for example, palmitoyl group) are activated when they bind to CoA and enter metabolic pathways such as TCA cycles and β-oxidation to perform their metabolism. In other words, CoA plays the role of a carrier in the metabolic process of acyl groups of organic acids and fatty acids. The total concentration of CoA in animal tissues ranges from 20 to 140 μmol/L, and is high in the liver and myocardium, about 30% of which is acyl-CoA. The concentration of total carnitine in the liver and the heart is 2 mM and 4 mM, respectively, and the total carnitine concentration is about 30–100 times higher than that of total CoA. CoA is present in high concentrations of 2–5 mM in the mitochondria and 0.7 mM in peroxisomes, and it is extremely low at 0.05–0.14 mM in the cytoplasm [2–4].

Pantothenic acid (vitamin B_5), which forms the central backbone of CoA, is biosynthesized from 2-oxoisovaleric acid in a three-step reaction in bacteria, but mammals lack these biosynthetic enzyme systems. Therefore, it is thought that mammals ingest pantothenic acid as vitamin B_5 from food, and a very small amount from intestinal bacteria. Since pantothenic acid is present in a wide variety of foods, it is rarely deficient, but intake from food is limited to a few mg/day. The

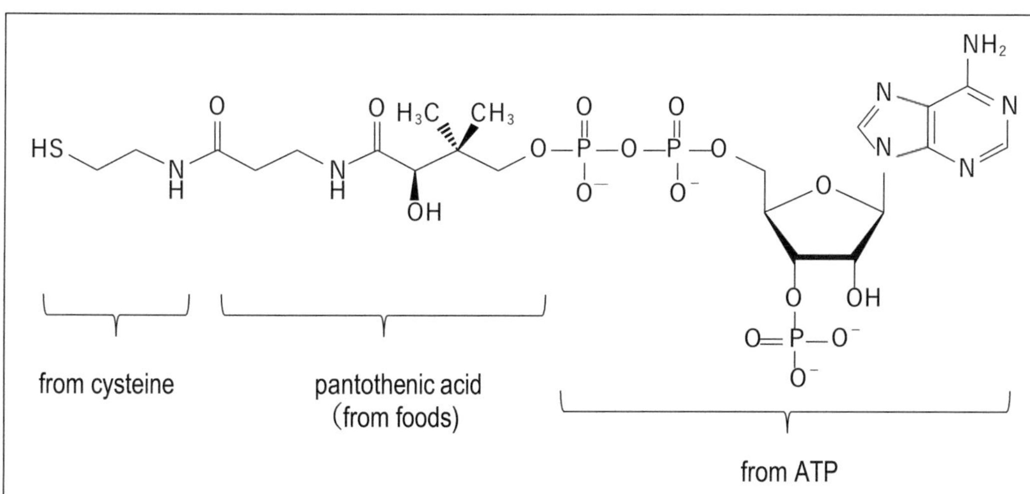

- A very important coenzyme in various energy metabolism and fatty acid metabolism
- A substance consists of cysteine, pantothenic acid and ADP (mol. wt.:767.5)
- Discovered by Lipmann in 1945 who received Nobel prize with Krebs in 1953

Fig. 3.4 Chemical structure of coenzyme A. Coenzyme A is biosynthesized from cysteine, pantothenic acid (vitamin B_5), and ATP. Coenzyme A is usually called as CoA, and the left-side SH- group is highly reactive and bound with acyl group forming acyl-CoA that acts as an acyl group carrier. So it is described as CoASH in the chemical reaction formula

CoA biosynthesis process in mammals involves five steps starting from pantothenic acid, and it requires four molecules of ATP and cysteine for biosynthesis, which is an energy-consuming process [3] (Fig. 3.5). In the first place, when acyl-CoA accumulates in the mitochondria and free CoA is insufficient, energy production is stagnant due to mitochondrial dysfunction, so it is not possible to biosynthesize CoA immediately in the required amount. Conversely, it is thought that all organisms have a regulatory mechanism that avoids excessive CoA biosynthesis. The rate-limiting step of the CoA biosynthesis system is the reaction of pantothenate kinase (PanK), which phosphorylates the first pantothenic acid, but this reaction is feedback inhibited by CoA or acetyl-CoA. In mammals, PanK has four isoforms: 1α, 1β, 2, and 3. PanK1α is expressed in the heart and kidneys, and PanK1β is expressed in the liver and kidneys [5, 6]. PanK activity is not only inhibited by CoA and acetyl-CoA, but also, interestingly, it is activated by carnitine [7, 8]. PanK2 is localized in the intermembrane space between the outer and inner membranes of mitochondria [4], but PanK2 is strongly inhibited by acetyl-CoA (IC50 < 1 μM), and it is inversely activated by palmitoyl carnitine [4]. It is known that in pathological conditions in which long-chain acylcarnitines such as palmitoylcarnitine accumulate, free CoA decreases and acyl-CoA increases, metabolic reactions requiring free CoA stagnate, and a variety of enzymes are inhibited. Therefore, it is considered that PanK2 is localized in mitochondria and plays the role of a sensor to promote the biosynthesis of free CoA when acyl-CoA accumulates and free CoA decreases relatively, and it is impossible to proceed smooth β-oxidation [4].

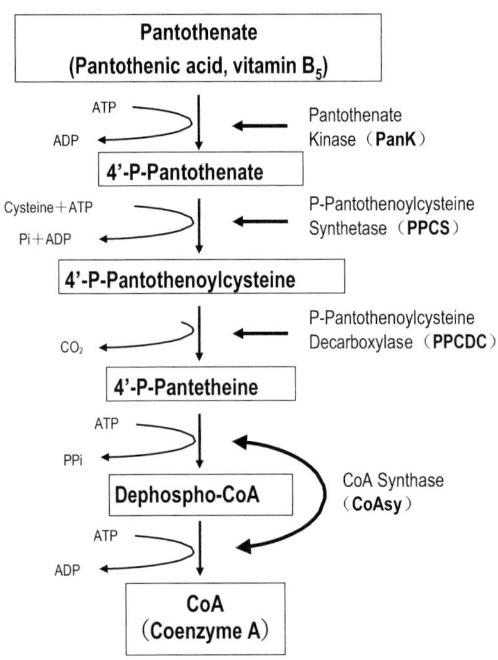

Fig. 3.5 Biosynthesis pathway of coenzyme A (CoA) in humans. Coenzyme A is biosynthesized from pantothenic acid with five steps of reactions that require four molecules of ATP and cysteine. The first pantothenic acid kinase (PanK) reaction is the rate-limiting step. The last two steps are processed multifunctionally by one CoA synthase, addition of nucleotide part, and phosphorylation reaction. (From Ref. [3])

In the CoA biosynthesis system, each of the three enzymes that perform the first three steps is located in the cytoplasm. The last two steps are performed by CoA synthase, a single enzyme that performs two different reactions in a multifunctional manner. Its localization is controversial, and there are reports that it is in the mitochondrial matrix, though other reports claim the localization in the nucleus, while some reports claim it to be in the outside of the outer mitochondrial membrane [9]. When CoA synthesis is required, it is thought that these four enzymes form complexes on the cytoplasmic side of the outer mitochondrial membrane to biosynthesize CoA [3].

More recently, CoA has been shown not only to play a classical role in mediating energy, fatty acid, and amino acid metabolism as described in biochemistry textbooks, but also to regulate the TCA cycle, fatty acid, and amino acid metabolism by acetylating lysine residues of the enzymes involved in these processes. In addition, it has been reported that CoA protects enzymes by acetylating thiol groups (-SH groups) of cysteine residues in various enzymes when oxidative and metabolic stresses are increased, which has attracted much attention (this action is called "CoAlation") [10–13]. The biosynthesis of CoA is regulated by nutrients, hormones, metabolites, cellular stress, and oxidative stress. Specifically,

CoA biosynthesis is promoted by fasting, glucagon, glucocorticoids, and hyperlipidemia treatments, and biosynthesis is suppressed by insulin, glucose, fatty acids, and pyruvate. Disruption of homeostasis of CoA is thought to lead to pathologies such as cancer, diabetes, myopathy, cardiomyopathy and cardiac hypertrophy, Reye-like syndrome, and neurodegenerative diseases [3]. In addition, neurodevelopmental and neurodegenerative diseases due to abnormalities in CoA biosynthesis have been discovered [14]. Gout et al. summarize the effects of CoA not only in classical metabolism and but also the new effects of CoA that have recently been elucidated (Fig. 3.6) [13]. Theodoulou et al. in the same group also position it as a "renaissance" in which a new role of CoA was discovered [10].

CoA plays an extremely important role in energy metabolism and lipid metabolism. CoA has a charged moiety so it cannot move freely across biological membranes such as the inner mitochondrial membrane which is composed of hydrophobic phospholipid bilayers, and its supply to the mitochondria, which is the main site of its existence, is quite limited. When organic acids and long-chain fatty acids accumulate in the mitochondria due to metabolic abnormalities, their acyl groups bind to free CoA forming acyl-CoA, so the amount of free CoA, which plays an important role in energy metabolism such as β-oxidation in mitochondria and the TCA cycle, becomes relatively deficient, and these metabolisms that require free CoA become stagnant. At this time, the ratio of acyl-CoA to free CoA (acyl-CoA/free CoA ratio) becomes high. If free carnitine is abundant in the mitochondria, carnitine receives the accumulated acyl group of acyl-CoA by the action of carnitine acyltransferases such as CPT2 and becomes acylcarnitine, and this acylcarnitine can be excreted outside the mitochondria via CACT. However, if the amount of free carnitine is not enough to buffering acyl-CoA toxicity and releasing free CoA, the amount of free CoA, which is very important in many metabolisms, might decrease, resulting in metabolic disorders (see "carnitine insufficiency" described below).

Thus, CoA not only has an effect in many classical metabolic reactions, but also new effects have been discovered, and the activation of CoA biosynthesis by carnitine and acylcarnitine has been discovered, and future research on the relationship between CoA biosynthesis and carnitine or acylcarnitine is expected.

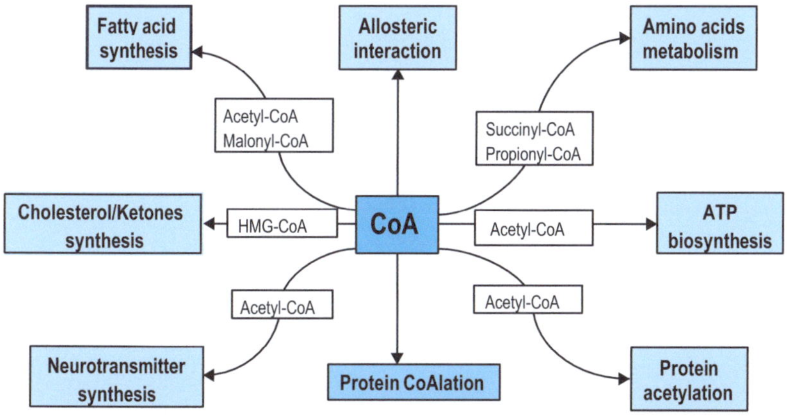

Fig. 3.6 Cellular functions of CoA and its derivatives. CoA thioester derivatives are Implicated in diverse cellular functions, including Krebs cycle, ketogenesis, biosynthesis of cholesterol or acetylcholine, degradation of amino acids, the synthesis and oxidation of fatty acid, biosynthesis of neurotransmitters, and the regulation of gene expression. Protein CoAlation is a novel, unconventional function of CoA in redox regulation and antioxidant defense. These metabolic functions of CoA are described in textbooks of biochemistry. However, it is discovered recently that CoAs not only have these functions but also act as gene expression modulator by acetylation of nuclear protein such as histone and act as metabolic modulator by acetylation of various metabolic enzymes. Furthermore, CoA acts as a defense function from oxidative stress by protein CoAlation. (From Ref. [13])

3.8 Biochemical and Physiological Actions of Carnitine

Numerous studies have shown that L-carnitine in vivo has various biochemical and physiological actions. They are summarized in Table 3.3.

The first of these biochemical and physiological actions of carnitine is to transport acyl groups of long-chain fatty acids into mitochondria through carnitine cycle (also called carnitine shuttle) and produce energy through β-oxidation (influx). The second is that when acyl groups (acyl CoA) accumulate in mitochondria, carnitine receives the acyl groups, converts them to acylcarnitine, and transports them out of mitochondria through the carnitine cycle (efflux). The third is modulation of acyl-CoA/free CoA ratio in the mitochondria to maintain free CoA by the second excreting action of acyl group. Acyl groups accumulate in mitochondria in various pathological conditions. The acyl groups accumulated in mitochondria combine with free CoA to form acyl CoA, which inhibits various important enzymes and deprives free CoA, which is important for energy metabolism, etc., resulting in an increase in the acyl CoA/free CoA ratio. If free carnitine is abundant, the harmful acyl groups can be received by carnitine and excreted out of mitochondria and eventually excreted in the urine. Therefore, the third action can be said to be a biochemical action derived from the second action. Thus, the first action of carnitine as an inward carrier of the acyl group and the second and third as an outward carrier, are the most important biochemical and physiological actions classically known.

Table 3.3 Biochemical and physiological functions of L-carnitine

1. Transport of long-chain fatty acid into mitochondria Stimulation of ATP production by β-oxidation
2. Excretion of accumulated harmful acyl-compounds from mitochondria or cell Important for removal of toxic acyl substances
3. Modulation of acyl-CoA/CoA ratio in mitochondria Important to maintain free CoA, which is important for energy metabolism
4. Repairing and stabilizing functions of membranes of red blood cells, etc. Important for turn-over and life-expansion of red blood cells, etc.
5. Reduction of oxidative stress and anti-inflammation effect Potentiate of anti-oxidative and anti-inflammatory function Anti-atherosclerotic action, cell protective action
6. Suppression of apoptosis Stimulation of red cell proliferation, suppression of muscle atrophy, fibrosis of cardiomyocytes and fibrosis of hepatocytes, etc.
7. Appetite enhancement and digestive enzymes secretion promoting effects

The fourth action is repairing and stabilizing of membrane phospholipids. There are no mitochondria in red blood cells, but there are enzymes called carnitine palmitoyltransferase (CPT) and lysophospholipid acyl CoA transferase (LAT), and acylcarnitines such as palmitoyl carnitine are utilized as acyl donors. Under oxidative stress, membrane phospholipids are peroxidized and cleaved, so that lysophospholipids may increase. Normal phospholipids have two fatty acids, but lysophospholipids have only one fatty acid, and act as a surfactant and destabilize membranes. Carnitine and palmitoyl carnitine play a role in transferring the palmitoyl group and repairing lysophospholipids to normal phospholipids. In addition, carnitine is said to enhance the packing of phospholipids in the erythrocyte membrane and contribute to membrane stability [15] (For details, see Chap. 14, Sect. 14.4.6. "Metabolic Actions of Carnitine"). The fifth and sixth actions are relatively new actions of carnitine, which have been known since around the year 2000, and will be described in detail separately in Chap. 7. Carnitine and carnosine in meat, and electrophilic substances in plants such as sulforaphane in broccoli and curcumin in curry activate a stress-responsive transcription factor called Nrf2 (nuclear factor erythroid-derived 2-related factor 2, or NF-E2-related factor 2). It binds to the antioxidant responsive element (ARE) or electrophile responsive element (EpRE) in the nucleus and enhances the expression of downstream antioxidant enzymes or cytoprotective proteins such as superoxide dismutase (SOD), catalase, glutathione peroxidase and heme oxygenase-1 (HO-1), and heat shock protein 70 (HSP70), and then these molecules exert anti-inflammatory, antioxidant, and anti-apoptotic actions. In other words, anti-inflammatory, antioxidant, and anti-apoptotic actions are exerted by an Nrf2-mediated mechanism completely different from the effect of carnitine as a carrier of the acyl group (the first, the second, and the third actions) [16].

The seventh action of carnitine, which is to stimulate appetite and digestive enzyme secretion, is the oldest known physiological action of carnitine and was clinically applied to improve anorexia in postoperative children and adults in Germany, France, and Belgium in the 1950s and 1960s. This is also discussed in detail in Chap. 6.

3.9 Free Carnitine and Acylcarnitine

Carnitine does not act as a coenzyme, but as a carrier, and what is carried is the acyl group of organic acids and fatty acids. So what is the difference between free carnitine and acylcarnitine? I would like to introduce an easy-to-understand illustration by Dr. Akiko Wakisaka, Department of Pediatrics, Iou Hospital, National Hospital Organization, in Kanazawa, Japan (Fig. 3.7). Free carnitine means it has not yet loaded acyl groups, so you can compare a free carnitine to an empty truck, which has carrying capacity. Acylcarnitine has already loaded acyl groups on the loading platform of a truck, so acylcarnitine cannot load acyl groups further. In other words, free carnitine is considered to be the sum of empty trucks which have carrying capacity, and acylcarnitine is the sum of the trucks which are already loaded with acyl-cargo (or acyl-rubble) and without carrying capacity. The sum of free carnitine and acylcarnitine is called "total carnitine," which means total numbers of trucks, empty and loaded (Fig. 3.7). The first task of free carnitine is to load long-chain fatty acids, which are fuel, and transport them to the mitochondria, the energy powerplants in cells (the upper part of Fig. 3.7, the first action in Table 3.3). An enzyme called carnitine acyltransferase helps to load the acyl group on the loading platform of the truck, and in the case of long-chain fatty acids, it is CPT1 of carnitine palmitoyltransferase (CPT). The acyl group of the long-chain fatty acid is loaded onto the truck by the action of CPT1 to become acylcarnitine, which is transported into the mitochondria through CACT, and then

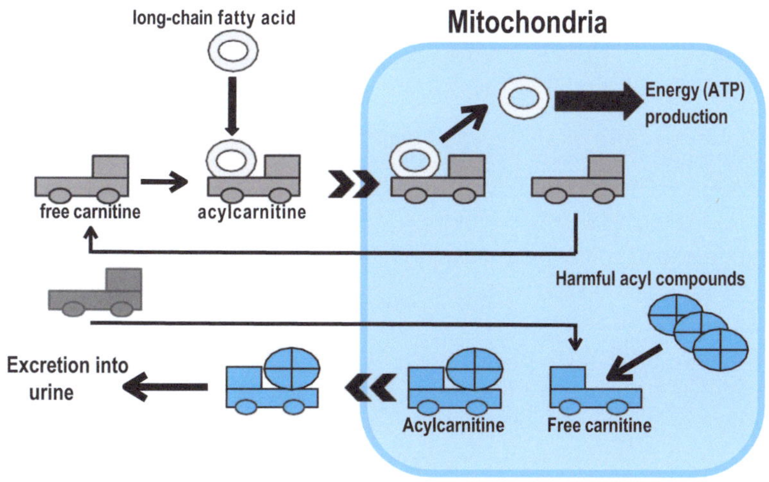

Fig. 3.7 An easy-to-understand schematic diagram showing the role of carnitine compared to a cargo truck. This figure explains the three functions of carnitine listed within Table 3.3: first, transport long-chain fatty acid into mitochondria; second, excrete harmful acyl compounds that accumulate in mitochondria; this excretion function results in the third function, maintenance of acyl-CoA/CoA ratio constant. (With permission from: Dr. Akiko Wakisaka, Pediatrics, NHO Iou National Hospital, Kanazawa, Japan)

inversely the acyl group is transferred to CoA by the action of CPT2 to become acyl-CoA (+ free carnitine) again, so CPT1 acts to load cargo on the truck, and CPT2 acts to unload the cargo. Acyl-CoA enters the β-oxidation system and is metabolized one by one of each two carbon atoms, to produce acetyl-CoA.

In addition, free carnitine, this empty truck, is also responsible for loading acyl groups of acyl-CoA accumulated in mitochondria in various pathological conditions and excreting them outside the mitochondria, outside the cells, and finally in the urine (Fig. 3.7, bottom). There is no problem with the presence of acyl-CoA in small quantities as an intermediate in metabolism such as β-oxidation of long-chain fatty acids, but when acyl groups accumulate in mitochondria due to various pathological conditions such as inherited metabolic disorders, liver cirrhosis, and heart failure, acyl groups bind to free CoA, which is limited in mitochondria but necessary for energy metabolism, become deficient relatively, and metabolic disorders occur. The total amount of CoA in the mitochondria is limited, and if acyl-CoA increases, the amount of free CoA decreases relatively (as a result, the acyl-CoA/free CoA ratio increases), and normal energy metabolism using free CoA becomes stagnant. In addition, acyl-CoA is important as an intermediate in metabolism, but when it accumulates, it inhibits various enzymes and causes mitochondrial dysfunction. So, in these cases, the accumulated acyl groups act as rubble, not useful fuel. Acylcarnitine is a truck that cannot load rubble (acyl group). Free carnitine is loadable because it is empty trucks. If free carnitine is abundant, carnitine can receive the accumulated harmful acyl groups as the buffering action, load them on the loading platform of the trucks, become acylcarnitines, carry them out of mitochondria and cells through the carnitine cycle, and finally excrete it in the urine. In the situation of carnitine deficiency, it is difficult to excrete these harmful acyl groups outside the mitochondria because free carnitine (trucks) is relatively deficient compared to the rubble (acyl-CoA). Therefore, the biochemical and physiological significance of carnitine supplementation therapy in carnitine deficiency is to administer carnitine preparations from outside and send free carnitine, which is an empty truck, to promote the excretion of harmful acyl groups outside the mitochondria (and eventually outside the body), thereby removing the toxicity of the acyl groups (actually the toxicity of the acyl-CoA) and normalizing the acyl-CoA/free CoA ratio.

3.10 Transport of Long-Chain Fatty Acids into Mitochondria

Next, let us take a closer look at the significance of the transport of long-chain fatty acids into the mitochondria, which is the first biochemical and physiological action in Table 3.3.

3.10.1 Carnitine Cycle and Energy Metabolism

As described in Chap. 2, the most important physiological function of carnitine, which was established in the research of Fritz and Bremer, is its role as an acyl group carrier that transports the acyl group of long-chain fatty acids into the inner mitochondrial membrane. Long-chain fatty acids are extremely important for energy metabolism in the heart, skeletal muscles, and liver. Therefore, carnitine is closely linked to energy metabolism. In Fig. 3.7, the function of carnitine was explained as trucks, but the carnitine cycle, which is the passage of carnitine and acylcarnitine, will be described in a little more detail (Fig. 3.8).

Long-chain acyl-CoA, in which the acyl group of a long-chain fatty acid is activated by the reaction of acyl-CoA synthetase, can pass through the outer membrane of mitochondria. This is because the outer membrane of mitochondria contains channel proteins called porins, which can pass through large molecules with a molecular weight of up to 5000. However, acyl-CoA cannot pass through the mitochondrial inner membrane as it is. In order to convert long-chain fatty acids into energy (ATP), the acyl groups of long-chain fatty acids must be converted to acetyl-CoA, NADH

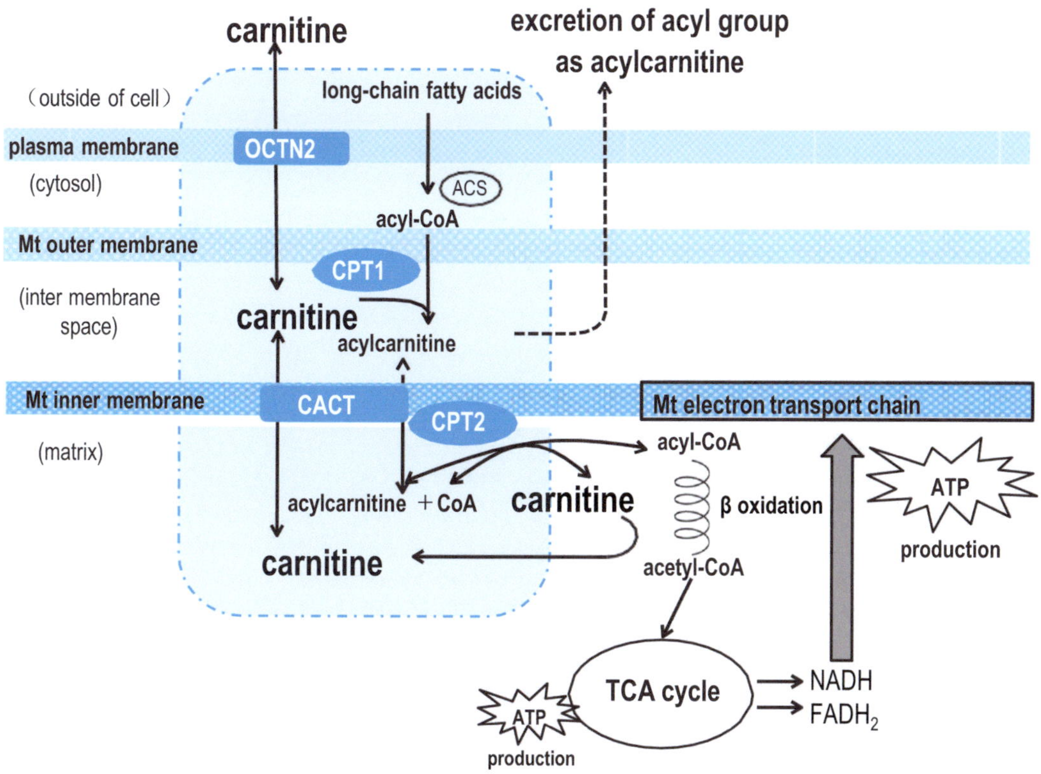

Fig. 3.8 Relationship of carnitine cycle and energy metabolism. The blue area surrounded with dashed-dotted line is called carnitine cycle (or carnitine shuttle), which is closely linked to mitochondrial energy metabolism such as fatty acid β-oxidation and TCA cycle. *ACS* acyl-CoA synthetase, *CACT* carnitine acylcarnitine translocase, *CPT1* carnitine palmitoyltransferase 1, *CPT2* carnitine palmitoyltransferase 2, *FADH*$_2$ reduced flavin adenine dinucleotide, *Mt* mitochondria, *NADH* reduced nicotinamide adenine dinucleotide, *OCTN2* carnitine transporter, *TCA* tricarboxylic acid

(reduced nicotinamide adenine dinucleotide), and $FADH_2$ (reduced flavin adenine dinucleotide) by the β-oxidation system in the inner mitochondrial membrane and these high-energy compounds are metabolized in the TCA cycle and electron transport system on the mitochondrial inner membrane. Therefore, CPT1 localized inside of the outer mitochondrial membrane transfers the acyl group to carnitine and becomes acylcarnitine and free CoA. Then, acylcarnitine is taken up into mitochondria through CACT, an exchange transporter of free carnitine and acylcarnitine in the inner mitochondrial membrane, and at the same time, free carnitine is transported in the opposite direction from the inside to the outside of inner mitochondrial membrane (so called "antiporter"). In other words, acylcarnitine is taken inside the inner mitochondrial membrane and free carnitine is transported to the outside of the inner mitochondria in the opposite direction, and these exchange reactions occur at the same time. Acylcarnitine transported into the mitochondria is converted back to acyl-CoA and free carnitine by carnitine palmitoyltransferase 2 (CPT2) located inside the inner mitochondrial membrane, and acyl-CoA is oxidized by β-oxidation to form acetyl-CoA. This acetyl-CoA enters the TCA cycle and produces NADH and $FADH_2$. NADH and $FADH_2$ enter into the complex of the electron transport system of the inner mitochondrial membrane, Complex I (NADH oxidase complex) and Complex II (succinate dehydrogenase complex), respectively, and produce large amounts of ATP (Fig. 3.8).

The dashed-dotted line area in Fig. 3.8 is called as carnitine cycle (also known as carnitine shuttle) consisting of four components—the carnitine transporter, CPT1, CPT2, and CACT—through which carnitine and acylcarnitine transport the acyl group into and out of the mitochondria. This carnitine cycle is linked to metabolic systems that are important for energy metabolism, such as β-oxidation that is localized inside soluble space of the inner mitochondrial membrane (called the matrix), TCA cycle, and electron transport system on the inner mitochondrial membrane.

In addition, the carnitine cycle plays an extremely important role in the excretion of acyl-CoA which accumulate in the inner membrane of the mitochondria in various pathologies, as opposed to the uptake of the acyl groups of long-chain fatty acids into the mitochondria. Therefore, when the acyl group is transported into the inner mitochondrial membrane and conversely, when the acyl group is transported out of the mitochondria, the exchange reaction between free carnitine and acyl carnitine through CACT is working, and free carnitine plays a role like a priming water for a hand-pump. For this priming action, adequate amount of free carnitine is needed inside and outside of the mitochondria.

Peroxisomes also have an oxidation system of β fatty acids, but they are not directly connected to the energy metabolism system and are closely related to lipid synthesis such as bile acid synthesis, ether-type phospholipid synthesis, and fatty acid remanufacturing [17]. Carnitine is used to transfer β-oxidation products in peroxisomes outside the peroxisome, but carnitine is not required for the incorporation of long-chain fatty acids into peroxisomes [18].

3.10.2 How Much ATP Do Humans Consume?

Carnitine promotes energy metabolism using long-chain fatty acids. Let us look back again the energy metabolism from the basics. The life activities of all living organisms, including humans, use the energy stored in ATP for their vital activities. How much ATP does a human consume in a day? According to Dr. Amagai et al., if the energy requirement of a 20-year-old woman is 2000 kcal, the amount of ATP is about 100 kg [19] (Table 3.4). This means that she is consuming about twice her body weight in ATP in a day! Of course, when ATP is used, it becomes ADP, and that ADP is soon converted back to ATP, so the body weight does not decrease, but this amount of energy is consumed under normal conditions. When there is an increased demand for energy, such as during strenuous exercise, fever, infection, fasting, or starvation, even more ATP is consumed. Therefore, it is very important for the metabolic system to be able to supply sufficient energy (ATP) promptly when energy demand increases.

Then, what organs consume the most ATP? According to Dr. Amagai et al., the liver consumes the largest amount of ATP (29%) per resting energy expenditure [19] (Table 3.5). It is clear that the liver, which is a silent organ, is actually actively working as a chemical factory for living organisms such as gluconeogenesis, lipid synthesis, amino acid metabolism, ammonia metabolism, and albumin synthesis. The heart consumes the largest amount of energy per organ weight, and since it repeatedly contracts and relaxes 100,000 times a day to pump blood throughout the body, it is extremely important to secure an energy source for this organ.

Table 3.4 Estimation of the amount of ATP consumption in human

The energy requirement for resting female aged 20 years is 2000 kcal/day
It is roughly 75% of the energy consumption ⇒ 1500 kcal/day
1 mole of ATP can produce 7.6 kcal (at pH 7.0, 25 °C)
It means 1500 ÷ 7.6 = 197.3 moles/day of ATP are consumed in a day
The molecular weight of ATP is 507.2 ⇒ 1 moles of ATP is 507.2 g
197.3 モル × 507.2 g = 100,070.56 g ≒ 100 kg
About 100 kg of ATP is consumed in a resting young female!

From Ref. [19]

Table 3.5 Energy consumption of various organs

Organs	% REE	% weight	% REE/% weight
Liver	29	2	14.5
Brain	19	2	9.5
Heart	10	0.4	25.0
Kidney	7	0.4	17.5
Muscle	18	40	0.45
Total	83	44.8	66.95

From Ref. [19]
REE resting energy expenditure

3.10.3 Numbers of ATP Produced from Glucose and Long-Chain Fatty Acids

The energy (ATP) necessary for these organisms is produced by metabolizing dietary carbohydrates and fats. When glucose is metabolized anaerobically in glycolysis, two molecules of ATP are produced. In aerobic condition, when glucose energy metabolism is extended to the mitochondrial TCA cycle and oxidative phosphorylation, a total of 32 molecules of ATP are produced (Table 3.6).

How many molecules of ATP are produced when long-chain fatty acids are metabolized? As an example, let us take the C_{16} saturated fatty acid palmitic acid, which is a major component of dairy products such as milk and butter, animal fats such as lard, and vegetable oils such as coconut oil. When palmitic acid is transported into the mitochondria by carnitine and undergoes β-oxidation, 8 molecules of acetyl-CoA are produced as shown in Fig. 3.9, and when it enters the TCA cycle and electron transport system, a total of 106 molecules of ATP are produced (assuming a P/O ratio of 2.5 for NADH and 1.5 for $FADH_2$. If a P/O ratio is 3.0 for NADH and 2.0 for $FADH_2$, a total of 129 molecules of ATP are produced). Thus, the β-oxidation of long-chain fatty acids is a very efficient and important energy metabolism process for organs that require a large amount of energy, such as the liver, heart, and skeletal muscles. Therefore, in severe carnitine deficiency, cardiac symptoms such as cardiomyopathy and cardiac dysfunction, and muscular symptoms such as hypotonia, limpness, and myalgia are often observed because energy production from long-chain fatty acids is impaired due to carnitine deficiency. Fatty acid energy metabolism is also required for the liver to supply sugar from amino acids and other substances through the glycogenic system during hypoglycemia, when food deprivation is prolonged, and for the liver to produce ketone bodies to supply to the brain. When energy production by fatty acid oxidation is impaired due to carnitine deficiency, clinical symptoms such as hypoglycemic attacks, cardiac symptoms such as

Table 3.6 Number of ATP generated in glycolysis[a]

Pathway	Reaction	ATP synthesis	No. of ATP
Glycolysis	Glucose → 2 pyruvate	Substrate-level phosphorylation	2
		Oxidative phosphorylation of NADH transported through malate-aspartate shuttle[b]	5
PDH	2 pyruvate → 2 acetyl-CoA	Oxidative phosphorylation of 2 NADH	5
TCA cycle	Oxidative degradation of 2 acetyl-CoA	From 2 GTP derived from substrate-level phosphorylation	2
		Oxidative phosphorylation of 6 NADH and 2 $FADH_2$	18
Total	4 ATP from substrate-level phosphorylation, 28 ATP from oxidative phosphorylation		32

PDH pyruvate dehydrogenase
[a]In the case of P/O ratio, it is assumed: for NADH 2.5, for $FADH_2$ 1.5. Some textbooks describe P/O ratio for NADH 3.0, for $FADH_2$ 2.0. In this case, total 38 ATP are generated from glycolysis and oxidative phosphorylation
[b]In the case of liver, heart, and kidney, cytosol NADH derived from glycolysis cannot pass mitochondrial inner membrane. So NADH is transported into mitochondria through malate-aspartate shuttle, and 1 NADH generates 2.5 ATP by the mitochondrial electron transport chain. In the case of brain and muscle, NADH is transported into mitochondria through glycerol-3-phosphate shuttle, and NADH is transferred to $FADH_2$, so that 1 NADH generates 1.5 ATP

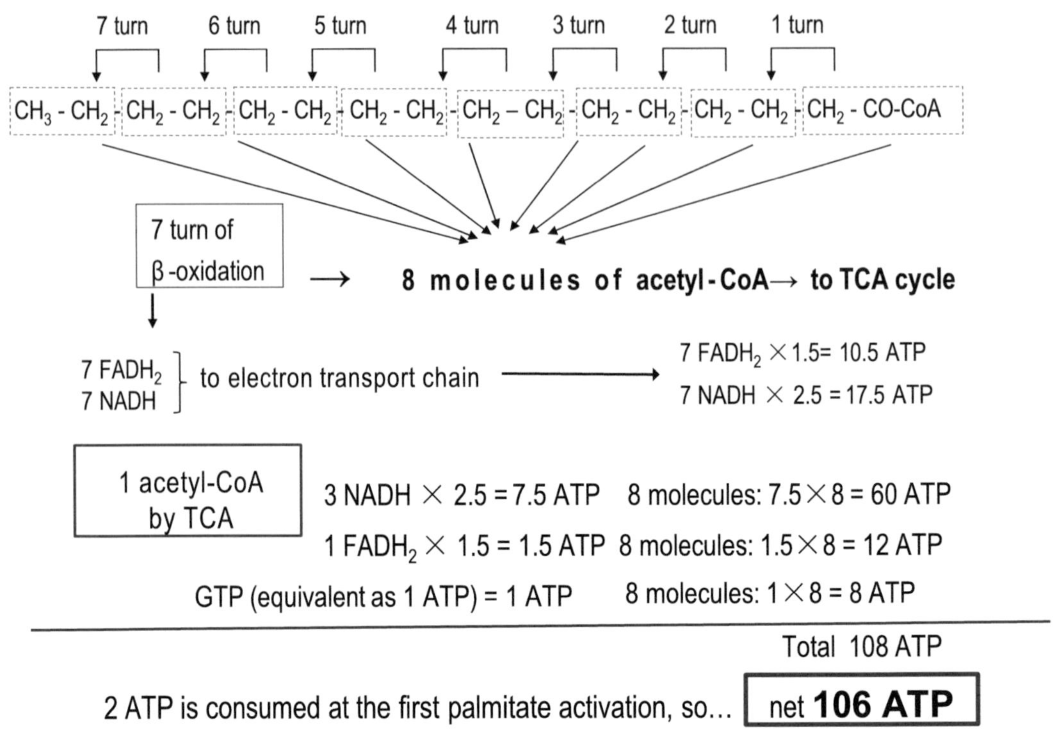

Fig. 3.9 Number of ATP generated by β-oxidation of one molecule of palmitate is shown when P/O ratio is assumed for NADH 2.5, and for $FADH_2$ 1.5. (In the case of P/O ratio is assumed for NADH 3.0 and $FADH_2$ 2.0, total 129 ATP is generated by β-oxidation of 1 molecule of palmitate)

cardiomyopathy and heart failure, and muscle symptoms such as muscle weakness are likely to occur. In addition, since a large amount of ATP is required for the conversion of ammonia to urea and albumin synthesis in the urea cycle, hyperammonemia and hypoalbuminemia are likely to occur when ATP production is impaired due to carnitine deficiency.

Thus, efficient utilization of long-chain fatty acids as an energy source is extremely important for maintaining blood glucose levels and producing ketone bodies, especially when energy demand increases due to fasting, starvation, infection, or fever, because it enables the production of large amounts of energy (ATP). A large amount of ATP is also required for ammonia detoxication and albumin synthesis, so if carnitine is deficient and long-chain fatty acids are not available for energy metabolism, hyperammonemia and hypoalbuminemia are likely to occur.

3.11 Excretion of Harmful Acyl Compound and Modulation of Acyl-CoA/Free CoA Ratio

3.11.1 The Concept of Carnitine Insufficiency

The second biochemical and physiological action of carnitine is the excretion of harmful acyl compounds accumulated in the mitochondria due to metabolic abnormalities. In the 1960s and 1970s, when the biochemical and physiological effects of carnitine in mammals were studied, the focus was only on the role of the inward transport of long-chain fatty acids to mitochondria, which is the first biochemical/physiological action described above. However, in the 1980s, it was reported that when carnitine is administered in propionic acidemia and methylmalonic acidemia, the accumulated harmful intermediate metabolite organic acids such as propionic acid and methylmalonic acid are excreted outside the mitochondria as acylcarnitine, suggesting that carnitine is also involved in the transport of toxic acyl groups out of mitochondria [20–22]. Furthermore, it was proposed for the first time that the essence of carnitine supplement therapy in organic acidemia is to normalize the acyl-CoA/free CoA ratio in the mitochondria by promoting the excretion of acyl groups of harmful organic acid metabolites out of the mitochondria, such as propionylcarnitine and methylmalonylcarnitine, and to maintain the amount of free CoA required for metabolism within a normal range. This situation in which acyl-CoA/free CoA ratio is abnormally high is named "carnitine insufficiency" [23, 24]. In this carnitine insufficiency, due to relative deficiency of free carnitine to buffer the accumulated harmful acyl-CoA, mitochondrial dysfunction might occur. The third biochemical and physiological action of carnitine, "modulation of the acyl-CoA/free CoA ratio in mitochondria," is the buffering action of carnitine to maintain free CoA levels. In other words, to put it simply, "modulation of acyl-CoA/free CoA ratio" can be simply expressed as a buffering action, or the action of regenerating free CoA, which is important for various metabolism, by removing the acyl group from acyl-CoA as acylcarnitine outside the mitochondria.

In other words, the pathophysiology of carnitine insufficiency is a condition in which free CoA, which is important for various metabolism, decreases and acyl-CoA increases relatively. Therefore, the essence of the pathology is that a deficiency of free CoA or an increase in acyl-CoA causes metabolic disorders. So it should be originally called "CoA insufficiency." There are many methods for quantifying CoA and acyl-CoA, such as enzyme method, mass spectrometry, and high-performance liquid chromatography, and all of them require deproteinization and specialized analytical techniques, and are not easy [25, 26]. On the other hand, carnitine measurement is not so difficult compared to that of CoA or acyl-CoA. So it is called "carnitine insufficiency" because it has been diagnosed based on the ratio of acylcarnitine and free carnitine (AC/FC) in blood by appropriate measurement. When acyl-CoA accumulates in the mitochondria, carnitine receives the acyl group of acyl-CoA to buffer (detoxify) it and become acylcarnitine. Since the free carnitine content in the mitochondria is much higher than the CoA content, a buffering effect is exerted. In this situation, which is a kind of steady state, acylcarnitine content reflects acyl-CoA content and free carnitine content reflects free CoA contents. However, in the pathogenesis of CoA insufficiency (≈carnitine insufficiency), the amount of free carnitine relative to acyl-CoA is insufficient, so its buffering action cannot be performed (carnitine insufficiency). Namely, AC/FC ratio is thought to reflect acyl-CoA/free CoA ratio. In normal condition, the free carnitine content in mitochondria is much higher than the CoA content, thus exerting a buffering action. However, in carnitine insufficiency, the amount of free carnitine is insufficient relative to accumulated acyl CoA, so insufficient amount of carnitine cannot exert the buffering action. In this situation, external supplementation with free carnitine causes the acyl groups of toxic acyl CoA to be excreted out of the mitochondria as acylcarnitine due to the increased free carnitine in the mitochondria, and also promotes the biosynthesis of free CoA, and the acyl CoA/free CoA ratio is modulated within the normal range, and mitochondrial energy metabolism and other processes proceed smoothly. Acyl-CoA compounds themselves are important as metabolic intermediates, but the amount of free CoA in mitochondria, which is important for energy metabolism, is limited. Therefore, when excess amount of the acyl group of organic acids or fatty acids accumulates in the mitochondria, the acyl group combines with free CoA to form acyl-CoA, and the free CoA decreases relatively, resulting in mitochondrial dysfunction.

3.11.2 Enzymes Inhibited by Acyl-CoA

Excessive accumulation of acyl CoA in mitochondria inhibits various enzymes, causing mitochondrial dysfunction and other clinical symptoms such as hypoglycemic attacks and hyperammonemia [27–37] (Table 3.7).

Table 3.7 Enzymes inhibited by various acyl-CoA

Enzymes	Acyl-CoA	Clinical significance and implications	Refs.
Adenine nucleotide translocase	Long-chain acyl-CoA (e.g., oleyl-CoA)	Rate-limiting enzymes of oxidative phosphorylation are inhibited, so ATP cannot be synthesized	[28]
Citrate synthase	Long-chain acyl-CoA (e.g., palmitoyl-CoA)	Ketone body such as acetoacetate cannot be synthesized, so might be prone to hypoglycemic attack	[29] [30]
Glutamate dehydrogenase	Long-chain acyl-CoA (e.g., palmitoyl-CoA)	Gluconeogenesis from glucogenic amino acids such as glutamate cannot progress, so might be prone to hypoglycemic attack	[31]
Hormone-sensitive lipase	Palmitoyl-CoA, oleyl-CoA, etc.	Lipolysis is suppressed and energy production or ketogenesis in starvation cannot progress, so might be prone to hypoglycemic attack	[32]
Malate dehydrogenase	Long-chain acyl-CoA (e.g., palmitoyl-CoA)	TCA cycle is inhibited, and gluconeogenesis from malate ⇔ oxaloacetate ⇔ phosphoenolpyruvate cannot progress	[31]
N-acetylglutamate synthase	Propionyl-CoA, methylmalonyl-CoA	N-acetylglutamate is an important cofactor of CPS-1 in the urea cycle, so might be prone to hyperammonemia	[33]
3-oxo-acyl-CoA thiolase	Pent-4-enoyl-CoA	Metabolites of pent-4-enolic acid, which structurally resemble hypoglycin, can cause Jamaican vomiting disease, inhibit β-oxidation, and cause hypoglycemia	[34]
Pyruvate dehydrogenase complex	Propionyl-CoA (low concentration)	Very important pyruvate dehydrogenase, located at the junction position of glycolysis and TCA cycle, and having central role in energy metabolism, which might be inhibited	[35]
Pyruvate dehydrogenase kinase	Propionyl-CoA (high concentration)	Pyruvate dehydrogenase activity is activated (conserved) because high concentration of propionate inhibits pyruvate dehydrogenase kinase	[35]
Pyruvate carboxylase	Malonyl-CoA, methylmalonyl-CoA, succinyl-CoA, etc.	Pyruvate carboxylase, which supplies oxaloacetate, is inhibited and gluconeogenesis in the liver can cause trouble	[36]
Succinate: CoA ligase (GTP)	Propionyl-CoA, glutaryl-CoA, methylmalonyl-CoA, etc.	Acyl-CoA increases and free CoA decreases, and succinate: CoA ligase (GTP) in TCA cycle is inhibited, hence might be prone to metabolic abnormalities such as Reye's syndrome	[37]

Created with reference to Refs. [27–37]
CoA coenzyme A, *CPS-1* carbamoyl phosphate synthetase-1, *OAA* oxaloacetic acid, *PEP* phosphoenolpyruvate, *TCA* tricarboxylic acid

Specifically, it inhibits enzymes that are members of the TCA cycle, such as malate dehydrogenase, citrate synthase, and succinate, CoA ligase (also called succinyl-CoA synthase), and pyruvate dehydrogenase complexes located at the junction between glycolysis and TCA cycles, causing serious disturbances in intracellular energy metabolism. The liver produces energy (ATP) by β-oxidation of fatty acids during fasting and uses that energy for gluconeogenesis system to produce glucose from pyruvate and glycogenic amino acids by the action of pyruvate carboxylase and glutamate dehydrogenase, and supplies glucose to the brain. In addition, long-chain fatty acids released by the action of hormone-sensitive lipase are oxidized by β-oxidation, producing acetyl-CoA which is used for ketone biosynthesis. Then ketone bodies are released into the bloodstream and used by the brain and heart during hypoglycemia. When the activity of citrate synthase and hormone-sensitive lipase is inhibited by acyl-CoA, ketone production is suppressed, and hypoketotic hypoglycemic attacks are more likely to occur. In addition, N-acetylglutamate synthase, which synthesizes N-acetylglutamate, an essential cofactor for maintaining activity of carbamoyl phosphate synthetase-1 (CPS-1), the first enzyme of the urea cycle, is also inhibited by acyl CoA, making hyperammonemia more likely to occur.

However, the metabolism of carnitine and CoA varies from tissue to tissue, and thus different tissues and organs are affected in various diseases. For example, propionic acid inhibits oxidative phosphorylation of mitochondria in the liver, but not in muscle mitochondria [38]. Although there are differences in the toxicity of acyl-CoA depending on these tissues, the highly toxic acyl-CoA is usually converted to acylcarnitine, which is not toxic. In addition, since the equilibrium between acyl-CoA and acyl-carnitine is rapid, it is effective in reducing the acyl-CoA concentration by excreting the acyl group outside the mitochondria and further into the urine as acylcarnitine as long as the carnitine concentration is sufficient. With the exception of primary carnitine deficiency (also known as systemic carnitine deficiency) due to genetic abnormalities in the carnitine transporter (OCTN2), the essence of most carnitine deficiency is not an absolute deficiency of free carnitine, but a relative deficiency of free carnitine to the accumulation of harmful acyl groups.

Many of these enzymes inhibited by acyl-CoA are localized in mitochondria and are involved in energy metabolism, lipid metabolism, and ammonia metabolism such as the TCA cycle, and various clinical symptoms such as hypoglycemia and hyperammonemia occur when inhibited (see Chap. 5, Sect. 5.5). If there is a sufficient amount of carnitine, it receives an acyl group, becomes an acylcarnitine such as propionylcarnitine, excretes the acyl group outside the mitochondria, and finally excretes it in the urine. As a result, metabolic disorders caused by propionic acid (or propionyl-CoA) are canceled and clinical symptoms are significantly improved. However, if this condition continues for a long time, a large amount of carnitine is excreted in the urine in the form of propionylcarnitine, etc., and especially in children, the amount of carnitine intake from the diet and the amount of biosynthesis in the body are limited, so the carnitine in the body is deficient and secondary carnitine deficiency occurs. Therefore, by administering carnitine, clinical symptoms such as hypoglycemic attacks and hyperammonemia caused by secondary carnitine deficiency can be improved, and metabolic disorders caused by acyl-CoA can be relieved. This is the significance of carnitine supplementation therapy.

Although there are many papers that define carnitine deficiency as the absolute value of free carnitine concentration in the blood and the total concentration of carnitine in tissues, the essence of carnitine deficiency is whether there is enough free carnitine to buffer (or detoxify) the toxicity of the acyl group (acyl-CoA) accumulated in the mitochondria, and hence the amount of free carnitine relative to the amount of acyl group, not absolute values, is the essential concern [27]. In conditions such as organic acidemia, where a large amount of harmful acyl groups accumulate, a suitable amount of free carnitine is required to detoxify it.

3.11.3 Types and Roles of Carnitine Acyltransferase

So far, carnitine and CoA have been discussed. Next the types and roles of carnitine acyltransferases are summarized and explained. Carnitine acyltransferases have often appeared in the pathophysiology of carnitine insufficiency, and have the action of maintaining the amount of free CoA, which is important for metabolism, by transferring the acyl group of acyl—CoA to carnitine and excreting it outside the cell as acylcarnitine, and by adjusting the acyl—CoA/free CoA ratio. Carnitine acyltransferases carry out the following reactions:

$$\text{carnitine} + \text{acyl-CoA} \underset{\text{carnitine acyltransferase}}{\rightleftarrows} \text{acylcarnitine} + \text{CoA}$$

This reaction is reversible, and, for example, when acyl—CoA increases, carnitine receives an acyl group and becomes acylcarnitine, which is excreted outside the mitochondria through CACT, and this rightward reaction proceeds to produce free CoA. In addition, after the acyl group of a long-chain fatty acid is transported into the mitochondria as acylcarnitine, it is converted back to acyl—CoA by CPT2 (leftward reaction), one of the carnitine acyltransferases, and this acyl—CoA is subjected to β-oxidation. The reaction between carnitine and carnitine acyltransferase can be regarded as a modulating reaction of the ratio of acyl—CoA to free CoA as follows (the third action of the biochemical and physiological actions of carnitine) [39].

$$\frac{\text{acyl-CoA}}{\text{CoA}} \Longleftrightarrow \frac{\text{acylcarnitine}}{\text{free carnitine}}$$

carnitine acyltransferases

Normally, the free carnitine content in the mitochondria is much higher than the CoA content, so it can exert the buffering action. However, in the condition of carnitine deficiency or carnitine insufficiency, the amount of carnitine is deficient or insufficient and this buffering effect cannot be exerted. Due to the action of carnitine and carnitine acyltransferase, the acyl—CoA/free CoA ratio is kept constant, and energy metabolism in mitochondria using free CoA proceeds smoothly.

In cells, several types of carnitine acyltransferases, whose specificities are different depending on the length of the carbon chain of the acyl group, are present in the outer and inner membranes of the mitochondria, peroxisomes, endoplasmic reticulum, cytoplasm, etc., and exert their functions. In addition, carnitine acyltransferase is also present in the erythrocyte cell membrane and the sarcoplasmic reticulum of the heart of muscle cells, and when the phospholipid of the cell membrane is peroxidized and cleaved by phospholipase to become lysophospholipid with a strong surfactant effect, it is thought that these acyltransferases act to repair the lysophospholipid to the original phospholipid and to regulate the microenvironment of Ca^{2+} transport of the sarcoplasmic reticulum of cardiomyocytes. The types and localizations of these carnitine acyltransferases are summarized in Table 3.8 [39].

CPT1, which is present in the outer mitochondrial membrane, usually exerts rightward reaction, and CPT2, which is present inside the inner mitochondrial membrane, promotes the leftward reaction as follows.

$$\text{carnitine} + \text{long-chain acyl-CoA} \underset{\text{CPT2}}{\overset{\text{CPT1}}{\rightleftarrows}} \text{long-chain acylcarnitine} + \text{CoA}$$

The acyl groups of long-chain fatty acids such as palmitic acid are used in these reactions. The actual incorporation of long-chain fatty acids into the mitochondria is mediated by long-chain acyl—CoA synthetase (ACS), CPT1, and porin in the mitochondrial outer membrane, and CPT2 and CACT in the mitochondrial inner membrane, which form a cluster at the site where the outer membrane and inner membrane are close together. CPT2 and CACT in the mitochondrial inner membrane are thought to efficiently convert long-chain fatty acids to acyl—CoA by forming clus-

Table 3.8 Acronyms, names, subcellular localization, acyl-chain specificity, and malonyl-CoA sensitivity of various carnitine acyltransferases

Acronym	Name	Subcellular localization	Acyl-chain specificity	Malonyl-CoA sensitivity
CrAT	Carnitine acetyltransferase	Mitochondria, peroxisome, ER (soluble membrane-bound)	$C_2 \sim C_6$	No
CPT1	Outer palmitoyltransferase	Mitochondria facing the cytosol (probably outer membrane)	$C_6 \sim C_{18}$	Yes
CPT2	Inner palmitoyltransferase	Mitochondrial inner membrane	$C_8 \sim C_{18}$	No
COT	Peroxisomal carnitine octanoyl-transferase	Peroxisome (membrane-bound)	$C_6 \sim C_{22}$	Yes
CPT (ER)	Microsomal carnitine acyltransferase	ER	$C_8 \sim C_{22}$	Yes
CPT (RBC)		Plasma membrane of RBC	$C_8 \sim C_{22}$	Yes
CPT (SR)		SR		Yes

From Ref. [39]

ER endoplasmic reticulum, *RBC* red blood cell, *SR* sarcoplasmic reticulum

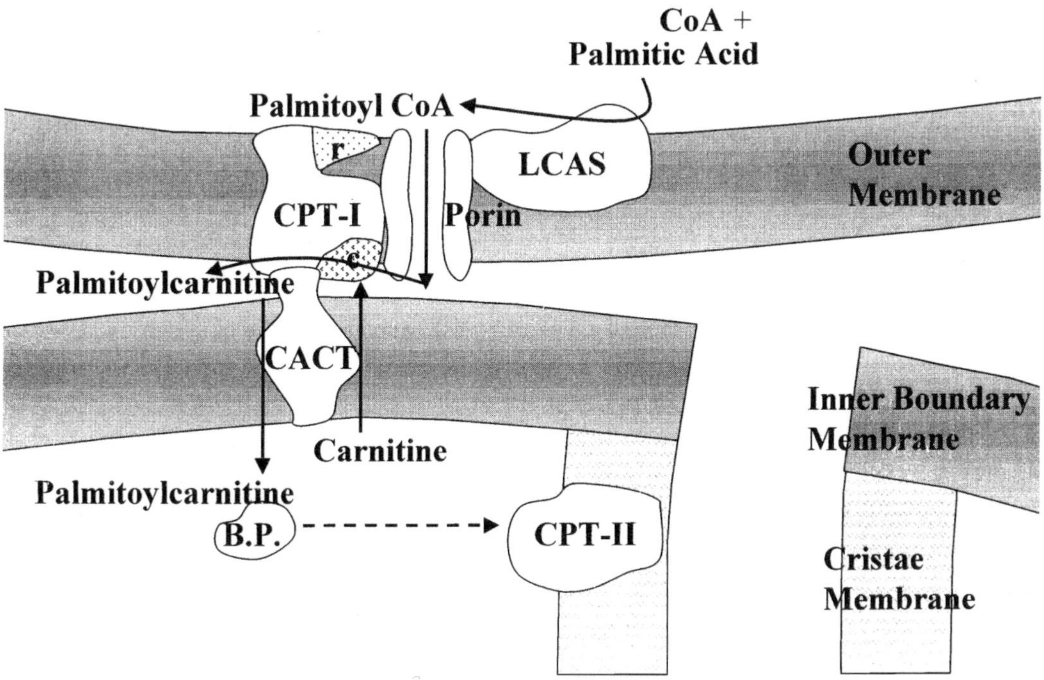

Fig. 3.10 Proposed model of mitochondrial fatty acid uptake with the malonyl-CoA binding site (r) of CPT-I exposed to the cytosol and catalytic site (c) exposed to the intermembrane face of the outer membrane. *BP* postulated matrix binding protein for palmitoylcarnitine, *CACT* carnitine acylcarnitine translocase, *CPT-I* malonyl-CoA sensitive carnitine palmitoyltransferase, *CPT-II* malonyl-CoA insensitive carnitine palmitoyltransferase, *LCAS* long-chain acyl-CoA synthase. (From Ref. [40])

ters at the site where the outer and inner membranes are close together [40] (Fig. 3.10).

In addition to acyltransferases such as CPT1 and CPT2, which transfer long-chain acyl group to carnitine (or remove acyl groups), carnitine acetyltransferases (hereinafter abbreviated as CrAT; there is also some literature that refer it as CRAT and CAT), which have high specificity for short-chain acyl groups, are also found in cells, along with carnitine octanoyltransferase (COT), which is highly specific for medium-chain acyl groups such as octanoic acid [41, 42].

CrAT catalyzes the following reaction.

$$\text{carnitine} + \text{acetyl-CoA} \underset{\text{CrAT}}{\rightleftarrows} \text{acetylcarnitine} + \text{CoA}$$

It is a reaction in which the acetyl group ($CH3CO^{-}$), which is the acyl group of acetic acid, is transferred to carnitine, but also acyl groups of short-chain organic acids of $C_2 \sim C_6$ is transferred. It is thought that CrAT converts acetyl CoA produced in excess by β-oxidation, etc. into acetylcarnitine, which is transported to other intracellular organelles and extracellular space for effective utilization in other metabolic processes, such as fatty acid synthesis. Therefore, acetylcarnitine and CrAT are extremely important in maintaining mitochondrial function and cooperative metabolism with other organelles such as peroxisomes [41, 42]. Acetyl—CoA is the end product of β-oxidation and is an important substance that enters the TCA cycle and is converted into energy, but since it is one of the acyl—CoA, if it accumulates in the mitochondria in excess, it will also cause carnitine insufficiency, and free CoA, which is important for energy metabolism, will decrease, causing mitochondrial dysfunction. Excess acetyl—CoA allosterically inhibits pyruvate dehydrogenase (PDH). PDH is an important enzyme at the junction of the glycolysis pathway and TCA cycles, and its inhibition leads to a stagnation of glucose metabolism [43].

This condition is called "metabolic inflexibility" and is considered to be one of the causes of conditions such as insulin resistance in diabetes and obesity, diabetic cardiomyopathy, liver diseases such as non-alcoholic steato hepatitis (NASH) and non-alcoholic fatty liver disease (NAFLD), and heart failure [44–50]. These kinds of metabolic disturbances are also thought to occur in intermittent claudication (IC). If there is sufficient free carnitine, the acetyl group of acetyl-CoA is transferred to carnitine by the action of CrAT, and it becomes acetylcarnitine, and the inhibition of PDH is canceled.

The COT catalyzes transfer of acyl group of C_6-C_{10} medium-chain fatty acids such as octanoic acid (C_8), but is found only in peroxisomes in the cell [41, 42].

$$\text{carnitine} + \text{octanoyl-CoA} \underset{\text{COT}}{\rightleftarrows} \text{octanoylcarnitine} + \text{CoA}$$

CrAT and COT are also thought to be used to transport short- and medium-chain acyl groups produced when medium-chain fatty acids and branched-chain fatty acids are oxidized by β- or ω-oxidation of peroxisomes to mitochondria as acylcarnitine. Crystal structure analysis has already been performed on CrAT and COT, and detailed structures of the sites where carnitine, acetyl-CoA, or octanoyl-CoA bind have been reported [51, 52].

The subcellular localization and function of CrAT and COT are shown in Fig. 3.11, focusing mainly on acetyl CoA and CrAT.

Thus, carnitine is not only an inward carrier of long-chain acyl groups, but also contributes to various metabolic activities by expelling various acyl groups accumulated in mitochondria out of mitochondria, transporting short-chain (mainly acetyl groups) and medium-chain acyl groups between organelles such as peroxisomes and mitochondria, and in some cases transporting them out of cells, contributing to various metabolic processes.

As described so far, carnitine and CoA have a very deep relationship. The central function of actual metabolism is performed by CoA, but carnitine always supports mitochondrial metabolism by CoA, such as passing an acyl group to the CoA to promote energy production by β-oxidation, and conversely, receiving the acyl group and excreting it outside the mitochondria when acyl—CoA accumulates. Both carnitine and CoA are classical molecules in biochemistry, but as mentioned above, new effects and physiological significance are being discovered. Abnormalities in homeostasis of carnitine and CoA are closely related to insulin resistance, chronic inflammation in lifestyle-related diseases, and aging, and are expected to be the focus of new research in the future.

3.12 L-Carnitine in Pharmaceuticals and Supplements

In Japan, a pharmaceutical L-carnitine preparation, L-Cartin® tablets, was approved and launched for the treatment of carnitine deficiency in methylmalonic acidemia and propionic acidemia in 1990. On March 10, 2011, the indication was expanded to carnitine deficiency by a public knowledge-based application, and the drug is now widely used not only for carnitine deficiency in methylmalonic acidemia and propionic acidemia but also for carnitine deficiency due to inherited metabolic diseases other than methylmalonic acidemia and propionic acidemia, as well as carnitine deficiency due to dialysis and valproic acid.

In Europe and the United States, pharmaceutical carnitine preparations were approved for the treatment of carnitine deficiency, while the dietary supplement L-carnitine was used as an ingredient in nutritional supplements, foods for infants and young children, and sports drinks. Against this unique Japanese background, where only pharmaceutical L-carnitine was approved, on March 19, 2001, the Office of Trade and Investment Ombudsman (OTO) of the Cabinet Office received a complaint from a Swiss chemical manufacturer through the Swiss Embassy in Tokyo as a petitioner, requesting that "The classification of L-carnitine as a food and drug should be reviewed and that supplemental L-carnitine

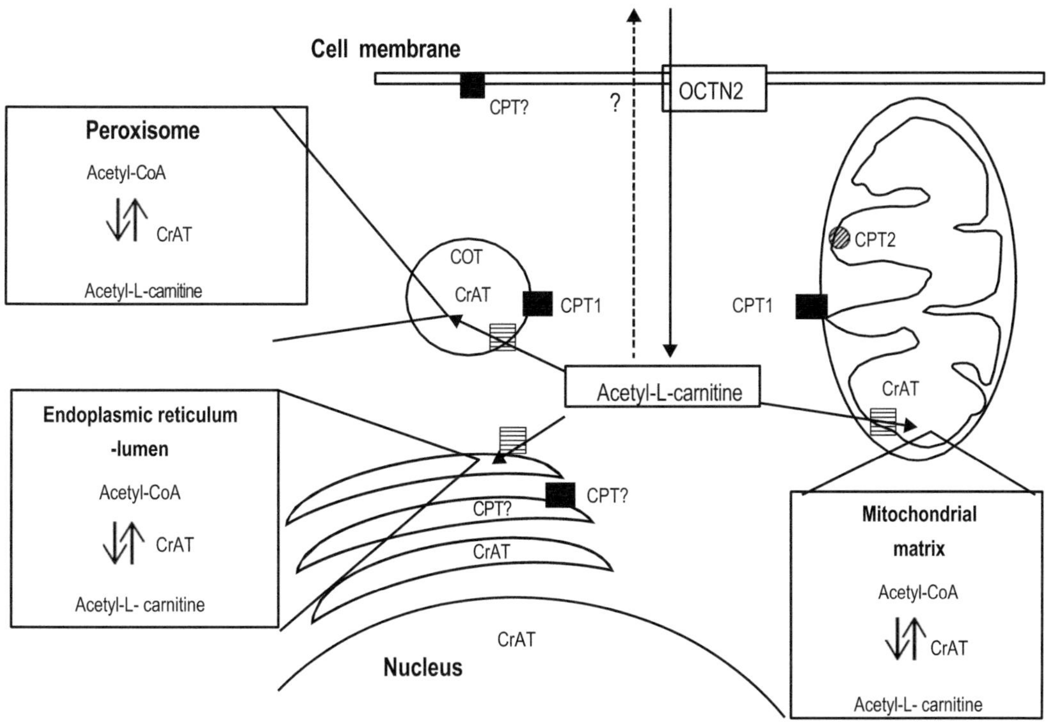

Fig. 3.11 Proteins of the carnitine system connect pools of acetyl-CoA. Malonyl-CoA sensitive enzymes such as CPT 1, shown as black squares, act on cytosolic long-chain substrate. CrAT is found only inside the organelles using matrix acetyl-CoA and acetyl-carnitine. The mitochondrial exchange carrier (CACT, striped squares) transfers carnitine and its esters across the membranes and similar transport proteins are suggested of other organelles. The sodium-dependent organic cation transporter (OCTN2) is the high-affinity carnitine transporter for uptake of carnitine into the cell but the mechanism for export from the cell is not well characterized. *CACT* (☒) carnitine acylcarnitine translocase, *COT* carnitine octanoyltransferase, *CPT 1* (■) carnitine palmitoyltransferase 1, *CPT 2* (⊘) carnitine palmitoyltransferase 2, *CrAT* carnitine acetyltransferase, *OCTN2* carnitine transporter. (From Ref. [41])

products should be allowed, same as in European countries and the United States." In other words, L-carnitine could not be imported into Japan for use other than pharmaceuticals. At the same time, the Ministry of Health, Labour and Welfare (MHLW) issued a draft judgment standard "Handling of Ingredient Essence (Raw Materials) in the Food and Drug Classification" in August 2000, in response to requests that many vitamins, amino acids, and minerals should be allowed to be used as food products in Japan. and studies were underway to review the food and drug classification. In this trend, L-carnitine was also included in the item to be reviewed in the food and drug classification along with other vitamins, etc., and after consultation with experts, the MHLW issued a notice on December 25, 2002 (Food and Drug Administration No. 1225001, December 25, 2002), stating that "L-carnitine shall not be classified as a pharmaceutical product unless it claims a pharmaceutical indication. L-carnitine can be used as a food additive without having to be designated as a food additive and can be immediately distributed as food." Regarding the "Revision of food-drug classification" including carnitine products, on June 1, 2004, MHLW issued the "Revision of the handling under the Foods Sanitation Law of ingredient essence (raw material), not deemed to be a drug unless it claims a drug indication," and the past Director's Notification was repealed. In the revised law, it is stated that "the acceptable daily intake of L-carnitine products is 20 mg/kg/day in the United States and 1000 mg/day is the intake requirement in Switzerland." As a result, many L-carnitine supplements and drinks containing L-carnitine have been sold in Japan since around 2003 [53].

Most of the carnitine products in supplements are tartrate and fumarate levocarnitine, and most use L-carnitine produced by the fermentation method of Lonza in Switzerland but switched to chemical synthesis using asymmetric synthesis method since 2016 (there is also a free levocarnitine product). L-carnitine in pharmaceuticals is a chemically synthesized DL-carnitine (racemic body) that is separated to the L-form by optical resolution. Although there is basically no difference in biochemical and physiological effects between L-carnitine preparations in pharmaceuticals and in supplements, L-carnitine preparations in supplements are sold as "fat burning supplements" for diet and cannot be used for "carnitine deficiency." In addition, L-carnitine preparations for pharmaceuticals are manufactured and approved under strict quality control in terms of purity, impurities, manufacturing processes, etc., and detailed information necessary for use as a pharmaceutical product can be found in the package insert and interview form of each product. However, there is no detailed package insert and it is not clear about the quality for the supplement product. Quality control is the responsibility of each manufacturer. According to Matsushima et al., when the carnitine content of 22 L-carnitine products sold as health foods was examined, the content fluctuated greatly from 79 to 145% in the analysis value relative to the labeled value, but D-carnitine, which antagonizes the action of L-carnitine, was not detected [54].

3.13 Acetyl-L-Carnitine and Propionyl-L-Carnitine

When searching for "L-carnitine" as a search term in PubMed, etc., be aware that in addition to information on L-carnitine, information on acetyl-L-carnitine (hereinafter abbreviated as ALC, and in some literature abbreviated as ALCAR) and propionyl-L-carnitine (hereinafter abbreviated as PLC) are often included. Both compounds were developed by Sigma Tau S.p.A. (now Alpha Sigma S.p.A.) of Italy, a pioneer in carnitine preparations, and are approved as drugs in many countries including Italy. Figure 3.12 shows the chemical structure of L-carnitine, ALC, and PLC chloride.

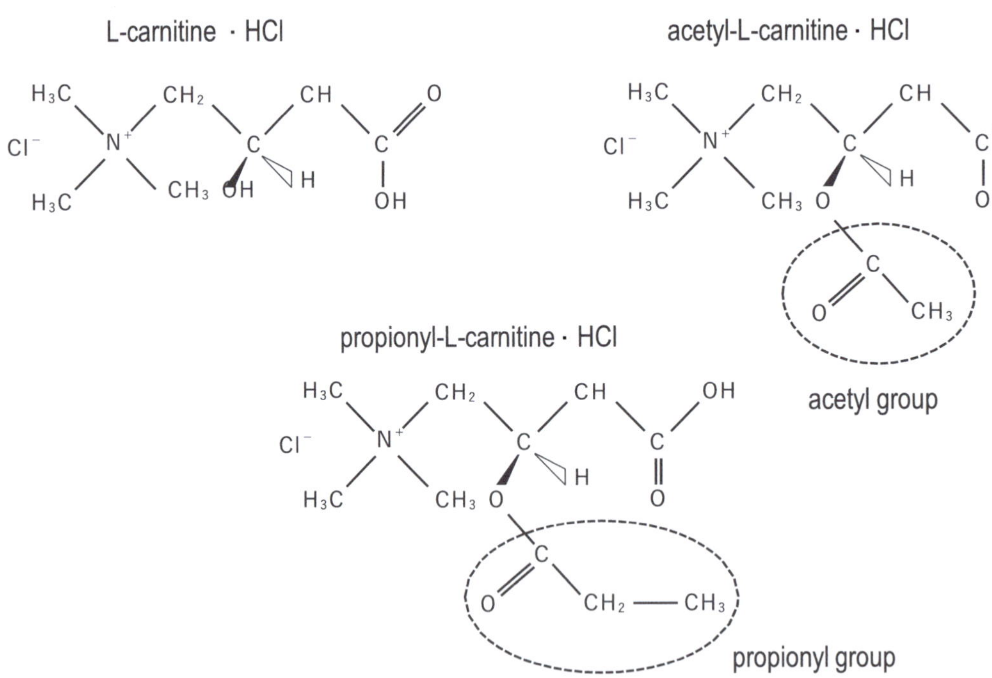

Fig. 3.12 Chemical structures of hydrochloride of L-carnitine, acetyl-L-carnitine, and propionyl-L-carnitine

ALC is sold as a supplement in Europe and the United States. On the other hand, neither compound is marketed as a drug nor as a dietary supplement in Japan, but imported products are available on the Internet.

ALC and PLC are also present in small amounts in the body as described below. Although the biochemical and physiological actions of ALC and PLC are similar to those of L-carnitine, it is unclear whether ALC and PLC act as L-carnitine in the body when administered externally as a drug, or whether ALC and PLC have their own actions, because there are very few studies that have directly compared the effects of L-carnitine, ALC, and PLC. Many meta-analyses and systematic reviews on the evaluation of carnitine in various diseases search for "L-carnitine" as a search term and analyze clinical studies in which L-carnitine is found, as well as ALC and PLC. Since ALC and PLC are analogous compounds similar to L-carnitine but having structures different from L-carnitine as compounds, this document is limited to summarize and review the findings of different compounds. It should be noted that it is considered to be not correct research to compare clinical studies of different compounds together. This book will only provide an overview of the findings of both compounds and attach the latest review article. If you are interested, please refer to those references.

3.13.1 Acetyl-L-Carnitine (ALC)

ALC is the shortest acylcarnitine, a compound in which the acetyl group of acetic acid (CH_3COOH) is esterified to L-carnitine. About 83% of the total carnitine in the blood of healthy adults is free carnitine and about 17% is acylcarnitine, of which about 75% is ALC, the most abundant acylcarnitine [55].

According to studies using animals, ALC administered has been reported to cross the blood brain barrier and enter the brain, increasing acetylcholine production, improving cognitive function in aging rats, improving energy metabolism and mitochondrial function, and having antioxidant and inhibitory effects on apoptosis [56–61].

For this reason, clinical studies are being actively conducted for dementia in the elderly [62], diabetic neuropathy, neuropathy caused by anticancer agents [63–66], and depression [67]. In particular, there are many clinical studies on dementia, but as a result of a meta-analysis of these studies, there are reports that L-carnitine is effective [68] and reports that its effectiveness is not clear [69]. In addition, even if L-carnitine is administered, about 10% is converted to ALC in the body, so carnitine administration is also thought to be effective [70].

3.13.2 Propionyl-L-Carnitine (PLC)

PLC is a compound in which L-carnitine is esterified with a propionyl group of propionic acid (CH_3CH_2COOH), which is also present in the body, but less than ALC. It is thought to be derived from the metabolic process of branched-chain amino acids and propionic acid produced by intestinal bacteria. PLC is easy to migrate to skeletal muscle and myocardium, and there are many clinical studies in heart disease due to ischemic myocardial protective action, and it is also said to have the effect of improving vascular endothelial disorders, and there are many clinical studies in intermittent claudication (IC) of peripheral angiopathy [71–75]. In a study by Brevetti and Hiatt, metabolic disorders due to carnitine deficiency and accumulation of short- and long-chain acylcarnitine occur in the skeletal muscles of IC patients, and it is thought that carnitine and PLC cause improvement in IC walking difficulties and pain symptoms by removing the harmful acyl groups of these acylcarnitines [76–81]. L-carnitine is also somewhat weaker than PLC, but it is effective [78], and it is not clear whether PLC acts as free carnitine after administration or whether PLC has its own effect.

References

1. Interview form of Levocarnitine preparations, “L-Cartin® FF Tablets 100 mg”, “L-Cartin® FF Tablets 250 mg”, “L-Cartin® FF Oral Solution 10%”, “L-Cartin® FF Oral Solution 10% Packet 5 mL”, “L-Cartin® FF Oral Solution 10% Packet 10 mL”. Otsuka Pharmaceutical Co., Ltd., https://www.otsuka-elibrary.jp/pdf_viewer/?f=/file/1032/info_if.pdf#page=1. Article in Japanese.
2. Leonardi R, Zhang YM, Rock CO, et al. Coenzyme A: back in action. Prog Lipid Res. 2005;44(2–3):125–53. https://doi.org/10.1016/j.plipres.2005.04.001.
3. Baković J, Martínez DL, Nicolaou S, et al. Regulation of the CoA biosynthetic complex assembly in mammalian cells. Int J Mol Sci. 2021;22(3):1131. https://doi.org/10.3390/ijms22031131.
4. Leonardi R, Rock CO, Jackowski S, et al. Activation of human mitochondrial pantothenate kinase 2 by palmitoylcarnitine. Proc Natl Acad Sci USA. 2007;104(5):1494–9. https://doi.org/10.1073/pnas.0607621104.
5. Rock CO, Calder RB, Karim MA, et al. Pantothenate kinase regulation of the intracellular concentration of coenzyme A. J Biol Chem. 2000;275(2):1377–83.
6. Dansie LE, Reeves S, Miller K, et al. Physiological roles of the pantothenate kinases. Biochem Soc Trans. 2014;42(4):1033–6. https://doi.org/10.1042/BST20140096.
7. Fisher MN, Neely JR. Regulation of pantothenate kinase from various tissues of the rat. FEBS Lett. 1985;190(2):293–6. https://doi.org/10.1016/0014-5793(85)81303-x.
8. Fisher MN, Robishaw JD, Neely JR. The properties and regulation of pantothenate kinase from rat heart. J Biol Chem. 1985;260(29):15745–51.
9. Zhyvoloup A, Nemazanyy I, Panashuk G, et al. Subcellular localization and regulation of coenzyme A synthase. J Biol Chem. 2003;278(50):50316–21. https://doi.org/10.1074/jbc.M307763200.
10. Theodoulou FL, Sibon OCM, Jackowski S, et al. Coenzyme A and its derivatives: renaissance of a textbook classic. Biochem Soc Trans. 2014;42(4):1025–32. https://doi.org/10.1042/BST20140176.
11. Tsuchiya Y, Peak-Chew SY, Newell C, et al. Protein CoAlation: a redox-regulated protein modification by coenzyme A in mammalian cells. Biochem J. 2017;474(14):2489–508. https://doi.org/10.1042/BCJ20170129.
12. Tsuchiya Y, Zhyvoloup A, Baković J, et al. Protein CoAlation and antioxidant function of coenzyme A in prokaryotic cells. Biochem J. 2018;475(11):1909–37. https://doi.org/10.1042/BCJ20180043.
13. Gout I. Coenzyme A, protein CoAlation and redox regulation in mammalian cells. Biochem Soc Trans. 2018;46(3):721–8. https://doi.org/10.1042/BST20170506.
14. Mignani L, Gnutti B, Zizioli D, et al. Coenzyme a biochemistry: from neurodevelopment to neurodegeneration. Brain Sci. 2021;11(8):1031. https://doi.org/10.3390/brainsci11081031.
15. Bonomini M, Zammit V, Pusey CD, et al. Pharmacological use of L-carnitine in uremic anemia: has its full potential been exploited? Pharmacol Res. 2011;63(3):157–64. https://doi.org/10.1016/j.phrs.2010.11.006.
16. Calabrese V, Cornelius C, Dinkova-Kostova AT, et al. Cellular stress responses, hormetic phytochemicals and vitagenes in aging and longevity. Biochim Biophys Acta. 2012;1822(5):753–83. https://doi.org/10.1016/j.bbadis.2011.11.002.
17. Tanaka T, Shimozawa N. Fatty acid metabolism in peroxisomes and diseases. J Jpn Biochem Soc. 2018;90(1):14–20. Article in Japanese.
18. Antonenkov VD, Hiltunen JK. Transfer of metabolites across the peroxisomal membrane. Biochim Biophys Acta. 2012;1822(9):1374–86. https://doi.org/10.1016/j.bbadis.2011.12.011.
19. Amagai T, Kaneko M, Tamura T, et al. Chapter 1. Significance and problems of energy consumption. An introduction to energy metabolism. In: Hosoya N, editor. Why energy metabolism now: for the prevention of lifestyle-related diseases. Tokyo: Daiichi Shuppan Co., Ltd.; 2000. p. 13–52. Book in Japanese, ISBN:4-8041-0899-8.
20. Roe CR, Bohan TP. L-carnitine therapy in propionicacidaemia. Lancet. 1982;1(8286):1411–2. https://doi.org/10.1016/s0140-6736(82)92524-7.
21. Roe CR, Hoppel CL, Stacey TE, et al. Metabolic response to carnitine in methylmalonic aciduria. An effective strategy for elimination of propionyl groups. Arch Dis Child. 1983;58(11):916–20. https://doi.org/10.1136/adc.58.11.916.
22. Sugiyama N, Morishita H, Nagaya S, et al. Biochemical evidence of carnitine effect on propionate elimination. J Inherit Metab Dis. 1984;7(3):137–8. https://doi.org/10.1007/BF01801775.
23. Chalmers RA, Roe CR, Tracey BM, et al. Secondary carnitine insufficiency in disorders of organic acid metabolism: modulation of acyl-CoA/CoA ratios by L-carnitine in vivo. Biochem Soc Trans. 1983;11:724–5. Abstract.
24. Chalmers RA, Roe CA, Stacey TE, et al. Urinary excretion of l-carnitine and acylcarnitines by patients with disorders of organic acid metabolism: evidence for secondary insufficiency of l-carnitine. Pediatr Res. 1984;18(12):1325–8. https://doi.org/10.1203/00006450-198412000-00021.
25. Tsuchiya Y, Pham Y. Gout I: methods for measuring CoA and CoA derivatives in biological samples. Biochem Soc Trans. 2014;42(4):1107–11. https://doi.org/10.1042/BST20140123.
26. Shurubor YI, D’Aurelio M, Clark-Matott J, et al. Determination of coenzyme A and acetyl-coenzyme A in biological samples using HPLC with UV detection.

Molecules. 2017;22(9):1388. https://doi.org/10.3390/molecules22091388.
27. Stumpf DA, Parker WD Jr, Angelini C. Carnitine deficiency, organic acidemias, and Reye's syndrome. Neurology. 1985;35(7):1041–5. https://doi.org/10.1212/wnl.35.7.1041.
28. Woldegiorgis G, Shrago E, Gipp J, et al. Fatty acyl coenzyme A-sensitive adenine nucleotide transport in a reconstituted liposome system. J Biol Chem. 1981;256(23):12297–300.
29. Wieland O, Weiss L. Inhibition of citrate-synthase by palmityl-coenzyme A. Biochem Biophys Res Commun. 1963;13:26–31. https://doi.org/10.1016/0006-291x(63)90156-6.
30. Mita M, Yasumasu I. Regulation of the tricarboxylic acid cycle in sea urchin eggs and embryos. J Exp Zool. 1983;228(1):71–7. https://doi.org/10.1002/jez.1402280107.
31. Kawaguchi A, Bloch K. Inhibition of glutamate dehydrogenase and malate dehydrogenases by palmitoyl coenzyme a. J Biol Chem. 1976;251(5):1406–12.
32. Severson DL, Hurley B. Inhibition of the hormone-sensitive lipase in adipose tissue by long-chain fatty acyl coenzyme A. Lipids. 1984;19(2):134–8. https://doi.org/10.1007/BF02534504.
33. Coude FX, Sweetman L, Nyhan WL. Inhibition by propionyl-coenzyme A of N-acetylglutamate synthetase in rat liver mitochondria. J Clin Invest. 1979;64(6):1544–51. https://doi.org/10.1172/JCI109614.
34. Holland PC, Senior AE, Sherratt HSA. Biochemical effects of the hypoglycaemic compound pent-4-enoic acid and related non-hypoglycaemic fatty acids. Effects of their coenzyme A esters on enzymes of fatty acid oxidation. Biochem J. 1973;136(1):173–84. https://doi.org/10.1042/bj1360173.
35. Patel TB, Debuysere MS, Olson MS. The effect of propionate on the regulation of the pyruvate dehydrogenase complex in the rat liver. Arch Biochem Biophys. 1983;220(2):405–14. https://doi.org/10.1016/0003-9861(83)90430-7.
36. Utter MF, Scrutton MC. Pyruvate carboxylase. Curr Top Cell Regul. 1969;1:253–96.
37. Stumpf DA, McAfee J, Parks JK, et al. Propionate inhibition of succinate: CoA ligase (GDP) and the citric acid cycle in mitochondria. Pediatr Res. 1980;14(10):1127–31. https://doi.org/10.1203/00006450-198010000-00008.
38. Matsuishi T, Stumpf DA, Seliem M, et al. Propionate mitochondrial toxicity in liver and skeletal muscle: Acyl CoA levels. Biochem Med Metab Biol. 1991;45(2):244–53. https://doi.org/10.1016/0885-4505(91)90027-i.
39. Ramsay RR, Arduini A. The carnitine acyltransferases and their role in modulating acyl-CoA Pools. Arch Biochem Biophys. 1993;302(2):307–14. https://doi.org/10.1006/abbi.1993.1216.
40. Kerner J, Hoppel C. Fatty acid import into mitochondria. Biochim Biophys Acta. 2000;1486(1):1–17. https://doi.org/10.1016/s1388-1981(00)00044-5.
41. Ramsay RR, Zammit VA. Carnitine acyltransferases and their influence on CoA pools in health and disease. Mol Aspect Med. 2004;25(5–6):475–93. https://doi.org/10.1016/j.mam.2004.06.002.
42. Zammit VA, Ramsay RR, Bonomini M, et al. Carnitine, mitochondrial function and therapy. Adv Drug Deliv Rev. 2009;61(14):1353–62. https://doi.org/10.1016/j.addr.2009.04.024.
43. Miyata Y, Shimomura I. Metabolic flexibility and carnitine flux: the role of carnitine acyltransferase in glucose homeostasis. J Diabetes Invest. 2013;4(3):247–9. https://doi.org/10.1111/jdi.12064.
44. Noland RC, Kovas TR, Seiler SE, et al. Carnitine insufficiency caused by aging and overnutrition compromises mitochondrial performance and metabolic control. J Biol Chem. 2009;284(34):22840–52. https://doi.org/10.1074/jbc.M109.032888.
45. Muoio DM, Noland RC, Kovalik JP, et al. Muscle-specific deletion of carnitine acetyltransferase compromises glucose tolerance and metabolic flexibility. Cell Metab. 2012;15(5):764–77. https://doi.org/10.1016/j.cmet.2012.04.005.
46. Smith RL, Soeters MR, Wüst RCI, et al. Metabolic flexibility as an adaptation to energy resources and requirements in health and disease. Endocr Rev. 2018;39(4):489–517. https://doi.org/10.1210/er.2017-00211.
47. Athithan L, Gulsin GS, McCann GP, et al. Diabetic cardiomyopathy: pathophysiology, theories and evidence to date. World J Diabetes. 2019;10(10):490–510. https://doi.org/10.4239/wjd.v10.i10.490.
48. Chakravarthy MV, Neuschwander-Tetri BA. The metabolic basis of nonalcoholic steatohepatitis. Endocrinol Diabetes Metab. 2020;3(4):e00112. https://doi.org/10.1002/edm2.112.
49. da Silva Rosa SC, Nayak N, Caymo AM, et al. Mechanisms of muscle insulin resistance and the cross-talk with liver and adipose tissue. Physiol Rep. 2020;8(19):e14607. https://doi.org/10.14814/phy2.14607.
50. Gilbert M. Role of skeletal muscle lipids in the pathogenesis of insulin resistance of obesity and type 2 diabetes. J Diabetes Investig. 2021;12(11):1934–41. https://doi.org/10.1111/jdi.13614.
51. Jogl G, Tong L. Crystal structure of carnitine acetyltransferase and implications for the catalytic mechanism and fatty acid transport. Cell. 2003;112(1):113–22. https://doi.org/10.1016/s0092-8674(02)01228-x.
52. Jogl G, Hsiao YS, Tong L. Crystal structure of mouse carnitine octanoyltransferase and molecular determinants of substrate selectivity. J Biol Chem. 2005;280(1):738–44. https://doi.org/10.1074/jbc.M409894200.
53. Mitsumoto Y. Food and Drug Classification and "Carnitine". Non-profit organization Institute of Medical Education. Alternative Medicine Information. https://www.ime.or.jp/daitai/daitai29.html. Article in Japanese.

54. Matsushima Y, Monma K, Oishi M. Content of pharmaceutical ingredients (gamma aminobutyric acid, curcumin, and carnitine) in health foods. Tokyo Inst Health Saf Annu Rep. 2013;64:113–8. Article in Japanese.
55. Reuter SE, Evans AM, Chase DH, et al. Determination of the reference range of endogenous plasma carnitines in healthy adults. Ann Clin Biochem. 2008;45(Pt 6):585–92. https://doi.org/10.1258/acb.2008.008045.
56. Ando S, Tadenuma T, Tanaka Y, et al. Enhancement of learning capacity and cholinergic synaptic function by carnitine in aging rats. J Neurosci Res. 2001;66(2):266–71. https://doi.org/10.1002/jnr.1220.
57. Rosca MG, Lemieux H, Hoppel CL. Mitochondria in the elderly: is acetylcarnitine a rejuvenator? Adv Drug Deliv Rev. 2009;61(14):1332–42. https://doi.org/10.1016/j.addr.2009.06.009.
58. Jones LL, McDonald DA, Borum PR. Acylcarnitines: role in brain. Prog Lipid Res. 2010;49(1):61–75. https://doi.org/10.1016/j.plipres.2009.08.004.
59. Scafidi S, Fiskum G, Lindauer SL, et al. Metabolism of acetyl-L-carnitine for energy and neurotransmitter synthesis in the immature rat brain. J Neurochem. 2010;114:820–31.
60. Traina G. The neurobiology of acetyl-L-carnitine. Front Biosci (Landmark Ed). 2016;21(7):1314–29. https://doi.org/10.2741/4459.
61. Ferreira GC, McKenna MC. L-Carnitine and acetyl-L-carnitine roles and neuroprotection in developing brain. Neurochem Res. 2017;42(6):1661–75. https://doi.org/10.1007/s11064-017-2288-7.
62. Pennishi M, Lanza G, Cantone M, et al. Acetyl-L-carnitine in dementia and other cognitive disorders: a critical update. Nutrients. 2021;12(5):1389. https://doi.org/10.3390/nu12051389.
63. Zaheer A, Zaheer F, Saeed H, et al. A review of alternative treatment options in diabetic polyneuropathy. Cureus. 2021;13(4):e14600. https://doi.org/10.7759/cureus.14600.
64. Li S, Li Q, Li Y, et al. Acetyl-L-carnitine in the treatment of peripheral neuropathic pain: a systematic review and meta-analysis of randomized controlled trials. PLoS One. 2015;10(3):e0119479. https://doi.org/10.1371/journal.pone.0119479.
65. Avan A, Postma TJ, Ceresa C, et al. Platinum-induced neurotoxicity and preventive strategies: past, present, and future. Oncologist. 2015;20(4):411–32. https://doi.org/10.1634/theoncologist.2014-0044.
66. Dinicola S, Fuso A, Cucina A, et al. Natural products —alpha-lipoic acid and acetyl-L-carnitine— in the treatment of chemotherapy-induced peripheral neuropathy. Eur Rev Med Pharmacol Sci. 2018;22(14):4739–54. https://doi.org/10.26355/eurrev_201807_15534.
67. Chiechio S, Canonico PL, Grilli M. L-acetylcarnitine: a mechanistically distinctive and potentially rapid-acting antidepressant drug. Int J Mol Sci. 2018;19(1):11. https://doi.org/10.3390/ijms19010011.
68. Montgomery SA, Thal LJ, Amrein R. Meta-analysis of double blind randomized controlled clinical trials of acetyl-L-carnitine versus placebo in the treatment of mild cognitive impairment and mild Alzheimer's disease. Int Clin Psychopharmacol. 2003;18(2):61–71. https://doi.org/10.1097/00004850-200303000-00001.
69. Hudson SA, Tabet N. Acetyl-l-carnitine for dementia. Cochrane Database Syst Rev. 2003;2003(2):CD003158. https://doi.org/10.1002/14651858.CD003158.
70. Ando S. Study of the lipid-trophic effect of carnitine and its anti-aging effect on the brain. J Lipid Nutr. 2010;19(1):19–23. Article in Japanese.
71. Mingorance C, Rodríguez-Rodríguez R, Justo ML, et al. Critical update for the clinical use of L-carnitine analogs in cardiometabolic disorders. Vasc Health Risk Manag. 2011;7:169–76. https://doi.org/10.2147/VHRM.S14356.
72. Mingorance C, Rodríguez-Rodríguez R, Justo ML, et al. Pharmacological effects and clinical applications of propionyl-L-carnitine. Nutr Rev. 2011;69(5):279–90. https://doi.org/10.1111/j.1753-4887.2011.00387.x.
73. Delaney CL, Spark JI, Thomas J, et al. A systematic review to evaluate the effectiveness of carnitine supplementation in improving walking performance among individuals with intermittent claudication. Atherosclerosis. 2013;229(1):1–9. https://doi.org/10.1016/j.atherosclerosis.2013.03.004.
74. Brass EP, Koster D, Hiatt WR, et al. A systematic review and meta-analysis of propionyl-L-carnitine effects on exercise performance in patients with claudication. Vasc Med. 2013;18(1):3–12. https://doi.org/10.1177/1358863X12467491.
75. Tama B, Fabara SP, Zarrate D, et al. Effectiveness of propionyl-L-carnitine supplementation on exercise performance in intermittent claudication: a systematic review. Cureus. 2021;13(8):e17592. https://doi.org/10.7759/cureus.17592.
76. Brevetti G, Chiariello M, Ferulano G, et al. Increases in walking distance in patients with peripheral vascular disease treated with L-carnitine: a double-blind, cross-over study. Circulation. 1988;77(4):767–73. https://doi.org/10.1161/01.cir.77.4.767.
77. Brevetti G, Angelini C, Rosa M, et al. Muscle carnitine deficiency in patients with severe peripheral vascular disease. Circulation. 1991;84(4):1490–5. https://doi.org/10.1161/01.cir.84.4.1490.
78. Brevetti G, Perna S, Sabbà C, et al. Superiority of L-propionylcarnitine vs L-carnitine in improving walking capacity in patients with peripheral vascular disease: an acute, intravenous, double-blind, cross-over study. Eur Heart J. 1992;13(2):251–5. https://doi.org/10.1093/oxfordjournals.eurheartj.a060155.
79. Brevetti G, di Lisa F, Perna S, et al. Carnitine-related alterations in patients with intermittent claudication. Indication for a focused carnitine therapy. Circulation. 1996;93(9):1685–9. https://doi.org/10.1161/01.cir.93.9.1685.

80. Hiatt WR, Wolfel EE, Regensteiner JG, et al. Skeletal muscle carnitine metabolism in patients with unilateral peripheral arterial disease. J Appl Physiol. 1992;73(1):346–53. https://doi.org/10.1152/jappl.1992.73.1.346.

81. Hiatt WR, Regensteiner JG, Creager MA, et al. Propionyl-L-carnitine improves exercise performance and functional status in patients with claudication. Am J Med. 2001;110(8):616–22. https://doi.org/10.1016/s0002-9343(01)00704-5.

4 Homeostasis of Carnitine

4.1 Introduction

Carnitine is not a vitamin, but it is an essential biological substance that plays an extremely important role in energy metabolism and other processes in mammals. Therefore, in order to maintain its biochemical and physiological functions, there is a mechanism to maintain the concentration of carnitine in the intracellular and extracellular fluids within a certain range and to maintain the homeostasis of carnitine. A correct understanding of the mechanism of carnitine homeostasis is very helpful in understanding "carnitine deficiency," which is a breakdown of this homeostasis.

4.2 Overview of Carnitine Distribution and Homeostasis

Figure 4.1 shows the overall picture of the factors that determine the distribution of carnitine and the homeostasis of carnitine in the body. The amount of carnitine stored in the body is 50 mmol (about 8 g) in children and 100–130 mmol (about 16–21 g) in 70 kg adults [1, 2].

It is said that dietary intake is about 75% of the daily requirement of adults, and biosynthesis is about 25% of the daily requirement of adults. Carnitine, ingested by the diet or biosynthesized, enters the compartment of the blood and extracellular fluid and is transported to each tissue. However, the amount of carnitine in the extracellular fluid including plasma is only about 0.5–0.6% of the body's total stores (of which 0.1% is in plasma), and most of it is distributed in tissues such as skeletal muscles (97%), myocardium (0.4%), liver (1%), and kidney (0.2%) [2]. Carnitine is filtered in the renal glomeruli, but 90–99% of free carnitine (FC) is reabsorbed in the renal tubules. Acylcarnitine (AC) is poorly reabsorbed and excreted in the urine. Therefore, dietary intake and biosynthesis determine the amount of inflow into the body of carnitine, and the efficiency of reabsorption of free carnitine from the renal tubules and the amount of renal excretion as acylcarnitine determine the amount of carnitine lost, and the balance between the amount of inflow and the amount of carnitine loss determines the amount of carnitine in the body.

Thus, factors affecting carnitine homeostasis and turnover of carnitine are:

1. Intake from the diet (usually about 75% of the daily requirement for adults; Fig. 4.1①)
2. Biosynthesis in the body (usually about 25% of daily requirement for adults; Fig. 4.1②)
3. Excretion from the kidneys (mainly as acylcarnitines; Fig. 4.1③)
4. Reabsorption from renal tubules (mainly free carnitine; Fig. 4.1④)
5. Tissue distributed via bloodstream (Fig. 4.4)

N. Nakayama, *Carnitine Deficiency and Insufficiency in Medicine*,
https://doi.org/10.1007/978-981-95-7797-2_4

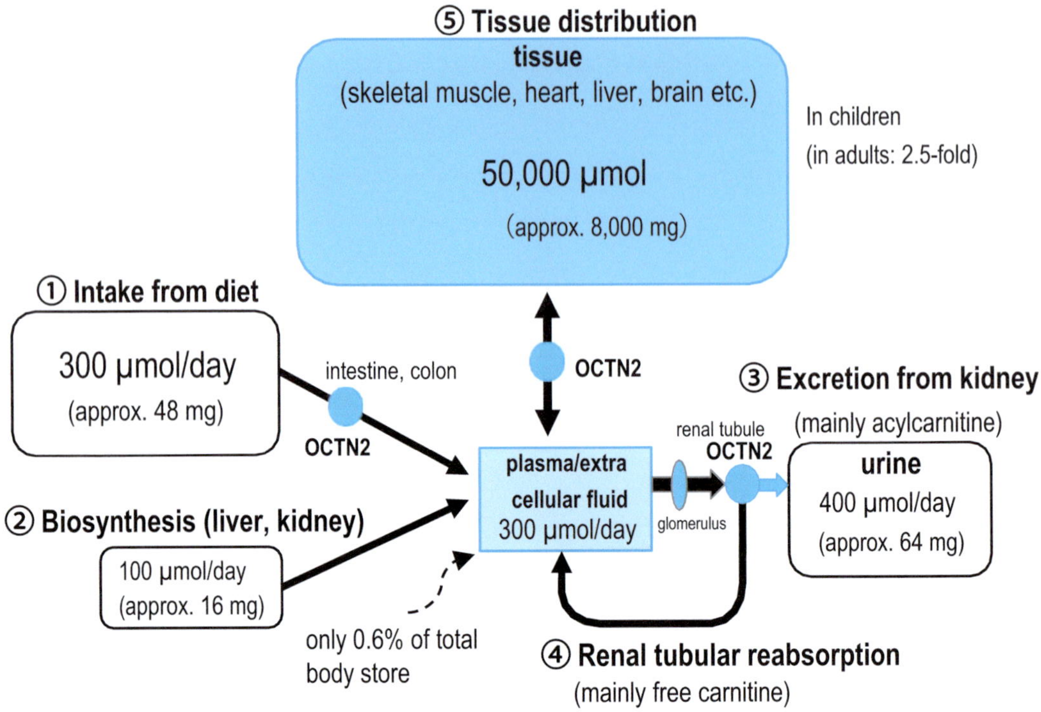

Fig. 4.1 Distribution of carnitine in the body and factors that influence carnitine homeostasis. *OCTN2* carnitine transporter. (Modified from Ref. [1])

The carnitine remaining in the plasma is transported to various tissues and used in concentrations several to several tens of times higher than the plasma concentration [1, 2]. When one or more of these five factors are significantly affected by a disease or factors, carnitine homeostasis will be disrupted, and carnitine deficiency may occur.

4.3 Tissue Content of Carnitine

The distribution of carnitine in the body is explained in more detail. Carnitine that enters the bloodstream is not metabolized or degraded in the liver or other organs and is transported to various tissues without protein binding. Carnitine, which enters the body through diet and biosynthesis, is thus taken up by tissues such as skeletal muscle, myocardium, and liver through the bloodstream (Fig. 4.1⑤).

The carnitine content in the body is about 130 mmol (about 21 g) in a healthy adult male weighing 70 kg. Table 4.1 shows the carnitine content of plasma, extracellular fluid including plasma, skeletal muscle, heart, kidney, liver, and brain; total carnitine content; percentage to the total body pool; affinity for carnitine uptake (Km value); and turnover time (time required for replacement) [2].

Since the concentration of carnitine in plasma and extracellular fluid is 40–50 μmol/L, and the volume of plasma and extracellular fluid in a 70 kg adult male is about 13 L, this pool size is 0.5–0.7 mmol. Of these, the volume of plasma is about 3 L, which means that there is only 0.2 mmol in plasma. The carnitine pool of plasma and extracellular fluid is an important compartment as a transport route for carnitine entering into the body, but the amount of carnitine itself is extremely low at around 0.5–0.6% of the entire body (0.1% in plasma alone). This is a particular point to keep in mind when diagnosing carnitine deficiency. Blood concentration of carnitine measurement is used in the diagnosis of carnitine deficiency on the premise that an equilibrium

Table 4.1 The distribution of carnitine into the primary sites of carnitine storage in human body for a 70 kg adult

Tissue compartment	Compartment size (plasma/extracellular fluid: L; others: kg)	Carnitine concentration (plasma/extracellular fluid: μmol/L; others: μmol/kg ww)	Carnitine content (mmol)	Total body carnitine (%)	Km (μmol/L)	Turnover time (h)
Plasma	3	50	0.2	0.1	–	–
Extracellular fluid including plasma	13	50	0.7	0.5	–	–
Skeletal muscle	30	4200	126	97	2 ~ 60	105
Heart	0.35	1300	0.5	0.4	2 ~ 60	21
Kidney	0.35	600	0.2	0.2	10 ~ 200	0.4
Liver	1.5	1000	1.5	1	500	1.3
Brain	–	–	–	–	>1000	223

ww wet weight, *Km* Michaelis-Menten constant for affinity of carnitine transporter
Modified from Ref. [2]

relationship is established over a long period of time between tissue concentration such as muscle and blood concentration. However, in acute onset carnitine deficiency that is deficient in tissues but has not yet decreased in the blood, or conversely, due to the decay of tissue cells in rhabdomyolysis or cirrhosis, etc., tissue carnitine leaks into the blood and the blood concentration may be high, so it is necessary to be careful in interpreting the data of blood concentration measurement. Also, in monitoring after administration of a carnitine preparation, the blood concentration may appear to be extremely high due to the administered carnitine and cannot be determined unless samples are taken after enough withdrawal period after carnitine administration (see "Chap. 18: "Guidelines for Diagnosis and Treatment of Carnitine Deficiency"). In addition, the fact that it is "transported to tissues via the bloodstream" means that carnitine is not sufficiently replenished in tissues with impaired blood flow, such as myocardial ischemia and intermittent claudication (especially myocardium and skeletal muscles that cannot biosynthesize carnitine on their own).

Among the organs, skeletal muscle has the highest carnitine content, followed by the heart, liver, and kidneys. These organs are rich in mitochondria, with active fatty acid metabolism and energy metabolism. Skeletal muscle and myocardium have extremely high carnitine content but do not have the last enzyme of the carnitine biosynthesis system. This means that carnitine biosynthesized in the liver and kidneys as well as carnitine ingested from food, must be taken up to the skeletal muscle or myocardium through the blood. The uptake of carnitine into skeletal muscle and myocardium is concentrated and taken up by a high-affinity uptake mechanism to several tens of times the blood concentration (Km value 2–60 μmol/L, probably via carnitine transporter [OCTN2]), and is thought to be in equilibrium with the blood pool. The carnitine content of skeletal muscle is approximately 4200 μmol/kg wet weight, and a 70 kg adult male has approximately 30 kg of skeletal muscle, resulting in a carnitine content of approximately 126 mmol. This is equivalent to about 97% of the entire body carnitine. The carnitine turnover in skeletal muscle is extremely slow at 105 h. Differences in blood-free carnitine concentrations between children and adults, or by sex, are thought to reflect differences in skeletal muscle mass. In other words, skeletal muscle is the organ that needs carnitine the most, and at the same time it is thought to play the role of a reservoir of carnitine in the body. Therefore, infants, women, the elderly, and patients with sarcopenia and cachexia with low skeletal muscle mass are prone to carnitine deficiency due to low carnitine stores in the body.

The carnitine content of the heart is reported to be 610–1300 μmol/kg wet weight, and the cardiac weight of 70 kg adult male is about 350 g, so it is considered to be a carnitine content of 0.2–0.5 mmol (0.4% of the total). Carnitine uptake into the heart is also carried out by a high-affinity uptake mechanism of 2–60 μmol/L, as is the case with skeletal muscle.

The liver is also an important organ for homeostasis of carnitine, and it has been reported to have a wet weight of 500–1000 μmol/kg, and if the liver weight of a 70 kg adult male is 1.5 kg, the carnitine content is 0.6–1.5 mmol (about 1% of the body's total pool). Unlike skeletal muscle and cardiac muscle, carnitine turnover in the liver is extremely fast at 1.3 h. The uptake of carnitine in the liver is said to be a low-affinity mechanism (Km = 500 μmol/L), but there are also reports that there is a high-affinity uptake [2].

There is little research on carnitine content in the brain and not much information. The uptake of carnitine into the brain is a low-affinity mechanism (Km value >1000 μmol/L), and the turnover is slow at 223 h.

The uptake of carnitine with high affinity is thought to be mediated by the carnitine transporter (OCTN2) (Km value 3–5 μmol/L), which is distributed in skeletal muscle, myocardium, and kidneys. The mechanism of low-affinity carnitine uptake in the liver, etc. is not well understood. OCTN1 has a relatively low affinity (Km value to 100 μmol/L); it is expressed in the liver in rats but not in humans [2]. Other low-affinity amino acid transporters for carnitine, $ATB^{0,+}$ expressed in the small intestine, lung, and mammary gland and OAT9 expressed in the liver and kidney in rats, may also be involved [2].

4.4 Dietary Intake of Carnitine

4.4.1 Absorption of Carnitine from the Gastrointestinal Tract

Normally, about 75% of the amount of carnitine needed by adults comes from the diet. Carnitine in food, pharmaceuticals, and supplement is absorbed by the carnitine transporter (OCTN2) present in intestinal epithelial cells, enters the bloodstream, and is transported to various tissues (Fig. 4.1①). Carnitine that cannot be fully absorbed is excreted in the feces or broken down by intestinal bacteria. It should be noted that in books and papers on carnitine, the term "abnormality of carnitine metabolism" is sometimes used, but this refers to an abnormality in which free carnitine becomes acylcarnitine and goes into fatty acid metabolism, or when the normal function of carnitine is not exerted. Except the esterification with the acyl group to form acylcarnitine, it is not metabolized or degraded and converted to other compounds in tissues such as the liver.

In humans, OCTN2 is thought to be widely distributed from the duodenum to the large intestine, but in a study investigating the mRNA expression of the human carnitine transporter (hOCTN2), it was reported that it was expressed more abundantly in the colon such as the ascending colon [12] (see Fig. 4.2), and that it was expressed more in the ileum and small intestine than in the colon [13]. In studies using mice, OCTN2 is also expressed in the duodenum, jejunum, ileum, large intestine, and even stomach [14].

4.4.2 Content of Carnitine in Foods

Since about 75% of the amount of carnitine necessary for humans is ingested from the diet, it is important to know the content of carnitine in foods as a nutritional science. Fraenkel et al., who studied the effects of carnitine in insects in the 1950s, reported carnitine content in many types of microorganisms, marine organisms, animals, and plants [15]. According to the study, carnitine is widely distributed in most organisms and is more abundant in animals than in microorganisms and plants, and the highest content is in muscles in both vertebrates and invertebrates. It is especially common in red meat and dairy products. *Escherichia coli*, *Tetrahymena* (*Tetrahymena geleii*), corn seeds, and pre-embryonic chicken eggs showed almost no carnitine. However, the method for measuring the carnitine content in biological samples reported by Fraenkel was calculated by comparing the growth rate and viability of the larvae of the mealworm (*Tenebrio molitor*) when they metamorphose into a pupae,

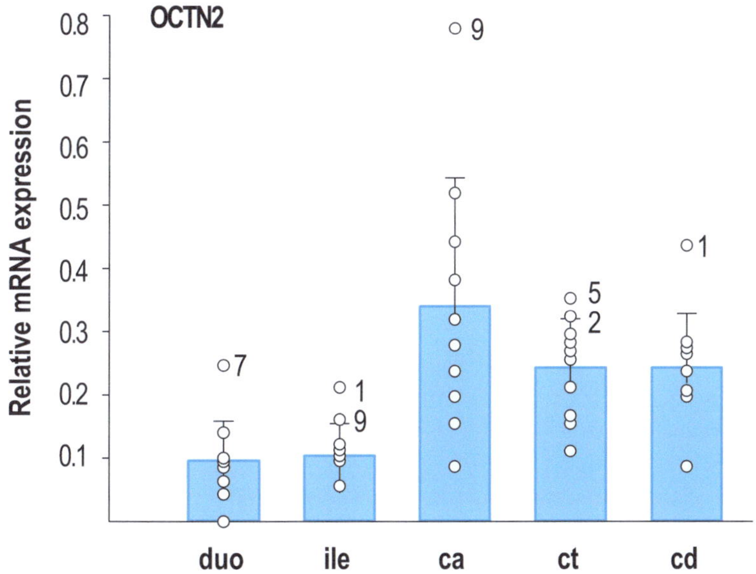

Fig. 4.2 Expression of OCTN2 mRNA along the human intestinal tract. The columns represent relative mRNA expression levels (OCTN2 cDNA copy numbers normalized to villin cDNA copy numbers). Data points (○) indicate the mRNA expression levels in individual patients. Individuals with OCTN2 mRNA expression level greater than mean + 1 SD (standard deviation) are identified by their identification numbers. *ca* colon ascendens, *cd* colon descendens, *ct* colon transversum, *duo* duodenum, *ile* ileum. (Modified from Ref. [12])

and was a bioassay method (see "Chap. 1: "The Discovery of Carnitine"). Since then, there have been studies on the carnitine content in some foods, but many of them are difficult to utilize; some studies express carnitine content as per dry weight of food.

4.4.2.1 Content of Carnitine in Foods Ingested on a Daily Basis

Tada et al. reported the content of free carnitine, short-chain acylcarnitine, long-chain acylcarnitine, and total carnitine in various daily foods we eat, using the enzyme cycling method [16] (Tables 4.2, 4.3, and 4.4). This data shows the carnitine content per 100 g of edible portion and can be utilized to estimate the amount of carnitine in cooked meals.

As shown in Table 4.2, the carnitine content is extremely low in plant vegetables and potatoes, except for avocados. In seafood, it is abundant in the muscular parts of scallops, red mussels, and blue squid (Table 4.3). However, there are fluctuations depending on the part and season.

Table 4.2 Carnitine contents in vegetables, potatoes, and avocado (mg/100 g of eatable parts)

Name of food	Free carnitine	Total carnitine
Cabbage	0	0
Spinach	0	0
Carrot	0	0
Broccoli	0.3	0.3
Potato	0	0
Sweet potato	0.9	0.9
Avocado	7.7	12.2

Note: The original paper describes the values of short-chain and long-chain acyl carnitine, but here shows only the value of total carnitine. The "0" in carrot and potato represent very small amounts
Modified from Ref. [16]

As shown in Table 4.4, carnitine is especially abundant in red meat and dairy products. Among bovine meat, pig meat, and chicken, bovine meat has the highest content. It is almost not contained in avian eggs. Considering that about 87% of milk is water, it can be said that it contains much of carnitine. However, there is almost no carnitine in soy milk derived from soybeans.

With recent trends of healthy gibier foods (wild game meat), there have been many studies

Table 4.3 Carnitine contents in seafood (mg/100 g of eatable part)

Name of food	Free carnitine	Total carnitine
Clams	8.6	23.8
Freshwater clams	1.3	23.9
Shrimps (frozen, August)	2.9	6.1
Shrimps (frozen, October)	3.9	6.5
Prawns (August)	8.5	9.3
Sea bream (August, back)	9.4	19.6
Sea bream (August, stomach)	10.8	22.1
Sea bream (October, back)	8.1	11.4
Sea bream (October, stomach)	10.7	13.0
Scallop (August, muscle)	47.8	90.3
Red clam (mixture of string and scallop)	21.7	82.7
Red clam (muscle)	134.4	175.4
Red clam (intestinal organs)	48.0	66.9
Octopus (boiled, legs, and muscle)	27.0	29.6
Octopus (boiled, tip part of legs, and muscle)	17.0	20.6
Oyster (include intestinal organs)	10.6	23.1
Tuna (cut out)	1.9	3.7
Tuna (red meat)	2.9	4.5
Tuna (medium tuna)	3.5	5.9
Horse mackerel (three piece grated)	6.0	14.3
Pacific saury (three piece grated)	12.1	16.6
Great squid (muscle part without legs and head)	48.1	58.4
Baked chikuwa	1.2	1.8
Hanpen	0.5	2.5

Note: The original paper describes the values of short-chain and long-chain acyl carnitine, but here shows only the value of total carnitine
Modified from Ref. [16]

Table 4.4 Carnitine contents in birds, animal meat, egg, cow milk, and soymilk (mg/100 g of eatable parts)

Name of food	Free carnitine	Total carnitine
Beef (rump, slice)	115.2	130.7
Pork (loin, thick slice)	48.6	69.6
Chicken (young Nagoya-cochin, thigh)	19.2	32.8
Chicken (young Nagoya-cochin, wing)	6.5	16.6
Ham (boneless, roast hams)	19.4	26.6
Ham (pressed hams)	53.3	66.2
Sausage (coarsely ground sausages)	29.6	43.2
Sausage (bologna sausages)	19.8	33.3
Sausage (pork cocktail sausages)	15.1	23.4
Sausage (wiener sausages)	8.7	12.9
Salami sausages	81.4	106.3
Egg (whole)	0	0
Cow milk (4 commercial products)	3.1	3.4
Soymilk (2 commercial products)	0	0

Note: The original paper describes the values of short-chain and long-chain acyl carnitine, but here shows only the value of total carnitine. The "0" in soymilk represent very small amounts
Modified from Ref. [16]

on carnitine content in deer meat, etc. [17–20]. The carnitine content of these various animal meats is summarized in Table 4.5. Carnitine is abundant in goat meat and sheep meat (mutton), followed by horse meat, deer meat, lamb meat, etc., and then cattle meat, pig meat, and chicken meat, in that order.

Carnitine deficiency is more likely to occur in patients with chronic renal failure and cirrhosis who restrict protein intake with low daily intake of carnitine-rich meat, seafood, and dairy products, as well as in patients with anorexia and the elderly who do not eat much meat. In addition, it should be noted that dietary carnitine intake decreases in patients with chronic renal failure and cirrhosis with hepatic encephalopathy when restricting protein intake for a long time.

In addition, many tube feedings and central venous feedings do not contain carnitine; besides, special formulas such as powdered formula without cow's milk allergen for children with cow's milk allergy also do not contain carnitine, and hence patients (especially children) who are managed with these nutritional supplements alone for a long time are prone to carnitine deficiency, and therefore require attention (see "Chap. 11: "Tube Feeding, TPN, Special Milk, and Carnitine Deficiency").

Table 4.5 Carnitine contents in various animal meat (mg/100 g of eatable parts)

Animal species	Free carnitine	References
Goat	183.1	Shimada [17]
Ram (tenderloin)	176.5	Erfle [18]
Ram (rump)	168.6	Erfle [18]
Sheep	158.9	Snoswell [19]
Horse (Hokkaido native horse)	79.8	Shimada [17]
Deer (Hokkaido native deer)	73.7	Shimada [17]
Deer (Japanese deer)	64.2	Yamashita [20]
Lamb	57.9	Snoswell [19]
Beef cattle (Japanese black)	57.5	Shimada [17]
Horse (Thoroughbred)	45.0	Shimada [17]
Calf (Holstein, steer)	30.0	Shimada [17]
Pig	17.6	Shimada [17]
Chicken (Broiler)	11.1	Shimada [17]

4.4.2.2 Carnitine Intake in Vegetarians

Among foods, carnitine is rich in red meat and dairy products. Therefore, what about carnitine intake in the case of vegetarians who eat little meat? Reuter et al. estimate that the omnivorous diet consumes 6–15 μmol/kg/day of carnitine, compared with vegetarian (semi-vegan diet) who eats chicken and fish but does not eat red meat, 1–8 μmol/kg/day, and if you are on a strict vegan diet only <1 μmol/kg/day of carnitine is ingested [2]. Lombard et al. have actually investigated and reported plasma carnitine concentrations and urinary excretion of carnitine in adult and pediatric groups of normal and vegetarian groups [21]. Plasma free carnitine and total carnitine concentrations, and urinary free carnitine and total carnitine excretion were examined in people who eat meat and dairy products normally (normal diet group), vegetarians who do not eat meat and fish but eat dairy products and eggs (lacto-ovo vegetarian group), and strict vegetarians who do not eat dairy products and eggs (strict vegetarian group). It was found that the plasma free carnitine and total carnitine concentrations were significantly lower in the lacto-ovo vegetarian group and the strict vegetarian group than in the normal diet group, and the urinary free carnitine and total carnitine concentrations were also lower than in the normal diet group. However, the decrease in plasma free carnitine concentration in these vegetarian groups was 10–20%, and the decrease in the strict vegetarian group of children who had the lowest concentration was about 30%, which was not low enough to cause carnitine deficiency. On the other hand, urinary excretion decreased significantly to 10–15% in the lacto-ovo vegetarian group and several percent in the strict vegetarian group. In other words, it is thought that a mechanism to prevent carnitine deficiency by increasing the reabsorption of carnitine in the renal tubules in vegetarians minimizes renal excretion as much as possible to maintain homeostasis and prevent carnitine deficiency. In particular, in children, there was a significant decrease in plasma free carnitine and total carnitine concentrations and urinary free carnitine concentrations in the complete vegetarian group compared to the lacto-ovo vegetarian group. It has also been reported that the biosynthetic capacity of carnitine is not yet fully developed in children [3, 4, 22], and it can be said that carnitine intake from breast milk and baby food is more important for infants and children.

4.5 Biosynthesis of Carnitine

4.5.1 Characteristics of Carnitine Biosynthesis

Approximately 25% of the adult daily requirement for carnitine is biosynthesized [1, 5, 23] (Fig. 4.1②). Therefore, when dietary intake decreases, carnitine supplied from biosynthesis becomes important. It is often said that carnitine is biosynthesized from lysine and methionine. However, it is important to note that lysine is not the same lysine that is produced when dietary proteins are broken down in the gastrointestinal tract. Carnitine is biosynthesized in several steps, starting from trimethyllysine, which is released by lysosomal degradation of trimethyllysine in several biological proteins [3, 5–9, 23]. The release of trimethyllysine in the body is said to be the rate-limiting step of carnitine biosynthesis [9]. In other words, carnitine is not simply biosynthesized from free amino acids lysine and methionine derived from degradation of the dietary proteins. There are some proteins such as actin, myosin, calmodulin, cytochrome *c*, and histones whose lysine residues are post-translationally trimethylated endogenously in mammals [24, 25]. Thus, skeletal muscle mass rich in actin and myosin are critical for maintaining carnitine biosynthesis.

The intermediates of carnitine biosynthesis are biosynthesized in various organs and tissues, including skeletal muscle. However, the γ-butyrobetaine hydroxylase, the last biosynthetic step, is mainly expressed only in the kidney, liver, and brain and its activity is highest in the kidney. Therefore, carnitine is basically biosynthesized in the kidney and liver and transported again to each organ such as skeletal muscle through the bloodstream. Not surprisingly, in patients with chronic renal failure and cirrhosis, carnitine biosynthesis is thought to be reduced.

To date, no cases of carnitine deficiency due to genetic abnormalities in enzymes in the carnitine biosynthesis system have been reported. However, it is proposed that limited carnitine deficiency due to defects of carnitine biosynthesis in brain during early developmental stage causes attention-deficit hyperactivity disorder (ADHD) and autism (see Chap. 17).

4.5.2 Biosynthetic Pathway of Carnitine

Details of the carnitine biosynthesis pathway are shown in Fig. 4.3.

1. The first step in the biosynthesis of carnitine is the trimethylation of lysine residues of histone, actin, myosin, calmodulin, cytochrome *c*, and other proteins by S-adenosylmethionine, which in turn donates a methyl group to produce 6-N-trimethyllysine (TML) of these proteins, and lysine provides the carbon backbone [26–28] and methionine provides the methyl group of the trimethyl group. It is important that this first step is not simply a reaction in which the amino acids lysine and methionine are combined, but a complex process. The lysine residues of these proteins are modified to 6-N-trimethyllysine by a post-translational reaction of a special methyltransferase with a methyl group supplied from S-adenosylmethionine [10, 29]. These processes are dependent on protein turnover, which is very active in muscles but not in kidney and liver. S-adenosylmethionine is produced by the reaction of methionine with ATP. Thus, three molecules of ATP are consumed for trimethylation of one lysine residue of these proteins.
2. Next, when these proteins are turned over, they are hydrolyzed in the lysosomes in the cell and 6-N-trimethyllysine is released into the bloodstream. This 6-N-trimethyllysine is practically the starting material for carnitine biosynthesis, and the stage in which this

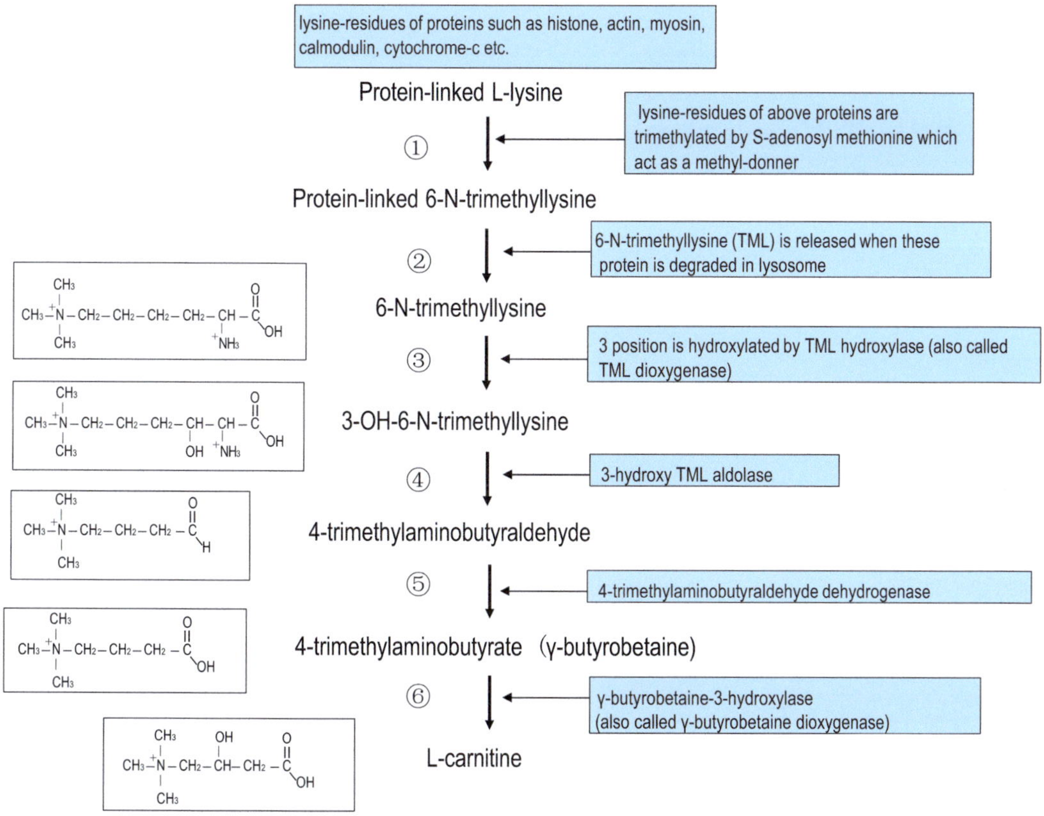

Fig. 4.3 Biosynthesis pathway of L-carnitine. (Created with reference to Refs. [4, 6, 9])

protein-binding 6-N-trimethyllysine is released is said to be the rate-limiting step of carnitine biosynthesis [9]. Therefore, the release of 6-N-trimethyllysine also occurs in various tissues such as the liver, but skeletal muscle, which is rich in actin and myosin, is considered to be the most important source of 6-N-trimethyllysine [8, 9].

3. Next, trimethyllysine hydroxylase (also called trimethyllysine dioxygenase), which exists in kidney and liver mitochondria, hydroxylates the 3-position of trimethyllysine to form 3-OH-6-N-trimethyllysine [29]. This reaction requires oxygen, divalent iron ion (Fe^{2+}), ascorbic acid, and α-ketoglutaric acid.
4. Trimethyllysine (3-OH-6-N-trimethyllysine), whose 3-position is hydroxylated, is next converted to 4-trimethylaminobutyraldehyde and glycine by the aldol cleavage reaction with specific aldolase. This reaction requires vitamin B_6 (pyridoxal phosphate) as a co-factor.
5. 4-Trimethylaminobutyraldehyde is oxidized by a specific nicotinamide adenine dinucleotide (NAD)-dependent aldehyde dehydrogenase to trimethylaminobutyrate (γ-butyrobetaine).
6. L-carnitine is formed by hydroxylation of γ-butyrobetaine at position 3 by γ-butyrobetaine dioxygenase in the cytoplasm. This reaction also requires divalent iron ions (Fe_{2+}) and ascorbic acid.

Thus, carnitine is biosynthesized in the body; it depends on skeletal muscle mass rich in actin and myosin, which supply the starting material 6-N-trimethyllysine, so the amount of biosynthesis is thought to be reduced in elderly people with sarcopenia. In addition, the liver and kidneys are important for the last step, and biosynthesis may be reduced in patients with cirrhosis or chronic renal failure.

4.5.3 Organ Distribution of Carnitine Biosynthetic Enzymes and Organ–Organ Interactions

Lysine residues in proteins such as calmodulin, myosin, actin, cytochrome *c*, and histone are trimethylated post-translationally, and these proteins are degraded by lysosomes to release 6-N-trimethyllysine into the bloodstream, which becomes the starting material for carnitine biosynthesis in each organ, and L-carnitine is biosynthesized through a four-step reaction (Fig. 4.3). Although the organ distribution of each of these enzymes varies greatly depending on the animal species [4], the distribution of these enzyme activities in humans by Rebouche et al. is shown in Table 4.6 [3].

The activity of 6-N-trimethyllysine hydroxylase in the biosynthetic pathway (step 3) is distributed in the mitochondrial fraction of various organs, but the activity in the kidneys is overwhelmingly strong, followed by the liver, heart, and brain. The next biosynthetic pathway (step 4), 3-OH-trimethyllysine aldolase activity, is overwhelmingly active in the liver. The activity of 4-trimethylaminobutyraldehyde dehydrogenase in the biosynthetic pathway (step 5) is overwhelmingly high in the liver, and the activity of the kidneys is less than 1/3 that of the liver, and almost no activity is observed in other organs. The activity of butyrobetaine-3-hydroxylase in the last biosynthetic pathway (step 6) was overwhelmingly high in the kidneys, followed by liver activity equaling about 30% of the kidneys, and in the brain about 10% of the kidneys, but almost no activity in other organs. In studies of Englard's on butyrobetaine-3-hydroxylase activity in humans and monkeys, it was reported that this enzyme activity in the kidney was more than 13 times higher in humans than in the liver [30]. In a study using cDNA of butyrobetaine-3-hydroxylase in humans by Vaz et al., northern blot showed that the highest expression was observed in the kidney, moderate expression in the liver, and extremely low expression in the brain [4]. These results suggest that intermediates of carnitine biosynthesis are biosynthesized in organs other than the kidney and liver, but they are transported to the kidneys and liver via the bloodstream, and eventually carnitine is biosynthesized from γ-butyrobetaine mainly in the kidney and liver. Carnitine synthesized in the kidney and liver is then transported through the bloodstream to the skeletal and heart muscles that need it most.

Figure 4.4 shows the interaction between tissues related to the homeostasis and biosynthesis of carnitine and the relationship between biosynthesis sites [2–11]. In other words, the role of the kidneys and liver in the biosynthesis of carnitine is extremely important. Concomitant liver and kidney diseases such as cirrhosis and chronic renal failure are said to be one of the factors that make carnitine deficiency more likely to occur [9, 31]. The reason for this is that the biosynthesis of carnitine is greatly reduced in patients with renal insufficiency and hepatic insufficiency due to the deterioration of parenchymal cells due to fibrosis,

Table 4.6 Organ distribution of activities of four biosynthesis enzymes of carnitine

Organ	6-N-trimethyllysine hydroxylase(③)	3-OH-trimethyllysine aldolase(④)	4-Trimethylaminobutyraldehyde dehydrogenase(⑤)	γ-Butyrobetaine-3-hydroxylase(⑥)
Liver	45.6 ± 15.5(7)	38.70 ± 48.0(6)	46,800 ± 9200(7)	369 ± 95 (8)
Kidney	181.0 ± 59.5(3)	2.70 ± 1.94(3)	15,700 ± 4900(3)	1350 ± 230(3)
Brain	36.6 ± 10.2(3)	5.23 ± 1.92(3)	<500 (3)	140 ± 56 (4)
Heart	38.2 ± 14.1(4)	2.19 ± 1.98(4)	<500 (4)	0.00 (4)
Skeletal muscle	23.6 ± 6.5 (4)	0.81 ± 0.37(4)	<500 (4)	0.00 (4)

Note: Adult human tissues obtained at autopsy were used. Values show enzyme activity per mg protein (Mean ± SD), and the numbers in parenthesis indicate numbers of cases. The circled number indicates the biosynthesis steps in Fig. 4.3. The enzyme activities of 6-N-trimethyllysine hydroxylase were the activities in the mitochondrial fractions; the other enzyme activities were those in the 105,000× *g* supernatants

Modified from Ref. [3]

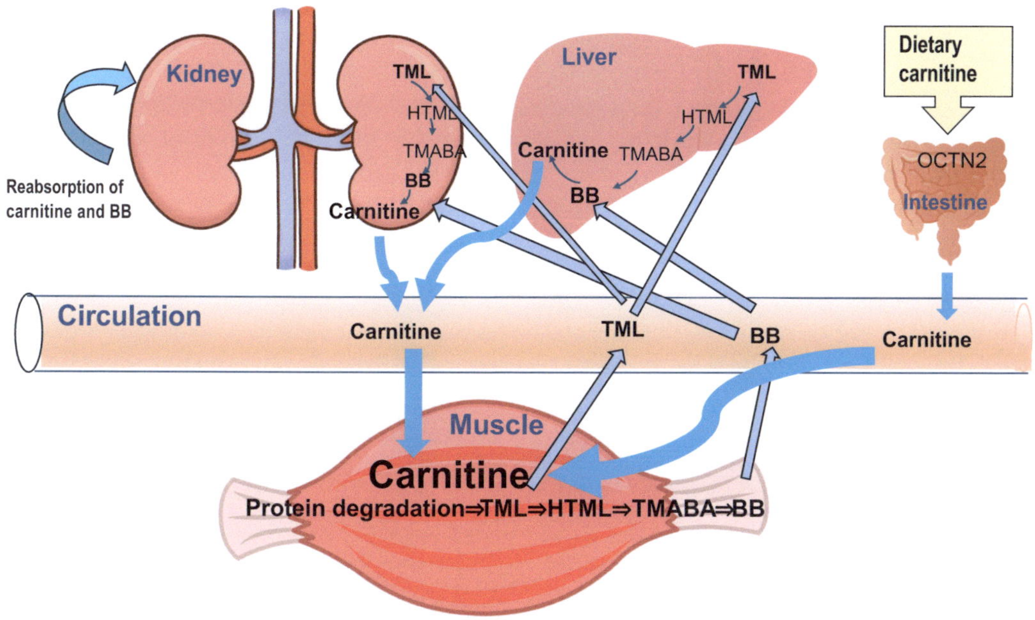

Fig. 4.4 Schematic representation of carnitine homeostasis and Inter-organ interaction in carnitine biosynthesis. About 75% of body carnitine comes from dietary carnitine absorbed through intestinal OCTN2 (carnitine transporter). About 25% of body carnitine comes from biosynthesis mainly in liver and kidney. These tissues have all four biosynthesis enzymes of carnitine (TML ⇒ HTML ⇒TMABA ⇒BB ⇒ carnitine). Muscles are important tissues for carnitine homeostasis because it contains more than 97% of body carnitine. However, muscle does not have a final biosynthesis enzyme which convert BB into carnitine. Muscle is considered as a major supply site of TML and BB to liver and kidney through blood circulation. TML is also derived from endogenous degradation of proteins such as actin, myosin, and histones. Kidney is an important tissue to maintain carnitine homeostasis because it reabsorbs most free carnitine and BB through OCTN2. *BB* trimethylaminobutyrate (butyrobetaine), *HTML* 3-OH-trimethyllysine, *OCTN2* carnitine transporter, *TMABA* 4-trimethylaminobutyraldehyde, *TML* 6-N-trimethyllysine. (Created using Refs. [2–11])

etc., and in patients with these diseases, the intake of meat and other foods is reduced due to protein intake restriction, and thus the intake of carnitine from the diet is also greatly reduced. In addition, it has been reported that the amount of carnitine in muscles decreases with aging [32], and the relationship between sarcopenia progression and carnitine deficiency in the elderly and the effects of carnitine supplementation have been investigated. Muscle does not have the final enzyme for biosynthesis of carnitine, but it could synthesize trimethylaminobutyrate (γ-butyrobetaine). So, muscle is considered as a major supply site of TML and BB (butyrobetaine) for carnitine biosynthesis to liver and kidney through blood circulation. Carnitine deficiency due to abnormalities in enzymes involved in carnitine biosynthesis is currently unknown [5, 33].

Human liver γ-butyrobetaine hydroxylase is age-dependent, and its activity is about 10% in infancy, about 30% in 3-year-old children, and reaches the adult level at 15 years of age, assuming that the normal adult biosynthesis is 100%. Therefore, the amount of biosynthesis in infancy and childhood is low, and carnitine deficiency is likely to occur when dietary intake is reduced. γ-Butyrobetaine hydroxylase in the kidneys is as active in newborns as it is in adults [3, 4, 22].

4.6 Excretion and Reabsorption of Carnitine from the Kidney

In carnitine homeostasis, dietary intake and biosynthesis are responsible for the influx of carnitine into the body, while renal excretion and reabsorption are responsible for the amount lost from the body (Fig. 4.1).

4.6.1 Handling of Free Carnitine and Acylcarnitine in the Kidney

The molecular weight of free carnitine is 161.2 and there is no protein binding, so it is 100% filtered from the renal glomeruli (Fig. 4.1③), but in healthy humans, 90–99% of free carnitine is reabsorbed by the carnitine transporter (OCTN2) in the renal tubules [1, 34–36] (Fig. 4.1④). Carnitine reabsorption by the renal tubules is thought to be a mechanism to maintain carnitine homeostasis by preventing the loss of free carnitine, which is important for energy metabolism. Since the reabsorption of free carnitine is carried out through a transporter, saturation dynamics is shown when the free carnitine concentration is extremely high. In other words, in individuals with very low dietary carnitine intake, such as vegetarians, the expression of carnitine transporters in the renal tubules is increased and most carnitine is reabsorbed. On the other hand, when urinary free carnitine concentrations exceed 60 μmol/L, expression of the transporter is reduced and reabsorption is diminished, with substantial amounts excreted in the urine [11, 37]. This mechanism also works when carnitine preparations are administered, and blood concentrations become very high, and excess carnitine is excreted in the urine.

On the other hand, in the case of acylcarnitines, not only short- and medium-chain acylcarnitines, but also long-chain acylcarnitines such as C16 palmitoylcarnitine with a molecular weight of 436.07 are 100% filtered in the glomerulus but are reabsorbed by the renal tubular carnitine transporter (OCTN2) same as free carnitine. However, free carnitine is preferentially reabsorbed, probably due to its higher affinity for OCTN2, resulting in a renal clearance of acylcarnitines that is four to eight times higher than that of free carnitine, and much of them are excreted into the urine [2, 7, 23]. In healthy subjects, the concentration of acylcarnitine in plasma is usually 6–23 μmol/L, which is lower than the free carnitine concentration of 36–74 μmol/L, and the concentration ratio (AC/FC) is $\leq$0.25, which means overwhelmingly higher plasma concentration for free carnitine than that of acylcarnitines. However, in urinary concentrations, acylcarnitines are several times higher than that of free carnitine.

The urinary excretion of acylcarnitine is considered to be a detoxification mechanism that is inherent in the body. In other words, carnitine binds to long-chain acyl groups that are toxic when accumulated in mitochondria and short- and medium-chain acyl groups of organic acids such as propionic acid and glutaric acid that accumulate in large amounts in inherited metabolic diseases such as organic acidemia and is excreted in the urine in the form of acylcarnitines. This reduces the toxicity caused by the accumulation of acyl groups in the mitochondria. In conditions such as propionic acidemia and methylmalonic acidemia, a large amount of propionyl-CoA and methylmalonyl-CoA accumulates in the liver and other organ mitochondria, and impairs carbamoyl phosphate synthase 1 (CPS1) enzyme in the urea cycle, glycine cleavage, and ATP production in energy metabolism, resulting in development of hyperammonemia, severe ketosis, metabolic acidosis, etc. Carnitine receives acyl groups such as propionyl and methylmalonyl groups and excretes their harmful acyl groups in the urine as propionylcarnitine and methylmalonylcarnitine. However, if this urinary excretion as acylcarnitine continues for a long time, a large amount of free carnitine itself in the body is used for excretion as acylcarnitine, and eventually carnitine deficiency occurs (secondary carnitine deficiency). In these conditions, the acylcarnitine/free carnitine ratio in the blood is often higher than normal (usually less than 0.25, and 0.4 or more is considered clearly abnormal). This is thought to be the mechanism of development of secondary carnitine deficiency in organic acidemia. In other words, most of free carnitine is reabsorbed in the renal tubules, so there is no carnitine deficiency due to increased urinary excretion of free carnitine, except for extremely rare special pathologies such as primary carnitine deficiency (genetic abnormality of the carnitine transporter) and Fanconi syndrome. When acyl-CoA accumulates in the mitochondria, a large amount of carnitine in the form of acylcarnitine is excreted from the kidneys into the urine, and the homeostasis of carnitine is disrupted. The mechanism of carnitine deficiency in inherited metabolic diseases is not only a simple and direct mechanism in which carnitine receives the acyl group of acyl-CoA of organic acids and fatty acids accumulated due to metabolic disorders and becomes acylcarnitine and is excreted in large quantities, but also it has been pointed out that carnitine deficiency is accelerated by an indirect mechanism in which accumulated acylcarnitines themselves inhibit the OCTN2 activity in the renal tubules and the threshold for carnitine reabsorption is lowered, reabsorption is reduced, and more free carnitine is excreted in the urine [38].

As already mentioned, the urinary excretion rate of carnitine in strict vegetarians is lower than that of normal eaters, and it is thought that a mechanism that minimizes the waste of carnitine in the body is working [21]. In a study of rats with drug-induced carnitine deficiency, the renal expression of the carnitine transporter (OCTN2) was increased 1.7-fold [39]. It is thought that the expression of the carnitine transporter in the renal tubules was increased to compensate for carnitine deficiency as much as possible. Thus, the reabsorption process of carnitine in the renal tubules is considered to play an extremely important role in maintaining carnitine homeostasis. Carnitine transporters in renal tubules are expressed in the apical membrane side of epithelial cells in the proximal and distal tubules and reabsorb free carnitine [40, 41]. The mechanism of carnitine reabsorption and excretion in the renal tubules plays a very important role in maintaining homeostasis of carnitine. As will be described later, inhibition of free carnitine reabsorption or increased excretion of free carnitine in the renal tubules due to congenital, acquired, or iatrogenic causes may cause primary and secondary carnitine deficiency, so it needs to be paid special attention [42].

4.7 Disruption of Carnitine Homeostasis in Various Diseases

When the carnitine homeostasis maintenance mechanism shown in Fig. 4.1①–⑤ is abnormal in various diseases, carnitine homeostasis is disrupted, and carnitine deficiency occurs. Typical examples are listed below.

4.7.1 Inherited Metabolic Disorders

In primary carnitine deficiency, the activity of the carnitine transporter (OCTN2) is reduced due to congenital abnormalities, and not only the absorption of carnitine from the intestinal tract and the uptake of carnitine into cardiomyocytes and skeletal muscle cells are reduced, but also the reabsorption of free carnitine from the renal tubules is reduced. Carnitine is excreted in the urine, resulting in extreme carnitine deficiency. In addition, in inherited metabolic diseases such as organic acid metabolism disorders, fatty acid metabolism disorders, and mitochondrial disorders, a defense mechanism works to detoxify accumulated intermediate metabolites and toxic acyl groups by excreting a large amount of acylcarnitines into the urine, and the accumulated acylcarnitines inhibit OCTN2 in the renal tubules and inhibit the reabsorption of free carnitine. As a result, the patients are prone to carnitine deficiency [38, 43, 44].

4.7.2 Liver Diseases

Liver is an important organ for carnitine biosynthesis, so in patients with chronic liver diseases such as cirrhosis, liver cancer, and hepatitis B, carnitine biosynthesis and intake are often decreased due to the dysfunction of the liver. Also, in patients with cirrhosis, especially patients who have suffered hepatic encephalopathy, protein intake restriction therapy might cause reduction of carnitine intake. In addition, accumulation of long-chain acylcarnitines has been reported due to mitochondrial dysfunction of the liver [45, 46].

4.7.3 Heart Failure

It was already reported in 1982 that myocardial free carnitine decreases and short- and long-chain acylcarnitine accumulates in patients with heart failure [47]. Recently, the association between the decrease in free carnitine and the increase in long-chain acylcarnitine in patients with heart failure and clinical symptoms and prognosis has been actively studied again [48–52].

4.7.4 Dialysis Patients

In dialysis patients, 100% of the low-molecular-weight free carnitine is filtered through the dialysis membrane and eliminated in the dialysis effluent, thus decreasing the free carnitine concentration in the blood. In addition, long-chain acylcarnitines accumulate, and the relationship between the accumulation of long-chain acylcarnitines and prognosis and severity of clinical symptoms such as muscle symptoms and erythropoietin-resistant anemia has been studied [53–55]. Carnitine deficiency is thought to affect treatment-resistant anemia and cardiac dysfunction, and it has been reported that carnitine supplementation therapy improves hematocrit, reduces the required dose of erythropoietin, and improves left ventricular ejection fraction in patients with heart failure [56]. The cause of acylcarnitines' accumulation in dialysis patients is not well understood. I would like to suggest following possibilities:

1. Long-chain acylcarnitine has a hydrophobic moiety and a hydrophilic moiety, and is amphiphilic, so it forms micelles and gains an apparent large molecular weight and is difficult to remove by dialysis. For example, if 100 of palmitoylcarnitines make micelle, its molecular weight will be more than 43,600 and will not be removed by dialysis.
2. The affinity between the dialysis membrane, especially the case of hydrophobic membrane, and the long-chain acylcarnitine is high and it is difficult to get out of the dialysis membrane.
3. It is conceivable that long-chain acylcarnitine causes protein binding due to its hydrophobic nature, and its apparent molecular weight becomes large, making it difficult to be dialyzed.

In any case, long-chain acylcarnitine is difficult to remove and accumulates in the body (especially long-chain acylcarnitine with a carbon chain of C12 or more). Thus, carnitine deficiency in dialysis patients is considered to be a special condition that combines two aspects: primary carnitine deficiency, in which free carnitine decreases significantly, and carnitine insufficiency, in which acylcarnitine increases relative to free carnitine and the acylcarnitine/free carnitine ratio increases.

4.7.5 Fanconi Syndrome

In Fanconi syndrome, a renal tubular dysfunction of various causes, carnitine deficiency is likely to occur due to impaired reabsorption of free carnitine as well as impaired reabsorption of phosphate, bicarbonate, potassium, and glucose [57–61]. In patients with such tubular disorders, attention should be paid to carnitine deficiency in addition to treatment of the disease.

4.7.6 Patients Receiving Valproate

Carnitine deficiency has been reported as a cause of valproate-induced hyperammonemic encephalopathy and hepatotoxicity, and one of the mechanisms is that valproic acid and its metabolites inhibit renal tubular reabsorption of free carnitine [62–68]. However, some studies have denied that valproic acid and its metabolites inhibit carnitine reabsorption, and no conclusion has been reached (for details, see "Chap. 12, "Carnitine Deficiency Caused by Sodium Valproate"). Such tubular dysfunction in patients taking valproate may be considered a drug-induced Fanconi's syndrome [69–71].

In any case, carnitine deficiency is more likely to occur when carnitine homeostasis is severely disrupted due to disease-specific causes.

References

1. Stanley CA. Carnitine deficiency disorders in children. Ann N Y Acad Sci. 2004;1033:42–51. https://doi.org/10.1196/annals.1320.004.
2. Reuter SE, Evans AM. Carnitine and acylcarnitines. Pharmacokinetic, pharmacological and clinical aspects. Clin Pharmacokinet. 2012;51(9):553–72. https://doi.org/10.1007/BF03261931.
3. Rebouche CJ, Engel AG. Tissue distribution of carnitine biosynthetic enzymes in man. Biochim Biophys Acta. 1980;630(1):22–9. https://doi.org/10.1016/0304-4165(80)90133-6.
4. Vaz FM, Wanders RJA. Carnitine biosynthesis in mammals. Biochem J. 2002;361(Pt 3):417–29. https://doi.org/10.1042/0264-6021:3610417.
5. Steiber A, Kerner J, Hoppel CL. Carnitine: a nutritional, biosynthetic, and functional perspective. Mol Asp Med. 2004;25(5–6):455–73. https://doi.org/10.1016/j.mam.2004.06.006.
6. Broquist HP, Borum PR. Carnitine biosynthesis: nutritional implications. Adv Nutr Res. 1982;4:181–204. https://doi.org/10.1007/978-1-4613-9934-6_7.
7. Bremer J. Carnitine —metabolism and functions—. Physiol Rev. 1983;63(4):1420–80. https://doi.org/10.1152/physrev.1983.63.4.1420.
8. Hoppel CL, Davis AT. Inter-tissue relationships in the synthesis and distribution of carnitine. Biochem Soc Trans. 1986;14(4):673–4. https://doi.org/10.1042/bst0140673.
9. Krähenbühl S. Carnitine metabolism in chronic liver disease. Life Sci. 1996;59(19):1579–99. https://doi.org/10.1016/0024-3205(96)00343-8.
10. Paik WK, Kim S. Protein methylation: chemical, enzymological, and biological significance. Adv Enzymol Relat Areas Mol Biol. 1975;42:227–86. https://doi.org/10.1002/9780470122877.ch5.
11. Rebouche CJ, Mack DL. Sodium gradient-stimulated transport of L-carnitine into renal brush border membrane vesicles: kinetics, specificity, and regulation by dietary carnitine. Arch Biochem Biophys. 1984;235(2):393–402. https://doi.org/10.1016/0003-9861(84)90212-1.
12. Meier Y, Eloranta JJ, Darimont J, et al. Regional distribution of solute carrier mRNA expression along the human intestinal tract. Drug Metab Dispos. 2007;35(4):590–4. https://doi.org/10.1124/dmd.106.013342.
13. Grigat S, Fork C, Bach M, et al. The carnitine transporter SLC22A5 is not a general drug transporter, but it efficiently translocates mildronate. Drug Metab Dispos. 2009;37(2):330–7. https://doi.org/10.1124/dmd.108.023929.
14. Alnouti Y, Petrick JS, Klaassen CD. Tissue distribution and ontogeny of organic cation transporters in mice. Drug Metab Dispos. 2006;34(3):477–82. https://doi.org/10.1124/dmd.105.006932.
15. Fraenkel G, Friedman S. Carnitine. Vitam Horm. 1957;15:73–118. https://doi.org/10.1016/s0083-6729(08)60508-7.
16. Tada M, Sugiyama S, Ozawa T. Investigation of carnitine content in food by enzymatic method. J Jpn Soc Nutr Food Sci. 1984;37:13–7. (Article in Japanese).
17. Shimada K, Sakuma Y, Wakamatsu J, et al. Species and muscle differences in L-carnitine levels in skeletal muscles based on a new simple assay. Meat Sci. 2004;68(3):357–62. https://doi.org/10.1016/j.meatsci.2004.04.003.
18. Erfle JD, Sauer FD, Fisher LJ. Interrelationships between milk carnitine and blood and milk components and tissue carnitine in normal and ketotic cows. J Dairy Sci. 1974;57(6):671–6. https://doi.org/10.3168/jds.S0022-0302(74)84945-3.
19. Snoswell AM, Koundakjian PP. Relationships between carnitine and coenzyme a esters in tissues of normal and alloxan-diabetic sheep. Biochem J. 1972;127(1):133–41. https://doi.org/10.1042/bj1270133.
20. Yamashita A, Kato Y, Yoshimura M. Effects of heat treatment on free carnitine and acylcarnitine contents in venison. J Jpn Soc Food Sci Technol. 2014;61:178–81. (Article in Japanese).
21. Lombard KA, Olson AL, Nelson SE, et al. Carnitine status of lactoovovegetarians and strict vegetarian adults and children. Am J Clin Nutr. 1989;50(2):301–6. https://doi.org/10.1093/ajcn/50.2.301.
22. Olson AL, Rebouche CJ. γ-butyrobetaine hydroxylase activity is not rate limiting for carnitine biosynthesis in the human infant. J Nutr. 1987;117(6):1024–31. https://doi.org/10.1093/jn/117.6.1024.
23. Rebouche CJ, Seim H. Carnitine metabolism and its regulation in microorganisms and mammals. Annu Rev Nutr. 1998;18:39–61. https://doi.org/10.1146/annurev.nutr.18.1.39.
24. Morse RK, Vergnes JP, Malloy J, et al. Sites of biological methylation of proteins in cultured chick muscle cells. Biochemistry. 1975;14(19):4316–25. https://doi.org/10.1021/bi00690a028.
25. Huszar G. Tissue-specific biosynthesis of epsilon-N-monomethyllysine and epsilon-N-trimethyllysine in skeletal and cardiac muscle myosin: a model for the cell-free study of post-translational amino acid modifications in proteins. J Mol Biol. 1975;94(3):311–26. https://doi.org/10.1016/0022-2836(75)90205-3.
26. Tanphaichitr V, Horne DW, Broquist HP. Lysine, a precursor of carnitine in the rat. J Biol Chem. 1971;246(20):6364–6.
27. Tanphaichitr V, Broquist HP. Role of lysine and ε-N-trimethyllysine in carnitine biosynthesis. II. Studies in the rat. J Biol Chem. 1973;248(6):2176–81.
28. Horne DW. Broquist HP: role of lysine and ε-N-trimethyllysine in carnitine biosynthesis. I. Studies in *neurospora crassa*. J Biol Chem. 1973;248(6):2170–5.
29. Henderson LM, Nelson PJ, Henderson L. Mammalian enzymes of trimethyllysine conversion to trimethylaminobutyrate. Fed Proc. 1982;41(12):2843–7.
30. Englard S. Hydroxylation of γ-butyrobetaine to carnitine in human and monkey tissues. FEBS Lett. 1979;102(2):297–300. https://doi.org/10.1016/0014-5793(79)80022-8.

31. Wanner C, Hörl WH. Carnitine abnormalities in patients with renal insufficiency. Pathophysiological and therapeutical aspects. Nephron. 1988;50(2):89–102. https://doi.org/10.1159/000185137.
32. Costell M, O'Connor JE, Grisolia S. Age-dependent decrease of carnitine content in muscle of mice and humans. Biochem Biophys Res Commun. 1989;161(3):1135–43. https://doi.org/10.1016/0006-291x(89)91360-0.
33. Kerner J, Hoppel C. Genetic disorders of carnitine metabolism and their nutritional management. Annu Rev Nutr. 1998;18:179–206. https://doi.org/10.1146/annurev.nutr.18.1.179.
34. Evans AM, Fornasini G. Pharmacokinetics of L-carnitine. Clin Pharmacokinet. 2003;42(11):941–67. https://doi.org/10.2165/00003088-200342110-00002.
35. Borum PR, Bennett SG. Carnitine as an essential nutrient. J Am Coll Nutr. 1986;5(2):177–82. https://doi.org/10.1080/07315724.1986.10720124.
36. Rebouche CJ. Kinetics, pharmacokinetics, and regulation of L-carnitine and acetyl-L-carnitine metabolism. Ann N Y Acad Sci. 2004;1033:30–41. https://doi.org/10.1196/annals.1320.003.
37. Huth PJ, Shug AL. Properties of carnitine transport in rat kidney cortex slices. Biochim Biophys Acta. 1980;602(3):621–34.
38. Stanley CA, Berry GT, Bennett MJ, et al. Renal handling of carnitine in secondary carnitine deficiency disorders. Pediatr Res. 1993;34(1):89–97. https://doi.org/10.1203/00006450-199307000-00021.
39. Schürch R, Todesco L, Novakova K, et al. The plasma carnitine concentration regulates renal OCTN2 expression and carnitine transport in rats. Eur J Pharmacol. 2010;635(1–3):171–6. https://doi.org/10.1016/j.ejphar.2010.02.045.
40. Tamai I, China K, Sai Y, et al. Na+-coupled transport of L-carnitine via high-affinity carnitine transporter OCTN2 and its subcellular localization in kidney. Biochim Biophys Acta. 2001;1512(2):273–84. https://doi.org/10.1016/s0005-2736(01)00328-5.
41. Wu X, Huang W, Prasad PD, et al. Functional characteristics and tissue distribution pattern of organic cation transporter 2 (OCTN2), an organic cation/carnitine transporter. J Pharmacol Exp Ther. 1999;290(3):1482–92.
42. Ohashi R, Tamai I, Yabuuchi H, et al. Na^+-dependent carnitine transport by organic cation transporter (OCTN2): its pharmacological and toxicological relevance. J Pharmacol Exp Ther. 1999;291(2):778–84.
43. Chalmers RA, Roe CR, Stacey CE, et al. Urinary excretion of L-carnitine and acylcarnitines by patients with disorders of organic acid metabolism: evidence for secondary insufficiency of L-carnitine. Pediatr Res. 1984;18(12):1325–8. https://doi.org/10.1203/00006450-198412000-00021.
44. Shekhawat PS, Matern D, Strauss AW. Fetal fatty acid oxidation disorders, their effect on maternal health and neonatal outcome: impact of expanded newborn screening on their diagnosis and management. Pediatr Res. 2005;57(5 Pt 2):76R–86R. https://doi.org/10.1203/01.PDR.0000159631.63843.3E.
45. Zhou L, Wang Q, Yin P, et al. Serum metabolomics reveals the deregulation of fatty acids metabolism in hepatocellular carcinoma and chronic liver diseases. Anal Bioanal Chem. 2012;403(1):203–13. https://doi.org/10.1007/s00216-012-5782-4.
46. Miyaaki H, Kobayashi H, Miuma S, et al. Blood carnitine profiling on tandem mass spectrometry in liver cirrhotic patients. BMC Gastroenterol. 2020;20(1):41. https://doi.org/10.1186/s12876-020-01190-6.
47. Suzuki Y, Masumura Y, Kobayashi A, et al. Myocardial carnitine deficiency in chronic heart failure. Lancet. 1982;1(8263):116. https://doi.org/10.1016/s0140-6736(82)90263-x.
48. Ahmad T, Kelly JP, McGarrah RW, et al. Prognostic implications of long-chain acylcarnitines in heart failure and reversibility with mechanical circulatory support. J Am Coll Cardiol. 2016;67(3):291–9. https://doi.org/10.1016/j.jacc.2015.10.079.
49. Masumura Y, Kobayashi A, Yamazaki N. Myocardial free carnitine and fatty acylcarnitine levels in patients with chronic heart failure. Jpn Circ J. 1990;54(12):1471–6. https://doi.org/10.1253/jcj.54.12_1471.
50. Ruiz M, Labarthe F, Fortier A, et al. Circulating acylcarnitine profile in human heart failure: a surrogate of fatty acid metabolic dysregulation in mitochondria and beyond. Am J Physiol Heart Circ Physiol. 2017;313(4):H768–81. https://doi.org/10.1152/ajpheart.00820.2016.
51. Ueland T, Svardal A, Øie E, et al. Disturbed carnitine regulation in chronic heart failure —increased plasma levels of palmitoyl-carnitine are associated with poor prognosis. Int J Cardiol. 2013;167(5):1892–9. https://doi.org/10.1016/j.ijcard.2012.04.150.
52. Yoshihisa A, Watanabe S, Yokokawa T, et al. Associations between acylcarnitine to free carnitine ratio and adverse prognosis in heart failure patients with reduced or preserved ejection fraction. ESC Heart Fail. 2017;4(3):360–4. https://doi.org/10.1002/ehf2.12176.
53. Kalim S, Clish CB, Wenger J, et al. A plasma long-chain acylcarnitine predicts cardiovascular mortality in incident dialysis patients. J Am Heart Assoc. 2013;2(6):e000542. https://doi.org/10.1161/JAHA.113.000542.
54. Kamei Y, Kamei D, Tsuchiya K, et al. Association between 4-year all-cause mortality and carnitine profile in maintenance hemodialysis patients. PLoS One. 2018;13(8):e0201591. https://doi.org/10.1371/journal.pone.0201591.
55. Murphy WJA, Steiber A, Connery GC, et al. Altered carnitine metabolism in dialysis patients with reduced physical function may be due to dysfunctional fatty acid oxidation. Nephrol Dial Transplant.

2012;27(1):304–10. https://doi.org/10.1093/ndt/gfr334.
56. Schreiber B. Levocarnitine and dialysis: a review. Nutr Clin Pract. 2005;20(2):218–43. https://doi.org/10.1177/0115426505020002218.
57. Ohtani Y, Nishiyama S, Matsuda I. Renal handling of free and acylcarnitine in secondary carnitine deficiency. Neurology. 1984;34(7):977–9. https://doi.org/10.1212/wnl.34.7.977.
58. Manz F, Waldherr R, Fritz HP, et al. Idiopathic de Toni-Debré-Fanconi syndrome with absence of proximal tubular brush border. Clin Nephrol. 1984;22(3):149–57.
59. Bernardini I, Rizzo WB, Dalakas M, et al. Plasma and muscle free carnitine deficiency due to renal Fanconi syndrome. J Clin Invest. 1985;75(4):1124–30. https://doi.org/10.1172/JCI111806.
60. Gahl WA, Bernardini I, Dalakas M, et al. Oral carnitine therapy in children with cystinosis and renal Fanconi syndrome. J Clin Invest. 1988;81(2):549–60. https://doi.org/10.1172/JCI113353.
61. Gahl WA, Bernardini IM, Dalakas MC, et al. Muscle carnitine repletion by long-term carnitine supplementation in nephropathic cystinosis. Pediatr Res. 1993;34(2):115–9. https://doi.org/10.1203/00006450-199308000-00001.
62. Matsuda I, Ohtani Y, Ninomiya N. Renal handling of carnitine in children with carnitine deficiency and hyperammonemia associated with valproate therapy. J Pediatr. 1986;109(1):131–4. https://doi.org/10.1016/s0022-3476(86)80592-3.
63. Tein I, DiMauro S, Xie ZW, et al. Valproic acid impairs carnitine uptake in cultured human skin fibroblasts. An in vitro model for the pathogenesis of valproic acid-associated carnitine deficiency. Pediatr Res. 1993;34(3):281–7. https://doi.org/10.1203/00006450-199309000-00008.
64. Lheureux PER, Hantson P. Carnitine in the treatment of valproic acid-induced toxicity. Clin Toxicol. 2009;47(2):101–11. https://doi.org/10.1080/15563650902752376.
65. Sztajnkrycer MD. Valproic acid toxicity: overview and management. J Toxicol Clin Toxicol. 2002;40(6):789–801. https://doi.org/10.1081/clt-120014645.
66. Verrotti A, Greco R, Morgese G, et al. Carnitine deficiency and hyperammonemia in children receiving valproic acid with and without other anticonvulsant drugs. Int J Clin Lab Res. 1999;29(1):36–40. https://doi.org/10.1007/s005990050060.
67. Segura-Bruna N, Rodriguez-Campello A, Puente V, et al. Valproate-induced hyperammonemic encephalopathy. Acta Neurol Scand. 2006;114(1):1–7. https://doi.org/10.1111/j.1600-0404.2006.00655.x.
68. Silva MFB, Aires CCP, Luis PBM, et al. Valproic acid metabolism and its effects on mitochondrial fatty acid oxidation: a review. J Inherit Metab Dis. 2008;31(2):205–16. https://doi.org/10.1007/s10545-008-0841-x.
69. Yoshikawa H, Watanabe T, Abe T. Tubulo-interstitial nephritis caused by sodium valproate. Brain Dev. 2002;24(2):102–5. https://doi.org/10.1016/s0387-7604(02)00007-4.
70. Hawkins E, Brewer E. Renal toxicity induced by valproic acid (Depakene). Pediatr Pathol. 1993;13(6):863–8. https://doi.org/10.3109/15513819309048273.
71. Watanabe T, Yoshikawa H, Yamazaki S, et al. Secondary renal Fanconi syndrome caused by valproate therapy. Pediatr Nephrol. 2005;20(6):814–7. https://doi.org/10.1007/s00467-005-1827-7.

5 What Are Carnitine Deficiency and Carnitine Insufficiency?

5.1 Definition of Carnitine Deficiency and Carnitine Insufficiency

According to Pons and De Vivo, carnitine deficiency is defined as "a condition in which the concentration of carnitine in plasma or tissues does not meet the requirements of the body to perform normal functions." [1] "Normal function" refers to energy metabolism such as fatty acid β-oxidation, ammonia metabolism such as the urea cycle, excretion of toxic long-chain acyl groups from the body, regulation of the acyl-CoA (coenzyme A)/free CoA ratio, and other functions of carnitine to maintain energy metabolism and mitochondrial function [1–3].

In addition, Stumpf et al. stated that the essence of carnitine deficiency is not the absolute value of free carnitine concentration or total carnitine concentration in the blood or tissues, but whether there is a sufficient amount of free carnitine to buffer (or detoxify) the toxicity of acyl groups (= acyl-CoA) accumulated in the mitochondria. The question is whether there is enough free carnitine to buffer (or detoxify) the toxicity of the accumulated acyl groups (= acyl-CoA) in the mitochondria, and the "relative" amount of free carnitine to the accumulated acyl groups (acyl-CoA) is the issue [4]. In other words, if acyl-CoA is too high even with a normal level of free carnitine, carnitine deficiency may develop.

5.2 Two Types of Carnitine Deficiency: Carnitine Deficiency and Carnitine Insufficiency

Carnitine deficiency develops in various etiologies, but if we summarize which fraction of free carnitine, acylcarnitine, and the total of carnitine in the blood is reduced from the definition of carnitine deficiency described above, it can be divided into the following two types of carnitine deficiency, A and B.

A. Absolute lack of free carnitine prevents the classical carnitine action of transporting long-chain fatty acids into the mitochondria for energy production (carnitine deficiency in the narrow sense).
B. Insufficient free carnitine relative to acyl-CoA accumulation, resulting in failure of acyl-CoA to excrete from the mitochondria (called "carnitine insufficiency").

"A" mainly refers to the inability of carnitine to carry long-chain fatty acids to the mitochondria, while "B" refers to the outward action of carnitine, which carnitines carry harmful acyl groups out of the mitochondria. In fact, these two types of "A" and "B" are usually treated together as "carnitine deficiency." The two types are often mixed in clinical cases. It is easy to understand

N. Nakayama, *Carnitine Deficiency and Insufficiency in Medicine*,
https://doi.org/10.1007/978-981-95-7797-2_5

the absolute deficiency of type "A" carnitine because it is considered in the same way as vitamin deficiency. However, it should be noted that carnitine is mostly present in tissues, and the pool size of plasma and extracellular fluid is about 0.5%, and the true deficiency state cannot be judged by blood concentration alone. Carnitine insufficiency (type "B") relative to the accumulation of acyl-CoA is described in detail in Chap. 3: "What Is Carnitine?" and it is an important point for understanding the action of carnitine and carnitine deficiency. It is also a concept that is difficult to understand for those who have come to view carnitine deficiency as similar to vitamin deficiency.

Carnitine deficiency due to absolute deficiency of "A" is caused by an absolute shortage of carnitine in the body due to reduced dietary intake of carnitine or great reduction of renal tubular reabsorption of free carnitine, resulting in carnitine deficiency. The most typical is primary carnitine deficiency (systemic carnitine deficiency) due to a genetic abnormality of the carnitine transporter (OCTN2), with extremely low levels of both free carnitine and acylcarnitines (free carnitine levels <10 μmol/L) and therefore very low total carnitine in blood. This type of deficiency includes carnitine deficiency in Fanconi syndrome and carnitine deficiency in patients managed with tube feedings that do not contain carnitine or formula that eliminates cow's milk allergens. Carnitine deficiency due to antibacterial agents containing a pivoxil group in drug-induced carnitine deficiency also causes an extreme decrease in free carnitine and a decrease in total carnitine.

On the other hand, the pathophysiology of carnitine insufficiency of "B" is due to the accumulation of acyl-CoA, a metabolite of organic acids and fatty acids, in the mitochondria in various pathologies such as inherited metabolic diseases, cirrhosis, and heart failure. The acyl group binds to free CoA to form acyl-CoA. It is a condition in which free CoA, which is important for mitochondrial function such as energy metabolism, is relatively reduced, and mitochondrial function is impaired. In these conditions, acyl-CoA accumulated in the mitochondria is excreted outside the mitochondria as acylcarnitines and eventually excreted in the urine, reducing the toxicity of the acyl group if there are enough free carnitines. However, if this continues for a long time, the amount of free carnitine in the body decreases, resulting in secondary carnitine deficiency. In type "B" carnitine deficiency, the concentration of free carnitine in the blood does not decrease significantly in the early stages of the disease, but high levels of acylcarnitines peculiar to the disease are observed in the blood and urine, and as the disease progresses, the concentration of free carnitine in the blood decreases, and if this continues, secondary carnitine deficiency develops. In this way, the accumulation of acyl-CoA in the mitochondria reduces the relative amount of free CoA, which is important for various metabolisms, so the acyl-CoA/free CoA ratio increases, and free CoA cannot be used for metabolism. Furthermore, these accumulated acyl-CoA inhibit various enzymes that are important for mitochondrial energy metabolism, gluconeogenesis, and ammonia metabolism.

While the classical physiological action of carnitine to transport the acyl group of a long-chain fatty acid into the mitochondria has been established, a new effect of regulating the ratio of acyl-CoA/free CoA to maintain the amount of free CoA, which is important for metabolism, has been found in cases such as organic acidemia. Until then, only the role of carnitine in the inward transport of long-chain fatty acids into the mitochondria had been considered, but the action of transporting acyl groups outward from the mitochondria was proposed. This new effect of carnitine was first reported in organic acidemia in the 1980s and is called carnitine insufficiency [5]. Free CoA is extremely important for various metabolisms such as energy metabolism, but the amount of free CoA in mitochondria is limited and cannot be synthesized and supplied quickly (see "Sect. 3.7: What Is Coenzyme A (CoA)" in "Chap. 3: What Is Carnitine?"). This concept of carnitine insufficiency was initially established and used in inherited metabolic diseases such as organic acidemia and fatty acid oxidation disorders, mitochondrial myopathy, pediatric cardiomyopathy patients, and dialysis patients, but is now used in the diagnosis and pathophysiology of carnitine deficiency due to a variety of causes [5–14]. It has also been reported that in adult chronic diseases such as chronic heart failure, renal failure, and liver cirrhosis, acyl groups accumulate in mitochondria, resulting in the same pathophysiology as that of carnitine insufficiency. [15–24]

Using the easy-to-understand illustration by Dr. Akiko Wakisaka, Department of Pediatrics, Iou Hospital, National Hospital Organization, in Kanazawa, Japan, in the same way as Fig. 3.7 in Chap. 3, Fig. 5.1a shows that the function of the powerplant stagnates due to the insufficient absolute amounts of trucks required for inward fuel transportation to the mitochondria. Figure 5.1b shows a shortage of trucks to carry out a large amount of rubble generated in the city due to the earthquake. If there is a large amount of rubble, even if there is a normal number of trucks, it will not be possible to keep up with the disposal of rubble. It will not be possible to restore the city systems unless a large number of trucks from all over the country are collected and sent to the affected areas. There is a relative shortage of trucks (free carnitine) necessary to excrete acyl-CoA accumulated in the mitochondria as acylcarnitines, and various symptoms and pathophysiology develop due to mitochondrial dysfunction. At first, the blood acylcarnitines level is high, and a large amount of acylcarnitines is also excreted in the urine. Therefore, the ratio of acylcarnitine/free carnitine (AC/FC) in the blood is high. As the disease progresses, blood free carnitine also decreases because it is used to excrete the acyl group as acylcarnitines, but it is rare to see an extreme decrease (free carnitine level <10 μmol/L) as in carnitine deficiency of A. Even if the free carnitine concentration is normal, if the acylcarnitine level (actually acyl-CoA accumulation in the mitochondria) is high, it is considered to be a state of relative shortage of free carnitine, and carnitine deficiency develops because free carnitine cannot detoxify the toxicity of accumulated acyl-CoA and secure the free CoA necessary for metabolism.

In type B carnitine deficiency, which is most of secondary carnitine deficiencies, the problem is not the absolute value of free carnitine itself, but whether there is enough free carnitine to bind to the accumulated acyl group and excrete the acyl group in the urine [4, 5]. Essentially, carnitine insufficiency refers to a condition in which acyl-CoA in tissues is too high relative to free CoA and metabolic failure occurs, so the diagnosis should be to quantify the amount of acyl-CoA and free CoA in the mitochondria, but it is not easy to quantify acyl-CoA and free CoA. When the acyl group accumulates in the mitochondria, it becomes acyl-CoA, but if the amount of carnitine is sufficient, the acyl group is received by free carnitine to become acylcarnitines, and it is excreted outside the mitochondria, where free CoA, which is important for metabolism, is regenerated. Over time, acyl-CoA/free CoA in the mitochondria is thought to be in equilibrium with the acylcarnitine/free carnitine ratio (AC/FC) in the blood. Therefore, acylcarnitine and free carnitine concentrations in the blood are measured to diagnose carnitine insufficiency based on AC/FC. Based on the findings of various carnitine deficiency cases, such as inherited metabolic diseases and dialysis patients, AC/FC in healthy people is usually less than 0.25, and when it exceeds 0.4, it is considered to be a state of carnitine insufficiency, that is, acyl-CoA is relatively more accumulating in the mitochondria than free CoA, and free carnitine is also insufficient. In the state of carnitine insufficiency, mitochondria fall into metabolic dysfunction, and various clinical symptoms such as fatigue, myasthenia, hypoglycemia, and hyperammonemia are likely to occur. Of course, this AC/FC > 0.4 is not absolute value, and carnitine deficiency can develop even if it is lower than that.

5.3 Significance of Carnitine Supplementation Therapy

Carnitine supplementation therapy is effective for both types of carnitine deficiency. As illustrated in Fig. 5.1, it is necessary to send a large number of trucks from outside to facilitate the transportation of fuel to the powerplant (Fig. 5.1a) or to facilitate the removal of rubble (Fig. 5.1b) in

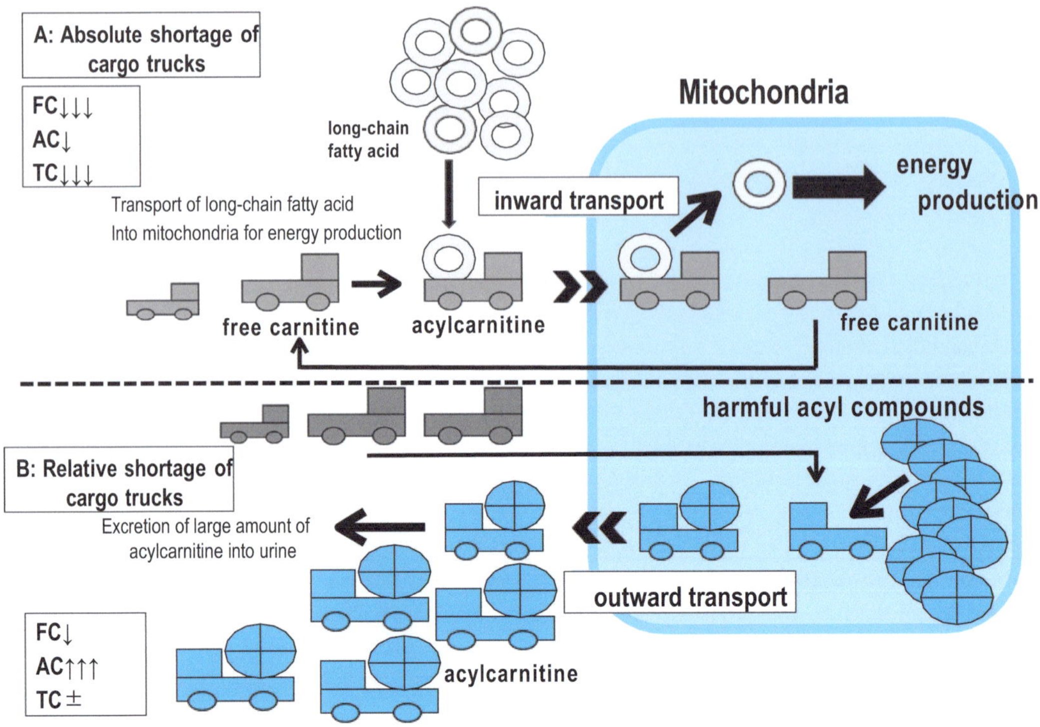

Fig. 5.1 An easy-to-understand illustration explaining the role of carnitine compared to cargo trucks in two types of carnitine deficiency. (**a**) (upper part of dotted line): Absolute deficiency of free carnitine (cargo trucks) causes trouble for energy production because fuel (long-chain fatty acid) cannot be transported into the powerplant (mitochondrial β-oxidation). (**b**) (lower part of dotted line): Relative deficiency of free carnitine (cargo trucks) in mitochondria causes trouble to excrete and remove the toxicity of accumulated harmful acyl-compound in mitochondria. The amount of free carnitine needed to remove acyl-compound is dependent on the amount of acyl-compound (relative). In other words, **a** is inward-transport of acyl group in long-chain fatty acid for energy production (classical role of carnitine). The deficiency of carnitine causes disturbance for energy production. **b** is a kind of new role of carnitine. The accumulated harmful acyl-compound(s) in various pathophysiology disturbs mitochondrial metabolism such as energy production, etc., so relative deficiency of carnitine causes trouble in outward transportation of acyl-group to excrete and detoxify them. *AC* acylcarnitine, *FC* free carnitine, *TC* total carnitine (=free carnitine + acylcarnitine). (With permission from Dr. Akiko Wakisaka, Pediatrics, NHO Iou National Hospital, Kanazawa, Japan)

order to normalize the disease. This is carnitine supplementation therapy. It is often said that "carnitine is just one of the nutrients," but among the many nutrients and pharmaceuticals, there is no substance other than carnitine that has the effect of transporting long-chain acyl groups to the mitochondria and excreting acyl groups accumulated in the mitochondria from the mitochondria. For carnitine deficiency, only "carnitine" can be clinically applied as a therapeutic or prophylactic agent. This is a very important point why it is necessary to have carnitine as a drug.

Primary carnitine deficiency, which is a typical example of type A deficiency, develops cardiomyopathy and sudden death by the age of 2–3 years, but a large dose of carnitine allows the patient to lead a normal life. Until then, cardiomyopathy was considered untreatable, but carnitine supplementation therapy for those cases was reported as "treatable" [25–30]. When large doses of carnitine are administered to patients with primary carnitine deficiency, the concentration of free carnitine in the blood often returns to nearly half of the normal value, and the concentration of carnitine in the liver is also restored to near normal, but the concentration in the tissues of skeletal muscle increases only slightly. Even so, it has been reported that exercise tolerance and cardiac function are dramatically improved, and findings due to cardiomyopathy such as cardiac hypertrophy are also improved [31–35].

A typical example of a deficiency of type B is organic acidemia, such as propionic acidemia. In the 1980s, it was reported that in propionic acidemia and methylmalonic acidemia, the accumulated harmful organic acids of intermediate metabolites such as propionic acid and methylmalonic acid were excreted from the mitochondria as acylcarnitine when carnitine was administered, and clinical symptoms were greatly improved [36–38]. When carnitine is administered, the toxicity of the acyl group is released by the excretion of the acyl group in the accumulated mitochondria, and at the same time, free CoA, which is important for metabolism, is regenerated (regulation of acyl-CoA/free CoA ratio [5, 6]; see Sect. 3.11 "Excretion of Harmful Acyl Compounds and Regulation of Acyl-CoA/Free CoA Ratio" in Chap. 3).

In both A and B types of carnitine deficiency, carnitine supplementation is the same as supplementation as a carrier of acyl groups, but the precise mechanism of action is different in that it promotes inward transport into the mitochondria in the case of A, and outward transport outside the mitochondria in the case of B. However, if a large amount of acyl groups accumulates in type B for some reasons and are excreted as acylcarnitines, the free carnitine will eventually decrease considerably and energy metabolism will stagnate, and a carnitine deficiency of type A will also be mixed.

As for the diagnosis of the two types of carnitine deficiency, based on the accumulation of numerous clinical studies to date, it is clear that there is an abnormality in the homeostasis of carnitine in the case of free carnitine concentrations <20 μmol/L, acylcarnitine ≥20 μmol/L, and AC/FC >0.4, and the causes of these diseases need to be further investigated [1, 3, 7, 10–12, 39–41] (see "Chap. 18: Guidelines for the Diagnosis and Treatment of Carnitine Deficiency," for details). However, the reference values of carnitine concentration and AC/FC in the blood are only used for diagnosis before administration of carnitine preparations, and after administration of carnitine preparations, the blood free carnitine levels are apparently high due to the effects of administration, so it may not be possible to diagnose correctly. At present, there is no established method for diagnosing and monitoring carnitine deficiency after administration of carnitine preparations (see "Chap. 18: Guidelines for the Diagnosis and Treatment of Carnitine Deficiency").

5.4 Classification of Carnitine Deficiency by Etiology

Carnitine deficiency is widely classified according to the cause of the disease. Carnitine deficiency is known to be caused by a variety of factors and is not necessarily a rare condition.

In the 1980s, carnitine deficiency disorders that developed in infancy due to inherited metabolic diseases were classified according to clinical presentation and tissue carnitine concentrations into: (1) muscle type, in which carnitine is reduced only in skeletal muscle; (2) systemic type, in which carnitine is reduced not only in muscle but also in liver and blood; and (3) mixed type, in which carnitine is reduced in a combination of both types and there is no fixed trend [42, 43]. However, due to the accumulation of subsequent cases and the progress of genetic research, the investigation of the mechanism that causes carnitine deficiency and the identification of mutation sites of specific genes have progressed dramatically. In addition, it has been found that carnitine deficiency can be caused by acquired or medical causes, or by various iatrogenic causes such as drug administration and dialysis. Therefore, carnitine deficiency is now broadly classified into primary carnitine deficiency (also called systemic carnitine deficiency), which develops due to an extreme decrease in systemic carnitine due to genetic abnormalities of the carnitine transporter (OCTN2), and secondary carnitine deficiency, which develops due to various other causes [1, 6,

41, 44–48]. Secondary carnitine deficiency is further caused by inborn metabolic disorders such as organic acid metabolism abnormalities, fatty acid metabolism disorders, mitochondrial disorders, cirrhosis of the liver, renal failure, pregnancy and malnutrition, long-term central venous nutrition (total parenteral nutrition [TPN]) treatment, or long-term hemodialysis. It is classified as iatrogenic factors due to the administration of antiepileptic drugs such as valproic acid and pivoxil-containing antimicrobials. Table 5.1 shows the classification of carnitine deficiency according to etiology. The characteristics of each pathology are outlined below, and clinical cases and treatments are explained in more detail in this book.

Each carnitine deficiency is outlined below.

5.4.1 Primary Carnitine Deficiency

It is also called systemic carnitine deficiency, carnitine transporter disorder, etc. In the 1970s and 1980s, there were cases of fat deposits in muscle cells among myopathy patients (lipid-storage myopathy), and the amount of carnitine in skeletal muscle alone was reduced (muscular-type carnitine deficiency). The name systemic carnitine deficiency was used for cases in which the carnitine content of the whole body, such as the liver, was extremely low. Numerous subsequent studies have shown that such systemic carnitine

Table 5.1 Classification of carnitine deficiency by etiology

Primary carnitine deficiency	
Carnitine transporter abnormality (also called systemic carnitine deficiency)	Genetic functional defects or decreased activity of carnitine transporter in intestinal tract, renal tubule, skeletal muscle, cardiac muscle, etc.
Secondary carnitine deficiency	
Inherited metabolic diseases	**Organic acidemias**: Propionic acidemia, methylmalonic acidemia, isovaleric acidemia, 3-methylcrotonylglycinuria, 3-hydroxy-3-methylglutaric acidemia, multiple carboxylase deficiency, β-ketothiolase deficiency, glutaric acidemia type 1, methylglutaconic aciduria, etc. **Fatty acid oxidation disorders**: Glutaric acidemia type 2, very long-chain acyl-CoA dehydrogenase deficiency, trifunctional protein (TFP) deficiency, medium-chain acyl-CoA dehydrogenase deficiency, CPT2 deficiency, CACT deficiency, etc. **Mitochondrial disorders**: Mitochondrial respiratory chain disorders (ComplexI~V), etc.
Acquired medical conditions	**Decreased biosynthesis**: Cirrhosis, chronic renal diseases **Decreased intake**: Chronic TPN, malnutrition, strict vegetarians, soy-protein infant formula without added L-carnitine, anorexia, etc. **Decreased absorption**: Short-gut syndrome, celiac disease, Crohn's disease, cystic fibrosis, etc. **Decreased body store/increased requirement:** Pregnant and lactating women, extreme prematurity, intrauterine growth retardation, infant of carnitine deficient mother, critically ill patients (sepsis and septic shock, etc.), acquired immune deficiency syndrome, etc. **Increased loss**: Fanconi syndrome, renal tubular acidosis, etc.
Iatrogenic factors	**Renal replacement therapy**: Hemodialysis, peritoneal dialysis, continuous renal replacement therapy (CRRT) **Drug induced**: Antiepileptics (valproate, phenobarbital, phenytoin, carbamazepine, etc.), pivoxil group containing antibiotics (cefcapene pivoxil, cefteram pivoxil, cefditoren pivoxil, etc.), anticancer drugs (platinum-containing drugs such as cisplatin, anthracycline drugs such as doxorubicin, alkylating agents such as ifosfamide), anesthetics (bupivacaine, propofol), AIDS drugs (zidovudine, etc.), sodium benzoate (for treatment of hyperammonemia in urea cycle disorders), etc.

Created using Refs. [1, 6, 41, 44–48]

CACT carnitine-acylcarnitine translocase, *CoA* coenzyme A, *CPT2* carnitine palmitoyltransferase 2, *TPN* total parenteral nutrition deficiency is caused by a genetic defect in the carnitine transporter (OCTN2), and it is now called primary carnitine deficiency or carnitine transporter disorder. Muscle-type carnitine deficiency syndromes described in those days are now thought to be fatty acid oxidation disorders such as medium-chain fatty acid dehydrogenase deficiency, based on the subsequent progress of neonatal mass screening and genetic research.

The carnitine transporter (OCTN2) is expressed in the intestinal tract, renal tubules, cardiac and skeletal muscles, etc. The congenital abnormality in OCTN2 gene interferes with the absorption of carnitine from the diet, reabsorption in the renal tubules to prevent the leakage, and uptake into cardiac and skeletal muscles, which require large amounts of carnitine, resulting in an extreme decrease in blood and tissue carnitine levels. Free carnitine levels in the blood are less than 10 μmol/L in most cases, and acylcarnitine levels are also decreased. Therefore, the total carnitine level, the sum of the free carnitine and acylcarnitine levels, is also reduced.

Homozygous mutations in OCTN2, which cause primary carnitine deficiency, are said to incidence 1 in 40,000 newborns in Japan, making it one of the most frequently inherited metabolic diseases [49, 50]. As of 2016, more than 138 genetic mutations in the carnitine transporter are known [51], and the mutations significantly affect the activity of the transporter, resulting in death from cardiomyopathy or hypoglycemic attacks in newborns and infants in severe cases. In the case of mildly reduced activity, a genetic abnormality of OCTN2 may be found after a close examination of a patient with heart failure in adulthood. In the daily care of patients with heart failure and cardiac hypertrophy, it is possible that cases of mildly reduced transporter activity or heterozygous cases may be found [52–54].

In patients with primary carnitine deficiency, OCTN2 in intestinal epithelial cells may also be mutated, resulting in decreased carnitine absorption. However, it is thought that large doses of carnitine are absorbed by pharmacological doses of carnitine replacement therapy (100–200 mg/kg/day) via the amino acid transporter $B^{0,+}$ ($ATB^{0,+}$), which has low affinity but high capacity for carnitine [51, 55]. However, when tissue distribution of this amino acid transporter $ATB^{0,+}$ was examined by Northern blotting, it was strongly expressed in the lung, trachea, and salivary glands, weakly expressed in the mammary gland, stomach, and pituitary gland, and hardly expressed in cardiac and skeletal muscles [56]. Therefore, the mechanism by which absorbed carnitine is taken up by cardiac and skeletal muscles is unknown, but it is thought to be through other low-affinity transporters or by passive transport into the cells. In any case, it has been established in domestic and foreign case reports that progressive cardiomyopathy in primary carnitine deficiency, which is fatal in the natural course of the disease, can be dramatically improved by carnitine supplementation therapy, and can be even prevented by early administration. So, in the *Clinical Practice Guidelines for Neonatal Mass Screening Diseases 2019*, edited by the Japanese Society for Inherited Metabolic Diseases, large doses of L-carnitine (100 to 100–400 mg/kg/day) are described [57].

5.4.2 Secondary Carnitine Deficiency

Secondary carnitine deficiency is broadly classified into those caused by inherited metabolic diseases, acquired medical conditions, and iatrogenic causes.

5.4.2.1 Due to Inherited Metabolic Diseases (Excluding Primary Carnitine Deficiency)

Since 1973, when Engel et al. reported a decrease in long-chain fatty acid oxidative capacity and carnitine deficiency in myopathy patients with sudanophilic lipid droplets in skeletal muscle [58], a large number of carnitine deficiency cases have been reported due to the excretion of a large amount of accumulated intermediate metabolites and free carnitine in the urine in the form of acylcarnitine due to metabolic abnormalities in various organic acid metabolism disorders, fatty acid oxidation disorders, and other inherited metabolic diseases. This measurement of acylcarnitine in blood and urine has been used in tandem mass newborn screening for inherited metabolic diseases and in the diagnosis of inherited metabolic diseases. Although carnitine administration to patients with these inherited metabolic diseases cannot improve the inherited metabolic diseases themselves, there have been numerous reports of marked improvement in clinical symptoms such as hypoglycemic attacks and myasthenia caused by carnitine deficiency that develop due to these inherited metabolic diseases [39, 41, 58–71]. In this way, clinicians and researchers in Japan and abroad have examined the effects of various treatments and made efforts to accumulate the number of cases, and the standard treatment methods and precautions for each disease have been clarified. For this reason, the Japanese Society for Inherited Metabolic Diseases has published the *Clinical Practice Guidelines for Diseases Subject to Mass Screening for Newborns 2019*. In it, the diagnosis and treatment methods for each disease are described in detail, and carnitine supplementation therapy is also described in detail in the acute and chronic phases. However, since inherited metabolic diseases are a rare disease with an extremely small number of cases, and because the patient background of each case differs greatly due to differences in genetic mutation sites, there have been no high-quality clinical trials such as placebo-controlled double-blind studies of carnitine supplementation therapy for these inherited metabolic diseases [72, 73]. Evidence-based medicine (EBM) is important, but it is not practical to demand the same level of evidence for rare diseases as for diseases such as hypertension, diabetes, and heart failure, which have a very large number of patients. Winter et al. summarized the results of carnitine supplementation therapy for patients with various inherited metabolic diseases in the United States and negotiated tenaciously with the Food and Drug Administration (FDA), and the FDA finally approved carnitine replacement therapy for carnitine deficiency in these inherited metabolic diseases with the accumulation of many individual case reports [41, 61].

There are various types of inherited metabolic diseases even with the same disease name, and diagnosis and treatment should be carried out in cooperation with specialists in each disease (for requests for detailed examinations for each disease in Japan, see the "List of Detailed Examination Facilities," http://jsimd.net/iof.html, the Japanese Society for Inherited Metabolic Diseases).

5.4.2.2 Due to Acquired Medical Conditions

Reduction of Biosynthesis

Carnitine is biosynthesized mainly in the liver and kidneys, and its biosynthesis is decreased in liver cirrhosis and chronic kidney disease, which can easily cause carnitine deficiency [15, 74–76]. The activity of γ-butyrobetaine hydroxylase in the carnitine biosynthetic pathway is not yet fully developed in infancy, and insufficient intake from breast milk or food can lead to carnitine deficiency [77–79]. However, no cases of carnitine deficiency due to genetic abnormalities in enzymes involved in carnitine biosynthesis have been reported so far [80, 81].

Decreased Intake

Patients who have been managed for a prolonged period of time with tube feedings that do not contain carnitine, high-calorie infusions used for central venous nutrition, and some powdered infant formulas that eliminate milk allergens are prone to carnitine deficiency. Carnitine deficiency is also more likely to occur in the elderly, anorexic patients, strict vegetarians, and others who do not eat much meat or fish rich in carnitine [82]. In addition, patients fed a diet based on soy protein are more likely to develop carnitine deficiency. Infants who are fed a diet based on soy protein are also at risk for carnitine deficiency [83]. Carnitine intake from breast milk is extremely important during infancy, when the biosynthesis of carnitine in the liver and kidneys is not fully developed [84–89]. Therefore, carnitine deficiency should also be considered when using special infant formula that does not contain carnitine or contains little carnitine for a long period of time, such as in the treatment of special diseases in newborns and infants, or when central venous nutrition with high-calorie infusions is used in infant surgery, etc. [88–100] Carnitine deficiency with fatty liver or abnormal liver function has been reported in adults and children who were managed with enteral nutrition alone without carnitine for a long period of time [101–103].

Decreased Absorption

Carnitine in foods such as meat and dairy products is absorbed by carnitine transporters (OCTN2), which are widely distributed in the intestinal tract. Therefore, in patients with short bowel syndrome seen when the small intestine is extensively resected due to intestinal diseases such as necrotizing enterocolitis, midgut volvulus, and small bowel atresia, intestinal insufficiency diseases such as Crohn's disease, and diseases such as follicular fibrosis, carnitine deficiency is likely to develop due to a decrease in the absorption of carnitine from the intestine along with other nutrient malabsorption [104–106]. When carnitine replacement therapy is used to treat carnitine deficiency in these diseases, care should be taken, such as administering carnitine intravenously, as oral administration may not be sufficient [104]. Patients with celiac disease, which is common among Westerners whose intestinal tracts are damaged by an excessive immune response to gluten, a protein found in wheat, may develop idiopathic dilated cardiomyopathy, which is thought to be caused by carnitine deficiency [107, 108].

Decrease in Body Storage/Increase in Requirement

Pregnant and lactating women are thought to be more susceptible to carnitine deficiency because of increased carnitine requirements [109]. It has been reported that free carnitine levels in the blood gradually decrease to 50% of control levels during pregnancy and recover during childbirth and lactation [110–115]. Angelini et al. found that the carnitine content of skeletal muscle was decreased in young women who developed generalized myasthenia after delivery. Carnitine supplementation therapy improved this symptom [116].

Carnitine deficiency is more likely to occur in patients with low muscle mass, the storage site of carnitine, such as extremely underweight infants, children of carnitine-deficient mothers, and AIDS patients. Patients with critical medical conditions such as surgery, sepsis, or septic shock have increased energy demands and an increased dependence on energy production by fatty acids and ketone bodies, requiring adequate amounts of carnitine [95, 100, 104].

Increased Losses

The mechanism of reabsorption of free carnitine in the renal tubules is extremely important in the maintenance of carnitine homeostasis. Therefore, in patients with severe tubular disorders, such as Fanconi syndrome, cystinosis, renal tubular acidosis, lysinuric protein intolerance, and Lowe syndrome, carnitine deficiency is likely to occur due to large amounts of carnitine leakage into the urine due to impaired reabsorption of carnitine from the renal tubules [117–124].

Among epidermolysis bullosa patients with recurrent blisters, erosions, and ulcers on the skin and mucous membranes throughout the body, the severe form of latent (recessive) nutritionally impaired epidermolysis bullosa can cause various trace element deficiencies, and complications of dilated cardiomyopathy due to selenium deficiency and carnitine deficiency have been reported. It is recommended to supplement selenium and carnitine as well as other trace elements such as iron and zinc [125–127].

5.4.2.3 Carnitine Deficiency Caused by Iatrogenic Treatment

Dialysis

It has been reported that patients with conservative renal failure have reduced oral carnitine intake due to protein restriction and impaired carnitine biosynthesis as renal failure progresses, and renal carnitine excretion is also reduced as glomerular function declines, and hence free carnitine in blood and muscle are all reported to be elevated [128–131]. On the other hand, in patients with advanced renal failure undergoing hemodialysis, carnitine is removed during dialysis, and free carnitine in the blood decreases, while acylcarnitine increases, as reported in many studies [132–137]. Carnitine deficiency also occurs in peritoneal dialysis patients after prolonged treatment [138, 139]. Pediatric renal failure patients often undergo peritoneal dialysis because it is difficult to secure blood vessels for shunt creation, but children are prone to carnitine deficiency due to underdevelopment of carnitine biosynthesis and small muscle mass [140], and the carnitine deficiency may lead to cardiomyopathy and cardiac dysfunction [141].

Of course, dialysis patients have a variety of clinical symptoms that are unique to dialysis patients because of their very specific pathological conditions, such as repeated and rapid water withdrawal and fluid retention, progressive sarcopenia due to increased oxidative stress and inflammatory response, and accelerated atherosclerosis and calcification of blood vessels and heart valves due to abnormalities in phosphorus and calcium metabolism. The development of these clinical manifestations is thought to be complicated by multiple factors, and in particular, the development of muscle manifestations (cramps, muscle weakness, extreme fatigue, etc.), cardiac manifestations (cardiac hypertrophy, cardiomyopathy, cardiac dysfunction, arrhythmia, hypotension during dialysis, sudden death, etc.), erythropoietin-resistant anemia, sarcopenia, anorexia, and other symptoms unique to hemodialysis patients. Carnitine deficiency is thought to be one of the causes of the development of these symptoms. Carnitine deficiency due to dialysis is called "dialysis-related carnitine deficiency" or "dialysis-related carnitine disorder" (DCD), and many clinical studies of carnitine supplementation therapy for these clinical symptoms have been reported [142–150].

In pediatric patients with severe diseases state (e.g. bone marrow transplantation for blood cancer, congenital heart diseases and sepsis, etc.), kidney function deteriorate and continuous renal replacement therapy (CRRT) is performed for a long time, resulting carnitine deficiency. It was reported that carnitine supplementation therapy improved [151, 152].

Carnitine deficiency in dialysis patients is described in detail in Chap. 14, "Carnitine Deficiency in Patients with Renal Failure."

Drug-Induced Carnitine Deficiency

Sodium valproate, which is widely used for the prevention and treatment of various types of epileptic seizures, mood stabilizers for manic and manic-depressive patients, and suppression of migraine attack, is widely known to cause carnitine deficiency [153–156].

Although valproic acid was first used as a well-tolerated antiepileptic drug, its rare fatal hepatotoxicity and hyperammonemic encephalopathy have become problems, and basic and clinical studies on its pathogenesis have been reported (see "Chap. 12: Carnitine Deficiency Caused by Sodium Valproate," for details). Among them, valproic acid and its metabolites cause carnitine deficiency, and inhibition of fatty acid oxidation in the mitochondria of the liver and the urea cycle is thought to cause lethal hepatotoxicity and hyperammonemic encephalopathy in some patients. Although the incidence of serious adverse reactions has decreased, a relatively high incidence of mild to moderate hyperammonemia has been observed in younger children and severely disabled children, who are more prone to carnitine deficiency. Particular attention should be paid to carnitine deficiency in younger children and children (and persons) with severe mental or physical disabilities who are prone to it.

In addition, in order to improve absorption, many prodrugs of antimicrobials in which a pivoxil group is attached to the main body of a cephem antibiotic are used in daily clinical practice in Japan. In addition, the carbapenem antimicrobial drug tebipenem pivoxil (trade name: Orapenem), which was developed and approved in Japan for the first time in the world, is also a prodrug that combines a pivoxil group and a carbapenem antimicrobial. When these antibiotics are absorbed in the intestinal tract, pivalic acid, cleaved by esterase, binds to free carnitine and is excreted as pivaloylcarnitine. Although it is rare for carnitine deficiency to progress to symptomatic severe carnitine deficiency in adults with sufficient skeletal muscle, which is the storage site of carnitine, there have been many reports of hypoglycemic attacks due to carnitine deficiency when prescribed for a long time for refractory acute otitis media and pharyngitis in children, which has become a problem [157–166]. For this reason, the use of pivoxil-containing antibiotics is extremely limited in overseas, but in Japan they are widely used in pediatrics and otolaryngology, and care should be paid to carnitine deficiency, especially when administered for a long period of time, such as carnitine administration along with antibiotic administration. Clinical guidelines for respiratory tract infections and pediatric acute otitis media include warnings for secondary carnitine deficiency caused by antibiotics containing the pivoxil group [167, 168]. Although it is not an antimicrobial drug, Sivelestat Sodium Hydrate, a neutrophil elastase inhibitor used for acute lung injury, also has a pivoxil group (2,2-dimethylpropanyl group), which has been reported to cause an increase in C5-acylcarnitine and hypocarnitinemia [169, 170]. Caution should also be exercised in infectious diseases of the elderly with carnitine deficiency due to the use of antibacterial drugs containing the pivoxil group.

Among anticancer drugs, it has also been reported that platinum agents such as cisplatin, anthracycline preparations such as doxorubicin, and alkylating agents such as cyclophosphamide and ifosfamide cause carnitine deficiency [171–180].

According to a basic study, cisplatin suppresses the expression of the carnitine transporter (OCTN2) in the renal tubules [172]. Many anticancer drugs, such as cisplatin and etoposide, inhibit OCTN2 activity [181]. Inhibition of OCTN2 expression and activity increases the excretion of carnitine from the renal tubules [182, 183]. Therefore, caution should be paid to carnitine deficiency in patients with poor nutrition and advanced cachexia, patients who are unable to eat adequately due to loss of appetite due to anticancer drugs treatment or radiation therapy in combination, and patients who have undergone repeated chemotherapy for a long time. The association between carnitine deficiency caused by anticancer drugs and cardiomyopathy, fatigue, anorexia, and neuropathy has been studied [173, 184, 185]. However, it has been reported that carnitine deficiency was not observed in terminally ill patients receiving palliative care alone [186].

Bupivacaine, a local anesthetic, has not been reported to cause carnitine deficiency by itself, but has been reported to cause severe arrhythmias and cardiac arrest when administered to patients of isovaleric acidemia or patients taking valproic acid who are carnitine deficient [187, 188]. Basic studies have reported that bupivacaine inhibits carnitine-acylcarnitine translocase (CACT) in myocardial mitochondria [189], or that bupivacaine cardiotoxicity is enhanced in the presence of carnitine deficiency [190].

Propofol is often used as a general anesthesia or sedative in children, but a condition called propofol syndrome (or propofol injection syndrome) might occur [191–193]. In this propofol infusion syndrome, mitochondrial fatty acid oxidation and respiratory chain disturbances have been found [194], and carnitine deficiency is thought to be related to the underlying cause [195–197]. Propofol is an extremely useful drug in clinical practice, and further research is required to investigate the acylcarnitine profile at the onset of propofol injection syndrome, its association with carnitine deficiency, and the usefulness of carnitine administration in order to use propofol more safely.

Many patients with AIDS also have decreased carnitine levels in blood and peripheral blood mononuclear cells, and it has been reported that carnitine supplementation enhances the proliferative response of lymphocytes stimulated by cell division [198–201]. It has also been pointed out that nucleoside reverse transcriptase inhibitors such as zidovudine, a drug used in the treatment of AIDS, may cause carnitine deficiency in muscle, possibly resulting in lipid-storage myopathy [202, 203]. In addition, protease inhibitors such as ritonavir and saquinavir and integrase inhibitors such as elvitegravir have been reported to inhibit OCTN2 activity in human placenta in basic studies [204].

Sodium benzoate, which is used for hyperammonemia in urea cycle disorders, may also lead to carnitine deficiency because carnitine is excreted in the urine as benzoylcarnitine, and treatment such as carnitine supplementation is necessary when blood carnitine is low [205, 206].

There have been clinical reports that emetine, a component of tocon syrup used as an emetic or antiprotozoal agent, and pyrimethamine and sulfadiazine, one of the sulfa drugs, also caused carnitine deficiency [207, 208].

Those mentioned above are the drugs that have been clinically reported to be associated with carnitine deficiency, but basic studies have reported that a number of drugs, including the calcium channel inhibitor verapamil and the antiarrhythmic agent quinidine, inhibit OCTN2 activity in vitro, so caution is needed [209, 210].

Among these drug-induced carnitine deficiencies, valproic acid and pivoxil-containing antibiotics are described in more detail in Chaps. 12 and 13.

5.5 Clinical Symptoms of Carnitine Deficiency

When carnitine deficiency occurs, various clinical symptoms develop. Since the main biochemical and physiological function of carnitine is to transport long-chain fatty acids into the mitochondria and promote β-oxidation, its clinical symptoms are almost identical to those observed in the case of abnormal fatty acid oxidation (Table 5.2).

These clinical symptoms and laboratory abnormalities are not always present in carnitine deficiency, and the affected organs may vary depending on the nutritional status and dietary pattern of the patient, the severity and pathophysiology of the underlying disease, and the presence or absence of concomitant medications.

These clinical manifestations are described in detail below.

Table 5.2 Common features of fatty acid oxidation disorders

Metabolic decompensation during fasting (starvation), infection, prolonged exercise, cold exposure, and stress
decreased oral intake, increased energy expenditure
progressive obtundation leading to coma
Reye-like syndrome
SIDS or "near miss" SIDS
recurrence, familial occurrence
Involvement of fatty acid-dependent tissues.
muscle: Myalgia, exercise intolerance, myoglobinuria, hypotonia
heart: Cardiac hypertrophy/dilatation, endocardial fibroelastosis, arrhythmias
liver: Hepatomegaly, hepatic dysfunction
others: Fatty infiltration of tissues on biopsy or autopsy
Hypoketotic hypoglycemia
elevated serum FFA (FFA/ketone body >2:1)
Hyperinsulinism and hypopituitarism ruled out (mandatory)
Alteration in plasma or tissue carnitine levels
decreased total carnitine (10–50% of normal) with increased acylcarnitine in intramitochondrial fatty acid oxidation defects
marked decrease in total carnitine (<5% of normal) and normal acylcarnitine (in the case of carnitine transporter defect)
CPT 1 deficiency is exception with high to normal carnitine and low acylcarnitine
Clinical laboratory abnormal findings.
dicarboxylic aciduria
Hyperammonemia
acidosis
during myoglobinuria, increased serum CK, hyperkalemia, hyperphosphatemia, hypocalcemia, hyperuricemia, and possible increased creatinine (with renal failure)
Increased AST and ALT

From Ref. [39]

ALT alanine aminotransferase, *AST* aspartate aminotransferase, *CK* creatine kinase, *CPT* carnitine palmitoyltransferase, *FFA* free fatty acid, *SIDS* sudden infant death syndrome

5.5.1 Acute Metabolic Compensatory Failure Associated with Fasting

Acute metabolic compensation failure is more likely to occur during fasting (starvation) or infection, when glucose and glycogen reserves are depleted, and energy production due to fatty acid oxidation is more burdened. Even in the case of carnitine deficiency due to inherited metabolic diseases, it is often reported that hypoglycemic attacks and hyperammonemia develop due to the burden of catching a cold, having a high fever, and going to bed without eating dinner. Burdens on energy metabolism include fasting (starvation), mild to moderate aerobic exercise for more than 1 h, infections and fever accompanied by vomiting, and tremor and heat production due to cold exposure. Tremor is an involuntary muscle activity that depends on the oxidation of long-chain fatty acids. These carnitine deficiency patients are often found in coma early in the morning. Infection with viral diseases is often accompanied by vomiting, decreased feeding, tremor, etc., and is likely to be a trigger for metabolic decompensation such as coma. Infants and toddlers are more likely to have hypoglycemic attacks because they have a lower ability to cope with fasting and starvation than older children and adults (fasting time: 6–10 h for children under 1 year of age, 12 h for 1–4 years old).

5.5.2 Recurrent Hypoketotic Hypoglycemic Attacks

Hypoketotic hypoglycemic attacks occur when glucose utilization is increased in situations where long-chain fatty acids cannot be used as an energy source, such as fatty acid oxidation disorder or carnitine deficiency due to various causes, or when ketone bodies cannot be produced. An elevated ratio of free fatty acids to ketone bodies in serum is also a clue to detect fatty acid β-oxidation disorder (normal 1:1; 2:1 or higher is highly suspicious for β-oxidation disorder). However, conditions that cause impaired fat transfer, such as hyperinsulinemia, should be excluded because of elevated free fatty acids.

The mechanism of blood glucose maintenance and hypoketotic hypoglycemic attacks in fatty acid oxidation disorders and carnitine deficiency is shown in Fig. 5.2 [211]. In normal subjects (children), glucose from the digestion, absorption, and breakdown of dietary carbohydrates is first used until about 2 or 3 h after a meal. During the next 4–5 h, after the glucose has decreased, glycogen stored in the liver and muscles is decomposed into glucose, which is released into the blood and used by the brain and other organs. In addition, during the night, 7–8 h after a meal or during fasting (starvation) until the next morning's meal, the glycogen in the liver is also depleted. In this state, fatty acids are mobilized from adipose tissue and acetyl-CoA is produced by fatty acid β-oxidation in the liver, which enters the tricarboxylic acid (TCA) cycle to produce ATP, and oxidative phosphorylation in the electron transfer system is also mobilized to produce a large amount of ATP (see in Chap. 3, "What Is Carnitine?" Sect. 3.10.3 "Number of ATP Produced from Glucose and Long-Chain Fatty Acids"). This ATP is used to produce glucose from pyruvate and glycogenic amino acids such as alanine and aspartate through gluconeogenesis. In gluconeogenesis, six molecules of ATP are required to produce one molecule of glucose, and efficient energy production by fatty acid β-oxidation is extremely useful. In addition, the brain can use ketone bodies instead of glucose when energy needs increase due to prolonged fasting, or fever caused by infection. Ketone bodies such as acetoacetic acid and β-hydroxybutyric acid are biosynthesized from acetyl-CoA, which is produced by fatty acid oxidation in the liver and supplied to the brain. Ketone bodies can pass through the blood-brain barrier, enter the brain, and are converted back to acetyl-CoA in the mitochondrial TCA cycle of brain cells for use in energy production. In other words, the fatty acid β-oxidation system in the liver has two important roles: one is to supply ATP for gluconeogenesis and the other is to be involved in ketone body synthesis [212]. Ketones are also used in skeletal muscle, cardiac muscle, and kidneys. However, patients with fatty acid oxidation disorders or carnitine deficiency are unable to generate glucose by glycogenesis using energy production from β-oxidation or to produce ketone bodies from acetyl-CoA produced by fatty acid β-oxidation [211]. Even if a person with fatty acid oxidation disorder or carnitine deficiency can manage to live a normal life with low energy demand, he or she may not be able to eat enough for a long time due to diarrhea or vomiting caused by infection, fever due to the common cold, tremor during cold stress, strenuous exercise at sports day or other events that are not normally performed, etc. Hypoketotic hypoglycemia often occurs when energy demands increase, such as during fevers associated with the common cold, tremors during cold stress, or during strenuous

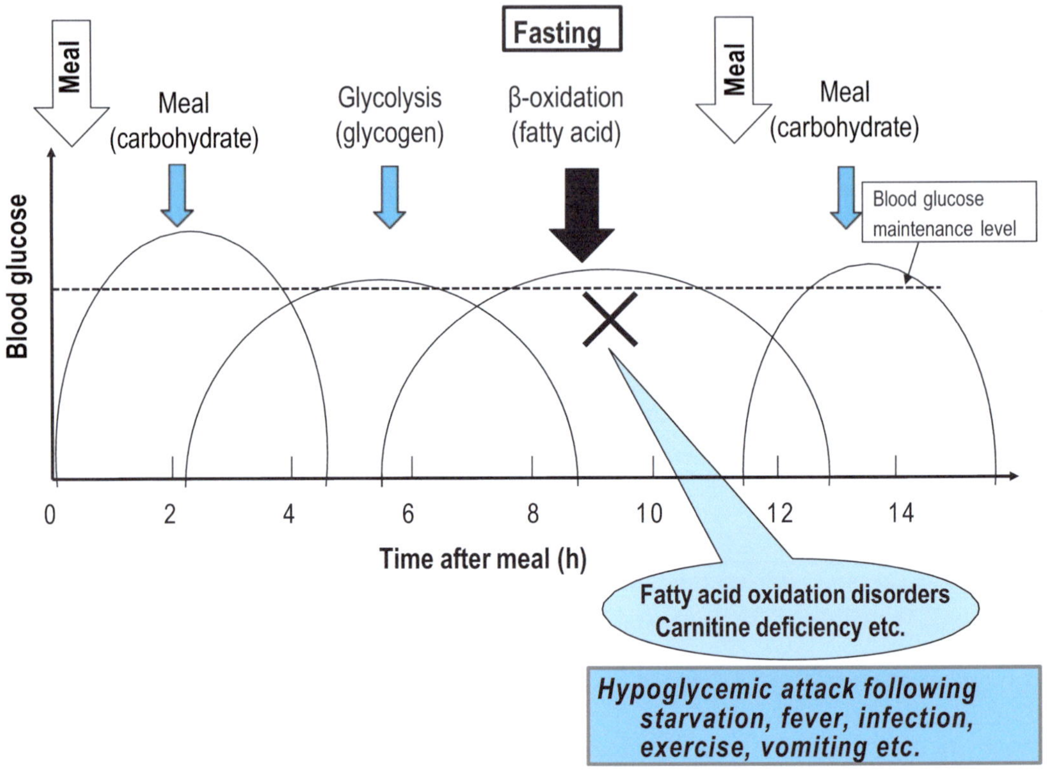

Fig. 5.2 Schematic diagram of the mechanisms to maintain blood glucose level. Disruption of these mechanisms due to unavailability of fatty acids in fasting situation in fatty acid oxidation disorders and carnitine deficiency due to the disruption of this system results in acute metabolic dysfunctions such as hypoketotic hypoglycemia. In healthy subjects, blood glucose level is maintained utilizing carbohydrates (glucose) from meal and then switched to glycolysis utilizing glycogen stored in the liver or muscle along with time course after meal. During fasting condition after glycogen is depleted, β-oxidation of fatty acids can be utilized to provide energy (the bold black arrow). Acetyl-CoA produced by β-oxidation of fatty acids flows into TCA cycle and electron transfer chain in mitochondria and ATP is produced. Using this ATP, glucose or ketone bodies are synthesized by gluconeogenesis and ketogenesis, and the blood glucose levels are maintained, and brain can utilize ketone bodies. In the case of patients with fatty acid oxidation disorders or carnitine deficiency (**X**), this ATP production by fatty acid oxidation cannot be utilized in fasting state, so they are prone to hypoketotic hypoglycemia. Especially these acute metabolic dysfunctions occur in increased energy-demand situations such as starvation, fever, infection, vigorous exercise, and vomiting. (Modified from Ref. [211])

exercise at athletic events, which are not usually performed. If the hypoglycemia is prolonged, it may cause sequelae in the brain, so special caution is required.

5.5.3 Tissues That Are Highly Dependent on Efficient Fatty Acid Oxidation at All Times

Organs with high energy demands, such as skeletal muscle, heart, and liver, depend on efficient fatty acid oxidation. When the production of ketone bodies in the liver is stagnant, glucose becomes the only energy source, and glucose and glycogen are depleted, and fatty acid oxidation is loaded, resulting in a rate-limiting step. As a result, the free fatty acids released during fasting are not fully metabolized and are stored as triglycerides in the cytoplasm. This leads to the development of lipid-storage myopathy accompanied by muscle pain, hypotonia, and weakness, and in severe cases, rhabdomyolysis and myoglobinuria. Cardiac myocytes, which constantly repeat contraction and relaxation, depend on fatty acid oxidation rather than glycolysis for energy production, and fatty acid oxidation abnormalities and carnitine deficiency can cause cardiomyopathy and heart failure with cardiac hypertrophy and dilation, and sometimes sudden death due to arrhythmias. In the liver, symptoms and pathological conditions similar to those seen in congenital fatty acid metabolism disorders, such as fatty liver, may occur. In the case of carnitine deficiency due to valproic acid, fat deposits are also observed in the liver. Increased concentrations of short- and medium-chain fatty acids and their dicarboxylic acids due to compensatory ω-oxidation have been observed in Reye-like syndrome and its model animals, resulting in secondary metabolic disorders such as impaired gluconeogenesis, hyperammonemia due to disturbances in the urea cycle, impaired β-oxidation, and impaired TCA cycles [213–215]. As a result, the concentration of ATP in the cells is further reduced.

Long-chain fatty acid oxidation disorders such as carnitine palmitoyltransferase 2 (CPT2) deficiency, long-chain acyl-CoA dehydrogenase deficiency (LCAD), very long-chain acyl-CoA dehydrogenase deficiency (VLCAD), and trifunctional protein deficiency often cause repeated attacks of acute muscle collapse and myoglobinuria. It has been suggested that accumulated long-chain fatty acids and long-chain acylcarnitines exert a surfactant-like effect on muscle cell membranes. Palmitoyl CoA and palmitoylcarnitine have surfactant effects and have been reported to exacerbate the peroxidation of membrane lipids by free radicals during ischemia although this is only at concentrations as high as 50 μM in isolated canine cardiomyocytes [216]. In addition, it has been reported that long-chain acylcarnitine has an activating effect on calcium channels in cardiomyocytes and smooth muscle cells, and it has been pointed out that it may have a proarrhythmic effect during ischemia [217, 218].

5.5.4 Changes in Serum and Tissue Carnitine Concentrations

In most cases of β-oxidation defects in the mitochondria, the total carnitine concentration decreased to 50% or less of normal, and acylcarnitine increased to 50% or more (normal range: 10–25% at feeding, 30–40% on fasting) [39]. Due to a defect in the mitochondrial β-oxidation process, the specific acyl-CoA produced before (proximal to) the defective site accumulates, and the acyl groups are transferred to carnitine to form specific acylcarnitines, which are transported out of the mitochondria and ultimately excreted in the urine. However, long-chain acylcarnitines such as palmitoylcarnitine (palmitoyl: C_{16}), myristoylcarnitine (myristoyl: C_{14}), and lauroylcarnitine (lauroyl: C_{12}) inhibit the carnitine transporter (OCTN2), which is involved in carnitine reabsorption in renal tubules, and free carnitine is excreted and decreases [210, 219]. Figure 5.3 shows the results of an experiment by Ohashi et al. on the inhibitory effect of various acylcarnitines on carnitine uptake in HEK293 cells expressing human OCTN2 [210]. It can be seen that the inhibition of carnitine uptake of long-chain acylcarnitines such as palmitoylcarnitine rather than short-chain acylcarnitine such as acetylcarnitine and propionylcarnitine is strong. The carnitine transporter in the renal tubules is involved in the reabsorption of carnitine, and when this activity is inhibited, a large amount of carnitine leaks into the urine, leading to carnitine deficiency. In primary carnitine deficiency, which is a carnitine transporter disorder, the total carnitine and free carnitine concentrations are extremely low because free carnitine is not reabsorbed in the renal tubules. In carnitine palmitoyltransferase 1 (CPT1) deficiency, carnitine is not taken up by mitochondria, so the concentration of free carnitine in the blood is conversely high.

5.5.5 Laboratory Abnormalities Commonly Seen in Carnitine Deficiency

Acute-onset catabolic disorders such as Reye-like syndrome often have moderate elevations in blood ammonia (100–200 μM/L) and three- to fivefold elevations in hepatic transaminases.

In patients with acute myoglobinuria, creatinine kinase activity is usually 250 U/L, but often exceeds 100,000 U/L. Amino acids such as serum

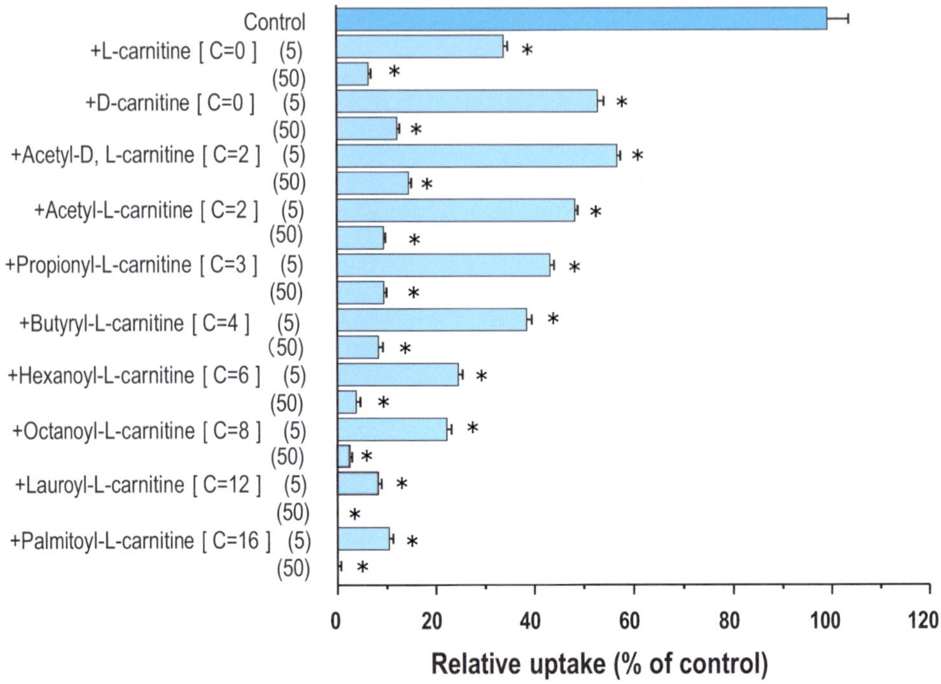

Fig. 5.3 Inhibitory activity of various acylcarnitines on carnitine uptake in the HEK293 cell expressed human OCTN2. The uptake of L-[^{3}H]-carnitine (10 nM) is studied in the presence of 5 or 50 μM of various acylcarnitines (37 °C, 3 min). Data shows Mean ± SE of three experiments. *Significant differences between control. *SE* standard error. (Modified from Ref. [210])

creatinine, potassium, phosphorus, uric acid, and taurine may also be elevated. Lactic acidosis is also common.

Hyperammonemia is also often seen as a laboratory abnormality for carnitine deficiency. As described in detail in "Chap. 12: Carnitine Deficiency Due to Sodium Valproate," the main part of the urea cycle that converts ammonia into urea and detoxifies it is in the liver. The reasons why hyperammonemia is more likely to develop due to carnitine deficiency are summarized below:

1. The urea cycle is an extremely energy-intensive metabolic system that requires three molecules of ATP to convert one molecule of ammonia into urea.
2. N-acetylglutamate (NAG) is required as a cofactor for the activity of carbamoyl phosphate synthase 1 (CPS1), the primary enzyme of the urea cycle.
3. NAG is synthesized from acetyl-CoA and glutamate by N-acetylglutamate synthase (NAGS).
4. Intermediate metabolites acyl-CoA and valproyl-CoA that accumulate in organic acidemia inhibit NAGS.

Half of the urea cycle takes place in the hepatic mitochondria. In the urea cycle, two molecules of ATP are required for the CPS1 reaction to synthesize the first carbamoyl phosphate, and one molecule of ATP is also required for argininosuccinate synthetase (ASS) to form argininosuccinate from citrulline. In other words, three molecules of ATP are required to convert one molecule of ammonia to one molecule of urea in the urea cycle, which is an extremely energy-demanding metabolic process. However, according to Stumpf et al. at the University of Colorado, who have been treating and researching urea cycle disorders for many years, when nitrogen from endogenous amino

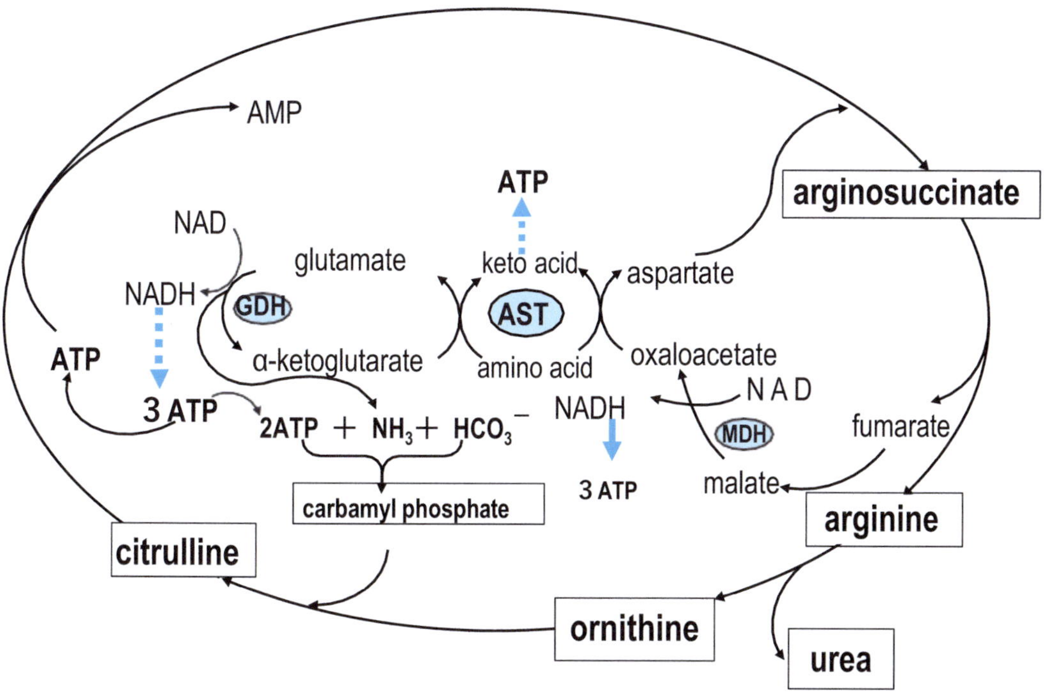

Fig. 5.4 The "low cost" of the urea production by urea cycle and the related amino acid metabolic systems. *AST* aspartate aminotransferase, *GDH* glutamate dehydrogenase, *MDH* malate dehydrogenase. (Modified from Ref. [220])

acids is converted to urea, urea is produced in a way that does not consume much energy (ATP) and, in some cases, acquires energy [220]. Stumpf et al. called this "low-cost urea production" (Fig. 5.4).

Furthermore, Stumpf et al. reported that the conversion process of ammonia to urea is not a single step in the urea cycle, but a complex process involving ornithine: α-ketoglutarate transaminase (OKT), glutamate semialdehyde dehydrogenase (GSDH), glutamate dehydrogenase (GDH), and nicotinamide adenine dinucleotide (NADH) and flavin adenine dinucleotide (FADH2) produced in the ornithine metabolic pathway are transferred to the oxidative phosphorylation system to produce ATP, and α-ketoglutarate enters the TCA cycle to efficiently produce ATP [220] (Fig. 5.5).

In other words, the urea cycle is not established in isolation, but is closely linked to the TCA cycle in the mitochondria, the oxidative phosphorylation of the electron transport system, ornithine, α-ketoglutarate, glutamic acid, etc., and is closely linked to the metabolic system conjugated to it, to process ammonia at a "low cost" by reducing net energy consumption. Originally, the urea cycle is an energy-consuming metabolic system that requires a total of three molecules of ATP: two molecules of CPS1 to convert one ammonia molecule into urea, and one molecule of ASS to produce argininosuccinic acid from aspartic acid. As discussed in Sect. 5.5, "Clinical Symptoms of Carnitine Deficiency," similar to the "hypoketotic hypoglycemia" discussed in Sect. 5.2, "Recurrent Hypoketotic Hypoglycemic Attacks," ammonia detoxification metabolism is also an energy-consuming metabolic system. Conversely, if the energy supply is disrupted, ammonia cannot be fixed (processed) by urea, and hyperammonemia and hyperammonemic encephalopathy are likely to develop. Reye-like syndrome and hepatic encephalopathy in children are examples. Ammonia produced in the amino acid metabolic system can be converted to urea at low cost, but when it comes to the processing of ammonia produced and absorbed outside the body, such as ammonia produced in the intestine, three molecules of ATP are required to

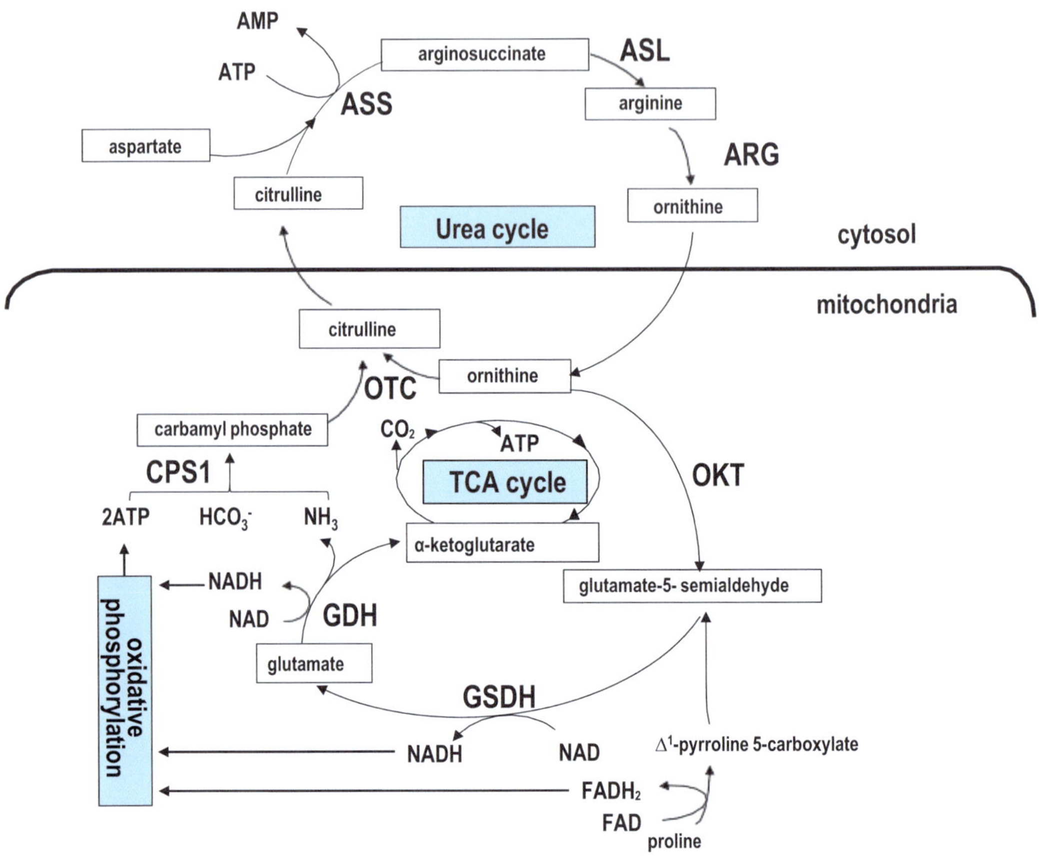

Fig. 5.5 Coupling of urea cycle, ornithine metabolism, and energy production systems. The urea cycle is not working alone but coupled and associated with ornithine metabolism and energy production systems such as TCA cycle and oxidative phosphorylation in mitochondria. *ARG* arginase, *ASL* arginosuccinate lyase, *ASS* arginosuccinate synthetase, *CPS1* carbamyl phosphate synthetase, *GDH* glutamate dehydrogenase, *GSDH* glutamate-5-semialdehyde dehydrogenase, *OKT* ornithine α-ketoglutarate transaminase, *OTC* ornithine transcarbamylase. (From Ref. [220])

convert one molecule of ammonia into one molecule of urea. Therefore, hyperammonemia is often associated with Reye-like syndrome, mitochondrial disorders, carnitine deficiency, patients receiving valproate, and cirrhosis of the liver, which often cause an energy crisis in which mitochondrial function is suppressed and ATP production capacity is reduced. In such a low-energy state, ornithine flows to another metabolic system at the expense of ammonia processing, and the electron transport system works to produce ATP. In other words, although ornithine is a member of the urea cycle, ornithine is used to maintain ATP levels when mitochondrial energy is deficient, resulting in a vicious cycle in which the flow to the urea cycle decreases, the activity of the urea cycle decreases, and ammonia production increases further. Considering this pathophysiological mechanism, hyperammonemia is a consequence of energy deficiency. Thus, in their paper, Stumpf et al. discussed that in the treatment of hyperammonemia such as mitochondrial disorders, Reye-like syndromes, or carnitine deficiency, "treatments that improve energy deficiency will be more beneficial than efforts to reduce ammonia." [220]

So, what is the relationship between carnitine and the urea cycle? The relationship between the carnitine cycle and the urea cycle is summarized in Fig. 5.6. Carnitine promotes long-chain fatty

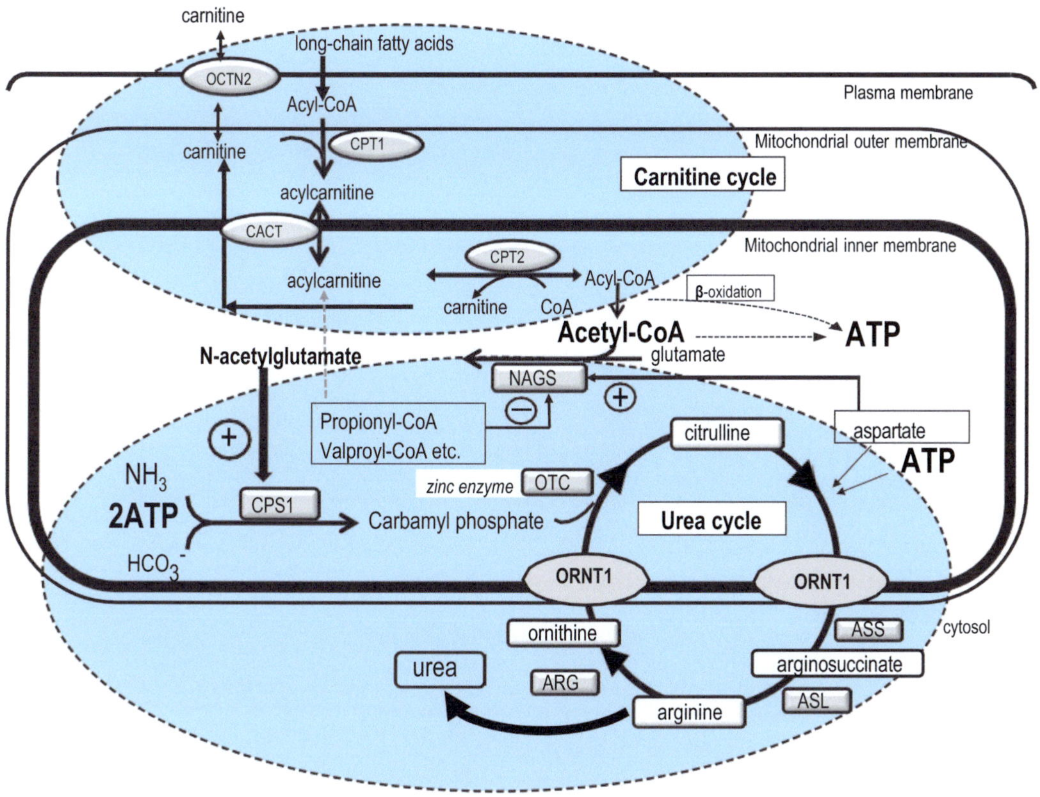

Fig. 5.6 Relationship between carnitine cycle and urea cycle. Long-chain fatty acids are transported into mitochondria by carnitine cycle and acetyl-CoA is produced by β-oxidation. N-acetylglutamate (NAG), a positive cofactor for the first enzyme of urea cycle, is produced from acetyl-CoA and glutamate by NAG synthase (NAGS). CPS1 converts ammonia to carbamyl phosphate with the aid of NAG. This CPS1 reaction needs two ATPs and also ASS reaction needs one ATP, so total three ATPs are required to convert one NH3 into urea by urea cycle. Acetyl-CoA flows into TCA cycle and ATP is produced and this ATP is provided to urea cycle. Propionyl-CoA or valproyl-CoA inhibits NAGS activity and suppresses the urea cycle and induces hyperammonemia. Free carnitine acts as an antidote of these harmful acyl-CoA by excreting these acyl groups into urine as acylcarnitines. In the case of carnitine deficiency or fatty acid oxidation disorders, the coordination of carnitine cycle and urea cycle is disturbed, and this lack of coordination between carnitine cycle and urea cycle is supposed to be leading to hyperammonemia. *ARG* arginase, *ASL* arginosuccinate lyase, *ASS* arginosuccinate synthetase, *CACT* carnitine acylcarnitine translocase, *CPS1* carbamyl phosphate synthetase 1, *CPT* carnitine palmitoyltransferase, *NAGS* N-acetylglutamate synthase, *OCTN2* carnitine transporter, *ORNT1* ornithine citrulline antiporter (also called as mitochondrial ornithine transporter), *OTC* ornithine transcarbamylase

acid oxidation, contributes to efficient energy (ATP) production, and promotes the supply of N-acetylglutamate by supplying acetyl-CoA by fatty acid β-oxidation. Therefore, it is easy to understand that when carnitine is deficient, the urea cycle becomes difficult to turn due to a decrease in ATP production and a decrease in the supply of N-acetylglutamate. The active expression of CPS1, which is the first starting enzyme of the urea cycle and fixes ammonia to carbamoyl phosphate, requires NAG as a positive cofactor. If there is not enough NAG, the urea cycle will not turn. However, propionyl-CoA, which accumulates in organic acid metabolism disorders such as propionic acidemia, and acyl-CoA, such as valproyl-CoA, which is produced when valproic acid is administered, inhibit NAGS, which synthesizes NAG, and suppress the urea cycle, which increases the occurrence of hyperammonemia. In addition, NAGS deficiency is also known among urea cycle abnormalities [221], and hyperammonemia naturally develops in this disease. Since it has been reported that long-chain acyl-CoA accumulates in patients with hepatic encephalopathy with cirrhosis, I suppose that such long-chain acyl-CoA might inhibit NAGS and make it easier to develop hyperammonemia, but there are no such reports yet. This is an interesting research topic and further research is required. However, ornithine transcarbamylase (OTC) in the urea cycle is a zinc enzyme, and it should be noted that carnitine supplementation alone may not be sufficient to reduce ammonia in patients with cirrhosis or tube feeding who might have also zinc deficiency.

5.6 Animal Models of Carnitine Deficiency

In order to study the pathophysiology of various diseases and examine the effects of treatments, it is useful to create and examine animal models of the diseases. In the case of conventional drug development, the mechanism of action and dosage are first examined as basic pharmacological research in an animal model of the target disease, and then it is moved on to clinical research. However, it is not easy to create an animal model of carnitine deficiency because carnitine is biosynthesized in the body and ingested from daily meals. The physiological role of carnitine in fatty acid metabolism and energy metabolism is extremely important in higher animals, including humans, because there is a mechanism by which homeostasis is maintained (see "Chap. 4: Homeostasis of Carnitine"). In addition, since the 1960s, carnitine has been used as a pharmaceutical product, as a nutritional agent with stunted growth and appetite-stimulating effects, and when the DL-form is used, it was known that there was no safety problem except for the problem that the D-body may antagonize the L-body and cause myasthenia-like symptoms. In the 1980s, clinical applications were made for organic acidemia and carnitine deficiency with valproic acid. Therefore, unlike other medicines, only a few pharmacological studies have been conducted on the efficacy of carnitine replacement therapy in carnitine deficiency model animals.

The following animal models of carnitine deficiency have been reported to date.

1. Juvenile visceral steatosis (JVS)mouse.
2. Secondary carnitine deficiency model by valproic acid administration (rats).
3. Secondary carnitine deficiency model by pivalate administration (rats).
4. Secondary carnitine deficiency model by D-carnitine administration (rats).
5. Secondary carnitine deficiency model by meldonium (trade name: mildronate; also called as MET-88, N-trimethyl-hydrazine-3-propionate [THP]) administration (rats).
6. Others

The following are the main reports of the effects of carnitine administration in these carnitine deficiency model animals.

1. JVS Mouse

JVS mice discovered by Koizumi et al. at Kanazawa University showed fat deposits in the liver, hypoglycemia, and hyperammonemia from the fifth day after birth [222] and significant hypocarnitinemia [223]. Subsequently, it was shown that the reabsorption process of carnitine filtered by the glomeruli was impaired due to abnormalities in the carnitine transporter in the renal tubules [224].

In JVS mice (homozygotes), weight gain stopped from the 15th day of life, and from the 22nd day, the body weight decreased significantly compared to the control mice, and growth was inhibited, but the growth disorder was normalized by intraperitoneal administration of 5 μmol of L-carnitine from the tenth day of life [225]. In addition, when the enzymatic activity of CPS1, OTC, and ASS, which are urea cycle enzymes associated with hyperammonemia, was examined, it was found that CPS1, OTC, and ASS were significantly reduced compared to control mice. It also prevented a decrease in the expression of CPS1 and ASS, which were also examined at the mRNA level. However, fat accumulation in the liver was not reduced by carnitine supplementation. These results indicate that hyperammonemia observed in JVS mice is strongly related to carnitine deficiency in abnormal expression and decreased activity of urea cycle enzymes [225].

JVS mice exhibit cardiac hypertrophy, and about 50% of mice die from cardiomyopathy at 1 month, and all mice die from cardiomyopathy at 3–4 months. Therefore, Horiuchi et al. investigated the effect of carnitine administration on cardiac hypertrophy in JVS mice [226]. The serum carnitine concentration of 2.3 ± 0.8 nmol/mL was significantly reduced in the JVS mouse group (homozygotes) that survived for 2–3 months without carnitine supplementation compared to the control group (45.7 ± 9.2 nmol/mL), and the cardiac wet weight increased by about 2 times (0.11 ± 0.02 g in the control group and 0.23 in the JVS mouse group ±0.03 g, $P < 0.01$). In JVS mice (homozygotes), the non-collagen protein content of the heart increased by about 2.5 times and the DNA content increased by about 1.7 times, and histological evaluation showed that individual cardiomyocytes were enlarged, and nucleus deformed and enlarged. When carnitine (5 μmol) is administered from 11 to 20 days of age, and carnitine (10 μmol) is administered twice a day until 30–90 days, it is recommended that carnitine (10 μmol) is administered twice a day from 11 to 20 days after birth. Compared to the JVS mouse group that received carnitine only on day 20 and did not receive carnitine thereafter, cardiac weight increase was suppressed (0.27 ± 0.05 g and 0.15 ± 0.03 g, $P < 0.01$), but cardiac hypertrophy remained slightly higher compared to the control group (control group 0.10 ± 0.01 g, JVS mouse group 0.15 ± 0.03 g, $P < 0.05$), though there was no significant difference in heart weight/body weight compared with the control group. These results indicate that carnitine deficiency plays an important role in the formation of cardiac hypertrophy. Kuwajima et al. investigated the effects of carnitine administration (intraperitoneal) on cardiac free carnitine, short-chain acylcarnitine, long-chain acylcarnitine, intracardiac lipids, and cardiac weight in JVS mice [227]. Comparisons were made between carnitine treatment and non-carnitine and control groups at 2, 4, and 8 weeks. In JVS mice, most mice die within 5 weeks if carnitine is not administered. Therefore, in the untreated 8-week group, 1 μmol/g body weight of carnitine was administered from the tenth day of life to 20–25 days, and no subsequent doses were administered. In the carnitine group, carnitine was administered continuously from the tenth day of life to the fourth or eighth week after birth. In the 2- and 8-week-old non-treated JVS mice, cardiac weight was significantly increased compared to the control group ($P < 0.01$ and $P < 0.001$, respectively). In JVS mice treated with carnitine, this cardiac hypertrophy was suppressed, and there was no significant difference at 4 weeks compared to the control group, and it was significant at 8 weeks ($P < 0.01$), but it was not as pronounced as in the non-treatment group. In JVS mice, body weight increased significantly in the carnitine group compared to the non-treatment group ($P < 0.05$), and cardiac weight/body weight, which is an indicator of cardiac hypertrophy, was significantly higher in the 2-, 4-, and 8-week non-treatment group than in the control group ($P < 0.001$), and in the 4- and 8-week carnitine group compared to the non-treatment group ($P < 0.001$). The cardiac content of total lipids, triglycerides, total phospholipids, phosphatidylinositol, phosphatidylethanolamine, and cardiolipin was significantly higher than that of the control group of 8-week-old JVS mice ($P < 0.05$–0.001). Administration of L-carnitine reduced these lipid contents, and there was no significant difference from that of the control group. On the other hand, when comparing the carnitine treatment group and the non-carnitine group, there was no significant difference in phosphatidylinositol, but there was a significant difference in the other five parameters ($P < 0.05$). When L-carnitine was administered intraperitoneally, the concentration of free carnitine in the heart remained constant, but the concentration of free carnitine in serum increased by up to 80 times. On the other hand, the short-chain acylcarnitine in the heart increased markedly. These results suggest that the increase in the concentration of short-chain acylcarnitine rather than free carnitine is the key to the protective effect of L-carnitine administration to JVS mice.

In 1998, the gene site of the human carnitine transporter (OCTN2) was found to be present on chromosome 5q31, and it was cloned [228, 229]. Since human chromosome 5q31 is considered to be synteny with chromosome 11 in mice, Lu et al. cloned OCTN2 on chromosome 11 in JVS mice and reported that the 352nd codon, which corresponds to the sixth transmembrane region, mutates from CTG (Leu) to CGG (Arg) in JVS mice, causing a missense mutation. [230] These studies suggest that JVS mice are extremely useful as a model of primary carnitine deficiency in humans [231, 232].

In addition, it has been reported that atrophy of the mucosal epithelium of the small intestine and colon and infiltrated inflammatory images of lymphocytes were observed in JVS mice, and pathological findings similar to those of human inflammatory bowel disease were observed in JVS mice [233].

2. Secondary Carnitine Deficiency Model by Valproate Administration (Rats)

Valproic acid is widely used worldwide as an antiepileptic drug, a treatment for mania and manic states, and a treatment for migraine, but in rare cases, it can cause serious lethal hepatotoxicity and hyperammonemic encephalopathy, including Reye-like syndrome, and carnitine deficiency is thought to be involved in the underlying cause (see "Chap. 12: Carnitine Deficiency Caused by Sodium Valproate"). As a basic study of the mechanism, carnitine dynamics and lesions in animals treated with valproic acid have been studied.

Nishida et al. examined the concentration of carnitine in serum, red blood cells, muscles, liver, and urine by administering valproic acid alone (500 mg/kg/day) or valproic acid (500 mg/kg/day) and L-carnitine (200 mg/kg/day) to rats for 7 days [234]. As a result, the serum and muscle free carnitine concentrations of the valproate group decreased compared to the control group, while the acylcarnitine concentration and the acylcarnitine/free carnitine ratio increased. Concentrations of free carnitine in serum, muscle, and liver were increased in the valproate group and the acylcarnitine/free carnitine ratio decreased in all tissues compared to the valproate group treated with L-carnitine alone ($P < 0.05$). The mean urinary free carnitine concentration in the valproate group did not change compared to the control group, but the urinary acylcarnitine concentration was elevated, and the acylcarnitine concentration on days 4 and 7 decreased after L-carnitine supplementation compared to the valproate group ($P < 0.05$). The serum concentration of β-hydroxybutyrate in the valproate group was lower than in the control group and the valproate + L-carnitine group. Thus, in a model of secondary carnitine deficiency in rats treated with valproate, free carnitine in serum and muscle decreased and acylcarnitine and acylcarnitine/free carnitine increased.

Murakami et al. measured urinary acetylcarnitine in rats treated with valproic acid by high-performance liquid chromatography [235]. In the urine of mature rats (180–200 g) treated with valproate (500 mg/kg/day), acetylcarnitine concentrations were higher than in the control group on days 4 and 7, but in rats that received L-carnitine supplementation (200 mg/kg/day) at the same time as valproate, it was lower than the acetylcarnitine concentration in valproate-treated rats. In the urine of rats treated with valproate, it increased on day 4 (0.60 ± 0.06 mmol/g creatinine) and day 7 (0.55 ± 0.1 mmol/g creatinine) compared to day 1 (0.2 ± 60.06 mmol/g creatinine). On the other hand, in the valproate group treated with L-carnitine supplementation, acetylcarnitine increased on day 4 (0.40 ± 0.07 mmol/g creatinine) and day 7 (0.34 ± 0.04 mmol/g creatinine) but was significantly lower than that of valproate alone ($P < 0.001$). The acetylcarnitine/acy lcarnitine ratio increased in the valproate group after the fourth day, but there was no significant change in the L-carnitine group compared to the control group. In the urine of immature rats (80–90 g) treated with valproate (50 mg/kg/day), acetylcarnitine was elevated after day 14. These results indicate that even small amounts of valproic acid administered over a long period of time cause elevation of urinary acetylcarnitine, and Murakami et al. suggested that acetylcarnitine is not a product of mitochondrial β-oxidation, since concurrent L-carnitine administration decreases acetylcarnitine levels that are elevated by valproic acid administration. L-carnitine administration may activate native mitochondrial β-oxidation and decrease acetylcarnitine production and urinary acetylcarnitine excretion by peroxisomes and other systems, thereby maintaining a normal acetylcarnitine/acylcarnitine ratio.

3. Secondary Carnitine Deficiency Model by Pivalate Administration (Rats)

Pivalic acid is often used in prodrugs such as pivampicillin and pivmecillinam to promote intestinal absorption of antibiotics and increase their oral bioavailability. However, when pivalate is absorbed, pivalic acid (pivoxil group) is cleaved and released from the drug by esterase in the intestinal tract, and it immediately forms an ester with carnitine in the blood, which is excreted as pivaloyl carnitine. When a pivoxil-containing antibiotic is administered over a long period of time, carnitine is persistently excreted in the form of this ester, resulting in a decrease in carnitine content in plasma and tissues, which may lead to carnitine deficiency (see "Chap. 13: Carnitine Deficiency Caused by Pivoxil-Containing Antibiotics").

Bianchi et al. examined the effect of carnitine administration on fat metabolism in a secondary carnitine deficiency model in which rats were administered sodium pivalate [236]. Weaned male rats were mixed with 20 mmol/L sodium pivalate or 20 mmol/L sodium bicarbonate in drinking water and administered for 2 weeks. Thereafter, in order to maximize fatty acid oxidation, they fasted for 24 h and exposed to cold stimuli for 4 h. In Experiment 1, group 1 was treated with sodium bicarbonate, group 2 was treated with sodium pivalate, group 3 was treated with sodium pivalate + L-carnitine (0.46 mmol/day feeding), and group 4 was treated with sodium pivalate + L-carnitine (0.95 mmol/day feeding). In Experiment 2, group 1 was a group of drinking water without additives, group 2 was a group of sodium bicarbonate, group 3 was treated with sodium pivalate, and group 4 was treated with sodium pivalate + L-carnitine (0.95 mmol/day feeding). Rats treated with sodium pivalate fed on a low-carnitine diet showed significant increases in triglycerides in plasma and liver (Experiment 1), plasma β-hydroxybutyrate concentrations (Experiments 1 and 2), and urinary dicarboxylic acid excretion (Experiment 2) compared to the control group ($P < 0.05$). Supplementation with L-carnitine (0.95 ± 0.06 mmol/day) significantly increased plasma and tissue carnitine concentrations, and plasma β-hydroxybutyrate and liver triglyceride concentrations decreased to concentrations similar to those in the control group. In addition, carnitine administration reduced fat accumulation in the liver and ketosis during starvation due to sodium pivalate. Patients with human disorders of organic acid metabolism also observed a decrease in the content of carnitine in tissues, ketosis during starvation, and fat accumulation in the liver. Thus, in a model of secondary carnitine deficiency in rats treated with pivalic acid, total carnitine in plasma, myocardium, and skeletal muscle were significantly reduced, and triglycerides, plasma β-hydroxybutyrate, and urinary dicarboxylic acid excretion in plasma and liver were increased.

4. Secondary Carnitine Deficiency Model by D-Carnitine Administration (Rats)

D-carnitine does not exist in nature, and it competitively antagonizes L-carnitine on carnitine transporters, which reduces the absorption of L-bodies, intracellular uptake, and reabsorption in renal tubules. Several models of secondary carnitine deficiency by D-carnitine administration have been reported using this mechanism.

Paulson et al. reported that intraperitoneal administration of D-carnitine (750 mg/kg) to rats for 4 days reduced the L-carnitine content in myocardial and skeletal muscle tissues to 31.8% and 48.2%, respectively, for controls [237]. However, the L-carnitine content of the kidneys decreased slightly, and the L-carnitine content of the liver, brain, and plasma did not decrease significantly. Thus, it was found that D-carnitine competes with its transporters in the heart and skeletal muscles that actively uptake the L-body and causes a significant decrease in carnitine content.

In this carnitine deficiency model by D-carnitine administration, a model in which meldonium, an inhibitor of γ-butyrobetaine hydroxylase, is also administered in addition to D-carnitine, which is described in the next section, has been reported. Tsoko et al. examined the effects on tissue carnitine content and lipid metabolism by administering D-carnitine for 2 days (18 mg/100 g body weight) and meldonium for 10 days (20 mg/100 g body weight) to rats, and reported that carnitine content in the heart, skeletal muscle, and kidneys was significantly reduced and triglycerides in the liver were significantly increased [238]. Fatani et al. investigated the development of cardiomyopathy caused by cyclophosphamide in cancer patients using a carnitine-deficient animal model in which rats were given D-carnitine (250 mg/kg/day) and meldonium (200 mg/kg/day) for 10 days and reported that carnitine deficiency was a risk factor for the development of cardiomyopathy caused by cyclophosphamide [239].

5. Secondary Carnitine Deficiency Model by Meldonium Administration (Rats)

Meldonium was developed in Latvia in the 1970s as a drug to speed up the growth of livestock. Also, it has been recognized as having the effect of improving cardiac function and blood flow and is used as a treatment for ischemic heart diseases such as angina pectoris and myocardial infarction in former Eastern European countries such as Latvia and Russia under the trade name Mildronate. It is also called MET-88 or N-trimethyl-hydrazine-3-propionate (THP). Because meldonium is very similar in chemical structure to γ-butyrobetaine, it acts as a competitive inhibitor of γ-butyrobetaine hydroxylase, which is involved in the reaction of the last step of the carnitine biosynthesis system, causing carnitine deficiency (Fig. 5.7). While meldonium is approved as a drug in a small number of countries, the World Doping Agency added it to the list of prohibited drugs in January 2016. It is a drug that became a hot topic used by professional tennis player Sharapova. Although it is still unclear why lowering tissue carnitine leads to ischemic cardioprotective effects and improvement of cardiac function, it is thought that it may exert a cardiac function protection effect by promoting the utilization of glucose in tissues by causing a decrease in tissue carnitine content and decreasing lipid utilization [240].

Spaniol et al. administered meldonium to rats by feeding them for 3 weeks (vegetarian diet, 20 mg/100 g body weight/day) and found that renal excretion of total carnitine was 96 ± 36 μmol/day, a significant increase compared to 5.3 ± 1.2 μmol/day for the control, and the total plasma carnitine concentration was 8.8 ± 2.1 μmol/L (control: 52.8 ± 11.4). Since the total concentration of carnitine in muscles decreased significantly to 0.24 ± 0.04 μmol/g (control: 1.07 ± 0.13) and the total concentration of carnitine in the liver decreased significantly to 0.19 ± 0.03 μmol/g (control: 0.59 ± 0.08), it was reported that it may be possible to use it as an animal model of carnitine deficiency [241]. Spaniol et al. examined changes in lipid metabolism in a similar carnitine-deficient animal model and found that 6 weeks of meldonium reduced the carnitine pool in the liver by about 75%, increased the long-chain acylcarnitine/free carnitine ratio, and caused steatosis [242]. Using a secondary carnitine-deficient animal model treated with meldonium, this group showed that a decrease in plasma carnitine concentrations was associated with a carnitine transporter (OCTN2) in the renal tubules [243], and a 25% decrease in CoA utilization in skeletal muscle, a 44% decrease in peak tension generation, and a 52% increase in glycogen hydrolysis were reported [244].

Thus, meldonium-treated animal models of secondary carnitine deficiency not only suppress carnitine biosynthesis, but also antagonize L-carnitine in the renal tubule carnitine transporter (OCTN2) and compete at other enzyme levels.

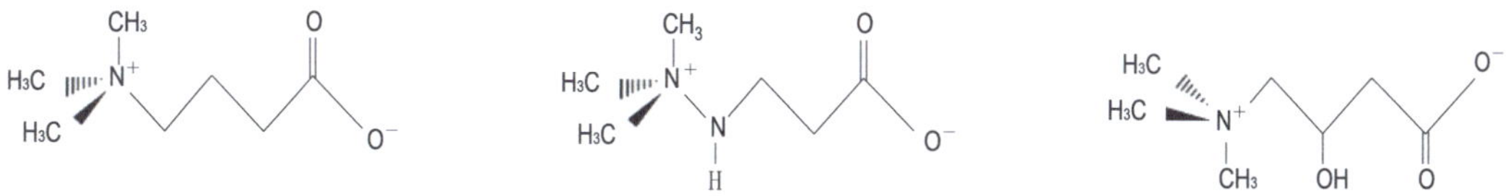

Fig. 5.7 Chemical structures of γ-butyrobetaine, N-trimethyl-hydrazine-3-propionate (THP, meldonium, MET-88, mildronate), and L-carnitine

6. Others

Severe carnitine deficiency in humans presents with cardiomyopathy and heart failure, but in Syrian hamsters, which have been actively studied as experimental animal models that cause cardiomyopathy, it was initially thought that abnormalities in catecholamine metabolism and intracellular calcium handling were the causes [245, 246]. However, there have been some reports that carnitine deficiency may be the cause of cardiomyopathy.

York et al. examined carnitine in Syrian hamsters (BIO 14.6 strain) that cause cardiomyopathy and Syrian hamsters with normal myocardium (F1B strain). The concentration of carnitine in plasma and liver was about 2.4 times higher than in the control, and was almost the same in skeletal muscle, but the carnitine content in myocardium decreased by almost half regardless of age [247]. This indicates that carnitine deficiency in the myocardium of this cardiomyopathy hamster is not secondary. They hypothesized that the affinity of carnitine-binding protein for carnitine in cardiomyopathy hamsters was lower than that of control hamsters, and that the inability to take up sufficient carnitine into the myocardium may be the cause of the development of cardiomyopathy.

Yamashita et al. also examined that the free carnitine content of the liver and skeletal muscle was unchanged in the cardiomyopathy Syrian hamster (BIO 14.6 strain) compared to the control, but the free carnitine content, short-chain acylcarnitine, and total carnitine in the myocardium were significantly reduced, and pathologically, necrosis with the melting of cardiomyocytes was more common in the left ventricle than in the right ventricle in addition to coagulation necrosis. They also examined diabetic KK mice as an animal model of cardiomyopathy but did not exhibit a decrease in carnitine and its derivatives as observed in Syrian hamsters, and pathologically, diabetic KK mice showed a large number of necrosis images in the right ventricle and just below the epicardium [248]. Yamashita et al. divided these cardiomyopathy hamsters into three groups at 20 days of age—a control group (intraperitoneal administration of saline), an L-carnitine group (300 mg/kg/day, intraperitoneal administration), and a verapamil group (0.5 mg, intraperitoneal administration)—and examined the carnitine content and pathological findings of tissues at 90 days of age. [249] As a result, the total disability area, fibrosis area, and calcification area of pathologic findings were significantly lower with L-carnitine or verapamil. These results suggest that carnitine deficiency may be related to the mechanism of cardiomyopathy in hamsters.

Whitmer examined the high-energy phosphate compound and carnitine content of Syrian hamsters with BIO 53.58 strain and control hamsters with dilated cardiomyopathy in addition to Syrian hamsters with BIO 14.6 strain, which cause hypertrophic cardiomyopathy, and reported that myocardial carnitine content was reduced in cardiomyopathy hamsters [250]. It was reported that administration of high doses of L-carnitine (1 g/kg/day, intraperitoneally for 6 months) improved cardiac function decline and reduction of high-energy phosphate compounds [251].

References

1. Pons R, De Vivo DC. Primary and secondary carnitine deficiency syndromes. J Child Neurol. 1995;10(Suppl 2):S8–S24.
2. Ramsay RR, Zammit VA. Carnitine acyltransferases and their influence on CoA pools in health and disease. Mol Aspect Med. 2004;25(5–6):475–93. https://doi.org/10.1016/j.mam.2004.06.002.
3. Reuter SE, Evans AM. Carnitine and acylcarnitines. Pharmacokinetic, pharmacological and clinical aspects. Clin Pharmacokinet. 2012;51(9):553–72. https://doi.org/10.1007/BF03261931.
4. Stumpf DA, Parker WD, Angelini C. Carnitine deficiency, organic acidemias, and Reye's syndrome. Neurology. 1985;35(7):1041–5. https://doi.org/10.1212/wnl.35.7.1041.
5. Chalmers RA, Roe CR, Stacey TE, et al. Urinary excretion of l-carnitine and acylcarnitines by patients with disorders of organic acid metabolism: evidence for secondary insufficiency of L-carnitine. Pediatr Res. 1984;18(12):1325–8. https://doi.org/10.1203/00006450-198412000-00021.
6. Sugiyama S. Carnitine deficiency. Jpn J Pediatr Med. 1996;28(Suppl):284–9. (Article in Japanese).
7. Winter SC, Zorn EM, Vance WH. Carnitine deficiency. Lancet. 1990;335(8695):981–2.
8. Angelini C, Vergani L, Martinuzzi A. Clinical and biochemical aspects of carnitine deficiency and insufficiency: transport defects and inborn errors of β-oxidation. Crit Rev Clin Lab Sci. 1992;29(3–4):217–42. https://doi.org/10.3109/10408369209114601.
9. Hoppel C. The role of carnitine in normal and altered fatty acid metabolism. Am J Kidney Dis. 2003;41(4 Suppl 4):S4–S12. https://doi.org/10.1016/s0272-6386(03)00112-4.
10. Campos Y, Huertas R, Lorenzo G, et al. Plasma carnitine insufficiency and effectiveness of L-carnitine therapy in patients with mitochondrial myopathy. Muscle Nerve. 1993;16(2):150–3. https://doi.org/10.1002/mus.880160205.
11. Böhles H, Evangeliou A, Bervoets K, et al. Carnitine esters in metabolic disease. Eur J Pediatr. 1994; 153(7 Suppl 1):S57–61. https://doi.org/10.1007/BF02138779.
12. Carter AL, Abney TO, Lapp DF. Biosynthesis and metabolism of carnitine. J Child Neurol. 1995;10(Suppl 2):S3–7.
13. Riley S, Rutherford S, Rutherford PA. Low carnitine levels in hemodialysis patients: relationship with functional activity status and intra-dialytic hypotension. Clin Nephrol. 1997;48(6):392–3.
14. Marin VB, Azocar M, Molina M, et al. Total carnitine and acylated carnitine ratio: relationship of free carnitine with lipid parameters in pediatric dialysis patients. Adv Perit Dial. 2006;22:130–5.
15. Krähenbühl S. Carnitine metabolism in chronic liver disease. Life Sci. 1996;59(19):1579–99. https://doi.org/10.1016/0024-3205(96)00343-8.
16. Zhou L, Wang Q, Yin P, et al. Serum metabolomics reveals the deregulation of fatty acids metabolism in hepatocellular carcinoma and chronic liver diseases. Anal Bioanal Chem. 2012;403(1):203–13. https://doi.org/10.1007/s00216-012-5782-4.
17. Miyaaki H, Kobayashi H, Miuma S, et al. Blood carnitine profiling on tandem mass spectrometry in liver cirrhotic patients. BMC Gastroenterol. 2020;20(1):41. https://doi.org/10.1186/s12876-020-01190-6.
18. Suzuki Y, Masumura Y, Kobayashi A, et al. Myocardial carnitine deficiency in chronic heart failure. Lancet. 1982;1(8263):116. https://doi.org/10.1016/s0140-6736(82)90263-x.
19. Hunter WG, Kelly JP, McGarrah RW, et al. Metabolomic profiling identifies novel circulating biomarkers of mitochondrial dysfunction differentially elevated in heart failure with preserved versus reduced ejection fraction: evidence for shared metabolic impairments in clinical heart failure. J Am Heart Assoc. 2016;5(8):e003190. https://doi.org/10.1161/JAHA.115.003190.
20. Ahmad T, Kelly JP, McGarrah RW, et al. Prognostic implication of long-chain acylcarnitines in heart failure and reversibility with mechanical circulatory support. J Am Coll Cardiol. 2016;67(3):291–9. https://doi.org/10.1016/j.jacc.2015.10.079.
21. Kalim S, Clish CB, Wenger J, et al. A plasma long-chain acylcarnitine predicts cardiovascular mortality in incident dialysis patients. J Am Heart Assoc. 2013;2(6):e000542. https://doi.org/10.1161/JAHA.113.000542.
22. Ruiz M, Labarthe F, Fortier A, et al. Circulating acylcarnitine profile in human heart failure: a surrogate of fatty acid metabolic dysregulation in mitochondria and beyond. Am J Physiol Heart Circ Physiol. 2017;313(4):H768–81. https://doi.org/10.1152/ajpheart.00820.2016.
23. Ueland T, Svardal A, Øie E, et al. Disturbed carnitine regulation in chronic heart failure – increased plasma levels of palmitoyl-carnitine are associated with poor prognosis. Int J Cardiol. 2013;167(5):1892–9. https://doi.org/10.1016/j.ijcard.2012.04.150.
24. Yoshihisa A, Watanabe S, Yokokawa T, et al. Associations between acylcarnitine to free carnitine ratio and adverse prognosis in heart failure patients with reduced or preserved ejection fraction. ESC Heart Fail. 2017;4(3):360–4. https://doi.org/10.1002/ehf2.12176.
25. Chapoy PR, Angelini C, Brown WJ, et al. Systemic carnitine deficiency – a treatable inherited lipid-storage disease presenting as Reye's syndrome. N Engl J Med. 1980;303(24):1389–94. https://doi.org/10.1056/NEJM198012113032403.

26. Tripp ME, Katcher ML, Peters HA, et al. Systemic carnitine deficiency presenting as familial endocardial fibroelastosis. A treatable cardiomyopathy. N Engl J Med. 1981;305(7):385–90. https://doi.org/10.1056/NEJM198108133050707.
27. Waber LJ, Valle D, Neill C, et al. Carnitine deficiency presenting as familial cardiomyopathy: a treatable defect in carnitine transport. J Pediatr. 1982;101(5):700–5. https://doi.org/10.1016/s0022-3476(82)80294-1.
28. Hirata K, Yoshioka F, Eto Y, et al. Carnitine deficiency: a treatable cardiomyopathy. J Cardiogr. 1986;16(1):217–25. (Article in Japanese).
29. Pierpont ME, Breningstall GN, Stanley CA, et al. Familial carnitine transporter defect: a treatable cause of cardiomyopathy in children. Am Heart J. 2000;139(2 Pt 3):S96–S106. https://doi.org/10.1067/mhj.2000.103921.
30. Cano A, Ovaert C, Vianey-Saban C, et al. Carnitine membrane transporter deficiency: a rare treatable cause of cardiomyopathy and anemia. Pediatr Cardiol. 2008;29(1):163–5. https://doi.org/10.1007/s00246-007-9051-9.
31. Karpati G, Carpenter S, Engel AG, et al. The syndrome of systemic carnitine deficiency. Clinical, morphologic, biochemical, and pathophysiologic features. Neurology. 1975;25(1):16–24. https://doi.org/10.1212/wnl.25.1.16.
32. Di Donato S, Peluchetti D, Rimoldi M, et al. Systemic carnitine deficiency: clinical, biochemical, and morphological cure with L-carnitine. Neurology. 1984;34(2):157–62. https://doi.org/10.1212/wnl.34.2.157.
33. von Petrykowski W, Ketelsen UP, Schmidt-Sommerfeld E, et al. Primary systemic carnitine deficiency under successful therapy: clinical, biochemical, ultrahistochemical and renal clearance studies. Clin Neuropathol. 1985;4(2):63–71.
34. Matsuishi T, Hirata K, Terasawa K, et al. Successful carnitine treatment in two siblings having lipid storage myopathy with hypertrophic cardiomyopathy. Neuropediatrics. 1985;16(1):6–12. https://doi.org/10.1055/s-2008-1052536.
35. Treem WR, Stanley CA, Finegold DN, et al. Primary carnitine deficiency due to a failure of carnitine transport in kidney, muscle, and fibroblasts. N Engl J Med. 1988;319(20):1331–6. https://doi.org/10.1056/NEJM198811173192006.
36. Roe CR, Bohan TP. L-carnitine therapy in propionicacidaemia. Lancet. 1982;1(8286):1411–2. https://doi.org/10.1016/s0140-6736(82)92524-7.
37. Roe CR, Hoppel CL, Stacey TE, et al. Metabolic response to carnitine in methylmalonic aciduria. An effective strategy for elimination of propionyl groups. Arch Dis Child. 1983;58(11):916–20. https://doi.org/10.1136/adc.58.11.916.
38. Sugiyama N, Morishita H, Nagaya S, et al. Biochemical evidence of carnitine effect on propionate elimination. J Inherit Metab Dis. 1984;7(3):137–8. https://doi.org/10.1007/BF01801775.
39. Tein I. Role of carnitine and fatty acid oxidation and its defects in infantile epilepsy. J Child Neurol. 2002;17(Suppl 3):3S57-3S83.
40. De Vivo DC, Bohan TP, Coulter DL, et al. L-carnitine supplementation in childhood epilepsy: current perspectives. Epilepsia. 1998;39(11):1216–25. https://doi.org/10.1111/j.1528-1157.1998.tb01315.x.
41. Winter SC, Szabo-Aczel S, Curry CJR, et al. Plasma carnitine deficiency. Clinical observations in 51 pediatric patients. Am J Dis Child. 1987;141(6):660–5. https://doi.org/10.1001/archpedi.1987.04460060076039.
42. Engel AG. Chapter 57. Carnitine deficiency syndromes and lipid storage myopathies. In: Engel AG, Banker BQ, editors. Myology: basic and clinical, vol. 2. New York: McGraw-Hill; 1986. p. 1663–96.
43. Di Mauro S, Trevisan C, Hays A. Disorders of lipid metabolism in muscle. Muscle Nerve. 1980;3(5):369–88. https://doi.org/10.1002/mus.880030502.
44. Giovannini M, Agostoni C, Salari PC. Is carnitine essential in children? J Int Med Res. 1991;19(2):88–102. https://doi.org/10.1177/030006059101900202.
45. Takayanagi M. Carnitine deficiency and secondary carnitine deficiency. Shounika. 2012;9(10):1271–9. (Article in Japanese).
46. Kuwashima M. Secondary carnitine deficiency. Jpn J Clin Med. 2002;60(Suppl 4):706–9. (Article in Japanese).
47. Sugiyama N, Wada YB. Carnitine metabolism disorder – carnitine deficiency. In: Kobayashi N, editor. New pediatric college. Advances in pediatric medicine '89C, annual edition. Tokyo: Nakayama Shoten Co., Ltd.; 1989. p. 155–69. (Book in Japanese).
48. Scaglia F, Longo N. Primary and secondary alterations of neonatal carnitine metabolism. Semin Perinatol. 1999;23(2):152–61. https://doi.org/10.1016/s0146-0005(99)80047-0.
49. Koizumi A, Nozaki J, Ohura T, et al. Genetic epidemiology of the carnitine transporter OCTN2 gene in a Japanese population and phenotypic characterization in Japanese pedigrees with primary systemic carnitine deficiency. Hum Mol Genet. 1999;8(12):2247–54. https://doi.org/10.1093/hmg/8.12.2247.
50. Koizumi A. Toward a more rational field genetic epidemiology. Nippon Eiseigaku Zasshi (Jpn J Hyg). 2010;65(1):37–47. https://doi.org/10.1265/jjh.65.37. (Article in Japanese).
51. Longo N, Frigeni M, Pasquali M. Carnitine transport and fatty acid oxidation. Biochim Biophys Acta. 2016;1863(10):2422–35. https://doi.org/10.1016/j.bbamcr.2016.01.023.
52. Longo N. Primary carnitine deficiency and newborn screening for disorders of the carnitine cycle.

Ann Nutr Metab. 2016;68(Suppl 3):5–9. https://doi.org/10.1159/000448321.
53. Rose EC, di San Filippo CA, Ndukwe Erlingsson UC, et al. Genotype-phenotype correlation in primary carnitine deficiency. Hum Mutat. 2012;33(1):118–23. https://doi.org/10.1002/humu.21607.
54. Chen YC, Chien YH, Chen PW, et al. Carnitine uptake defect (primary carnitine deficiency): risk in genotype-phenotype correlation. Hum Mutat. 2013;34(4):655. https://doi.org/10.1002/humu.22286.
55. Nakanishi T, Hatanaka T, Huang W, et al. Na^+- and Cl^--coupled active transport of carnitine by the amino acid transporter $ATB^{0,+}$ from mouse colon expressed in HRPE cells and Xenopus oocytes. J Physiol. 2001;532(Pt 2):297–304. https://doi.org/10.1111/j.1469-7793.2001.0297f.x.
56. Sloan JL, Mager S. Cloning and functional expression of a human Na^+ and Cl^--dependent neutral and cationic amino acid transporter B^{0+}. J Biol Chem. 1999;274(34):23740–5. https://doi.org/10.1074/jbc.274.34.23740.
57. Chapter 24. Systemic Carnitine Deficiency (OCTN2 Abnormality), In: The Japanese Society for Inherited Metabolic Diseases ed. Clinical practice guidelines for neonatal mass screening diseases 2019, p. 228–39, Tokyo: SHINDAN TO CHIRYO SHA, Inc. 2019. (Book in Japanese).
58. Engel AG, Angelini C. Carnitine deficiency of human skeletal muscle with associated lipid storage myopathy: a new syndrome. Science. 1973;179(4076):899–902. https://doi.org/10.1126/science.179.4076.899.
59. Engel AG, Rebouche CJ. Carnitine metabolism and inborn errors. J Inherit Metab Dis. 1984;7(Suppl 1):38–43. https://doi.org/10.1007/BF03047372.
60. Olpin SE. Fatty acid oxidation defects as a cause of neuromyopathic disease in infants and adults. Clin Lab. 2005;51(5–6):289–306.
61. Winter SC. Treatment of carnitine deficiency. J Inherit Metab Dis. 2003;26(2–3):171–80. https://doi.org/10.1023/a:1024433100257.
62. Hörster F, Hoffmann GF. Pathophysiology, diagnosis, and treatment of methylmalonic aciduria – recent advances and new challenges. Pediatr Nephrol. 2004;19(10):1071–4. https://doi.org/10.1007/s00467-004-1572-3.
63. Gordon N. Glutaric aciduria types I and II. Brain Dev. 2006;28(3):136–40. https://doi.org/10.1016/j.braindev.2005.06.010.
64. Deodato F, Boenzi S, Santorelli FM, et al. Methylmalonic and propionic aciduria. Am J Med Genet C Semin Med Genet. 2006;142C(2):104–12. https://doi.org/10.1002/ajmg.c.30090.
65. Kölker S, Christensen E, Leonard JV, et al. Diagnosis and management of glutaric aciduria type I – revised recommendations. J Inherit Metab Dis. 2011;34(3):677–94. https://doi.org/10.1007/s10545-011-9289-5.
66. Larson A, Goodman S. Glutaric Acidemia Type 1. In: Adam MP, Ardinger HH, Pagon RA, et al., editors. GeneReviews® [Internet]. Seatle: University of Washington; 2019.
67. Manoli I, Sloan JL, Venditti CP. Isolated methylmalonic acidemia. In: Adam MP, Ardinger HH, Pagon RA, et al., editors. GeneReviews® [Internet]. Seattle: University of Washington; 2005.
68. Longo N, Di San Filippo CA, Pasquali M. Disorders of carnitine transport and the carnitine cycle. Am J Med Genet C Semin Med Genet. 2006;142C(2):77–85. https://doi.org/10.1002/ajmg.c.30087.
69. Shigematsu Y, Hata I. Abnormal metabolism of organic acids. Jpn J Pediatr. 2006;59(4):635–41. (Article in Japanese).
70. Yamaguchi S. Overview of abnormal in organic acid metabolism. Jpn J Clin Med. 1998;18(Suppl 1):261–6. (Article in Japanese).
71. Kitagawa T, Matsuda I, Tada K, et al. Report of the expert committee on examination of treatment guidelines for disorders of organic acid metabolism and fatty acid oxidation. Specific Milk Information. 2003;39:58–67. (Article in Japanese).
72. Alfadhel M, Al-Thihli K, Moubayed H, et al. Drug treatment of inborn errors of metabolism: a systematic review. Arch Dis Child. 2013;98(6):454–61. https://doi.org/10.1136/archdischild-2012-303131.
73. Walter JH. L-carnitine in inborn errors of metabolism: what is the evidence? J Inherit Metab Dis. 2003;26(2–3):181–8. https://doi.org/10.1023/a:1024485117095.
74. Rudman D, Sewell CW, Ansley JD. Deficiency of carnitine in cachectic cirrhotic patients. J Clin Invest. 1997;60(3):716–23. https://doi.org/10.1172/JCI108824.
75. Evans AM, Fornasini G. Pharmacokinetics of L-carnitine. Clin Pharmacokinet. 2003;42(11):941–67. https://doi.org/10.2165/00003088-200342110-00002.
76. Broquist HP, Borum PR. Carnitine biosynthesis. Nutritional implications. Adv Nutr Res. 1982;4:181–204. https://doi.org/10.1007/978-1-4613-9934-6_7.
77. Rebouche CJ, Engel AG. Tissue distribution of carnitine biosynthetic enzymes in man. Biochim Biophys Acta. 1980;630(1):22–9. https://doi.org/10.1016/0304-4165(80)90133-6.
78. Vaz FM, Wanders RJA. Carnitine biosynthesis in mammals. Biochem J. 2002;361(Pt 3):417–29. https://doi.org/10.1042/0264-6021:3610417.
79. Olson AL, Rebouche CJ. γ-Butyrobetaine hydroxylase activity is not rate limiting for carnitine biosynthesis in the human infant. J Nutr. 1987;117(6):1024–31. https://doi.org/10.1093/jn/117.6.1024.
80. Kerner J, Hoppel C. Genetic disorders of carnitine metabolism and their nutritional management. Ann Rev Nutr. 1998;18:179–206. https://doi.org/10.1146/annurev.nutr.18.1.179.

81. Steiber A, Kerner J, Hoppel CL. Carnitine: a nutritional, biosynthetic, and functional perspective. Mol Asp Med. 2004;25(5–6):455–73. https://doi.org/10.1016/j.mam.2004.06.006.
82. Rebouche CJ, Paulson DJ. Carnitine metabolism and function in humans. Annu Rev Nutr. 1986;6:41–66. https://doi.org/10.1146/annurev.nu.06.070186.000353.
83. Novak M. Carnitine supplementation in soy-based formula-fed infants. Biol Neonate. 1990;58(Suppl 1):S89–92. https://doi.org/10.1159/000243303.
84. Ohtani Y, Higashi A, Matsuda I. On the content of carnitine in breast milk and various artificial milks. J Jpn Soc Pediatr. 1984;88(9):1943–9. (Article in Japanese).
85. Penn D, Dolderer M, Schmidt-Sommerfeld E. Carnitine concentrations in the milk of different species and infant formulas. Biol Neonate. 1987;52(2):70–9. https://doi.org/10.1159/000242686.
86. Tokuriki S, Hayashi H, Okuno T, et al. Biotin and carnitine profiles in preterm infants in Japan. Pediatr Int. 2013;55(3):342–5. https://doi.org/10.1111/ped.12053.
87. Takayanagi M, Okimoto Y, Kakuta H. Causes and responses to vitamin deficiency in Japan. Jpn J Pediatr Med. 2009;41(9):1289–92. (Article in Japanese).
88. Kodama H, Shimizu T, Takitani K, et al. Pitfalls of using of some types of milk and enteral nutrient formulas. J Jpn Soc Pediatr. 2012;116(4):637–54. (Article in Japanese).
89. Takayanagi M. Application and problems of special formula milk powder for medical use for infant. Jpn J Pediatr Med. 2014;46(8):1063–6. (Article in Japanese).
90. Kodama H. Deficiency of some nutrients due to use of enteral and special milk formulas. Brain Dev. 2014;46(1):5–9. (Article in Japanese).
91. Crill CM, Helms RA. The use of carnitine in pediatric nutrition. Nutr Clin Pract. 2007;22(2):204–13. https://doi.org/10.1177/0115426507022002204.
92. Penn D, Schmidt-Sommerfeld E, Wolf H. Carnitine deficiency in premature infants receiving total parenteral nutrition. Early Hum Dev. 1980;4(1):23–34. https://doi.org/10.1016/0378-3782(80)90005-5.
93. Nanni G, Pittiruti M, Castagneto M. Carnitine plasma levels during total parenteral nutrition. Am J Clin Nutr. 1983;38(3):339–41. https://doi.org/10.1093/ajcn/38.2.339.
94. Bowyer BA, Fleming CR, Ilstrup D, et al. Plasma carnitine levels in patients receiving home parenteral nutrition. Am J Clin Nutr. 1986;43(1):85–91. https://doi.org/10.1093/ajcn/43.1.85.
95. Borum PR, Bennett SG. Carnitine as an essential nutrient. J Am Coll Nutr. 1986;5(2):177–82. https://doi.org/10.1080/07315724.1986.10720124.
96. Bowyer BA, Fleming CR, Haymond MW, et al. L-carnitine: effect of intravenous administration on fuel homeostasis in normal subjects and home-parenteral-nutrition patients with low plasma carnitine concentrations. Am J Clin Nutr. 1989;49(4):618–23. https://doi.org/10.1093/ajcn/49.4.618.
97. Tibboel D, Delemarre FMC, Przyrembel H, et al. Carnitine deficiency in surgical neonates receiving total parenteral nutrition. J Pediatr Surg. 1990;25(4):418–21. https://doi.org/10.1016/0022-3468(90)90384-l.
98. Schmidt-Sommerfeld E, Penn D. Carnitine and total parenteral nutrition of the neonate. Biol Neonate. 1990;58(Suppl 1):S81–8. https://doi.org/10.1159/000243302.
99. Schmidt-Sommerfeld E, Penn D, Bieber LL, et al. Carnitine ester excretion in pediatric patients receiving parenteral nutrition. Pediatr Res. 1990;28(2):158–65. https://doi.org/10.1203/00006450-199008000-00016.
100. Ida S, Hebiguchi T. Nutritional management at home for intestinal failure and severe motor and intellectual disabilities children. J Jpn Soc Pediatr Gastroenterol Hepatol Nutr. 2013;27(1):1–5. (Article in Japanese).
101. Btaiche IF, Khalidi N. Parenteral nutrition-associated liver complications in children. Pharmacotherapy. 2002;22(2):188–211. https://doi.org/10.1592/phco.22.3.188.33553.
102. Tanaka S, Miki T, Hsieh ST, et al. A case of severe hyperlipidemia caused by long-term tube feedings. J Atheroscler Thromb. 2003;10(5):321–4. https://doi.org/10.5551/jat.10.321.
103. Kumpf VJ. Parenteral nutrition-associated liver disease in adult and pediatric patients. Nutr Clin Pract. 2006;21(3):279–90. https://doi.org/10.1177/0115426506021003279.
104. Hirose S, Hirata M, Azuma N, et al. Carnitine depletion during total parenteral nutrition despite oral L-carnitine supplementation. Acta Paediatr Jpn. 1997;39(2):194–200. https://doi.org/10.1111/j.1442-200x.1997.tb03580.x.
105. Nozaki A, Kusunoki T, Yoda A. A case of Crohn's disease in which carnitine and selenium deficiency was observed due to long-term use of Ehrenthal ®. J Jpn Soc Pediatr Gastroenterol Hepatol Nutr. 2015;29(1):8–12. (Article in Japanese).
106. Treem WR, Stanley CA. Massive hepatomegaly, steatosis, and secondary plasma carnitine deficiency in an infant with cystic fibrosis. Pediatrics. 1989;83(6):993–7.
107. Curione M, Danese C, Viola F, et al. Carnitine deficiency in patients with coeliac disease and idiopathic dilated cardiomyopathy. Nutr Metab Cardiovasc Dis. 2005;15(4):279–83. https://doi.org/10.1016/j.numecd.2005.01.002.
108. Uslu N, Demir H, Karagöz T, et al. Dilated cardiomyopathy in celiac disease: role of carnitine deficiency. Acta Gastroenterol Belg. 2010;73(4):530–1.
109. Tanphaichitr V, Leelahagul P. Carnitine metabolism and human carnitine deficiency. Nutrition. 1993;9(3):246–54.

110. Scholte HR, Stinis JT, Jennekens FGI. Low carnitine levels in serum of pregnant women. N Engl J Med. 1978;299(19):1079–80. https://doi.org/10.1056/nejm197811092991922.
111. Novak M, Monkus EF, Chung D, et al. Carnitine in the perinatal metabolism of lipids. I. Relationship between maternal and fetal plasma levels of carnitine and acylcarnitines. Pediatrics. 1981;67(1):95–100.
112. Cederblad G, Niklasson A, Rydgren B, et al. Carnitine in maternal and neonatal plasma. Acta Paediatr Scand. 1985;74(4):500–4. https://doi.org/10.1111/j.1651-2227.1985.tb11017.x.
113. Cederblad G, Fåhraeus L, Lindgren K. Plasma carnitine and renal-carnitine clearance during pregnancy. Am J Clin Nutr. 1986;44(3):379–83. https://doi.org/10.1093/ajcn/44.3.379.
114. Bargen-Locker C, Hahn P, Wittmann B. Plasma carnitine in pregnancy. Am J Obstet Gynecol. 1981;140(4):412–4. https://doi.org/10.1016/0002-9378(81)90035-1.
115. Marzo A, Cardace G, Corbelletta C, et al. Plasma concentration, urinary excretion and renal clearance of L-carnitine during pregnancy: a reversible secondary L-carnitine deficiency. Gynecol Endocrinol. 1994;8(2):115–20. https://doi.org/10.3109/09513599409058032.
116. Angelini C, Govoni E, Bragaglia MM, et al. Carnitine deficiency: acute postpartum crisis. Ann Neurol. 1978;4(6):558–61. https://doi.org/10.1002/ana.410040616.
117. Bernardini I, Rizzo WB, Dalakas M, et al. Plasma and muscle free carnitine deficiency due to renal Fanconi syndrome. J Clin Invest. 1985;75(4):1124–30. https://doi.org/10.1172/JCI111806.
118. Gahl WA, Bernardini I, Dalakas M, et al. Oral carnitine therapy in children with cystinosis and renal Fanconi syndrome. J Clin Invest. 1988;81(2):549–60. https://doi.org/10.1172/JCI113353.
119. Gahl WA, Bernardini IM, Dalakas MC, et al. Muscle carnitine repletion by long-term carnitine supplementation in nephropathic cystinosis. Pediatr Res. 1993;34(2):115–9. https://doi.org/10.1203/00006450-199308000-00001.
120. Besouw M, Cornelissen E, Cassiman D, et al. Carnitine profile and effect of suppletion in children with renal Fanconi syndrome due to cystinosis. JIMD Rep. 2014;16:25–30. https://doi.org/10.1007/8904_2014_312.
121. Gahl WA, Thoene JG, Schneider JA. Chapter 199: Cystinosis: a disorder of lysosomal membrane transport. In: Scriver CR, Beaudet AL, Sly WS, et al., editors. The metabolic & molecular bases of inherited disease. 8th ed. New York: McGraw-Hill; 2001. p. 5085–108.
122. Tanner LM, Näntö-Salonen K, Niinikoski H, et al. Nephropathy advancing to end-stage renal disease: a novel complication of lysinuric protein intolerance. J Pediatr. 2007;150(6):631–4. https://doi.org/10.1016/j.jpeds.2007.01.043.
123. Tanner LM, Näntö-Salonen K, Rashed MS, et al. Carnitine deficiency and L-carnitine supplementation in lysinuric protein intolerance. Metabolism. 2008;57(4):549–54. https://doi.org/10.1016/j.metabol.2007.11.019.
124. Nussbaum RL, Suchy SF. Chapter 252: The oculocerebrorenal Syndrome of Lowe (Lowe Syndrome). In: Scriver CR, Beaudet AL, Sly WS, et al., editors. The metabolic & molecular bases of inherited disease. 8th ed. New York: McGraw-Hill; 2001. p. 6257–66.
125. Lara-Corrales I, Mellerio JE, Martinez AE, et al. Dilated cardiomyopathy in epidermolysis bullosa: a retrospective, multicenter study. Pediatr Dermatol. 2010;27(3):238–43. https://doi.org/10.1111/j.1525-1470.2010.01127.x.
126. Sidwell RU, Yates R, Atherton D. Dilated cardiomyopathy in dystrophic epidermolysis bullosa. Arch Dis Child. 2000;83(1):59–63. https://doi.org/10.1136/adc.83.1.59.
127. Pfendner EG, Lucky AW. Dystrophic Epidermolysis Bullosa. 2006 Aug 21 [updated 2018 Sep 13]. In: Adam MP, Ardinger HH, Pagon RA, et al., editors. GeneReviews® [Internet]. Seattle: University of Washington; 1993–2022.
128. Bertoli M, Battistella PA, Vergani L, et al. Carnitine deficiency induced during hemodialysis and hyperlipidemia: effect of replacement therapy. Am J Clin Nutr. 1981;34(8):1496–500. https://doi.org/10.1093/ajcn/34.8.1496.
129. Bartel LL, Hussey JL, Shrago E. Perturbation of serum carnitine levels in human adults by chronic renal disease and dialysis therapy. Am J Clin Nutr. 1981;34(7):1314–20. https://doi.org/10.1093/ajcn/34.7.1314.
130. Novoa D, Rodriguez-Segade S, Romero R, et al. Carnitine levels and hypertriglyceridemia in undialyzed patients. Nephron. 1987;47(3):231. https://doi.org/10.1159/000184499.
131. Wanner C, Förstner-Wanner S, Rössle C, et al. Carnitine metabolism in patients with chronic renal failure: effect of L-carnitine supplementation. Kidney Int. 1987;37(Suppl 22):S132–5.
132. Bohmer T, Bergrem H, Eiklid K. Carnitine deficiency induced during intermittent haemodialysis for renal failure. Lancet. 1978;1(8056):126–8. https://doi.org/10.1016/s0140-6736(78)90422-1.
133. Battistella PA, Angelini C, Vergani L, et al. Carnitine deficiency induced during haemodialysis. Lancet. 1978;1(8070):939. https://doi.org/10.1016/s0140-6736(78)90714-6.
134. Leschke M, Rumpf KW, Eisenhauer T, et al. Quantitative assessment of carnitine loss during hemodialysis and hemofiltration. Kidney Int. 1983;24(Suppl 16):S143–6.
135. Bellinghieri G, Savica V, Mallamace A, et al. Correlation between increased serum and tissue L-carnitine levels and improved muscle symptoms in hemodialyzed patients. Am J Clin Nutr.

1983;38(4):523–31. https://doi.org/10.1093/ajcn/38.4.523.
136. Guarnieri G, Toigo G, Crapesi L, et al. Carnitine metabolism in chronic renal failure. Kidney Int. 1987;32(Suppl 22):S116–27.
137. Hiatt WR, Koziol BJ, Shapiro JI, et al. Carnitine metabolism during exercise in patients on chronic hemodialysis. Kidney Int. 1992;41(6):1613–9. https://doi.org/10.1038/ki.1992.233.
138. Guarnieri G, Situlin R, Biolo G. Carnitine metabolism in uremia. Am J Kidney Dis. 2001;38(4Suppl 1):S63–7. https://doi.org/10.1053/ajkd.2001.27408.
139. Guarnieri G, Biolo G, Vinci P, et al. Advances in carnitine in chronic uremia. J Ren Nutr. 2007;17(1):23–9. https://doi.org/10.1053/j.jrn.2006.10.006.
140. Aguilar-Kitsu A, Ibarra-Cazares P, Mendoza-Guevara L, et al. Frequency of low carnitine levels in children on dialysis. Adv Perit Dial. 2006;22:208–10.
141. Kawahara K, Sato T, Kawamura K, et al. Effect of carnitine administration on cardiomyopathy in children on peritoneal dialysis. J Pediatr Ren Fail. 1995;15:224–7. (Article in Japanese).
142. Evans AM, Faull R, Fornasini G, et al. Pharmacokinetics of L-carnitine in patients with end-stage renal disease undergoing long-term hemodialysis. Clin Pharmacol Ther. 2000;68(3):238–49. https://doi.org/10.1067/mcp.2000.108850.
143. Evans AM, Faull RJ, Nation RL, et al. Impact of hemodialysis on endogenous plasma and muscle carnitine levels in patients with end-stage renal disease. Kidney Int. 2004;66(4):1527–34. https://doi.org/10.1111/j.1523-1755.2004.00916.x.
144. Golper TA, Goral S, Becker BN, et al. L-carnitine treatment of anemia. Am J Kidney Dis. 2003;41(4 Suppl 4):S27–34. https://doi.org/10.1016/s0272-6386(03)00114-8.
145. Schreiber B. Levocarnitine and dialysis: a review. Nutr Clin Pract. 2005;20(2):218–43. https://doi.org/10.1177/0115426505020002218.
146. Hedayati S. Dialysis-related carnitine disorder. Semin Dial. 2006;19(4):323–8. https://doi.org/10.1111/j.1525-139X.2006.00180.x.
147. Belay B, Esteban-Cruciani N, Walsh CA, et al. The use of levo-carnitine in children with renal disease: a review and a call for future studies. Pediatr Nephrol. 2006;21(3):308–17. https://doi.org/10.1007/s00467-005-2085-4.
148. Hurot JM, Cucherat M, Haugh M, et al. Effects of L-carnitine supplementation in maintenance hemodialysis patients: a systematic review. J Am Soc Nephrol. 2002;13(3):708–14. https://doi.org/10.1681/ASN.V133708.
149. Lynch KE, Feldman HI, Berlin JA, et al. Effects of L-carnitine on dialysis-related hypotension and muscle cramps: a meta-analysis. Am J Kidney Dis. 2008;52(5):962–71. https://doi.org/10.1053/j.ajkd.2008.05.031.
150. Miller B, Ahmad S. A review of the impact of L-carnitine therapy on patient functionality in maintenance hemodialysis. Am J Kidney Dis. 2003;41(4 Suppl 4):S44–8. https://doi.org/10.1016/s0272-6386(03)00116-1.
151. Sgambat K, Moudgil A. Carnitine deficiency in children receiving continuous renal replacement therapy. Hemodial Int. 2016;20(1):63–7. https://doi.org/10.1111/hdi.12341.
152. Sgambat K, Clauss S, Moudgil A. Effect of levocarnitine supplementation on myocardial strain in children with acute kidney injury receiving continuous kidney replacement therapy: a pilot study. Pediatr Nephrol. 2021;36(6):1607–16. https://doi.org/10.1007/s00467-020-04862-3.
153. Böhles H, Richter K, Wagner-Thiessen E, et al. Decreased serum carnitine in valproate induced Reye syndrome. Eur J Pediatr. 1982;139(3):185–6. https://doi.org/10.1007/BF01377353.
154. Coulter DL. Carnitine deficiency: a possible mechanism for valproate hepatotoxicity. Lancet. 1984;1(8378):689. https://doi.org/10.1016/s0140-6736(84)92209-8.
155. Murphy JV, Marquardt KM, Shug AI. Valproic acid associated abnormalities of carnitine metabolism. Lancet. 1985;1(8432):820–1. https://doi.org/10.1016/s0140-6736(85)91481-3.
156. Coulter DL. Carnitine, valproate, and toxicity. J Child Neurol. 1991;6(1):7–14. https://doi.org/10.1177/088307389100600102.
157. Melegh B, Kerner J, Bieber LL. Pivampicillin-promoted excretion of pivaloylcarnitine in humans. Biochem Pharmacol. 1987;36(20):3405–9. https://doi.org/10.1016/0006-2952(87)90318-2.
158. Holme E, Greter J, Jacobson CE, et al. Carnitine deficiency induced by pivampicillin and pivmecillinam therapy. Lancet. 1989;2(8661):469–73. https://doi.org/10.1016/s0140-6736(89)92086-2.
159. Brass EP. Pivalate-generating prodrugs and carnitine homeostasis in man. Pharmacol Rev. 2002;54(4):589–98. https://doi.org/10.1124/pr.54.4.589.
160. Abrahamsson K, Eriksson BO, Holme E, et al. Impaired ketogenesis in carnitine depletion caused by short-term administration of pivalic acid prodrug. Biochem Med Metab Biol. 1994;52(1):18–21. https://doi.org/10.1006/bmmb.1994.1028.
161. Abrahamsson K, Mellander M, Eriksson BO, et al. Transient reduction of human left ventricular mass in carnitine depletion induced by antibiotics containing pivalic acid. Br Heart J. 1995;74(6):656–9. https://doi.org/10.1136/hrt.74.6.656.
162. Abrahamsson K, Eriksson BO, Holme E, et al. Pivalic acid-induced carnitine deficiency and physical exercise in humans. Metabolism. 1996;45(12):1501–7. https://doi.org/10.1016/s0026-0495(96)90179-1.
163. Ito T, Sugiyama N, Kobayashi M, et al. Alteration of ammonia and carnitine levels in short-term treatment with pivalic acid-containing prodrug. Tohoku J Exp Med. 1995;175(1):43–53. https://doi.org/10.1620/tjem.175.43.
164. Pap M, Kopcsányi G, Bieber LL, et al. Cefetamet pivoxil treatment causes loss of carnitine reserves

that can be prevented by exogenous carnitine administration. J Nutr Biochem. 1999;10(11):670–3. https://doi.org/10.1016/s0955-2863(99)00051-0.
165. Ito S, Yoshikawa T, Itabashi K, et al. Attention to secondary carnitine deficiency due to administration of antibiotics containing a Pivoxil Group (Report of the Pharmaceutical Affairs Committee of the Japan Pediatric Society). J Jpn Soc Pediatr. 2012;116(4):804–6. (Article in Japanese).
166. Pharmaceuticals and Medical Devices Agency: Request for proper use of medicines from the PMDA. Severe hypocarnitinemia and hypoglycemia in children due to administration of antibiotics with a pivoxil group. April 2012, No. 8. (Article in Japanese). https://www.pmda.go.jp/files/000143929.pdf
167. Japan Otological Society, Japan Society for Pediatric ORL, Japan Society for Infection and Aerosol in Otorhinolaryngology. Clinical practice guidelines for the diagnosis and management of acute otitis media (AOM) in children in Japan – 2018. Tokyo: Kanehara & Co., Ltd.; 2018. (Article in Japanese).
168. Hotomi M, Yanagihara K, Ishiwada N, Ito M, et al. Recommendations on the appropriate use of antimicrobial agents for airway infections (Report of the Committee on the appropriate use of antimicrobial agents for airway infections, Japanese Society of Infectious Diseases). J Japanese Soc Infect Dis. 2019;93(5):623–42. (Article in Japanese).
169. Akiyama S, Fujinaga H, Kaneshige M, et al. Neonatal cases of hypocarnitinemia caused by sivelestat administration. J Japan Soc Perinat Neonatal Med. 2013;49:695. (Abstract O-330, Article in Japanese).
170. Yamada K, Kobayashi H, Bo R, et al. Elevation of pivaloylcarnitine by sivelestat sodium in two children. Mol Genet Metab. 2015;116(3):192–4. https://doi.org/10.1016/j.ymgme.2015.09.009.
171. Sayed-Ahmed MM. Role of carnitine in cancer chemotherapy-induced multiple organ toxicity. Saudi Pharm J. 2010;18(4):195–206. https://doi.org/10.1016/j.jsps.2010.07.008.
172. Lancaster CS, Hu C, Franke RM, et al. Cisplatin-induced downregulation of OCTN2 affects carnitine wasting. Clin Cancer Res. 2010;16(19):4789–99. https://doi.org/10.1158/1078-0432.CCR-10-1239.
173. Hockenberry MJ, Hooke MC, Gregurich M, et al. Carnitine plasma levels and fatigue in children/adolescents receiving cisplatin, ifosfamide, or doxorubicin. J Pediatr Hematol Oncol. 2009;31(9):664–9. https://doi.org/10.1097/MPH.0b013e3181b259a7.
174. Kawai A, Matsumoto H, Endou Y, et al. Repeated combined chemotherapy with cisplatin lowers carnitine levels in gastric cancer patients. Ann Nutr Metab. 2017;71(3–4):261–5. https://doi.org/10.1159/000485808.
175. Graziano F, Bisonni R, Catalano V, et al. Potential role of levocarnitine supplementation for the treatment of chemotherapy-induced fatigue in non-anaemic cancer patients. Br J Cancer. 2002;86(12):1854–7. https://doi.org/10.1038/sj.bjc.6600413.
176. Marthaler NP, Visarius T, Küpfer A, et al. Increased urinary losses of carnitine during ifosfamide chemotherapy. Cancer Chemother Pharmacol. 1999;44(2):170–2. https://doi.org/10.1007/s002800050963.
177. Armenian SH. Anthracycline-induced cardiotoxicity in young cancer patients: the role of carnitine. Ann Nurt Metab. 2016;68(Suppl 3):S10–4. https://doi.org/10.1159/000448322.
178. Yaris N, Ceviz N, Coskun T, et al. Serum carnitine levels during the doxorubicin therapy. Its role in cardiotoxicity. J Exp Clin Cancer Res. 2002;21(2):165–70.
179. De Leonardis V, Neri B, Bacalli S, et al. Reduction of cardiac toxicity of anthracyclines by L-carnitine: preliminary overview of clinical data. Int J Clin Pharm Res. 1985;5(2):137–42.
180. Matsui H, Einama T, Shichi S, et al. L-carnitine supplementation reduces the general fatigue of cancer patients during chemotherapy. Mol Clin Oncol. 2018;8(3):413–6. https://doi.org/10.3892/mco.2018.1557.
181. Hu C, Lancaster CS, Zuo Z, et al. Inhibition of OCTN2-mediated transport of carnitine by etoposide. Mol Cancer Ther. 2012;11(4):921–9. https://doi.org/10.1158/1535-7163.MCT-11-0980.
182. Heuberger W, Berardi S, Jacky E, et al. Increased urinary excretion of carnitine in patients treated with cisplatin. Eur J Clin Pharmacol. 1998;54(7):503–8. https://doi.org/10.1007/s002280050504.
183. Haschke M, Vitins T, Lüde S, et al. Urinary excretion of carnitine as a marker of proximal tubular damage associated with platin-based antineoplastic drugs. Nephrol Dial Transplant. 2010;25(2):426–33. https://doi.org/10.1093/ndt/gfp456.
184. Endo K, Tsuji A, Kondo S, et al. Carnitine is associated with fatigue following chemoradiotherapy for head and neck cancer. Acta Otolaryngol. 2015;135(8):846–52. https://doi.org/10.3109/00016489.2015.1030769.
185. Endo K, Ueno T, Ishikawa K, et al. Effects of L-carnitine administration on health-related quality of life during cisplatin-based chemoradiotherapy in patients with head and neck squamous cell carcinoma. Auris Nasus Larynx. 2019;46(3):431–6. https://doi.org/10.1016/j.anl.2018.10.007.
186. Sakamoto A, Tsukahara Y, Gomi D, et al. Plasma L-carnitine levels in terminally ill cancer patients receiving only palliative care. Ann Palliat Med. 2017;6(Suppl 1):S52–7. https://doi.org/10.21037/apm.2017.06.08.
187. Weinberg GL, Laurito CE, Geldner P, et al. Malignant ventricular dysrhythmias in a patient with isovaleric acidemia receiving general and local anesthesia for suction lipectomy. J Clin Anesth. 1997;9(8):668–70. https://doi.org/10.1016/s0952-8180(97)00187-6.
188. Wong GK, Joo DT, McDonnell C. Lipid resuscitation in a carnitine deficient child following intravascular migration of an epidural catheter. Anaesthesia. 2010;65(2):192–5. https://doi.org/10.1111/j.1365-2044.2009.06131.x.

189. Weinberg GL, Palmer JW, VadeBoncouer TR, et al. Bupivacaine inhibits acylcarnitine exchange in cardiac mitochondria. Anesthesiology. 2000;92(2):523–8. https://doi.org/10.1097/00000542-200002000-00036.
190. Wong GK, Crawford MW. Carnitine deficiency increases susceptibility to bupivacaine-induced cardiotoxicity in rats. Anesthesiology. 2011;114:1417–24.
191. Bray RJ. Propofol infusion syndrome in children. Paediatr Anesth. 1998;8(6):491–9. https://doi.org/10.1046/j.1460-9592.1998.00282.x.
192. Wolf AR, Potter F. Propofol infusion in children: when does an anesthetic tool become an intensive care liability? Pediatr Anesth. 2004;14(6):435–8. https://doi.org/10.1111/j.1460-9592.2004.01332.x.
193. Fudickar A, Bein B. Propofol infusion syndrome: update of clinical manifestation and pathophysiology. Minerva Anestesiol. 2009;75(5):339–44.
194. Wolf A, Weir P, Segar P, et al. Impaired fatty acid oxidation in propofol infusion syndrome. Lancet. 2001;357(9256):606–7. https://doi.org/10.1016/S0140-6736(00)04064-2.
195. Uezono S, Hotta Y, Takakuwa Y, et al. Acquired carnitine deficiency: a clinical model for propofol infusion syndrome? Anesthesiology. 2005;103(4):909. https://doi.org/10.1097/00000542-200510000-00046.
196. Weinberg GL, Baughman V. Carnitine deficiency, mitochondrial metabolism, and abnormal response to anesthetics. Anesthesiology. 2006;104(6):1343. https://doi.org/10.1097/00000542-200606000-00036.
197. Okazaki K. Propofol infusion syndrome. J Clin Anesth (Japan). 2009;33(Suppl):329–48. (Article in Japanese).
198. De Simone C, Ferrari M, Lozzi A, et al. Vitamins and immunity: II. Influence of L-carnitine on the immune system. Acta Vitaminol Enzymol. 1982;4(1–2):135–40.
199. De Simone C, Tzantzoglou S, Jirillo E, et al. L-carnitine deficiency in AIDS patients. AIDS. 1992;6(2):203–5. https://doi.org/10.1097/00002030-199202000-00011.
200. De Simone C, Tzantzoglou S, Famularo G, et al. High dose L-carnitine improves immunologic and metabolic parameters in aids patients. Immunopharmacol Immunotoxicol. 1993;15(1):1–12. https://doi.org/10.3109/08923979309066930.
201. De Simone C, Famularo G, Tzantzoglou S, et al. Carnitine depletion in peripheral blood mononuclear cells from patients with AIDS: effect of oral L-carnitine. AIDS. 1994;8(5):655–60. https://doi.org/10.1097/00002030-199405000-00012.
202. Campos Y, Arenas J. Muscle carnitine deficiency associated with zidovudine-induced mitochondrial myopathy. Ann Neurol. 1994;36(4):680–1. https://doi.org/10.1002/ana.410360427.
203. Dalakas MC, Leon-Monzon ME, Bernardini I, et al. Zidovudine-induced mitochondrial myopathy is associated with muscle carnitine deficiency and lipid storage. Ann Neurol. 1994;35(4):482–7. https://doi.org/10.1002/ana.410350418.
204. Karahoda R, Ceckova M, Staud F. The inhibitory effect of antiretroviral drugs on the L-carnitine uptake in human placenta. Toxicol Appl Pharmacol. 2019;368:18–25. https://doi.org/10.1016/j.taap.2019.02.002.
205. Sakuma T, Asai K, Ichiki T, et al. Identification of benzoylcarnitine in the urine of a patient of hyperammonemia. Tohoku J Exp Med. 1989;159(2):147–51. https://doi.org/10.1620/tjem.159.147.
206. Sakuma T. Alteration of urinary carnitine profile induced by benzoate administration. Arch Dis Child. 1991;66(7):873–5. https://doi.org/10.1136/adc.66.7.873.
207. Kuntzer T, Reichmann H, Bogousslavsky J, et al. Emetine-induced myopathy and carnitine deficiency. J Neurol. 1990;237(8):495–6. https://doi.org/10.1007/BF00314772.
208. Sekas G, Paul HS. Hyperammonemia and carnitine deficiency in a patient receiving sulfadiazine and pyrimethamine. Am J Med. 1993;95(1):112–3. https://doi.org/10.1016/0002-9343(93)90240-p.
209. Wagner CA, Lükewille U, Kaltenbach S, et al. Functional and pharmacological characterization of human Na+-carnitine cotransporter hOCTN2. Am J Physiol Renal Physiol. 2000;279(3):F584–91. https://doi.org/10.1152/ajprenal.2000.279.3.F584.
210. Ohashi R, Tamai I, Yabuuchi H, et al. Na+-dependent carnitine transport by organic cation transporter (OCTN2): its pharmacological and toxicological relevance. J Pharmacol Exp Ther. 1999;291(2):778–84.
211. Yamaguchi S. Sudden death syndrome (SIDS) and inherited metabolic diseases. Mater Child Health Information. 2006;53:39–45. (Article in Japanese).
212. Horiike K. Giving a lecture on metabolic biochemistry. J Jpn Biochem Soc. 2012;84:623. (Article in Japanese).
213. Corkey BE, Hale DE, Glennon MC, et al. Relationship between unusual hepatic acyl coenzyme A profiles and the pathogenesis of Reye syndrome. J Clin Invest. 1988;82(3):782–8. https://doi.org/10.1172/JCI113679.
214. Tonsgard JH, Getz GS. Effect of Reye's syndrome serum on isolated chinchilla liver mitochondria. J Clin Invest. 1985;76:816–25.
215. Tonsgard JH. Serum dicarboxylic acids in patients with Reye syndrome. J Pediatr. 1986;109(3):440–5. https://doi.org/10.1016/s0022-3476(86)80114-7.
216. Mak IT, Kramer JH, Weglicki WB. Potentiation of free radical-induced lipid peroxidative injury to sarcolemmal membranes by lipid amphiphiles. J Biol Chem. 1986;261(3):1153–7.
217. Inoue D, Pappano AJ. L-palmitylcarnitine and calcium ions act similarly on excitatory ionic currents in avian ventricular muscle. Circ Res. 1983;52(6):625–34. https://doi.org/10.1161/01.res.52.6.625.

218. Spedding M, Mir AK. Direct activation of Ca2+ channels by palmitoyl carnitine, a putative endogenous ligand. Br J Pharmacol. 1987;92(2):457–68. https://doi.org/10.1111/j.1476-5381.1987.tb11343.x.
219. Stanley CA, DeLeeuw S, Coates PM, et al. Chronic cardiomyopathy and weakness or acute coma in children with a defect in carnitine uptake. Ann Neurol. 1991;30(5):709–16. https://doi.org/10.1002/ana.410300512.
220. Stumpf DA, Parks JK. Urea cycle regulation: I. Coupling of ornithine metabolism to mitochondrial oxidative phosphorylation. Neurology. 1980;30(2):178–83. https://doi.org/10.1212/wnl.30.2.178.
221. Ah Mew N, Caldovic L. N-acetylglutamate synthase deficiency: an insight into the genetics, epidemiology, pathophysiology, and treatment. Appl Clin Genet. 2011;4:127–35. https://doi.org/10.2147/TACG.S12702.
222. Koizumi T, Nikaido H, Hayakawa J, et al. Infantile disease with microvesicular fatty infiltration of viscera spontaneously occurring in the C3H-H-2° strain of mouse with similarities to Reye's syndrome. Lab Anim. 1988;22(1):83–7. https://doi.org/10.1258/002367788780746511.
223. Kuwajima M, Kono N, Horiuchi M, et al. Animal model of systemic carnitine deficiency: analysis in C3H-H-2° strain of mouse associated with juvenile visceral steatosis. Biochem Biophys Res Commun. 1991;174(3):1090–4. https://doi.org/10.1016/0006-291x(91)91532-h.
224. Horiuchi M, Kobayashi K, Yamaguchi S, et al. Primary defect of juvenile visceral steatosis (jvs) mouse with systemic carnitine deficiency is probably in renal carnitine transport system. Biochim Biophys Acta. 1994;1226(1):25–30. https://doi.org/10.1016/0925-4439(94)90054-x.
225. Horiuchi M, Kobayashi K, Tomomura M, et al. Carnitine administration to juvenile visceral steatosis mice corrects the suppressed expression of urea cycle enzymes by normalizing their transcription. J Biol Chem. 1992;267(8):5032–5.
226. Horiuchi M, Yoshida H, Kobayashi K, et al. Cardiac hypertrophy in juvenile visceral steatosis (jvs) mice with systemic carnitine deficiency. FEBS Lett. 1993;326(1–3):267–71. https://doi.org/10.1016/0014-5793(93)81805-a.
227. Kuwajima M, Horiuchi M, Harashima H, et al. Cardiomegaly in the juvenile visceral steatosis (JVS) mouse is reduced with acute elevation of heart short-chain acyl-carnitine level after L-carnitine injection. FEBS Lett. 1999;443(3):261–6. https://doi.org/10.1016/s0014-5793(98)01732-3.
228. Wu X, Prasad PD, Leibach FH, et al. cDNA sequence, transport function, and genomic organization of human OCTN2, a new member of the organic cation transporter family. Biochem Biophys Res Commun. 1998;246(3):589–95. https://doi.org/10.1006/bbrc.1998.8669.
229. Tamai I, Ohashi R, Nezu J, et al. Molecular and functional identification of sodium ion-dependent, high affinity human carnitine transporter OCTN2. J Biol Chem. 1998;273(32):20378–82. https://doi.org/10.1074/jbc.273.32.20378.
230. Lu K, Nishimori H, Nakamura Y, et al. A missense mutation of mouse OCTN2, a sodium-dependent carnitine cotransporter, in the juvenile visceral steatosis mouse. Biochem Biophys Res Commun. 1998;252(3):590–4. https://doi.org/10.1006/bbrc.1998.9708.
231. Lahjouji K, Mitchell GA, Qureshi IA. Carnitine transport by organic cation transporters and systemic carnitine deficiency. Mol Genet Metab. 2001;73(4):287–97. https://doi.org/10.1006/mgme.2001.3207.
232. Tein I. Carnitine transport: pathophysiology and metabolism of known molecular defects. J Inherit Metab Dis. 2003;26(2–3):147–69. https://doi.org/10.1023/a:1024481016187.
233. Shekhawat PS, Sonne S, Matern D, et al. Spontaneous development of intestinal and colonic atrophy and inflammation in the carnitine-deficient jvs(OCTN2−/−)mice. Mol Genet Metab. 2007;92(4):315–24. https://doi.org/10.1016/j.ymgme.2007.08.002.
234. Nishida N, Sugimoto T, Araki A, et al. Carnitine metabolism in valproate-treated rats: the effect of L-carnitine supplementation. Pediatr Res. 1987;22(5):500–3. https://doi.org/10.1203/00006450-198711000-00003.
235. Murakami K, Sugimoto T, Nishida N, et al. Alterations of urinary acetylcarnitine in valproate-treated rats: the effect of L-carnitine supplementation. J Child Neurol. 1992;7(4):404–7. https://doi.org/10.1177/088307389200700414.
236. Bianchi PB, Lehotay DC, Davis AT. Carnitine supplementation ameliorates the steatosis and ketosis induced by pivalate in rats. J Nutr. 1996;126(11):2873–9. https://doi.org/10.1093/jn/126.11.2873.
237. Paulson DJ, Shug AL. Tissue specific depletion of L-carnitine in rat heart and skeletal muscle by D-carnitine. Life Sci. 1981;28(26):2931–8. https://doi.org/10.1016/0024-3205(81)90269-1.
238. Tsoko M, Beauseigneur F, Gresti J, et al. Enhancement of activities relative to fatty acid oxidation in the liver of rats depleted of L-carnitine by D-carnitine and a γ-butyrobetaine hydroxylase inhibitor. Biochem Pharmacol. 1995;49(10):1403–10. https://doi.org/10.1016/0006-2952(95)00019-v.
239. Fatani AG, Darweesh AQ, Rizwan L, et al. Carnitine deficiency aggravates cyclophosphamide-induced cardiotoxicity in rats. Chemotherapy. 2010;56(1):71–81. https://doi.org/10.1159/000298822.
240. Dambrova M, Makrecka-Kuka M, Vilskersts R, et al. Pharmacological effects of meldonium: biochemical mechanisms and biomarkers of cardiometabolic

activity. Pharmacol Res. 2016;113(Pt B):771–80. https://doi.org/10.1016/j.phrs.2016.01.019.
241. Spaniol M, Brooks H, Auer L, et al. Development and characterization of an animal model of carnitine deficiency. Eur J Biochem. 2001;268(6):1876–87.
242. Spaniol M, Kaufmann P, Beier K, et al. Mechanisms of liver steatosis in rats with systemic carnitine deficiency due to treatment with trimethylhydraziniumpropionate. J Lipid Res. 2003;44(1):144–53. https://doi.org/10.1194/jlr.m200200-jlr200.
243. Schürch R, Todesco L, Novakova K, et al. The plasma carnitine concentration regulates renal OCTN2 expression and carnitine transport in rats. Eur J Pharmacol. 2010;635(1–3):171–6. https://doi.org/10.1016/j.ejphar.2010.02.045.
244. Roberts PA, Bouitbir J, Bonifacio A, et al. Contractile function and energy metabolism of skeletal muscle in rats with secondary carnitine deficiency. Am J Physiol Endocrinol Metab. 2015;309(3):E265–74. https://doi.org/10.1152/ajpendo.00001.2015.
245. Sole MJ, Liew CC. Catecholamines, calcium and cardiomyopathy. Am J Cardiol. 1988;62(11):20G–4G. https://doi.org/10.1016/0002-9149(88)90027-6.
246. Opie LH, Walpoth B, Barsacchi R. Calcium and catecholamines: relevance to cardiomyopathies and significance in therapeutic strategies. J Mol Cell Cardiol. 1985;17(Suppl 2):S21–34. https://doi.org/10.1016/0022-2828(85)90005-7.
247. York CM, Cantrell CR, Borum PR. Cardiac carnitine deficiency and altered carnitine transport in cardiomyopathic hamsters. Arch Biochem Biophys. 1983;221(2):526–33. https://doi.org/10.1016/0003-9861(83)90171-6.
248. Yamashita T, Hayashi H, Kaneko M, et al. Carnitine derivatives in hereditary cardiomyopathic animals. Jpn Heart J. 1985;26(5):833–44. https://doi.org/10.1536/ihj.26.833.
249. Yamashita T, Kobayashi A, Yamazaki N, et al. Effects of L-carnitine and verapamil on myocardial carnitine concentration and histopathology of Syrian hamster BIO 14.6. Cardiovasc Res. 1986;20(8):614–20. https://doi.org/10.1093/cvr/20.8.614.
250. Whitmer JT. Energy metabolism and mechanical function in perfused hearts of Syrian hamsters with dilated or hypertrophic cardiomyopathy. J Mol Cell Cardiol. 1986;18(3):307–17. https://doi.org/10.1016/s0022-2828(86)80413-8.
251. Whitmer JT. L-carnitine treatment improves cardiac performance and restores high-energy phosphate pools in cardiomyopathic Syrian hamster. Circ Res. 1987;61(3):396–408. https://doi.org/10.1161/01.res.61.3.396.

6 Appetite Enhancement and Digestive Enzyme Secretion Promoting Effects of Carnitine

6.1 Introduction

Among the physiological and biochemical effects of carnitine, the effects of carnitine on appetite and digestive enzyme secretion were the first to be clinically applied earlier than carnitine deficiency, although these effects have been forgotten and little attention has been paid to them. In the 1950s, it was actively studied in France, Belgium, Germany, and other countries, and was the earliest clinical application to anorexia in children and postoperative patients. Not only that, but in Japan, tablets, liquids, and injections of DL-carnitine chloride were launched in the 1960s and sold until around 2000. Unfortunately, most of the research on appetite enhancement and digestive enzyme secretion promotion is old research papers written in German and French, but I would like to introduce them in this chapter because I feel that some of them are still applicable to the current clinical practice where the indications for carnitine have been expanded, and some issues should be re-examined in the current science.

6.2 Digestive Enzyme Secretion Stimulating Effects of DL-Carnitine Preparation (Bicarnesine)

As described in Chap. 1: The Discovery of Carnitine, the discovery of acetylcholine in the 1920s and 1930s led to extensive physiological research on carnitine, which has the same quaternary amine structure. In a review article on carnitine written by Fraenkel himself in 1957, an overview of carnitine research at the time was described (following) [1].

In 1926, Komaroff et al. reported that carnitine stimulated intestinal fluid secretion. In 1927, Korchov et al. confirmed this intestinal fluid secretion stimulating action and reported that it also stimulated gastric fluid secretion via parasympathomimetic effects. However, these studies were conducted using "crude carnitine" extracted from muscle, and there was a possibility that impurities such as carnosine, methylguanidine, and betaine crotonate were contaminated, and the effects were observed at very high doses

N. Nakayama, *Carnitine Deficiency and Insufficiency in Medicine*,
https://doi.org/10.1007/978-981-95-7797-2_6

(80–160 mg/kg), so the study was neglected without much attention.

However, in 1948, the entomologist Fraenkel et al. named a new essential factor as vitamin B_T and began insect nutritional studies, and when it was found to be identical to carnitine in 1952, there was renewed interest in the physiological and nutritional actions of carnitine. Dechamps and Binon of the Belgian pharmaceutical company Laboratoires Labaz synthesized a molecule called bicarnesine, in which two carnitine chloride molecules are ester-bonded with each other's hydroxyl groups and carboxylic acids (Fig. 6.1). This bicarnesine is relatively stable and is also called dicarnitine or carnitine carnitate and was used in various physiological studies at the time. Since bicarnesine is hydrolyzed in vivo to form two molecules of carnitine chloride (racemic form), its action is considered to be a physiological effect of carnitine.

Binon et al. reported in 1956 that the pancreatic exocrine fluid contains carnitine and that intravenous administration of this bicarnesine to dogs not only enhanced pancreatic exocrine secretion but also contained more carnitine in the exocrine fluid than when secretin was administered [2].

Charlier has also reported that bicarnesine stimulates the secretion of saliva, gastric juice, pancreatic juice, and bile through parasympathetic stimulation [3–5].

Thus, it was successively reported in animal experiments that bicarnesine, which is hydrolyzed to carnitine when administered in vivo, enhances the secretion of digestive fluids such as saliva, gastric juice, intestinal juice, pancreatic juice, and bile [1].

Unfortunately, most of the research papers on the appetite-stimulating effects of carnitine in the 1950s were in old French or German literature and are not often seen today, but I had the opportunity to read some of them in translation. It seems to be a very useful study for the clinical application of carnitine's digestive enzyme secretion-promoting and appetite-stimulating effects, so it is introduced as below.

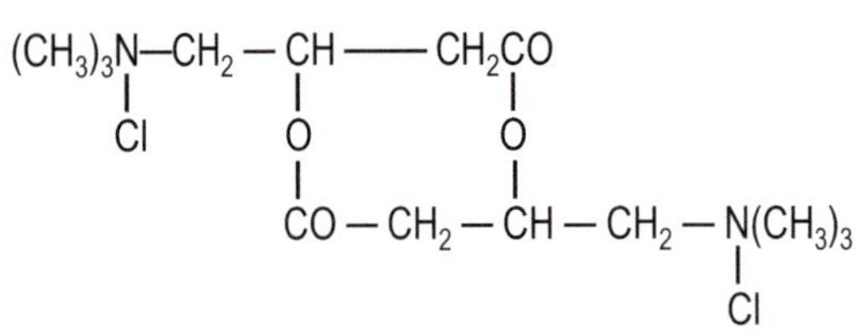

Fig. 6.1 Chemical structure of bicarnesine (dicarnitine, carnitine carnitate). (Modified from Ref. [1])

6.3 Appetite Enhancement and Weight Gain Effects of Bicarnesine

Canlorbe et al. found that bicarnesine stimulated digestive enzyme secretion and showed no toxicity, so they investigated the effects of bicarnesine on anorexia and weight gain in 46 infants, young children, and preterm infants [6]. The results showed that bicarnesine was effective in increasing appetite and weight gain.

Anorexia, in which psychological factors were the definitive cause, was excluded from this study.

6.3.1 Group I: Infants

First, 22 infant subjects (14 girls and 8 boys) between 1 month and 2 years of age were treated with bicarnesine. These subjects were infants hospitalized for infections of various causes and treated with conventional acute care. The subsequent clinical challenge was to promote digestive function, act on post-infectious anorexia, and shorten this period of somewhat prolonged stagnation of weight gain. The dose was 10 mg/kg/day and the duration of administration ranged from 7 days to 2 months. The drug was perfectly tolerated, and there were no cases of apparent side effects that necessitated discontinuation of treatment. A rapid increase in appetite was observed in all 22 patients after bicarnesine administration. As a result, weight gain resumed and accelerated, and the weight curve increased several days after administration in 21 infants who could be followed. This acceleration of weight gain persisted long after administration was discontinued. Table 6.1 summarizes the results for the 22 infant subjects.

Table 6.1 Effects of bicarnesine on 22 cases of infant subjects

No.	Name	Gender	Age (month)	Duration (day)	Dose (mg/day)	Tolerability	Appetite	Action on body weight	Effects
1	Mar.., M.	F	3.5	11	50	Good	Improved	5.175 kg → 5.275 kg	+
2	Les.., J.	F	7	7	60	>	>	6 kg → 6.470 kg	+
3	Le Man.., M.	F	1.5	13	100	>	>	Accelerate BW increase	+
4	Lab.., G.	M	9.5	10	80	>	>	8.400 kg → 8.550 kg	+
5	Bon.., D.	F	0.5	60	40	>	>	3.010 kg → 4.100 kg	+
6	Bon.., M.	F	2.5	17	40	>	>	During study 4.440 kg → 4.780 kg, later 5.130 kg	+
7	Tab.., M.	F	0.2	15	45	>	>	During study 4.300 kg → 4.400 kg, later 5.080 kg	+
8	Por.., G. C.	M	1	35	30	>	>	3.330 kg → 4.330 kg	+
9	Pev.., A.	F	3	38	40	>	>	4.620 kg → 6.020 kg	+
10	Duq.., G.	M	0.5	60	30	>	>	3.225 kg → 4.740 kg	+
11	Gruq.., G.	M	4.5	24	50	>	>	4.500 kg → 5.075 kg	+
12	Dal.., V.	F	7	22	50	>	>	5.750 kg → 6.370 kg	+
13	Ruf.., M. G.	F	6	29	50	>	>	5.390 kg → 5.970 kg	+
14	Ran.., G.	M	8.5	16	50	>	>	6.700 kg → 6.930 kg	+
15	Gaz.., C.	F	7	20	50	>	>	5.300 kg → 5.580 kg	+
16	Bug.., M.	F	3	15	50	>	>	Accelerate BW curve from final 3 days	±
17	Dou.., M.	M	12	25	60	>	>	5.800 kg → 6.270 kg continuous BW increase after discontinuation of administration	+
18	Le Van.., B.	F	2.5	30	50	>	>	BW increase 3.850 kg → 4.200 kg	+
19	Abd.., M. G.	M	3.5	36	50	>	>	5.400 kg → 5.930 kg	+
20	Cel.., L.	F	1	30	50	>	>	Enough BW increase 5.350 kg → 6.740 kg	+
21	Caz..,	F	1	33	35	>	>	Enough BW increase	+
22	Man..,	M	12	30	60	>	>	Significant acceleration of BW increase	+

Modified from Ref. [6]

Notes: It is not clear the meaning of ">" in "Tolerability" and "Appetite," but maybe same as Case No. 1's "good" and "improved"

M male, *F* female, *BW* body weight

6.3.2 Group II: Children 2 Years and Older

Next, 19 subjects (11 boys and 8 girls) aged 2 years and older were examined. These subjects were children who had been followed up due to growth retardation and were underweight for their height. The mean dose of bicarnesine was 10–15 mg/kg/day, and the duration of treatment ranged from 2 to 7.5 months, depending on the case. Again, bicarnesine was extremely well tolerated, with only two cases of mild, transient disturbances of dizziness, headache, or nausea. Increased appetite was particularly marked in 13 of the 19 patients, and accelerated weight gain was observed in almost half (9 patients). Furthermore, the daily dose of bicarnesine was probably inadequate in those cases in which weight gain was not clearly observed. In fact, during the course of the study, a total of 200 mg/day was considered inadequate in certain cases, and 400 mg/day of bicarnesine was able to induce weight gain. Table 6.2 summarizes the results in 19 pediatric subjects aged 2 years and older.

6.3.3 Group III: Preterm Infants

Finally, bicarnesine was administered to five preterm infants with birth weights of 1.320, 1.690, 1.800, 0.920, and 0.980 kg, respectively (see Table 6.3).

An average dose of 10 mg/kg/day was started at 1 month of age and continued for 10–15 days. These preterm infants showed no signs of tolerability problems: appetite was restored in four of five cases and weight gain was accelerated in three cases. Of course, the limited number of cases of preterm infants and the short observation period did not allow definitive conclusions to be drawn, but it did suggest the possibility of clinical application of bicarnesine. Canlorbe's group then studied the effects of bicarnesine in 12 malnourished children (1–8 years old) after infection [7]. All of these patients were hypoalbuminemic and had significant weight loss, but after 2–5 months of treatment with bicarnesine solution, albumin levels increased, and weight gain resumed.

Alexander selected 65 children of similar age, size, and nutritional status from more than 200 children living at the Le Coq-sur-Mer Marin Tuberculosis Prevention Clinic and divided them into two groups: a bicarnesine-treated group and a bicarnesine-naive (control) group and studied the effect of bicarnesine on body weight gain [8]. The two groups were further classified into the following two groups.

1. Severe weakness group: Severe weakness or severe anorexia (excluding "nervous" anorexia).
2. Early tuberculosis group: Cases of early tuberculosis. Children with tracheobronchial lymphadenopathy, cervical adenitis, erythema nodosum, and other symptoms of primary pulmonary tuberculosis infiltration.

Children in the treatment group received bicarnesine 300 mg/day in two doses two minutes before each meal (30 drops of 20% commercial solution diluted with a small amount of water). Table 6.4 shows the classification of each group and the number of cases per age group. Although the number of patients in the control and bicarnesine groups was not necessarily homogeneous, a total of 65 patients were examined for changes in body weight over a 4-month period. Figure 6.2 shows the 4-month weight change for each age group in the severely frail (left side) and early tuberculosis (right side) bicarnesine and control groups, respectively. There were no problems with tolerability. Therefore, the dose could have been increased for the 6-year-olds, but all were given 300 mg/day. The difference in weight gain between the bicarnesine and control groups observed in this study was striking and quite surprising [8]. Furthermore, these results are consistent with those of Canlorbe et al. in infants, children, and preterm infants.

Based on these results, Alexander considered that bicarnesine solution (carnitine chloride) is extremely easy to administer to children, even at

Table 6.2 Effects of bicarnesine on 19 infant subjects older than 2 years old

No.	Name	Gender	Age	Duration (month)	Dose (mg/day)	Tolerability	Appetite	Action on BW	Effects
1	Nor.., G.	M	14 y	4	300	Good	Almost no change	No change in BW curve 40 kg → 40.550 kg	–
2	Per.., G.	M	9 y	2 1/2	200	>	Improved	21.930 kg → 22.500 kg Namely +500 g	+
3	Bra.., G.	M	14 y, 10 m	3	300	>	Improved	34.800 kg → 36.400 kg	+
4	Ste.., G.	M	8 y, 9 m	5	200, then 100	>	No change	20.250 kg → 21.550 kg	±
5	Fou.., M.	M	15 y, 5 m	3.5	300	>	Improved	Not confirmed	±
6	Hus.., M.	F	11 y	2.5	250	>	No change	No BW increase	–
7	Con.., O.	F	13 y	4	300	>	Increased	Significant action on BW: BW increase to height + 5% → +12%	+
8	Lo Cos.., F.	M	11 y, 9 m	5	200, then 300	>	Increased	Upward change in BW curve:20.400 kg → 24.300 kg	+
9	Mas..,	M	12 y 6 m	21 (days)	300	>	Increased	Not obvious	±
10	Let..,	M	14 y	5	300	>	No change	35.700 kg → 38.850 kg: BW increase to height + 6% → +12%	+
11	Paq..,	M	2 y, 6 m	7.5	100, then 200	>	Increased	No change in BW increase rate	–
12	Kub..,	M	6 y	6	200, then 400	>	Increased	Accelerated BW curve, deviation of BW to height – 17% → -6%	+
13	Lef..,	F	9 y, 6 m	3.5	300, then 400	>	Increased	No change in BW increase rate	±
14	Gos..,	M	4 y, 6 m	2	200	>	No change	No steady action on BW	–
15	Meg..,	F	6 y	2	500	>	Increased	Mean BW increase 19.500 kg → 20.100 kg	+
16	Mel..,	F	9 y	2	300	>	Increased	No change in BW increase rate	–
17	Boa..,	F	16 y, 7 m	2	300	>	No change	No action on BW	–
18	Ron..,	F	12 y, 6 m	3	300	>	Increased	Slight BW increase 27 kg → 27.700 kg	±
19	Bue..,	M	10 y	4	300	>	Increased	No change in BW curve	±

Modified from Ref. [6]

Note: It is not clear the meaning of ">" in "Tolerability" and "Appetite," but maybe same as the case No. 1's "good" and "improved"

M male, *F* female, *y* year, *m* month, *BW* body weight

Table 6.3 Effects of bicarnesine on preterm infants

No	Name	Gender	BW at birth	Age (month)	Duration (day)	Dose (mg/day)	Tolerability	Appetite	Actions on BW curve	Effect
1	Le Rol..,	M	1.380 kg	1	25	25	Good	Improved	No acceleration in BW increase	–
2	Gar..,	M	1.690 kg	1.5	27	50	>	Increase	Quite significant acceleration of BW increase	+
3	Mol..,	M	1.800 kg	1	10	25	>	Increase	Quite significant acceleration of BW increase	+
4	Zam..,	F	920 g	1.5	45	20, then 40	>	Enough	Quite significant acceleration of BW increase	+
5	Ver..,	M	980 g	1	33	25	>	Good	No change in acceleration of BW increase	–

Modified from Ref. [6]

Note 1: It is not clear the meaning of ">" in "Tolerability," but maybe same as Case No. 1's "good"

Note 2: The birth weight of case No. 1 is 1.380 kg in Table 6.3, but it is described in the text as 1.320 kg

M male, *F* female, *BW* body weight

Table 6.4 Classification of infant cases and number of cases by age

	Severe frail		Early stage of tuberculosis	
Age	Group I (Bicarnesine)	Group II (Control)	Group III (Bicarnesine)	Group IV (Control)
6 years	–	–	1	1
5 years	1	3	3	4
4 years	2	4	10	8
3 years	3	3	7	6
2 years	–	–	5	4

Modified from Ref. [8]

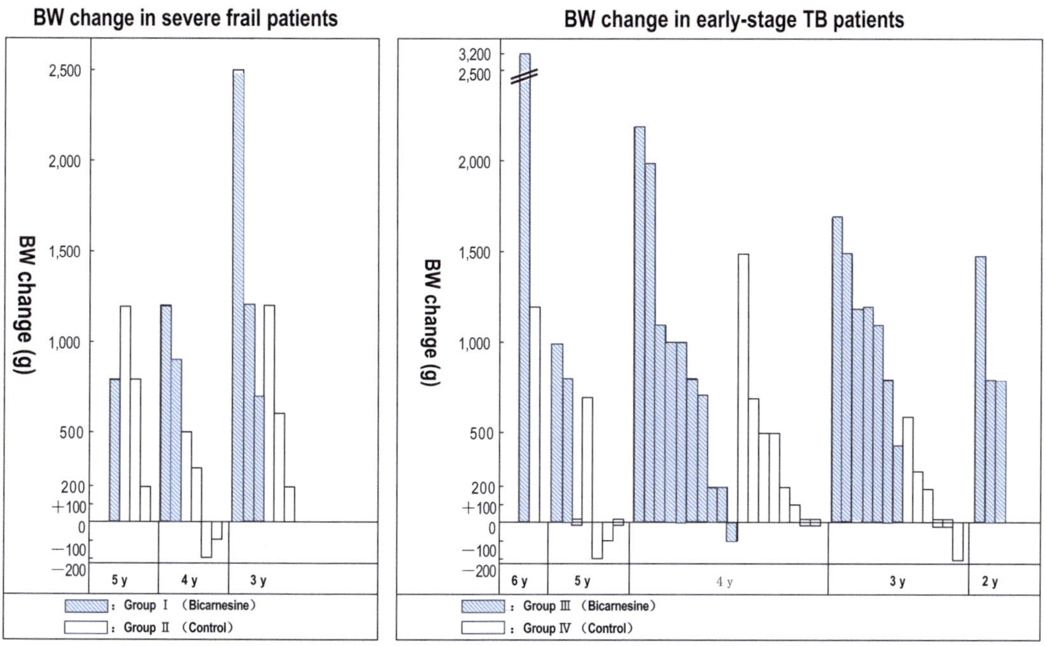

Fig. 6.2 Body weight changes during 4 months in severe frail patients (left) and early-stage TB patients (right) after bicarnesine administration or control. *BW: body weight, y: years, TB: tuberculosis*. (Modified from Ref. [8])

very young ages, due to its stability, tolerability, and organoleptic characteristics, and that it has not reported any side effects and is significantly superior to many other neurostimulants and other preparations that do not have long-lasting effects.

Bicarnesine was positioned as a drug that stimulates digestive function and accelerates weight gain in children and infants with anorexia, stunting of growth, and malnutrition.

6.4 Digestive Secretion Promotion Effect of Bicarnesine, D-Carnitine and L-Carnitine

Aubert et al. ligated the gastric pylorus of anesthetized dogs, administered bicarnesine intravenously, and collected gastric juice using a sonde to determine gastric juice volume and acidity, finding that gastric juice secretion was markedly increased at 5 mg/kg bicarnesine [9]. Many investigators have reported that all of these gastric juice secretion promotions by bicarnesine were inhibited by prior atropine administration. Therefore, these gastric juice secretory effects have parasympathetic aspects, strongly suggesting the involvement of vagal nerve stimulation. Next, they compared the effects of bicarnesine (10 mg/kg, administered by gastric sonde) on gastric juice secretion with those of histamine (1 mg, administered subcutaneously) in 10 healthy subjects, and found that bicarnesine markedly increased gastric juice secretion, but slightly less than that of histamine 1 mg (Table 6.5).

In addition, 12 volunteer subjects without gastrointestinal diseases were administered 200 mg of 10% bicarnesine solution from a duodenal sonde, and the effects on the increase in pancreatic juice volume and lipase and trypsin activity in pancreatic juice were examined. In 11 of the 12 patients, a significant increase in pancreatic exocrine was observed, trypsin activity was increased in 6 patients, and lipase activity was increased in 7 patients, not related to each other (Table 6.6). In five other subjects, the action of bicarnesine was compared with that of secretin, and the activity of 200 mg of bicarnesine transduodenally showed almost the same activity as five units of secretin injected intravenously. Bicarnesine was not thought to enhance the action of secretin, nor was it thought to act through its intermediates, and Aubert et al. considered that the explanation for the mechanism of nutritional effects such as increased appetite and promotion of weight gain in infants and children reported by Canlorbe et al. may be due to the secretion of digestive juices such as gastric juice and pancreatic exocrine juice of bicarnesine, which have been reported by many researchers and confirmed by themselves. He also suggested that bicarnesine may have clinical applications in gastrointestinal diseases and nutritional disorders [9].

When many French and Belgian clinicians reported that bicarnesine had a hypersecretory effect on digestive juices and nutritional effects such as increased appetite and weight gain, the question arose as to whether D-carnitine chloride or L-carnitine chloride, a component of bicarnesine, was responsible for these effects.

Kunze et al. at the Institute for Physiological Sciences, Karl Marx University in Leipzig compared and reported the effects of synthesized pure D- and L-carnitine (0.2 g/kg) on pancreatic juice secretion in rabbits [10]. They placed a transperitoneal sonde 5 cm distal to the pancreatic duct opening of the duodenum of rabbits, administered the test drug intravenously, and collected duodenal secretions over time. As a positive control, Cecekin (0.8 mg/kg, Vitrum, Sweden) containing cholecystokinin, secretin, and pancreozymin was used.

As a result, it was found that D- and L-carnitine enhanced the secretion of duodenal fluid, although slightly weaker than that of Cecekin, and that the response of the L-body was later than that of the D-body (Fig. 6.3). Furthermore, when the lipase, trypsin, and amylase activities of duodenal fluid were examined, the L-form (solid line in the figure) was weaker than that of Cecekin (dotted line in the figure), but the lipase activity increased by about 14 times and the trypsin activity increased by about 7 times, but only a slight increase was observed in the D-form (dashed line in the figure). In the case of amylase, both D- and L-form carnitine showed a fivefold increase in activity. The secretion-promoting effect of digestive enzymes in the L-body appeared about 30 min later than that of the D-body in both enzymes (Fig. 6.4). Kunze considered that the reason why the action of the L-body is delayed is that L-carnitine does not directly promote the secretion of digestive enzymes but may increase the secretion of digestive enzymes through metabolic promoting effects. Although there is not much difference between the D- and L-forms of

Table 6.5 Changes in gastric fluid volume, free acidity, and total acidity after administration of histamine or bicarnesine in 10 subjects without gastric disorders

	Histamine						Bicarnesine					
	Before administration			After administration			Before administration			After administration		
Subjects	Volume	Free acidity	Total acidity	Volume	Free acidity	Total acidity	Volume	Free acidity	Total acidity	Volume	Free acidity	Total acidity
I	10	25	36	107	++	++	25	40	71	130	+	—
II	15	0	11	131	+++	++	17	0	1	117	++	+++
III	10	0	40	109	++	++	18	6	31	73	+++	+++
IV	7	34	42	40	+++	+++	2	—	—	16	—	—
V	5	0	2	32	0	+	5	0	5	16	0	++
VI	28	0	13	200	+++	+++	20	20	25	127	++	++
VII	10	0	2	56	+++	+++	5	28	31	107	++	+
VII	30	70	90	230	+++	+++	20	37	42	77	++	++
IX	8	0	15	162	+++	+++	12	0	7	85	+++	+++
X	15	0	20	233	+++	++	9	9	17	128	++	++

Modified from Ref. [9]

Table 6.6 Changes in pancreatic juice volume, pancreatic lipase and trypsin activity in healthy subjects administered bicarnesine

Subjects	Change in pancreatic juice volume	Increase in pancreatic lipase activity	Increase in pancreatic trypsin activity
I	+++	+++	+
II	0	0	0
III	++	0	++
IV	+++	++	0
V	+	0	+
VI	++	++	0
VII	++	++	+
VII	+	0	+
IX	+++	+++	0
X	+++	+++	0
XI	++	++	0
XII	++	0	++

Modified from Ref. [9]

amylase, this is the first time that L-carnitine has been proven to have a strong effect on the secretion of lipase and trypsin.

In addition, Kunze et al. administered L-carnitine and D-carnitine (2 g, intraduodenum) to 13 patients with gastrointestinal disorders who did not have pancreatic disease but had various symptoms in the abdomen and examined the amount of duodenal fluid secretion and the lipase, amylase, and trypsin in them [11] (Fig. 6.5). As a result, the volume of duodenal fluid increased by 3 times on average after administration of 2 g of L-carnitine, but the enzyme activity in the solution was basically the same, and therefore the absolute release of the enzyme was significantly increased. D-carnitine showed only a slight stimulatory effect on secretion.

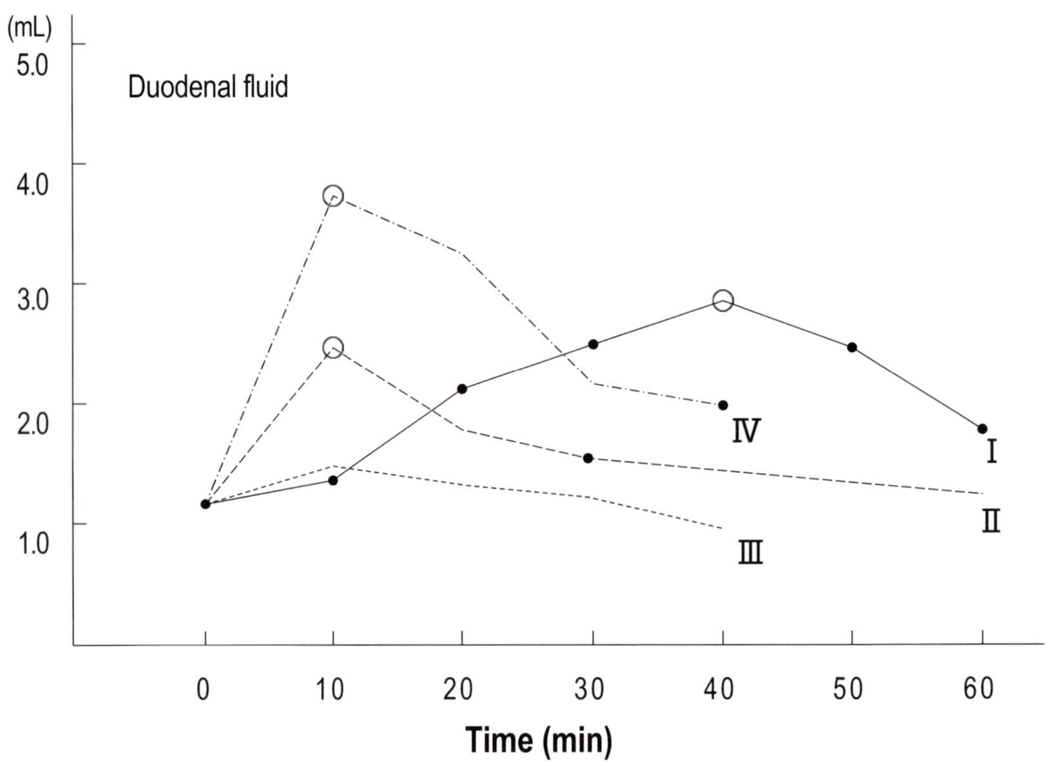

Fig. 6.3 Changes in duodenal fluid volume after intravenous administration of L-carnitine (0.2 g/kg, I: solid line), D-carnitine (0.2 g/kg, II: broken line), NaCl solution (0.9%, III: dotted line) and Cecekin (0.8 mg/kg, IV: dash-dotted line). (Modified from Ref. [10])

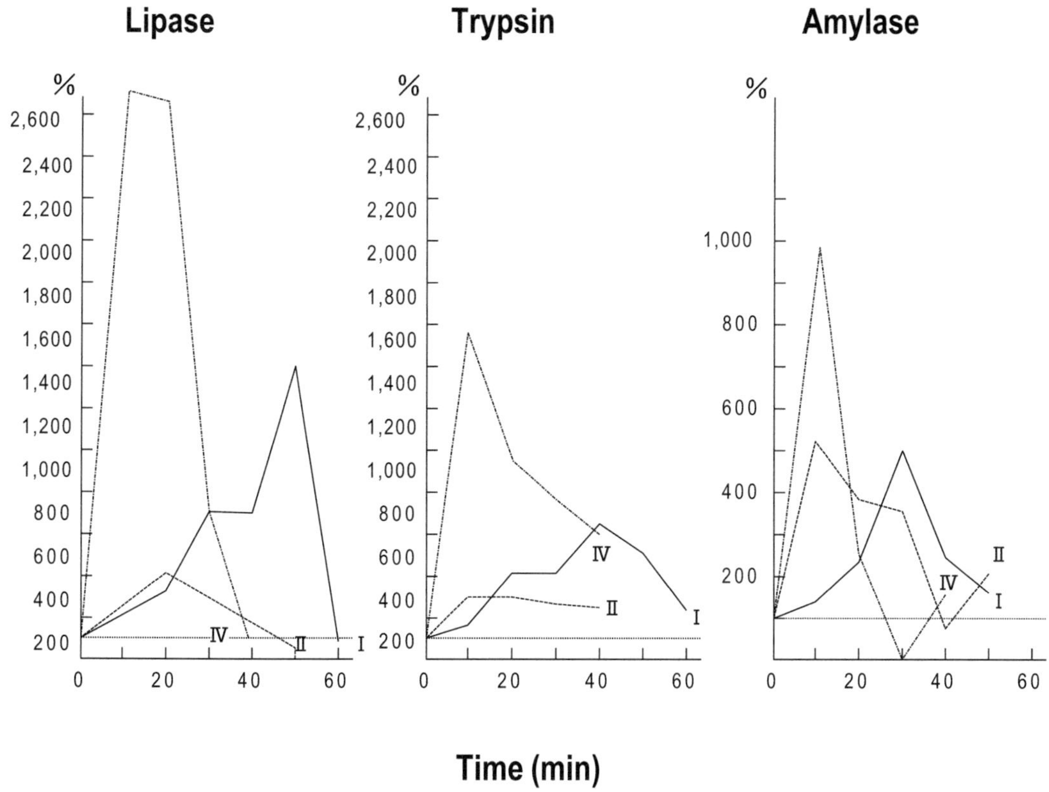

Fig. 6.4 Changes of enzyme secretion in duodenum fluid after administration of L-carnitine (I), D-carnitine (II), Cecekin (IV). The horizontal dotted lines indicate the levels of 0.9% NaCl solution (100%). (Modified from Ref. [10])

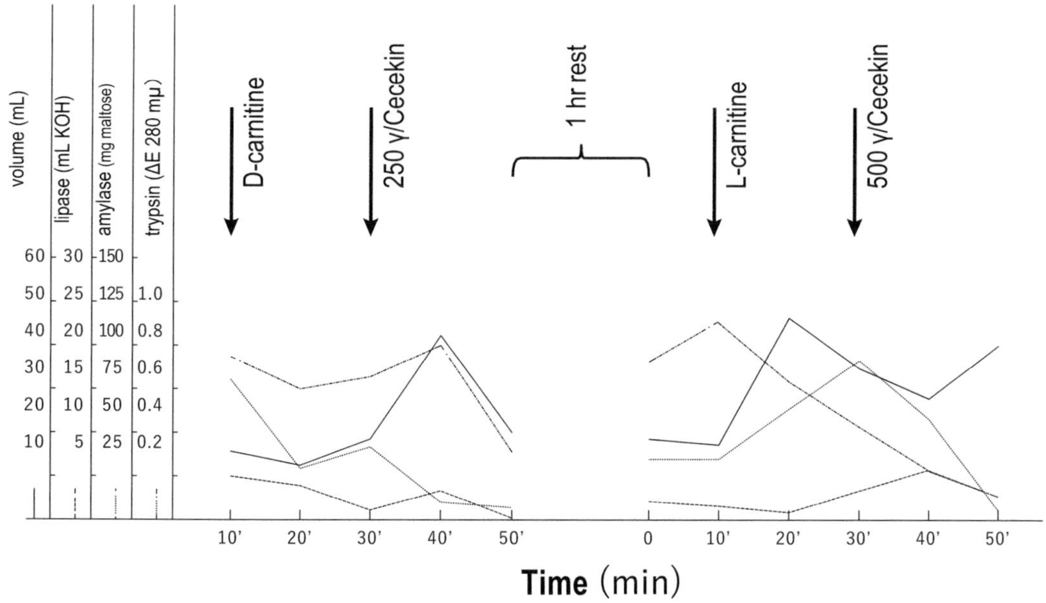

Fig. 6.5 Changes in duodenal fluid volume, activities of lipase, amylase and trypsin after administration of L-carnitine D-carnitine (2 g, intra-duodenum) in 13 digestive diseases patients without pancreatic diseases. Solid line: duodenum fluid volume, broken line: lipase, dotted line: amylase, dash-dotted line: trypsin. (Modified from Ref. [11])

However, Bender et al. reported that these effects of carnitine on growth and stimulation of gastric enzymes secretion were not confirmed in rat and toad [12].

Thus, carnitine's parasympathomimetic action, which promotes digestive fluid secretion and gastrointestinal motility, was reported to be useful in pediatric patients with anorexia in France, Belgium, and Germany in the 1950s and 1960s, and was applied clinically. The Italian company Sigma Tau (now Alphasigma S.p.A.), the world leader in the development of carnitine products for carnitine deficiency, marketed a syrup formulation of DL-carnitine chloride from 1961 to 1981 for the treatment of stunted growth and anorexia in children. In 1981, the U.S. Food and Drug Administration (FDA) banned carnitine products containing the D-body due to side effects of the D-body, and the company switched to levocarnitine chloride, the L-body.

According to data obtained by the author in 2010 from Sigma Tau S.p.A. of Italy, which markets pharmaceutical carnitine products in 38 countries worldwide, most of the indications are primary and secondary carnitine deficiency, and no indications related to improving anorexia or promoting digestive enzyme secretion were identified in any of these countries. I cannot find out the reason for the indications in European countries for appetite enhancement and digestive enzyme secretion promoting effects of carnitine had disappeared.

6.5 DL-Carnitine Chloride Preparation in Japan

In this way, the clinical application of DL-carnitine chloride to promote digestive enzyme secretion and improve appetite has been reported from France, Belgium, and Germany, and several similar studies have been reported in Japan.

In 1961, Abe et al. administered carnitine chloride (600 mg/day, 3–4 weeks) to 14 patients (18–57 years old, mean age 40 years) with anorexia, pulmonary tuberculosis, gastritis and gastroptosis, and examined its effects on anorexia and weight gain [13]. As a result, anorexia returned to normal in 5 patients, improved in 3 patients, and showed little improvement in 6 patients. In addition to these anorexic patients, the effects on weight gain were also examined in a total of 20 patients, including patients who had lost weight, but no significant effects were observed except for a weight gain of about 1 kg in 9 patients.

Thus, the gastroprokinetic action of DL-carnitine has been reported in Japan, and although the details and basis of the clinical data submitted for approval are not well known, a number of DL-carnitine products have been marketed in Japan since around 1962, including Avedine tablets (100 mg) and its liquid formulation by Nippon Zoki Seiyaku, and Monokamine tablets (100 mg) by Tanabe Seiyaku Co. DL-carnitine preparations were marketed in Japan around 1962 for such indications as anorexia, gastric juice secretion disorder, constipation, and acid-free and low-acidity (Table 6.7).

According to the 1974 edition of the Japanese Pharmacopoeia (Ethical Drugs), the "action" of DL-carnitine chloride preparations is described as "stimulating the parasympathetic nervous system, increasing the concentration of digestive enzymes and acids in the digestive juices, and promoting gastric juice secretion." In the 1984 re-evaluation of pharmaceuticals, a total of 23 items, including DL-carnitine chloride Entomin (10 items such as powders, tablets, liquids, injections, and granules), Avedaine (tablets, liquids, and granules), Monocamine (tablets, powders, granules, liquids, syrups), and Monocamine granules of pyrophosphate were indicated as "chronic gastritis with decreased gastrointestinal function" could be presumed to be "effective" and were judged to be "those that are recognized as useful" in the overall evaluation judgment [14]. However, demand for Monocamine declined around 1996 and Abedyne around 2004,

Table 6.7 Pharmaceutical products of DL-carnitine sold in Japan in the past

Pharmaceutical products of DL-carnitine	Pharmaceutical company
Abedine (tablets, liquid, granules)	Nippon Zoki
Monocamin (tablets, powder, granules, liquid, syrup)	Tanabe
Entomin (tablets, powder, granules, liquid, injection)	Maruko (now Nichi-Iko)
Carnitan (granules)	Kaken Chemical
Vitaminbete (tablets, granules)	Meiji
Rhythm (tablets, liquid)	Toyama
Pasron N (granules), Pannitin (tablets)	Choseido
Abedamine (tablets)	Hokuriku
Admoresan (syrup, injection)	Dojin
Polycartine (tablets, granules, injection)	Daigo Nutrition
Beten (tablets)	Homei (oversea)
Carnitine chloride SS (syrup)	SSP
Carnitine chloride tablets (tablets)	Nihon Vitamin Chemical

Modified from Ref. [9]
These products were classified as stomach digestive aids, and their indications were loss of appetite, gastric fluid secretion disorders, constipation, hypochlorhydria, or achlorhydria. Most of these DL-carnitine chloride products were launched in 1960s, but the sales were discontinued (Monocamine in 1996, and Abedine in 2004). In the case of Entomin, most of the product sales were discontinued in 1998 except intravenous formulation. So only Entomin injection (200 mg) is still on the market as digestive aid for chronic gastritis patients with decreased digestive functions
Note: The spelling of these products is uncertain except Monocamine, Abedine, and Entomin, because they are sold only in Japan and English name is not available now

and most of the products were discontinued and drug prices were deleted. All products are DL-form (racemic) carnitine chloride preparations. Among them, Marco Pharmaceutical Co., Ltd. had been selling DL-carnitine chloride in many dosage forms such as powders, tablets, liquids, injections, and granules as a digestive function stimulant since 1962, and it passed the re-evaluation of drugs in 1984, but around 1998 the drug price was removed except for the injectable drug, and for some reason only the injectable drug still remains. This product was marketed by Nichi-Iko Co., Ltd. in 2004, and the product name was changed to "Entomin Injection 200 mg" in 2007 [15]. The indication and effect of this Entomin are "chronic gastritis with decreased gastrointestinal function." In the world, the use of DL-form (racemic) carnitine preparations containing D-carnitine was banned in 1981 due to side effects such as myasthenia caused by the D-body, but inexplicably DL-form carnitine preparations (and injections that enter the entire amount into the blood) are still approved and sold in Japan.

6.6 Gastrointestinal Motility Disorders, Constipation and Carnitine Deficiency

Severely mentally and physically disabled patients are characterized by decreased peristalsis of the intestinal tract due to bedridden living and other factors and are prone to severe constipation and ileus [16].

Weaver et al. examined a child on a soy-based liquid diet due to growth retardation and peripheral myopathy who presented with symptoms of postprandial vomiting, delayed gastric contents evacuation, and decreased intestinal motility by esophageal manometry, and found decreased gastrointestinal motility and decreased blood carnitine concentration (9.9 μ mol/L). When carnitine was administered to the patient, gastrointestinal motility improved markedly [17].

Murata et al. divided 27 severely mentally and physically handicapped patients with severe intellectual and motor disabilities into a severe constipation group (14 patients) and a non-constipation group (13 patients) and measured

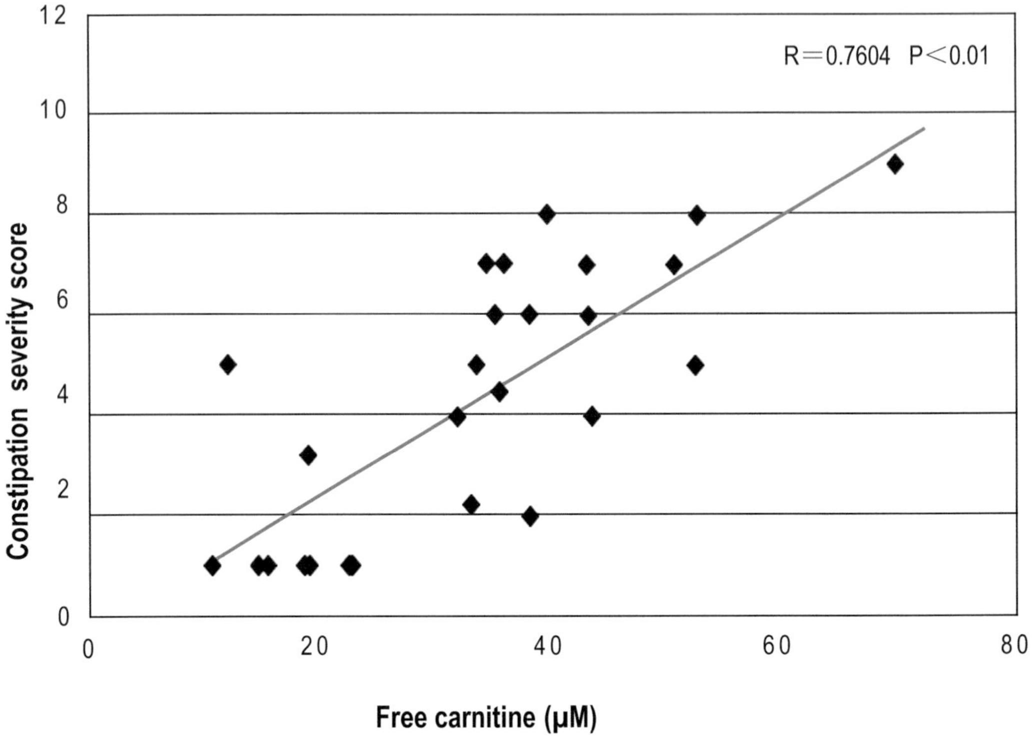

Fig. 6.6 Correlation between constipation severity score and blood free carnitine concentration in patients with severe motor and intellectual disabilities. Constipation severity score: frequency of defecation/a day +Bristol stool score. (From Ref. [18])

blood carnitine levels and patient background and nutritional indices [18]. Of the 14 patients with constipation, 11 were taking valproate, 9 were receiving nutritional management by tube feeding, and 6 were prone to carnitine deficiency. They found a positive correlation when plotting the relationship between blood carnitine concentration and constipation severity score (number of bowel movements per day + Bristol stool scale) in 27 patients ($R = 0.7604$, $P < 0.01$, Fig. 6.6). Therefore, when carnitine (10–50 mg/kg/day) was administered to 10 patients in the constipation group, the number of bowel movements per day and the Bristol stool scale were significantly improved ($P < 0.001$, Fig. 6.7). These results suggest that carnitine deficiency is one of the causes of severe defecation disorders in people with severe physical and mental disabilities. Murata et al. concluded that various factors can be considered as the cause of severe constipation in people with severe physical and mental disabilities, but that carnitine deficiency should also be considered, and that carnitine deficiency may also be a cause of laxative-resistant defecation in the elderly and bedridden patients, and that future studies are needed [18]. Also, it is necessary to further investigate whether the effect of carnitine on gastrointestinal motility is due to the improvement in fatty acid metabolism or the effect of enhancing gastrointestinal motility and promoting the secretion of digestive enzymes.

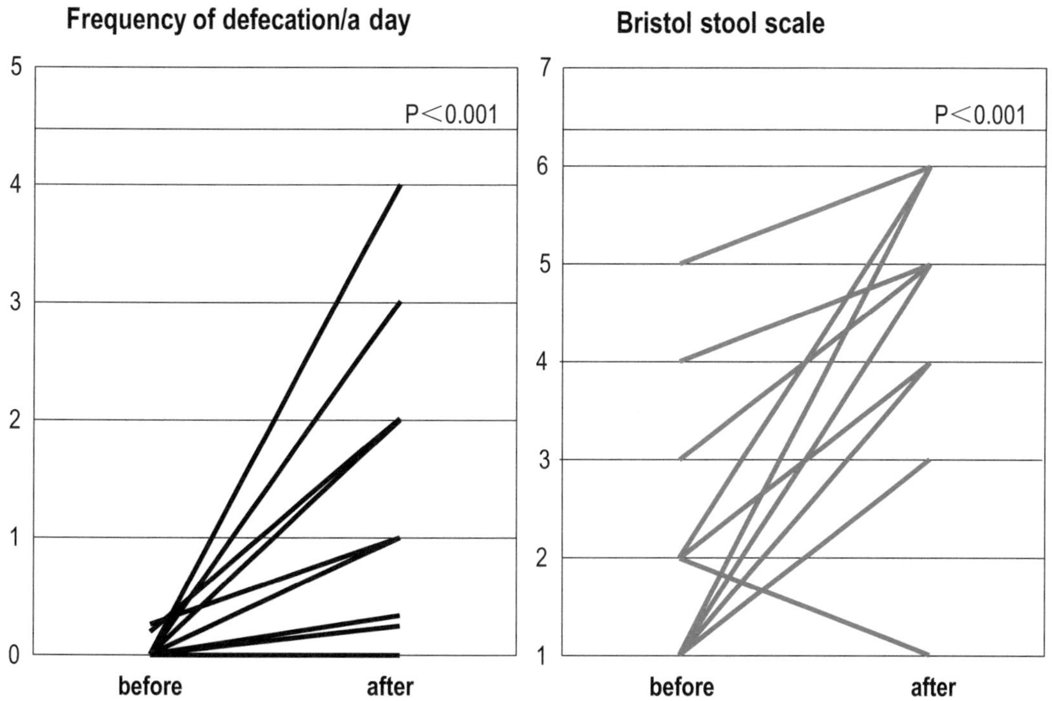

Fig. 6.7 Frequency of defecation/a day (left) and Bristol stool score (right) before and after supplementation of L-carnitine (10–50 mg/kg/day) in patients with severe motor and intellectual disabilities. (From Ref. [18])

6.7 Orexin and Carnitine

Orexin has recently been discovered as a neurotransmitter in the brain involved in the regulation of appetite and feeding activity, as well as the regulation of wakefulness and sleep [19]. In JVS mice, an animal model of systemic carnitine deficiency, orexin neuronal activity is decreased during weaning and a high percentage of the animals die due to hypoglycemia, etc. However, it has been reported that rearing with carnitine-rich feed releases the decreased orexin expression and improves death due to hypoglycemia [20, 21]. Although it is unclear whether the decrease in orexin expression affects the awake time and reduces feeding activity and hypoglycemia, it has been observed that JVS mice with carnitine deficiency do not eat much even when fed.

Orexins are also involved in the regulation of wakefulness and sleep. Miyagawa et al. identified CPT1B as a susceptibility gene based on genome-wide association analysis of abnormal carnitine metabolism in narcoleptic patients and a randomized, double-blind, placebo-controlled study of L-carnitine administration in patients with narcolepsy and reported a significant decrease in daytime dozing time in narcoleptic patients [22–27]. Although carnitine's appetite-enhancing effect has been thought to be due to its ability to stimulate digestive enzyme secretion, new research on the relationship between carnitine deficiency and orexin and on the regulatory mechanisms of sleep and wakefulness in anorexia, narcolepsy, and other difficult-to-treat diseases awaits further development.

6.8 Remaining Issues in Appetite Enhancement and Digestive Enzyme Secretion Promoting Effects of Carnitine

Although it has been shown in animal experiments and clinical studies by Kunze et al. that L-carnitine itself has a digestive enzyme secretion promoting effect, most of the overseas clinical reports on appetite-stimulating and weight-gaining effects in the 1960s are based on bicarnesine, which is a two-molecule combination of DL-carnitine chloride. It is not clear whether L-carnitine itself has a digestive secretion-enhancing effect, an appetite-promoting effect, or a weight-gain promotion effect on stunted growth, as there are few clinical studies other than the study of Kunze et al.

Domestic clinical studies are also based on DL-carnitine chloride, and there is no data on L-carnitine chloride or L-carnitine in free form. According to a study by Kunze et al., D-carnitine and L-carnitine differ in the onset time of digestive enzyme secretion, and in the former case, the effect appears immediately after administration, while in the case of L-carnitine, it is observed after 30 to 40 min. The reason for this difference in action remains to be clarified.

References

1. Fraenkel G, Friedman S. Carnitine. Vitam Horm. 1957;15:73–118. https://doi.org/10.1016/s0083-6729(08)60508-7.
2. Binon F, Deltour G. Pancreatropic action of carnitine. Experientia. 1956;12(9):357. https://doi.org/10.1007/BF02165354. (Article in French).
3. Charlier R. Action of dicarnitine on the cardiovascular system on some digestive secretions. Arch Int Pharmacodyn Ther. 1954;98(2):251–4. (Article in French).
4. Charlier R. Effects of bicarnesine on vagal excitability. CR Seances Soc Biol Fil. 1955;149(9–10):934–7. (Article in French).
5. Charlier R. About the pharmacology of bicarnesine. Arch Int Pharmacodyn Ther. 1956;106(1–2):184–98. (Article in French).
6. Canlorbe P, Deltour G, Borniche P, et al. Therapeutic trials of a stable carnitine derivative. Its action on digestive functions and weight growth. Ann Pédiatr Semaine Hospitaux. 1956;32:276–9. (Article in French).
7. Borniche P, Brandel M, Canlorbe P, et al. Effect of a synthetic carnitine compound on the blood during therapy of post-infectious anorexias in childhood. Sem Hop. 1958;34(49/10):2515–7. (Article in French).
8. Alexander F. Effect of a carnitine derivative on digestive functions and weight increase in young children in preventoria. Acta Paediatr Belg. 1956;10(4):170–5. (Article in French).
9. Aubert P, Andre J, Dallemagne MJ, et al. Action of a carnitine derivative on digestive secretions. Therapie. 1957;12(3):430–7. (Article in French).
10. Kunze D, Rotzsch W, Strack E. On the effect of carnitine on pancreatic secretion. Dtsch Z Verdau Stoffwechselkr. 1962;22:97–101. (Article in German).
11. Kunze D, Drechsler R, Rotzsch R, et al. On the secretory effect of (−)-carnitine on the human pancreas after enteral application. Dtsch Z Verdau Stoffwechselkr. 1963;23:137–42. (Article in German).
12. Bender AE, Adams EP. An investigation of suggested physiological function of carnitine. Biochem J. 1962;82(2):232–6. https://doi.org/10.1042/bj0820232.
13. Abe T, Abei T, Abei R. Usage experience of Monocamine (carnitine chloride). J New Remedies Clin. 1961;10(3):197–9. (Article in Japanese).
14. Ministry of Health, Labour and Welfare: Pharmaceutical Bulletin No. 385. Results of Drug Reevaluation and Measures Based on the Results – Part 22. Evaluation results of gastrointestinal agents 9,2. Carnitine salts. June 1, 1959, http://www.fpmaj-saihyoka.com/efficacy/data/05/05_19840601-385.pdf (Article in Japanese).
15. Nichi-Iko Co., Ltd.: Entomin ® Note 200 mg Interview Form 10th Edition, Revised March 2019. Japan Standard Product Classification Number 872336. https://www.nichiiko.co.jp/medicine/file/45500/interview/45500_interview.pdf. (Article in Japanese).
16. Tokumitsu A. Proper use of enteral nutrition in consideration of the characteristics of severely mentally and physically handicapped children. J Soc Severe Mental Phys Disabil. 2017;42(1):35–43. (Article in Japanese).
17. Weaver LT, Rosenthal SR, Gladstone W, et al. Carnitine deficiency: a possible cause of gastrointestinal dysmotility. Acta Paediatr. 1992;81(1):79–81. https://doi.org/10.1111/j.1651-2227.1992.tb12085.x.
18. Murata S, Inoue K, Aomatsu T, et al. Supplementation with carnitine reduces the severity of constipation: a retrospective study of patients with severe motor and intellectual disabilities. J Clin Biochem Nutr. 2017;60(2):121–4. https://doi.org/10.3164/jcbn.16-52.
19. Sakurai T, Amemiya A, Ishii M, et al. Orexins and orexin receptors: a family of hypothalamic neuropeptides and G protein-coupled receptors that regulate feeding behavior. Cell. 1998;92(4):573–85. https://doi.org/10.1016/s0092-8674(00)80949-6.

20. Yoshida G, Li MX, Horiuchi M, et al. Fasting-induced reduction in locomotor activity and reduced response of orexin neurons in carnitine-deficient mice. Neurosci Res. 2006;55(1):78–86. https://doi.org/10.1016/j.neures.2006.02.003.
21. Kuwajima M, Fujihara H, Sei H, et al. Reduced carnitine level causes death from hypoglycemia: possible involvement of suppression of hypothalamic orexin expression during weaning period. Endocr J. 2007;54(6):911–25. https://doi.org/10.1507/endocrj.k07-044.
22. Miyagawa T, Kawashima M, Nishida N, et al. Variant between CPT1B and CHKB associated with susceptibility to narcolepsy. Nat Genet. 2008;40(11):1324–8. https://doi.org/10.1038/ng.231.
23. Miyagawa T, Honda M, Kawashima M, et al. Polymorphism located between CPT1B and CHKB, and HLA-DRB1*1501-DQB1*0602 haplotype confer susceptibility to CNS hypersomnias (essential hypersomnia). PLoS One. 2009;4(4):e5394. https://doi.org/10.1371/journal.pone.0005394.
24. Miyagawa T. Narcolepsy and fatty acid metabolism. J Clin Exp Med. 2011;238(2):199–200. (Article in Japanese).
25. Miyagawa T, Miyadera H, Tanaka S, et al. Abnormally low serum acylcarnitine levels in narcolepsy patients. Sleep. 2011;34(3):349–53. https://doi.org/10.1093/sleep/34.3.349.
26. Miyagawa T, Kawamura H, Obuchi M, et al. Effects of oral L-carnitine administration in narcolepsy patients: a randomized, double-blind, cross-over and placebo-controlled trial. PLoS One. 2013;8(1):e53707. https://doi.org/10.1371/journal.pone.0053707.
27. Miyagawa T, Tokunaga K. Genetics of narcolepsy. Hum Genome Var. 2019;6:4. https://doi.org/10.1038/s41439-018-0033-7.

Anti-inflammatory, Antioxidant, and Anti-apoptotic Effects of Carnitine

7

7.1 Introduction

Most people think of carnitine's physiological and biochemical actions as transporting long-chain fatty acids to the mitochondria. However, since around 2000, it has been known that carnitine has anti-inflammatory, antioxidant, and anti-apoptotic effects by a mechanism completely different from the transport of acyl groups of long-chain fatty acids. In this chapter, the findings of basic and clinical studies on the anti-inflammatory, antioxidant, and anti-apoptotic effects of carnitine will be described.

7.2 Role of the Transcription Factor NF-κB in the Inflammatory Response

The inflammatory response is considered to be a protective response in the event of infection or biological damage. Redness, fever, radical production, and cytokine release at the site of infection due to the inflammatory response are defense reactions of immune system cells against foreign substances that have invaded the body. However, it is known that chronic inflammatory responses are deeply involved in the deterioration of pathological conditions in various diseases. In addition, a weak and persistent inflammatory response is considered to be the basis of the pathogenesis of diseases and conditions that progress over a long period of time, such as arteriosclerosis, dementia, neurodegenerative diseases, heart failure, renal failure, and liver cirrhosis. Therefore, the control of inflammation is a crucial therapeutic target [1].

The transcription factor nuclear factor-κB (NF-κB) plays a central role in the inflammatory response, which normally exists in the cytoplasm in an inactive state bound to NF-κB inhibitor protein(IκB). NF-κB regulates not only inflammatory responses but also genes which control cell proliferation and cell survival. Once foreign substances such as bacteria and viruses invade the body, activated macrophages and B lymphocytes release cytokines such as tumor necrosis factor (TNF-α) and lipopolysaccharides (LPS) from dead bacterial cell walls. When these cytokines and LPS bind to specific receptors, the IκB kinase (IKK) complex, which consists of heterodimers of IKKα and IKKβ, and regulators such as NEMO (NF-κB essential modulator, or NF-κB essential modifier, also called IKKγ), is activated by phosphorylation. The activated IKK complex phosphorylates IκB due to its kinase activity. Then, IκB is ubiquitinated and degraded by the proteasome. Free NF-κB (p65 + p50) migrates to the nucleus and binds to target genes such as inflammatory cytokines such as interleukin-1 (IL-1), IL-6, IL-8, and TNF-α, as well as adhesion factors such as VCAM-1 and ICAM-1, and enhances their expression, triggering a sus-

N. Nakayama, *Carnitine Deficiency and Insufficiency in Medicine*,
https://doi.org/10.1007/978-981-95-7797-2_7

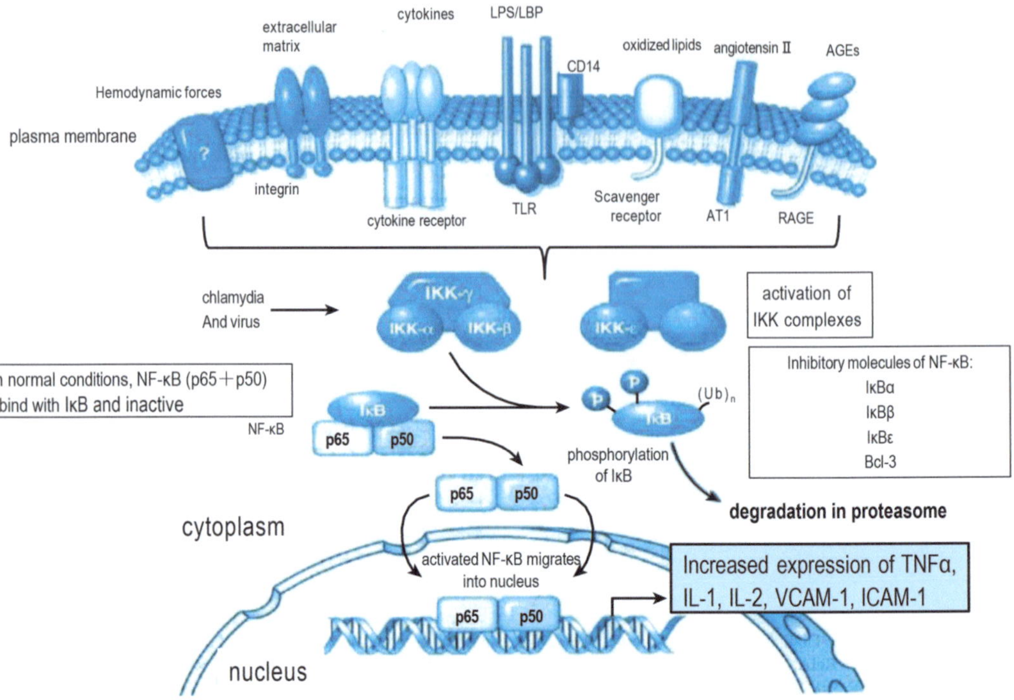

Fig. 7.1 General concept of NF-κB pathway and mechanisms of increased expression of inflammatory cytokines and adhesion molecules. In normal conditions, NF-κB is localized in cytoplasm and binds with IκB and inactivated. However, various inflammatory ligands bind with their receptors, IKK complex is activated, and then IκB is ubiquitinated and degraded in proteasome. Released from inhibitory molecules, free NF-κB migrates into nucleus and increases the expression of target molecules such as inflammatory cytokines or adhesion molecules. Among the increased expression molecules, there is IκB, and this IκB binds to NF-κB and inactivates it. NF-κB signal is oscillatory system. *LPS* lipopolysaccharide, *LBP* LPS binding protein, AGEs advanced glycation end products, *RAGE* AGE receptor, *AT1* type 1 angiotensin IIreceptor, *TLR* toll-like receptor, *IKK* IκB kinase, *IκB* NF-κB inhibitory molecule, *Bcl-3* B cell lymphoma/leukemia 3, *TNF-α* tumor necrolysis factor α, *IL-1* interleukin 1, *IL-2* interleukin 2, *VCAM-1* vascular cell adhesion molecule 1, *ICAM-1* intercellular adhesion molecule 1. (Modified from ref. [1])

tained inflammatory response. Figure 7.1 summarizes the mechanism of NF-κB-mediated hyperexpression of inflammatory cytokines and adhesion factors in the inflammatory response. For more information on the target genes of NF-κB, please refer to the following sites (NF-κB Target Genes) NF-κB Transcription Factors (https://www.bu.edu/nf-kb/gene-resources/target-genes/) and Rel/NF-kappaB target genes (https://bioinfo.lifl.fr/NF- KB/).

7.3 Stress Response and Nrf2-Keap1 System

Although the transcription factor NF-κB plays a central role in the response to induce inflammation, it has been found that there is a stress response in vivo that protects cells in response to oxidative stress caused by inflammatory responses and maintains homeostasis. Transcription factors (nuclear factor erythroid 2-related factor 2 (Nrf2) play a central role.

Nrf2 was discovered in 1994 by Moi et al. as a new transcriptional activator of the leucine zipper type in the study of nuclear factor erythroid 2 (NF-E2). [2] and in 1995, Itoh et al. also studied the gene expression regulation region of avian β-globin. They discovered a new transcription factor that forms heterodimers with small Maf proteins (sMaf) and named erythroid cell-derived protein with CNC homology (ECH. [3] It was later shown that Nrf2 and ECH are orthologous genes that arose from a common ancestral gene during the divergence of species. Furthermore, in 1999, Itoh et al. discovered Kelch-like

ECH-associated protein 1 (Keap1), which binds to Nrf2 and suppresses its activity [4]. Subsequent studies using Nrf2 and Keap1-deficient mice have shown that Nrf2 is a transcription factor that regulates the expression of genes related to biological defense in response to various environmental stresses, and that it is usually bound to Keap1 and is ubiquitinated and degraded by proteasomes, but there are many highly reactive cysteine residues in the Keap1 molecule (27 in humans). It has been found that when specific residues of these cysteine residues are oxidized or alkylated by reactive oxygen species or various electrophilic chemicals, Keap1 undergoes conformational changes and the bond with Nrf2 is broken, and Nrf2 escapes degradation and migrates to the nucleus. Nrf2 translocated to the nucleus forms a heterodimer with sMaf and binds to the antioxidant/electrophile responsive element (ARE/EpRE or CNC-sMaf binding element: CsMBE) and enhances the expression of downstream oxidative stress protection genes such as superoxide dismutase (SOD), catalase, glutathione peroxidase, heme oxygenase-1 (HO-1), and drug detoxification and metabolism enzymes (so-called phase II enzymes). In addition, Nrf2 has been found to promote stress responses by uniformly inducing the expression of enzymes such as glutathione S-transferase (GST) and NAD(P)H quinone reductase (NQO1) which are detoxification enzymes of electrophilic substances, anti-inflammatory genes, heat shock proteins, and drug transporter genes [5–12]. Figure 7.2 shows a schematic diagram of the simplified Nrf2-Keap1 system and the stress response of cells.

Activation of Nrf2 enhances transcription of hundreds of gene groups, which vary depending on animal species, tissues, and cells, but it enhances transcription of reactive oxygen species degrading enzymes such as SOD, which is involved in antioxidant defense, and detoxification of metabolic enzymes such as GST, which is involved in the detoxification of chemical sub-

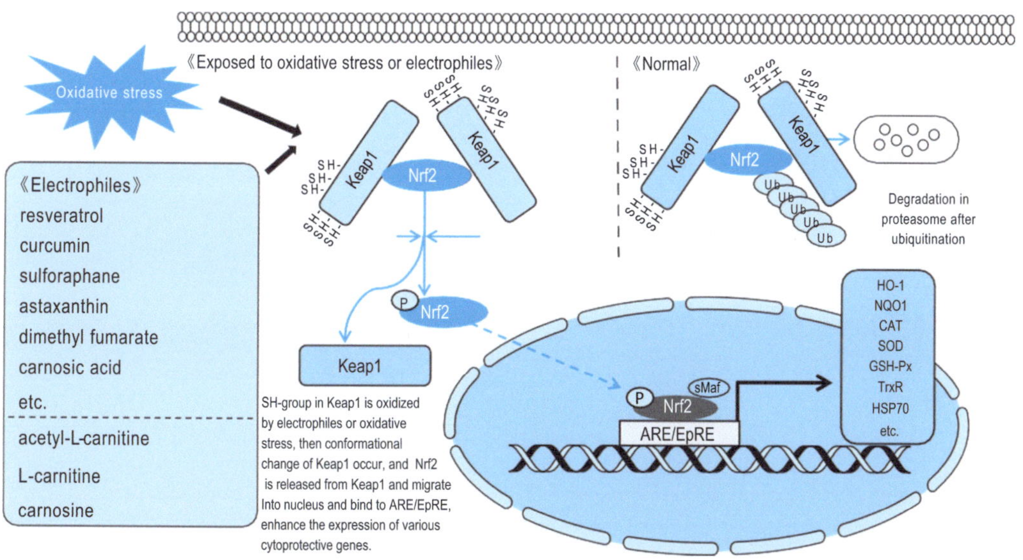

Fig. 7.2 Simplified schematic diagram of Nrf2-Keap1 system and stress response. In normal conditions, transcription factor Nrf2 is bind with Keap1 and degraded in proteasome after ubiquitination. However, when cells are exposed to oxidative stress or electrophiles, these substances react with highly reactive SH-groups on Keap1, then conformational change of Keap1 occurs, and Nrf2 is released and migrated into nucleus, binds with sMaf and forms heterodimer, and binds to ARE/EpRE. Then the expression of downstream antioxidant enzymes genes, detoxification enzyme genes, and cytoprotective molecule genes are enhanced. *Nrf2* nuclear factor erythroid 2-related factor 2, *Keap1* Kelch-like ECH associated protein 1, *ARE/EpRE* antioxidant/electrophile responsive element, *HO-1* heme oxygenase 1, *NQO1* NAD(P)H quinone reductase, *CAT* catalase, *SOD* superoxide dismutase, *GSH-Px* glutathione peroxidase, *TrxR* thioredoxin reductase, *HSP70* heat shock protein 70

stances. Nrf2 is also involved in cross-talk with genes involved in autophagy, mitochondrial biosynthesis, matrix remodeling, cell proliferation, anabolic metabolism, and signaling systems such as MAP kinase (MAPK) and protein kinase C (PKC). The cytoprotective effect of Nrf2 is shown in Fig. 7.3 in the case of the heart [7].

For more information on the genes involved in oxidative stress defense and redox signaling regulated by Nrf2, please refer to the following site (https://www.annualreviews.org/doi/suppl/10.1146/annurev-pharmtox-011112-140320/suppl_file/pa_53_ma_supmat.pdf).

Substances that ultimately activate Nrf2 by reacting with the SH-group of Keap1 and oxidizing or alkylating it include small amounts of reactive oxygen species, curcumin (a polyphenol found in turmeric), resveratrol (a stilbenoid polyphenol found in red wine), sulforaphane (an isothiocyanate compound found in broccoli sprouts), carnosic acid (a diterpene compound found in rosemary and sage), flavonoids and flavonols, and many phytochemicals found in plants, such as ginseng and many components of Chinese herbal medicines. It is now known that many phytochemicals, such as flavonoids, flavo-

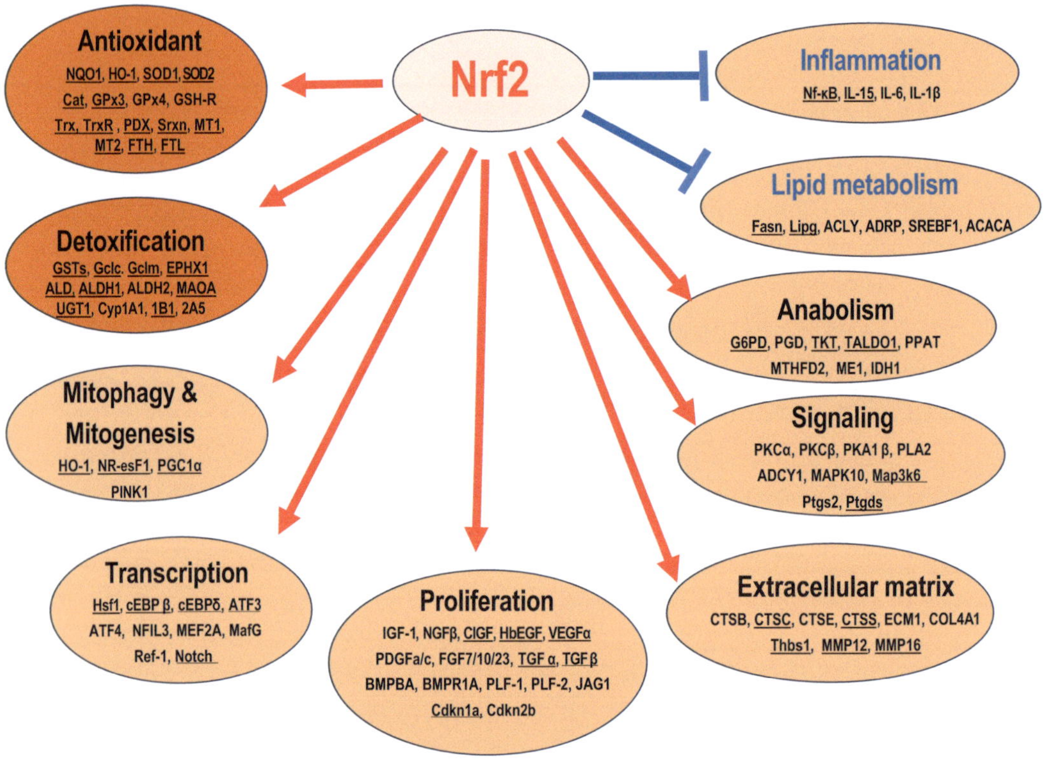

Fig. 7.3 Function of Nrf2 protein after its activation and binding to ARE/EpRE based on their downstream cardioprotective genes and cross talk. The downstream genes were revealed by microarray or RNA-seq as transcripts increased or reduced due to Nrf2 activation of deficiency or are targets of Nrf2 binding as determined by DNA sequencing following ChIP. Underlines indicate the genes, expression of which has been found in cardiomyocytes in culture or in mouse myocardium related to Nrf2 status. Color shade reflects the ratio of genes in the functional group under control of Nrf2 in the myocardium and the amount of literature validating the function by Nrf2 inducers or Nrf2 knockout in experimental animals. *ARE* antioxidant response element, *EpRE* electrophile responsive element, *ChIP* chromatin-immunoprecipitation, *Cat* catalase, *c/EBP* CCAAT-enhancer-binding protein, *GPx* glutathione peroxidase, *GSH-R* glutathione reductase, *GST* glutathione S transferase, *HbEGF* heparin-binding epidermal growth factor-like growth factor, *HO-1* heme oxygenase 1, *MT* metallothionein, *NF-κB* nuclear factor kappa B, *NO1* NAD(P)H quinone oxidoreductase 1, *Pdx* peroxiredoxin, *PGC1α* peroxisome proliferator-activated receptor gamma activator 1 alpha, *PINK1* PTEN-induced kinase 1, *SOD* superoxide dismutase, *Srxn* sulfiredoxin, *TGF* transforming growth factor, *Trx* thioredoxin, *TrxR* thioredoxin reductase, *VEGF* vascular endothelial growth factor. (From ref. [7])

nols, ginseng (Chinese ginseng), and components of many Chinese medicines, exhibit anti-inflammatory and antioxidant effects by activating Nrf2 [8–13]. Many of these substances are considered to be beneficial to health, and some are used as dietary supplements. Not only phytochemicals, but also astaxanthin, the red pigment of salmon and shrimp [12], L-carnitine, acetyl-L-carnitine and carnosine [9, 10] (dipeptides composed of β-alanine and histidine, similar in name but different from carnosic acid). Talalay et al. started a project around 1980 to search for compounds which suppress carcinogenesis in foods such as vegetables and fruits. Although initially met with little success and ridicule from other researchers, their research has yielded significant results, including the discovery that sulforaphane in broccoli sprouts is a potent inducer of the drug-metabolizing phase II enzyme, and is now an extremely important area of research. In his memoirs celebrating the 100th anniversary of the Journal of Biological Chemistry, Talalay described the background and philosophy of his exploratory studies, which is very interesting [14].

Sulforaphane, curcumin, and resveratrol have been used not only as supplements but also as medicines for clinical applications in various diseases, but the long-term safety of high doses as medicines is not yet clear. In addition, there are many poorly soluble substances that are difficult to be used as pharmaceuticals, and there are problems such as bioavailability [15–19]. Many attempts have been made to improve the bioavailability of curcumin, quercetin, resveratrol, and other substances through various formulation technologies [18, 19].

In the field of drug development, there is bardoxolone methyl (RTA402), a small-molecule compound that activates Nrf2. Clinical trials were conducted overseas to administer the drug to patients with stage IV renal failure with type II diabetes mellitus, but it was discontinued due to side effects of heart failure due to increased fluid retention [20–22]. In Japan, Kyowa Kirin Co., Ltd. has introduced and developed this compound, and it has been designated as an orphan drug for the indication of "improvement of renal function in Alport syndrome" and is in the process of applying for approval, and clinical trials for diabetic nephropathy [23, 24] and autosomal (dominant) polycystic kidney disease (ADPKD) is being investigated in a global Phase III study (FALCON trial). However, the clinical development of bardoxolone methyl was discontinued in May 2023.

The structure of these main Nrf2-activated compounds is shown in Fig. 7.4.

Although research on Nrf2 activators has only just begun, new nutritional, biochemical and pharmacological reviews and clinical applications of vegetables, fruits, and medicinal herbs that have long been considered to be good for health have begun. Classical nutrition has dealt with macronutrients such as carbohydrates, lipids, and proteins, and micronutrients such as vitamins and minerals. In recent years, it has become clear that trace substances such as sulforaphane, curcumin, and resveratrol, which are not originally considered nutrients, exert antioxidant, anti-inflammatory and anticancer effects by regulating the expression of various genes, and this has led to rapid development of the new field called "nutritional genomics" (nutrigenomics). Although not derived from plants, many new studies on the effects of carnitine and acetylcarnitine have emerged from this perspective (see below). Supplements of resveratrol and sulforaphane are already available, and these substances are also understood by the general public to be beneficial to health. In addition, many nephrologists and diabetologists know about bardoxolone methyl. However, it is currently known only to a small number of researchers that carnitine and acetylcarnitine also have such as Nrf2 activation effects, which exert anti-inflammatory, antioxidant, and anti-apoptotic effects.

In cancer, activation of Nrf2 may promote the elimination of carcinogens, oxidative stress, and drugs that damage DNA, and prevent carcinogenesis. However, it has been found that Nrf2 is constitutively activated in certain cancer cells by mutation, etc., and that they acquire resistance to radiation and anticancer drugs, and Nrf2 inhibitors are attracting attention as therapeutic targets. In other words, Nrf2 can act as a "double-edged

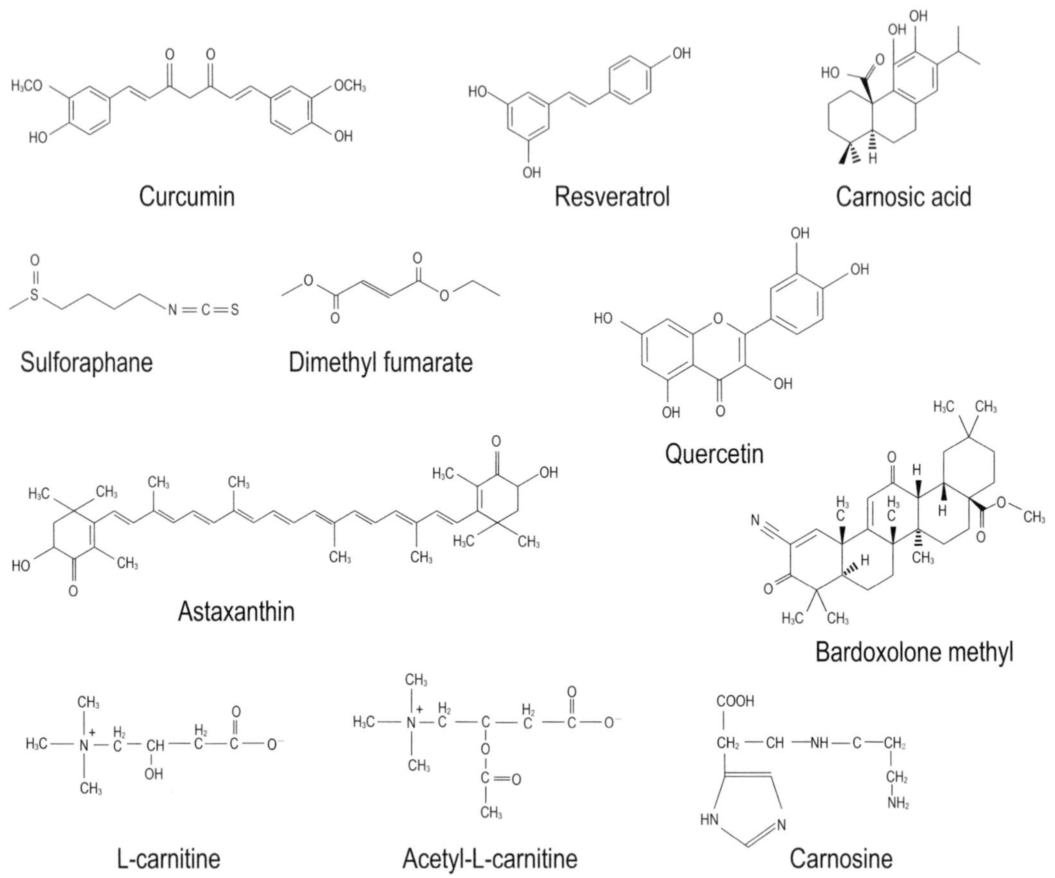

Fig. 7.4 Chemical structures of main compounds which activate Nrf2. Among them, there are many electrophilic substances from plant origin such as curcumin in turmeric, resveratrol in red wine, carnosic acid in rosemary, sulforaphane in broccoli sprouts, methyl fumarate in shepherd's purse, quercetin in onion/apple/buck wheat. However, it is proved that astaxanthin in salmon, carnitine/acetylcarnitine/carnosine in meat also activates Nrf2. As others, there is bardoxolone methyl which is a drug candidate compound developing as a Nrf2-activator

sword" in cancer, which is complicated. This is beyond the scope of this book, so please refer to the detailed review article [25–27]. It should be noted, however, that there have been no reports of Nrf2 activators such as those shown in Fig. 7.4 promoting carcinogenesis or having an adverse effect on cancer therapy. Rather, it is thought to have the effect of suppressing carcinogenesis by detoxifying carcinogens and inducing enzymes that reduce oxidative stress, and protecting normal cells from damage caused by anticancer drugs and irradiation.

For more information on the Nrf2-KEAP1 system and its pathological control, please refer to other excellent review articles and explanations [28–31].

7.4 Anti-inflammatory, Antioxidant, and Anti-apoptotic Effects of Carnitine

7.4.1 Interaction Between the Nrf2-Keap1 and NF-κB System

Nrf2 acts in an anti-inflammatory manner by suppressing the NF-κB system, which plays a central role in inflammatory responses. Conversely, the

NF-κB system acts inhibiting the Nrf2 system. Nrf2 and NF-κB systems are thought to mutually cross-talk and regulate inflammatory responses and carcinogenesis in a balanced manner between promotion and inhibition [32, 33] (Fig. 7.5).

For example, it is known that in animals knocked out of Nrf2, NF-κB is activated, inflammatory cytokine production is enhanced, and cell damage progresses [33]. Therefore, there are high expectations for safe Nrf2 activators that can be used as pharmaceuticals.

7.4.2 Anti-inflammatory, Antioxidant and Anti-apoptotic Effects of Carnitine (Cellular-Level Studies)

Cao et al. examined the effects of L-carnitine on protein expression of Nrf2, Keap1, HO-1, and γ-glutamyl cysteine synthetase (γ-GCS) by Western blotting in cultured retinal ganglion cells in a cytotoxic experimental system with high glucose (30 mM)-induced oxidative stress [34]. As a result, when comparing the carnitine group (LC group: 50, 100, and 200 μM of L-carnitine + high glucose) with the only high glucose group (HG group), Nrf2 was 0.71 ± 0.04, 0.89 ± 0.05, 1.24 ± 0.05 in the LC group, and 0.56 ± 0.03 in HG group ($P < 0.05$), significantly upregulated in the LC group. In the case of HO-1, 0.58 ± 0.04, 0.76 ± 0.06, 0.89 ± 0.07 in the LC group, and 0.250 ± 03 in the HG group ($P < 0.01$). In the case of γ-GCS, 0.66 ± 0.03, 0.79 ± 0.05, 0.84 ± 0.08 in the LC group, and 0.42 ± 0.08* in the HG group ($P < 0.01$) (*Note: In the paper, the HG group is 0.84 ± 0.08, but in the figure, it is around 0.48, which is clearly low and seems to be a mistake). Thus, Nrf2, HO-1, and γ-GCS were significantly upregulated in the LC group, and Keap1 expression was significantly decreased (LC group: 0.50 ± 0.03, 0.45 ± 0.02, 0.53 ± 0.03, HG group: 0.86 ± 0.05, $P < 0.01$) (Fig. 7.6).

Increased expression of HO-1 works to suppress the increase in ROS, and γ-GCS, the rate-limiting enzyme for glutathione biosynthesis, also promotes the production of glutathione, a reducing force, in response to oxidative stress.

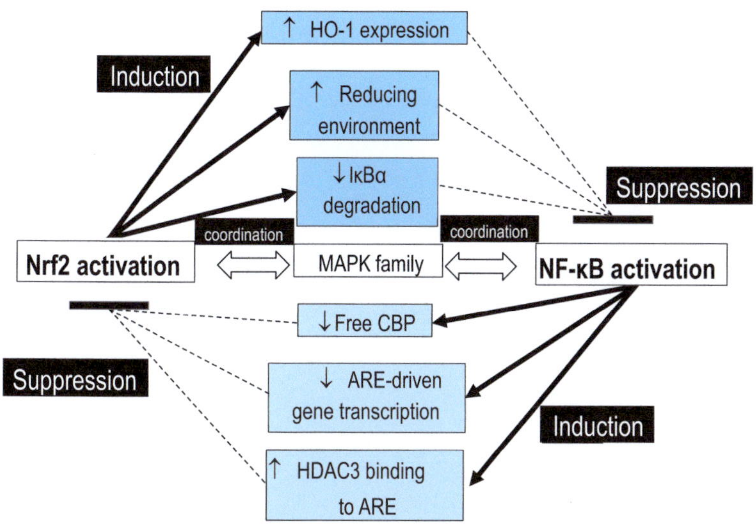

Fig. 7.5 Outline of proposed cross-talk between Nrf2 and NF-κB systems. Nrf2 activation induces intracellular events that concur to NF-κB suppression and vice versa. Namely, Nrf2 activation induces enhanced HO-1 expression and reduced environment and inhibits breakdown of IκBα, an inhibitory protein of NF-κB, resulting in suppression of NF-κB system. On the other hand, NF-κB activation reduces CBP, suppresses gene expression of cytoprotective genes located in downstream of ARE, enhances the binding of HDAC3 to ARE, and acetylation of ARE is also reduced by interaction of HDAC3 with CBP or MafK, resulting in reduction in gene expression of the downstream of ARE. *HO-1* heme oxygenase 1, *MAPK* mitogen-activated protein kinase, *CBP* CREB binding protein, *ARE* antioxidant responsive element. *HDAC3* histone deacetylase 3. (Modified from ref. [32])

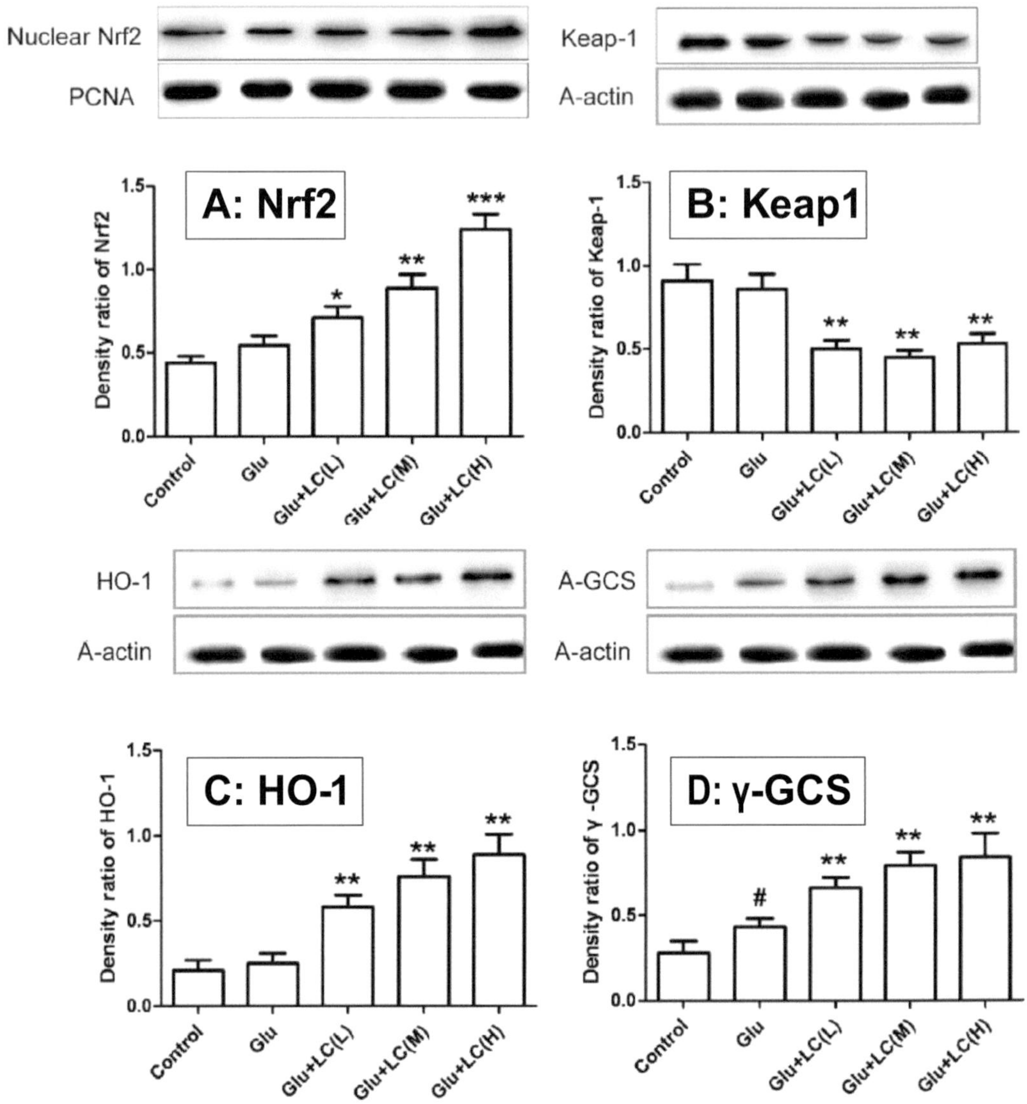

Fig. 7.6 Effects of L-carnitine on protein expression of Nrf2, Keap1, HO-1, and γ-GCS in cultured retinal ganglion cells under high glucose conditions. Comparison of expression of Nrf2 (**a**), Keap1 (**b**), HO-1 (**c**) and γ-glutamyl cysteine synthetase (γ-GCS, D) proteins in the control, high glucose stimulation (Glu, 30 mM) and L-carnitine (LC) treatment group under high glucose stimulation (L): 50 μM, LC(M): 100 μM, LC(H): 200 μM) in cultured retinal ganglion cells. The vertical axes show relative expression ratio to proliferating cell nuclear antigen (PNCA) in the case of Nrf2, and to β-actin in the cases of Keap1, HO-1 and γ-GCS. (From ref. [34])

Thus, carnitine may protect cells from oxidative stress by enhancing the expression of HO-1 and γ-GCS through the activation of Nrf2.

Montesano et al. found that high fructose intake increases the incidence of non-alcoholic fatty liver disease (NAFLD), so they cultured HepG2 cells derived from human liver cancer under high fructose (5 mM) as a model of NAFLD and examined the effect of carnitine addition (5 mM) [35]. As a result, it was found that the fat accumulation stained with oil red O increased by about 4 times due to fructose loading, but the increase was suppressed by the addition of carnitine by about 2.5 times, and that the effect of carnitine on preventing fat accumulation was canceled by compound C (dorsomorphin, 2 μM), which is an inhibitor of AMP-activated protein kinase (AMPK). Furthermore, fructose administration enhances the generation of reactive oxygen species, while carnitine administration increases the expression of peroxisome proliferator-activated receptor gamma coactivator 1α (PGC1α), promotes mitochondrial production, and enhances the expression of Nrf2 and SOD2. It was found that fructose suppresses the expression of PGC1α and the generation of reactive oxygen species (Fig. 7.7, PGC1α is not shown). In other words, carnitine activates Nrf2 through the AMPK system, enhances the expression of antioxidant enzymes such as SOD2, and decreases reactive oxygen. Therefore, carnitine may be effective in inhibiting the pathological progression of NAFLD.

Angiotensin II (Ang II) increases oxidative stress through activation of NADPH oxidase and aggravates pathological conditions such as hypertension. Blanca et al. stimulated NRK-52E cells derived from normal rat kidney with Ang II and investigated the effects of carnitine on oxidative stress production [36]. In NRK-52E cells, Ang II. stimulation increased reactive oxygen species such as superoxide anion (O^{2-}), H_2O_2, and nitrotyrosine by activation of NADPH oxidase, while carnitine (0.1–10 mM) suppressed the increase in the activity of NADPH oxidase and suppressed O^{2-} production (Fig. 7.8). The mechanism of action of carnitine on Ang II signaling was investigated, and it did not affect the phosphorylation of Akt, p38MAPK, or ERK1/2 by Ang II. Ang II phosphorylates the inhibitory protein IκB by activating PKC, deviates from IκB and promotes the transfer of NF-κB to the nucleus, while carnitine inhibits PKC activity and suppresses the phosphorylation of IκB, thereby suppressing the transfer of NF-κB to the nucleus (Fig. 7.9).

Calò et al. investigated the effects of L-carnitine, acetyl-L-carnitine, and propionyl-L-carnitine on the expression of HO-1 and ecNOS (NO synthase constitutively present in endothelial cells) when cultured human vascular endothelial cells were subjected to oxidative stress injury induced by H_2O_2 [37]. As a result, L-carnitine, acetyl-L-carnitine, and propionyl-L-carnitine increased mRNA and protein expression of HO-1 and ecNOS (Table 7.1). Both HO-1 and NO produced by ecNOS are expected to act as antioxidants and anti-inflammatory, reduce oxidative stress, inhibit cell proliferation, and prevent the progression of arteriosclerosis.

Ye et al. examined the protective effect of carnitine on cell damage caused by H_2O_2 in HK-2 cells, a human proximal tubular epithelial cell [38]. The amount of H_2O_2 added and the treatment time were examined, and the cells were moderately damaged after 30 min of treatment with H_2O_2 (500 μM). Therefore, the effect of carnitine was examined under this condition. As a result, carnitine (10, 50, and 100 μM) suppressed reactive oxygen species and lipid peroxide caused by H_2O_2, and its strength was almost the same as that of N-acetyl-L-cysteine (5 mM), which is known as an antioxidant. SOD, catalase, glutathione peroxidase, and total antioxidant capacity (T-AOC) were greatly reduced by H_2O_2 but were restored by carnitine (Fig. 7.10).

During H_2O_2-induced cell damage, apoptosis-promoting events such as DNA fragmentation, mitochondrial membrane potential decrease, and cytochrome c release occur, but carnitine addition suppressed these events. Furthermore, during cell damage induced by H_2O_2, the expression of apoptosis-promoting factor Bax was increased, caspase-3 was activated, and the expression of apoptosis-suppressing factor Bcl-2 was decreased. Carnitine addition reduced the Bax/Bcl-2 ratio, which increased about fivefold after H_2O_2 treatment compared to controls and acted to prevent apoptosis (Fig. 7.11).

Thus, it was found that carnitine suppresses oxidative stress damage caused by H_2O_2 in HK-2 cells and suppresses cell apoptosis by suppressing mitochondrial disorders. Carnitine may be useful in the treatment of cellular damage caused by oxidative stress, such as renal tubular damage.

Wang et al. examined the effect of carnitine in a cytotoxic model of 1-methyl-4-phenylpyridinium ion (MPP+), an inhibitor of mitochondrial complex I used in Parkinson's disease models, to primary cultured cells of the

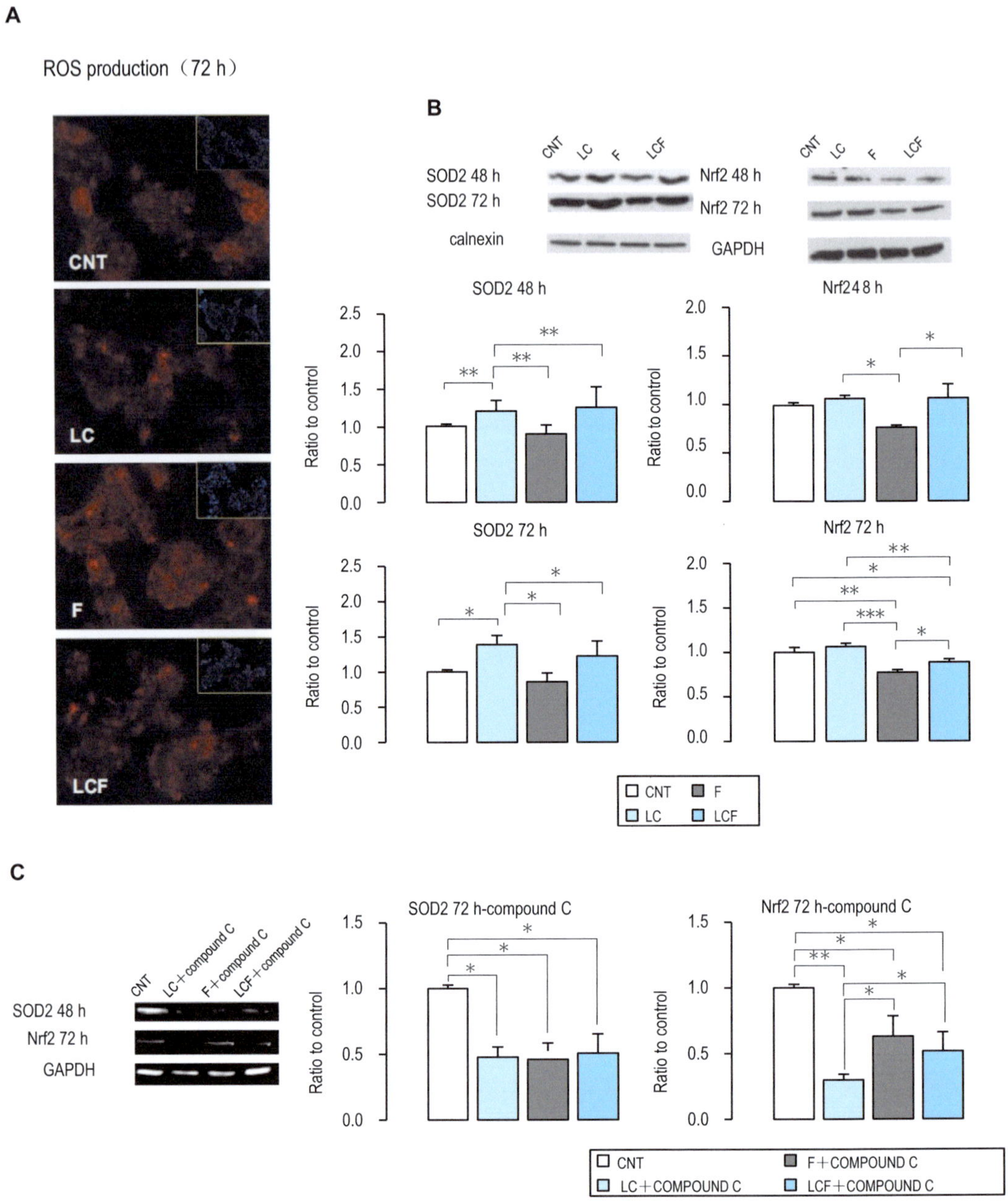

Fig. 7.7 Effects of L-carnitine (5 mM) on reactive oxygen species (ROS) production and expression of anti-oxidant enzymes in human hepatoma HepG2 cells under high fructose condition (5 mM). (**a**): H_2O_2 stimulation of HepG2 cells enhanced ROS production (CNT), and addition of L-carnitine lowered ROS production (LC). High fructose treatment further enhanced ROS production (F), but addition of L-carnitine suppressed the ROS production (LCF). (**b**): Expression of SOD2 and Nrf2 in HepG2 cells. *CNT* control, *F* high fructose (5 mM), *LC* L-carnitine (5 mM), *LCF* high fructose (5 mM) + L-carnitine (5 mM). (**c**) Effect of AMPK inhibitor, compound C on the action of L-carnitine. Addition of compound C cancelled the action of L-carnitine. *: $P \leqq 0.05$, **: $P \leqq 0.01$, ***: $P \leqq 0.001$. (From ref. [35])

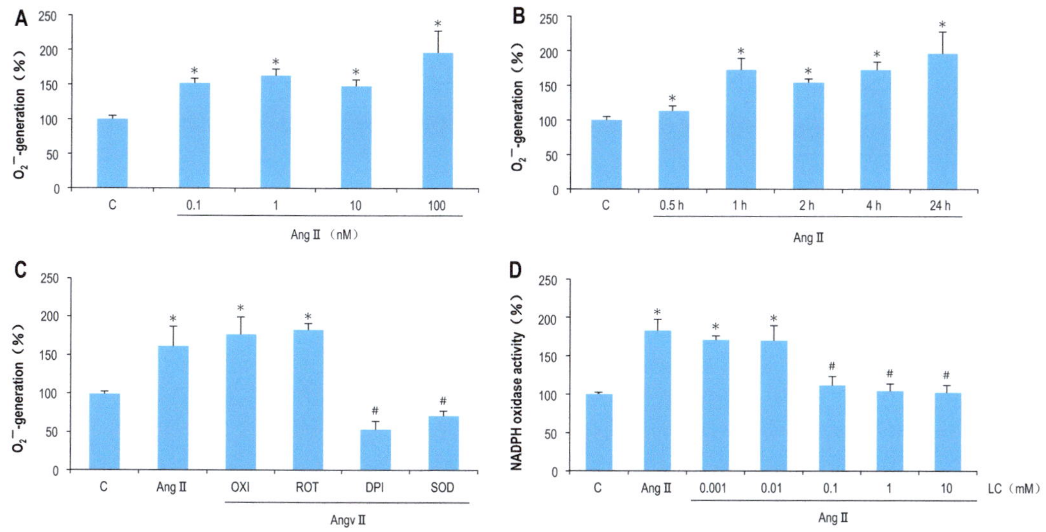

Fig. 7.8 Effects of L-carnitine on angiotensin II (AngII) induced superoxide anion (O_2^-) production and NADPH oxidase activity in NRK-52E cells derived from normal rat kidney. (**a**) Dose-responsive curve for Ang II-dependent O_2^- production. NRK-52E cells were incubated with AngII (0.1–100 nM) for 24 h. (**b**): Time course of O_2^- production induced by AngII (100 nM). (**c**): Effects of oxypurinol (OXI, an inhibitor of xanthine oxidase), rotenone (ROT, an inhibitor of mitochondrial electron transport system), diphenyleneiodonium (DPI, an inhibitor of flavoprotein), and superoxide dismutase (SOD, 100 U/mL, scavenger enzyme of O_2^-) on AngII-induced O_2^- production in NRK-52E cells. (**d**) Effects of L-carnitine (0.001–10 mM) on NADPH oxidase activity. *: $P < 0.05$ vs control, #: $P < 0.05$ vs AngII. (From ref. [36])

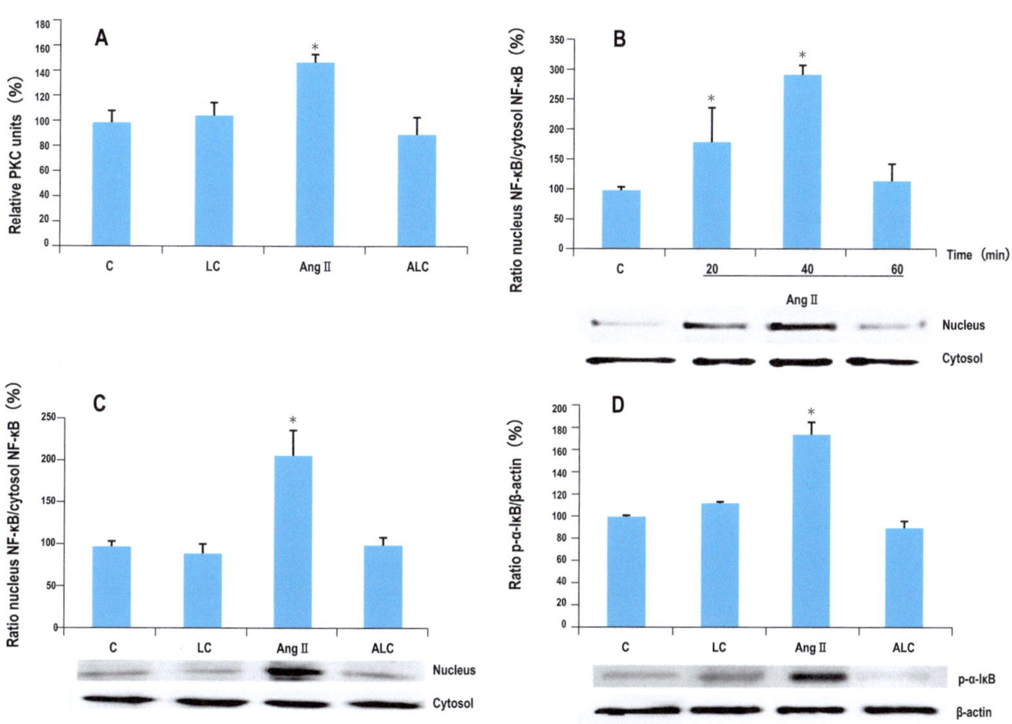

Fig. 7.9 Effects of L-carnitine on Ang II-induced PKC activation in NRK-52E cells. (**a**): Activation of PKC by Ang II (100 nM). This activation of PKC was suppressed by L-carnitine (1 mM, ALC). (**b**): The time course of migration of NF-κB into nucleus by Ang II. (**c**): Effects of L-carnitine (ALC) on migration of NF-κB into nucleus by Ang II. (**d**): Effects of L-carnitine (ALC) on phosphorylation of I κB induced by Ang II. *: $P < 0.05$ vs other groups. PKC: protein kinase C. (From ref. [36])

Table 7.1 ecNOS and HO-1 gene (A) and protein expression (B) in human umbilical endothelial cells incubated with H_2O_2 alone and with H_2O_2 and L-carnitine, or acetyl-L-carnitine, or propionyl-L-carnitine

			L-carnitine			Acetyl-L-carnitine			Propionyl-L-carnitine		
	Basal	H_2O_2	0.5 mM	1.0 mM	2.0 mM	0.1 mM	0.2 mM	0.4 mM	0.05 mM	0.1 mM	0.2 mM
(A) mRNA											
eNOS	0.75 ± 0.04	0.39 ± 0.08[a]	0.82 ± 0.06[b]	1.12 ± 0.07[b]	1.02 ± 0.08[b]	0.85 ± 0.03[b]	0.82 ± 0.03[b]	0.80 ± 0.02[b]	1.11 ± 0.04[b]	1.17 ± 0.05[b]	1.06 ± 0.03[b]
HO-1	0.48 ± 0.05	0.61 ± 0.02[a]	0.72 ± 0.03[c]	0.75 ± 0.02[c]	0.76 ± 0.03[c]	0.72 ± 0.03[c]	0.75 ± 0.02[c]	0.76 ± 0.03[c]	0.67 ± 0.02[c]	0.68 ± 0.02[c]	0.69 ± 0.02[c]
(B) Protein											
eNOS	0.85 ± 0.02	0.79 ± 0.02[a]	0.92 ± 0.01[b]	1.03 ± 0.03[b]	1.03 ± 0.02[b]	1.40 ± 0.06[b]	1.36 ± 0.05[b]	1.22 ± 0.02[b]	1.07 ± 0.05[b]	1.23 ± 0.09[b]	1.08 ± 0.06[b]
HO-1	0.63 ± 0.05	0.76 ± 0.03[a]	1.02 ± 0.01[b]	1.02 ± 0.02[b]	0.98 ± 0.02[b]	1.13 ± 0.05[b]	1.08 ± 0.04[b]	1.02 ± 0.03[b]	0.87 ± 0.04[c]	0.99 ± 0.05[b]	0.86 ± 0.03[c]

From ref. 37

[a]:$P<0.001$ vs. Basal, [b]:$P < 0.001$, [c]:$P<0.01$ vs. H_2O_2

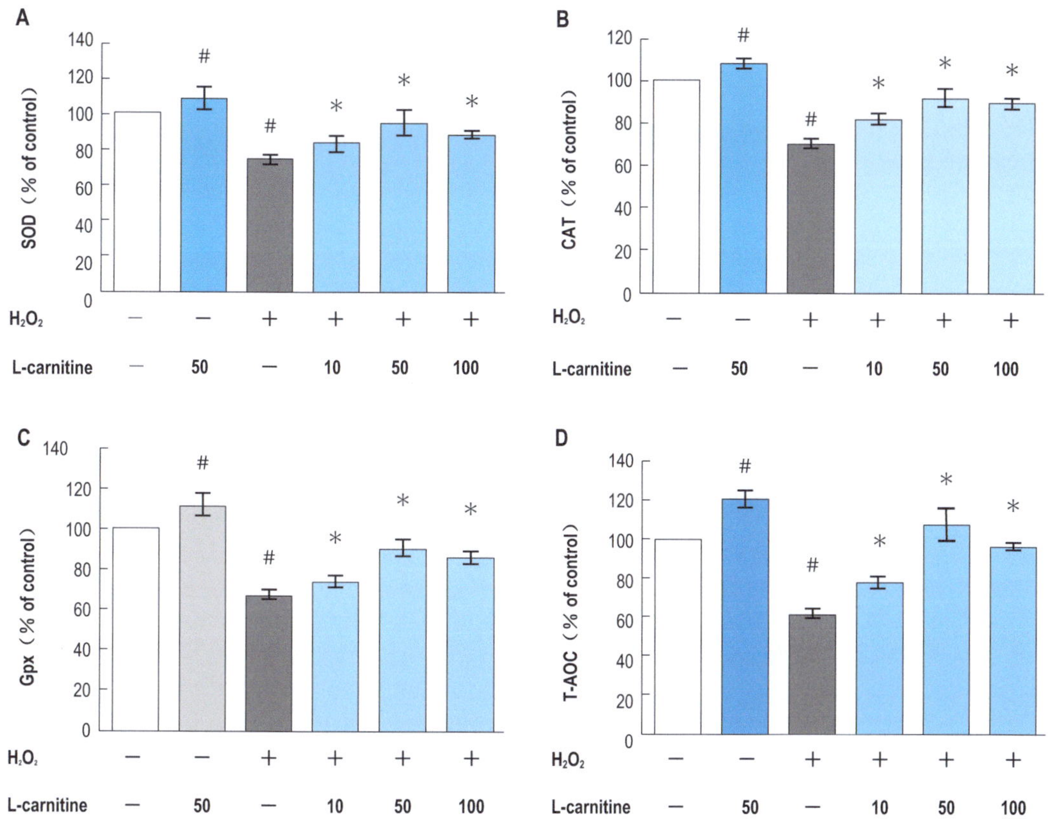

Fig. 7.10 Reduction of SOD (**a**), catalase (**b**), glutathione peroxidase (Gpx, **c**), and total anti-oxidant capacity (TOC, **d**) by H_2O_2 in HK-2 cells, and effects of L-carnitine (10, 50, 100 μM) on them. #: $P < 0.05$ vs non-treated cells, *: $P < 0.05$ vs H_2O_2 alone-treated cells. *SOD* superoxide dismutase, *CAT* catalase, *Gpx* glutathione peroxidase, *TOC* total anti-oxidant capacity. (From ref. [38])

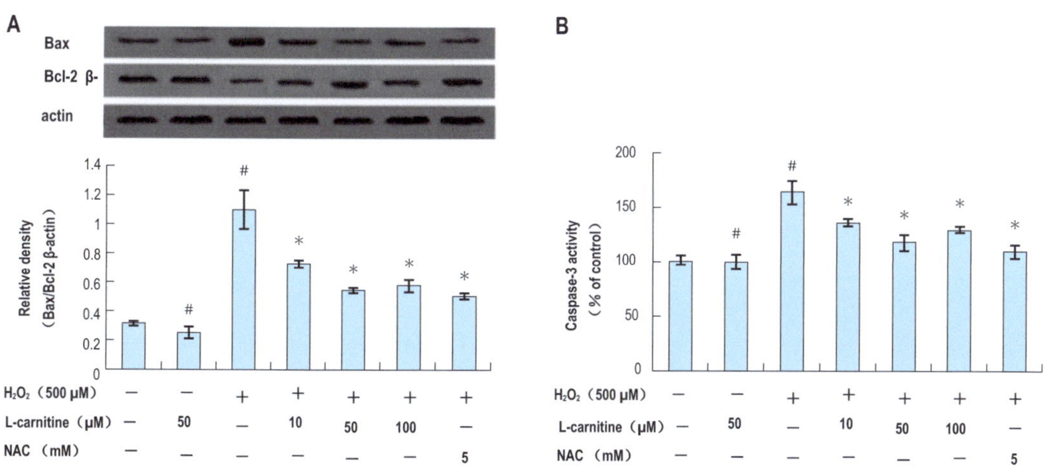

Fig. 7.11 Effects of L-carnitine on expression of Bax, Bcl-2 and Bax/Bcl-2 ratio (**a**) and caspase 3 activity (**b**) in HK-2 cells during H_2O_2-induced cellular injury. #: $P < 0.05$ vs non treated cells, *: $P < 0.05$ vs H_2O_2 alone-treated cells. *NAC* N-acetylcysteine. (From ref. [38])

rat forebrain [39]. The results showed that MPP^+ (500 μM) reduced the number of viable cells by about half and increased the number of dead cells, while carnitine (30 and 100 μM) almost completely inhibited the cytotoxic effect of MPP^+ (Fig. 7.12).

This MPP^+-induced cytotoxicity is indicated by a fivefold increase in the number of cells positive for TUNEL staining, a staining method for apoptotic cells, indicating apoptotic cell death, and carnitine (30 μM) almost completely inhibited apoptosis (Fig. 7.13). Furthermore, carnitine (30 μM) suppressed the expression of Bax, an apoptosis-promoting factor, while it enhanced the expression of BCL-XL, an apoptosis-inhibiting factor, and restored the MPP^+-induced decrease in the BCL-XL/Bax ratio (Fig. 7.14).

Thus, carnitine may be an effective treatment for neurodegenerative diseases such as Parkinson's disease, Alzheimer's disease, and other mitochondrial diseases.

Ye et al. also reported that the addition of carnitine (100 μM) prevented cellular apoptosis in the same way as N-acetylcysteine (5 mM), which is known as an antioxidant, in H_2O_2 stimulation or tunicamycin-induced cytotoxic model of endoplasmic reticulum stress in neuroblastoma-derived SH-SY5Y cells [40]. The mechanism is that carnitine suppresses endoplasmic reticulum stress caused by H_2O_2 stimulation.

Doxorubicin (adriamycin) is a useful anticancer drug for various solid and hematologic cancers, but its use is limited because it causes cardiotoxicity. Chao et al. studied the protective effect of carnitine in a model of myocardial cell injury by doxorubicin (1 μM) in neonatal rat myocardial cells [41]. The results showed that carnitine (1–30 mM) inhibited oxidative stress induced by doxorubicin, activation of NADPH oxidase, and apoptosis induced by caspase-3 activation. The mechanism of the inhibition of apoptosis was the upregulation of BCL-XL (apoptosis inhibitory factor) and the activation of signaling pathways such as prostacyclin (PGI_2) and peroxisome proliferator-activated receptor alpha (PPARα).

Normal apoptosis is extremely important in organogenesis and cell metabolism during development and differentiation, whereas essentially unnecessary apoptosis of normal cells in pathological conditions and aging leads to organ dysfunction and fibrosis. As described above, carnitine is thought to suppress apoptosis by reducing oxidative stress, enhancing the expression of genes involved in Nrf2-mediated cytoprotection, and mediating various signal transduction systems. In addition to Nrf2-mediated suppression of oxidative stress and apoptosis, carnitine has also been reported to suppress the apoptotic mechanism involving long-chain fatty acids. In apoptosis, cytochrome c is released from mitochondria into the cytoplasm, apoptosomes are formed, and caspase-3 activation triggers DNA fragmentation and promotes apoptosis. During the release of cytochrome c from the mitochondrial inner membrane, the membrane potential and permeability of the mitochondrial inner

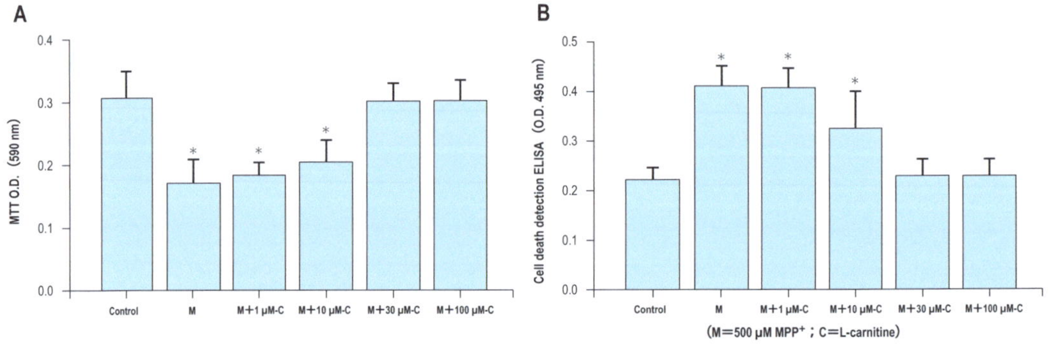

Fig. 7.12 Inhibition of MPP^+ (500 μM)-induced neurotoxicity by L-carnitine and dose-response studies as measured by MTT assay (**a**) and cell death ELISA that measures nucleosomes (**b**). *: $P < 0.05$ vs control. *M: 500 μM MPP⁺, C: L-carnitine* (From ref. [39])

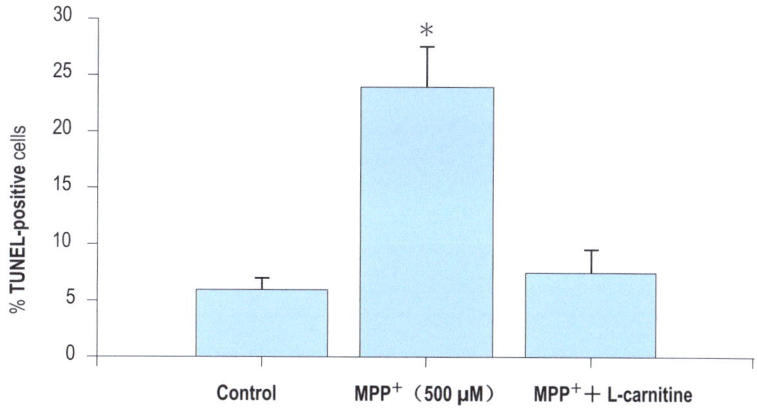

Fig. 7.13 Quantitative assessment of TUNEL-positive cells (apoptosis) in the forebrain cultures following control, 500 μM MPP^+, and 500 μM MPP^+ + 30 μM L-carnitine for 24 h. *: $P < 0.05$ vs control. (From ref. [39])

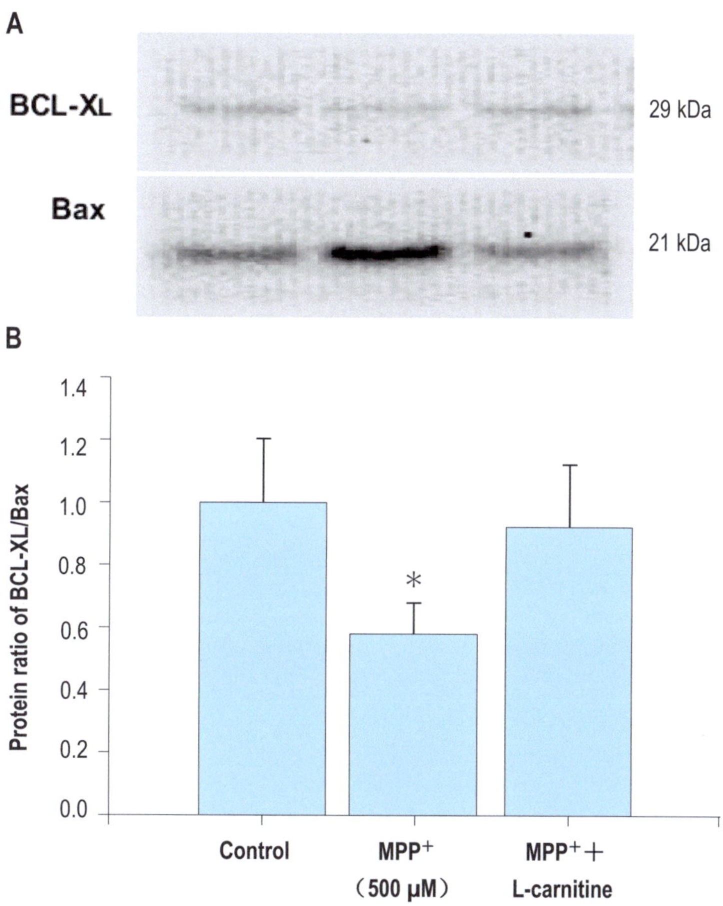

Fig. 7.14 Western blot analysis of the effect of MPP^+ and L-carnitine on the regulation of BCL-XL and Bax protein expression (**a**) and a ratio of BCL-XL to Bax by densitometry measurements (**b**). *: $P < 0.05$ vs control. (From ref. [39])

membrane change, a phenomenon called "mitochondrial membrane permeability transition" (MPT). This process is thought to be caused by the opening of voltage-dependent anion channels (VDACs), but recently, a more complex mechanism of pore opening has been proposed: when cells are exposed to oxidative stress, the opening of the fusion zone of the inner membrane cristae and the formation of Bax and Bak pore formation in the outer membrane and cytochrome c are released into the cytoplasm [42].

Furuno et al. reported that the addition of palmitoyl-CoA to isolated rat liver mitochondria induces cytochrome c release and apoptosis, and that carnitine inhibits the release of cytochrome c from the mitochondrial to the cytoplasm by palmitoyl-CoA [43] (Fig. 7.15). This may be a result of the inhibition of MPT by palmitoyl CoA.

Oyanagi et al. in the same group reported that incubation of isolated rat liver mitochondria with oleic acid causes mitochondrial swelling and cytochrome c release, and that carnitine (1 mM) inhibits cytochrome c release in mitochondria by promoting beta oxidation of oleic acid [44], Also, Oyanagi et al. reported that palmitoyl CoA promotes cytochrome c release in isolated rat cardiac mitochondria as well as in liver mitochondria, and carnitine prevents this release [45]. Nishimura et al. in the same group found that the addition of 3-nitropropionic acid (3NP), an inhibitor of mitochondrial electron transport chain complex II (succinate dehydrogenase), to isolated rat liver mitochondria caused the release of cytochrome c and suppressed mitochondrial respiratory function, and the addition of carnitine cancels these actions of 3NP [46]. 3NP is known to cause symptoms similar to Huntington's disease when administered to rats.

These results suggest that carnitine suppresses MPT caused by palmitoyl-CoA and oleic acid.

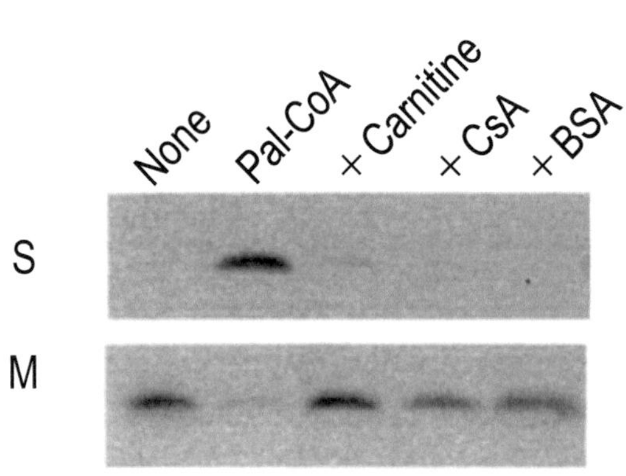

Fig. 7.15 Effects of various reagents on Pal-CoA-induced cytochrome c release (Western blot). Addition of palmitoyl-CoA (Pal-CoA, 5 μM) induces cytochrome c release from mitochondria (M) to soluble fraction (cytosol, S). Carnitine (1 mM), cyclosporin A (1 μM, CsA), BSA (100 μg/mL) suppressed the cytochrome release. (From ref. [43])

Thus, it is interesting that long-chain acyl CoA and fatty acids such as palmitoyl CoA and oleic acid, inhibitors of mitochondrial electron transfer system, promote cytochrome c release from mitochondria, and then trigger apoptosis. In recent years, it has become clear that long-chain acyl CoA accumulates in mitochondria in heart failure, chronic liver disease, and chronic renal failure, causing impaired energy metabolism. Free carnitine promotes β-oxidation of long-chain fatty acids in these conditions and improves mitochondrial metabolic disorders by eliminating acyl CoA accumulated in mitochondria as acylcarnitine in the condition of carnitine insufficiency (see Chap. 3) In other words, carnitine not only enhances the expression of Nrf2-mediated antioxidant enzymes and anti-apoptotic factors such as Bcl-2 and BCL-XL, or suppresses the expression of caspase-3, but also promotes the efflux of long-chain acyl CoA from mitochondria and β-oxidation of carnitine, thereby suppressing MPT may be one of the mechanisms by which carnitine suppresses apoptosis.

In addition to these reports, there are several reports about carnitine as following:

Carnitine (0.5 or 5 mM) suppressed oxidative stress damage in cultured rat cardiomyocytes (H9c2 cells) induced by high glucose [47]. Carnitine addition increased the expression and activity of SOD, catalase, etc. in a model of H_2O_2-stimulated cell damage in human hepatocytes (HL7702 cells) [48]. Carnitine (1 or 5 mM) increased the expression and activity of podocin and protects podocytes (glomerular peduncle cells) in the kidney from damage caused by hyperglycemia by increasing the expression of podocin and nephrin [49]. Carnitine protects against the cellular damage of neuroblastoma cells (SH-SY5Y cells) and cortical neurons caused by glyceraldehyde, an inhibitor of glycolysis, by suppressing ROS, maintaining mitochondrial membrane potential, maintaining ATP synthesis, and decreasing Tau protein, suggesting that carnitine may also be effective against mitochondrial dysfunction caused by Tau protein accumulation in Alzheimer's disease [50].

Mutomba et al. reported that carnitine suppresses apoptosis induced by anti-Fas antibodies

Jurkat cells derived from human leukemia T lymphocytes [51]. They also found that carnitine (5 mM) suppressed the recombinant caspase 3, 7, and 8, and palmitoyl carnitine cancelled carnitine-induced inhibition of caspase. The activity of caspase, which is important as a trigger for apoptosis, may be regulated by the balance of carnitine and palmitoyl carnitine.

Andrieu-Abadie et al. found that apoptosis caused by doxorubicin (0.5 μM) in rat isolated cardiomyocytes was accompanied by a decrease in sphingomyelin and an increase in ceramide due to activation of intracellular sphingomyelinase, and that carnitine (200 μg/mL) suppressed sphingomyelinase activity in a dose-dependent manner. It was reported to inhibit ceramide elevation and subsequent cell death [52].

7.4.3 Anti-inflammatory, Antioxidant, and Anti-apoptotic Effects of Carnitine (Animal-Level Studies)

Zhao et al. investigated the effects of carnitine administration (150 mg/kg/day, intraperitoneally, 1 month before ischemia–reperfusion) in a rat cardiac ischemia–reperfusion model (30 min ischemia, 4 h reperfusion) [53]. Ischemia–reperfusion significantly increased malondialdehyde (MDA) in myocardial tissue, lactate dehydrogenase (LDH), and creatine kinase (CK) in serum, and significantly decreased SOD, Na-K-ATPase, and Ca-Mg-ATPase activities in myocardial tissue, which are indicators of oxidative stress. In the carnitine-treated group, myocardial cell damage due to ischemia–reperfusion was prevented and was almost the same as that in the control group. In addition, when the mRNA expression of antioxidative enzymes SOD1, SOD2, and apoptosis-related genes was studied, caspase-3, caspase-8, and Bax increased two- to threefold, oxidative stress and apoptotic cells increased two- to threefold, and Bcl-2 mRNA expression was significantly decreased by ischemia–reperfusion, whereas these changes were prevented in the carnitine-treated group (Fig. 7.16).

In addition to this in vivo ischemia–reperfusion experiment, Zhao et al. performed a similar evaluation in an in vitro model in which cultured rat cardiomyocytes (H9c2) were placed in a hypoxic (95% N2) CO_2 incubator for 4 h and then reoxygenated, showing that carnitine enhances Nrf2 and HO-1 expression (Fig. 7.17).

Thus, in the myocardial ischemia–reperfusion model, carnitine administration was shown to activate the Nrf2-HO-1 system and exert antioxidant and cytoprotective effects.

Fan et al. reported the effects of L-carnitine administration in a model of cardiac injury in mice with radiation [54]. Gamma irradiation (3 Gy/day) of mice with cobalt-60 for 5 days resulted in myocardial damage, including decreased left ventricular contractility, a fourfold increase in TUNEL-stain positive apoptotic cells, and a marked increase in the expression of active caspase-3 and caspase-3, while the intraperitoneally administered carnitine before gamma irradiation However, in the groups treated with carnitine intraperitoneally before gamma irradiation (Treat 1: 100 mg/kg, Treat 2: 200 mg/kg), cardiac dysfunction and changes associated with apoptosis were suppressed. However, administration of SB203580 (15 mg/kg), an inhibitor of p38 MAPK, together with carnitine (200 mg/kg) (Treat 3) attenuated the effect of carnitine (Fig. 7.18).

In order to investigate the mechanism of action of carnitine, they examined the ratio of phosphorylated Nrf2, which is the active form, to Nrf2 (p-Nrf2/Nrf2), NQO1, HO-1, active phosphorylated p38 MAPK (p-p38 MAPK). NQO1, HO-1, p-Nrf2/Nrf2, and p-p38 MAPK were greatly reduced by gamma irradiation and recovered by carnitine treatment, while the protective effect of carnitine was attenuated in the presence of p38 MAPK inhibitors (Fig. 7.19). Taken together, carnitine reduces gamma-irradiation-induced myocardial damage and preserves cardiac function by enhancing antioxidant capacity and anti-apoptotic effects through signaling of the p38 MAPK/Nrf2 system.

Zambrano et al. reported the protective effect of carnitine in a model of hypertension created by NG-nitro-L-arginine methyl ester (L-NAME), an

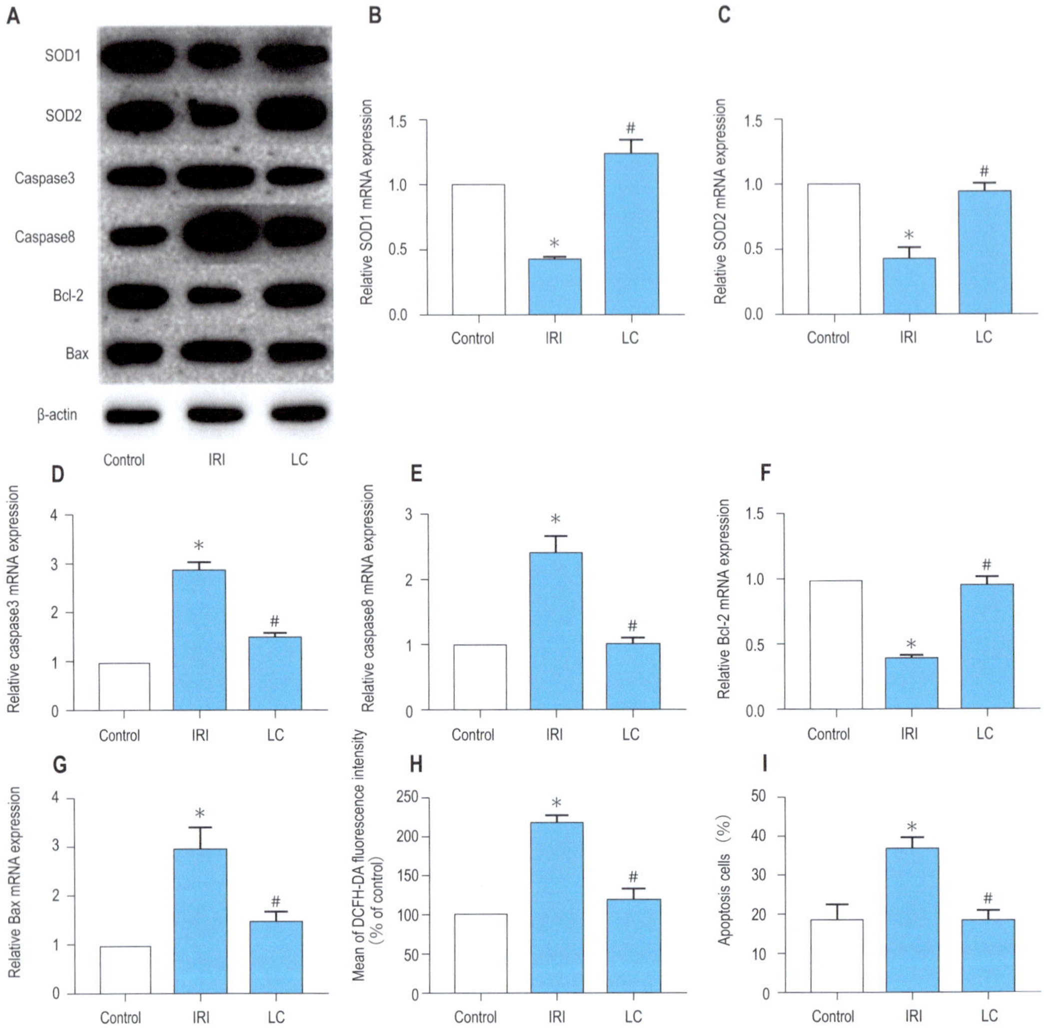

Fig. 7.16 L-carnitine reduced oxidative stress and apoptosis of rat cardiomyocytes in vivo. Expression of SOD1, SOD2, caspase8, Bcl-2, and Bax in the 3 groups was detected by Western blot (**a**) and qRT-PCR (**b**–**g**). ROS level (**h**) and cell apoptosis level (**i**) induced by DCFH-DA were determined by flow cytometry. *Control* control group, *IRI* ischemia–reperfusion group, *LC* L-carnitine administration group (150 mg/kg/day for 1 month) with ischemia–reperfusion. *: $P < 0.05$ vs Control, #:$P < 0.05$ vs IRI *DCFH-DA: 2′, 7′-dichlorodihydrofluorescein diacetate, an oxidant-sensing probe.* (Modified from ref. [53])

NO synthase inhibitor (25 mg/kg in drinking water for 10 weeks) to rats [55]. Compared to controls, rats treated with L-NAME for 10 weeks had significantly higher renal weight/body weight ratio and urine protein/100 g BW/24 h ($P < 0.001$ and $P < 0.01$), significantly lower creatinine clearance ($P < 0.05$), and significantly higher diastolic and systolic blood pressure ($P < 0.001$), indicating kidney impairment. Carnitine administration (400 mg/kg, drinking water for 12 weeks from 2 weeks before L-NAME) normalized the renal weight/body weight ratio and creatinine clearance, and significantly improved proteinuria and blood pressure. In addition, L-NAME administration significantly reduced NOx in the renal cortex and significantly increased lipid peroxide ($P < 0.001$), but carnitine administration normalized these levels. In addition, in rats treated with L-NAME, SOD activity, glutathione peroxidase activity, and their protein expression were signifi-

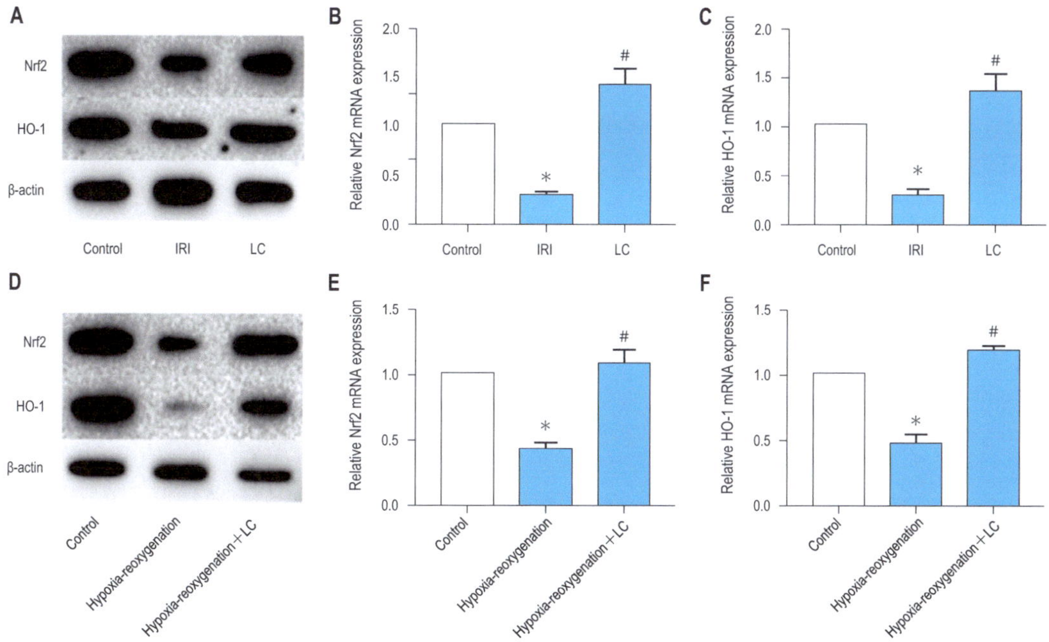

Fig. 7.17 L-carnitine increased the activity of Nrf2/HO-1 signaling pathway in rat myocardial tissue and H9c2 cells. A-C: Expression of Nrf2 and HO-1 in rat myocardial tissue was determined by Western blot (**a**) and qRT-PCR (**b**, **c**). D-F: Expression of Nrf2 and HO-1 in H9c2 cells was determined by Western blot (**d**) and qRT-PCR (**e**, **f**). *: $P < 0.05$ vs control, #: $P < 0.05$ vs IRI group or hypoxia-reoxygenation group. *IRI* ischemia–reperfusion injury, *HO-1* heme oxygenase 1, *LC* L-carnitine. (From ref. [53])

cantly reduced, and glutathione reductase activity and protein expression were increased. NADPH oxidase activity and the expression of NOX4 protein in it were enhanced, and superoxide was elevated, but carnitine administration normalized these changes. In addition, mRNA expression of endothelial nitric oxide synthase (eNOS) was significantly reduced by L-NAME administration, and mRNA expression of angiotensin-converting enzyme and angiotensin II receptor type 1 was significantly enhanced, but carnitine administration normalized these changes. In order to investigate the mechanism of the nephroprotective effects of carnitine, they examined the mRNA expression of PPARα, Nrf2, and NF-κB, and found that the mRNA expression of PPARα and Nrf2 was significantly reduced by L-NAME administration, but these changes were normalized by carnitine administration (Fig. 7.20).

Thus, carnitine administration was thought to exert its nephroprotective effect by activating Nrf2 and PPARα and decreasing the activity of NF-κB, causing a decrease in the activity of NADPH oxidase, which in turn decreases superoxide and other reactive oxygen species.

In addition to these studies, Salama et al. reported that carnitine alleviated lung injury caused by potassium dichromate, a hexavalent chromium, by activating the Nrf2-HO-1 signaling pathway [56], and Liu et al. reported that carnitine administration enhanced HO-1 and Bcl-2 expression in a rat middle cerebral artery ischemia/reperfusion model by activating the signaling pathway of the Nrf2/ARE system and exerted antioxidant and anti-apoptotic effects [57].

Blanca et al. reported the mechanism of cardiotoxicity of the anticancer drug sunitinib, an inhibitor of various receptor-type tyrosine kinases, and the effect of carnitine administration. [58] Rats were administered 25 mg/kg/day of sunitinib in drinking water for 8 weeks. Carnitine was administered 2 weeks earlier (400 mg/kg/day, drinking water, for total of 10 weeks). The amount of drug in the drinking water was adjusted as body weight increased. Sunitinib significantly increased blood pressure, heart rate, cardiac weight, and

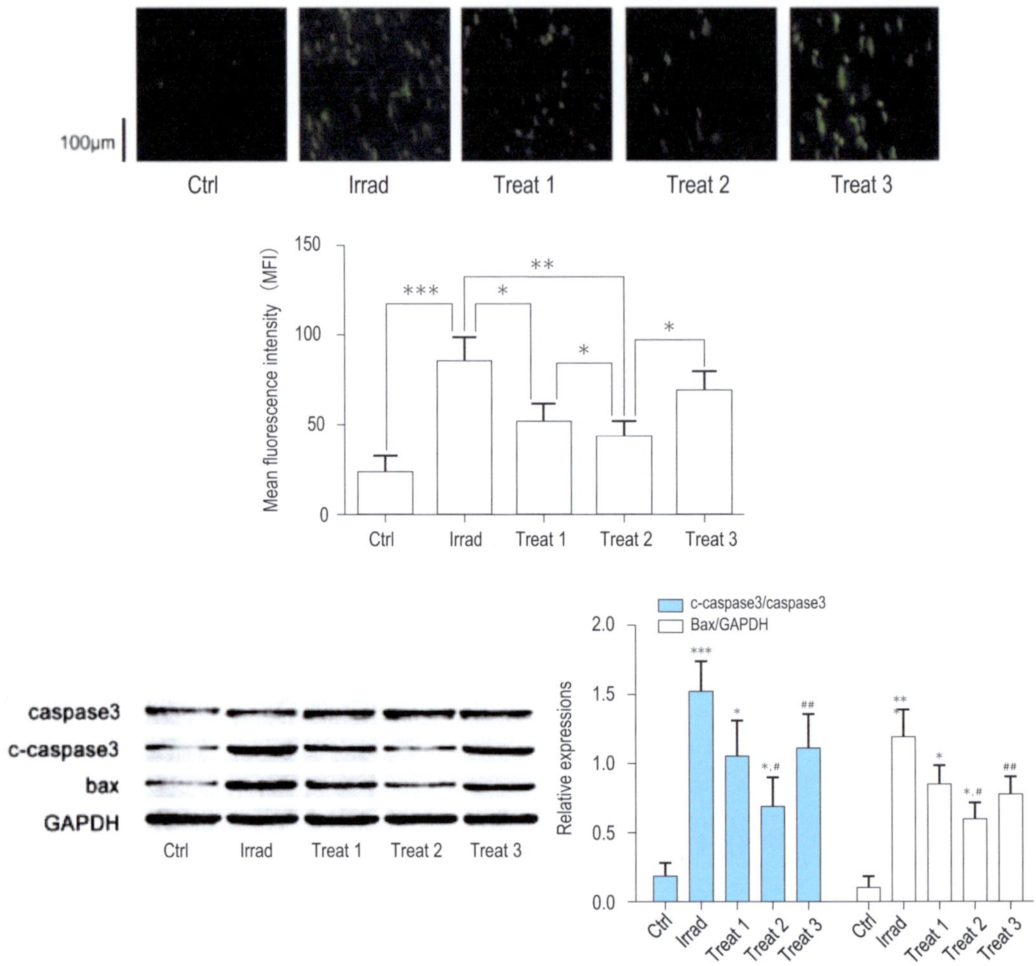

Fig. 7.18 Protective effects of l-carnitine in cardiomyocytes injury model irradiated with γ-ray. (**a**): The captured images from the TUNEL assay of cardiac tissue from Ctrl, Irrad, Treat 1, Treat 2, and Treat 3, respectively, and the quantification results. *: $P < 0.05$, **: $P < 0.01$ and ***: $P < 0.001$. (**b**): The expression of apoptosis facilitating factor Bax, active caspase3 (cleaved-caspase3 or c-caspase3) and caspase3. Ctrl: control (without irradiation), Irrad: irradiation (3 Gy/day, for 5 days), Treat 1: before irradiation, L-carnitine administration (100 mg/kg/day, i. p.), then irradiation. Treat 2: before irradiation, L-carnitine administration (200 mg/kg/day, i. p.), then irradiation. Treat 3: before irradiation, L-carnitine administration (200 mg/kg/day, i. p.) and p38 MAPK inhibitor SB203580 (15 mg/kg, i. p), then irradiation. *GAPDH* glycelaldehyde-3-phosphate dehydrogenase. (Modified from ref. [54])

increased mRNA expression of IL-1β, IL-6, IL-10, and NF-κB, but these changes were suppressed in the carnitine group. In addition, administration of sunitinib significantly increased mRNA expression of factors that promote tissue fibrosis, such as collagen type 1 (Col 1), TGF-β1, connective tissue growth factor (CTGF), and tissue inhibitor matrix metalloproteinase 1 (TIMP1), while mRNA expression of matrix metalloproteinase 9 (MMP9), which degrades collagen, was markedly suppressed. Carnitine treatment normalized these changes. Further examination of NF-κB translocation to the nucleus revealed that sunitinib treatment enhanced NF-κB translocation to the nucleus, whereas carnitine treatment prevented it (Fig. 7.21). Histologically, carnitine inhibited the increase in myocardial fibrosis induced by sunitinib administration.

In the field of liver diseases, cirrhosis and carcinogenesis caused by hepatitis C are becoming

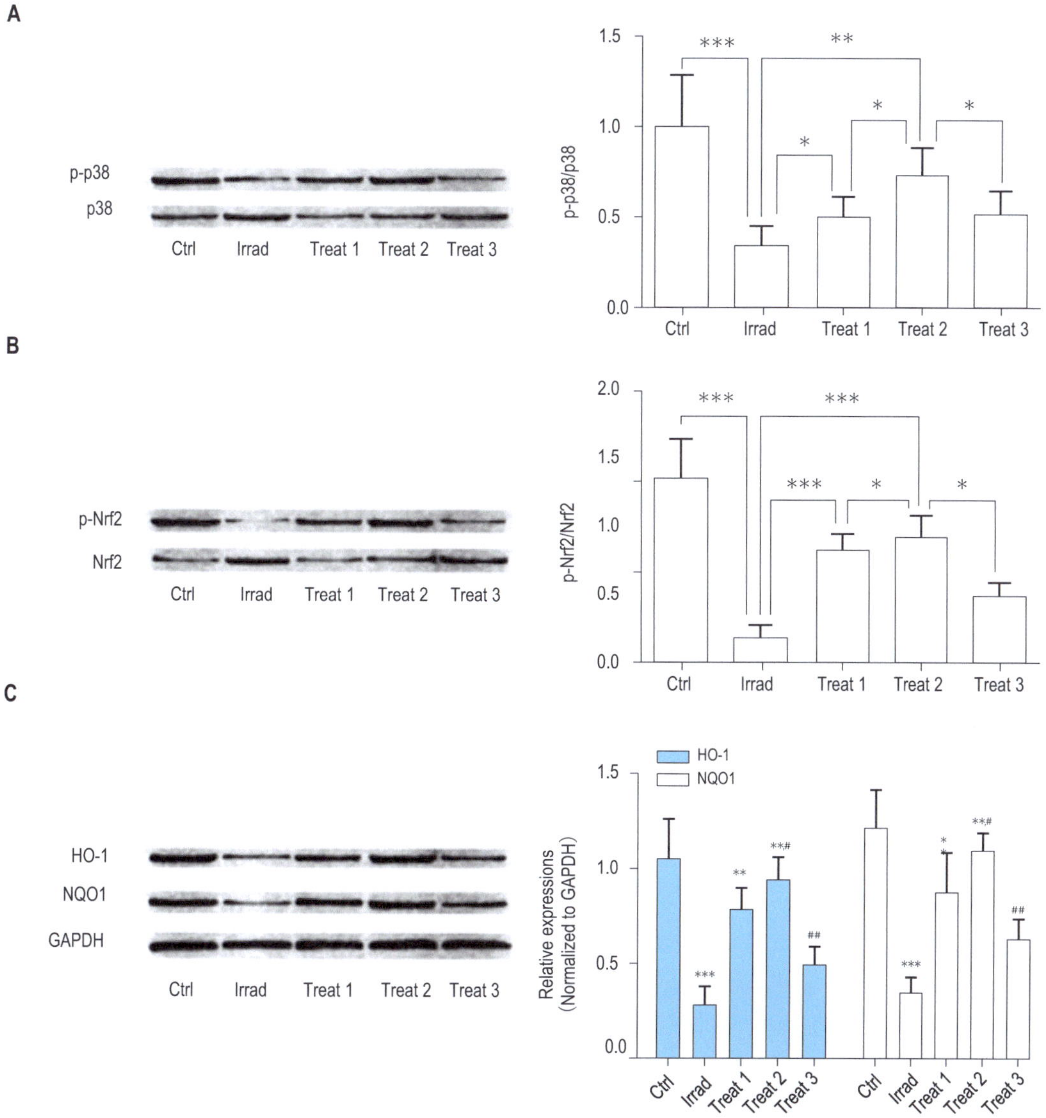

Fig. 7.19 Expression of NQO1, HO-1, pNrf2, Nrf2 and p-p38 MAPK in irradiated rat cardiomyocyte, and effects of L-carnitine and p38 MAPK inhibitor on them. Each group of horizontal axes are the same as Fig. 7.18a, b *: $P < 0.05$, **: $P < 0.01$, ***: $P < 0.001$. C: ***: $P < 0.001$ vs Ctrl, **: $P < 0.01$ vs Irrad, #: $P < 0.05$ vs treated 1, ##: $P < 0.01$ vs Treated 2. (From ref. [54])

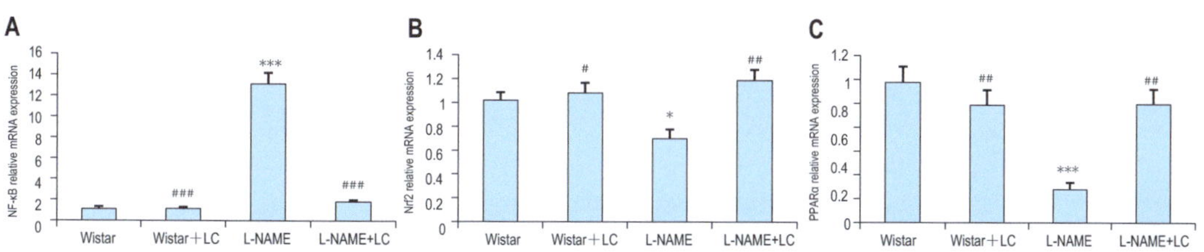

Fig. 7.20 The mRNA expression of NF-κB (**a**), Nrf2 (**b**) and PPARα (**c**) in Wistar rat kidney cortex after administration of L-carnitine, L-NAME and L-NAME + L-carnitine. L-NAME: NG-nitro-L-arginine methyl ester (25 mg/kg, 10 weeks, in drinking water), LC: L-carnitine (400 mg/kg, in drinking water, 2 weeks before L-NAME administration and total 10 weeks). *: $P < 0.05$ vs Wistar, ***: $P < 0.001$ vs Wistar, $^{\#}$: $P < 0.05$ vs L-NAME, $^{\#\#}$: $P < 0.01$ vs L-NAME, $^{\#\#\#}$: $P < 0.001$ vs L-NAME. (From ref. [55])

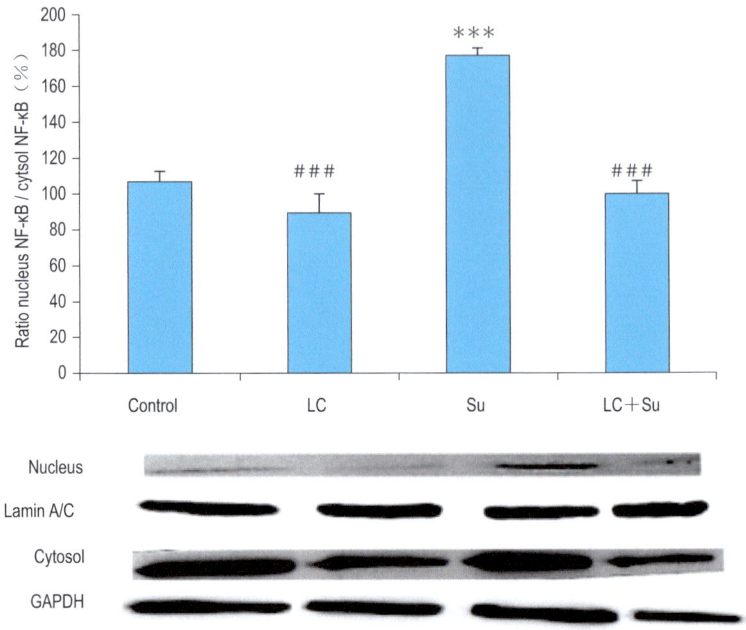

Fig. 7.21 Activation of NF-κB and migration into nucleus after sunitinib administration suppress effects of L-carnitine on it. *LC* L-carnitine (400 mg/kg, in drinking water, 10 weeks), *Su* sunitinib (25 mg/kg, in drinking water, 8 weeks), LC + Su: L-carnitine, in drinking water, before 2 weeks of sunitinib administration and more 8 weeks, and sunitinib for 8 weeks). ***: $P < 0.001$ vs Control, ###: $P < 0.001$ vs Su. (From ref. [58])

treatable, but the number of cases of cirrhosis and liver cancer due to NASH/NAFLD is increasing, and the establishment of a treatment method is an issue. Ishikawa et al. compared the effects of carnitine with α-tocopherol in STAM mice, a NASH model animal in which fat accumulation leads to carcinogenesis [59]. They found that carnitine treatment suppressed fatty liver formation, enhanced the expression of transporters, enzymes and antioxidant enzymes involved in long-chain fatty acid transport and mitochondrial β-oxidation, and strongly suppressed hepatic oxidative stress and proinflammatory cytokines. Carnitine administration significantly reduced the number of hepatocellular carcinomas (see Chap. 15 for details).

Canbaz et al. examined the effects of carnitine administration in an ischemia–reperfusion model of rat liver [60]. Ischemia–reperfusion increased ALT, AST, MDA, and myeloperoxidase (MPO) in plasma, and increased MDA and MPO and decreased T-AOC in liver tissue, but these changes were suppressed in the group of carnitine administration 2 h before ischemia–reperfusion (200 mg/kg, intravenously). Carnitine administration reduced the leukocyte fraction of plasma and liver tissue and increased T-AOC.

There are also numerous studies on renal impairment models. Liu et al. also reported that phospholipase A_2 activity and MDA levels were elevated in an ischemia–reperfusion model of rat kidneys, but these changes were suppressed when carnitine (500 mg/kg, intravenously) was administered 30 min before ischemia–reperfusion [61].

Myoglobin released from the muscles in rhabdomyolysis may cause acute kidney injury. Aydogdu et al. reported that in a model of myoglobin acute kidney injury induced by intramuscular glycerol injection, MDA is increased, and glutathione, SOD, catalase, and glutathione peroxidase are markedly decreased, but carnitine administration (200 mg/kg, intraperitoneal) suppressed these changes [62].

Kunak et al. reported that in a contrast nephropathy model in which rats were given glycerol (10 mL/kg of 25% glycerol, intramuscularly) and then contrast medium (iohexol, 8 mL/kg), blood BUN and creatinine increased, renal TNF-α, TGFβ, IL-1β, caspase-3, and MDA are markedly elevated, and SOD and glutathione are decreased, but carnitine administration (200 mg/kg and 400 mg/kg, intraperitoneally) suppressed these changes [63].

Boyacioglu et al. also observed a decrease in SOD, catalase, and glutathione in renal tissues and an increase in MDA in a rat model of contrast-induced nephropathy in which contrast media (iohexol, 10 mL/kg, intravenous) was administered, but carnitine administration (500 mg/kg, intraperitoneal) prevented these changes64) Carnitine administration (500 mg/kg, intraperitoneal) prevented these changes [64].

The anticancer drug cisplatin may cause kidney damage and small intestine damage. Chang et al. examined the effects of carnitine administration (200 mg/kg/day, administered subcutaneously 30 min before cisplatin administration) on pathological changes in the kidney and small intestine, lipid peroxide (TBARS), and renal cortical mitochondrial function when cisplatin (5 mg/kg, intraperitoneal administration) was administered to rats [65]. As a result, cisplatin administration increased plasma BUN and creatinine, lipid peroxide in the renal cortex and small intestine mucosal cells significantly increased, mitochondrial respiratory control index deteriorated, and apoptotic cells increased (TUNEL staining), but carnitine administration suppressed these changes.

Vescovo et al. established a model of right ventricular failure due to pulmonary hypertension in rats treated with monocrotaline (30 mg/kg, intraperitoneal) and reported the effect of carnitine on apoptosis of skeletal muscle in their heart failure rats [66]. In heart failure rats, Bcl-2 expression in anterior tibial myocytes was decreased, caspase-3 and -9 expression was increased, and apoptotic cells increased, but these changes were suppressed in the monocrotaline + carnitine group (50 mg/kg, drinking water administration). These findings suggest that carnitine may be effective in treating myopathy in patients with heart failure.

Busquets et al. studied the effects of carnitine administration (1 g/kg) in rats with AH-130 Yoshida ascites hepatoma as a model of cachexia in cancer patients [67]. The carnitine-treated rats had increased food intake and gastrocnemius muscle mass, and increased activity, locomotor speed, and walking distance. Carnitine administration decreased proteasome activity and related gene expression (ubiquitin, C8 proteasome subunit, MuRF-1). Carnitine treatment also decreased caspase-3 mRNA expression and inhibited apoptosis. In the gastrocnemius muscle extracted from these carcinoma rats, the addition of 50 μM carnitine directly inhibited proteolysis. These findings suggest that carnitine administration may be useful in the treatment of cancer-related cachexia.

These effects of carnitine on antioxidant, anti-apoptotic, and mitochondrial function are reviewed in a list of cellular and animal studies on L-carnitine, acetyl-L-carnitine, and propionyl-L-carnitine by Modanloo et al. [68] Moghaddas et al. also reviewed and tabulated the effects of L-carnitine (including acetyl-L-carnitine) on ischemia–reperfusion injury in the central nervous system and muscle [69]. Also, Ucar et al. reviewed the concept and issues of treating ischemia–reperfusion injury with Nrf2 activators, including L-carnitine [70].

7.4.4 Anti-inflammatory, Antioxidant, and Anti-apoptotic Effects of Carnitine (Clinical Studies)

As described above, studies at the cellular and animal levels have shown that carnitine has antioxidant, anti-inflammatory, anti-apoptotic effects, mitochondrial function-improvement, and mitochondrial growth effects, and carnitine is an extremely safe substance, so there are many studies at clinical level. Some of them are introduced below.

It is known that increased oxidative stress is involved in the deterioration of the disease in cardiovascular disease. Lee et al. conducted a randomized, placebo-controlled trial in patients with coronary artery disease with ≥50% stenosis of the major coronary arteries and evaluated MDA, SOD, catalase, and glutathione peroxidase before and after administration of placebo (19 patients) and carnitine (20 patients, 1000 mg/day, 12 weeks, orally. [71] As a result, MDA in plasma, which is an indicator of oxidative stress, was significantly reduced in the carnitine group, and SOD, catalase, and glutathione peroxidase, which are responsible for eliminating oxidative stress in red blood cells, were significantly increased (Fig. 7.22).

Lee et al. also conducted a similar study in patients with coronary artery disease, and examined plasma MDA, C-reactive protein (CRP), IL-6, and TNF-α in the placebo group (19 patients) and the carnitine group (20 patients, 1000 mg/day, 12 weeks), and found significant reductions in CRP, IL-6, and TNF-α in the carnitine group compared to placebo [72].

Pertosa et al. report that when white blood cells come into contact with the dialysis membrane during dialysis procedures, white blood cells are activated via the Jun N-terminal kinase (JUNK) system, resulting in elevated inflammatory cytokines such as IL-6 [73]. In addition, Pertosa et al.

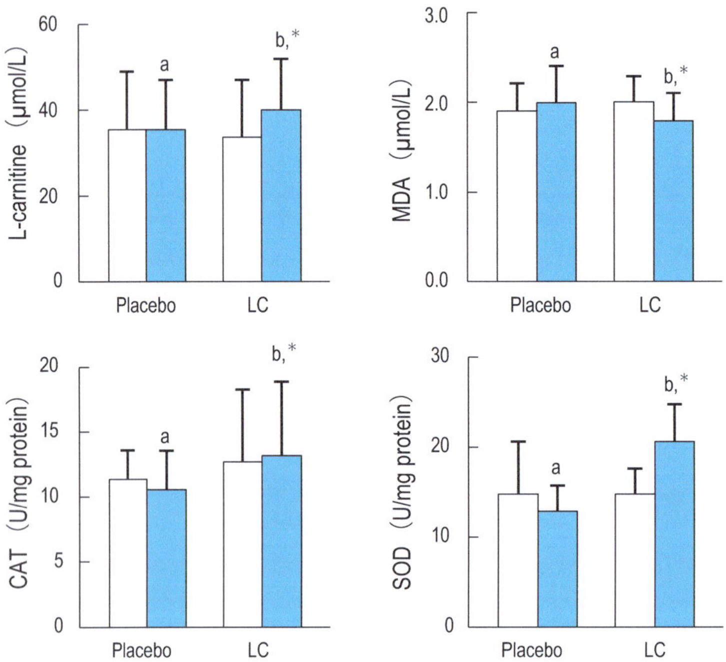

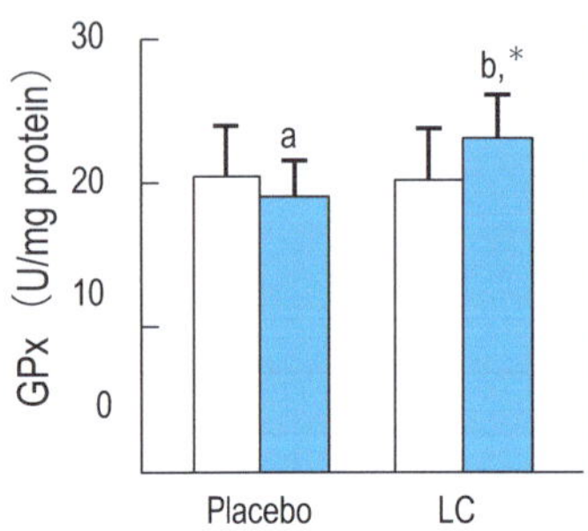

	Placebo (n=19)	LC (n=20)	P values#
LC (μmol/L)	−1.2±20.0 (0.8)	7.6±15.2 (7.2)	0.04
MDA (μmol/L)	0.1±0.5 (0.0)	−0.2±0.5 (−0.1)	0.03
CAT (U/mg protein)	−0.9± 3.4 (- 1.0)	2.0±6.9 (1.8)	0.02
SOD (U/mg protein)	−1.9±5.7 (- 1.2)	5.9±4.9 (6.8)	<0.01
GPx (U/mg protein)	−1.4±3.5 (- 1.3)	2.7±4.4 (2.4)	<0.01

mean ± SD (median).
values are significantly different between the placebo and LC groups. CAT : catalase, GPx : glutathione peroxidase, LC : L-carnitine, MDA : malondialdehyde, SOD : superoxide dismutase

Fig. 7.22 Change in L-carnitine levels, oxidative stress marker, and antioxidant enzymes activities after L-carnitine supplementation in patients with coronary artery disease. L-carnitine (LC, 1000 mg) was administered for 12 weeks. White bar: 0 week, Blue bar 12 weeks. *: $P < 0.05$ in same group after intervention, #: $P < 0.05$ between placebo and LC groups, [a, b]: $P < 0.05$ between different two groups. *MDA* malondialdehyde, *CAT* catalase, *SOD* superoxide dismutase, *GPx* glutathione peroxidase. (From ref. 71, modified) reported that carnitine administration (2 g intravenously after dialysis, 3 times a week for 3 months) significantly reduced JUNK activity in dialysis patients [74]. There are many reports of significant reductions in CRP and hsCRP with carnitine administration [75–79]. In addition, a decrease in serum amyloid A (SAA) [80], a decrease in advanced glycation end products (AGEs) [81], and a decrease in IL-6 have been reported (see Chap. 14 for details). CRP and SAA are produced in the liver, but their production is regulated by IL-6. Therefore, the decrease of CRP and SAA by carnitine is thought to be due to the suppression of JUNK activity and the decrease in the production of IL-6. Fatouros et al. also reported a decrease in MDA, a decrease in carbonylated proteins, an increase in glutathione, and a decrease in oxidized glutathione (GSSG) in the carnitine group (20 mg/kg, intravenously administered three times a week after dialysis) when dialysis patients were subjected to exercise load [82].

Koeth et al. of the Hazen group [83] claimed that carnitine and choline are degraded by the intestinal microbiota to become trimethylamine-N-oxide (TMAO) in the liver, and this TMAO promotes arteriosclerosis, which is a big debate (see Chap. 9 for details). Fukami et al., in a study on TMAO concentrations when carnitine was administered to dialysis patients, reported that carnitine administration increased TMAO and TMA concentrations, but significantly reduced MDA, an index of oxidative stress, and sVCAM-1 and sICAM-1, which are arteriosclerosis promoting factors [84] (see Chap. 9 for details). In addition, Higuchi et al. examined the effect of carnitine administration (20 mg/kg/day, oral administration, 12 months) on brachial-ankle pulse wave velocity (baPWV), which is an indicator of arteriosclerosis itself, and reported that carnitine administration significantly reduced baPWV [85] (for details, see Chap. 14).

The effects of inflammation and carnitine in dialysis patients are described in detail in a review article by Khalatbari-Soltani et al. [86] Haghighatdoost et al. also conducted a systematic review and meta-analysis of 13 trials that reported the effects of carnitine administration on inflammatory markers in patients with various diseases and healthy subjects and reported that carnitine administration showed a significant reduction in CRP, IL-6, and TNF-α [87] (Figs. 7.23, 7.24, and 7.25). Further research is needed to determine how this decrease in inflammatory mediators affects clinical outcomes.

Intravenous iron administration is often used to treat anemia in patients with chronic renal failure and dialysis patient and may increase oxidative stress and exacerbate the condition. Armaly et al. found that administration of carnitine (20 mg/kg, intravenously) before intravenous iron suppressed increases in advanced oxidative protein products (AOPP), a marker of oxidative stress and a factor that promotes atherosclerosis, and neutrophil gelatinase-associated lipocalin (NGAL), a marker of renal damage, and that oxidative stress during intravenous iron infusion was associated with a higher risk in patients with a haptoglobin phenotype of Hp2–2 [88].

Elevations of SOD, catalase, glutathione peroxidase, and T-AOC by carnitine administration were also observed in healthy subjects. Cao et al. administered carnitine (liquid, 2 g, single oral dose) to 12 healthy subjects and measured blood carnitine concentration and antioxidant enzyme activity over time and found that the carnitine concentration reached Cmax in 3.4 ± 0.46 h and gradually decreased. Glutathione peroxidase and T-AOC increased by 1.84-fold, 1.65-fold, 1.45-fold, and 1.52-fold, respectively [89].

In an unusual case, there is a clinical study in which carnitine was added to cardioplegia during cardiac surgery. Li et al. divided the 129 patients with rheumatic valvular heart disease into three groups of each 43 patients: control group with cardioplegic solution (Thomas' II. cardioplegic solution), a group of the cardioplegic solution with 6 g/L carnitine (group 1), and a group of the cardioplegic solution with 12 g/L carnitine (group 2), and heart valve replacement surgery with a heart-lung machine was performed. [90] There were no significant differences in age, sex, weight, left ventricular ejection fraction, cardiopulmonary device bypass time (about 80 h) and clamping time (about 70 min) between the three groups, and the results were the same in 32 mitral valve replacements, 6 aortic valve replacements, and 5

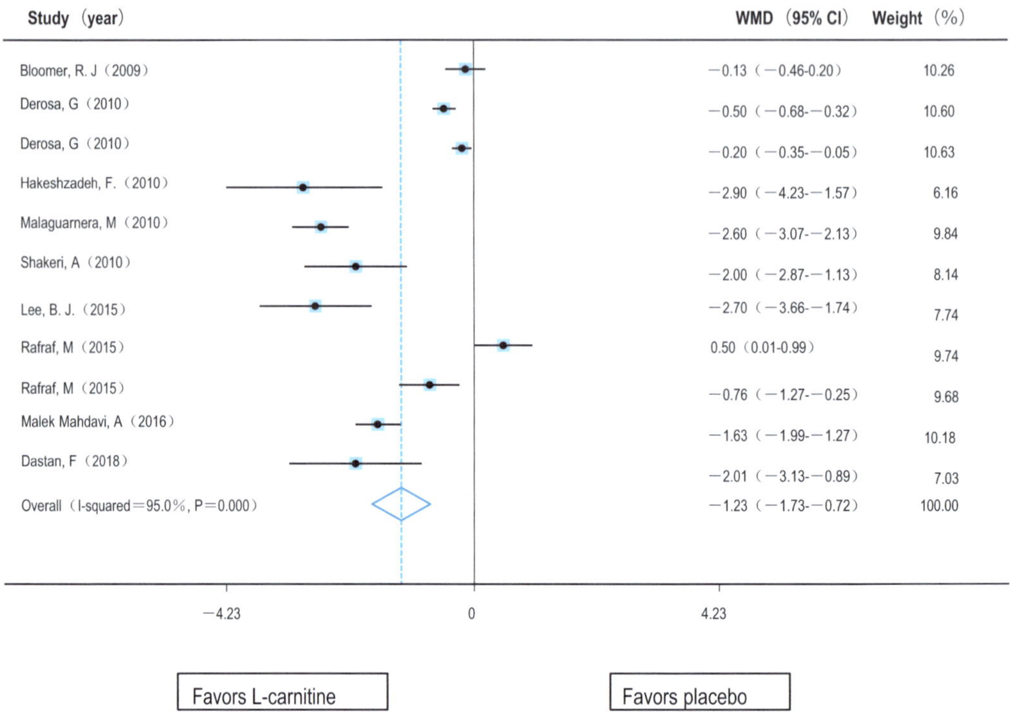

Fig. 7.23 Forest plot of the effect of L-carnitine supplementation on serum CRP. *WMD* weighted mean difference, *CI* confidence interval. (Modified from ref. [87])

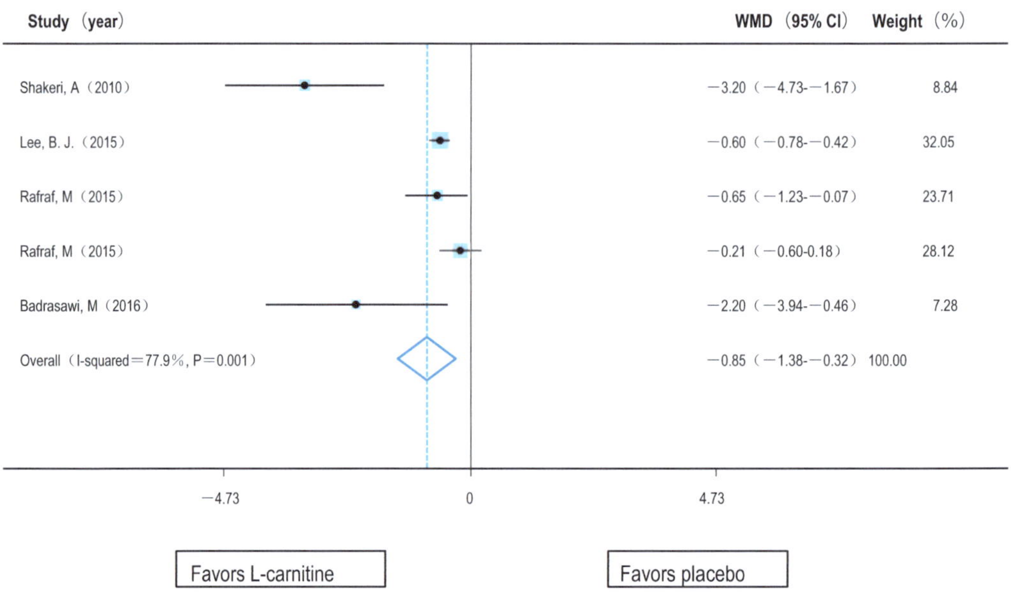

Fig. 7.24 Forest plot of effect of L-carnitine supplementation on serum IL-6. *WMD* weighted mean difference, *CI* confidence interval. (Modified from ref. [87])

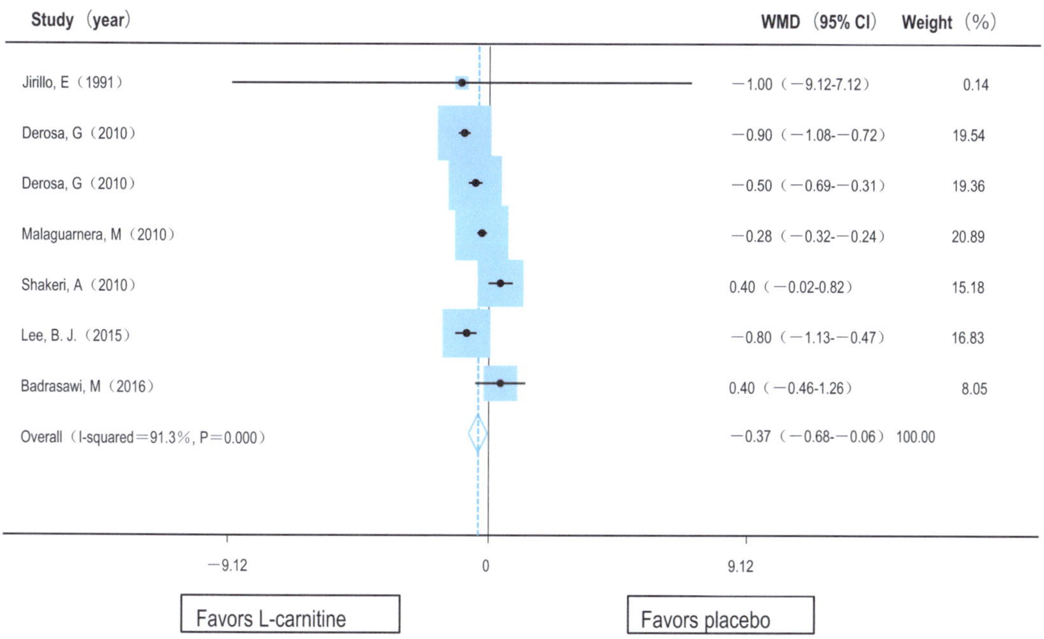

Fig. 7.25 Forest plot of the effect of L-carnitine supplementation on serum TNF-α. *WMD* weighted mean difference, *CI* confidence interval. (Modified from ref. [87])

aortic and mitral valve replacements, respectively. Serum and right atrial muscle before and after valve replacement surgery were collected, and various markers and histopathological images were compared. As a result, histological scores such as cardiomyocyte swelling and muscle fiber degeneration, as well as myeloperoxidase in myocardial tissue, which represents the infiltration of neutrophils into myocardial tissue, increased significantly after surgery, but the elevation was significantly suppressed in groups 1 and 2 to which carnitine was added. In addition, markers of cardiomyopathy disorders such as serum CK-MB, troponin C, high-sensitivity troponin C, and ischemic degenerative albumin were elevated 2, 6, 24, and 48 h after surgery, but the increase was significantly suppressed in the carnitine groups, and the left ventricular ejection fraction after surgery was significantly improved compared to the control. In addition, the postoperative enhancement of the expression of TNF-α, IL-6, and IL-1β proteins and RNA in myocardial tissue were significantly suppressed in the carnitine groups. In addition, SOD and catalase in myocardial tissue decreased significantly after surgery, and oxidative stress markers such as MDA and protein carbonyl increased, but the addition of carnitine significantly suppressed these changes. The expression of IκB decreased significantly and the expression of NF-κB (p65) increased after surgery, but the addition of carnitine significantly suppressed these changes (Fig. 7.26). In addition, the addition of carnitine enhanced the mRNA expression of Nrf2, which in turn increased the expression of HO-1 and NQO1 (Fig. 7.27).

From the above, the antioxidant effect of carnitine may be useful as an addition to cardioprotective fluids during cardiac surgery, and further studies are desired in the future.

7.4.5 Sarcopenia-MIA Syndrome and Carnitine

The treatment of sarcopenia and frailty has become a major issue in our aging society. There is also a condition called MIA syndrome, which is composed of three components: malnutrition, chronic inflammation, and atherosclerosis, and is also called protein-energy malnutrition (PEM).

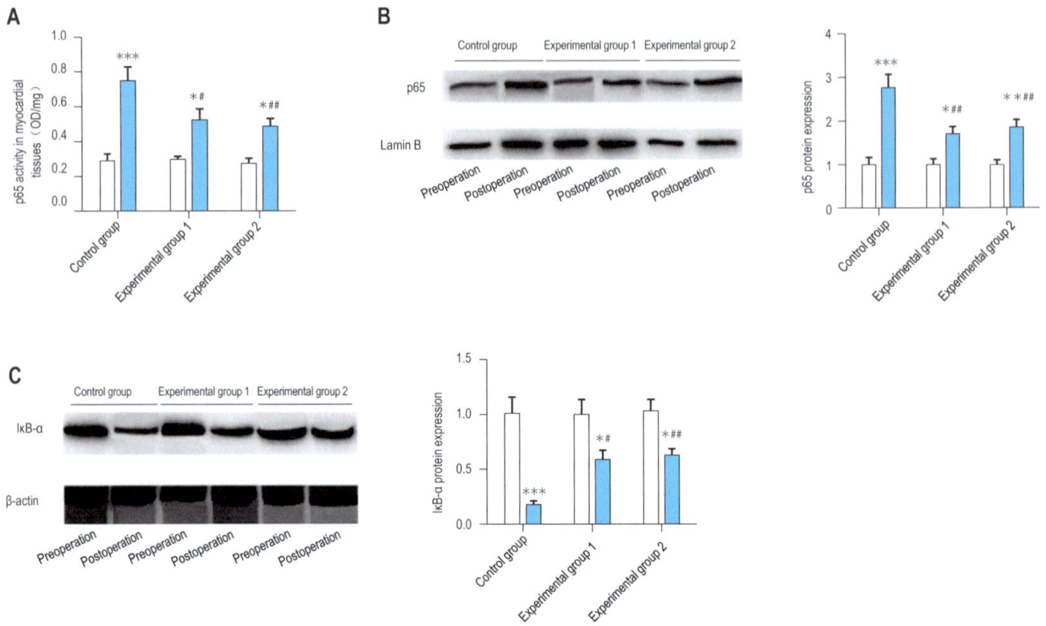

Fig. 7.26 L-carnitine suppressed NF-κB pathway in patients with rheumatic valvular heart disease (RVHD) undergoing valve replacement surgery. (**a**): p65 activity in the nuclear extract of heart tissue was determined. (**b**, **c**): Nuclear p65 and cytosolic IκB-α protein in heart tissue was detected by Western blot analysis. Control: St. Thomas'IIcardioplegic solution only, Experimental group 1: same cardioplegic solution +6 g/L of L-carnitine, Experimental group 2: same cardioplegic solution +12 g/L of L-carnitine. Data were presented as the mean ± SD ($n = 43$). *: $P < 0.05$, ***: $P < 0.001$ vs preoperation of the same group, #: $P < 0.05$, ##: $P < 0.01$ vs the corresponding control group, preoperation; , postoperation; . (From ref. [90]

The same disease concept is also referred to as protein-energy malnutrition (PEM) syndrome or protein-energy wasting (PEW) syndrome. Sarcopenia and MIA syndromes may occur early in some elderly patients [91, 92] and are especially prevalent in patients with cirrhosis [93–95], chronic kidney disease and dialysis patients [96, 97], heart failure patients [98, 99], and cancer patients [100, 101]. Although the pathogenesis of sarcopenia has not yet been fully elucidated, it is thought to be caused by a vicious cycle of aging, chronic low-level inflammation, increased muscle cell catabolism and apoptosis due to oxidative stress, and undernutrition due to anorexia.

Crentsil proposed the hypothesis that "carnitine deficiency may be the basis of the pathogenesis of sarcopenia, muscle weakness, and fatigue in the elderly and these diseases" [102] (Fig. 7.28).

Genetic mutations of the carnitine transporter (OCTN2) and CPT2 include severe mutations that develop immediately after birth, as well as weak and heterozygous mutations. It is hypothesized that these genetic mutations and age-related functional decline cause carnitine deficiency, which in turn causes mitochondrial dysfunction, which, in combination with apoptosis and increased catabolism of muscle cells due to chronic inflammation and oxidative stress, may lead to sarcopenia and frailty.

Ubiquitin ligase, the rate-limiting enzyme in the ubiquitin-dependent proteolytic pathway, is known to be an important causative enzyme in the disuse muscle atrophy seen in sarcopenia and other disorders [103].

Keller et al. found that when piglets were administered carnitine (500 mg/kg/day, feeding administration, 3 weeks), genes for ubiquitin-dependent proteolysis pathways in skeletal muscle and liver (atrogin-1, MuRF1, E214k, Psma1, Psmb1, ubiquitin) mRNA expression were significantly suppressed compared to the control

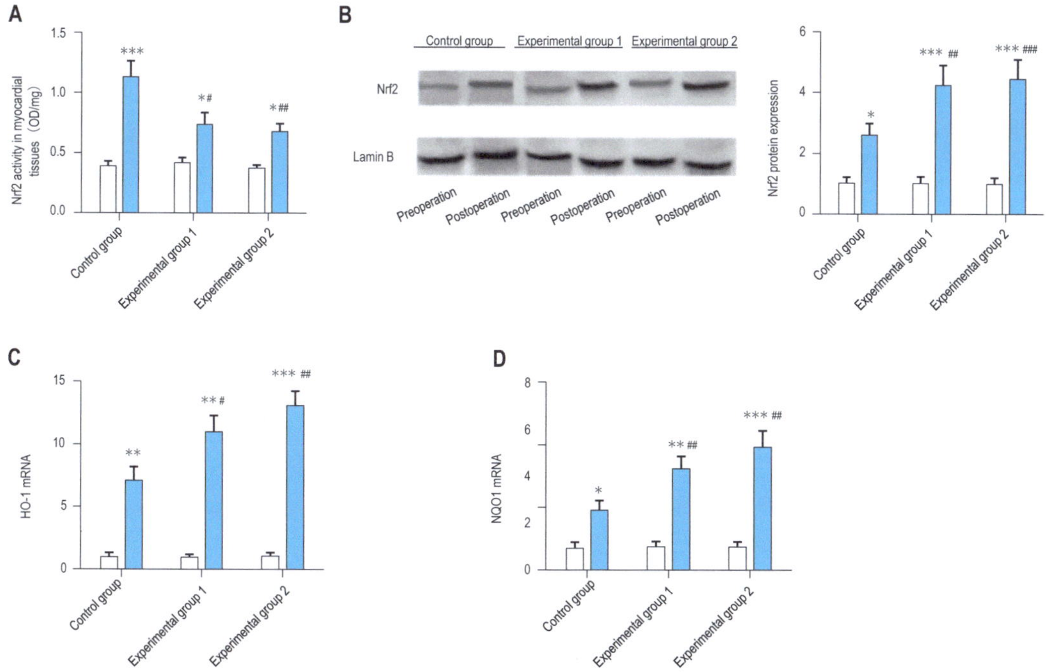

Fig. 7.27 L-carnitine activated Nrf2 pathway in patients with rheumatic valvular heart disease (RVHD) undergoing valve replacement surgery. (**a**): Nrf2 activity in the nuclear extract of heart tissue was determined. (**b**): Nuclear Nrf2 protein in heart tissue was detected by Western blot analysis. (**c, d**): Nrf2-regulated genes including HO-1 (**c**) and NOQ1 (**d**) mRNA in heart tissue were determined by real-time PCR. Control: St. Thomas'IIcardioplegic solution only, Experimental group 1: the same cardioplegic solution +6 g/L of L-carnitine, Experimental group 2: the same cardioplegic solution +12 g/L of L-carnitine. Data were presented as the mean ± SD (n = 43). *: $P < 0.05$, **: $P < 0.01$, ***: $P < 0.001$ vs preoperation of the same group, $^{\#}$: $P < 0.05$, $^{\#\#}$: $P < 0.01$, $^{\#\#\#}$: $P < 0.001$ vs the corresponding control group, preoperation; , postoperation; . *HO-1* heme oxygenase-1, *NOQ1* NAD(P)H quinone oxidoreductase. (From ref. [90])

group fed a low-carnitine diet [104], and when the mechanism was investigated in rat skeletal muscle, it was reported that carnitine administration activated the IGF-1/PI3K/Akt signaling system and downregulated E3 ligase MuRF1 [105].

Montesano et al. examined the effect of carnitine administration on myotube formation in C2C12 myoblasts, immortalized mouse skeletal muscle cells [106]. The results showed that carnitine promoted myotube formation and induced morphological changes seen in the early stages of muscle hypertrophy. Carnitine also enhanced IGF-1 receptor expression, AKT activation, and its downstream signaling.

Badrasawi et al. conducted a randomized, double-blind, placebo-controlled study in which 50 pre-frail elderly patients were divided into a carnitine group (26 patients, 1.5 g/day, orally administered, for 10 weeks) and a placebo group (24 patients). [107] The results showed that significantly improved frailty index and grip strength in the carnitine group. However, there was a trend toward a decrease in IL-6 and an increase in IGF-1 in the carnitine group; the differences were not significant.

Tokuchi et al. examined changes in psoas muscle mass index (PMI) by subsequent CT scan in patients with hepatitis C virus infection who were treated with direct acting antivirals (DAA) and showed a sustained virological response (SVR) [108]. The results of this study showed that the psoas muscle mass index (PMI) of patients who showed an SVR to DAA was significantly lower than that of patients who did not. They found that 41 of 58 patients (70.7%) showed an increase in

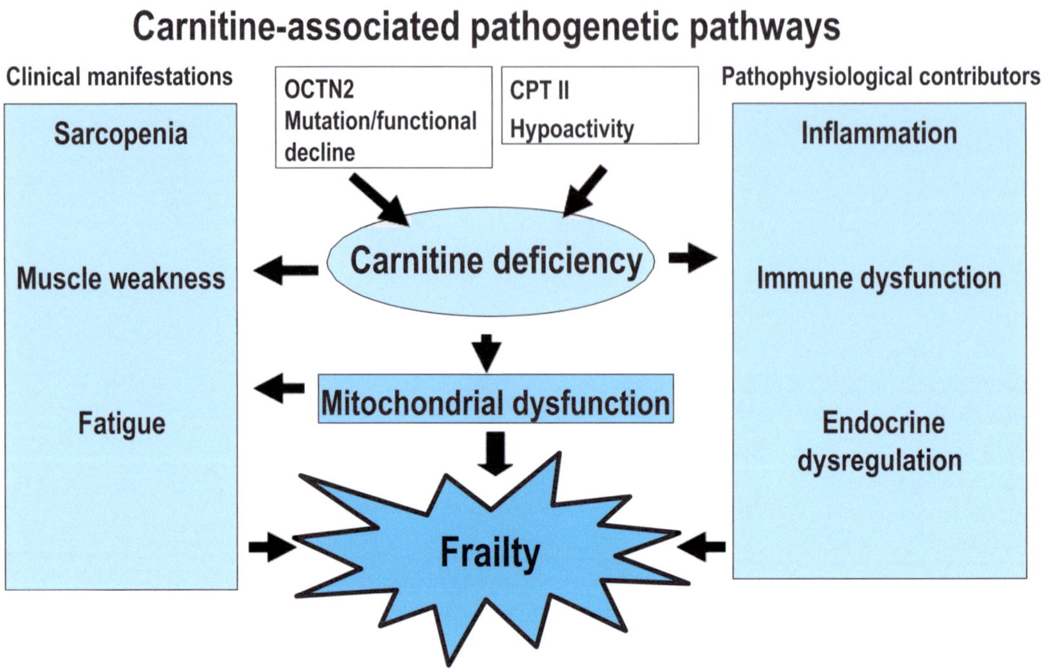

Fig. 7.28 Potential pathogenetic relationships between carnitine deficiency and geriatric frailty (Crentsil's hypothesis). *OCTN2* organic cation transporter 2, *CPT II* carnitine palmitoyltransferase II. (Modified from ref. [102]

PMI at 48 weeks after SVR. They examined clinical laboratory parameters that correlated with an increase in PMI, but there was no correlation between the parameters performed in routine tests, and only serum zinc, total carnitine, and free carnitine were elevated after DAA therapy, and the increase in free carnitine and PMI was significantly correlated ($r = 0.305$, $P = 0.02$). This suggests that carnitine supplementation may benefit from muscle mass loss in patients with hepatitis C who have undergone DAA therapy (see Chap. 15 for details on sarcopenia in liver disease).

Malaguarnera et al. investigated the relationship between serum carnitine concentration and frailty score and other parameters in 521 outpatients aged 65 years or older without serious diseases or cognitive impairment [109]. The results showed that plasma total carnitine and free carnitine concentrations in the frail (187 patients), pre-frail (166 patients) and healthy (168 patients) groups were significantly lower in the frail group than in the pre-frail and healthy groups, and the pre-frail group was significantly lower than the healthy group. Conversely, acylcarnitine concentrations were significantly higher in the frail group and significantly lower in the healthy group than in the other groups. The results suggest that maintaining high blood carnitine levels may have a positive effect on the maintenance of exercise function and may be a possible treatment for frailty in the elderly.

Ringseis et al. reviewed the complex mechanisms of carnitine replacement therapy in the treatment of muscle-wasting disorders such as sarcopenia and frailty, dividing it into basic and clinical studies [110]. The mechanism of muscle atrophy in pathologies such as sarcopenia is that the promotion of proteolysis, the decrease in protein synthesis, the increase in apoptosis of muscle cells, and the activation of NF-κB induce chronic inflammation, and the elevation of inflammatory cytokines and oxidative stress promote the induction of the ubiquitin-dependent proteolysis system. The combination of the fact that the decline in mitochondrial function induces a ubiquitin-dependent proteolytic system is working. There are many reports in basic and clinical studies that carnitine blocks these processes. As specific examples in clinical research, the effects of carnitine in AIDS patients, dialysis patients, chronic obstructive pulmonary disease (COPD) patients, etc. are summarized in six tables. Specifically, the effects on the proteolytic system (6 studies), the effects on protein synthesis (9 studies), the effects on lymphocyte apoptosis (6 studies), the effects on inflammatory parameters (12 studies), the effects on oxidative stress (7 studies), and the effects on mitochondrial function maintenance and mitochondrial biosynthesis (3 studies) are described. Not only from basic research, but also from clinical research, carnitine supplementation may be useful in the treatment of amyotrophic pathologies. In other words, carnitine suppresses the decline of IGF-1 and promotes the synthesis of muscle proteins, suppresses the production of inflammatory cytokines and reactive oxygen species (ROS), suppresses muscle protein degradation, apoptosis, inflammation and oxidative stress by suppressing casspace3, and further inhibits muscle atrophy by maintaining mitochondrial function and promoting mitochondrial biosynthesis (Fig. 7.29).

Cancer patients also often suffer from cachexia, a condition characterized by muscle atrophy and anorexia. Silvério et al. found that muscle atrophy and fatigue in cachexia in cancer patients were also associated with chronic inflammation and anorexia, similar to the mechanisms of muscle atrophy and fatigue in other pathological conditions such as sarcopenia. Silvério et al. reported that, similar to the mechanism of muscle atrophy in sarcopenia and other pathological conditions, chronic inflammation and increased production of oxidative stress and inflammatory cytokines due to anticancer drugs and radiotherapy cause a vicious cycle, and that carnitine is a safe drug with anti-inflammatory and antioxidant effects, and therefore, carnitine is a promising candidate for the treatment of muscle atrophy and fatigue in cancer patients [111] (Fig. 7.30).

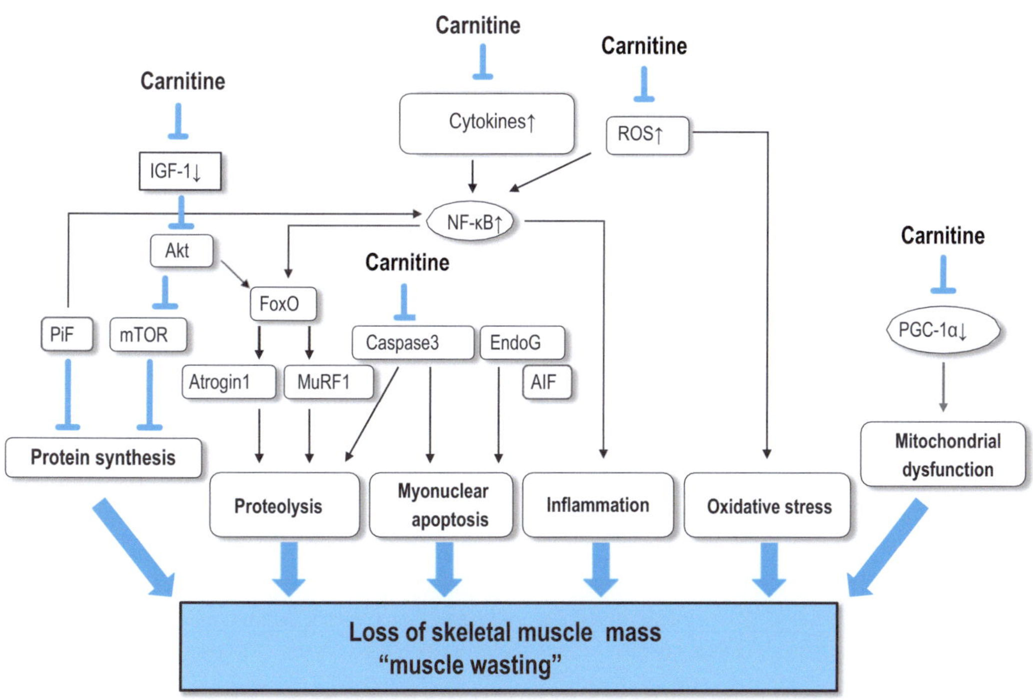

Fig. 7.29 Hypothetical model of mechanisms underlying the anti-wasting effects of carnitine supplementation under pathologic conditions. *IGF-1* insulin-like growth factor 1, *ROS* reactive oxygen species, *PGC-1α* peroxisome proliferator-activated receptor γ coactivator-1α, *NF-κB* nuclear factor κB, *PiF* proteolysis inducing factor, *mTOR* mechanistic target of rapamycin, *FoxO* forkhead box O transcription factor, *MuRF1* muscle-specific RING finger protein 1, *AIF* apoptosis-inducing factor, *EndoG* endonuclease G. (Modified from ref. [110])

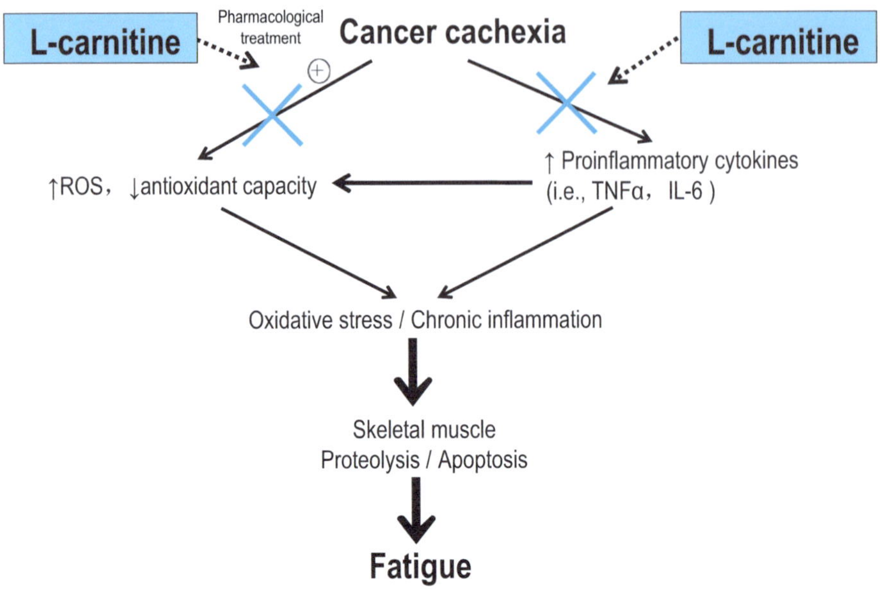

Fig. 7.30 Action mechanism of L-carnitine on cancer cachexia. (Modified from ref. [111]

Carnitine not only has the anti-inflammatory, antioxidant, and anti-apoptotic effects discussed in this chapter, but also maintains mitochondrial function, promotes mitochondrial growth, and stimulates appetite, and may be effective against cachexia in cancer patients. Further clinical studies are expected in the future.

7.4.6 Potential of Carnitine in Hereditary Neurometabolic Diseases and Neurodegenerative Diseases

Ribas et al. reported that hereditary neurometabolic diseases such as organic acidemia, maple syrup urine disease, and phenylketonuria are prone to carnitine deficiency due to restricted protein intake, and the condition is often exacerbated by increased oxidative stress, making carnitine supplementation with its antioxidant effects significan [112]. Other neurodegenerative diseases that occur in adults, such as amyotrophic lateral sclerosis (ALS), multiple sclerosis (MS), Parkinson's disease, and Alzheimer's disease, have been shown to be closely related to oxidative stress and increased inflammation. Guerreiro et al. reviewed the neuroprotective effects of carnitine toward oxidative stress and inflammatory processes, and these action and pointed out that carnitine might have potential as a therapeutic drug in some disorders [113]. Therefore, there is growing interest in the regulation of oxidative stress by Nrf2 activators [114, 115]. However, there are still issues to be solved in the wide use of phytochemicals such as sulforaphane as pharmaceuticals due to their bioavailability and their ability to cross the blood-brain barrier. In this sense, carnitine, which has already been used as a drug for various pathological conditions and has a high safety profile, is most promising for application as an Nrf2 activator in these neurodegenerative diseases.

In addition, in cerebral infarction, which occurs in about 290,000 people a year in Japan, oxidative stress and apoptosis have a significant impact on the progression and prognosis of the disease. Ueno et al. investigated the effects of carnitine in a chronic hypoperfusion model of in rats by ligation of the bilateral common carotid arteries (LBCCA) [116]. Carnitine (600 mg/kg/day in 0.7 mL saline, oral route, daily for 28 days) was continuously administrated after LBCCA. The escape latency in the Morris water maze test was 4–5 s before

LBCCA, but gradually increased to around 80 s at 28 days after LBCCA, indicating a decline in learning function and spatial memory. The number of apoptotic cells in the corpus callosum was significantly suppressed in the carnitine-treated group. This carnitine administration was not given before but after LBCCA. Therefore, they investigated the mechanism of the effect of carnitine administration on learning and cognitive function, and found that in the vehicle group, phosphorylated high molecular weight neurofilament (pNFH), a marker of axons, and myelin basic protein (MBP), a marker of mature oligodendrocytes, decreased in the vehicle group, but significantly increased in the carnitine group (Fig. 7.31), myelin sheath thickness also increased, and the number of pNFH positive + MBP positive nerve fibers increased significantly after carnitine treatment (Fig. 7.32).

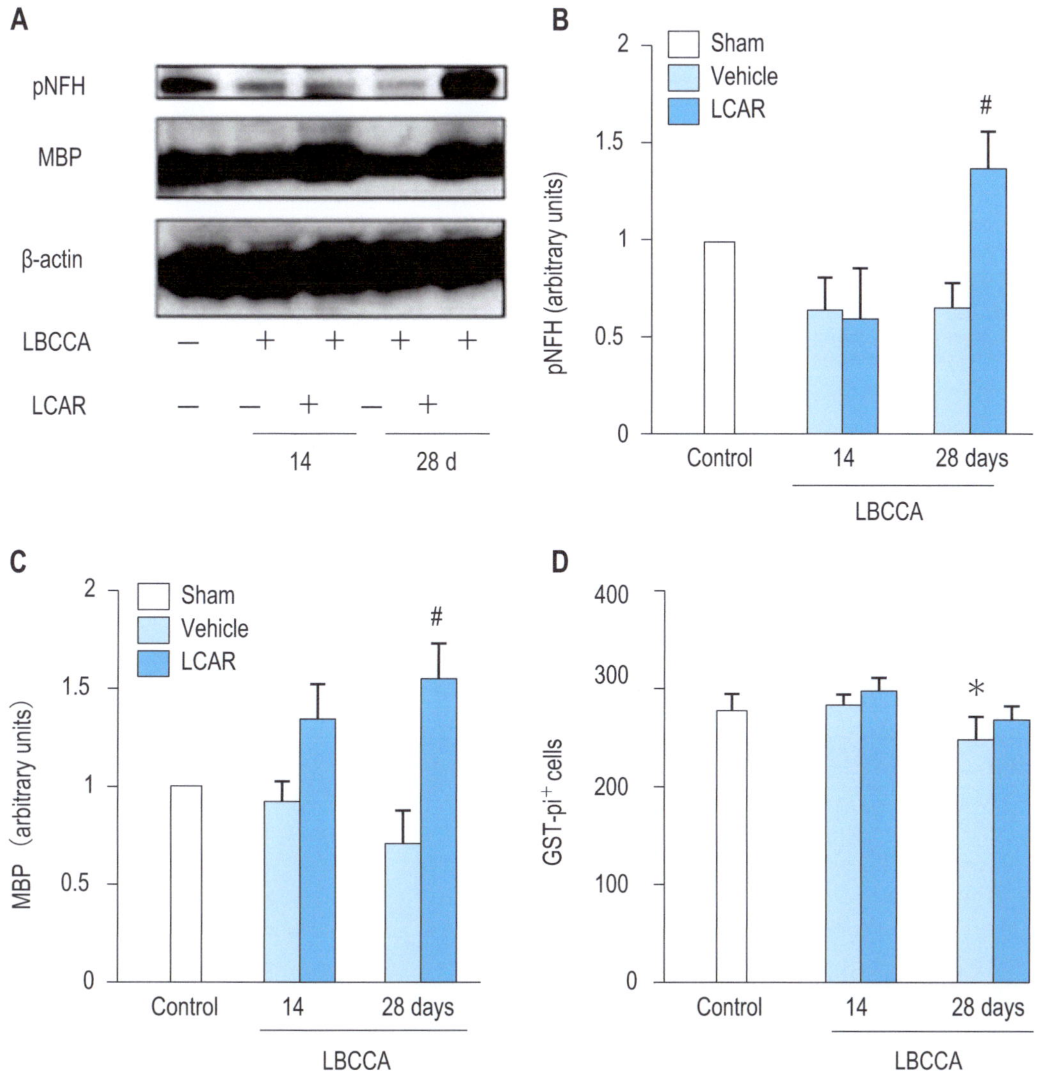

Fig. 7.31 L-carnitine increased phosphorylated high molecular weight neurofilament (pNFH) and myelin binding protein (MBP) after chronic hypoperfusion in rats with ligation of the bilateral common carotid artery (LBCCA). L-carnitine (LCAR, 600 mg/kg/day, oral route) was administration for 28 days after LBCCA. (**a**–**c**): pNFH and MBP levels were measured using Western bolts. β-actin was used as an internal control. N = 5/group. Values are expressed as Mean ± S.E., #: $P < 0.05$ vs vehicle-treated rats. (**d**): Quantitation of GST-pi+ cell numbers in the corpus collosum. *N* = 4/group. Values are expressed as Mean ± S.E., *: $P < 0.05$ vs sham-operated rats. *GST-pi*+ Glutathione-S-transferase pi (a marker of mature oligodendrocytes). (From ref. [116])

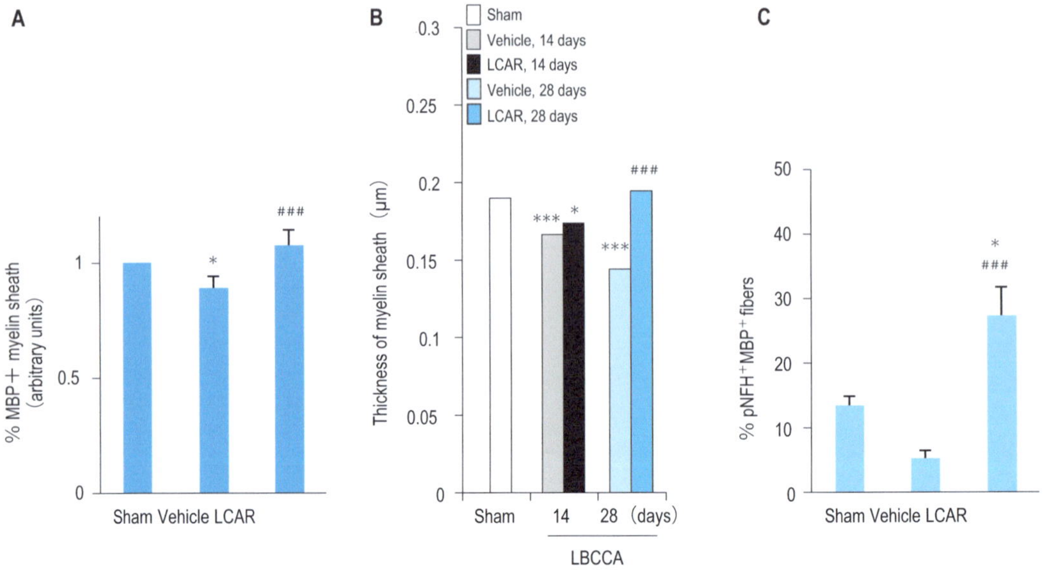

Fig. 7.32 L-carnitine increased myelin sheath thickness after chronic hypoperfusion by LBCCA. (**a**) Quantification of the diameter of MBP+ processes in the corpus collosum. $N = 4$/group. (**b**) Quantification of average myelin sheath thickness in sham-operated rats, and at 14 and 28 days after LBCCA in vehicle-treated and LCAR-treated rats. $N = 4$/group. (**c**): Quantification of pNFH+ axons colocalized with MBP+ processes in the corpus collosum. $N = 4$/group. Values are expressed as Mean ± S.E. *: $P < 0.05$, ***: $P < 0.001$ vs the sham-operated rats, and ###: $P < 0.001$ vs the vehicle-treated group. Abbreviations are same as Fig. 7.31. (Modified from ref. [116])

In the carnitine group, pPTEN/PTEN decreased, and pAkt/AKT and pmTOR/mTOR increased. In addition, 4-hydroxy-2-nonenal (HNE), a marker of lipid peroxidation, and 8-hydroxy-deoxyguanosine (8OHdG), a marker of oxidative DNA damage, were significantly lower in the carnitine group. In addition, mitochondrial CPT1 and CPT2 protein expression increased in the carnitine group. Thus, in the lesion formation of white matter due to chronic brain hypoperfusion, carnitine administration was thought to reduce phosphorylation of PTEN and increase phosphorylation of Akt and mTOR to reduce oxidative stress and promote the maintenance of mitochondrial function, and to protect nerves and improve cognitive impairment by promoting plasticity and myelin formation of axons.

To confirm the neuroprotective effect of carnitine in clinical practice, Ueno et al. compared 14 hemodialysis patients with carnitine deficiency to 7 patients who received no or short-term L-carnitine supplementation (NSTLC group: no or short-term L-carnitine treatment) and long-term L-carnitine treatment (LTLC group) and evaluated cognitive performance by blood biochemical tests and neuropsychological tests. Also, to evaluate micro-damage in the white matter of the brain in 7 patients who have undergone long-term carnitine replacement therapy (LTLC group), a new technique called a diffusion magnetic resonance imaging (dMRI), and the effect of carnitine was evaluated by comparing it with 10 age- and gender-matched healthy controls [117]. As a result, white matter microlesions were detected by dMRI more widely in dialysis patients than in healthy controls. However, in the LTLC group, white matter lesions by dMRI were reduced compared to the NSTLC group along with a significant decrease in hsCRP, and the achievement time was significantly shorter than in the NSTLC group in the neuropsychological trail making test A (76.2 ± 53.5 vs. 50.4 ± 21.3 sec, $P < 0.05$). Therefore, dMRI is a new method for evaluating white matter lesions in dialysis patients, and long-term carnitine supplementation in dialysis patients may reduce cerebral white matter lesions and maintain cognitive function.

7.5 Summary

As described above, it has been clarified that carnitine has anti-inflammatory, antioxidant, and anti-apoptotic effects at the cellular level, animal experiments, and clinical studies. Its mechanism of action is thought to be due to the fact that it acts as an Nrf2 activator in the Nrf2-Keap1 system in the same way as phytochemicals such as plant-derived sulforaphane, curcumin, and resveratrol, and increases the expression of antioxidant enzymes and proteins that work on cell protection. In addition to the suppression of the production of inflammatory cytokines and reactive oxygen species that induce apoptosis, carnitine directly suppresses caspase 3 and attenuates mitochondrial MPT induced by acyl groups of long-chain fatty acids by eliminating these acyl groups as acylcarnitine, as one of the well-known carnitine's actions. Further clinical studies are needed to determine whether the anti-inflammatory, antioxidant, and anti-apoptotic effects of carnitine are reduced in carnitine deficiency, and whether these effects may be exacerbated by carnitine deficiency, and in what conditions the anti-inflammatory, antioxidant, and anti-apoptotic effects should be applied, and if so, the dosage and administration of carnitine. Further clinical studies are desirable.

References

1. Hishikawa K, Nakaki T. NF-κB as a therapeutic drug target. Folia Pharmacol Jpn. 2001;118(3):197–202. (Article in Japanese)
2. Moi P, Chan K, Asunis I, et al. Isolation of NF-E2-related factor 2 (Nrf2), a NF-E2-like basic leucine zipper transcriptional activator that binds to the tandem NF-E2/AP1 repeat of the beta-globin locus control region. Proc Natl Acad Sci USA. 1994;91(21):9926–30. https://doi.org/10.1073/pnas.91.21.9926.
3. Itoh K, Igarashi K, Hayashi N, et al. Cloning and characterization of a novel erythroid cell-derived CNC family transcription factor heterodimerizing with the small Maf family proteins. Mol Cell Biol. 1995;15(8):4184–93. https://doi.org/10.1128/MCB.15.8.4184.
4. Itoh K, Wakabayashi N, Katoh Y, et al. Keap1 represses nuclear activation of antioxidant responsive elements by Nrf2 through binding to the amino-terminal Neh2 domain. Genes Dev. 1999;13(1):76–86. https://doi.org/10.1101/gad.13.1.76.
5. Itoh K, Wakabayashi N, Katoh Y, et al. Keap1 regulates both cytoplasmic-nuclear shuttling and degradation of Nrf2 in response to electrophiles. Genes Cells. 2003;8(4):379–91. https://doi.org/10.1046/j.1365-2443.2003.00640.x.
6. Kobayashi A, Kang MI, Okawa H, et al. Oxidative stress sensor Keap1 functions as an adaptor for Cul3-based E3 ligase to regulate proteasomal degradation of Nrf2. Mol Cell Biol. 2004;24(16):7130–9. https://doi.org/10.1128/MCB.24.16.7130-7139.2004.
7. Chen QM, Maltagliati AJ. Nrf2 at the heart of oxidative stress and cardiac protection. Physiol Genomics. 2018;50(2):77–97. https://doi.org/10.1152/physiolgenomics.00041.2017.
8. Shahcheraghi SH, Salemi F, Peirovi N, et al. Nrf2 regulation by curcumin: molecular aspects for therapeutic prospects. Molecules. 2022;27(1):167. https://doi.org/10.3390/molecules27010167.
9. Calabrese V, Cornelius C, Mancuso C, et al. Cellular stress response: a novel target for chemoprevention and nutritional neuroprotection in aging, neurodegenerative disorders and longevity. Neurochem Res. 2008;33(12):2444–71. https://doi.org/10.1007/s11064-008-9775-9.
10. Calabrese V, Cornelius C, Dinkova-Kostova AT, et al. Cellular stress responses, hormetic phytochemicals and vitagenes in aging and longevity. Biochim Biophys Acta. 2012;1822(5):753–83. https://doi.org/10.1016/j.bbadis.2011.11.002.
11. Calabrese EJ, Kozumbo WJ. The phytoprotective agent sulforaphane prevents inflammatory degenerative diseases and age-related pathologies via Nrf2-mediated hormesis. Pharmacol Res. 2021;163:105283. https://doi.org/10.1016/j.phrs.2020.105283.
12. Davinelli S, Saso L, D'Angeli F, et al. Astaxanthin as a modulator of Nrf2, NF-κB, and their crosstalk: molecular mechanisms and possible clinical applications. Molecules. 2022;27(2):502. https://doi.org/10.3390/molecules27020502.
13. Calabrese EJ. Hormesis and ginseng: ginseng mixtures and individual constituents commonly display hormesis dose responses, especially for neuroprotective effects. Molecules. 2020;25(11):2719. https://doi.org/10.3390/molecules25112719.
14. Talalay P. A fascination with enzymes: the journey not the arrival matters. J Biol Chem. 2005;280(32):28829–47. https://doi.org/10.1074/jbc.X500004200.
15. Yagishita Y, Fahey JW, Dinkova-Kostova AT, et al. Broccoli of sulforaphane: is it the source or dose that matters? Molecules. 2019;24(19):3593. https://doi.org/10.3390/molecules24193593.
16. Fahey JW, Holtzclaw WD, Wehage SL, et al. Sulforaphane bioavailability from glucoraphanin-rich broccoli: control by active endogenous myrosi-

nase. PLoS One. 2015;10(11):e0140963. https://doi.org/10.1371/journal.pone.0140963.
17. Houghton CA, Fassett RG, Coombes JS. Sulforaphane and other nutrigenomic Nrf2 activators: can the clinician's expectation be matched by the reality? Oxidative Med Cell Longev. 2016;2016:7857186. https://doi.org/10.1155/2016/7857186.
18. Grilc NK, Sova M, Kristl J. Drug delivery strategies for curcumin and other natural Nrf2 modulators of oxidative stress-related diseases. Pharmaceutics. 2021;13:2137.
19. Vesely O, Baldovska S, Kolesarova A. Enhancing bioavailability of nutraceutically used resveratrol and other stilbenoids. Nutrients. 2021;13(12):3095. https://doi.org/10.3390/pharmaceutics13122137.
20. Chin MP, Wrolstad D, Bakris GL, et al. Risk factors for heart failure in patients with type 2 diabetes mellitus and stage 4 chronic kidney disease treated with bardoxolone methyl. J Card Fail. 2014;20(12):953–8. https://doi.org/10.1016/j.cardfail.2014.10.001.
21. Chin MP, Reisman SA, Bakris GL, et al. Mechanisms contributing to adverse cardiovascular events in patients with type 2 diabetes mellitus and stage 4 chronic kidney disease treated with bardoxolone methyl. Am J Nephrol. 2014;39(6):499–508. https://doi.org/10.1159/000362906.
22. Chin MP, Bakris GL, Block GA, et al. Bardoxolone methyl improves kidney function in patients with chronic kidney disease stage 4 and type 2 diabetes: post-hoc analysis from bardoxolone methyl evaluation in patients with chronic kidney disease and type 2 diabetes study. Am J Nephrol. 2018;47(1):40–7. https://doi.org/10.1159/000486398.
23. Sakashita M, Tanaka T, Nangaku M. Effect of bardoxolone methyl on diabetic kidney disease. Jpn J Nephrol. 2019;61(4):482–9. (Article in Japanese)
24. Kanda H, Yamawaki K. Bardoxolone methyl: drug development for diabetic kidney disease. Clin Exp Nephrol. 2020;24(10):857–64. https://doi.org/10.1007/s10157-020-01917-5.
25. Milkovic L, Zarkovic N, Saso L. Controversy about pharmacological modulation of Nrf2 for cancer therapy. Redox Biol. 2017;12:727–32. https://doi.org/10.1016/j.redox.2017.04.013.
26. Geismann C, Arlt A, Sebens S, et al. Cytoprotection "gone astray": Nrf2 and its role in cancer. Onco Targets Ther. 2014;7:1497–518. https://doi.org/10.2147/OTT.S36624.
27. Wu S, Lu H, Bai Y. Nrf2 in cancers: a double-edged sword. Cancer Med. 2019;8(5):2252–67. https://doi.org/10.1002/cam4.2101.
28. Ito K. Pathological control by Nrf2 oxidative stress response system. J Jpn Biochem Soc. 2009;81(6):447–55. (Article in Japanese)
29. Ito K, Ebina M. Intracellular signaling cascade for the protection against oxidative stress: Nrf2/KEAP1 system. Kagaku to Seibutsu. 2012;50(6):423–9. (Article in Japanese)
30. Taguchi K. Integrative regulation of Nrf2 by two proteolytic systems. J Jpn Biochem Soc. 2016;88(5):630–8. (Article in Japanese)
31. Tonelli C, Chio IIC, Tuveson DA. Transcriptional regulation by Nrf2. Antioxid Redox Signal. 2018;29(17):1727–45. https://doi.org/10.1089/ars.2017.7342.
32. Bellezza I, Mierla AL, Minelli A. Nrf2 and NF-κB and their concerted modulation in cancer pathogenesis and progression. Cancers(Basel). 2010;2(2):483–97. https://doi.org/10.3390/cancers2020483.
33. Wardyn JD, Ponsford AH, Sanderson CM. Dissecting molecular cross-talk between Nrf2 and NF-κB response pathways. Biochem Soc Trans. 2015;43(4):621–6. https://doi.org/10.1042/BST20150014.
34. Cao Y, Li X, Wang CJ, et al. Role of NF-E2-related factor 2 in neuroprotective effect of L-carnitine against high glucose-induced oxidative stress in the retinal ganglion cells. Biomed Pharmacother. 2015;69:345–8. https://doi.org/10.1016/j.biopha.2014.12.030.
35. Montesano A, Senesi P, Vacante F, et al. L-carnitine counteracts in vitro fructose-induced hepatic steatosis through targeting oxidative stress markers. J Endocrinol Investig. 2020;43(4):493–503. https://doi.org/10.1007/s40618-019-01134-2.
36. Blanca AJ, Ruiz-Armenta MV, Zambrano S, et al. L-carnitine ameliorates the oxidative stress response to angiotensin II by modulating NADPH oxidase through a reduction in protein kinase c activity and NF-κB translocation to the nucleus. Food Chem. 2017;228:356–66. https://doi.org/10.1016/j.foodchem.2017.02.011.
37. Calò LA, Pagnin E, Davis PA, et al. Antioxidant effect of L-carnitine and its short chain esters: relevance for the protection from oxidative stress related cardiovascular damage. Int J Cardiol. 2006;107(1):54–60. https://doi.org/10.1016/j.ijcard.2005.02.053.
38. Ye J, Li J, Yu Y, et al. L-carnitine attenuates oxidant injury in HK-2 cells via ROS-mitochondria pathway. Regul Pept. 2010;161(1–3):58–66. https://doi.org/10.1016/j.regpep.2009.12.024.
39. Wang C, Sadovova N, Ali HK, et al. L-carnitine protects neurons from 1-methyl-4-phenylpyridinium-induced neuronal apoptosis in rat forebrain culture. Neuroscience. 2007;144(1):46–55. https://doi.org/10.1016/j.neuroscience.2006.08.083.
40. Ye J, Han Y, Chen X, et al. L-carnitine attenuates H_2O_2-induced neuron apoptosis via inhibition of endoplasmic reticulum stress. Neurochem Int. 2014;78:86–95. https://doi.org/10.1016/j.neuint.2014.08.009.
41. Chao HH, Liu JC, Hong HJ, et al. L-carnitine reduces doxorubicin-induced apoptosis through a prostacyclin-mediated pathway in neonatal rat cardiomyocytes. Int J Cardiol. 2011;146(2):145–52. https://doi.org/10.1016/j.ijcard.2009.06.010.

42. Wu CC, Bratton SB. Regulation of the intrinsic apoptosis pathway by reactive oxygen species. Antioxid Redox Signal. 2013;19(6):546–58. https://doi.org/10.1089/ars.2012.4905.
43. Furuno T, Kanno T, Arita K, et al. Roles of long chain fatty acids and carnitine in mitochondrial membrane permeability transition. Biochem Pharmacol. 2001;62(8):1037–46. https://doi.org/10.1016/s0006-2952(01)00745-6.
44. Oyanagi E, Yano H, Kato Y, et al. L-carnitine suppresses oleic acid-induced membrane permeability transition of mitochondria. Cell Biochem Funct. 2008;26(7):778–86. https://doi.org/10.1002/cbf.1506.
45. Oyanagi E, Yano H, Uchida M, et al. Protective action of L-carnitine on cardiac mitochondrial function and structure against fatty acid stress. Biochem Biophys Res Commun. 2011;412(1):61–7. https://doi.org/10.1016/j.bbrc.2011.07.039.
46. Nishimura M, Okimura Y, Fujita H, et al. Mechanism of 3-nitropropionic acid-induced membrane permeability transition of isolated mitochondria and its suppression by L-carnitine. Cell Biochem Funct. 2008;26(8):881–91. https://doi.org/10.1002/cbf.1521.
47. Vacante F, Senesi P, Montesano A, et al. L-carnitine: an antioxidant remedy for the survival of cardiomyocytes under hyperglycemic condition. J Diabetes Res. 2018;2018:4028297. https://doi.org/10.1155/2018/4028297.
48. Li JL, Wang QY, Luan HY, et al. Effects of L-carnitine against oxidative stress in human hepatocytes: involvement of peroxisome proliferator-activated receptor alpha. J Biomed Sci. 2012;19(1):32. https://doi.org/10.1186/1423-0127-19-32.
49. Fan JP, Kim D, Kawachi H, et al. Ameliorating effects of L-carnitine on diabetic podocyte injury. J Med Food. 2010;13(6):1324–30. https://doi.org/10.1089/jmf.2010.1079.
50. Magi S, Preziuso A, Piccirillo S, et al. The neuroprotective effect of L-carnitine against glyceraldehyde-induced metabolic impairment: possible implications in Alzheimer's disease. Cells. 2021;10(8):2109. https://doi.org/10.3390/cells10082109.
51. Mutomba MC, Yuan H, Konyavko M, et al. Regulation of the activity of caspases by L-carnitine and palmitoylcarnitine. FEBS Lett. 2000;478(1–2):19–25. https://doi.org/10.1016/s0014-5793(00)01817-2.
52. Andrieu-Abadie N, Jaffrézou JP, Hatem S, et al. L-carnitine prevents doxorubicin-induced apoptosis of cardiac myocytes: role of inhibition of ceramide generation. FASEB J. 1999;13(12):1501–10. https://doi.org/10.1096/fasebj.13.12.1501.
53. Zhao T, Chen S, Wang B, et al. L-carnitine reduces myocardial oxidative stress and alleviates myocardial ischemia-reperfusion injury by activating nuclear transcription-related factor 2 (Nrf2)/heme oxygenase-1 (HO-1) signaling pathway. Med Sci Monit. 2020;26:e923251. https://doi.org/10.12659/MSM.923251.
54. Fan Z, Han Y, Ye Y, et al. L-carnitine preserves cardiac function by activating p38 MAPK/Nrf2 signalling in hearts exposed to irradiation. Eur J Pharmacol. 2017;804:7–12.
55. Zambrano S, Blanca AJ, Ruiz-Armenta MV, et al. The renoprotective effect of L-carnitine in hypertensive rats is mediated by modulation of oxidative stress-related gene expression. Eur J Nutr. 2013;52(6):1649–59. https://doi.org/10.1007/s00394-012-0470-x.
56. Salama A, Fayed HM, Elgohary R. L-carnitine alleviated acute lung injuries induced by potassium dichromate in rats: involvement of Nrf2/HO-1 signaling pathway. Heliyon. 2021;7(6):e07207. https://doi.org/10.1016/j.heliyon.2021.e07207.
57. Liu SH, Zhang YC. Effect of levocarnitine on cerebral ischemia-reperfusion rats via activating Nrf2/ARE signaling pathway. Eur Rev Med Pharmacol Sci. 2019;23(23):8168–74. https://doi.org/10.26355/eurrev_201912_19650.
58. Blanca AJ, Ruiz-Armenta MV, Zambrano S, et al. Inflammatory and fibrotic processes are involved in the cardiotoxic effect of sunitinib: protective role of L-carnitine. Toxicol Lett. 2016;241:9–18. https://doi.org/10.1016/j.toxlet.2015.11.007.
59. Ishikawa H, Takaki A, Tsuzaki R, et al. L-carnitine prevents progression of non-alcoholic steatohepatitis in a mouse model with upregulation of mitochondrial pathway. PLoS One. 2014;9(7):e100627. https://doi.org/10.1371/journal.pone.0100627.
60. Canbaz H, Akca T, Tataroglu C, et al. The effects of exogenous L-carnitine on lipid peroxidation and tissue damage in an experimental warm hepatic ischemia-reperfusion injury model. Curr Ther Res Clin Exp. 2007;68(1):32–46. https://doi.org/10.1016/j.curtheres.2007.02.002.
61. Liu Y, Yan S, Ji C, et al. Metabolomic changes and protective effect of L-carnitine in rat kidney ischemia/reperfusion injury. Kidney Blood Press Res. 2012;35(5):373–81. https://doi.org/10.1159/000336171.
62. Aydogdu N, Atmaca G, Yalcin O, et al. Protective effects of L-carnitine on myoglobinuric acute renal failure in rats. Clin Exp Pharmacol Physiol. 2006;33(1–2):119–24. https://doi.org/10.1111/j.1440-1681.2006.04336.x.
63. Kunak CS, Ugan RA, Cadirci E, et al. Nephroprotective potential of carnitine against glycerol and contrast-induced kidney injury in rats through modulation of oxidative stress, proinflammatory cytokines, and apoptosis. Br J Radiol. 2016;89(1958):20140724. https://doi.org/10.1259/bjr.20140724.
64. Boyacioglu M, Turgut H, Akgullu C, et al. The effect of L-carnitine on oxidative stress responses of experimental contrast-induced nephropathy in rats. J Vet Med Sci. 2014;76(1):1–8. https://doi.org/10.1292/jvms.13-0202.

65. Chang B, Nishikawa M, Sato E, et al. L-carnitine inhibits cisplatin-induced injury of the kidney and small intestine. Arch Biochem Biophys. 2002;405(1):55–64. https://doi.org/10.1016/s0003-9861(02)00342-9.
66. Vescovo G, Ravara B, Gobbo V, et al. L-carnitine: a potential treatment for blocking apoptosis and preventing skeletal muscle myopathy in heart failure. Am J Physiol Cell Physiol. 2002;283(3):C802–10. https://doi.org/10.1152/ajpcell.00046.2002.
67. Busquets S, Serpe R, Toledo M, et al. L-carnitine: an adequate supplement for a multi-targeted anti-wasting therapy in cancer. Clin Nutr. 2012;31(6):889–95. https://doi.org/10.1016/j.clnu.2012.03.005.
68. Modanloo M, Shokrzadeh M. Analyzing mitochondrial dysfunction, oxidative stress, and apoptosis: potential role of L-carnitine. Iran J Kidney Dis. 2019;13(2):74–86.
69. Moghaddas A, Dashti-Khavidaki S. Potential protective effects of l-carnitine against neuromuscular ischemia-reperfusion injury: from experimental data to potential clinical applications. Clin Nutr. 2016;35(4):783–90. https://doi.org/10.1016/j.clnu.2015.07.001.
70. Ucar BI, Ucar G, Saha S, et al. Pharmacological protection against ischemia-reperfusion injury by regulating the Nrf2-Keap1-ARE signaling pathway. Antioxidants(Basel). 2021;10(6):823. https://doi.org/10.3390/antiox10060823.
71. Lee BJ, Lin JS, Lin YC, et al. Effects of L-carnitine supplementation on oxidative stress and antioxidant enzymes activities in patients with coronary artery disease: a randomized, placebo-controlled trial. Nutr J. 2014;13:79. https://doi.org/10.1186/1475-2891-13-79.
72. Lee BJ, Lin JS, Lin YC, et al. Antiinflammatory effects of L-carnitine supplementation (1000 mg/d) in coronary artery disease patients. Nutrition. 2015;31(3):475–9. https://doi.org/10.1016/j.nut.2014.10.001.
73. Pertosa G, Grandaliano G, Soccio M, et al. Vitamin E-modified filters modulate Jun N-terminal kinase activation in peripheral blood mononuclear cells. Kidney Int. 2002;62(2):602–10. https://doi.org/10.1046/j.1523-1755.2002.00458.x.
74. Pertosa G, Grandaliano G, Simone S, et al. Inflammation and carnitine in hemodialysis patients. J Ren Nutr. 2005;15(1):8–12. https://doi.org/10.1053/j.jrn.2004.09.031.
75. Savica V, Santoro D, Mazzaglia G, et al. L-carnitine infusions may suppress serum C-reactive protein and improve nutritional status in maintenance hemodialysis patients. J Ren Nutr. 2005;15(2):225–30. https://doi.org/10.1053/j.jrn.2004.10.002.
76. Duranay M, Akay H, Yilmaz FM, et al. Effects of L-carnitine infusions on inflammatory and nutritional markers in haemodialysis patients. Nephrol Dial Transplant. 2006;21(11):3211–4. https://doi.org/10.1093/ndt/gfl356.
77. Suchitra MM, Ashalatha VL, Sailaja E, et al. The effect of L-carnitine supplementation on lipid parameters, inflammatory and nutritional markers in maintenance hemodialysis patients. Saudi J Kidney Dis Transpl. 2011;22(6):1155–9.
78. Hakeshzadeh F, Tabibi H, Ahmadinejad M, et al. Effects of L-carnitine supplement on plasma coagulation and anticoagulation factors in hemodialysis patients. Ren Fail. 2010;32(9):1109–14. https://doi.org/10.3109/0886022X.2010.510617.
79. Shakeri A, Tabibi H, Hedayati M. Effects of L-carnitine supplement on serum inflammatory cytokines, C-reactive protein, lipoprotein (a), and oxidative stress in hemodialysis patients with Lp (a) hyperlipoproteinemia. Hemodial Int. 2010;14(4):498–504. https://doi.org/10.1111/j.1542-4758.2010.00476.x.
80. Tabibi H, Hakeshzadeh F, Hedayati M, et al. Effects of L-carnitine supplement on serum amyloid A and vascular inflammation markers in hemodialysis patients: a randomized controlled trial. J Ren Nutr. 2011;21(6):485–91. https://doi.org/10.1053/j.jrn.2011.01.001.
81. Fukami K, Yamagishi S, Sakai K, et al. Potential inhibitory effects of L-carnitine supplementation on tissue advanced glycation end products in patients with hemodialysis. Rejuvenation Res. 2013;16(6):460–6. https://doi.org/10.1089/rej.2013.1459.
82. Fatouros IG, Douroudos I, Panagoutsos S, et al. Effects of L-carnitine on oxidative stress responses in patients with renal disease. Med Sci Sports Exerc. 2010;42(10):1809–18. https://doi.org/10.1249/MSS.0b013e3181dbacab.
83. Koeth RA, Wang Z, Levison BS, et al. Intestinal microbiota metabolism of L-carnitine, a nutrient in red meat, promotes atherosclerosis. Nat Med. 2013;19(5):576–85. https://doi.org/10.1038/nm.3145.
84. Fukami K, Yamagishi S, Sakai K, et al. Oral L-carnitine supplementation increases trimethylamine-N-oxide but reduces markers of vascular injury in hemodialysis patients. J Cardiovasc Pharmacol. 2015;65(3):289–95. https://doi.org/10.1097/FJC.0000000000000197.
85. Higuchi T, Abe M, Yamazaki T, et al. Effects of levocarnitine on brachial-ankle pulse wave velocity in hemodialysis patients: a randomized controlled trial. Nutrients. 2014;6(12):5992–6004. https://doi.org/10.3390/nu6125992.
86. Khalatbari-Soltani S, Tabibi H. Inflammation and L-carnitine therapy in hemodialysis patients: a review. Clin Exp Nephrol. 2015;19(3):331–5. https://doi.org/10.1007/s10157-014-1061-3.
87. Haghighatdoost F, Jabbari M, Hariri M. The effect of L-carnitine on inflammatory mediators: a systematic review and meta-analysis of randomized clinical trials. Eur J Clin Pharmacol. 2019;75(8):1037–46. https://doi.org/10.1007/s00228-019-02666-5.

88. Armaly Z, Abd El Qader A, Jabbour A, et al. Effects of carnitine on oxidative stress response to intravenous iron administration to patients with CKD: impact of haptoglobin phenotype. BMC Nephrol. 2015;16:135. https://doi.org/10.1186/s12882-015-0119-0.
89. Cao Y, Qu HJ, Li P, et al. Single dose administration of L-carnitine improves antioxidant activities in healthy subjects. Tohoku J Exp Med. 2011;224(3):209–13. https://doi.org/10.1620/tjem.224.209.
90. Li M, Xu S, Geng Y, et al. The protective effects of L-carnitine on myocardial ischaemia-reperfusion injury in patients with rheumatic valvular heart disease undergoing CPB surgery are associated with the suppression of NF-κB pathway and the activation of Nrf2 pathway. Clin Exp Pharmacol Physiol. 2019;46(11):1001–12. https://doi.org/10.1111/1440-1681.13155.
91. Watanabe K. Mutritional management in the elderly. J Nihon Univ Med Ass. 2019;78(4):215–21. (Article in Japanese)
92. Mathewson SL, Azevedo PS, Gordon AL, et al. Overcoming protein-energy malnutrition in older adults in the residential care setting: a narrative review of causes and interventions. Ageing Res Rev. 2021;70:101401. https://doi.org/10.1016/j.arr.2021.101401.
93. Shiraki M, Hanai T. Epidemiology and clinical significance of sarcopenia in liver disease. J Japanese Soc Gastroenterol. 2018;115(5):424–9. (Article in Japanese)
94. Saeki C, Tsubota A. Influencing factors and molecular pathogenesis of sarcopenia and osteosarcopenia in chronic liver disease. Life (Basel). 2021;11(9):899. https://doi.org/10.3390/life11090899.
95. Traub J, Reiss L, Aliwa B, et al. Malnutrition in patients with liver cirrhosis. Nutrients. 2021;13(2):540. https://doi.org/10.3390/nu13020540.
96. Honda H, Akizawa T. MIA syndrome in dialysis patients. Jpn J Clin Dial. 2013;29(8):1175–80. (Article in Japanese)
97. Mori K. Maintenance of skeletal muscle to counteract sarcopenia in patients with advanced chronic kidney disease and especially those undergoing hemodialysis. Nutrients. 2021;13(5):1538. https://doi.org/10.3390/nu13051538.
98. Yamada S, Adachi T. Mechanism of frailty generation in heart failure. Jpn J Clin Physiol. 2019;49(3):125–9. (Article in Japanese)
99. Fernández-Pombo A, Rodríguez-Carnero G, Castro AI, et al. Relevance of nutritional assessment and treatment to counteract cardiac cachexia and sarcopenia in chronic heart failure. Clin Nutr. 2021;40(9):5141–55. https://doi.org/10.1016/j.clnu.2021.07.027.
100. Ohsawa M, Mori N, Kawamura K, et al. Recent molecular understanding in pathophysiology of cachexia its application for pharmacotherapy. Jpn J Pharm Palliat Care Sci. 2012;5:31–7. (Article in Japanese)
101. Amano K, Ishiki H. Pathophysiology of cancer cachexia and significance of nutritional support during the treatment in palliative care. Palliat Care Res. 2017;12(2):401–7. (Article in Japanese)
102. Crentsil V. Mechanistic contribution of carnitine deficiency to geriatric frailty. Ageing Res Rev. 2010;9(3):265–8. https://doi.org/10.1016/j.arr.2010.02.005.
103. Hirasaka K. Molecular nutritional studies on the prevention of muscle atrophy. J Jpn Soc Nutr Food Sci. 2014;67:291–7. (Article in Japanese)
104. Keller J, Ringseis R, Koc A, et al. Supplementation with L-carnitine downregulates genes of the ubiquitin proteasome system in the skeletal muscle and liver of piglets. Animal. 2012;6(1):70–8. https://doi.org/10.1017/S1751731111001327.
105. Keller J, Couturier A, Haferkamp M, et al. Supplementation of carnitine leads to an activation of the IGF-1/PI3K/Akt signalling pathway and down regulates the E3 ligase MuRF1 in skeletal muscle of rats. Forum Nutr Metab (Lond). 2013;10(1):28. https://doi.org/10.1186/1743-7075-10-28.
106. Montesano A, Senesi P, Luzi L, et al. Potential therapeutic role of L-carnitine in skeletal muscle oxidative stress and atrophy conditions. Oxid Med Cell Longev. 2015;2015:646171.
107. Badrasawi M, Shahar S, Zahara AM, et al. Efficacy of L-carnitine supplementation on frailty status and its biomarkers, nutritional status, and physical and cognitive function among prefrail older adults: a double-blind, randomized, placebo-controlled clinical trial. Clin Interv Aging. 2016;11:1675–86. https://doi.org/10.2147/CIA.S113287.
108. Tokuchi Y, Suda G, Kimura M, et al. Possible correlation between increased serum free carnitine levels and increased skeletal muscle mass following HCV eradication by direct acting antivirals. Sci Rep. 2021;11(1):16616. https://doi.org/10.1038/s41598-021-96203-z.
109. Malaguarnera G, Catania VE, Bonfiglio C, et al. Carnitine serum levels in frail older subjects. Nutrients. 2020;12(12):3887. https://doi.org/10.3390/nu12123887.
110. Ringseis R, Keller J, Eder K. Mechanisms underlying the anti-wasting effect of L-carnitine supplementation under pathologic conditions: evidence from experimental and clinical studies. Eur J Nutr. 2013;52:1421–42.
111. Silvério R, Laviano A, Fanelli FR, et al. L-carnitine and cancer cachexia: clinical and experimental aspects. J Cachexia Sarcopenia Muscle. 2011;2(1):37–44. https://doi.org/10.1007/s13539-011-0017-7.
112. Ribas GS, Vargas CR, Wajner M. L-carnitine supplementation as a potential antioxidant therapy for inherited neurometabolic disorders. Gene. 2014;533(2):469–76. https://doi.org/10.1016/j.gene.2013.10.017.

113. Guerreiro G, Deon M, Becker GS, et al. Neuroprotective effects of l-carnitine towards oxidative stress and inflammatory processes: a review of its importance as a therapeutic drug in some disorders. Metab Brain Dis. 2015;40(2):127. https://doi.org/10.1007/s11011-025-01545-6.
114. Sivandzade F, Bhalerao A, Cucullo L. Cerebrovascular and neurological disorders: protective role of NRF2. Int J Mol Sci. 2019;20(14):3433. https://doi.org/10.3390/ijms20143433.
115. Boas SM, Joyce KL, Cowell RM. The NRF2-dependent transcriptional regulation of antioxidant defense pathways: relevance for cell type-specific vulnerability to neurodegeneration and therapeutic intervention. Antioxidants(Basel). 2022;11(1):8. https://doi.org/10.3390/antiox11010008.
116. Ueno Y, Koike M, Shimada Y, et al. L-carnitine enhances axonal plasticity and improves white-matter lesions after chronic hypoperfusion in rat brain. J Cereb Blood Flow Metab. 2015;35(3):382–91. https://doi.org/10.1038/jcbfm.2014.210.
117. Ueno Y, Saito A, Nakata J, et al. Possible neuroprotective effects of L-carnitine on white-matter microstructural damage and cognitive decline in hemodialysis patients. Nutrients. 2021;13(4):1292. https://doi.org/10.3390/nu13041292.

8 Quantification and Pharmacokinetics of Carnitine

8.1 Introduction

Carnitine may be administered externally as a drug or taken as a supplement and understanding the pharmacokinetics (PK) of these externally administered carnitine products is important in the clinical application. PK characteristics usually include absorption, distribution, metabolism, and excretion, also known as Absorption, Distribution, Metabolism, Excretion (ADME) for short. In addition, drug–drug interactions and protein binding are also important characteristics of PK.

For PK of ordinary drugs, it is sufficient to consider externally administered substances only. However, in the case of carnitine, it is necessary to consider the presence of endogenous substances because carnitine is contained in the daily foods, including meat and dairy products, and is also biosynthesized in the body. Usually, to simplify the interpretation of measurement results, various parameters are calculated by subtracting the baseline carnitine concentration before external administration of carnitine. In the case of pharmaceutical carnitine, there are two routes of administration: oral administration in the form of tablets or oral solution, and intravenous administration used in dialysis patients, etc. These two routes of administration need to be organized and considered separately.

In this book, we would like to avoid describing the specialized PK of carnitine, but only describe the essence of PK as a basis for the concept of dosage and administration in clinical applications of carnitine and in clinical research. Those interested in specialized carnitine PK studies are referred to the following excellent reviews [1–3].

8.2 Overview of Old Analytical Measurement Methods for Carnitine

To study the PK of carnitine, the carnitine content in biological samples such as blood (usually serum or plasma), urine, milk, and tissues must be measured. Since the beginning of research on the physiological and biochemical role of carnitine in vivo in the 1950s, methods for the determination of carnitine based on various principles have been developed and there have been many reports of how they have been further improved. Therefore, it is not easy to understand the differences in past carnitine assay methods and the differences in PK caused by them. In particular, unless the researcher conducts comparative research following past measurement methods, it is sufficient to understand the general principles and advantages and disadvantages of the carnitine measurement methods such as the radioenzyme method used at that time. This is because most of the clinical studies of carnitine deficiency due to various causes described in this book have measured free carnitine and acylcarnitine levels

N. Nakayama, *Carnitine Deficiency and Insufficiency in Medicine*,
https://doi.org/10.1007/978-981-95-7797-2_8

in plasma, urine, and tissues by radioenzyme method. However, this radioenzyme method is currently rarely used in Japan.

8.2.1 Principle of Enzyme Method and Radioenzyme Method

In 1964, Marquis et al. reported a method for the determination of free carnitine using the reaction of carnitine acetyltransferase (CAT) in purified porcine myocardium [4]. The method is based on the reaction of CoASH with 5,5′-dithiobis-2-nitrobenzoic acid (DTNB). When CoASH produced in the reaction shown in Fig. 8.1 (I) is reacted with DTNB, thiophenolic acid with an absorption peak at 412 nm is produced and color develops, which is measured spectrophotometrically (Fig. 8.1, II).

Pearson et al. also reported a method using the same CAT enzyme reaction to react CoASH with sorbate and measure the absorption at 300 nm of the sorboyl-CoA produced [5], which was later improved to a method using DTNB [6]. Although the sensitivity of these enzyme methods was improved over the earlier carnitine measurement methods, they still lacked sensitivity for the measurement of trace amounts of carnitine and had various limitations in measurement, such as being easily affected by foreign substances in the sample.

Therefore, in 1972, Cederblad et al. reported a radioenzyme method that utilized a similar CAT reaction and replaced acetyl CoA with a radiolabeled one (Fig. 8.2) [7].

Carnitine in the sample is converted to ^{14}C-acetylcarnitine. Thereafter, the unreacted ^{14}C-acetyl-CoA remaining in the reaction solution is removed by passing it through an anion exchange resin column (Dowex AG2-X8 or Dowex 1-X10, etc., activated in advance to the Cl-form), and ^{14}C-acetylcarnitine is measured from the radioactive activity in the eluate, which in turn quantifies L-carnitine. This method is nearly 100 times more sensitive than the colorimetric quantification method used so far, and can quantify 10–20 pmoles of carnitine, even with 20 μL of plasma samples. Usually, plasma treated with the anticoagulant heparin is used, but serum can also be measured. It can also be measured with 10–20 mg of muscle and other tissue extracts. However, since this reaction is a reversible reaction, it is necessary to add a large excess amount of ^{14}C-acetyl-CoA in order to advance the reaction to the right, and there is a problem that the linearity of the standard curve is not good. Böhmer et al. devised an improved method to trap CoASH produced by adding DTNB to suppress reverse reactions and measured and reported blood carnitine concentrations in patients on dialysis, Crohn's disease, and anorexia [8]. The improved method by Böhmer et al. also had problems such as the fact that not only free carnitine but also acetylcarnitine in the sample was measured because CAT reacted with acetyl-L-carnitine, and DTNB slowly inactivated CAT. Therefore, McGarry et al. reported an

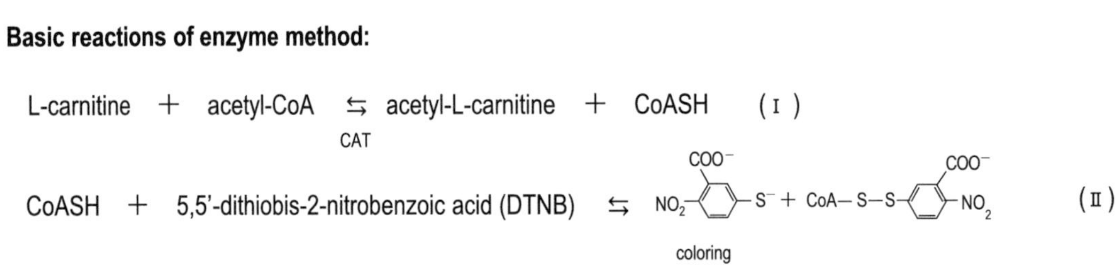

Fig. 8.1 Basic reactions of enzyme method. (I) Carnitine in the sample is reacted with acetyl-CoA in the presence of carnitine acyltransferase (CAT) purified from porcine heart and producing acetylcarnitine and CoASH. (II) CoASH react with DTNB and produce thiophenolic acid which has an absorption peak at 412 nm and is measured by spectrophotometrically. Notes: "CoASH" means free CoA. In reaction formula, sometimes "CoASH" is used instead "CoA" to show the presence of highly reactive SH-group in CoA molecule. *CAT* carnitine acetyltransferase, *DNTB* 5′,5′-dithiobis-nitrobenzoic acid

Basic reaction of radioenzyme method:

$$\text{L-carnitine} + {}^{14}\text{C-acetyl-CoA} \underset{\text{CAT}}{\leftrightarrows} {}^{14}\text{C-acetyl-L-carnitine} + \text{CoASH}$$

Fig. 8.2 Basic reaction of radioenzyme method. Carnitine in the sample reacts with ^{14}C-acethyl-CoA in the presence of CAT as Fig. 8.1 and produces ^{14}C-acetyl-L-carnitine and CoASH. Remaining un-react ^{14}C-acethyl-CoA is removed by anion exchange resin column, and the eluent radioactivity is measured. *CAT* carnitine acetyltransferase

improved method using sodium tetrathionate instead of DTNB as a trapping agent for the generated CoASH [9]. This radioenzyme method was superior to the original method of Cederblad et al. and the improved method of Böhmer et al. in the following respects.

1. The standard curve of free carnitine shows a wide linearity.
2. There is no interference effect of short-chain acylcarnitine or long-chain acylcarnitine such as acetylcarnitine.
3. Acylcarnitine in the sample can be hydrolyzed with alkali to quantify as total carnitine (free carnitine + acylcarnitine).
4. The acylcarnitine concentration can be measured by subtracting the free carnitine concentration (the value when the sample is not treated with alkali) from the total carnitine concentration.
5. The work was simplified by incubating a certain amount of anion exchange resin in a test tube without using an anion exchange resin column and measuring the radioactivity of the supernatant.
6. Sodium tetrathionate is used instead of DTNB, which has an enzyme inactivation effect, as a trapping agent for CoASH.

The advent of the radioenzyme method, which combines the enzyme and radioisotope methods, established a highly sensitive and rapid method for the determination of carnitine and was used in many subsequent clinical studies. This method has excellent sensitivity and does not require expensive instruments such as high-performance liquid chromatography or tandem-mass spectrometry. However, the disadvantages of this method are that it requires an isotope laboratory and is difficult to automate due to complicated operations such as alkaline treatment. Therefore, it is only for research purposes and is not suitable for clinical testing in hospitals, so it has been used only by some researchers in Japan.

8.2.2 Short-Chain Acylcarnitine and Long-Chain Acylcarnitine

As described in Chap. 3 carnitine includes free carnitine and acylcarnitine, which need to be measured separately. Acylcarnitine is a generic term, and its acyl groups can be as short as acetyl group (2 carbons) or as long as palmitoyl group (18 carbons): acylcarnitine (SCAC) and long-chain acylcarnitine (LCAC). Especially in studies of carnitine content in tissues and clinical studies of free carnitine and acylcarnitine in various diseases, these SCAC and LCAC are often measured separately. The principle of the measurement of SCAC and LCAC is difficult to understand, so the radioenzyme method of McGarry et al. that combines the most widely used enzyme method and radioisotope method is used here as an example (Fig. 8.3).

Free carnitine in plasma or serum can be measured directly by this method. To measure carnitine in muscle, liver, or other tissues, the tissue is chopped into small pieces under ice-cold conditions, buffer solution is added, and carnitine is extracted from the cells using a homogenizer. The free carnitine content in the extract (supernatant of centrifugation) can be determined from the radioactivity of ^{14}C-acetyl-L-carnitine. Next, 0.2 N NaOH is added to the sample to hydrolyze acylcarnitines in the sample in alkaline condition, then neutralize the solution. In this solution, there is the original free carnitine and the free carnitine

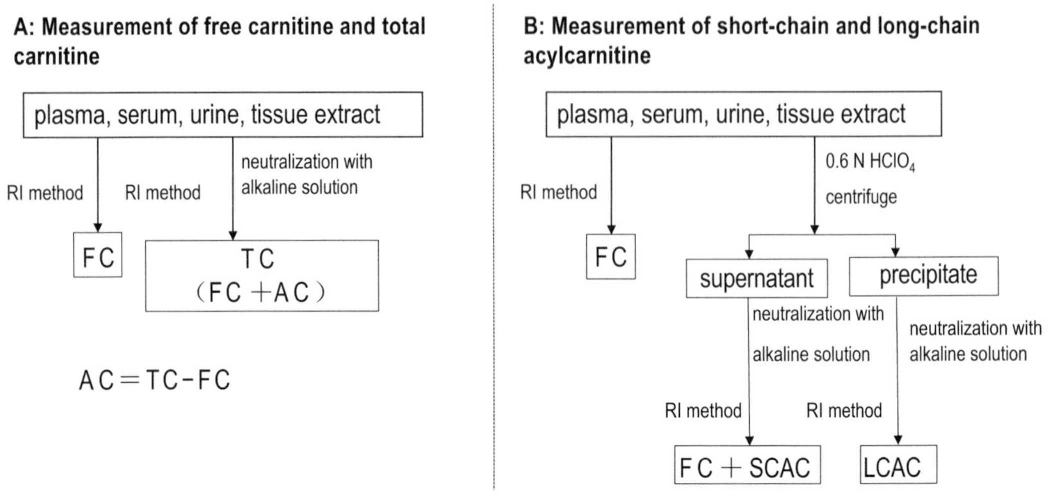

Fig. 8.3 Measurement method of carnitine fractions with radioenzyme method. *RI method* radioenzyme method, *FC* free carnitine, *AC* acylcarnitine, *TC* total carnitine, *SCAC* short-chain acylcarnitine, *LCAC* long-chain acylcarnitine

with acyl groups removed from the acylcarnitine by alkaline hydrolysis. When this is quantified by the free carnitine measurement method described above, total carnitine can be determined (total carnitine = original free carnitine + free carnitine derived from acylcarnitine by alkaline treatment). Therefore, the acylcarnitine value can be obtained by subtracting the free carnitine value from the total carnitine value (Fig. 8.3a).

When measuring carnitine in tissue extracts, since the presence of various enzymes and acetyl CoA in the extract may interfere with the measurement, the protein is usually removed with 0.6 N perchloric acid. Since the addition of perchloric acid precipitates proteins, etc., the supernatant and precipitate are separated by centrifugation. The supernatant is neutralized with alkaline solution and measured by the radioenzyme method, etc. The supernatant actually contains free carnitine and SCAC (C2–C5 acylcarnitines such as acetylcarnitine and propionylcarnitine). When the precipitate is dissolved and neutralized with alkali, it contains LCAC (C10 or higher), because LCAC is precipitated by strong acid (Fig. 8.3b). Thus, LCAC precipitates during perchloric acid treatment, while SCAC and free carnitine remain in the supernatant, and can be separated into LCAC and SCAC. However, the recovery of LCAC remains problematic. Furthermore, it has not been clarified whether medium-chain acylcarnitines (C6–C10) go to the supernatant or to precipitation upon perchloric acid treatment [10].

8.3 Current Carnitine Measurement Methods

Currently, there are three main methods for measuring carnitine in Japan.

(1) Tandem-mass spectrometry

Applications: Neonatal mass screening and definitive diagnosis of inherited metabolic diseases.

(2) Enzyme cycling method

Applications: Diagnosis of carnitine deficiency due to causes other than inherited metabolic diseases, follow-up after confirmation of inherited metabolic diseases, etc. This enzyme cycling method is covered by insurance in Japan.

(3) High-performance liquid chromatography: HPLC

These carnitine measurement methods will be described below.

8.3.1 Determination of Acylcarnitine by Tandem Mass Spectrometer

To make a definitive diagnosis of inherited metabolic diseases, it is necessary to identify and quantify the molecular species of acylcarnitine that accumulate in the blood and urine due to metabolic disorders. An expensive and large tandem-mass spectrometer (MS/MS) or a high-performance liquid chromatography-tandem-mass spectrometer (LC/MS/MS) is required for this measurement. In Japan, Dr. Yosuke Shigematsu of the Department of Pediatrics, University of Fukui, started pilot screening using tandem mass in 1997, and since 2004, Dr. Yosuke Shigematsu and Dr. Seiji Yamaguchi of the Department of Pediatrics, Shimane University have been conducting "Research on the Establishment of a New Newborn Mass Screening System Introducing New Technologies such as Tandem Mass" under the "Grant-in-Aid for Research on Children and Families under the Grant-in-Aid for Research on Children and Families." Based on the results, neonatal mass screening has been introduced nationwide since 2014, and tests are being conducted for 20 diseases. As of 2020, it has been conducted at 36 facilities nationwide (see the website of the Japan Society of Mass Screening: https://www.jsms.gr.jp/). In the actual screening, filter paper blood collected from newborns is used to identify and quantify acylcarnitine as an internal standard by contrasting the peaks of various acylcarnitine labeled with deuterium. Free carnitine is C0, acetylcarnitine is C2, palmitoyl carnitine is C16, and so on. Further details are beyond the scope of this book, so you can deepen your understanding by referring to technical books [11, 12], by contacting the Japanese Society for Neonatal Screening (https://www.jsms.gr.jp/) and the Japanese Society for Inherited Metabolic Diseases (https://jsimd.net/).

In addition, although contract laboratories overseas are commissioned to perform acylcarnitine analysis using a tandem-mass spectrometer, to the best of the author's knowledge, among laboratories that perform newborn mass screening, there are no laboratories that perform acylcarnitine analysis using a tandem-mass spectrometer for clinical research purposes other than for neonatal metabolic disorders on a commercial basis. Therefore, there are no facilities that contract the analysis of acylcarnitines using a tandem-mass spectrometer on a commercial basis. Therefore, if you want to analyze and quantify acylcarnitines down to their molecular species using a tandem-mass spectrometer, you should consult with the pediatric department or collaborative research facility dealing with inherited metabolic diseases at the nearest university medical school, the pharmacokinetic laboratory of the pharmacology department, or contact a major contract analysis company. In this sense, acylcarnitine analysis using a tandem-mass spectrometer is not an analytical method that can be used immediately in fields other than inherited metabolic diseases.

8.3.2 Principle of Determination of Free and Acylcarnitine by Enzyme Cycling Method

The radioenzyme method has high sensitivity and has been used in many clinical studies. However, it has disadvantages such as the need for a radioisotope laboratory, complicated operation, and the inability to measure a large number of specimens in a short period of time. Therefore, many methods using spectrophotometers have been developed, but all of them have problems such as sensitivity and strong influence of contaminants.

In 1994, Takahashi (Toyo Brewery, later Asahi Kasei Corporation) and his colleagues developed a highly sensitive and simple enzyme cycling method [13]. This is a method using carnitine dehydrogenase (CDH) isolated and purified from soil bacteria of the genus Alcalygenes (*Alcaligenes* sp. *No. 981*) isolated from the soil of a potato field in Gojo City, Nara Prefecture. Free carnitine is converted to dehydrocarnitine by CDH, which specifically reacts with L-carnitine, and at the same time as the reaction, thio-NAD^+, a coenzyme of CDH, changes from the oxidized form to the reduced thio-nicotinamide adenine dinucleotide (NADH). When NADH is added to the measurement system, NADH is oxidized by dehydrocarnitine to NAD+, dehydrocarnitine becomes L-carnitine again, and thio-NAD+ becomes thio-NADH by the reaction of CDH. Each time this cycling reaction is repeated, the amount of thio-NADH produced is amplified and thio-NADH accumulates. Since the amount of thio-NADH produced is proportional to the concentration of free carnitine in the added sample, the free carnitine concentration can be measured with high sensitivity by measuring the change in UV absorption (415 nm) of thio-NADH. Total carnitine (free carnitine + acylcarnitine) is measured by cleaving acylcarnitine in the sample into free carnitine using acylcarnitine esterase (ACE), which was also discovered in the same soil bacteria by Takahashi et al.

This reaction solution contains a combination of free carnitine and free carnitine formed by cleavage of acylcarnitine with ACE, which was originally in the sample, and is called "total carnitine." The total carnitine content can be measured by measuring the combined free carnitine amount by the method using CDH as described above. Acylcarnitine can be calculated by subtracting the free carnitine value obtained without reacting ACE from the total carnitine value [14].

The CDH and ACE enzymes discovered by Takahashi et al. are extremely stable compared to similar enzymes isolated from other bacteria and can be stored at 2–10 °C in aqueous solution for 24 months (1 month after opening). ACE can also efficiently cleave short-, medium-, and long-chain acylcarnitines. These enzymes are of bacterial origin and are hardly affected by various substances and enzymes in vivo. Therefore, free carnitine, total carnitine, and calculated acylcarnitine in the sample solution can be measured without the complicated and hazardous alkaline treatment and perchloric acid solution deproteinization that are required by the radioenzyme method (Fig. 8.4). These assay kits are supplied with a standard solution of L-carnitine (50 μmol/L), which is used to quantitate from the obtained standard straight line. The kits can be applied to automatic analyzers for biochemical tests commonly used in clinical laboratories in hospitals, or the kits can be purchased in laboratories and samples can be measured manually using a spectrophotometer. The normal refer-

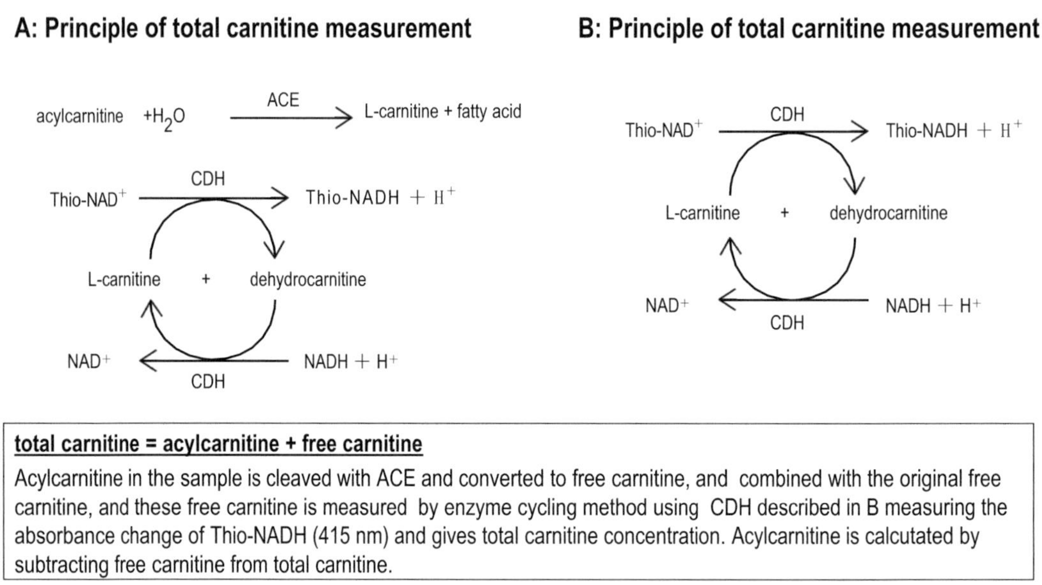

Fig. 8.4 Principle of enzyme cycling method for free and total carnitine measurement. *ACE: acylcarnitine esterase, CDH: carnitine dehydrogenase, Thio-NAD+* thionicotinamide adenine dinucleotide, *Thio-NADH* thionicotinamide adenine dinucleotide reduced form. (Modified from refs. [13, 16])

ence values are based on the report by Tsushima et al. [14] This enzyme cycling method of measuring carnitine correlates extremely well with conventional enzyme methods, radioenzyme methods, and high-performance liquid chromatography-tandem-mass spectrometry (LC/MS/MS) [13, 15, 16]. Unfortunately, this enzyme cycling method is rarely used overseas due to its limited export volume. Unlike the tandem-mass spectrometry method, the molecular species of acylcarnitines cannot be identified, but it is an excellent carnitine quantification method that can easily quantify free carnitine and acylcarnitines, respectively.

These carnitine measurement kits ("Free Carnitine Kainos" and "Total Carnitine Kainos") by enzyme cycling method were released by Kainos Laboratories Inc. as research reagents in 1993, and clinical testing laboratories such as SRL and BML started to accept them as clinical research tests in 1994. It has been used for diagnosis of carnitine deficiency and clinical research in various clinical settings but was not covered by insurance. The Diagnostic Division of Otsuka Pharmaceutical Co., Ltd. cooperated with Kainos in the clinical performance test, and the drug was approved as an in-vitro diagnostic reagent on February 26, 2014, and launched by Kainos as an in-vitro diagnostic reagent on December 7, 2015. The current sales names of the two kits are as follows:

1. For total carnitine measurement: "Total Carnitine Kit T-Carnitine Reagent Kainos".
   http://www.kainos.co.jp/jp/products/pdf/TKA4710.pdf
2. For free carnitine measurement: "Free Carnitine Kit F-Carnitine Reagent Kainos".
   http://www.kainos.co.jp/jp/products/pdf/TKA4610.pdf

However, it was still difficult to get insurance coverage, and from February 2018, insurance coverage was finally applied to the diagnosis of major carnitine deficiency. The carnitine measurement method covered in the "Guidelines for the Diagnosis and Treatment of Carnitine Deficiency 2018" formulated mainly by the Nutrition Committee of the Japan Pediatric Society, specialists in hepatology and nephrology, etc., is this enzyme cycling method (https://www.jpeds.or.jp/uploads/files/20181207_shishin.pdf, for details, see Chap. 18).

When measuring using an automatic analyzer for biochemical testing in a clinical laboratory of a hospital, the parameter setting and the reagent liquid volume setting differ depending on the model of the instrument, so it is necessary to contact Kainos. In addition, if the number of samples is small or there is no automatic analytical equipment for biochemical testing, it can be outsourced to a clinical testing laboratory as before. This enzyme cycling method is used in the carnitine measurement method of most clinical testing laboratory companies in Japan.

I deeply hope this enzyme cycling method will be introduced worldwide because it is very useful to diagnose carnitine deficiency and insufficiency in clinical practice of various medical fields explained in this book.

8.3.3 Determination of Carnitine by HPLC Method

High-performance liquid chromatography (HPLC) is also used to analyze free carnitine and acylcarnitine in blood (plasma or serum) and urine in PK studies, but it requires advanced analytical techniques and expensive equipment. Analysis of carnitine by HPLC is usually performed in the following four steps:

1. Solid-phase extraction: An operation in which carnitine compounds in a sample are adsorbed on a column and proteins and fat-soluble substances are washed, then carnitine was eluted and concentrated.
2. Precolumn derivatization: A highly polar carnitine compound is reacted with 4′-bromophenacyl-trifluoro-methane sulfonate and N,N-diisopropylethylamine, etc., to improve the efficiency of subsequent separation and detection by the column.
3. Reversed-phase separation: Separation operation by reversed-phase column.

4. UV or fluorescence detection: Carnitine derivatives are detected by UV or fluorescence and quantified by calibration curves prepared with various carnitine standards.

Minkler et al. have been researching HPLC-based analytical and quantification methods for carnitine for more than 30 years and have analyzed and reported urine samples from healthy individuals, methylmalonic acidemia, isovaleric acidemia, urine from patients with medium-chain acyl-dehydrogenase deficiency, short-chain and medium-chain carnitine in plasma, and tissue carnitine during exercise [17–21]. In recent years, high-performance liquid chromatography-tandem-mass spectrometry (LC/MS/MS), which separates by HPLC and then detects and quantifies by MS/MS, has also been reported [22, 23]. When analyzing carnitine compounds using HPLC, please refer to these references.

8.3.4 Selecting the Right Measurement Method for the Purpose of the Research

For details on the characteristics and problems of each of the carnitine analytical methods, including chromatographic, enzyme, and radioenzyme methods, please refer to the reviews by Marzo et al. and Evans et al. [1, 10, 24, 25] If the carnitine fractions of interest are free carnitine and acylcarnitine only, the enzyme cycling method is sufficient. On the other hand, if the molecular species of acylcarnitines are to be identified and quantified, LC/MS/MS is the most suitable method. In particular, LC/MS/MS should be used to identify and quantify the specific molecular species of acylcarnitines that increase with disease in order to establish the name of the disease, especially in inherited metabolic diseases. Normal reference values and cut-off values for acylcarnitines have been established for each disease. In addition, HPLC methods should also be considered if research focusing on propionyl carnitine and acetylcarnitine is to be conducted. When re-examining the long-chain acylcarnitine and short-chain acylcarnitine fractions of old research papers, the radioenzyme method should be used, but it is not suitable if the medium-chain acylcarnitine is to be focused on [10]. The LC/MS/MS method can identify and quantify even the molecular species of medium-chain acylcarnitine.

Regarding the handling of specimens, there is no significant difference between plasma and serum blood samples, but since acylcarnitine is susceptible to hydrolysis, it is necessary to separate it into plasma or serum immediately after blood collection and store it at −20 °C or lower (preferably −70 °C) until just before measurement (stable for 6 months). However, according to Fornasini et al., a 20–30% reduction in acetylcarnitine degradation was observed even when plasma samples were frozen-stored at –20ºC [25]. Although free carnitine is extremely stable as a substance, bacterial degradation should be considered. Urine samples are susceptible to bacterial contamination, so sodium azide (NaN_3, 0.05–0.1% of urine) or thymol (thymol, 10% (w/v) solution of isopropyl alcohol in 5 mL/L urine) should be added and stored below –20ºC.[10.]

8.4 Absorption and Bioavailability of Orally Administered Carnitine

For information on the intake of carnitine from the diet, its biosynthesis in the body, and its excretion and reabsorption by the kidneys, please refer to Chap. 4. This chapter describes only the absorption of carnitine administered from outside the body as a drug or supplement.

The oral absorption rate of a drug is calculated from the area under the blood concentration-time curve (AUC) of a fixed dose of the drug administered intravenously as 100%, and the AUC ratio of the same dose of the drug administered orally. In PK, this is called biological availability (BA).

Bioavailability (%) = [AUC for oral administration/AUC for intravenous administration] × 100

Bioavailability is defined as the percentage of drug that enters the bloodstream by oral administration, as opposed to 100% by intravenous administration. Bioavailability is mainly determined by the absorption rate of the administered drug in the gastrointestinal tract and the metabolism called the first-pass effect in the liver, but in the case of carnitine preparations, there is no metabolism in the liver (described later), so the bioavailability is considered to represent the absorption rate in the gastrointestinal tract.

Sahajwalla et al. conducted an equivalence study of three levocarnitine formulations (tablets, oral solutions, and chewables) in 15 healthy adults and reported bioavailability as one of the parameters [26]. In the study of three oral preparations, the pharmacokinetics were examined by repeated administration of 2 g twice daily (6 tablets of 330 mg in two doses; 1980 mg) for 4 days, and the pharmacokinetics were calculated by a single intravenous dose of 20 mg/kg. As a result, the pharmacokinetics of the three oral formulations (oral solution, tablet, and chewable) were similar in steady state, and the bioavailability corrected for endogenous free carnitine was 15.1 ± 5.3% for tablets, 14.8 ± 5.1% for chewables, and 15.9 ± 4.9% for oral solutions.

Segre et al. reported the pharmacokinetics of carnitine in six healthy adults. Levocarnitine was administered as followings: First, a single intravenous dose of 30 mg/kg of, then after washout, followed by a single oral dose of 100 mg/kg, and after washout again, a subsequent single oral dose of 30 mg/kg in three subjects [27]. The bioavailability at oral administration was 18%, corrected by subtracting the baseline value from the blood concentration at all timepoints, considering the baseline free carnitine concentration to be the endogenous free carnitine value. They also stated that an absorption saturation phenomenon may be occurring, but more precise studies are needed. In 1992, Rizza et al. examined the pharmacokinetics of levocarnitine 30 mg/kg and 100 mg/kg single oral and intravenous doses in six healthy adults. In the analysis corrected for baseline endogenous free carnitine, the bioavailability at oral administration was still 13–21%, although it was reported that the concentration of free carnitine in plasma was almost proportional to the dose administered [28]. On the other hand, Harper et al. also examined the pharmacokinetics of 2 and 6 g of L-carnitine in six healthy adults by administering them orally and intravenously [29]. As a result, the AUC at a single dose of 2 g and the AUC at a single dose of 6 g were almost the same, and the bioavailability rate was 16% at a single dose of 2 g, but it was 5% at a single dose of 6 g, indicating saturation of absorption from the gastrointestinal tract. Bain et al. studied changes in blood concentrations of carnitine and its metabolites caused by intestinal bacteria in seven healthy adult subjects after administration of 0, 0.5, 1, and 2 g three times daily for 7 days (daily doses of 0, 1.5, 3, and 6 g. [30] In all doses, the blood concentrations of carnitine and its metabolites increased approximately 4 h after administration and reached a steady state on the third day. However, the AUC of free carnitine concentration was not significantly different among the three doses. Bain et al. considered the possibility that not only saturation of absorption in the gastrointestinal mucosa but also saturation of reabsorption of carnitine in the renal tubules occurred and the AUC did not increase even if the dosage was increased beyond 1.5 g. The reason for this is that when the carnitine dosage was increased from 0.5 to 2 g (daily dosage from 1.5 to 6 g), renal clearance of carnitine increased approximately threefold from 11.1 ± 3.7 mL/min to 29.2 ± 12.6 mL/min.

The absorption rate of carnitine preparations when administered orally is 10–20%, which is considerably lower than the more than 75% absorption rate reported by Rebouche et al., in their study of trace amounts of radiolabeled carnitine added to a normal diet [31]. These results suggest that the low bioavailability of carnitine in oral formulations and the decrease in AUC at high doses are due to saturation of the carnitine transporter (OCTN2), which is involved in the in the mucous membrane which are the sites of gastrointestinal absorption and reabsorption process in the renal tubules. As a result, it is considered that this problem cannot be solved by devising a formulation design for the purpose of improving absorption. Therefore, when high doses of oral carnitine are administered in severe carnitine deficiency such as primary carnitine deficiency, divided small doses are administered as frequently as possible.

Although there are no detailed literature reports on the effects of meals on oral administration of carnitine, it has been reported that in a phase 1 study of levocarnitine chloride conducted by Earth Pharmaceutical Co., Ltd., there were comparative data on fasting and postprandial administration of levocarnitine chloride administered to five healthy adults when a single dose of 800 mg was administered orally and concluded "no dietary effects were observed" [32].

Table 8.1 Carnitine content and distribution of various tissues and organs of the human body

Compartments	Concentration (μmol/L or μmol/kg)	Estimated content[a] (mmol)	Distribution (%)
Plasma/extracellular fluid	40–50	0.5	~ 0.4
Liver	500–1000	1.3	~ 1
Kidney	330–600	0.2	~ 0.2
Skeletal muscle	Up to 4000	126.4	> 98
Total of listed compartments		128.4 (about 21 g)	

(From ref. [1])

[a]Estimated content in a 70 kg healthy male

8.5 Distribution of Carnitine in the Body

The distribution of carnitine ingested from normal meals and carnitine biosynthesized in the body has already been shown in Table 4.1 of Chap. 4 but data from other literature and pool size are shown in Table 8.1 [1, 33, 34].

More than 98% of the carnitine content of the whole body is distributed in muscle tissue such as skeletal muscle, less than 1% is distributed in the liver, 0.2% is distributed in the kidneys, and less than 0.4% is distributed in blood and extracellular fluid. The concentration in skeletal muscle (2000–4000 μmol/kg wet weight) is more than 100 times higher than the concentration in plasma. Tissues such as liver, kidney, and brain have intermediate concentrations of 300–1000 μmol /kg wet weight. Carnitine concentrations in the heart range from 3500 to 6000 μmol/kg wet weight, which are equal to or higher than those of skeletal muscle. Carnitine concentrations in the epididymis are highest (53 mM) and 2000 times higher than blood levels [35]. The total amount of carnitine in the body is 128 mmol, which corresponds to about 21 g (Table 8.1). It is important to note that the carnitine content in the blood is less than 1% of the total carnitine in the body. This means that the measurement of carnitine levels in the blood may not necessarily reflect the status of carnitine in the body as a whole. However, when the input of carnitine into the body (dietary ingestion and biosynthesis) and excretion are steady-state, the blood concentration and tissue concentration of carnitine are considered to be correlated [36]. It has been reported that blood and muscle levels are correlated in healthy people, but not in dialysis patients [37]. It should be noted that in the case of dialysis patients, the concentration of carnitine in the blood fluctuates greatly, and the data varies greatly depending on when the blood was collected (before or after dialysis, etc.).

When considering the distribution of carnitine to tissues, it should always be taken into account that carnitine is a low molecular weight, highly polar substance that exists in vivo and is concentrated intracellularly via carnitine transporters in the cell membrane.

When the blood concentration of carnitine is measured when administered externally, the value is the sum of the endogenous carnitine concentration and the externally administered and absorbed carnitine. It should also be noted that endogenous carnitine concentrations vary greatly from person to person. Although the presence of endogenous carnitine is a troublesome problem when calculating the pharmacokinetic parameters of externally administered carnitine, it is usually simply assumed that the diurnal variation of endogenous carnitine is small, and the blood concentration before administration of carnitine from the outside is subtracted from the blood concentration data at all time points to draw a blood concentration curve and calculate the parameter [1].

Protein binding is usually a factor that greatly affects the distribution of drugs in the body, especially their presence in the blood. However, L-carnitine and SCAC (acetyl-L-carnitine and propionyl-L-carnitine) have been found to be completely free of protein binding [38].

When carnitine is administered intravenously, the initial distribution volume is 0.1–0.2 L/kg, approximately equivalent to the extracellular fluid [27, 39–41]. When carnitine is administered intravenously, the blood level immediately reaches a sharp peak and then shows an exponential decline in two (or three) phases. The initial half-life is 0.5–1 h, and the final half-life is 3–12 h [1]. By 12–24 h after administration, the blood level of carnitine is almost the same as the level before administration. This does not mean that most of the dose has been excreted. A certain amount is excreted in the urine, but the rest is incorporated into the body's carnitine pool. For example, when 1 g of carnitine is administered intravenously, the carnitine pool in the body is about 20 g, so 1 g increases by only about 5%. And if we assume that half of it is excreted, the amount of pool in the body will increase by only 2.5%. The amount of change is within the range of diurnal variation and measurement error, and it is difficult to analyze pharmacokinetics by ordinary means. In particular, more than 98% of the carnitine pool in the body is skeletal muscle, and the rate of uptake of carnitine from the central compartment (= blood and extracellular fluid) into skeletal muscle is extremely slow, so pharmacokinetic analysis from blood concentration measurement is impossible. Therefore, Rebouche et al. reported a method in which trace amounts of L-[methyl-^{3}H]-carnitine were administered intravenously to six healthy adults and analyzed the radioactivity and carnitine content in the blood and urine for 29 days [42]. The original purpose of Rebouche et al.'s study was to determine whether carnitine deficiency in patients with systemic carnitine deficiency or medium-chain fatty acid dehydrogenase deficiency was caused by a defect in carnitine transport, and in doing so, they examined in detail the dynamics of intravenous carnitine in six healthy subjects. As a result of the analysis, it was found that the three compartments of (1) extracellular fluid (initial volume of distribution), (2) liver and kidney (early equilibrium related tissues), and (3) skeletal muscle and myocardium (slow equilibrium related tissues) were considered as the best fit. Evans et al. used the pool size of each compartment of Rebouche et al.'s three-compartment model as an image to make it easier to understand (Fig. 8.5) [1].

The arrow numbers in the figure indicate how much of the total pool moves in the direction of the arrow per hour. For example, 0.3/h of the arrow from the "central pool (extracellular fluid pool)" to the "muscle pool" means that 0.3 (i.e.,

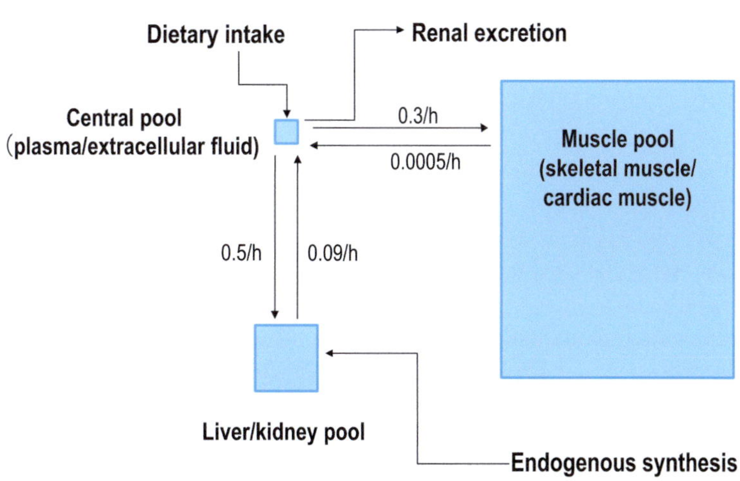

Fig. 8.5 Pharmacokinetic 3 compartment model of disposition of L-carnitine and rate constants. (From ref. [1])

30%) of the total pool moves per hour. Conversely, 0.005/h of the arrow from the muscle pool to the central pool means that 0.005 (0.5%) of the muscle pool is transferred, and the half-life of this process is calculated to be 5–6 days. The turnover time with these three compartments was 1, 12, and 191 h, respectively, and 66 days for the whole body. These data are close to the results of Rebouche et al. when they studied the kinetics of L-[methyl-^{3}H]-carnitine in dogs prior to their human study (62.9 ± 5.6 days), with no species differences [43].

Since the apparent volume of distribution (Vss) in steady state is the value obtained by dividing the amount in the body by the concentration in the plasma, in the case of carnitine, dividing 128 mmol by 40–50 μmol/L yields a large figure of about 3000 L. Significant discrepancies with the reported 20–50 L of intravenous carnitine preparations administered to humans [27, 39–41]. This shows the difficulty of analyzing carnitine, which is mostly distributed in tissues that are distributed in very slow equilibrium, such as skeletal muscle. Acute or short-term continuous administration of carnitine does not change the carnitine content in the muscles [39]. This is because the entry and exit from the central pool to the skeletal muscle pool is an extremely slow process, and the carnitine administered during that time is eliminated from the body by urinary excretion. However, chronic administration is a different story, and the carnitine content in the muscles gradually increases with sufficiently long-term administration [33].

8.6 Metabolism of Carnitine

It is believed that carnitine is not metabolized in the intestinal mucosa or liver. This is based on a report by Yue et al. that no radioactive metabolites were detected when ^{3}H-labeled carnitine was administered to dogs and rats [44]. Rebouche et al.'s study also found no urinary metabolites of intravenously administered L-[methyl-^{3}H]-carnitine [43]. However, it has been reported that when DL-carnitine labeled with a methyl group of ^{14}C was administered to rats, 2–5% of the dose was metabolized to β-methylcholine by carnitine decarboxylase and excreted in the urine [45]. Since little research has been done on the metabolism of carnitine decarboxylase since then, it is not known whether this enzyme is also active in human.

As described in the Sect. 8.4, 10–20% of externally administered carnitine is absorbed. Carnitine that is not absorbed is partly excreted in the feces and partly metabolized by intestinal bacteria into trimethylamine (TMA) and γ-butyrobetaine, which are transported to the liver via bloodstream and are mostly metabolized by the flavin-containing enzyme monooxygenase (flavin-containing monooxygenase form 3), to trimethylamine-N-oxide (TMAO), which is then excreted into the bloodstream and in the urine. This conversion of carnitine to TMA and TMAO by intestinal bacteria is discussed in detail in Chap. 9.

8.7 Excretion of Carnitine

As described in Chap. 4 carnitine is 100% filtered by the renal glomeruli, but 90–99% is reabsorbed by the carnitine transporter (OCTN2) in the renal tubules [1, 46–48]. The amount filtered by the glomeruli is 8–9 mmol, most of which is reabsorbed by the carnitine transporter in the renal tubules, and excreted is less than 0.4 mmol. On the other hand, biosynthesis and dietary intake are 0.1–0.3 mmol, so homeostasis is maintained. If, for some reason, carnitine filtered by the glomeruli is not efficiently reabsorbed, carnitine deficiency can easily develop [1]. The administration of drugs that affect reabsorption and the genetic impairment of the carnitine transporter (OCTN2) have a significant impact on carnitine homeostasis (see Chap. 4). In patients with systemic carnitine deficiency caused by a decrease in the function of the carnitine transporter due to genetic abnormalities in OCTN2, the reabsorption threshold in the renal tubules drops to 2 μmol/L or less, resulting in the excretion of carnitine in the urine [46]. It has been found that carnitine deficiency caused by valproic acid is also enhanced in urinary excretion due to decreased reabsorption of free carnitine in the renal tubules (see Chap. 12). Many anticancer drugs, such as cisplatin and etoposide, suppress the expression and activity of the carnitine transporter (OCTN2) in the renal tubules, resulting in increased excretion of carnitine from the renal tubules [49–52]. In dialysis patients, there is no reabsorption process in the renal tubules, and free carnitine leaks into the dialysate three times a week, causing carnitine deficiency [2, 3, 53, 54].

Carnitine administered from the outside is rapidly excreted by the kidneys. This is because the threshold for reabsorption of carnitine in the renal tubules is 40–60 μmol/L, which is almost the same as the normal value of the blood concentration of free carnitine, and if the concentration in the blood is much higher than that, it is no longer necessary for reabsorption and is excreted. In other words, the efficient carnitine reabsorption mechanism of the renal tubules is purposeful to prevent carnitine deficiency, and when the blood concentration becomes extremely low, it works to reabsorb and maintain carnitine as much as possible, but when the blood concentration rises above the normal value, the reabsorption mechanism is no longer necessary and does not work.

As described in the Sect. 8.4 when Bain et al. administered 0.5, 1, and 2 g three times a day for 7 days (daily doses of 1.5, 3, and 6 g), the AUC of the blood concentration of free carnitine was not significantly different between the three doses, and together with the phenomenon of saturation of absorption in the gastrointestinal mucosa, it is considered that saturation of the reabsorption of carnitine in the renal tubules may also have occurred. In fact, when the carnitine dosage was increased from 0.5 to 2 g (from 1.5 to 6 g in daily dosage), renal carnitine clearance increased approximately threefold from 11.1 ± 3.7 mL/min to 29.2 ± 12.6 mL/min [30]. Thus, an increase in renal clearance has been observed even in oral administration, but a more pronounced renal clearance value has been reported in intravenous administration. Harper et al. reported 2 and 6 g of L-carnitine intravenously, the renal clearance was 78 and 100 mL/min [29], respectively, and Sahajwalla et al. reported 50 mL/min renal clearance at 20 mg/kg intravenously [40]. Thus, renal clearance values with intravenous carnitine administration are close to the glomerular filtration rate of 100–120 mL/min, which is thought to indicate near saturation of reabsorption in the renal tubules. Therefore, intravenous administration is not necessarily an efficient route of administration for patients with normal renal function, although high blood concentrations can be obtained. Intravenous administration is useful in cases of acute phase of carnitine deficiency due to inherited metabolic diseases or hyperammonemia due to carnitine deficiency caused by valproic acid toxicity, where increase in blood concentration is desired as possible as we can. On the other hand, in dialysis patients with impaired renal function, carnitine is only removed from the dialysis membrane during dialysis three times a week without renal excretion, so intravenous administration after dialysis can maintain a high blood carnitine concentration until the next dialysis therapy and efficiently replenish skeletal muscle, cardiac muscle, and other tissues. It can be said that it is an extremely useful route of administration.

References

1. Evans AM, Fornasini G. Pharmacokinetics of L-carnitine. Clin Pharmacokinet. 2003;42(11):941–67. https://doi.org/10.2165/00003088-200342110-00002.
2. Reuter SE, Evans AM. Carnitine and acylcarnitines: pharmacokinetic, pharmacological and clinical aspects. Clin Pharmacokinet. 2012;51(9):553–72. https://doi.org/10.1007/BF03261931.
3. Fornasini G, Upton RN, Evans AM. A pharmacokinetic model for L-carnitine in patients receiving haemodialysis. Br J Clin Pharmacol. 2007;64(3):335–45. https://doi.org/10.1111/j.1365-2125.2007.02926.x.
4. Marquis NR, Fritz IB. Enzymological determination of free carnitine concentrations in rat tissues. J Lipid Res. 1964;5:184–7.
5. Pearson DJ, Tubbs PK. A sensitive enzymic assay for carnitine. Biochem J. 1964;91(1):2C–4C.
6. Pearson DJ, Tubbs PK, Chase JF. Carnitine and acylcarnitines. In: Bergmeyer M, editor. Methods of enzyme analysis. 2nd ed. New York: Academic; 1974. p. 1758–71.

7. Cederblad G, Lindstedt S. A method for the determination of carnitine in the picomole range. Clin Chim Acta. 1972;37:235–43. https://doi.org/10.1016/0009-8981(72)90438-x.
8. Böhmer T, Rydning A, Solberg HE. Carnitine levels in human serum in health and disease. Clin Chim Acta. 1974;57(1):55–61. https://doi.org/10.1016/0009-8981(74)90177-6.
9. McGarry JD, Foster DW. An improved and simplified radioisotopic assay for the determination of free and esterified carnitine. J Lipid Res. 1976;17(3):277–81.
10. Marzo A, Cardace G, Monti N, et al. Chromatographic and non-chromatographic assay of L-carnitine family components. J Chromatogr. 1990;527(2):247–58. https://doi.org/10.1016/s0378-4347(00)82109-x.
11. Yamaguchi S. A well-understood guidebook for newborn mass screening. Tokyo: SHINDAN TO CHIRYO SHA Inc; 2019. (Book in Japanese)
12. Japanese Society for Inherited Metabolic Diseases ed.: Clinical practice guidelines for diseases subject to mass screening of newborns 2019. SHINDAN TO CHIRYO SHA Inc, Tokyo, 2019. (Book in Japanese).
13. Takahashi M, Ueda S, Misaki H, et al. Carnitine determination by an enzymatic cycling method with carnitine dehydrogenase. Clin Chem. 1994;40(5):817–21.
14. Tsushima K, Kase N. Methods for the determination of carnitine in human serum. Jpn J Clin Dial. 2000;16(2):167–73. (Article in Japanese)
15. Sugiyama N. Carnitine. Jpn J Pediatr. 1996;59(Suppl):181–5. (Article in Japanese)
16. Itami N, Aoyama S, Yasuo Kudoh Y, et al. Clinical performance of KDK-1201, the carnitine measurement reagent based on the principle of enzyme cycling method. Rinsho Kensa (J Med Technol). 2014;58(5):651–8. (Article in Japanese)
17. Minkler PE, Ingalls ST, Hoppel CL. High-performance liquid chromatographic separation of acylcarnitines following derivatization with 4′-bromophenacyl trifluoromethanesulfonate. Anal Biochem. 1990;185(1):29–35. https://doi.org/10.1016/0003-2697(90)90250-d.
18. Minkler PE, Hoppel CL. Determination of free carnitine and 'total' carnitine in human urine: derivatization with 4′-bromophenacyl trifluoromethanesulfonate and high performance liquid chromatography. Clin Chim Acta. 1992;212(1–2):55–64. https://doi.org/10.1016/0009-8981(92)90137-f.
19. Minkler PE, Hoppel CL. Quantification of carnitine and specific acylcarnitines by high-performance liquid chromatography: application to normal human urine and urine from patients with methylmalonic aciduria, isovaleric acidemia or medium-chain acyl-CoA dehydrogenase deficiency. J Chromatogr. 1993;613(2):203–21. https://doi.org/10.1016/0378-4347(93)80135-q.
20. Minkler PE, Hoppel CL. Quantification of free carnitine, individual short- and medium-chain acylcarnitines, and total carnitine in plasma by high-performance liquid chromatography. Anal Biochem. 1993;212(2):510–8. https://doi.org/10.1006/abio.1993.1361.
21. Minkler PE, Brass EP, Hiatt WR, et al. Quantification of carnitine, acetylcarnitine, and total carnitine in tissues by high-performance liquid chromatography: the effect of exercise on carnitine homeostasis in man. Anal Biochem. 1995;231(2):315–22. https://doi.org/10.1006/abio.1995.0057.
22. Minkler PE, Stoll MS, Ingalls ST, et al. Quantification of carnitine and acylcarnitines in biological matrices by HPLC electrospray ionization-mass spectrometry. Clin Chem. 2008;54(9):1451–62. https://doi.org/10.1373/clinchem.2007.099226.
23. Minkler PE, Stoll MS, Ingalls ST, et al. Validated method for the quantification of free and total carnitine, butyrobetaine, and acylcarnitines in biological samples. Anal Chem. 2015;87(17):8994–9001. https://doi.org/10.1021/acs.analchem.5b02198.
24. Marzo A, Curti S. L-carnitine moiety assay: an up-to-date reappraisal covering the commonest methods for various applications. J Chromatogr B Biomed Sci Appl. 1997;702(1–2):1–20. https://doi.org/10.1016/s0378-4347(97)00376-9.
25. Fornasini G, Evans AM. Analysis of free carnitine in plasma of ESRD patients undergoing chronic dialysis: what is the best assay? Dial Transplant. 2003;32(6 Suppl 1):S1–S12.
26. Sahajwalla CG, Helton ED, Purich ED, et al. Multiple-dose pharmacokinetics and bioequivalence of L-carnitine 330-mg tablet versus 1-g chewable tablet versus enteral solution in healthy adult male volunteers. J Pharm Sci. 1995;84(5):627–33. https://doi.org/10.1002/jps.2600840520.
27. Segre G, Bianchi E, Corsi M, et al. Plasma and urine pharmacokinetics of free and of short-chain carnitine after administration of carnitine in man. Arzneimittelforschung. 1988;38(12):1830–4.
28. Rizza V, Lorefice N, Rizza N, et al. Chapter 4. Pharmacokinetics of L-carnitine in human subjects. In: Ferrari R, DiMauro S, Sherwood G, et al., editors. L-carnitine and its role in medicine: from function to therapy. New York: Academic; 1992. p. 63–77.
29. Harper P, Elwin CE, Cederblad G. Pharmacokinetics of intravenous and oral bolus doses of L-carnitine in healthy subjects. Eur J Clin Pharmacol. 1988;35(1):69–75. https://doi.org/10.1007/BF00555510.
30. Bain MA, Milne RW, Evans AM. Disposition and metabolite kinetics of oral L-carnitine in humans. J Clin Pharmacol. 2006;46(10):1163–70. https://doi.org/10.1177/0091270006292851.
31. Rebouche CJ, Chenard CA. Metabolic fate of dietary carnitine in human adults: identification and quantification of urinary and fecal metabolites. J Nutr. 1991;121(4):539–46. https://doi.org/10.1093/jn/121.4.539.
32. Nakashima M, Takiguchi Y, Mizuno A, et al.: Earth Pharmaceutical Co., Ltd. Internal Report, Phase I study of L-carnitine chloride (LC-80), 1986. (Article in Japanese).

33. Brass EP. Pharmacokinetic considerations for the therapeutic use of carnitine in hemodialysis patients. Clin Ther. 1995;17(2):176–85. https://doi.org/10.1016/0149-2918(95)80017-4.
34. Angelini C, Vergani L, Martinuzzi A. Clinical and biochemical aspects of carnitine deficiency and insufficiency: transport defects and inborn errors of β-oxidation. Crit Rev Clin Lab Sci. 1992;29(3–4):217–42. https://doi.org/10.3109/10408369209114601.
35. Hinton BT, Snoswell AM, Setchell BP. The concentration of carnitine in the luminal fluid of the testis and epididymis of the rat and some other mammals. J Reprod Fertil. 1979;56(1):105–11. https://doi.org/10.1530/jrf.0.0560105.
36. Savica V, Bellinghieri G, Di Stefano C, et al. Plasma and muscle carnitine levels in haemodialysis patients with morphological-ultrastructural examination of muscle samples. Nephron. 1983;35(4):232–6. https://doi.org/10.1159/000183087.
37. Debska-Slizień A, Kawecka A, Wojnarowski K, et al. Correlation between plasma carnitine, muscle carnitine and glycogen levels in maintenance hemodialysis patients. Int J Artif Organs. 2000;23(2):90–6.
38. Marzo A, Martelli EA, Mancinelli A, et al. Protein binding of L-carnitine family components. Eur J Drug Metab Pharmacokinet Spec No. 1991;3:364–8.
39. Brass EP, Hoppel CL, Hiatt WR. Effect of intravenous L-carnitine on carnitine homeostasis and fuel metabolism during exercise in humans. Clin Pharmacol Ther. 1994;55(6):681–92. https://doi.org/10.1038/clpt.1994.85.
40. Sahajwalla CG, Helton ED, Purich ED, et al. Comparison of L-carnitine pharmacokinetics with and without baseline correction following administration of single 20-mg/kg intravenous dose. J Pharm Sci. 1995;84(5):634–9. https://doi.org/10.1002/jps.2600840521.
41. Uematsu T, Itaya T, Nishimoto M, et al. Pharmacokinetics and safety of l-carnitine infused i.v. in healthy subjects. Eur J Clin Pharmacol. 1988;34(2):213–6. https://doi.org/10.1007/BF00614562.
42. Rebouche CJ, Engel AG. Kinetic compartmental analysis of carnitine metabolism in the human carnitine deficiency syndromes. Evidence for alterations in tissue carnitine transport. J Clin Invest. 1984;73(3):857–67. https://doi.org/10.1172/JCI111281.
43. Rebouche CJ, Engel AG. Kinetic compartmental analysis of carnitine metabolism in the dog. Arch Biochem Biophys. 1983;220(1):60–70. https://doi.org/10.1016/0003-9861(83)90387-9.
44. Yue KT, Fritz IB. Fate of tritium-labeled carnitine administered to dogs and rats. Am J Phys. 1962;202:122–8. https://doi.org/10.1152/ajplegacy.1962.202.1.122.
45. Khairallah EA, Wolf G. Carnitine decarboxylase. The conversion of carnitine to β-methylcholine. J Biol Chem. 1967;242(1):32–9.
46. Stanley CA. Carnitine deficiency disorders in children. Ann N Y Acad Sci. 2004;1033:42–51. https://doi.org/10.1196/annals.1320.004.
47. Borum PR, Bennett SG. Carnitine as an essential nutrient. J Am Coll Nutr. 1986;5(2):177–82. https://doi.org/10.1080/07315724.1986.10720124.
48. Rebouche CJ. Kinetics, pharmacokinetics, and regulation of L-carnitine and acetyl-L-carnitine metabolism. Ann N Y Acad Sci. 2004;1033:30–41. https://doi.org/10.1196/annals.1320.003.
49. Lancaster CS, Hu C, Franke RM, et al. Cisplatin-induced downregulation of OCTN2 affects carnitine wasting. Clin Cancer Res. 2010;16(19):4789–99. https://doi.org/10.1158/1078-0432.CCR-10-1239.
50. Hu C, Lancaster CS, Zuo Z, et al. Inhibition of OCTN2-mediated transport of carnitine by etoposide. Mol Cancer Ther. 2012;11(4):921–9. https://doi.org/10.1158/1535-7163.MCT-11-0980.
51. Heuberger W, Berardi S, Jacky E, et al. Increased urinary excretion of carnitine in patients treated with cisplatin. Eur J Clin Pharmacol. 1998;54(7):503–8. https://doi.org/10.1007/s002280050504.
52. Haschke M, Vitins T, Lüde S, et al. Urinary excretion of carnitine as a marker of proximal tubular damage associated with platin-based antineoplastic drugs. Nephrol Dial Transplant. 2010;25(2):426–33. https://doi.org/10.1093/ndt/gfp456.
53. Borum PR, Taggart EM. Carnitine nutriture of dialysis patients. J Am Diet Assoc. 1986;86(5):644–7.
54. Evans AM, Faull R, Fornasini G, et al. Pharmacokinetics of L-carnitine in patients with end-stage renal disease undergoing long-term hemodialysis. Clin Pharmacol Ther. 2000;68(3):238–49. https://doi.org/10.1067/mcp.2000.108850.

Safety of Carnitine

9

9.1 Introduction

Carnitine is a nutrient that is biosynthesized in the body and is abundant in foods such as dairy products and meat, as well as breast milk, so it is considered to be extremely safe. However, there is a new topic about the safety of carnitine, so I would like to describe the safety issues and summarize up-to-date information and data.

9.2 Daily Intake of Carnitine as a Food

In Japan, the use of carnitine as a food has been approved since December 2002, and L-carnitine has been added to diet supplements and beverages that claim to burn fat and can be sold (for details, see Sect. 3.12 in Chap. 3). At that time, regarding the general use of this ingredient as a food, cautionary statement as "In the use of this ingredient, it is necessary to take care not to overdose and provide information to consumers appropriately, given that the allowable daily intake (ADI) is evaluated at 20 mg/kg/day in the United States and 1,000 mg/day is a condition for intake in Switzerland" was described, and has since been partially revised. [1]

9.3 Risk Assessment of Carnitine

In the safety evaluation of compounds such as pharmaceuticals, a half-lethal dose (LD50) is used for acute toxicity, and a no observed effect level (NOEL) or maximum no observed adverse effect level (NOAEL) is used for chronic toxicity. On the other hand, since many nutrients do not cause adverse events or toxicity even when administered at high doses, a method is used to determine the acceptable daily intake based on experience from clinical trials in which high doses were administered. Carnitine is a nutrient similar to vitamins and minerals, and its safety is evaluated in the same way as vitamins and minerals. Carnitine risk assessments by experts in the field using this method include clinical trials of L-carnitine, acetyl-L-carnitine (ALCAR) and propionyl-L-carnitine (PLCAR) at relatively high doses for various diseases (2000 mg/day or more: 12 studies, 2000 mg/day or less, 16 studies or less), in addition safety was also evaluated in animal studies [2]. In clinical studies using ALCAR and PLCAR, the safety of carnitine was evaluated as 72.3 and 68.8% of ALCAR and PLCAR, respectively, calculated from the molecular weight, because the acetyl and propionyl groups are cleaved after being absorbed into the

N. Nakayama, *Carnitine Deficiency and Insufficiency in Medicine*,
https://doi.org/10.1007/978-981-95-7797-2_9

body [3]. As a result, there were no safety problems even when administered at high doses except for mild symptoms such as fish odor and diarrhea, and the observed safe level (OSL) of chronic administration was said to be no problem up to 2000 mg/day. No adverse events have been reported at higher doses, but it was decided that future studies should be waited for long-term safety [2]. The clinical studies used to assess the risk of carnitine were conducted in patients with acute myocardial infarction (437 patients, 9 g/day intravenously for 5 days, followed by 6000 mg/day orally for 1 year) [4], ADHD patients (10 patients, 100 mg/kg/day, 6 months) [5], patients with Rett syndrome (35 patients, 100 mg/kg/day, 8 weeks) [6], and hyperthyroidism patients (10 patients, 4000 mg/day orally for 3 months) [7] and other special diseases for which carnitine supplementation therapy is not currently widely used. Carnitine as a drug is mostly applied to carnitine deficiency in Japan and other countries, and carnitine administration is used as a supplementation therapy, intended to "bring low carnitine back to normal," and the risk of overdosage is considered to be extremely low. The highest dose of carnitine is administered in primary carnitine deficiency (systemic carnitine deficiency), which is an abnormality of the carnitine transporter, and the maintenance dose in the chronic phase is 100–400 mg/kg/day, and the dose should be adjusted so that the free carnitine concentration does not drop below 20 μmol/L in the guideline for this disease [8]. Other than primary carnitine deficiency, there is no other disease for which such high doses are administered chronically and for a long period of time.

Since the active ingredient is the same as that of levocarnitine chloride (L-carnitine chloride), the data from the toxicity test conducted on levocarnitine chloride are also used in the application for approval. In a single-dose toxicity study in mice and rats, LD50 was >3000 and > 1000 mg/kg for intravenous administration in both mice and rats. In addition, the non-toxic doses were 450 and 200 mg/kg/day, respectively, in a 13-week repeated oral toxicity study in rats (SD series) and dogs (beagles), and in a 12-month oral toxicity study, the non-toxic doses were 272 and 200 mg/kg/day, respectively, indicating that the compound is highly safe over a long period of time up to extremely high doses [9]. In addition, there were no notable problems in genotoxicity tests, carcinogenicity tests, reproductive and developmental toxicity tests, and antigenicity tests [9]. In addition, single-dose toxicity tests, repeated dose toxicity tests, genotoxicity tests, and tests to investigate the effects on the cardiovascular system and respiratory system were conducted for crotonoylbetaine, which is contained in very small amounts as an impurity in free-body intravenous preparations, and it was considered that there was no risk of toxicity at the clinical dose [9].

9.4 Carnitine and Fish Odor Syndrome

Although carnitine is an extremely safe compound, a problem that is very rarely observed when administered orally as a drug is a fish odor (a foul smell of rotting fish on the breath or in the sweat). Even if fish odor is observed in the early stages of administration, in most cases it is usually alleviated by reducing the dosage or administering an antibacterial agent such as metronidazole. Carnitine administered orally is not metabolized in the liver after absorption. However, unabsorbed carnitine is metabolized by intestinal bacteria to trimethylamine (TMA) and gamma-butyrobetaine. Normally, the TMA is absorbed and converted into trimethylamine-N-oxide (TMAO) by the liver's flavin-containing monooxygenase form 3 (FMO3) and excreted in the bloodstream then in urine. However, in patients with reduced enzyme activity due to congenital mutations in the FMO3 gene, which encodes this hepatic monooxygenase, TMA is not sufficiently converted to TMAO and remains in the bloodstream. In other words, it is TMA that emits a foul odor, and when metabolized to TMAO, it does not smell. And because TMA is volatile, it emits a strong foul odor from exhaled breath, urine, sweat, etc., like rotten fish. This is one of the congenital metabolic disorders called fish-odor syndrome (trimethylamineurea), but it does not cause serious life-threatening disorders such as organic acidemia [10–12]. However, because of body odor, patients with fish-odor syndrome may be bullied, marginalized or withdrawn, which can lead to schooling, psychological, or social problems. Although the number of patients overseas is relatively large, it was said that there are almost no patients with fish odor disease in Japan. However, even in Japan, there have been cases of children with fish odor becoming targets of bullying or committing suicide due to concerns during adolescence, and they are sometimes featured on TV programs. In Japan, Associate Professor Makiko Shimizu and Professor Hiroshi Yamazaki of the Department of Pharmacokinetics, Showa Pharmaceutical University, are working on clinical collaboration on urinary TMA analysis and FMO genotype in patients with fish-odor syndrome (http://hiroshi-yamazaki.my.coocan.jp/). Fish-odor syndrome occurs when TMA is produced by intestinal bacteria from trimethyl groups, which are tertiary amines such as choline and lecithin (phosphatidylcholine) contained in foods such as eggs, meat, and legumes, and carnitine, which are abundant in meat and dairy products, and cannot be sufficiently converted to TMAO in the liver. In addition, fish such as cod and sea bream also contain a large amount of TMA, and when people with fish odor eat these seafoods, they emit a body odor like rotten fish. At present, the only measure to prevent the occurrence of fish odor as much as possible is to avoid foods high in choline, lecithin, carnitine, or TMA itself as much as possible, but in the case of children, it is necessary to be aware of nutritional imbalances in terms of growth and development, and consultation with a specialist such as a pediatrician should be consulted. For the content of carnitine in various foods, please refer to Chap. 3. Table 9.1 shows the choline content in the main foods, and Table 9.2 shows the total TMA contained in fish and the urinary excretion rate after 8 h (both from the website of the Department of Pharmacokinetics, Showa Pharmaceutical University, http://hiroshi-yamazaki.my.coocan.jp/).

Table 9.1 Total choline content (mg) in 100 g of various foods

Whole egg	250
Meats	5 ~ 140
Organ meats such as liver	130 ~ 400
Processed meats	30 ~ 70
Fishes	10 ~ 90
Crops	2 ~ 80
Beans	20 ~ 190
Fruits	1 ~ 20
Vegetables	0.4 ~ 40
Milk, yogurt	10 ~ 20

Choline is rich in egg, meats, liver and beans

From home page of the Laboratory of Drug Metabolism and Pharmacokinetics (Dr. Hiroshi Yamazaki), Showa Pharmaceutical University; http://hiroshi-yamazaki.my.coocan.jp

Table 9.2 Total trimethylamine (TMA) contents in various fishes and urinary excretion rate

Fish species	TMA content in 100 g μmol, (mg)	Urinary excretion rate in 8 h., %[a]
Bonito	100(5.9)	87
Salmon	600(34.5)	
Tuna	1300(76.8)	43
Sea bream	3700(218.7)	
Cod	7400(437.4)	38

From home page of the Laboratory of Drug Metabolism and Pharmacokinetics (Dr. Hiroshi Yamazaki), Showa Pharmaceutical University; http://hiroshi-yamazaki.my.coocan.jp

[a] Urinary excretion rate in 8 h when eat 200 g of each fish

9.5 Carnitine and Atherosclerosis

As described above, carnitine is a nutrient that is biosynthesized in vivo and is contained in large amounts in meat, dairy products, and breast milk, and has been used as a pharmaceutical product for more than 60 years, but there have been no particular safety problems. Therefore, it is considered to be an extremely safe compound. However, in 2013, Koeth et al. in the group of Prof. Hazen at the Cleveland Clinic published in Nature Medicine that "TMAO is a causative compound of atherosclerosis," arising a major controversy regarding carnitine's merits and demerits, which continues to present [13]. The following is a summary of the study by Koeth et al. and the subsequent controversy.

9.5.1 Overview of the Study by Koeth et al.

Hazen et al. reported that phosphatidylcholine, which is abundant in eggs and other foods, is metabolized by the intestinal microflora to TMA, which is further oxidized in the liver to TMAO, which is related to the risk of developing cardiovascular disease [14, 15] before Koeth et al. reported in 2013. Although there are various studies on the relationship between cardiovascular disease and diet, and studies have shown that meat intake as a source of protein correlates with the risk of developing cardiovascular disease [16], a meta-analysis has shown that the amount of saturated fatty acids and cholesterol contained in meat alone does not explain the increased risk of cardiovascular disease [17], the focus turned to choline and phosphatidylcholine, which are abundant in meat. In this series of study flows, Koeth et al.'s research focused on carnitine, which has a tertiary amine structure same as choline and phosphatidylcholine and also abundant in red meat.

Koeth et al. first administered deuterium-labeled L-carnitine to healthy volunteers, and at the same time fed steak, TMAO was detected in plasma, but when a broad-spectrum antibiotic was administered, the TMAO was not recognized, and TMAO was detected when antibiotic administration was stopped for more than 3 weeks, and the same carnitine and steak tolerance test was performed again. It was shown that carnitine is metabolized by the intestinal microflora to become TMAO. Next, it was shown that carnitine was produced only slightly when carnitine was administered to vegans and vegetarians, and that the intestinal microbiota was different between vegans and normal eaters. He also reported that long-term administration of carnitine to mice (1.3%) changed their gut microbiota to produce more TMAO. It has been widely known that carnitine is metabolized and degraded into TMA by intestinal bacteria and converted into TMAO in liver, so that it is not a new discovery, but it is interesting to note that the intestinal microbiota changes toward the production of TMA and TMAO when a diet rich in carnitine is consumed on a regular basis.

Next, Koeth et al. examined the association between blood carnitine concentration quartiles and the prevalence of coronary artery disease (CAD), peripheral artery disease (PAD), and total cardiovascular disease (CVD) in a study cohort (2595 patients) examining long-term cardiac disease outcomes. The higher the plasma carnitine concentration, the higher the prevalence ($P < 0.05$). Logistic analyses adjusted for factors such as age, sex, diabetes history, smoking history, systolic blood pressure, LDL-cholesterol, and HDL-cholesterol related to well-known arteriosclerosis also showed significant associations (Fig. 9.1a–c). Regardless of whether patients had 1-, 2-, or 3-vessel disease, blood carnitine levels were significantly higher in patients with CAD lesions than in those without ($P < 0.001$, Fig. 9.1d). Hazard ratios (HRs) for the composite endpoint of major adverse cardiac events (MACE) of cardiac death, myocardial infarction, stroke, and revascularization also increased with increasing carnitine concentration, but did not differ when adjusted for traditional atherosclerosis-promoting factors (Fig. 9.1e). Then, a Kaplan-Meier plot of 3-year event-free survival showed that only the group with both high blood carnitine and TMAO concentration quartiles had significantly lower survival rates after adjustment for atherosclerosis-promoting factors (Fig. 9.1f). Finally, the female Apoe−/− When L-carnitine was administered to mice from 4 to 19 weeks of age (1.3%), and the arteriosclerotic lesion area at the base of the aorta was stained with oil Red O and quantitatively evaluated, it was reported that the arteriosclerotic lesions were about 1.8 times higher in the carnitine-loaded group, and that the suppression of intestinal bacteria with an antimicrobial cocktail did not promote arteriosclerosis. As a mechanism for promoting atherosclerosis in TMAO, it was shown that TMAO suppresses the reverse transfer of cholesterol (about 28% suppression

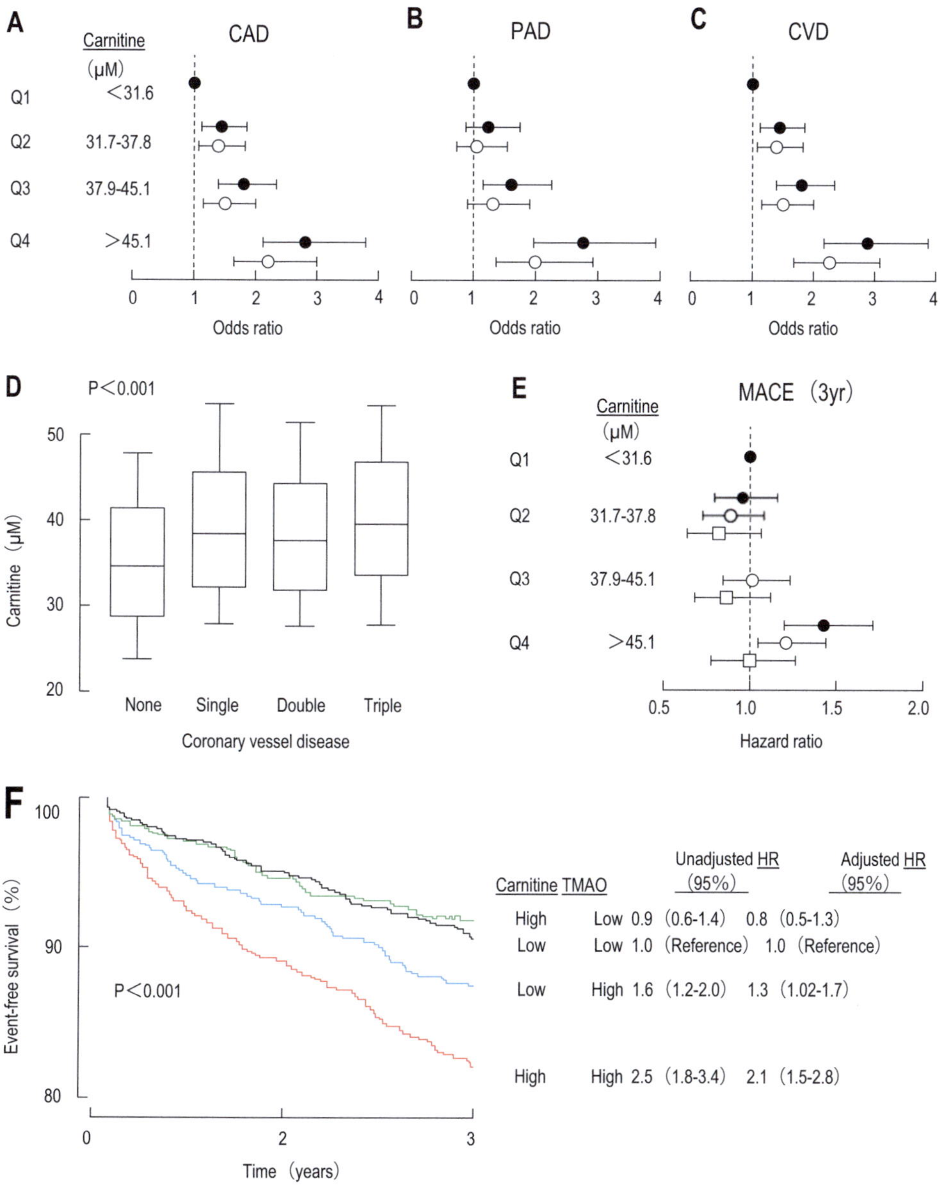

Fig. 9.1 Relationship between plasma carnitine concentration and CVD risk. (**a–c**): Forest plots of Relationship between odds ratios of CAD, PAD, CVD and quartiles of plasma carnitine concentrations before (closed circles) and after (open circles) logistic regression adjustments with traditional cardiovascular risk factors, including age, sex, history of diabetes mellitus, smoking, systolic blood pressure. Bars represent 95% confidence intervals. (**d**): Relationship of fasting plasma carnitine concentrations and angiographic evidence of CAD. Boxes represent the 25th, 50th and 75th percentile of plasma carnitine concentrations, and bars represent the tenth and 90th percentiles. (**e**): Forest plot of the hazard ratio of MACE and quartiles of carnitine unadjusted (closed circle) and after adjusting for traditional cardiovascular risk factors (open circles), or traditional cardiac risk factors plus creatinine clearance, history of myocardial infarction, history of CAD, burden of CAD (one-, two- or three-vessel disease), left ventricular ejection fraction, baseline medications (angiotensin-converting enzyme inhibitors, statins, beta blockers and aspirin) and TMAO levels (open squares). Bars represent 95% confidence intervals. (**f**): Kaplan-Meire plot and hazard ratios with 95% confidence intervals for unadjusted model, or following adjustments for traditional risk factors as in E. Median plasma concentration of carnitine (46.8 μM) and TMAO (4.6 μ) within the cohort were used to stratify subjects as having 'high' (≥median) or 'low' (<median) values. *CVD* cardiovascular diseases, *CAD* coronary artery diseases, *PAD* peripheral artery diseases, *MACE* major adverse cardiac events, *TMAO* trimethylamine-N-oxide. (From ref. [13])

by carnitine administration and about 31% suppression by choline), reduces the expression of cytochrome P7a1 and P27a1, which are involved in the synthesis of bile acids, and changes sterol metabolism. In summary, as shown in Fig. 9.2, they claimed that the hypothesis that the intake of L-carnitine and choline increases with the intake of red meat, and that they become TMA in the intestinal microbiota and are oxidized to TMAO in the liver, and that this TMAO promotes arteriosclerosis has been proven [13].

Subsequently, Koeth et al. orally administered deuterium-labeled carnitine to vegetarians and normal eaters and investigated in detail the process of carnitine conversion to TMAO, and reported that carnitine is first converted to γ-butyrobetaine by the action of multiple intestinal bacterial flora in both vegetarians and normal eaters, and then γ- butyrobetaine is converted to TMA by the action of multiple intestinal microflora, and the only process for conversion of γ-butyrobetaine to TMA is performed by a bacterium of the order Clostridiales, *Emergencia timonensis* [18].

In summary, Hazen's group claims that when people consume foods rich in tertiary amines such as carnitine, choline, and phosphatidylcholine, which are abundant in meat and eggs, the intestinal microflora changes and the number of bacteria that metabolize them into TMA increases, leading to the production of more TMA, which is then TMA is metabolized by

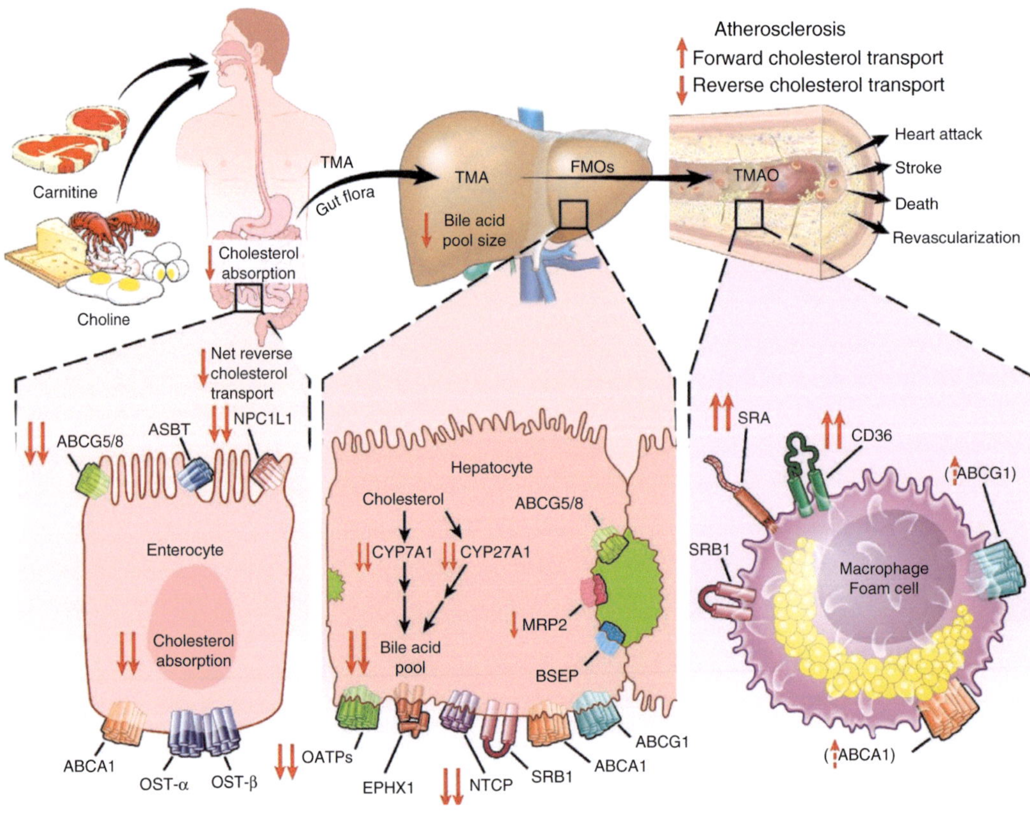

Fig. 9.2 A working hypothesis by Koeth et al. Carnitine and choline in foods are metabolized and degraded to γBB and further converted into TMA by bacterial flora in the intestine of Clostridium order. TMA go to liver via blood stream and converted into TMAO by flavin-containing monooxygenase (FMO). Their hypothesis claims that this TMAO induces atherosclerosis, and causes heart failure, heart attack, stroke, etc. Koeth et al. describes that vegetarian who don't eat meats have little intestinal bacterial flora which convert γBB into TMA, so that TMA is very low. *γBB* γ-butyrobetaine, *TMA* trimethylamine, *TMAO* trimethylamine-N-oxide, *FMO* flavin-containing monooxygenase. (From ref. [13])

FMO in the liver to TMAO. This TMA is metabolized by FMO in the liver to TMAO, which is hypothesized to be the substance responsible for causing atherosclerosis.

In the United States, partly due to the high cost of medical care, many people take choline, phosphatidylcholine (lecithin) and carnitine, etc. as supplements to maintain their health. In particular, with regard to the prevention of cardiovascular disease, Dr. Stephen T. Sinatra's book "The Sinatra Solution: Metabolic Cardiology," [19, 20] which suggests that CoQ10, L-carnitine and D-ribose are effective in preventing heart failure and other heart diseases, became a bestseller and carnitine was widely used as a supplement. So that the paper by Koeth et al. claiming that carnitine increases atherosclerosis was picked up and caused a sensation, not only in academic journals but also on TV programs and the Internet.

9.5.2 Rebuttal to the Research of the Hazen'S Group

These studies by the Hazen group on TMAO causing atherosclerosis have been uncritically praised by arteriosclerosis researchers and others as a unique study that links arteriosclerosis promotion and intestinal microbiota, and many reviews had appeared. On the other hand, there have also been a number of reports refuting the reproducibility of the Hazen group's findings and the hypothesis that carnitine, choline, and TMAO induce atherosclerosis. The main ones are described below.

1. Research on animal arteriosclerosis models

Koeth et al. reported that L-carnitine was administered to female Apoe−/− mice for 15 weeks and their atherosclerosis increased by 1.8-fold, and the dose of carnitine was 1.3% when administered with drinking water. An adult weighing 50–60 kg has a daily water intake of 2000–2500 mL, of which about 2000 mL comes from food and drink [21]. Assuming that an adult drinks approximately 2 L of water per day, the 1.3% carnitine dosage of Koeth et al. in drinking water is a tremendous dosage of 26 g/day in humans. Carnitine in food is not abundant so 54–87% is absorbed, but the remainder is broken down by intestinal bacteria. [22] It is natural that if a large amount of carnitine is administered and remains in the intestine without being absorbed, the number of bacteria that utilize it will increase. In addition, the TMAO concentration in mice used by Wang and Koeth et al. for the study of choline and carnitine was 2.5 ± 0.1 μM in the control group of male Apoe−/− mice and 28.3 ± 2.4 μM in the high-choline diet group. In mice, it was 158.6 ± 32.9 μM, and the highest quartile of TMAO levels in the clinical research cohort of Koeth et al. was 6.18 μM at the highest and 3.67 μM at the median, indicating that the TMAO concentration in the female Apoe−/− mouse model used by Koeth et al. is at least one order of magnitude or more away from the clinical level. [23]

Collins et al. orally administered L-carnitine to male Apoe−/− mice expressing human cholesteryl ester transfer protein (hCETP) for 12 weeks (87 and 352 mg/kg, corresponding to 500 mg/day and 2000 mg/day in humans, respectively) [24]. For some unknown reason, female Apoe−/− mice were used in the Koeth et al. study, whereas male Apoe−/− mice are usually used for atherosclerosis studies. Collins et al. used male Apoe−/− mice because the expression of flavin monooxygenase (FMO) isoforms is sex-dependent and influenced by hormones, and TMAO in female mice can be 10 times higher than in male mice, resulting in greater variability. The reason for expressing hCETP was to study the development of atherosclerosis under conditions more similar to those in humans, since Apoe−/− mice do not express CETP, although CETP plays an important role in the reverse transfer of cholesterol. TMAO levels increased approximately twofold in the high-dose carnitine group, but atherosclerotic lesions decreased in the opposite direction; correlation analysis of TMAO levels and aortic atherosclerotic lesion area revealed that TMAO levels of 0.05–0.1 ppm (low), 0.1—0.2 ppm (moderate), and > 0.2 ppm (high) were significantly higher

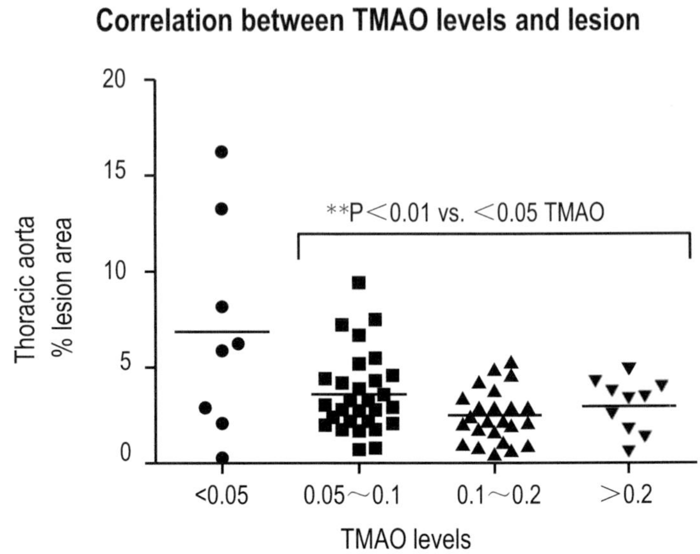

Fig. 9.3 A correlation analysis between the thoracic aorta % lesion area and very low (<0.05 ppm), low (0.1 ppm), moderate (0.1–0.2 ppm), and high (>0.2 ppm) TMAO levels was conducted in hCETP expressed male *Apoe*$^{-/-}$ mouse administered L-carnitine for 12 weeks. The results were rather low thoracic aorta % lesion area in low, moderate and high groups compared to very low group ($P < 0.01$). (From ref. [24])

than those of <0.05 ppm (very low) group had significantly lower atherosclerotic lesion area [24] ($P < 0.01$, Fig. 9.3).

In addition, when the FMO inhibitor methimazole was administered and TMAO production was suppressed, arteriosclerosis increased. Furthermore, in in vitro foam cell formation experiments using mouse macrophages, adding TMAO at a concentration 10 times higher than the TMAO concentration reported in humans had no effect on foam cell formation. Therefore, Collins et al. concluded that TMAO does not promote atherosclerosis, but rather acts in the direction of blocking arteriosclerosis in vivo. Collins et al. questioned the fact that only 3 out of 11 mice showed atherosclerosis in Koeth et al.'s experiment (as described in detail when looking closely at the supplement data in Koeth et al.'s paper) and that they did not examine dose dependence.

Aldana-Hernández et al. also tested the hypothesis of Koeth et al. that TMAO is a causative agent of accelerated atherosclerosis by loading male *Apoe−/−* mice and LDL-receptor knockout male *Ldlr−/−* mice were choline-loaded (1.0%), betaine-loaded (0.9%), and TMAO-loaded (0.12 and 0.2%) for 16–28 weeks to examine the effects on aortic atherosclerotic lesions and lipid profiles [25]. The results showed that TMAO levels in the blood of *Ldlr−/−* mice were approximately doubled by TMAO loading after 16 weeks, and in *Apoe−/−* mice they were 1.8- and 1.5-fold higher after 28 weeks, but not in no difference in aortic atherosclerotic lesions was observed compared to controls (choline content 0.1%). Furthermore, in the same male *Ldlr−/−* mice, on the basis of a high calorie diet (0.5% cholesterol), a control group (0.1% choline), a choline loading group (0.4% choline), and a phosphatidylcholine (PC) loading group (0.1% choline + The PC-loading group had a twofold increase in TMAO concentration compared with the control group and the choline-loading group, but atherosclerotic lesions were significantly lower) [26]. In the PC-loading group, VLDL cholesterol and apoprotein B48 were lower, and HDL cholesterol and HDL cholesterol B48 were decreased, and HDL cholesterol was increased in the PC-loading group, but VLDL secretion was unchanged. Chemokines that promote atherogenesis were also decreased in the PC-loaded group. Thus, the Hazen group reported that choline and TMAO produced from PC did not promote atherosclerosis as they claimed. [26]

Jonsson et al. also examined the effects of choline loading and changes in the intestinal microbiota in sterile or normally bred Apoe−/− mice [27]. First, in preliminary experiments, atherosclerotic lesions in female Apoe−/− mice were compared with those in male mice, but since there were no significant sex differences, only male Apoe−/− mice were used. Apoe −/− mice on the diet for Apoe−/− mice, a choline-loaded (1.2%) group on the diet for Apoe−/− mice, a Western diet group and a Western diet plus choline (1.0%) group were reared for 12 weeks, and TMAO levels, atherosclerotic lesions, and lipid profiles were examined. Normal-reared mice had fewer atherosclerotic lesions and lower plasma cholesterol than aseptic-reared mice. In aseptically bred mice, TMAO concentrations were barely detectable even after choline loading, while in normally bred Apoe−/− mice, TMAO concentrations increased markedly in the choline-loaded group, both in the diet for Apoe−/− mice and in the Western diet. However, the area of atherosclerotic lesions in the aorta was not increased by choline loading, and there was no correlation with TMAO concentration. Choline loading did not cause significant changes in the intestinal microbiota, but bacteria belonging to the order Clostridiales were relatively abundant [27].

Seyed-Ahmed et al. reported that in a 2% cholesterol-loaded rabbit atherosclerosis model, L-carnitine administration (250 mg/kg, intraperitoneal, for 28-days) completely prevented atherosclerotic lesions in the aorta and coronary arteries, and reduced glutathione in the aorta and heart in 2% cholesterol loaded group compared to the control group increased [28]. Conversely, carnitine deficiency in rabbits treated with D-carnitine (250 mg/kg, intraperitoneal, for 28 days) exacerbated atherosclerotic lesions. However, malondialdehyde, an oxidative stress marker, was reduced by about 40% in the aorta and by about 85% in the heart under both L-carnitine and D-carnitine. These findings suggest that L-carnitine may prevent the progression of atherosclerosis by another mechanism in addition to its antioxidant and lipid-lowering effects [28].

Thus, many experimental results in animal models of atherosclerosis do not fit the hypothesis of the Hazen Group that TMAO causes atherosclerosis.

2. Carnitine and risk of developing cardiovascular disease

A large number of clinical studies have been conducted to verify the Hazen group's assertion that TMAO is a causative agent of atherosclerosis, and numerous systematic reviews and meta-analyses of these studies have been reported. Many of them found significant associations between TMAO concentrations and cardiovascular events and all-cause mortality [29–33].

On the other hand, TMAO concentrations were significantly higher in patients with diabetes mellitus and metabolic syndrome, and TMAO and choline concentrations increased with decreased renal function, but TMAO and betaine concentrations were not associated with CVD [34], and the results of a clinical study on the association between carotid artery atherosclerosis and γ-butyrobetaine, carnitine, trimethyllysine, and TMAO concentrations showed that γ-butyrobetaine, trimethyllysine, and TMAO concentrations were associated with cardiovascular death, and γ-butyrobetaine and trimethyllysine were associated with cardiovascular death when adjusted for age and eGFR, but not significantly associated with TMAO [35]. In addition, TMAO concentrations were not independently associated with atherosclerotic cardiovascular diseases such as acute myocardial infarction or cardiogenic stroke, but were independently associated with obesity and diabetes, and were independent predictors of long-term mortality in patients with coronary artery disease [36], and TMAO concentrations were associated with the prevalence of cardiovascular disease but not with the carnitine concentration [37]. Thus, when the clinical studies of TMAO and cardiovascular disease risk are reviewed in detail, the conclusions regarding the association of TMAO and carnitine concentrations with cardiovascular disease are slightly different from those of the Hazen group.

In addition, a meta-analysis of clinical studies has also reported results that contradict the hypothesis that carnitine increases TMAO, causes arteriosclerosis, and increases the risk of developing cardiovascular disease. DiNicolantonio et al. reported a systematic review and meta-analysis of clinical trials of carnitine on cardiovascular disease risk [38]. Among 153 reports, 13 studies were selected and analyzed in which adults (18 years of age and older) were given carnitine (duration of treatment: 0.7–12 months, median 2 months) compared with placebo or control, and showing results of all-cause mortality, cardiovascular events (including re-infarction), heart failure, and ventricular arrhythmias. All of these trials were compared to carnitine versus placebo or control in acute myocardial infarction. All-cause mortality was reported in 11 trials (n = 3579), and L-carnitine was associated with a 27% reduction in all-cause mortality compared to placebo or controls (odds ratio: 0.73, 95% confidence interval [CI]: 0.54–0.99, P = 0.05, Fig. 9.4). Although the results are not shown, five of the 11 trials were small-scale, but analysis of five of these small trials showed that L-carnitine administration reduced the odds ratio of all-cause mortality by 34%.

The risk of developing heart failure after acute myocardial infarction was reported in six studies (n = 3214), and no increased risk of heart failure was observed with L-carnitine administration (relative risk [RR]: 0.85, 95% CI: 0.67–1.09, P = 0.21, Fig. 9.5).

The risk of developing ventricular arrhythmias was also examined in five studies (n = 229), and L-carnitine was associated with a 65% reduction in risk (RR = 0.35, 95% CI: 0.21–0.58, $P < 0.0001$, Fig. 9.6).

In addition, the risk of developing reinfarction was examined in four studies (n = 829), and no significant increase in the risk of developing reinfarction was observed with L-carnitine administration (RR = 0.78, 95% CI: 0.41–1.48, $P < 0.45$). The risk of developing angina pectoris was examined in two studies (n = 261), and L-carnitine administration reduced the risk of angina pectoris by 40% (RR = 0.60, 95% CI: 0.5–0.72, $P < 0.00001$).

Contrary to the Hazen group's claims, many clinical studies cited in the meta-analysis by DiNicolantonio et al. did not confirm that carnitine administration promoted arteriosclerosis and worsened the condition even when administered to patients with heart disease after acute myocardial infarction for a long time, but rather reduced the risk of developing cardiovascular disease. Among the many papers on TMAO and arterio-

Study or subgroup	L-carnitine (No.)		Control (No.)		Weight (%)	Odds ratio IV, random (95% CI)
	Events	Total	Events	Total		
De Pasquale et al, 1990	0	49	18	97	1.2	0.04 (0.00-0.74)
Davini et al, 1992	1	81	10	79	2.1	0.09 (0.01-0.69)
Jacoba et al, 1996	0	22	1	17	0.9	0.24 (0.01-6.39)
Rebuzzi et al, 1984	0	12	1	10	0.8	0.25 (0.01-6.94)
Xue et al, 2007	0	48	1	48	0.9	0.33 (0.01-8.22)
Iyer et al, 1999	1	30	2	30	1.5	0.48 (0.04-5.63)
Singh et al, 1996	4	51	6	50	5.1	0.62 (0.17-2.36)
Kobulia et al, 2002	4	45	7	53	5.4	0.64 (0.17-2.35)
Iliceto et al, 1995	21	233	27	239	22.8	0.78 (0.43-1.42)
Tarantini et al, 2006	67	1,168	75	1,161	57.0	0.88 (0.63-1.24)
Rizzon et al, 1989	2	28	2	28	2.2	1.00 (0.13-7.64)
Total	100	1,767	150	1,812	100	0.73 (0.54-0.99)

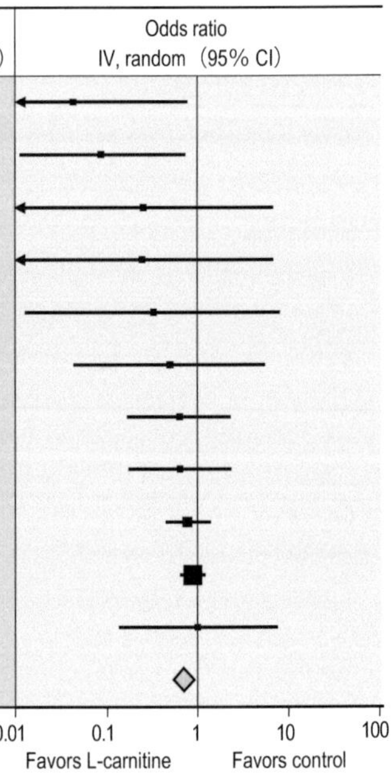

Heterogeneity: $Tau^2=0.01$; $\chi^2=10.36$, df=10 (P=0.41); $I^2=4\%$
Test for overall effect: Z=2.00 (P=0.05), **IV = inverse variance**

Fig. 9.4 Forest plot of odds ratios for all-cause mortality. *IV* inverse variance, *CI* confidence interval. (From ref. [38])

Study or subgroup	L-carnitine (No.)		Control (No.)		Weight (%)	Risk ratio IV, random (95% CI)
	Events	Total	Events	Total		
Davini et al, 1992	0	81	2	79	0.7	0.20 (0.01-4.00)
Rizzon et al, 1989	1	28	5	28	1.4	0.20 (0.02-1.60)
Xue et al, 2007	0	48	1	48	0.6	0.33 (0.01-7.98)
Singh et al, 1996	12	51	18	50	15.8	0.65 (0.35-1.21)
Tarantini et al, 2006	41	1,168	46	1,161	35.4	0.89 (0.59-1.34)
Iliceto et al, 1995	46	233	48	239	46.1	0.98 (0.68-1.41)
Total	100	1,609	120	1,605	100	0.85 (0.67-1.09)

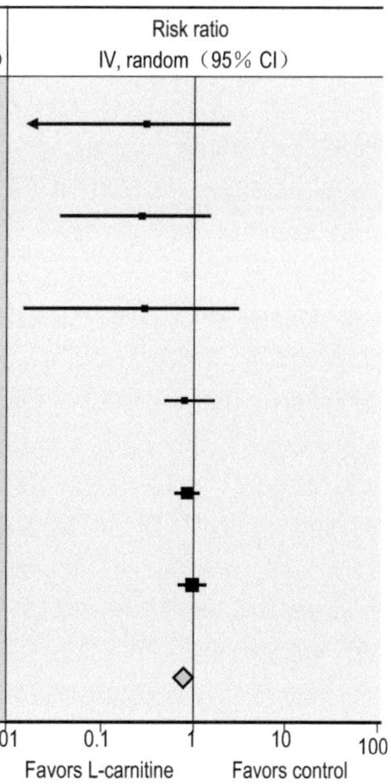

Heterogeneity:Tau^2=0.00; χ^2=4.45, df=5 (P=0.49); I^2=0%
Test for overall effect: Z=1.26 (P=0.21), *IV*=inverse variance, CI: confidence interval

Fig. 9.5 Forest plot of risk ratios for the development of heart failure (6 studies, $n = 3214$). *IV* inverse variance, *CI* confidence interval. (From ref. [38])

Study or subgroup	L-carnitine (No.)		Control (No.)		Weight (%)	Risk ratio IV, random (95% CI)
	Events	Total	Events	Total		
Rizzon et al, 1989	2	28	13	28	13.6	0.15 (0.04-0.62)
Pehlivanoglu et al, 1996	1	12	9	18	7.1	0.17 (0.02-1.15)
Rebuzzi et al, 1984	0	12	1	10	2.8	0.28 (0.01-6.25)
Martina et al, 1992	4	12	7	8	37.2	0.38 (0.16-0.88)
Singh et al, 1996	7	51	14	50	39.3	0.49 (0.22-1.11)
Total	14	115	44	114	100	0.35 (0.21-0.58)

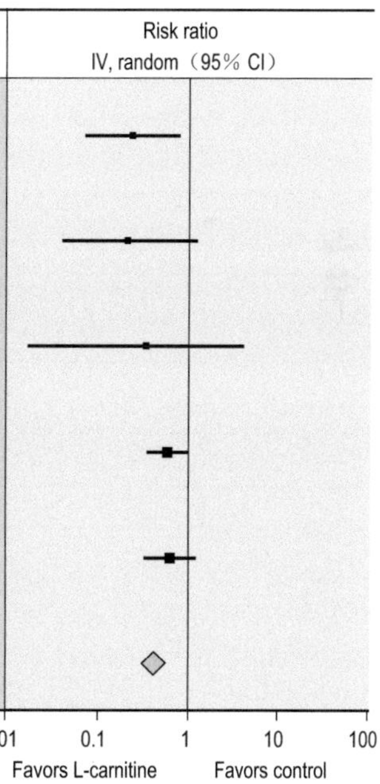

Heterogeneity: Tau^2=0.00; χ^2=2.61, df=4 (P=0.63); I^2=0%
Test for overall effect: Z=4.03(P<0.0001), IV= inverse variance.

Fig. 9.6 Forest plot of risk ratios for ventricular arrhythmia (5 studies, $n = 229$). *IV* inverse variance, *CI* confidence *interval.* (From ref. [38])

sclerosis, there is an argument that the intake of choline, phosphatidylcholine, and carnitine, which are the precursors of TMAO, should be avoided as much as possible. However, there are many nutritional epidemiological studies that have failed to provide clear data on whether carnitine, as well as choline and phosphatidylcholine, increase TMAO levels or increase the incidence of cardiovascular disease. For example, there have been studies that examined TMAO concentrations by eating eggs rich in phosphatidylcholine (1–3 eggs per day), and studies that examined TMAO concentrations by eating a Mediterranean diet rich in choline, phosphatidylcholine, and carnitine, but no increase in TMAO concentrations has been confirmed [39].

Since TMAO produced from carnitine and choline is excreted by the kidneys, the concentration may be high, especially in patients with chronic renal failure and dialysis. Bain et al. first administered 1.5 g/day (0.5 g × 3 times), 3.0 g/day (1.0 g × 3 times), and 6.0 g/day (2.0 g × 3 times) of L-carnitine for 7 days, and examined the blood carnitine, TMA, and TMAO concentrations on the seventh day, and found that the TMAO concentration was 34.5 ± 2.0 μM at baseline, 32.4 ± 21.1 μM in the 1.5 g group, and 35.2 ± 18.5 μM in the 3 g group. It was the same as the baseline, but it was significantly increased to 149 ± 145 μM in the 6.0 g group [40]. In other words, up to an oral dose of about 3 g of carnitine per day, TMAO is eliminated by renal excretion in healthy subjects, but at a high dose of about 6 g per day, the amount produced exceeds that of renal excretion, and the blood concentration become higher.

Next, Bain et al. compared pre- and post-dialysis TMA and TMAO concentrations in hemodialysis patients on a regular diet with those of healthy controls [41]. TMA and TMAO concentrations were 0.418 ± 0.124 and 37.8 ± 20.4 μM, respectively, in healthy subjects, but were significantly higher ($P < 0.05$) in dialysis patients, 1.39 ± 0.483 and 99.9 ± 31.9 μM, before dialysis. After dialysis, however, the concentrations were 0.484 ± 0.164 and 41.3 ± 18.8 μM, not different from those in healthy subjects. This indicates that TMA and TMAO are efficiently removed with each dialysis session.

In addition, Bain et al. orally administered L-carnitine (1 g/day, evening) daily for 12 consecutive days to hemodialysis patients and examined plasma carnitine, TMA, and TMAO concentrations before and after dialysis [42]. The results showed that the pre-dialysis concentrations of carnitine and TMAO increased significantly from baseline after 12 days (carnitine: 32.4 ± 6.1 ⇒ 61.9 ± 21.8 μM, TMAO: 289.1 ± 236.1 ⇒ 529.0 ± 237.9 μM, both $P < 0.0001$), while TMA concentration did not change significantly (1.2 ± 0.2 to 1.5 ± 0.5 μM). After dialysis, carnitine, TMA, and TMAO were efficiently removed, and carnitine, TMA, and TMAO concentrations did not increase significantly after 12 days of carnitine administration compared to baseline. This indicates that TMA and TMAO do not accumulate and are efficiently removed by dialysis at a carnitine dosage of approximately 1 g/day when administered orally [42].

Ozasa et al. administered L-carnitine intravenously (1 g three times a week after dialysis) or orally (300 mg daily) to hemodialysis patients and compared plasma TMA and TMAO concentrations at 0, 2, 4, and 8 weeks with those of non-treated patients. The results showed that carnitine administration significantly increased TMA concentration compared to the non-treated group, but the change in TMAO concentration was not significantly different within the same level of daily fluctuation as that of healthy Japanese people consuming a normal diet [43]. Thus, it is unlikely that carnitine administered orally to end-stage renal failure patients undergoing hemodialysis would result in a significant increase in TMAO concentrations. In Japan, the oral formulation of L-carnitine was used for dialysis patients when the indication was expanded in March 2011, but the intravenous formulation was launched in February 2013, and now the oral formulation is rarely used in the dialysis field. It has been shown that neither TMA nor TMAO increases with intravenous L-carnitine administration due to the lack of gut microbiota involvement [43].

So, what is the relationship between the risk of developing cardiovascular disease in patients with end-stage renal disease and TMAO concentration? Kim et al. examined the association between cardiovascular events and TMAO concentrations in patients with stage 3b and 4 chronic renal failure and reported that TMAO levels were independently associated with the onset of cardiovascular events even after adjusting for well-known cardiovascular risk factors [44].

On the other hand, Kaysen et al. examined TMAO concentrations in hemodialysis patients, which were markedly elevated compared to controls (median 43 μmol/L, compared to 1.41 ± 0.49 μmol/L in controls), but there was no association between TMAO concentration and all-cause mortality, cardiovascular death, or cardiovascular-related hospitalizations, and it was positively correlated with nutritional factors such as albumin, prealbumin, and creatinine, and inversely correlated with log CRP [45]. These findings suggest that the negative effects of TMAO on blood vessels may be balanced by the positive effects on nutrition and inflammatory status.

Higuchi et al. divided 176 hemodialysis patients into two groups, a carnitine group (oral administration, 20 mg/kg, 12 months) and a control group, and examined the effect on brachial-ankle pulse wave velocity (baPWV) [46]. As a result, carnitine administration did not promote arteriosclerosis, but on the contrary, it significantly reduced baPWV (see Chap. 14, Fig. 14.36).

Similarly, Lai et al. measured carnitine concentrations and baPWV in patients who underwent kidney transplantation with end-stage renal failure and examined the association between various factors and peripheral arteriosclerosis [47]. As a result, it was found that 55.4% of patients had peripheral arterial stiffness (PAS) when the cut-off value was 14.0 m/s > baPWV. Diabetes mellitus, hypertension, metabolic syndrome, and hypertriglyceridemia were significantly more common in patients with PAS, and significantly more common in patients with low serum carnitine levels ($P < 0.001$). The diagnostic power of the presence or absence of PAS based on serum carnitine concentration in ROC analysis was as high as 0.789 (95% CI: 0.670–0.881, $P < 0.001$).

Higuchi et al. conducted a randomized controlled trial in which 148 hemodialysis patients were divided into two groups: L-carnitine treatment (75 patients, 20 mg/kg/day, oral administration) and control group (73 patients) and investigated the effects on left ventricular ejection fraction and left ventricular myocardial weight coefficient [48]. The results showed that carnitine treatment significantly increased left ventricular ejection fraction from 6 months after treatment and significantly decreased left ventricular myocardial weight coefficient, an index of cardiac hypertrophy, in the carnitine-treated group (see Fig. 14.27 in Chap. 14).

Fukami et al. administered L-carnitine (900 mg/day) orally to 31 hemodialysis patients for 6 months, and measured plasma TMA and TMAO concentrations and AGE, sICAM-1, sVCAM-1, and MDA, markers of vascular damage and oxidative stress, to examine their changes [49]. The results showed that TMA and TMAO increased significantly from baseline after 6 months of carnitine treatment (TMA: 39.1 ± 15.3 ⇒ 85.1 ± 36.5 μmol/L, TMAO: 222.5 ± 111.7 ⇒ 548.4 ± 206.4 μmol/L, all $P < 0.001$), but sICAM −1, sVCAM-1, and logMDA were significantly decreased by carnitine administration. Serum AGEs showed a decreasing trend, but not significantly.

Thus, it has been reported that carnitine administration reduces arterial stiffness in end-stage renal failure patients and dialysis patients with advanced arterial stiffness and a high incidence of cardiovascular disease. In addition to these data, many papers have shown that carnitine administration in dialysis patients does not promote arteriosclerosis, but rather provides benefits such as improved cardiac function (see Chap. 14).

3. Fish diet and TMAO

The greatest weakness of the Hazen group's hypothesis that TMAO is a causative agent of atherosclerosis is that the large amounts of TMA and TMAO contained in fish and marine products that are considered healthy have not been examined, as described in Sect. 9.4.

He et al. conducted a meta-analysis of 11 studies (13 cohorts, 222,364 cases, mean follow-up: 11.8 years) reporting fish consumption or frequent fish consumption and mortality from coronary artery disease [50]. Compared to those who do not eat fish at all or eat less than once a month, the mortality rate from coronary artery disease is lower in people who eat fish 1 to 3 times a month. Analyzing the pooled data, those who eat fish 1 to 3 times a month have a relative risk of 0.89 (95% CI: 0.79–1.01), while those who eat fish once a week have a relative risk of 0.85 (95% CI: 0.76–0.96). Those who ate fish 2–4 times a were 0.77 (95% CI: 0.66–0.89) and those who ate fish 5 or more times a week were 0.62 (95% CI: 0.46–0.82). For every 20 g/day of fish intake, there is a 7% reduction in coronary artery disease mortality (*P* for trend = 0.03). In addition to this meta-analysis, there are many clinical studies and meta-analyses on the risk of developing cardiovascular disease with fish intake, and all of them show that consistently high fish intake significantly reduces the risk of cardiovascular disease.

Japanese eat fish more often than Westerners, but what about that? Iso et al. conducted a large-scale cohort study to investigate whether people who frequently eat fish in Japan have less coronary artery disease [51]. From 1990–1992 to 2001, a total of 41,578 Japanese men and women (40–59 years old) were followed up with a questionnaire to determine their dietary intake, including fish. A total of 258 cases of coronary artery disease developed during this period (196 were nonfatal and 62 were fatal). Multivariate analysis showed that the quartile with the highest frequency of fish consumption (8 times/week, median daily intake 180 g) and the quartile with the lowest frequency of fish consumption (once/week, median daily intake 23 g) had a hazard ratio of 0.63 (95% CI: 0.38–1.04) for developing total coronary artery disease and 0.44 (0.44) for definite myocardial infarction 0.44 (0.24–0.81) and 1.14 (0.36–3.63) for sudden death, respectively. The hazard ratios were significantly lower in the group of frequent fish eaters except for sudden death. A reduction in risk was seen for nonfatal events (HR = 0.43, 0.23–0.81) but not for fatal coronary events (HR = 1.08, 0.42–2.76). There was also a strong inverse association between n3 fatty acid intake and risk of incident events (HR = 0.35, 0.18–0.66) for definite myocardial infarction and HR = 0.33, 0.17–0.63 for non-fatal coronary events.

ISO et al. suggest that the effect of fish eating on coronary artery disease is due to the effects of ω-3 polyunsaturated fatty acids EPA, DHA, and DPA contained in fish [52, 53].

The Hazen group emphasizes that carnitine, choline, and phosphatidylcholine in food are precursors of TMA, but fish diets should consume large amounts of TMA and TMAO itself. Zhang et al. reported 8-h urinary excretion of TMA and TMAO when humans were fed 227 g of 46 foods (fruits, vegetables, meat, grains, dairy products, and seafood), but there was no significant increase in carnitine-rich meat and dairy products compared to the control background values, and only when fish and shellfish were fed [54]. The increase is 9 times higher for sardines, 10 times for crabs, 26 times for haddock, 28 times for herring, 34 times for cod, and 54 times for halibut. Naturally, one would expect these TMA and TMAO to remain at high blood concentrations until excreted in the urine.

Then, is the anti-arteriosclerotic effect of ω-3 polyunsaturated fatty acids by fish eating a decrease in arteriosclerotic disease because it overcomes the arteriosclerosis promotion effect of high TMAO levels?

DiNicolantonio et al. suggest that, although there are several meta-analyses on the cardiovascular disease risk of ω-3 polyunsaturated fatty acids, the result that ω-3 polyunsaturated fatty acids had no coronary artery disease-lowering effect or, if any, it is considered a weak effect, suggests that if TMAO is a strong atherosclerosis promoter, as claimed by Hazen's group, then ω-3 polyunsaturated fatty acids may not have a countervailing effect [55]. In fact, studies showing that ω-3 polyunsaturated fatty acids were effective against cardiovascular diseases such as coronary artery disease [56, 57], meta-analyses that showed that they were not effective in all-cause mortality and the onset of cerebral infarction, but had a significant protective effect on cardiac death, sudden death, and myocardial infarction [58], and reports that they were considered to be effective but weak [59, 60], and there are clinical studies and meta-analyses that show no effect [61, 62]. Even if it is effective, it is considered that it is not very powerful.

The diet of northern Portugal in the northwestern part of the Iberian Peninsula and the Galicia region in northwestern Spain is called the Southern European Atlantic Diet (SEAD), and in addition to seafood such as cod, halibut, crab, herring, and clams, meat such as pork and beef represented by Iberian piglet, and flaxseed oil are attracting attention. Oliveira et al. examined the detailed dietary content of patients hospitalized with acute myocardial infarction (820 patients) and healthy controls (2196 patients) using a questionnaire and examined the association with the onset of acute myocardial infarction using an index called SEAD score [63]. The odds ratio of acute myocardial infarction decreased by 10% for each 1-point increase in SEAD score, and the odds ratio decreased by 33% for those in the highest SEAD score quartile compared to those in the lowest SEAD score quartile (odds ratio 0.67, 95% CI: 0.51–0.88, *P* for trend = 0.003). Interestingly, when looking at adherence to the nine foods individually, the odds ratios increased for red meat, pork products, and potatoes, while the odds ratios decreased for cod, dairy products, beans and vegetables, vegetable soups, whole grain bread, and wine. Carnitine is more common in red meat and dairy products, but results are different in red meat and dairy products. In addition, SEAD diets should include a high intake of fish and meat, and an increase in the intake of TMA, TMAO, and carnitine, but the incidence of acute myocardial infarction is rather low.

Norway and other Nordic countries are also areas where fish high in TMA and TMAO, such as cod and halibut, are consumed in very high amounts. Norwegian fisheries scientist Valeur and colleagues wrote, "The Hazen group is focusing on the choline and carnitine in meat and eggs and claiming that these diets are unhealthy because they promote atherosclerosis, but are they forgetting about fish, which is a source of large amounts of TMA and TMAO itself?" and commented in the Journal of the American College of Cardiology, which was published along with a response from the Hazen group [64]. Tang and Hazen's response to this question was a vague one, stating that TMAO is excreted by the kidney, that some fish species contain less TMAO than others, and that the interaction between the host and intestinal bacteria promotes atherosclerosis, not attacking specific foods [64]. The answer is a dodge, such as, "We do not intend to attack any specific food." This makes one wonder if the much smaller amounts of TMAO produced from choline and carnitine are also excreted by the kidneys and are not a problem. To this response, Landfald of the same group as Valeur responded, "The Hazen group has called meat high in carnitine and eggs high in choline as unhealthy foods that promote atherosclerosis, while fish, which is the highest in TMAO, is widely known to lower cardiovascular disease, a paradox, but they did not give straight answer rather ignored. The TMAO theory is a "red herring [65]." In fact, a literature review on the intake of various foods and TMAO concentrations has shown that the consumption of saltwater fish and shellfish, but not freshwater fish, increases TMAO in plasma and urine [66].

Meat, eggs, dairy products, and seafood are nutritionally important foods as protein sources, and it is a very important question whether consuming a large amount of such foods really promotes arteriosclerosis and increases the incidence of cardiovascular disease, and carnitine is caught up in the fray.

Papandreou et al. provide a comprehensive list of clinical and basic studies on the TMAO issue, including studies that support, deny, or are neutral to the claims of the Hazen group et al. [67] Furthermore, with regard to the interpretation of the data from Apoe−/− mice, they found that Apoe−/− mice are quite different from human atherosclerosis, and that the model in which Apoe−/− mice express CETP is closer to the human pathology. He also noted that the increase in TMAO in cardiovascular disease may be a consequence and not a cause of the disease. For example, many animal studies that have shown accelerated atherosclerosis have used models in which animals are subjected to intense stress but stress itself has also been shown to increase TMAO. In addition, in the animal studies by the Hazen group, extremely high doses of precursors such as choline and carnitine, as well as TMAO itself, were administered, and at levels equivalent to those administered to humans, atherosclerosis protective effects have been reported. The fact that TMAO is elevated in cardiovascular disease and type II diabetes mellitus may be due to some compensatory mechanism or dysbiosis (intestinal toxemia) caused by disease [67]. Similarly, Velasquez et al. suggested that TMAO levels are elevated in various diseases because cells adapt to stress, and therefore TMAO may be a marker rather than a mediator of disease [68]. Cho et al. have also reported that high TMAO levels are associated with cardiovascular disease, but it is inconclusive whether TMAO is the cause or merely a marker of disease, and that foods such as meat and eggs that are said to increase TMAO are nutritionally very important, so restricting these foods is not justified at this time [69].

Wang et al. have listed clinical studies, meta-analyses, and animal studies on TMAO, but there are positive and negative reports, and there is still debate about whether TMAO is really a causative agent that promotes arteriosclerosis, and further research is needed in the future [70].

Thus, carnitine, which has been used safely in clinical practice for many years as an appetite suppressant and as a treatment for carnitine deficiency, is now facing the brunt of the Hazen Group's TMAO atherosclerosis-promoting theory. However, there are many clinical studies and meta-analyses that show that the intake of processed meat increases the risk of arteriosclerosis. Processed meat contains a large amount of carnitine, but also salt and preservatives contained in ham and sausages, and many studies suggest that salt and preservatives may be the cause of arteriosclerosis, but the debate is still ongoing [71–75]. Studying the causal relationship between daily food intake and the risk of developing diseases over decades is not an easy task.

Taking the above into consideration, the use of L-carnitine as a supplementation therapy for carnitine deficiency or the intake of up to 2 g/day of L-carnitine as a dietary supplement do not seem to pose any risk of promoting atherosclerosis.

9.6 Other Safety Considerations Carnitine

Although the safety of carnitine is not expected to be a problem in daily clinical practice, the following two points should be noted.

9.6.1 Aggravation of Symptoms Due to Carnitine's Stimulatory Effect on Digestive Enzyme Secretion

As described in Chap. 6, carnitine has been used to improve appetite loss since early times, because it enhances digestive enzyme secretion in addition to its effect on fatty acid oxidation. In Japan, carnitine chloride preparations were marketed from the 1960s to the 1980s, albeit in the DL-form. Most of them have already had their prices removed and are no longer on the market, but only one drug, Entomin Injection 200 mg, is still on the market in Japan. Its indication is for "chronic gastritis with decreased gastrointestinal function" and is said to be a digestive function stimulant. And the "contraindications" in the package insert are described as follows [76].

[Contraindications (do not administer to the following patients)]

1. Patients with hyperacidity [Increased gastric juice secretion]
2. Patients with acute pancreatitis or acute exacerbation of chronic pancreatitis [Increased pancreatic fluid secretion]

Otsuka Pharmaceutical's L-carnitine tablets, oral solution, and injection are L-carnitine products (the trade name: L-Cartin®), and their active ingredients are different from those of Entomin Injection 200 mg which is DL-carnitine. Therefore, when discussing the contents of the package insert with the PMDA at the time of approval, the contraindication of Entomin Injection in the package insert was not discussed. In addition, the use of DL-carnitine in carnitine preparations overseas was banned in the 1980s, and the indication of L-carnitine preparation is "primary and secondary carnitine deficiency" and not related to gastrointestinal hyperactivity, so there are no "contraindications" as described in the package insert for Entomin Injection. However, as described in Chap. 6, although it is an old literature, it has been shown that DL-carnitine chloride has the effect of stimulating the secretion of gastric juice and pancreatic juice. Entomin was launched in 1962, and it is unclear whether there was a description of contraindications from the beginning, but in the 1974 Japan Pharmaceutical Collection (prescription drugs), the product names of 11 "carnitine chloride" preparations are listed as a healthy stomach digestive agent, and as a "caution," "Do not use if you have gastrointestinal pain due to pronounced chronic recurrent pancreatitis or acute pancreatitis, etc., because it may sometimes worsen." [77] Furthermore, in the 1993 Japan Pharmaceutical Collection (prescription drugs), there is a description of three products and five items of carnitine chloride preparations (Abedyne tablet and liquid, Entomin tablet and injection, and Monocamine tablet), and as "cautions," "Contraindications: a. patients with hyperacidity, b. patients with acute exacerbation of acute pancreatitis or chronic pancreatitis" were described [78].

L-Cartin® (Otsuka Pharmaceutical Co., Ltd., tablets, oral solution and intravenous preparation), which is now a free-form L-carnitine preparation currently used for carnitine deficiency, tablets, and intravenous preparations of generic drugs, and levocarnitine chloride preparations (tablets) do not contain D-form, but L-carnitine is also thought to have a digestive enzyme secretion stimulating effect. Therefore, it is considered that there is a possibility that symptoms may develop or worsen in patients with hyperacidity or acute pancreatitis due to excessive gastric acid or increased pancreatic fluid secretion. In particular, when the blood concentration becomes high in a short period of time, such as intravenous preparations, it is better to pay attention to patients with hyperacidity, acute pancreatitis, and chronic pancreatitis.

9.6.2 Side Effects of D-Carnitine

As described in Chap. 6, the appetite enhancement and digestive enzyme secretion-stimulating effects of DL-carnitine were reported in Belgium, France, Germany, and other countries in the 1950s, and DL-carnitine preparations were used for stunted growth in children and anorexia in postoperative patients. In addition, a large number of DL-carnitine preparations were sold in Japan, and the classification was healthy gastric digestive, and the indications were anorexia, gastric secretion disorder, constipation, and achlorhydria/hypoacidity. Although the digestive enzyme secretion stimulating effect of carnitine has not been denied, it has been found that the D-form of the DL-body (racemic body) causes myasthenia-like symptoms, and the use of the DL-form is prohibited in the world. The background of the ban is as followings;

Bazzato et al. administered DL-carnitine (2 g) intravenously to 15 patients undergoing hemodialysis 3 times a week for 45 days after the completion of hemodialysis, with the expectation of improving hypertriglyceridemia in patients undergoing hemodialysis 3 times a week [79]. Since carnitine deficiency in hemodialysis patients is said to be associated with muscle weakness and other symptoms, they further investigated muscle symptoms. Conversely, 3 of 15 patients developed asthenia of the muscles of the lower limbs and jaw 2 to 3 weeks after starting intravenous DL-carnitine. The tensilon test (edrophonium test) was positive in all three patients, and electromyography showed a decrease in muscle action potentials characteristic of myasthenic patients. This symptom was relieved by the dialysis procedure, but reappeared between dialyses.

Subsequently, Bazzato et al. found that these myasthenia-like symptoms did not occur when intravenous L-carnitine was administered but were reproduced when DL-carnitine was administered [80].

Paulson et al. reported that intraperitoneal administration of D-carnitine in rats (750 g/day for 4 days) depleted L-carnitine in cardiac and skeletal muscle [81]. The heart of rats treated with D-carnitine was excised and perfused as a working heart specimen, and the function of cardiac contractility was significantly reduced compared to a control using palmitic acid as a substrate.

As a result of several reports of such clinical and animal studies, the FDA notified the pharmaceutical companies concerned in 1981 that the use of DL-carnitine preparations would henceforth be prohibited. Sigma Tau S.p.A. of Italy (now Alphasigma S.p.A.), the world's leading manufacturer of carnitine products for carnitine deficiency, marketed a syrup formulation of DL-carnitine chloride from 1961 to 1981 for the treatment of stunting growth and anorexia in children, but later switched to levocarnitine chloride, which is the L-body. The company has since switched to L-body of carnitine chloride. Sigma Tau also exported the product to the United States as a drug and dietary supplement, but the FDA notified the company in 1984 that it considered DL-carnitine and D-carnitine to be illegal food additives and prohibited their advertising and importation.

In this way, DL-carnitine preparations have been previously marketed in Japan and have been used as a "healthy stomach digestive" for patients with symptoms such as anorexia, gastric secretion disorders, constipation, and achlorhydria/hypoacidity. In addition, a disease called carnitine deficiency has been reported overseas, and it has been used in clinical studies of carnitine deficiency in patients treated with sodium valproate. Currently, only 200 mg of Entomin remains on the market in Japan. It is an injectable and is "intramuscular or intravenous," but when it is injected intravenously, the entire amount enters the blood, so it is considered necessary to be careful.

Since D-carnitine act antagonistic to L-carnitine, such as D-carnitine, is sometimes used to create carnitine deficiency as animal model, it is thought that the concentration of L-carnitine in tissues decreases, and mitochondrial dysfunction might occur.

On May 1, 1992, the FDA published a general policy on optical isomers of drugs, not only for carnitine preparations [82]. The underlying principle is that drugs such as the DL-form (racemic form) should be considered to contain 50% impurities, since the L-form may be active and the D-form inactive but antagonistic on receptors and enzymes. In some cases, as has been reported for some calcium channel blockers, one optical isomer has calcium antagonist activity while the other optical isomer has calcium agonist activity [83] and the opposite physiological activity may be exhibited. Therefore, the FDA's policy is that compounds that are scientifically capable of optical resolution should be developed separately. Although the optical separation of some compounds with many optical isomers is very labor intensive and costly, the basic principle is that compounds should be optically separated, and preclinical studies should be conducted before proceeding to clinical development.

References

1. Director, Standards Review Division, Food Safety Department 0601001, Pharmaceuticals and Food Bureau, Ministry of Health, Labour and Welfare: Revision of the handling under the Food Sanitation Law of "Ingredient essence (raw material) that is not judged to be a pharmaceutical unless it claims to have a pharmaceutical efficacy effect" (June 1, 2004), and Shokuanki No. 1105001 "Essence of ingredients (raw materials) that are not judged to be pharmaceuticals unless they claim pharmaceutical efficacy effects" Partial revision of "Revision of Handling under the Food Sanitation Law" (November 5, 2004) (in Japanese).
2. Hathcock JN, Shao A. Risk assessment for carnitine. Regul Toxicol Pharmacol. 2006;46(1):23–8. https://doi.org/10.1016/j.yrtph.2006.06.007.

3. Siliprandi N, Di Lisa F, Menabó R, et al. Transport and functions of carnitine in muscles. J Clin Chem Clin Biochem. 1990;28(5):303–6.
4. Iliceto S, Scrutinio D, Bruzzi P, et al. Effects of L-carnitine administration on left ventricular remodeling after acute anterior myocardial infarction: the L-carnitine Ecocardiografia Digitalizzata Infarto Miocardico (CEDIM) trial. J Am Coll Cardiol. 1995;26(2):380–7. https://doi.org/10.1016/0735-1097(95)80010-e.
5. Oudheusden LJ, Scholte HR. Efficacy of carnitine in the treatment of children with attention-deficit hyperactivity disorder. Prostagrandins Leukot Essent Fatty Acids. 2002;67(1):33–8. https://doi.org/10.1054/plef.2002.0378.
6. Ellaway C, Williams K, Leonard H, et al. Rett syndrome: randomized controlled trial of L-carnitine. J Child Neurol. 1999;14(3):162–7. https://doi.org/10.1177/088307389901400306.
7. Benvenga S, Ruggeri RM, Russo A, et al. Usefulness of L-carnitine, a naturally occurring peripheral antagonist of thyroid hormone action, in iatrogenic hyperthyroidism: a randomized, double-blind, placebo-controlled clinical trial. J Clin Endocrinol Metab. 2001;86(8):3579–94. https://doi.org/10.1210/jcem.86.8.7747.
8. Chapter 24. Systemic carnitine deficiency (OCTN2 Abnormality), In: The Japanese Society for Inherited Metabolic Diseases ed. Clinical practice guidelines for neonatal mass screening diseases 2019, Tokyo, SHINDAN TO CHIRYO SHA, Inc.; 2019. p. 228–239. (Book in Japanese).
9. Interview Form of Levocarnitine preparations, "L-Cartin® FF Tablets 100 mg", "L-Cartin® FF Tablets 250 mg", "L-Cartin® FF Oral Solution 10%", "L-Cartin® FF Oral Solution 10% Packet 5 mL", "L-Cartin® FF Oral Solution 10% Packet 10 mL" (Otsuka Pharmaceutical Co., Ltd.) https://www.otsuka-elibrary.jp/pdf_viewer/index.html?f=/file/1032/lfo_if.pdf#page=1 (Article in Japanese)
10. Bain MA, Fornasini G, Evans AM. Trimethylamine: metabolic, pharmacokinetic and safety aspects. Curr Drug Metab. 2005;6(3):227–40. https://doi.org/10.2174/1389200054021807.
11. Schmidt AC, Leroux JC. Treatments of trimethylaminuria: where we are and where we might be heading. Drug Discov Today. 2020;25(9):1710–7. https://doi.org/10.1016/j.drudis.2020.06.026.
12. Messenger J, Clark S, Massick S, et al. A review of trimethylaminuria (fish odor syndrome). J Clin Aesthet Dermatol. 2013;6(11):45–8.
13. Koeth RA, Wang Z, Levison BS, et al. Intestinal microbiota metabolism of L-carnitine, a nutrient in red meat, promotes atherosclerosis. Nat Med. 2013;19(5):576–85. https://doi.org/10.1038/nm.3145.
14. Wang Z, Klipfell E, Bennett BJ, et al. Gut flora metabolism of phosphatidylcholine promotes cardiovascular disease. Nature. 2011;472(7341):57–63. https://doi.org/10.1038/nature09922.
15. Tang WH, Wang Z, Levison BS, et al. Intestinal microbial metabolism of phosphatidylcholine and cardiovascular risk. N Engl J Med. 2013;368(17):1575–84. https://doi.org/10.1056/NEJMoa1109400.
16. Bernstein AM, Sun Q, Hu FB, et al. Major dietary protein sources and the risk of coronary heart disease in women. Circulation. 2010;122(9):876–83. https://doi.org/10.1161/CIRCULATIONAHA.109.915165.
17. Siri-Tarino PW, Sun Q, Hu FB, et al. Meta-analysis of prospective cohort studies evaluating the association of saturated fat with cardiovascular disease. Am J Clin Nutr. 2010;91(3):535–46. https://doi.org/10.3945/ajcn.2009.27725.
18. Koeth RA, Lam-Galvez BR, Kirsop J, et al. L-carnitine in omnivorous diets induces an atherogenic gut microbial pathway in humans. J Clin Invest. 2019;129(1):373–87. https://doi.org/10.1172/JCI94601.
19. Sinatra ST. The Sinatra solution: metabolic cardiology. North Bergen: Basic Health Publications; 2008.
20. Sinatra ST. Metabolic cardiology: an integrative strategy in the treatment of congestive heart failure. Altern Ther Health Med. 2009;15(3):44–52.
21. Baba A, Goda N, Shirai K, et al. A survey study on daily water intake and urination volume by eating and drinking in female nursing students. J Nutr. 2001;59(1):19–25. (Article in Japanese)
22. Rebouche CJ, Chenard CA. Metabolic fate of dietary carnitine in human adults: identification and quantification of urinary and fecal metabolites. J Nutr. 1991;121(4):539–46. https://doi.org/10.1093/jn/121.4.539.
23. MacCarty MF. L-carnitine consumption, its metabolism by intestinal microbiota, and cardiovascular health. Mayo Clin Proc. 2013;88(8):786–9. https://doi.org/10.1016/j.mayocp.2013.06.004.
24. Collins HL, Drazul-Schrader D, Sulpizio AC, et al. L-carnitine intake and high trimethylamine N-oxide plasma levels correlate with low aortic lesions in Apoe−/− transgenic mice expressing CETP. Atherosclerosis. 2016;244:29–37. https://doi.org/10.1016/j.atherosclerosis.2015.10.108.
25. Aldana-Hernández P, Leonard KA, Zhao YY, et al. Dietary choline or trimethylamine N-oxide supplementation does not influence atherosclerosis development in *Ldlr−/−* and *Apoe−/−* male mice. J Nutr. 2020;150(2):249–55. https://doi.org/10.1093/jn/nxz214.
26. Aldana-Hernández P, Azarcoya-Barrera J, van der Veen JN, et al. Dietary phosphatidylcholine supplementation reduces atherosclerosis in Apoe−/− male mice. J Nutr Biochem. 2021;92:108617. https://doi.org/10.1016/j.jnutbio.2021.108617.
27. Jonsson AL, Caesar R, Akrami R, et al. Impact of gut microbiota and diet on the development of atherosclerosis in Apoe−/− mice. Arterioscler Thromb Vasc Biol. 2018;38(10):2318–26. https://doi.org/10.1161/ATVBAHA.118.311233.

28. Sayed-Ahmed MM, Khattab MM, Gad MZ, et al. L-carnitine prevents the progression of atherosclerotic lesions in hypercholesterolaemic rabbits. Pharmacol Res. 2001;44:235–42.
29. Heianza Y, Ma W, Manson JE, et al. Gut microbiota metabolites and risk of major adverse cardiovascular disease events and death: a systematic review and meta-analysis of prospective studies. J Am Heart Assoc. 2017;6(7):e004947. https://doi.org/10.1161/JAHA.116.004947.
30. Schiattarella GG, Sannino A, Toscano E, et al. Gut microbe-generated metabolite trimethylamine-N-oxide as cardiovascular risk biomarker: a systematic review and dose-response meta-analysis. Eur Heart J. 2017;38(39):2948–56. https://doi.org/10.1093/eurheartj/ehx342.
31. Yao ME, Liao PD, Zhao XJ, et al. Trimethylamine-N-oxide has prognostic value in coronary heart disease: a meta-analysis and dose-response analysis. BMC Cardiovasc Disord. 2020;20(1):7. https://doi.org/10.1186/s12872-019-01310-5.
32. Farhangi MA, Vajdi M, Asghari-Jafarabadi M. Gut microbiota-associated metabolite trimethylamine N-oxide and the risk of stroke: a systematic review and dose-response meta-analysis. Nutr J. 2020;19(1):76. https://doi.org/10.1186/s12937-020-00592-2.
33. Farhangi MA. Gut microbiota-dependent trimethylamine N-oxide and all-cause mortality: findings from an updated systematic review and meta-analysis. Nutrition. 2020;78:110856. https://doi.org/10.1016/j.nut.2020.110856.
34. Mueller DM, Allenspach M, Othman A, et al. Plasma levels of trimethylamine-N-oxide are confounded by impaired kidney function and poor metabolic control. Atherosclerosis. 2015;243(2):638–44. https://doi.org/10.1016/j.atherosclerosis.2015.10.091.
35. Skagen K, Trøseid M, Ueland T, et al. The carnitine-butyrobetaine-trimethylamine-N-oxide pathway and its association with cardiovascular mortality in patients with carotid atherosclerosis. Atherosclerosis. 2016;247:64–9. https://doi.org/10.1016/j.atherosclerosis.2016.01.033.
36. Ringel C, Dittrich J, Gaudl A, et al. Association of plasma trimethylamine N-oxide levels with atherosclerotic cardiovascular disease and factors of the metabolic syndrome. Atherosclerosis. 2021;335:62–7. https://doi.org/10.1016/j.atherosclerosis.2021.09.026.
37. Mente A, Chalcraft K, Ak H, et al. The relationship between trimethylamine-N-oxide and prevalent cardiovascular disease in a multiethnic population living in Canada. Can J Cardiol. 2015;31:1189–94.
38. DiNicolantonio JJ, Lavie CJ, Fares H, et al. L-carnitine in the secondary prevention of cardiovascular disease: systematic review and meta-analysis. Mayo Clin Proc. 2013;88(6):544–51. https://doi.org/10.1016/j.mayocp.2013.02.007.
39. Thomas MS, Fernandez ML. Trimethylamine N-oxide (TMAO), diet and cardiovascular disease. Curr Atheroscler Rep. 2021;23(4):12. https://doi.org/10.1007/s11883-021-00910-x.
40. Bain MA, Milne RW, Evans AM. Disposition and metabolite kinetics of oral L-carnitine in humans. J Clin Pharmacol. 2006;46(10):1163–70. https://doi.org/10.1177/0091270006292851.
41. Bain MA, Faull R, Fornasini G, et al. Accumulation of trimethylamine and trimethylamine-N-oxide in end-stage renal disease patients undergoing haemodialysis. Nephrol Dial Transplant. 2006;21(5):1300–4. https://doi.org/10.1093/ndt/gfk056.
42. Bain MA, Faull R, Milne RW, et al. Oral L-carnitine: metabolite formation and hemodialysis. Curr Drug Metab. 2006;7(7):811–6. https://doi.org/10.2174/138920006778520561.
43. Ozasa H, Shimizu M, Koizumi A, et al. Trimethylamine generation in patients receiving hemodialysis treatment with L-carnitine. Clin Kidney J. 2014;7(3):329. https://doi.org/10.1093/ckj/sfu045.
44. Kim RB, Morse BL, Djurdjev O, et al. Advanced chronic kidney disease populations have elevated trimethylamine N-oxide levels associated with increased cardiovascular events. Kidney Int. 2016;89(5):1144–52. https://doi.org/10.1016/j.kint.2016.01.014.
45. Kaysen GA, Johansen KL, Chertow GM, et al. Associations of trimethylamine N-oxide with nutritional and inflammatory biomarkers and cardiovascular outcomes in patients new to dialysis. J Ren Nutr. 2015;25(4):351–6. https://doi.org/10.1053/j.jrn.2015.02.006.
46. Higuchi T, Abe M, Yamazaki T, et al. Effects of levocarnitine on brachial-ankle pulse wave velocity in hemodialysis patients: a randomized controlled trial. Nutrients. 2014;6(12):5992–6004. https://doi.org/10.3390/nu6125992.
47. Lai YH, Lee MC, Ho GJ, et al. Association of low serum L-carnitine levels with peripheral arterial stiffness in patients who undergo kidney transplantation. Nutrients. 2019;11(9):2000. https://doi.org/10.3390/nu11092000.
48. Higuchi T, Abe M, Yamazaki T, et al. Levocarnitine improves cardiac function in hemodialysis patients with left ventricular hypertrophy: a randomized controlled trial. Am J Kidney Dis. 2016;67(2):260–70. https://doi.org/10.1053/j.ajkd.2015.09.010.
49. Fukami K, Yamagishi S, Sakai K, et al. Oral L-carnitine supplementation increases trimethylamine-N-oxide but reduces markers of vascular injury in hemodialysis patients. J Cardiovasc Pharmacol. 2015;65(3):289–95. https://doi.org/10.1097/FJC.0000000000000197.
50. He K, Song Y, Daviglus ML, et al. Accumulated evidence on fish consumption and coronary heart disease mortality: a meta-analysis of cohort studies. Circulation. 2004;109(22):2705–11. https://doi.org/10.1161/01.CIR.0000132503.19410.6B.
51. Iso H, Kobayashi M, Ishihara J, et al. Intake of fish and n3 fatty acids and risk of coronary heart disease among Japanese: the Japan Public Health Center-Based (JPHC) study Cohort I. Circulation. 2006;113(2):195–202. https://doi.org/10.1161/CIRCULATIONAHA.105.581355.

52. Yamagishi K, Iso H, Date C, et al. Fish, ω-3 polyunsaturated fatty acids, and mortality from cardiovascular diseases in a nationwide community-based cohort of Japanese men and women: the JACC (Japan Collaborative Cohort Study for Evaluation of Cancer Risk) study. J Am Coll Cardiol. 2008;52(12):988–96. https://doi.org/10.1016/j.jacc.2008.06.018.
53. Yamagishi K, Hori M, Iso H. Fish and ω-3 polyunsaturated fatty acids in relation to risk of cardiovascular disease. Nippon Rinsho. 2013;71(9):1552–7. (Article in Japanese)
54. Zhang AQ, Mitchell SC, Smith RL. Dietary precursors of trimethylamine in man: a pilot study. Food Chem Toxicol. 1999;37(5):515–20. https://doi.org/10.1016/s0278-6915(99)00028-9.
55. DiNicolantonio JJ, McCarty M, OKeefe J. Association of moderately elevated trimethylamine N-oxide with cardiovascular risk: is TMAO serving as a marker for hepatic insulin resistance. Open Heart. 2019;6(1):e000890. https://doi.org/10.1136/openhrt-2018-000890.
56. de Oliveira Otto MC, Wu JH, Baylin A, et al. Circulating and dietary omega-3 and omega-6 polyunsaturated fatty acids and incidence of CVD in the multi-ethnic study of atherosclerosis. J Am Heart Assoc. 2013;2:e000506.
57. Bhatt DL, Steg PG, Miller M, et al. Cardiovascular risk reduction with icosapent ethyl for hypertriglyceridemia. N Engl J Med. 2019;380(1):11–22. https://doi.org/10.1056/NEJMoa1812792.
58. Casula M, Soranna D, Catapano AL, et al. Long-term effect of high dose omega-3 fatty acid supplementation for secondary prevention of cardiovascular outcomes: a meta-analysis of randomized, placebo controlled trials [corrected]. Atheroscler Suppl. 2013;14(2):243–51. https://doi.org/10.1016/S1567-5688(13)70005-9.
59. Filion KB, El Khoury F, Bielinski M, et al. Omega-3 fatty acids in high-risk cardiovascular patients: a meta-analysis of randomized controlled trials. BMC Cardiovasc Disord. 2010;10:24. https://doi.org/10.1186/1471-2261-10-24.
60. Innes JK, Calder PC. Marine omega-3 (N-3) fatty acids for cardiovascular health: an update for 2020. Int J Mol Sci. 2020;21(4):1362. https://doi.org/10.3390/ijms21041362.
61. Rizos EC, Ntzani EE, Bika E, et al. Association between omega-3 fatty acid supplementation and risk of major cardiovascular disease events: a systematic review and meta-analysis. JAMA. 2012;308(10):1024–33. https://doi.org/10.1001/2012.jama.11374.
62. Khoueiry G, Rafeh NA, Sullivan E, et al. Do omega-3 polyunsaturated fatty acids reduce risk of sudden cardiac death and ventricular arrhythmias? A meta-analysis of randomized trials. Heart Lung. 2013;42(4):251–6. https://doi.org/10.1016/j.hrtlng.2013.03.006.
63. Oliveira A, Lopes C, Rodríguez-Artalejo F. Adherence to the Southern European Atlantic Diet and occurrence of nonfatal acute myocardial infarction. Am J Clin Nutr. 2010;92(1):211–7. https://doi.org/10.3945/ajcn.2009.29075.
64. Valeur J, Landfald B, Berstad A, et al. Trimethylamine N-oxide in seafood: rotten or forgotten? J Am Coll Cardiol. 2016;68(25):2916–7. https://doi.org/10.1016/j.jacc.2016.08.077.
65. Landfald B, Valeur J, Berstad A, et al. Microbial trimethylamine-N-oxide as a disease marker: something fishy? Microb Ecol Health Dis. 2017;28(1):1327309. https://doi.org/10.1080/16512235.2017.1327309.
66. Lombardo M, Aulisa G, Marcon D, et al. Association of urinary and plasma levels of trimethylamine N-oxide (TMAO) with foods. Nutrients. 2021;13(5):1426. https://doi.org/10.3390/nu13051426.
67. Papandreou C, Moré M, Bellamine A. Trimethylamine N-oxide in relation to cardiometabolic health-cause of effect? Nutrients. 2020;12(5):1330. https://doi.org/10.3390/nu12051330.
68. Velasquez MT, Ramezani A, Manal A, et al. Trimethylamine N-oxide: the good, the bad and the unknown. Toxins(Basel). 2016;8(11):326. https://doi.org/10.3390/toxins8110326.
69. Cho CE, Caudill MA. Trimethylamine-N-oxide: friend, foe, or simply caught in the cross-fire? Trends Endocrinol Metab. 2017;28(2):121–30. https://doi.org/10.1016/j.tem.2016.10.005.
70. Wang BY, Qiu J, Lian JF, et al. Gut metabolite trimethylamine-N-oxide in atherosclerosis: from mechanism to therapy. Front Cardiovasc Med. 2021;8:723886. https://doi.org/10.3389/fcvm.2021.723886.
71. Micha R, Wallace SK, Mozaffarian D. Red and processed meat consumption and risk of incident coronary heart disease, stroke, and diabetes: a systematic review and meta-analysis. Circulation. 2010;121(21):2271–83. https://doi.org/10.1161/CIRCULATIONAHA.109.924977.
72. Micha R, Michas G, Mozaffarian D. Unprocessed red and processed meats and risk of coronary artery disease and type 2 diabetes — an updated review of the evidence. Curr Atheroscler Rep. 2012;14(6):515–24. https://doi.org/10.1007/s11883-012-0282-8.
73. Rohrmann S, Overvad K, Bueno-de-Mesquita HB, et al. Meat consumption and mortality — results from the European prospective investigation into cancer and nutrition. BMC Med. 2013;11:63.
74. Händel MN, Cardoso I, Rasmussen KM, et al. Processed meat intake and chronic disease morbidity and mortality: an overview of systematic reviews and meta-analyses. PLoS One. 2019;14(10):e0223883. https://doi.org/10.1371/journal.pone.0223883.
75. Zhong VW, Van Horn L, Greenland P, et al. Associations of processed meat, unprocessed red meat, poultry, or fish intake with incident cardiovascular disease and all-cause mortality. JAMA Intern Med. 2020;180(4):503–12. https://doi.org/10.1001/jamainternmed.2019.6969.

76. Nichi-Iko Co., Ltd.: Entomin note 200 mg package insert 9th Edition (Revised in March 2019)(Article in Japanese).
77. Japan Pharmaceutical Information Center: Japan Pharmaceutical Collection (Ethical Drugs) 1974 edition. Carnitine chloride, p 96. (Article in Japanese).
78. Japan Pharmaceutical Information Center. Carnitine chloride. In: Japan Pharmaceutical Collection (Pharmaceuticals) 1993 edition. p. 294. (Article in Japanese).
79. Bazzato G, Mezzina C, Ciman M, et al. Myasthenia-like syndrome associated with carnitine in patients on long-term haemodialysis. Lancet. 1979;313:1041–2.
80. Bazzato G, Coli U, Landini S, et al. Myasthenia-like syndrome after D, L- but not L-carnitine. Lancet. 1981;1(8231):1209. https://doi.org/10.1016/s0140-6736(81)92372-2.
81. Paulson DJ, Shug AL. Tissue specific depletion of L-carnitine in rat heart and skeletal muscle by D-carnitine. Life Sci. 1981;28(26):2931–8. https://doi.org/10.1016/0024-3205(81)90269-1.
82. U.S Food & Drug Administration, Center for Drug Evaluation and Research: Development of new stereoisomeric drugs (May 1,1992, https://www.fda.gov/regulatory--information/search-fda-guidance-documents/development-new-stereoisomeric-drugs).
83. Hof RP, Rüegg UT, Hof A, et al. Stereoselectivity at the calcium channel: opposite action of the enantiomers of a 1,4-dihydropyridine. J Cardiovasc Pharmacol. 1985;7(4):689–93. https://doi.org/10.1097/00005344-198507000-00012.

Carnitine Deficiency in Inherited Metabolic Diseases

10

10.1 Introduction

As already described in Chap. 5, "What is Carnitine Deficiency?," it is widely known that carnitine deficiency can occur due to primary carnitine deficiency (systemic carnitine deficiency) caused by genetic abnormalities of the carnitine transporter (OCTN2), or secondary carnitine deficiency caused by inherited metabolic diseases, including abnormal organic acid metabolism, fatty acid metabolism, and mitochondrial abnormalities. Carnitine deficiency may also occur in urea cycle disorders and other inherited metabolic diseases due to drug administration or restricted protein intake.

Japanese Society for Inherited Metabolic Diseases published "Guidelines for the Diagnosis and Treatment of Diseases Subject to Mass Screening in Newborns 2015," which was revised in 2019 ("Guidelines for the Diagnosis and Treatment of Diseases Subject to Mass Screening in Newborns 2019," hereafter "Guidelines 2019") [1]. In it, there are detailed descriptions of the dosage and administration of carnitine in the acute and chronic phases, as well as drugs and dietary therapy for the diagnosis and treatment of each disease covered in newborn mass screening. In addition, there are many detailed specialized manuals [2–5]. Therefore, this book is limited to the causes of carnitine deficiency in inherited metabolic diseases, an overview of carnitine replacement therapy described in "Guidelines 2019," and other cases of carnitine deficiency in inherited metabolic diseases, which are not described in "Guidelines 2019."

10.2 Etiology and Pathophysiology of Carnitine Deficiency in Organic Acidemia

Primary carnitine deficiency (also known as systemic carnitine deficiency) due to genetic abnormalities in the carnitine transporter (OCTN2) is characterized by a decrease in the absorption of carnitine from the diet, the uptake of carnitine into cardiomyocytes and skeletal muscle cells, and the reabsorption of carnitine in the renal tubules, resulting in an extreme decrease in carnitine throughout the body (Chap. 5: "What is carnitine deficiency?").

In organic acidemia, due to deficiency or reduced activity of disease-specific metabolic enzymes, organic acid metabolites upstream of their metabolic processes accumulate intracellularly, causing metabolic disorders and tissue damage. Secondary carnitine deficiency occurs because carnitine is utilized in their detoxification and is excreted in the urine as acylcarnitine (Fig. 10.1) [6]. For example, in propionic acidemia, propionyl CoA upstream of propionyl CoA carboxylase (② in Fig. 10.1) accumulates intracellularly and becomes toxic due to its deficiency

N. Nakayama, *Carnitine Deficiency and Insufficiency in Medicine*,
https://doi.org/10.1007/978-981-95-7797-2_10

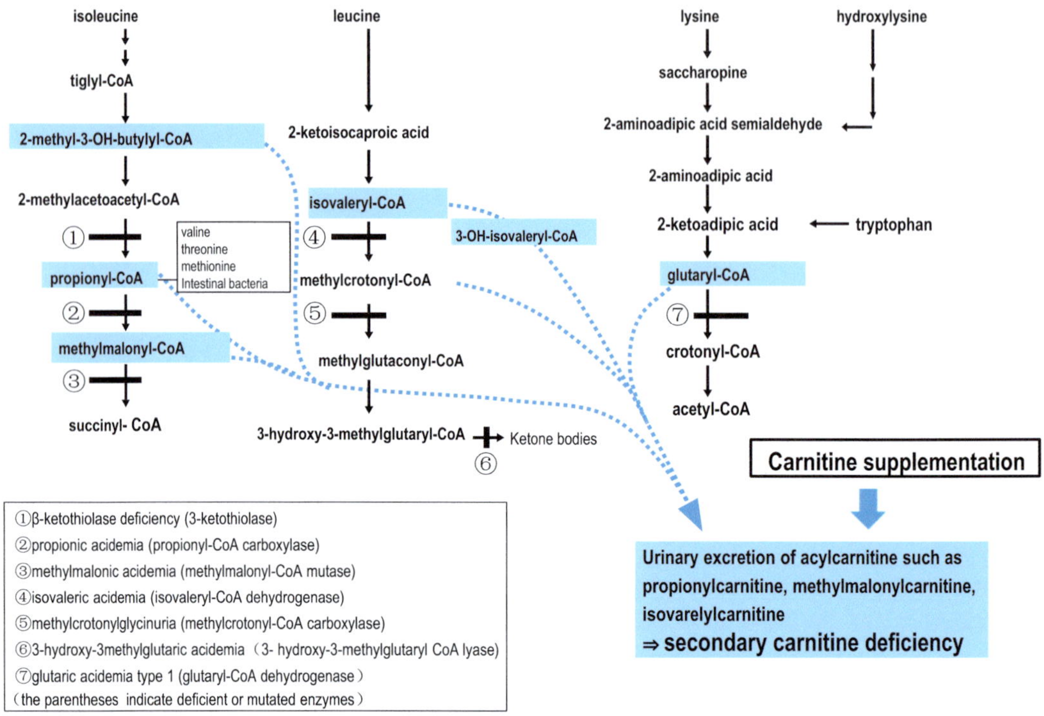

Fig. 10.1 Pathogenic mechanisms of main organic acidemias and role of carnitine supplementation. Deficiency or mutation of amino acid (isoleucine, leucine, lysine, etc.) metabolism enzymes cause accumulation of metabolic intermediates at the upstream and causes cell damage. Free carnitine is utilized to detoxify these intermediate toxicity and large amount of these disease specific acylcarnitine is excreted as related acylcarnitines and causes secondary carnitine deficiency. Carnitine supplementation therapy prevents, treats and normalizes these abnormalities

or reduced activity. Free carnitine is consumed to detoxify the propionyl CoA by urinary excretion as propionylcarnitine, resulting in secondary carnitine deficiency. Depending on the toxicity of the accumulated organic acids and their CoA, different organs and symptoms may develop, including hypercapnia, ketoacidosis, hyperammonemia, hyperlactatemia, and hypoglycemia.

Figures 10.2, 10.3, 10.4, and 10.5 show an overview of the etiology, pathogenesis, symptoms, diagnosis, treatment, and prognosis of propionic acidemia, methylmalonic acidemia, isovaleric acidemia, and glutaric acidemia type [8].

Summary of propionic acidemia

Propionic acidemia

1. **Etiology, pathology**
   Propionic acidemia is an autosomal latent (recessive) genetic disease in which a large amount of propionic acid and propionyl-CoA metabolites accumulate in the body due to a decrease in the activity of propionyl-CoA carboxylase present in the intermediate metabolic pathway of four amino acids (valine, isoleucine, threonine, and methionine), cholesterol, and odd-chain fatty acids. In tandem mass screening in Japan, the frequency of mild cases is about 1 in 30,000 people, and the frequency of severe cases is 1 in 400,000 people.
2. **Symptoms**
   Severe cases: It develops from the early stages of newborns and infants due to poor feeding ability, vomiting, hyperpnea, and impaired consciousness, and causes growth and developmental delay due to recurrence of ketoacidosis attack. **Mild cases**: Majority of patients were "most mild cases" discovered by tandem mass screening without ketoacidosis attack in early stage, in Japan.
3. **Diagnosis**
   Metabolic ketoacidosis, hyperammonemia, and sometimes hypoglycemia are observed. High levels of glycine in the blood and urine, and increased excretion of methylcitric acid, 3-OH-propionic acid and propionylglycine in the urine by GC-MS are detected. Increased propionylcarnitine (C3) and C3/C2 ratio is observed by tandem mass analysis. The diagnosis is confirmed by a decrease in the activity of propionyl-CoA carboxylase in leukocytes and cultured skin fibroblasts.
4. **Treatment, prognosis**
   In the acute phase of metabolic acidosis, correction by alkaline therapy, and in severe cases, exchange blood transfusion, peritoneal dialysis, hemodialysis, etc. are performed. Give a restricted diet of valine, isoleucine, methionine, and threonine, and carnitine administration to simulate the excretion of methylmalonic acid (in severe cases, intravenous is preferable). Although a small number of effective cases of liver transplantation therapy have been reported. In general, growth and developmental delays are observed, and the prognosis is poor.

Fig. 10.2 Summary of propionic acidemia. (Created using as Refs. [2, 6, 7])

Summary of methylmalonic acidemia

Methylmalonic acidemia

1. **Etiology, pathology**
   Methylmalonic acidemia is an autosomal latent (recessive) genetic disease in which a large amount methylmalonic acid and methylmalonyl-CoA metabolites accumulate in the body due to a decrease in the activity of methylmalonyl-CoA mutase present in the intermediate metabolic pathway of four amino acids (valine, isoleucine, threonine, and methionine), cholesterol, and odd-chain fatty acids. The causes of this decrease in enzyme activity are divided into abnormalities in methylmalonyl-CoA mutase and abnormalities in the metabolic pathway of coenzyme cobalamin (vitamin B_{12}). The former is abbreviated as *mut*, the latter as *cbl*, and are further classified as a subtype by cell-cell complementarity testing. Many of the latter are resistant to high-dose vitamin B_{12} therapy, and some forms of the disease are accompanied by homocystinuria. There is also benign methylmalonic acidemia of unknown pathology. It is an autosomal recessive inheritance, and the frequency in Japan is 1 in 80,000.
2. **Symptoms**
   It has symptoms similar to propionic acidemia.
3. **Diagnosis**
   Metabolic ketoacidosis, hyperammonemia, high levels of glycine in the blood and urine, and increased excretion of methylmalonic acid in the urine are detected by GC-MS. Increased propionylcarnitine (C3) and C3/C2 ratio is observed by tandem mass analysis same as propionic acidemia. The diagnosis is confirmed by a decrease in the activity of methylmalonyl-CoA mutase in leukocytes and cultured skin fibroblasts.
4. **Treatment, prognosis**
   Equivalent to propionic acidemia. Try high-dose vitamin B_{12} therapy to see if it responds. In many cases, the prognosis is poor and are accompanied by developmental delays.

Fig. 10.3 Summary of methylmalonic acidemia. (Created using as Refs. [2, 6, 7])

Summary of isovaleric acidemia

Isovaleric acidemia

1. **Etiology, pathology**
   Isovaleric acidemia is the first organic acidemia discovered by K. Tanaka using GC-MS. It is an autosomal latent (recessive) genetic disorder in which metabolites of isovaleryl CoA accumulate in the body due to deficiency of isovaleryl-CoA dehydrogenase in the metabolic process of leucine. In tandem mass screening in Japan, the frequency is 1 in 520,000.
2. **Symptoms**
   There are acute and chronic intermittent types. In the acute form, poor feeding capacity, vomiting, polypnea, convulsions, and a peculiar body odor like swollen feet (due to isovaleric acid) are observed from early infancy. In the chronic intermittent type, vomiting and lethargic attacks occur when there is hypercatabolism during protein overdose or infection.
3. **Diagnosis**
   Metabolic ketoacidosis, peculiar urine odor, and increased excretion of isovaleryl glycine and 3-OH-isovaleric acid (in acute phase) in the urine are detected by GC-MS. A definitive diagnosis is made by measuring the activity of isovaleryl CoA dehydrogenase in leukocytes and cultured skin fibroblasts.
4. **Treatment, prognosis**
   The acute phase is equivalent to propionic acidemia. As a maintenance therapy, a leucine-restricted diet is given, and carnitine is administered. The prognosis is relatively good if maintenance therapy is started after the acute phase is overcome.

Fig. 10.4 Summary of isovaleric acidemia. (Created using as Refs. [2, 6, 7] and others)

Summary of glutaric acidemia type 1

Glutaric acidemia type 1

1. **Etiology, pathophysiology**
   An autosomal latent (recessive) genetic disorder due to deficiency of glutaryl-CoA dehydrogenase in the lysine, OH-lysine and tryptophan metabolism in which glutaric acid accumulates in the body. Glutaryl-CoA dehydrogenase uses FAD (flavin adenine dinucleotide, vitamin B_2) as a coenzyme, and the electrons from the reaction are transmitted to the respiratory chain via electron transfer flavin protein (ETF) and ETF-ubiquinone oxidoreductase (ETF-QO). In the tandem mass screening in Japan, the frequency is 1 in 280,000.
2. **Symptoms**
   In infancy, hypotonia, ankylosis, posterior arch tension, dystonia, convulsions, disturbed consciousness, and encephalopathy appear. Symptoms temporarily relieved, but involuntary movements such as tonia and dystonia progressed, and psychomotor retardation became apparent.
3. **Diagnosis**
   Symptoms such as progressive involuntary movements and muscle tone abnormalities, as well as atrophy of the frontal and temporal lobes on brain CT and MRI. Characteristic images such as enlargement of the Sylvius fissure are present, and an increase in glutaric acid excretion in the urine is detected by GC-MS. A definitive diagnosis is made by a decrease in glutaryl-CoA dehydrogenase activity in cultured skin fibroblasts.
4. **Treatment, prognosis**
   In the acute phase, alkaline therapy and high-calorie infusion are administered. In the intermittent period, a lysine and tryptophan restricted diet are given, and carnitine, riboflavin, and GABA analogues are administered. Involuntary movements and psychomotor retardation appear, and the prognosis is poor.

Fig. 10.5 Summary of glutaric acidemia type 1. (Created using as Refs. [2, 6, 7] and others)

10.3 Overview of Fatty Acid β Oxidation and Its Abnormalities

In fatty acid β-oxidation disorders, genetic defects or reduced activity of specific metabolic enzymes cause accumulation of upstream fatty acid metabolite intermediates, resulting in mitochondrial dysfunction and abnormal energy metabolism (Fig. 10.6). It is thought that genetic abnormalities in specific metabolic enzymes involved in fatty acid β-oxidation cause deficiency or reduced activity, leading to accumulation of upstream acyl CoA, resulting in a state of so-called carnitine insufficiency. Carnitine replacement therapy is indicated when blood carnitine levels are low, or the acylcarnitine/free carnitine ratio is high.

Linear long-chain fatty acids such as palmitic acid (C = 16) and stearic acid (C = 18) are metabolized by mitochondrial β-oxidation to acetyl CoA, a unit of two carbon atoms each, for ATP production (see "Chap. 3: What is Carnitine," Fig. 10.9). Conventional textbooks have suggested that this long-chain fatty acid β-oxidation produces long-chain acyl CoA→ enoyl CoA→ 3-hydroxyacyl CoA→ acyl CoA in the matrix, the soluble fraction of mitochondria, and that this cycle is repeated several times to form a spiral (Fig. 10.7).

Subsequent studies, however, have revealed that there are twoβ-oxidation systems: the long-chain fatty acid β-oxidation system in the mitochondrial inner membrane and the β-oxidation system that degrades medium- and short-chain fatty acids that are shortened to some extent in

Summary of fatty acid β oxidation disorders

Fatty acid β oxidation disorders

1. **Etiology, pathophysiology**
   An autosomal latent (recessive) genetic disorder in which the utilization of fatty acids is impaired, and fatty acids and their metabolites accumulate due to deficiencies in short-chain acyl-CoA dehydrogenase (SCAD),medium-chain acyl-CoA dehydrogenase (MCAD), long-chain acyl-CoA dehydrogenase (LCAD), and very-long-chain acyl-CoA dehydrogenase (VLCAD), which is involved in the β oxidation of fatty acids in mitochondria. In tandem mass screening in Japan the frequency is 1 in 130,000 in MCAD deficiency, and 1 in 93,000. Most of the cases of SCAD deficiency were asymptomatic and excluded from newborn screening. of LCAD deficiency was reported in 1984 but later research revealed that the role of LCAD was degradation of branched long chain fatty acids, and the presence disease of the deficiency become unclear.
2. **Symptoms**
   During the neonatal period, prolonged starvation, and increased catabolism, such as infection disease, poor feeding capacity, vomiting, impaired consciousness, and sudden death may occur. Hyperammonemia, metabolic acidosis, and hypoglycemia were observed.
3. **Diagnosis**
   Hyperammonemia, metabolic acidosis, hypoglycemia, and urinary excretion of organic acids by GC-MS Increased excretion of organic acids is noted. Defects in each enzyme in leukocytes and cultured skin fibroblasts confirm the diagnosis.
4. **Treatment, prognosis**
   In acute phase, glucose infusion and carnitine is administered. Avoid prolonged starvation (frequent food), and give them a low-fat, high-carbohydrate diet. If treated, the prognosis is good.

SCAD: short chain acyl-CoA dehydrogenase, MCAD: medium chain acyl-CoA dehydrogenase,
LCAD: long chain acyl-CoA dehydrogenase, VLCAD: very long chain acyl-CoA dehydrogenase

Fig. 10.6 Summary of fatty acid β oxidation disorders. *SCAD short chain acyl-CoA dehydrogenase, MCAD medium chain acyl-CoA dehydrogenase, LCAD long chain acyl-CoA dehydrogenase, VLCAD very long chain acyl-CoA dehydrogenase.* (Created using as Refs. [2, 6, 7])

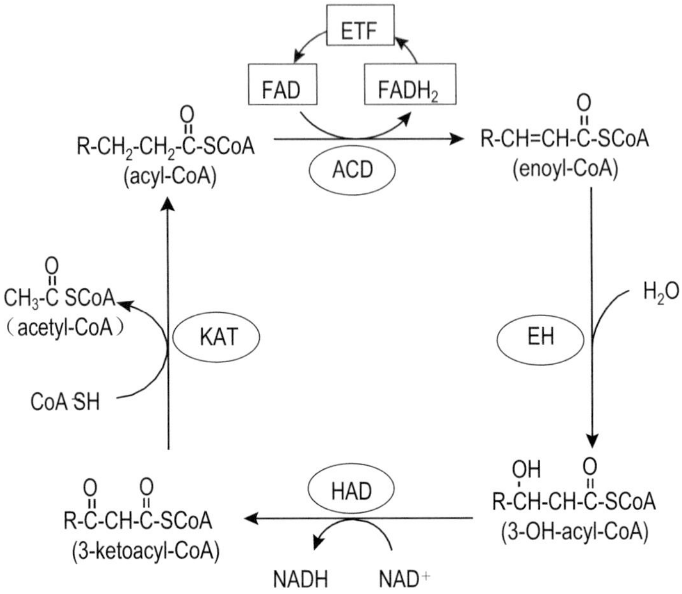

Fig. 10.7 Conventional spiral schema of long chain fatty acid β-oxidation. Acyl-CoA of a long-β chain fatty acid, becomes enoyl-CoA by dehydrogenation by acyl-CoA dehydrogenase (ACD), and then by the reaction of enoyl-CoA hydratase (EH) to 3-hydroxyacyl-CoA, and by the reaction of hydroxyacyl-CoA dehydrogenase (HAD) to produce 3-ketoacyl-CoA. Finally, the reaction of 3-ketoacyl-CoA thiolase (KAT) results in a long-chain acyl-CoA with two fewer carbons than at the start. These reactions are repeated in a spiral shape, and acetyl-CoA is produced and enters the TCA cycle, and the H^+ generated by the ACD reaction enters the electron transport chain via electron transfer flavoprotein (ETF), and NADH produced by the HAD reaction also enters the electron transport chain and energy is produced. These reactions were supposed to take place in the mitochondrial matrix, but now it was revealed that the long chain acyl-CoA dehydrogenase reactions take place in mitochondrial inner membrane (See, text). (From Ref. [10])

the matrix [10]. The former is a very-long-chain acyl-CoA dehydrogenase (VLCAD) and the mitochondrial trifunctional protein (MTP or TFP), which performs the EH, HAD, and KAT reactions shown in Fig. 10.7 in a single enzyme. The latter is a system that degrades the acyl CoA of the shorter fatty acids in the inner membrane long-chain fatty acid degradation system by two carbon atoms, with medium-chain acyl CoA dehydrogenase (MCAD) and short-chain acyl CoA dehydrogenase responsible for the initial acyl CoA dehydrogenation reaction. As the name suggests, MTP performs three functions in a single molecule. The three reactions are (1) reaction of long-chain enoyl CoA hydratase (LCEH), (2) reaction of long-chain hydroxyacyl CoA dehydrogenase (LCHAD), and (3) reaction of long-chain ketoacyl CoA thiolase (LCKT). In the β-oxidation system of medium- and short-chain acyl CoA, after the reaction of medium- and short-chain acyl CoA dehydrogenases (MCAD and SCAD), the reactions of (1) short-chain enoyl CoA hydratase (SCEH, also called crotonase), (2) short-chain 3-hydroxyacyl CoA dehydrogenase (SCHAD), (3) medium-chain ketoacyl CoA thiolase (MCKAT), which performs each reaction. Thus, the long-chain fatty acid β-oxidation system in the inner membrane and the medium-/short-chain fatty acid β-oxidation system in the matrix, which is the soluble fraction, are responsible for the two systems.

As shown in Table 10.1, there are four types of acyl CoA dehydrogenases responsible for fatty acid β-oxidation: VLCAD, MCAD, and SCAD, and LCAD, and the enzyme that acts depends on the number of the carbon of the fatty acid [11].

VLCAD is an enzyme discovered by Prof. Takashi Hashimoto of Shinshu University and Izai et al. of Gifu University group, which exists in the inner mitochondrial membrane and oxi-

Table 10.1 Subcellular localization, substrate specificity, and structure of acyl-CoA dehydrogenases in mitochondrial β-oxidation of long chain fatty acids

	SCAD	MCAD	VLCAD	LCAD[a]
Subcellular localization	Matrix	Matrix	Inner membrane	Matrix
Substrate specificity (carbon chain-length)	C4	C4–C16	C12–C22	C8–C18
Subunit molecule, kDa (precursor)	41 (44)	42 (45)	66 (71)	44 (48)
Protein structure	Homotetramer	Homotetramer	Homodimer	Homotetramer

Long chain fatty acids are degraded each 2 carbon atoms by VLCAD localized inner membrane, then oxidized by MCAD, and SCAD

SCAD short-chain acyl-CoA dehydrogenase, MCAD medium-chain acyl-CoA dehydrogenase, VLCAD very long-chain acyl-CoA dehydrogenase, LCAD long-chain acyl-CoA dehydrogenase

Modified from Ref. [11]

[a]The later research revealed that the role of LCAD was degradation of branched long chain fatty acids

dizes long-chain fatty acids with carbon chains of 12 to 22, carbon 2 units each [12]. The mitochondrial trifunctional protein (MTP) was also discovered by Uchida et al. in the same group [13]. Until then, it had been thought that the β oxidation system of long-chain fatty acids existed in the mitochondrial matrix, but the presence of VLCAD and MTP in the inner mitochondrial membrane was a very important discovery [14]. MCAD oxidizes medium-chain fatty acids with 12 carbons by the action of VLCAD, and SCAD oxidizes short-chain fatty acids with 4 carbons or less by TFP [14]. MCAD and SCAD are localized in the mitochondrial matrix. The later research revealed that the role of LCAD was degradation of branched long chain fatty acids [11].

Figure 10.8 shows a schematic diagram of the overall β-oxidation system for long-chain fatty acids and medium- and short-chain fatty acids in mitochondria as currently known.

Acyl CoA of long-chain fatty acids that enters the mitochondria through the carnitine cycle (CPT1, CACT, and CPT2) is oxidized by VLCAD and MTP, the β-oxidation systems for long-chain fatty acids in the inner mitochondrial membrane, by two each atom at a time. Medium-chain acyl CoA and short-chain acyl CoA with carbon chains of 12 or less are oxidized by the β-oxidation system of medium- and short-chain fatty acids in the matrix two carbon atoms at a time to acetyl CoA, which enters the TCA cycle and is used for energy production. VLCAD or MCAD/SCAD is a flavin enzyme that receives a proton (H^+) in the dehydrogenation reaction of fatty acids to form $FADH_2$, which is re-oxidized by ETFα/ETFβ to form FAD, and the next dehydrogenation reaction proceeds. ETFα/ETFβ receive electrons from $FADH_2$ of VLCAD or MCAD/SCAD, which are further transferred via ETFDH to the electron-transfer system of the oxidative phosphorylation system in the mitochondrial inner membrane. In addition, LCAD, which had previously been reported to have LCAD deficiency, was initially not confirmed the deficiency itself in detail, but subsequent studies revealed that it is a dehydrogenase of the branched chain long-chain fatty acids [14]. Thus, the β-oxidation of long-chain fatty acids involves a much greater number of enzymes than previously thought, and many β-oxidation disorders due to genetic abnormalities of these enzymes have been reported [15–17].

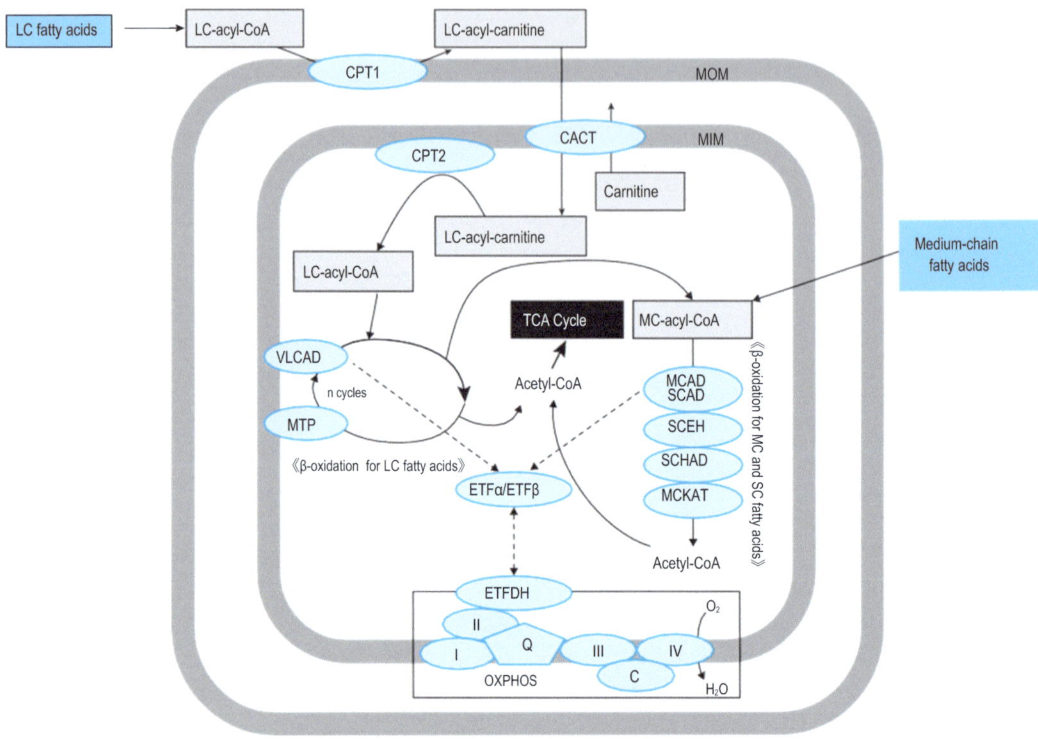

Fig. 10.8 Enzymology of mitochondrial fatty acid oxidation. Detailed overview of the enzymes involved in the sequential reactions that compose mitochondrial fatty acid oxidation. *LC long-chain, MC middle-chain, SC short-chain, CPT carnitine palmitoyltransferase, CACT carnitine acylcarnitine translocase, MOM mitochondrial outer membrane, MIM mitochondrial inner membrane, TCA tricarboxylic acid, VLCAD very-long-chain acyl-CoA dehydrogenase, MTP mitochondrial trifunctional protein, ETF electron transfer flavoprotein, EFTDH electron transfer flavoprotein dehydrogenase, MCAD middle-chain acyl-CoA dehydrogenase, SCAD short-chain acyl-CoA dehydrogenase, SCEH short-chain enoyl-CoA hydratase (crotonase), SCHAD short-chain 3-hydroxyacyl-CoA dehydrogenase, MCKAT middle-chain 3-ketoacyl-CoA thiolase, OXPHOS mitochondrial oxidative phosphorylation system.* (From Ref. [15], with permission from The Korean Society of Lipid and Atherosclerosis)

10.4 Urea Cycle Disorders and Carnitine Deficiency

The urea cycle, which converts ammonia produced by nitrogen metabolism of proteins and amino acids and by intestinal bacteria into urea for urinary excretion, is localized in the liver. Figure 10.9 shows the overall picture of the urea cycle [9].

Defects or reduced activity of enzymes and transporters of this urea cycle due to genetic abnormalities result in urea cycle disorders, as shown in Table 10.2 [8].

In these urea cycle disorders, carnitine deficiency is often caused by restricted protein intake and sodium benzoate administration, and national and international guidelines state that carnitine should be administered by measuring blood carnitine concentration and targeting 50 μmol/L (see below) [1, 18].

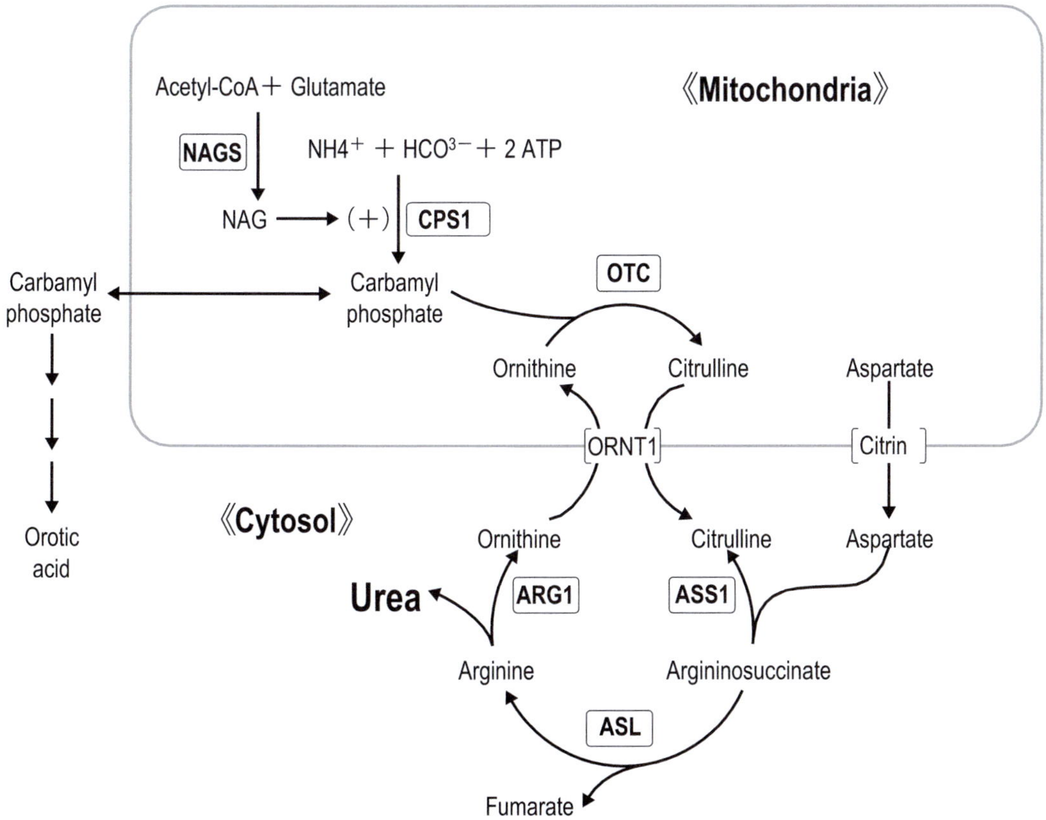

Fig. 10.9 Overall scheme of urea cycle. Urea cycle localize in liver and consist of enzymes such as NAGS, CPS1, OTC, ASS1, ARG1, and transporter/carrier protein such as ORNT1, citrin (aspartic acid/glutamic acid carrier). NAGS, CPS1 and OTC localize in mitochondria, and ARG1, ASS1 and ASL localize in cytosol. Ammonia is converted to urea by urea cycle and detoxified. *NAGS N-acetylglutamate synthase, CPS1 carbamyl phosphate synthetase 1, OTC ornithine transcarbamylase, ORNT1 mitochondrial ornithine/citrulline antiporter, ASS1 argininosuccinate synthetase, ASL argininosuccinate lyase, ARG1 arginase 1.* (Modified from Ref. [9])

Table 10.2 Characteristics of main urea cycle disorders

Disease	Clinical features	Specific changes in amino acids		Urinary orotic acid	Inheritance
		Plasma	Urine		
CPS1 deficiency	Hyperammonemia	Glutamine ↑, glutamate ↑ citrulline ↓		—	AR
OTC deficiency	Hyperammonemia	Glutamine ↑, glutamate ↑ citrulline ↓		++	XLR
ASS deficiency (citrullinemia type 1)	Hyperammonemia	Citrulline ↓	Citrulline ↑	++	AR
ASL deficiency (argininosuccinic aciduria)	Hyperammonemia hepatomegaly,hair abnormality	Argininosuccinate ↑ citrulline ↓	Argininosuccinate ↑	+	AR
ARG1 deficiency (argininemia)	Hyperammonemia	Arginine ↑↑	Arginine ↑, lysine ↑ cystine ↑	++	AR

(continued)

Table 10.2 (continued)

Disease	Clinical features	Specific changes in amino acids		Urinary orotic acid	Inheritance
		Plasma	Urine		
NAGS deficiency	Hyperammonemia	Glutamine ↑, glutamate ↑		+ ~ ±	AR
Citrin deficiency	Hyperammonemialiver dysfunction	Citrulline ↑		±	AR
HHH syndrome (ornithine transporter 1 deficiency)	Hyperammonemia	Citrulline ↑, ornithine ↑	Homocysteine ↑	+	AR

Citrin: glutamate aspartate carrier localized in liver mitochondria membrane
CPS1 carbamyl phosphate synthetase 1, OTC ornithine transcarbamylase, ASS argininosuccinate synthetase, ASL argininosuccinate lyase, ARG1 arginase 1, NAGS N-acetyl glutamate synthase, HHH hyperornithinemia-hyperammonemia-homocitrullinuria, AR autosomal recessive, LR X-linked recessive
Modified from Refs. [1, 8]

Table 10.3 Main points of carnitine supplementation in metabolic emergency

Case examples	Carnitine formulations	Dose
Metabolic acidosis suspected inherited metabolic diseases	L-carnitine (intravenous (i.v.): L-Cartin FF injection 1000 mg, oral medicine: L-Cartin FF)	100–150 mg/kg/day
Example administration to patients of body weight 3 kg	Side tube (L-Cartin FF injection 1000 mg)	Initially 1.5 mL i.v. 100 mg/kg, then 0.75 mL × 3 times i.v., 150 mg/kg/day
Hyperammonemia + metabolic acidosis (NH_3: newborn >250 μg/dL, infancy >170 μg/dL, anion gap >20 mEq/L)	Respiratory and blood glucose managements, alkalizing agents, blood purifications with vitamin cocktail (in the case of suspected organic aciduria, carnitine supplementation is indispensable)	Side tube Initial dose: 100 mg/kg, i.v. Later: 150 mg/kg/day, divided 3 times, i.v.

The above description shows case examples of carnitine supplementation methods in the case of metabolic emergency at the first visit and definitive diagnosis is not yet done, but it is necessary to do differential diagnosis and initiation of treatment
Modified from Ref. [1]

10.5 Dosage of Carnitine for Inherited Metabolic Diseases

Usage and dose of carnitine supplementation in inherited metabolic diseases are described in "Guidelines 2019" and elsewhere. Even for the same disease, there are a wide range of dosages depending on the individual patient's condition, concomitant medications, or the severity of the mutation.

Tables 10.3, 10.4, 10.5, and 10.6 summarize the main points of carnitine usage/dosage and precautions in organic acid metabolism disorders, fatty acid metabolism disorders, and urea cycle disorders, as well as guidelines for the treatment of metabolic emergencies, which are mainly listed in "Guidelines 2019."

In fatty acid metabolism disorders such as VLCAD, carnitine administration is described as "controversial," but the details of its controversy are described by Van Calcar et al. [19]

Table 10.4 Main points of L-carnitine supplementation in inherited metabolic diseases (1)

Disorders	Treatment after definitive diagnosis with newborn mass screening (without symptoms)	Treatment in acute attack onset	Management in chronic phase
Urea cycle disorders	Treatment for urea cycle disorders (right)	Hyperammonemic crisis and definitive diagnosis is not yet done: 100 mg/kg/day. Intravenous administration of LC should be considered.	In the case of carnitine deficiency due to protein restriction, LC should be administered as target blood concentration as 50 μmol/L. Also, prophylactic administration is necessary (20–50 mg/kg/day) To maintain mitochondrial function, LC administration is required. Carnitine deficiency is prone in the case of sodium benzoate administration.
Methylmalonic acidemia Propionic acidemia	LC (oral): 50–150 mg/kg/day divided 3 times Keep serum FC (or filter paper blood) above 50 μmol/L	LC administration LC (i.v.): initial dose: 100 mg/kg (bolus, or 1–2 h infusion) Maintenance dose: 100–200 mg/kg/day In the case of i.v. formulation is not available: LC (oral): 100–150 mg/kg/day	LC (oral): 100–200 mg/kg/day divided 3 times Keep serum FC (or filter paper blood) above 50 μmol/L.
Isovaleric acidemia	LC (oral): 100–200 mg/kg/day, divided 3 times Keep serum FC (or filter paper blood) above 50 μmol/L		
HMG-CoA lyase deficiency		In the case of HMG-CoA lyase deficiency: Acute attack: 50–100 mg/kg i.v., 3 times/day In the case of i.v. formulation is not available: LC (oral): 100–150 mg/kg/day	To prevent secondary carnitine deficiency, LC administration: 30–100 mg/kg/day
Methylcrotonylglycinuria	LC (oral): 50–150 mg/kg/day Keep serum FC (or filter paper blood) above 50 μmol/L.		LC (oral): 100–200 mg/kg/day divided 3 times Keep serum FC (or filter paper blood) above 50 μmol/L.
	Issues in adulthood Symptoms due to original disorders are observed; LC supplementation is strongly recommended regardless of the blood concentrations. In the case of asymptomatic but low FC concentration, LC supplementation is recommended because there is a possibility to cause complications such as myocardial disorders. In the case of asymptomatic and normal FC concentration, there is no study to show benefits of the supplementation and there is no consensus to recommend the supplementation.		

LC L-carnitine, FC free carnitine, i.v. intravenous
Modified from Ref. [1]

Table 10.5 Main points of L-carnitine supplementation in inherited metabolic diseases (2)

Disorders	Treatment after definitive diagnosis with newborn mass screening (without symptoms)	Treatment in acute attack onset	Management in chronic phase
Multiple carboxylase deficiency		LC (i.v.) Initial dose: 100 mg/kg (bolus, or 1–2 h infusion) maintenance dose: 100–200 mg/kg/day In the case of intravenous formulation is not available: LC (oral): 100–150 mg/kg/day	LC (oral): 100–200 mg/kg/day divided 3 times Keep serum (or filter paper blood) FC above 50 μmol/L
β-ketothiolase deficiency			To prevent secondary carnitine deficiency: LC 30 ~ 100 mg/kg/day
Glutaric acidemia type 1	LC (oral): 100 mg/kg/day Keep FC concentration higher around 60–100 μmol/L	LC (i.v.) Initial dose: 100 mg/kg (bolus), maintenance dose: 100–200 mg/kg/day In the case of intravenous formulation is not available: LC (oral): 100–150 mg/kg/day	LC (oral): 100 mg/kg/day divided 3 times Keep FC around 60–100 μmol/L
Glutaric acidemia type 2		Enough glucose supplementation is considered useful in lipid and protein restriction, although no enough evidences. There is no specific useful drugs for lipid oxidation disorders. LC administration is controversial, but normal i.v. dose in acute phase is not contraindications. After stable condition, diets mainly special milk and carbohydrates should be started	It is not conclusive, but there are many case reports of LC supplementation in symptomatic infants, and at least it is not contraindications in acute phase administration Risk of carnitine deficiency due to increased short or medium-chain acylcarnitine excretion, controlled administration to keep above 20 μmol/L
Very long chain acyl-CoA dehydrogenase deficiency (VLCAD deficiency)		LC administration is controversial, and it is considered LC (i.v.) in acute phase is contraindications	LC supplementation is controversial. There is possibility to aggravate cardiomyopathy and muscle symptoms, and there is a report of brother cases of rhabdomyolysis and not recommended When LC supplementation, start from low dose, and keep FC around 20 μmol/L.

LC L-carnitine, FC free carnitine, i.v. intravenous
Modified from Ref. [1]

Table 10.6 Main points of L-carnitine supplementation in inherited metabolic disorders (3)

Disorders	Treatment after definitive diagnosis with newborn mass screening (without symptoms)	Treatment in acute attack onset	Management in chronic phase
Trifunctional protein(TFP) deficiency		It is controversial about LC administration and i.v. administration is contraindication. After stable condition, initiate diet of special milk and carbohydrates	LC administration is controversial. There is no evidence of efficacy in long term LC supplementation, and it is said to avoid administration during severe attack As general, consider supplementation if FC is low (to keep FC above 20 μmol/L)
Medium chain acyl-CoA dehydrogenase (MCAD) deficiency		At present, there is no specifically effective drugs, and dietary lifestyle guidance is reported quite good result. It is controversial about LC administration and i.v. administration is contraindication. After stable condition, initiate diet of carbohydrates. Avoid starvation, drug treatment, after stable state treatment is same as the case of newborn mass screening cases	It is unnecessary to MCAD deficiency However, in the case of low carnitine by nutritional condition, LC supplementation to keep FC around normal lower limit (20 μmol/L) is recommended
Systemic carnitine deficiency, or organic carnitine transporter 2 (OCTN2) deficiency, carnitine uptake defect	There is no definitive opinion for treatment before definitive diagnosis, but start treatment is not denied if necessary	Continuous i.v. administration (300 mg/kg/day) control FC keeps above 20 μmol/L If oral administration is possible, 100–400 mg/kg/day, divided 4 time If oral intake is not possible, i.v. (dose is not established), start i.v. administration from a half of normal or normal doses	The most and the only effective treatment is a large amount of LC supplementation 100–400 mg/kg/day divided 4 times in infants, divided 3 times in adults. In some cases, FC are still below normal limits. Periodical monitoring is required (up to 4 years: every 2 months, later: every 3–6 months. The LC dose is controlled as target above 20 μmol/L
Carnitine cycle abnormality (CPT1 deficiency, CACT deficiency, CPT2 deficiency)		LC supplementation for CPT2 or CACT deficiency is controversial. At least, i.v. administration in acute phase is contraindications	LC supplementation is not done in CPT1 deficiency. LC supplementation is controversial in CPT2 and CACT deficiencies, but in low FC cases(<15–20 μmol/L) usually LC administered to keep FC above 20 μmol/L. In CACT deficiency, efficacy is reported in large dose(50–150 mg/kg/day)

LC L-carnitine, FC free carnitine, i.v. intravenous
Modified from Ref. [1]

10.6 Mitochondrial Diseases

Mitochondrial diseases is one of the most common inherited metabolic diseases. Mitochondrial diseases is a general term for inherited metabolic diseases caused by mitochondrial dysfunction, such as abnormal energy metabolism due to abnormalities in the mitochondrial respiratory chain. It is sometimes referred to as mitochondrial disorders. It is also called mitochondrial encephalomyopathy because the symptoms often appear in the muscles and brain. It is said to occur in 1 in 5000 live births, and cases that occur in infancy are often severe and have a high mortality rate. The diseases have an extremely wide variety of symptoms, manifests as abnormalities in various organs, and have a wide age range of onset. Currently, there is no fundamental treatment for the disease, which is limited to symptomatic therapy. Since sperm contains only a few dozen mitochondria, which are eliminated at fertilization, it is believed that the traits of the mitochondria in the mother's egg cells (approximately 200,000 mitochondria) are mainly inherited. Table 10.7 summarizes the classification of the broad sense of mitochondrial abnormalities by function, etc. [20].

In a narrow sense, mitochondrial diseases may refer to an abnormality in the mitochondrial electron transport system, and Goto has also tried to classify it according to biochemical, morphological, DNA abnormalities, and clinical symptoms [21] and clinical phenotype and onset time [22]. In addition, Koga summarizes the classification of mitochondrial encephalomyopathy according to clinical pathotype, biochemical classification, and genetic abnormality [23].

Treatment for mitochondrial diseases is only symptomatic treatment, and cocktails containing

Table 10.7 Classification of mitochondrial abnormalities by functions

I. *Abnormalities in proteins related to energy metabolisms*	II. *Disorders in enzymes in reactive oxygen metabolism* (a) Abnormality in *SOD2* (animal model only) III. *Abnormalities in proteins related to apoptosis* IV. *Abnormalities in proteins related to calcium dynamics* V. *Abnormalities in proteins of unknown function* VI. *Abnormalities in proteins related to biogenesis of mitochondria* 1. Abnormalities in proteins related to mitochondrial DNA replication/maintenance (a) Adenine nucleotide translocase 1 disorder (b) *Twinkle* disorder (c) DNA polymerase γ disorder (d) Thymidine phosphorylase deficiency (e) Thymidine kinase deficiency (f) Deoxyguanosine kinase deficiency 2. Abnormalities in proteins related to mitochondrial DNA transcription/translation (a) *Tfam1* disorder (animal model only) (b) *RMPP* disorder (cartilage/hair hypoplasia) 3. Abnormalities in proteins related to transport into mitochondrial (a) DFN1 disorder 4. Abnormalities in proteins related to processing of mitochondrial protein 5. Abnormalities in proteins related to fusion/separation of mitochondria 6. Abnormalities in proteins related to intracellular migration of mitochondria
1. Disorders of substrate transport (a) CPT2 deficiency (b) Carnitine deficiency	
2. Disorders of substrate availability (a) Pyruvate carboxylase deficiency (b) Pyruvate dehydrogenase complex deficiency (c) Disorders of β-oxidation enzymes (d) Electron transport flavoprotein deficiency (e) Trifunctional enzyme deficiency	
3. Disorders in TCA cycle (a) Fumarase deficiency (b) α-ketoglutarate dehydrogenase deficiency (c) Aconitase deficiency	
4. Disorders of the coupling in oxidative phosphorylation (a) Luft disease	
5. Disorders of enzymes in electron transport system (a) Complex I deficiency (b) Complex II deficiency (c) Complex III deficiency (d) Complex IV deficiency (e) Complex V deficiency (f) Multiple complexes deficiency	
6. Disorders in urea cycle	
7. Disorders in heme, iron metabolism	

From Ref. [20]

riboflavin (vitamin B_2), thiamine (vitamin B_1), biotin, CoQ10, L-carnitine, creatine, α-lipoic acid, etc., which are involved in maintaining mitochondrial function, are administered [24–31]. Campos et al. report that many patients with mitochondrial myopathy often have secondary carnitine deficiency and a condition of carnitine insufficiency [32].

Koga et al. conducted an investigator-initiated clinical trial of L-arginine for mitochondrial myopathy, encephalopathy, lactic acidosis, and stroke-like episodes (MELAS), and found that it could extend the life of MELAS patients, and that activities of daily living (activities of daily life: ADL), but it has not yet been approved [23, 33–35]. (https://www.kurume-u.ac.jp/site/backno/20181002-02.html). However, arginine hydrochloride preparation (granules and injection) was exceptionally covered by insurance in Japan in October 2022.

On January 31, 2019, a large dose of taurine was approved as an orphan drug for the indication of "suppression of stroke-like seizures in MELAS" and was covered by insurance. This was the first drug to be covered by insurance for mitochondrial disease in Japan [36] (Taisho Pharmaceutical Co., Ltd., Taurine powder, https://www.amed.go.jp/news/release_20190219-01.html).

Regarding the diagnosis and treatment of mitochondrial diseases, the Japan Mitochondrial Society has published the "Mitochondrial Diseases Clinical Treatment Manual 2017" [37]. Currently, treatment methods are extremely limited, but further drug development for mitochondrial diseases is desired in the future.

10.7 Other Inherited Metabolic Diseases and Carnitine

Carnitine supplementation is also effective in cases of secondary carnitine deficiency or carnitine insufficiency in inherited metabolic diseases other than those described so far.

In phenylketonuria (PKU), the use of a strict protein-restricted diet and long-term use of special milk has been reported to result in decreased levels of vitamins and trace elements such as biotin, iodine, and selenium [38], and supplementation with appropriate tests is recommended in the "Guidelines 20,191," [1] and supplementation is recommended for selenium deficiency in particular, but there is no description for carnitine deficiency. Overseas studies have reported that PKU patients have a decrease in blood free carnitine [39], especially in patients who strictly adhere to a protein-restricted diet [40]. In addition, it has been reported that carnitine deficiency may occur in PKU patients due to insufficient intake of iron, which is necessary for carnitine biosynthesis [41]. Sitta et al. found that blood carnitine levels were about half of the control level in PKU patients, and that plasma phenylalanine levels were positively correlated with free carnitine levels, that is, patients who strictly followed a protein-restricted diet had lower blood free carnitine levels [42]. In these patients, thiobarbituric acid-reactive species (TBARS) measured were significantly elevated compared to controls, and the blood-free carnitine concentration was negatively correlated with oxidative stress and the total antioxidant capacity was positively correlated. When L-carnitine (98 mg/day) and selenium (31.5 μg/day) were administered to these PKU patients for at least 6 months, the TBARS, which had increased by about 3 times, almost normalized, and the reduced sulfhydryl (SH) content was restored [43] (Fig. 10.10). At that time, glutathione peroxidase activity, which had been significantly reduced, was restored (SOD activity was not restored).

It has been reported that maple syrup urine disease (MSUD) also has a decrease in blood carnitine compared to controls [44]. The pathogenesis of MSUD is not yet fully understood, but it is thought to be related to oxidative stress. Mescka et al. examined the effect of carnitine (100 mg/kg, intraperitoneal, 1 day before branched-chain amino acid administration, intraperitoneally) on lipid peroxidation and oxidative stress in the cerebral cortex in a chemically induced MSUD model in which a branched-chain amino acid pool was injected subcutaneously into rats [45]. As a result, glutathione peroxidase activity in the cerebral cortex was significantly reduced com-

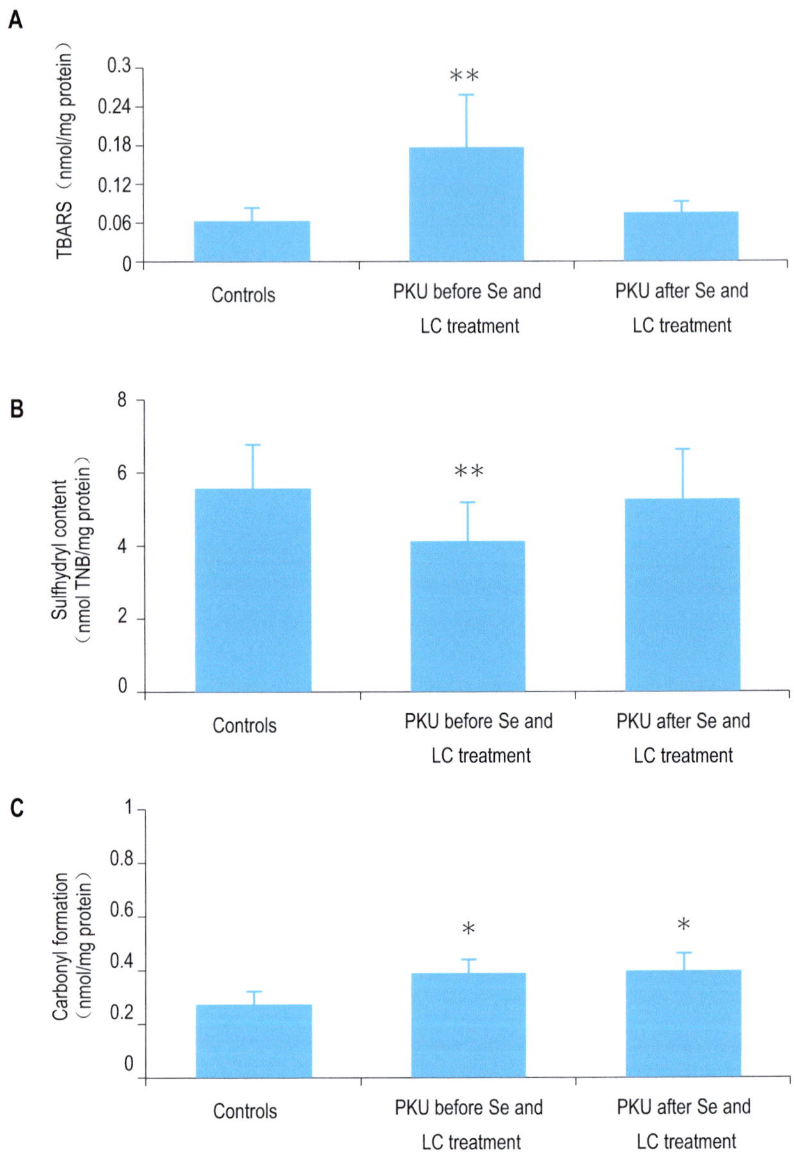

Fig. 10.10 Lipid and protein oxidative damage in plasma from phenylketonuric (PKU) patients ($n = 18$) and effects of L-carnitine (LC) and selenium (Se) treatment on them. (**a**) Thiobarbituric acid-reactive species (TBARS). (**b**) Sulfhydryl content. TNB: A yellow derivative which is generated when 5, 5′-dithio-bis (2-nitrobenzoic acid) (DTNB) is reduced by sulfhydryl-group and has an absorption peak at 412 nm. (**c**) Carbonyl formation. Data represent mean ± S.D., *: $P < 0.05$, **: $P < 0.01$ vs controls. (From Ref. [43])

pared to control in MSUD model rats, and significant increases in TBARS and protein carbonyl and sulfhydryl content were observed, but carnitine administration completely suppressed these fluctuations. In a follow-up report, Meskka et al. conducted a similar experiment in the MSUD model rat, and when they examined changes in the cerebellum in addition to the cerebral cortex, they found that TBARS was significantly increased in the cerebral cortex and cerebellum, and that carbonyls were significantly increased in the cerebellum in the MSUD model rats [46]. In addition, carnitine administration suppressed the elevation of TBARS in the cerebral cortex and cerebellum, inhibited the elevation of catalase in the cerebral cortex, and suppressed the decrease in SOD, glutathione peroxidase, and glucose 6-phosphate dehydrogenase activity. Therefore, it is possible that the antioxidant effect of carnitine may be effective in inhibiting the onset and progression of MSUD [46]. Ribas et al. introduced research on carnitine supplementation in organic acidemia, PKU, MSUD, etc., and discussed that carnitine supplementation with antioxidant effects is significant because these inherited neurometabolic diseases are prone to carnitine deficiency due to protein intake restriction and the condition is often worsened by increased oxidative stress [47]. (Chap. 7: “Anti-inflammatory, Antioxidant, and Anti-apoptotic Effects of Carnitine”).

In Rett syndrome, carnitine deficiency may occur due to nutritional management by valproic acid administration or tube administration for epileptic seizures. Basic studies have reported abnormalities in the carnitine cycle [48]. Ellaway et al. also conducted an open-label study of 21 patients with Rett syndrome treated with carnitine (100 mg/kg/day, min. 2.6 months) and 62 patients in the non-treatment group (control group) [49]. As a result, carnitine administration showed significant improvement in measures such as sleep efficiency, energy level, communication skills, and expressive speech.

Thus, it is important to note that carnitine deficiency may also occur in inherited metabolic diseases not listed in the “Guidelines 2019.” In the presence of unexplained hypoglycemic attacks, hyperammonemia, cardiomyopathy or heart failure, and muscle weakness, blood carnitine levels should be measured to diagnose the presence of carnitine deficiency.

References

1. The Japanese Society for Inherited Metabolic Diseases, editor. Clinical practice guidelines for neonatal mass screening diseases 2019. Tokyo: SHINDAN TO CHIRYO SHA, Inc.; 2019. (Book in Japanese).
2. Yamaguchi S, editor. A well-understood newborn mass screening guidebook. Tokyo: SHINDAN TO CHIRYO SHA, Inc.; 2019. (Book in Japanese).
3. The Japanese Society for Inherited Metabolic Diseases, editor. Learning from cases of inherited metabolic diseases: an approach from everyday practice. Tokyo: SHINDAN TO CHIRYO SHA, Inc.; 2019. (Book in Japanese).
4. The Japanese Society for Inherited Metabolic Diseases, editor. Inherited metabolic diseases. Diagnosis at a glance. Tokyo: SHINDAN TO CHIRYO SHA, Inc.; 2011. (Book in Japanese).
5. Ida S, Sakai N, Shintaku H, et al., editors. Practice! Inherited metabolic diseases. Nutrition and dietary guidance casebook. Tokyo: SHINDAN TO CHIRYO SHA, Inc.; 2019. (Book in Japanese).
6. Special Milk Joint Safety Development Committee, editor. Treatment guidelines for new diseases subject to screening associated with the introduction of tandem mass. Special milk information 42: November 2006 (Suppl.), Social Welfare Corporation, Imperial Gift Foundation Boshi-Aiiku-Kai, Tokyo; 2007. (Booklet in Japanese).
7. Abe T, Iinuma K, Yoshioka H, editors. Textbook of pediatrics and neonatology, revised 4th Edition. Tokyo: SHINDAN TO CHIRYO SHA, Inc.; 2003. p. 204–26. (Book in Japanese).
8. Matsumoto S, Häberle J, Kido J, et al. Urea cycle disorders-update. J Hum Genet. 2019;64(9):833–47. https://doi.org/10.1038/s10038-019-0614-4.
9. Kenneson A, Singh RH. Presentation and management of N-acetylglutamate synthase deficiency: a review of the literature. Orphanet J Rare Dis. 2020;15(1):279. https://doi.org/10.1186/s13023-020-01560-z.
10. Yamaguchi S. Overview of enzymes of mitochondrial β oxidation system. Jpn J Clin Med. 2002;60(Special Issue 4):88–93. (Article in Japanese).
11. Yamaguchi S. Mitochondrial acyl-CoA dehydrogenases (VLCAD, LCAD, MCAD, SCAD). Jpn J Clin Med. 2002;60(Suppl 4):94–7. (Article in Japanese).
12. Izai K, Uchida Y, Orii T, et al. Novel fatty acid β-oxidation enzymes in rat liver mitochondria. I. Purification and properties of very-long-chain acyl-coenzyme A dehydrogenase. J Biol Chem. 1992;267(2):1027–33.
13. Uchida Y, Izai K, Orii T, et al. Novel fatty acid β-oxidation enzymes in rat liver mitochondria. II. Purification and properties of enoyl-coenzyme A (CoA) hydratase/3-hydroxyacyl-CoA dehydrogenase/3-ketoacyl-CoA thiolase trifunctional protein. J Biol Chem. 1992;267(2):1034–41.
14. Yamaguchi S. Progress in research on abnormal metabolism of organic acids and fatty acids. Jpn J Inherit Metab Dis. 2004;20(1):26–36. (Article in Japanese).
15. Wanders RJA, Visser G, Ferdinandusse S, et al. Mitochondrial fatty acid oxidation disorders: laboratory diagnosis, pathogenesis, and the complicated route to treatment. J Lipid Atheroscler. 2020;9(3):313–33. https://doi.org/10.12997/jla.2020.9.3.313.
16. Merritt JL 2nd, Norris M, Kanungo S. Fatty acid oxidation disorders. Ann Transl Med. 2018;6(24):473. https://doi.org/10.21037/atm.2018.10.57.
17. Ruiz-Sala P, Peña-Quintana L. Biochemical markers for the diagnosis of mitochondrial fatty acid oxidation diseases. J Clin Med. 2021;10(21):4855. https://doi.org/10.3390/jcm10214855.
18. Häberle J, Burlina A, Chakrapani A, et al. Suggested guidelines for the diagnosis and management of urea cycle disorders: first revision. J Inherit Metab Dis. 2019;42:1192–230.

19. Van Calcar SC, Sowa M, Rohr F, et al. Nutrition management guideline for very-long chain acyl-CoA dehydrogenase deficiency (VLCAD): an evidence- and consensus-based approach. Mol Genet Metab. 2020;131(1–2):23–37. https://doi.org/10.1016/j.ymgme.2020.10.001.
20. Goto Y. Concept and classification of mitochondrial diseases (broad sense). Jpn J Clin Med. 2002;60(Special Issue 4):213–7. (Article in Japanese).
21. Goto Y. Problems related to the concept and classification of mitochondrial diseases (narrowly defined). Jpn J Clin Med. 2002;60(Special Issue 4):221–3. (Article in Japanese).
22. Goto Y. Classification of mitochondrial diseases (narrowly defined) by clinical phenotype. Jpn J Clin Med. 2002;Special Issue 4:499–500. (Article in Japanese).
23. Koga Y. Diagnosis and treatment of mitochondrial diseases. Brain Dev. 2010;42:124–9. (Article in Japanese).
24. Bernsen PL, Gabreëls FJ, Ruitenbeek W, et al. Successful treatment of pure myopathy, associated with complex I deficiency, with riboflavin and carnitine. Arch Neurol. 1991;48(3):334–8. https://doi.org/10.1001/archneur.1991.00530150106028.
25. Marriage BJ, Clandinin MT, Macdonald IM, et al. Cofactor treatment improves ATP synthetic capacity in patients with oxidative phosphorylation disorders. Mol Genet Metab. 2004;81(4):263–72. https://doi.org/10.1016/j.ymgme.2003.12.008.
26. Scaglia F, Northrop JL. The mitochondrial myopathy encephalopathy, lactic acidosis with stroke-like episodes (MELAS) syndrome: a review of treatment options. CNS Drugs. 2006;20(4):443–64. https://doi.org/10.2165/00023210-200620060-00002.
27. DiMauro S, Hirano M, Schon EA. Approaches to the treatment of mitochondrial diseases. Muscle Nerve. 2006;34(3):265–83. https://doi.org/10.1002/mus.20598.
28. DiMauro S, Mancuso M. Mitochondrial diseases: therapeutic approaches. Biosci Rep. 2007;27(1–3):125–37. https://doi.org/10.1007/s10540-007-9041-4.
29. Sproule DM, Dyme J, Coku J, et al. Pulmonary artery hypertension in a child with MELAS due to a point mutation of the mitochondrial tRNA(Leu) gene (m.3243A>G). J Inherit Metab Dis. 2008;31(suppl 3):497–503. https://doi.org/10.1007/s10545-007-0735-3.
30. Orsucci D, Filosto M, Siciliano G, et al. Electron transfer mediators and other metabolites and cofactors in the treatment of mitochondrial dysfunction. Nutr Rev. 2009;67(8):427–38. https://doi.org/10.1111/j.1753-4887.2009.00221.x.
31. Pfeffer G, Majamaa K, Turnbull DM, et al. Treatment for mitochondrial disorders. Cochrane Database Syst Rev. 2012;4:CD004426. https://doi.org/10.1002/14651858.CD004426.pub3.
32. Campos Y, Huertas R, Lorenzo G, et al. Plasma carnitine insufficiency and effectiveness of L-carnitine therapy in patients with mitochondrial myopathy. Muscle Nerve. 1993;16(2):150–3. https://doi.org/10.1002/mus.880160205.
33. Koga Y. L-arginine therapy on MELAS. Rinsho Shinkeigaku. 2008;48(11):1010–2. https://doi.org/10.5692/clinicalneurol.48.1010. (Article in Japanese).
34. Koga Y, Povalko N, Inoue E, et al. Therapeutic regimen of L-arginine for MELAS: 9-year, prospective, multicenter, clinical research. J Neurol. 2018;265(12):2861–74. https://doi.org/10.1007/s00415-018-9057-7.
35. Ikawa M, Povalko N, Koga Y. Arginine therapy in mitochondrial myopathy, encephalopathy, lactic acidosis, and stroke-like episodes. Curr Opin Clin Nutr Metab Care. 2020;23(1):17–22. https://doi.org/10.1097/MCO.0000000000000610.
36. Ohsawa Y, Hagiwara H, Nishimatsu S, et al. Taurine supplementation for prevention of stroke-like episodes in MELAS: a multicentre, open-label, 52-week phase III trial. J Neurol Neurosurg Psychiatry. 2019;90(5):529–36. https://doi.org/10.1136/jnnp-2018-317964.
37. Japan Mitochondrial Society, editor. Mitochondrial disease treatment manual 2017. Tokyo: SHINDAN TO CHIRYO SHA, Inc.; 2016. (Book in Japanese).
38. Okano Y, Hattori T, Fujimoto H, et al. Nutritional status of patients with phenylketonuria in Japan. Mol Genet Metab Rep. 2016;8:103–10.
39. Schulpis KH, Nounopoulos C, Scarpalezou A, et al. Serum carnitine level in phenylketonuric children under dietary control in Greece. Acta Paediatr Scand. 1990;79(10):930–4. https://doi.org/10.1111/j.1651-2227.1990.tb11354.x.
40. Vilaseca MA, Briones P, Ferrer I, et al. Controlled diet in phenylketonuria may cause serum carnitine deficiency. J Inherit Metab Dis. 1993;16:101–4.
41. Böhles H, Ullrich K, Endres W, et al. Inadequate iron availability as a possible cause of low serum carnitine concentrations in patients with phenylketonuria. Eur J Pediatr. 1991;150(6):425–8. https://doi.org/10.1007/BF02093725.
42. Sitta A, Barschak AG, Deon M, et al. L-carnitine blood levels and oxidative stress in treated phenylketonuric patients. Cell Mol Neurobiol. 2009;29:211–8.
43. Sitta A, Vanzin CS, Biancini GB, et al. Evidence that L-carnitine and selenium supplementation reduces oxidative stress in phenylketonuric patients. Cell Mol Neurobiol. 2011;31(3):429–36. https://doi.org/10.1007/s10571-010-9636-3.
44. Kumru B, Hismi BO. Investigation of L-carnitine concentrations in treated patients with maple syrup urine disease. J Pediatr Genet. 2019;8(3):133–6. https://doi.org/10.1055/s-0039-1691789.
45. Mescka C, Moraes T, Rosa A, et al. In vivo neuroprotective effect of L-carnitine against oxidative stress in maple syrup urine disease. Metab Brain Dis. 2011;26:21–8.
46. Mescka CP, Rosa AP, Schirmbeck G, et al. L-carnitine prevents oxidative stress in the brains of rats sub-

jected to a chemically induced chronic model of MSUD. Mol Neurobiol. 2016;53(9):6007–17. https://doi.org/10.1007/s12035-015-9500-z.

47. Ribas GS, Vargas CR, Wajner M. L-carnitine supplementation as a potential antioxidant therapy for inherited neurometabolic disorders. Gene. 2014;533(2):469–76. https://doi.org/10.1016/j.gene.2013.10.017.

48. Mucerino S, Di Salle A, Alessio N, et al. Alterations in the carnitine cycle in a mouse model of Rett syndrome. Sci Rep. 2017;7:41824. https://doi.org/10.1038/srep41824.

49. Ellaway CJ, Peat J, Williams K, et al. Medium-term open label trial of L-carnitine in Rett syndrome. Brain Dev. 2001;23(Suppl 1):S85–9. https://doi.org/10.1016/s0387-7604(01)00346-1.

11 Tube Feeding, Total Parenteral Nutrition (TPN), Specific Milk Formula, and Carnitine Deficiency

11.1 Tube Feeding, TPN, Specific Milk, and Carnitine Deficiency

It is estimated that 75% of the daily requirement of carnitine comes from dietary intake, and 25% is biosynthesized in the body [1, 2]. Moreover, infants and young children have not yet developed biosynthesis (see "Chap. 4: Homeostasis of Carnitine" for details). Therefore, the effect of diet on homeostasis of carnitine is very large, and there is a high risk of carnitine deficiency in children with severe physical and mental disabilities and children who have undergone surgery if they are nutritionally managed for a long time with enteral nutrition preparations or total parenteral nutrition (TPN) that do not contain carnitine. In addition, carnitine deficiency has been reported in infants who are allergic to milk and are fed with some allergen-free formula that does not contain carnitine. Carnitine deficiency has also been reported in infants who take only soybean protein-based infant food, strict vegetarians, and patients with anorexia. As described in Chap. 5 "IV: Classification of carnitine deficiency by etiology," these belong to secondary carnitine deficiency caused by acquired medical conditions and decreased intake [3, 4].

As early as 1979, Borum et al. reported the carnitine content of 42 liquid specific nutrients used in infant nutrition at the time [5]. When the proteins of these nutrients were derived from soybeans, casein, and egg whites, the carnitine content was less than 4 nmol/mL, and most of them were below the detection limit, whereas products derived from cow's milk or beef had a higher content of 50–656 nmol/mL. Therefore, the authors cautioned against carnitine deficiency in the long-term use of these soy, casein, and egg white derived specific nutritional products in the nutritional management of infants and young children. Carnitine supplementation in such cases should be tailored to the individual patient's condition and growth, with 2–5 mg/kg/day in nutritional formulas and 50–100 mg/kg/day in carnitine formulations recommended [6].

In Japan, Kodama et al., in their review of nutritional management by tube feeding and specific milk formula, reported cases of fatty acid, biotin, zinc, and carnitine deficiencies and alerted to these deficiencies [7]. Among them, carnitine deficiency was most frequently observed in infants, but also in elderly adults, and various clinical manifestations of carnitine deficiency were observed, including hypercapnia, vomiting, hypoglycemia, hepatic dysfunction, hyperCKemia, hyperlipidemia, and dilated cardiomyopathy (Table 11.1).

Carnitine deficiency in children (and persons) with severe mental and physical disabilities is not simple because, in addition to the problem of carnitine content in enteral nutritional products used for tube feeding, there are additional problems such as carnitine deficiency caused by antiepilep-

N. Nakayama, *Carnitine Deficiency and Insufficiency in Medicine*,
https://doi.org/10.1007/978-981-95-7797-2_11

Table 11.1 Case reports of carnitine deficiency with specific milk and tube feeding

Age	Gender	Underlying disease	Symptoms of carnitine deficiency	Carnitine deficiency laboratory findings	Nutritional supplement	Duration	Reporter	Publication year
Infant	Male	Milk allergy	Polypnea	Free carnitine 0 μmol/L	Specific milk (Meiji 605Z)	3 months	Masaki Takayanagi	2009
Infant	Male	Congenital myopathy	Vomiting, hypoglycemia	Free carnitine Less than measurement sensitivity	Enteral nutrition (Elental P)	1 year, 2 months	Koji Nakajima	2009
Infant	Female	Severely handicapped child	No clinical symptoms	Free carnitine 8.7 μmol/L	Enteral nutrition (Elental P)	11 months	Masaki Takayanagi[a]	Not reported
Infant	Male	Sequelae of hypoxic encephalopathy	Dilated cardiomyopathy	Marked decline in free carnitine[b]	Enteral nutrition	—	Toru Yoneda	2008
Infant	Male	Very low birth weight infant	Hepatic dysfunction hyperCKemia	Free carnitine 13.8 μmol/L	Specific milk (MCT milk)	2 months	Atsuko Goto	1992
78 years	Female	Diabetes, hyperlipidemia	Hyperlipidemia	Free carnitine 22 μmol/L	Enteral nutrition (Ensure Liquid)	2 years, 6 months	Sin-ichiro Tanaka	2000

Modified from Ref. [7]

[a]Self-studied case, and not reported, [b]Later, it turned out as organic carnitine transporter 2 (OCTN2) deficiency

tic drugs such as valproic acid and antibacterial drugs containing pivoxil groups, and muscle loss due to bedridden or neuromuscular diseases.

11.2 Species of Tube Feeding Products and Carnitine Contents

Enteral nutrition is divided into semi-digestible nutrition, digestive nutrition, and component nutrition, and is further divided into pharmaceutical and food products. Table 11.2 shows the main types of enteral nutrition and their carnitine content in Japan [8, 9].

Pharmaceutical enteral formulas containing carnitine currently include Enevo® and Enoras® as semi-digestible formulas. There are a large number of food-based nutritional products containing carnitine, such as Isocal® 1.0 Junior (20 mg/100 kcal) and Sanet K2 (20 mg/100 kcal) (Table 11.2).

In nutritional management with enteral nutrition and TPN, it is natural to consider the calorific value and nutrients contained according to the patient's condition, but especially when enteral nutrition is used for long periods of time in pediatric patients, the non-protein calorie/nitrogen ratio (NPC/N) and other factors must be carefully considered [8, 10, 11]. In addition to the carnitine content, there are other complex issues that need to be considered when selecting an enteral formula. Regarding the patient's responsibility for enteral nutritional supplements, some families prefer to use products treated as pharmaceutical drugs for economic reasons (Table 11.3), because enteral nutritional supplements treated as pharmaceutical drugs are covered by insurance, while nutritional supplements treated as food are, in principle, paid for by the patient [12]. Although there are a number of enteral nutritional products containing carnitine that are treated as food, there are some problems, such as low carnitine content in some products or lack of another nutrient, such as chlorine, which may cause dietary chloride deficiency syndrome in patients taking high doses of valproic acid [13, 14]. (Table 11.3).

Table 11.2 Main products of enteral nutrition in Japan and their carnitine content

Classification	Product name	Carnitine content (mg/100 kcal)
Ingredients nutrition	Elental, Elental P	0
	Hepan ED	0
Digestive nutrition	Twinline NF	0
Semi-digestible nutrition	Ensure liquid, Ensure H	0
	Enevo	10.7
	Racol NF	0
	Enoras	16.7
	Aminoleban EN	0
Digestive nutrition	Entemeal R	0
	Peptino	0
	Peptamen intense	40.0
	Hinex E-gel	0
Semi-digestible nutrition	Isocal 1.0 Junior	20.0
	Tapion α	20.0
	Terumeal Uplead	12.5
	PG Soft (EJ, Ace, Ace MP)	3.75
	F2 shot (EJ, Light, Light 55, Light MP)	3.75
	Mermed (One, Plus)	4.0
	Glucerna REX	8.0
	Pulmocare EX	10.1
	Meiji Main	15.0
	Meiji Renalen MP	25.0
	Meiji MeiBalance RehaSupport Mini	15.0
	Prona	20.0
	Renargy U	12.7
	Hepas	5.0
	Sanet K2	20.0
	Isocal Glucopal TF	45.0

Modified from Refs. [8, 9]

Mixer meals are also very useful because they can be administered by boiling meat and other foods high in carnitine, and it is recommended that mixer meals be given even once a day [8]. However, it is very labor intensive to prepare mixer meals for individual patients in a facility with many patients. In short, it is impossible to provide long-term nutritional management with only one type of tube feedings, and nutritional management should be as varied as possible, just as we usually take a very wide variety of foods.

Table 11.3 Types of concentrated liquid foods and patient burden in Japan

		Pharmaceuticals	Foods (concentrated liquid food)
Law		Law for quality and efficacy of pharmaceuticals, medical devices, etc. and the act on ensuring safety	Food sanitation law, etc.
Manufacturing and marketing authorization		Acquisition of pharmaceutical manufacturing and marketing approval. It is used for the diagnosis, treatment, and prevention of diseases, and has been reviewed and approved for its efficacy and safety.	Unnecessary However, if you want to obtain a labeling permit as a comprehensive nutritional food for special purposes, It is necessary to obtain a labeling permission from the Consumer Affairs Agency.
Doctor's prescription		Necessity	Unnecessary
Insurance coverage		Yes	No
Patient burden	At the time of admission	Bear the legal burden rate as a drug cost	Self-payment for meals expenses at the time of hospitalization
	Outpatient and home-based	Bear the legal burden rate as a drug cost	Full self-payment
Treatment of medical fees		Pharmaceuticals	Foodstuff
Private purchases		Impossible	Possible

From Ref. [12]

Depending on the patient's condition and the degree of risk, it may be advisable to use different carnitine oral solution formulas or carnitine-containing enteral feedings, and to incorporate a mixer diet as much as possible.

11.3 Case Reports of Carnitine Deficiency in Severely Handicapped Children (Patients) Undergoing Tube Feeding

Ochi et al. measured serum carnitine concentration, blood glucose level, ammonia level, and other parameters to investigate secondary carnitine deficiency in severely mentally and physically disabled children (13 cases, 1 year 2 months to 9 years 4 months, median 4 years 8 months) undergoing tube feeding [15]. The patient background is shown below (Table 11.4). Serum carnitine concentrations of total carnitine and free carnitine were decreased in all 13 patients, indicating carnitine deficiency, and also found hypoglycemia in 5 patients, high triglyceride (TG) levels in 5 patients, abnormal free fatty acid (FFA) levels in 5 patients, abnormal *creatine kinase* (CK) levels in 5 patients, hyperammonemia in 9 patients, and abnormal findings on echocardiography in 3 patients (Table 11.5).

When carnitine (30 mg/kg/day) was administered to 6 patients (cases 1–4, 6, and 7), serum carnitine concentration increased, CK, TG, total cholesterol, and FFA remained unchanged, but blood glucose levels ranged from 78 to 108 mg/dL and in patients with high ammonia levels (cases 1, 4, and 6) normalized. However, case 1, diagnosed with acute heart failure due to dilated cardiomyopathy, showed no marked improvement in cardiac function after 1 month despite administration of selenium, carnitine, diuretics, dopamine, and milrinone.

Ohtaki et al. investigated serum total carnitine concentrations in 21 hypoalbuminemia (≤3.4 g/dL) severely disabled children (persons) undergoing tube feeding [16]. The results showed that 19 patients were below the reference value (45–91 μmol/L), and carnitine was significantly lower in the valproic acid group (10 patients, 14.5 ± 7.6 μmol/L) than in the nonmedicated group (11 patients, 35.7 ± 12.7 μmol/L, $P = 0.0002$).

Ozawa of the same group described that in carnitine deficiency in severely mentally and

Table 11.4 Patient background of carnitine deficiency of 13 infants and children with severe physical and mental disabilities undergoing long-term tube feeding

Cases	Age/gender	Underlying diseases	RDS	Height/ weight	Nutrient route	Nutrient	Duration	Antimicrobials	VPA
1	Infant/M	Autosomal anomalies, slippery syndrome, difficult to treat sexuality	39	105 cm/16 kg	ED	Elental P → Racol	11 m → 3 m	EM	–
2	Infant/F	Severe neonatal asphyxia, HIE, refractory epilepsy	42	92 cm/19 kg	Gastric fistula	Elental P	4y	EM	–
3	Infant/F	Very low birth weight infants and HIE	47	70 cm/11 kg	ED	Elental P	1 y 4 m	EM + ST	–
4	Pediatric/M	Congenital CMV infections, refractory epilepsy	32	104 cm/18 kg	ED	Elental P → Racol	5 y → 1 y	EM	+
5	Infant/F	13 trisomy/refractory epilepsy	27	78 cm/11 kg	Nasogastric tube	Follow-up milk → Racol	1 y → 1 y	No	+
6	Infant/F	Severe neonatal asphyxia, HIE, refractory epilepsy	35	73 cm/13 kg	Gastric fistula	Follow-up milk + Elental P	5 m	EM	+
7	Pediatric/F	HIE and refractory epilepsy	32	110 cm/22 kg	Nasogastric tube	Ensure liquid	6 y	No	+
8	Infant/F	Arnold–Chiari malformation type II	42	88 cm/13 kg	ED	Elental P	1 y → 3 m	EM + ST	–
9	Infant/M	HIE and refractory epilepsy	35	74 cm/11 kg	Gastric fistula	Elental P → Racol	5 m → 2 m	EM	+
10	Pediatric/M	Congenital anomaly syndrome, HIE	32	123 cm/26 kg	Nasogastric tube	Racol	7 y	CAM	–
11	Infant/F	Nagar syndrome	27	82 cm/11 kg	Nasogastric tube	Follow-up milk + liquid diet	3 y	No	–
12	Infant/M	Sequelae after head injury, intractable epilepsy	27	103 cm/15 kg	ED	Ensure liquid	1 m	No	+
13	Pediatric/F	Very low birth weight infants,cerebral palsy, and intractable epilepsy	27	125 cm/26 kg	Gastric fistula	Racol	9 y	No	+

CMV cytomegalovirus, HIE hypoxic-ischemic encephalopathy, RDS very critically ill infant score, ED duodenal tube, EM erythromycin, ST sulfamethoxazole-trimethoprim mixing drug, CAM clarithromycin, VPA valproic acid
From Ref. [15]

Table 11.5 Laboratory findings at the time of diagnosis of carnitine deficiency

Cases	STC	SFC	SAC	Glc	HbA1c	TG	T.cho	FFA	CK	NH_3	Echocardiography (EF/% FS/Tei)
1	28.3	25.1	3.2	33	3.6	81	78	0.13	478	185	54.7/23.2/0.36
2	16.3	13.8	2.5	16	4.4	372	245	0.35	28	45	72/34/0.51
3	9.8	7.8	2.0	31	3.7	338	189	0.43	69	47	69/32/0.40
4	9.2	7.0	2.2	87	3.7	123	148	0.65	33	140	69/32/0.53
5	36.1	25.3	10.8	69	4.0	243	148	0.30	281	79	80/42/ND
6	10.3	7.8	2.5	82	4.4	196	124	0.21	32	133	ND
7	5.2	4.3	0.9	102	4.8	47	138	1.23	134	82	66/35/0.41
8	37.1	31.0	6.1	111	4.4	109	222	0.09	132	45	80/42/0.22
9	17.7	14.2	3.5	84	4.5	191	153	0.26	575	67	ND
10	38.8	34.0	4.8	87	5.0	103	146	0.33	45	92	74/38/0.41
11	30.4	17.9	12.5	65	4.4	91	134	1.20	29	ND	ND
12	36.1	29.3	6.8	88	5.4	38	99	0.06	28	83	ND
13	36.8	25.3	11.5	103	5.0	89	183	1.00	103	221	ND

All patients had carnitine deficiency, and hypoglycemia (5 patients), high triglyceride levels (5 patients), abnormal FFA levels (5 patients), abnormal CK levels (5 patients), hyperammonemia (9 cases), and abnormal echocardiography findings (3 cases) were observed. Underlines indicate outliers. Normal values: STC (45.0–91.0 μmol/L), SFC (36.0–74.0 μmol/L), SAC (6.0–23.0 μmol/L), Glc (70–109 mg/dL), HbA1c (4.3–5.8%), TG (50–149 mg/dL), T.cho (90–229 mg/dL), FFA (0.10–0.81 mEq/L), CK (50–230 IU/ L), NH_3 (12–66 μg/dL), EF (55% or more), %FS (30–50%), Tei(0.36 ± 0.07).Note: The CK value of 69 IU/L in Case 3 was an outlier in the original report, but it was judged to be incorrect

STC Serum total carnitine, SFC serum free carnitine, SAC serum acylcarnitine, Glc blood glucose, HbA1c hemoglobin A1c, TG triglycerides, T.cho total cholesterol, FFA free fatty acids, CK creatine kinase, NH_3 ammonia, EF left ventricular ejection fraction, %FS left chamber diameter shortening, Tei Tei index, ND not determined

From Ref. [15]

physically disabled children (persons), carnitine preparations (30 mg/kg/day) were administered for 1 month to 15 patients with low carnitine levels, and when administration was discontinued because an increase in carnitine levels was observed, the levels again decreased to the same levels as before administration 5 months later [17]. The authors stated that further studies are needed to determine the dosage, how to continue administration, and how to supplement carnitine from food.

Ozawa et al. in the same group subsequently administered a low-dose carnitine preparation (L-Cartin® FF Oral Solution 10%, 2 mg/kg/day: 3 cases and 4 mg/kg/day: 9 cases) to 12 severely mentally and physically disabled children (persons) who were being managed with enteral nutrition without carnitine and compared serum-free carnitine concentrations and biochemical tests after 1 month. The serum-free carnitine concentration and biochemical tests were compared at 1 and 3 months [18]. The results showed that serum-free carnitine levels were 9.3 μmol/L before treatment, 31.4 μmol/L at 1 month, and 48.1 μmol/L at 3 months after treatment, which were significantly increased. There were no significant differences in serum-free carnitine concentrations between patients taking valproic acid and those not taking valproic acid. Among the two patients receiving antibiotics containing a pivoxil group, one patient received antibiotics for 1 week up to 2 days before the start of carnitine administration, and the serum-free carnitine concentration was extremely low at 2.39 μmol/L but increased markedly to 67 μmol/L after administration. In the other case, patients received a pivoxil-containing antibiotic for 5 days from 4 days before the 3rd month of carnitine administration, and the serum-free carnitine concentration at the 3rd month was halved from 51.74 μmol/L at the first month to 25.16 μmol/L. There was no significant change in biochemical tests after carnitine administration, probably because there were no patients with large abnormal values prior to administration, but in patients with high total ketone levels

(814 μmol/L) prior to administration, there was a decrease to 390 μmol/L after 1 month of carnitine administration and to 190 μmol/L after 3 months.

Wakisaka et al. investigated in detail the effects of enteral nutrition and antiepileptic drugs on serum carnitine concentrations in 78 children (persons) with severe mental and physical disabilities [19, 20]. The patients were divided into three groups according to carnitine intake. Group A: enteral feedings without carnitine (36 cases), group B: enteral feedings with carnitine (8 cases), and group C: oral feedings (34 cases). The mean serum total carnitine concentration (reference value 46–91 μmol/L) was significantly decreased to 16.8 ± 8.9 μmol/L in all patients in group A and 38.2 ± 12.1 μmol/L in 24 patients (about 70%) in group C, but 64.4 ± 23.5 μmol/L in group B with only one patient having a mild decrease (Fig. 11.1a). Free carnitine levels (reference value 36–74 μmol/L) were similar, with 13.9 ± 7.5 μmol/L in group A, 48.2 ± 16.0 μmol/L in group B, and 32.1 ± 10.7 μmol/L in group C (Fig. 11.1b).

Next, a comparison of abnormal laboratory findings in the low-carnitine group (free carnitine concentrations <20 μmol/L, 42 patients) and the non-low-carnitine group (36 patients) showed a trend toward more cases of hypoglycemia and hyperammonemia in the low-carnitine group (Table 11.6). A negative correlation between blood ammonia concentration and serum-free carnitine concentration was also observed in those taking valproic acid (32 cases, $r = -0.495$).

Oral antiepileptic drugs are a risk factor for low serum carnitine levels. In group A (enteral feedings without carnitine), free carnitine levels were compared with oral administration of valproate and phenobarbital (PB) and without oral antiepileptic drugs administration (Fig. 11.2). The patients taking valproate and PB had significantly lower blood carnitine levels than those in the monotherapy or no medication groups (Fig. 11.2).

In group C (oral food intake), those taking multiple antiepileptic drugs also had significantly lower free carnitine levels than those taking single agents or no antiepileptic drugs (Fig. 11.3a). In particular, those taking valproate and PB had significantly lower serum-free carnitine than those taking valproate alone, PB alone, or no antiepileptic drugs (Fig. 11.3).

Based on these results, Wakisaka et al. identified the following risk factors for carnitine deficiency in patients with severe mental and physical disabilities: Use of enteral nutritional supple-

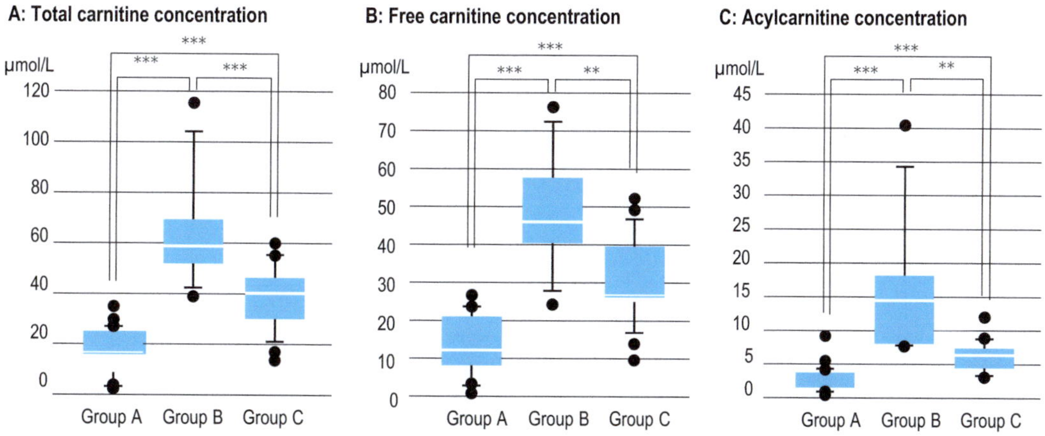

Fig. 11.1 Serum total carnitine concentration (**a**), serum-free carnitine concentration (**b**), and serum acylcarnitine concentrations (**c**) in enteral nutrition users without carnitine (Group A: 36 cases), carnitine-supplemented enteral nutrition users(Group B: 8 cases), and oral foods users(Group C: 34 cases). (**a**) Total carnitine concentration (normal range: 46–91 μmol/L), Group A: 16.8 ± 8.9, Group B: 64.2 ± 23.5 μmol/L), Group C: 38.2 ± 12.1 μmol/L. (**b**) Free carnitine concentration (normal range: 36–71 μmol/L), Group A: 13.9 ± 7.5, Group B: 48.2 ± 15.9 μmol/L), Group C: 32.1 ± 10.7 μmol/L. (**c**) Acylcarnitine concentration (normal range: 6–23 μmol/L), Group A: 2.9 ± 1.6, Group B: 16.2 ± 11.0 μmol/L), Group C: 6.1 ± 1.9 μmol/L. **: *P* 0.01, ***: *P* 0.001. [20])

Table 11.6 Abnormal laboratory findings due to hypocarnitinemia

Laboratory findings	Hypocarnitinemia group (42 cases)	Non-Hypocarnitinemia group (36 cases)
Hypoglycemia	7 case (Fasting blood glucose 61–68 mg/dL)	0 case
Low HbA1c	4 cases (HbA1c 4–4.1%)	5 cases
Impaired liver function	4 cases (AST 46–62， ALT 56–74 U/L)	6 cases
Hyperammonemia	23 cases (68–216 mg/dL)	9 cases
Decreased cardiac function(EF% 55 or less)	0 case	ND

Serum-free carnitine concentrations of <20 μmol/L were used in the hypocarnitinemia group (42 patients) and ≥20 μmol/L were used in the non-hypocarnitinemia group (36 patients), and the number of abnormal laboratory findings was compared in both groups

ND not determined

From Ref. [20]

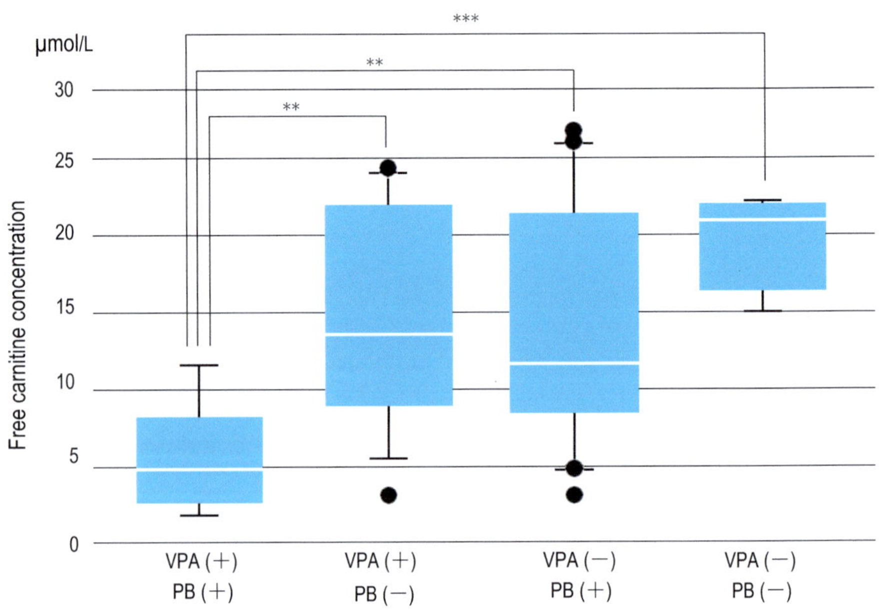

Fig. 11.2 Comparison of free carnitine concentration by valproic acid (VPA) and phenobarbital (PB) administration in enteral nutrition users without carnitine (Group A). **: $P < 0.01$, ***: $P < 0.001$. (From Ref. [20])

ments that do not contain carnitine, consumption of hospital diets with low carnitine content, and patients taking multiple antiepileptic drugs, mainly valproic acid.

Kato also divided 45 children (and persons) with severe mental and physical disabilities who were taking valproic acid into 31 children who fed normal food and 14 children who were tube fed (14 cases, including 10 Racol®, 3 patients who were Liferon®, and 1 patient who were subjected to Termir®), and also compared the blood-free carnitine concentrations of children who were tube fed without valproate (7 patients) [21]. As a result, the blood-free carnitine concentration was below normal (36–74 μmol/L) in 15 out of 31 children and 13 out of 14 children on enteral nutrition taking valproic acid, and the blood-free carnitine concentration was 36.1 ± 12.0 μmol/L for the former and 14.6 ± 10.0 μmol/L for the latter. It was significantly lower in enterally fed

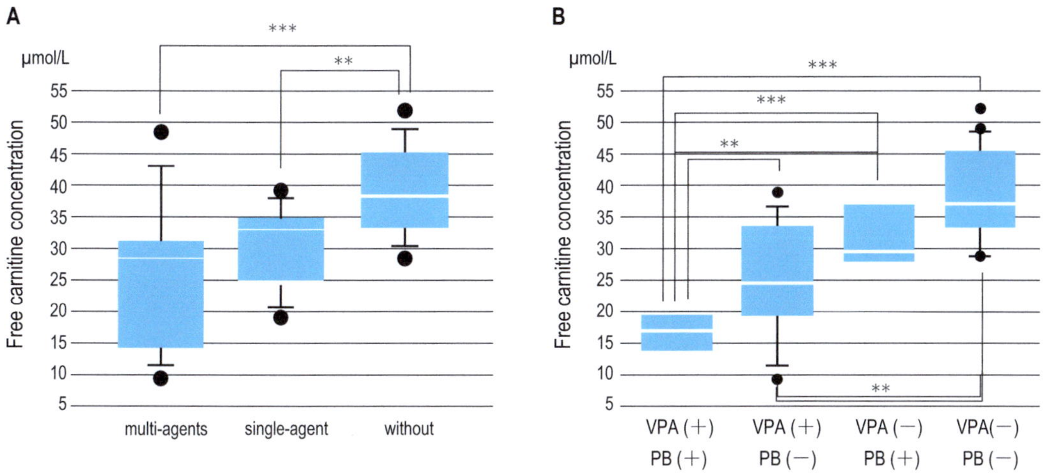

Fig. 11.3 Comparison of serum-free carnitine concentration by the species of anti-epileptics in oral foods users(Group C). (**a**) Comparison by multi-agents, single agent and without anti-epileptics. (**b**) Comparison by with or without valproic acid (VPA) and phenobarbital (PB). **: $P < 0.01$, ***: $P < 0.001$. (From Ref. [20])

children taking valproate ($P < 0.001$). In seven enteral nutrition patients who did not receive valproic acid, the blood-free carnitine concentration was 31.0 ± 9.8 μmol/L, which was only a slight decrease. In addition, among the normal food patients taking valproic acid, the concentration of free carnitine in the blood of the valproate group was significantly lower in the multidrug group compared to valproic alone group ($P < 0.05$).

Takeda et al. administered levocarnitine (30 mg/kg/day, for 2 weeks) to 26 children (persons) with severe mental and physical disabilities, and continued administration for another 10 months to 20 patients who were at high risk for decline if carnitine administration was discontinued and compared biochemical laboratory values before and after administration [22]. The results showed that blood carnitine levels were low before treatment but normalized after 2 weeks of treatment (median and range of serum-free carnitine levels (28.6 [5.1–50.7] before treatment, 50.0 [37.4–111.5] after treatment; μmol/L), uric acid and N-Terminal pro-Brain Natriutetic Peptide (NT-proBNP) improved significantly and ammonia levels decreased in 8 of the 9 patients with elevated levels before treatment (not significant difference).

Patients with neurological intractable diseases such as muscular dystrophy and amyotrophic lateral sclerosis (ALS) are also often managed by tube feeding. Ueno et al. examined serum-free carnitine concentrations in 29 patients with muscular dystrophy, ALS, and other neurological intractable diseases [23]. Table 11.7 shows the patient background of the included patients.

Serum-free carnitine concentrations were lower than the standard values in the muscular dystrophy group (16 patients) < the ALS group (10 patients) < the neurological disease group, in that order (Table 11.8), and the mean values in all three groups were significantly lower than those in the oral food intake group (50.4 ± 3.8 μmol/L). By case, 14 out of 17 patients with muscular dystrophy, 9 out of 12 patients with ALS, and 6 out of 8 patients with neurological disorders were below the standard value, but none of the patients in the oral food intake group (7 patients) were below the standard value. In the muscular dystrophy and ALS groups, 3 of the 5 cases in which serum-free carnitine concentrations were within the reference value were used with enteral nutrition with carnitine (carnitine amount: 10–62.5 mg/day). Excluding these three cases, serum-free carnitine concentrations were 21.0 ± 10.9 μmol/L in the muscular dystrophy group and 24.4 ± 12.8 μmol/L in the ALS group. A correlation analysis between the duration of tube feeding and serum-free carnitine concentrations in these patients showed a negative correlation with

Table 11.7 Background of 29 patients with intractable neurological diseases such as muscular dystrophy and ALS

	Tube feeding			Oral intake
	With muscle atrophy		Without muscle atrophy	With muscle atrophy
	Muscular dystrophy group	ALS group	Neurological disorders group	Oral intake group
Number of cases	17 cases (12 males, 5 females)	12 cases (9 males, 3 females)	8 cases (6 males, 2 females)	7 cases (4 males and 3 females)
Diseases	Muscular dystrophy (Duchenne type, Fukuyama type, oculopharyngeal type, myotonic type)	ALS	Parkinson's disease, spinocerebellar degeneration, basal cortical degeneration, progressive supranuclear paralysis	Muscular dystrophy (Duchenne type, myotonic type, Becker type)
Age (years)	42.7 ± 15.1	62.4 ± 12.8	61.9 ± 18.2	62.6 ± 3.0

ALS amyotrophic lateral sclerosis
From Ref. [23]

Table 11.8 Duration of tube feeding and serum-free carnitine concentrations in 29 patients with intractable neurological diseases such as muscular dystrophy and ALS

	With muscle atrophy		No muscle atrophy
	Muscular dystrophy group	ALS group	Neurological disorders group
Number of cases	16 cases (11 males and 5 females)	10 cases (7 males, 3 females)	8 cases (6 males, 2 females)
Duration of tube feeding (months)	44.9 ± 34.0	30.3 ± 23.4	67.0 ± 72.5
Serum-free carnitine concentration (μmol/L)	21.0 ± 10.9	24.4 ± 12.8	30.9 ± 9.3

Normal serum free carnitine concentration: 36–74 μmol/L
From Ref. [23]

the muscular dystrophy group ($r = -0.7049$, $P = 0.0006$) and the ALS group ($r = -0.8333$, $P = 0.0275$). In particular, it tended to decline faster in the ALS group. No correlation was observed in the neuromuscular diseases group ($r = -0.2156$, $P = 0.5684$). Seven patients in the muscular dystrophy group who consented were administered 1500 mg/day with carnitine, but there was no significant improvement in left ventricular ejection fraction after 12 months, but the frequency of bowel movements tended to improve.

Muguruma et al. examined serum carnitine concentrations and trace elements such as zinc and selenium in 24 patients with intractable neurological diseases (10 patients with Parkinson's disease, 8 ALS, 2 patients with spinocerebellar degeneration, 2 patients with progressive supranuclear palsy, and 1 each with multiple system atrophy and multiple sclerosis) undergoing long-term tube feeding [24]. As a result, serum concentrations of zinc, selenium, and free carnitine decreased in 17/24 patients (70.8%), 14/24 patients (58.3%), and 21/24 patients (87.5%), respectively. The carnitine content of these enteral nutritional supplements treated as pharmaceuticals (Ensure Liquid®, Elental®, Racol®) and treated as food (The carnitine content of CZ-Hi®, CZ-Hi1.5®, Termir Mini®, Medief Push Care®, Medief Soy Bag®, Calm Solid®, E-7II®), which are commonly used in these patients with neurological disorders, were below the measurement limit. Paroxysmal tachycardia and frequent ventricular extrasystoles were observed in 4 (19%) of 21 patients with low serum-free carnitine, and elevated BNP was observed in 9 (42.9%) patients. Echocardiography showed no obvious left ventricular dysfunction, but 4 patients (19%) had diastolic dysfunction, and chest X-ray and computed tomography (CT) scans showed cardiac enlargement in 4 patients (19%). In addition, 15 patients (71.4%) had decreased natural killer cell activity. In patients with neurological intractable disease undergoing long-term tube feeding, periodic testing of trace element and carnitine levels should be performed, and enteral nutritional supplements should be re-evaluated, and carnitine formulation should be administered.

Death due to respiratory failure was common in patients with muscular dystrophy, but recent advances in medical technology, including advances in ventilatory management and efforts toward standardization, have greatly improved patient prognosis [25–27]. On the other hand, cardiomyopathy and heart failure are now the most common causes of death [28–31]. Although the pathogenesis of cardiomyopathy and heart failure in patients with muscular dystrophy has not yet been fully elucidated, carnitine deficiency and carnitine insufficiency are now known to occur in general heart failure (see "Chap. 16: Cardiovascular Diseases and Carnitine Deficiency"). Further studies are needed in patients with muscular dystrophy and other disorders from this perspective.

Le Borgne et al. found that carnitine content, carnitine uptake activity, and lipid metabolism in muscle cells of muscular dystrophy patients were reduced compared to controls and were restored by the addition of L-carnitine at the cellular level, suggesting that carnitine supplementation may also have positive effects in patients [32]. Watanabe provides an easy-to-understand explanation of carnitine deficiency in neuromuscular diseases, including precautions and dosage and administration, based on the role of carnitine in muscles and nerves [33]. Patients with neuromuscular disease who are managed with tube feeding are at high risk for carnitine deficiency due to the addition of carnitine-free tube feedings and decreased skeletal muscle mass, which is a storage site for carnitine.

11.4 Carnitine Deficiency in Intravenous Nutrition

Penn et al. examined serum carnitine levels in 12 low-birthweight infants receiving nutritional management with TPN (group A) and 8 infants who were fed with milk containing carnitine of approximately the same gestational age and weight (group B) and found that serum carnitine in group A was significantly lower than in group B at 5 days after birth [34]. After that, when the infants were switched to milk or breast milk containing carnitine, their serum carnitine levels recovered to the same levels as those in group B. In addition, when the carnitine content of tissues of infants with low birth weight who died after receiving TPN nutritional management was examined, it was reported that the total carnitine content of liver and heart tissues was significantly lower in low birth weight infants who were on TPN for more than 15 days after birth, compared to infants who were not on TPN [35] (Table 11.1). As a result of these reports, Borum et al. warned that there is a risk of carnitine deficiency because it does not contain carnitine at all, especially when TPN is used for postoperative nutrition management of newborns [36, 37].

Tibboel et al. reported a case in which a neonate born at 37.5 weeks gestational age, undergoing surgery for meconium peritonitis, and receiving nutritional management by TPN, presented with hypotonia, hyperbilirubinemia, and heart failure, and the symptoms disappeared after starting carnitine supplementation (100 mg/kg/day) after 4 weeks [38]. They further investigated plasma carnitine levels and symptoms possibly caused by carnitine deficiency in neonates (13 cases) who underwent surgery. The mean plasma-free carnitine concentration was extremely low (7.9 ± 2.2 μmol/L), three patients had hypotonia and two had cardiac hypertrophy, and eight patients were treated with carnitine supplementation therapy at 50–100 mg/kg/day.

Schmidt-Sommerfeld et al. reviewed these studies and found that although evidence is accumulating that carnitine is an essential nutrient in neonates, further study is needed to determine whether carnitine supplementation should be routinely given to infants who are nutritionally managed on TPN [39].

In Japan, there have been reports of cases in which hypoglycemic symptoms appeared after prolonged nutritional management with TPN due to surgery for intestinal necrosis in very low birth weight infants, and of mild hepatic dysfunction that appeared to be caused by carnitine deficiency in a boy (8 months old) with short bowel syndrome after massive bowel resection due to midgut axis distortion [40, 41]. In addition, Hirose et al. reported a case of carnitine deficiency in a 3-year-old boy with short bowel syndrome who had been on TPN for a long period of time, and his blood carnitine level was low, and although oral doses of 60, 90, and 120 mg/kg/day were tried, the level did not reach normal levels, probably due to his short intestinal tract, especially when he developed sepsis. They also reported that the intravenous formulation was not available at that time and that early approval of an intravenous formulation would be desirable [42].

Okanari et al. also reported a case of a 1-year and 11-month-old girl with brain damage who died of heart failure 9 days after vaccination due to measles vaccine-related febrile disease [43]. Serum total carnitine was 7.4 μmol/L, free carnitine was extremely low at 5.6 μmol/L, and the patient had nonketotic hypoglycemia (blood glucose level 13 mg/dL). The patient had difficulty eating orally and had been on intravenous nutrition for a long period of time, beginning in early infancy. Although the patient was not taking valproic acid, it was thought that the patient was suffering from secondary carnitine deficiency, the fever induced by the vaccine caused cardiac failure related to nonketotic hypoglycemia.

What about carnitine deficiency in patients nutritionally controlled with TPN in adult patients? Hahn et al. examined blood total carnitine levels in 47 adult patients undergoing surgery for cancer or Crohn's disease and being managed with TPN [44]. Unlike in pediatric patients, total carnitine levels remained constant without significant change for up to about 15 days, then gradually decreased, averaging 32.6 ± 5.46% after 40 days, but began to rise again within a few days after oral intake was initiated.

In response to Hahn's report, Nanni et al. discussed in a "Letter to the Editor" that blood carnitine levels in postoperative patients managed with TPN do not decrease for about 20 days in their study, but that urinary free carnitine excretion is increased in patients with invasive procedures or disease such as surgery or sepsis, and the degree of tissue depletion is not determined from blood levels alone. In the "Letter to the Editor," they discussed that these patients have severe abnormal energy metabolism, and that carnitine supplementation is recommended [45].

Berner et al. reported plasma carnitine concentrations in 17 adult patients under long-term (12–196 months, mean 69 ± 11 months) nutritional management with TPN due to malabsorption disorders caused by enteritis, short bowel syndrome, or inflammatory bowel disease due to radiation therapy for cancer [46]. In 9 patients, lysine and methionine, precursors of carnitine biosynthesis, were measured. There was no significant difference in the methionine concentration; however, the lysine concentration was reduced to 58% of the control level. Plasma carnitine concentration showed a positive correlation with serum albumin ($r = 0.62$, $P < 0.05$) and a negative correlation with serum alkaline phosphatase ($r = -0.64$, $P < 0.05$).

11.5 Hepatobiliary Dysfunction Due to TPN

Hepatic dysfunction is also a problem in nutritional management with TPN and is divided into neonatal/infantile type with biliary stasis as the main pathology and adult type with fatty liver, which is thought to be caused by various factors, including secondary carnitine deficiency [47–50].

Worthley et al. found that a patient (41 years old, male) undergoing long-term TPN nutrition management complained of continuous elevation of bilirubin, hypotonia, reactive hypoglycemia, etc., and that the blood carnitine concentration was decreased. Therefore, carnitine was administered (intravenous, 400 mg/day for 7 days, then 40 mg/day for 3 weeks, later 60 mg/day), and then bilirubin and blood glucose were normalized and reactive hypoglycemia also disappeared [51]. The effect of carnitine supplementation was subsequently reported in a patient who complained of fatigue and asthenia due to long-term TPN [52].

Palombo et al. also reported a case of a patient (37-year-old female) undergoing TPN at home with elevated bilirubin, AST, ALT, and AP, who was hospitalized for catheter sepsis and was given carnitine (1 g/day, intravenous, for 12 days), which significantly improved her bilirubin level and liver function enzymes [53].

In an article on surgical and nutritional management of patients who undergo multiple surgeries due to cancer recurrence, Ottery also described a case in which a 33-year-old female patient with rectal cancer underwent a total pelvic exenteration due to pelvic metastasis after low anterior rectal resection surgery, followed by further bowel resection and other operations, which resulted in short bowel syndrome and TPN management [54]. Approximately 6 months after the last surgery, the patient presented with progressive jaundice, abnormal liver function, hypertriglyceridemia, and progressive muscle fatigue. The blood carnitine level was found to be low (free carnitine 11.8 μmol/L), and after the addition of intravenous carnitine supplementation (1 g/day), liver function abnormalities and hypertriglyceridemia normalized within 2 weeks, and muscle fatigue resolved. This is a successful case in which the diagnosis of carnitine deficiency and treatment before irreversible liver damage occur.

Forchielli et al. reported the case of a patient (41-year-old male) with short bowel syndrome due to surgery for midgut axis torsion since the age of 15, who became home TPN management (HPN) and has lived a normal life for 26 years [55]. The patient was hospitalized 8 times during the first decade due to catheter-related sepsis, etc., and was treated with vitamin E and iron supplementation; during the next decade, abnormal liver function (normal bilirubin) occurred, and carnitine supplementation was started (100 mg/day, then continued) and ursodeoxycholic acid (oral, 300 mg, 3 times/day) was administered, and the patient was able to live on HPN nutritional management for as long as 26 years.

Thus, it is known that cholestasis, fatty liver, and steatohepatitis occur frequently in patients who receive long-term nutritional management with TPN, and carnitine deficiency is considered to be one of the factors. Bowyer et al. examined plasma carnitine concentrations in 37 patients receiving long-term HPN nutrition and found that total carnitine and free carnitine concentrations were reduced in 13 patients [56]. Plasma carnitine concentrations were compared in 15 of the 37 patients and in 15 age- and sex-matched patients with Crohn's disease who did not require HPN. Total carnitine and free carnitine were lower in the former group than in the patients who did not require HPN (total carnitine: 32.2 ± 11.9 μmol/L, free carnitine: 28.4 ± 10.8 μmol/L in the former group; total carnitine: 49.1 ± 10.9 μmol/L, free carnitine: 46.4 ± 11.5 μmol/L in the latter group, $P < 0.001$ for both). However, there was no correlation between carnitine concentration and liver function abnormalities such as bilirubin level and AST. Furthermore, Bowyer et al. treated HPN patients (female, 4 patients, average 53 months of HPN) with carnitine (1 g/day, intravenous administration, 1 month) who developed hepatic dysfunction and fatty liver with low carnitine level, but despite increased carnitine concentrations in blood and liver tissue, there was no improvement in hepatic dysfunction or other abnormalities. Therefore, they concluded that carnitine deficiency is not the cause of abnormal liver function or fatty liver in HPN patients [57]. However, since the number of patients was only 4 and the duration of carnitine administration was only 1 month, it cannot be simply concluded that carnitine deficiency is not the cause of hepatic dysfunction in TPN.

11.6 Celiac Disease and Cardiomyopathy

Curione et al. reported that among patients with idiopathic dilated cardiomyopathy (52 patients), there were three patients with celiac disease, an autoimmune disease triggered by an immune response to gluten, a protein found in wheat, barley, and rye [58]. Conversely, an increasing number of patients with idiopathic dilated cardiomyopathy were found among those with celiac disease [59]. Therefore, Curione et al.

compared the total serum carnitine concentrations of patients with idiopathic dilated cardiomyopathy (Group A, 10 cases), patients with idiopathic dilated cardiomyopathy + celiac disease (Group B, 3 cases), and healthy individuals (Group C, 10 cases), and reported that total carnitine concentrations in group A were 26.8 ± 7.35 μmol/L, in group B 16.7 ± 2.86 μmol/L, in group C, significant decrease in group B [60]. When the patients in group B were given a gluten-free diet for 2 years, the total serum carnitine concentration increased to 27.2 ± 1.73 μmol/L, which was equivalent to that of group A (27.7 ± 7.89 μmol/L). In addition, three patients with celiac disease with idiopathic dilated cardiomyopathy were treated with a gluten-free diet, and two patients strictly switched to a gluten-free diet, and after 28 months, echocardiographic parameters and QOL improved, but one patient did not adhere to a gluten-free diet and echocardiography parameters deteriorated [61].

On the other hand, Bene et al. measured the plasma carnitine profile of 33 adult celiac disease patients who had been on a gluten-free diet for a long period of time by tandem mass analysis and compared it with healthy controls and found that the free carnitine concentration did not change, but acetylcarnitine was 45% of the control level, propionylcarnitine 61.5%, and butyrylcarnitine 56.9%, and each acylcarnitine from short chain to long chain was low ($P < 0.01$ for all) [62]. These celiac disease patients did not develop cardiomyopathy, but it was thought that an abnormality in the metabolism of acylcarnitine was occurring.

Uslu et al. found that a 6-year-old girl was diagnosed with celiac disease and switched to a gluten-free diet, and although weight gain (24.2%) was observed after 2 months, she exhibited symptoms of dilated cardiomyopathy such as dyspnea, tachycardia, cyanosis, ventricular dilation, and decreased cardiac ejection fraction (46%). After carnitine administration (dose unknown), cardiac contractile function was normalized after 7 months [63].

Patients with celiac disease are at risk of carnitine deficiency due to short bowel syndrome and TPN caused by intestinal resection in severe cases, but even nonsurgical patients may develop carnitine deficiency due to absorption disorders caused by intestinal mucosal damage and nutritional imbalance caused by gluten-free diet, and regular monitoring of blood carnitine levels may be necessary. Carnitine supplementation may be worthwhile in patients with low blood carnitine level with cardiomyopathy.

11.7 Carnitine Deficiency Caused by Specific Milk

Specific milks include a wide variety of products used in the treatment of metabolic diseases, milk allergen-removing formula for children suffering from milk allergies, lactose-degraded milk for lactose intolerance, and middle-chain triglyceride (MCT) milk for fat malabsorption, and some products are treated as pharmaceuticals in Japan (Table 11.9).

Many of these specific milks do not contain micronutrients such as carnitine, biotin, and selenium, and many cases of carnitine deficiency and biotin deficiency due to these specific milks have been reported [7, 64–66].

The "Guidelines for the Diagnosis and Treatment of Carnitine Deficiency 2018" (hereinafter referred to as "Guidelines for Diagnosis and Treatment") introduces cases of carnitine deficiency caused by mainly specific milks and milk for allergies, which have been reported in papers and academic mmeetings [67]. These case reports of carnitine deficiency caused by milk for allergies and specific milk are summarized in Table 11.10 [65, 68–75].

However, since there have been many case reports of deficiency due to the lack of biotin, carnitine, and selenium in specific milk, biotin and carnitine have been added to many products as of July 2021 due to the efforts of dairy companies

Table 11.9 Classification of specific milk in Japan

Classification	Pharmaceuticals	Registered items	Non-registered items	Commercial items
Contents	Prescription drugs (use requires a doctor's prescription)	Items whose quality, ingredients, and usage methods have been examined by the Joint Safety Development Committee for Special Milk based on certain standards	It is manufactured at the expense of dairy companies. Items whose quality, ingredients, and usage methods have been examined based on certain standards	It is sold by dairy companies
Expenses	Health insurance coverage. "Pediatric Chronic Disease Treatment Research Project" covers part of medical expenses (under 20 years old)	Free of charge at public expense or at the expense of the dairy company (under 20 years old)	Free of charge at the expense of the dairy company	Pay
How to get started	The doctor instructs the pharmacy with a prescription (you can't buy it at the pharmacy without a prescription)	The doctor addresses the Special Milk Secretariat Request and receive approval by e-mail using the "Application for Special Milk Supply"	The doctor addresses the Special Milk Secretariat Request by e-mail the "Special Milk Supply Application Form"	Inquire with the branch of each dairy company (some pharmacies may be located)
Applicable conditions	Use for indications	(1) It is an inherited metabolic(2) Be less than 20 years of age	As a rule, it is an inherited metabolic disorders. Please contact each dairy company	Use for indications

From Specific Milk Office, Boshi-Aiiku-Kai, Social Welfare Corporation, Imperial Gift Foundation: http://www.boshi-aiikukai.jp/milk02_01.html, and modified

(gray area in Table 11.11). However, many of them have not yet been added selenium. In addition, biotin and carnitine have not yet been added to other Snow Brand Megmilk products that use the same production line as the pharmaceutical product, and this situation is likely to continue for the time being. It is necessary for doctors and nutritionists to pay attention to the deficiency of biotin and carnitine when using these products. The Specific Milk Office has published "Precautions for the Use of Specific Milk for the Treatment of Inherited Metabolic Diseases" (http://www.boshiaiikukai.jp/img/milk/R03/20210714chuui.pdf, Specific Milk Office, Boshi-Aiiku-Kai, Social Welfare Corporation, Imperial Gift Foundation, in Japanese).

Table 11.10 Case reports of carnitine deficiency caused by Specific milks such as milk for allergies in Japan

Cases	Gender	Underlying diseases	Clinical manifestations, etc.	Nutritional supplements used	Duration of use	Blood Carnitine (μmol/L)	Supplemented carnitine (mg/kg/day)	Reporter
0 year(Day 8)	Unknown	At the 32nd week of fatal distress, the emperor cut the low birth weight	Changed to allergic milk due to milk allergy and frequently vomiting, hyperpnea, retractive breathing, anemia, enlarged liver, inguinal hernia, atopic dermatitis, hair loss, conjunctival hyperemia, lip fissure, genital erosion, hypotonia, high lactate/pyruvate	Artificial Milk⇒Milk protein degradation adjustment powder (706A) ⇒ Amino acid preparation powder(605Z)	3 months	0	100 (500 mg/day) Biotin 10 mg/day	Abe et al. (1991)(Ref. [68])
3 months old	Male	Very low birth weight infants	Hepatic dysfunction and hyperCKemia	MCT milk	2 months	13.8	80	Goto et al. (1992) (Ref. [69]) Ishida et al. (1994) (Ref. [70])
4–17 months old(average 8.0 months old)	8 cases(unknown)	Cow's milk allergy	Refractory eczema, hair loss, poor weight gain, developmental delay (secondary multiple carboxylase deficiency due to biotin deficiency)	Special milk for allergies	3–13 カ month(average 6.4 month)	Tandem mass analyzed 6 cases, all deficient	Along with biotin, carnitine supplementation relieves symptoms (dosage unknown)	Mushimoto et al. (2008) (Ref. [71])
3 months old	Male	Cow's milk allergy	Hyperpnea	Meiji 605Z	3 months	0	With biotin supplementation (dosage of carnitine unknown)	Takayanagi (2009)(Ref. [65])
8 months old	Male	Treated as refractory atopic dermatitis	Eyelids, lips, anal refractory dermatitis, hair loss	Milfy HP, New MA-1, Elemental formula	8 months	8.8 (C5OH high: 3.83)	Carnitine approx. 39(300 mg/day), improved with biotin 1 mg/day supplementation	Mushimoto et al. (2010) (Ref. [72])

8 months old	Male	Cow's milk allergy	Persistent loose stools, diarrhea, vomiting, hypoglycemia, general weakness, impaired consciousness, metabolic acidosis, high lactate level, brown hair change	Milfy HP	7 months	4.62 (C5OH high)	200 mg per day,Biotin 1 mg/day(later 0.5 mg), with ordinary milk symptom improvement	Komatsu et al. (2012) (Ref. [73])
1 year, 0 months	Female	Very low birth weight infant/chronic lung disease, neonatal - infant gastrointestinal allergy	Hepatic dysfunction, hyperammonemia, hyperlactatemia, hair loss	Milfy HP	6 months	13.8	200 mg/day, 1 mg/day of biotin, 25 μg of selenium symptom improvement	Nozaki et al. (2014)(Ref. [74])
Gestation 205.8 ± 27.1 days	Male 3, female 3	6 cases of premature infants diagnosed with cow's milk allergy	After being diagnosed with milk allergy due to bile vomiting, bloody stools, stunted growth, etc., switched to allergy milk, and developed erythema with clear boundaries, hair loss, and hypopigmentation	Amino acid formula, hydration solution formula(product name unknown)	11–76 days	5 cases <20 μmol/L,1 case, normal range	Biotin (0.5–5 mg/day) Carnitine (15–99 mg/kg/day)	Hayashi ら (2014)(Ref. [75])

Modified from Ref. [67]

Table 11.11 Micronutrient contents of special formulae in Japan (as of August 2024)

Company name	Code	Product name	Biotin (μg)		Carnitine (mg))		Selenium (μg))		Iodine (μg))	
			per 100 g	per 100 kcal	per 100 g	per 100 kcal	per 100 g	per 100 kcal	per 100 g	per 100 kcal
Meiji	110	Galactose-free formula	11.6	2.4	9.6	2.0	13.5	2.8	50	10.4
Meiji	GSD-D	Lactose and fructose-free low-fat formula (milk protein, for daytime) for GSD	10.3	2.4	8.6	2.0	12.0	2.8	30	7.0
Meiji	GSD-N	Lactose and fructose-free fat-free formula (milk protein, for night) for GSD	9.2	2.4	7.7	2.0	10.8	2.8	40	10.4
Meiji	8007	Lactose and fructose-free low-fat formula (soy protein, for lunch) for GSD	10.3	2.4	8.6	2.0	12.0	2.8	40	9.3
Meiji	8009	Lactose and fructose-free fat-free formula (soy protein, for night) for GSD	9.2	2.4	7.7	2.0	10.8	2.8	60	15.6
Meiji	7925-A	Low protein & arginine enriched formula	14.0	2.4	11.6	2.0	16.2	2.8	60	10.4
Meiji	8003	Leucine-free formula	11.0	2.4	9.4	2.0	13.1	2.8	30	6.4
Meiji	721	Essential fatty acid enriched MCT formula	12.1	2.4	10.1	2.0	14.1	2.8	40	7.9
Meiji	817-B	Ketone formulas	17.8	2.4	14.8	2.0	20.7	2.8	100	13.5
Snow Brand	A-1	Phenylalanine-free powder of amino acid mixture	0	0	0	0	2	0.5	0	0
Snow Brand	S-26	Methionine removal milk formula	0	0	0	0	1	0.2	32	7.0
Snow Brand	S-1	Phenylalanine and tyrosine removal milk formula	0	0	0	0	1	0.2	33	7.2
Snow Brand	S-23	Protein removal milk formula	0	0	0	0	0	0	41	8.5
Snow Brand	S-22	Isoleucine, valine, methionine, threonine, glycine removal milk formula	0	0	0	0	1	0.2	33	7.2
Snow Brand	S-30	Lysine and tryptophan removal milk formula	0	0	0	0	1	0.2	33	7.2
Morinaga	MP-11	Low phenylalanine peptide powder	<0.1	<0.03	1.1	0.3	15.1	4.6	150	46
Morinaga	MM-5	Low phosphorus formula	15	2.9	12	2.3	6.7	1.3	<10	<2
Morinaga	ML-3	Protein hydrolysis MCT formula	15	3.2	12	2.6	7.9	1.7	<10	<3

Non-registered milk supplied by provision business of special formula

Company name	Code	Product name	Biotin (μg)		Carnitine (mg))		Selenium (μg))		Iodine (μg))	
			per 100 g	per 100 kcal	per 100 g	per 100 kcal	per 100 g	per 100 kcal	per 100 g	per 100 kcal
Meiji	8103	Low protein, essential amino acid enriched, arginine removal formula	12.0	2.4	10.0	2.0	14.0	2.8	90	18.0
Meiji	502	Medium protein and low sodium formula	11.5	2.4	9.6	2.0	13.4	2.8	90	18.8
Meiji	8806H	Low potassium and medium phosphorus formula	11.3	2.4	9.4	2.0	13.2	2.8	80	17.0
Meiji	603	Sugar-free MCT formula	15.0	2.4	12.5	2.0	17.6	2.8	180	28.7
Meiji	817-B	Ketone formulas	17.8	2.4	14.8	2.0	20.7	2.8	100	13.5
Morinaga	MM-2	Low potassium milk	15	2.9	12	2.3	6.4	1.2	<10	<2
Morinaga	MM-4	Low calcium formula	15	2.9	12	2.3	6.9	1.3	<10	<2
Morinaga	MM-5	Low phosphorus formula	15	2.9	12	2.3	6.7	1.3	<10	<2
Morinaga	ML-1	Low fat formula	15	4.0	12	3.2	9.3	2.5	<10	<3

Special formula covered by health insurance

Company name	Product name	Biotin (μg)		Carnitine (mg))		Selenium (μg))		Iodine (μg))	
		per 100 g	per 100 kcal	per 100 g	per 100 kcal	per 100 g	per 100 kcal	per 100 g	per 100 kcal
Snow Brand	Phenylalanine-free milk blend	0	0	0	0	1	0.2	32	7.0
Snow Brand	Leucine, isoleucine and valine-free milk blend	0	0	0	0	1	0.2	33	7.2

Commercially available special formula

Company name	Product name	Biotin (μg)		Carnitine (mg))		Selenium (μg))		Iodine (μg))	
		per 100 g	per 100 kcal	per 100 g	per 100 kcal	per 100 g	per 100 kcal	per 100 g	per 100 kcal
Meiji	Essential fatty acid enriched MCT formula	12.1	2.4	10.1	2.0	14.1	2.8	40	7.9
Meiji	MCT formula	12.1	2.4	10.1	2.0	14.1	2.8	40	7.9
Meiji	Milfy HP	11.1	2.4	9.2	2.0	12.9	2.8	0	0
Meiji	Elemental formula	9.4	2.4	7.8	2.0	10.9	2.8	0	0
Morinaga	New MA-1	15	3.2	12	2.6	7	1.5	<10	<3
Morinaga	AR Milk	15	3.0	12	2.4	6.8	1.4	32	6
Morinaga	Nonlact	15	3.1	12	2.5	7	1.5	60	12.5

0 means below detection limit, The gray area indicate display values

From Website of Imperial Gift Foundation Boshi Aiiku Kai: http://www.boshiaiikukai.jp/milk02_01.html

11.8 Monitoring of Carnitine Deficiency in Patients Nutritionally Managed by Tube Feeding

Regarding the diagnosis and risk judgment of carnitine deficiency in patients who are fed by tube feeding, it can be diagnosed by blood carnitine 2-fractions test (using enzymatic cycling method, see Chap. 8) as described in the "Guidelines for Diagnosis and Treatment" before taking carnitine preparations or carnitine-containing enteral nutrition. The method of monitoring after administration of carnitine products has not been established at present, since the effect of administered carnitine makes it impossible to make a correct diagnosis due to an apparent increase in blood concentration. This is because only 0.5% or less of the carnitine pool in the body is in the blood and extracellular fluid pool, where the administered carnitine enters and is absorbed very slowly by the tissues over time (see "Chap. 4: Carnitine Homeostasis"). In many of the clinical studies described in this book, they described as "carnitine administration normalized blood carnitine levels" or "blood carnitine levels were high," but these results indicate an apparent increase in blood carnitine levels due to carnitine administration.

How can we determine whether the dosage and type of carnitine or carnitine-containing enteral nutrition is appropriate for a patient who is already receiving carnitine or carnitine-containing enteral nutrition? The author believes that in patients at low risk of developing carnitine deficiency immediately, administration of carnitine products or carnitine-containing enteral feeds should be stopped for about a week, and the diagnosis should be made by blood carnitine test. The blood concentration of orally administered carnitine decreases exponentially, with an initial half-life of 0.5–1.0 h and a final half-life of 3–12 h, and 12–24 h after administration, the blood concentration is almost the same as that before administration. Therefore, the effect of the carnitine administration is completely negligible after 1-week withdrawal (theoretically 3 days may be enough). However, in patients who are at high risk of developing carnitine deficiency by withdrawal of carnitine administration, such as those who are taking multidrug antiepileptic drugs containing valproic acid, or who have experienced frequent hypoglycemic attacks or hyperammonemia due to carnitine deficiency in the past, it is not recommended to suspend the administration of carnitine preparations or enteral nutrition containing carnitine for about 1 week, because there is a risk of developing carnitine deficiency. In patients with stable conditions, such as hemodialysis patients, who are at low risk of developing carnitine deficiency immediately, it is possible to avoid the effects of the apparent increase in blood concentration due to carnitine administration by suspending treatment for 1 week and performing a blood carnitine test, and to make a diagnosis according to the "Guidelines for Diagnosis and Treatment," and consider continuation of enteral nutrition containing carnitine, change of the type of nutrient, and additional administration of carnitine preparations (for details, see "Chap. 18: Guidelines for Diagnosis and Treatment of Carnitine Deficiency"). However, at present, there are no reports of post-administration monitoring using this proposed method. The author looks forward to future studies.

Carnitine deficiency due to tube feeding, intravenous nutrition, and specific milk formulas is described above. Humans live on dietary intake of nutrients. "It is only nutrition, but it is nutrition." Tube feeding, intravenous nutrition, and specific milk formulas are life-sustaining nutrition for low birthweight infants, postoperative patients, children with inherited metabolic diseases, severely mentally or physically disabled children (and persons), and patients who must consume allergic milk due to allergies. In these nutritional formulas, it is necessary to pay close attention to the deficiency of micronutrients such as biotin and selenium, including carnitine.

References

1. Rebouche CJ. Kinetics, pharmacokinetics, and regulation of L-carnitine and acetyl-L-carnitine metabolism. Ann N Y Acad Sci. 2004;1033:30–41. https://doi.org/10.1196/annals.1320.003.
2. Stanley CA. Carnitine deficiency disorders in children. Ann N Y Acad Sci. 2004;1033:42–51. https://doi.org/10.1196/annals.1320.004.

3. Rebouche CJ, Paulson DJ. Carnitine metabolism and function in humans. Annu Rev Nutr. 1986;6:41–66. https://doi.org/10.1146/annurev.nu.06.070186.000353.
4. Pons R, De Vivo DC. Primary and secondary carnitine deficiency syndromes. J Child Neurol. 1995;10(Suppl 2):S8–S24.
5. Borum PR, York CM, Broquist HP. Carnitine content of liquid formulas and specific diets. Am J Clin Nutr. 1979;32(11):2272–6. https://doi.org/10.1093/ajcn/32.11.2272.
6. Borum PR. Carnitine in parenteral nutrition. Gastroenterology. 2009;137(5 Suppl):S129–34. https://doi.org/10.1053/j.gastro.2009.08.016.
7. Kodama H, Shimizu T, Takitani K, et al. Pitfalls of using some types of milk and enteral nutrient formulas. J Japanese Soc Pediatr. 2012;116(4):637–54. (Article in Japanese).
8. Keitani Y. Practice of enteral nutrition management in children with severe physical and mental disabilities. J Child Health. 2020;79(1):10–9. (Article in Japanese).
9. Ogino A. Carnitine deficiency during intravenous and enteral nutrition and its therapy. Jpn J Surg Metab Nutr. 2020;54(2):71–6. (Article in Japanese).
10. Tokumitsu A. Proper use of enteral nutrition in consideration of the characteristics of severely mentally and physically handicapped children. J Severe Motor Intellect Disabil. 2017;42(1):35–43. (Article in Japanese).
11. Soh H. The significance and practice of intravenous nutrition in pediatric nutrition management. Jpn J Soc Parenter Enteral Nutr. 2018;33(3):831–4. (Article in Japanese).
12. Japanese Industry Association for Medical Nutrition: I. The species of concentrated liquid food. https://www.ryudoshoku.org/howto/use1. (Article in Japanese).
13. Signorelli GC, Bianchetti MG, Jermini LMM, et al. Dietary chloride deficiency syndrome: pathophysiology, history, and systematic literature review. Nutrients. 2020;12(11):3436. https://doi.org/10.3390/nu12113436.
14. Miyahara J, Aramaki S, Yokochi K. Dietary chloride deficiency due to new liquid nutritional products. Pediatr Int. 2009;51(2):197–200. https://doi.org/10.1111/j.1442-200X.2008.02670.x.
15. Ochi F, Ohmori H, Hino K, et al. Secondary carnitine deficiency in tube-fed children with severe motor and intellectual disabilities. J Jpn Pediatr Soc. 2011;115(8):1314–20. (Article in Japanese).
16. Otaki U, Ozawa H, Ishizuka T. Evaluation of total carnitine values in persons with severe motor and intellectual disabilities with enteral (tube) feeding. Brain Dev. 2012;44(5):374–7. (Article in Japanese).
17. Ozawa H. Nutritional management in children (and persons) with severe motor and intellectual disabilities: focusing on carnitine. J Severe Motor Intellect Disabil. 2012;37(1):101–6. (Article in Japanese).
18. Ozawa Y, Nomura Y, Amemiya K, et al. Low-dosage of L-carnitine supplementation to the severe motor and intellectual disabilities with secondary carnitine deficiency. J Severe Mot Intellect Disabil. 2016;41(3):357–62. (Article in Japanese).
19. Wakisaka A, Niida Y, Yamada S, et al. The association of hypocarnitinemia with enteral diets and antiepileptic drugs in children and adults with severe physical and mental disabilities. Brain Dev. 2013;45(6):445–50. (Article in Japanese).
20. Wakisaka A. Risk management of hypocarnitinemia in children with severe mental and physical disabilities: from the standpoint of pediatrics. J Severe Mot Intellect Disabil. 2015;40(1):103–8. (Article in Japanese).
21. Kato A. Carnitine deficiency with valproate sodium therapy. the difference by normal diets and enteral nutrition. Brain Dev. 2013;45(1):17–20. (Article in Japanese).
22. Takeda Y, Tomiwa K, Kin H, et al. Changes in carnitine and biochemical variables after levocarnitine administration in patients with severe motor intellectual disabilities. J Jpn Pediatr Soc. 2014;118(8):1208–12. (Article in Japanese).
23. Ueno K, Tanabe K, Toh N, et al. Current status of carnitine deficiency in patients with neuromuscular disease during long-term tube feeding. Jpn J Clin Nutr. 2014;125(5):635–9. (Article in Japanese).
24. Muguruma T, Omori H, Takeda I, et al. Nutritional evaluation in patients with intractable neurological diseases undergoing long-term tube feeding. in point of trace elements and carnitine. J Jpn Soc Clin Nutr. 2015;37(1):36–43. (Article in Japanese).
25. Tatara K, Ishikawa Y, Ichihara N, et al. Survey of patients with neuromuscular disorders requiring long-term ventilator-assistance in Hospitals of Japanese National Hospital Organizations: a trial for the standardization of the ventilator. Iryo (Official Publication of Japanese Society of National Medical Services). 2010;64(4):257–64. (Article in Japanese).
26. Mikami J, Higuchi K, Nakai K, et al. Evaluation of the severity and nursing requirements of patients in a muscular dystrophy ward. Iryo (Official Publication of Japanese Society of National Medical Services). 2010;64(5):322–7. (Article in Japanese).
27. Tatara K, Ishikawa Y, Takashi Imai, et al. Survey on long-term ventilator-assisted individuals in Japanese National Hospital Organization 3rd report. J Jpn Soc Respir Care Rehabil. 2009;19(2):151–5. (Article in Japanese).
28. Tatara K, Shinno S. Management of mechanical ventilation and prognosis in Duchenne muscular dystrophy. Iryo (Official Publication of Japanese Society of National Medical Services). 2008;62(10):566–71. (Article in Japanese).
29. Tatara K, Fukunaga H, Kawai M. Clinical survey of muscular dystrophy in hospitals of National Hospital Organization. Iryo (Official Publication of Japanese Society of National Medical Services). 2006;60(2):112–8. (Article in Japanese).
30. Matsumura T, Saito T, Fujimura H, et al. A longitudinal cause-of death analysis of patients with

Duchenne muscular dystrophy. Clin Neurol. 2011;51(10):743–50. (Article in Japanese).
31. Komaki H. Best supportive Care for Muscular Dystrophy. J Jpn Pediatr Soc. 2012;116(4):655–65. (Article in Japanese).
32. Le Borgne F, Guyot S, Logerot M, et al. Exploration of lipid metabolism in relation with plasma membrane properties of Duchenne muscular dystrophy cells: influence of L-carnitine. PLoS One. 2012;7(11):e49346. https://doi.org/10.1371/journal.pone.0049346.
33. Watanabe S. L-carnitine deficiency and treatment in neuromuscular diseases. Jpn J Surg Metab Nutr. 2020;54(2):77–80. (Article in Japanese).
34. Penn D, Schmidt-Sommerfeld E, Wolf H. Carnitine deficiency in premature infants receiving total parenteral nutrition. Early Hum Dev. 1980;4(1):23–34. https://doi.org/10.1016/0378-3782(80)90005-5.
35. Penn D, Schmidt-Sommerfeld E, Pascu F. Decreased tissue carnitine concentrations in newborn infants receiving total parenteral nutrition. J Pediatr. 1981;98:976–8.
36. Borum PR, Bennett SG. Carnitine as an essential nutrient. J Am Coll Nutr. 1986;5(2):177–82. https://doi.org/10.1080/07315724.1986.10720124.
37. Borum PR. Carnitine in neonatal nutrition. J Child Neurol. 1995;10(Suppl 2):2S25–31.
38. Tibboel D, Delemarre FM, Przyrembel H, et al. Carnitine deficiency in surgical neonates receiving total parenteral nutrition. J Pediatr Surg. 1990;25(4):418–21. https://doi.org/10.1016/0022-3468(90)90384-l.
39. Schmidt-Sommerfeld E, Penn D. Carnitine and total parenteral nutrition of neonate. Biol Neonate. 1990;58(Suppl 1):81–8. https://doi.org/10.1159/000243302.
40. Ohashi Y, Gomi A, Sugiyama A, et al. A case of very low birth weight infants with secondary carnitine deficiency symptoms during long-term TPN management. J Japanese Soc Pediatr Surg. 2009;45(3):595. (P-144, meeting abstract. Article in Japanese).
41. Tazuke Y, Sawai T, Soh H, et al. On the necessity of L-carnitine administration in long-term TPN patients - based on the experience of treating patients with short bowel syndrome. J Japanese Soc Pediatr Surg. 2009;45(3):595. (P-145, meeting abstract. Article in Japanese)
42. Hirose S, Hirata M, Azuma N, et al. Carnitine depletion during total parenteral nutrition despite oral L-carnitine supplementation. Acta Paediatr Jpn. 1997;39(2):194–200. https://doi.org/10.1111/j.1442-200x.1997.tb03580.x.
43. Okanari K, Takahashi M, Maeda T, et al. Hypocarnitinemic hypoglycemia and heart failure in an infant with a constant parenteral elementary nutrition during measles vaccination-related febrile illness. Neuropediatrics. 2007;38(6):317–9. https://doi.org/10.1055/s-2008-1062714.
44. Hahn P, Allardyce DB, Frohlich J. Plasma carnitine levels during total parenteral nutrition of adult surgical patients. Am J Clin Nutr. 1982;36(4):569–72. https://doi.org/10.1093/ajcn/36.4.569.
45. Nanni G, Pittiruti M, Castagneto M. Carnitine plasma levels during total parenteral nutrition. Am J Clin Nutr. 1983;38(2):339–41. https://doi.org/10.1093/ajcn/38.2.339.
46. Berner YN, Larchian WA, Lowry SF, et al. Low plasma carnitine in patients on prolonged total parenteral nutrition: association with low plasma lysine. JPEN J Parenter Enteral Nutr. 1990;14(3):255–8. https://doi.org/10.1177/0148607190014003255.
47. Btaiche IF, Khalidi N. Parenteral nutrition-associated liver complications in children. Pharmacotherapy. 2002;22(2):188–211. https://doi.org/10.1592/phco.22.3.188.33553.
48. Tanaka Y, Yagi M. Pathogenesis of hepatobiliary dysfunction caused by total parenteral nutrition. J Clin Exp Med. 2006;218(5):409–12. (Article in Japanese).
49. Kumpf VJ. Parenteral nutrition-associated liver disease in adult and pediatric patients. Nutr Clin Pract. 2006;21(3):279–90. https://doi.org/10.1177/0115426506021003279.
50. Hisamatsu C, Okata Y, Nishijima E. Nutritional management for the parenteral nutrition associated liver disease. Jpn J Pediatr Surg. 2011;43(4):403–7. (Article in Japanese).
51. Worthley LI, Fishlock RC, Snoswell AM. Carnitine deficiency with hyperbilirubinemia, generalized skeletal muscle weakness and reactive hypoglycemia in a patient on long-term total parenteral nutrition: treatment with intravenous L-carnitine. JPEN J Parenter Enteral Nutr. 1983;7(2):176–80. https://doi.org/10.1177/0148607183007002176.
52. Worthley LI, Fishlock RC, Snoswell AM. carnitine balance and effects of intravenous L-carnitine in two patients receiving long-term total parenteral nutrition. JPEN J Parenter Enteral Nutr. 1984;8(6):717–9. https://doi.org/10.1177/0148607184008006717.
53. Palombo JD, Schnure F, Bistrian BR, et al. Improvement of liver function tests by administration of L-carnitine to a carnitine-deficient patient receiving home parenteral nutrition: a case report. JPEN J Parenter Enteral Nutr. 1987;11(1):88–92. https://doi.org/10.1177/014860718701100188.
54. Ottery FD. Nutritional consequences of reoperative surgery in recurrent malignancy. Semin Oncol. 1993;20(5):528–37.
55. Forchielli ML, Richardson D, Folkman J, et al. Better living through chemistry, constant monitoring, and prompt interventions: 26 years on home parenteral nutrition without major complications. Nutrition. 2008;24(1):103–7. https://doi.org/10.1016/j.nut.2007.10.001.
56. Bowyer BA, Fleming CR, Ilstrup D, et al. Plasma carnitine levels in patients receiving home parenteral nutrition. Am J Clin Nutr. 1986;43(1):85–91. https://doi.org/10.1093/ajcn/43.1.85.
57. Bowyer BA, Miles JM, Haymond MW, et al. L-carnitine therapy in home parenteral nutrition patients with abnormal liver

tests and low plasma carnitine concentrations. Gastroenterology. 1988;94(2):434–8. https://doi.org/10.1016/0016-5085(88)90433-7.
58. Curione M, Barbato M, De Biase L, et al. Prevalence of coeliac disease in idiopathic dilated cardiomyopathy. Lancet. 1999;354(9174):222–3. https://doi.org/10.1016/s0140-6736(99)01501-9.
59. Fonager K, Sørensen HT, Nørgård B, et al. Cardiomyopathy in Danish patients with coeliac disease. Lancet. 1999;354(9189):1561. https://doi.org/10.1016/S0140-6736(05)76595-8.
60. Curione M, Danese C, Viola F, et al. Carnitine deficiency in patients with coeliac disease and idiopathic dilated cardiomyopathy. Nutr Metab Cardiovasc Dis. 2005;15(4):279–83. https://doi.org/10.1016/j.numecd.2005.01.002.
61. Curione M, Barbato M, Viola F, et al. Idiopathic dilated cardiomyopathy associated with coeliac disease: the effect of a gluten-free diet on cardiac performance. Dig Liver Dis. 2002;34(12):866–9. https://doi.org/10.1016/s1590-8658(02)80258-4.
62. Bene J, Komlósi K, Gasztonyi B, et al. Plasma carnitine ester profile in adult celiac disease patients maintained on long-term gluten free diet. World J Gastroenterol. 2005;11(42):6671–5. https://doi.org/10.3748/wjg.v11.i42.6671.
63. Uslu N, Demir H, Karagöz T, et al. Dilated cardiomyopathy in celiac disease: role of carnitine deficiency. Acta Gastroenterol Belg. 2010;73(4):530–1.
64. Yamamoto S, Ohtake M, Takayanagi M, et al. Carnitine deficiency in infants using therapeutic specific milk: a study based on the measurement of plasma free carnitine levels and the ability to produce ketone bodies from triglycerides. J Jpn Pediatr Soc. 1985;89(11):2488–94. (Article in Japanese).
65. Takayanagi M, Okimoto Y, Kakuta H. Causes and treatment to vitamin deficiency in Japan. Jpn J Pediatr Med. 2009;41(9):1289–92. (Article in Japanese).
66. Takayanagi M. Indications and problems of specific medical formula (specific milk) for infants. Jpn J Pediatr Med. 2014;46(8):1063–6. (Article in Japanese).
67. Ida S, Iwasaki Y, Uchida K, et al. Guideline for diagnosis and treatment of carnitine deficiency 2018. https://www.jpeds.or.jp/uploads/files/20181207_shishin.pdf. (Article in Japanese).
68. Abe H, Kanazawa M, Ohtake A, et al. A case of biotin deficiency thought to be caused by amino acid conditioning formula (605Z). Jpn J Inherit Metab Dis. 1991;7:172. (P-44, meeting abstract. Article in Japanese).
69. Goto A, Ishida A, Arai H, et al. A case of extremely low birth weight infants with dicarboxylic aciduria and carnitine deficiency caused by MCT milk. Acta Neonatologica Japonica. 1992;28(4):710. (202, meeting abstract. Article in Japanese).
70. Ishida A, Goto A, Takahashi Y, et al. A preterm infant with secondary carnitine deficiency due to MCT formula effective treatment of L-carnitine. Tohoku J Exp Med. 1994;172(1):59–64. https://doi.org/10.1620/tjem.172.59.
71. Mushimoto Y, Taketani T, Kobayashi H, et al. Examination of 8 cases of refractory eczema due to biotin deficiency in specific milk for allergies. Jpn J Pediatr Allergy Clin Immunol. 2008;22:661. (18, meeting abstract. Article in Japanese).
72. Mushimoto Y, Taketani T, Hasegawa Y, et al. Biotin deficiency misdiagnosed as steroid-resistant atopic dermatitis. Allergy Pract. 2010;30(6):543–7. (Article in Japanese).
73. Komatsu J, Ohura T, Kitamura T, et al. A case of biotin and carnitine deficiency due to long-term use of milk for allergies. J Sendai Hosp. 2012;32:43–8. (Article in Japanese).
74. Nozaki A, Kusunoki T. A case in which biotin, carnitine, and selenium deficiency was observed by long-term use of whey protein hydrolyzed milk. Jpn J Pediatr Gastroenterol Nutr. 2014;28(2):69–76. (Article in Japanese).
75. Hayashi H, Tokuriki S, Okuno T, et al. Biotin and carnitine deficiency due to hypoallergenic formula nutrition in infants with milk allergy. Pediatr Int. 2014;56(2):286–8.

Carnitine Deficiency Due to Sodium Valproate

12

12.1 Outline of Sodium Valproate

12.1.1 Chemical Structure of Sodium Valproate

Valproic acid was synthesized in the United States by Burton in 1882 as a chemical solvent. Valproic acid is a branched-chain lower fatty acid with the structural formula shown in Fig. 12.1, and sodium valproate is its sodium salt.

Although not marketed in Japan, divalproex sodium (trade name: Depakote), a coordination compound consisting of two molecules of valproic acid and sodium, is also available overseas and is widely used. It should be noted that clinical studies, including case reports of prescriptions of divalproex sodium (Depakote), can be found in the overseas literature. The indications, pharmacological effects, metabolism, and toxicity of divalproex sodium are basically the same as those of sodium valproate.

In this book, the description of sodium valproate is simplified to "valproic acid" or "VPA."

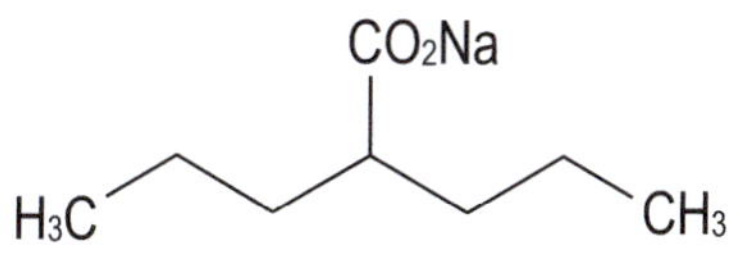

Chemical formula: $C_8H_{15}NaO_2$
Chemical name: monosodium 2-propyl pentanoate
sodium 2-propyl pentanoate
sodium dipropyl acetate
sodium propyl valerate
Trade name: Depaken (Kyowa Kirin), Selenica R (Kowa, Mitsubishi Tanabe, Yoshitomi), many generic products etc.

Fig. 12.1 Chemical structure and chemical name of sodium valproate

12.1.2 Discovery of the Antiepileptic Effect of Valproic Acid

In 1962, Eymard in Carraz's laboratory at the University of Medicine and Pharmacy of Grenoble, France, synthesized a khelline derivative and evaluated its anticonvulsant activity. Meunier et al., who were researchers at a small pharmaceutical company in Grenoble, tried to evaluate the experimental anticonvulsant activity of a series of compounts of khelline (also spelled as khelline). In this period valproic acid was used as a solvent agent. Accidentally they discoverd that valproic acid itself had anticonvulsant activity regardless of the khelline evaluated, using pentylenetetrazole-induced convulsions model in rabbits. They reported this finding at academic conferences and papers, obtained a npatent, and began research on valproic acid as an antiepileptic drug [1–3]. It was approved as an antiepileptic drug in France in 1967, Germany in 1973, Japan in 1975, and the United States in 1978. Sodium

N. Nakayama, *Carnitine Deficiency and Insufficiency in Medicine*,
https://doi.org/10.1007/978-981-95-7797-2_12

valproate is one of the first-line drugs for the treatment of epilepsy because it has a broad therapeutic spectrum for the treatment of various types of epilepsy (small seizures, focal seizures, psychomotor seizures, and mixed seizures) and personality behavioral disorders associated with epilepsy (moodiness, irritability, etc.). Since the mechanism of action is different from other antiepileptic drugs, it is often used in combination with other antiepileptic drugs. It is also used to prevent convulsive seizures after brain surgery. In addition, it is indicated for "treatment of manic and manic-depressive manic states" in the psychiatric field and is widely used as a so-called mood stabilizer along with lithium preparations, carbamazepine, and lamotrigine. In 2011, the indication for "migraine attack suppression" was also added.

12.1.3 Current Indications for Valproic Acid

The approved indications for valproic acid as of April 2019 are as follows:

1. Treatment of various types of epilepsy (small, focal, psychomotor, and mixed seizures) and personality and behavioral disorders associated with epilepsy (mood swings, irritability, etc.)
2. Treatment of manic states in mania and manic-depression
3. Suppression of migraine attack onset

The second of these indications, "treatment of manic states," was approved in Japan in 2002 without comparative studies, on the grounds that its usefulness for manic states was "well known in the medical and pharmaceutical literature." The approval of the drug was triggered by a request submitted to the Ministry of Health, Labour and Welfare by The Japanese Society of Neuropsychopharmacology and The Japanese Society of Clinical Neuropsychopharmacology in 2000 [4]. Due to this indication, it is widely used as a so-called mood stabilizer for bipolar disorder in psychiatric patients, along with lithium carbonate, carbamazepine, and lamotrigine. In addition, as in the case of the expansion of the indication for L-carnitine (L-Cartin® Tablets), the indication for the third migraine was added in June 2011 at the "The Evaluation Committee on Unapproved or Off-label Drugs with High Medical Needs" in October 2010, stating that the usefulness of valproic acid in suppressing the onset of migraine attacks was already "known in medical and pharmaceutical sciences" ("Kouchi-Shinsei" in Japanese, meaning public knowledge-based application).

Furthermore, although it is not considered to be an official indication, the use of valproate in dementia has been increasing rapidly in recent years, because it is recommended that it is effective for convulsive seizures in elderly dementia patients and for irritability, aggression, and mood disorders that are peripheral symptoms of dementia [5].

Because of this wide range of indications, it is thought to be prescribed to more than 1 million people in Japan, including adults as well as pediatric epilepsy patients.

12.1.4 History of Clinical Application of Valproic Acid

Table 12.1 summarizes the history of clinical application and approval status of sodium valproate. As can be seen, it was first approved and used in clinical practice as an antiepileptic drug, followed 20–30 years later by approval for use in psychiatry. Therefore, clinical experience with the adverse effects of valproic acid has been accumulated in the fields of pediatric neurology and the treatment of severely mentally and physically disabled patients with epilepsy. In the fields of pediatric neurology and severe mental and physical disabilities, it is widely known that when valproic acid is used to treat or prevent epilepsy, carnitine deficiency should be noted and carnitine supplementation

Table 12.1 History of clinical application of valproic acid

1882	Synthesized in the United States as a solvent for chemicals
1963	Meunier in France reported an effect on pentetrazol (pentylenetetrazol) convulsions induced in rabbits
1967	Approved in France as an antiepileptic drug
1973	Approved in Germany (antiepileptic drug)
1975	Approved in Japan (antiepileptic drug)
1978	Approved in the United States (antiepileptic drug)
1995	Approved for bipolar disorder in the United States
1996	Approved for the treatment of migraine in the United States
2000	Approved for the treatment of bipolar disorder in France
2001	Approved for the treatment of bipolar disorder in the United Kingdom
2002	Approved in Japan for the treatment of manic and manic-depressive disorders (public knowledge-based application based on requests from the Japan Society of Neuropsychopharmacology and the Japan Society of Clinical Neuropsychopharmacology)
2003	Approved for the treatment of bipolar disorder in Germany
2011	Approval for migraine in Japan (public knowledge-based application)

therapy should be given, when necessary, but this is still not well known in the field of psychiatry.

12.1.5 Side Effects of Valproic Acid

Valproic acid is usually a well-tolerated drug with a wide margin of safety and is widely used. Common adverse drug reactions observed at the time of approval and in adverse drug reaction surveys through the end of December 1977 included somnolence (5.5%), ataxia and lightheadedness (3.6%), nausea, nausea and vomiting (2.6%), anorexia (1.7%), gastrointestinal disturbances (1.5%), and general malaise (0.7%). In the case of use for "manic and manic-depressive states of mania and manic-depressive disorder" and "suppression of the onset of migraine," no domestic survey has been conducted to clarify the frequency of side effects [2]. In addition, in the section "Serious Side Effects and Initial Symptoms in the Drug Interview Form [2], there are descriptions as followings: (1) serious liver damage such as fulminant hepatitis, jaundice, and fatty liver; (2) impaired consciousness with hyperammonemia; (3) hemolytic anemia, erythroblastosis, pancytopenia, serious thrombocytopenia, granulocytopenia; (4) acute pancreatitis; and (5) interstitial nephritis and Fanconi syndrome.

Among these, valproate-induced hepatotoxicity (VHT) and valproate-induced hyperammonemic encephalopathy (VHE), although rare, have been reported. Understanding and prevention of these VHT and VHE are essential for the safe use of valproic acid, which is extremely useful in the treatment of epilepsy and bipolar disorder. The following is an overview of the findings to date regarding these two serious side effects and their relationship with carnitine deficiency.

12.2 Valproate-Induced Hepatotoxicity (VHT)

After valproic acid was first launched, it was initially emphasized only for its usefulness in the treatment of epilepsy and was said to have few side effects. Abnormal liver functions such as elevated serum glutamate oxaloacetate transaminase (SGOT) and serum glutamate pyruvate transaminase (SGPT) have been reported as side effects, but these liver function abnormalities were initially thought to be reversed by reducing the dose [6].

12.2.1 Early VHT Case Reports

However, since 1979, a series of VHT cases have been reported. In 1979, Donat et al. reported the first death from valproic acid hepatotoxicity in the United States (status epilepticus, 17-month-old boy), although it was short and simple one [7].

Suchy et al. also reported details from the onset of two VHT cases to the autopsy results in the United States [8]. One was a 5-year-old boy who had moderate psychomotor retardation and was prescribed 650 mg/day of valproate 3 months before hospitalization because phenytoin was ineffective. One month before hospitalization, the patient became weak, with somnolence, loss of appetite, and facial edema. Since the blood concentration of valproic acid was 166 μg/mL, which exceeded the normal range (50–100 μg/mL), the dose was reduced to 500 mg/day. One week before hospitalization, peripheral edema, abdominal distention, and scleral icterus were observed. At the time of admission, somnolence, hypotonia, generalized edema, abdominal prominence, ascites (no hepatic hypertrophy or spleen enlargement), SGOT 106 International Units (IU)/L, SGPT 58 IU/L, bilirubin 7 mg/dL, and serum ammonia 36.4 μg/dL. Supportive care was provided, but he died of pulmonary hemorrhage and sepsis on the 36th day of hospitalization. The second case was an 11.5-year-old boy who was initially prescribed 250 mg/day of valproic acid in addition to phenytoin (300 mg/day), primidone (400 mg/day), and clonazepam (3.5 mg/day), and was eventually increased to 1000 mg/day. After 1 week, the SGOT was 54 IU/L and the SGPT was 131 IU/L, and after 6 weeks, the patient developed weakness and ataxia and died on the 10th day. In this report, Suchy et al. criticized the lack of information at the beginning of the launch, saying, "Valproic acid is said to have few side effects, but the details have not been properly reported as case reports, even though there have been at least six deaths in the United Kingdom and the United States." Thus, it is undeniable that when valproic acid was first introduced, only its usefulness in the treatment of epilepsy was emphasized, and serious side effects were neglected.

When Bowdle et al. conducted a pharmacokinetic study of valproate in six healthy subjects, they reported that all subjects who received 1000 or 1500 mg/day of valproate experienced symptoms that appeared to be central, such as unexpected sedation and nausea [9]. Specifically, daytime sleepiness (3 cases), lack of concentration (3 cases), fatigue (2 cases), speech difficulties (stuttering and slurred speech, 2 cases), and frequent nightmares (1 case) were common. At the beginning of the report, Bowdle et al. disputed the widespread belief at the time that "valproic acid has almost no sedative or central effects" by quoting the opinion of Mr. Temple, then the director general of the U.S. Food and Drug Administration (FDA), that "case reports on post-marketing side effects are indispensable for assessing the toxicity of new drugs introduced to the pharmaceutical market" [10].

Mathis et al. also experienced two cases very similar to Suchy et al., but instead of the damage to the bile secretory apparatus such as capillary bile ducts and herring ducts pointed out as the cause of liver damage by Suchy et al., when examined with an electron microscope, swollen hepatocyte mitochondria, decreased matrix density, deformed cristae, and decreased dense bodies, suggesting parenchymal organelle injury observed in Reye syndrome [11]. They speculated that 4-pentenoic acid and 2-propylpentanoic acid, which are metabolites of valproic acid, which have been reported to cause mitochondrial oxidative phosphorylation decoupling in rat studies, may cause mitochondrial damage.

However, Partin et al. in Suchy's group found mitochondrial degeneration in two of their own cases, but it was not a lesion seen in Reye syndrome or a lesion in rat mitochondria treated with hypoglycin or 4-pentenoic acid, but a lesion preceded by parenchymal damage [12].

In 1979, Gerber et al. reported a case in which a 12-year-old girl with epilepsy with mental retardation who took valproic acid along with phenytoin and phenobarbital developed the so-called Reye syndrome and died 2 months later after taking valproic acid. In this study, they discussed the structural similarity between valproic acid and its metabolites, which have a

2-propylpentanoic acid structure, and hypoglycin, which is abundant in an immature fruit native to west Africa called ackee, which causes Jamaican vomiting disease, and 4-pentenoic acid, which is known to cause hepatotoxicity when administered to animals. At the end of the paper, it is stated that by January 1, 1979, the pharmaceutical company had identified 14 VHT cases, six in the United States and eight in Europe [13].

Jacobi et al. also reported a VHT case of valproic acid and phenobarbital in West Germany (at that time) (a boy aged 6 years and 11 months with psychomotor retardation), and described, "Valproic acid is said to be a safe antiepileptic drug, and it is said that there is only a transient asymptomatic abnormality of liver function, but it should be noted that it has serious lethal hepatotoxicity. In addition to monitoring liver function, regular examinations of the coagulation system are also necessary." [14]

Ware et al. also reported acute liver injury due to valproic acid [15]. Of the five cases, four were children and three died. All of them developed 3–6 months after administration and were accompanied by symptoms such as thrombocytopenia and hair loss. At the time, it was not generally known that valproic acid could cause serious hepatotoxicity, but the attending physician was very concerned about the hepatotoxicity of valproic acid. The clinical course of five patients was described, and the structural similarity between valproic acid and hypoglycin was found to indicate that hypoglycemia caused by hypoglycin did not occur in all cases of hepatotoxicity caused by valproic acid. The mechanism is unknown, but it is thought to be idiosyncrasy or hypersensitivity reaction.

12.2.2 Clinical Characteristics of VHT Cases

Many VHT cases have been reported, and several research reports have been published that collect these VHT cases and examine the similarities and differences.

In 1982, Zimmerman et al. analyzed 23 previously reported VHT cases in detail, including their clinical characteristics and course of onset [16]. A summary is given below.

12.2.2.1 Age of Onset and Clinical Signs

Of the 23 patients, 20 were under 20 years of age, and 19 patients took valproic acid for more than 1 month. Prominent clinical signs included convulsions, facial edema, fatigue, and nausea. Hypoglycemia was observed in 6 cases. The elevation of transaminases of liver function enzymes was moderate, with SGOT less than 500 IU and SGPT less than 200 IU. In most patients, autopsy or biopsy results show lesions similar to Jamaican vomiting disease, such as droplet fat deposits, lesions with hepatocellular necrosis, necrosis of the central venous region, microvesicular fat deposition in hepatocytes, swelling of the liver mitochondria, and deformation of cristae. In four cases, fibrosis of hepatocytes was also observed.

12.2.2.2 Estimation of the Pathogenic Mechanism

These cases of lethal hepatotoxicity are considered to be extremely rare idiosyncratic reactions, but they may be specific manifestations of common side effects such as hyperammonemia and abnormal liver function, which occur at a higher rate. The specific constitution is metabolic and not immune, and it is speculated that the metabolite of valproic acid may be the cause of steatosis.

Zafrani et al. also reported a brief review of 31 VHT cases similar to that of Zimmerman et al. in terms of clinical signs, biochemical and histological characteristics [17].

Furthermore, in 1984, Powell-Jackson et al. conducted a detailed analysis of 42 VHT cases from 1978 to 1984 in terms of age of onset, sex, clinical signs, clinical course, and concomitant drugs other than valproic acid [18]. The following is a summary of the report.

Clinical Signs

Abnormal liver enzymes were found in 11% of the 1197 patients monitored. Only 2 of 57 patients with elevated liver enzymes were symptomatic with discomfort, anorexia, or coma, which resolved with dose reduction or withdrawal. Reye-like syndrome was present in 3 of 42 VHT cases, and 22 patients had hyperammonemia. These cases of hyperammonemia had no obvious signs of liver damage but were often confused with fulminant liver failure. The age of onset of VHT in the 42 cases was 0–5 years in 16 cases (38%), 6–10 years in 13 cases (31%), 11–15 years in 3 cases (7.1%), and 16–20 years in 5 cases (11.9%) were over 20 years of age. More than two-thirds of the VHT patients had prodromal symptoms such as anorexia, nausea, decreased epileptic control, decreased level of consciousness, and ataxia. One third had signs of liver failure such as fever, jaundice, ascites, peripheral edema, and bruising. Finally, hepatic coma occurred, and 30% of the patients had complications such as hypoglycemia, uremia, pneumonia, and sepsis. Blood bilirubin, alkaline phosphatase, aspartate transaminase (AST), and prothrombin time were elevated in 29, 5, 33, and 19 patients, respectively, and in 30 of 42 patients, AST was elevated above 100 IU/L and prothrombin time was above 18 seconds, indicating severe hepatocellular damage. The three patients with Reye-like syndrome were 3-, 8-, and 14-year-old patients with mental retardation who developed anorexia, nausea, and fever during treatment of epileptic seizures, followed by loss of consciousness and onset of symptoms of cerebral edema within a few days, metabolic acidosis, elevated ammonia (155, 120, and 208 μmol/L, respectively), prolonged prothrombin time, and AST The patient had metabolic acidosis, elevated ammonia (155, 120, and 208 μmol/L, respectively), prolonged prothrombin time, and extremely elevated AST (4500, 133, and 564 IU/L), but the bilirubin level was normal. All three patients died within 3 weeks of onset (two from cerebral edema and one from pneumonia). Symptomatic hyperammonemia occurred in 22 of 42 patients without overt liver disease, 17 of whom were children aged 9 months to 14 years, and 5 of whom were adults aged 20–51 years. Disorders of consciousness and ataxia such as confusion, paralysis, and coma preceded gastrointestinal disturbances such as nausea, vomiting, anorexia, and diarrhea. Valproic acid treatment was associated with moderate AST elevation in 7 patients, but no liver function abnormalities were observed in the remaining 15 patients. After withdrawal of the drug, the biochemical abnormalities normalized within 1–3 days and the patients became asymptomatic. Because hyperammonemia can occur in asymptomatic patients with unchanging electroencephalogram (EEG) and no clinical signs or biochemical changes, it is difficult to diagnose the cause of decreased level of consciousness when epileptic seizures are no longer controlled in the absence of hepatic dysfunction.

Pathological and Histological Characteristics

Histological abnormalities in these VHT deaths included fatty deposits in the hepatic microtubules in 10 of 26 autopsied patients, 4 patients had only this abnormality, 8 patients had massive lobular centrilobular necrosis, and the remaining 16 patients had lobular centrilobular necrosis without fatty deposits. In the remaining 16 cases, lobular centrilobular necrosis without fatty deposits was the predominant finding. Fatty degeneration of the microtubules was observed in three patients with Reye-like syndrome, and similar changes were seen in the proximal renal tubules in two patients. In one case with electron microscopic findings, a large number of mitochondrial abnormalities with distortion of the cristae and residual bodies were noted.

Dosage of Valproic Acid and Other Concomitant Medications

Data on the dose of valproic acid were available for 22 of 42 patients, and in 9 cases the dose was higher than the recommended dose of 30 mg/kg/day. The time of onset of hepatotoxicity varied from 1 week to 2 years after initiation of valproic acid. However, one-third of the 36 patients with data on onset from initiation to onset developed hepatotoxicity within 1–2 months, and only 2 patients developed hepatotoxicity more than 5 months after initiation. Thirty-six patients were prescribed concomitant medications such as phenobarbital, phenytoin, carbamazepine, and primidone, which induce hepatic metabolizing enzymes. Twelve patients were prescribed one concomitant drug with valproic acid, 12 patients were prescribed two concomitant drugs, and 8 patients were prescribed more than three drugs.

Necessity of Liver Function Monitoring

In patients with mental retardation, structural brain damage, or metabolic disorders, liver function monitoring during the first 3 months of valproic acid administration is mandatory. Particular attention should be paid when multiple medications are administered, and dose reductions may be necessary when abnormalities are observed. Even so, it is difficult to prevent the development of idiosyncratic hepatotoxicity.

12.3 Valproate-Induced Hyperammonemic Encephalopathy (VHE)

As mentioned in the VHT section above, many patients who showed hepatotoxicity presented with hyperammonemia and Reye-like syndrome. In the aforementioned review by Powell-Jackson et al., 22 of 42 patients presented with hyperammonemia [18].

Sackellares et al. reported that confusion (stupor) developed in four patients who were prescribed valproate and other antiepileptic drugs such as phenobarbital, phenytoin, carbamazepine and primidone, and that symptoms disappeared when valproate or other antiepileptic drugs were discontinued [19]. In this study, they reported that generalized high-amplitude rhythmic bisynchronous delta activity was detected, and that the onset of stupor may be due to drug interactions caused by the concomitant use of valproic acid and other antiepileptic drugs.

Coulter et al. summarized the experience of using valproate alone or in combination with about 100 patients with epilepsy in the early stages of valproate and reported that although it was a useful antiepileptic drug with an 82% efficacy rate for epileptic seizures, serious side effects such as coma were observed in 7 cases, and 3 of them died [20].

Coulter et al. found that three pediatric epilepsy patients with consent to take 500 mg of valproate resulted in severe vomiting, somnolence, and coma, as well as significant increases in blood glycine and propionate levels [21]. In addition, when a 4-year-old girl with epilepsy taking multiple drugs was given 750 mg of valproate, clinical symptoms such as somnolence, vomiting, fever, and bluntness, and the ammonia level was 80 μmol/L, and it was reported that hyperammonemia developed with valproate administration even if there was no abnormal liver function [21].

In addition, Coulter et al. identified 11 patients with symptoms of somnolence, confusion, and coma from hundreds of patients receiving valproate, and reported that eight of them had high blood ammonia levels [22]. The ammonia values in eight patients ranged from 58 to 488 μmol/L (mean 118.2 μmol/L). Two of the eight patients had signs of abnormal liver function, but six had no abnormal liver function or other disease. Although hyperammonemia with hepatic dysfunction or Reye-like syndrome has been reported in valproic acid-induced hyperammonemia, he cautioned that hyperammonemia without hepatic dysfunction may lead to somnolence or coma in some cases. Thus, it was found that hyperammonemia caused by valproic acid is not necessarily due to abnormal liver function or Reye-like syndrome, and that ammonia levels of 60 μmol/L or more are often symptomatic, and measures such as reducing the dose of valproic acid are necessary when ammonia levels are 60 μmol/L (about 102 μg/dL) or more. In addition, although the mechanism of the development of VHE in liver damage, Reye syndrome, gastrointestinal bleeding, and other conditions is unknown, hyperammonemia may be more likely to occur by mechanism proposed by Gruskay et al. that propionyl-CoA, an acyl-CoA ester of propionic acid produced by the metabolism of valproic acid, inhibits carbamoyl phosphate synthase 1 (CPS1), an important enzyme in the first stage of the urea cycle. In addition, it was suggested that propionyl-CoA inhibits synthase of N-acetylglutamate which is a positive cofactor of

CPS1, and decreases CPS1 activity, which may be one of the factors that secondary hyperammonemia occur due to valproic acid [21, 22].

Young et al. also reported cases of Reye-like syndrome caused by valproic acid [23]. There are several patterns of liver damage caused by valproic acid, such as (1) elevation of liver function enzymes (response to a 10% dose reduction), (2) hepatocellular necrosis from cholestatic hepatitis, and (3) Reye-like syndrome (common in children). They experienced a case of an 8-year-old boy with moderate mental retardation who developed Reye-like syndrome after taking valproic acid. Three weeks before admission, he was started on 375 mg valproate twice daily. The patient presented with fever of 42 °C, hyperpnea, and decreased responsiveness. The patient had chronic asynchronous activity modulation with aimless wiggling, blood ammonia of 354 μg/dL, SGOT of 564 U/mL, lactate dehydrogenase (LDH) of 1490 U/mL, and blood glucose below detection limits. Postmortem examination revealed panlobular microvesicular fatty degeneration of the liver and renal tubules, and a slightly distended brain.

Keene et al. also reported a case of Reye-like syndrome after taking valproic acid [24]. The patient was a 40-month-old boy with mental retardation and severe refractory multifocal seizure disorder from the age of 5.5 months but was frequently hospitalized for generalized tonic–clonic convulsive status and was treated with drugs such as carbamazepine, primidone, ethosuximide, phenytoin, and valproic acid, but did not achieve adequate seizure control, and tried a ketogenic diet but had no effect. Eventually, he was treated with phenytoin, ethosuximide, and valproic acid, and died of Reye-like syndrome. Up to 2 weeks before hospitalization, SGOT was in the normal range, and at the time of admission, it was about twice the normal range (30 IU or more). After 36 hours of hospitalization, the patient became unresponsive and the SGOT level rose to 4500 IU, but on the third day, it dropped to 219 IU, and the ammonia concentration was 264 (unit unknown). In this discussion, they described that cases of fatal liver failure can be divided into two types: (1) a type that progresses slowly over 4–6 weeks with clinical symptoms such as somnolence, vomiting, nausea, anorexia, jaundice, and edema, and (2) a type that rapidly develops fulminant. In the former case, recovery occurs if valproic acid is discontinued early, and histologically, there are no characteristic lesions other than cholestasis, and it is common in patients in a relatively older age group. The age of onset of the latter acute-onset type is often 12 years old or younger, and it often develops rapidly without signs of coma, increased brain pressure, or jaundice. Abnormalities in the coagulation system, elevated SGOT, and elevated ammonia levels occur at the onset of symptoms, but not earlier. This type of liver biopsy shows droplet fat deposits and mitochondrial lysis, and all reported patients have died. In addition, Keene et al. found that both diseases are common in children with fulminant liver failure due to valproic acid and Reye syndrome, and that some metabolic system failures occur, coma is caused by increased brain pressure, but there is no jaundice, and liver biopsies show droplet fat deposition and mitochondrial lysis. Furthermore, due to the structural similarity between valproic acid and hypoglycin, which causes Jamaican vomiting disease, and the report that pentanoic acid inhibits mitochondrial β oxidation, taking valproic acid slowly suppresses the oxidative function of mitochondria, and SGOT and ammonia rise. They hypothesized that a small proportion of predisposing factors, such as patients with congenital abnormalities in CPS1 or pyruvate dehydrogenase, patients with viral infections, and patients taking other antiepileptic drugs metabolized by the mitochondria, may experience rapid mitochondrial dysfunction and lead to a rapid decline in liver function. In other words, the hypothesis is that the suppression of mild mitochondrial function and the suppression of ammonia metabolism by valproic acid may be the basis of most cases.

12.4 Epidemiological Studies of VHT and VHE

Determining the causes of VHT and VHE and their prevention/prediction is essential for the safer use of valproic acid, which is useful in the treatment of epilepsy. Therefore, epidemiological studies and studies to determine risk factors for VHT were conducted in the United States and West Germany (at that time).

Dreifuss et al. conducted a detailed epidemiological study of VHT cases in the United States between 1978 and the end of December 1984 [25]. As a method, they obtained a report of adverse reactions that appeared to be VHT from Abbott, a manufacturer of valproic acid preparations in the United States and tried to contact the doctor in charge of each case to obtain more detailed information and cross-checked the side effect information on the FDA's computer and determined that 37 of the 47 cases reported to Abbott were VHT cases. The age of onset, the dose and duration of valproic acid, concomitant medications, and the medical characteristics of the cases were analyzed. The age of onset of 37 patients (23 males, 14 females) ranged from 5 months to 71 years, but the most common cases were 0–2 years old (16 cases), and children under 10 years old were 27 cases, accounting for 73% of the total. In 24 patients, the dose of valproate ranged from 13 to 120 mg/kg/day, and in 4 patients, the recommended dose of 60 mg/kg/day was exceeded. In addition to epilepsy, 36 patients had mental retardation, developmental retardation, birth defects, and other neurological disorders. Phenytoin, phenobarbital, clonazepam, and carbamazepine were the most common combination drugs, and although only 5 VHT cases were treated with valproate alone, there were a total of 32 VHT cases, including 22 children under 10 years of age. Calculated from the number of patients prescribed valproic acid in the United States, the incidence rate of valproic acid monotherapy in all age groups is 1/37,000, and the incidence rate of monotherapy for 0–2 years old is 1/7000. The incidence rate was 1/500 in multidrug treatment at 2 years of age, and extremely high in the combination of polypharmacy at a young age. Based on these findings, they called attention to the risk factors for the development of VHT, such as children under 2 years of age, concomitant use of polypharmacy drugs, abnormal liver function, metabolic disorders, and neurological disorders.

Following this study, Dreifuss et al. conducted a similar additional study in 1985–1986 [26]. As a result, four VHT cases were reported in children under the age of 10 in each of the past two years, and the incidence rate decreased sharply from 0.93 to 0.2 per 10,000 people, about one-fifth, compared to the previous survey from 1978 to 1984, probably because the risk factor was widely recognized.

Scheffner reported that in 1985 alone, five cases of VHT occurred in southern Germany alone and the incidence of this side effect was estimated to be at least 1/5000, not 1/20,000 as reported by Jeavon [27]; therefore, Scheffner questioned that valproic acid was considered the first-line drug for generalized epilepsy in children [28].

In addition, Scheffner et al. reported a detailed analysis of 16 cases of VHT in West Germany between 1977 and 1986 [29]. During this period, 24 cases of death due to valproic acid were investigated in West Germany, and 16 cases were judged to be VHT cases, excluding 8 cases in which death was thought to be due to other causes. Contrary to the findings of Dreifuss et al., 5 of the 16 cases were developed in children with normal development, 5 cases were also caused by valproate monotherapy, and only 2 cases were developed before 3 years of age. The first clinical signs observed in these VHT cases were nausea, vomiting, and apathy, and abnormalities in liver function enzymes indicated the progression of liver failure. Pathological findings included microvesicular steatosis, cell necrosis, and various degrees of bile duct proliferation. An abnormal metabolite, 4-en-VPA was observed in 6 of the 6 patients for which metabolites of valproic acid were examined, and this metabolite was detected even after discontinuation of treatment. In this study, 112 cases of VHT have been reported in the world so far, of which 95 cases have been ruled out as having other causes. Of these 95 cases, 90% of the patients were under the age of 20, 95% were within 6 months of the start of treatment, and 16 were treated with valproic acid as a single agent. It is stated that it is difficult to identify the risk of developing VHT, and that it is essential to pay attention to changes such as nausea, vomiting, and emotional blunting as clinical signs, and to perform clinical tests of liver function enzymes and coagulation systems during the first 6 months, and they described that in some cases it is necessary to re-evaluate the indications for valproate [29].

In addition, König et al. in Scheffner's group reported that 12 cases of VHT have been reported worldwide since the 198, and other 8 cases of VHT in Germany and Switzerland, so that for a total of 132 cases of VHT were developed since the introduction of valproic acid worldwide [30]. Compared to the analysis by Scheffner et al. in 1988, the characteristics of VHT cases after 1988 are described in detail. According to the report, two-thirds of the onset ages were children over 2 years old and young adults (13–19 years), and one 21-year-old adult died, which was significantly different from the case until 1988. In addition, 33.3% of patients were normally developed, and there were 4 cases in which the disease developed with valproic acid monotherapy. In the survey up to 1988, only 6% of patients had the onset of valproate after 6 months, but after 1988, 33% of patients had developed the disease after 6 months. The reason for these changes is that in the first 6 months after the start of valproic acid administration, the attention and visits of doctors and patients increased, and regular laboratory examinations became sufficient, but after 6 months, attention to hepatotoxicity gradually became neglected. Early clinical signs of patients with VHT included nausea, vomiting, loss of appetite, emotional blunting, coma, increased seizures, status epilepticus, and fever in 7 out of 8 patients. In addition, there were 20 cases of severe hepatotoxicity that did not die by 1988 and 6 cases after that, and there are many cases that are not reported in the paper. In addition, Scheffner et al. reported in 1988 that 4-en-VPA, an unsaturated metabolite of valproate, was detected in most of the cases examined [29], but many subsequent studies did not necessarily detect 4-en-VPA, and the search for metabolites of valproate was no longer valuable to prevent or treat VHT.

12.5 Overview of Studies on the Causes of VHT and VHE

As described above, as many cases of VHT and VHE have been reported, many clinical and basic studies have been reported on the causes of VHT and VHE. According to Kaneko et al., they can be broadly classified as follows [31].

1. Hyperammonemia hypothesis
2. Carnitine deficiency hypothesis
3. Inherited metabolic disorders hypothesis
4. Unsaturated metabolites of valproic acid hypothesis
5. Reduced free radical scavenger activity hypothesis

1. The hyperammonemia hypothesis is an interesting hypothesis because it is gradually becoming clear that hyperammonemia occurs in nearly half of VHT cases, and that valproic acid and its metabolites inhibit the detoxification and metabolic system of ammonia such as the urea cycle. However, hyperammonemia does not precede VHT in all cases.
3. In the inherited metabolic disorders hypothesis, Lenn et al. found that in the case of a brother and sister with progressive encephalopathy with hepatic damage presumed to be an autosomal latent (recessive) genetic disease, the younger sister developed VHT-like hepatotoxicity even though she did not take valproate [32], and Appleton et al. reported that two boys from different families with presumed progressive neuronal degeneration of childhood (PNDC) received valproic acid and developed VHT, but two sisters and one sister in each case developed VHT and died even though they had not received valproic acid,

and all three cases had steatosis, leading us to speculate that the two boys may have also developed VHT without valproic acid [33]. This hypothesis is also interesting, but since many of the cases reported by Lenn and Appleton have developmental delay and poor growth, and VHT also occurs in normally developing children who have been confirmed to have no inherited metabolic disorders. So that this hypothesis may be true in some VHT cases, but inherited metabolic disorders cannot be judged as the cause of all VHT cases.

4. Regarding the unsaturated metabolite hypothesis of valproate, the association between the concentration of unsaturated metabolites such as 2-en-VPA and 4-en-VPA and the development of VHT and VHE was also vigorously studied [34–38]. Studies of these valproate metabolites have also been discussed in review articles [31, 39]. Initially, it was thought that specific metabolites such as 4-en-VPA might be the cause of VHT and VHE but based on the results of many subsequent studies on metabolites of valproic acid, it is now difficult to predict the onset of VHT and VHE based only on the concentration of these unsaturated metabolites [30, 31]. For example, with regard to the concentration of 4-en-VPA, which was considered to be the most toxic, the results of studies in a wide age range of 0–60 years showed that the concentration of 4-en-VPA was higher in younger age groups [35], but when the age range was limited to children, the concentration of 4-en-VPA in infants was rather lower than that in older children, and 4-en-VPA levels do not correlate with liver function abnormalities or blood ammonia levels [31, 35]. Based on these findings, it is thought that valproic acid itself, rather than metabolites such as 4-en-VPA, may be an important factor in the toxicity of VHT [31].

On the other hand, among these various hypotheses regarding the cause of VHT and VHE, the carnitine deficiency hypothesis of (2) is described in detail below.

12.5.1 Carnitine Deficiency Due to Valproic Acid

The development of VHT and VHE caused by this valproic acid has been found to be closely related to carnitine deficiency and abnormal carnitine metabolism.

12.5.1.1 Early Clinical Studies Such as Blood Carnitine Concentrations in Patients Receiving Valproate

In 1982, Ohtani et al. reported the concentration of free carnitine and ammonia in the blood of 25 children with severe physical and mental disabilities, 14 treated with antiepileptic drugs containing valproic acid and 11 treated with antiepileptic drugs without valproic acid [40]. In 14 patients treated with valproate, blood concentrations of valproate and blood-free carnitine concentrations were inversely correlated ($r = -0.81$, $P < 0.005$, Fig. 12.2a), and in all 25 patients, the blood-free carnitine concentration and blood ammonia level were also inversely correlated ($r = -0.69$, $P < 0.005$, Fig. 12.2b), and the ammonia level was higher in the case with low blood-free carnitine. There was no significant difference in SGOT and SGPT in both groups. The blood ammonia levels in the 14 patients treated with valproate were high at 143.8 ± 42.4 μg/dL compared to 46.7 ± 13.9 μg/dL for the control, but when 50 mg/kg/day DL-carnitine (L-carnitine was not available at that time, and only DL-form was available) was administered, it was significantly reduced to 70.8 ± 45.8 μg/dL. All of the patients in this study consumed a diet rich in carnitine, and the blood-free carnitine levels of the 11 patients treated with antiepileptic drugs without valproate were not significantly different from those of controls. Therefore, the decrease in blood-free carnitine concentration and hyperammonemia observed in 14 patients treated with valproate were considered to be side effects of valproic acid administration. This study was the first report in the world to show the association between hyperammonemia and carnitine defi-

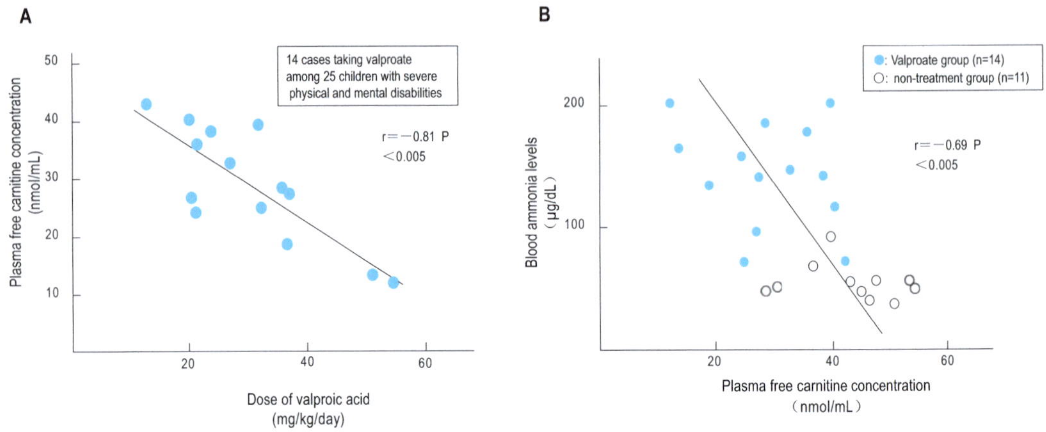

Fig. 12.2 Relationship of valproate administration on plasma-free carnitine and blood ammonia concentrations in 25 children with severe physical and mental disabilities. (**a**) Relationship of plasma-free carnitine concentration and valproate dose in 14 children with severe physical and mental disabilities who received valproate administration. (**b**) Relationship of blood ammonia and plasma-free carnitine concentrations in 25 children with severe physical and mental disabilities. (Modified from Ref. [40])

ciency caused by valproic acid, and this pioneering study has led to the publication of a number of studies on blood carnitine concentrations in patients with epilepsy treated with valproate.

Following this report by Ohtani et al., Böhles et al. reported a decrease in blood-free carnitine concentration in a 3 year 6 month-old girl who developed Reye-like syndroment receiving valproate for 6 months (450 mg/day) [41]. In the second month of treatment with valproic acid, the patient vomited and stopped taking valproic acid. The parents stopped taking valproic acid for 2 weeks and the vomiting disappeared, but the seizures increased, so valproic acid was restarted and the patient was finally treated at a dose of 600 mg/day for 4 and 3 weeks before admission. He was hospitalized after losing consciousness following an epileptic seizure but died on the 10th day of hospitalization due to jaundice, hyperammonemia, and abnormal liver function. At that time, his blood ammonia level was 176 μg/dL, SGOT was 329 IU/L, SGPT was 82 IU/L, and LDH was 2000 IU/L. He showed symptoms of liver failure. The blood-free carnitine concentration was extremely low at 7.0 ng/mL (normal value was 42.3 ± 7.2 ng/mL). In the discussion, it is known that a decrease in free carnitine decreases β-oxidation and causes small-drop fatty liver, but since only free carnitine was measured in this study, we cannot say for sure. However, it is possible that valproic acid and its metabolites trap free carnitine and decrease free carnitine levels, and further studies are needed to investigate the changes in carnitine metabolism induced by valproic acid.

In response to these studies, Coulter proposed the carnitine deficiency hypothesis that VHT and VHE may be caused by carnitine deficiency, which is caused by inhibition of carnitine metabolism by valproic acid and its metabolites [42]. Children with hepatic impairment and children with severe physical and mental disabilities have a high risk of carnitine deficiency due to reduced carnitine biosynthesis in the liver and decreased carnitine intake due to poor nutrition, and he thought that valproic acid administration might further exacerbate carnitine deficiency.

Murphy et al. also supported Coulter's hypothesis that valproic acid may cause carnitine deficiency and thereby develop Reye-like syndrome. They examined blood-free carnitine levels in 13 asymptomatic patients under 10 years of age receiving treatment with valproate and one infant showing signs of hepatotoxicity due to valproate [43]. Of the 13 asymptomatic patients, 7 were treated with polypharmacy and 5 were treated with valproate within 4 months. The concentration of free carnitine was

39.6 ± 5.6 μmol/L in 13 patients receiving valproic acid compared to the normal value of 50 ± 10 μmol/L. In addition, 6-month-old infants with developmental delays who showed signs of hepatotoxicity started receiving valproate at 3 months of age and were hospitalized at 6 months of age due to decreased responsiveness. At that time, the ammonia level was 100 μmol/L (normal value 35 or less), the AST value was 7080 IU (normal value 35 or less), and the ALT value was 9690 IU (normal value 35 or less). The patient's blood-free carnitine concentration was extremely low at 5.3 μmol/L but normalized when valproate was discontinued. In other words, the concentration of free carnitine in the blood was slightly reduced in asymptomatic patients taking valproate, and it was extremely low in symptomatic patients. Although Reye-like syndrome is a rare disease, it is a disease that has been experienced in two or three cases in every children‘s hospital, and carnitine supplementation therapy may be effective [43].

Morita et al. also conducted a multivariate analysis of blood carnitine levels and the factors influencing them in 25 severely mentally and physically handicapped children aged 1–13 years who were receiving polypharmacy of antiepileptic drugs [44]. Twelve patients were treated with multidrug antiepileptic drugs containing valproic acid, and the remaining 13 patients were treated with multidrug antiepileptic drugs without valproic acid. In addition, 32 age- and sex-matched patients aged 2 months to 30 years were included in the control group. The 25 patients treated with multidrug antiepileptic drugs had significantly lower total carnitine and free carnitine levels than those in the control group but were particularly significantly lower in the multidrug antiepileptic drug treatment group containing valproate. There was no significant change in blood acylcarnitine levels. Multivariate analysis showed that the factors associated with total blood carnitine and free carnitine concentrations were positive factors for upper arm circumference (i.e., muscle mass) and negative factors for valproic acid doses. Doses of phenytoin and nitrazepam also had a weak effect.

Rodriguez-Segade et al. compared blood carnitine levels in adult patients with epilepsy in a group of 34 patients treated with antiepileptic drugs including valproate and in a group of 149 patients treated with antiepileptic drugs other than valproate such as carbamazepine, phenobarbital, and diphenylhydantoin [45]. In the valproate-treated group, 76.5% had free carnitine levels below the lower limit of normal, and acylcarnitine levels were significantly higher. In the group treated with antiepileptic drugs other than valproate, free carnitine was below the lower limit of normal in only 21.5%.

Shapira et al. reported muscular-free carnitine concentrations in biopsies of the vastus lateralis muscle in seven children with severe epilepsy who had been treated with valproate for a long period of time [46]. The concentration of free carnitine in the blood was 29.3 ± 8.7 nmol/mL compared to the normal value of 37 ± 6 nmol/mL. The total carnitine concentration in the muscle was 7.2 ± ±4.9 nmol/mg NCP compared to the normal value of 226 nmol/mg NCP, which was significantly lower even in the muscle. Although there was no obvious decrease in blood-free carnitine concentrations of 31, 35, 43, and 32 nmol/mL in 4 of the 7 patients, the total carnitine concentrations in the muscles were 8.4, 7.8, 5.6, and 17.1 nmol/mg NCP, respectively, and 3 out of 4 patients were extremely low compared to control. Of the six patients who could be observed with an electron microscope, a large number of fat droplets were observed in the muscles, and abnormal mitochondria were observed in one case. When carnitine supplementation was performed in patients 3 and 6, an improvement in muscle tone was observed. The study showed that valproic acid administration reduced carnitine levels in tissues in some cases, even though blood carnitine levels were normal.

Matsuda et al. examined blood and urinary carnitine levels in five severely mentally and physically handicapped children (three of whom had Reye-like syndrome) who presented with hyperammonemia after receiving valproate [47]. The mean ammonia levels in the five patients were high at 196.5 ± 22.9 μg/dL compared to the normal value of 46.7 ± 2.8 μg/dL, the free carnitine concentration was low at 19.4 ± 4.6 nmol/mL (normal value 56.0 ± 2.1 nmol/mL), and the acylcarnitine concentration was high at 12.7 ± 4.2 nmol/mL (normal value 6.0 ± 0.3 nmol/mL). The acylcarnitine/free carnitine ratio was 0.62 ± 0.1, compared with normal values of 0.11 ± 0.02. Urinary-free carnitine was 431.1 ± 88.7 nmol/mg creatinine, much higher than the normal value of 51.5 ± 8.3 nmol/mg creatinine, and urinary acylcarnitine was 841.8 ± 329.0 nmol/mg creatinine, higher than the normal value of 200.2 ± 39.2 nmol/mg creatinine. Originally, free carnitine is reabsorbed in the renal tubules, and acylcarnitine is excreted in the urine without much reabsorption. However, the authors reported that in patients receiving valproic acid, the reabsorption of free carnitine in the renal tubules is inhibited and the clearance of acylcarnitine is also reduced, thereby lowering the blood-free carnitine and increasing blood acylcarnitine and blood acylcarnitine/ free carnitine ratio. This may be one of the mechanisms of carnitine deficiency induced by valproic acid.

It was also shown that valproic acid inhibits carnitine uptake by carnitine transporter in later basic study conducted by Tein et al. They investigated the effect of valproic acid on carnitine uptake in cultured human skin fibroblasts and found that valproic acid suppresses carnitine uptake in fibroblasts in a dose-dependent and processing time-dependent manner [48], and that the inhibitory effect of carnitine uptake by valproic acid is canceled by washing out the cells in valproate-free culture medium for more than 5 days [49]. Inhibition of free carnitine reabsorption in renal tubules and inhibition of carnitine transporters in muscle cells by valproic acid may be one of the mechanisms for the development of carnitine deficiency. However, when Wagner et al. subsequently expressed the human carnitine transporter (hOCTN2) in xenopus oocytes and examined its functional and pharmacological properties, they reported that hOCTN2-mediated carnitine uptake was inhibited by emetine, quinidine, and verapamil, but not directly by valproic acid [50]. Wagner et al. suggest that Tein et al. incubated fibroblasts and valproate for 24 hours, which suppressed the expression of hOCTN2, or that a metabolite of valproate inhibited carnitine uptake, by either mechanism other than direct carnitine transporter inhibition by valproate.

On the other hand, Melegh et al. examined blood and urinary carnitine concentrations in 11 severely mentally and physically handicapped children who were treated with valproic acid monotherapy [51]. Blood-free carnitine concentrations were significantly lower at 16.8 ± 1.8 nmol/mL in the valproate treatment group compared to 26.5 ± 2.1 nmol/mL in the control group, but there was no significant change in blood acylcarnitine concentrations, and urinary-free carnitine excretion was 21.2 ± 5.8 μmol/day compared to 154.3 ± 33.6 μmol/day in the control group. The concentration of β-hydroxybutyrate in the blood, one of the ketone bodies, decreased to 23.2 ± 2.5 nmol/mL compared to 81.9 ± 7.8 nmol/mL in the control group, and the urinary carnitine concentration was measured continuously for 10 days in one patient with infantile spasm. Since urinary carnitine excretion decreased continuously, and the blood-free carnitine concentration decreased from 44.7 nmol/mL on the first day to 10.4 nmol/mL on the 10th day, the mechanism of carnitine deficiency caused by valproic acid may not be due to an increase in carnitine excretion in the urine, but to a decrease in carnitine biosynthesis in liver [51].

Krähenbühl et al. also reported free carnitine and acylcarnitine levels in blood and liver tissue of a 39-year-old woman with VHT [52]. The patient had been suffering from bipolar disorder since the age of 20 years but was untreated. She developed frontal status epilepticus 5 months earlier and was prescribed 2400 mg/day of valproic acid and 6 mg/day of clonazepam. Initially, liver function was normal and blood levels of valproic acid and clonazepam were within therapeutic limits. Four months later, she developed ataxia and apathy and was prescribed carbamazepine 400–600 mg/day, but this did not improve. She developed leukopenia, progressive anemia, thrombocytopenia, etc., and was hospitalized with elevated bilirubin and AST. Five days later, she died of multiple organ failure due to hepatorenal syndrome. The patient's blood-free carnitine and total carnitine concentrations were 18.8 and 27.4 μmol/L (controls were 35.8 and 44.7 μmol/L), and free and total carnitine concentrations in liver tissue were decreased to 2.98 and 4.97 nmol/mg non-collagen protein. The free CoA, acetyl CoA, and total CoA contents of the liver tissue were extremely decreased to 101, 20, and 320 pmol/mg non-collagen protein (controls were 1530, 89, and 2060 pmol/mg non-collagen protein, respectively), while propionyl CoA and succinyl CoA were significantly increased to 32 and 89, respectively, compared with 4 and 36 in the control group. In other words, the decrease in free carnitine in the blood may reflect changes in hepatic tissue. They hypothesized that the decrease in free carnitine and the depletion of CoA in the liver tissue caused a decrease in liver β-oxidation and other mitochondrial functions, leading to rapid liver failure. Based on this, they concluded that valproic acid should not be prescribed for epilepsy in patients with signs of abnormal mitochondrial energy metabolism, such as ptosis or ocular paralysis, or in patients with abnormal fatty acid oxidation or inherited metabolic diseases.

Laub et al. also measured blood carnitine levels in 21 patients each taking valproate, other antiepileptic drugs, and a control group and reported that blood-free carnitine levels were significantly lower in the valproate group than in the other groups, that short-chain acylcarnitine levels and short-chain acylcarnitine/free carnitine and short-chain acylcarnitine and short-chain acylcarnitine/free carnitine ratios were significantly higher in the valproate group than in the other groups, and there was no correlation between blood-free carnitine or total carnitine concentration and ammonia [53]. However, Laub et al. experienced a case of hepatotoxicity in a 3.5-year-old girl who developed hepatotoxicity despite normal blood carnitine levels and died after carnitine supplementation therapy at 100 mg/kg/day was ineffective, suggesting that carnitine deficiency may not be the only cause of valproic acid-induced hepatotoxicity.

Murphy et al. also reported a case in which a 16-month-old child with mild developmental delay began to exhibit periodic vomiting, ataxia, chorea-like symptoms that constantly moved the legs, and peculiar eye movements, and was considered to be a general kinesthetic neuropathy based on electromyography [54]. Prior to valproic acid administration, the concentration of free carnitine in the blood at 16 months of age was 21.3 nmol/mL, the lower limit of normal (19–50 nmol/mL). At 17 months of age, the child started receiving valproate at 15 mg/kg/day and carnitine at 90 mg/kg/day but died due to hepatotoxicity. They described that carnitine is not a panacea and should be used for side effects other than valproic acid-induced hepatotoxicity.

12.5.1.2 Basic and Clinical Research on Carnitine Deficiency Caused by Valproic Acid

Coulter reviewed basic and clinical studies on valproic acid use and carnitine deficiency in such epileptic patients, including his own cases, and again proposed the carnitine deficiency hypothesis as a cause of VHT and VHE [55, 56]. Raskind et al. also extracted only clinical articles on pediatric subjects from the literature hit by "valproic acid" and "carnitine" from 1966 to December 1998 and scrutinized their contents [57]. The contents of these three reviews and the results of numerous studies up to around 2000 are summarized as follows.

Suppression of Mitochondrial Function by Valproic Acid

Many basic studies have shown that valproic acid and its metabolites inhibit mitochondrial fatty acid β oxidation and oxidative phosphorylation. Valproic acid is a branched fatty acid that is metabolized by the β oxidation system when incubated with isolated rat liver mitochondria. It has been reported that valproic acid and its unsaturated metabolite such as 2-n-propyl-4-pentenoic acid inhibit β oxidation [58, 59], and deprive free CoA, which is important for metabolism [60, 61]. Valproic acid also inhibits oxidative phosphorylation of isolated rat liver mitochondria [62]. In vivo administration of valproic acid or valproic acid and phenobarbital to rats caused a decrease in cristae and granules in the rat liver mitochondria, and the deposition of fat droplets in hepatocytes, resulting in a decrease in free carnitine in the blood and liver [63], and carnitine supplementation prevented these mitochondrial disorders [64]. Incubation of rat liver cells with 0.1–5 mM valproic acid significantly suppressed the oxidation of ^{14}C-palmitic acid [65]. In addition, administration of valproic acid to mice not only reduced free carnitine but also decreased CoA in hepatocytes, and β oxidation and ketone production were suppressed [66]. It is thought that valproyl CoA increases valproyl-CoA and consequently decreases free CoA, which is important for metabolism, and that the metabolite 4-en-VPA directly inhibits β oxidation [67, 68]. It has been speculated that carnitine supplementation prevents these mitochondrial dysfunctions induced by valproic acid because valproic acid and its metabolites are eliminated from the mitochondria as carnitine esters.

These mitochondrial dysfunctions and inhibition of fatty acid oxidation by valproic acid have been shown in clinical studies too. Compared to patients treated with other antiepileptic drugs, patients with epilepsy receiving valproate had higher levels of lactate, pyruvate, and alanine in their blood and suppressed mitochondrial function, and the combination of valproic acid and other antiepileptic drugs such as valproate and phenobarbital tended to have higher ammonia levels and lower blood-free carnitine [69]. In addition, patients taking valproic acid for a long time had a decrease in fat burning and an increase in carbohydrate utilization, resulting in a significantly higher respiratory quotient (0.91 ± 0.01 vs. 0.87 ± 0.01, $P < 0.05$), and these changes reversed after 1 month of carnitine supplementation therapy [70]. Arimoto et al. examined the effect of valproic acid on fatty acid β oxidation from the amount of $^{13}CO_2$ excreted in exhaled $^{13}CO_2$ by administering octanoic acid or palmitic acid labeled with the stable isotope of carbon ^{13}C in pediatric epilepsy patients taking valproate and children not taking valproate as a control group [71, 72]. All of the subjects taking valproate were taking other antiepileptic drugs, and laboratory test values such as ammonia levels were normal and asymptomatic. The amount of $^{13}CO_2$ excreted in exhaled air (i.e., the β oxidative capacity of fatty acids) was greatly suppressed in the valproic acid group, and the exhaled excretion was inversely correlated with the dose and blood concentration of valproate. In this way, it has been shown that mitochondrial β oxidation is actually suppressed even in asymptomatic conditions. Many other studies have reported that valproic acid inhibits mitochondrial functions such as fatty acid β oxidation, and it is thought that valproic acid inhibits fatty acid oxidation, which in turn causes weight gain and steatosis.

Baldwin et al. reported that one of the mechanisms for inhibiting fatty acid oxidation by valproic acid is that the metabolite 2,4-dien-valproate and its CoA ester reversibly bind to the α subunit of mitochondrial trifunctional proteins and inhibit their activity [73]. Aires et al. also reported that the inhibition of hepatic CPT1A activity by valproyl-CoA in vitro may be a mechanism for weight gain, fatty liver, and hepatotoxicity caused by valproic acid [74].

In this way, it has been reported that valproic acid and its metabolites suppress various steps such as mitochondrial β oxidation and oxidative phosphorylation. However, in most cases, these mitochondrial function suppressions are compensated for by biological defense mechanisms, and hepatotoxicity and encephalopathy are avoided. However, it is thought that in a small percentage of cases, the mitochondrial suppression may not be fully compensated and disrupted due to idiosyncratic reaction, etc. [68, 75].

Blood Carnitine Concentration in Epilepsy Patients Treated with Valproate

As have already shown in many studies, valproic acid is associated with carnitine deficiency (decrease in free carnitine and total carnitine) and carnitine insufficiency (increase in acylcarnitine/free carnitine ratio, acyl-CoA increase, indicating a relative decrease in free CoA, which is essential for various metabolisms, see Chap. 5), which has been shown in many clinical stud-

Table 12.2 Clinical studies on the concentrations of total carnitine, free carnitine, and acylcarnitine in patients treated with valproate

Subjects taking VPA						Control				
Source[a]	Description	*n*	TC	FC	AC	Description	*n*	TC	FC	AC
Ohtani et al. [2]	Age 3~21 yr	14	32.6[b, c]	28.6[c]	–	Epilepsy, other drugs	11	48.6	43.0	–
						Healthy	27	49.0	44.2	–
Laub et al. [3]	VPA alone; age 3~21 yr	14	46.4	34.7	11.7	Epilepsy, other drugs	21	47.0	39.9	7.0
	VPA plus other drugs	7	36.4	28.9	7.5	Healthy	21	49.5	41.2	8.0
Morita et al. [4]	Age 1–30 yr	12	33.4[c]	21.5[c]	12.0	Epilepsy, other drugs	13	43.6	31.5	12.0
						Healthy	32	60.3	51.7	9.7
Melegh et al. [5]	Children	11	24.3[c]	16.8[c]	–	Healthy	11	34.9	26.5	–
Rodriguez-Segade et al. [6]	Age 17~65 yr	34	36.9	26.4	10.5	Epilepsy, other drugs	149	48.1	41.2	6.9
						Healthy	49	53.3	47.1	6.2
Beghi et al. [7]	VPA alone; age 1~39 yr	54	49.4	36.2	13.1	Epilepsy, other drugs	51	46.9	37.0	10.1
	VPA plus other drugs	55	44.2	33.0	11.2	Epilepsy, no drugs	53	50.8	41.4	9.2
Hug et al. [8]	VPA alone	53	35.6[c]	27.0[c]	8.6[c]	Phenobarbital	119	32.7[c]	24.6[c]	8.1[c]
	VPA plus other drugs	18	30.1[c]	23.2[c]	6.9[c]	Healthy	32	57.8	42.5	15.3
Opala et al. [9]	VPA alone	43	40.8	29.9[c]	10.9	Epilepsy, other drugs	43	48.1	36.7	8.9
	VPA plus other drugs	91	29.3[c]	21.4[c]	8.0	Healthy	89	44.2	36.8	8.9
Riva et al. [10]	Taking VPA for 45 days	22	50.0[c]	35.0[c]	15.0[c]	Same patients before VPA	22	60.0	49.0	11.0
Toksoy et al. [11]	Children	24	–	33.5[c]	–	Healthy age/sex matched	24	–	50.8	–
Zelnik et al. [12]	Taking VPA	14	–	29.1[c]	–	Same patients before VPA	14	–	37.6	–
Murphy et al. [13]	Under age 10 yr	13	39.6	–	–	Not stated	–	50.0	–	–

From Ref. [56]

VPA valproate, *TC* total carnitine, *FC* free carnitine, *AC* acylcarnitine

[a]The reference numbers in this table is the reference number in the original Coulter's review (Ref. [56] in this chapter)

[b]All carnitine values are mean levels in μmol/L

[c]Value is significantly different ($P < 0.05$) from respective value of the control group shown in italics in the same study

ies (Table 12.2). There are differences in whether or not the patients in the study include patients with mental retardation or severe disorders, and some of them may not be significantly different from normal values, but in general, it can be seen that patients taking valproate have lower blood-free carnitine and total carnitine concentrations and the acylcarnitine/free carnitine ratio is high. A comparison before and after valproic acid showed that the carnitine concentration was lower after taking it. In addition, the multidrug antiepileptic drug treatment group containing valproate tended to be lower than the valproic acid monotherapy group. Patients with epilepsy aged 1–10 years have lower blood levels of free carnitine than patients with epilepsy aged 10–18 years (Bohan TP. Ann Neurol 1991; 30: 491, Conference Abstracts). When the blood levels of 18 newborns were examined, 13 had low free carnitine levels (Coulter DL. Ann Neurol 1991; 36: 510–11, Conference Abstracts). In this way, it has been reported that blood carnitine concentrations decrease at a younger age, but there are also reports that blood-free carnitine concentrations decrease in adult patients taking valproic acid [45].

Rationality for the Diagnosis of Carnitine Deficiency

The rationality and justification of diagnosing and assessing the state of carnitine in the entire body by measuring blood carnitine levels are debatable. For example, Shapira et al. reported a case in which blood carnitine levels were normal, but biopsy showed a lower level of carnitine in the muscle than in the control group [46]. This is because more than 90% of carnitine in the body is present in the muscles, and the concentration of carnitine in the muscles is said to be more than 10 times higher than in the blood. In other words, when the blood carnitine concentration is low, it is expected that carnitine is also low in tissues such as muscles and liver, and it is always necessary to keep in mind that the carnitine concentration may be low in tissues even if the blood carnitine concentration is normal.

Many of the studies in Table 12.2 found significantly lower carnitine levels in patients taking valproate, while others found no significant difference. It has been reported that 4–76% of patients in the valproic acid group and 8–36% of patients treated with other antiepileptic acid drugs had carnitine deficiency, and although it tends to be more common in the valproic acid group, it can be said that the carnitine concentration is originally highly variable. A decrease in 2 SD or more from the normal mean value is considered a clear decrease. From an ethical point of view, it is unrealistic to measure the concentration in tissues by biopsy in actual clinical practice, and the only way to diagnose carnitine deficiency is to measure the blood concentration and comprehensively diagnose carnitine deficiency based on other patient background and clinical signs. However, not all patients diagnosed with carnitine deficiency show clinical symptoms or laboratory abnormalities due to carnitine deficiency. For example, Igarashi et al. reported blood carnitine levels in 40 severely mentally and physically handicapped patients aged 2–36 years and 69 controls, but no clinical signs such as non-ketotic hypoglycemia, hepatic hypertrophy, hypotonia, abnormal cardiac function, or abnormal liver function were observed in six patients with hypocarnitinemia (Note: Although the author states that ammonia levels were also normal, five of the six hypocarnitinemic patients had ammonia levels of 80 μg/dL or higher, which are elevated considering the reference value of 40–80 μg/dL) [76]. However, the ketone formation reaction was impaired when fat emulsion was loaded, and it was restored by carnitine supplementation. In addition, there are reports that blood carnitine levels and blood ammonia levels are inversely correlated, as in the study of Ohtani et al., but there are reports that no association was observed, so it is not necessarily correlated. Hirose et al. examined the blood-free carnitine and ammonia levels of 45 healthy valproate patients aged 6.3–21.7 years with no neurological abnormalities or malnutrition except for epilepsy, and age-matched 45 control patients who did not take valproate and did not have epilepsy. Although a weak inverse correlation was observed in which blood-free carnitine and total carnitine decreased as blood concentrations of valproate increased in the valproate-treated group ($r = -0.40$; $P < 0.01$ and $r = -0.34$; $P < 0.05$, respectively), these values were in the normal range, and there was no significant difference in the blood-free carnitine, total carnitine, and ammonia values between the valproate-treated epilepsy patients and the control group [77]. Thus, blood carnitine concentrations are more dependent on nutritional conditions than valproate concentrations, and carnitine supplementation should be reconsidered in otherwise healthy epileptic patients without neurological abnormalities or nutritional disorders.

Fung et al. also examined blood-free carnitine and acylcarnitine in 43 epilepsy patients under 19 years of age (0.9–18.7 years, mean 9.1 years) receiving valproate therapy and found that only 2 patients had free carnitine levels below the normal range, raising questions about routine carnitine measurement and carnitine supplementation [78]. Of the 43 patients, 40% were polypharmacy, 49% had mental retardation, 26% had difficulty walking, 28% had multiple neurological abnormalities, 23% had eating problems, total 83.7% had one or more risk factors for carnitine deficiency, and 23.2% had three or more risk factors. So, not all patients are healthy except for epilepsy. However, since the acylcarnitine level is not described in this report, it is not known whether the patient was in a state of carnitine insufficiency, and there is no clinical data such as clinical symptoms and ammonia level, so it is not possible to discuss it enough. Fung et al. also suggest that carnitine supplementation should be considered in patients with hyperammonemia, anorexia, constipation, nausea, vomiting, weakness, and hypotonia.

Abnormal laboratory values and clinical symptoms due to carnitine deficiency in patients with epilepsy are thought to develop not only from carnitine levels in blood and tissues but also in association with other factors when energy demand increases, such as fever and infection, and carnitine deficiency or carnitine insufficiency. The state of carnitine deficiency or carnitine insufficiency itself should be considered a dangerous state, regardless of the presence or absence of clinical symptoms such as hyperammonemia.

Association Between VHT and Carnitine Deficiency

In most cases of VHT, liver function abnormalities such as mild high SGOT levels were observed in the early stages. Although it is a basic study, Takeuchi et al. found that when valproic acid was incubated with rat liver primary cultured cells in vitro, intrahepatocellular enzymes such as lactate dehydrogenase (LDH), glutamate oxaloacetate transaminase (GOT), and glutamate pyruvate transaminase (GPT) leaked into the culture medium in a concentration-dependent manner due to the cytotoxicity of valproate. It was reported that administration of DL-carnitine suppressed these leakage [79]. In addition, microvesicular steatosis was observed in most cases of VHT, suggesting that fatty acid metabolism was suppressed. These facts raise the possibility that carnitine deficiency may be associated with VHT. However, it is unclear whether carnitine deficiency was present in all cases of VHT. As mentioned above, there have been reports of decreased blood-free carnitine in these patients [41, 43, 47], while others have reported no decrease in blood-free carnitine [53]. In addition, it has been reported that carnitine supplementation therapy was given to patients who developed hepatotoxicity, but in some cases, it was ineffective, and they died [53, 54, 80]. Carnitine deficiency is known to occur in many inherited metabolic disorders, and it is thought that valproic acid administration may worsen it and cause Reye-like syndrome and VHT. In other words, some of the cases that developed these VHTs may have had undiagnosed hidden inherited metabolic abnormalities such as fatty acid oxidation disorders and urea cycle disorders. Therefore, when administering valproic acid to patients who are suspicious of inherited metabolic disorders, urinary organic acid, blood gas, lactic acid, pyruvate, carnitine, ammonia, etc. should be screened to see whether there are any hidden inherited metabolic disorders or not. Coulter reported at the conference that in his several own cases in which the absence of inherited metabolic disorders was confirmed, patients who had previously developed hepatotoxicity with multidrug antiepileptic drug treatment containing valproic acid and had recovered once were again successfully administered valproate alone with carnitine supplementation therapy to prevent recurrence of hepatotoxicity [81]. However, Coulter described, "Clinicians should not rely on carnitine supplementation therapy to prevent hepatotoxicity, because there are not sufficient controlled trials" [56]. In 1995, the differences between fatality and survivors of VHT were unknown, and clinical studies of carnitine supplementation therapy were limited to a small number of cases, and the evidence was considered insufficient at that time.

VHE and Carnitine Deficiency

It is common for epileptic patients receiving valproic acid to exhibit hyperammonemia in the absence of signs of liver failure. These patients often present with clinical symptoms such as somnolence and hypotonia [21, 22, 82–85], but they may also be asymptomatic [86]. If patients receiving valproate show clinical signs such as altered levels of consciousness, hypotonia, asterixis (flapping tremors), ataxia, nausea or vomiting, and loss of appetite, blood ammonia should be measured. If ammonia is higher than the reference value, measures should be taken, such as switching valproic acid to another antiepileptic drug, if other antiepileptic drugs such as phenytoin or phenobarbital are also used change to valproic acid alone, or adding carnitine supplementation therapy to valproic acid, if carnitine is measured and reduced.

Risk Factors for Carnitine Deficiency

Carnitine deficiency is not uncommon in patients with epilepsy, and the etiology of carnitine deficiency includes younger patients such as neonates (undeveloped carnitine biosynthetic capacity), patients with poor nutritional status (reduced carnitine intake), patients with metabolic disorders and neuropathy, and patients treated with multiple antiepileptic drugs, including valproic acid, and these are risk factors for carnitine deficiency (Table 12.3).

Table 12.3 Risk factors for carnitine deficiency

Young age (less than 10 years old)
Multiple neurological disabilities (mental retardation, cerebral palsy, blindness/blindness, microcephaly, etc.)
Nonambulatory status
Underweight (decreased weight for height)
Diet low in meat and dairy products
On tube feeding or intravenous hyperalimentation
Taking multiple anticonvulsant drugs including valproate
Hyperammonemia
Hypoglycemia
Metabolic acidosis

From Ref. [56]

Younger Age Younger infants and children are also listed as high-risk groups for carnitine deficiency, but these young children may not be able to adequately compensate for carnitine reduction caused by valproic acid due to their low ability to biosynthesize carnitine.

Undernutrition Carnitine is abundant in meat and milk, and patients who consume low intake of these foods, and children who are fed by tube feeding or high-calorie intravenous infusion are more likely to develop carnitine deficiency because many of these nutrients do not contain carnitine. It can be said that children with severe mental and physical disabilities are also prone to carnitine deficiency due to these factors.

Inherited Metabolic Diseases Many cases have been reported in which patients with underlying hidden inherited metabolic diseases were treated with valproic acid and developed coma and other serious symptoms [33, 87–93]. Inherited metabolic diseases include several fatty acid oxidation disorders, mitochondrial diseases such as MELAS and Leigh syndrome, glutaric acidemia type 2, other organic acidemias, CPT deficiency, and urea cycle disorders. In urea cycle disorders, hyperammonemia often develops immediately after birth, and is diagnosed and treated, but in the case of heterozygous mutations, hyperammonemia may manifest itself after taking valproic acid because it has a weak ammonia processing capacity after growth [94, 95]. Therefore, even now that tandem mass screening of newborns is widespread, it is necessary to keep in mind the possibility that there are patients with inherited metabolic diseases that are difficult to find with weak mutations and heterozygous mutations in daily clinical practice. These patients with inherited metabolic diseases are prone to carnitine deficiency or carnitine insufficiency, since large amounts of their acyl groups are excreted in the urine as acylcarnitines to detoxify the toxicity of fatty acids and organic acids, which are metabolic intermediates accumulated as a result of metabolic abnormalities.

In clinical practice, moderate carnitine deficiency in tissues causes mild reversible hepatotoxicity, while more severe carnitine deficiency is thought to cause severe conditions such as Reye-like syndrome. Tissue carnitine levels are said to be much higher than blood carnitine levels. In cases of hepatotoxicity in patients with normal blood carnitine levels, it may be due to carnitine insufficiency or relative deficiency in the tissues. When valproic acid causes liver mitochondrial dysfunction, ATP production is reduced due to impaired energy metabolism, and metabolic systems that consume large amounts of ATP, such as albumin synthesis and ammonia detoxification, are suppressed. Itokazu et al. reported two cases of hypoalbuminemia in severely disabled patients with hypoalbuminemia who were receiving valproic acid and promptly improved after valproic acid administration was discontinued [96].

12.6 Carnitine Supplementation Therapy

12.6.1 The Ideas of Initial Carnitine Supplementation Therapy

Around 1991 and 1995, when Coulter proposed the carnitine deficiency hypothesis as a cause of VHT and VHE, there was not enough evidence for carnitine supplementation therapy for these rare but serious side effects of valproate. Therefore, in his 1995 review, Coulter concluded that carnitine supplementation therapy should not be indicated for all patients taking valproate but should be used in patients with risk factors for carnitine deficiency such as those listed in Table 12.3, with blood carnitine concentrations measured and either decrease or when carnitine insufficiency is detected [56]. The risk factors for carnitine deficiency in Table 12.3 coincide with those for VHT and VHE. The greater the number of risk factors, the more severe carnitine deficiency is expected [97]. On the other hand, Kelly pointed out that carnitine supplementation therapy for patients with epilepsy lacks sufficient scientific data and requires careful controlled clinical trials [98]. Freeman et al. conducted a double-blind, placebo-controlled crossover study in which 47 epilepsy patients were randomized to 17 to valproate, 5 to valproate plus other antiepileptic drugs, and 15 to carbamazepine, and their parents were evaluated for well-being [99]. However, the patients in this study were asymptomatic, had normal carnitine levels, and were not at any risk for carnitine deficiency. The "well-being" endpoint is also subtle and ambiguous. Conversely, carnitine supplementation therapy in asymptomatic patients with normal carnitine levels is not supported unless there are several risk factors. In addition, the results of this clinical study should not be misinterpreted to mean that carnitine supplementation therapy is ineffective for patients with symptomatic risk factors. Coulter reported an improvement in clinical symptoms before and after carnitine supplementation therapy in 20 patients with epilepsy with two or more risk factors for carnitine deficiency (Table 12.4) [56].

Carnitine is a safe drug with only mild side effects such as diarrhea. In extremely rare cases, fish odor may appear. In children with severe physical and mental disabilities, severe constipation often leads to serious conditions such as ileus, and carnitine also has the effect of improving these symptoms by improving gastrointestinal motility [100, 101].

Table 12.4 Results before and after carnitine therapy for 20 patients with two or more risk factors for carnitine deficiency

Symptom or sign examined	Present before treatment No.	Improved after treatment No. (%)[a]
Muscle weakness	13	2 (15)
Muscle pain or tenderness	0	– (–)
Hypotonia	7	2 (29)
Lncoordination	12	2 (17)
Frequent seizures	16	5 (31)
Lethargy	11	7 (64)
Listlessness	11	6 (55)
Apathy	13	7 (54)
Poor concentration	7	1 (14)
Headaches	0	– (–)
Loss of appetite	8	7 (88)
Nausea or vomiting	6	5 (83)
Constipation	4	3 (75)

From Ref. [56]
[a]One child was worse because of irritability and agitation

12.6.2 Effect of Carnitine Supplementation Therapy for VHT on Liver Survival Rate

Thus, there were only individual case reports of carnitine supplementation therapy in VHT, and there were several reports of deaths with no effect of carnitine supplementation. However, in 2001, an international research group consisting of pediatricians from the United States, Germany, and Japan led by Bohan and others conducted a retrospective inception cohort study that examined the effects of carnitine supplementation therapy on liver survival [102]. Bohan et al. investigated the effect of carnitine supplementation therapy on hepatic survival in patients with acute and severe symptomatic VHT with valproate, inspired by the fact that carnitine supplementation therapy improves survival in inherited metabolic diseases that cause secondary carnitine deficiency. The following is an overview of the report.

Methods Deaths and cases of liver transplantation were combined as liver failure cases. Of the 123 cases of hepatotoxicity registered in the International Registry of Adverse Effects of Valproic Acid, 92 patients were selected who met the inclusion criteria by reviewing their medical records. Ninety patients were patients with epilepsy and two were patients with bipolar disorder. All patients treated with carnitine were included, regardless of the turning point. Patients treated with carnitine to prevent hepatotoxicity were not excluded. No patients were treated with carnitine at the time of hepatotoxicity, but three patients had received carnitine therapy, and carnitine supplementation had been stopped more than 50 days before the onset of hepatotoxicity in all cases. Forty-two patients were in the carnitine-treated group and 50 were in the untreated group. Carnitine was administered intravenously in 21 cases and orally in 21 cases, at doses ranging from 50 to 100 mg/kg/day, for varying durations.

Results There were no significant differences in clinical characteristics such as country distribution, sex, cognitive function, single or multidrug valproate treatment, age distribution, weight percentile, last year hospital stay, and early symptoms between the two groups, and only in the calendar year when hepatotoxicity developed.

Of the 92 VHT cases overall, 42 were treated with carnitine supplementation and 50 with no carnitine. Table 12.5 shows their liver survival rates. Liver survival was significantly higher in the carnitine-treated group ($P = 0.001$). That is, liver survival was 20 (47%) among 42 patients in the carnitine-treated group and 5 (10%) among 50 patients in the untreated group. The effect of carnitine treatment on liver survival was evident in all subgroups: by country, sex, cognitive function, multiple or single drug administration, age at onset, age at onset, and hospitalization in the past year.

In 42 patients treated with carnitine, intravenous and oral preparations were used in 21 cases each, and the liver survival rates were 14/21 (67%) and 6/21 (29%), respectively. In addition, the period from the onset of hepatotoxicity to discontinuation of valproic acid was divided into three periods: 0–5 days, 6–10 days, and 10 days or more. The effect was high at 5 days (12 out of

Table 12.5 Hepatic survival in the L-carnitine-treated and untreated groups

Variable	L-carnitine treated cases/ total (%)	Untreated cases/total (%)
All patients	20/42 (47)	5/50 (10)
Country		
United States	13/26 (50)	3/30 (10)
Germany[a]	5/13 (38)	2/16 (13)
Other	2/3 (67)	0/4 (0)
Sex	14/25 (54)	5/30 (17)
M	7/17 (39)	0/20 (0)
F		
Cognitive abilities	9/22 (41)	1/26 (4)
Normal	11/20 (55)	4/24 (17)
Mentally retarded		
Valproate therapy	17/32 (54)	3/40 (8)
Polytherapy	3/10 (30)	2/10 (20)
Monotherapy		
Year of event		
1977~1987	2/11 (18)	1/22 (5)
1988~1992	7/14 (50)	4/20 (20)
1993~1997	11/17 (65)	0/8 (0)
Age, y		
<2	4/9 (44)	2/11 (18)
2~5	9/19 (47)	3/18 (17)
5~10	5/7 (71)	0/8 (0)
>10	2/7 (29)	0/13 (0)
Hospitalizations in last year		
None	8/10 (80)	1/6 (17)
One	6/14 (43)	3/25 (12)
Two or more	5/13 (38)	1/13 (8)

From Ref. [102]

Overall, hepatic survival was 20 of 42 patients (47%) in the L-carnitine treated group and 5 of 50 patients in the untreated group (10%)

[a]For L-carnitine-treated patients, the dissimilar survival of American and German patients may be related to differing use of intravenous carnitine use (60% of American, but only 40% of German patients) during the crucial first 5 days of the hepatotoxic episode

17 patients in the carnitine treatment group survived, and only 1 out of 13 patients in the non-treatment group survived), and the survival rate tended to decrease in the carnitine treatment group as the discontinuation of valproate and the start of carnitine treatment were delayed (Fig. 12.3a, blue bar graph). However, in the non-treatment group, the survival rate was always low regardless of the timing of discontinuation of valproate (Fig. 12.3a, white bar graph). The carnitine treatment group was divided into an intravenous formulation group (Fig. 12.3b, blue bar graph) and an oral preparation group (Fig. 12.3b, white bar graph), and the liver survival rate was compared by the time of discontinuation of valproic acid. The liver viability rate decreased as the number of days increased, from 1/2 (50%) at 10 days to 3/9 (33%) at 10 days or more. On the other hand, the oral formulation yielded only a liver survival rate of around 30% regardless of the number of days until discontinuation of valproate.

Liver survivors survived for more than 1 year after fully recovering not only their livers but also their neurologically recovering to their original state. According to the exploratory logistic analysis model, the liver survival rate was significantly higher when treated with intravenous preparations than oral formulations ($P = 0.002$) and conventional supportive care ($P = 0.001$). Treatment with oral formulations was only suggestive compared to supportive care ($P = 0.110$).

Looking at the liver survival rate of carnitine-treated patients by age group, 2/11 (18%) in 1977–1987, 7/14 (50%) in 1988–1992, and 1993–1997 in 11/17 (65%), as the year progresses, the number of patients treated with carnitine and the liver survival rate are increasing. This may be due to the growing recognition of the carnitine deficiency hypothesis in VHT and the widespread use of intravenous preparations. Incidentally, in the United States, an oral preparation of carnitine was approved in 1985 and an intravenous preparation in 1992 for side effects caused by valproic acid.

Discussion The typical treatment for drug-induced hepatotoxicity is only the removal of toxic components and supportive care, and the mortality rate is more than 80%, and progress has not been made for a long time. The same was true in the case of hepatotoxicity caused by valproic acid, and the situation was getting worse year by year. In this situation, it is surprising that carnitine treatment has a liver survival rate of nearly 50% compared to 10% of supportive care, even if it is not a panacea. The mechanism is that patients taking valproic acid are prone to secondary carnitine deficiency because carnitine is used in the

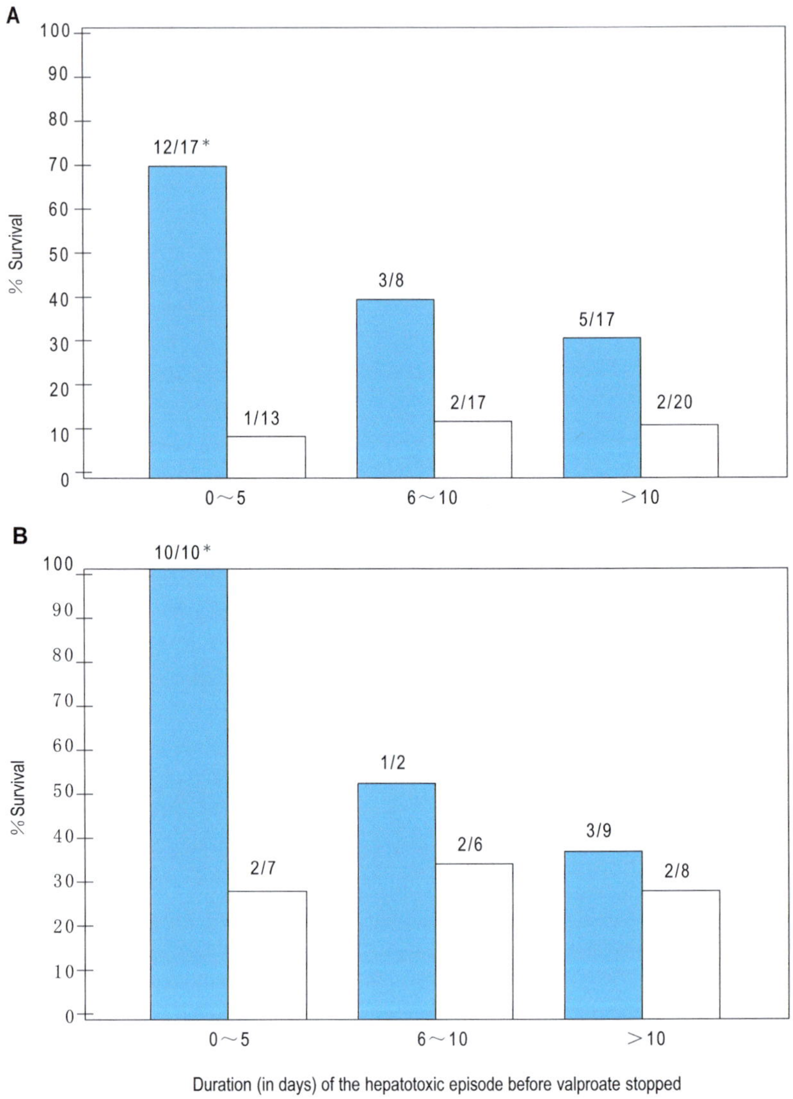

Fig. 12.3 Relationship between hepatic survival rate and time of carnitine administration (**a**), and comparison of efficacy of intravenous or oral carnitine on hepatic survival (**b**) in VHT cases. (**a**) Non-carnitine-supplemented group (white bar) showed lower hepatic survival compared to carnitine-supplemented group (blue bar) regardless of time of carnitine administration. Among carnitine supplemented group, administration of carnitine within 0–5 days after onset of symptoms showed high efficacy. *$P < 0.001$ vs. non carnitine-supplemented group. (**b**) Intravenous carnitine-supplemented group (blue bar) showed high efficacy on hepatic survival compared to oral carnitine-supplemented group (white bar), but it is important that carnitine is administered as soon as possible same as (**a**). *$P < 0.001$ vs. oral carnitine-supplemented group. (From Ref. [102])

process of detoxification of its toxicity, especially those who are underweight, young, polypharmacy, mentally retarded, and frequent hospitalization. When infection or severe epileptic seizures are added to this, stress-induced catabolism may occur, and VHT may develop in some patients (i.e., carnitine deficiency is the base of VHT). The high effect of intravenous injections may be due to the low bioavailability of oral administration (absorption rate in the intestinal tract of 15% or less), while the high blood concentration of intravenous injectables can be maintained and can be efficiently replenished in tissues. Whether or not to treat all patients with hepatotoxicity with carnitine depends on the thinking and preferences of the doctors at the advanced medical institutions that accept these patients and does not depend on the quality of the medical institutions. In any case, if VHT develops, valproate should be discontinued as soon as possible, and intravenous carnitine treatment should be given in addition to supportive care.

12.7 Recommendation of Carnitine Supplementation Therapy for Children with Epilepsy

12.7.1 Background

In August 1989, an advisory committee of pediatric neurologists recommended (1) prophylactic carnitine administration to epilepsy patients taking valproate under 2 years of age, and (2) carnitine supplementation therapy for children 1 week of age or older with carnitine deficiency. Carnitine deficiency was defined as a plasma-free carnitine concentration of 20 μM or less or a plasma acylcarnitine/free carnitine ratio greater than 0.4. However, these two recommendations for carnitine supplementation therapy were somewhat controversial. Therefore, in order to update this recommendation, a panel of nine eminent pediatric neurologists and inherited metabolic diseases specialists (hereinafter referred to as the "Expert Committee") was convened on November 23, 1996, including De Vivo, Coulter, Bohan, and Dreifuss, and the contents of the panel were published in 1998 [103].

The contents of the report include epilepsy patients treated with valproic acid or other antiepileptic drugs and epilepsy patients on a ketogenic diet, and it can be said that carnitine supplementation therapy is summarized not only for epilepsy patients but also for carnitine deficiency caused by inherited metabolic diseases and other causes. The physiological effects and functions of carnitine, the association between valproic acid and other antiepileptic drugs and carnitine deficiency, the ketogenic diet and carnitine deficiency, carnitine deficiency and its diagnosis, indications for carnitine supplementation therapy and the dosage of carnitine preparations in pediatric epilepsy patients, and patients who should receive carnitine supplementation therapy. The following is a summary of the recommendations of the expert committee.

12.7.2 Recommendations for Carnitine Supplementation Therapy by the Expert Committee

12.7.2.1 Functions of Carnitine in the Metabolism

The physiological functions of carnitine in metabolism in vivo can be summarized in Table 12.6.

12.7.2.2 Carnitine Deficiency in Patients with Epilepsy

Clinicians should consider the items listed in Table 12.7 when determining whether patients with epileptic encephalopathy or difficult-to-treat epileptic seizures have hidden metabolic disor-

Table 12.6 Metabolic functions of carnitine

Transport of long-chain fatty acids across the inner mitochondrial membrane
Facilitation of branched-chain α-ketoacid oxidation
Shuttling of acyl-CoA products of peroxisomal β oxidation to mitochondrial matrix in the liver
Modulation of the acyl CoA/CoA ratio in mammalian cells
Esterification and excretion of potentially toxic acyl-CoA metabolites that impair the citric acid cycle, urea cycle, and the pathways for gluconeogenesis and fatty acid oxidation during acute clinical crises

Modified from Ref. [103]

Table 12.7 Clues favoring an underlying metabolic abnormality in patients with epileptic encephalopathies or seizures

Family history of unusual neurologic problems or deaths in infancy or childhood
Episodic metabolic derangement or progressive epileptic disorder
Intractable early-onset seizures
Failure to thrive
Ataxia
Strokes at an early age
Seizures with minor motor manifestations
Extrapyramidal syndrome
Peripheral neuropathy

From Ref. [103]

ders or carnitine deficiency. Alternatively, metabolic disorders associated with a patient's specific epileptic symptoms can be scrutinized. For example, in the case of West syndrome (frontal status epilepticus), metabolic disorders such as phenylketonuria, non-ketotic hyperglycinemia, hyperornithinemia-hyperammonemia homocitrullinuria (known as HHH syndrome) should be scrutinized. Unfortunately, it is often difficult to identify specific epilepsy syndromes in infants and younger children with inherited metabolic diseases.

12.7.2.3 Carnitine Deficiency with Valproic Acid and Other Antiepileptic Drugs

As previously described, many studies have already shown that total carnitine and free carnitine are reduced in patients taking valproate. Carnitine deficiency is more likely to occur in patients who have been treated with valproic acid for a long time, in patients who have undergone polydrug treatment with valproate, and in patients who are younger (especially under 10 years of age). In addition, carnitine deficiency is more likely to occur in patients with severe disabilities such as cerebral palsy than in healthy children other than in children with epilepsy. A decrease in blood carnitine means a decrease in carnitine in tissues such as muscles.

On the other hand, studies on the administration of antiepileptic drugs other than valproic acid and blood carnitine concentrations are limited. According to the study by Hug et al, low carnitine levels (blood total carnitine concentrations more than 2 SD lower than the mean control level) was observed 23% in patients taking valproate (53 patients), 36% in patients taking phenobarbital (119 patients), 12% in patients taking phenytoin (92 patients), 8% of patients on carbamazepine (141 patients), 37% of patients on phenobarbital and phenytoin (19 patients), 18% of patients on phenobarbital and carbamazepine (17 patients), and 44% of patients on valproic acid and carbamazepine (18 patients). None of the 12 patients taking carbamazepine and phenytoin had reduced total carnitine levels. However, this study did not measure blood carnitine levels before taking antiepileptic drugs. In a study of 183 outpatient adult patients by Rodoriguez-Segade et al., 76.5% of patients taking valproate (34 patients), 26.8% of patients taking phenytoin and phenobarbital (41 patients), 22.8% of patients taking carbamazepine (57 patients), and 16% of patients taking phenytoin (25 patients) had a decrease of 2SD or more below the mean value of the control [45]. In addition, Zelnik et al. examined blood carnitine concentrations before and after antiepileptic treatment and reported that total carnitine levels and free carnitine levels were significantly reduced after administration in only valproate users compared to before treatment, but no significant reductions were observed in carbamazepine and phenobarbital [105]. Thus, among antiepileptic drugs, carnitine deficiency due to valproic acid is the most prominent, and it is thought that it is the basis for the development of VHT and VHE.

12.7.2.4 Ketogenic Diet and Carnitine Deficiency

In recent years, the ketogenic diet has been reviewed for children with refractory epilepsy and children who cannot use antiepileptic drugs due to unacceptable side effects. The brain normally oxidizes glucose aerobically for energy, but during starvation or high-fat diets, the brain uses ketone bodies as an energy source. The ketogenic diet consists of a high-fat diet, low carbohydrate, and protein (usually 3:1 or 4:1 by weight) diet that mimic starvation to create a state of ketosis. Then, the brain's energy source switches from carbohydrates to fats. This shift in metabolism raises the epilepsy threshold and makes epilepsy less likely to occur. Since biblical times, it has been well known that hunger suppresses epilepsy. It was actively studied in the 1920s, before the development of many antiepileptic drugs.

Although the mechanism by which the ketogenic diet suppresses epilepsy is not yet fully understood, ketosis is thought to improve energy metabolism in brain cells and increase tissue ATP levels [106]. Although the effect of the ketogenic diet on suppressing epilepsy was higher in younger children and also differed depending on the type of epilepsy, the results of many trials showed that about one-third of the patients with the ketogenic diet had a positive response, such as a reduction in seizures of more than 50%, a decrease in the amount of antiepileptic drugs, and an increase in alertness. In the remaining one-third, the epileptic seizure reduction was less than 50%, and in the remaining one-third, there was no effect. The ketogenic diet depletes the carnitine pool in the body by lowering the intake of carnitine, increasing the demand for carnitine use, and increasing the urinary excretion of acylcarnitine. Based on the data presented at academic conferences at this time, it is considered that patients on the ketogenic diet have a decrease in free carnitine and a very high acylcarnitine/free carnitine ratio, indicating a state of carnitine insufficiency. Carnitine supplementation in these patients may further promote ketone production and contribute to antiepileptic effects.

The panel believes that patients who are scheduled to follow a ketogenic diet should be screened for organic acids in advance to check for hidden metabolic disorders. It is also useful to check the urinary organic acid and acylcarnitine profiles at the time of introductory fasting. Patients with epilepsy may have been prescribed many antiepileptic drugs over the years and may have developed a carnitine deficiency. If the free carnitine in the blood is low, the side effects of hypoglycemia are more likely to occur when starting a ketogenic diet. While some doctors may prescribe a combination of valproate and ketogenic diets, the panel expressed serious concerns. This is because valproic acid is an inhibitor of fatty acid oxidation, and administration of valproic acid to patients on a ketogenic diet may cause mitochondrial dysfunction. At this time, there is no report on the effects of carnitine supplementation in patients who use a combination of valproate and ketogenic diets.

12.7.2.5 Risk Factors for Carnitine Deficiency in Patients with Epilepsy

This panel of experts reviewed a number of clinical studies and identified mental developmental delay, neurological disorders, poor nutrition, multiple drug treatment for epilepsy, and hyperammonemia as risk factors for carnitine deficiency. For example, Hunt et al. studied 640 teenagers and adults with mental retardation at a facility and reported that almost half of the subjects had abnormal carnitine concentrations, including low levels of total and free carnitine or high acylcarnitine/free carnitine ratios [107]. Most of these patients were taking antiepileptic drugs, most of them other than valproate. The results of this study suggest that carnitine deficiency correlates with multiple severe neurological disorders that require institutionalization. In addition, small children and undernourished children are especially at risk of carnitine deficiency, and studies have shown that upper arm circumference correlates with blood carnitine concentration as an indicator of nutritional status [44]. Ohtani et al. were the first to show a relationship between valproic acid, hyperammonemia, and carnitine deficiency [40]. Hyperammonemia is associated with somnolence, but ammonia itself may not be causative toxin. Rather, it may be a marker of other metabolic abnormalities, such as free fatty acids. Not all cases of coma in patients taking valproate are due to hyperammonemia.

12.7.2.6 Clinical Perspectives of Carnitine Supplementation Therapy

Unfortunately, there is limited scientific data on the effects of carnitine supplementation therapy in people with epilepsy. There is only one controlled crossover trial by Freeman et al. in which 47 pediatric epilepsy patients aged 3–17 years, 13 of whom had developmental delays, were treated with carnitine supplementation therapy [99]. Carnitine (100 mg/kg divided into three daily doses up to a maximum of 3 g/day) or placebo was administered for 4 weeks, and each week parents were evaluated with scores on a questionnaire about well-being (the patient's energy level, physical activity, mental alertness, appetite, and school performance). After a 1-week wash out, the carnitine treatment group switched to placebo and the placebo group switched to carnitine treatment. The results were analyzed in 37 patients who participated in the trial until the end, and it was reported that carnitine supplementation had no effect on well-being.

However, the expert committee believes that this result should be interpreted with caution. It is unclear how many risk factors were associated with carnitine deficiency in these subjects, and baseline carnitine levels were not low. Coulter conducted a study to examine the effects of carnitine administration in 20 epilepsy patients with two or more risk factors for carnitine deficiency and found that more than half of the subjects improved gastrointestinal symptoms such as apathy, low energy, somnolence, loss of appetite, nausea, vomiting, and constipation [97]. In some cases, muscle weakness and the frequency of epilepsy have improved. However, this test is also a non-blinded study with no controls, so there is a limit to the evaluation.

12.7.2.7 Summary: Carnitine Supplementation Therapy for Pediatric Epilepsy Patients

There is a need for well-designed clinical studies of special or general carnitine supplementation therapy in children with epilepsy, but it would not be possible to conduct statistical analyses of such studies due to the rare characteristics of hepatotoxicity. In such a situation, clinical practice must be based on empirical observation, clinical experience, and theory. Taken together, the currently available evidence provides a reasonable argument for the role of carnitine supplementation therapy in cases of childhood-onset epilepsy. The panel summarizes the key points for the diagnosis and treatment of carnitine deficiency in pediatric epilepsy patients as follows:

Diagnosis

- Carnitine deficiency is defined as a free blood carnitine value of <20 μM.
- Carnitine insufficiency is defined as an acylcarnitine/free carnitine ratio >0.4.
- A beneficial clinical response to carnitine supplementation supports the diagnosis of symptomatic carnitine deficiency.
- Muscle carnitine levels can be measured to assess tissue carnitine levels and are more accurate measure of carnitine deficiency.
- Carnitine deficiency can be suspected clinically and confirmed biochemically.

Rational Use of Carnitine Supplementation Therapy

- The benefits of carnitine for patients with some inherited metabolic diseases, such as primary carnitine deficiency due to defects in the carnitine transporter of cell membranes, have been documented.
- Carnitine supplementation for most patients with epilepsy has not been studied satisfactorily, and the possible value of this treatment is based on an understanding of the pathophysiological processes associated with carnitine deficiency.
- Intravenous, high-dose carnitine treatment may be life-saving in cases of VHT.
- The judicious use of carnitine supplementation requires careful consideration of the formulation, dosage, route of administration, and measure of response.

12.7.2.8 Recommendation of Carnitine Treatment

The panel of experts summarized the recommendations in Table 12.8 that there is, or will likely be, clinically supported benefits. This panelist consensus is based on previously published critical reviews of published research findings and the panelists' own clinical experiences.

The panel considers the risks and benefits of carnitine supplementation to be positive. The main side effects of carnitine are nausea, diarrhea, and fishy odor at high doses, but decreasing the oral dosage can reduce the complaints of these gastrointestinal symptoms. In addition, insurance reimbursement of medical expenses is permitted. The panel believes that if the indications for carnitine supplementation therapy are more widely disseminated, insurance reimbursement will be promoted. The panel also discussed the need for further research to clarify the risks and benefits of carnitine in the treatment of special fatty acid oxidation disorders.

12.7.2.9 Carnitine Dosage Recommendation

The literature on the optimal dosage of carnitine is limited. Therefore, the doses described here

Table 12.8 Recommendations for carnitine administration

Carnitine clearly indicated	Carnitine strongly recommended
VPA-induced hepatotoxicity (intravenous administration) VPA-overdose of valproic acid (intravenous administration) Primary carnitine deficiency (carnitine transporter dysfunction of cell membranes)	Inherited metabolic diseases which cause specific secondary carnitine deficiency syndromes (except fatty acid oxidation disorders) Symptomatic VPA-associated hyperammonemia Renal associated syndromes Infants and children receiving VPA, especially those younger than 2 years with a complex neurologic disorder who are receiving multiple anticonvulsants Patients who have multiple risk factors for hepatotoxicity (e.g., neurologic impairments, poor nutrition, failure to thrive, chronic illness, multiple anticonvulsants) Epilepsy patients on the ketogenic diets who have hypocarnitinemia Premature infants who are receiving total parenteral nutrition (TPN) Patients who are receiving dialysis

Modified from Ref. [103]
VPA valproic acid

should be regarded as general dosage guidelines of the expert committee.

In general, in the case of oral L-carnitine administration, it should be divided into 3–4 doses. The daily dose should be 100 mg/kg/day or 2 g/day, whichever is less. These doses fall within the dosage range described in the package insert for carnitine: "50 to 100 mg/kg/day, up to 3 g per day."

When used to save the lives of patients in the acute phase, it is recommended to administer intravenous preparations in high doses of 150–500 mg/kg/day.

These dosing guidelines are based on current knowledge, including previous reports, clinical experience, and the apparently safe nature of carnitine.

12.8 Asymptomatic Hyperammonemia Caused by Valproic Acid

So far, I described case reports, reviews, pathogenic mechanisms, and relationships with carnitine deficiency. regarding VHT and VHE. Such rare but life-threatening hepatotoxicity and encephalopathy need to be dealt with the utmost care. However, some case reports of adverse reactions to valproic acid showed asymptomatic hyperammonemia [56, 84, 108]. Murphy et al. describe that these cases of asymptomatic hyperammonemia did not require reduction or discontinuation of valproate [86]. Coulter himself, alerted to the existence of hyperammonemia in the absence of valproate-induced hepatotoxicity, stated, "Hyperammonemia is certainly not good, but it is not clear whether it is bad enough to warrant a change in therapy. I would not change treatment for asymptomatic hyperammonemia if it is within twice the normal range" [56].

So, how often does asymptomatic hyperammonemia caused by valproic acid occur? In a study of 157 outpatients with epilepsy aged 2 years and older by Haidukewych et al., there were no patients with hyperammonemia in the valproate single-agent group. Asymptomatic hyperammonemia was observed in 45.5% of patients with valproate plus phenytoin and phenobarbital, 33.3% in the valproic acid plus phenobarbital group, and 15.4% in the valproic acid plus phenytoin group, and ammonia levels correlated with blood levels of valproate and other antiepileptic drugs [109]. Ratnaike et al. reported that in a study of 81 adults with epilepsy, 8% of patients treated with multidrug antiepileptic drugs without valproate, 52.2% in the valproic acid monotherapy group, and 54.6% in the combination of valproate with phenytoin, carbamazepine, phenobarbital, or primidone had asymptomatic hyperammonemia [110]. Verrotti et al. also reported that in a study of 84 patients with epilepsy aged 5.6–19.1 years, ammonia was higher, and blood-free carnitine was lower in the valproate monotherapy group or in the valproate plus multidrug antiepileptic drug group, and that ammonia levels correlated with valproate dose, blood levels, and duration of treatment [111]. In this study, the proportion of patients with hyperammonemia was unknown, but it was described that patients who presented with hyperammonemia after receiving valproic acid did not have symptoms such as nausea, vomiting, somnolence, confusion, or coma. Although there are differences depending on the target population, asymptomatic hyperammonemia due to antiepileptic drug treatment containing valproic acid is quite frequent.

12.8.1 Systematic Review of Studies on Valproic Acid-Induced Hyperammonemia

Chicharro et al. searched the literature reported between 1980 and 2005 for "valproic acid" and "ammonia" or "hyperammonemia" and conducted a systematic review of clinical studies that met the following criteria [112].

1. Paper describing the prevalence and clinical importance of hyperammonemia in valproic acid-treated patients
2. Human clinical research
3. Research with 10 or more cases
4. Prospective or cross-sectional research
5. Reports written in English, German, French, and Spanish

This review excluded reports of patients diagnosed with urea cycle disorders, patients with hepatic failure with hyperammonemia, and patients with VHE. As a result, 24 reports were selected from 183 reports that met the above criteria. Seven of the studies included prospective studies, seventeen cross-sectional studies, and two included components of both studies.

First, they reviewed four studies that prospectively examined how much blood ammonia levels increased before and after valproic acid administration (two of the four studies were in French and Spanish). The number of patients in these four studies was small (12–25), most were treated with valproate monotherapy (only one study reported multiple agents), and the duration of treatment with valproate varied: two studies reported 4 days, one 3 months, and one 12 months, but all reported significantly higher blood ammonia levels after valproate administration compared to before the start of valproate administration and almost twice as high after treatment compared to before valproate initiation. Most of the 14 cross-sectional studies also found significantly higher ammonia levels in the valproate-treated group compared with controls. Next, the prevalence of hyperammonemia in valproate-treated patients in the three prospective studies and 14 cross-sectional studies in which ammonia values for individual cases were available was 90.3% (range: 71–100%) in the prospective studies and 47.4% (range: 16.2–100%) in the cross-sectional studies, suggesting quite high rate of hyperammonemia (Table 12.9).

The remaining studies were unable to calculate prevalence because they described comparisons based solely on data from of means. Chicharro et al. discussed that hyperammonemia caused by valproic acid administration is quite common. So that they questioned the Physician's Desk Reference description that "If hyperammonemia occurs with valproic acid administration, careful monitoring should be performed, and if high ammonia levels persist, discontinuation of valproic acid should be considered." They described that ammonia should be measured only when there are signs of acute encephalopathy and questioned the significance of blood ammonia monitoring in patients receiving valproic acid [112]. However, as discussed later, asymptomatic hyperammonemia is also dangerous, and I do not agree with Chicharro's statement that conventional hyperammonemia should be left untreated, but it is meaningful that their systematic review confirmed that asymptomatic hyperammonemia occurs at a high rate with valproic acid.

Table 12.9 Prevalence of valproate-treated hyperammonemia in prospective studies (3 reports) and cross-sectional studies (14 studies)

Reference	Prevalence	Venous/arterial	Fasting conditions
Prospective studies			
Warter et al. [3][a]	20/20 (100%)	Arterial	Variable
Zaccara et al. [7][b]	17/24 (71%)	Venous	Overnight
Navarro-Quesada et al. [11][a, c]	25/25 (100%)	Venous	9–12 h
Cross-sectional studies			
Warter et al. [3][a]	10/10 (100%)	Arterial	Variable
Navarro-Quezada et al. [11][a]	49/56 (80%)	Venous	2 h
Murphy and Marquardt [14]	29/55 (53%)	Venous	Nonfasting
Wyllie et al. [16]	41/56 (73%)	Venous	Not reported
Williams et al. [17]	12/19 (63%)	Venous	3 h
Haidukewych et al. [18]	26/125 (21%)	Venous	Variable
Laub et al. [19]	8/30 (27%)	Venous	Variable
Ettlin et al. [20][d]	44/69 (63%)	Venous	2–4 h
Ratnaike et al. [21]	30/56 (54%)	Not reported	Not reported
Iinuma et al. [22]	8/41 (20%)	Venous	4–5 h
Thom et al. [23]	6/37 (16.2%)	Not reported	Not reported
Altunbasak et al. [25]	14/68 (20.6%)	Venous	2 h
Raja and Azzoni [27]	62/123 (51%)	Not reported	Not reported
Kugoh et al. [28]	46/193 (23.8%)	Venous	2–4 h

Modified from Ref. [112]

Notes: The studies listed in this table are those that provided the individual data for patients. The ref. numbers are those of the original review (Ref. [112]), not of the reference number of this chapter

[a]Studies with two components, prospective and cross-sectional

[b]In Ref. [112], it is Zaccara et al. [4], but it seems to be an error of [7]

[c]In Ref. [112], it is Navarro-Quesada et al. [8], but it seems to be an error of [11]

[d]In this study, hyperammonemia was also detected in 45% of the control group

12.8.2 Studies on the Incidence of Hyperammonemia in Epilepsy Patients in Japan

Yamamoto et al. reported the incidence of hyperammonemia in a large cohort study of risk factors for valproate-induced hyperammonemia in adults with epilepsy in 2012 and in pediatric patients with epilepsy in 2013 [113, 114].

In a cohort study of adults (2724 patients (1217 males, 1507 females, age: 16–76 years old) treated with valproate from January 2006 to December 2010), a retrospective survey of blood ammonia levels was conducted [113]. Figure 12.4 shows the distribution of blood ammonia levels in 2724 patients. The mean ammonia value was 80.7 ± 1.0 μg/dL (facility reference value: 16–76 μg/dL). About 1171 cases (43.0%) exceeded the facility's ammonia standards. The data were not much different from the incidence rates of many of the reports covered in the systematic review by Chicharro et al. Eighty-one patients (3.0%) showed very dangerous levels of ammonia above 200 μg/dL, and 33 of them were treated with carnitine supplementation therapy, intravenous drip therapy, or valproic acid reduction. Of the 81 patients who exceeded 200 μg/dL, 30 (37.0%) patients reported adverse events such as somnolence, vomiting, loss of appetite, ataxia, and increased frequency of epilepsy.

Blood ammonia concentrations increased as the number of antiepileptic drugs in addition to valproic acid increased (Fig. 12.5). Multiple regression analysis using a definition of hyperammonemia of >200 μg/dL showed that the following combinations of drugs were associated with hyperammonemia: valproic acid at ≥20 mg/kg/day (odds ratio: 4.1), phenytoin (odds ratio: 11.0), phenobarbital (odds ratio: 4.3), carbamazepine (odds ratio: 2.8), topiramate (odds ratio: 2.8), and valproic acid in combination with phenytoin (odds ratio: 1.0).

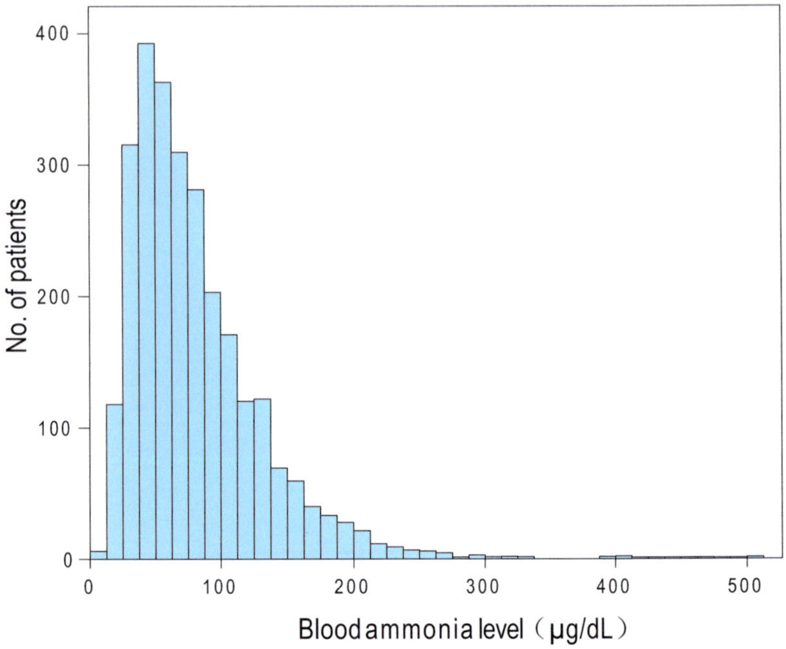

Fig. 12.4 Ammonia levels in 2724 adult epilepsy patients receiving VPA. (From Ref. [113])

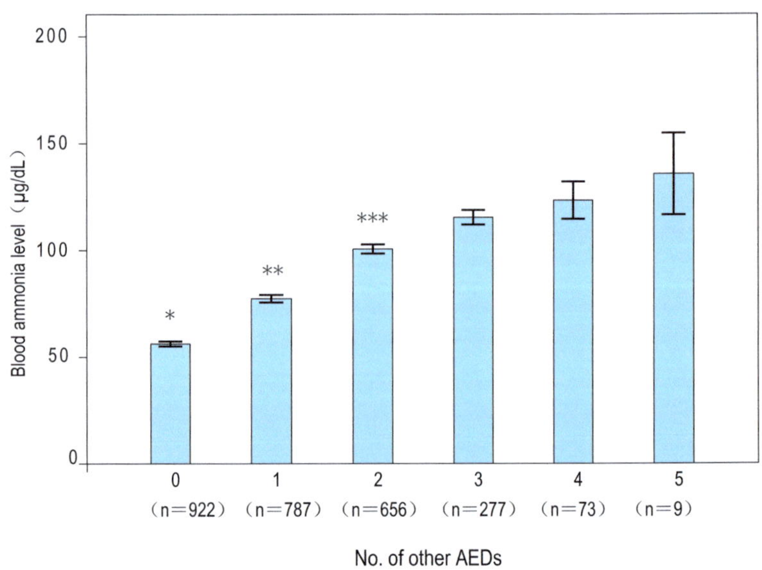

Fig. 12.5 Blood ammonia level stratified by the number of combined anti-epileptic drugs (AEDs) other than valproate. *$P < 0.001$ vs. 1, 2. 3, 4, and 5 AEDs. **$P < 0.001$ vs. 2, 3 and 4 AEDs; $P < 0.05$ vs. 5 AEDs, ***$P < 0.001$ vs. 3 AEDs; $P < 0.01$ vs. 4 AEDs (analysis of variance (ANOVA) and Scheffe's multiple comparison test). (From Ref. [113])

Concomitant use of hepatic enzyme-inducing antiepileptic drugs was a risk factor for valproic acid-induced hyperammonemia.

In a large cohort study of pediatric epilepsy patients by Yamamoto et al., 2944 pediatric epilepsy patients (1610 males, 1334 females, age: 0–15 years of age) were divided into patients who had not yet received antiepileptic acid treatment (Group I: 445 patients), patients who were treated with antiepileptic drugs other than valproate (Group II: 673 patients), and patients who were receiving valproate treatment (Group III: 1826 patients), and the definition of hyperammonemia was defined as >100 μg/dL, and risk factors for hyperammonemia were examined [114]. Hyperammonemia (>100 μg/dL) was extremely high in patients taking valproate, with 7 patients (1.6%) in Group I, 52 patients (7.7%) in Group II, and 578 patients (31.7%) in Group III. Of the patients with group III hyperammonemia, 110 (19.0%) needed carnitine supplementation therapy. Although the study did not systematically investigate the symptoms of patients, 37 of the 166 patients with ammonia levels above 150 μg/dL had clinical symptoms such as somnolence, lethargy, nausea, vomiting, and loss of appetite. In all groups, patients aged 3 years or younger had significantly higher ammonia levels than patients aged 4–15 years.

From these two large cohort studies by Yamamoto et al., it was shown that about 30% of pediatric valproate patients and about 26% of adult valproate patients in Japan exhibited hyperammonemia with blood ammonia concentrations of 100 μg/dL or more, which is not much different from the results of the systematic review by Chicharro et al.

Further clinical studies are needed to determine how to prevent and treat asymptomatic hyperammonemia in pediatric and adult patients receiving valproic acid, or whether it can be left untreated. There is particular concern about the effects of ammonia in pediatric patients with immature brain development.

12.9 Hyperammonemic Encephalopathy Caused by Valproic Acid in Psychiatry

As mentioned in Sects. 12.1.3 and 12.1.4, the indication for bipolar disorder in the psychiatric field was approved in 1995 in the United States, 17 years after the use of valproic acid in the field of epilepsy began. In Japan, after a delay of 27 years, in 2002, the indication for the treatment of manic and manic-depressive disorders was approved. Therefore, clinical experience in the field of psychiatry is still insufficient and awareness of VHT and VHE is not widely understood. In addition, pediatric neurologists in the field of epilepsy pay special attention to ammonia toxicity, especially since they are dealing with infants and children whose brains are still developing and pay attention to regular monitoring of ammonia levels and changes in clinical symptoms in patients taking valproate. On the other hand, in the field of psychiatry, where there are many adult patients, attention to hyperammonemia caused by valproic acid has not yet been sufficiently disseminated.

In 2001, Bohan et al. reported efficacy of carnitine supplementation therapy on hepatic survival in 92 cases of valproate hepatotoxicity, and described that 90 cases were epileptic patients and 2 cases were bipolar disorder patients. However more detail information about these 2 cases was not available in this article [102]. Also, a PubMed search of the literatures revealed only three cases of VHT with valproic acid in psychiatry in two reports [115, 116].

On the other hand, since its widespread use of valproic acid for treatment of bipolar disorder in the United States since 1995, reports of hyperammonemia and VHE caused by valproic acid in psychiatry have increased rapidly. The following are the main case reports.

12.9.1 Case Reports of VHE in Psychiatry

12.9.1.1 Overseas Case Reports

Settle reported a case in which a 57-year-old female patient with recurrent depression and hypomania was hospitalized for mania, started valproic acid treatment, and added clonazepam on the 6th day of hospitalization, and the next day she had slurred speech and mild ataxia, and the drug administration was discontinued, but somnolence and sluggishness were also observed, and she suddenly became comatose and was assisted with artificial respiration in the intensive care unit (ICU) [117]. Based on the results of computed tomography (CT) scans, it was diagnosed that encephalopathy had developed due to diffuse cerebral edema. The blood ammonia level was 134 μg/dL. Later, the condition improved, and clonazepam was resumed, but valproic acid was not re-administered.

Raby reported cases in which two vegetarian female psychiatric patients developed symptomatic hyperammonemia due to valproic acid treatment and recovered with carnitine treatment [118]. The first case was a 24-year-old patient with borderline personality disorder who was hospitalized due to depressive symptoms and a risk of suicide attempt. The patient was a strict vegetarian. After 2 weeks of washout, tranylcypromine was started and increased to 40 mg/day. Other drugs were administered with liothyronine 50 μg/day, thioridazine 10 mg 3 times a day, and clonazepam 0.5 mg 3 times a day. After 50 days, valproic acid 250 mg twice a day started, and after 3 days the dose was increased to 250 mg 3 times a day, and after 5 days to 500 mg twice a day. On the 60th day of hospitalization, the patient complained of fatigue, somnolence, and persistent nausea, and valproate was discontinued. The blood ammonia concentration was 101.5 μmol/L (normal value 11–35), and the dose of valproic acid was 250 mg in the morning and 500 mg in the evening. Despite the dose adjustment, somnolence persisted. Liver function tests were normal, and the blood concentration of valproic acid was 89.9 μmol/L and the blood ammonia concentration was 51 μmol/L. Carnitine administration (1 g, 2 times a day) was started. After 10 days, somnolence was resolved, and the ammonia concentration was 28 μmol/L. The second patient was a 38-year-old female with bipolar disorder type I since the age of 16 and rapid cycling for the past 2 years. She had a nearly vegetarian diet due to cholelithiasis. Treatment included venlafaxine 300 mg/day, L-thyroxine 0.15 mg/day, lithium 900 mg/day, and valproic acid 1000 mg/day. At 6 months, the patient complained of cognitive decline and somnolence, which interfered with her work. The blood ammonia concentration was 101 μmol/L, the blood concentration of valproic acid was 73 μmol/L, and liver function was normal. After starting carnitine (1 g, twice daily), somnolence disappeared after 2 weeks, and the ammonia level was 22 μmol/L. Raby described that carnitine supplementation therapy is also useful in the psychiatric field for symptoms of hyperammonemia such as somnolence, cognitive decline, and fatigue due to valproate administration, and that it may be possible to continue valproate treatment.

In 1998, Eze et al. reported a case in which a 69-year-old female patient with major affective disorder, bipolar disorder type II, and benzodiazepine dependence was admitted to the hospital and started receiving valproic acid 250 mg three times a day with trazodone and clonidine, and the staff judged that the hypomania had improved after 4 days, but the patient complained of fatigue the next day and gradually fell into a stupor and coma [119]. The blood valproic acid concentration was 107.2 μg/mL, and valproic acid was discontinued immediately. Liver function was normal, but the blood ammonia concentration was 143 μmol/L, more than three times the normal value. All psychotropic drugs were discontinued, and naloxone hydrochloride was slowly injected to attenuate the central action of valproic acid, supportive treatment, lactulose enema, and oxygen inhalation. The ammonia level gradually decreased, and the level of consciousness was restored to normal on the 5th day after the coma of level 3 of the Glasgow coma scale. In the discussion, the author stated that there is a danger of interpreting the change in the patient's condition due to hyperammonemia caused by valproic acid as an improvement in the symptoms of hypomania, and that it is necessary to educate staff about the possibility of hyperammonemia caused by valproic acid. Although it is costly to measure blood ammonia and valproic acid concentrations, considering the more expensive and time-consuming treatment of VHE, these monitoring should be performed in patients who are considered to have hyperammonemia due to valproic acid.

Panikkar et al. reported two cases of symptomatic hyperammonemia induced by valproic acid and noted that although there are several case reports of hyperammonemia caused by valproic acid, they are not widely known to psychiatric clinicians, and that the changes in mental status due to hyperammonemia are often overlooked because they are confused with exacerbation of mental disorders and mania [120]. The first case was a 53-year-old woman with bipolar disorder type I and alcoholism, who was admitted to a psychiatric hospital for substance abuse. The patient, who had not previously been admitted to a psychiatric hospital, was taken away by a police officer after a loud argument at a local hotel claiming that she was the personal assistant of a famous songwriter who was staying at the same hotel. After 5 days of abstinence, patient was started on divalproex 500 mg 3 times a day for the treatment of mania, haloperidol 5 mg twice a day for the treatment of acute psychiatric disorders, and benztropine 1 mg twice a day. On the fifth day after starting valproate, irritability and aggression decreased. The blood concentration of valproic acid at that time was 74 μg/mL. After a few days, the dose of valproic acid increased to 1750 mg/day, and the manic state disappeared. A few days later, somnolence, invading other patients' rooms, incomprehensible speech, and daytime drowsiness were observed, and the blood ammonia level was 79 μmol/L (normal value 11–35), but liver function was normal. Valproic acid was reduced by more than half to 250 mg three times a day, and lithium 300 mg twice a day was added to control mania, and the ammonia level dropped to 54, 30 after 1 week and was discharged from the hospital a few days later.

The second case was a 23-year-old man who was hospitalized for schizoaffective disorder and mania. He had progressive delusional thinking disorder and mania, as well as a long-term cocaine addiction and alcoholism. Olanzapine 10 mg 1 time per day and valproic acid (divalproex) 500 mg 3 times a day were started. During that time, blood levels of valproic acid were 70–80 μmol/L. Two weeks later, somnolence during the day and bizarre talk was observed, but it was initially thought to be due to a mental disorder. The blood ammonia concentration was 54 μmol/L and liver function was normal. In the following week, the patient's mental state became even more confused, and the ammonia level was 193 μmol/L, and the administration of valproate was immediately discontinued, and lactulose was administered instead of 300 mg lithium twice a day, and the ammonia level was 44 in the next 2 days. After that, treatment with olanzapine and lithium continued.

The authors cautioned that hyperammonemia caused by valproic acid causes changes in mental status such as somnolence, drowsiness, reversible cognitive decline, and delirium, but it is easily overlooked in psychiatric patients as a result of the original disease, and that if such symptoms occur in patients treated with valproate, blood ammonia measurement and liver function tests should be performed, and the dose of valproate should be reduced or discontinued. In the discussion, it is stated that carnitine supplementation therapy (2 g/day) should also be considered.

Barrueto et al. also reported how a 41-year-old male patient with bipolar disorder with mild psychotic disorder was brought to the emergency department with VHE [121]. The patient had a history of psoriasis, bipolar disorder, ventilation, hypertension, etc., and was in a nursing home, where staff were taking valproate and several other medications. On the day of the event, he was found in a state of somnolence and incontinence in the room and was taken to the hospital. The patient's Glasgow coma scale was level 8. He had vomiting during the examination and his oxygen saturation level was 85%, so he was intubated to secure his airway, gastric lavage, and 100 g of activated charcoal was added as intragastrical administration. Laboratory tests showed that the blood ammonia level was extremely high at 377 μM/L (normal value was <35 μM/L), and the patient was immediately administered intravenous carnitine (10 g, about 100 mg/kg with a body weight of 100 kg) for 1 hour and transported to the intensive care unit. The ammonia level dropped to 47 μM/L, the patient regained consciousness, and the tracheal intubation was removed 14 hours after being transferred to the emergency department. The patient denied attempting suicide and said he was fuzzy and could not remember the past 2 days. After that, no activated charcoal or carnitine treatment was given, and the ammonia level eventually decreased with lactulose treatment, and the patient returned to a psychiatric facility. The patient's urinalysis did not detect any abnormal amino acids as seen in urea cycle disorders. It is described that if the state of consciousness changes significantly during valproic acid treatment, monitoring of ammonia is necessary.

Kimmel et al. also reported that hyperammonemia developed in a 50-year-old female patient with schizoaffective disorder (bipolar type) who was administered valproic acid 1250 mg/day, lamotrigine 200 mg/day, quetiapine 200 mg/day, and lorazepam 1 mg/day [122]. Valproic acid was continuously administered at this dose for 4 years, and lithium and bupropion were initially used for depressive symptoms, but it could not be controlled well, so it was switched to lamotrigine, and this dose was increased over 4 months. The blood level of valproate was 88 mg/L 5 months before the onset of hyperammonemia and 1 month before the start of lamotrigine, liver function was normal, and there was no history of epilepsy, dementia, liver disease, hyperammonemia, or substance dependence. He was receiving levothyroxine 50 μg/day for hypothyroidism. Four months after starting lamotrigine, the patient was transferred to the emergency department due to loss of consciousness, loss of time, incontinence, tremor, etc. He denied taking a drug overdose and had no head injuries. Tremor of the head and bilateral upper limbs was not noted at the previous examination. Blood levels of valproic acid were 101 mg/L and liver function was normal. On the other hand, the blood ammonia level was as high as 242 μg/dL (normal value 10–40), and when valproic acid was immediately discontinued and lactulose administration was started, after 48 hours, the ammonia level was 40 μg/dL, convulsions, delirium, and loss of consciousness resolved, and the patient was discharged from the hospital without problems after 72 hours of hospitalization. Thereafter, there was no tremor at follow-up, and the doses of lamotrigine, quetiapine, and lorazepam remained unchanged and the manic-depressive remission phase was maintained. Based on this case, the authors concluded that ammonia measurement should be performed when new neurological symptoms or changes in mental status are observed, even if the blood concentration of valproate is in the normal range, the dose is constant, and liver function is normal during long-term valproate treatment. He also cautioned that although valproic acid is widely used in the psychiatric field for bipolar disorder, symptoms such as coma, altered state of consciousness, delirium, violent impulses, delusional disorder, and hypothermia are often overlooked, and that VHE itself is not widely recognized in the psychiatric field.

Yehya et al. reported cases of VHE in childhood psychiatric disorders [123]. He pointed out that child psychiatrists pay less attention to VHE than pediatric neurologists who treat epilepsy. The use of valproic acid for bipolar disorder in children is widespread, and its side effects are said to be the same as those in adults, but VHE has only been reported in adults, and a case of intermittent explosive disorder in a 9-year-old boy was presented. The patient was admitted to the hospital with increased aggression, such as biting and throwing furniture at home and school. He did not have epilepsy, and although the details of his life were unknown, he was suspected of being exposed to drugs during the fetal period. Behavioral therapy was performed with an IQ of 70, but it was ineffective, and valproic acid 1500 mg/day had been administered for 7 months to stabilize mood, but the blood level was 107 μg/mL, so the dose was reduced to 1000 mg/day and the blood level downed to 99 μg/mL. Against aggression quetiapine

125 mg 2 times a day was prescribed. A week later, he became increasingly irritable and rebellious, disobeying orders, and was forced to quarantine due to violent tantrums, including kicking adult staff. Lithium 300 mg twice a day was added, but there was no improvement, and the patient continued to behave violently on the door, such as banging his head and body violently. Since the near-delirious behavior was thought to be due to drug toxicity, valproate was further reduced to 500 mg/day. At that time, the blood lithium concentration was 0.9 mmol/L, the blood concentration of valproic acid was 113 μg/mL, the blood ammonia concentration was 127 μg/dL, and liver function was normal. Administration of valproic acid discontinued. There was tremor in the upper extremities, which was considered a side effect of lithium. Electroencephalogram (EEG) showed symmetrical 5–6 Hz waves, which were considered to be diffuse encephalopathy, and were judged to be hyperammonemic encephalopathy. Lithium was also discontinued, and the tremor was resolved. On the fourth day after discontinuation of valproate, the blood ammonia level was 55 μg/dL. Chlorpromazine 75 mg/day was administered in three divided doses for aggression, and although it was partially effective, mild dystonia was reported twice. The addition of diazepam 5 mg three times a day improved dystonia and reduced isolation by 70%. Carbamazepine 150 mg twice daily was added 2 weeks before discharge from the hospital for mood stabilization. At the time of discharge, the diagnosis was intermittent explosive disorder, oppositional defiant disorder, and borderline mental retardation. In this case, liver function was normal, but valproic acid increased ammonia levels, which may have increased violence and aggression. It was described that it is important to measure ammonia and grasp electroencephalogram in consideration of hyperammonemic encephalopathy when managing acute drug treatment with valproic acid.

Carlson et al. also reported two pediatric psychiatric cases with VHE [124]. In the first case, an 11-year-old boy with Asperger's syndrome and attention deficit hyperactivity disorder was treated with valproic acid along with risperidone and lithium, and although his liver function was normal, he repeatedly vomited and became abnormally manic. When the ammonia concentration reached a high level of 213 μmol/L and valproic acid administration was discontinued, the ammonia concentration normalized on the third day, and the abnormal manic state disappeared. The second case was an 11-year-old boy with absence epilepsy and attention deficit hyperactivity disorder, who was prescribed methylphenidate for attention deficit hyperactivity disorder and valproic acid and ethosuximide for absence epilepsy. When risperidone was started, the manic state worsened and became aggressive, and the ammonia level was 113 μmol/L. Risperidone was discontinued, and valproic acid was discontinued, and the ammonia level was in the normal range, and the manic state disappeared. The author cautioned that hyperammonemia caused by valproic acid can aggravate the manic state, and that ammonia measurement is essential in such cases.

In addition, cases of VHE in the field of psychiatry have been reported in the Department of Critical Care and Critical Care. Nicolai et al. reported a case of a 33-year-old man who was admitted to a psychiatric hospital for 3 months, and after discharge, his family found him in a comatose state after 5 days of decreased wakefulness and finally transferred him to the emergency room [125]. The patient had been prescribed clozapine 450 mg/day for the diagnosis of schizophrenia for 5 years and also 500 mg valproate twice a day to prevent epilepsy. The patient had a mild fever of 37.6 °C and hyperventilation (36 times/min), the blood ammonia concentration was 582 μmol/L (normal value <33), the free carnitine concentration was as low as 17 μmol/L, and the acylcarnitine/free carnitine ratio was as high as 0.65. In addition, ketone bodies were detected in the urine, and ketoacidosis caused by clozapine and hyperammonemic encephalopathy caused by valproic acid were diagnosed, and the patient was transported to the intensive care unit for tracheal intubation and sedation. Valproic acid administration was discontinued, and 20 g of lactulose and 1 g of neomycin were administered

6 times and 4 doses of 1 g to lower ammonia, and after 10 hours, the ammonia decreased. The next day, he was treated for respiratory distress syndrome and disseminated intravascular coagulation syndrome 3 days later and was discharged from the hospital in good condition after 7 weeks. Instead of clozapine prescribed haloperidol (2.5 mg/day). In this way, even if valproic acid has been prescribed for several years and is stable, hyperammonemic encephalopathy may develop suddenly, and caution should be exercised.

Elgudin et al. also reported a case in which a psychiatric patient prescribed valproate became comatose and was transported to the critical care department [126]. A 35-year-old male who had been diagnosed with bipolar disorder type I (rapid cycling), anxiety disorder, and alcoholism for 5 years was prescribed valproic acid 500 mg 3 times a day, temazepam 30 mg, lorazepam 1 mg 3 times a day, and 2 mg before bedtime. The patient's liver function was normal, the blood valproate level was 100 μg/mL, and the Glasgow coma scale was 3. The next day, the patient was somnolent and somewhat confused, and was able to recognize himself but not the time or place. Blood ammonia levels on the second day were 83 mmol/L, which was higher than normal. All psychotropic medications were discontinued, and the patient gradually improved. On the third day, he complained of insomnia and anxiety, so zolpidem 10 mg and lorazepam 1 mg 3 times a day were administered 3 times a day for anti-anxiety. By the time he was discharged from the hospital, his cognition returned to normal, and he was prescribed 300 mg lithium twice a day instead of valproic acid for bipolar disorder. In the discussion, the author discusses the pathogenesis of VHT and hyperammonemia in pediatric neurological areas and carnitine deficiency. In addition, it was emphasized that when patients receiving valproic acid change their mental status or become coma, it is necessary to measure their blood ammonia, because there is a possibility of hyperammonemia.

Thus, since the indication of valproic acid for bipolar disorder was approved in the United States in 1995, VHE has been reported one after another in the psychiatric field.

Carr et al. found that a 45-year-old female with schizoaffective disorder and bipolar disorder was successfully treated with risperidone and carbamazepine, but her mental state deteriorated after her mother's death, and she had auditory hallucinations, unstable emotions, and confused thoughts, and when valproic acid (initially 1000 mg, 1500 mg after 3 days) was started, after 6 days, agitation, confusion, decreased level of consciousness, incontinence, and hyperammonemia (445 μg/dL), and was treated with lactulose [127]. In addition, in their report Carr et al. reviewed VHE cases in psychiatry area as described in this book above (14 cases in 11 reports) and described that it is often difficult to distinguish the change in level of consciousness and mental status caused by hyperammonemia due to valproate from the symptoms of the primary illness, and that life-saving measures such as intubation are necessary if the patient becomes comatose too late to notice (5 of 14 cases resulted in coma and 4 cases were intubated). The treatment of VHE includes discontinuation or reduction in valproic acid, administration of lactulose, and administration of carnitine.

Chopra et al. discussed 5 own cases and 30 psychiatric VHE cases from the previous literature and summarized their risk factors, clinical correlates, and treatment [128]. Case 1 was a 19-year-old female patient who had been diagnosed with bipolar affective disorder 1 week earlier, and she had been given valproic acid 1500 mg/day and clonazepam 0.5 mg twice a day for anxiety symptoms, but complained of hyposensitivity, ataxia, somnolence, drowsiness, nausea, and vomiting. The blood valproic acid concentration was in the normal range of 87 μg/mL, but the blood ammonia concentration was 124 μmol/L (normal value 9–35), and these symptoms recovered when valproic acid was discontinued. Case 2 was a 36-year-old male, schizophrenic bipolar type, alcohol and cannabis dependence, and hepatitis C; he started valproic acid 1000 mg/day, risperidone 3 mg/day, clozapine 300 mg/day, clonazepam 3 mg/day, disulfiram 250 mg/day, topiramate 300 mg/day, lithium 450 mg/day, and the patient was taken to the ICU due to slurred speech, drowsiness, and hypoesthesia. The blood concentration of valproic acid was 114 μg/mL, liver function was normal, but the blood ammonia level was high at 111 μmol/L, and generalized rhythm abnormal waves were observed in electroencephalography, so when valproic acid was discontinued, the ammonia level normalized and the symptoms disappeared in 3 days. Case 3 was a 53-year-old female patient with bipolar disorder type II and alcoholism, who was taking valproic acid 750 mg/day and quetiapine 300 mg/day, but nausea, vomiting, awkwardness, and coordination disorders were observed. The blood level of valproic acid was 52 μg/mL and the blood ammonia level was 97 μmol/L, and liver function was normal. Symptoms improved 3 days after discontinuation of valproate. Case 4 was a 41-year-old female patient with mild mental retardation with bipolar affective disorder who had repeated symptoms of rapid alternation for 10 days for the past 5 months. When the dose of valproic acid was increased from 1000 to 1250 mg/day and olanzapine was started to treat manic symptoms, strange behaviors such as taking off clothes and drawing pictures on the wall with crayons occurred 1 day later, and isolation and physical restraint in two parts were required. When valproic acid was discontinued at a high level of 133 μmol/L in the blood, the concentration of ammonia was 12 μmol/L the next day, but aggression and manic states remained, so paliperidone was prescribed and the symptoms improved after 2 weeks. Case 5 was a 71-year-old female patient with schizoaffective disorder (bipolar type) with mild mental retardation, and for 3 weeks, a decrease in attention and functionality was observed compared to the previous one. Patients were treated with valproic acid 1500 mg/day and lithium 300 mg/day (later increased to 300 + 450 mg/day). Blood valproate concentrations were 26 μg/mL and blood ammonia concentrations were 133 μmol/L, respectively, and liver function was at the upper limit of normal. When severe generalized slow-wave three-phase waves were observed on electroencephalography, which was diagnosed as suspected encephalopathy, valproic acid administration was discontinued, and the symptoms improved within 2 days, and the ammonia level decreased to 45 μmol/L.

Chopra et al. compiled a list of these five self-reported cases and 30 previously reported cases of VHE in the psychiatric field (Table 12.10), and updated their risk factors, clinical correlations, and treatments. These results showed that there was no correlation between the dose or blood concentration of valproate and the severity of hyperammonia encephalopathy; that drug interactions, carnitine deficiency, urea cycle disorder, and mental retardation due to multidrug administration were risk factors; and that carnitine administration was effective in preventing and treating VHE in six cases. If left untreated, VHE can lead to life-threatening coma, and it is important to fully understand its pathophysiology and treatment. Asymptomatic hyperammonemia (subclinical VHE) may also affect cognitive function if left untreated, and if in doubt, neuropsychological tests [129] and neuroimaging tests [130] used to detect latent hepatic encephalopathy in hepatic encephalopathy should be performed to actively detect abnormalities and take countermeasures. Symptomatic hyperammonemia typically causes sudden onset of somnolence, disorientation, hypoesthesia, and focal neuropathy. In addition, nausea and vomiting, flapping tremors, ataxia, epilepsy and its frequency increase, changes in the level of consciousness such as coma, and triphasic EEG of theta and δ waves were observed. Symptomatic hyperammonemia may occur within a few days of starting valproate treatment, or it may appear suddenly several months later, even if blood levels of valproate are within the normal range. The symptoms of symptomatic hyperammonemia are mild and difficult to understand, and it is not easy, especially in daily clinical practice in psychiatry, where it is often difficult to distinguish them from the original changes in the symptoms of psychiatric disorders. To know the severity of encephalopathy, one should know quantitative rating scales such as the Glasgow coma scale. As a treatment for hyperammonemia, enema administration, or neomycin administration with a nasal tube or enema of lactulose, citing a review article on the treatment of hepatic encephalopathy, dialysis is the most efficient treatment for cases of 400 μmol/L or more, and carnitine is recom-

Table 12.10 Clinical characteristics of VHE cases in psychiatric patients

Author (Ref.[a])	Age, gender	Diagnoses	Additional medications	Daily VPA dosage (mg)	Serum VPA levels (g/dL)	Serum ammonia levels (mol/L)	Time to onset of VHE	VHE symptoms	Rx and time to recovery
Settle (1995) [4]	57, F	BD II	Clonazepam	Unavailable	67	79	Day 7	Slurred speech, mild ataxia, obtundation followed by coma	VPA discontinuation; <5 days
Raby (1997) [5]	24, F	BPD, MDD-atypical	Tranylcypromine, liothyronine, thioridazine, clonazepam	1000	Unknown	101.5	Day 10	Fatigue, lethargy, persistent nausea	VPA discontinuation L-carnitine; 10 days
Raby (1997) [5]	38, F	BD I	Venlafaxine, L-thyroxine, lithium	1000	73	101	Month 6	Cognitive changes and lethargy	VPA discontinuation L-carnitine; 14 days
Eze et al. (1998) [6]	69, F	BD II	Sertraline, alprazolam, trimipramine, meclizine hydrochloride, nabumentone, trazodone, clonidine	750	107.2	143	Day 4	Disorientation to time, feeling slowed followed by coma	VPA discontinuation, naloxone hydrochloride; 5 days
Panikkar and Gilman (1999) [7]	53, F	BD I, AD	Haloperidol, benztropine	1750	107	79	Week 2	Lethargic, wandering, illogical speech	VPA discontinuation; 3 days
Panikkar and Gilman (1999) [7]	23, M	Schizoaffective disorder—bipolar type, AD	Olanzapine	1500	75	54	Week 2	Lethargic, bizarre speech	VPA discontinuation; 2 days
Rottach et al. (2000) [8]	28, M	BD I	Doxepin, lorazepam, risperidone, biperiden	600	72	Normal	Week 2	Tiredness, tremor, generalized epileptic seizure	VPA discontinuation; 4 days
Barrueto and Hack (2001) [9]	41, M	BD I, MR	Acebutolol, hydrochlorothiazide, ibuprofen, haloperidol, benztropine, gemfibrozil, lorazepam, multivitamins	Unavailable	73.5	377	Year 3	Increasing lethargy, inappropriate behavior	VPA discontinuation, L-carnitine + lactulose; 1 day

Nicolai et al. (2001) [10]	33, M	Schizophrenia; diabetic ketoacidosis	Clozapine	1000	Unavailable	582	Year 4	Unconscious	VPA discontinuation, lactulose, neomycin, treatment of diabetic ketoacidosis; day 26
Feil et al. (2002) [11]	62, F	Generalized tonic clonic seizures, mood and anxiety disorders	Estradiol, levothyroxine, diazepam, cyclobenzaprine, trazodone, acetaminophen with codeine and sulindac	1000	75	99	Unavailable	Somnolence, confusion, unresponsive	VPA discontinuation; 3 days
McCall and Bourgeois (2004) [12]	88, M	Unclear, possible dementia	None	1000	48	836	Unclear	Confusion	VPA discontinuation; 2 days
Elgudin et al. (2003) [13]	35, M	BD I, GAD	Temazepam, lorazepam	1500	100	83	Year 5	Unresponsive	VPA discontinuation; >3 days
Reif et al. (2004) [14]	42, M	MDD with psychotic features	Lithium, risperidone, flupentixol, clomipramine	1500	102	57	Day 7	Sedation, tremor, myoclonus	VPA discontinuation; 2 days
Yehya et al. (2004) [15]	9, M	Intermittent explosive disorder, MR	Quetiapine	1500	113	75	Month 7	Aggressive behavior	VPA discontinuation; >4 days
Ricard et al.(2005) [16]	58, M	BD I	Amisulpride, liothyronine, ciprofibrate, benfluorex trihexyphenidyl	1000	79	124	Month 7	Cognitive changes	VPA discontinuation; <30 days
Stewart (2005) [17]	79, M	Schizophrenia—paranoid type	Olanzapine, phenytoin, amlodipine, finasteride, risedronate, aspirin	2250	48.5	89	Day 2	Lethargy, confusion, disorientation	VPA discontinuation, lactulose and protein restriction; >/2 days
Kimmel et al. (2005) [18]	50, F	Schizoaffective disorder—bipolar type	Lamotrigine, quetiapine, lorazepam	1250	101	242 μg/dL	Year 4	Loss of consciousness, bladder incontinence and cognitive changes	VPA discontinuation, lactulose; 3 days

(continued)

Table 12.10 (continued)

Author (Ref.[a])	Age, gender	Diagnoses	Additional medications	Daily VPA dosage (mg)	Serum VPA levels (g/dL)	Serum ammonia levels (mol/L)	Time to onset of VHE	VHE symptoms	Rx and time to recovery
Carlson et al. (2007) [19]	11, M	Asperger's syndrome, ADHD	Lithium, risperidone	500	90	213 (units not given)	<7 days	Agitation, disinhibition and manic behavior	VPA discontinuation, 3 days
Carlson et al. (2007) [19]	11, M	Absence seizures, ADHD	Risperidone, ethosuximide	Unavailable	71	113 (units not given)	3~4 weeks after initiation of risperidone	Aggression, disinhibition, and manic behavior	VPA and risperidone discontinuation, <2 weeks
Wadzinski et al. (2007) [20]	51, F	PTSD	Topiramate, quetiapine	1000	145	232	Day 7	Nonresponsive	VPA discontinuation, L-carnitine; 4 days
Wadzinski et al. (2007) [20]	29, F	BD, OCD	Fluvoxamine, clonazepam	1500	113	182	Day 17 since dose increase from 1000 to 1500	Confusion, disorientation, hypersomnia and memory loss	VPA discontinuation; 2 days
Stewart (2008) [21]	76, M	BD	Quetiapine, primidone, metoprolol, atorvastatin, lisinopril, levothyroxine and aspirin	3000	73	214	Year 11	Confusion, lethargy and near mute	VPA discontinuation, lactulose; 3 days
Fan et al. (2008) [22]	72, F	BD I with psychosis	Clozapine, lamotrigine	900	85.9	101	Week 6	Weakness, hand tremor, lethargy and asterixis	VPA discontinuation; 7 days
Dealberto and Sarazin (2008) [23]	41, F	BD I with psychosis	Olanzapine	1000	131	61	Day 1	Disorientation, drowsiness and agitation	VPA discontinuation; 2 days
Eubanks et al. (2008) [24]	33, F	BD, PTSD, substance abuse, BPD	Clonazepam, venlafaxine- XR, bupropion, hydroxyzine, mirtazapine	1500	120	283	Day 3 since start of VPA	Altered mental status and minimally responsive	VPA discontinuation, L-carnitine and lactulose; 1 day
Deutsch et al. (2009) [25]	31, F	BD I, BPD	Citalopram, quetiapine, topiramate	1000	104	41	Week 2	Somnolence, ataxia, dysarthria, memory problems	VPA discontinuation; 1 day

Young and Coffey (2010) [26]	15, M	BD I, diabetes	Haloperidol, clonazepam	1750	100	96	Unknown	Excessive drowsiness	VPA discontinuation after failed trial of l-carnitine; unkown
Sunkavalli et al. (2011) [27]	52, F	Schizoaffective disorder, remote seizures	Risperidone, phenytoin, haloperidol, lorazepam, ciprofloxacin	Unavailable (recent increase in dosage)	81	218	Day 2	Altered mental status, suicidal ideation, low- grade temperature, worsening seizures	VPA discontinuation, lactulose
Trojak et al. (2011) [28]	36, F	BD I	Diazepam, olanzapine	1500	120	Normal	Day 4	Drowsy, comatose, required intubation	VPA and diazepam discontinuation; 1 day
Abreu et al. (2009) [29]	47, F	BD I	None	2500	83.5	48	7 months	Apathy, confusion, memory loss, Parkinsonism, catatonia	VPA and diazepam discontinuation; 7 days

From Ref. [128]

VHE valproate-induced hyperammonemic encephalopathy, *ADHD* attention-deficit/hyperactivity disorder, *BD* bipolar disorder, *BPD* borderline personality disorder, *GAD* generalized anxiety disorder, *MDD* major depressive disorder, *MR* mental retardation, *OCD* obsessive compulsive disorder, *PTSD* post-traumatic stress disorder, Rx prescription

[a]The reference numbers in this table are the reference numbers in the original Chopra's review (Ref. [128] in this chapter)

mended if there are signs of carnitine deficiency such as hypotonia or somnolence, or if there is a decrease in blood-free carnitine, decreased liver function, or hyperammonemia. Outcomes of VHE may be completely reversible, or serious neurological impairment may remain, or death may occur.

Brown et al. summarized three self-reported cases (one of which was symptomatic) and six literature cases (five of which were symptomatic) on carnitine supplementation therapy for hyperammonemia with valproate in the psychiatric field [131]. The clinical course of the three self-study cases is as follows (see also Table 12.11).

Case 1 A 65-year-old man with bipolar disorder type I was admitted to a psychiatric inpatient facility due to an acute manic episode. The medication prescribed by the outpatient psychiatric department was discontinued, and valproic acid solution started at 500 mg in the morning, 1000 mg before bedtime, and risperidone 1 mg twice a day. Valproic acid and risperidone gradually increased to 3000 and 6 mg/day over 5 days. On day 7, the patient was drowsy, slurred speech, and abnormal mental state, and the blood valproate concentration was 68 μg/mL (normal value 50–100), but the blood ammonia level was 56 μmol/L (normal value 16–53), and the symptoms were resolved by discontinuing valproic acid administration. However, the manic state, including aggression and hyper-libido behavior, recurred, and on the 11th day of hospitalization, lithium 900 mg/day and valproate solution 1000 mg/day were started, and the valproic acid was finally increased to 1500 mg/day (divided into 2 doses per day). When the dose of valproic acid was increased, carnitine 990 mg/day was administered to prevent hyperammonemic encephalopathy, and after 3 days, the dose was increased to 990 mg twice a day. The ammonia level was 23 μmol/L, and no symptoms of encephalopathy were reported when carnitine was also administered during hospitalization. However, after being transferred to the outpatient clinic, valproic acid administration was continued, but carnitine administration was discontinued for unknown reasons, and about 3 months after being discharged from the hospital, he was hospitalized again for encephalopathy. Blood ammonia levels were as high as 66 μmol/L and valproic acid levels were as low as 19 μg/mL. Valproate was discontinued and not re-administered.

Case 2 A 64-year-old man with schizoaffective disorder and bipolar type was hospitalized due to disrupted behavioral and psychiatric symptoms. He was prescribed clonazepam, risperidone, and valproic acid for outpatient treatment, but gradually changed to valproic acid solution (1250 mg in the morning, 1500 mg before bedtime), haloperidol (5 mg, twice a day), and clonazepam (1 mg, before bedtime) for 8 weeks of inpatient treatment. The valproic acid blood concentrations were 141 μg/mL at a steady state of 7.5 hours post-dose, with calculated true trough values of ~88 μg/mL and blood ammonia levels of 63 μmol/L, elevated from 47 to 49 μmol/L at admission, but no change in mental status. After a single dose of 990 mg of carnitine, the ammonia level decreased to 44 μmol/L after 10 hours. The patient was discharged from the hospital and received 990 mg/day of carnitine for 2 days and then 990 mg twice daily to prevent hyperammonemia due to valproic acid. Valproic acid solution was switched to divalproex ER 2500 mg/day, which is a valproic acid derivative. After discharge, blood ammonia levels were elevated twice to 67 and 71 μmol/L, although it was not known whether the patient had been taking carnitine since discharge. The patient was readmitted several months later with a valproic acid blood level of 75 μg/mL and an ammonia level of 75 μmol/L. Under supervision, the patient was given carnitine 330 mg three times a day, properly administered, and his ammonia dropped to 29 μmol/L and his valproic acid concentration was 25 μg/mL. The patient was discharged and given 990 mg of carnitine twice daily to continue divalproex.

Case 3 A 53-year-old man with bipolar disorder type I and substance use disorder who developed acute mania was transferred from an external hospital. He was treated with buspirone, hydroxy-

Table 12.11 L-carnitine supplementation therapy for patients with VPA-induced hyperammonemia in psychiatric setting

Author	Age	Sex	DSM	Clinical picture	VPA dose, mg/d	VPA, μg/mL	NH_3, μmol/L	Treatment	Outcome	VPA D/C?
Case series										
Patient 1	65	M	BD	AMS resolved after VPA D/C then restarted	3000 then ↓ to 1500	68	56	L-carnitine 990 mg PO daily × 3 d then 990 mg PO BID	sxs did not recur, NH_3 level 50	No
Patient 2	64	M	SAD	No sxs	2750	141	63	L-carnitine 990 mg PO daily × 3 d then 990 mg PO BID	No sxs emerged, NH_3 level 44 within 10 h	No
Patient 3	53	M	BD	No sxs	2000	104	82	L-carnitine 1 g PO daily × 3 d then 1 g PO BID × 2 wk then 1 g QAM, 1.5 g QHS	No sxs emerged, NH_3 level ↓ to WNL over 3 wk	No
Literature review (Ref.[a])										
Aiyer et al. [9]	58	F	BD	No sxs	2000	117	225	Lactulose 20 g PO BID, L-carnitine 990 mg PO TID	NH_3 level ↓ to 163 then 27 Within 2 wk	No
Barrueto and Hack [15]	41	M	BD	Lethargy, AMS, vomiting, coma	NR, same dose × 3y	73.5	377	Activated charcoal, L-carnitine 100 mg/kg IV × 1	NH_3 level ↓ to 47 within 7 h, sxs resolved within 24 h	Yes
Eubanks et al. [16]	33	F	BD PTSD BPD	Lethargy, AMS	1500	120	283	L-carnitine 3 g IV followed by 990 mg PO BID, lactulose 30 mL PO every 6 h × 7 d	NH_3 level ↓ to 41, sxs resolved within 6 h, 66, 30, 25 on next d	Yes
Raby [17]	24	F	BPD MDD	Lethargy, fatigue, persistent nausea	1000 then ↓ 750	89.9[b]	102 then 51	L-carnitine 1 g PO BID	NH_3 level ↓ to 28, sxs resolved within 10 d	No
Raby [17]	38	F	BD	Congnitive slowing, lethargy	1000	73[b]	101	L-carnitine 1 g PO BID	NH_3 ↓ to 22, sxs resolved within 2 wk	No
Young and Coffey [18]	15	M	BD	Sedation, lethargy	1750 then ↓ to 1500	100	96	L-carnitine 500 mg PO BID then ↑ to 750 mg PO BID	Persistent NH_3 elevations, VPA eventually D/C after failed trial of levocarnitine	No

From Ref. [131]

AMS altered mental status, *BD* bipolar disorder, *BID* twice daily, *BPD* borderline personality disorder, *D/C* discontinued, *DMS* Diagnostic and Statistical Manual of Mental Disorders, 5th edition (DMS-5), *F* female, *IV* intravenously, *M* male, *MDD* major depressive disorder, *NH3* ammonia, *NR* not reported, *PO* by mouth, *PTSD* post-traumatic stress disorder, *QAM* every morning, *QHS* at bedtime, *QPM* every evening, *SAD* schizoaffective disorder, *sxs* symptoms, *TID* 3 times daily, *VPA* valproic acid, *WNL* within normal limits

[a]The reference numbers in this table are the reference numbers in the original Brown's review (Ref. [131] in this chapter)

[b]Reported in μmol/L in case report, unclear if units are correct given that conversion puts levels at 12.9 and 10.5 μg/mL (below therapeutic range)

zine, lithium, etc., but at the time of hospitalization, he discontinued lithium due to hypernatremia caused by lithium and started valproic acid 1000 mg twice a day. Blood ammonia levels for the next 4 days were 59, 65, and 82 μmol/L, and blood levels of valproic acid were 104 μg/mL. Manic symptoms persisted, but there was no change in mental status. However, lactulose was administered for 3 days to lower the ammonia. In addition, carnitine was administered 1000 mg/day and later 1000 mg twice a day from 2 days after the start of lactulose. Over the next 2 weeks, the ammonia levels decreased slightly after carnitine administration, but were higher than normal, at 70, 58, 67, and 68 μmol/L. In order to reduce the ammonia level from the second week, the dose of carnitine was increased to 1000 mg in the morning and 1500 mg in the evening. After 3 weeks, the ammonia levels were checked every 8–10 days, and were within the normal range or slightly higher at 46, 46, 59, and 49 μmol/L. The patient was discharged from the hospital and continued to take valproate and carnitine. As in Case 2, it is unclear whether the patient was adhering to carnitine, but the patient was readmitted to the hospital with a valproic acid concentration of 115 μg/mL and an ammonia level of 74 μmol/L. It was reduced to 30 μmol/L by carnitine administration.

Thus, the usefulness of carnitine supplementation therapy has been shown in VHE cases in the psychiatric field.

12.9.1.2 Case Reports in Japan

In Japan, sodium valproate was approved in 2002 for the treatment of manic and manic-depressive disorders, and it is widely used in the psychiatric field. It is also used in patients with aggressive dementia and in the prevention of epilepsy after brain surgery. In addition, in Japan, carnitine preparations were only indicated for carnitine deficiency caused by two inherited metabolic diseases, methylmalonic acidemia and propionic acidemia, but in March 2011, the indication for carnitine deficiency was approved by a public knowledge application, and it can now be used for hyperammonemia caused by valproic acid.

Aoki et al. reported for the first time in Japan that 12 out of 26 elderly patients aged 65 years and older taking valproic acid for the treatment of psychiatric disorders and dementia had asymptomatic hyperammonemia (>80 μg/dL), and that carnitine administration improved hyperammonemia (Table 12.12) [132].

The dose of carnitine required to normalize ammonia levels was positively correlated with blood ammonia levels ($r = 0.916$), with higher doses required at higher blood ammonia levels. In their discussion, the authors stated that it is clear that exposure to high ammonia, even if asymptomatic, continues to damage brain neurons through destruction of brain mitochondria and astrocytes, and that carnitine administration and other measures are necessary not only in cases of symptomatic hyperammonemia but also in cases of asymptomatic hyperammonemia.

Ito et al. also measured the blood ammonia concentrations of 66 psychiatric patients taking valproic acid and found that 14 patients (21.2%) had hyperammonemia of 80 μg/dL or more while asymptomatic, and when carnitine was administered to 10 patients who agreed (900 or 900 mg → 1800 mg/day) (Table 12.13), ammonia levels normalized in 9 patients (Fig. 12.6) [133]. The authors concluded that persistent hyperammonemia may cause cognitive decline, cerebral edema, and cerebral atrophy, and that regular blood ammonia measurement is necessary in patients receiving valproic acid in the psychiatric field, and that carnitine administration is useful for the treatment of hyperammonemia.

12.9.2 Prevalence of Valproic Acid-Induced Hyperammonemia in Psychiatry

Raja et al. examined the incidence of hyperammonemia in patients admitted to psychiatric hospitals and reported that 63 out of 123 patients (51.2%) in the valproic acid treatment group had hyperammonemia (>97 μg/dL), compared with 5 out of 23 patients (21.7%) who were prescribed mood stabilizing drugs other than valproic acid [134].

Table 12.12 Effect of carnitine on valproate-induced asymptomatic hyperammonemia in 12 patients with psychiatric disorder or dementia aged 65 years and older

				VPA dose (mg)				
No.	Age	Gender	Psychiatric diagnosis	Blood VPA concentration (og/mL)	NH_3 before LC supplementation (bg/dL)	LC dose (mg)	NH_3 after LC supplementation (after	Concomitant medications
1	65	Male	Schizophrenia	1000 64	135	2700	67	Haloperidol, levomepromazine, risperidone
2	65	Male	Schizophrenia	600 86	118	1800	74	Biperiden, chlorpromazine,etizolam, haloperidol
3	75	Male	Dementia	600 68	106	1800	68	Donepezil hydrochloride, Yokukansan extract
4	76	Male	Schizophrenia, dementia	600 68	88	600	68	Lormetazepam, risperidone
5	77	Male	Organic mental disorders	1000 66	93	1800	74	Levomepromazine, ramelteon
6	65	Female	Organic mental disorders	800 75	175	3600	79	Flunitrazepam, levomepromazine, phenytoin, risperidone
7	69	Female	Schizophrenia, epilepsy	1200 86	154	2700	69	Carbamazepine, levomepromazine, nitrazepam
8	76	Female	Dementia	600 50	120	2700	59	Nitrazepam, olanzapine, remelteon, sulpiride, zopiclone
9	77	Female	Dementia	1000 83	116	1800	55	Levomepromazine, zopiclone
10	79	Female	Dementia	600 61	88	600	66	Tiapride hydrochloride, zopiclone
11	80	Female	Dementia	600 60	139	2700	53	Brotizolam
12	85	Female	Bipolar disorder	600 72	89	600	58	Flunitrazepam, quetiapine fumarate

From Ref. [132]
LC L-carnitine chloride, *VPA* sodium valproate

Table 12.13 Patients background of 10 psychiatric patients with hyperammonemia during valproic acid administration

Case	Gender	Age	Inpatient/ Outpatient	VPA Dose (mg/ day)	VPA concentration in blood (μg/mL)	Before carnitine administration: Blood ammonia concentration (μg/ dL)	Carnitine dose (mg/ day)
1	Male	73	Hospitalization	600	68	104	900
2	Female	67	Hospitalization	600	72	81	900
3	Male	63	Hospitalization	600	66	108	900
4	Male	36	Hospitalization	1000	95	101	900
5	Female	29	Outpatient	600	41	83	900
6	Male	45	Hospitalization	600	66	130	900
7	Male	40	Hospitalization	1000	45	108	900–1800
8	Male	31	Hospitalization	1000	74	101	900–1800
9	Male	41	Hospitalization	1000	75	80	900
10	Male	56	Hospitalization	600	42	88	900–1800

From Ref. [133]

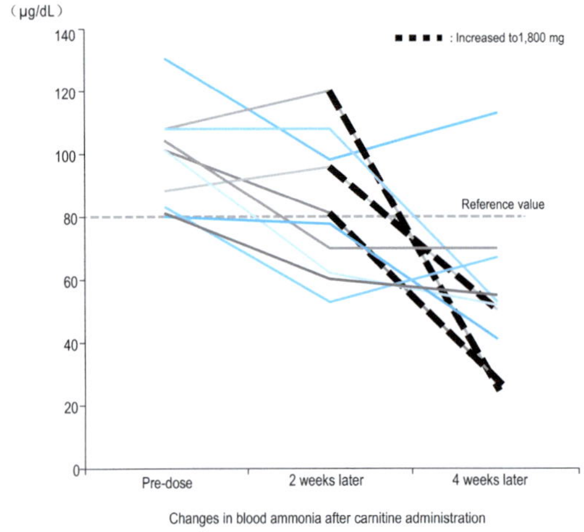

Fig. 12.6 Changes of ammonia levels after carnitine administration in hyperammonemic psychiatric patients receiving VPA. L-carnitine was administered 900 mg/day at first, then increased to 1800 mg/day after 2 weeks in the 3 cases (bold dashed lines). (From Ref. [133])

In addition, Baddour et al. reported the prevalence of hyperammonemia by extracting patients who had been admitted to a psychiatric ward and receiving valproic acid and measuring their blood ammonia levels using electronic medical records [135]. Of the 357 patients screened, 10 had concomitant cirrhosis, and 347 were excluded and 347 were investigated. Hyperammonemia was defined as 47 μmol/L (80 μg/dL) or more. There were 232 male patients (66.9%) with a range of ages 18–97 (median 47 years). There were 84 patients with intellectual disability (24.2%), 2 patients with topiramate (0.6%), 111 patients (32%) with schizophrenia, 106 patients (30.6%) with schizoaffective disorder, 78 patients with dementia (22.5%), 74 patients with bipolar disorder type I (21.3%), 47 patients with substance use disorder (13.6%), and 28 patients with major depression (8.1%) (overlap). The prevalence of hyperammonemia was 36% in 125 of 347 cases.

Of the 125 cases of hyperammonemia, 54 (43.2%) were symptomatic hyperammonemia. However, in addition to somnolence, disorientation, and vomiting, changes in mental status, confusion, stupor, delirium, and epileptic seizures were added as symptoms of hyperammonemia, and the prevalence of symptom was 144 out of 222 patients (64.9%) even in the group with normal blood ammonia concentrations. The treatment of 113 cases of hyperammonemia is as follows: 32 patients (28.3%) discontinued valproate, 38 patients (33.6%) received carnitine, 19 patients (16.8%) received carnitine + lactulose, 55 patients (48.7%) received lactulose, 25 patients (22.1%) reduced valproate dose, and 24 patients (21.2%) received no treatment, and the success rates of ammonia levels in the normal range were 56.3, 50.0, 47.4, 41.8, 40.0, and 29.2%, respectively. Thus, in the United States, it seems that carnitine supplementation therapy is becoming popular in the psychiatric field for hyperammonemia caused by valproic acid.

Nakamura et al. found that psychiatric patients receiving valproate had significantly higher blood ammonia levels than those receiving non-valproate (100.4 ± 24.8 vs. 70.1 ± 11.1 μg/dL, $P = 0.0006$, normal value of 30–86 μg/dL), and a high acylcarnitine/free carnitine ratio (0.364 ± 0.089 vs. 0.267 ± 0.071, $P = 0.0027$) [136]. In addition, when carnitine (30 mg/kg/day) was administered for 3 months to 22 patients with high ammonia levels (>86 μg/dL), about 55% of patients showed decrease of ammonia levels [137]. Nagamine et al. reported that a 59-year-old male schizophrenic inpatient treated with valproic acid (900 mg/day) and risperidone (2 mg/day) had weight loss and recurrent hypoglycemic attacks (fasting blood glucose: 21 mg/dL), which were judged to indicate carnitine deficiency. They reported that the patient no longer had hypoglycemic attacks or weight loss after starting carnitine administration (900 mg/day) and alerted the patient to the possibility of carnitine deficiency in psychiatric patients receiving valproic acid therapy [138, 139].

Thus, it can be seen that hyperammonemia develops at a high rate of 36% in psychiatry, mainly in adult patients, as in patients with epilepsy. So, can asymptomatic hyperammonemia be left untreated?

12.10 Central Toxicity of Ammonia

Although it is not the purpose of this book to go into detail on the toxicity of ammonia, especially its central toxicity, the recent basic and clinical data and concept on ammonia toxicity are described in relation to valproic acid-induced hyperammonemia and hepatic encephalopathy (Chap. 15).

As described in Sect. 12.8.1, Chicharro et al. questioned the significance of blood ammonia monitoring in patients treated with valproate, stating that hyperammonemia with valproate is "quite common" and that ammonia should be measured only in the suspected cases of signs of acute encephalopathy [112]. However, in the pediatric neurological field, where epilepsy is treated in children whose brains are still developing, close attention is paid to hyperammonemia even if it is asymptomatic. In their discussion of hyperammonemia in epileptic patients, Yamamoto et al. cited Lichter-Konecki's study of astrocytes in mouse brains to explain why chronic hyperammonemia causes brain edema and encephalopathy [140] and concluded that long-term exposure of the brain to high ammonia, whether symptomatic or asymptomatic, is itself extremely dangerous and that ammonia toxicity can cause a risk of brain function disorders such as mental retardation and intellectual disabilities due to ammonia toxicity [113]. Regarding the brain toxicity of ammonia, there are theories such as increased glutamine in cells and vascular disorders. In a new study by Lichter-Konecki, water and potassium homeostasis is disrupted in astrocytes in mice exposed to hyperammonemia, resulting in astrocyte edema [140]. In any case, such prolonged chronic hyperammonemia can eventually lead to major brain damage, even if it is asymptomatic. In the field of epilepsy, valproate-induced hyperammonemia occurs in 30–40% of patients treated with valproate, and in psychiatry, where many adult patients are treated with valproate, the incidence of asymptomatic hyperammonemia is 30–50% of patients treated with valproate, although studies vary [134, 135].

Hyperammonemia may be caused by various causes such as inherited metabolic diseases such as urea cycle disorders, hepatic encephalopathy in patients with cirrhosis, valproic acid administration, and anticancer drugs, and appropriate treatment is necessary. In the case of urea cycle disorder with extremely high ammonia levels, drugs such as sodium benzoate and sodium phenylbutyrate (Buphenyl) to excrete glycine and glutamine in the urine as hippuric acid or phenylacetylglutamine, were used to reduce ammonia levels through alternative routes that do not involve in urea cycle, or L-arginine hydrochloride (intravenous Algi-U infusion) to maximize urea cycle. Treatment for other hepatic encephalopathy and mild-to-moderate drug-induced hyperammonemia is essentially the same, regardless of the cause. These include protein restriction, suppression of ammonia production by intestinal bacteria by administration of hard-to-absorb antimicrobials such as kanamycin and rifaximin, suppression of ammonia absorption by administration of lactulose, a non-absorbable synthetic disaccharide, and administration of carnitine. There are few books on ammonia medicine, and although it is an old book, "Medical Ammoniaology" covers a wide range of topics from basic to clinical and industrial fields [141]. In addition, the clinicians who are most knowledgeable about the treatment of hyperammonemia are hepatologists who treat patients with cirrhosis of the liver and a small number of specialists who study inherited metabolic diseases such as urea cycle disorders. Currently, The Japan Society for Portal Hypertension is active in Japan as an internationally unique society specializing in the research and treatment of hepatic encephalopathy. Internationally, there is a small international society called the International Society for Hepatic Encephalopathy and Nitrogen Metabolism (ISHEN), which brings together about 200 researchers on ammonia, including the toxicity mechanism of ammonia and the clinical development of drugs for the treatment of hyperammonemia, stays in a limited accommodation for about 4 days and discusses on ammonia from morning to night (http://www.ishen.org/). In 2014, the author had the opportunity to participate in the 16th ISHEN held in Surrey, a suburb of London, and I was able to learn a lot.

The normal value (reference value) of blood ammonia level is 30–80 µg/dL, but it varies slightly depending on each institution. In addition, it is said that the ammonia level itself tends to fluctuate depending on the contents of the previous day's diet and the method of blood collection. However, it is said that no matter how much meat you ate the day before, a blood ammonia level exceeding 100 µg/dL is impossible for a healthy person. In other words, the body is usually equipped with a metabolic system that strictly prevents hyperammonemia. Conversely, ammonia is an extremely toxic substance for living organisms, and fish can drip it into the water, but birds and mammals that have come to land have developed a metabolic system that excretes ammonia in the form of uric acid and urea. The breakdown of this metabolic system causes hyperammonemia, still which is extremely dangerous even asymptomatic conditions.

Hepatic encephalopathy in patients with cirrhosis is described in detail in Chap. 15, but basic and clinical research on ammonia toxicity is most active in the field of hepatic encephalopathy, which presents with hyperammonia encephalopathy due to decreased liver function and portal hypertension with progression of cirrhosis. According to these studies, it is clear that ammonia plays a central role in the development of encephalopathy, but it is also believed that many encephalopathy-causing factors, such as the ratio of branched amino acids to aromatic amino acids and the composition of fatty acids, work together with ammonia toxicity to cause encephalopathy. In addition to overt hepatic encephalopathy with clinical symptoms such as hyperammonemia, minimal hepatic encephalopathy (MHE), which is not yet symptomatic and can be diagnosed only by special tests, is also considered important, and actively being studied how to diagnose and start to treat MHE as early as possible [142, 143].

According to Dr. Akinobu Kato, a hepatologist and director of Morioka City Hospital, who is familiar with hepatic encephalopathy, states that "There are patients who have ammonia levels above 150 µg/dL but do not have abnormal symptoms, and there are patients who have 80–90 µg/dL, some patients develop symptoms such as somnolence, vomiting, and flapping tremors, so that ammonia level itself is not absolute. Brains exposed to ammonia for a long period of time even at low concentrations are transformed into sensitive brains [144]; ammonia toxicity suppresses mitochondrial function and reduces energy stores such as ATP and glucose, and coma-inducing factors (methionine and other amino acids, abnormal tryptophan metabolites, indoles, skatoles, short-chain fatty acids, and ammonia), etc., and there is a risk of developing encephalopathy rapidly. That is why it is necessary to pay attention from the stage of MHE." In other words, even if it is asymptomatic, it is a subtle and extremely dangerous condition in which the development of encephalopathy may or may not occur depending on a slight imbalance, such as the amount and combination of ammonia and other encephalopathy-causing factors, the energy state of astrocytes and nerve cells in the brain, and the resistance to oxidative stress. It must be said that it is an extremely important question how to deal with asymptomatic hyperammonemia not only in pediatric patients whose brains are still developing but also in elderly dementia patients. In the future, further research on the risk of asymptomatic hyperammonemia is expected to progress, research on supplementation therapy with carnitine preparations, and the development of new treatments and therapeutic agents for hyperammonemia.

12.11 Carnitine Supplementation Therapy for Valproate Poisoning

12.11.1 Valproic Acid Poisoning

Numerous cases of acute valproic acid poisoning have also been reported in Japan and abroad, such as children accidentally ingesting large doses of valproic acid or psychiatric patients taking large doses of valproic acid in suicide attempts [145–150].

According to an annual report of the American Association of Poison Control Centers' National Poison Data System, in 2016, among those transported for poison poisoning or overdose were 7763 cases of overdose of several drugs including valproic acid, and 3060 cases of valproic acid alone, making it the third most common drug after GABA (γ-aminobutyric acid) and its analogues (23,390 cases) and lamotrigine (10,197 cases), and almost as frequent as carbamazepine and its analogues (7614 cases). In addition, 316 cases in children under 5 years of age and 492 cases of children under 12 years of age were reported [151]. Of these 7763 cases, 1270 were unintentional overdoses and 1215 were intentional overdoses, resulting in 2 deaths and 71 serious outcomes. Thus, there are many cases of poisoning due to overdose of valproic acid in the United States, and it is thought that there are a considerable number of cases in Japan where people are transported to the emergency department due to an overdose of valproic acid.

Sztajnkrycer summarized the details of the acute toxicity of valproic acid overdose in his review [152]. It does not inhibit sodium channels or act as a direct GABA agonist as do other antiepileptic drugs but indirectly increases local GABA levels in the brain. In overdose, central nervous system depression is the most frequently observed symptom, leading to coma and respiratory depression. Cerebral edema is also frequently observed. In acute poisoning, hepatotoxicity is rare, but pancreatitis, hyperammonemia, metabolic and hematologic abnormalities are common. Treatment includes administration of activated charcoal or hemodialysis to remove excessive amounts of valproic acid from the body and administration of carnitine.

12.11.2 Hemodialysis/Direct Hemoperfusion Therapy for Valproic Acid Poisoning

If a comatose patient is found to have taken a large dose of valproate by the time he or she is transported to the critical care department, or if a high concentration of valproic acid is detected in the blood drug analysis, the usual treatment for drug addiction is activated charcoal, gastric lavage, etc., and if necessary, the airway is secured by tracheal intubation, continuous hemofiltration dialysis therapy (continuous hemodiafiltration: CHDF) or hemodialysis therapy should be performed.

Hicks et al. reported a case of successful lifesaving treatment of a patient with bipolar affective disorder who was overdosed with valproate and summarized 11 cases of valproic acid poisoning treated by hemodialysis or hemoperfusion [153]. Their own case is a 56-year-old female patient with bipolar affective disorder attempted suicide three times in the past, and this time too, when she was found unconscious, she was littered with tablets of divalproex sodium (brand name: Epival), a valproic acid preparation. Blood levels of valproic acid were extremely high at 998 μg/mL. Naloxone administration, activated charcoal administration through a nasal tube, and artificial respiration by tracheal intubation, etc. were performed, but there was no improvement in the condition. After 7.7 hours of hemodialysis from a femoral vein catheter, the blood level of valproic acid rapidly decreased to the therapeutic range, and although there was a slight rebound after discontinuation of dialysis, the patient began to move her limbs during dialysis, opened her eyes spontaneously, woke up after dialysis, and recovered to the point where she could communicate using a notepad. Valproic acid has a high protein binding ratio of 90%, and the volume of distribution is 0.1–0.4 L/kg according to the package insert, but in the case of poisoning, the blood concentration at the usual therapeutic dose (50–100 μg/mL) is far exceeded. It is considered to be a condition in which a large amount of free valproic acid is present in extracellular fluids such as blood [147, 153–155]. Therefore, hemodialysis therapy is considered to be one of the effective therapeutic measures. In addition, it has been reported that a direct hemoperfusion method using an activated charcoal column is also effective [146]. The patient was a 19-year-old female epilepsy patient with chronic tonic seizures who was transported to the hospital with 18 g (90 tablets) of valproate in an attempt to commit suicide. The blood concentration of valproic acid was 800 μg/mL, while the blood ammonia level was 72 μg/dL. Direct hemoperfusion was performed using a column containing 200 g of activated charcoal for 6 hours, and leakage of valproic acid from the column was observed in the 3rd hour, and the column was replaced with a new one at the 3rd hour, and the valproate concentration decreased from 404 μg/mL at the start of hemoperfusion to 45 μg/mL, and the patient recovered.

12.11.3 Carnitine Supplementation Therapy for Patients with Valproic Acid Poisoning

There have been many reports of successful life-saving cases in which carnitine supplementation therapy has been performed in patients with valproic acid poisoning.

Murakami et al. reported a 15-month-old boy who accidentally ingested 4000 mg (400 mg/kg) of valproate prescribed for his mother during epilepsy treatment [149]. The blood level of valproate at admission was 1316 μg/mL and the blood ammonia level was 69 μg/dL. Carnitine was administered through a nasal tube for 3 days (100 mg/kg/day) along with gastric lavage and infusion, and the patient was discharged from the hospital on day 8 without any abnormalities in liver function. Initially, the metabolite 4-en-VPA of valproic acid was significantly elevated, and the β oxidation metabolite of valproic acid decreased markedly, but these levels normalized after carnitine administration. Ishikura et al. reported the same case in detail separately from the standpoint of the emergency department [148].

Auinger et al. reported a 19-year-old male case with near-brain death, in which the brain stem response was reduced to the point where the pupillary response to light was not seen, but he was saved by carnitine (dose and route unknown) and CHDF [156]. The blood valproic acid concentration was 12,430 μmol/L (normal value: 350–700 μmol/L) and the blood ammonia concentration was 500 μmol/L (normal value: <30 μmol/L), but no cerebral edema was observed on CT examination.

Mitani et al. reported a case in which a 22-year-old manic-depressive female patient was transported to an advanced critical care center after taking a large amount of lithium carbonate and valproic acid in a suicide attempt [147]. In this case, CHDF was performed because it was necessary to remove lithium, which has a low protein binding rate, and since the prescribed valproic acid preparation was a sustained-release formulation, it may remain in the gastrointestinal tract, and lactulose and carnitine were administered for hyperammonemia. In the discussion, they described that the therapeutic range of valproic acid is 40–120 μg/mL and the poisoning dose is >200 μg/mL, very narrow safety margin, and it is prescribed as a mood stabilizer with various psychotropic drugs, so there is a possibility that the number of cases of valproic acid poisoning will increase in the future.

Perrott et al. conducted a systematic review of case reports of carnitine administration in patients with valproate poisoning and summarized 8 cases in 7 reports [157] (Note: The aforementioned reports of Murakami and Ishikura were reporting a same case, but Perrott et al. taken the case separately, so there are really 7 cases). Although the number of cases is small and the evidence is limited, these case reports suggest that patients with valproic acid poisoning who have a decrease in consciousness or an increase in ammonia should be administered 100 mg/kg of carnitine intravenously, followed by continuous intravenous infusion (50 mg/kg) every 8 hours until consciousness is restored or ammonia is reduced.

LoVecchio et al. extracted 674 cases of valproic acid poisoning from a 3-year database of more than 300,000 cases at the Poison Control Center in Arizona and examined the safety of carnitine in patients administered with it [158]. There were 19 cases of overdose of valproate alone, and 55 doses of carnitine were administered. Carnitine was administered 196 times to patients who were poisoned with a mixture of valproate and other drugs, but there were no adverse events such as allergic reactions or hypotension in any of the cases, so there was no safety problem in administering carnitine to patients poisoned with valproate. No efficacy was assessed in this study.

Mock et al. also reviewed the usefulness and safety of carnitine supplementation therapy for VHE, including cases of valproate overdose and poisoning [159]. Although there were no double-blind comparative trials, it was concluded that intravenous administration of carnitine was recommended for patients with acute poisoning with valproic acid.

Russell summarized a review titled "Carnitine as an 'antidote' to the acute toxicity of valproate in children" and found that many case reports and carnitine administration increased the survival rate of patients with hepatotoxicity caused by valproate, and that children overdose on valproate, resulting in coma, hyperammonemia, and valproate concentration of >450 mg/L, carnitine should be administered to these patients [160]. The usage and dosage are not established; for patients with acute toxicity of valproic acid who do not have hepatotoxicity or hyperammonemia, 100 mg/kg/day should be administered every 6 hours up to a maximum of 3 g per day as a prophylactic dose, and for patients with acute toxicity of valproic acid with symptomatic hyperammonemia or hepatotoxicity, 100 mg/kg should be administered intravenously for 30 minutes as an initial loading dose for up to 6 g/day, followed by a maintenance dose of 15 mg/kg/day every 4 hours over 1–30 minutes.

As described above, there are many cases of accidental ingestion and suicide attempts in cases of overdose and poisoning of valproic acid, and advanced evidence such as double-blind comparative trials is not desirable. There are many reports that carnitine supplementation therapy is extremely effective and safe.

Carnitine supplementation therapy for VHT and hyperammonemic encephalopathy was approved in the United States in 1992, and intravenous preparation was approved in 1996. In Japan, carnitine preparations were approved in 1990, but they were only indicated for carnitine deficiency due to methylmalonic acidemia and propionic acidemia and could not be officially used for side effects caused by valproic acid, but on March 11, 2011, through a public knowledge application led by the author's efforts, an application for the indication of "carnitine deficiency" in general was approved, and it can now be used for side effects caused by valproic acid. However,

Table 12.14 Description of carnitine administration for hyperammonemia due to valproic acid poisoning in the Washington Manual

Supplementation of L-carnitine for hyperammonemia due to valproic acid poisoning
Most poisoning symptoms improve with symptomatic treatment. If there is no impairment of consciousness and the airway is secured, activated charcoal administration is recommended.
If hyperammonemia (>35 mmol/L or >80 μg/dL) is observed, L-carnitine is administered. In the absence of impaired consciousness, oral administration of carnitine is desirable. 50~100 mg/kg per day every 6 hours (up to a maximum of 3 g per day).
If oral administration is not possible, an initial dose of 100 mg/kg (up to a maximum of 6 g) is administered intravenously, followed by an additional 15 mg/kg every 4 hours.
If the blood ammonia concentration drops below 35 mmol/L, discontinue treatment.

Modified from Ref. [161]

the intravenous preparations had not yet been released, and in cases of overdose of valproic acid, intragastric administration using a nasal tube with tablets as a suspension was performed. However, from February 26, 2013, intravenous carnitine preparations (free form) were launched in Japan and can be used. At the time of its launch, it was ample preparation, but since December 20, 2017, it has been a prefilled syringe (L-Cartin® FF intravenous injection 1000 mg syringe). Hospitals with critical care departments should always have intravenous carnitine preparations on hand in case such patients are transported.

Based on these backgrounds, the "Washington Manual, 13th edition," which is said to be the bible of residents, the carnitine supplementation therapy for valproic acid poisoning is summarized in the section on "Valproic acid" in Chapter 28, "Toxicology" of hyperammonemia caused by valproic acid as shown in Table 12.14 [161].

12.12 Pathogenic Mechanism of VHT and VHE

As described above, the pathogenic mechanism of VHT and VHE has not yet been completely elucidated, although various basic and clinical studies have been reported. It is also undeniable that the pathogenic mechanism may differ from case to case. However, it is thought that mitochondrial dysfunction due to carnitine deficiency is the underlying cause, and that valproic acid itself and its metabolites affect the urea cycle, carnitine biosynthesis, and reabsorption. The following is a summary of the current understanding of the pathogenic mechanism of VHT and VHE.

12.12.1 Carnitine Deficiency Due to Valproic Acid

Why is valproic acid administration more likely to cause carnitine deficiency and hyperammonemia? In their 2005 and 2009 review articles, Lheureux et al. summarized the pathogenic mechanism of VHT and VHE [162, 163]. Nakajima et al. also summarized the mechanism of carnitine deficiency in a report that examined changes in acylcarnitine in the blood of epilepsy children receiving valproate using LC/MS/MS [164]. In addition, there are many review papers on the relationship between valproic acid and its metabolites and fatty acid metabolism, and the pathogenic mechanism of VHT and VHE has been speculated [152, 165–167]. Summary of them, although somewhat complicated, is shown in Fig. 12.7.

The overview is given below (the numbers correspond to the circled number in Fig. 12.7).

❶ Urinary excretion in the form of valproylcarnitine causes carnitine deficiency.

About 17% of valproic acid enters the blood as unchanged form, and 70% of it binds to free carnitine and is transported into the mitochondria as valproylcarnitine. Initially, there was a theory that carnitine was excreted in the urine in the form of valproylcarnitine, causing decrease of carnitine, but unlike the case of carnitine deficiency caused by antibiotics containing a pivoxil group, it is now thought to be less than 1% [168, 169]. However, since valproic acid administration produces substantial amounts of 3-hydroxyisovalerylcarnitine (C5OH) and

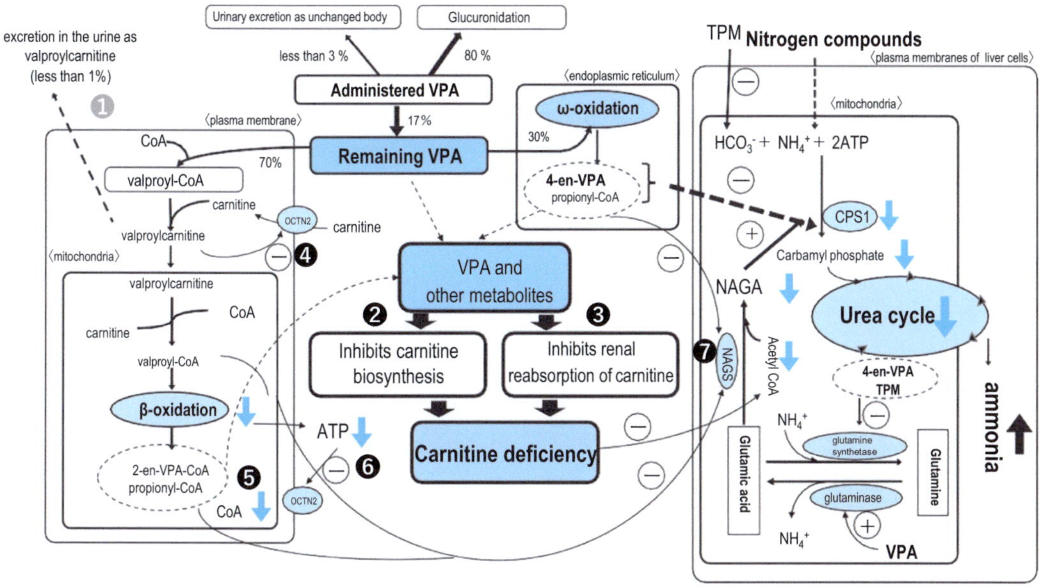

Fig. 12.7 Speculated mechanisms of carnitine deficiency and hyperammonemia by VPA. ⊖ inhibition, ⊕ activation, CPS1 carbamyl phosphate synthetase, NAGS N-acetylglutamate synthase, VPA valproic acid, TPM topiramate, OCTN2 carnitine transporter. (Created from Refs. [162, 164–167])

2-propylglutarylcarnitine, it cannot be ruled out that the urinary excretion of acylcarnitine is related to carnitine deficiency [169, 170].

❷ Valproic acid reduces the biosynthesis of carnitine and leads to carnitine deficiency.

When valproic acid is administered to rats, the biosynthesis of carnitine decreases, resulting in a decrease in the content of carnitine in the liver and serum. The mechanism is that valproic acid does not directly inhibit biosynthetic enzymes, but that valproic acid or its metabolites reduce α-ketoglutaric acid, which is a cofactor of the enzyme γ-butyrobetaine hydroxylase in the final step of carnitine biosynthesis, thereby reducing the biosynthesis of carnitine in rats [171], which has been shown in the vivo study.

❸ Valproic acid and its metabolites inhibit the reabsorption of free carnitine in the renal tubules and carnitine transporter activity in the cell membrane, resulting in carnitine deficiency.

Matsuda et al. examined serum and urinary carnitine in pediatric patients with Reye-like syndrome receiving valproate who presented with hyperammonemia and carnitine deficiency and reported that renal reabsorption of free carnitine and acylcarnitine was reduced [47]. Camiña et al. also reported that in a study in mice, valproate administration at a clinical dose for 7 days inhibited the reabsorption of acylcarnitine in the kidneys but did not significantly affect the reabsorption of free carnitine, and that administration of carbamazepine, phenytoin, and phenobarbital suppressed the reabsorption of free carnitine in the kidneys [172]. In addition, basic studies have shown by Tein et al. that valproic acid inhibits the uptake of carnitine in cultured fibroblasts [48, 49]. Ohashi et al. reported inhibition of carnitine transporter activity by valproic acid in hOCTN2-expressing HEK293 cells, and Wu et al. reported inhibition of carnitine transporter activity by valproic acid in hOCTN2 in human placenta [173, 174]. However, no direct inhibition of carnitine uptake by valproic acid was observed in the system in which human carnitine transporter was expressed in *Xenopus* oocytes by Wagner et al. Wagner et al. suggested that metabolites of valproic acid may have inhibited uptake, or the expression of carnitine transporters may have been suppressed in Tein's study where fibroblast was incubated with valproic acid for 24 hours [50]. Further studies are needed.

❹ Intracellular carnitine is deficient because valproic acid or valproylcarnitine inhibits the carnitine transporter (OCTN2) in the cell membrane.

Valproic acid and its metabolites, as well as valproylcarnitine, inhibit the carnitine transporter of the cell membrane, leading to carnitine deficiency in the cell.

❺ The decrease in free CoA makes it difficult for free carnitine to be regenerated from valproylcarnitine, and the free carnitine decreases.

It has been reported that short-chain acyl-CoA and medium-chain acyl-CoA were increased and free CoA were significantly decreased in the liver of patients with Reye syndrome [175]. Even in the case of valproic acid therapy, valproyl CoA and mitochondrial metabolites of valproic acid suppress free CoA-dependent metabolic processes by seizing a limited amount of free CoA in the mitochondria [61, 176]. Valproic acid is a branched chain fatty acid, so that usually valproyl-CoA reacts with carnitine, and free CoA is regenerated and reused by following reaction:

$$\text{valproyl} - \text{CoA} + \text{carnitine} \qquad \text{valproylcarnitine} + \text{free CoA}$$

However, the reaction of branched chain valproyl-CoA is much slower than that of linear fatty acid acyl-CoA, and free CoA is sequestrate [177–179]. It has been reported that in experiments with rats treated with valproate, free CoA decreased, while the total amount of CoA remained the same [68]. The acylcarnitine/free carnitine ratio in the blood is thought to reflect the acyl-CoA/free CoA ratio in the mitochondria. This may be the reason why the acylcarnitine/free carnitine ratio has increased in a study that examined blood carnitine concentrations in patients receiving valproate, resulting in a state of carnitine insufficiency. Furthermore, it has been shown that valproyl-CoA is produced outside the mitochondria, and the β oxidation of long-chain fatty acids, which are important for energy metabolism, is suppressed [166, 180].

❻ When β oxidation is suppressed, efficient ATP production is not possible, the activity of ATP-dependent carnitine transporters is reduced, and the reabsorption of free carnitine in the renal tubules is reduced, resulting in carnitine deficiency.

❼ Valproyl-CoA and propionyl-CoA inhibit the synthase of N-acetylglutamate, a cofactor of the primary enzyme CPS1 in the urea cycle, causing hyperammonemia [181, 182].

In this way, carnitine deficiency caused by valproic acid is thought to be caused by a combination of various mechanisms, and it is thought that the genetic background, clinical background, and concomitant medications of each individual case have a further influence. For example, hyperammonemia in patients who used valproate in combination with topiramate has been reported in many cases in the fields of pediatric neurology and psychiatry in recent years [135, 183]. As shown in the upper right part of Fig. 12.7, topiramate reduces bicarbonate ions (HCO_3^-), which is one of the substrates of CPS1, by inhibiting carbonic anhydrase, and reduces the activity of the urea cycle. Therefore, caution is required when used in combination with valproic acid, which may cause hyperammonemia even if the concentration of carnitine in the blood is normal [167].

12.12.2 Mechanism of Hyperammonemia Induced by Valproic Acid

I have described here and there why ammonia rises due to valproic acid, but I would like to summarize the mechanism again.

In order to understand hyperammonemia caused by valproic acid, it is necessary to understand the relationship between the urea cycle and carnitine. As described in detail in Chap. 5, carnitine is an essential substance for transporting long-chain fatty acids to the inner membrane of the mitochondria, and the long-chain fatty acids transported to the inner mitochondrial membrane undergo β oxidation and eventually produce acetyl-CoA. This acetyl-CoA enters the TCA cycle and produces NADH and $FADH_2$, which enter in the electron transport chain of the inner mitochondrial membrane and produce a large amount of ATP (see Sect. 3.10 in Chap. 3). In order to efficiently produce energy, β oxidation of long-chain fatty acids using carnitine is important.

Next, let's take a closer look at the urea cycle (see Chap. 5, Figs. 5.5 and 5.6). Ammonia is produced when protein and other nitrogen compounds in the body are metabolized, or when protein derived from the diet is decomposed by intestinal bacteria, and the concentration of ammonia in the blood increases. In the first step of the urea cycle, ammonia (ammonium ions) and two molecules of ATP and HCO_3^- are converted to carbamyl phosphate (also called carbamoyl phosphate) by the action of CPS1. The presence of a cofactor called N-acetylglutamic acid (NAGA) is essential for the activity of CPS1. This NAGA is synthesized from acetyl-CoA, the end products of β-oxidation, and glutamic acid, by the action of N-acetylglutamate synthase (NAGS, also called glutamate N-acetyltransferase). This amount of NAGA is said to be the rate-limiting step of the urea cycle. Therefore, when the biosynthesis and reabsorption of carnitine are inhibited by valproic acid or its metabolites, carnitine deficiency occurs. As a result, the transport of long-chain fatty acids into mitochondria is suppressed, and the free CoA decreases, resulting in suppression of β-oxidation. In addition, when the β-oxidation of fatty acids is directly suppressed by valproic acid or its metabolites, the production of acetyl-CoA decreases, and NAGA, synthesized from glutamic acid and acetyl-CoA, decreases, and CPS1 activity that requires NAGA as a cofactor decreases (NAGS deficiency is one of the inherited urea cycle abnormalities, which causes hyperammonemia).

As a point of action where valproic acid and its metabolites inhibit the β-oxidation of long-chain fatty acids, the α subunit of trifunctional protein [73] and carnitine palmitoyltransferase 1 (CPT1) [74] are reported. Aires et al. also reported that valproyl-CoA directly inhibits NAGS [182]. Thus, valproic acid, valproyl-CoA, and metabolites of valproic acid are thought to inhibit fatty acid β oxidation and urea cycles in various ways, causing hepatotoxicity and hyperammonemia.

In summary, the urea cycle is an energy-requiring metabolic system in which a total of 3 molecules of ATP are used to convert 1 molecule of ammonia into urea, and if the β oxidation of long-chain fatty acids is suppressed by carnitine deficiency due to valproic acid administration, the urea cycle becomes difficult to turn, and the supply of acetyl-CoA decreases due to carnitine deficiency. When NAGS is inhibited by propionyl-CoA or valproyl-CoA, the synthesis of NAGA, a cofactor of CPS1, is suppressed and the urea cycle is further disrupted, resulting in the development of hyperammonemia.

12.13 Summary

It is clear from many basic and clinical studies that carnitine deficiency is underlying of VHT and VHE. There are no biological substances or medicines that can replace the action of carnitine. Therefore, carnitine supplementation therapy is the single and most effective preventive or therapeutic measure for carnitine deficiency. Therefore, carnitine supplementation for VHT and VHE ameliorates the pathogenesis of carnitine insufficiency due to acyl-CoA accumulation such as valproyl-CoA, restores ATP production by β oxidation, and increases the supply of acetyl-

CoA required for the synthesis of NAGA. Furthermore, by promoting the excretion of valproyl-CoA and propionyl-CoA from the mitochondria, it is thought that the suppression of NAGS by valproyl-CoA and propionyl-CoA is canceled, and the urea cycle proceeds smoothly.

Valproic acid is a highly useful drug in both the treatment of epilepsy and psychiatric disorders, and it is expected that further research on the carnitine supplementation will be conducted for safer use of valproic acid in clinical practice.

References

1. López-Muñoz F, Baumeister AA, Hawkins MF, et al. The role of serendipity in the discovery of the clinical effects of psychotropic drugs: beyond of the myth. Actas Esp Psiquiatr. 2012;40(1):34–42.
2. Kyowa Kirin Co., Ltd.. Depaken Interview form revised January 2024 (5th Edition). 2024. https://medical.kyowakirin.co.jp/site/drugpdf/interv/dpa_in.pdf. (Article in Japanese).
3. Löscher W. The discovery of valproate. In: Löscher W, editor. Valproate. Basel: Springer Basel; 1999. p. 13–5. https://play.google.com/books/reader?id=WjUDCAAAQBAJ&hl=ja&pg=GBS.PA3.w.1.0.0.
4. Saio T, Kurihara M. Expansion of indications not based on clinical trials. The impact of the "Second section manager's notice": focusing on the case of sodium valproate. Clin Eval. 2004;31(3):587–601. (Article in Japanese).
5. Guidelines for the Practice of Dementia Committee, Japanese Society of Neurology. Chapter 3. Treatment CQ 3C-2. How to deal with epilepsy including convulsions in people with dementia. In: Guidelines for the practice of dementia diseases 2017. Tokyo: IGAKU-SHOIN Co., Ltd.; 2017. p. 94–6. (Book in Japanese).
6. Willmore LJ, Wilder BJ, Bruni J, et al. Effect of valproic acid on hepatic function. Neurology. 1978;28(9 Pt 1):961–4. https://doi.org/10.1212/wnl.28.9.961.
7. Donat JF, Bocchini JA Jr, Gonzalez E, et al. Valproic acid and fatal hepatitis. Neurology. 1979;29(2):273–4.
8. Suchy FJ, Balistreri WF, Buchino JJ, et al. Acute hepatic failure associated with the use of sodium valproate – report of two fatal cases. N Engl J Med. 1979;300(17):962–6. https://doi.org/10.1056/NEJM197904263001706.
9. Bowdle TA, Patel IH, Wilensky AJ, et al. Hepatic failure from valproic acid. N Engl J Med. 1979;301(8):435–7. https://doi.org/10.1056/nejm197908233010817.
10. Temple RJ, Jones JK, Crout JR. Adverse effects of newly marketed drugs. N Engl J Med. 1979;300(18):1046–7. https://doi.org/10.1056/NEJM197905033001810.
11. Mathis RK, Hanson RF, Sharp HL. Hepatic failure from valproic acid. N Engl J Med. 1979;301(8):436.
12. Partin JS, McAdams AJ, Suchy FJ, et al. Hepatic failure from valproic acid. N Engl J Med. 1979;301(8):436–7.
13. Gerber N, Dickinson RG, Harland RC, et al. Reye-like syndrome associated with valproic acid therapy. J Pediatr. 1979;95(1):142–4. https://doi.org/10.1016/s0022-3476(79)80110-9.
14. Jacobi G, Thorbeck R, Ritz A, et al. Fatal hepatotoxicity in child on phenobarbitone and sodium valproate. Lancet. 1980;1(8170):712–3.
15. Ware S, Millward-Sadler GH. Acute liver disease associated with sodium valproate. Lancet. 1980;2(8204):1110–3. https://doi.org/10.1016/s0140-6736(80)92543-x.
16. Zimmerman HJ, Ishak KG. Valproate-induced hepatic injury: analyses of 23 fatal cases. Hepatology. 1982;2(5):591–7. https://doi.org/10.1002/hep.1840020513.
17. Zafrani ES, Berthelot P. Sodium valproate in the induction of unusual hepatotoxicity. Hepatology. 1982;2(5):648–9. https://doi.org/10.1002/hep.1840020520.
18. Powell-Jackson PR, Tredger JM, Williams R. Hepatotoxicity to sodium valproate: a review. Gut. 1984;25(6):673–81. https://doi.org/10.1136/gut.25.6.673.
19. Sackellares JC, Lee SI, Dreifuss FE. Stupor following administration of valproic acid to patients receiving other antiepileptic drugs. Epilepsia. 1979;20(6):697–703. https://doi.org/10.1111/j.1528-1157.1979.tb04853.x.
20. Coulter DL, Wu H, Allen RJ. Valproic acid therapy in childhood epilepsy. JAMA. 1980;244(8):785–8.
21. Coulter DL, Allen RJ. Secondary hyperammonaemia: a possible mechanism for valproate encephalopathy. Lancet. 1980;1(8181):1310–1. https://doi.org/10.1016/s0140-6736(80)91773-0.
22. Coulter DL, Allen RJ. Hyperammonemia with valproic acid therapy. J Pediatr. 1981;99(2):317–9. https://doi.org/10.1016/s0022-3476(81)80489-1.
23. Young RS, Bergman I, Gang DL, et al. Fatal Reye-like syndrome associated with valproic acid. Ann Neurol. 1980;7(4):389. https://doi.org/10.1002/ana.410070423.
24. Keene DL, Humphreys P, Carpenter B, et al. Valproic acid producing a Reye-like syndrome. Can J Neurol Sci. 1982;9(4):435–7. https://doi.org/10.1017/s0317167100044358.
25. Dreifuss FE, Santilli N, Langer DH, et al. Valproic acid hepatic fatalities: a retrospective review. Neurology. 1987;37(3):379–85. https://doi.org/10.1212/wnl.37.3.379.
26. Dreifuss FE, Langer DH, Moline KA, et al. Valproic acid hepatic fatalities. II. US experience since 1984. Neurology. 1989;39(2 Pt 1):201–7. https://doi.org/10.1212/wnl.39.2.201.

27. Jeavons PM. Non-dose-related side effects of valproate. Epilepsia. 1984;25(Suppl 1):S50–5. https://doi.org/10.1111/j.1528-1157.1984.tb05638.x.
28. Scheffner D. Fatal liver failure in children on valproate. Lancet. 1986;2(8505):511. https://doi.org/10.1016/s0140-6736(86)90376-4.
29. Scheffner D, König S, Rauterberg-Ruland I, et al. Fatal liver failure in 16 children with valproate therapy. Epilepsia. 1988;29(5):530–42. https://doi.org/10.1111/j.1528-1157.1988.tb03757.x.
30. König SA, Siemes H, Bläker F, et al. Severe hepatotoxicity during valproate therapy: an update and report of eight new fatalities. Epilepsia. 1994;35(5):1005–15. https://doi.org/10.1111/j.1528-1157.1994.tb02546.x.
31. Kaneko S, Nau H. Chapter 9. Hepatotoxicity and metabolites of valproic acid. In: Clinical pharmacology of valproic acid, revised ed. Tokyo: Life Science Co., Ltd.; 2006. p. 86–99. (Book in Japanese).
32. Lenn NJ, Ellis WG, Washburn ER, et al. Fatal hepatocerebral syndrome in siblings discordant for exposure to valproate. Epilepsia. 1990;31(5):578–83. https://doi.org/10.1111/j.1528-1157.1990.tb06109.x.
33. Appleton RE, Farrell K, Applegarth DA, et al. The high incidence of valproate hepatotoxicity in infants may relate to familial metabolic defects. Can J Neurol Sci. 1990;17(2):145–8. https://doi.org/10.1017/s0317167100030353.
34. Kondo T, Kaneko S, Otani K, et al. Associations between risk factors for valproate hepatotoxicity and altered valproate metabolism. Epilepsia. 1992;33(1):172–7. https://doi.org/10.1111/j.1528-1157.1992.tb02302.x.
35. Kondo T, Ishida M, Kaneko S, et al. Is 2-propyl-4-pentenoic acid, a hepatotoxic metabolite of valproate, responsible for valproate-induced hyperammonemia? Epilepsia. 1992;33(3):550–4. https://doi.org/10.1111/j.1528-1157.1992.tb01708.x.
36. Siemes H, Nau H, Schultze K, et al. Valproate (VPA) metabolites in various clinical conditions of probable VPA-associated hepatotoxicity. Epilepsia. 1993;34(2):332–46. https://doi.org/10.1111/j.1528-1157.1993.tb02419.x.
37. Baillie TA. Metabolic activation of valproic acid and drug-mediated hepatotoxicity. Role of the terminal olefin, 2-n-propyl-4-pentenoic acid. Chem Res Toxicol. 1988;1(4):195–9. https://doi.org/10.1021/tx00004a001.
38. McLaughlin DB, Eadie MJ, Parker-Scott SL, et al. Valproate metabolism during valproate-associated hepatotoxicity in a surviving adult patient. Epilepsy Res. 2000;41(3):259–68. https://doi.org/10.1016/s0920-1211(00)00151-0.
39. Cotariu D, Zaidman JL. Valproic acid and the liver. Clin Chem. 1988;34:890–7.
40. Ohtani Y, Endo F, Matsuda I. Carnitine deficiency and hyperammonemia associated with valproic acid therapy. J Pediatr. 1982;101(5):782–5. https://doi.org/10.1016/s0022-3476(82)80320-x.
41. Böhles H, Richter K, Wagner-Thiessen E, et al. Decreased serum carnitine in valproate induced Reye syndrome. Eur J Pediatr. 1982;139(3):185–6. https://doi.org/10.1007/BF01377353.
42. Coulter DL. Carnitine deficiency: a possible mechanism for valproate hepatotoxicity. Lancet. 1984;1(8378):689. https://doi.org/10.1016/s0140-6736(84)92209-8.
43. Murphy JV, Marquardt KM, Shug AL. Valproic acid associated abnormalities of carnitine metabolism. Lancet. 1985;1(8432):820–1. https://doi.org/10.1016/s0140-6736(85)91481-3.
44. Morita J, Yuge K, Yoshino M. Hypocarnitinemia in the handicapped individuals who receive a polypharmacy of antiepileptic drugs. Neuropediatrics. 1986;17(4):203–5. https://doi.org/10.1055/s-2008-1052530.
45. Rodriguez-Segade S, de la Peña CA, Tutor JC, et al. Carnitine deficiency associated with anticonvulsant therapy. Clin Chim Acta. 1989;181(2):175–81. https://doi.org/10.1016/0009-8981(89)90185-x.
46. Shapira Y, Gutman A. Muscle carnitine deficiency in patients using valproic acid. J Pediatr. 1991;118(4 Pt 1):646–9. https://doi.org/10.1016/s0022-3476(05)83396-7.
47. Matsuda I, Ohtani Y, Ninomiya N. Renal handling of carnitine in children with carnitine deficiency and hyperammonemia associated with valproate therapy. J Pediatr. 1986;109(1):131–4. https://doi.org/10.1016/s0022-3476(86)80592-3.
48. Tein I, DiMauro S, Xie ZW, et al. Valproic acid impairs carnitine uptake in cultured human skin fibroblasts. An in vitro model for the pathogenesis of valproic acid-associated carnitine deficiency. Pediatr Res. 1993;34(3):281–7. https://doi.org/10.1203/00006450-199309000-00008.
49. Tein I, Xie ZW. Reversal of valproic acid-associated impairment of carnitine uptake in cultured human skin fibroblasts. Biochem Biophys Res Commun. 1994;204(2):753–8. https://doi.org/10.1006/bbrc.1994.2523.
50. Wagner CA, Lükewille U, Kaltenbach S, et al. Functional and pharmacological characterization of human Na+-carnitine cotransporter hOCTN2. Am J Physiol Renal Physiol. 2000;279(3):F584–91. https://doi.org/10.1152/ajprenal.2000.279.3.F584.
51. Melegh B, Kerner J, Kispál G, et al. Effect of chronic valproic acid treatment on plasma and urine carnitine levels in children: decreased urinary excretion. Acta Paediatr Hung. 1987;28(2):137–42.
52. Krähenbühl S, Mang G, Kupferschmidt H, et al. Plasma and hepatic carnitine and coenzyme A pools in patient with fatal, valproate induced hepatotoxicity. Gut. 1995;37(1):140–3. https://doi.org/10.1136/gut.37.1.140.
53. Laub MC, Paetzke-Brunner I, Jaeger G. Serum carnitine during valproic acid therapy. Epilepsia. 1986;27(5):559–62. https://doi.org/10.1111/j.1528-1157.1986.tb03584.x.

54. Murphy JV, Groover RV, Hodge C. Hepatotoxic effects in a child receiving valproate and carnitine. J Pediatr. 1993;123(3):318–20. https://doi.org/10.1016/s0022-3476(05)81711-1.
55. Coulter DL. Carnitine, valproate, and toxicity. J Child Neurol. 1991;6(1):7–14. https://doi.org/10.1177/088307389100600102.
56. Coulter DL. Carnitine deficiency in epilepsy: risk factors and treatment. J Child Neurol. 1995;10(Suppl 2):S32–9.
57. Raskind JY, El-Chaar GM. The role of carnitine supplementation during valproic acid therapy. Ann Pharmacother. 2000;34(5):630–8. https://doi.org/10.1345/aph.19242.
58. Bjorge SM, Baillie TA. Inhibition of medium-chain fatty acid β-oxidation in vitro by valproic acid and its unsaturated metabolite, 2-n-propyl-4-pentenoic acid. Biochem Biophys Res Commun. 1985;132(1):245–52. https://doi.org/10.1016/0006-291x(85)91014-9.
59. Bjorge SM, Baillie TA. Studies on the beta-oxidation of valproic acid in rat liver mitochondrial preparations. Drug Metab Dispos. 1991;19(4):823–9.
60. Ponchaut S, van Hoof F, Veitch K. In vitro effects of valproate and valproate metabolites on mitochondrial oxidations: relevance of CoA sequestration to the observed inhibitions. Biochem Pharmacol. 1992;43(11):2435–42. https://doi.org/10.1016/0006-2952(92)90324-c.
61. Silva MF, Ruiter JP, IJlst L, et al. Differential effect of valproate and its Δ2- and Δ4-unsaturated metabolites, on the β-oxidation rate of long-chain and medium-chain fatty acids. Chem Biol Interact. 2001;137(3):203–12. https://doi.org/10.1016/s0009-2797(01)00234-4.
62. Haas R, Chir B, Stumpf DA, et al. Inhibitory effects of sodium valproate on oxidative phosphorylation. Neurology (NY). 1981;31(11):1473–6. https://doi.org/10.1212/wnl.31.11.1473.
63. Sugimoto T, Woo M, Nishida N, et al. Hepatotoxicity in rat following administration of valproic acid. Epilepsia. 1987;28(2):142–6. https://doi.org/10.1111/j.1528-1157.1987.tb03640.x.
64. Sugimoto T, Araki A, Nishida N, et al. Hepatotoxicity in rat following administration of valproic acid: effect of L-carnitine supplementation. Epilepsia. 1987;28(4):373–7. https://doi.org/10.1111/j.1528-1157.1987.tb03660.x.
65. Coudé FX, Grimber G, Pelet A, et al. Action of the antiepileptic drug, valproic acid, on fatty acid oxidation in isolated rat hepatocytes. Biochem Biophys Res Commun. 1983;115(2):730–6. https://doi.org/10.1016/s0006-291x(83)80205-8.
66. Thurston JH, Carroll JE, Hauhart RE, et al. A single therapeutic dose of valproate affects liver carbohydrate, fat, adenylate, amino acid, coenzyme A, and carnitine metabolism in infant mice: possible clinical significance. Life Sci. 1985;36(17):1643–51. https://doi.org/10.1016/0024-3205(85)90367-4.
67. Rozas I, Camiña MF, Paz JM, et al. Effects of acute valproate administration on carnitine metabolism in mouse serum and tissues. Biochem Pharmacol. 1990;39(1):181–5. https://doi.org/10.1016/0006-2952(90)90663-6.
68. Kesterson JW, Granneman GR, Machinist JM. The hepatotoxicity of valproic acid and its metabolites in rats. I. Toxicologic, biochemical and histopathologic studies. Hepatology. 1984;4(6):1143–52. https://doi.org/10.1002/hep.1840040609.
69. Kanayama M, Sugiyama N, Morishita H, et al. Effects on mitochondrial function in patients taking valproate. Brain Dev. 1985;17(6):507–13. (Article in Japanese).
70. Melegh B, Pap M, Morava E, et al. Carnitine-dependent changes of metabolic fuel consumption during long-term treatment with valproic acid. J Pediatr. 1994;125(2):317–21. https://doi.org/10.1016/s0022-3476(94)70218-7.
71. Arimoto K, Sakuragawa N, Suehiro M, et al. Abnormalities in the ^{13}C-palmitic acid breath test in children taking valproic acid. Brain Dev. 1986;18(5):354–9. (Article in Japanese).
72. Arimoto K, Sakuragawa N, Suehiro M, et al. Abnormal ^{13}C-fatty acid breath tests in patients treated with valproic acid. J Child Neurol. 1988;3(4):250–7. https://doi.org/10.1177/088307388800300404.
73. Baldwin GS, Abbott FS, Nau H. Binding of a valproate metabolite to the trifunctional protein of fatty acid oxidation. FEBS Lett. 1996;384(1):58–60. https://doi.org/10.1016/0014-5793(96)00267-0.
74. Aires CCP, IJlst L, Stet F, et al. Inhibition of hepatic carnitine palmitoyl-transferase I (CPT IA) by valproyl-CoA as a possible mechanism of valproate-induced steatosis. Biochem Pharmacol. 2010;79(5):792–9. https://doi.org/10.1016/j.bcp.2009.10.011.
75. Granneman GR, Wang SI, Kesterson JW, et al. The hepatotoxicity of valproic acid and its metabolites in rats. II. Intermediary and valproic acid metabolism. Hepatology. 1984;4(6):1153–8. https://doi.org/10.1002/hep.1840040610.
76. Igarashi N, Sato T, Kyouya S. Secondary carnitine deficiency in handicapped patients receiving valproic acid and /or elemental diet. Acta Paediatr Jpn. 1990;32(2):139–45. https://doi.org/10.1111/j.1442-200x.1990.tb00799.x.
77. Hirose S, Mitsudome A, Yasumoto S, et al. Valproate therapy does not deplete carnitine levels in otherwise healthy children. Pediatrics. 1998;101(5):E9. https://doi.org/10.1542/peds.101.5.e9.
78. Fung EL, Tang NL, Ho CS, et al. Carnitine level in Chinese epileptic patients taking sodium valproate. Pediatr Neurol. 2003;28(1):24–7. https://doi.org/10.1016/s0887-8994(02)00460-5.
79. Takeuchi T, Sugimoto T, Nishida N, et al. Evaluation of the cytotoxicity of sodium valproate on primary cultured rat hepatocytes. Neuropediatrics. 1988;19(3):158–61. https://doi.org/10.1055/s-2008-1052421.

80. Siemes H, Nau H, Seidel U, et al. Irreversible valproate-associated liver failure. Monatsschr Kinderheilkd. 1992;140(12):869–75. (Article in German).
81. Coulter DL. Prevention of hepatotoxicity recurrence with valproate monotherapy and carnitine. Ann Neurol. 1988;24:301. (Abstract).
82. Rawat S, Borkowski WJ Jr, Swick HM. Valproic acid and secondary hyperammonemia. Neurology. 1981;31(9):1173–4. https://doi.org/10.1212/wnl.31.9.1173.
83. Zaret BS, Beckner RR, Marini AM, et al. Sodium valproate-induced hyperammonemia without clinical hepatic dysfunction. Neurology. 1982;32(2):206–8. https://doi.org/10.1212/wnl.32.2.206.
84. Williams CA, Tiefenbach S, McReynolds JW. Valproic acid-induced hyperammonemia in mentally retarded adults. Neurology. 1984;34(4):550–3. https://doi.org/10.1212/wnl.34.4.550.
85. Kulick SK, Kramer DA. Hyperammonemia secondary to valproic acid as a cause of lethargy in a postictal patient. Ann Emerg Med. 1993;22(3):610–2. https://doi.org/10.1016/s0196-0644(05)81955-9.
86. Murphy JV, Marquardt K. Asymptomatic hyperammonemia in patients receiving valproic acid. Arch Neurol. 1982;39(9):591–2. https://doi.org/10.1001/archneur.1982.00510210061016.
87. Triggs WJ, Bohan TP, Lin SN, et al. Valproate-induced coma with ketosis and carnitine insufficiency. Arch Neurol. 1990;47(10):1131–3. https://doi.org/10.1001/archneur.1990.00530100101021.
88. Papadimitriou A, Servidei S. Late onset lipid storage myopathy due to multiple acyl CoA dehydrogenase deficiency triggered by valproate. Neuromuscul Disord. 1991;1(4):247–52. https://doi.org/10.1016/0960-8966(91)90097-c.
89. Kai Y, Taku K, Ohtani Y, et al. A case of ornithine transcarbamylase deficiency with minor epileptic status-like symptoms. Brain Dev. 1987;19:239–43. (Article in Japanese).
90. Batshaw ML, Brusilow SW. Valproate-induced hyperammonemia. Ann Neurol. 1982;11(3):319–21. https://doi.org/10.1002/ana.410110315.
91. Tripp JH, Hargreaves T, Anthony PP, et al. Sodium valproate and ornithine carbamyl transferase deficiency. Lancet. 1981;1(8230):1165–6. https://doi.org/10.1016/s0140-6736(81)92338-2.
92. Kay JD, Hilton-Jones D, Hyman N. Valproate toxicity and ornithine carbamoyltransferase deficiency. Lancet. 1986;2(8518):1283–4. https://doi.org/10.1016/s0140-6736(86)92714-5.
93. Ellaway CJ, Bennetts B, Tuck RR, et al. Clumsiness, confusion, coma, and valproate. Lancet. 1999;353(9162):1408. https://doi.org/10.1016/s0140-6736(99)01433-6.
94. Leão M. Valproate as a cause of hyperammonemia in heterozygotes with ornithine-transcarbamylase deficiency. Neurology. 1995;45(3 Pt 1):593–4. https://doi.org/10.1212/wnl.45.3.593.
95. Honeycutt D, Callahan K, Rutledge L, et al. Heterozygote ornithine transcarbamylase deficiency presenting as symptomatic hyperammonemia during initiation of valproate therapy. Neurology. 1992;42(3 Pt 1):666–8. https://doi.org/10.1212/wnl.42.3.666.
96. Itokazu N, Sawada K, Maruyama R, et al. Possibility of hypoalbuminemia induced by sodium valproate in severely mentally and physically disabled people. Pediatr Jpn. 2013;54(7):1047–51. (Article in Japanese).
97. Coulter DL. Carnitine and anticonvulsant drugs. In: Perruca E, editor. L-carnitine in pharmacotherapy in epilepsy. Milan: Adis International; 1994. p. 9–15.
98. Kelly RI. The role of carnitine supplementation in valproic acid therapy. Pediatrics. 1994;93(6 Pt 1):891–2.
99. Freeman JM, Vining EP, Cost S, et al. Does carnitine administration improve the symptoms attributed to anticonvulsant medications?: a double-blinded, crossover study. Pediatrics. 1994;93(6 Pt 1):893–5.
100. Weaber LT, Rosenthal SR, Gladstone W, et al. Carnitine deficiency: a possible cause of gastrointestinal dysmotility. Acta Paediatr. 1992;81(1):79–81. https://doi.org/10.1111/j.1651-2227.1992.tb12085.x.
101. Murata S, Inoue K, Aomatsu T, et al. Supplementation with carnitine reduces the severity of constipation: a retrospective study of patients with severe motor and intellectual disabilities. J Clin Biochem Nutr. 2017;60(2):121–4. https://doi.org/10.3164/jcbn.16-52.
102. Bohan TP, Helton E, McDonald I, et al. Effect of L-carnitine treatment for valproate-induced hepatotoxicity. Neurology. 2001;56(10):1405–9. https://doi.org/10.1212/wnl.56.10.1405.
103. De Vivo DC, Bohan TP, Coulter DL, et al. L-carnitine supplementation in childhood epilepsy: current perspectives. Epilepsia. 1998;39(11):1216–25. https://doi.org/10.1111/j.1528-1157.1998.tb01315.x.
104. Hug G, McGraw CA, Bates SR, et al. Reduction of serum carnitine concentrations during anticonvulsant therapy with phenobarbital, valproic acid, phenytoin, and carbamazepine in children. J Pediatr. 1991;119(5):799–802. https://doi.org/10.1016/s0022-3476(05)80306-3.
105. Zelnik N, Fridkis I, Gruener N. Reduced carnitine and antiepileptic drugs: cause relationship or co-existence? Acta Paediatr. 1995;84(1):93–5. https://doi.org/10.1111/j.1651-2227.1995.tb13494.x.
106. De Vivo DC, Leckie MP, Ferrendelli JS, et al. Chronic ketosis and cerebral metabolism. Ann Neurol. 1978;3(4):331–7. https://doi.org/10.1002/ana.410030410.
107. Hunt PA, Pellock JM. Carnitine deficiency. Pediatrics. 1995;96(6):1175.
108. Altunbaşak Ş, Baytok V, Tasouji M, et al. Asymptomatic hyperammonemia in children treated with valproic acid. J Child Neurol. 1997;12(7):461–3. https://doi.org/10.1177/088307389701200709.

109. Haidukewych D, John G, Zielinski JJ, et al. Chronic valproic acid therapy and incidence of increases in venous plasma ammonia. Ther Drug Monit. 1985;7(3):290–4. https://doi.org/10.1097/00007691-198507030-00009.
110. Ratnaike RN, Schapel GJ, Purdie G, et al. Hyperammonaemia and hepatotoxicity during chronic valproate therapy: enhancement by combination with other antiepileptic drugs. Br J Clin Pharmacol. 1986;22(1):100–3.
111. Verrotti A, Greco R, Morgese G, et al. Carnitine deficiency and hyperammonemia in children receiving valproic acid with and without other anticonvulsant drugs. Int J Clin Lab Res. 1999;29(1):36–40. https://doi.org/10.1007/s005990050060.
112. Chicharro A, de Marinis AJ, Kanner AM. The measurement of ammonia blood levels in patients taking valproic acid: looking for problems where they do not exist? Epilepsy Behav. 2007;11(3):361–6. https://doi.org/10.1016/j.yebeh.2007.06.015.
113. Yamamoto Y, Takahashi Y, Suzuki E, et al. Risk factors for hyperammonemia associated with valproic acid therapy in adult epilepsy patients. Epilepsy Res. 2012;101(3):202–9. https://doi.org/10.1016/j.eplepsyres.2012.04.001.
114. Yamamoto Y, Takahashi Y, Imai K, et al. Risk factors for hyperammonemia in pediatric patients with epilepsy. Epilepsia. 2013;54(6):983–9. https://doi.org/10.1111/epi.12125.
115. Gayam V, Mandal AK, Khalid M, et al. Valproic acid induced acute liver injury resulting in hepatic encephalopathy – a case report and literature review. J Community Hosp Intern Med Perspect. 2018;8(5):311–4. https://doi.org/10.1080/20009666.2018.1514933.
116. Walker CP, Deb S. Rhabdomyolysis and hepatotoxicity from valproic acid: case reports. J Pharm Pract. 2021;34(4):648–52. https://doi.org/10.1177/0897190019882880.
117. Settle EC Jr. Valproic acid-associated encephalopathy with coma. Am J Psychiatry. 1995;152(8):1236–7. https://doi.org/10.1176/ajp.152.8.1236b.
118. Raby WN. Carnitine for valproic acid-induced hyperammonemia. Am J Psychiatry. 1997;154(8):1168–9. https://doi.org/10.1176/ajp.154.8.1168.
119. Eze E, Workman M, Donley B. Hyperammonemia and coma developed by a woman treated with valproic acid for affective disorder. Psychiatr Serv. 1998;49(10):1358–9. https://doi.org/10.1176/ps.49.10.1358.
120. Panikkar GP, Gilman SM. Valproate-induced hyperammonemia in the psychiatric setting: 2 cases. J Clin Psychiatry. 1999;60(8):557–9. https://doi.org/10.4088/jcp.v60n0809g.
121. Barrueto F, Hack JB. Hyperammonemia and coma without hepatic dysfunction induced by valproate therapy. Acad Emerg Med. 2001;8(10):999–1001. https://doi.org/10.1111/j.1553-2712.2001.tb01102.x.
122. Kimmel RJ, Irwin SA, Meyer JM. Valproic acid-associated hyperammonemic encephalopathy: a case report from the psychiatric setting. Int Clin Psychopharmacol. 2005;20(1):57–8. https://doi.org/10.1097/00004850-200501000-00012.
123. Yehya N, Saldarini CT, Koski ME, et al. Valproate-induced hyperammonemic encephalopathy. J Am Acad Child Adolesc Psychiatry. 2004;43(8):926–7. https://doi.org/10.1097/01.chi.0000129217.79887.c8.
124. Carlson T, Reynolds CA, Caplan R. Case report: valproic acid and risperidone treatment leading to development of hyperammonemia and mania. J Am Acad Child Adolesc Psychiatry. 2007;46(3):356–61. https://doi.org/10.1097/chi.0b013e31802ed8b2.
125. Nicolai J, Smith SJ, Keunen RW. Simultaneous side effects of both clozapine and valproate. Intensive Care Med. 2001;27(5):943. https://doi.org/10.1007/s001340100937.
126. Elgudin L, Hall Y, Schubert D. Ammonia induced encephalopathy from valproic acid in a bipolar patient: case report. Int J Psychiatry Med. 2003;33(1):91–6. https://doi.org/10.2190/XW8E-Q283-4429-QL71.
127. Carr RB, Shrewsbury K. Hyperammonemia due to valproic acid in the psychiatric setting. Am J Psychiatry. 2007;164(7):1020–7. https://doi.org/10.1176/ajp.2007.164.7.1020.
128. Chopra A, Kolla BP, Mansukhani MP, et al. Valproate-induced hyperammonemic encephalopathy: an update on risk factors, clinical correlates and management. Gen Hosp Psychiatry. 2012;34(3):290–8. https://doi.org/10.1016/j.genhosppsych.2011.12.009.
129. Dealberto MJ. Valproate-induced hyperammonaemic encephalopathy: review of 14 cases in the psychiatric setting. Int Clin Psychopharmacol. 2007;22(6):330–7. https://doi.org/10.1097/YIC.0b013e3281c61b28.
130. O'Carroll R. Neuropsychological and neuroimaging aspects of latent hepatic encephalopathy (LHE). Alcohol Alcohol Suppl. 1993;2:191–5.
131. Brown LM, Cupples N, Moore TA. Levocarnitine for valproate-induced hyperammonemia in the psychiatric setting: a case series and literature review. Ment Health Clin. 2018;8(3):148–54. https://doi.org/10.9740/mhc.2018.05.148.
132. Aoki T, Nakashima M, Tokura A, et al. Effect of levocarnitine chloride on sodium valproate-induced hyperammonemia in elderly patients in a psychiatry setting. Clin Psychiatry. 2013;55(5):441–8. (Article in Japanese).
133. Ito K, Takaya A. Benefits of administering L-carnitine in asymptomatic hyperammonemia induced by sodium valproate used for psychiatric treatment. Jpn J Psychiatry. 2015;20(4):1–7. (Article in Japanese).
134. Raja M, Azzoni A. Valproate-induced hyperammonaemia. J Clin Psychopharmacol.

2002;22(6):631–3. https://doi.org/10.1097/00004714-200212000-00019.
135. Baddour E, Tewksbury A, Stauner N. Valproic acid-induced hyperammonemia: incidence, clinical significance, and treatment management. Ment Health Clin. 2018;8(2):73–7. https://doi.org/10.9740/mhc.2018.03.073.
136. Nakamura M, Nagamine T. Hyperammonemia and carnitine deficiency treated with sodium valproate in psychiatric setting. Int Med J. 2015;22(3):132–5.
137. Nakamura M, Nagamine T. The effect of carnitine supplementation on hyperammonemia and carnitine deficiency treated with valproic acid in a psychiatric setting. Innov Clin Neurosci. 2015;12(9–10):18–24.
138. Nagamine T. Carnitine deficiency and severe hypoglycemia associated with valproic acid. Clin Neuropsychopharmacol Ther. 2012;3:37–9.
139. Nagamine T, Nakamura M. Carnitine deficiency in psychiatric practice. Jpn J Clin Psychopharmacol. 2013;16(3):429–31. (Article in Japanese).
140. Lichter-Konecki U. Profiling of astrocyte properties in the hyperammonaemic brain: shedding new light on the pathophysiology of the brain damage in hyperammonaemia. J Inherit Metab Dis. 2008;31(4):492–502. https://doi.org/10.1007/s10545-008-0834-9.
141. Watanabe M, Saeki T. Medical science of ammoniaology. Tokyo: Medical Review Co., Ltd; 1995. (Book in Japanese).
142. Kato A, Suzuki K. Diagnosis of hepatic encephalopathy. J Japanese Soc Gastroenterol. 2007;104(3):344–51. (Article in Japanese).
143. Watanabe M. Hepatic encephalopathy. Acta Hepatol Jpn. 2002;43(2):77–83. (Article in Japanese).
144. Walker CO, Schenker S. Pathogenesis of hepatic encephalopathy with special reference to the role of ammonia. Am J Clin Nutr. 1970;23(5):619–32. https://doi.org/10.1093/ajcn/23.5.619.
145. Andersen GO, Ritland S. Life threatening intoxication with sodium valproate. J Toxicol Clin Toxicol. 1995;33(3):279–84. https://doi.org/10.3109/15563659509018000.
146. Matsumoto J, Ogawa H, Maeyama R, et al. Successful treatment by direct hemoperfusion of coma possibly resulting from mitochondrial dysfunction in acute valproate intoxication. Epilepsia. 1997;38(8):950–3. https://doi.org/10.1111/j.1528-1157.1997.tb01263.x.
147. Mitani T, Mato T, Matsueda H, et al. A patient who was successfully saved with continuous hemofiltration after massive dose of lithium carbonate and sustained-release sodium valproate. J Jpn Assoc Acute Med. 2013;24:425–30. (Article in Japanese).
148. Ishikura H, Matsuo N, Matsubara M, et al. Valproic acid overdose and L-carnitine therapy. J Anal Toxicol. 1996;20(1):55–8. https://doi.org/10.1093/jat/20.1.55.
149. Murakami K, Sugimoto T, Woo M, et al. Effect of L-carnitine supplementation on acute valproate intoxication. Epilepsia. 1996;37(7):687–9. https://doi.org/10.1111/j.1528-1157.1996.tb00634.x.
150. Itoh T, Konda A, Yasuda S, et al. A case of valproate overdose complicated by severe hyperammonemia that was ameliorated with time concomitant with decline in serum valproate concentration. Jpn J Clin Pharmacol Ther. 2015;46(2):77–9.
151. Gummin DD, Mowry JB, Spyker DA, et al. 2016 annual report of the American Association of Poison Control Centers' National Poison Data System (NPDS): 34th annual report. Clin Toxicol Phila. 2017;55(10):1072–254. https://doi.org/10.1080/15563650.2017.1388087.
152. Sztajnkrycer MD. Valproic acid toxicity: overview and management. J Toxicol Clin Toxicol. 2002;40(6):789–801. https://doi.org/10.1081/clt-120014645.
153. Hicks LK, McFarlane PA. Valproic acid overdose and haemodialysis. Nephrol Dial Transplant. 2001;16(7):1483–6. https://doi.org/10.1093/ndt/16.7.1483.
154. Gómez Bellver MJ, García Sánchez MJ, Alonso González AC, et al. Plasma protein binding kinetics of valproic acid over a broad dosage range: therapeutic implications. J Clin Pharm Ther. 1993;18(3):191–7. https://doi.org/10.1111/j.1365-2710.1993.tb00612.x.
155. Zaccara G, Messori A, Moroni F. Clinical pharmacokinetics of valproic acid 1988. Clin Pharmacokinet. 1988;15(6):367–89. https://doi.org/10.2165/00003088-198815060-00002.
156. Auinger K, Müller V, Rudiger A, et al. Valproic acid intoxication imitating brain death. Am J Emerg Med. 2009;27(9):1177.e5–6. https://doi.org/10.1016/j.ajem.2009.02.019.
157. Perrott J, Murphy NG, Zed PJ. L-carnitine for acute valproic acid overdose: a systematic review of published cases. Ann Pharmacother. 2010;44(7–8):1287–93. https://doi.org/10.1345/aph.1P135.
158. LoVecchio F, Shriki J, Samaddar R. L-carnitine was safely administered in the setting of valproate toxicity. Am J Emerg Med. 2005;23(3):321–2. https://doi.org/10.1016/j.ajem.2005.02.015.
159. Mock CM, Schwetschenau KH. Levocarnitine for valproic-acid-induced hyperammonemic encephalopathy. Am J Health Syst Pharm. 2012;69:35–9.
160. Russell S. Carnitine as an antidote for acute valproate toxicity in children. Curr Opin Pediatr. 2007;19(2):206–10. https://doi.org/10.1097/MOP.0b013e32805e879a.
161. Godara H, Hirbe A, Nassif M, et al. Chapter 28: Toxicology. In: The Washington Manual™ of medical therapeutics. 34th edition, Japanese edition, Takaku F and Wada O supervised translation: Washington Manual 13th Edition. Tokyo: Medical Sciences International LTD; 2015. p. 1040–1.
162. Lheureux PE, Penaloza A, Zahir S, et al. Science review: carnitine in the treatment of valproic acid-

induced toxicity – what is the evidence? Crit Care. 2005;9(5):431–40. https://doi.org/10.1186/cc3742.
163. Lheureux PE, Hantson P. Carnitine in the treatment of valproic acid-induced toxicity. Clin Toxicol (Phila). 2009;47(2):101–11. https://doi.org/10.1080/15563650902752376.
164. Nakajima Y, Ito T, Maeda Y, et al. Evaluation of valproate effects on acylcarnitine in epileptic children by LC-MS/MS. Brain Dev. 2011;33(10):816–23. https://doi.org/10.1016/j.braindev.2010.12.003.
165. Segura-Bruna N, Rodriguez-Campello A, Puente V, et al. Valproate-induced hyperammonemic encephalopathy. Acta Neurol Scand. 2006;114(1):1–7. https://doi.org/10.1111/j.1600-0404.2006.00655.x.
166. Silva MF, Aires CC, Luis PB, et al. Valproic acid metabolism and its effects on mitochondrial fatty acid oxidation: a review. J Inherit Metab Dis. 2008;31(2):205–16. https://doi.org/10.1007/s10545-008-0841-x.
167. Blackford MG, Do ST, Enlow TC, et al. Valproic acid and topiramate induced hyperammonemic encephalopathy in a patient with normal serum carnitine. J Pediatr Pharmacol Ther. 2013;18(2):128–36. https://doi.org/10.5863/1551-6776-18.2.128.
168. Sugimoto T, Muro H, Woo M, et al. Valproate metabolites in high-dose valproate plus phenytoin therapy. Epilepsia. 1996;37(12):1200–3. https://doi.org/10.1111/j.1528-1157.1996.tb00553.x.
169. Silva MF, Selhorst J, Overmars H, et al. Characterization of plasma acylcarnitines in patients under valproate monotherapy using ESI-MS/MS. Clin Biochem. 2001;34(8):635–8. https://doi.org/10.1016/s0009-9120(01)00272-7.
170. Werner T, Treiss I, Kohlmueller D, et al. Effects of valproate on acylcarnitines in children with epilepsy using ESI-MS/MS. Epilepsia. 2007;48(1):72–6. https://doi.org/10.1111/j.1528-1167.2006.00833.x.
171. Farkas V, Bock I, Cseko J, et al. Inhibition of carnitine biosynthesis by valproic acid in rats – the biochemical mechanism of inhibition. Biochem Pharmacol. 1996;52(9):1429–33. https://doi.org/10.1016/s0006-2952(96)00507-2.
172. Camiña MF, Rozas I, Castro-Gago M, et al. Alteration of renal carnitine metabolism by anticonvulsant treatment. Neurology. 1991;41(9):1444–8. https://doi.org/10.1212/wnl.41.9.1444.
173. Ohashi R, Tamai I, Yabuuchi H, et al. Na+-dependent carnitine transport by organic cation transporter (OCTN2): its pharmacological and toxicological relevance. J Pharmacol Exp Ther. 1999;291(2):778–84.
174. Wu SP, Shyu MK, Liou HH, et al. Interaction between anticonvulsants and human placental carnitine transporter. Epilepsia. 2004;45(3):204–10. https://doi.org/10.1111/j.0013-9580.2004.29603.x.
175. Corkey BE, Hale DE, Glennon MC, et al. Relationship between unusual hepatic acyl coenzyme A profiles and the pathogenesis of Reye syndrome. J Clin Invest. 1988;82(3):782–8. https://doi.org/10.1172/JCI113679.
176. Becker CM, Harris RA. Influence of valproic acid on hepatic carbohydrate and lipid metabolism. Arch Biochem Biophys. 1983;223(2):381–92. https://doi.org/10.1016/0003-9861(83)90602-1.
177. Li J, Norwood DL, Mao LF, et al. Mitochondrial metabolism of valproic acid. Biochemistry. 1991;30(2):388–94. https://doi.org/10.1021/bi00216a012.
178. Moore KH, Decker BP, Schreefel FP. Hepatic hydrolysis of octanoyl-CoA and valproyl-CoA in control and valproate-fed animals. Int J Biochem. 1988;20(2):175–8. https://doi.org/10.1016/0020-711x(88)90482-x.
179. Silva MF, Ruiter JP, Overmars H, et al. Complete β-oxidation of valproate: cleavage of 3-oxovalproyl-CoA by a mitochondrial 3-oxoacyl-CoA thiolase. Biochem J. 2002;362(Pt 3):755–60. https://doi.org/10.1042/0264-6021:3620755.
180. Aires CC, Ruiter JP, Luís PB, et al. Studies on the extra-mitochondrial CoA-ester formation of valproic and delta4-valproic acids. Biochim Biophys Acta. 2007;1771(4):533–43. https://doi.org/10.1016/j.bbalip.2007.01.010.
181. Coude FX, Sweetman L, Nyhan WL. Inhibition by propionyl-coenzyme A of N-acetylglutamate synthetase in rat liver mitochondria. A possible explanation for hyperammonemia in propionic and methylmalonic acidemia. J Clin Invest. 1979;64(6):1544–51. https://doi.org/10.1172/JCI109614.
182. Aires CC, van Cruchten A, IJlst L, et al. New insights on the mechanisms of valproate-induced hyperammonemia: inhibition of hepatic N-acetylglutamate synthase activity by valproyl-CoA. J Hepatol. 2011;55(2):426–34. https://doi.org/10.1016/j.jhep.2010.11.031.
183. Latour P, Biraben A, Polard E, et al. Drug induced encephalopathy in six epileptic patients: topiramate? valproate? or both? Hum Psychopharmacol. 2004;19(3):193–203. https://doi.org/10.1002/hup.575.
184. Cheung E, Wong V, Fung CW. Topiramate-valproate-induced hyperammonemic encephalopathy syndrome: case report. J Child Neurol. 2005;20:157–60.
185. Deutsch SI, Burket JA, Rosse RB. Valproate-induced hyperammonemic encephalopathy and normal liver functions: possible synergism with topiramate. Clin Neuropharmacol. 2009;32(6):350–2. https://doi.org/10.1097/WNF.0b013e3181ac3615.
186. Vivekanandan S, Nayak SD. Valproate-induced hyperammonemic encephalopathy enhanced by topiramate and phenobarbitone: a case report and an update. Ann Indian Acad Neurol. 2010;13(2):145–7. https://doi.org/10.4103/0972-2327.64638.
187. Twilla JD, Pierce AS. Hyperammonemic encephalopathy due to valproic acid and topiramate interaction. Case Rep Psychiatry. 2014;2014:410403. https://doi.org/10.1155/2014/410403.
188. Raru Y, Zeid F. Hypoxic respiratory failure due to hyperammonemic encephalopathy induced by concurrent use of valproic acid and topiramate, a case report and review of the literature. Respir Med Case Rep. 2018;25:1–3. https://doi.org/10.1016/j.rmcr.2018.05.026.

13 Carnitine Deficiency Caused by Antibiotics Containing Pivoxil Group

13.1 What Are Antibiotics Containing Pivoxil Group?

The pivoxil group (also called pivalic acid, pivalate, 2,2-dimethylpropanoic acid, 2,2-dimethylpropionic acid, trimethylacetic acid, and neopentanoic acid, dotted circles in Fig. 13.1) is a simple branched-chain carboxylic acid with a tert-butyl group that has been applied to the design of pharmaceutical prodrugs. If there is a polar group such as a -OH group in the body of an antibiotic such as a cephem antibiotic, absorption from the intestinal mucosa is reduced, but when the pivoxil group is bonded, the polar group is masked and the fat solubility increases, so the absorption of the antibiotic in the intestinal tract is promoted and the bioavailability at the time of oral administration increases. For this purpose, a pivoxil group or a similar chemical group is attached to the body of the antibiotic, and it has been used in many antimicrobial drugs such as pivampicillin and pivmecillinam.

Until now, five types of antibiotics containing a pivoxil group have been launched in Japan, but pivmecillinam hydrochloride (trade name: Melisin) has been discontinued, and as of 2022, four types of antimicrobials are on the market by various companies (Table 13.1). These antimicrobials containing pivoxil group are often used in children and the elderly in Japan, and products in various dosage forms are distributed. However, as will be described later, only a few cases of carnitine deficiency in chil-

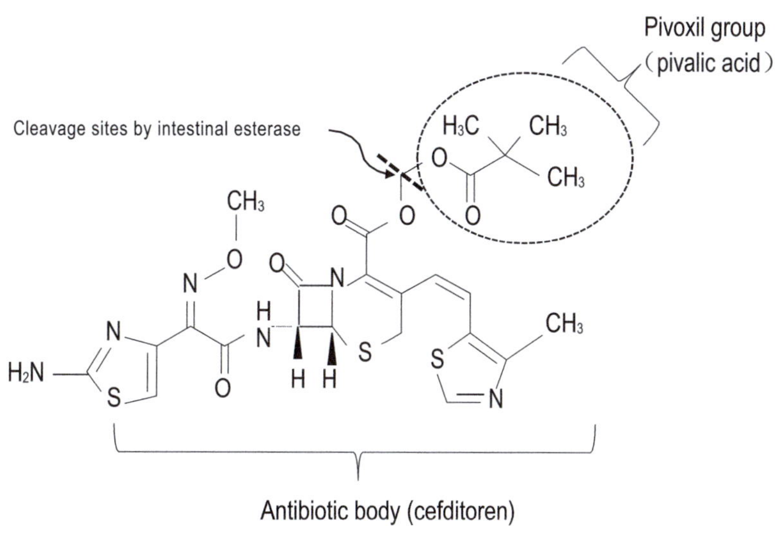

Fig. 13.1 Chemical structure of pivoxil group containing antibiotics (example of cefditoren pivoxil)

N. Nakayama, *Carnitine Deficiency and Insufficiency in Medicine*,
https://doi.org/10.1007/978-981-95-7797-2_13

Table 13.1 Pivoxil-containing antibiotics current and past on the market in Japan (as of April 2022)

Classification	Ingredient	Abbreviation	Brand name	Sales company
Currently available pivoxyl-containing antibiotics				
Third-generation cephem antibiotics	Cefcapene pivoxil hydrochloride hydrate	CFPN-PI	Flomox tablets 75 mg/100 mg Flomox pediatric granules 100 mg	Shionogi Pharmaceutical Co., Ltd.
			Cefcapene pivoxil hydrochloride tablets 75 mg/100 mg 「SW」 Cefcapene pivoxil hydrochloride fine granules pediatric use 10% 「SW」	Sawai Pharmaceutical Co., Ltd.
			Cefcapene pivoxil hydrochloride tablets 75 mg/100 mg 「Tōwa」	C.H.O. New Pharmaceutical Co., Ltd.
			Cefcapene pivoxil hydrochloride tablets75 mg/100 mg 「CH」 Cefcapene Pivoxil hydrochloride fine granules pediatric use 10% 「CH」	Choseido Pharmaceutical Co., Ltd. Japan Generic Co., Ltd.
			Cefcapene pivoxil hydrochloride fine grain pediatric 10% 「Tōwa」	Towa Pharmaceutical Co., Ltd.
			Cefcapene pivoxil hydrochloride tablets 75 mg/100 mg 「Nichi-Iko」 Cefcapene pivoxil hydrochloride fine granules pediatric use 10% 「Nichi-Iko」	Nichi-Iko Pharma Co., Ltd.
			Cefcapene pivoxil hydrochloride tablets75 mg/100 mg 「Pfizer」 Cefcapene pivoxil hydrochloride fine granules pediatric use 10% 「Pfizer」	Viatoris Pharmaceutical Co., Ltd.
			Cefcapene pivoxil hydrochloride tablets 75 mg/100 mg 「YD」 Cefcapene pivoxil hydrochloride fine granules pediatric use 10% 「YD」	Yoshindo Corporation
	Cefditoren pivoxil	CDTR-PI	Meiact MS tablets100 mg Meiact MS fine granules pediatric use 10%	Meiji Seika Pharma Co., Ltd.
			Cefditoren pivoxil tablets 100 mg 「SW」 Cefditoren pivoxil fine granules pediatric use 10% 「SW」	Sawai Pharmaceutical Co., Ltd.
			Cefditoren pivoxil tablets 100 mg 「CH」 Cefditoren pivoxil fine granules pediatric use 10% 「CH」	Choseido Pharmaceutical Co., Ltd.
			Cefditoren pivoxil tablets100 mg 「Tōwa」 Cefditoren pivoxil fine granules pediatric use 10% 「Tōwa」	Towa Pharmaceutical Co., Ltd.
			Cefditoren pivoxil tablets 100 mg 「OK」 Cefditoren pivoxil fine granules pediatric use 10% 「OK」	Okura Pharmaceutical Co., Ltd.
			Cefditoren pivoxil tablets100 mg 「Nichi-Iko」 Cefditoren pivoxil fine granules pediatric use 10% 「Nichi-Iko」	Nichi-Iko Pharma Co., Ltd.
			Cefditoren pivoxil fine granules pediatric use 10% 「EMEC」	Medisa Shinyaku Co., Ltd., Elmed Co., Ltd., Nichi-Iko Co., Ltd.
	Cefteram pivoxil	CFTM-PI	Tomiron tablets 50/100 Tomiron fine granules pediatric use 20% (Changed from 10%, at 2016)	Toyama Chemical Industry Co., Ltd. (now FUJIFILM Toyama Chemical Co., Ltd.)

Carbapenem antimicrobials	Tebipenem pivoxil	TBPM-PI	Orapenem fine granules pediatric use 10%	Meiji Seika Pharma Co., Ltd.	
Discontinued pivoxyl-containing antibiotics in Japan					
Classification	Ingredient	Abbreviation	Brand name	Sales Company	Remarks
Third-generation cephem antibiotics	Cefetamet pivoxil	CEMT-PI	Cefir tablets	KYORIN Pharmaceutical Co., Ltd.	Discontinued in 2002
	Cefteram pivoxil	CFTM-PI	Cetlat tablets 100 mg fine granules pediatric use 10%	Sawai Pharmaceutical Co., Ltd.	Discontinued in 2012
			Somatron fine granules pediatric use 100	Teva Pharmaceutical Co., Ltd.	Discontinued in 2012
			Teramiron fine granules pediatric use 10%	Towa Pharmaceutical Co., Ltd.	Discontinued in 2017
			Cefteram pivoxil fine granules pediatric use 10% 「Nichi-Iko」 (December 2013, renamed from Teracefron fine granules 100 mg pediatric)	Nichi-Iko Corporation	Discontinued in 2017
Penicillin antimicrobials	Pivmecillinam hydrochloride	PMPC	Melisin tablets 50 mg	Takeda Pharmaceutical Company Limited	Discontinued in 2013

CFPN-PI cefcapene pivoxil, CDTR-PI cefditoren pivoxil, CFTM-PI cefteram pivoxil, TBPM-PI tebipenem pivoxil, RSV respiratory syncytial virus
Unit of clinical laboratory test blood glucose: mg/dL, ammonia: μg/dL, carnitine: nmol/mL
Note: Case 9 is same patient as case 3

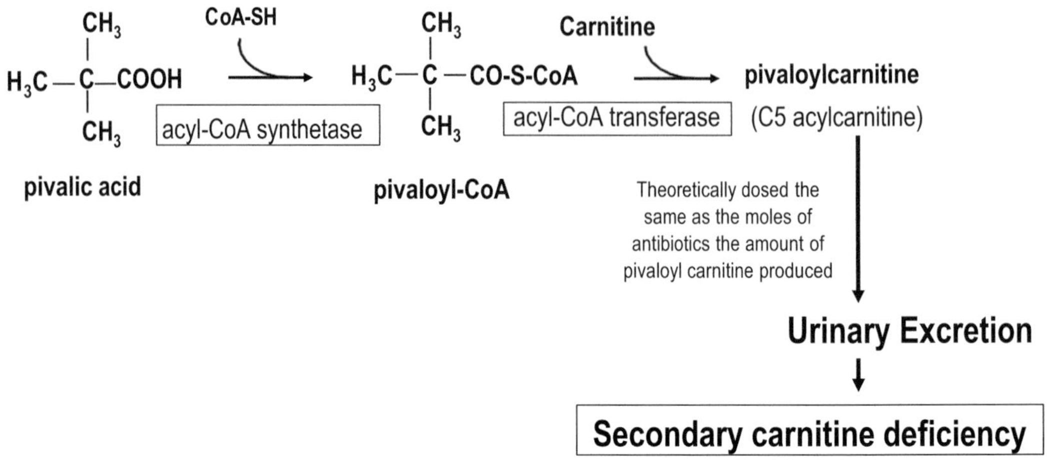

Fig. 13.2 Metabolism of pivalic acid and mechanism of the carnitine deficiency with pivoxil-containing antibiotics. Note: Molecular weight of pivaloyl-carnitine (245.3) is same as isovalerylcarnitine which is observed in high concentration in the urine in isovaleric acidemia, one of the inherited metabolic diseases, so that it is not possible to distinguish between pivaloyl-carnitine and isovaleryl-carnitine by mass spectrometer. In the case of carnitine deficiency due to pivoxil-containing antibiotics, C5 carnitine increases in the urine but it is necessary to confirm that this C5 carnitine is isovalerylcarnitine or not by GC/MS. (Modified from Ref. [1, 2])

dren due to antibiotics containing the pivoxil group have been reported overseas.

13.2 Metabolism of Pivalic Acid

Pivalic acid, cleaved with esterases in the intestinal tract, is converted to pivaloyl-CoA by the action of acyl-CoA synthetase. However, in mammalian cells, pivaloyl-CoA is not further metabolized to carbon dioxide and accumulates in the cell, and pivaloyl-CoA becomes a substrate for various acyl-CoA transferases. In humans, the main elimination route of pivalic acid is that the pivoxil group is transferred to carnitine and becomes pivaloylcarnitine, which is excreted from the kidneys [1, 2] (Fig. 13.2). Theoretically, an amount of carnitine equal to the number of moles of the administered pivoxil-containing antibacterial drug is excreted in the urine.

13.3 Clinical Pharmacological Study of Carnitine Deficiency with Pivoxil-Containing Antibiotics

Melegh et al. reported that seven pediatric patients were administered 2 g/day of pivampicillin (500 mg × 4) for 7 days and measured carnitine in plasma and urine, and that free carnitine and long-chain acylcarnitine in plasma decreased, and pivaloylcarnitine, which accounts for most of the acylcarnitine in urine, was excreted in large quantities in the urine without being reabsorbed [3]. Thus, when a pivoxil-containing antimicrobial drug is administered, pivalic acid (pivoxil group) is cleaved and released from the body of the antibiotic by esterases in the intestinal tract when it is absorbed, and immediately forms an ester with free carnitine in the blood and is excreted in the urine as pivaloylcarnitine.

In fact, Holme et al. reported that when pivampicillin was administered at 1000 mg twice a day for 6 days, followed by 400 mg pivmecillinam three times a day for 9 days to 41-year-old men and 42-year-old women and examined blood and urinary carnitine, the total carnitine reduced to half, and the urinary acylcarnitine excretion increased drastically, and these changes persisted 10 days after antibiotic discontinuation [4]. Theye also examined seven young girls who were receiving long-term treatment of pivampicillin and pivmecillinam. Five patients aged 8 months to 6 years and 5 months at the start of treatment received a daily dose of a mixture of 50 mg of pivmecillinam and 62.5 mg of pivampicillin, which corresponds to 250 μmol of pivarate moles daily. Two patients, aged 8 and 15, received the drug twice daily. In these seven patients, the concentration of total carnitine in plasma decreased to an average of 15% (7–27%) before administration, and acyl carnitine increased from about 2% of total carnitine before administration to 11–57%. The concentration of carnitine in the muscles of the two children who had been treated with these antibiotics for 22 and 30 months had decreased to 10% of the pre-dose level, indicating that serious carnitine deficiency could develop at any time. Treatment was discontinued due to aggression and nausea in a 15-year-old child, and in an 8-month-old child due to malaise and frequent screaming adverse events. As a result, Holme et al. warned that long-term administration of antibiotics containing a pivoxil group could cause severe carnitine deficiency [4].

Abrahamsson et al. in the same group examined blood carnitine concentrations and urinary carnitine in 44 patients with urinary tract infections in the group treated with pivmecillinam (200 mg × 3 days, 7 days) and norfloxacin (200 mg × 2, 7 days) [5]. As a result, the plasma-free carnitine concentration decreased from 44.6 to 12.9 μmol/L even after standard 7 day administration in the pivmecillinam group, while the free carnitine concentration did not decrease from 40.0 to 40.5 μmol/L in the group treated with norfloxacin, a new quinolone antibiotic that does not contain a pivoxil group. Plasma-free carnitine concentrations of 10 μmol/L or less are concentrations that interfere with ketone production, and it was noted that carnitine deficiency occurred in the pivmecillinam group regardless of age [5].

Abrahamsson et al. also administered 1200 mg/day of pivmecillinam to seven healthy volunteers (five females, two males, ages 17–54 years). After 8 weeks, plasma and muscle carnitine concentrations and echocardiographic changes in the heart were examined [6]. As a result, the plasma-free carnitine concentration decreased to 7% before administration, and the total carnitine concentration in the muscle decreased to 46% before administration. Except for cases that showed extreme values in echocardiography, the diastolic septum thickness decreased from 10.9 (7.0–15.3) mm to 7.3 (6.5–9.1) mm and the left ventricular weight decreased 185.1 (99.0–237.3) to 156.2 (73.5–237.3), significantly. Fifteen months after discontinuation of pivmecillinam, these changes were almost reversed. Abrahamsson et al. pointed out that carnitine deficiency caused by antibiotics containing the pivoxil group affects the dimensions of the heart even in healthy individuals and may leave irreversible damage in patients with pre-existing abnormalities in coronary blood flow [6].

Abrahamsson et al. examined 13 more healthy volunteers by dividing them into two groups [7]. First, 6 patients were given 1200 mg of pivmecillinam for 10 days, and 7 patients were given 1200 mg of pivmecillinam for 54 days to measure blood carnitine and muscle carnitine. Serum-free carnitine was reduced from 35 to 3.5 μmol/L after 54 days of administration, and muscle-free carnitine was reduced to 10.0 NCP (non-collagen protein)–4.3 μmol/g NCP. The ATP content and phosphocreatine content of muscles during exercise did not change, but the amount of work at maximal oxygen uptake and maximum heart rate was significantly reduced. Since maximal oxygen uptake is known to depend on cardiac function, it was thought that carnitine deficiency may affect cardiac function. Five out of seven participants who received pivmecillinam for 54 days complained of fatigue during walking exercises on an uphill slope, and 6 out of 7 complained of pain in their legs during a 60-min bicycle exercise test with a reduced load and could not sustain it until the end [7].

Melegh et al. examined the effects on energy metabolism in 10 children receiving pivampicillin (4 × 500 mg/day, 8 days) in addition to pivampicillin (4 × 1 g/day) for the last 4 days [8]. After 4 days of pivampicillin, the respiratory quotient increased from 0.86 ± 0.01 to 0.96 ± 0.01, fat oxidation decreased from 1.27 ± 0.17 to 0.31 ± 0.17 g/kg/day, and carbohydrate oxidation increased from 4.00 ± 0.50 to 6.20 ± 0.51 g/kg/day. In other words, it was thought that carnitine deficiency occurred due to the administration of pivampicillin, and that there was a shift from fatty acid oxidation to carbohydrate oxidation-based energy metabolism. Carnitine supplementation reduced the respiratory quotient to

0.90 ± 0.01, increased fat oxidation, and decreased carbohydrate oxidation. In addition, carnitine administration promoted the excretion of the pivoxil group in the body [8].

Pap et al. in the same group divided 14 pediatric patients receiving cefetamet pivoxil due to infection into two groups, group A (7 patients) received cefetamet pivoxil only (3 × 500 mg/day, 7 days), group B (7 patients) received cefetamet pivoxil (3 × 500 mg/day, 7 days), and an excessive amount of carnitine (3 × 1 g) was administered as a molar ratio. Blood carnitine concentrations and urinary carnitine excretion were examined [9]. As a result, the plasma-free carnitine concentration decreased from 32.7 ± 2.87 to 11.9 ± 1.36 μmol/L after 7-day administration of cefetamet pivoxil in group A, but there was no decrease in plasma-free carnitine concentration in group B, and urinary excretion of pivaloylcarnitine was promoted. In other words, it was thought that the externally supplemented carnitine was used to excrete pivarate in group B, and that the decrease in plasma-free carnitine observed in group A did not occur. Carnitine supplementation is only used to excrete pivalic acid produced by cleavage with esterases and does not affect the medicinal efficacy of the antimicrobial itself. Therefore, it can be said that carnitine supplementation is also effective when using a pivoxil-containing antibiotic for patients at risk of carnitine deficiency.

Ito et al. examined cefteram pivoxil (3 × 200 mg/day, 7 days) after meals in six of eight healthy adult male volunteers and placebo in two of them and examined the kinetics of carnitine in the blood and urine and pivaloylcarnitine in the urine [10]. Serum-free carnitine concentrations decreased sustainably from the second day of cefteram pivoxil administration and were 18.7 ± 4.4 nmol/mL (50.8 ± 4.5 nmol/mL for placebo) on the seventh day of treatment but recovered to the previous level on the third day after discontinuation of antibiotics. On the second day of administration, excretion of pivaloylcarnitine in the urine was characteristic. In addition, a negative correlation was observed between serum-free carnitine concentration and blood ammonia concentration ($r = -0.51$, $P < 0.001$). Ito et al. conducted a similar study in 17 pediatric patients receiving cefteram-pivoxil for the prevention of infections such as upper respiratory tract infections, anaphylactoid purpura, and surgery. The dose of cefteram pivoxil is 8.5–16.8 mg/kg/day, and the duration of administration is 3–14 days (mean 6.2 ± 3.1 days). In these patients, cefteram pivoxil administration significantly reduced serum-free carnitine concentrations, and a significant negative correlation was observed between blood ammonia concentrations and serum-free carnitine concentrations ($r = -0.60$, $P < 0.05$, $n = 16$) [10].

Nakajima et al. divided healthy adult male volunteers into two groups and administered cefetamet pivoxil (2 × 500 mg/day, 8 days) to group A (5 people), and cefetamet pivoxil (2 × 500 mg/day, 8 days) and carnitine preparations (L-Cartin®, 2 × 200 mg/day, 8 days) to reduce plasma and urinary carnitine concentrations. Detailed clinical pharmacological studies including pharmacokinetic analysis of cefetamet pivoxil were reported [11]. The dosage of this carnitine preparation is approximately equimolar with the dose of cefetamet pivoxil. As a result, the plasma-free carnitine concentration decreased to 17.7 μmol/L (37% of the pre-dose value) in group A and 29.2 μmol/L (58% of the pre-dose value) in group B on the 5th day after cefetamet pivoxil administration for 8 days. Nakajima et al. noted that carnitine was decreased even in Group B, which received approximately the same molar carnitine supplementation, because the bioavailability of carnitine was low (16%) when administered orally, compared with 55% of bioavailability for cefetamet pivoxil, which may have prevented sufficiently carnitine depletion [11].

The findings from these clinical pharmacological studies are summarized below.

1. Pivalic acid is released according to the number of moles of the administered pivoxil group-containing antimicrobial drug.
2. The number of pivaloylcarnitine, which is almost the same as the number of moles of pivalic acid, is excreted in the urine.

3. Even in healthy men, the administration of antibiotics containing the pivoxil group rapidly decreases free carnitine in the blood, but there are no symptoms of carnitine deficiency due to the large amount of skeletal muscle mass.
4. The risk of developing carnitine deficiency is higher in younger children and patients with low skeletal muscle mass.
5. If a pivoxil-containing antimicrobial drug is administered for a long period of time, carnitine supplementation therapy should be considered as necessary. In this case, since the bioavailability of oral carnitine preparations is low, it is not sufficient to use equimolar drugs with antimicrobials (Note: If the bioavailability of oral carnitine is 20%, it is considered that the dose of about 5 times that of antimicrobials is required). In addition, in the case of oral administration, it is not necessary to increase the dose of carnitine because there is a phenomenon of saturation of absorption when the dose of carnitine increases. Switching to antibiotics that do not contain pivarate is the best way to avoid risks).

13.4 Case Report of Carnitine Deficiency Caused by Pivoxil-Containing Antibiotics in Japan

Thus, it has been reported that blood-free carnitine is decreased and urinary pivaloylcarnitine excretion is increased in healthy subjects and patients treated with pivoxil-containing antimicrobial agents. In 2013, Ohura reported detailed clinical course of a case of carnitine deficiency caused by a pivoxil-containing antimicrobial agent and summarized 17 previously reported cases [12]. The majority of the cases were infants between 11 months and 4 years of age, especially 1 year of age; most of the cases were acute upper respiratory tract infection and acute otitis media; 10 cases had received these antimicrobial agents for more than 1 month, and some had received them for 8–12 months to prevent infection. Clinical and laboratory abnormalities included hypoglycemia, impaired consciousness, convulsions, and hypocarnitinemia, and there were cases in which encephalopathy was suspected on admission due to superimposed convulsions and cases in which patients were left with sequelae such as paraplegia upon discharge. Even after antimicrobial therapy was discontinued, there were cases in which the recovery of free carnitine in the blood was slow, and carnitine preparations were necessary.

However, carnitine deficiency itself is not well recognized in clinical practice, and hypoglycemia and hyperammonemia due to carnitine deficiency caused by antibiotics containing a pivoxil group have been reported frequently in daily clinical practice in pediatrics and otolaryngology in Japan. In the "Guidelines for the Diagnosis and Treatment of Carnitine Deficiency 2018" prepared mainly by the Japan Pediatric Society and others, 19 cases of carnitine deficiency caused by antibiotics containing a pivoxil group are mainly in children and 1 case in the elderly are listed in a table [13].

In addition to the cases listed in this table, the following is a summary of the cases reported in Japan so far on carnitine deficiency caused by antibiotics containing the pivoxil group, as well as the reports of academic conferences (Table 13.2) [12, 14–39]. Many of these reports are from the Department of Pediatrics and the Department of Pediatric Emergency Medicine at university hospitals, which have a deep understanding of carnitine deficiency, and it is likely that there are many cases that have not been reported in actual clinical practice.

From these case reports, the following points can be drawn.

1. Many practitioners of pediatricians and otolaryngologists easily prescribe these antibiotics for cold symptoms and otitis media for a long time.
2. Although there are cases in which the antibacterial agent is switched to another drug, there are many cases in which the same antibacterial agent containing a pivoxil group is used, resulting in prolonged administration.

Table 13.2 Case reports of carnitine deficiency caused by pivoxil-containing antibiotics in Japan

Paper report					
Case	Age, gender	Symptoms, laboratory tests, treatments, etc.	Antibiotics, duration	Carnitine, duration, prognosis, etc.	Reporter
1	1 year, 3 months old female	At the time of the first seizure; vomiting, hollow eyes, ankylosis in the limbs, hypoglycemia10, ammonia 114 → glucose drip regained consciousness and went home. At the second attack; limb ankylosis, impaired consciousness, blood glucose 12, ammonia 319 → consciousness regained by glucose drip. Both are hypoketotic hypoglycemia, free carnitine 9.6	She had been prescribed CFPN-PI and CDTR-PI for cold symptoms at another hospital. 19 days until the first seizure (intermittent) and 31 days until the second seizure (intermittent)	Even 1 month after the second seizure, it was still low carnitine, so 60 mg/kg/day was taken orally for 1 week.	Teraoka (2004) (Ref. [14])
2	4 years old male	Hypoketotic hypoglycemia 26, impaired consciousness, tonic–clonic convulsions. Symptoms improved by intravenous glucose injection. Marked free carnitine decline (4.4)	Patient with a cold and were treated with CFPN-PI, CDTR-PI, or CFTM-PI for 52 of the 72 days (intermittent)	Since the blood carnitine level did not improve even after 2 months of natural follow-up, 40 mg/kg/day was administered for 10 days and improved.	Nagasaki et al. (2005) (Ref. [15])
3	18 months old male	Loss of appetite, repeated vomiting, tonic convulsions, hypoglycemia 11, blood-free carnitine 6.2. Immunoglobulin, dexamethasone, and acyclovir were administered on suspicion of encephalopathy without improvement in consciousness impairment even after glucose infusion.	After the procedure, when the parent checked the medication history, it was found that he had been receiving CDTR-PI for 6 months due to refractory otitis media	Intravenous infusion of carnitine 100 mg/kg/day for 3 days. Acidosis improves in 1 h after the start of intravenous infusion, and conversation is possible after half a day	Makino et al. (2007) (Ref. [16])

4	11 months old male	Sudden impairment of consciousness, hypoketotic hypoglycemia 11, tonic convulsions, free carnitine 4.7. Intravenous glucose, intravenous diazepam, hospitalization. Ammonia 124, metabolic acidosis	Interviews with parent after the onset of the disease revealed that the following antibiotics had been used intermittently in the 2 months before the onset of the disease at pediatrics and otolaryngology clinics for frequent pharyngitis and otitis media from around 5 months of age. Received CDTR-PI 13–15 mg/kg for 17 days and CFPN-PI 11–15 mg/kg for 10 days, with onset on day 9 after discontinuation of medication	No carnitine supplementation	Kanda et al. (2008) (Ref. [17])
5	2 years, 1 month old male	Fever, generalized clonic convulsions, hypoketotic hypoglycemia 12, and a marked decrease of 0.73 free carnitine. There is an elevation of C5 acylcarnitine, but isovaleric acidemia is ruled out by urinary organic acid analysis. Adenovirus-positive interstitial pneumonia diagnosed → prednisolone administered for 4 days	Due to upper respiratory tract inflammation, he had been prescribed CDTR-PI for 4 days and CFPN-PI for 7 days and 14 days before the onset of the disease at his local doctor	On the 11th day after the onset of the disease, the free carnitine level was 1.5 and had not recovered, and L-carnitine (20 mg/kg/day) was started. On the 7th day after the start of oral administration, the free carnitine recovered to 13.94. Administered for 17 days	Yamada et al. (2008) (Ref. [18])
6	3 years old female	Both limb tremor, somnolence, impaired consciousness, hypoketotic hypoglycemia 16, ammonia 81, free carnitine 12.3	Due to cough and nasal discharge, the patient was administered CFPN-PI for 15 days and CDTR-PI for 7 days out of 27 days at the family pediatrician and otolaryngology department just before the onset of the disease (intermittent)	Free carnitine levels normalized after 19 days after stop antibiotics containing the pivoxil group	Nakatani et al. (2009) (Ref. [19])

(continued)

Table 13.2 (continued)

Paper report					
Case	Age, gender	Symptoms, laboratory tests, treatments, etc.	Antibiotics, duration	Carnitine, duration, prognosis, etc.	Reporter
7	1 year, 9 months old male	Fever, intermittent limb tonic convulsions, hypoglycemia 21 (blood ketones were high in this case). After continuous infusion, hypoglycemia improved, but consciousness did not return. Mannitol infusion for cerebral edema relief, phenobarbital suppositories for brain protection, steroid pulse for suspected encephalopathy. Free carnitine 7.4 and extreme drop.	Due to otitis media, CDTR-PI 13.5 mg/kg was administered at the otolaryngologist for 10 days (intermittent), and the dose was increased to 18 mg for 2 days without effect	Levocarnitine chloride 20 mg/kg/day for 17 days improved and was discharged from the hospital, but mild paralysis of the left side of the body remained.	Hanayama et al. (2010) (Ref. [20])
8	16 months old male	In the morning, he was taken to the hospital by ambulance due to tremors in his hands and feet, generalized convulsions. Hypoketotic hypoglycemia 18 → Intravenous glucose improves convulsions	After interviewing his parents, it was found that he had been administering CDTR-PI, CFPN-PI, and CFTM-PI intermittently for 87 days at 6 months due to otitis media and upper respiratory tract infections.	Normalized after 3 months without supplementation, adequate diet	Nakajima et al. (2010) (Ref. [21])
9	18 months old male	Loss of appetite, repeated vomiting, and tonic convulsions were taken to the hospital by ambulance. Hypoketotic hypoglycemia 11. Intravenous glucose infusion improved hypoglycemia, but convulsions and impaired consciousness did not improve. Immunoglobulin, dexamethasone, and acyclovir on suspicion of encephalitis/encephalopathy	After first aid, checked the medication history with his parents, and it was found that I had been receiving CDTR-PI for 6 months due to refractory otitis media	Intravenous carnitine injection 100 mg/kg/day for 3 days. Acidosis improves in 1 h after the start of intravenous infusion, and conversation is possible after half a day	Nakajima et al. (2010) (Ref. [21])

10	3 years, 11 months old male	Lumbar meningocele with ventriculoperitoneal shunt placement. 2 weeks before hospitalization, hypoketotic hypoglycemia 12, impaired consciousness. At the time of hospitalization, hypoketotic hypoglycemia 16, blood glucose drip improved blood sugar, but consciousness disorder and convulsions in the right arm continued. Methylprednisolone pulse therapy, intravenous midazolam. He regained consciousness on the 11th day. Discharged after 26 days. After discharge from the hospital, there was a hypoglycemic attack again, and the suspected carnitine deficiency in non-ketotic dicarboxylic acid urine → free carnitine 0.37 and high levels of C5 acylcarnitine in the blood.	Cefteram pivoxil 100 mg for 12 months to prevent urinary tract infections.	Replace CFTM-PI with cefaclor and oral carnitine supplementation. After 3 months, free carnitine 58.65, no hypoglycemic attacks	Okumura et al. (2011) (Ref. [22])
11	1 year old female	First seizure; vomiting after ingesting a small amount of dinner, going to bed as it was, and being transported to the emergency hospital the next morning in unconscious state. Limb ankylosis, ketotic hypoglycemia (10), ammonia 114. Second seizure: One and a half months later, he went to the emergency room with stiffness in his upper limbs and impaired consciousness. Hypoglycemia (12), ammonia 319.free carnitine decreased (9.6), C5 acylcarnitine elevated	Interviews with the mother revealed that she had been taking CFPN-PI intermittently for cold symptoms for 4 months	After administration of L-Cartin, the level of free carnitine normalized (dose and duration of administration unknown), and there was no decrease thereafter	Ohura et al. (2013) (Ref. [12])

(continued)

Table 13.2 (continued)

Paper report					
Case	Age, gender	Symptoms, laboratory tests, treatments, etc.	Antibiotics, duration	Carnitine, duration, prognosis, etc.	Reporter
12	1 year, 2 months old female	Emergency transport with fever and generalized tonic convulsions. Extremely low levels of hypoglycemia 19, free carnitine 1.42, and high levels of C5 acylcarnitine, but isovaleric acidemia was denied by urinary organic acid analysis.	From 32 days before the onset of the disease, CDTR-PI 15 mg/kg/day was administered by a nearby doctor for about 25 days due to cold symptoms	Vitamins and carnitine supplementation (dose unknown), impaired consciousness from the 6th day of hospitalization, tendency to improve motor function, discharged from the hospital on the 39th day, no hypoglycemic attack even after discontinuation of carnitine administration	Kajimoto et al. (2013) (Ref. [23])
13	1 year, 7 months old female	Fever, clonic convulsions, impaired consciousness, hypoglycemia 15, glucose solution administration improved blood sugar, but convulsive seizures did not improve, and she was emergency transported. Extremely low with free carnitine 5.1. High levels of C5 acylcarnitine (no abnormalities in neonatal filter blood acylcarnitine analysis), ammonia 183 (previous doctor) → 80, metabolic acidosis. Diagnosed as acute encephalopathy and cerebral reception was started with artificial respiration, continuous EEG monitoring, and barbiturate coma therapy.	Participants had received CDTR-PI for 15 days 3 months prior to onset and TBPM-PI for 3 days immediately prior to onset of disease	Carnitine 30 mg/kg/day, vitamins B1, B6, B12, vitamin C, vitamin E, biotin, CoQ10 supplementation started. On the fifth day of illness, the impairment of consciousness was prolonged even after the end of the brain warmth therapy. He was discharged from the hospital on the 28th day, but left upper and lower limb paralysis and severe developmental delay remained as sequelae	Nishiyama et al. (2014) (Ref. [24])
14	1 year, 6 months old female	He had no appetite for 2 days before his visit, and on the day of his visit, he was taken to the hospital by ambulance due to impaired consciousness and convulsions. Significant hypoglycemia was observed, and consciousness was restored by intravenous glucose injection.	Four days before coming to the hospital, a nearby doctor prescribed cefditoren pivoxil for a cold.	Diagnosis of carnitine deficiency due to pivoxil-containing antibiotics with low C0 and high C5 levels in filter paper blood at the time of consultation (systemic carnitine deficiency and isovaleric acidemia were later denied). The tandem mass method is also useful in the emergency room.	Kajiwara et al. (2017) (Ref. [38])

15	1 year, 11 months old male	Hospitalized with anorexia, fever, cough, tendency to somnolence, positive RSV antigen test. *B*lood sugar 38 and high ketone bodies. After intravenous glucose injection, he recovered and was discharged from the hospital on the 5th day. Free carnitine 10.8 and low, C5 acylcarnitine 1.4 and high. Isovaleric acidemia is ruled out	Nasal discharge and cough symptoms, fever (38 °C) 3 days before hospitalization due to upper respiratory tract infection, and a local doctor prescribed CDTR-PI at 6 mg/kg/day.	Carnitine supplementation was not performed due to the short duration of administration of the pivoxil-containing antibiotic. 30 days after the attack, blood carnitine normalized	Nakazaki et al. (2018) (Ref. [25])
16	1 year, 5 months old male	Undernourished infants. Hospitalized for loss of appetite, convulsions accompanied by fever, impaired consciousness. Hypoglycemia 26, hyperammonemia 440, metabolic acidosis, free carnitine low 8.5, C5 acylcarnitine (pivaloyl carnitine) 3.30.	Upper respiratory tract infection on CFTM-PI for 6 days.	Hypothermia therapy, intravenous methylprednisolone (30 mg/kg/day, 3 days), carnitine (80 mg/kg/day). Cerebral atrophy on MRI on 30th day, low blood flow to bilateral frontal lobe and left parietal lobe on SPECT.	Shimizu et al. (2019) (Ref. [26])
Case reports of carnitine deficiency caused by pivoxil-containing antibiotics in Japan (continued)					
Paper report Cases in elderly patients					
17	85 years old female	Patient with nasal tube feeding (gastrostomy), fever, sputum, dyspnea, decreased urine output. Suspected hypoglycemia due to profuse sweating and cold sensation in the extremities→42. Glucose administration	CFPN-PI 100 mg × 2 administered for 2 days in the diagnosis of acute bronchitis	No administration of carnitine	Tanigawa et al. (2013) (Ref. [27])
18	85 years old male	Gastrectomy for gastric cancer, pharyngeal resection for pharyngeal chondrosarcoma. I have no teeth and no longer eat meat or fish. Antimicrobial medication is administered on suspicion of urinary tract infection. He was taken to the emergency room due to an intermittent hypoglycemic attack. Hypoglycemia 58 (normal ketones, low for hypoglycemic status), low at 12.6 free carnitine	She had been prescribed CFPN-PI for 3 days at a nearby urology clinic due to a suspected urinary tract infection	Oral 1, 500 mg/day (1 week), then 750 mg/day (1 week)	Hanai et al. (2019) (Ref. [28])

(continued)

Table 13.2 (continued)

Paper report					
Case	Age, gender	Symptoms, laboratory tests, treatments, etc.	Antibiotics, duration	Carnitine, duration, prognosis, etc.	Reporter
A case of administration of a pivoxil-containing antibiotic to a patient with muscular dystrophy					
19	6 years old girl, Fukuyama-type muscular dystrophy patient	Impaired consciousness, hypoglycemia31, hypocarnitinemia (free carnitine 6.2, 29.0 and normal range before CDTR-PI administration). Recovery by intravenous glucose injection. CT showed significant muscle atrophy compared to healthy children, and it was determined that carnitine deficiency developed even after short-term administration of CDTR-PI	I oral administration of CDTR-P for 3 days in pharyngitis (8.5 mg/kg/day)	Carnitine supplementation (dose unknown)—free carnitine 56.7 after 1 month	Ito et al. (2017) (Ref. [29])
Cases reports at academic conferences (Abstracts)					
Case	*Age, gender*	*Symptoms, laboratory tests, treatments, etc.*	*Antibiotics, duration*	*Carnitine, duration, prognosis, etc.*	*Reporter*
20	1 year old male	Impaired consciousness, convulsions, hypoglycemia, mild elevation of ammonia, marked decrease in carnitine	Repeated upper respiratory tract inflammation and oral CDTR-PI for about 1 month (intermittent).		Ito et al. (2003) (Ref. [30])
21	11 months old female	Hypoglycemia, hyperammonemia, convulsions, impaired consciousness, low free carnitine, high C5 acylcarnitine (pivaloyl carnitine)	Multiple pivoxil-containing antibiotics were administered to a nearby doctor for otitis media (administration period unknown)		Noguchi et al. (2009) (Ref. [31])
22	1 year old male	At dawn on the day of hospitalization, afebrile generalized tonic convulsions, hypoglycemia22, and intravenous glucose infusion restored the level of consciousness. Low levels of free carnitine (14.67), mild elevation of C5/C2 and C5/C3	A nearby doctor administered CFPN-P for slight fever and cold symptoms 6 days before hospitalization, and CDTR-PI for otitis media from 2 days before hospitalization		Kasai (2010) (Ref. [32])

23	1 year and 11 months old male	The day before, he went to another hospital for vomiting. In the early morning of the day, a single point stared appeared and was transported to an ambulance. Clonic convulsions, hypoglycemia 14, hyperammonemia 186, urine ketone negative. Suspected of a congenital error of metabolism, administration of glucose infusion, arginine, and carnitine was started. Marked free carnitine decline and C5 acylcarnitine elevation	At the age of 9 months, he was hospitalized and treated at another hospital for a bacterial infection. After discharge from the hospital, prophylactic administration of antibiotics, bacterial infection developed at 1 year and 1 year and 3 months, and TBPM-PI was changed to TBPM-PI (normal dose). If discontinued, there will be a fever, so continuous administration for 8 months	Carnitine supplementation for 4 days (dose unknown)	Hamahira (2011) (Ref. [33])
24	2 years old male	From the day before hospitalization, he was transported to the emergency hospital due to poor vitality, poor oral intake, and generalized tonic convulsions while sleeping in the middle of the night. Persistent convulsions, hypoglycemia 11, ammonia >400, dehydration. After glucose administration, his consciousness disorder prolonged and hospitalized. Decrease in total carnitine levels, elevation of C5 acylcarnitine. Urinary organic acid analysis denies isovaleric acidemia	He was receiving CDTR-PI or CFPN-PI almost every day for 8 months at the Department of Pediatrics	There is supplementation, but the dose is unknown	Uozumi et al. (2011) (Ref. [34])
25	1 year old female	She did not eat anything after eating at 6 p.m. the day before and was taken to the hospital by ambulance in the early morning due to impaired consciousness and convulsions. Extremely low levels of hypoketotic hypoglycemia 29 (urine ketone negative) and free carnitine 3.5	From around 5 months of age, CDTR-PI was administered intermittently for about 6 months by a nearby physician for cold symptoms		Murayama et al. (2012) (Ref. [35])

(continued)

Table 13.2 (continued)

Paper report					
Case	Age, gender	Symptoms, laboratory tests, treatments, etc.	Antibiotics, duration	Carnitine, duration, prognosis, etc.	Reporter
26	1 year old female	She was admitted to the emergency department for impaired consciousness and loss of appetite and was hospitalized for further examination. Low blood sugar 20, free carnitine 9.0	For cold symptoms such as fever, cough, and nasal discharge, he was administered CFPN-PI, CFTM-PI, and CDTR-PI by his doctor for 40 days	Discontinuation of antibiotics, improvement with glucose administration and discharge	Inoue et al. (2012) (Ref. [36])
27	1 year and 5 months old male	On the 7th day after the start of oral administration, after a fever of 38 °C, he underwent repeated generalized tonic–clonic seizures. Intravenous glucose administration for hypoglycemia. Administration of diazepam and fosphenytoin eliminated convulsions, but prolonged impairment of consciousness. Pharyngeal adenovirus positive, hyperammonemia. Acute encephalopathy was considered and started steroid pulse therapy. Decrease in free carnitine (8.5) and an increase in C5 acylcarnitine, and isovaleric acidemia was ruled out by urinary organic acid analysis, and an increase in C5 was judged to be pivaloyl carnitine	He was prescribed CFTM-PI for 6 days at a local doctor due to cold symptoms, and it developed on the 7th day	Suspected carnitine deficiency by hypoglycemia and hyperammonemia and started intravenous carnitine. Decreased diffusion in the bilateral subcortical white matter was observed on MRI, cerebral cryotherapy was initiated, and MRI after reheating showed decreased diffusion of the bilateral cerebral cortex, followed by cerebral atrophy and severe neurological sequelae	Minamikawa (2014) (Ref. [37])
28	Newborn at 4 days of age (gender unknown)	Neonatal mass screening of infants born at 39 weeks gestational age showed low levels of C0 (free carnitine) (6.8) and high levels of C5 acylcarnitine (4.44). Weight gain and feeding were good, and there were no abnormalities in physical findings	When interviewed mother in detail, it was found that the patient had been taking CDTR-PI for 2 weeks from prenatal to the first day of the puerperium due to sinusitis. The drug may have affected the baby trans-placentally	Oral use of carnitine preparation for 1 month, no abnormalities	Endo et al. (2019) (Ref. [39])

CFPN-PI cefcapene pivoxil, CDTR-PI cefditoren pivoxil, CFTM-PI cefteram pivoxil, TBPM-PI tebipenem pivoxil
Unit of clinical laboratory test: blood glucose: mg/dL, ammonia: μg/dL, carnitine: nmol/mL

3. Most cases occurred after long-term administration of 2 weeks or longer, but some cases occurred after short-term administration such as 3 days.
4. The main clinical symptoms and laboratory abnormalities are almost always hypoketotic hypoglycemia (normal ketones and hyperketones are also possible), convulsions, impaired consciousness, and hyperammonemia with a marked decrease in blood-free carnitine levels.
5. In many of these cases, symptoms have recovered by intravenous glucose injection and carnitine supplementation, but in some cases, sequelae such as hemiplegia remain due to cerebral atrophy.
6. The age of onset is mostly young infants and young children, but there are reports in the elderly, and special attention should be paid to patients with low muscle mass such as children with severe mental and physical disabilities and patients with low intake of carnitine from the diet.

13.5 Examination of Carnitine Deficiency Caused by a Pivoxil-Containing Antibiotics with a Tandem Mass Analysis

Kobayashi et al. retrospectively analyzed data of 13,451 patients from 2004 to 2012 who underwent tandem mass spectrometry analysis for suspected fatty acid metabolism disorders or organic acidemia. Total 22 patients were supposed as carnitine deficiency due to pivoxil-containing antimicrobial agents, whose free carnitine level was low (C0 ≤20 μmol/L) and high C5 acylcarnitine. All 22 patients were confirmed to have received pivoxil-containing antimicrobial agents. These patients ranged in age from 2 months to 42 years (median: 1 year 11 months, 1 case <1 year, 10 cases 1 year, 1 case 2 years, and 10 cases >3 years). Nine patients had known disease. Fourteen patients developed acute encephalopathy, 13 patients hypoglycemia, and 4 patients hypoglycemia without signs of encephalopathy. C0 values (free carnitine, cut-off value ≤20 μmol/L) ranged from 0.25 to 19.66 μmol/L (median 1.31 μmol/L) and C5 acylcarnitine (cut-off ≥0.6 μmol/L) ranged from 0.43 to 11.92 μmol/L (median 3.23 μmol/L). There was no correlation between duration of antimicrobial therapy and free carnitine levels, and 10 patients developed the disease after 14 or more days of therapy, while 6 patients developed the disease within 14 days. These results indicate that carnitine deficiency caused by pivoxil-containing antimicrobial agents often develops as hypoglycemic seizures and/or acute encephalopathy and is more common in infants around 1 year of age, and in some cases occurs within 14 days of administration [40].

Thus, C5 acylcarnitine (i.e., pivaloylcarnitine) is elevated in cases of neonatal mass screening in which the patient was taking antimicrobial agents containing pivoxil groups, but C5 acylcarnitine (i.e., isovalerylcarnitine) is also elevated in cases of isovaleric acidemia. Other C5 acylcarnitines include 2-methylbutyrylcarnitine, a metabolite related to metabolic diseases, and n-valerylcarnitine, which is rarely found in vivo, and these four C5 acylcarnitines have exactly the same mass and cannot be distinguished by mass spectrometry. Therefore, special tests such as urinary organic acid analysis using GC/MS are necessary. Kajiwara et al. reported that tandem mass spectrometry was useful in differentiating carnitine deficiency caused by pivoxil-containing antimicrobial agents from systemic carnitine deficiency or isovaleric acidemia in pediatric emergency departments, although its use is not common except for mass screening of inherited metabolic diseases in newborns [38].

Recently, Maeda et al. reported a simple method for identifying pivaloylcarnitine in these cases of high C5 acylcarnitine levels [41]. Acylcarnitine was eluted from dried filter paper blood used for tandem mass screening with water, and then reacted with acylcarnitine esterase (ACE)) in the diagnostic kit for total carnitine determination described in Sect. 8.3.2. This ACE isolated from the soil bacterium completely degraded isovalerylcarnitine and n-valerylcarnitine to free carnitine and the corresponding organic acids with high affinity, and almost completely degraded 2-methylbutyrylcarnitine, although a little remain. However, in the case of pivaloylcarnitine, the affinity for this ACE was low, probably due to steric hindrance, and the peak of C5 acylcarnitines did not decrease when reacted with this ACE. Therefore, when C5 acylcarnitines are high in newborn mass screening, it is possible to determine whether they are pivaloylcarnitine, isovalerylcarnitine, or other acylcarnitines by reacting with this ACE and seeing if the peak decreases. If the possibility of pivaloylcarnitine is ruled out, the name of the inherited metabolic diseases is determined by analyzing organic acids in urine to determine whether it is isovalerylcarnitine or 2-methylbutyrylcarnitine, if necessary.

13.6 Issues from a Regulatory Science Perspective

13.6.1 Approval Status of Pivoxil-Containing Antibiotics in Japan and Overseas

According to the review by Brass on the effects of pivalic acid-containing prodrugs on carnitine homeostasis, cefditoren pivoxil was the only pivoxil-containing antibacterial agent approved in the United States as of 2002 [11]. In comparison, in Japan, a very large number of pediatric fine granules of original and generic antibiotics containing pivoxil groups are on launched, and carnitine deficiency and hypoglycemic attacks caused by pivoxil-containing antibiotics have been reported one after another. In 2012, when the Japanese Society of Pediatrics issued the "Warning for Secondary Carnitine Deficiency Caused by Antibiotics Containing Pivoxil Groups" [42], a total of 24 products were on the market, including 10 cefcapene pivoxil including Flomox, 7 cefditoren pivoxil, 5 cefteram pivoxil, tebipenem pivoxil (trade name: Orapenem), and a pivmecillinam (trade name: Melicin). Although some antimicrobials have been discontinued in recent years, such as pivmecillinam in 2013 (see the bottom of Table 13.1), a very large number of antimicrobials containing the pivoxil group are still on the market compared to other countries and are especially widely used in pediatrics and otolaryngology. In 2012, in response to a warning from the Japan Pediatric Society, the Pharmaceuticals and Medical Devices Agency (PMDA) issued a "Request for Proper Use of Medicines from the PMDA" [43]. This alert summarizes 38 cases in which adverse reactions have been reported so far, and also describes the clinical course of 4 representative cases and calls for attention. In one of these cases, the expectant mother received cefcapene pivoxil at 27 weeks of gestation due to pyelonephritis and gave birth at 39 weeks with normal delivery, but the newborn was found to have hypocarnitinemia by mass screening for inherited metabolic diseases (PMDA alert case 4) [43]. Subsequently, cases of carnitine deficiency in infants via the placenta have been reported (Table 13.2, case 28). If pregnant women take these antibiotics, there is a risk that the newborn child will also develop hypocarnitinemia, and caution should be exercised.

According to Brass, the package inserts for these antimicrobials in other countries state that they are contraindicated in patients with carnitine deficiency or metabolic diseases that cause carnitine deficiency. And because these antimicrobials are intended for short-term use, there is no mention in the package inserts of blood carnitine measurement or carnitine replacement therapy. The German label for cefetamet pivoxil states that it is contraindicated in patients with primary carnitine deficiency, dialysis patients, and diabetics, and also includes a warning about patients receiving valproic acid. In Sweden, the pivampicillin package inserts a similar warning for unstable diabetics, stunted children, patients on valproic acid, and patients with extremely low muscle mass. On the other hand, the Canadian pivampicillin package insert states that short-term administration of pivampicillin causes a decrease in carnitine but no adverse events, and that neither blood carnitine monitoring nor carnitine replacement therapy is required. The UK pivampicillin package insert also includes warnings about carnitine deficiency and patients receiving valproic acid.

In Japan, in April 2012, cautions regarding hypoglycemia due to hypocarnitinemia were included in the "Precautions" section of the package inserts for cefteram pivoxil, cefcapene pivoxil, cefditoren pivoxil, and tebipenem pivoxil [44]. The package insert for tebipenem pivoxil also states as a contraindication, "Patients receiving sodium valproate [may experience recurrent epileptic seizures]". It is widely known that valproic acid administration can cause hypocarnitinemia and hyperammonemic encephalopathy (see "Chap. 12: Carnitine Deficiency Caused by Sodium Valproate"), and in fact, hyperammonemic encephalopathy has been reported overseas in patients receiving valproic acid in combination with a pivoxil-based antibacterial drug, pivmecillinam [45]. Why is there no mention of sodium valproate in the package inserts of cefteram pivoxil, cefcapene pivoxil, and cefditoren pivoxil, which are also pivoxil-containing antibacterial agents, as there is in the package insert for tebipenem pivoxil? The PMDA and related pharmaceutical companies should have consistency in the description about the pivoxil group-containing antimicrobial agents in the package inserts.

In any case, prescribing antimicrobial agents containing a pivoxil group when epileptic patients or severely mentally or physically handicapped children (persons) taking valproic acid develop an infection is thought to increase the risk of hypocarnitinemia, so switching to other antimicrobial agents such as new quinolones that do not contain a pivoxil group when possible, or supplementation with carnitine agents should be considered. Ito et al., who have led clinical research in this field in Japan, provide an easy-to-understand explanation of these drug-induced carnitine deficiencies and carnitine supplementation therapy [46].

13.6.2 Examination of Carnitine Deficiency in Clinical Trials at the Time of Application for Approval of Pivoxil-Containing Antibiotics

In fact, the risk of carnitine deficiency was also examined in detail in clinical trials at the time of approval of antibiotics containing the pivoxil group. For example, with regard to cefditoren pivoxil (CDTR-PI, brand name Meiact®), Fujii and 13 other pediatricians organized the "Cefditren pivoxil Carnitine Research Group" to investigate carnitine metabolism in 16 patients with pediatric infections [47]. In this clinical study, Group I: 3 mg/kg 3 times a day, Group II: 6 mg/kg 3 times a day after meals, 9–57 doses, the concentration of free carnitine in the blood and the concentration of pivaloylcarnitine in the urine before and after administration were examined in detail. As a result, the concentration of free carnitine in the blood decreased from 50.9 ± 15.0 nmol/mL before administration to 21.4 ± 9.3 nmol/mL (about 42%) in the I. group, and 40.5 ± 22.3 nmol/mL before administration decreased to 11.1 ± 4.5 nmol/mL after administration, about 1/3 in the II group. In the II group, in 3 out of 9 patients, the free carnitine level was less than 10 nmol/mL, which is said to be the cause of carnitine deficiency at any time, and one of the patients was administered 57 times (19 days), and the remaining two patients were 6.5 nmol/mL and trace level (<6.3 nmol/m) of free carnitine concentration, because the levels of before administration were low at 14.8 and 20.6 nmol/mL. In other words, carnitine deficiency caused by antibiotics containing pivoxil group, which had been pointed out in previous clinical pharmacological studies, had already been reproduced in clinical trial data at the time of application for approval. However, clinical symptoms such as hypoglycemic attacks and convulsions did not develop. Surprisingly, however, the conclusion of this paper is that "*the concentration of free carnitine in serum (plasma) recovered rapidly after discontinuation of CDTR-PI administration, and there were no side effects or abnormalities in laboratory values considered to be caused by a decrease in carnitine, so we believe that the safety of the effects of pivarate in children with general infectious diseases has been confirmed as long as the standard dosage and dose of this drug are observed.*" (underlined by the author). What was the purpose of this "Cefditren pivoxil Carnitine Research Group"? Since there was such a clear decrease in carnitine in 16 cases, it can be fully assumed from the data in this report that safety problems will occur when it is administered for a longer period of time or used in underweight children in actual clinical practice, and it is thought that further risk assessment and countermeasures should have been considered.

Subsequently, in the field of pediatric infectious diseases, *Streptococcus pneumoniae* and *Haemophilus influenzae*, which are the major causative organisms of acute otitis media, acute sinusitis, and bacterial pneumonia, the drug-resistance of *Streptococcus pneumoniae* and *Haemophilus influenzae* has progressed, and the intractable and prolonged infection has become a problem, and clinical trials of high dose (6 mg/kg, 3 times a day after meals) for pediatric infections have been conducted for CDTR-PI and CFTM-PI, and it became a kind of trend to use high doses [48, 49]. However, in clinical trials of CDTR-PI, blood carnitine levels have been evaluated, but CFTM-PI has not even been evaluated. Subsequently, in the field of pediatric infectious diseases, the carbapenem-based antimicrobial drug tebipenem pivoxil (TBPM-PI, trade name: Orapenem), which has a broader antimicrobial spectrum and contains a pivoxil group that exhibits stronger antimicrobial activity against various resistant bacteria, has been developed. Carnitine deficiency has been examined in the clinical evaluation, and a decrease in blood-free carnitine concentration has been clearly observed. In four patients whose carnitine concentration was measured 3 days after administration, the concentration decreased to 20–55% of the pre-dose level 7–8 days after administration but showed a tendency to recover 3 days after the end of treatment. However, it is not considered to be a serious problem because there were no adverse events such as hypoketotic hypoglycemia, which seemed to be due to carnitine deficiency, and recovery after discontinuation of treatment [50, 51]. A study of the use of TBPM-PI in 3547 patients from April 2010 to March 2013 showed that there were no hypoglycemic attacks due to carnitine deficiency [52]. However, in actual clinical practice, hypoglycemic attacks caused by TBPM-PI have also reported, as shown in Table 13.2. Clinical trials for approval range from a few dozen to several hundred patients at most, and post-marketing surveillance is in the thousands, which is an extremely limited number considering that it is widely used in clinical practice for pediatric upper respiratory tract infections, sinusitis, acute otitis media, etc. When hypoglycemic attacks or encephalopathy occur due to carnitine deficiency, it is necessary to pay attention to some cases in which serious sequelae remain.

13.6.3 Responses of Related Academic Societies

As mentioned above, the Pharmaceutical Affairs Committee of the Japan Pediatric Society issued a “Warning on Secondary Carnitine Deficiency Caused by Administration of Antibiotics Containing Pivoxil Groups” in 2012 [42]. In response to this, in April 2012, the Pharmaceuticals and Medical Devices Agency (PMDA) issued a revised Precautions for the Use of Antibiotics Containing Ivoxyl Groups [44] and a “Request for Proper Use of Pharmaceuticals” [43]. Subsequently, the Nutrition Committee in Child Health Consortium Japan, which consists of four organizations: Japan Pediatric Society, Japan Pediatric Association, The Japanese Society of Child Health, and the Japanese Association of Pediatric Surgery Related Societies, led the publication of the “Guidelines for the Diagnosis and Treatment of Carnitine Deficiency 2016” and the revised “Guidelines for the Diagnosis and Treatment of Carnitine Deficiency 2018” was published [13]. As a result, the understanding of carnitine deficiency has become more widely spread, but there is no end to the case reports of carnitine deficiency caused by antibiotics containing the pivoxil group. Therefore, in July 2019, the Pharmaceutical Affairs Committee of Japan Pediatric Society issued a “Warning on Hypocarnitinemia Associated with the Use of Antibiotics Containing Pivoxil Group” [53]. The committee investigated from the previous warning and the PMDA alert in 2012 to 2018 and identified 22 cases of hypocarnitinemia and hypoglycemia were reported even with short-term administration of these antibiotics for less than 7 days. The alert pointed out that hypocarnitinemia may occur even in a short period of time, in rare cases it becomes severe, and encephalopathy is caused, and sequelae remain, so that in suspicious cases, the blood carnitine concentration should be measured, and carnitine preparations should be administered (40–60 mg/kg/day), and intravenous administration may be useful in the cases of impaired consciousness. On August 30, 2019, the president of Japan Pediatric Society sent a request to the president of The Japan Society of Obstetrics and Gynecology and the president of The Japanese Dermatological Association, etc., requesting that the members of these societies be fully aware of the "Warning of hypocarnitinemia associated with the use of antibiotics containing the pivoxil group. " (https://www.jsog.or.jp/news/pdf/20190909_shuuchiirai.pdf, https://www.dermatol.or.jp/uploads/uploads/files/news/G20190911_JPEDS_Syuchi.pdf).

In the field of otolaryngology, Japan Otological Society, Japan Society for Pediatric Otorhynolaryngology, and Japan Society for Infectious Diseases in Otorhinolaryngology (Japan Society for Infection and Aerosol in Otorhinolaryngology since August 2013) organized a committee to create a clinical guideline for pediatric acute otitis media, and in 2006 the "Pediatric Acute Otitis Media Clinical Practice Guidelines, 2006 Edition" was published. Subsequently, in 2012, the Pharmaceutical Affairs Committee of the Japan Pediatric Society issued a "Warning for Secondary Carnitine Deficiency with Antibiotics Containing Pivoxil Groups." So that in the "Pediatric Acute Otitis Media Clinical Practice Guidelines, 2013 Edition," there is a description on the treatment algorithm for acute otitis media in children, as a note; "Caution should be taken to the development of secondary carnitine deficiency for long-term continuous administration of antibiotics containing a pivoxil group" [54, 55]. In the 2018 edition of the guideline on the treatment algorithm for mild, moderate, and severe cases, a cautionary statement was added as "antimicrobials with a pivoxil group should be carefully examined for the development of secondary hypocarnitinemia" [56]. In the 2010 supplementary edition of the Acute Rhinosinusitis Clinical Practice Guidelines, there is a note on the treatment algorithm stating that "there have been reports of hypoglycemia associated with hypocarnitinemia in long-term administration of drugs with the above pivoxil groups to infants and young children" [57]. In addition, the Committee on the Appropriate Use of Antimicrobials for Respiratory Tract Infections in The Japanese Association for Infectious Diseases has published a "Recommendation on the Appropriate Use of Antimicrobials for Respiratory Tract Infections," and as a precaution in the treatment algorithm for acute rhinosinusitis and acute pharyngitis/tonsillitis, "Regarding hypocarnitinemia and hypoglycemia caused by antibiotics with a pivoxil group: When antibiotics containing a pivoxil group are administered, carnitine excretion is enhanced. Severe hypocarnitinemia may be accompanied by hypoglycemia, convulsions, encephalopathy, etc., and there are cases of sequelae. Particular caution should be exercised in children (especially infants) due to low blood carnitine levels" [58].

13.7 Summary

In this way, in Japan, antibiotics containing pivoxil group are more easily used for the treatment of upper respiratory tract infections and otitis media than in other countries, and there have been many reports of hypoglycemic attacks due to carnitine deficiency. There are some cases where serious sequelae remain, and the situation has finally changed being understood that carnitine deficiency has serious consequences, but it must be said that it is still insufficient. Of course, these infections diseases need treatment, and it is understandable to use high doses of antibiotics, especially if they become intractable due to the spread of resistant bacteria. However, the risk of carnitine deficiency should always be considered, and alternatives to antibiotics containing the pivoxil group should be used whenever possible. Tanizaki et al. reported that a-69 years-old woman with non-dialysis chronic kidney disease and schizophrenia presented progression of painful involuntary nocturnal leg cramping for several months. She was prescribed cefcapene-pivoxil 100 mg 3 times for 6 months to treat pyorrhea [59]. After L-carnitine treatment, her symptom improved. This case shows that antibiotics containing the pivoxil group cause severe carnitine deficiency even in adults, especially in patients with chronic kidney disease or liver cirrhosis.

In addition, it has been reported that a new drug containing a pivoxil group, although not an antimicrobial drug, causes carnitine deficiency. It was reported that C5 acylcarnitine was elevated in two patients who were administered injectable sivelestat sodium hydrate (brand name: Elaspol®) for acute respiratory distress syndrome (ARDS) in the indication for "improvement of acute lung injury associated with systemic inflammatory response syndrome" [60]. In fact, it has been reported in the annual meeting of Japan Society of Perinatal and Neonatal Medicine that two children with severe respiratory disorders who received sivelestat for 14 days and severe hypocarnitinemia occurred although hypoglycemia and other symptoms were not observed [61]. In the use of drugs containing a pivoxil group, extreme caution should always be exercised against carnitine deficiency.

References

1. Brass EP. Pivalate-generating prodrugs and carnitine homeostasis in man. Pharmacol Rev. 2002;54(4):589–98. https://doi.org/10.1124/pr.54.4.589.
2. Oura T. Pivoxil-containing antibacterial. Prescr Inform. 2019;25(5):658–62. (Article in Japanese).
3. Melegh B, Kerner J, Bieber LL. Pivampicillin-promoted excretion of pivaloylcarnitine in humans. Biochem Pharmacol. 1987;36(20):3405–9. https://doi.org/10.1016/0006-2952(87)90318-2.
4. Holme E, Greter J, Jacobson CE, et al. Carnitine deficiency induced by pivampicillin and pivmecillinam therapy. Lancet. 1989;2(8661):469–73. https://doi.org/10.1016/s0140-6736(89)92086-2.
5. Abrahamsson K, Holme E, Jodal U, et al. Effect of short-term treatment with pivalic acid containing antibiotics on serum carnitine concentration — a risk irrespective of age. Biochem Mol Med. 1995;55(1):77–9. https://doi.org/10.1006/bmme.1995.1036.
6. Abrahamsson K, Mellander M, Eriksson BO, et al. Transient reduction of human left ventricular mass in carnitine depletion induced by antibiotics containing pivalic acid. Br Heart J. 1995;74(6):656–9. https://doi.org/10.1136/hrt.74.6.656.
7. Abrahamsson K, Eriksson BO, Holme E, et al. Pivalic acid-induced carnitine deficiency and physical exercise in humans. Metabolism. 1996;45(12):1501–7. https://doi.org/10.1016/s0026-0495(96)90179-1.
8. Melegh B, Pap M, Molnár D, et al. Carnitine administration ameliorates the changes in energy metabolism caused by short-term pivampicillin medication. Eur J Pediatr. 1997;156(10):795–9. https://doi.org/10.1007/s004310050716.
9. Pap M, Kopcsányi G, Bieber LL, et al. Cefetamet pivoxil treatment causes loss of carnitine reserves that can be prevented by exogenous carnitine administration. J Nutr Biochem. 1999;10(11):670–3. https://doi.org/10.1016/s0955-2863(99)00051-0.
10. Ito T, Sugiyama N, Kobayashi M, et al. Alteration of ammonia and carnitine levels in short-term treatment with pivalic acid-containing prodrug. Tohoku J Exp Med. 1995;175(1):43–53. https://doi.org/10.1620/tjem.175.43.
11. Nakashima M, Kosuge K, Ishii I, et al. Influence of multi-dose administration of cefetamet pivoxil on blood and urinary concentrations of carnitine and effects of simultaneous administration of carnitine with cefetamet pivoxil. Jpn J Antibiot. 1996;49(10):966–79. (Article in Japanese).
12. Ohura T. Secondary carnitine deficiency due to pivalate-conjugated antibiotics. J Japanese Soc Pediatr Pharmacol. 2013;26(1):14–7. (Article in Japanese).
13. Ida S, Iwasaki Y, Uchida K, et al. Guideline for diagnosis and treatment of carnitine deficiency 2018. https://www.jpeds.or.jp/uploads/files/20181207_shishin.pdf. (Article in Japanese).
14. Teraoka M, Wada T, Ogura K, et al. A case of hypocarnitinemia due to long-term antibiotic therapy. J Japan Pediatr Soc. 2004;108(8):1059–61. (Article in Japanese).
15. Nagasaki K, Hiura H, Kikuchi T, et al. A case of hypoglycemia with secondary hypocarnitinemia due to long-term administration of antibiotics. Jpn J Pediatr. 2005;58(9):1893–7. (Article in Japanese).
16. Makino Y, Sugiura T, Ito T, et al. Carnitine-associated encephalopathy caused by long-term treatment with an antibiotic containing pivalic acid. Pediatrics. 2007;120(3):e739–41. https://doi.org/10.1542/peds.2007-0339.
17. Kanda T, Horimukai K, Inoue T, et al. A case of hypocarnitinemia suspected to be associated with intermittent antibiotic administration. J Pediatr Pract. 2008;71(6):1053–6. (Article in Japanese).
18. Yamada K, Kobayashi H, Endo M, et al. A case of 2-year-old boy with the secondary carnitine deficiency after taking cephem antibiotics with pivoxil moiety of 11 days. J Jpn Pediatr Soc. 2008;112(9):1382–5. (Article in Japanese).
19. Nakatani K, Nakachi T, Hidaka F. A case of hypoglycemia with hypocarnitinemia due to continuous use of antibiotics containing pivoxil group. J Miyazaki Med Assoc. 2008;32(1):46–9. (Article in Japanese).
20. Hanayama R, Okida, Tsuboi A, et al. A case of suspected carnitine deficiency due to oral antibiotics for status epilepticus. Pediatr Jpn. 2010;51(8):1087–90. (Article in Japanese).

21. Nakajima Y, Ito T, Maeda Y, et al. Detection of pivaloylcarnitine in pediatric patients with hypocarnitinemia after long-term administration of pivalate-containing antibiotics. Tohoku J Exp Med. 2010;221(4):309–13. https://doi.org/10.1620/tjem.221.309.
22. Okumura A, Morita M, Ikeno M, et al. Acute encephalopathy in a child with secondary carnitine deficiency due to pivalate-conjugated antibiotics. Pediatr Infect Dis J. 2011;30(1):92. https://doi.org/10.1097/INF.0b013e3181fe3953.
23. Kajimoto M, Matsushige T, Yamada K, et al. A case of acute encephalopathy in which long-term administration of antimicrobials with a pivoxil group was associated with hypoglycemia. Brain Dev. 2013;45:150–2. (Article in Japanese).
24. Nishiyama M, Tanaka T, Fujita K, et al. A case of acute encephalopathy with secondary carnitine deficiency after taking pivalate-conjugated antibiotics for 3 days. J Japan Pediatr Soc. 2014;118(5):812–8. (Article in Japanese).
25. Nakazaki K, Ogawa E, Ishige M, et al. Hypocarnitinemia observed in an infant treated with short-term administration of antibiotic containing pivalic acid. Tohoku J Exp Med. 2018;244(4):279–82. https://doi.org/10.1620/tjem.244.279.
26. Minamikawa-Shimizu Y, Sassa K, Abe Y, et al. A case of acute encephalopathy caused by secondary carnitine deficiency due to Cefteram Pivoxil. J Jpn Soc Pediatr Radiol. 2019;35(1):56–60.
27. Tanikawa M, Honda T, Kaiyama H, et al. A case of hypoglycemia caused by cefcapene pivoxil hydrochloride in the elderly patient. Jpn J Med Technol. 2013;62(3):290–2. (Article in Japanese).
28. Hanai S, Iwata M, Terasawa T. Relapsing hypoglycemia associated with hypocarnitinemia following treatment with cefcapene pivoxil in an elderly man. Intern Med. 2019;58(19):2891–4. https://doi.org/10.2169/internalmedicine.2339-18.
29. Ito M, Fukuda M, Suzuki Y, et al. Carnitine-related hypoglycemia caused by 3 days of pivalate antibiotic therapy in a patient with severe muscular dystrophy: a case report. BMC Pediatr. 2017;17(1):73. https://doi.org/10.1186/s12887-017-0835-7.
30. Ito M, Kakinuma H, Takahashi H. A case of a boy who was thought to have hypoglycemia, convulsions, and loss of consciousness due to long-term antibiotic administration. J Jpn Pediatr Soc. 2003;107(3):566–7. (Meeting abstract, Article in Japanese).
31. Noguchi M, Toda S, Yano A, et al. A case with convulsions and impaired consciousness due to hypoglycemia and hyperammonemia relation to pivoxil-antibiotics. J Japan Pediatr Soc. 2009;113(6):1000. (Meeting abstract, Article in Japanese).
32. Kasai E, Komatsu N, Kamiya N, et al. A case of hypoglycemia with secondary hypocarnitinemia after administration of a pivoxil-containing antibiotic for 6 days. J Japan Pediatr Soc. 2010;114:319. (Meeting abstract, Article in Japanese)
33. Hamahira H. A case of secondary carnitine deficiency due to long-term oral administration of tebipenem pivoxil. J. Jpn. Soc. Emerg Pediatr. 2011;10:264 (PO-072). (Meeting abstract, Article in Japanese).
34. Uozumi K, Seino Y, Takeshita Y, et al. A case of repeated hypoglycemic attacks due to long-term administration of antibiotics containing a pivoxil group. J Jpn Soc Emerg Pediatr. 2011;10:265(PO-073). (Meeting abstract, Article in Japanese).
35. Murayama T, Sato H, Ogawa Y, et al. A case in which non-ketotic hypoglycemia, convulsions, and impaired consciousness due to long-term oral administration of cefditoren pivoxil. J Japan Pediatr Soc. 2012;116:887. (Meeting abstract, Article in Japanese).
36. Inoue M, Ito T, Kawamura M, et al. A 1-year-old girl who was considered to have hypocarnitinemia and hypoglycemia after administration of a cephem antibiotic containing pivoxil group. Akita Med J. 2012;62(2):138–9. (Meeting abstract, Article in Japanese).
37. Minamikawa Y, Abe Y, Morita E, et al. A case of acute encephalopathy suspected to be involved in hypocarnitinemia caused by antibiotics with pivoxil group. The 61st Annual Meeting of The Japanese Society of Child Neurology, Kanto Regional Meeting: 34 (No. 21), 2014 (Meeting abstract, Article in Japanese).
38. Kajiwara W, Tsuchiya S, Sasaki M, et al. Secondary carnitine deficiency due to pivalate-conjugated antibiotics diagnosed with tandem mass spectrometry: a case report. J Jpn Pediatr Soc. 2017;121(11):1862–6. (Article in Japanese).
39. Endo T, Arao M, Naomi Hotta, et al. A case of neonate with secondary carnitine deficiency due to maternal oral administration of pivoxil-containing antibiotics. J Japan Soc Perinat Neonatal Med. 2019;55(2):503 (O-313). (Meeting abstract, Article in Japanese).
40. Kobayashi H, Fukuda S, Yamada K, et al. Clinical features of carnitine deficiency secondary to pivalate-conjugated antibiotic therapy. J Pediatr. 2016;173:183–7. https://doi.org/10.1016/j.jpeds.2016.02.080.
41. Maeda M, Nakajima Y, Yamaguchi T, et al. A method for identifying pivaloylcarnitine by enzymatic reaction for secondary test of patients with high C5 acylcarnitine levels in tandem mass screening. J Jpn Soc Mass Screen. 2020;30(1):35–41. (Article in Japanese).
42. Ito S, Yoshikawa N, Itabashi K, et al. Information calling for attention to secondary carnitine deficiency due to administration of antibiotics containing pivoxil group. J Jpn Pediatr Soc. 2012;116(4):804–6. (Article in Japanese).
43. Pharmaceuticals and Medical Devices Agency (PMDA): PMDA request for proper use of medicines. No. 8 April 2012. https://www.pmda.go.jp/files/000143929.pdf (Article in Japanese).
44. Pharmaceuticals and Medical Devices Agency (PMDA): revision of "Precautions for Use" of antimicrobials with a Pivoxil Group. April 24, 2014. https://www.pmda.go.jp/files/000145832.pdf (Article in Japanese).

45. Lokrantz CM, Eriksson B, Rosén I, et al. Hyperammonemic encephalopathy induced by a combination of valproate and pivmecillinam. Acta Neurol Scand. 2004;109(4):297–301. https://doi.org/10.1046/j.1600-0404.2003.00227.x.
46. Ito T, Nakajima Y. Drug-induced L-carnitine deficiency and treatment. Jpn J Surg Metabol Nutr. 2020;54(2):57–61. (Article in Japanese).
47. Fujii Y, Chiba S, Numazaki K, et al. Effects of the oral cephem agent cefditoren pivoxil on pediatric carnitine metabolism. Jpn J Antibiot. 1993;46(10):926–37. (Article in Japanese).
48. Sunakawa K, Ouchi K, Suzuki K, et al. An open clinical study of high-dose cefditoren pivoxil in children with bacterial pneumonia, acute otitis media, or acute rhinosinusitis. Jpn J Chemother. 2012;60(4):478–91. (Article in Japanese).
49. Kuroki H, Sakata H, Sato Y, et al. The efficacy and safety of cefteram pivoxil (CFTM-PI) fine granules at the maximum approved dosage in pediatric with respiratory tract infections. Jpn J Chemother. 2008;56(4):453–61. (Article in Japanese).
50. Sunakawa K, Yamanaka N, Iwata S, et al. An open clinical study of tebipenem pivoxil in children with acute otitis media and acute upper respiratory tract infection (phase II). Jpn J Chemother. 2009;57(S-1):S115–24. (Article in Japanese).
51. Hori S, Sunakawa K. Safety profile of tebipenem pivoxil, a new oral carbapenem, in pediatric patients. Jpn J Chemother. 2009;57(S-1):S192–204. (Article in Japanese).
52. Kataoka H, Kasahara K, Sasagawa Y, et al. Evaluation of safety and efficacy of tebipenem pivoxil granules for pediatric in pneumonia, otitis media and sinusitis. Jpn J Antibiot. 2016;69(1):53–76. (Article in Japanese).
53. Pharmaceutical Affairs Committee of the Japan Pediatric Society. Information calling for attention to secondary carnitine deficiency due to administration of antibiotics containing pivoxyl group. J Jpn Pediatr Soc. 2019;123:1363–4. https://www.jpeds.or.jp/uploads/files/20190820pivoxil_chuikanki.pdf (Article in Japanese).
54. Japan Otological Society, Japan Pediatric Otorhinolaryngological Society, Japan Society of Otorhinolaryngological Infectious Diseases and Aerosols, editors. Pediatric acute otitis media clinical practice guidelines, 2013 Edition. Tokyo: Kanehara Publishing; 2013. (Book in Japanese).
55. Kudo F. Management of upper respiratory tract infection in children. clinical practice guidelines for the diagnosis and management of acute otitis media (AOM) in children and acute rhinosinusitis. Pediatr Otorhinolaryngol. 2013;34(3):306–11. (Article in Japanese).
56. Japan Otological Society, Japan Society for Pediatric ORL, Japan Society for Infection and Aerosol in otorhinolaryngology, editors. Clinical practice guidelines for the diagnosis and management of acute otitis media (AOM) In children in Japan –2018. Tokyo: Kanehara Publishing; 2018. (Book in Japanese).
57. Japan Rhinologic Society, editor. Clinical practice guidelines for acute sinusitis. 2010 Edition (Supplementary Edition). Jpn J Rhinol. 2014;53:103–60. (Article in Japanese).
58. Hotomi M, Yanagihara K, Ishiwada N, et al., editors. (The Committee for the Appropriate Use of Antibiotics in Acute Respiratory Tract Infection, The Japanese Association for Infectious Diseases): Recommendation for the appropriate use of antibiotics in acute respiratory tract infection. Jpn J Infect Dis. 2019;93(5):623–42. (Article in Japanese).
59. Tanizaki R, Miyamatsu Y. Nocturnal leg cramping caused by carnitine deficiency due to long-term pivalate antibiotics administration in patient with chronic kidney disease. Cureus. 2023;15(11):e48927. https://doi.org/10.7759/cureus.48927.
60. Yamada K, Kobayashi H, Bo R, et al. Elevation of pivaloylcarnitine by sivelestat sodium in two children. Mol Genet Metab. 2015;116(3):192–4. https://doi.org/10.1016/j.ymgme.2015.09.009.
61. Akiyama S, Fujinaga Y, Kaneshige M, et al. A case of neonatal hypocarnitinemia caused by ciberestat. J Japan Soc Perinat Neonatal Med. 2013;49(2):695. (Meeting abstract, article in Japanese).

14 Carnitine Deficiency in Patients with Renal Failure

14.1 Introduction

The kidney, along with the liver, is a major organ for the biosynthesis of carnitine, and is also greatly involved in the homeostasis of carnitine as a site for the reabsorption of free carnitine. In particular, it has been reported that when patients undergo hemodialysis (HD) or peritoneal dialysis (PD) due to end-stage renal failure, carnitine deficiency becomes apparent due to carnitine leakage from the dialysis membrane or peritoneum, and various clinical symptoms such as erythropoietin resistance (or hyporesponsive) anemia, muscular symptoms such as cramps and extreme fatigue, cardiac symptoms such as left ventricular dysfunction and arrhythmia, and hypotension during dialysis appear. Carnitine supplementation therapy has been reported in these hemodialysis patients [1–6]. In this chapter, I would like to introduce those reports, as well as understandings and controversies to date on carnitine deficiency in hemodialysis patients in Japan and overseas, to help clinicians who decide future treatment policies.

Note Erythropoietin (EPO) is technically described as human recombinant erythropoietin (rHuEPO), but with the advent of darbepoetin, which replaces some of the amino acids of EPO and loads it with sialic acid to extend its half-life, it has recently been used as a hematopoietic factor (erythropoiesis-stimulating agents: ESA). In this document, do not dare to unify them, and use "EPO" or "ESA" terms as necessary. In addition, terms such as "EPO hyporesponsive anemia" or "EPO-refractory anemia" are also used in the same sense as "EPO-resistant anemia," but in this book it is referred to as "EPO resistance" or "ESA resistance."

14.2 Carnitine Kinetics in Patients with Conservative Kidney Management

Currently, carnitine supplementation therapy is used in hemodialysis patients with end-stage renal disease and peritoneal dialysis patients, but the carnitine kinetics in patients with chronic

N. Nakayama, *Carnitine Deficiency and Insufficiency in Medicine*,
https://doi.org/10.1007/978-981-95-7797-2_14

renal failure who have not yet undergone dialysis (hereinafter referred to as patients with conservative kidney management), which is the previous stage, are first described.

In 1974, Böhmer et al. examined the blood carnitine concentration in patients with various diseases such as cirrhosis, right heart failure, diabetes, and cardiomyopathy, and reported that it was higher than the normal range in patients with severe renal insufficiency [7]. In dialysis patients, the blood concentration was slightly lower than normal before dialysis and well below the lower limit of normal after dialysis. Subsequently, Chen et al. examined the blood carnitine concentrations of patients with myocardial infarction, coronary artery disease, renal failure, cirrhosis, and malignant tumors, and reported that many patients with renal failure showed higher levels than the normal range (renal failure: 17 out of 26 patients, liver cirrhosis: 8 out of 11 cases), and that blood carnitine concentrations correlated with blood urea nitrogen (BUN) in patients with renal insufficiency [8].

Bartel et al. also examined serum carnitine concentrations in patients with severe chronic renal failure in dialysis patients (hemodialysis patients and peritoneal dialysis patients) as well as non-dialysis patients [9]. The serum carnitine concentrations of non-dialyzed chronic renal failure patients were 117.7 ± 6.7 μmol/L for males and 84.4 ± 10.0 μmol/L for females, compared to healthy control of 53.9 ± 12.2 μmol/L for males and 45.5 ± 9.9.9 μmol/L for females, which were 218% for males and 186% for females. Serum carnitine concentrations in these patients with non-dialyzed chronic renal failure were positively correlated with serum creatinine levels ($r = 0.734$, $P < 0.001$, Fig. 14.1). They concluded that just as renal excretion of waste products such as creatinine decreased with the decline of renal function, renal excretion of carnitine also decreased, and blood levels increased.

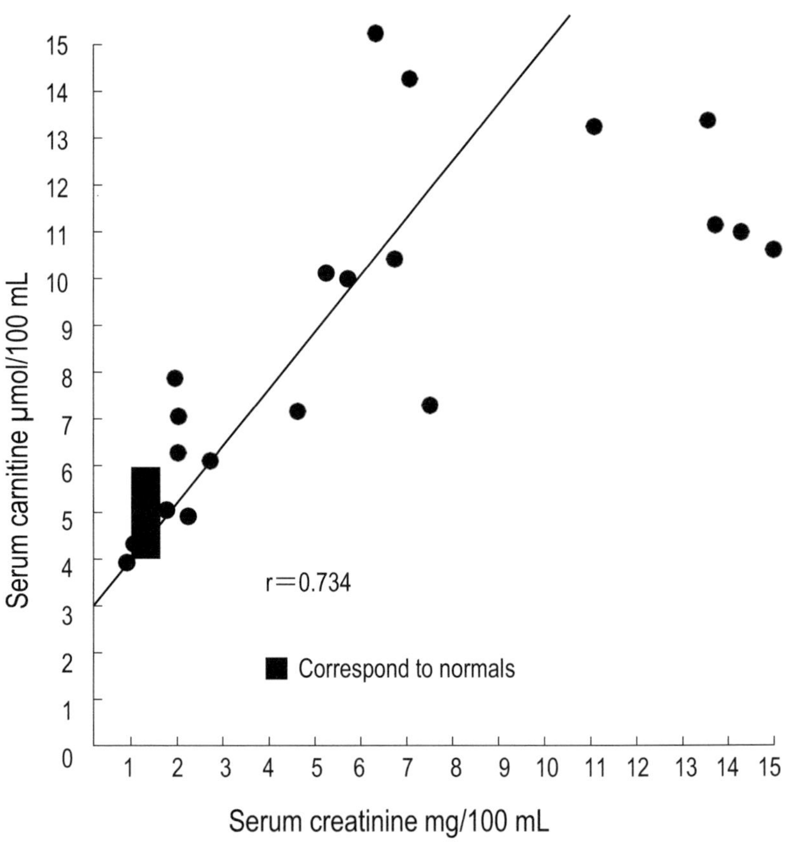

Fig. 14.1 Correlation between serum carnitine and serum creatinine in patients with non-dialyzed chronic renal disease. The black rectangle indicates normal range. The correlation coefficient: $r = 0.731$. (From Ref. [9])

In this study by Bartel et al., serum carnitine concentrations in dialysis patients were 72.8 ± 17.4 μmol/L in men and 72.6 ± 11.6 μmol/L in women, which were slightly higher than that of the control group, probably because the dialysis history was still short.

On the other hand, Rodriguez-Segade et al. compared the plasma total carnitine concentration, free carnitine concentration, and acylcarnitine concentration in patients with conservative period of renal failure (non-dialyzed, 34 patients) and in patients of dialyzed renal failure (54 patients) with the healthy control group (49 patients) (Table 14.1) [10].

In addition, when examined the plasma carnitine distribution of patients with conservative kidney management, many patients had high total carnitine levels and free carnitine levels, but a large number of patients had high acylcarnitine levels (Fig. 14.2). It is necessary to further investigate whether the pathology of carnitine insufficiency due to acyl group accumulation in dialysis patients may also occur in patients with conservative renal insufficiency in patients with acyl group accumulation in the conservative period.

In this study by Rodriguez-Segada et al., plasma total carnitine concentrations and free carnitine concentrations were weakly correlated with plasma urea and uric acid concentrations, but not with plasma creatinine. They discussed that free carnitine and acylcarnitine in patients with conservative kidney management may be different from the behavior of waste products such as urea, uric acid, and creatinine.

Bertoli et al. reported that the concentration of carnitine in muscles was 1.5 times higher in non-dialyzed end-stage renal failure patients than in the control group and lower in dialysis patients (14 patients, 6–61 months of dialysis history) than in the control group (14 patients, 6–61 months of dialysis history) (Fig. 14.3) [11].

In a recent study, Yano et al. examined the association between total carnitine, free carnitine, and acylcarnitine levels in the blood and urine in 75 patients with chronic kidney disease (CKD) not on dialysis (mean age 56.1 ± 20.0 years) and renal function [12]. As a result, the total serum carnitine and free carnitine concentrations were independent of the degree of progression (stage) of renal failure, and the urinary free carnitine concentration was an independent correlator of serum free carnitine, while serum acylcarnitine and acylcarnitine concentration/free carnitine concentration (AC/FC) of serum increased with the progression of renal failure, especially at the CKD5 stage, and BUN was an independent determinant of serum acylcarnitine (Fig. 14.4).

Thus, other research groups have reported that blood carnitine concentrations and muscle carnitine concentrations are higher than those in the control group in patients with conservative kidney management who do not receive dialysis therapy, and blood carnitine concentrations correlate with serum creatinine and urea nitrogen [13, 14]. In addition, elevated acylcarnitine levels in patients with conservative kidney management have been reported in many groups, suggesting that the accumulation of acyl groups with AC/FC greater than 0.4 causes carnitine insufficiency. However, there are few studies on acylcarnitine in patients with conservative kidney management. In addition, it is necessary to study in the future whether there are tissue carnitine deficiency and carnitine insufficiency, or whether high levels of AC/FC in patients with conservative renal failure in the

Table 14.1 Plasma carnitine concentration in patients with conserved period of renal failure, dialyzed patients with renal failure and healthy control group

	Plasma carnitine concentration(μmol/L)		
Carnitine fractions	Patients with conserved period of renal failure	Dialyzed patients with renal failure	Healthy control group
Total carnitine	92.5 ± 37.5	62.7 ± 22.3	53.3 ± 8.4
Free carnitine	63.8 ± 29.0	39.3 ± 16.5	47.1 ± 7.7
Acylcarnitine	28.7 ± 16.5	23.4 ± 10.9	6.2 ± 2.4

Modified from Ref. [10]

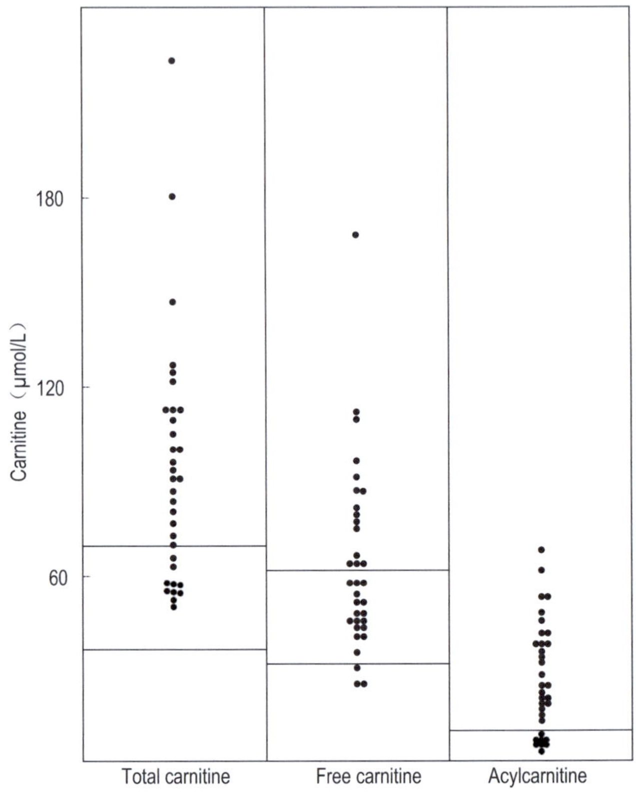

Fig. 14.2 Concentrations of plasma total carnitine, free carnitine, and acylcarnitine in patients with non-dialyzed chronic renal disease. Each point indicates the data of each patient. The dotted lines indicate the normal range of each carnitine fraction. (From Ref. [10])

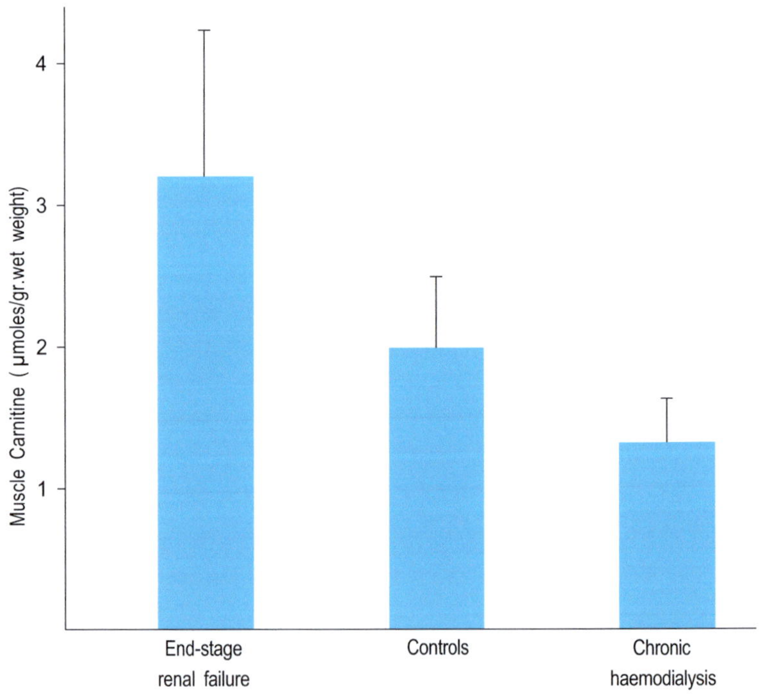

Fig. 14.3 Muscle free carnitine level in uremic patients, normal controls, and patients on maintenance hemodialysis. (From Ref. [11])

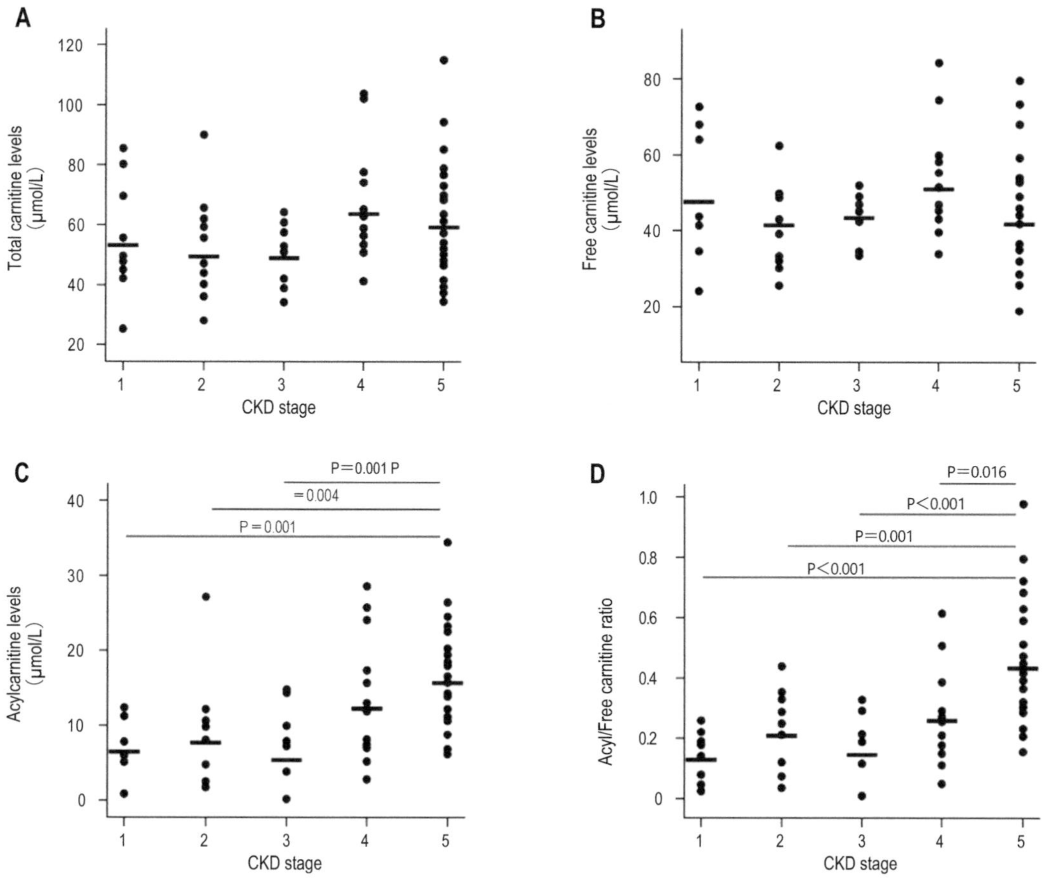

Fig. 14.4 Serum total carnitine (**a**), free carnitine (**b**), acylcarnitine (**c**), and acyl/free carnitine ratio (**d**) levels according to CKD stage in patients with CKD who were not on dialysis. eGFR: estimated glomerular filtration ratio, CKD: chronic kidney disease. (From Ref. [12])

elderly, which has been increasing in Japan in recent years. Because, this carnitine insufficiency, a relative lack of free carnitine and accumulation of acylcarnitine, might be involved in the progression of tubular disorders in the renal tubules, which are rich in mitochondria and play an important role in energy metabolism. Carnitine has been found to have new actions such as antioxidant, anti-inflammatory, and anti-apoptotic effects, and it has been reported that carnitine administration suppresses tubular disorders in animal models and clinical studies of hypermyoglobinemia, contrast agent nephropathy, and diabetic nephropathy [15–21].

Whether or not carnitine supplementation therapy improves carnitine deficiency and carnitine insufficiency in patients with pre-dialysis conservative renal failure and prevents the progression of diabetic nephropathy or contrast nephropathy has not yet been fully investigated in clinical practice, and it is an issue for the future.

14.3 Carnitine Kinetics in Dialysis Patients

14.3.1 Early Study on Carnitine Kinetics in Dialysis Patients

As described in the section on patients with conservative renal failure, Böhmer et al. found that blood carnitine levels were markedly decreased after dialysis in dialysis patients [7]. Subsequently, Böhmer et al. reported a more detailed study of carnitine concentrations in blood and muscle of nine dialysis patients [22].

The results showed that the concentration of carnitine in muscle of hemodialysis patients was significantly lower than that of the control group ($P < 0.05$), although there was a large variation among individual patients, and the blood carnitine concentration decreased significantly in 8 out of 9 patients due to dialysis (Fig. 14.5). The amount of carnitine filtered into the dialysate was found to be proportional to the plasma carnitine concentration by 10–15% of the plasma concentration. This finding raises the possibility that carnitine deficiency may be the cause of cardiomyopathy and heart failure often observed in patients on dialysis for long periods of time.

Battistella et al. also examined changes in blood carnitine concentrations over time from dialysis therapy to the end of the next dialysis treatment in eight dialysis patients (mean dialysis history 4.9 years, Fig. 14.6) [23]. The blood carnitine concentration before the start of dialysis was within the normal range in males, but in some females, it was lower than the normal range. In both sexes, blood carnitine levels decreased with dialysis treatment, recovered rapidly at first and then slowly with the end of dialysis, and then decreased again with the next dialysis treatment. Battistella et al. suggested that these dialysis patients with severely depleted carnitine levels may have a carnitine deficiency that could lead to irreversible cardiomyopathy. Battistella et al. suggested that carnitine supplementation therapy may be necessary to prevent irreversible cardiomyopathy in these hemodialysis patients with severely depleted carnitine levels [23].

Several studies have investigated how much carnitine leaks into the dialysate solution during dialysis therapy (Table 14.2) [24], which compared plasma carnitine concentrations, muscle carnitine concentrations, and carnitine levels in dialysate in hemodialysis, peritoneal dialysis, and continuous ambulatory peritoneal dialysis (CAPD). According to the study, hemodialysis patients had significantly lower plasma total carnitine concentrations ($P < 0.001$) and post-dialysis total carnitine concentrations in the muscles of the vastus lateralis were also reduced to about 30% of normal controls, while peritoneal dialysis did not significantly reduce post-dialysis plasma total carnitine concentrations, and no significant decrease in total carnitine concentration in muscle was also observed in CAPD patients. The total amount of carnitine lost in dialysate in hemodialysis three times a week is 1518 μmol, which is about 245 mg. On the other hand, peritoneal dialysis and CAPD lose about 190 mg of carnitine per week and CAPD lose about 174 mg of carnitine per week.

Not surprisingly, Bizzi et al. reported that oral administration of carnitine after dialysis (DL-carnitine, 3 g) or addition of L-carnitine to dialysate (65 μmol/L) can prevent the decline of blood carnitine after dialysis [25, 26].

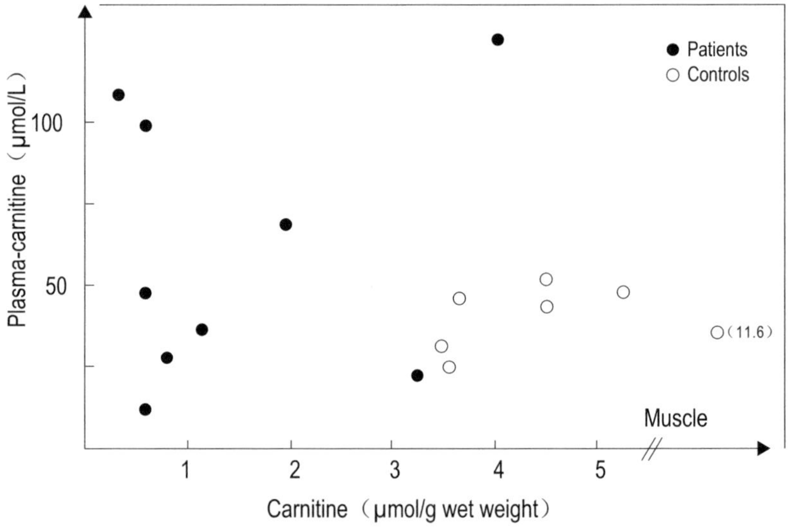

Fig. 14.5 Relation between concentrations of carnitine in plasma and muscle. Mean concentration of carnitine in muscle was significantly less in hemodialysis patients than controls ($P < 0.05$, Wilcoxon two-sample test). (From Ref. [22])

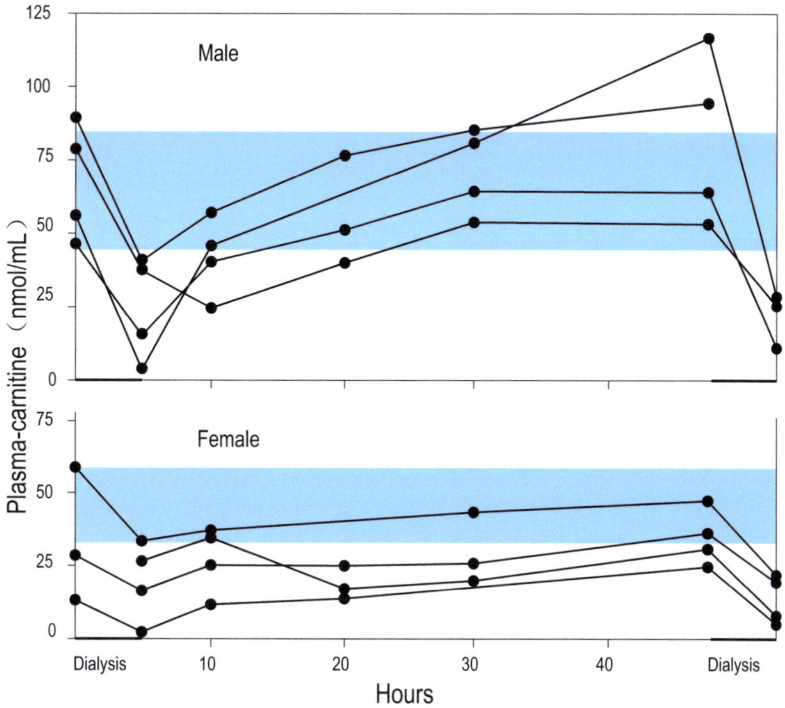

Fig. 14.6 Changes in plasma carnitine concentration from dialysis to next dialysis. Upper panel is male patients ($n = 4$) and lower panel is female patients ($n = 4$). The blue area indicates normal range. (From Ref. [23])

Table 14.2 Plasma and muscle carnitine concentration and dialysate loss in hemodialysis, peritoneal dialysis, and CAPD patients

Treatment	Number of subjects	Sample	Specimen collection time	Carnitine concentration[a]	Recovered carnitine concentration in dialysate[a]	Carnitine loss per week (μmol)
Hemodialysis (3 times/week)	6	Plasma	Before dialysis	46.2 ± 4.5	506 ± 91	1518 ± 273
			After dialysis	18.8 ± 2.7[b]		
	4	Muscle	After dialysis	1561.9 ± 124.4		
Peritoneal dialysis (3 times/week)	10	Plasma	Before dialysis	70.5 ± 12.2	391.5 ± 75.9	1174.5 ± 237.7
			After dialysis	67.9 ± 12.1[c]		
CAPD	12	Plasma	During dialysis	50.0 ± 5.2	154.3 ± 14.9	1078 ± 14.3
	4	Muscle	During dialysis	4983.2 ± 399.8		
Normal control	25	Plasma	–	47.4 ± 10.3	–	–
	25	Muscle	–	5230.5 ± 142.7		

From Ref. [24]

[a]Mean ± SEM, total carnitine concentration in plasma and dialysate: μmol/L, the amount of total carnitine in the muscle: nmol/g wet muscle weight

[b]$P < 0.001$

[c]*n.s.* not specified

CAPD continuous ambulatory peritoneal dialysis

Hemodialysis therapy was established in Europe and the United States in the 1960s and 1970s, and research on carnitine deficiency and carnitine supplementation therapy in dialysis patients began at a very early stage.

14.3.2 Carnitine Kinetics and Carnitine in Muscles in Dialysis Patients

Carnitine is a low molecular weight compound and has no protein binding, so it is 100% filtered out by the glomerulus. After that, more than 99% of carnitine is reabsorbed in the renal tubules in normal subjects. In hemodialysis patients, however, carnitine is not reabsorbed in the renal tubules, which is important for carnitine homeostasis, and is efficiently excreted through the dialysis membrane into the dialysate solution. As described above, there is no doubt that a single dialysis treatment can decrease carnitine concentrations in the blood, but it is important to consider in a long-term point of view that how long dialysis treatment continues, and carnitine deficiency becomes apparent.

Evans et al. examined in detail dialysis history and plasma and muscle carnitine concentrations [27]. First, plasma free carnitine, acetylcarnitine, and total carnitine concentrations were examined in 21 patients with end-stage renal failure newly started on dialysis to determine changes in plasma and muscle carnitine levels during the first 12 months of dialysis therapy, and in 8 of these patients, carnitine levels in the quadriceps muscle were also examined at 0, 6, and 12 months after the start of dialysis. In addition, blood carnitine concentrations and intramuscular carnitine concentrations of quadriceps muscle samples were also examined in all 20 patients who had been on dialysis for more than 12 months (average of 5.10 years on dialysis). Both plasma carnitine concentrations were from blood samples taken before dialysis on the day of dialysis. The relationship between plasma and muscle carnitine concentrations and dialysis history in these 41 hemodialysis patients was compared with data from 60 healthy adult controls [27]. Two of them were eliminated from the study due to kidney transplantation and one was due to death. The results showed that plasma carnitine concentrations were significantly higher than those in healthy adults before the start of dialysis therapy but declined significantly at 6 and 12 months after the start of dialysis therapy and were less than half those in healthy adults after 12 months or more of dialysis. Acetyl-L-carnitine levels, which were nearly three times higher than those of healthy adults before the start of treatment, decreased as dialysis therapy progressed, and after 12 months of dialysis, the levels were about half of those before the start of treatment (Table 14.3).

Total carnitine concentrations were 1.5 times higher than those of healthy adults before the start of treatment, but in the 21 patients who were examined for changes over time after the start of dialysis, carnitine concentrations decreased as dialysis therapy progressed and were slightly lower than those of healthy adults at 12 months of dialysis (Table 14.3). Blood carnitine concentrations in patients with end-stage renal failure who had been on dialysis for more than 12 months (mean 5.1 years) were lower in all carnitine fractions than blood carnitine concentrations at 12 months in newly initiated dialysis patients, but

Table 14.3 Change in plasma carnitine concentration in patients with end-stage renal disease on dialysis therapy (μmol/L)

		Duration of dialysis in patients with end-stage renal disease			
Analytical items	Healthy adult group	Before treatment	6 months	12 months	> 12 months
L-carnitine	43.3 ± 8.6[b]	50.6 ± 20.0[a]	34.6 ± 12.9[a,b]	29.7 ± 10.5[a,b]	22.0 ± 5.4[a,b]
Acetyl-L-carnitine	6.65 ± 2.11[b]	19.8 ± 8.9[a]	12.4 ± 4.7[a,b]	10.0 ± 2.7[a,b]	9.23 ± 3.47[a,b]
Total carnitine	49.2 ± 9.3[b]	75.3 ± 21.9[a]	48.9 ± 18.5[b]	44.7 ± 11.1[b]	42.6 ± 11.9[b]

Modified from Ref. [27]
[a]$P < 0.05$ vs. healthy adult group
[b]$P < 0.05$ vs. before treatment

the decrease was not significant. In other words, the decline in blood carnitine fraction due to dialysis therapy is more pronounced within the first 12 months of dialysis therapy (Fig. 14.7). As expected, plasma carnitine concentrations were below baseline levels early in the course of dialysis therapy and were below baseline levels after 1 year in most patients. The muscle carnitine concentrations in patients on dialysis therapy for more than 12 months were approximately 38% lower than those in patients not yet on dialysis therapy (Table 14.4) [27].

There have been many studies on muscle carnitine concentrations in hemodialysis patients, but they are not generally comparable because of differences in muscle sample collection sites, measurement methods, and data presentation (per g wet weight, per g dry weight, per g non-collagen protein, etc.) [24, 28–36]. However, all studies have shown a decrease in blood carnitine concentration with increasing duration of dialysis, followed by a decrease in muscle carnitine, as in the study by Evans et al. For example, Savica et al. reported that plasma carnitine concentration correlated with deltoid muscle carnitine concentration in 14 dialysis patients (Fig. 14.8), and that carnitine concentration in deltoid muscle tended to correlate inversely with dialysis duration, although the difference was not significant due to variability [35].

Hiatt et al. also measured plasma carnitine concentrations and muscle carnitine concentrations in the vastus lateralis muscle of eight hemodialysis patients (0.5–16 years of dialysis history, mean 7.86 years) and reported that total carnitine concentration in the muscle was inversely related to dialysis history (Fig. 14.9) [31].

Thus, carnitine in the blood is removed considerably after a single dialysis treatment, but carnitine levels in the blood recover at the end of dialysis. However, this recovery flows from the carnitine pool in muscle and other tissues, and the longer dialysis therapy continues, the more the carnitine content in muscle and other tissues gradually declines.

Fornasini et al. studied in detail the changes in blood L-carnitine concentrations in patients undergoing dialysis therapy by simulating the decrease in muscle carnitine concentrations using a three-compartment model [37]. Fornasini compared the simulation results with pharmacokinetic data obtained by Evans et al. in same group of plasma carnitine concentrations in 12 hemodialysis patients who underwent dialysis therapy (180–270 min, mean 230 min) three times a week, followed by intravenous carnitine administration (20 mg/kg) after the first dialysis treatment of the following week [38]. They also compared the simulated changes in carnitine concentration in muscle with the aforementioned data on carnitine concentration in muscle of hemodialysis patients by Evans et al. [27] The carnitine pool was 40 μmol/L in the central blood/extracellular fluid compartment 1, 900 μmol in compartment 2 (mainly liver and kidney), which is in relatively rapid equilibrium with compartment 1, and 128,000 μmol in compartment 3 (muscle), which is in very slow equilibrium with compartment 1. The measured baseline plasma carnitine concentrations (black circles) of the 12 hemodialysis patients were in excellent agreement with the results of the three-compartment model simulation (solid line) (Fig. 14.10).

What then are the changes in carnitine concentrations in the three compartments? In compartment 1, the plasma concentration of carnitine decreased by about 80% after each dialysis treatment, increased rapidly by nearly 25-fold after intravenous carnitine administration (20 mg/kg), and immediately fell rapidly (Figs. 14.10 and 14.11a). In compartment 2 (liver and kidney), the decrease in carnitine concentration due to dialysis treatment was about 30%, with an eightfold increase and decrease in tissue carnitine concentration due to intravenous administration, both of which were blunted compared to plasma changes (Fig. 14.11b). In compartment 3 (muscle), the decrease in tissue carnitine concentration due to dialysis treatment was almost negligible, and intravenous administration of carnitine only increased tissue concentration by about 20% (Fig. 14.11c).

Fornasini et al. further superimposed the three-compartment model simulation on the data from the aforementioned study on the duration of dialysis and muscle carnitine concentration in

Fig. 14.7 Mean ± SEM plasma concentrations of L-carnitine (upper panel), (middle panel), and total L-carnitine (lower panel) in patients with end-stage renal disease (ESRD) during the first 12 months of hemodialysis treatment. Mean ± SEM plasma concentrations of L-carnitine (upper panel), acetyl-L-carnitine (middle panel), and total L-carnitine (lower panel) in patients with end-stage renal disease during the first 12 months of hemodialysis treatment. *: Compared with baseline, $P < 0.05$. (From Ref. [27])

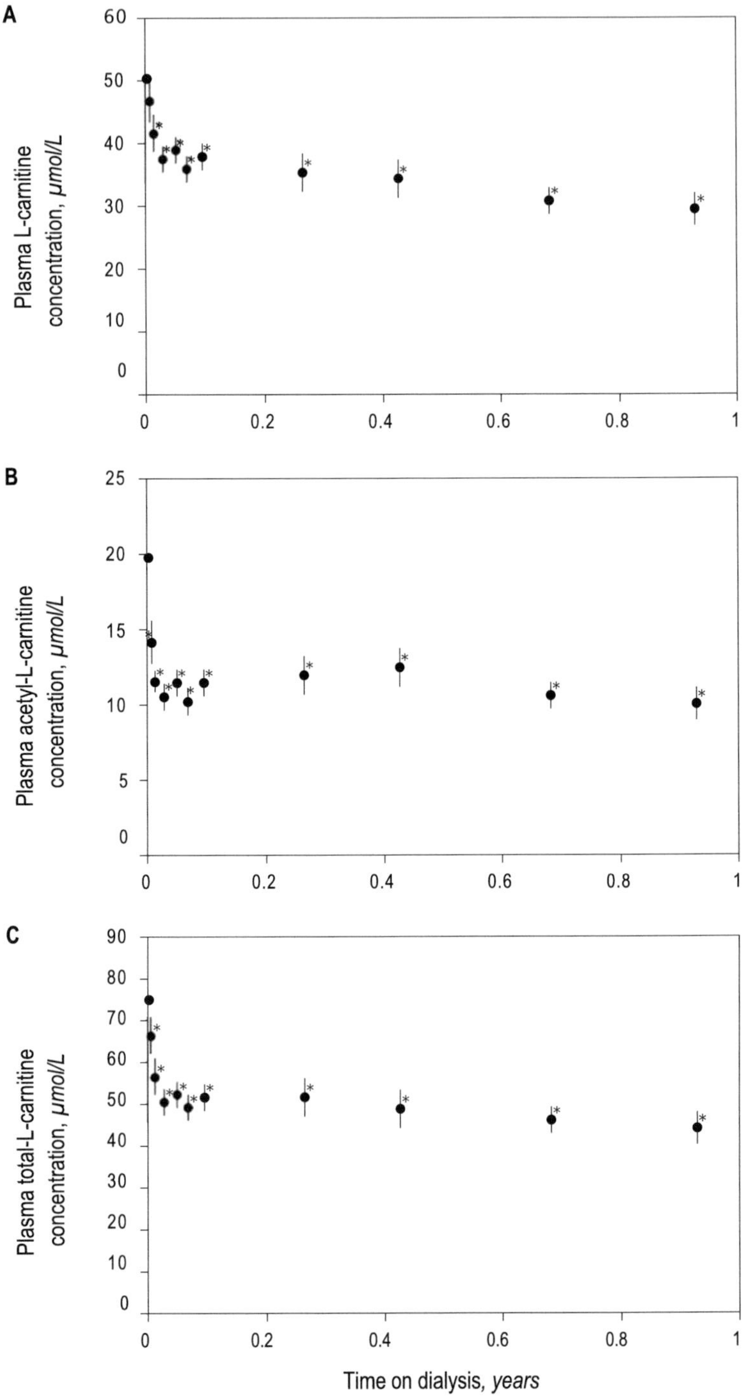

Mean ± SEM plasma concentrations of L-carnitine (upper panel), acetyl-L-carnitine (middle panel), and total L-carnitine (lower panel) in patients with end-stage renal disease (ESRD) during the first 12 months of hemodialysis treatment.
*: Compared with baseline, P<0.05.

Table 14.4 Change in muscular carnitine concentration in patients with end-stage renal disease on dialysis therapy (μmol/g)

	Duration of dialysis in patients with end-stage renal disease			
Analytical items	Before treatment	6 months	12 months	>12 months
L-carnitine	3.01 ± 0.47	2.60 ± 0.57	2.64 ± 0.52	1.87 ± 0.67[a]
Acetyl-L-carnitine	1.52 ± 0.64	1.71 ± 0.76	0.914 ± 0.377[a]	0.727 ± 0.297[a]
Total carnitine	4.54 ± 0.88	4.17 ± 1.01	3.56 ± 0.93	2.61 ± 0.91[a]

From Ref. [27]
[a]$P < 0.05$ vs. before treatment

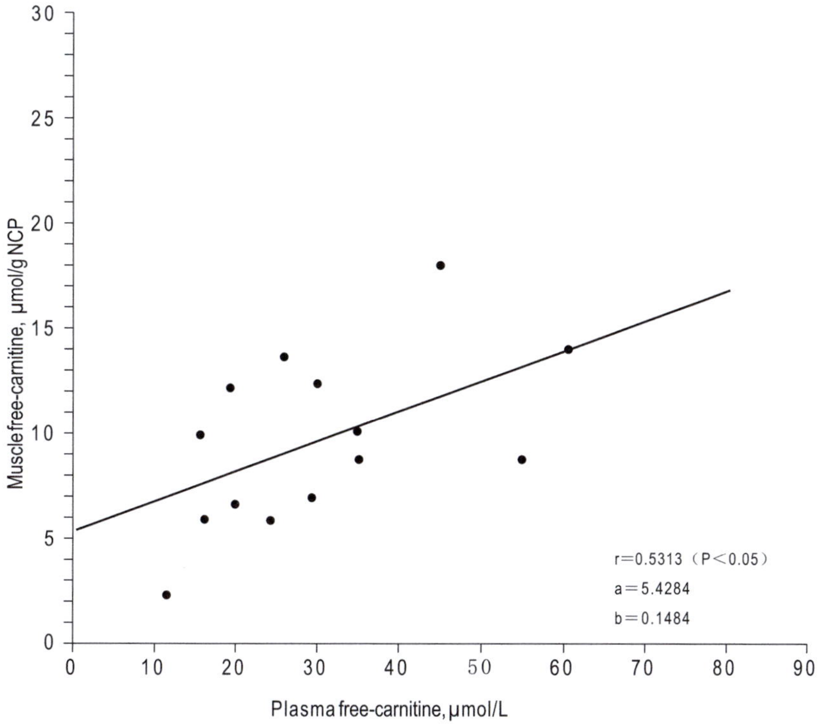

Fig. 14.8 Correlation between plasma and muscle free carnitine in a group of 14 patients on hemodialysis ($P < 0.05$). (From Ref. [35])

Fig. 14.9 Relationship between duration of hemodialysis in patients with chronic renal failure and total muscle carnitine content at rest in the vastus lateralis. Each point represents an individual patient. The duration of hemodialysis was inversely correlated with the muscle carnitine content ($y = -138x + 3409$; $R = -0.74$, $P < 0.05$). (Modified from Ref. [31])

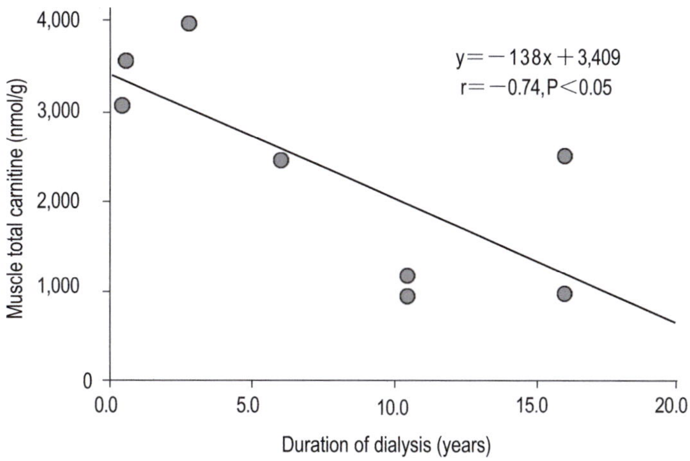

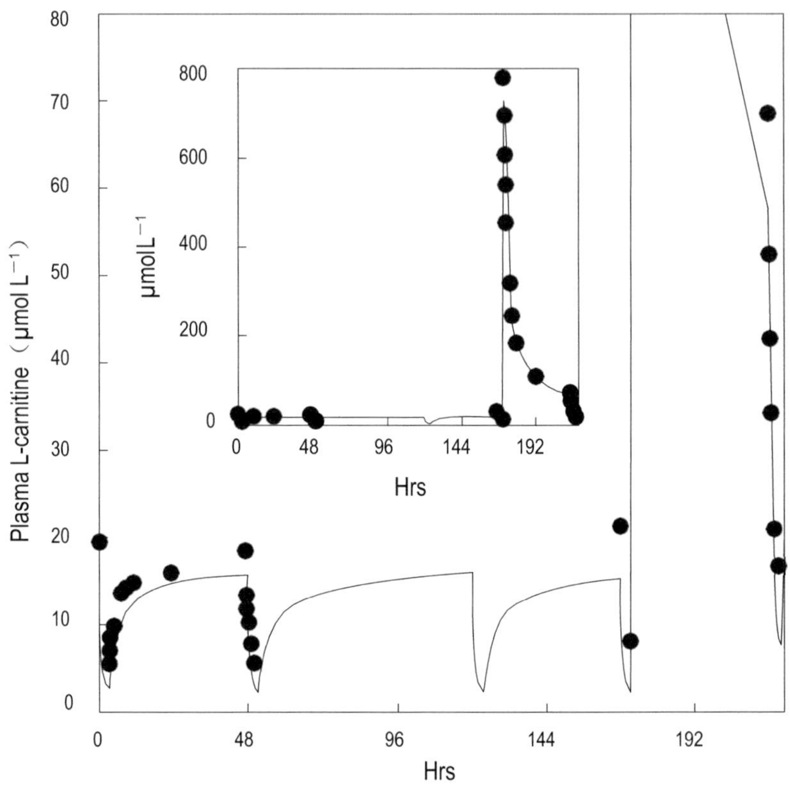

Fig. 14.10 Plasma concentration versus time profiles for L-carnitine in 12 end-stage renal disease patients under baseline conditions and after a single intravenous dose of L-carnitine (20 mg/kg). The symbols (●) represent pooled concentrations from 12 patients, and the continuous line represents the concentrations predicted by the fitted model with three compartment model simulation. The inserted figure indicates 10-folds scale of vertical axis. (Modified from Ref. [37])

dialysis patients [27] (Fig. 14.12). The data on the dialysis history (years) and muscle carnitine concentration in actual dialysis patients (black circles) and the simulated changes in the dialysis history and muscle carnitine concentration (solid lines) were presented, assuming four different endogenous carnitine supplementation levels (2, 4, 6, and 8 μmol/h), depending on the biosynthesis of carnitine and dietary intake. The data are presented.

For example, if 8 μmol/h of carnitine is supplemented by biosynthesis or dietary intake, carnitine levels in muscle would be expected to remain virtually unchanged after 18 years of dialysis and no carnitine deficiency would occur (Fig. 14.12, top solid line). This 8 μmol/h is approximately 30 mg/day. This means that if the blood is supplemented with a net daily supply of 30 mg of carnitine, carnitine deficiency will not occur even if carnitine is removed by dialysis three times a week. As shown in the bottom solid line, endogenous carnitine supplementation at 2 μmol/h would reduce the carnitine concentration in muscle by half in 4–5 years. As indicated by the black circles, the data on muscle carnitine concentration in actual dialysis patients also show that the muscle carnitine concentration in many patients declines with the dialysis history. The study by Fornasini et al. also shows that it is difficult to diagnose carnitine deficiency from blood carnitine concentration alone. Even if carnitine is administered intravenously and the blood concentration is extremely high, the muscle carnitine level may still be low, and even if the blood carnitine level is extremely low after dialysis, there may still be sufficient carnitine in the

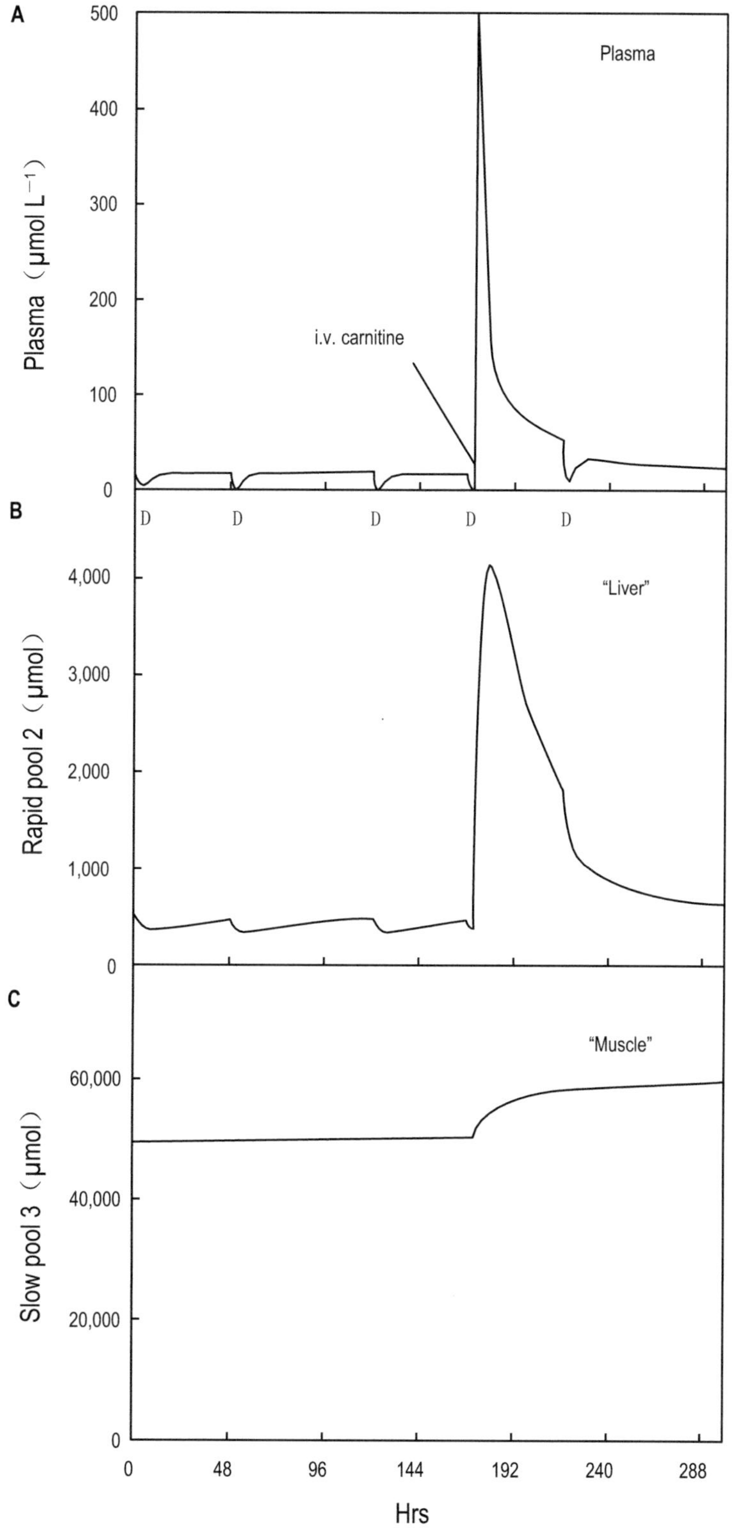

Fig. 14.11 Simulated amounts of L-carnitine in the three compartments (**a**: Plasma, **b**: liver, and **c**: muscle) of the pharmacokinetic model in patient 5 undergoing intermittent hemodialysis three times per week. (From Ref. [37])

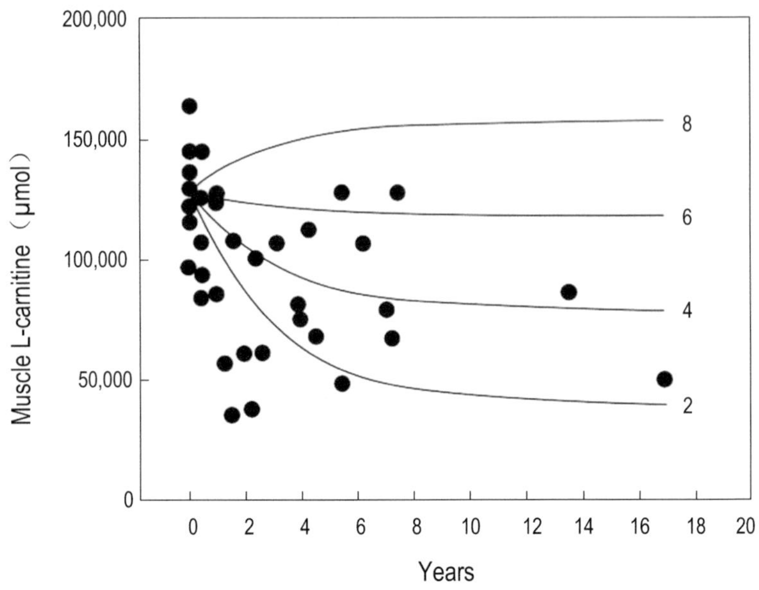

Fig. 14.12 The predicted and observed muscle contents of L-carnitine when an individual with normal carnitine pools is placed on dialysis. The symbols show the muscle carnitine contents estimated from measures of muscle carnitine concentration in dialysis patients before and during continuous dialysis for up to 17 years. The lines show the muscle content predicted using the model with initially normal carnitine pools and with parameter values and dialysis condition set at either 2, 4, 6, or 8 μmol h^{-1}. The extent of depletion of the muscle carnitine depended on the balance between the loss of carnitine via dialysis and the rate of inputs (from diet, endogenous synthesis, and supplement). (From Ref. [37])

muscle. The rate of carnitine outflow from the carnitine pool in muscle is extremely low (0.00014/h), and it takes approximately 200 days for the carnitine concentration in muscle to change by 50%. Therefore, it takes several years for carnitine deficiency in muscle to become obvious in hemodialysis patients. Compared with Westerners, who consume more carnitine-rich meat and dairy products, Japanese renal failure patients often receive dietary advice to reduce protein intake, and carnitine deficiency may develop earlier in many Japanese patients, who eat less meat as they get older.

14.3.3 Kinetics of Acylcarnitine in Dialysis Patients

Until now, it has been described that in dialysis patients, free carnitine is removed from the blood by dialysis treatment, and then carnitine is eluted from storage sites such as muscles, resulting in carnitine deficiency if dialysis therapy continues for a long time. In studies of carnitine deficiency in dialysis patients, it was initially thought that a decrease in free carnitine or total carnitine might be associated with cardiomyopathy and heart failure. Then, how about the changes in acylcarnitine fraction, which is considered as a problem in the concept of "carnitine insufficiency."

In the late 1980s, it was reported that the concentration of acylcarnitine was higher in hemodialysis and peritoneal dialysis patients than in healthy subjects [10, 14, 31, 39, 40].

Sakurabayashi et al. studied the relationship between free carnitine or acylcarnitine concentrations, AC/FC in blood and the duration of dialysis in 99 hemodialysis patients [41]. The plasma free carnitine decreased below the normal range with duration of dialysis, while AC/FC was increased above 0.4 with duration, especially 1–2 years after starting dialysis. The acylcarnitine concentrations were within the normal range or slightly higher in many cases [41].

This AC/FC is an indicator of carnitine insufficiency as described in detail in Chap. 5, "What Is Carnitine Deficiency?"; Carnitine insufficiency means that carnitine's function to regulate the acyl CoA/free CoA ratio is impaired due to shortage of free carnitine relative to the amount of accumulated acyl CoA in mitochondria. Free carnitine has buffering action to accept these harmful acyl group of acyl CoA as acylcarnitine, and release free CoA and maintain free CoA amount required for intracellular energy metabolism, etc. Based on clinical experience with carnitine deficiency in inherited metabolic diseases and mitochondrial diseases, normal AC/FC values are 0.2 to 0.25 or less, and at least >0.4 should be considered to indicate carnitine insufficiency [42–47].

Hatanaka et al. examined serum carnitine dynamics in 150 hemodialysis patients based on the diagnostic criteria in the "Diagnostic and Therapeutic Guidelines for Carnitine Deficiency 2018" (see Chap. 18) and found that about 25.3% of patients diagnosed as carnitine deficiency (<20 μmol/L), and 64.7% of patients had 20–36 μmol/L of free carnitine, below the normal range (36–74 μmol/L) and diagnosed as "high risk of carnitine deficiency," so total 90% of patients were considered as carnitine deficiency or high risk of carnitine deficiency [48] (Fig. 14.13a). On the other hand, the prevalence of carnitine insufficiency (AC/FC >0.4) was 86.7% (Fig. 14.13b). The pre- and post-dialysis carnitine reduction rates of 19 hemodialysis patients and 19 hemodiafiltration patients receiving carnitine supplementation therapy (1000 mg, intravenous) three times a week after dialysis were examined, and the reduction rates for all carnitine fractions were significantly higher in the hemodiafiltration patients.

In a study of muscle symptoms such as weakness, fatigue, cramps, and myalgia and carnitine deficiency in hemodialysis patients, Sakurauchi et al. also found that plasma free carnitine concentration decreased, and AC/FC increased with increasing duration of dialysis and that patients with muscle symptoms had significantly lower plasma free carnitine concentration and significantly higher AC/FC than patients without muscle symptoms [49].

Riley et al. reported that dialysis patients who develop hypotensive symptoms during dialysis tend to have significantly lower plasma free carnitine concentrations and higher AC/FC than those who do not, and that AC/FC in dialysis patients is inversely related to the Karnofsky activity scale score [50].

Ito et al. evaluated exercise parameters via cardiopulmonary testing in 82 of patients with incident dialysis who initiated on dialysis (20–90 years old, mean 62.1 ± 15.2 years old) [51]. As the results, fully-adjusted multivariate linear regression analysis showed that AC/FC was significantly associated with decreased peak VO_2 and with increased VE/VCO_2 slope, and FC or AC did not exhibit similar association

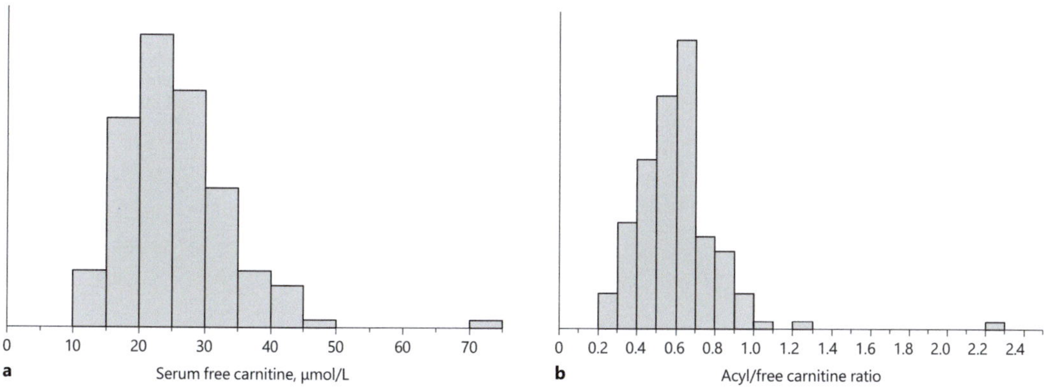

Fig. 14.13 (**a**) Histogram showing serum free carnitine concentrations in 150 patients on dialysis. (**b**) Histogram showing serum acyl/free carnitine ratio in these patients. (From Ref. [48])

Table 14.5 Simple linear regression analysis to determine the association of clinical parameters with primary outcomes

	Peak VO_2		% Peak VO_2		VE/VCO_2 slope	
Variables	β (95% CI)	*P*-value	β (95% CI)	*P*-value	β (95% CI)	*P*-value
Free carnitine (μmol/L)	0.04 (–0.03–0.10)	0.27	0.14 (–0.12–0.40)	0.29	–0.11 (–0.24–0.04)	0.15
Acylcarnitine (μmol/L)	–0.69 (–0.17–0.03)	0.18	–0.30 (–0.71–0.10)	0.14	0.12 (–0.10–0.35)	0.27
AC/FC	–7.18 (–12.11 to –2.24)	0.005	–30.23 (–49.65 to –10.82)	0.003	14.8 (4.14–25.47)	0.007

From Ref. [51]

VO_2: oxygen consumption; VE: minute ventilation: VCO_2: carbon dioxide uptake; 95% CI: 95% confidence interval; AC/FC: acylcarnitine/free carnitine ratio

with these parameters (Table 14.5). Moreover, only AC/FC was associated with decreased peak work rate (WR), percent-predicted WR, anaerobic threshold, delta VO_2/delta WR, and chronotropic index (Table 14.6). Thus, exercise parameters, including those related to both skeletal muscle and cardiac function, were strongly associated with AC/FC, a marker of carnitine insufficiency indicating altered fatty acid metabolism. Also, this result indicates that carnitine insufficiency is already developed with incident dialysis patients who just initiated dialysis treatment, suggesting that carnitine insufficiency is already developed in end-stage renal failure patients. Further studies might be needed to evaluate the prevalence of carnitine insufficiency in end-stage renal failure patients.

Thus, it has been reported that not only a decrease in free carnitine but also a relative decrease in free carnitine due to acyl group accumulation occurs in carnitine deficiency in dialysis patients, and that AC/FC, which is an indicator of carnitine insufficiency, is associated with the various clinical symptoms of dialysis patients.

14.3.4 Acylcarnitine Accumulation in Dialysis Patients and Pathophysiology

So far, I have described that carnitine deficiency in hemodialysis patients is associated with high acylcarnitine levels and high AC/FC as well as low free carnitine levels. The term "acylcarnitine" is a generic term, and there are various short- to long-chain acylcarnitines, ranging from C2 acetylcarnitine to C16 palmitoylcarnitine and C18 stearoylcarnitine (see Table 2, Chap. 2). In the carnitine assay using enzyme cycling method, total carnitine and free carnitine are measured, and the acylcarnitine concentration is calculated by subtracting the free carnitine concentration from the total carnitine concentration. In the radioisotope method, short-chain acylcarnitines and long-chain acylcarnitines are measured. In any case, these methods collectively measure many different acylcarnitine molecular species. The tandem mass spectrometer allows us to identify the molecular species of the various acylcarnitines. Recently, studies have been reported on the relationship between acyl group lengths and molecular species and clinical symptoms, and it has become clear that among acylcarnitines, high levels of long-chain acylcarnitines cause a variety of clinical symptoms.

Reuter et al. used LC/MS/MS to quantify serum carnitine in 87 dialysis patients by dividing it into free carnitine and individual acylcarnitine molecular species and examined the association with the erythropoietin resistance index (ERI) obtained by the following formula [52].

$$\mathrm{ERI} = \left[\mathrm{rHuEPO / kg / week}\right] / \left[\mathrm{g\,Hemoglobin}\right]$$

And when the cutoff value of ERI was set at 0.02 μg/kg/week/g Hb and patients with ERI ≥0.02 were defined as the high-ERI group and those with ERI <0.02 as the low-ERI group, the high-ERI group (n = 12) had significantly lower free carnitine levels than the low-ERI group

Table 14.6 Simple linear regression analysis to determine the association of clinical parameters with secondary outcomes

	Peak WR		% Peak WR		AT		Delta VO_2/delta WR		Chronotropic index	
Variables	β (95% CI)	*P*-value	β (95% CI)	*P*-value	β (95% CI)	*P*-value	β (95% CI)	*P*-value	β (95% CI)	*P*-value
Free carnitine (μmol/L)	0.24 (–0.17–0.65)	0.24	0.22 (–0.06–0.50)	0.11	0.02 (–0.02–0.06)	0.26	0.01 (–0.03–0.05)	0.57	0.06 (–0.34–0.45)	0.78
Acylcarnitine (μmol/L)	–0.17 (–0.81–0.46)	0.29	–0.26 (–0.70–0.17)	0.23	–0.03 (–0.08–0.03)	0.38	–0.03 (–0.08–0.03)	0.33	–0.39 (–1.00–0.22)	0.21
AC/FC	–34.03 (–64.91 to –3.15)	0.03	–36.78 (–57.22 to –16.34)	<0.001	–7.18 (–12.11 to –2.24)	0.005	–3.09 (–5.94 to –0.23)	0.03	–32.43 (–62.33 to –2.54)	0.03

From Ref. [51]

WR: work rate; AT: anaerobic threshold; VO_2: oxygen consumption; 95% CI: 95% confidence interval; AC/FC: acylcarnitine free carnitine ratio

(n = 75) (17.0 ± 5.47 vs. 23.8 ± 8.67 μM, $P < 0.05$). A significant negative correlation between free carnitine levels and ERI was found (P = p.0421). All patients categorized as high ERI exhibited subnormal free carnitine levels (<30 μM). Furthermore, the ratio of acylcarnitine concentration (AC-C2), which is all acylcarnitines minus C2 acylcarnitine (acetylcarnitine), to total carnitine (TC) [(AC-C2)/TC] was significantly higher in the high-ERI group (0.248 ± 0.0733 vs. 0.205 ± 0.0504, $P < 0.05$), and this ratio was significantly positively correlated with ERI ($P = 0.0062$). In other words, the higher the percentage of acylcarnitine, excluding acetylcarnitine, in the total carnitine, the higher the EPO resistance. Acetylcarnitine is a type of acylcarnitine, which is produced by combining carnitine with acetyl CoA, which is produced in large amounts when energy metabolism such as fatty acid metabolism and glycolysis proceeds smoothly and shows high values when mitochondrial function is normal. In addition, acylcarnitine in the blood of healthy individuals is mostly acetylcarnitine. Therefore, acetylcarnitine (C2) was subtracted from acylcarnitine.

Murphy et al. examined the relationship between C-reactive protein (CRP), a marker of inflammation, ESA dose, 6-min walking distance, number of 30-s sit-to-stand movements, and physical composite score by the Medical Outcomes Study Short Form 36 Item Instrument (SF-36) and the length of the acylcarnitine carbon chain was examined, and it was found that the more acylcarnitine of longer chain length, the higher the inflammation, the lower the physical activity level, and the higher the ESA dose (Fig. 14.14) [53].

Acylcarnitine in the blood is thought to reflect the accumulation of acyl-CoA in the mitochondria. These results suggest that dialysis patients with some kind of fatty acid oxidation disorder and a high amount of long-chain acylcarnitine, especially those with a long chain length of acylcarnitine, have a more promoting effects on inflammatory response, a higher dose of EPO, and a decrease in physical activity and quality of life (QOL).

Kamei et al. analyzed the serum carnitine profile of 79 dialysis patients using LC/MS/MS and examined the association with ERI and found that C5-OH acylcarnitine (3-hydroxy-isovaleryl-carnitine) showed a significant inverse correlation with ERI and a positive correlation with C18 acylcarnitine (stearoylcarnitine) [54].

In addition, it was reported that C2/(C16+C18:1) was found to be a significant independent factor when the serum carnitine profile of 111 dialysis patients was analyzed using LC/MS/MS and the association with 4-year all-cause mortality was examined from the same group [55]. C16 acylcarnitine is palmitoylcarnitine, and C18:1 acylcarnitine is octadecenoylcarnitine. C2/(C16 + C18:1), which indicates the efficacy of mitochondrial β-oxidation was negatively correlated with 4-year mortality. If the amount of such long-chain acylcarnitine is higher than that of acetylcarnitine, the mortality rate is high. As in the case of Reuter et al. [6], the acetyl group of C2 is a compound formed by the combination of acetyl-CoA and carnitine, which are produced at the end of fatty acid metabolism and are thought to reflect the good energy state in the body. On the other hand, if C16 + C18:1 is high, there is some kind of impairment of fatty acid oxidation, and not metabolized by β oxidation to acetyl-CoA, and such long-chain acyl-CoA accumulates in the mitochondria, which becomes these long-chain acylcarnitines. The relationship between long-chain acylcarnitine and the prognosis and QOL of patients with heart failure, not limited to dialysis patients, has been elucidated, and it is an important research topic for the future (see Chap. 16, "Cardiovascular Disease and Carnitine Deficiency").

Thus, it has been found that the accumulation of acylcarnitine, especially long-chain acylcarnitine, worsens the clinical condition of dialysis patients, such as a decrease in QOL and promotion of EPO-resistant anemia. In the future, it is necessary to investigate in more detail the relationship between the amount of acylcarnitine molecular species and clinical symptoms in dialysis patients using tandem mass spectrometry, and further research is required to determine how

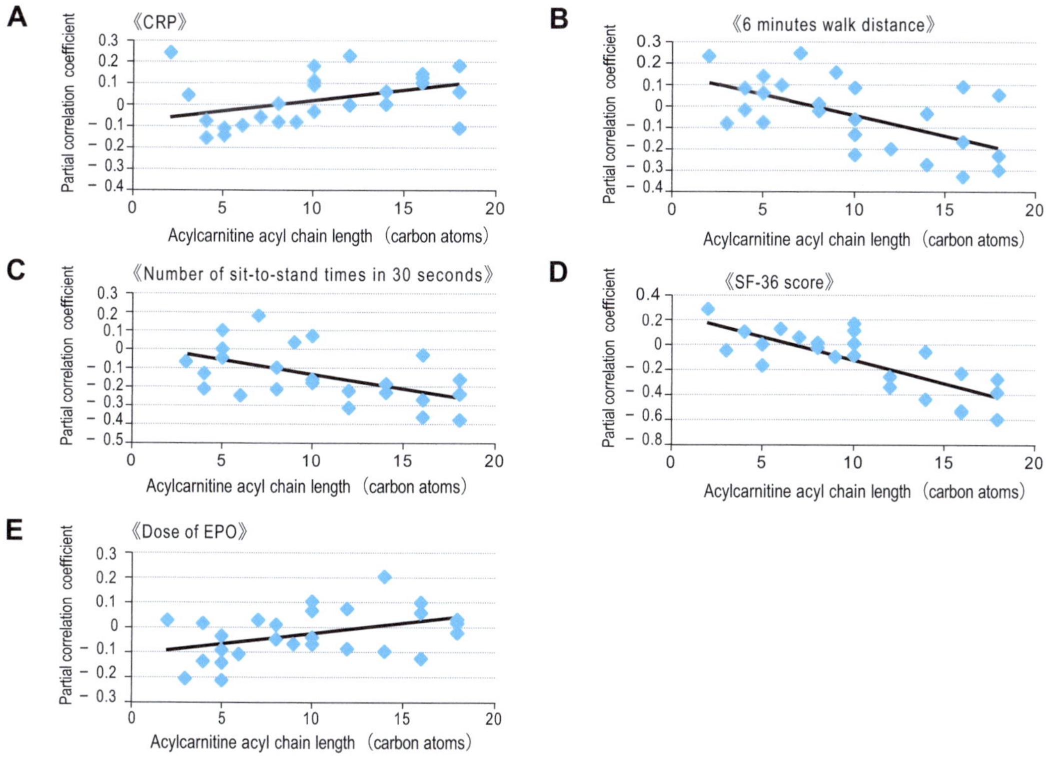

Fig. 14.14 Scatter plots with trend lines between acylcarnitine acyl chain length and strength of relationship (as represented by partial correlation coefficient) to outcome measures: (**a**) C-reactive protein, (**b**) 6-min walk distance, (**c**) sit-to-stand maneuver count, (**d**) Medical Outcomes Trust SF-36 PCS, and (**e**) ESA dose. Medical Outcomes Trust SF-36 PCS: Questionary of Medical Outcomes Trust short form physical composite score. (From Ref. [53])

carnitine supplementation therapy affects the accumulation of acylcarnitine and AC/FC, an indicator of carnitine insufficiency.

14.3.5 Dialysis Properties of Long-Chain Acylcarnitines

For acylcarnitine, which accumulates in dialysis patients, the relationship between the length of the carbon chain of the acyl group and dialysis has been investigated using a mass spectrometer. Reuter et al. analyzed the acyl groups of 50 patients with end-stage renal disease (56 ± 15 years old, 3.15 ± 3.07 years) undergoing long-term dialysis therapy using an electrospray ionization tandem mass spectrometer (ESI-MS/MS) and compared them with data from healthy subjects (39 ± 14 years, 60 patients [56]. Of the 31 acylcarnitines examined, 29 were significantly higher in patients with end-stage renal disease ($P < 0.05$). In particular, acylcarnitine of C5 and C8:1 was significantly increased. Acylcarnitines with a carbon chain number of 8 or less were significantly reduced by a single dialysis procedure, but their concentrations after dialysis were significantly higher than in healthy subjects. In addition, when the relationship between the number of carbon chains of acylcarnitine to the ratio of [post-dialysis concentration]/[pre-dialysis concentration] is plotted in a graph, the concentration of long-chain acylcarnitine with a carbon chain number of 8 or more was hardly removed by dialysis, and the ratio was close to 1 (Fig. 14.15). In other words, acylcarnitine in the blood of dialysis patients is less likely to be removed by dialysis than clearance from the kidneys in healthy subjects, and about 40% of acyl-

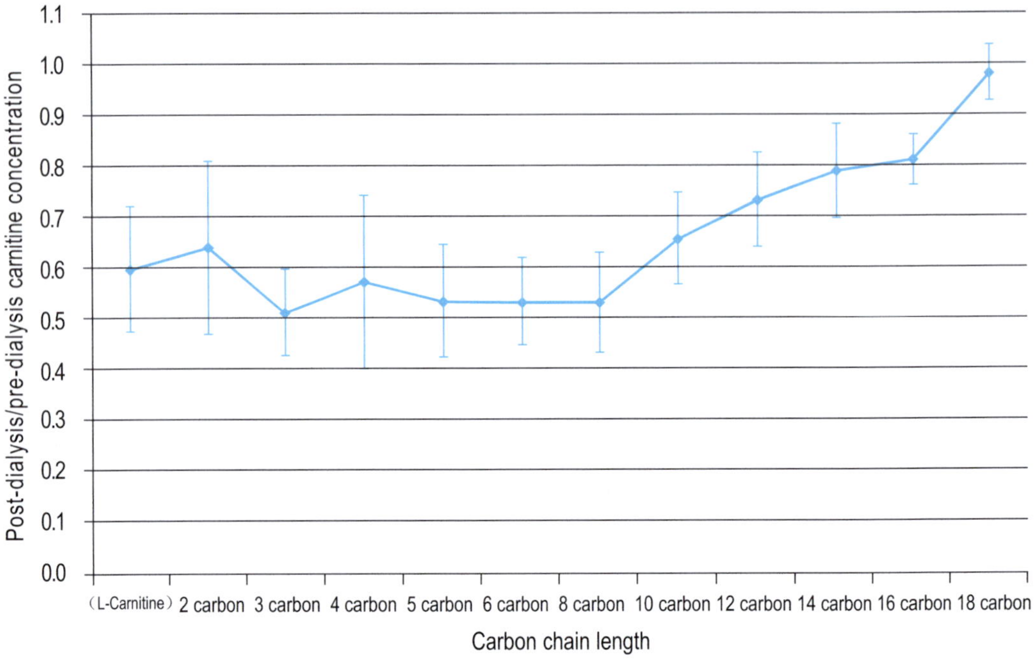

Fig. 14.15 Relationship between mean (±SEM) ratio of post-dialysis/pre-dialysis acylcarnitine concentration and carbon chain length of the acyl group. (From Ref. [56])

carnitine with a carbon chain number of 8 or less is removed by dialysis, but long-chain acylcarnitine with a carbon chain number of 8 or more is particularly difficult to remove from the dialysis membrane as the chain length increases.

Same as Reuter et al., Csiky et al. used ESI-MS/MS to investigate the change in concentration of various acylcarnitine molecular species in dialysis patients by a single dialysis and reported that long-chain acylcarnitine was difficult to remove by dialysis treatment [57].

Although the reasons why long-chain acylcarnitine is difficult to remove by dialysis are not described in detail in any of the papers, Reuter et al. consider two possibilities: (1) long-chain acylcarnitine has a long chain length and a large molecular weight, and (2) it may increase lipophilicity and make dialysis difficult. However, the molecular weight of C16 palmitoyl carnitine is 399.6 and that of C18 stearoyl carnitine is about 427.67; it is considered that they are still enough to remove by dialysis. Then, why does it become difficult to dialyze when the chain length of acylcarnitine becomes longer and the lipophilicity or hydrophobicity increases? Interaction with a highly hydrophobic dialysis membrane is also one of the possibilities but I hypothesize that micelle formation is the main cause. These long-chain acylcarnitines are amphiphilic substances with a hydrophobic fatty acid moiety and a hydrophilic carnitine moiety, and when the concentration exceeds the critical micelle concentration (CMC), these long-chain acylcarnitines surely form micelle. If about 100 molecules of long-chain acylcarnitine form micelle (hydrophilic quaternary amine moiety of carnitine oriented outside and covers the surface and the hydrophobic long-chain fatty acid moiety aggregates inside of the micelles), the apparent molecular weight will be as large as 40,000–50,000, like round green algae "marimo". It is difficult to removed these large molecule micelle by dialysis treatment making it difficult to be dialysis. Yalkowsky et al. reported that laurylcarnitine, myristylcarnitine, and palmitoylcarnitine form micelles in which 10–100 of molecules aggregate under various conditions [58]. According to the study, the CMC of palmitoylcarnitine in

0.2 M KCl solution is 8.5 μmol/L. The sum of the acylcarnitine concentrations of long-chain fatty acids of C8 and above in dialysis patients is at the level of several μmol/L [56, 59], and it is quite conceivable that various long-chain acylcarnitines with high hydrophobicity form mixed micelles. In addition, it is possible that long-chain acylcarnitine is difficult to dialyze due to protein binding. According to a study by Marzo et al., free carnitine and acetylcarnitine have no protein binding [60], and medium-chain octanoylcarnitine has 12~30% and 100% long-chain acylcarnitine are said to be protein binding. It is also possible that the hydrophobic part of the long-chain acylcarnitine may become difficult to dialyze due to the apparent large molecular weight when the hydrophobic part binds to the protein, and that the hydrophobic dialysis membrane and the hydrophobic long-chain acylcarnitine may interact with each other and be difficult to remove.

If micelle formation is the main reason why long-chain acylcarnitine is not removed by dialysis, carnitine should be administered from the conservative renal failure or the early stage of start of dialysis. Acylcarnitines can be removed as much as possible by residual renal function or dialysis treatment before the concentration exceeds CMC. Even if protein binding is the main cause, it is still important to remove long-chain acyl-CoA in the mitochondria as long-chain acylcarnitine as possible by replenishing carnitine as soon as possible to prevent the accumulation of long-chain acyl-CoA in the mitochondrial due to fatty acid oxidation impairment and to prevent mitochondrial dysfunction.

14.3.6 Carnitine Kinetics in Peritoneal Dialysis Patients

There are not as many studies on carnitine dynamics in peritoneal dialysis patients as in hemodialysis patients, but Albright et al. studied blood free carnitine levels in 9 CAPD patients (31–76 years old, 2 months to 2 years on dialysis) and found that free carnitine levels ranged from 33 to 95 μmol/L (mean 45), which is not decreased compared with normal levels of 30–70 μmol/L. In cases with high triglyceride levels, muscle biopsies were performed to measure carnitine concentrations in muscle, which were normal, so they concluded that there is no carnitine deficiency in peritoneal dialysis patients [61].

Leschke et al. also reported that plasma free carnitine concentrations were significantly lower than controls in hemodialysis/hemofiltration patients, but not significantly different from controls in CAPD patients [62].

Wanner et al. also measured blood carnitine levels in hemodialysis patients, CAPD patients, and intermittent peritoneal dialysis (IPD) patients, and reported that free carnitine levels were significantly lower in hemodialysis patients than controls, but not in CAPD and IPD patients [63]. In this report, hemodialysis patients with hemodialysis history were 6.1 ± 0.8 years, while dialysis history was 12.1 ± 3.0 months for patients with CAPD and 18.0 ± 6.9 months for IPD patients.

Subsequently, Constantin-Teodosiu et al. found that in eight patients with CAPD (mean age 43.6 ± 4.8 years-old, mean dialysis history of 27 ± 14 months), free carnitine concentrations were significantly lower (28.5 ± 1.4 vs. 36.2 ± 2.5 μmol/L, $P < 0.05$) and acylcarnitine/free carnitine ratio was significantly higher (0.52 ± 0.08 vs. 0.20 ± 0.05, $P < 0.01$), and carnitine excretion rates in urine and peritoneal dialysate were reported [40]. According to the report, the clearance of free carnitine was high and the clearance of acylcarnitine was low in patients with CAPD.

Grzegorzewska et al. also examined serum free carnitine concentrations in 23 patients with CAPD and found that they were normal in 17 patients but significantly decreased in 6 patients (25.4 ± 5.7 μmol/L), and there was no significant difference in dialysis history between normal and reduced groups (11.1 ± 9.6 vs. 9.7 ± 4.1 months), total creatinine clearance and dialysate volume were high in the reduced group, and carnitine deficiency was more likely to occur in peritoneal dialysis as well as adequate dialysis was performed [64].

Clinical studies in Japan have also reported that carnitine deficiency also occurs in peritoneal dialysis patients. Horie et al. examined the carnitine dynamics of 20 hemodialysis patients (170 months of dialysis history) and 13 patients with CAPD (43.8 months of dialysis history) and found that serum free carnitine concentrations were 21.5 ± 4.4 μmol/L and 31.3 ± 7.4 μmol/L, and AC/FC was 0.84 ± 0.21 and 0.46 ± 0.13. To a lesser extent, carnitine deficiency and carnitine insufficiency were observed in patients with CAPD as in hemodialysis patients [65].

In addition, Mochizuki et al. examined the relationship between carnitine dynamics and nutritional indices in 51 peritoneal dialysis patients and found that total carnitine and free carnitine in the blood decreased (37.9 ± 8.3 μmol/L and 24.9 ± 5.6 μmol/L), and residual renal function was negatively correlated with blood acylcarnitine concentration and AC/FC. The serum carnitine levels were found to be positively correlated with lean body mass (LBM), body mass index (BMI), serum creatinine, and albumin [66].

Ohashi et al. measured free carnitine, atrial natriuretic peptide (ANP), and brain natriuretic peptide (BNP) in 45 peritoneal dialysis patients, mainly CAPD. ^{123}I-labeled metaiodobenzylguanidine (MIBG) and β-methyl-p-iodophenyl-pentadecanoic acid (BMIPP) myocardial scintigraphy were performed to investigate the association with carnitine levels [67]. The results showed that myocardial scintigrams showed MIBG and BMIPP defects in 37.8 and 62.2% of patients, respectively, and that left ventricular systolic function was reduced in patients with extensive defects, especially in those with extensive BMIPP myocardial scintigrams, and serum free carnitine levels were reduced in those with extensive defects.

Shimizu et al. studied the free, acyl-, and total carnitine concentrations and AC/FC ratio in 34 PD patients and 34 age-, sex-, and dialysis duration-matched HD patients. Prevalence of carnitine deficiency was 8.8% in PD group and 17.7% in the HD group ($P = 0.283$). High risk of carnitine deficiency (FC: 20–36 μmole/L) was 73.5 and 76.4%, respectively. Carnitine insufficiency (AC/FC>0.4) was found 82.3 and 88.2% ($P = 0.733$), respectively. This result shows that PD patients have carnitine deficiency and carnitine insufficiency comparable to those in HD patients.

Kaneko et al. reported that when blood carnitine was measured in 60 peritoneal dialysis patients (mean age 62.5 years, mean dialysis history 24 months), 5 patients (8%) had 20 μmol/L < free carnitine and 56 patients (93%) had 0.4 > AC/FC (Fig. 14.16) [68].

In pediatric patients with renal failure, peritoneal dialysis is performed at a high rate due to various factors, such as the difficulty of securing blood vessels. Warady et al. reported that plasma carnitine concentrations measured in 12 pediatric CAPD patients were not significantly different from controls, and that carnitine supplementation (100 mg/kg/day, oral, 2 months) did not change serum triglyceride levels, which were elevated [69]. Koşan et al. examined plasma carnitine concentrations in 20 pediatric CAPD patients (mean ages 11.2 ± 4.3 years) and found that free carnitine concentrations were significantly lower (26.9 ± 4.9 vs. 32.8 ± 6.0 μmol/L, $P < 0.001$), and acylcarnitine concentrations and AC/FC were reported to be significantly higher in patients with CAPD (12.3 ± 3.1 vs. 8.4 ± 1.9 μmol/L, 0.46 ± 0.12 vs. 0.26 ± 0.08, both $P < 0.001$) [70]. Carnitine supplementation in pediatric CAPD patients (50 mg/kg/day, 30 days) showed that triglyceride levels remained unchanged, but high levels of apolipoprotein B were significantly reduced (146.6 ± 51.8 ⇒ 63.6 ± 22.2 mg/dL).

Kawahara et al. examined the effect of carnitine supplementation on cardiomyopathy in pediatric peritoneal dialysis patients [71]. Blood carnitine was measured in 36 pediatric peritoneal dialysis patients. Carnitine was administered to four patients with renal insufficiency who had not recovered cardiac function (30 mg/kg/day, increased by 10 mg/kg/day every month, maintained at 50 mg/kg/day and administered for 10 months). Serum free carnitine concentrations in pediatric peritoneal dialysis patients were 25.4 ± 7.3 μmol/L in all patients and 23.8 ± 10.0 μmol/L in four patients treated with carnitine, which was lower than normal values

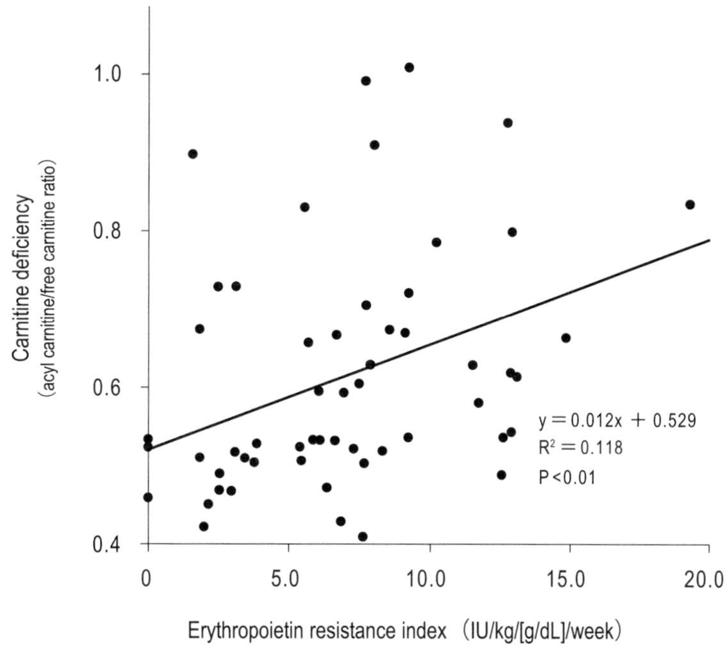

Fig. 14.16 Significant correlation between carnitine deficiency (acyl carnitine/free carnitine ratio) and the erythropoietin resistance index in patients ($n = 60$) who underwent peritoneal dialysis. (From Ref. [69])

(45.6 ± 11.0). Ejection fraction was examined after carnitine administration, and in two out of four patients, there was no improvement in ejection fraction itself, but symptoms of heart failure and pulmonary edema improved, and in the other case, dilated cardiomyopathy before carnitine administration and thickening of the ventricular septum, but the thickening became thinner after administration.

Furthermore, Uemura et al. in the same group reported that as a result of examining serum carnitine kinetics before and after living donor kidney transplantation in six pediatric peritoneal dialysis patients, serum free carnitine concentrations were low compared to normal values before renal transplantation but increased and became normal values in five out of six patients, and acylcarnitine concentrations, which had been high, decreased in all patients [72]. This suggests that the change in carnitine kinetics, such as a decrease in free carnitine and an increase in acylcarnitine, was caused by peritoneal dialysis treatment performed before kidney transplantation.

Nakano et al. discussed that the cause of low free carnitine and high levels of acylcarnitine in pediatric peritoneal dialysis patients is that free carnitine is reabsorbed in the kidneys and acylcarnitine is preferentially excreted, while free carnitine and acylcarnitine are excreted in the peritoneum, but acylcarnitine is excreted in small quantities due to its high molecular weight. It is considered that free carnitine is relatively likely to decrease [73]. It is said that the biosynthesis of carnitine is immature in children and reaches the same level as adults around the age of 10. Therefore, it is thought that carnitine deficiency and carnitine insufficiency are likely to occur in pediatric patients with end-stage renal failure due to low muscle mass. In pediatric peritoneal dialysis patients, it is important to suspect carnitine deficiency and investigate if there are clinical symptoms such as cardiomyopathy.

14.4 Clinical Manifestations of Carnitine Deficiency in Dialysis Patients and Carnitine Supplementation Therapy

What are the symptoms of carnitine deficiency in dialysis patients? There have been numerous studies on carnitine deficiency in dialysis patients. Taken together, the following clinical symptoms are thought to be associated with carnitine deficiency, and carnitine supplementation therapy has been attempted for them.

1. Lipid metabolism disorders: Hypertriglyceridemia, hypercholesterolemia, hyper-LDL-cholesterolemia, hypo-HDL-cholesterolemia, etc.
2. Muscle symptoms and decreased physical function: Muscle weakness, fatigue, muscle pain, cramps, decreased mobility, etc.
3. Cardiac symptoms: Left ventricular hypertrophy, decreased cardiac contractility, heart failure, cardiomyopathy, atherosclerosis progression, hypotension during dialysis, arrhythmia, etc.
4. Inflammation, atherosclerosis, undernutrition, sarcopenia, etc.
5. Anemia: EPO-resistant anemia, hemolysis (due to decreased stability of red blood cell membranes), etc.

The following is a summary of the findings so far.

14.4.1 Lipid Metabolism Disorders

Since the most important function of carnitine is the transport of long-chain fatty acids to mitochondria, it is thought that carnitine deficiency may be related to lipid metabolism abnormalities in dialysis patients, and many attempts at carnitine supplementation therapy have been reported. The effects of carnitine supplementation therapy on lipid metabolism disorders in dialysis patients were mixed, and the results of the systematic review did not show a consistent effect.

One of them was a systematic review by Hurot et al. [1] in 2002. They selected 163 of the 620 clinical studies of carnitine from 1978 to 1999 related to renal failure, excluded 84 because they did not receive carnitine, excluded 62 of the remaining 79 studies because they were not controlled trials, and contacted the authors and other work, and 18 of the studies were examined. A systematic review of lipid metabolism disorders and EPO-resistant anemia was conducted (in a total of 482 patients with renal failure). The results of the review on lipid profiles are described below. Hypertriglyceridemia is known to be common in dialysis patients. A meta-analysis of nine trials investigating the effects of carnitine on serum triglycerides showed no effect in five studies, and a positive significant difference in two trials and a negative significant difference in two studies compared to placebo or no carnitine in four trials. Summarizing the results of all trials, the effect of carnitine administration was not observed due to the heterogeneity between the studies and the lack of sufficient sample size (Fig. 14.17a).

Nine studies examined total serum cholesterol, and no effect of L-carnitine administration was observed overall (Fig. 14.17b). In addition, meta-analyses were conducted on three, five, and eight studies that examined the effects of carnitine administration on VLDL-cholesterol, LDL-cholesterol, and HDL-cholesterol. Again, no effect of carnitine administration was observed in any of the items. No heterogeneity was observed in total cholesterol and cholesterol fractions.

Although the trials included in the systematic review by Hurot et al. were limited to randomized controlled trials, the route of administration of carnitine varied in the nine trials that examined the effects on serum triglycerides, three of which were given orally, five were given intravenously, and one was given carnitine in dialysate. In addition, there is a large gap in the administration period from 3 weeks to 6 months, and it may be natural that there is a large heterogeneity in the administration effect. In any case, the results of this systematic review did not indicate that carnitine administration had an effect on serum triglycerides, total cholesterol, VLDL-cholesterol, LDL-cholesterol, and HDL-cholesterol in dialysis patients.

In addition, Reuter et al. reviewed 17 studies that examined the effects of carnitine on serum lipids in dialysis patients [6]. Seven of the 17 studies were also covered in a systematic review by Hurot et al., and 11 were open-label studies. Looking at the results of these 17 trials, they found that long-term carnitine supplementation therapy tended to significantly reduce serum tri-

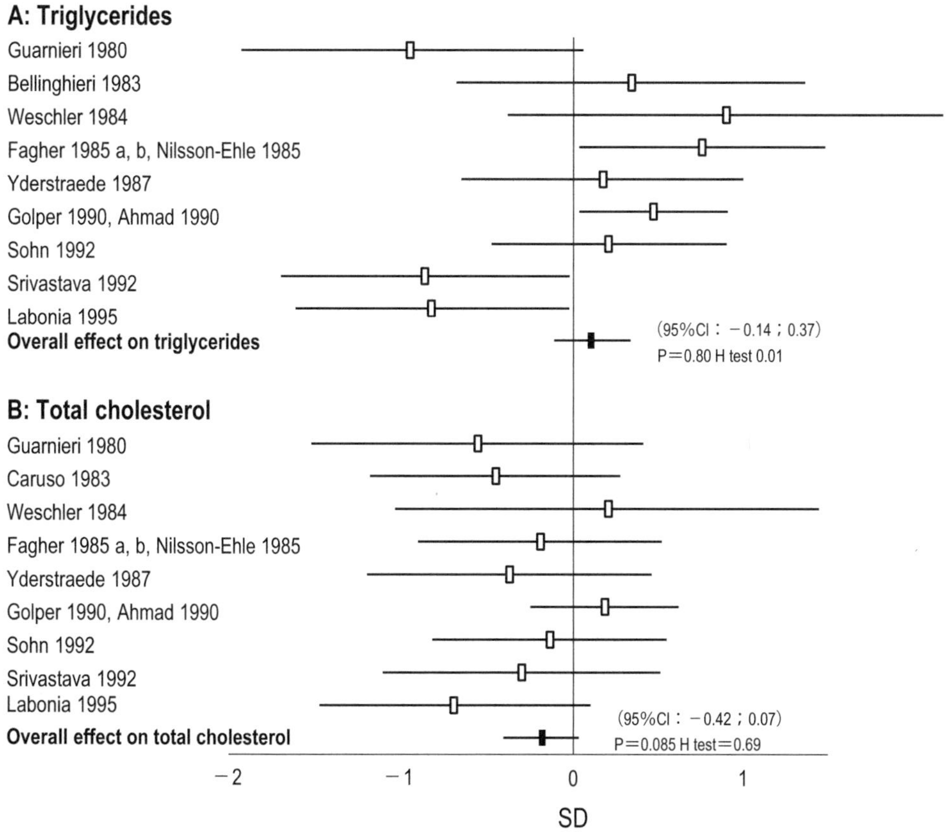

Fig. 14.17 The effect of L-carnitine on serum triglyceride (**a** random model analysis) and on total cholesterol (**b** fixed model analysis). Open boxes indicate the effects of each individual trial, and the closed boxes the overall effect of treatment. The *P* value is given for overall treatment effect and is significant for values ≦ 0.01. Heterogeneity of treatment effect among trials is present if H test value is <0.10. (From Ref. [1])

glyceride concentrations [11, 74–77] or significantly increase serum HDL-cholesterol [78–80]. In addition, Vacha et al. found that after at least 12 months of carnitine supplementation therapy, serum triglycerides were lowered and HDL-cholesterol was elevated, but when treatment was discontinued, it returned to normal, and when treatment was started again, these serum lipids improved [81]. On the other hand, many studies reported no improvement in serum lipid profiles [30, 82–85], and some reported negative results for serum triglycerides and HDL-cholesterol [63, 86]. Approximately 40% of dialysis patients are said to have lipid metabolism abnormalities [78], and most of them have type IV hyperlipidemia, high triglycerides, and low HDL-cholesterol, which are thought to promote atherosclerosis and increase the risk of cardiovascular disease. On the other hand, it has been reported that atorvastatin administration to dialysis patients with diabetes did not reduce the risk of cardiovascular events (cardiovascular death, non-fatal myocardial infarction, and cerebral infarction) [87], and a new therapeutic agent for lipid metabolism disorders in dialysis patients is desired, and many studies have been conducted on carnitine. Lipid metabolism disorders in dialysis patients may be partly due to carnitine deficiency, but many patients have complications such as diabetes, hypertension, and hyperlipidemia, and it is thought that they are caused by multiple very complex etiologies rather than simple etiology. Vacha et al. found that patients with HDL-cholesterol of <400 mg/L responded to 20 mg/kg carnitine supplementation therapy and HDL-cholesterol was elevated, but patients with ≧400 mg/L did not, and that non-responders were given higher doses of carnitine (60 mg/kg, 2 months) reported a decrease in serum triglycerides [88]. Wanner et al., who reported that carnitine supplementation therapy was negative for serum triglycerides and serum HDL-cholesterol, also reported that low-dose carnitine supplementation therapy (1 mg/kg, intravenously after dialysis three times a week for 3 months) lowered serum triglycerides in patients with high levels of long-chain acylcarnitine and short-chain acylcarnitine [89].

Thus, it is necessary to examine in more detail what kind of patients are responders of carnitine supplementation therapy for lipid metabolism disorders in dialysis patients, such as stratification by serum triglyceride and HDL-cholesterol concentration, and stratification by acylcarnitine concentration and acylcarnitine profiles.

14.4.2 Muscular Symptoms and Decreased Physical Function in Dialysis Patients

1. Carnitine supplementation therapy for muscular symptoms and decreased physical function in dialysis patients

There is no doubt that there has been significant progress in dialysis therapy and the prognosis of patients with renal failure has improved. Nonetheless, dialysis patients still have a high rate of muscle symptoms such as cramps, muscle pain, and psychological symptoms such as chronic fatigue and malaise, decreased activity in daily and social life, and decreased physical functionality. Carnitine deficiency has been considered to be one of the causes of various muscular symptoms that are common in these dialysis patients, as well as hypotension during dialysis [2, 3, 5].

Table 14.7 shows a list of major clinical studies that examine the effects of carnitine administration on muscle symptoms and QOL in dialysis patients. The effects of oral or intravenous administration of carnitine have been investigated, and some studies have been conducted by adding carnitine to dialysate. In addition, the dosage varies, but there are many studies that include 1 to 2 g/day for oral administration and 20 mg/kg for intravenous administration, after dialysis is completed. As a result, many studies have reported that the frequency and intensity of clinical symptoms such as cramps, muscle weakness, myalgia, and hypotension during dialysis have decreased. There have been studies that have evaluated QOL using SF-36, and it has been reported that physical functioning, vitality, and general health scores have improved, and that the total SF-36 score has improved [90, 91]. On the other hand, there are some reports that no effect was observed [84, 92, 93].

The study by Fagher et al. [92], which found no effect of carnitine administration on muscle symptoms, was conducted in patients who had been on dialysis for only about 36 months (about 3 years) on average and were still young, with an average age of 42–48 years, and it is questionable whether these dialysis patients had carnitine deficiency. The study by Bellinghieri et al. found an effect of carnitine on cramps and asthenia, but no morphological changes in muscles. This study was also conducted in relatively young patients with a mean history of dialysis of 23 ± 4 months and a mean age of 49 ± 4 years [32].

As shown in the study of Spagnoli et al. [94] and Giovenali et al. [95], it is reported that type I and type IIa muscle fibers, which mainly use fatty acid oxidation, were enlarged by carnitine administration, and type IIb muscle fibers, which mainly use glycogen, did not change. It is thought that carnitine deficiency caused a decrease in metabolism in muscle fibers dependent on fatty acid oxidation in skeletal muscle, which was improved by carnitine supplementation. The increase in fatty acid oxidative capacity is consistent with the results of studies by Ahmad et al. [96] and Brass et al. [97], which found that maximal oxygen uptake increased in the carnitine-treated group or that maximal oxygen uptake was not observed in the carnitine-treated group, as observed in the placebo group. Rocchi et al. also reported a decrease in the number of polyphasic muscle action potentials and an increase in total electromyography power with carnitine adminis-

Table 14.7 Major clinical studies investigating the effect of carnitine supplementation on muscle symptoms and QOL in dialysis patients

Oral administration					
Reference	Study design	Number of subjects	Dose of carnitine	Duration of administration	Result
Casciani, 1982 (Ref. 76)	Double-blind, cross-over, placebo-controlled trial	18 Group 1: 9 Group 2: 9	990 mg/day	60 days Wash out for 10 days	Decreased cramps, muscle weakness, and dyspnea
Bellinghieri, 1983 (Ref. 32)	Double-blind, cross-over, placebo-controlled trial	14 Group 1: 7 Group 2: 7	2000 mg/day (divided 2 times per day)	8 weeks Wash out for2 weeks	Reduction in cramps and asthenia No morphological changes in the muscles were noted
Rogerson, 1989 (Ref. 82)	Open study	6	2000 mg/day	6 weeks	No change in triceps thickness, middle arm circumference, total muscle mass, or lower limb muscle strength
Feinfeld, 1996 (Ref. 98)	Open study	6	1980 mg/day	8 weeks	Decreased stiffness, muscle weakness, and fatigue Decreased myoglobin levels that were high
Sakurauchi, 1998 (Ref. 48)	Placebo-controlled trial	51 With muscular symptoms:30 No muscular symptoms:21	500 mg/day	12 weeks	Decreased muscular symptoms. Patients with muscle symptoms were low plasma-free carnitine concentrations and high AC/FC than patients without muscle symptoms
Sloan, 1998 (Ref. 88)	Double-blind, parallel, placebo-controlled trial	101 Group 1 and 2: 63 Group 3: 38	1000 mg before and after dialysis 3 times a week	Groups 1 and 2: 6months, Group 3: 3 months	Groups 1 and 2: QOL assessed by SF-36(general health, physical function) improved at 1.5 and 3 months, no improvement after 4.5 months Group 3: Improvement of vitality and general health

Intravenous administration (some oral administration, some dialysate administration)					
Reference	Study design	Number of subjects	Dose of carnitine	Duration of administration	Result
Fagher, 1985 (Ref. 90)	Double-blind, parallel, placebo-controlled trial	28	3 times a week after dialysis 2 g intravenously	6 weeks	No effect on muscle metabolism, subjective symptoms, or nerve conduction velocity
Rocchi, 1986 (Ref. 96)	Open study	20	3 g intravenously after dialysis 3 times a week	7 months	The polyphasic muscle action potential of computerized electromyography decreases, and the total electromyography power increases

(continued)

Table 14.7 (continued)

Intravenous administration (some oral administration, some dialysate administration)					
Reference	Study design	Number of subjects	Dose of carnitine	Duration of administration	Result
Ahmad, 1990 (Ref. 94)	Double-blind, parallel, placebo-controlled trial	82 Carnitine group: 38 Placebo group: 44	3 times a week after dialysis 20 mg/kg intravenous administration	6 months	Reduction in cramps, increased brachial muscle area and maximal oxygen consumption, overall happiness improves, decrease in hypotension during dialysis
Spagnoli, 1990 (Ref. 92)	Open study	22	3 times a week after dialysis 2~4 g intravenously	12~16 months	Enlargement of type I muscle fibers
Siami, 1991 (Ref. 36)	Double-blind, parallel, placebo-controlled trial	14	3 times a week before and after dialysis 2 g intravenously	6 months	Reduced muscle weakness Increased physical activity
Sohn, 1992 (Ref. 80)	Double-blind, parallel, placebo-controlled trial	30	3 times a week after dialysis 1~1.5 g intravenously	2 months	Reduction in cramps. There was no change in triceps thickness, middle arm circumference, and brachial muscle mass
Giovenali, 1994 (Ref. 93)	Open study	Group 1: 11 Group 2: 6 Group 3: 9	Group 1: Addition in dialysate 2 g/170L Group 2: 2 g/day orally Group 3: 3 times per week after dialysis 20 mg/kg intravenous administration	6 months	Reduction in cramps (6 out of 10 cases), decrease of muscle pain (2 out of 5 cases). Type I and IIa muscle fiber diameter increased by 7% (muscle fibers that utilize fatty acid oxidation). Type II.b muscle fibers unchanged (muscle fibers that utilize glycogen)
Semeniuk, 2000 (Ref. 91)	Double-blind, cross-over, placebo-controlled trial	16	3 times a week after dialysis 20 mg/kg intravenous administration	6 months	There is no change in QOL evaluation by KDQ
Brass, 2001 (Ref. 95)	Double-blind, parallel, placebo-controlled trial	Study A: 30 Study B: 133	Study A:3 times a week after dialysis 20 mg/kg intravenous administration	6 months	In the carnitine group, there was no deterioration in maximal oxygen consumption, the placebo group decreased, Significant difference between both groups ($P = 0.009$) Fatigue reduced by KDQ evaluation
Rathod, 2006 (Ref. 89)	Single-blind, placebo-controlled trial	20 Carnitine group: 10 Placebo group: 10	3 times a week after dialysis 20 mg/kg intravenous administration	8 weeks	QOL assessment at SF-36. Physical functioning, general health, vitality, social functioning, mental health were improved. Overall physical health and overall mental health were improved, and total SF-36 score was improved

Created from Refs. [2, 3, 5]

AC: acylcarnitine concentration, FC: free carnitine concentration, SF-36: Questionnaire of Medical Outcomes Study Short Form 36 Item Instrument

KDQ Questionnaire of Kidney Disease Questionary, *QOL* quality of life tration on computerized electromyography [98]. In this way, it is interesting to note that the effects of carnitine administration have been verified not only in indicators that tend to be influenced by subjectivity, such as "reduction in fatigue" in questionnaires, but also in objective indicators using measuring instruments.

Hypermyoglobinemia and myoglobinuria in rhabdomyolysis and other conditions can cause acute renal failure, but mild-to-moderate myoglobinuria is also observed in patients with chronic renal failure [99]. It is well known that myoglobin causes renal tubular injury by its oxidative stress, etc., and it is desirable to lower as much as possible, but it will not be lowered by dialysis treatment. Feinfeld et al. treated dialysis patients with carnitine (1.98 g/day, orally, for 8 weeks), who had hypermyoglobinemia, and found that carnitine significantly reduced blood myoglobinuria [100]. It is possible that the "trophic effect" on skeletal muscle, as described in the studies by Spagnoli et al. [94] and Giovenali et al. [95], prevents muscle damage, or that carnitine's antioxidant and anti-oxidant effects are responsible for the decrease in blood myoglobin concentration, and anti-apoptotic effects on skeletal muscle prevent muscle damage.

Sakurauchi et al. examined the effect of oral carnitine (500 mg/day, 12 weeks) in a group of 30 dialysis patients (Group A, mean age 61.9 ± 10.7 years, dialysis history 138 ± 85 months) with muscle symptoms during dialysis such as cramps, muscle weakness, and fatigue [49]. They also examined plasma free carnitine concentrations and AC/FC in 21 dialysis patients (Group B, mean age 57.8 ± 10.2 years, dialysis history 102 ± 70 months) without muscular symptoms. As a result, the plasma free carnitine concentration was significantly lower than that of the group B patient group without muscular symptoms, and the AC/FC was significantly higher (Fig. 14.18).

When carnitine was administered to Group A patients with muscular symptoms, about two-thirds of the patients showed some improvement over time. Figure 14.19 shows the rate of improvement scores for muscle weakness, muscle fatigue, and muscle cramps/myalgia after 12 weeks of administration, and the evolution of the improvement index calculated from percentages over time.

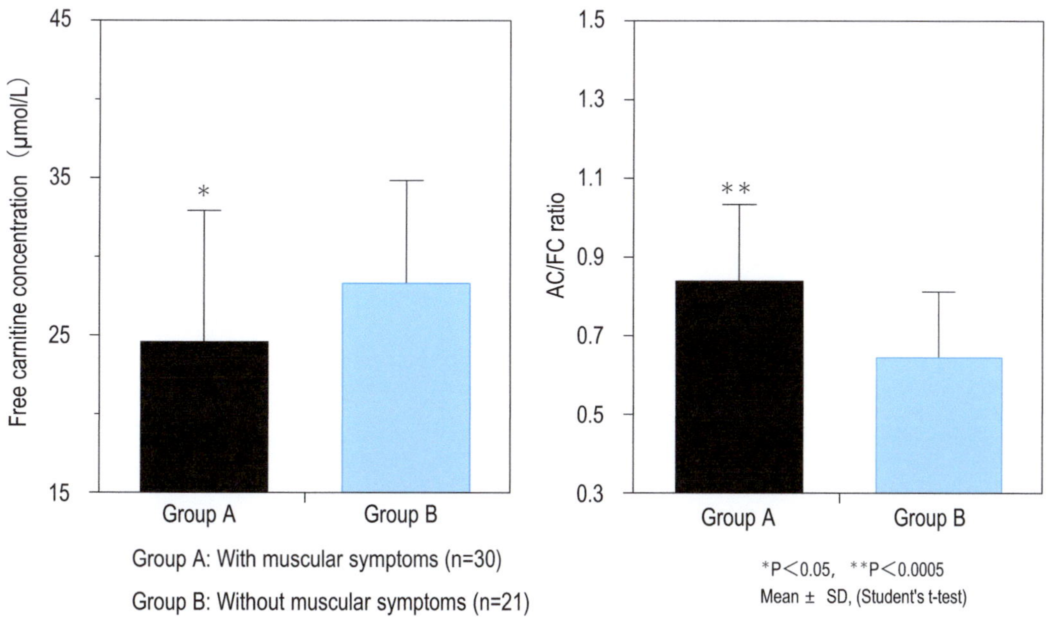

Fig. 14.18 Difference of free carnitine levels (left panel) and AC/FC ratio (right panel) between patients with symptoms (group A) and patients without symptoms (group B). This figure shows the mean ± SD of data in each group. *: denotes $P < 0.05$ for comparison with group B. **: denotes $P < 0.0005$ for comparison with group B. Statistical significance was determined be student's t-test. (From Ref. [49])

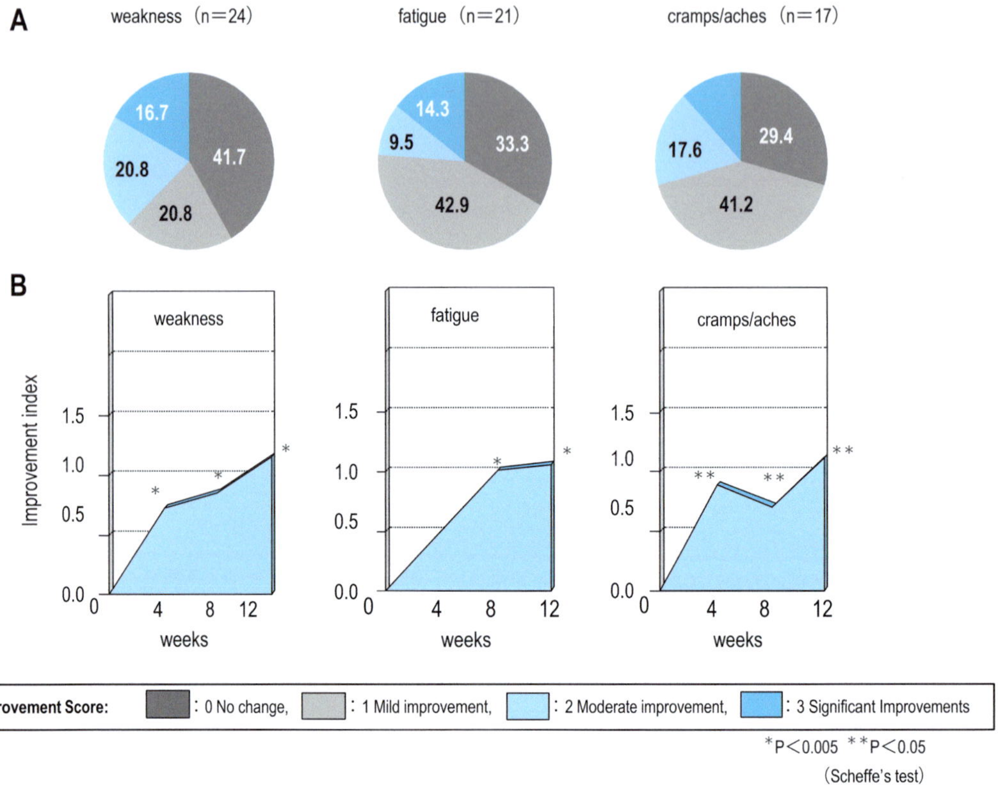

Fig. 14.19 Improvement in dialysis-associated muscle symptoms by oral L-carnitine substitution. (**a**) Effects on muscle weakness, fatigue, and cramps/aches after 12 weeks of treatment. (**b**). Improvement index in muscular symptoms during the course of the study. Approximately two-thirds of patients showed some improvement in each symptom. *: $P < 0.005$ for comparison with the baseline; **: $P < 0.05$ for comparison with the baseline. Statistical significance was determined by Scheffe's test. (From Ref. [49])

Table 14.8 Effect of intravenous carnitine supplementation on hypotension, cramps, and asthenia during dialysis

	Hypotension		Cramps		Asthenia	
	Carnitine	Placebo	Carnitine	Placebo	Carnitine	Placebo
Baseline[a]	17	9	14	10	15	18
Treatment[b]	7	11	5	9	7	10
P(Chi^2)	0.02	NS	0.02	NS	0.04	0.05

Modified from Ref. [96]

[a]Data were collected from 38 patients receiving carnitine (20 mg/kg intravenously after dialysis 3 times a week for 6 months) and 44 patients receiving placebo before and after treatment (There was no significant difference between the two groups at baseline)

[b]Post-treatment data were obtained during the last 2 months

Ahmad et al. reported that intravenous administration of carnitine (20 mg/kg, 24 weeks) after dialysis three times a week in long-term dialysis patients was investigated for the effects on hypotension, cramps, and myasthenia during dialysis, as well as on maximal oxygen uptake in exercise tolerance tests (Table 14.8) [96].

As a result, intravenous carnitine administration significantly reduced hypotension and cramps during dialysis, but there was no change in the placebo group. Myasthenia was significantly reduced after treatment in both groups compared to before treatment. Assessments of asthenia tend to be subjective and may have been reflected by the placebo effect.

In addition, an exercise tolerance test using a bicycle ergometer was conducted, and the maximum oxygen uptake was measured at that time, and there was no change in the placebo group, but a significant increase in maximal oxygen uptake was observed in the carnitine group after 24 weeks of administration (Fig. 14.20). In addition, in the carnitine group, the arm muscle area (AMA), calculated from the circumference of the upper arm and the subcutaneous fat thickness of the triceps, was significantly increased, and the overall sense of well-being and exercise tolerance were improved.

Takeuchi et al. studied carnitine supplementation therapy in hemodialysis patients with ESA-resistant anemia, as described later, and also studied and reported muscle symptoms such as asthenia and muscle cramps [101]. Levocarnitine chloride supplementation therapy was started in hemodialysis patients (600 mg/day, divided 2 times, daily oral administration for 24 weeks), and the effect was evaluated by scoring asthenia (fatigue and malaise) and muscle cramps during dialysis. The results showed that "no symptoms" for asthenia was 18.8% before the start of treatment, 43.8% at 12 weeks, and 78.1% at 24 weeks after the start of treatment, and that several percent of the patients no longer had severe symptoms. Muscle cramps improved significantly, from 34.4% of patients with no symptoms before treatment to 75% after 12 weeks, and from 21.9% of patients who had muscle cramps "every time" before treatment to 0% after 12 weeks. The effect of intravenous levocarnitine on muscle symptoms was also examined in 43 patients who switched from oral to intravenous levocarnitine and in 89 patients who were newly started on intravenous levocarnitine. In some cases, patients with inadequate response to the oral formulation improved after switching to the intravenous formulation [102]. In this paper, Takeuchi et al. presented data from their own study and explain the classification and characteristics of skeletal muscle myofiber types and the relationship between the energy-producing system in skeletal muscle and carnitine deficiency, referring to studies by Giovenali et al. [95] and others.

Decreased exercise tolerance in dialysis patients leads to poor prognosis and ability to perform daily activities. Muto et al. examined 13 dialysis patients (mean dialysis history 5.5 ± 4.9 years) who were receiving exercise therapy using a bicycle ergometer in the carnitine supplementation group (levocarnitine chloride 300 mg/day, min 3 and 6 patients, mean age 69.3 ± 8.4 years, mean dialysis history

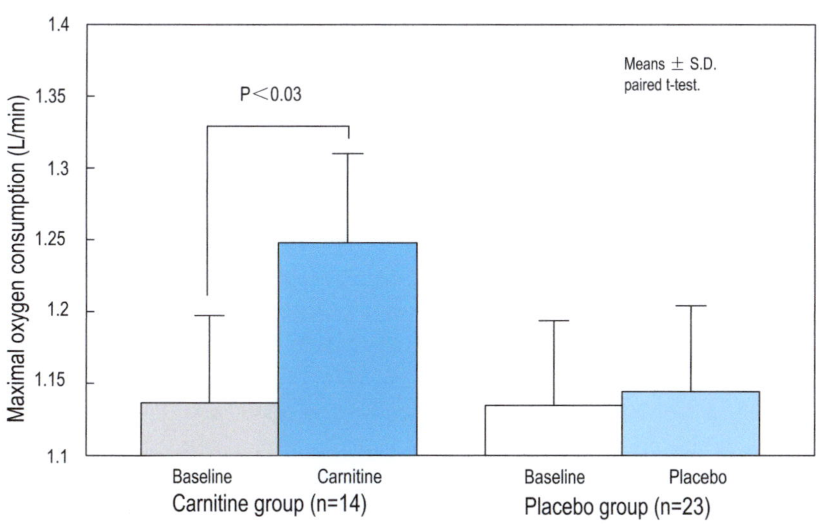

Fig. 14.20 Maximal oxygen consumption during progressive work exercise test at baseline and repeated at 24 weeks in patients receiving L-carnitine (N = 14) and those receiving placebo (N = 23). The comparisons between baseline and 24 weeks were made by paired t-test. Values are mean (SEM). (From Ref. [96])

6.8 ± 5.4 years) and non-supplemented group (7 patients, mean age 75.0 ± 10.4 years, mean dialysis history 4.5 ± 4.5 years), and it was reported that the maximum oxygen uptake (15.4 ± 3.9 ⇒ 16.2 ± 4.3 mL/kg/min, $P < 0.05$) and anaerobic metabolic threshold (9.8 ± 1.8 ⇒ 11.6 ± 2.4 mL/kg/min) were significantly improved in the carnitine supplementation therapy group [103].

Kuwasawa-Iwasaki et al. administered carnitine supplementation therapy to 80 dialysis patients (62 on hemodialysis and 18 on peritoneal dialysis) for 1 year (600 mg/day, orally) and intravenous carnitine supplementation therapy for hemodialysis patients for an additional 1 year (1 g intravenously injected after dialysis, 3 times a week). It was reported that 31% of patients who had been on hemodialysis for more than 4 years showed a decrease in muscle cramps and a significant increase in hemoglobin (Hb) levels [104].

Maruyama et al. conducted a randomized controlled trial in which 91 dialysis patients with carnitine deficiency were divided into a carnitine group (1000 mg intravenous injection three times a week after dialysis for 12 months) and a non-carnitine group and examined the effects on body composition and physical function [105]. As a result, there was no significant change in AMA in the carnitine group, but a significant decrease was observed in the non-carnitine group, with a difference of 6.22% (95% confidence interval (CI): 1.90–10.5, $P = 0.037$) between the two groups, and no significant change in grip strength in the carnitine group, but significantly lower in the non-treatment group. The difference between the two groups was 4.27% (95% confidence interval: 0.42–8.12, $P = 0.030$), and there was no significant change in LBM in the carnitine group, but a significant decrease was observed in the non-treatment group, and the difference between the two groups was 2.92% (95% confidence interval: 1.28–4.61, $P = 0.0007$). In other words, carnitine supplementation therapy can be expected to prevent the decline in muscle mass and muscle strength in dialysis patients with carnitine deficiency.

Matsufuji et al. reported that intravenous carnitine supplementation for 1 years is associated with change in knee extensor strength in patients undergoing hemodialysis [263].

2. Mechanism of onset of muscle cramps in hemodialysis patients

Muscle cramps, which are common in dialysis patients, often occur mainly in the gastrocnemius muscles of the lower extremities in the latter half of dialysis treatment and may also develop in the upper limbs and abdomen. In addition, it may occur not only during dialysis but also while going out or at night, and it reduces QOL and sleep quality. Takahashi examines the pathogenesis of cramps in dialysis patients from his own research. According to the report, various factors such as excessive dehydration, use of dialysate with low Na or inappropriate osmotic pressure, and hypocalcemia were cited, and carnitine deficiency was considered to be one of them [106, 107] (Fig. 14.21).

The intracellular calcium ion concentration in muscle cells is kept very low at resting state, and calcium ions are stored in the sarcoplasmic reticulum. Depolarizing stimuli at the neuromuscular junction induce an influx of very small amounts of calcium ions from outside the cell into the cell through dihydropyridine receptors (voltage-dependent L-type calcium channels) in the T-tubules, which are membrane structures with a deeply invaginated plasma membrane, and this triggers the release of calcium ions from ion channels formed by ryanodine receptors in the sarcoplasmic reticulum, increasing the intracellular calcium ion concentration drastically to 2.5×10^{-6} to 10^{-5} M (this phenomenon called as "calcium-induced calcium release"). Released calcium ions bind to troponin, which is bound to actin, and the interaction between actin and myosin causes muscle contraction. Conversely, when muscle relaxes, intracellular calcium ions are taken up again into the sarcoplasmic reticulum and the intracellular calcium ion concentration drops to 5×10^{-8} to 10^{-7} M, calcium ions are removed from troponin, and the interaction between actin and myosin is released. This causes relaxation. The reuptake of calcium ions into the

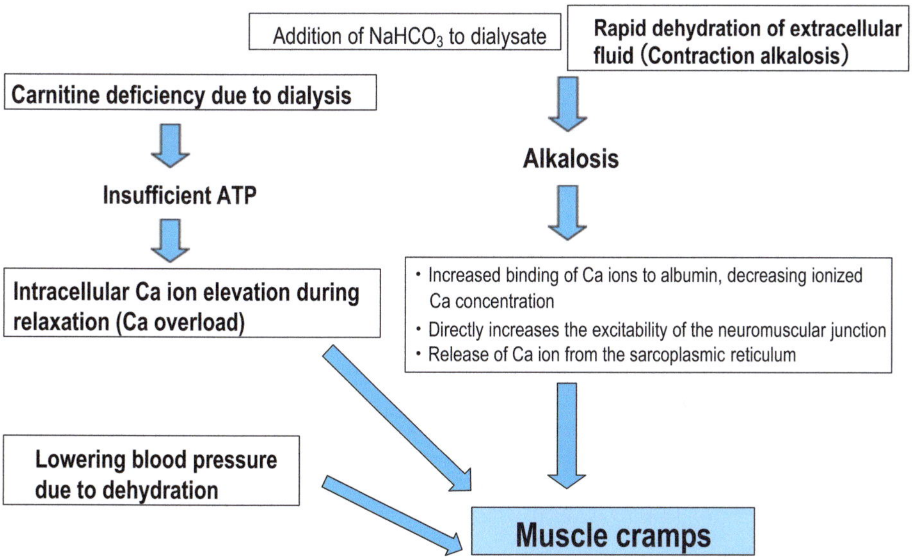

Fig. 14.21 Proposed pathogenic mechanism of muscle cramps in dialysis patients. (From Refs. [106, 107])

sarcoplasmic reticulum during muscle relaxation is carried out by Ca-ATPase, which requires ATP. Muscle is said to consume more ATP during relaxation than during contraction. In skeletal and cardiac muscle cells, ATP is produced mainly by fatty acid oxidation, so that intracellular calcium ion concentrations cannot be reduced sufficiently when carnitine is deficient and fatty acid oxidation is reduced and ATP cannot be supplied sufficiently. This phenomenon is called calcium overload and is thought to be the cause of myalgia and cramps due to skeletal muscle overstrain, as well as elevated end-diastolic pressure and diastolic failure in heart failure.

Cramps caused by carnitine deficiency are mainly reported in dialysis patients and cirrhosis patients. In particular, on March 10, 2011, carnitine administration for carnitine deficiency was expanded in Japan, and it was used for muscular symptoms such as cramps in dialysis patients, as well as hyperammonemia and severe cramps in patients with cirrhosis, and it was reported that carnitine administration was significantly effective in reflexing (see Chap. 15, "Liver Disease and Carnitine Deficiency").

Dialysis patients are often treated with Shakuyakukanzoto (Peony-Licorice Decoction), but there are patients who are ineffective, or due to side-effect such as hypokalemia and heart failure, and the package insert describes that the elderly are "cautiously administered," and there are patients who should not use Shakuyakukanzoto in dialysis patients, who are often complicated by heart failure. For patients who are suspected of having carnitine deficiency, such as those who have been on dialysis for a long time and have low muscle mass, it is worth trying a safer carnitine administration.

14.4.3 Cardiac Symptoms

1. "Dialysis heart" and carnitine

The myocardium requires a large amount of energy for its constant pumping function, and 60–70% of its ATP production depends on the β oxidation of long-chain fatty acids. Therefore, cardiomyopathy and heart failure are frequent clinical manifestations in severe carnitine deficiency due to inherited metabolic diseases. The leading cause of death in dialysis patients is heart failure, and cardiac symptoms such as cardiac hypertrophy, cardiomyopathy, decreased cardiac function, arrhythmia, and hypotension during dialysis are often present in Japan without obvious organic heart disease, or even after ischemia is treated, and are sometimes referred to as "dialysis heart" in Japan [108–113]. The term "dialysis heart" is close to uremic cardiomyopathy used overseas, but it is not an official term in cardiology, and is simply treated as "cardiovascular complications of dialysis patients" in overseas. The concept of cardiorenal linkage has been established since around 2000, and its mechanism has been actively studied, and the cardiac symptoms peculiar to dialysis patients are a typical example. In addition to rapid dehydration at dialysis, abnormal electrolyte balance, renal anemia, hypervolemia, secondary hyperthyroidism, abnormal calcium and phosphorus metabolism, arteriovenous shunting, and poor nutrition, there are many other factors such as diabetes, hypertension, and progression of atherosclerosis due to hyperlipidemia. So that the etiology of heart symptoms such as heart failure and cardiac hypertrophy in dialysis patients is not simple, but carnitine deficiency is thought to be one of the factors.

For example, in 1983, Kudoh et al. reported that the cardiothoracic ratio of 33 hemodialysis patients (mean age 42.7 ± 13.1 years) was inversely correlated with plasma free carnitine concentration, suggesting that carnitine deficiency is one of the causes of cardiac hypertrophy commonly seen in "dialysis heart" (Fig. 14.22) [114, 127].

2. Carnitine supplementation therapy for cardiac symptoms in dialysis patients

There have been many clinical studies to date, and many of them have reported that administering carnitine to dialysis patients has improved clinical symptoms such as hypotension, palpitations, shortness of breath, and chest pain during dialysis, as well as improvements in cardiothoracic ratio, left ventricular ejection fraction (LVEF), left ventricular end-diastolic volume, and other indicators of left ventricular contractile and diastolic capacity (Table 14.9) [78, 96, 114–125]. Maruyama et al. also reviewed 12 clinical trials of carnitine treatment in dialysis patients up to 2018 for cardiac symptoms and hypotension during dialysis [115]. Although some of these reports indicate that carnitine administration was ineffective, it is thought that clinical trials in which many of the subjects were not carnitine deficient or had no clear cardiac symptoms (e.g., patients without cardiac hypertrophy) would not make a significant difference in the effect of carnitine administration. The main clinical studies shown in Table 14.9 are introduced below.

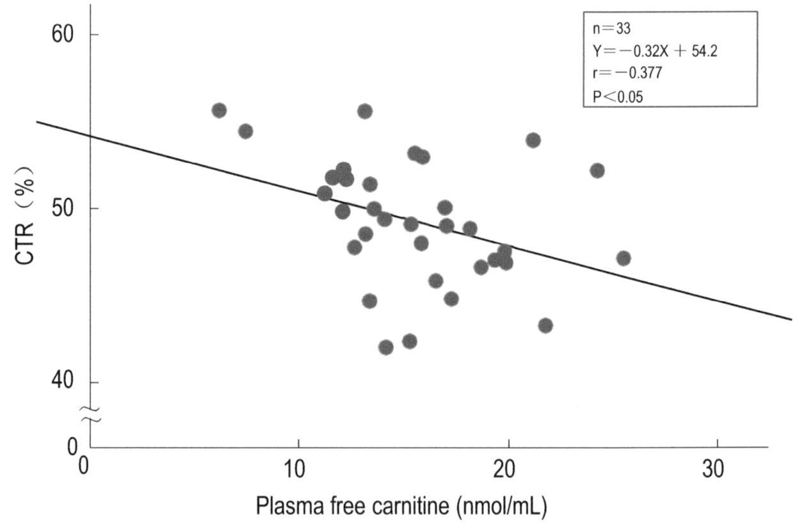

Fig. 14.22 Correlation between plasma carnitine levels and the cardiothoracic ratio (CTR). (From Ref. [127])

Table 14.9 Studies of effects of L-carnitine on cardiac function, arrhythmias, and hypotension in dialysis patients

Studies with oral administration					
References	Study design	Number of subjects	Dose of carnitine	Duration	Results
Casciani, 1982 (Ref. 76)	Double-blind, crossover, placebo-controlled trial	18 Group 1: 9 Group 2: 9	990 mg/day	60 days Wash out 10 days	Decreased hypotension during dialysis ($P < 0.001$)
Suzuki, 1982 (Ref. 114)	Open study	17	2000 mg	2 months	Decreased ventricular extrasystole or supraventricular tachycardia during dialysis ($P < 0.05$)
Fujita, 1988 (Ref. 115)	Open study	28 Carnitine group: 14 Control group: 14	1200 mg/day	3 months	Improved persistent hypotension before dialysis
Sakurabayashi, 1999 (Ref. 116)	Open study	11	1000 mg (First month) 500 mg (Next 1 month)	2 months	Improved BMIPP wash out rate ($P = 0.001$)
Matsumoto, 2000 (Ref. 117)	Open study	9	500 mg/day	6 months	Increased EF ($P = 0.005$) and Thoracic/Heart ratio($P = 0.042$). Reduction in symptoms such as dyspnea, palpitations, and chest pain during exercise

Studies with Intravenous administration					
References	Study design	Number of subjects	Dose of carnitine	Duration	Results
Fagher, 1985 (Ref. 118)	Double-blind, parallel, placebo-controlled trial	28	2 g after dialysis 3 times a week	1.5 months	No change in EF or cardiac dimension parameters
Fricke, 1988 (Ref. 119, in German)	Open study	16	20 mg/kg after dialysis 3 times a week	2 months	Reduced ventricular arrhythmias, no change in EF
Ahmad, 1990 (Ref. 94)	Double-blind, parallel, placebo-controlled trial	82 Carnitine group: 38 Placebo group: 44	20 mg/kg after dialysis 3 times a week	6 months	Decrease in hypotension during dialysis ($P = 0.02$), arrhythmia unchanged
van Es', 1992 (Ref.120)	Open study	16	1000 mg after dialysis 3 times a week	3 months	Elevated EF ($P < 0.05$) in all patients, elevated EF ($P < 0.02$) in patients with conditions, no significant difference in patients without symptoms
Záruba, 1994 (Ref. 121)	Open study	10	1000 mg after dialysis 3 times a week	1 month	In 6 of 10 cases, vascular refilling (refilling of circulating blood volume due to extravascular fluid inflow = an important factor in preventing hypotension during dialysis) increased.

(continued)

Table 14.9 (continued)

Studies with Intravenous administration					
References	Study design	Number of subjects	Dose of carnitine	Duration	Results
Trovato, 1998 (Ref. 122)	Open study	60	1000 mg intravenously after dialysis 3 times a week, and 1000 mg/day orally	More than 3 years	EF increased, EDV decreased ($P < 0.05$)
Romagnoli, 2002 (Ref. 123)	Open study	23	1000 mg after dialysis 3 times a week	8 months	EF increased, EPO requirements decreased ($P < 0.05$)

Modified from Ref. [3]
EF: ejection fraction, EDV: end diastolic volume, BMIPP: ^{123}I-labeled β-methyl-p-iodophenyl-pentadecanoic acid, EPO: erythropoietin

Suzuki et al. examined the effect of carnitine administration on the number of occurrences of ventricular premature contraction or supraventricular tachycardia by administering 2 g/day of carnitine orally for 8 weeks to eight patients who experienced ventricular premature contraction or supraventricular tachycardia during dialysis [116]. As a result, a significant suppression of the onset of arrhythmias was observed from 4 weeks after administration. The antiarrhythmic effects of carnitine are thought to be due to the effect of excreting the long-chain acyl group from the mitochondria as described in the section on arrhythmias in Chap. 16, "Cardiovascular Diseases and Carnitine Deficiency."

Fujita et al. examined the effects on blood pressure and lipid metabolism of 28 long-term hemodialysis patients with persistent hypotension by dividing them into a carnitine group (1200 mg/day, divided 3 times, oral administration, 12 weeks) and a control group [117]. As a result, the mean blood pressure before dialysis was 67.2 ± 13.9 mmHg, but it increased significantly to 74.2 ± 15.3 mmHg at 8 weeks and 80.0 ± 11.1 mmHg at 12 weeks, and there was no change in the control group. In addition, a significant increase in total ketone concentrations was observed in the carnitine group.

Sakurabayashi et al. performed ^{123}I-labeled BMIPP myocardial scintigraphy before and after oral carnitine administration (1 g/day for the first month and 500 mg/day for the next month) in 11 dialysis patients (mean age 60.7 ± 8.5 years, 21.1 ± 6.8 years on dialysis). Myocardial washout rate, an index of the rate of fatty acid metabolism in the myocardium, was decreased in dialysis patients compared to healthy controls (8 subjects, mean age 62.3 ± 9.4 years), but improved to the same level as healthy controls after 2 months of carnitine administration [118] (Fig. 14.23b). This may indicate that myocardial fatty acid metabolism, which had been decreased, was restored by carnitine administration.

Matsumoto et al. performed carnitine supplementation therapy (oral administration, 500 mg/day, 6 months) in 9 dialysis patients (mean age 62.1 ± 16.9 years, 107 ± 54 months of dialysis history) with cardiac symptoms and found that 2 out of 4 patients had dyspnea during exercise, 4 out of 4 patients had palpitations, and 4 out of 4 patients had chest pain. LVEF and left ventricular mass were significantly improved [119] (Fig. 14.24). Among the trials discussed in Table 14.9, Fagher et al. [120] and Fricke et al. [121] did not show any improvement in LVEF with carnitine administration, but the latter report showed a reduction in ventricular arrhythmias.

van Es' et al. examined the relationship between blood free carnitine concentration and ejection fraction in seven patients who developed hypotension during dialysis and nine patients who did not and found that blood free carnitine concentration and LVEF were positively correlated [122] (Fig. 14.25). Carnitine

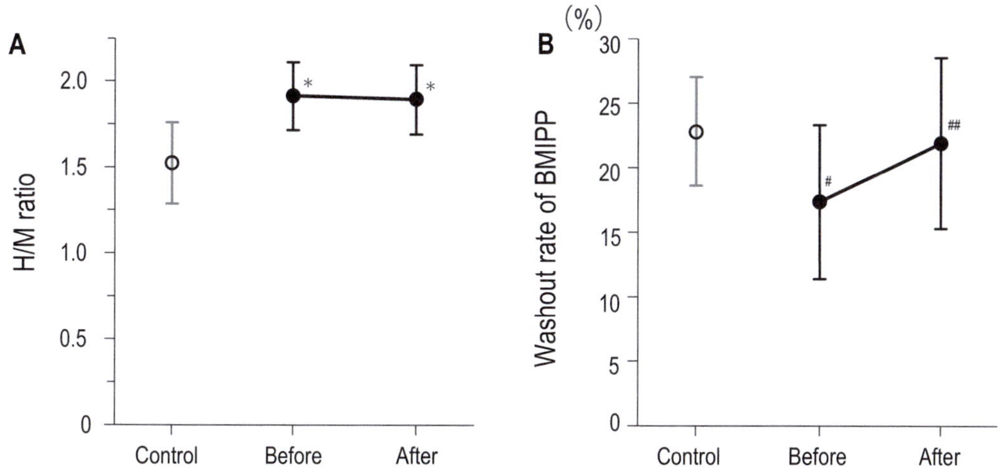

Subjects and Methods: Measurement of H/M ratio and myocardial washout rate by BMIPP myocardial scintigraphy before and after oral administration of L-carnitine 1 g/day for 1 month and 0.5 g/day for 1 month in 11 patients on hemodialysis

Fig. 14.23 Results of BMIPP myocardial scintigraphy of patients before and after carnitine supplementation compared with normal control. (**a**) H/M ratio, (**b**) myocardial washout rate of BMIPP. Data are expressed as means ± SD, $^{*}P < 0.005$ vs. control, $^{\#}P < 0.05$ vs. before, $^{\#\#}P < 0.001$ vs. before. H/M: Heart to mediastinal ratio, BMIPP: ^{125}I-labeled β-methyl-p-iodophenyl-pentadecanoic acid. (From Ref. [118])

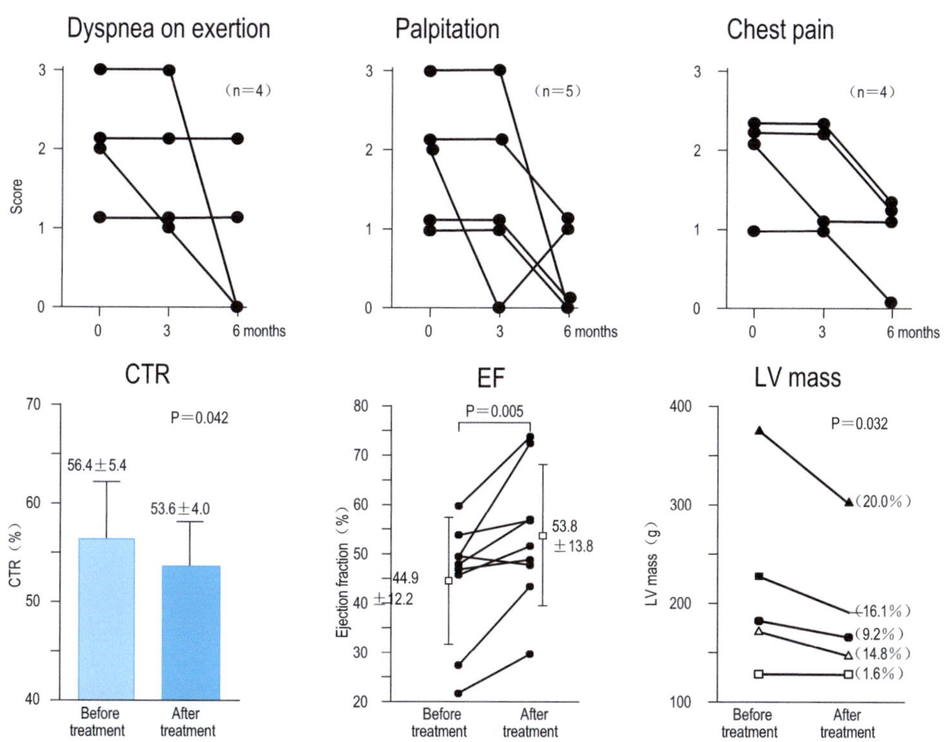

Fig. 14.24 Effect of carnitine supplementation on cardiac symptoms (dyspnea on exertion, palpitation, chest pain), cardiac hypertrophy (CTR and LV mass), and cardiac function (EF) in patients on hemodialysis. CTR: cardiothoracic ratio, EF: ejection fraction, LV mass: left ventricular mass. (From Ref. [119])

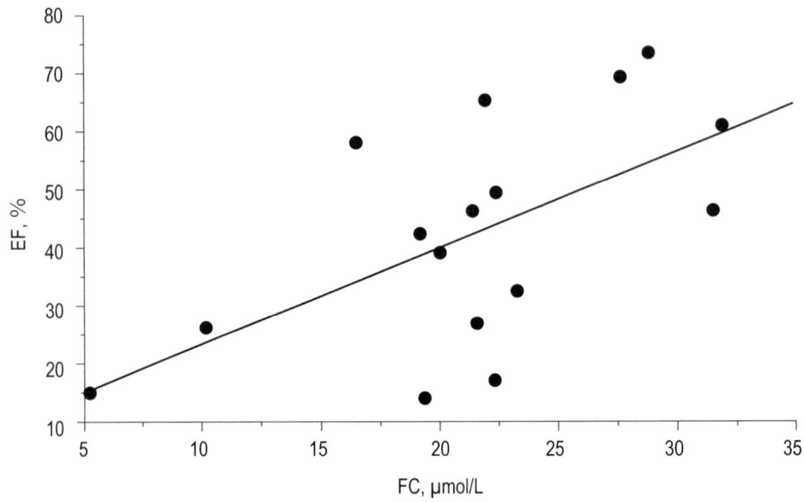

Fig. 14.25 The relationship between EF and FC before carnitine supplementation therapy in 16 dialysis patients ($P < 0.02$). EF: ejection fraction, FC: free carnitine concentration. (From Ref. [122])

supplementation therapy after dialysis three times a week (1000 mg, intravenous administration at the end of dialysis, 3 months) significantly increased LVEF from 30.4 ± 16.0% to 41.7 ± 15.9% before and after carnitine treatment in seven patients with symptoms of hypotension during dialysis ($P < 0.02$), but 51.8 ± 17.0% to 56.7 ± in nine patients without hypotensive symptoms, not a significant increase, and significant increase from 42.4 ± 19.4% to 48.6 ± 17.6% ($P < 0.05$) in all 16 cases [122]. In other words, the LVEF-improving effect of carnitine is remarkable in patients with low blood free carnitine concentrations and cardiac symptoms such as hypotension during dialysis, and it is considered that a larger number of cases are required to confirm the effect with significant differences in asymptomatic patients.

Romagnoli et al. also found that 1000 mg of carnitine intravenously administered after dialysis three times a week to 11 dialysis patients (mean age 63 ± 10 years, 20 ± 17 months of dialysis) with LVEF ≦45% showing a significant improvement in LVEF ($P < 0.05$) from 2 months after administration [125] (Fig. 14.26).

Among these clinical studies, most of the studies in Japan have used a supplement product of L-carnitine, but in March 2011, the indication for L-Cartin® tablets was expanded to "carnitine deficiency" by a public knowledge application, and it became possible to use it in dialysis patients, and clinical studies of carnitine supplementation therapy in dialysis patients have been published in Japan. Some of them are introduced below.

Kudoh et al. examined the LVEF and number of episodes of dialysis hypotension in 18 dialysis patients in a double-blind study (900 mg/day, 3 months) and no carnitine (900 mg/day, 3 months) [126, 127]. Although there was a large variation, LVEF was significantly increased by carnitine supplementation therapy (61.8 ± 16.0% ⇒ 64.4 ± 13.8%, $P < 0.05$), and hypotensive episodes during dialysis that were about 4 times per month were significantly reduced (4.0 ± 1.7 times/month ⇒ 1.3 ± 0.9 times/month, $P < 0.05$ Fig. 14.27).

Recently, Chewcharat A. et al. reported a systematic review, meta-analysis, and trial sequential analysis of the effect of levocarnitine supplementation on dialysis-related hypotension and concluded that carnitine supplementation helps prevent hypotension [264].

Hypotension during dialysis is caused by a variety of factors, but cardiac dysfunction is considered to be the main factor. It is of great importance to reduce the burden on patients and

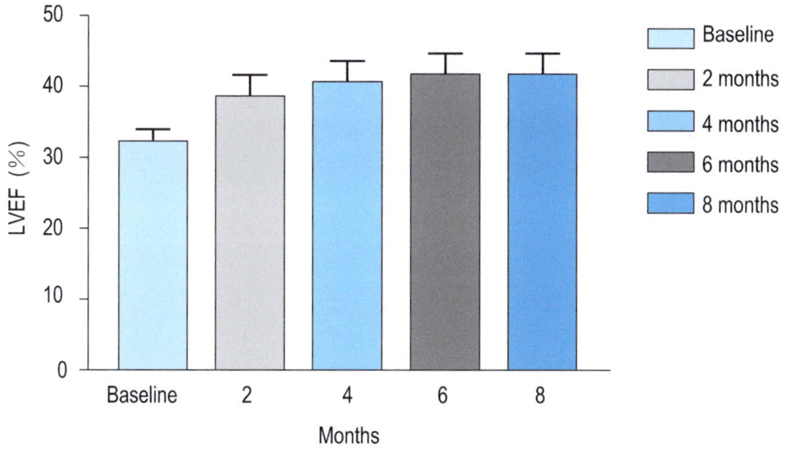

Fig. 14.26 Effect of carnitine supplementation (intravenously, 1000 mg/3 times a week, after dialysis session) in 11 dialysis patients whose LVEF were low. There was significant improvement from 2 months treatment ($P < 0.05$ vs. baseline). LVEF: left ventricular ejection fraction. (From Ref. [125])

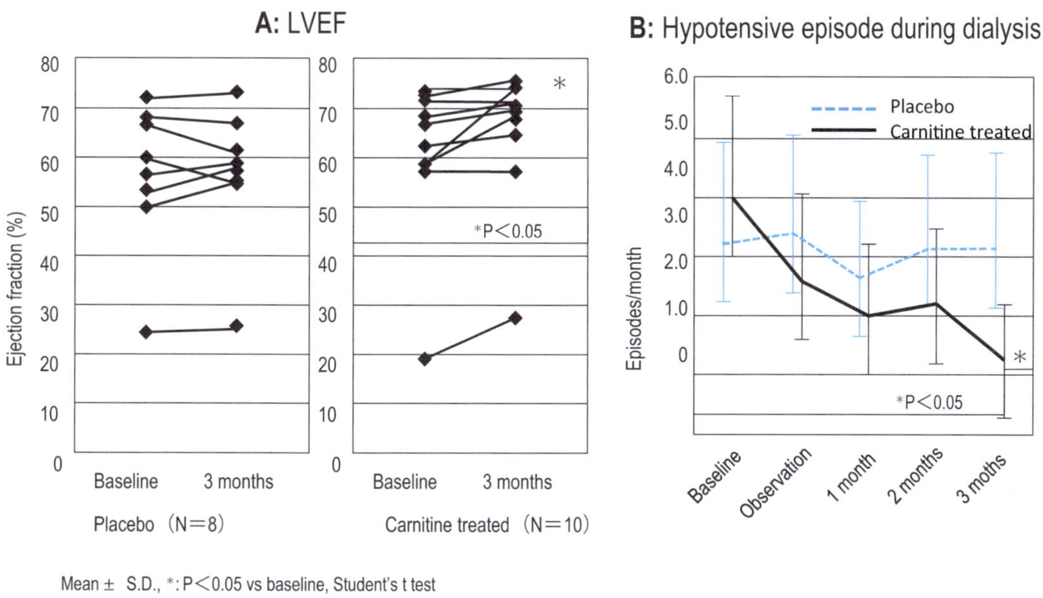

Fig. 14.27 Effect of carnitine supplementation (oral. 900 mg/day, for 3 months) on LVEF (**a**) and hypotensive episodes during dialysis (**b**) in 18 dialysis patients. LVEF: left ventricular ejection fraction. (From Refs. [126, 127])

clinic staffs because hypotension episode causes interruption of dialysis or the delay in the completion time of dialysis to deal with it.

Higuchi et al. conducted a randomized controlled trial of 148 dialysis patients in the carnitine treatment group (75 patients, 20 mg/kg/day, oral administration, 12 months) and the control group (73 patients) to investigate the effects on LVEF and left ventricular mass index (LVMI) [128]. As a result, there was no change in the placebo group after 6 and 12 months compared to before administration, but in the carnitine treatment group, LVEF increased significantly from 6 months after administration. LVMI, which is an indicator of left ventricular hypertrophy, was defined as >102 g/m^2 in men and >88 g/m^2 in women, and the carnitine treatment group was divided into 52 patients with left ventricular hypertrophy and 23 patients without left ventricular hypertrophy. As a result, a significant increase in LVEF and LVEF change was observed after 6 and 12 months of carnitine treatment in

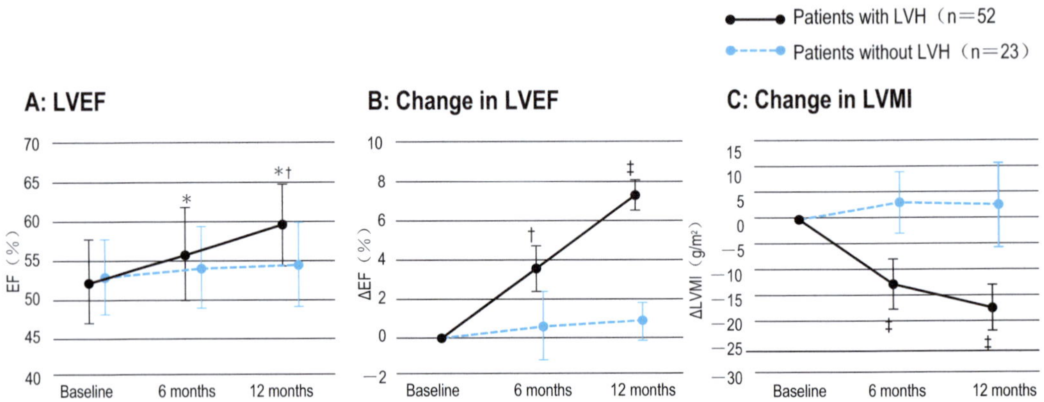

Fig. 14.28 Effect of carnitine supplementation (oral, 20 mg/kg/day, for 12 months) on LVEF (**a**), change in LVEF (**b**) and change in LVMI (**c**) in dialysis patients. (**a**) Ejection fraction(EF)data are expressed as mean ± SD, (**b**) Change in EF (ΔEF) is expressed as mean ± 95% confidence interval, (**c**) Change in left ventricular mass index (ΔLVMI) is expressed as mean ± 95% confidence interval. **P*<0.001 vs. baseline, †*P*<0.01, ‡*P*<0.001 vs. the levocarnitine subgroup of patients without left ventricular hypertrophy (LVH). LVEF: left ventricular ejection fraction, LVMI: left ventricular mass index. (From Ref. [128])

the group with left ventricular hypertrophy, and LVMI was significantly reduced, but no significant effect of carnitine administration on LVEF or LVMI was observed in the group without left ventricular hypertrophy (Fig. 14.28).

In addition, Higuchi et al. reported a case of hemodialysis patients with reduced cardiac function who were significantly effective in carnitine supplementation therapy, although it is only one case report [129]. In this case, a 58-year-old man underwent hemodialysis due to diabetic nephropathy, and an echocardiography showed an LVEF of 48.9%, E/e′ = 19.5, systolic and diastolic disorders, and an LVMI of 151 g/m^2 and left ventricular hypertrophy. Since EPO resistance was also observed, carnitine supplementation therapy (1200 mg/day, orally administered) was started, and after 1 year, there was a marked improvement in LVEF (48.9 ⇒ 72.7%), a marked improvement in LVMI (151 ⇒ 107 g/m^2), and an improvement in NT-proBNP (12,800 ⇒ 7850 pg/mL). Significant improvement in ERI (13.5 ⇒ 3.9), improvement in brachial-ankle pulse wave velocity (baPWV) (1832 ⇒ 1545 cm/s) which is an indicator of atherosclerosis, and improvement in brachial muscle area (32.9 ⇒ 39.3 cm^2) were observed, and alanine aminotransferase (ALT) and aspartate aminotransferase (AST) were also slightly improved.

Maruyama et al. reviewed 12 clinical studies that examined the effects of carnitine on cardiac symptoms and hypotension during dialysis as described in previous cases, including a recent report from Japan [115]. As a result, most of the studies showed that carnitine supplementation therapy reduced LVMI, increased LVEF, and decreased the frequency of hypotension during dialysis, but there were two reports that no effect was observed. Maruyama et al. argue that LVEF before carnitine in these trials was normal, and that caution should be exercised in interpreting the results of these trials. This is because, in a clinical study in Maruyama et al.'s group, the responders of carnitine administration were patients with left ventricular hypertrophy [128]. In other words, the effect of carnitine administration was observed in subjects with clinical symptoms such as decreased cardiac function, left ventricular hypertrophy, and hypotension during dialysis.

Kaneko et al. administered intravenous carnitine at 2000 mg/week to a female dialysis patient (51 years old, NYHA class III) with heart failure with dyspnea symptoms and found that a marked improvement was observed in the cardiothoracic ratio (71.3% ⇒ 51.4%), LVEF (50.6% ⇒ 71.2%), LVMI (214.0 g/m^2 ⇒ 144.0 g/m^2), and BNP (8257 pg/mL ⇒ 378 pg/mL) (Figs. 14.29 and 14.30) [130]. In addition, myocardial fatty acid metabolism assessed by BMIPP myocardial scintigraphy was significantly improved (BMIPP score 20 ⇒ 6).

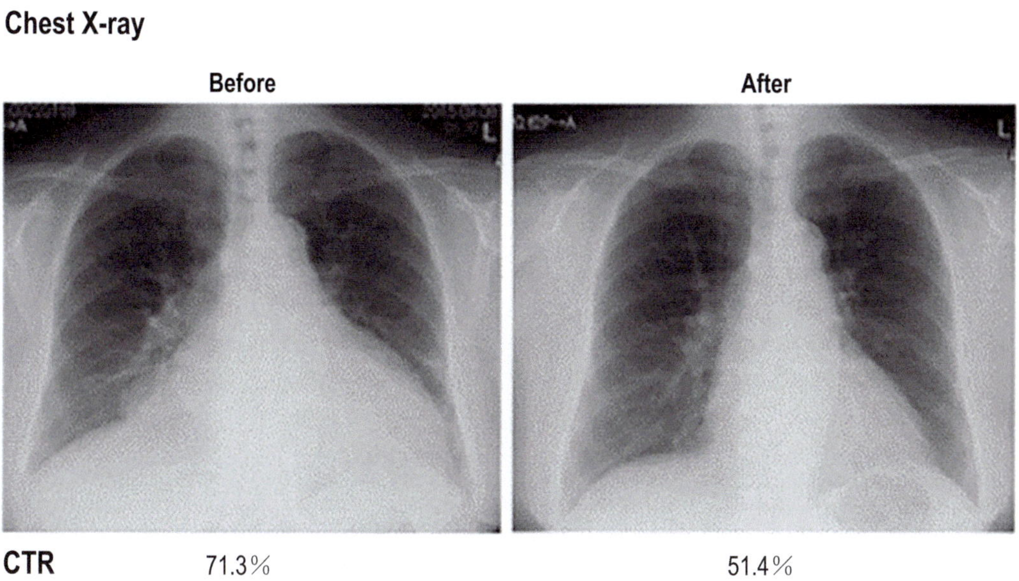

Fig. 14.29 Image of chest X-ray of a heart failure dialysis patient with significant effect after intravenous carnitine administration (intravenous, 1000 mg/after dialysis session, 2 times a week (total 2000 mg/week), for 1 year). After 1 year, CTR improved from 71.3 to 51.4%. CTR: cardiothoracic ratio. (Modified from Ref. [130])

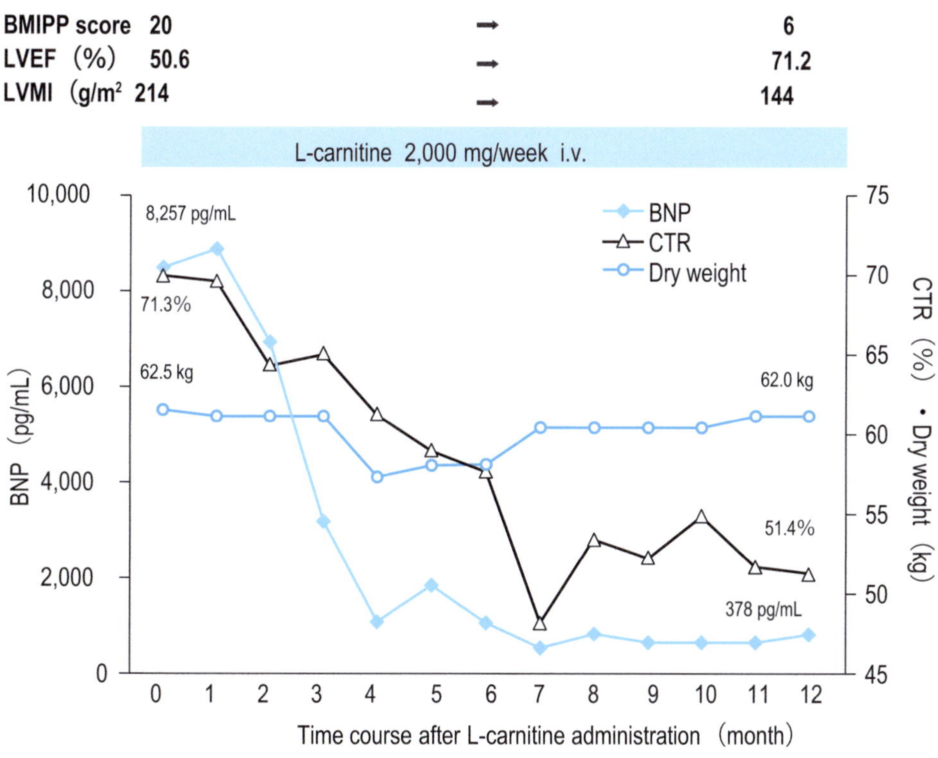

Fig. 14.30 Changes in BMIPP score, LVEF, LVMI, and BNP in heart failure dialysis patient with significant effect after intravenous carnitine administration (intravenous, 1000 mg/after dialysis session, 2 times a week (total 2000 mg/week), for 1 year). After carnitine supplementation for 1 year, BMIPP score decreased from 20 to 6 and BNP level from 8257 to 378 pg/mL, and LVEF improved from 50.6 to 71.2 and LVMI from 214 to 144 g/m^2. BMIPP: ^{125}I-labeled β-methyl-*p*-iodophenyl-pentadecanoic acid, LVEF: left ventricular ejection fraction, LVMI: left ventricular mass index, BNP: brain natriuretic peptide. (From Ref. [130])

Naito et al. administered intravenous carnitine to 88 hemodialysis patients (1000 mg three times a week for 12 months after completion of dialysis) and investigated its effect on cardiac function using echocardiography [131]. The early diastolic wave (E), atrial systolic wave (A), and early diastolic motion velocity (e′) of the left ventricular inflow blood velocity wave form were measured, and defined as indices of left ventricular diastolic dysfunction, E/A $\leq$ 0.8, E/e′ > 14, and left atrial volume index (LAVI) >34 mL/m,[2] and LVEF <55% as an indicator of left ventricular systolic dysfunction. They also evaluated the effect of carnitine administration on heart failure with preserved ejection fraction (HFpEF), a subset of heart failure. First, E/A, an index of left ventricular diastolic function, increased at 3, 6, 9, and 12 months after carnitine administration in patients with E/A $\leq$ 0.8, which is considered to indicate left ventricular diastolic dysfunction, and the increase at 6 months was significant compared to that before administration. The increase was significant at 6 months and even at 12 months. E/e′, another index of left ventricular diastolic dysfunction, significantly decreased at 3 months after treatment in the group of patients with E/e′>14 (24 patients before treatment, who were considered to have left ventricular diastolic dysfunction), and at 12 months after treatment, E/e′ was significantly decreased.. The decrease was still significant at 12 months post-dose when evaluated as corresponding data at 12 months post-dose. However, patients with E/e′ $\leq$ 14 did not show these changes with carnitine administration (Please see Fig. 1 in Ref. [131]).

When the effect of carnitine administration was examined in two groups of hemodialysis patients, one with LAVI >34 mL/m^2 and the other with LAVI $\leq$34 mL/m,[2] a significant decrease in LAVI was observed from the third month of carnitine administration in the group of patients with LAVI >34 mL/m,[2] who were considered to have left ventricular diastolic dysfunction, and a significant decrease at 12 months after administration (P < 0.001). However, no such changes were observed in the group of patients with LAVI $\leq$34 mL/m^2, considered to have normal left ventricular diastolic function. Next, LVEF, an indicator of left ventricular systolic dysfunction, was analyzed separately for patients with LVEF <55% which is considered to indicate left ventricular systolic dysfunction and LVEF $\geq$55%. The increase was significant when evaluated as paired data (P < 0.01). No such changes were observed in the group of patients with LVEF $\geq$55%, which is considered normal left ventricular systolic function (Please see Fig. 2 in Ref. [131]).

Next, they examined the effect of carnitine supplementation on heart failure (HFpEF) with preserved ejection fraction. HFpEF was defined as LVEF>50%, LAVI>34 mL/m2, E/e′>14, and e′<7 cm/s according to the Guidelines for the Practice of Acute and Chronic Heart Failure (Joint Guidelines of the Cardiovascular Society of Japan/Japan Heart FailureHeart failure Society, 2017 revision), and patients with no significant decrease in left ventricular systolic function (LVEF >50%), LAVI > 34 mL/m^2, E/e′ > 14, and e′ < 7 cm/s. As a result, a significant improvement in these indices was observed in the LVEF > 50%, LAVI > 34 mL/m^2, E/e′ > 14, and e′ < 7 cm/s groups after 12 months of carnitine supplementation (Please see Fig. 3 in Ref. [131]).

The effects of carnitine supplementation on LVMI, which is an indicator of cardiac hypertrophy, and BNP, which is also an indicator of myocardial hypertrophy and ventricular load, were examined in all patients; LVMI showed a significant decrease from 6 months after carnitine administration and a significant decrease at 12 months after administration compared to before administration (P < 0.001). A significant decrease in BNP was observed 9 months after administration, but there was a large variation at 12 months after administration, and no significant decrease was observed compared to before administration (Please see Fig. 4 in Ref. [131]).

Sada et al. administered carnitine (400 mg/day, 30–30 mg starting dose of 300 mg/day for 50 kg, 200 mg/day for less than 30 kg, oral administration) to 21 hemodialysis patients (mean ages 74 ± 11 years, mean dialysis history 60 ± 84 months), and blood levels measured in the first 3 months exceeded the upper limit of normal (≥70 μmol/L) 100 mg reduction for 6 months) [132]. As a result, nutritional indices such as serum transferrin, retinol-binding protein, and triglycerides, triceps subcutaneous fat thickness, and NK cell activity were significantly increased, left ventricular internal diameter shortening rate (% fractional shortening) was significantly increased (42.2 ± 11.4% ⇒ 46.1 ± 9.8%, $P < 0.01$), and BNP was significantly reduced (621.4 ± 666.8 ⇒ 412.0 ± 426.0, $P < 0.02$).

Nishimura et al. administered carnitine (1000 mg intravenously infusion 3 times a week after dialysis for 1 year) to half of 72 dialysis patients with myocardial fatty acid imaging disorders and left ventricular insufficiency, and performed pre- and post-BMIPP myocardial scintigraphy, and evaluated 53 patients (27 in the carnitine group and 26 in the control group) excluding dropouts [133]. As a result, although there was no significant difference in BMIPP summed score before treatment in both groups, it was reported that seven patients (25.9%) in the carnitine group showed an improvement in BMIPP summed score (<−20%), while only two patients (7.7%) in the control group showed improvement. Multivariate logistic regression analysis showed that intravenous carnitine may improve myocardial fatty acid imaging in dialysis patients with hypo-albuminemia (<35 g/L).

Thus, intravenous carnitine has been shown to be effective in the treatment of left ventricular diastolic failure and failure in hemodialysis patients. Since heart failure and other cardiac diseases are the leading cause of death in hemodialysis patients, it is important to understand and apply these studies to daily clinical practice.

Since children have little muscle mass and the carnitine biosynthesis system is still developing, pediatric renal failure patients undergoing dialysis therapy are likely to suffer from carnitine deficiency, and lowering cardiac function is common in pediatric dialysis patients. Sgambat et al. evaluated the myocardial strain rate by echocardiography with speckle-tracking method that appears earlier than decline in cardiac function and examined the effect of carnitine supplementation (20 mg/kg/day, intravenous administration after dialysis) [134]. As a result, the low plasma free carnitine concentration increased, AC/FC ratio decreased (0.73 ± 0.04 ⇒ 0.65 ± 0.05, $P = 0.02$), and the longitudinal strain rate was significantly improved (−1.48 ± 0.11 ⇒ −1.91 ± 0.12, $P = 0.017$). In addition, Sgambat et al. also reported that carnitine deficiency occurs more frequently in pediatric patients with acute renal failure, etc., undergoing continuous renal replacement therapy (CRRT) in the intensive care unit where the treatment is prolonged longer. [135] In a pilot controlled cohort study of carnitine supplementation therapy in pediatric patients undergoing CRRT for acute renal failure, they reported that longitudinal and circumferential myocardial strain improvement were observed in the carnitine replacement group [136].

Carnitine deficiency by CRRT is reported not only in children but also in adult patients [260, 261], and Berger et al sound the alarm to micronutrients deficiency including carnitine [262].

Kido et al. performed carnitine supplementation therapy (1 g, intravenously after dialysis 3 times a week for 1 year) in 31 elderly hemodialysis patients (mean age 72 years) and examined the effects on amino acid metabolism. It was reported that the valine, tyrosine, phenylalanine, citrulline levels in blood increased significantly, and left ventricular internal diameter shortening rate and LVEF were also significantly improved [137]. It is thought that carnitine supplementation in dialysis patients affects not only fatty acid metabolism but also amino acid metabolism and improves energy production in skeletal muscle and myocardium.

14.4.4 Inflammation, Atherosclerosis and Carnitine in Dialysis Patients

In dialysis patients, complications such as diabetes, hypertension, and hyperlipidemia are common, and arteriosclerotic disease is also very common due to abnormalities in calcium-phosphorus metabolism. In addition to these classic atherosclerosis promoters, increased oxidative stress, accumulation of advanced glycation end products (AGEs), and chronic inflammatory factors are associated with low nutritional state called protein-energy wasting (PEW) or protein-energy malnutrition (PEM) and calcification of blood vessels and heart valves, which are thought to increase cardiovascular events in dialysis patients [138–140]. In dialysis patients, it has been reported that cardiovascular events are more likely in patients with high levels of C-reactive protein (CRP), which is one of the indicators of chronic inflammation, and low albumin levels as nutritional state indicators [141]. Therefore, when considering the causes and countermeasures for cardiovascular death, which are extremely common in dialysis patients, it is necessary to take measures not only for diastolic and systolic insufficiency of the heart itself but also for the promotion of atherosclerosis in the large and microvasculature systems.

Savica et al. divided 113 dialysis patients into a carnitine-treated group (48 patients, 20 mg/kg intravenous infusion, 3 times a week after dialysis for 6 months) and a placebo group (65 patients) and examined changes in serum CRP, albumin, transferrin, and Hb [142]. As a result, CRP was significantly reduced by carnitine supplementation (1.8 ± 1.2 ⇒ 1.2 ± 0.2 mg/dL, $P < 0.002$), and there was no change in the placebo group. In addition, when patients with CRP≧3 mg/dL and CRP < 3 mg/dL were analyzed, carnitine supplementation significantly reduced CRP ≧ 3 mg/dL patients, but there was no significant change in the placebo group, and there was no significant change in both groups in patients with CRP < 3 mg/dL (Fig. 14.31).

In addition, serum albumin was significantly increased in the carnitine supplementation group (3.6 ± 0.3 ⇒ 3.9 ± 0.4 g/dL, $P < 0.0001$), but a significant decrease was observed in the placebo group (3.7 ± 0.2 ⇒ 3.5 ± 0.2 g/dL, $P < 0.0001$, Fig. 14.32). In addition, carnitine supplementation significantly increased serum transferrin, Hb levels, and BMI ($P < 0.0002$, $P < 0.0001$, and $P < 0.0001$).

Duranay et al. divided 42 dialysis patients into a carnitine supplementation group (intravenous infusion of 20 mg/kg after dialysis 3 times a week for 6 months) and a placebo group to examine changes in CRP and other factors, and as reported by Savica et al., CRP was significantly reduced, and transferrin, serum total protein, albumin, and BMI were significantly increased [143].

Suchitra et al. divided 35 dialysis patients into a carnitine supplementation group (intravenous injection of 1000 mg/week 3 times a week after dialysis, 6 months) and a control group, and examined changes in high-sensitivity CRP (hsCRP) and found that a significant decrease in hsCRP was observed in the carnitine supplementation group (2.1 ± 0.6 ⇒ 0.67 ± 0.1 mg/dL. $P = 0.02$), and hsCRP was significantly elevated in the control group (0.9 ± 0.4 ⇒ 1.4 ± 0.7 mg/dL, $P = 0.01$) [144].

Shakeri et al. examined the effects on inflammatory cytokines, CRP, Lp(a), and oxidative stress in 36 dialysis patients with high levels of lipoprotein(a)(Lp(a)) in the carnitine supplementation group (1000 mg/day, orally, 12 weeks) and placebo [145]. As a result, CRP decreased by 29% ($P < 0.05$) and interleukin-6 decreased by 29% ($P < 0.001$) in the carnitine supplementation group, but there was no change in the control group. In addition, there was no change in interleukin-1β, TNFα, Lp(a), and oxidized LDL with carnitine administration.

Tabibi et al. divided 36 dialysis patients into a carnitine supplementation group (1000 mg/day, oral administration, 12 weeks) and a placebo group and examined changes in serum amyloid A (SAA), a systemic inflammatory marker that is said to be a risk factor for cardiovascular disease, and ICAM-1, E-selectin, P-selectin, and oxidized LDL, which are inflammatory markers of blood vessels [146]. As a result, carnitine supplementation significantly reduced SAA by 32% ($P < 0.001$), but there was no change in other vascular inflammatory markers.

Adachi et al. found that skin AGEs levels were inversely correlated with blood carnitine levels in dialysis patients [147]. Fukami et al., in the same group, divided dialysis patients into a carnitine treatment group (oral, 900 mg/day, 6 months) and a placebo group to investigate its effect on skin AGEs [148]. The autofluorescence of AGEs accumulated in the skin was measured using an automatic AGE reader (DiagOptics BV,

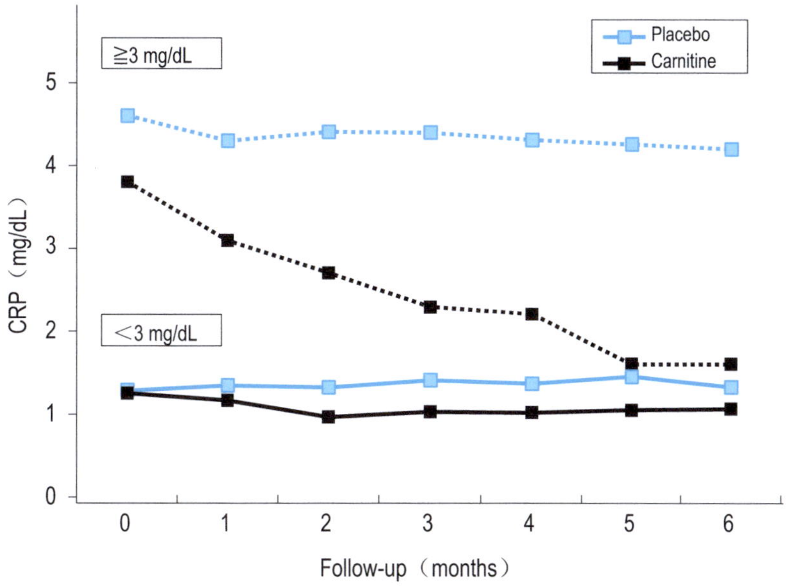

Fig. 14.31 Serum C-reactive protein (CRP) levels in maintenance hemodialysis patients with baseline levels <3 mg/dL (solid lines) or ≥3 mg/dL (dotted lines), treated with L-carnitine 20 mg/kg or placebo thrice weekly for 6 months. (Modified from Ref. [142])

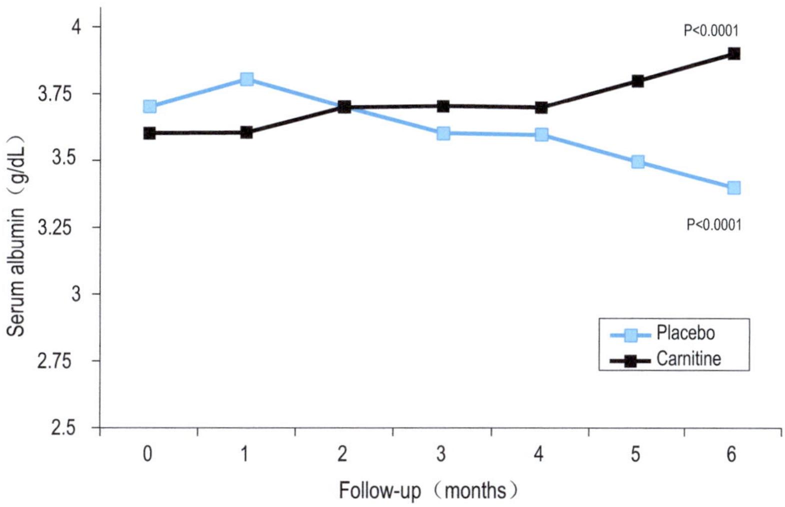

Fig. 14.32 Serum albumin concentrations (g/dL) in maintenance hemodialysis patients treated with L-carnitine or placebo for 6 months. (Modified from Ref. [142])

Groningen, Netherlands). As a result, total protein and albumin were significantly reduced in the placebo group, but there was no change in the carnitine group, while the concentration of skin AGEs and β2-microglobulin, which causes dialysis amyloidosis, was significantly reduced in the carnitine group. Multiple regression analysis revealed that the change in free carnitine concentration (Δfree carnitine concentration) before and after administration was an independent determinant of the change in AGE concentration (ΔAGEs) before and after administration (Fig. 14.33). This study shows that carnitine administration may be a novel treatment to reduce the accumulation of AGEs in tissues in dialysis patients with carnitine deficiency.

It has been reported that the level of skin AGEs, which is the cause of the unique darkening skin color of the elderly and dialysis patients, is not only a problem of appearance but also correlates with atherosclerosis indicators such as arterial plaque and coronary artery calcification [149], and it has also been shown that skin AGEs is a strong predictor of mortality and high levels of skin AGEs in hemodialysis patients reduce the 3-year survival rate [150]. Nagano et al. reported

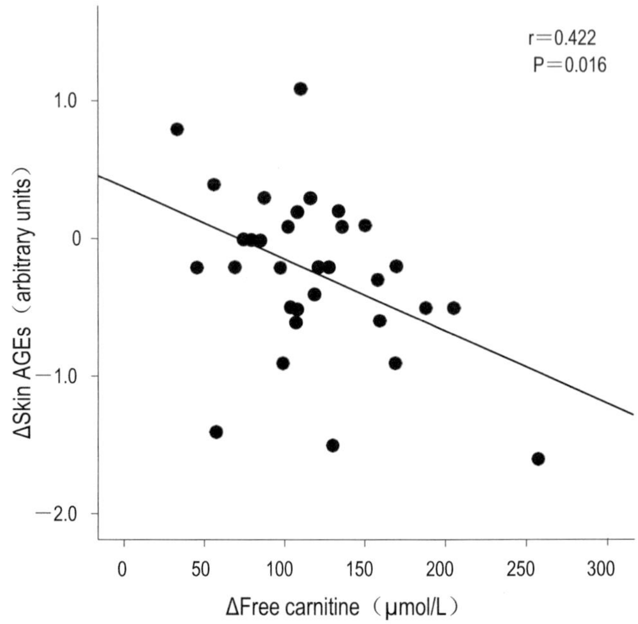

Fig. 14.33 Correlation between Δskin advanced glycation end products (AGEs) and Δfree carnitine in L-carnitine-treated hemodialysis patients (n = 32). (From Ref. [148])

that the level of skin AGEs in hemodialysis patients is an independent determinant of high-sensitivity CRP, and that high levels of skin AGEs and high-sensitivity CRP significantly increase the carotid atherosclerosis index [151] (Fig. 14.34). Fukami et al. explain the significance of carnitine supplementation therapy by focusing on carnitine deficiency and oxidative stress in hemodialysis patients, especially AGEs levels (Fig. 14.35) [152].

Although there are many studies that show that carnitine administration lowers inflammatory markers, oxidative stress factors, and AGEs, Higuchi et al. examined the effect of carnitine administration on brachial-ankle pulse wave velocity (baPWV), which is an indicator of atherosclerosis itself, in dialysis patients [153]. When the arteries become stiff due to fibrosis or calcification of atherosclerosis, the pulse wave velocity increases, and the progress of atherosclerosis can be measured. A total of 176 dialysis patients (mean age: 67.2 ± 10.3 years, mean dialysis history: 54 ± 51 months) with a free carnitine concentration of 40 μmol/L or less were randomly divided into a carnitine-treated group (oral dose of 20 mg/kg/day, 12 months) and placebo group, and baPWV was measured before, 6 months, and 12 months after administration. When 67 patients in the carnitine-treated group and 64 patients in the placebo group were excluded due to death or hospital transfer, a significant reduction in baPWV was observed in the carnitine-treated group from 6 months after administration ($P < 0.05$) and further decreased at 12 months ($P < 0.001$, Fig. 14.36).

These results suggest that carnitine supplementation therapy for dialysis patients reduces inflammatory markers such as CRP and SAA, AGEs, and reduces atherosclerosis itself, which was thought to progress irreversibly. In the future, further clinical studies are awaited to see whether the improvement in inflammatory markers and atherosclerosis indices by carnitine supplementation therapy actually leads to the suppression of cardiovascular events.

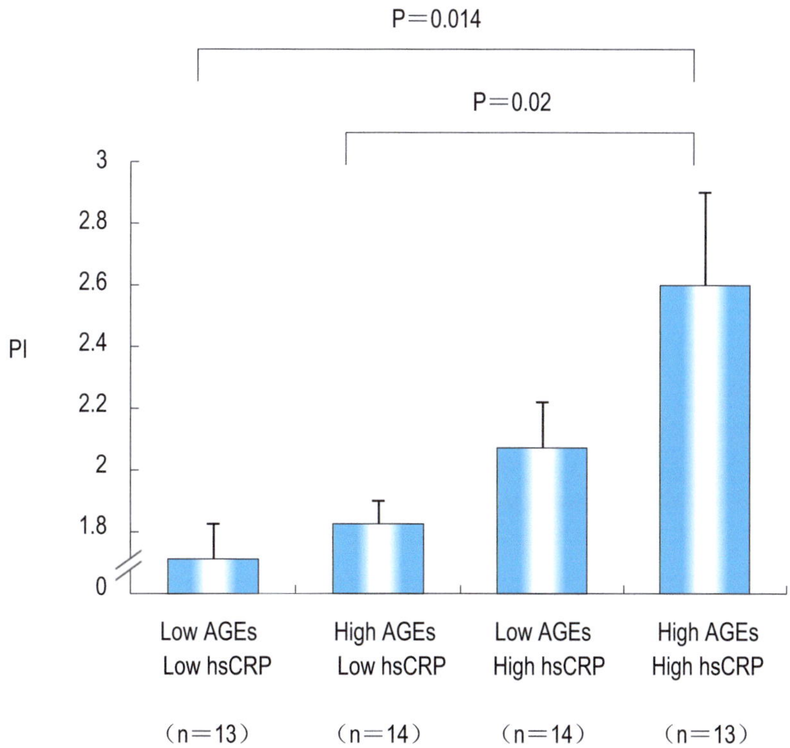

Fig. 14.34 Relationships among skin AGE, hsCRP, and carotid PI in hemodialysis patients. Low hsCRP: log ≦ 2.42, high hsCRP: log ≧ 2.43, low AGEs: ≦2.84, high AGEs: >2.85. AGE: advanced glycation end products, hsCRP: high-sensitive C-reactive protein, PI: pulsatility index. (From Ref. [151])

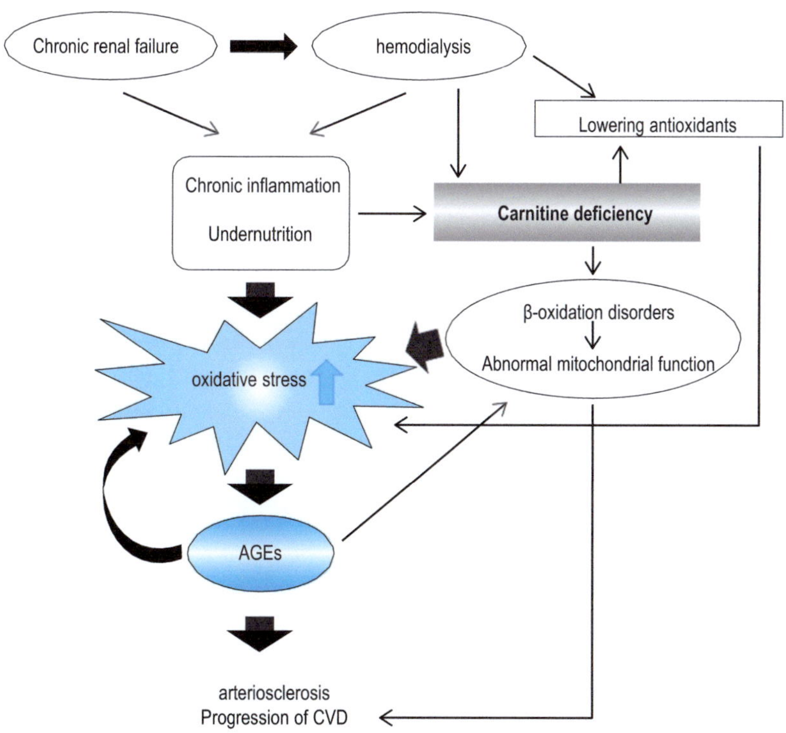

Fig. 14.35 Proposed mechanisms of progression of CVD via oxidative stress-AGEs system by carnitine deficiency. CVD: cardio-vascular diseases, AGEs: advanced glycation end products. (From Ref. [152])

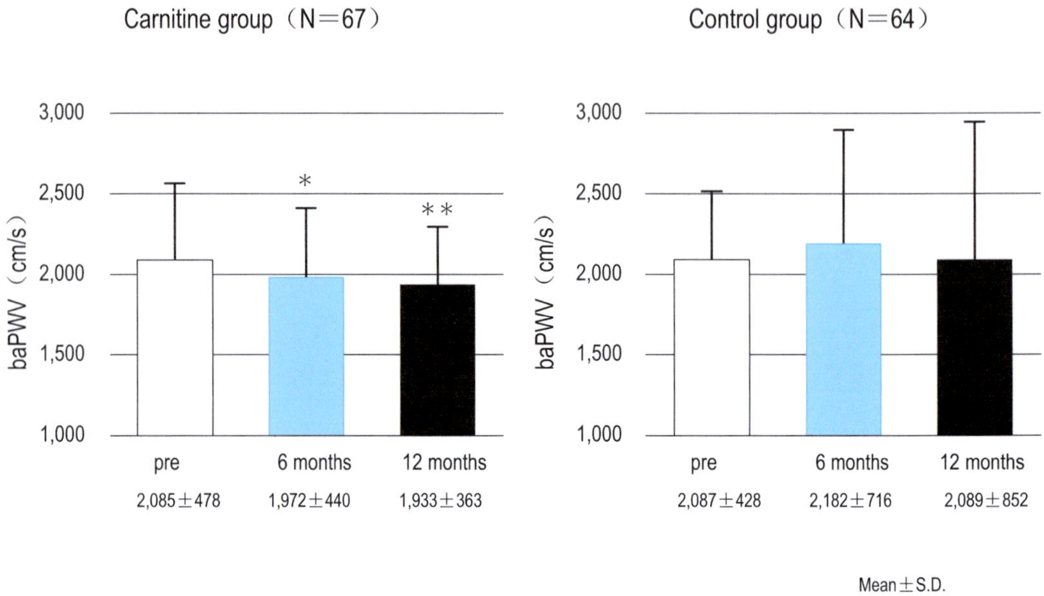

Fig. 14.36 Changes in baPWV following L-carnitine treatment (oral, 20 mg/kg/day, for 12 months) in hemodialysis patients. *$P < 0.05$, **$P < 0.001$ vs. pre value, baPWV: brachial-ankle pulse wave velocity. (From Ref. [153])

14.4.5 EPO-Resistant Anemia and Carnitine Supplementation Therapy

1. Early clinical study of carnitine supplementation therapy for EPO-resistant anemia

In dialysis patients, the secretion of EPO from the kidneys decreases as renal failure progresses, thereby reducing the hematopoietic effect in the bone marrow. In addition, hemolysis is promoted by destabilization of red blood cell membranes due to increased oxidative stress and uremic toxin accumulation, and it is thought that anemia progresses due to shortening the lifespan of red blood cells. Iron deficiency is the most common factor that causes anemia in dialysis patients, but micronutrient deficiencies such as folic acid, vitamin D_3, vitamin B_{12} and zinc, severe secondary hyperparathyroidism, inflammation, infections, and surgery can also affect individual patients. In addition, hemolysis, chronic blood loss, and shortened erythrocyte lifespan may cause reduced reactivity to genetically modified human erythropoietin (rHuEPO) and increased dose. EPO-resistant anemia is associated with a decrease in the quality of life of dialysis patients [154, 155] and is an independent risk factor for the progression of heart failure and a predictor of mortality [156]. Therefore, effective treatment for EPO-resistant anemia is crucial.

rHuEPO has been available overseas since 1989 (in Japan since 1991), and until then, blood transfusions were often performed for renal anemia in patients with end-stage renal disease who were undergoing dialysis therapy. During this period, there were many studies that examined the effects of carnitine supplementation therapy on Hb levels and hematocrit (Ht) in dialysis patients.

Trovato et al. divided 24 hemodialysis patients into a carnitine supplementation group (1.6 g/day orally, for 12 months) and a placebo group to examine the effects on Hb and Ht [157]. The results showed that carnitine supplementation significantly improved Ht (Fig. 14.37) and Hb. In the discussion, the possibility that carnitine improved anemia by increasing the uptake of cholesterol and phospholipids into the erythrocyte membrane and stabilizing the membrane is described.

Bellinghieri et al. also reported a significant increase in Ht ($25 \pm 2 \Rightarrow 31 \pm 2$ mL/dL, $P < 0.05$)

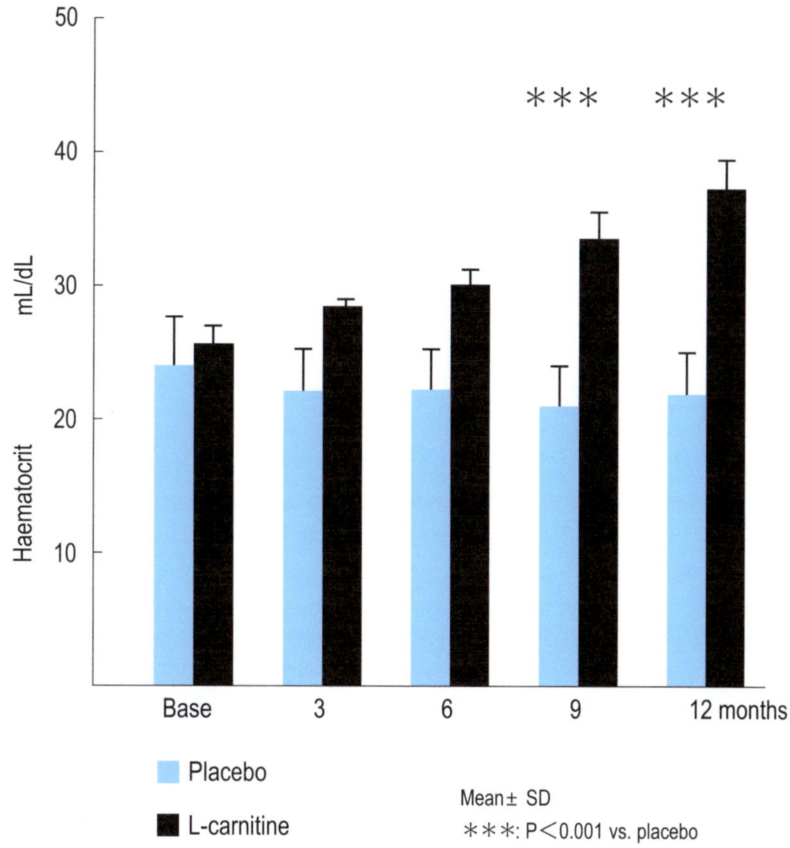

Fig. 14.37 Effect of carnitine supplementation (oral, 1.6 g/day, for 12 months) on hematocrit in dialysis patients. (Modified from Ref. [157])

in the carnitine supplementation group (7 patients, 2 g/day orally divided 2 times, 2 months) in a study to investigate the effect of carnitine supplementation on muscle symptoms [32].

Besides these two studies, Vacha (1983) [88] Albertazzi [158], and Donatelli [159] reported a significant increase in Ht in dialysis patients treated with carnitine.

Since 1989, rHuEPO products have been available in Europe and the United States, and the treatment of anemia in patients with end-stage renal failure has advanced dramatically. However, some patients require high doses of rHuEPO or do not respond adequately to rHuEPO, and this has become a new clinical issue for dialysis patients with EPO refractory anemia or EPO resistant anemia.

Kooistra et al. were the first to report on the relationship between carnitine and anemia [160]. They found that patients with severe anemia requiring rHuEPO had lower levels of carnitine than those with relatively mild or no anemia. Therefore, when the rHuEPO dose and serum carnitine concentration of 16 dialysis patients maintained at 30–35% Ht values were examined, the maintenance rHuEPO dose was inversely correlated with the total serum carnitine concentration ($r = -0.58$; $P < 0.05$, Fig. 14.38). However, there was no correlation between rHuEPO doses and serum free carnitine levels.

On the other hand, Matsumura et al. found an inverse correlation between the osmotic fragility of erythrocytes and serum total carnitine or free carnitine concentration in hemodialysis patients and further reported a significant negative correlation between the rHuEPO dose required to maintain the Ht target level and serum total carnitine concentration ($r = -0.54$; $P < 0.05$) or free carnitine concentration ($r = -0.50$; $P < 0.05$), respectively (Fig. 14.39a, b [161]. No correlation was found between serum acylcarnitine concentrations and rHuEPO doses (Fig. 14.39c).

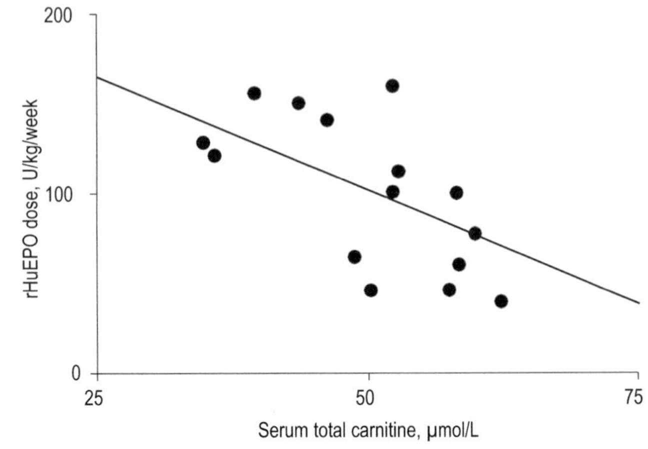

Fig. 14.38 Relationship between rHuEPO dose and serum total carnitine concentration in 16 hemodialysis patients. (From Ref. [160])

Thus, it was reported that carnitine is effective in improving Ht and Hb in dialysis patients, and after the advent of EPO, the relationship between carnitine concentration and rHuEPO requirement began to be studied, and carnitine deficiency, which is often observed in dialysis patients, was taken up as one of the causes of EPO-resistant anemia. In other words, the effectiveness of carnitine supplementation therapy in reducing the requirement of rHuEPO for EPO resistance has been actively studied, and there have been systematic reviews and review articles on the effects of carnitine administration on anemia [1, 6, 162]. The following is a summary of these research.

Labonia et al. divided 24 patients on hemodialysis for more than 1 year into a carnitine group (1000 mg intravenous, three times a week after dialysis for 6 months) and a placebo group and compared the amount of EPO use required to maintain Ht at 28–33% before and after treatment [163]. The results showed that the mean rHuEPO dose decreased by 38.1% from 102.2 ± 52.6 to 63.3 ± 37.8 U/kg/week in the carnitine group ($P < 0.02$), while the rHuEPO requirement was unchanged in the placebo group (Fig. 14.40 and Table 14.10). Of the 13 patients in the treatment group, rHuEPO requirement decreased in 7 patients, while it remained unchanged in the remaining 6 patients (Fig. 14.40). Responders tended to have higher mean rHuEPO requirements and higher endogenous EPO concentrations at baseline, but there was no correlation between serum carnitine concentration and response.

Kletzmayr et al. studied the effect of carnitine administration on rHuEPO dosage in 40 hemodialysis patients divided into a carnitine group (intravenous post-dialysis infusion 3 times a week, 5 mg/kg in 15 patients, 25 mg/kg in 5 patients, and 4 months with IV iron) and 20 placebo patients [164]. There was a 36.9% (183.7 ± 131.7 ⇒ 126.6 ± 127.9 U/kg/week, $P < 0.001$) decrease in rHuEPO dose in 8 of 19 patients (3 of 5 patients at 25 mg/kg and 5 of 14 patients at 5 mg/kg) who received IV carnitine. However, there was no significant change in rHuEPO dose in carnitine-treated patients as a whole. However, ERI was significantly reduced in the carnitine-

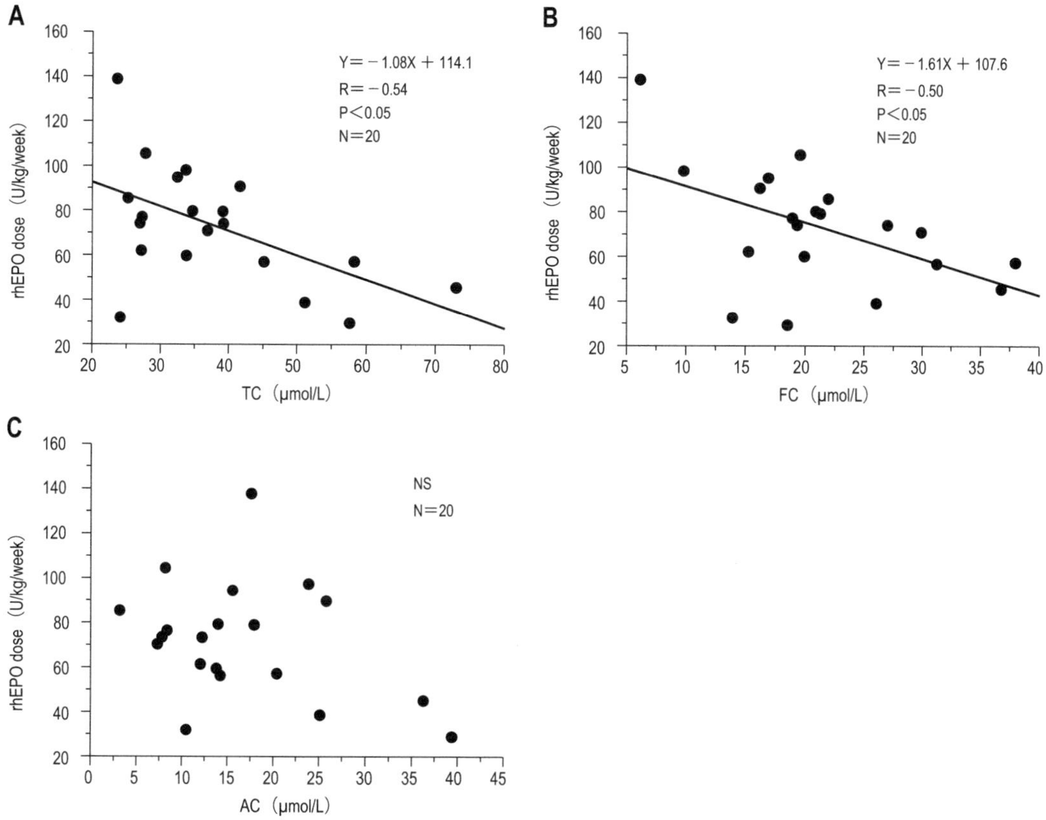

Fig. 14.39 Relation between maintenance rhEPO dose and serum carnitine levels. (**a** and **b**) Maintenance rhEPO dose was correlated with serum total carnitine (TC) and free carnitine (FC) levels. (**c**) Maintenance rhEPO dose was not correlated with serum acylcarnitine (AC) levels. (From Ref. [161])

treated group compared with the pre-treatment (16.0 ± 11.0 ⇒ 13.6 ± 10.5 U/kg/week/gHb, $P < 0.02$). Responders had higher serum total carnitine concentrations at baseline than non-responders ($P < 0.05$), but there was no correlation between individual patient rHuEPO requirements and carnitine concentration. Carnitine at 25 mg/kg increased serum carnitine concentrations to the normal range, but due to the small number of patients receiving the high dose ($n = 5$), they were unable to establish a clear benefit of low- or high-dose carnitine supplementation therapy.

Caruso et al. examined the effects of carnitine in a study in which 31 dialysis patients were randomly assigned to receive carnitine (15 patients, intravenous 1 g intravenously three times a week after dialysis for 6 months) and placebo (16 patients) and found that the rHuEPO requirement decreased after 6 months in elderly patients aged 65 years and older in the carnitine group (134.96 ± 79.41 ⇒ 118.34 ± 108.05 IU/week), and no effect of carnitine administration was observed in patients as a whole [165].

Thus, although it is not effective in all patients with EPO-resistant anemia, it has been found that at least some of them respond to carnitine supplementation therapy.

Hurot et al. conducted a meta-analysis of 18 randomized controlled clinical trials examining

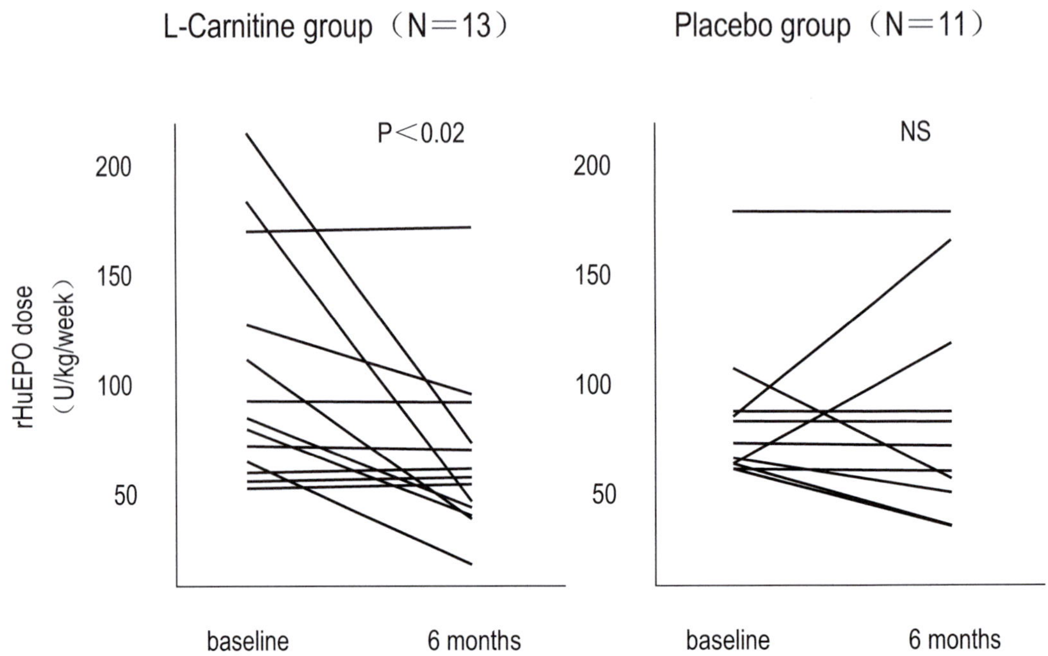

Fig. 14.40 Values for rHuEPO requirements in L-carnitine group (n = 13) and placebo group (n = 11). L-carnitine (intravenously, 1000 mg, 3 times a week) or placebo were administered for 6 months. (Modified from Ref. [163])

Table 14.10 rHuEPO dose, globular osmotic fragility, plasma erythropoietin, and carnitine levels in carnitine and placebo groups at the beginning (T0) and (T6) of the study

	Carnitine (N = 13)		Placebo (N = 11)	
	T0	T6	T0	T6
rHuEPO dose (U/week) (U/kg/week)	5615.4 ± 2631.2 102.2 ± 52.6	3538.5 ± 1898[a] 63.3 ± 37.8 *	4909.1 ± 2314.2 78.6 ± 32.1	4909.1 ± 2968.1 79.9 ± 46.7
Plasma carnitine Total (μmol/L) Free (μmol/L) Free/Total (%)	69.9 ± 11.4 41.8 ± 7.1 60 ± 7.7	394.6 ± 124.2[b] 247.7 ± 81.5[b] 61.6 ± 5.7	62.9 ± 18 36.3 ± 9.7 58.3 ± 6.6	71.7 ± 13.1 47 ± 11.3 64.6 ± 6.9
Globular osmotic fragility (% NaCl)	0.447 ± 0.031	0.452 ± 0.015	0.460 ± 0.015	0.453 ± 0.013
Plasma EPO (mU/mL)	33.1 ± 11.4	29.9 ± 15.2	40.1 ± 14.9	32.8 ± 13.4

From Ref. [163]

[a]$P < 0.02$ (T0 vs. T6), [b]$P < 0.001$ (T0 vs. after T6), Student's t-test

Twenty-four chronic maintenance dialysis patients with a dialysis period of more than 1 year, Carnitine group: 1 g of carnitine intravenously after dialysis 3 times weekly (for 6 months). rHuEPO: recombinant human erythropoietin

the effect of carnitine on anemia [1]. Prior to the introduction of rHuEPO, only three randomized clinical trials examined the effect of carnitine on Ht or Hb: the aforementioned studies by Trovato et al., Bellinghieri et al., and Nilsson-Ehle et al. [166] Pooling and analyzing these data, the conclusion was that carnitine administration significantly increased Ht or Hb (Fig. 14.41, above the dotted line). Pooling the data from the six randomized clinical trials introduced by Labonia, Kletzmayr, Caruso, and colleagues after 1989, when rHuEPO was introduced, revealed that carnitine combination therapy significantly decreased rHuEPO requirement (Fig. 14.41, below the dotted line). Furthermore, comparing the ERIs of placebo and carnitine-treated patients in four of the six studies for which ERIs could be calculated, the carnitine-treated group alone

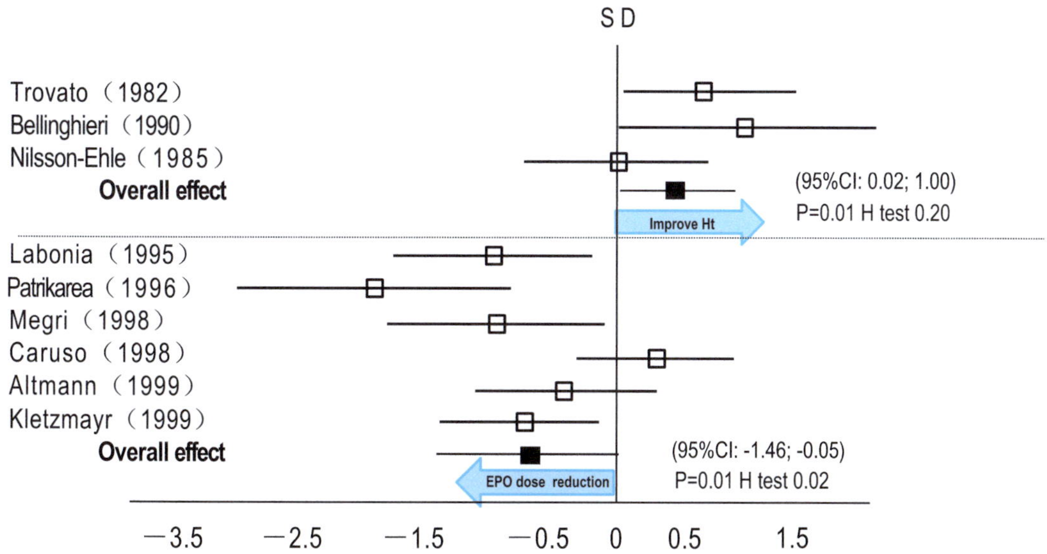

Fig. 14.41 The effect of L-carnitine on anemia (hematocrit: Ht) control (above the dotted line; fixed model analysis) and erythropoietin (EPO) dose reduction (below the dotted line; random model analysis). Open boxes indicate the effect of each individual trial, and the closed boxes indicate the overall effect of treatment. The *P* value is given for overall treatment effect and is significant for values ≤0.01. Heterogeneity of treatment effect among trials is present if H test value is <0.10. (Modified from Ref. [1])

showed a reduction in ERI in three of the four studies (Fig. 14.42) [1].

Although not a double-blind study, Matsumoto et al. divided 32 hemodialysis patients into three groups: rHuEPO-resistant patients (Group A: 14 patients) with Ht < 27.5% even with 9000 units of rHuEPO, patients who could maintain Ht > 30% with 9000 units of rHuEPO (Group B: 7 patients), and patients who could maintain Ht > 30% without 9000 units of rHuEPO (Group C: 11 patients). The carnitine concentration in serum or erythrocytes of the control group (15 healthy subjects) and the effect of carnitine administration (oral administration, 500 mg/day, for 3 months) were examined [167]. Serum free carnitine and total carnitine concentrations in Groups A and B were significantly lower than in the control group, while total carnitine concentrations in erythrocytes were not different from those in controls (Fig. 14.43).

When carnitine was orally administered to this group of rHuEPO-resistant patients (Group A: 14 patients) and Ht before and after administration, an increase in Ht was observed in 7 patients, and a significant increase in Ht was observed overall (Fig. 14.44).

The study by Matsumoto et al. a supplemental carnitine was used because at that time pharmaceutical carnitine has no indication for carnitine deficiency in dialysis patients, but the indication for L-Cartin® Tablets, the only pharmaceutical L-carnitine preparation in Japan, was changed to "carnitine deficiency" on March 10, 2011, through a public knowledge application, allowing its use for carnitine deficiency in dialysis patients too.

Takeuchi et al. examined the ESA dosage of carnitine tablets (L-Cartin® 600 mg/day orally, 2.24 weeks) and the effects of ERI and muscular symptoms in dialysis patients with ESA-resistant anemia who were receiving rHuEPO 9000 units/week or darbepoetin (DA) 40 μg/week or more [101]. As a result, Hb was significantly elevated in the carnitine group, and ESA dose and ERI were significantly reduced from 12 weeks of treatment. However, in the control group, there was no increase in Hb, ESA dose was significantly reduced but not significant, and ERI decreased only slightly at 16 and 20 weeks (Figs. 14.45 and 14.46). At the same time, nurses interviewed the patients to examine the improvement in dialysis-related muscle symptoms and

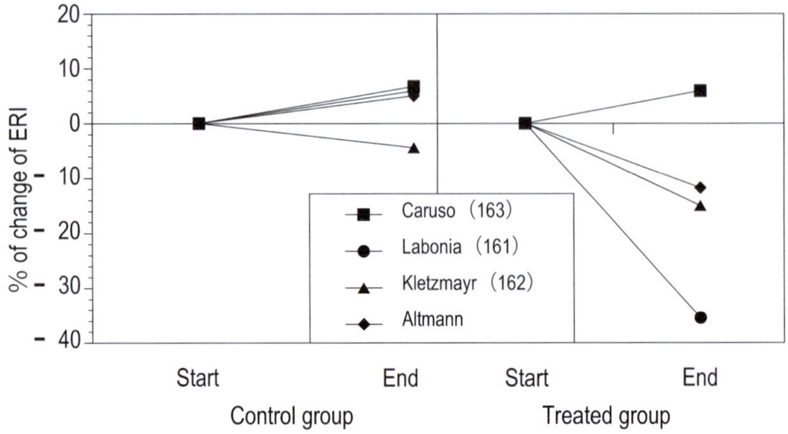

Fig. 14.42 The effect of L-carnitine on erythropoietin resistance index (ERI) in control (left) and L-carnitine-treated (right) groups. ERI was defined as the ratio of EPO dose divided by the patient's hemoglobin level. Data were available in four trials. The numbers in parenthesis indicate the reference numbers in Chap. 14 of this book, and Altmann's data are from a meeting abstract. (Modified from Ref. [1])

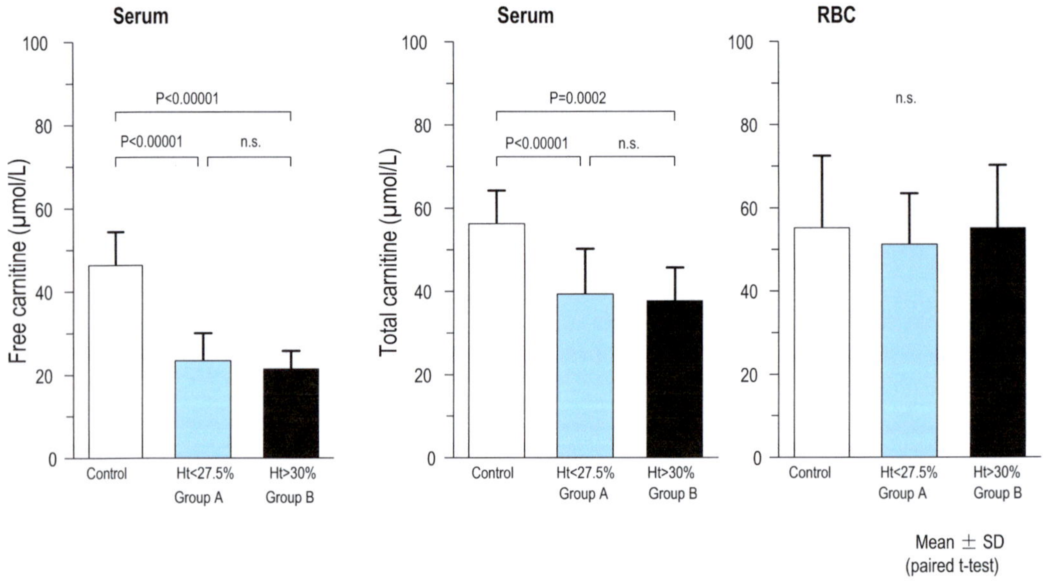

Fig. 14.43 Serum free carnitine (left), serum total carnitine (center), and RBC total carnitine (right) concentrations in normal controls and patients receiving 9000 U/week of EPO (group A and B). Results are expressed as mean ± SD. Serum carnitine levels in both group A and group B were significantly decreased compared to controls, while no difference was observed between group A and group B. RBC carnitine levels showed no obvious difference among three groups. Control: adults healthy subjects (n = 15), group A: patients with Ht < 27.5% when treated with EPO 9,000 U/week, group B: patients with Ht > 30 5 when treated with EPO 9000 U/week. Ht: hematocrit, RBC: red blood cells. (From Ref. [167])

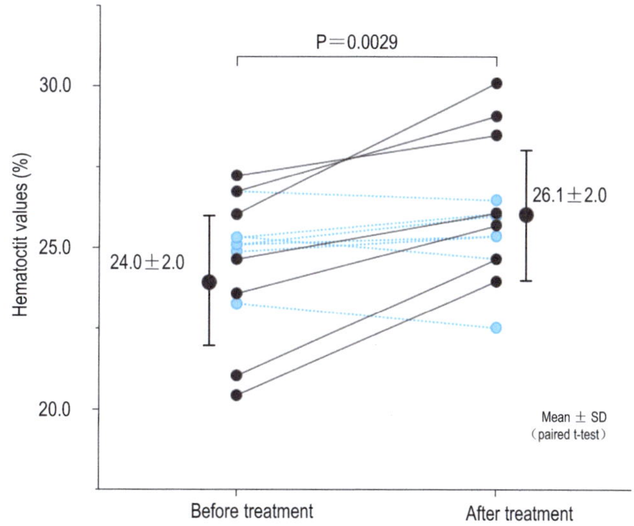

Fig. 14.44 Effect of L-carnitine treatment on hematocrit levels in the poor responders to EPO (group A). Hematocrit levels were measured before a dialysis session. The change of hematocrit levels in each patient of group A. Solid or dotted lines indicate patients with or without efficacy, respectively. (From Ref. [167])

found that asthenia (fatigue and malaise) improved significantly, from 18.8% of patients with no symptoms before treatment to 43.8% at 12 weeks and 78.1% at 24 weeks after treatment, and muscle cramps from 34.4% of patients with no symptoms before treatment to 75% at 12 weeks after treatment. Muscle cramps significantly improved from 34.4% of patients who had "no symptoms" before treatment to 75% at 12 weeks after treatment, and 21.9% of patients who had them every time before treatment to 0% at 12 weeks after treatment.

To investigate the effect of carnitine on renal anemia in hemodialysis patients, Higuchi et al. conducted a nonrandomized comparative study in which patients were divided into those who wanted to receive carnitine and those who did not [168]. In the carnitine group (113 patients, oral, 20 mg/kg/day, maximum 1200 mg/day, 6 months), there was a significant decrease in rHuEPO use by an average of about 31% at 6 months (Fig. 14.47) compared with the non-treated group (40 patients), and ERI was significantly reduced at 3 and 6 months of treatment compared to the control group (Fig. 14.48). No change in ESA use or ERI was observed in the control group.

Kudoh et al. reported that oral carnitine (L-Cartin® tablets, 900 mg/day for 12 months) significantly reduced ERI in 7 of 18 patients, but not in 11 patients, and that AC/FC was significantly lower in responders compared to non-responders (0.7 ± 0.09 vs. 0.83 ± 0.12) [169].

Murakami et al. also administered carnitine orally to dialysis patients with ESA-resistant anemia (L-Cartin® tablets, 600 mg/day, 7 months), and the Hb level increased significantly after 7 months ($9.5 \pm 1.0 \Rightarrow 10.9 \pm 1.3$ g/dL, $P < 0.01$), and the DA dose was 60 μg/week in 2/3 patients (22 patients) before carnitine treatment, but 48.2 ± 7.3 μg/week at 3 months and 45.0 ± 10.2 μg/week at 7 months after carnitine administration ($P < 0.01$). Also, significant improvement was observed in the nutritional indices albumin and protein catabolic rate (PCR) and the inflammatory marker CRP [170].

2. Intravenous L-Carnitine Preparation in Japan

Prior to 2011, studies on carnitine administration in dialysis patients in Japan had mainly used supplemental L-carnitine; on March 10, 2011, the indication for L-Cartin® Tablets (levocarnitine chloride) was changed to "carnitine deficiency," and pharmaceutical L-carnitine products became available for carnitine deficiency in dialysis patients. On February 26, 2013, "L-Cartin® FF Injection 1000 mg," an intravenous formulation of free L-carnitine (initially in ampule form), was launched, making the injectable form available to dialysis patients.

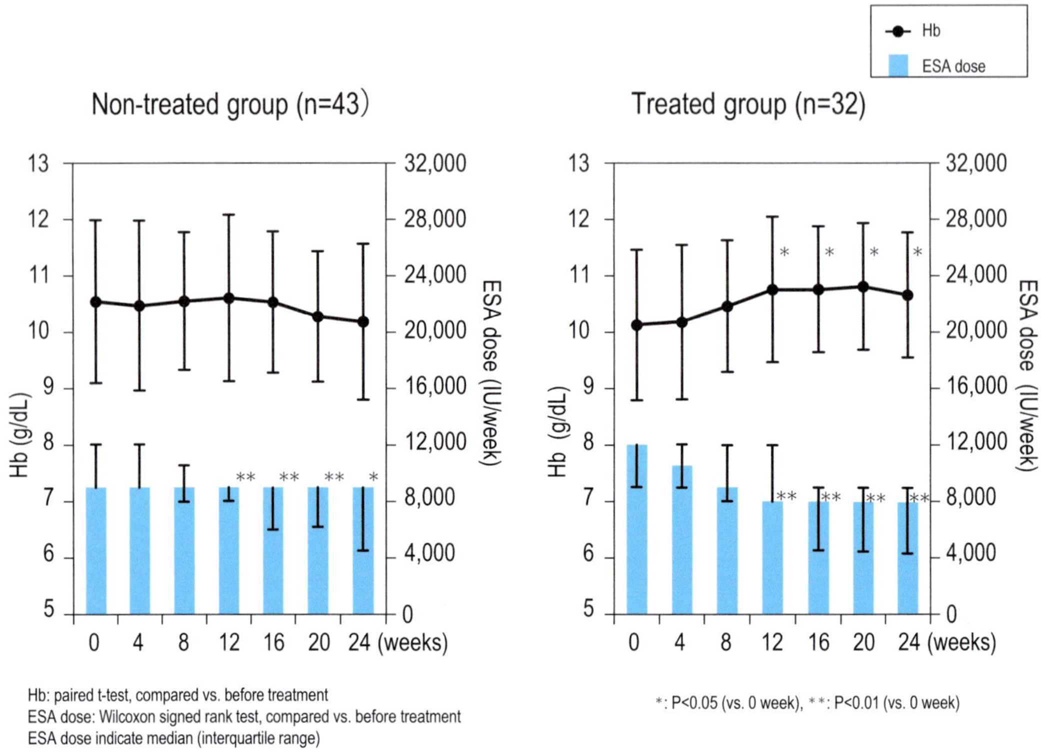

Fig. 14.45 Effect of L-carnitine supplementation (oral, 600 mg/day, divided 2 times, for 24 weeks) on hemoglobin (Hb) and on erythropoiesis stimulating agents (ESA) dose in dialysis patients who were low-responsive to ESA treatment. (From Ref. [101])

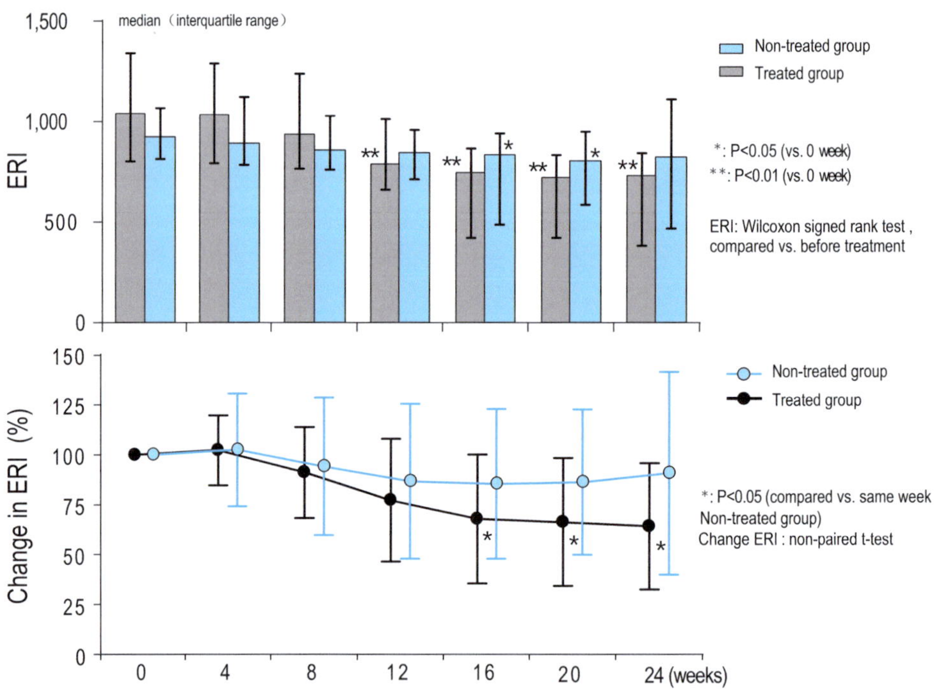

Fig. 14.46 Change in ERI after carnitine treatment in dialysis patients with ESA-resistant anemia. ERI: erythropoietin resistance index = ESA dose (units/week)/hemoglobin (g/dL). (Modified from Ref. [101])

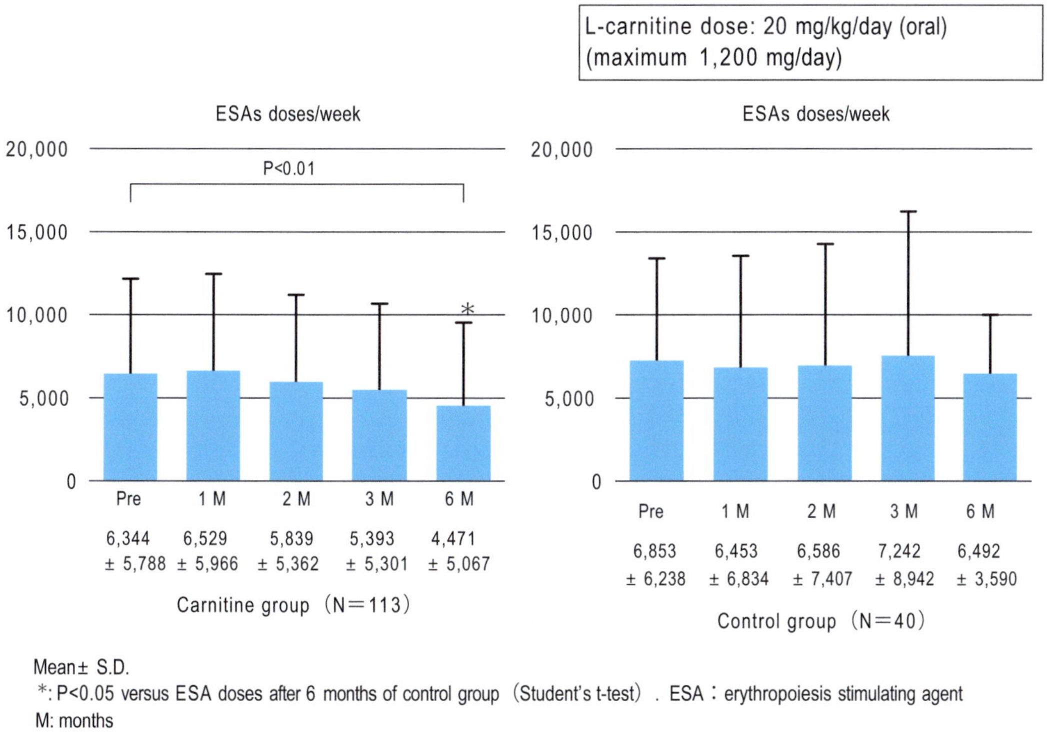

Fig. 14.47 Change of quantity of ESA from baseline to month 6 in carnitine-treated and control groups. In the carnitine-treated group, 20 mg/kg/day (oral, maximum 1200 mg/day) of carnitine was administered for 6 months. *: $P < 0.05$ vs. ESA doses after 6 months of control group, ESA: erythropoiesis stimulating agent. (Modified from Ref. [168])

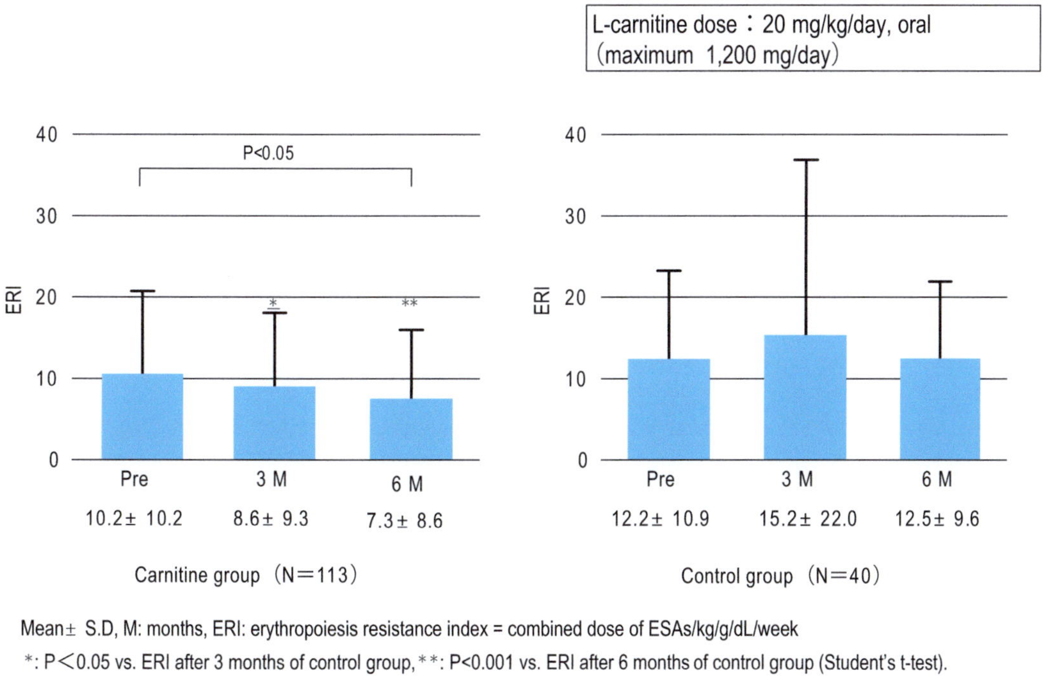

Fig. 14.48 Change of ERI from baseline to months 3 and 6 in treated and control groups and comparison between two groups. *: $P < 0.05$ vs. ERI after 3 months of control group, **: $P < 0.001$ vs. ERI after 6 months of control group. ERI: erythropoiesis resistance index = weekly ESAs dose/kg/g Hb/dL. (From Ref. [168])

This intravenous preparation of carnitine is indispensable for the acute treatment of inherited metabolic diseases and the treatment of patients with valproic acid poisoning, but it was not previously marketed in Japan, and when necessary, the intravenous preparation, it was unavoidable to dissolve the reagent L-carnitine in the pharmacy department of the hospital, sterilize it, and used as an in-hospital preparation. In addition to an oral solution that is easy to administer to pediatric patients and patients with severe mental and physical disabilities, Japanese Society for Inherited Metabolic Diseases and other organizations have strongly requested the early development of intravenous L-carnitine preparations to the "Evaluation Committee on Unapproved or Off-Label Drugs with High Medical Needs." In February 2013, Otsuka Pharmaceutical Co., Ltd. released "L-Cartin® FF Intravenous Injection 1000 mg" as a free L-carnitine preparation instead of the conventional levocarnitine chloride preparation. The expansion of the indication of L-Cartin® tablets was carried out based on the overseas approval status, papers, meta-analyses, and descriptions in textbooks and guidelines, without conducting clinical trials due to the system of public knowledge application. The development of the intravenous formulation that followed was selected because there was a request for early development from the academic society, the evidence on the efficacy and safety of the L-carnitine intravenous formulation in the free form was established overseas, and it does not contain chlorine like the levocarnitine chloride preparation, and side effects such as hyperchloremic acidosis are unlikely to occur even if it is injected intravenously. Since the 1970s, it has been involved in the manufacture and sale of carnitine preparations for carnitine deficiency, and it was introduced from an Italian pharmaceutical company with a proven track record. However, L-Cartin® (levocarnitine), which is a free form that differs from levocarnitine chloride, is judged to be a "new active ingredient drug" and requires several clinical trials in the patient phase from healthy subjects in phase I, phase II, and phase III. However, in consideration of the fact that the efficacy and safety of the free form have already been sufficiently established overseas, and the need to supply it to clinical sites in Japan as soon as possible, it was approved by conducting only clinical pharmacological studies (pharmacokinetic studies) using healthy subjects and comparing them with overseas data. As a result, intravenous preparations can be used not only for patients in the acute phase of inherited metabolic diseases in Japan but also for dialysis patients. Initially, ampoules were introduced, but a safer prefilled syringe formulation ("L-Cartin® FF Intravenous Injection 1000 mg Syringe") was launched in December 2017, and this formulation is now widely used. As described in Chap. 8, "Quantification and Pharmacokinetics of Carnitine," the bioavailability of orally administered carnitine is around 20%, which is not an efficient route of administration. On the other hand, in the case of intravenous preparations, 100% of the drug enters the blood, so it appears to be efficient, but in patients other than renal insufficiency with normal renal function, high concentrations of carnitine that exceed the threshold of reabsorption ($\leq$ the upper limit of normal) are not reabsorbed and are excreted in the urine. However, in patients with end-stage renal disease, fortunately or unfortunately, due to reduced renal function, it is not excreted, and the blood concentration is maintained until the next dialysis. On dialysis days three times a week, 70–80% of blood carnitine is removed in the dialysate during dialysis treatment, but by administering intravenous infusion after dialysis, the high blood concentration is maintained until the next dialysis and efficiently taken up by tissues such as muscles. For this reason, intravenous injection is considered to be the most suitable route of administration for carnitine deficiency in dialysis patients, and the U.S. FDA has approved only intravenous preparations for carnitine deficiency in dialysis patients.

3. Effect of Intravenous Carnitine on EPO-Resistant Anemia

Maruyama et al. examined the effect of intravenous carnitine preparations on EPO-resistant anemia in dialysis patients [171]. Sixty hemodialysis patients were randomized into two groups, one receiving carnitine (1000 mg intravenously after dialysis three times a week for 12 months) and the other untreated group, and the dosage of rHuEPO preparations required to maintain Hb at around 11 and its effect on ERI were examined. As a result, the ESA dose was significantly reduced from 6 months in the carnitine group, and a saving of about 40% was observed at 12 months, but the control group remained unchanged (Fig. 14.49). Similarly, ERI decreased significantly from 6 months after carnitine administration, and decreased by about 40% at 12 months (Fig. 14.50).

Funakoshi et al. administered intravenous carnitine (1000 mg) to 130 dialysis patients three times a week after dialysis and evaluated the reduction in rHuEPO after 6 months to determine whether to continue administration [172]. A saving effect was observed in about one-third of the cases, an undeterminable in one-third, and no effect in one-third. In patients with a saving effect and in cases undeterminable, intravenous carnitine was continued (reduced to twice a week). As a result, it was reported that the amount of rHuEPO used could be reduced by 15% across the entire facility, and the cost of intravenous carnitine injection could be recovered after 26 months.

Aoki et al. reported that 23 patients on hemodialysis were treated with carnitine supplementation therapy (2 g intravenously followed by dialysis 3 times a week for the first 10 months, followed by 1 g intravenously for 10 months) and that intravenous 2 g of carnitine reduced the DA dose required to maintain Hb at 10 to 11 g/dL by 90% and significantly improved LVEF and BNP levels [173]. This DA-reducing effect was greater with 2 g of intravenous carnitine than with 1 g of intravenous carnitine.

Suzuki et al. switched 17 hemodialysis patients who had received oral carnitine for at least 11 months (5 patients receiving 200 mg 3 times a day, 12 patients receiving 300 mg 3 times a day, mean age 61.6 ± 9.2 years, mean dialysis history 8.3 ± 6.2 years) to intravenous carnitine (1000 mg after dialysis 3 times a week) and discontinued carnitine administration 1 year after administration, and changes in blood carnitine levels were examined [174]. Total plasma carnitine, free carnitine, and acylcarnitine levels increased significantly for 3 months after switch-

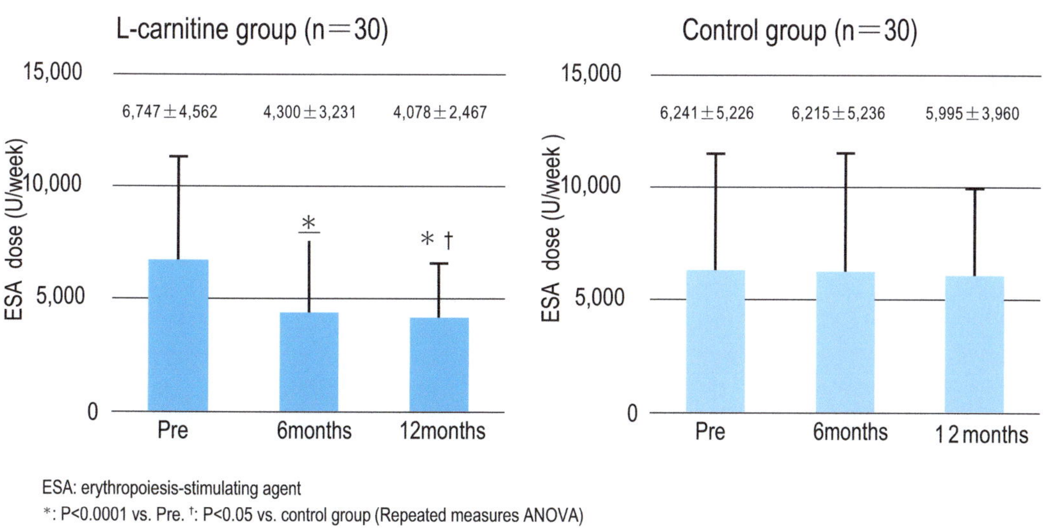

Fig. 14.49 Changes in the total ESA dose per week from baseline to 6 and 12 months in the L-carnitine- treated and control groups with comparison between the 2 groups. In the L-carnitine-treated group, L-carnitine was administered for 12 months (intravenous, 1000 mg, 3 times a week after each dialysis session). *$P < 0.0001$ vs. Pre, †$p < 0.05$ vs. ESA dose at the same time point in the control group. ESA: erythropoiesis-stimulating agent. (Modified from Ref. [171])

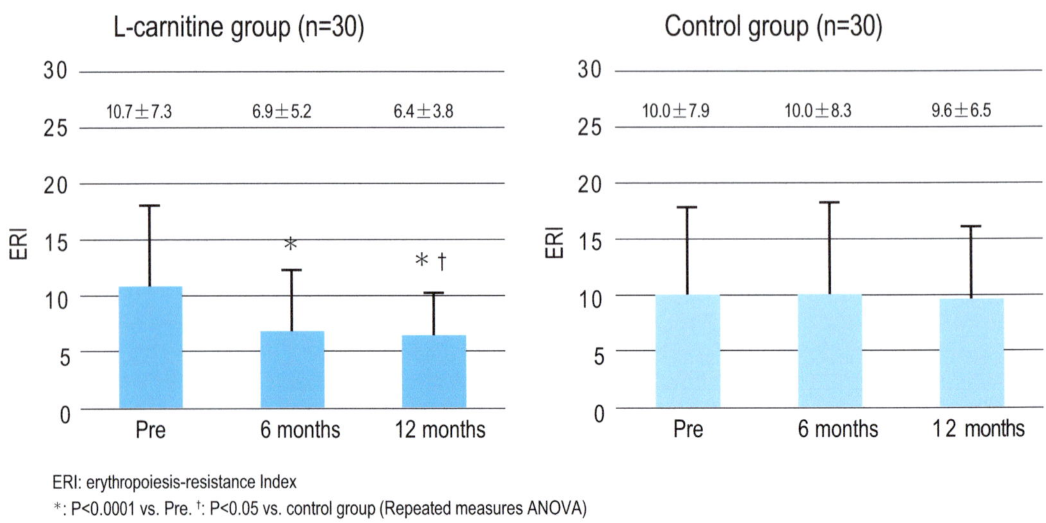

Fig. 14.50 Changes in ERI from baseline to 6 and 12 months in the L-carnitine treated and control groups with a comparison between the 2 groups. $*P < 0.0001$ vs. Pre, $^{\dagger}P < 0.05$ vs. ERI at the same time point in the control group. In the L-carnitine-treated group, L-carnitine was administered intravenously (1000 mg, 3 times a week after each dialysis session). ERI: erythropoiesis resistance index. (Modified from Ref. [169])

ing to intravenous carnitine but remained almost constant thereafter. AC/FC was consistently around 0.6 when switched to intravenous injections, and there was no significant change. When intravenous carnitine was discontinued, the concentration of each carnitine fraction decreased sharply over a period of 3 months, and thereafter it decreased slowly.

4. EPO-Resistant Anemia and Multiple Nutrient Deficiencies: Dobeneck's Barrel

As described above, in dialysis patients with EPO-resistant anemia, oral or intravenous carnitine administration may be effective in reducing rHuEPO use, at least in some patients. Currently considered causes of EPO resistance include hemolysis and hemorrhage, chronic hyperinflammation, inadequate dialysis, hyperparathyroidism, and nutrient deficiencies such as vitamin B_{12} and folate, which should be considered along with carnitine deficiency [175]. Patients undergoing dialysis therapy are often undernourished, with low intake of animal protein such as meat, due to dietary guidance to reduce protein intake in conservative renal failure and the fact that many of them are elderly, and in addition to carnitine deficiency, iron deficiency, zinc deficiency, and vitamin D_3 deficiency are also very common.

In 58 dialysis patients with renal anemia and low zinc concentrations than standard value (<80 μg/dL), Fukushima et al. treated them with two packets of polaprezinc (150 mg, twice daily, 34 mg/day as zinc) after breakfast and dinner, and found that Hb and Ht increased, and the rHuEPO dose was significantly reduced from the third month of treatment, and was the same as the rHuEPO dose in the zinc non-treated group (38 patients) who had normal zinc levels or could not give consent (Fig. 14.51). This finding indicates the importance of evaluating zinc levels in dialysis patients whose anemia did not improve with high-dose rHuEPO administration [176].

Kobayashi et al. also randomly divided 70 dialysis patients with serum zinc concentrations of <65 μg/dL into two groups, one of whom was administered polaprezinc (150 mg/day, 34 mg/day as zinc) for 12 months and compared the rHuEPO dose and ERI with the non-treated group [177]. As a result, the dose of rHuEPO decreased significantly from the 10th month and the ERI decreased significantly from the 9th month in the zinc group compared to the control group (Fig. 14.52).

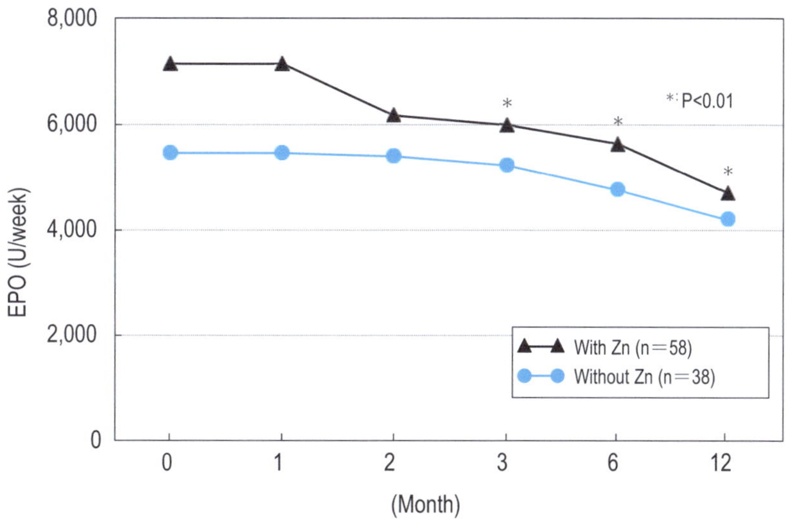

Fig. 14.51 Change in EPO dose with or without zinc supplementation in dialysis patients with ESA resistant anemia and zinc deficiency. Zinc supplementation: Polaprezinc 2 bags (containing 34 mg/day) for 12 months. (From Ref. [176])

Shimizu et al. in the same group examined the prevalence of zinc deficiency in 47 peritoneal dialysis patients and 47 hemodialysis patients and found that zinc deficiency with a serum zinc concentration of <60 μg/dL was found in 59.6% of peritoneal dialysis patients and 70.2% of hemodialysis patients, and latent zinc deficiency with serum zinc concentration of 60–80 μg/dL was 40.4% in the former and 29.8% in the latter, and there were no patients in either patient group with a serum zinc concentration above 80 μg/dL, the lower limit of the normal reference value [178]. This means that zinc deficiency, as well as carnitine deficiency, is extremely common in dialysis patients. Zinc is transported to tissues by binding to metallothionein, but it is still a small molecule and is easily removed by dialysis, and also dialysis patients tend to reduce the intake of animal proteins such as meat and seafood containing a lot of zinc, and other factors that reduce absorption from the stomach.

In addition, dialysis patients are more likely to have vitamin D_3 deficiency, and mortality increases by 2.76 times when there is a deficiency, and 5.88 times when it is accompanied by PEW [179]. In addition, serum 25(OH)D_3 concentrations were significantly correlated with Hb and ESA reactivity in maintenance dialysis patients, and vitamin D_3 deficiency is considered to be one of the factors of rHuEPO resistance [180]. Kumar et al. also examined 25(OH)D_3 concentrations in dialysis patients and reported that 89% of them had <30 ng/mL, and that when ergocalciferol was administered to these patients, rHuEPO doses were reduced in many patients [181]. Patel et al. also report that vitamin D_3 deficiency is associated with lower Hb levels [182]. It has also been suggested that 25(OH)D_3 deficiency is associated with vascular calcification in dialysis patients [183].

Vitamin D_3 acts on various organs as a hormone that regulates mineral metabolism, but it has been thought that it has an indirect effect of improving anemia by suppressing fibrosis in the bone marrow by suppressing the synthesis and secretion of parathyroid hormone (PTH) [175]. However, it has recently been reported that vitamin D_3 improves anemia by a route that does not involve PTH [184–187].

Against these backgrounds, Date et al. divided 73 dialysis patients into a vitamin D_3 pulse therapy group (VD_3-P group, 27 patients) and a non-vitamin D_3 pulse therapy group (non-VD_3-P group, 46 patients) and examined the dose of EPO preparations [188]. As a result, in the VD_3-P group, 52% of patients were able to maintain Hb at 10 to 12 g/dL without rHuEPO, but in the non-VD_3-P group, only 24% of patients were able to maintain Hb at 10 to 12 g/dL without rHuEPO. None of the patients were able to maintain 12 g/dL ($P < 0.01$, Fig. 14.53). To investigate the mechanism by which Hb could be maintained

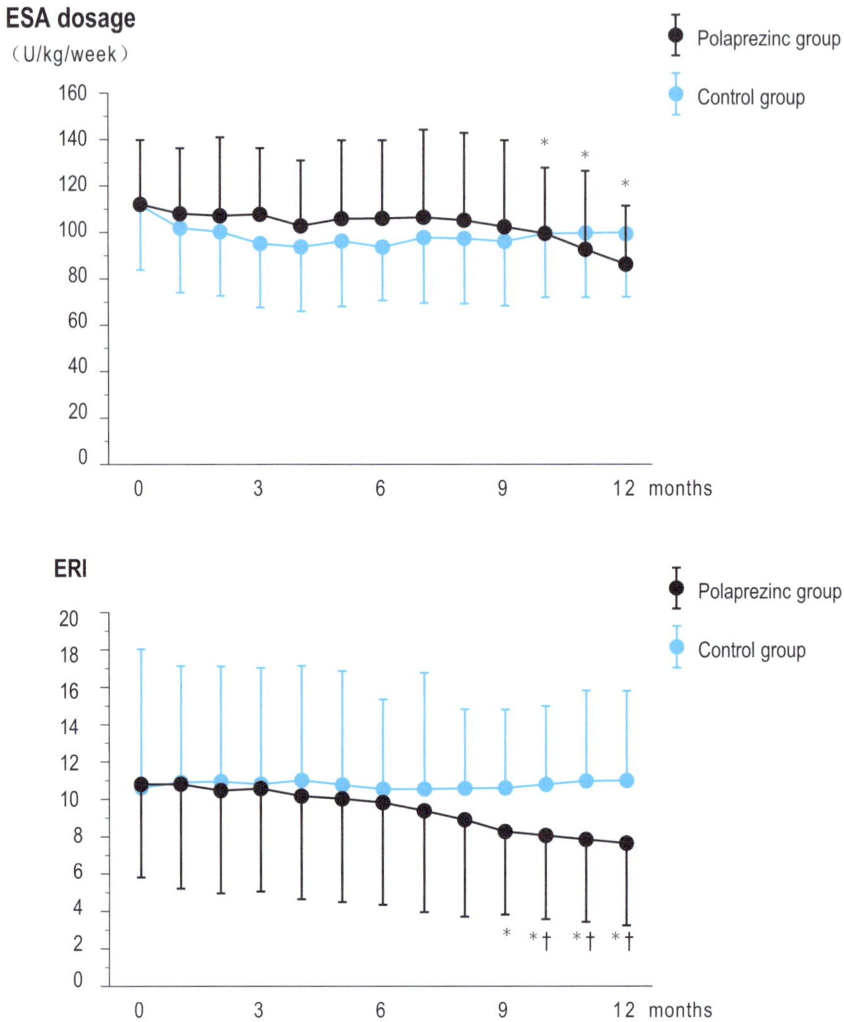

Fig. 14.52 Changes in ESA dosage (upper part) and erythropoietin responsiveness index (ERI, lower part) during the study period with or without zinc supplementation (34 mg/day). Data are expressed as mean ± SD. *P < 0.05 vs. baseline, †P < 0.05 vs. control group. (Modified from Ref. [177])

at the target level without rHuEPO in the VD_3-P group, they examined the kidneys by ultrasound echo and found that 85% of patients in the VD_3-P group had acquired renal cysts that may have formed after the patient was transferred to dialysis, while only 27% of the non-VD_3-P group had acquired renal cysts (Fig. 14.54). This suggests that VD_3-P therapy may have promoted the formation of acquired renal cysts, and that endogenous EPO may have been produced from these cysts, requiring further investigation.

In this way, in addition to strict dry weight management as a basic policy, the dialysis facility of Date et al. is based on the idea that hemodialysis patients are chronically deficient in vitamin D_3, and alfacalcidol to 0.25 to 0.75 μg/day or calcitriol to 0.25 to 0.250.5 μg/day was administered orally. Intravenous vitamin D_3 therapy (calcitriol 0.5 μg or 1.0 μg, 3 times a week or maxacalcitol 2.5 or 5.0 or 7.5 or 10.0 μg) was given to patients who could not maintain parathyroid hormone (iPTH) below 180 pg/mL. In other words, almost

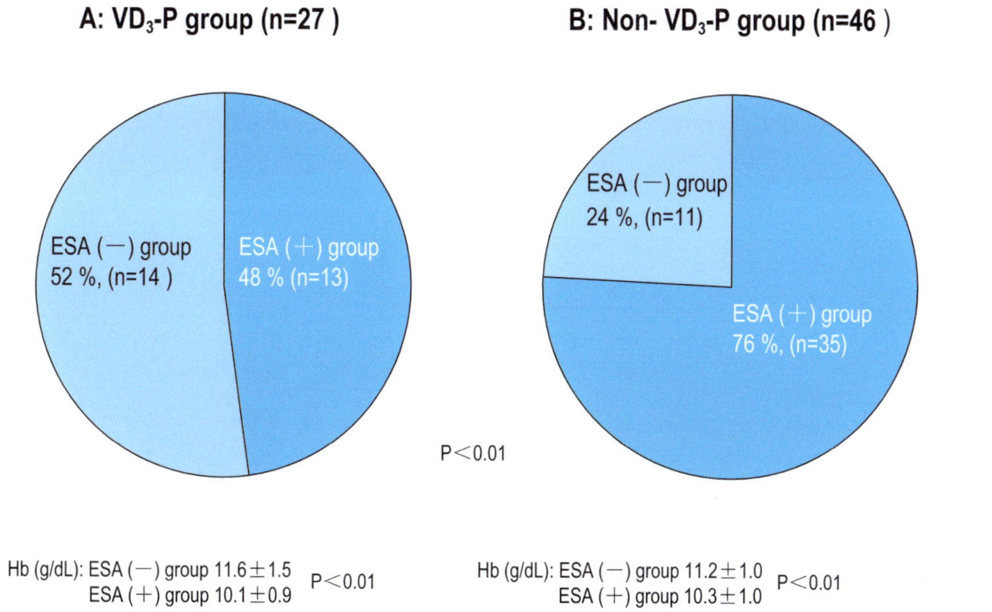

Fig. 14.53 Proportion of who could control hemoglobin (Hb) without rHuEPO between vitamin D$_3$ pulse therapy group (A, $n = 27$) and non-vitamin D$_3$ pulse therapy group (B, $n = 46$) in maintenance hemodialysis patients. VD$_3$-P: vitamin D$_3$ pulse therapy. (Modified from Ref. [188])

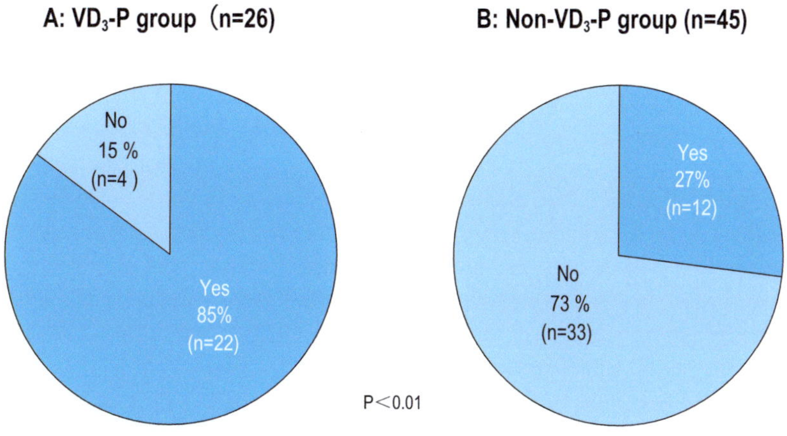

Fig. 14.54 Proportion of patients who had acquired renal cysts in vitamin D$_3$ pulse therapy group (**a**) and non-vitamin D$_3$ pulse therapy group (**b**) in maintenance hemodialysis patients. No: patients who had not acquired renal cysts, Yes: patients who had acquired renal cysts. VD$_3$-P: vitamin D$_3$ pulse therapy. (Modified from Ref. [188])

all patients received vitamin D$_3$ therapy. Carnitine supplementation was used in a small number of patients, but in March 2011, the indication for carnitine deficiency was approved through a public knowledge application, allowing its use in hemodialysis patients. In May 2012, almost all dialysis patients were finally treated with vitamin D$_3$ pulse therapy, 98% patients were treated with carnitine supplementation therapy who were diagnosed as carnitine deficiency, and zinc preparations (polaprezinc, 34 mg/day as zinc, trade name: Promac) were administered to 92% of the patients whose zinc level is below the standard (59–135 μg/dL). As a result, the percentage of patients who were able to maintain Hb at 10–12 g/dL without ESAs was initially in the 40% range but increased to around 60% 16 months after the start of carnitine and zinc administration. Date reported such interesting clinical experience at his institution [189] (Fig. 14.55).

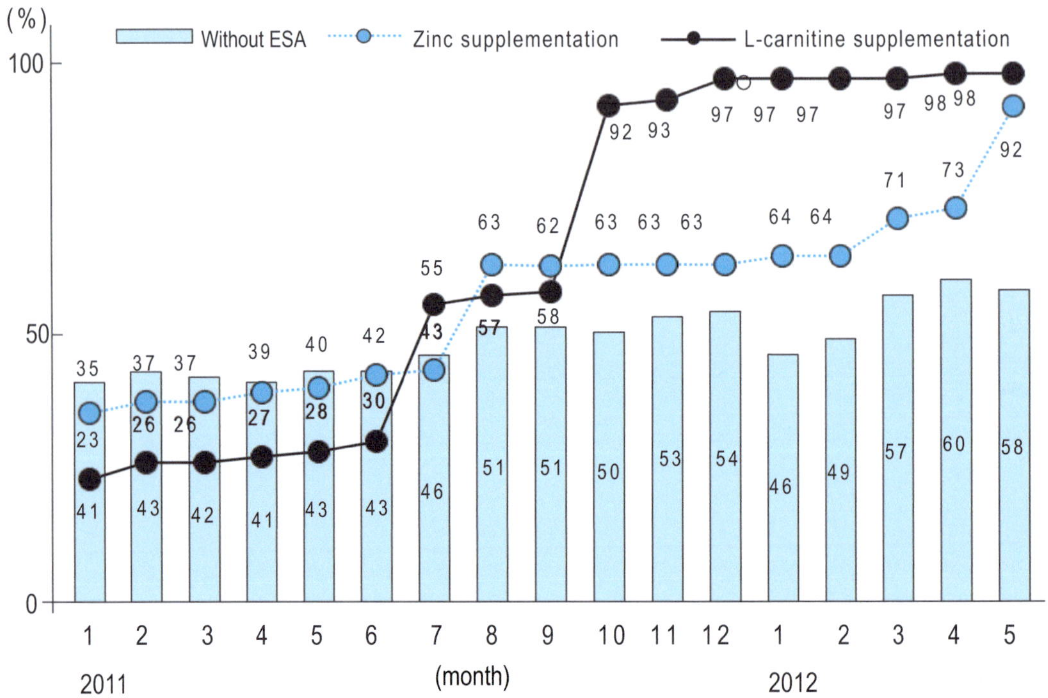

Fig. 14.55 Changes in monthly proportion of dialysis patients whose hemoglobin could control within target range without ESA (bar graph), and changes in monthly proportion of zinc supplementation (blue circle and dotted line) and L-carnitine supplementation (black circle and solid line) from January 2011 to May 2012. Most of patients received vitamin D_3 supplementation, and along with increase of the proportion of zinc and L-carnitine supplementation, the proportion of patients whose hemoglobin could control within target range without ESA gradually increased and finally reached around 60 %. The numbers indicate the percentage of each month. (Modified from Ref. [189])

It is surprising that about 60% of dialysis patients can maintain target Hb level without using ESA preparations, and it has a great impact for both health economics and hospital management.

How should we interpret these clinical experiences of Date et al. Although I have described carnitine deficiency in dialysis patients so far, it should be noted that deficiency of other nutrients and essential trace metals is also often observed in patients of carnitine deficiency, as described in the "Guideline for diagnosis and treatment of carnitine deficiency 2018." [190] It may be that carnitine deficiency, zinc deficiency, and vitamin D_3 deficiency are very common in hemodialysis patients, with two or three deficiencies in many cases (Fig. 14.56).

So, what role do zinc and vitamin D_3 play in the hematopoietic process? According to Deicher et al., various steps in the hematopoietic process involve co-hematopoietic factors such as vitamin D_3 (calcitriol), androgens, zinc, IGF-1, and EPO (Fig. 14.57) [191]. In other words, in order to produce sufficient mature red blood cells, not only rHuEPO but also many cofactors such as zinc and active vitamin D_3 are indispensable. Unfortunately, this Deicher review article does not mention about carnitine deficiency, but as mentioned above, it is necessary to consider not only carnitine deficiency but also zinc and vitamin D_3 deficiency in anemia in dialysis patients, and it is thought that hematopoiesis does not progress in the state of zinc and vitamin D_3 deficiency even with carnitine supplementation.

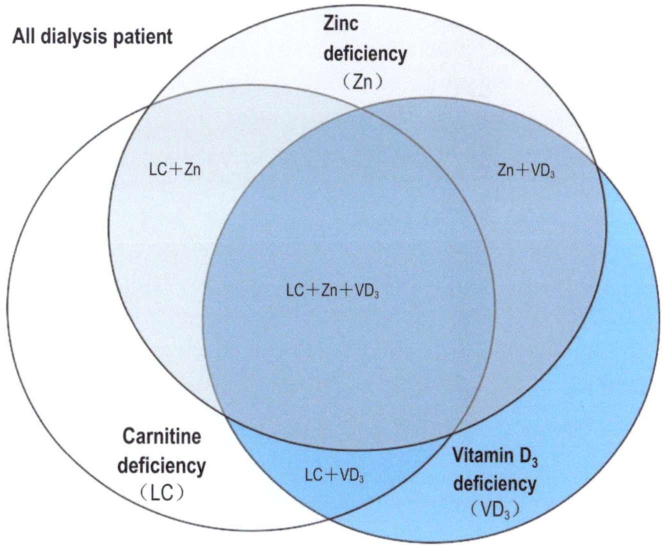

Fig. 14.56 An image figure for carnitine deficiency, zinc deficiency, and vitamin D_3 deficiency in maintenance dialysis patients. In dialysis patients, carnitine deficiency, zinc deficiency, and vitamin D_3 deficiency are common and sometimes they overlap each other, two or three of these deficiencies. LC: L-carnitine, Zn: zinc, VD_3: vitamin D_3

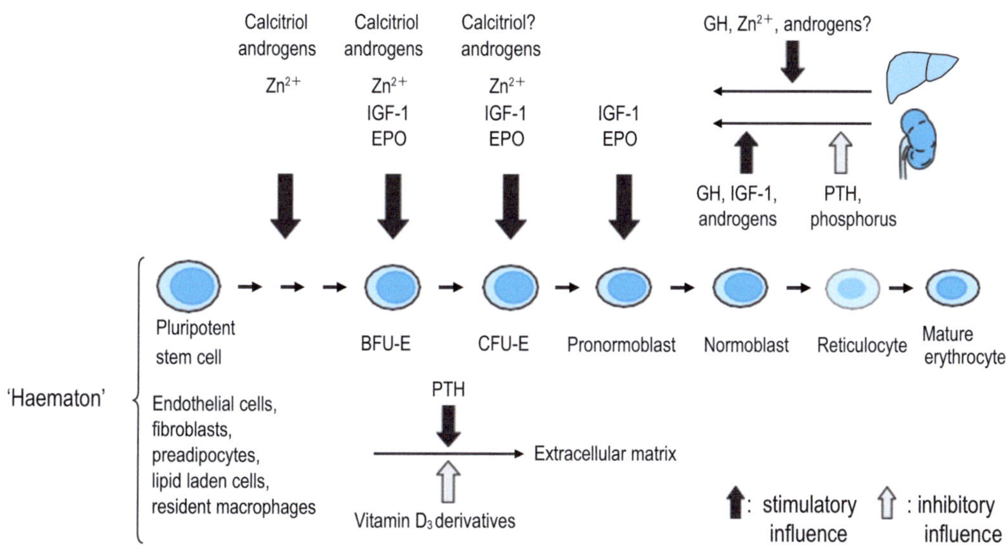

Fig. 14.57 Schematic synopsis of the mode of action of selected hormonal adjuvants that are potentially applicable to increase erythropoiesis in chronic kidney disease patients. Closed arrows represent stimulatory, open arrows inhibitory influence. Note that data on the effect of zinc largely stem from experimentally induced states of zinc deficiency. EPO: erythropoietin, GH: growth hormone, IGF-1: insulin-like growth factor-1, PTH: parathyroid hormone; Zn: zinc. (Modified from Ref. [191])

The foundation of crop nutrition is Liebig's law of the minimum. Plants need nitrogen, phosphate, and potassium for growth, as well as light, water, and temperature. If any one of these essential elements is lacking, plants cannot grow satisfactorily. If one of these essential elements is missing, such as lack of sunlight due to bad weather, cold temperatures due to cold damage, lack of water due to drought, or nitrogen, phosphate, or potassium in fertilizers, it becomes the rate-limiting factor. This Leibig's law of the minimum is often described by the analogy of a "Dobeneck's barrel (pale)" (Fig. 14.58).

Anemia and decreased cardiac function in dialysis patients also need to be dealt with by considering multiple nutrient deficiencies such as carnitine deficiency, zinc deficiency, and selenium deficiency. As shown in Fig. 14.57, multiple hormones and cofactors are required in the hematopoietic process, and sufficient hematopoiesis cannot be expected even if one of them is missing. When a clinical symptom develops due to a deficiency of multiple factors, replacement therapy for a single factor alone may not be sufficiently effective. The most rational method is to diagnose the degree of deficiency of each factor in each patient and replace all the deficient factors (Fig. 14.59). Carnitine supplementation therapy for "carnitine deficiency" in various pathologies described in this book, not limited to ESA-resistant anemia and heart failure in dialysis patients, may be partially or completely ineffective, and many clinicians have doubt about the efficacy of carnitine supplementation therapy. However, clinical trials of the effect of supplementation on deficiencies in multiple nutrients or factors are not easy, as it is necessary to evaluate the degree of deficiency of each relevant factor and then randomize patients. The clinical experience of Date et al. may be that supplementation

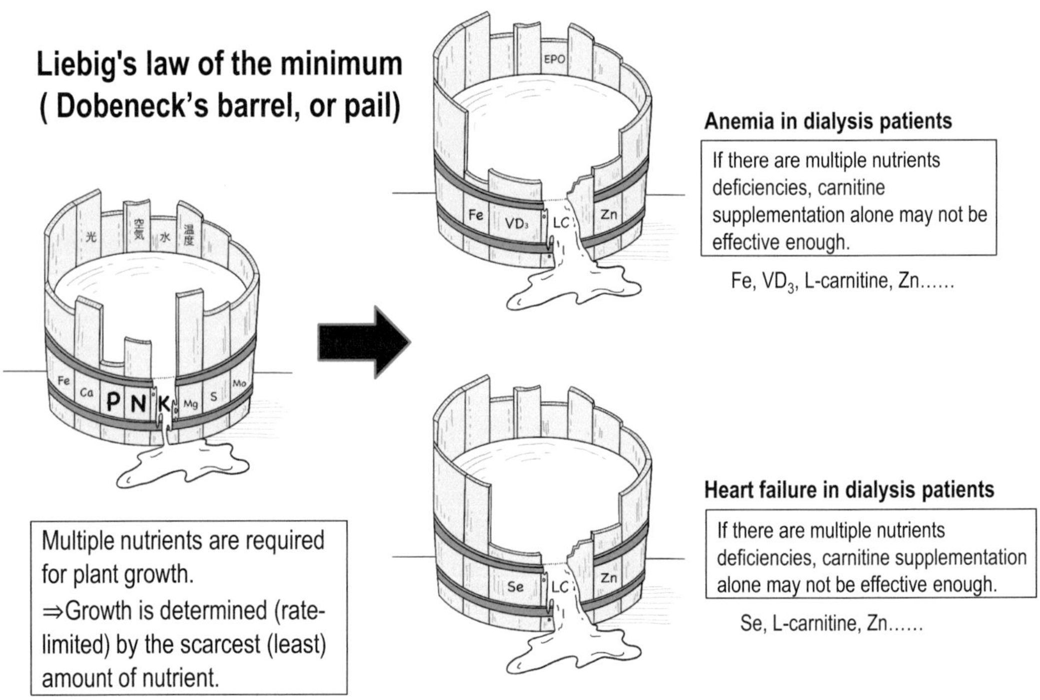

Fig. 14.58 Analogy of heart failure and anemia in dialysis patients between Liebig's law of the minimum. Liebig's law of the minimum (Dobeneck's barrel, or pail) says multiple nutrients are required for plant growth, but the growth is determined (rate-limited) by the scarcest (least) amount of nutrient. To treat pathophysiology such as anemia and heart failure in dialysis patients, we have to always consider multiple nutrients deficiencies and supplementation therapy for corresponding deficiencies. EPO: erythropoietin, Fe: iron, VD_3: vitamin D_3, Zn: zinc, Se: selenium, LC: L-carnitine

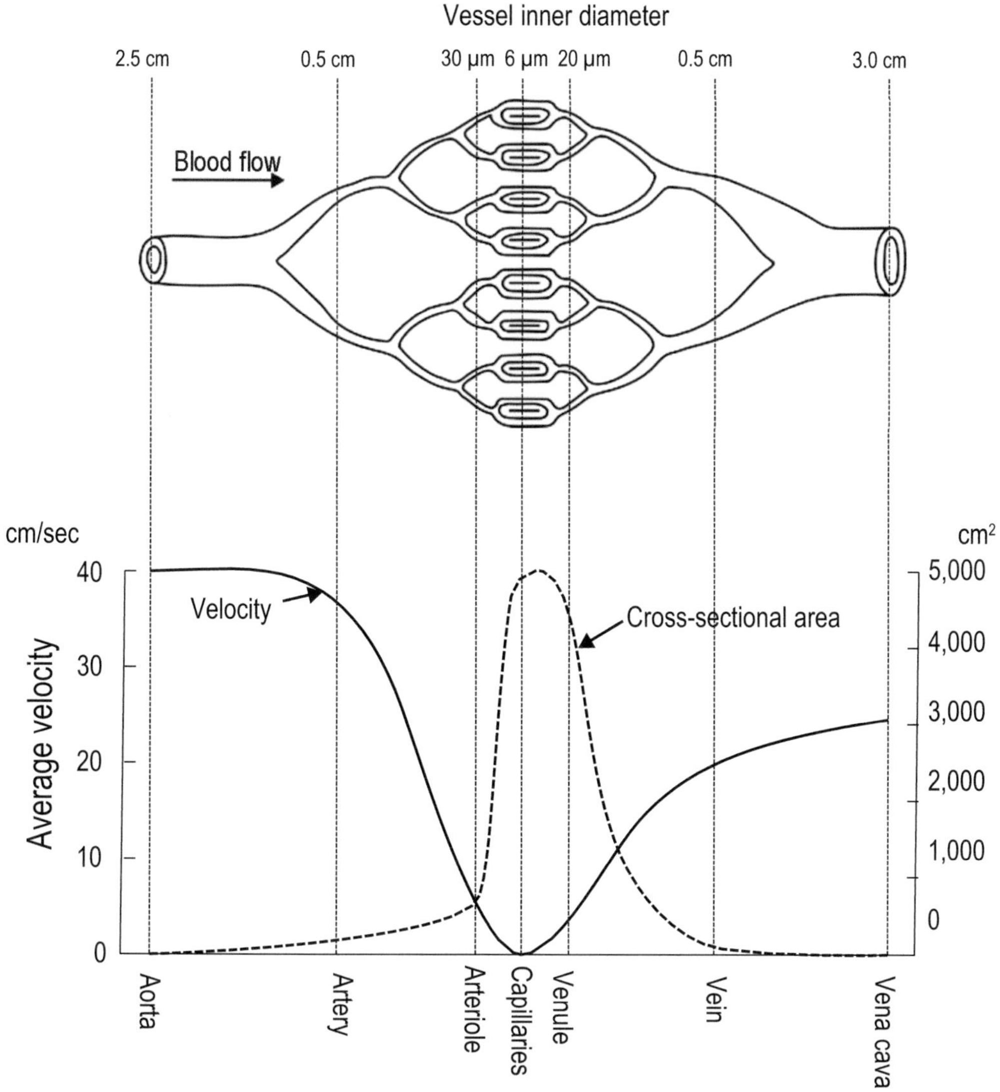

Fig. 14.59 Total cross-sectional area and average blood flow velocity of blood vessels in systemic circulation. Blood flow volume is equal in each blood vessel, and blood flow velocity is high in aorta and vena cava where the cross-sectional area is smallest. In the capillaries, the cross-sectional area is largest, but the blood flow velocity is smallest which contributes to exchange of substrates, waste substances, and gases between capillaries and the tissues. The dotted line: total cross-sectional area of blood vessels; The solid line: average blood flow velocity. (From Ref. [202])

of all three possible factors, conversely and simply, resulted in dramatic ESA non-use rates. In the treatment of carnitine deficiency in other areas, it is always necessary to consider the possibility of deficiencies in other multiple nutrients and factors, recalling Liebig's the law of minimum (Dobeneck's barrel).

5. Mechanism of the Anemia-Improving Effect of Carnitine

Mitochondria remain in reticulocytes until the final stage of erythropoiesis, when they are lost during enucleation. Why does carnitine have an effect on anemia when erythrocytes lack mitochondria, the primary site of action of carnitine? Although not discussed in detail, it is known that mature erythrocytes also contain significant amounts of free carnitine and acylcarnitine. Once carnitine is incorporated into mature erythrocytes during hematopoiesis, it is not transported to and from the plasma. Therefore, the carnitine content of erythrocytes probably depends on the plasma concentration during the hematopoietic process.

According to Bonomini et al., carnitine in erythroblasts and mature erythrocytes during the hematopoietic process has an anemia-improving effect mainly through the following three actions [192].

1. Metabolic effects
2. Biophysical action
3. Anti-apoptotic action

In the following section, I will describe detail.

14.4.6 Metabolic Actions of Carnitine

Patients with end-stage renal failure undergoing hemodialysis or peritoneal dialysis have increased oxidative stress, while their defense system against oxidative stress by antioxidant enzymes such as superoxide dismutase (SOD), catalase (CAT), and glutathione peroxidase is weakened, which may lead to peroxidation of phospholipids in the erythrocyte membrane. This is thought to contribute to the hemolysis and anemia seen in hemodialysis patients [193].

It is also known that the lifespan of preserved blood used for blood transfusion is shortened when the cytoskeleton is oxidized and changed by oxidative stress, etc., and drugs that act on it are required [194–196]. Arduini et al. compared the properties of preserved blood with 5 mM L-carnitine and found that preserved blood supplemented with carnitine had less hemolysis, higher ATP content, less damage due to oxidation of erythrocyte membrane phospholipids, and longer red blood cell lifespan when transfused into humans. [197] In addition, a series of studies by Arduini et al. revealed that although there are no mitochondria in red blood cells but carnitine, enzymes such as carnitine palmitoyltransferase (CPT), and lysophospholipid acyl-CoA transferase (LAT), which transfers the acyl group of acyl-CoA to lysophospholipids, are present. When phospholipids become phospholipid hydroperoxides by oxidative stress, the peroxidized fatty acid portion is cleaved by phospholipase A_2 to become lysophospholipids with only one fatty acid, and these lysophospholipids have a strong detergent effect and disturb the membrane structure and cause hemolysis. Carnitine, palmitoylcarnitine, and CPT in erythrocytes play a metabolic role in restoring the lysophospholipids to their original two long-chain fatty acid-linked phospholipids by adding acyl groups such as palmitoyl groups (Fig. 14.60a) [192, 198–200]. In addition, enzyme activity such as CPT and LAT are attenuated in erythrocytes from dialysis patients [193], and carnitine supplementation therapy partially reverses these disorders [192].

14.4.7 Biophysical Action

14.4.7.1 Rheology of Red Blood Cells

Unlike cells in solid tissues, red blood cells flow through blood vessels at extremely high velocities and are exposed to physical forces such as shear stress. However, the flow velocity is extremely slow in the capillaries beyond the small arteries. These capillaries are said to be 5–10 microns in diameter (average 7 microns), and a disc-shaped red blood cell with a dented center of about 8 microns in diameter can just barely pass through them. However, immediately

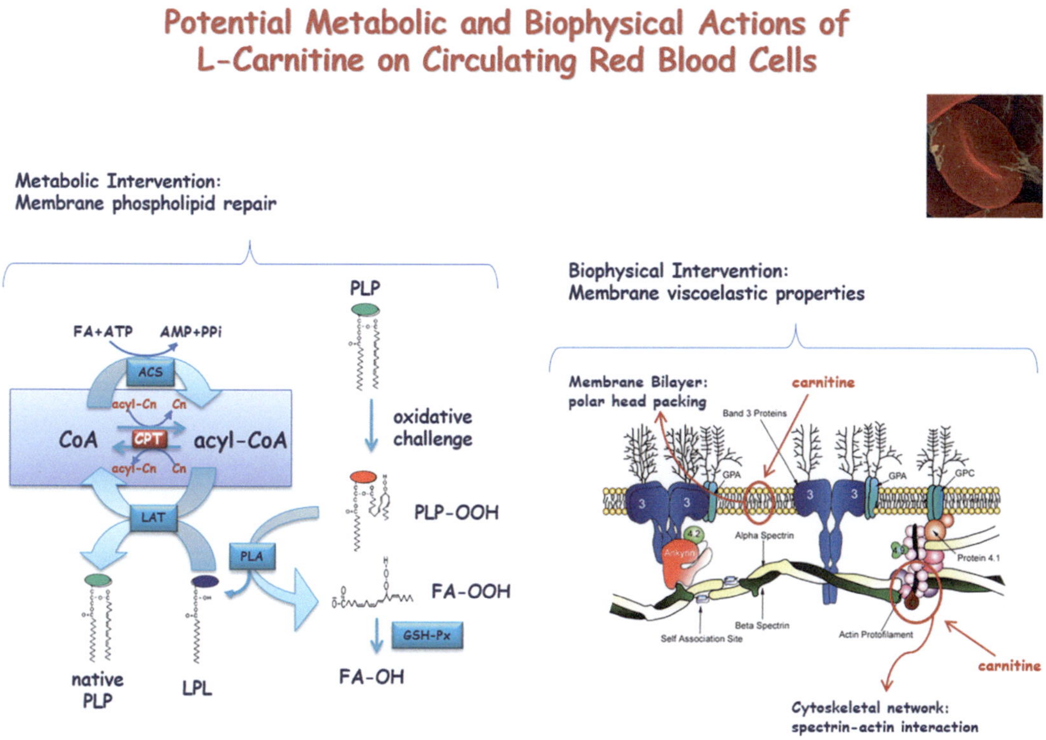

Fig. 14.60 Potential metabolic (**a**) and biophysical (**b**) interventions of L-carnitine on circulating red blood cells. *Membrane phospholipid repair (A)*: When oxidative challenge of a polyunsaturated fatty acid esterified in membrane phospholipid (PLP) generates a phospholipid hydroperoxide (PLP-OOH), the deacylating activity of a phospholipase A2 (PLA) hydrolyzes the fatty acid hydroperoxide (FA-OOH), which in turn may be further reduced to the corresponding non-reactive fatty acid alcohol (FA-OH) by a glutathione peroxidase (GSH-Px). This is followed by a parallel increase of lysophospholipid acyl-CoA transferase (LAT), which would regenerate the original PLP and limit the accumulation of the potentially toxic lysophospholipid (LPL). The increased LAT activity requires an increased supply of activated acyl units (acyl-CoA) from the activity of the ATP-dependent enzyme acyl-CoA synthetase (ACS). The combined activities of LAT and ACS generate a cycle, whose efficiency is strictly dependent on the extent of the enzymatic rates of LAT and ACS. Thus, the presence of a rate-coupler, carnitine palmitoyltransferase (CPT), would allow the ACS-LAT cycle to keep the acyl-CoA/free CoA ratio constant. Indeed, given the kinetic properties of CPT, any alteration of the acyl-CoA/free CoA ratio would be promptly buffered by the fully reversible and mass action-sensitive CPT reaction. The presence of supraphysiological concentration of carnitine (Cn) would support the buffering action of CPT. In addition, acyl-carnitine (acyl-Cn) represents a reservoir of acyl-CoA units at no ATP cost. *Membrane viscoelastic properties (B)*: Major determinants of the viscoelastic properties of RBC membrane are the cytoskeletal network and membrane phospholipid bilayer. The cytoskeletal network is an organization of supramolecular proteins lying beneath the inner monolayer of erythrocyte membrane bilayer. LC may improve the viscoelastic properties of RBC by exerting a stabilizing effect on the membrane via a specific interaction with certain cytoskeletal components (spectrin/actin as indicated in the figure by right red circle). LC would also affect the viscoelastic properties of RBC membrane by strengthening the polar head packing of membrane phospholipids (left red circle). (From Ref. [192])

in front of the capillaries, there is an anterior capillary sphincter, which contracts upon sympathetic nerve stimulation to regulate blood flow. Normal red blood cells can pass through capillaries as small as 2.7 microns by lowering their flow resistance through deformation [201]. This capillary network has a total length of 100,000 km and an area of 5000 cm^2. In other words, the blood flow velocity gradually decreases as it flows from arterioles to capillaries, and red blood cells pass through the capillaries very slowly, deforming as they do so, and after passing necessary oxygen and nutrients to each tissue and receiving waste products in the process, they gradually increase their speed to return to arterioles and veins [202] (Fig. 14.59).

The problem in ischemic heart disease and cerebral infarction is plaque and thickening of large blood vessels such as coronary arteries and cerebral arteries, but such large blood vessels have high blood flow velocity and high share stress. In these large blood vessels, the blood flow does not decrease until more than 90% of stenosis. However, arterioles and sphincters contract due to the slight influence of sympathetic nerves and smoking, and blood flow decreases. Naturally, the blood flow rate of the capillaries beyond it also decreases, and the exchange of oxygen and nutrients is also delayed. When we think of atherosclerosis, we tend to think of narrowing large blood vessels and plaques, but we have to consider the importance of arterioles, sphincters, and capillaries from Fig. 14.59. In particular, smoking dramatically reduces blood flow in arterioles and capillaries.

14.4.7.2 Effects of Carnitine on the Deformability of Red Blood Cells

Although erythrocytes are exposed to various chemical and physical attacks in the bloodstream, they survive through their deformability and damage repair functions and play a role in the transport of oxygen and waste products. In patients with end-stage renal failure, deformability is reduced due to increased oxidative stress caused by accumulation of urinary toxins, ATP depletion, and impaired ion homeostasis [192]. The deformability of erythrocytes is maintained by the cell membrane, which consists of a phospholipid bilayer, and the cytoskeleton, which supports the cell membrane on the underside (inside) of the membrane, but these are affected by changes such as lipid peroxidation, resulting in reduced deformability.

One of the simple methods for assessing the deformability of red blood cells is to investigate the osmotic fragility of red blood cells. Matsumura et al. examined the relationship between erythrocyte osmotic resistance and blood carnitine concentration in 26 dialysis patients using the coil planet centrifuge method [161, 203]. As a result, it was found that the point at which hemolysis ends (HEP) due to a decrease in osmotic pressure is inversely correlated with the total blood concentration of dialysis patients. In other words, it was found that patients with low blood carnitine levels were more likely to lyse at higher osmotic pressures.

Nikolaos et al. reported that when carnitine was administered to 15 dialysis patients for 3 months (intravenous infusion of 30 mg/kg/day, after dialysis 3 times a week) and examined the ability of red blood cells to deform by the filter method, the erythrocyte deformability of dialysis patients decreased and became rigid compared to normal control but was significantly improved by carnitine administration [204].

Ibrahim et al. also reported that the deformability of red blood cells was reduced in dialysis patients compared to healthy subjects, and as a result, the osmotic resistance of red blood cells was attenuated and hemolysis occurred at a higher osmotic pressure, and that osmotic fragility improved slightly after dialysis, but they were still susceptible to hemolysis [205]. This suggests that the presence of urine toxin is partly involved in the decrease in erythrocyte deformability in dialysis patients, but that there are other factors such as changes in the composition of red blood cell membranes.

Vlassopoulos et al. studied the effect of carnitine administration at 20 mg/kg three times a week for 1 year after dialysis on 13 patients with low erythrocyte osmotic resistance among 30 dialysis patients who had been on dialysis for more than 1 year, while 17 patients with normal osmotic resistance were treated as control group [206]. In the control group, free carnitine concentration in the blood decreased as the duration of dialysis increased, and osmotic resistance also decreased; however, in the 13 patients treated with carnitine, osmotic resistance improved in 5 patients, worsened in 7 patients, and remained unchanged in 1 patient.

On the other hand, Labonia administered carnitine to 13 of 24 dialysis patients (1000 mg intravenously 3 times a week after dialysis for 6 months) and compared the required amount of rHuEPO with the placebo group of 11 patients. They reported a 38.1% reduction in rHuEPO requirements in the carnitine group, but there was no significant change in osmotic vulnerability between the two groups before and after carnitine supplementation [163].

Thus, it was suggested that carnitine supplementation therapy may improve the osmotic fragility of red blood cells at least in some dialysis patients. According to Bonomini et al., the effect of carnitine supplementation is thought to be effective at a concentration above the physiological concentration (supra-physiological level of L-carnitine), such as 1 g administered intravenously [192]. Kobayashi et al. reported that in an in vitro experiment using 5-doxylstearic acid, in which isolated erythrocytes were spin-labeled, the addition of 100 μM palmitoyl carnitine resulted in a change in the fluidity of erythrocyte membranes, a morphological change from disc-shaped to spiked shape, and sphere shape, then hemolysis occurred, but the addition of 2 mM carnitine improved these changes [207]. Palmitoyl carnitine is said to increase during myocardial ischemia, and the concentration of such long-chain acylcarnitine also increases in dialysis patients. Long-chain acylcarnitine is an amphiphilic substance with hydrophobic and hydrophilic moieties and is thought to have a detergent effect that changes the fluidity of biological membranes and disrupts them. In addition, carnitine on the order of millimole has been shown in vitro studies to loosen the packing of phospholipids by attenuating the interaction of the polar head of the erythrocyte membrane toward the phospholipids [208]. In addition, carnitine strengthens the cytoskeletons (lining proteins), which is formed by a network of spectrin and actin directly under the erythrocyte membrane [209], thereby strengthening the cytoskeleton and maintaining the deformability of the erythrocyte membrane [210, 211]. In other words, it is thought that the erythrocyte membrane is softened, and the cytoskeleton, which is the lining proteins, is strengthened and the deformability is strengthened (Fig. 14.60b) [192]. In this way, by administering a high concentration of carnitine by intravenous administration, carnitine can exert physicochemical effects, and it is thought that the reduced deformability of red blood cells will be improved.

14.4.8 Anti-apoptotic Action of Carnitine

14.4.8.1 Apoptosis in the Production Process Inside the Red Blood Cells and Action of Carnitine

Red blood cells make up about 10% of the total cell volume of an adult. Red blood cells are produced in the bone marrow as following sequences: hematopoietic stem cells (HSCs) ⇒ multipotent progenitors (MPPs) ⇒ common myeloid progenitors (CMPs) ⇒ megakaryocyte-erythrocytes progenitor (MEP) ⇒ burst-forming unit-erythroid (BFU-E) ⇒ colony-forming unit-erythroid (CFU-E) ⇒ proerythroblast basophilic erythroblast ⇒ polychromatic erythroblast ⇒ orthochromatic erythroblasts erythroblast ⇒ reticulocyte, and finally mature erythrocytes are formed. When enucleation occurs during the process of forming reticulocytes from orthochromatic erythroblasts during this maturation process, and when Hb is stored in the process of becoming mature erythrocytes from reticulocytes, nucleic acid synthesis and protein synthesis systems are no longer needed, and intracellular organelles such as ribosomes, endoplasmic reticulum, and mitochondria disappear, resulting in highly streamlined cells specialized for oxygen transport [212].

Apoptosis is often studied as a programmed death of cells with nuclei (suicide death). In the hematopoietic process, there are nuclei up to orthochromatic erythroblasts, and mitochondria are also present in the process up to reticulocytes, and the apoptosis system is at work, and it is involved in the regulation of red blood cell production and the reduction in red blood cell production during bone marrow suppression during anticancer drugs and radiation [213].

The roles of apoptosis and carnitine in the production and decay processes of red blood cells are described below.

During hematopoiesis in the bone marrow, when EPO binds to the erythropoietin receptor (EPOR) on the erythroid progenitor cell membrane, JAK2 which bound to EPOR, was autophosphorylated, and phosphorylates tyrosine residues on the EPOR, which in turn causes phosphorylation of STAT5. STAT5 translocates in the nucleus and cooperates with GATA-1, which is the most important transcription factor for erythroblastic differentiation, to initiate gene expression and protein synthesis such as heme synthase and globin genes, which are important for the structure and function of erythrocytes. On the other hand, the signal from EPOR promotes the expression of apoptotic suppressor genes such as Bcl-XL and suppresses apoptosis [192, 214] (Fig. 14.61a, right side).

When an apoptotic signal is triggered by EPO deficiency or EPO resistance, or a cell death signal such as TNFα or Fas ligand, caspase 3 is eventually activated to degrade the transcription factor GATA-1, which is important for erythrocyte differentiation, and the maturation process of erythrocytes is blocked (maturation arrest, Fig. 14.61a, left side). The apoptotic system of this hematopoietic process

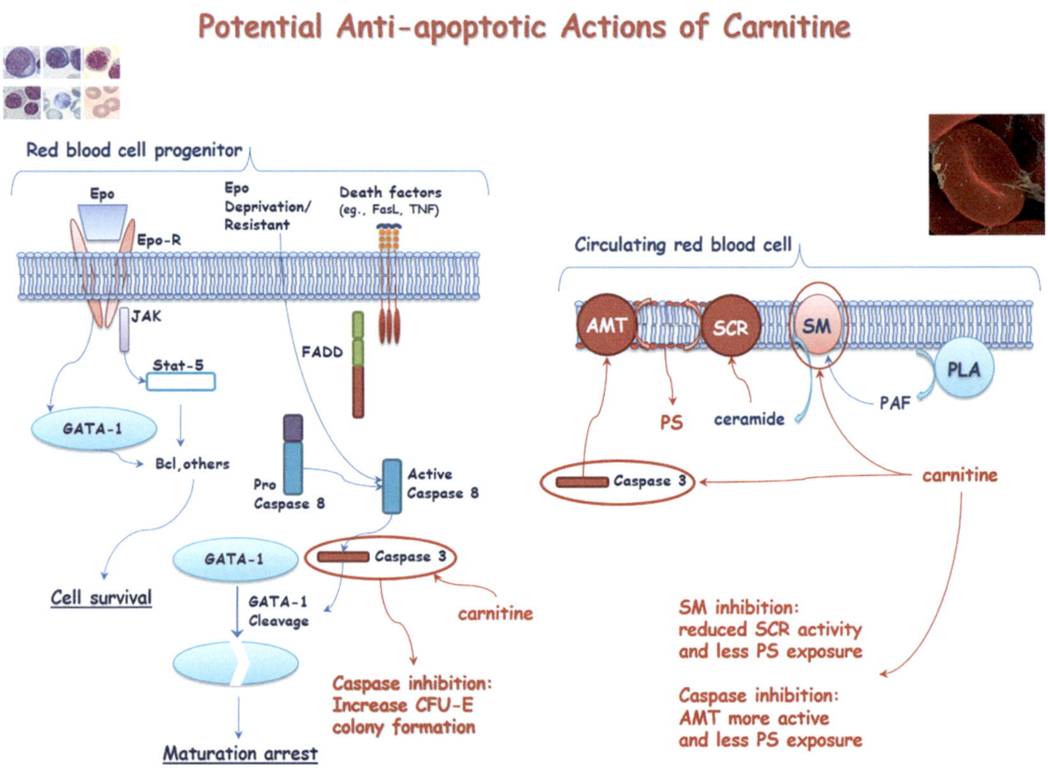

Fig. 14.61 Potential anti-apoptotic intervention of L-carnitine. (**a**) *Schematic representation of negative regulation of erythropoiesis triggered by death receptor activation or erythropoietin deprivation/resistance*: Both these conditions induce the apoptotic cascade, which the consequent activation of caspase 3 that cleaves the transcription factors GATA-1. The cleavage of GATA-1 is responsible for either the maturation arrest or apoptosis of erythroid cells. By inhibiting caspase 3, L-carnitine would favorably affect the maturation of erythroid cells. EPO: erythropoietin, FADD: FAS-associated death domain, Bcl: antiapoptotic protein, JAK: Janus kinase, CFU: colony forming unit. (**b**) *Ceramide- and caspase 3-dependent cell death pathways are also operative in circulating red blood cells and known as erythro apoptosis*: Indeed, activation of such cell death pathways leads to phosphatidylserine (PS) exposure at the surface of the erythrocyte membrane, a well-known signal for the elimination of RBCs by macrophages in vivo. L-carnitine may counteract PS exposure by either preventing ceramide production or inhibiting caspase-3. SCR: scramblase, AMT: aminophospholipid translocase, SM: sphingomyelinase, PLA: Phospholipase A_2, PAF: platelet activating factor. (From Ref. [192])

actually acts as a negative feedback mechanism mediated by the Fas–Fas ligand to regulate the amount of erythrocyte hematopoiesis [212, 213]. In other words, when EPO binds to the EPOR of undifferentiated erythroblasts, a proliferation signal acts and differentiates into mature erythrocytes, but the Fas expressed on the surface of mature erythrocyte cells binds to the Fas ligand on the surface of undifferentiated erythroblasts, and an apoptotic cascade signal enters, and finally caspase 3 is activated and GATA-1 is degraded, blocking the differentiation of undifferentiated erythroblasts. When the amount of EPO is normal or excessive, this apoptotic system is activated, and the maturation process of erythroblasts is stopped. In addition, if the amount of EPO is depleted or if there is resistance to EPO, proliferation does not proceed because the proliferation signal from EPOR does not work, and anemia progresses if the apoptotic system is activated in some way. It has been reported that L-carnitine suppresses apoptosis and morphological changes caused by caspase-3 in Jurkat cells derived from human T-cell leukemia and platelet cells of patients with end-stage renal disease and also suppresses caspase-3 activity produced by recombinants [215, 216].

Matsumura et al. reported that the addition of 400 μM carnitine to the medium of erythroblastic colonies derived from the liver of mouse fetuses, which is the main hematopoietic site in the fetal period, promoted erythroblastic colony formation by about 34%, palmitoyl carnitine by about 57% at 12.5 μM, and by about 91% at 100 μM. [217] In addition, Kitamura et al. cultured mouse bone marrow cells, the only hematopoietic site after birth, and examined the effect of carnitine on the cells, and reported that carnitine added to the medium promoted erythroblastic colonization by 21% at 200 μM and 32% at 400 μM at an EPO concentration of 0.5 IU/mL [218] When the EPO concentration was 1.0 IU/mL, the carnitine added to the medium promoted erythroblastic colonization by 12% at 200 μM and 23% at 400 μM. These results suggest that carnitine supplementation therapy for dialysis patients with rHuEPO-resistant anemia may be able to control anemia at a lower dose of rHuEPO. The data of Matsumura et al. and Kitamura et al. suggest that carnitine suppressed apoptosis during erythroblastic colony formation and promoted erythroblastic colony. The anti-apoptotic effect of carnitine has since been reported in various cells such as skeletal muscle cells, cardiomyocytes, neurons, and liver cells, and the clinical significance and utilization of its action are being investigated.

Calò et al. had been studying the antioxidant and anti-apoptotic effects of carnitine in vascular endothelial cells via its ability to enhance heme oxygenase-1 (HO-1) expression [219], and immediately commented upon reading the report by Kitamura et al., that “the hematopoietic effects of carnitine found by Kitamura et al. may be mediated by HO-1,” [220] and subsequently confirmed that carnitine improves EPO reactivity in humans and animals via enhancement of HO-1 expression. [221]

At present, it was found that carnitine exerts antioxidant, anti-inflammatory, anti-arteriosclerotic, and cytoprotective effects by enhancing the expression of antioxidant enzymes such as SOD, catalase, glutathione peroxidase, and cytoprotective factors of HO-1 and heat shock protein (HSP70, etc.) through the activation of Nrf2. In addition, it has been found that it exerts extremely multifaceted effects, such as suppression of apoptosis by reducing inflammatory cytokines and reactive oxygen species, and inhibition of apoptosis by direct suppression of caspace-3 (see Chap. 7, “Anti-inflammatory, Antioxidant, and Anti-apoptotic Effects of Carnitine”).

14.4.8.2 The Eryptosis and the Action of Carnitine in the Process of Decay of Erythrocytes in the Circulation

Red blood cells are formed in the bone marrow and mature into erythrocyte, where they perform the function of transporting oxygen and waste products such as CO_2, but as they age, they are processed by the reticuloendothelial system of the spleen and liver in a healthy adult with a lifespan of about 120 days [222]. It is thought that eryptosis similar to apoptosis also occurs during the disposal process of aged erythrocytes [223–225].

The lifespan of red blood cells in healthy individuals is around 120 days, but the lifespan of red blood cells in dialysis patients has been shortened to about 45–85 days [226, 227]. In addition, the lifespan of red blood cells in dialysis patients with ESA-resistant anemia is said to be 30–70% shorter than that of red blood cells in patients without ESA-resistant anemia [228]. ESA preparations such as rHuEPO can promote the hematopoietic process, but do not extend the shortened red blood cell life of dialysis patients. Conversely, it has been reported that the amount required for ESA preparations is inversely correlated with a decrease in the deformability of red blood cells [229] and a shortening of lifespan [228]. In other words, in order to consider measures to essentially treat ESA-resistant anemia, it is important to take measures to shorten the life of red blood cells [230]. So, what is the lifespan of red blood cells, that is, the mechanism by which aging red blood cells are recognized and processed by macrophages of the reticuloendothelial system?

Basically, biomembrane consists of phospholipid bilayer of outer and inner leaflet. From many studies using various cells including erythrocytes, it is well known that the composition of the phospholipids of the cell membrane is different between the outer leaflet and the inner leaflet. Under the normal state, the inner leaflet is rich with aminophospholipid such as phosphatidylserine (PS) and phosphatidylethanolamine (PE), and the outer leaflet is rich with phosphatidylcholine (PC), sphingomyelin, and glycosphingolipids. Especially, there is little PS in the outer leaflet, and PS is abundant in the inner leaflet. It is highly energy-requiring process to exchange lipids from outer to inner leaflet (vice versa). This exchange is called flip-flop. So, to keep asymmetrical distribution of phospholipids, there are three types of lipid transporters (enzymes) to maintain this asymmetry of phospholipids composition [231]. Using the energy of ATP, flippases and floppases translocate specific lipids from the outer to inner leaflet or from the inner to outer leaflet, respectively, against the concentration gradient, whereas scramblases, driven by the existing lipid gradient, nonspecifically and bidirectionally transport lipids between the inner and outer leaflet. Specifically, floppases transport phospholipids from the inside to the outside in a non-specific and slow manner. Also, the Ca^{2+}-dependent scramblases (SCR) transport phospholipids in both directions (inner ⇔ outer). When these enzymes are activated, PS is exposed to the outside of the cell, but usually the ATP-dependent aminophospholipid translocase (AMT, a kind of flippases) is working to return the exposed PS from the outer leaflet to the inner leaflet. Intracellular Ca^{2+} promotes SCR and inhibits AMT. In addition, when sphingomyelinase in the cell membrane is activated by platelet activating factor (PAF), the ceramide is produced and activates scramblase. As a result, the opportunity for PS exposure increases. Also, caspase 3 inhibits AMT and increases PS exposure. L-carnitine has property to inhibit SCR and caspase 3 and reduce PS exposure which result in increased lifespan of red blood cells (Fig. 14.61b) [192].

As shown in Fig. 14.62, L-carnitine has multiple mode of action which affects the number and quality of red blood cells. Inhibition of apoptosis by carnitine is expected to increase red blood cell number; on the other hand, by promoting phospholipid remodeling and improving viscoelastic properties, carnitine is expected to affect red blood cell quality and increase their lifespan [192].

14.4.9 Detail of the Studies on PS Exposure in Erythrocyte

In aging erythrocytes and erythrocytes damaged by oxidative stress, these repair mechanism is disrupted, and PS is exposed in outside leaflet of the bilayer cell membrane, and macrophages in the reticuloendothelial system recognize this exposed PS as an "eat-me signal" and phagocytose it in circulation [232] (Fig. 14.63). In other words, macrophages recognize PS exposed out-

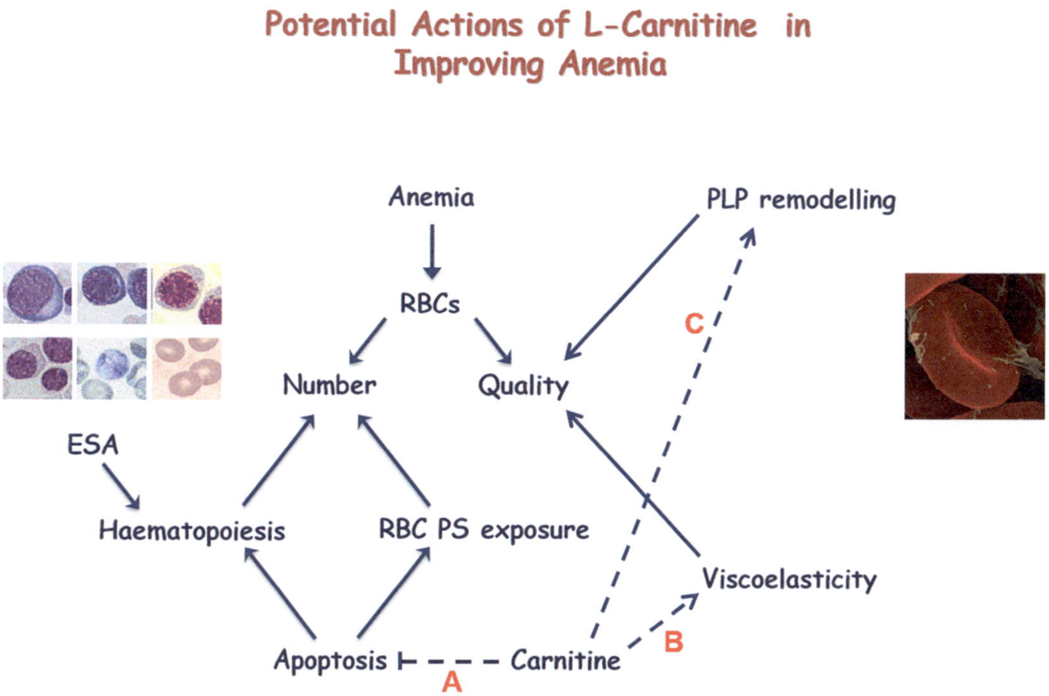

Fig. 14.62 Multiple mode of action through which L-carnitine may affect the number and quality of red blood cells in the circulation. Inhibition of apoptosis by L-carnitine (**a**) is expected to increase red cell number. On the other hand, by promoting phospholipid remodeling (**b**) and improving viscoelastic properties (**c**), L-carnitine is expected to affect red blood cells quality, and hence, increase their lifespan. EPO: erythropoietin, FADD: FAS-associated death domain, Bcl: antiapoptotic protein, JAK: Janus kinase, CFU: colony forming unit. (From Ref. [192])

side the cell membrane as a marker of senescent red blood cells, phagocytose, and process it. Namely, it is thought that the reason for the shortened lifespan of red blood cells in patients with end-stage renal disease is increased PS exposure due to oxidative stress and uretic toxins. These PS-exposed red blood cells are processed by macrophages faster than their normal original lifespan.

Bonomini et al. examined the amount of PS exposed to the outside of red blood cells in patients with end-stage renal disease and found that red blood cells in patients with non-dialysis end-stage renal failure, continuous peritoneal dialysis, and hemodialysis patients were 3.46, 3.15, and 4.54 times more exposed than that of healthy subjects, respectively [233] (Fig. 14.64).

Also, Bonomini et al. incubated red blood cells and monocyte-derived macrophages from hemodialysis patients with increased PS exposure and examined phagocytic red blood cells, which were significantly more phagocytic than red blood cells in healthy subjects (51.4 ± 1.6% vs. 13.5 ± 0.8%, $P < 0.001$) [234]. Therefore, it is thought that the recognition and phagocytosis of PS exposed on the surface of erythrocyte membranes by macrophages and the promotion of the disposal of red blood cells are one of the important factors in shortening the lifespan of red blood cells in patients with end-stage renal disease. Even if expensive EPO preparations are administered to promote hematopoiesis, the red blood cells produced are eaten by macrophages. Carnitine has the effect of inhibiting sphingomyelinase, which is activated when cultured cardiomyocytes are treated with doxorubicin, thereby reducing ceramide and attenuating the activity of scramblases and suppressing the apoptosis of cardiomyocytes caused by doxorubicin [235]. Carnitine also inhibits caspase3 in Jurkat cells [215]. Erythrocyte don't have mitochondria, but caspase 3 remains in mature erythrocytes, and when it is activated, AMT is suppressed and the action of AMT returning the PS from the outside

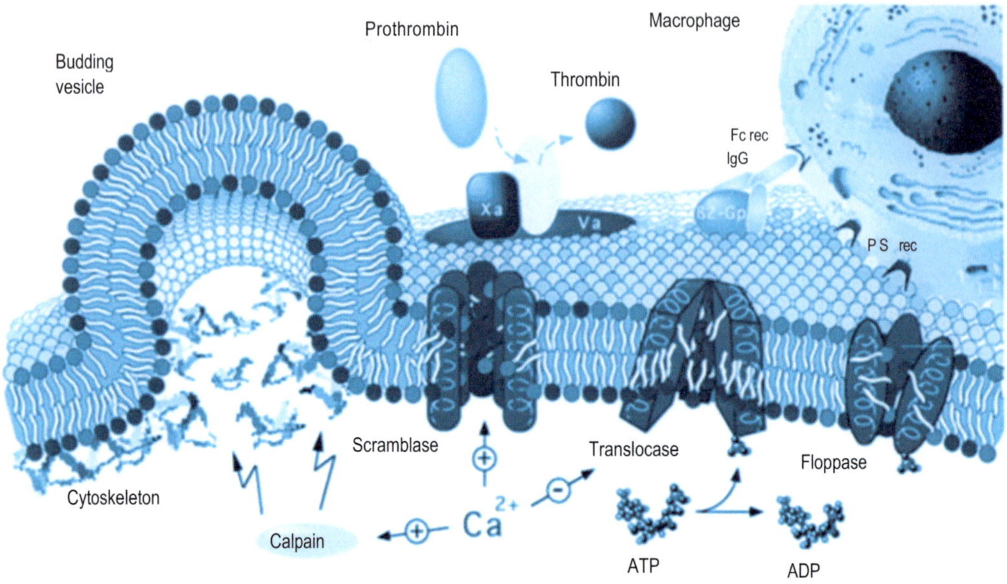

Fig. 14.63 The regulation and physiology of membrane phospholipid asymmetry. This model describes how membrane phospholipid asymmetry is generated, maintained, and perturbated as a prerequisite to various phosphatidylserine-related pathophysiologies. Membrane lipid asymmetry is regulated by the cooperative activities of three transporters. The ATP-dependent aminophospholipid-specific translocase, which rapidly transports PS and PE from the cell's outer-to-inner leaflet; the ATP-dependent nonspecific lipid floppase, which slowly transports lipids from the cell's inner-to-outer leaflet; and the Ca^{2+}-dependent nonspecific lipid scramblase, which allows lipids to move randomly between both leaflets. The model predicts that the translocases are targets for Ca^{2+} that directly regulate the transporter's activities. The figure shows that elevated intracellular Ca^{2+} induces PS randomization across the cell's plasma membrane by providing a stimulus that positively and negatively regulates scramblase and translocase activities, respectively. At physiologic Ca^{2+} concentrations, PS asymmetry is promoted because of an active translocase and floppase but inactive scramblase. Depending on the type of cell, elevated intracellular Ca^{2+} can be achieved by cellular activation that generally results in the concomitant influx and accumulation of extracellular Ca^{2+} and by its release from intracellular stores. Increased cytosolic Ca^{2+} can also result in calpain activation, which facilitates membrane blebbing and the release of PS-expressing procoagulant microvesicles. Exposure of PS at the cell's outer leaflet. The appearance of PS at the cell's outer leaflet promotes coagulation and thrombosis by providing a catalytic surface for the assembly of the prothrombinase and tenase (not shown) complexes and marks the cell as a pathologic target for elimination by phagocytes. Recognition of the PS-expressing targets can occur by both antibody-dependent and direct receptor-mediated pathways. β2-Gp: β2-glycoprotein-1, rec: receptor, PS: phosphatidylserine, PE: phosphatidylethanolamine. (From Ref. [232])

leaflet to the inside leaflet is weakened, but carnitine inhibits caspase 3 and maintains normal AMT function and reduces PS exposure (Fig. 14.61b) [192].

Arduini et al. studied the effect of carnitine on lifespan in dialysis patients who were given carnitine (20 mg/kg, intravenously, 3 times a week after dialysis for 24 weeks) and labeled them with chromium-51 (^{51}Cr). As a result, it was reported that there was no significant difference in the erythrocyte half-life of 39.1 ± 7.5 days and 40.2 ± 8.9 days at the beginning of the study between the carnitine and non-carnitine groups, but the half-life was 4.8 ± 11.1 days shorter in the non-carnitine group, and it was extended by 3.6 ± 10.6 days in the carnitine group, although there was no significant difference due to large variation ($P = 0.058$) [236]. In addition, Mizuguchi et al. had been studying the half-life of red blood cells in dialysis patients by the method based on the concentration of carbon monoxide in exhaled breath [226, 227]. to inves-

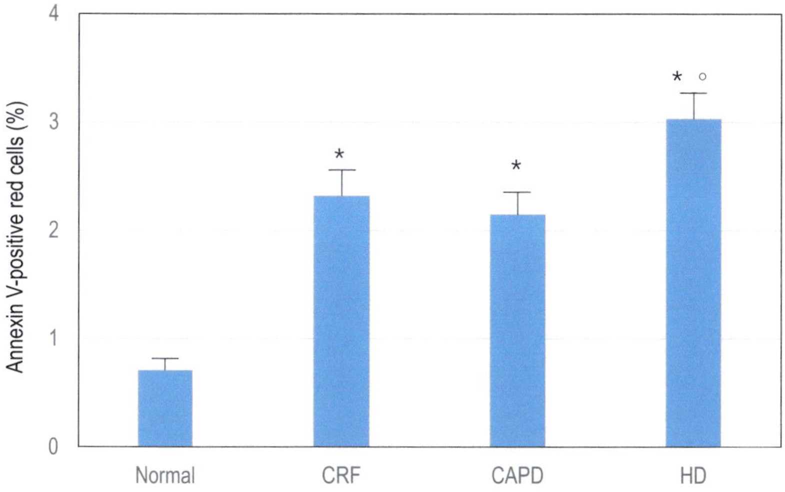

Fig. 14.64 PS exposure rates of RBC in chronic renal failure patients. PS exposure rates were studied using FITC-labeled Annexin V which bound to PS on RBC surface. Normal: healthy subjects, CRF: non-dialyzed chronic renal failure patients, CAPD; chronic renal failure patients treated with CAPD, HD: chronic renal failure patients treated with hemodialysis. PS: phosphatidyl serine, RBC: red blood cells, FITC: fluorescein isothiocyanate (a fluorescence probe). *: $P < 0.05$ vs. healthy subjects, °: $P < 0.05$ vs. CAPD patients. (From Ref. [233])

tigate the effect of levocarnitine chloride supplementation therapy (600 mg/day orally administered, 12 weeks) on red blood cell life in 13 hemodialysis patients with ESA-resistant anemia [237, 238]. The results showed that the Hb level increased by 1.0 g/dL or more in seven patients, and ESA was reduced by 25% or more in four patients, which was effective in a total of 9 patients, and the shortened red blood cell life was improved in three of the effective cases. Carnitine deficiency is one of the causes of ESA resistance, and carnitine supplementation therapy has been shown to be effective in at least some patients with ESA-resistant anemia in dialysis patients. Further studies are awaited.

I will not describe further details about this PS exposure, but PS exposure is an important element in the physiology of diverse range of developmental and pathogenic processes in multicellular organisms, and not only eryptosis but also various cell–cell interactions and apoptosis (please see Fig. 8 in Ref) [239]. Future studies about L-carnitine's actions on these PS-exposure-related pathophysiology are interesting and promising.

14.5 Hospitalization of Dialysis Patients and Carnitine

Kazmi et al. studied the effects of carnitine on hospitalization of dialysis patients [240]. Many dialysis patients have cardiovascular diseases and are said to be at high risk of hospitalization due to low nutrition, hyperinflammatory, atherosclerosis, and infections. According to 1998 statistics in the United States, the number of hospitalizations for the general population is 0.31 hospitalizations per year and the number of days of hospitalization is 1.9 days, but for patients with end-stage renal disease, the number of hospitalizations is dramatically higher at 1.9 hospitalizations per year and the number of days of hospitalization is 14 days, which accounts for 41% of the total medical cost of patients with end-stage renal disease. Therefore, Kazmi et al. used the Fresenius Medical Care North America Dialysis Database" to study the effects of carnitine supplementation therapy on the number of days of hospitalization. Total 2967 patients who were treated with carnitine for more than a month was retrospectively

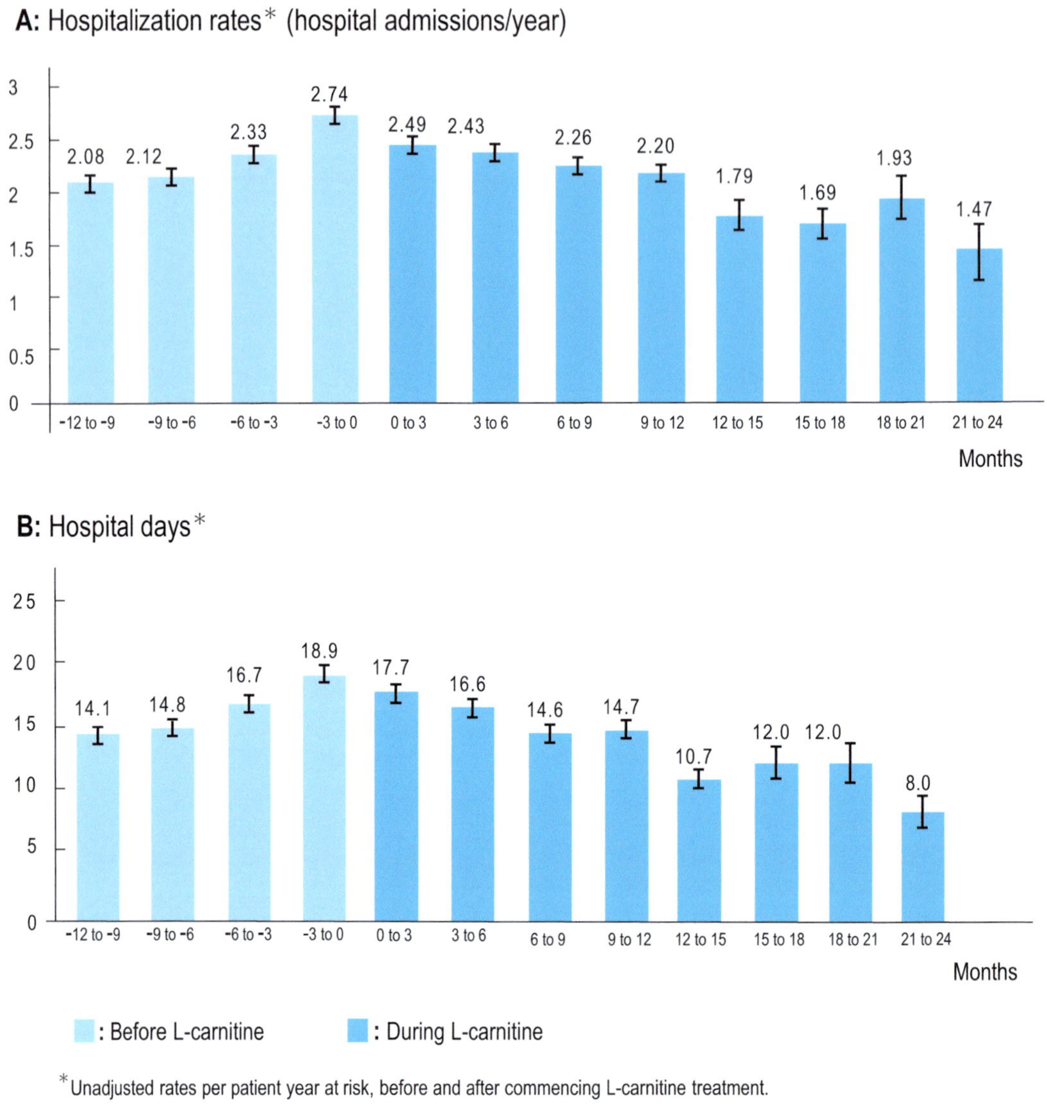

Fig. 14.65 Hospitalization rate (**a**) and hospital days (**b**) in dialysis patients (n = 2967) who received carnitine supplementation therapy more than 3 months. (From Ref. [240])

compared with the matched group of patients who did not receive carnitine treatment. As a result, carnitine supplementation therapy (mean dose of 1.5 ± 0.7 g, 9.7 ± 5.4 months) reduced the relative risk of hospitalization by 11, 11, and 15%, at 3, 6, and 9 months, respectively, compared to 3 months before carnitine administration. In patients with cardiovascular disease, the relative risk of hospitalization was reduced by 24, 31, and 34%, at 3, 6, and 9 months after treatment, respectively (Fig. 14.65).

Weinhandl et al. used the Centers for Medicare & Medicaid Services (CMS) database, a US health insurance payer, to retrospectively analyze the number of hospital days for dialysis patients on carnitine supplementation therapy (1 g or more intravenously administered at least 10 times/month) from 1998 to 2003 [241]. The number of dialysis patients was about 160,000 in 1998 and gradually increased to about 220,000 in 2003. From these patients, they excluded about 5–6% who had undergone renal transplantation and 20–25% for whom insurance claims could not be verified (116,534 patients/1998 to 153,081 patients/2003). Results showed that 3–7% of patients received carnitine replacement therapy for more than 1 month, and these patients were older, had more serious comorbidities, and required higher doses of ESA than patients not receiving carnitine. Patients receiving carnitine supplementation therapy had 10.8% fewer hospital days in the repeated measure model analysis ($P < 0.01$) and 21.7% fewer hospital days in the marginal structural model analysis ($P < 0.01$). In the United States, dialysis patients are hospitalized for an average of 15 days/year. Therefore, carnitine replacement therapy that reduces the number of hospitalization days may be very important for reducing medical costs.

14.6 Status of Carnitine Supplementation Therapy for Dialysis Patients in the Worldwide

14.6.1 Approval Status in Each Country

In Japan, the indication for L-Cartin® tablets was changed to carnitine deficiency in March 2011 by a public knowledge application, and carnitine can also be used for carnitine deficiency in dialysis patients. At that time, I investigated the approval status of carnitine preparations in countries around the world. As a result, carnitine supplementation therapy for carnitine deficiency has been approved in at least 38 countries worldwide, and carnitine supplementation therapy for dialysis patients has been approved in 26 countries worldwide. In addition, in major developed countries, the use of carnitine has been approved for carnitine deficiency in dialysis patients.

In Germany, oral and intravenous carnitine is approved for use in hemodialysis patients, and there are description in the package insert that the approval dosage of oral carnitine is 1 to 2 g/day, and the necessity of administration was judged based on clinical signs and symptoms. Also intravenous carnitine can be used in the case of oral carnitine was metabolized by intestinal bacteria to trimethylamineT and other substances ad caused fish odor. The efficacy of treatment should be evaluated every 3 months and determine whether the drug should be administered or not.

In the United Kingdom, carnitine supplementation therapy is recommended for dialysis patients with cramps, decreased energy, decreased quality of life, muscle weakness, myopathy, cardiomyopathy, EPO-resistant anemia, muscle mass loss, and undernutrition. Initially, intravenous administration is administered, and if the effect is confirmed, maintenance therapy is performed by oral administration (1 g/day). Intravenous injections are considered to be the most efficient formulation for dialysis patients, but in the United Kingdom, oral injections are recommended to control medical costs due to the high price of intravenous formulations.

In the United States, injectable carnitine was approved in December 1999 for the prevention and treatment of carnitine deficiency in patients with end-stage renal disease undergoing dialysis. The FDA's summary basis of approval (SBA) describes in detail the background and basis for approval, and in the "Information pertinent to the current sNDA" section of the medical review is described as the following:

"Carnitine deficiency has pronounced clinical manifestations only when the concentration of carnitine in tissues drops below 5–10%. There is ample evidence and clarity that a deficiency of an important molecule such as carnitine can lead to serious life-threatening consequences, and it is also clear that dialysis can cause a deficiency of carnitine in the blood, and a long dialysis period can lead to a deficiency of carnitine in the tissues. It is unethical to expose patients to risks and disadvantages in order to demonstrate the clinical benefit of carnitine supplementation therapy for carnitine deficiency in dialysis patients" (Carnitor injection, 020182, S006, https://www.accessdata.fda.gov/drugsatfda_docs/nda/99/20182S6_carnitor%20injection_medr.pdf).

In other words, the use of carnitine in dialysis patients was approved only for the review of the data from the clinical trial submitted in 1989 (which was not approved at that time) and the newly submitted pharmacokinetic study in dialysis patients, and no new clinical trial was demanded. This is very unusual case for the FDA's drug approval process. In the case of conventional medicines, the FDA demands two large-scale, placebo-controlled, double-blind trials (so-called pivotal clinical trials), but did not request in the case of carnitine. The FDA did not demand some pivotal studies because carnitine is a substance with important physiological functions in the body, and if it is deficient for any reason, it is clear that it will lead to serious clinical consequences. In the United States, drugs for patients with less than 200,000 patients are treated as orphan drugs, and intravenous carnitine injects for dialysis patients are also treated as orphan drugs. Intravenous injections are considered to be the most suitable route of administration for dialysis patients because they maintain a high concentration in the blood until the next dialysis and are efficiently taken up by tissues such as muscles, so only intravenous injections are approved in the United States. The dose is 10–20 mg/kg over 2–3 min at the end of each dialysis, usually 3 times a week. In addition, the package insert of the intravenous L-carnitine formulation in the United States describes that downward dose adjustment should be considered 3–4 weeks after the start of administration, and it is written that "for example, 5 mg/kg," but in reality, in most cases, 1 g/vial is injected intravenously after dialysis.

14.6.2 Status of Carnitine Supplementation in Hemodialysis Patients

Carnitine supplementation therapy has been approved for carnitine deficiency in dialysis patients in Europe and the United States, but the actual usage status is difficult to grasp due to the lack of statistical data. Hörl et al. summarize the use of adjunctive drugs such as vitamin B_{12}, vitamin E, vitamin C, vitamin D, and folic acid in the treatment of ESA-resistant anemia in each country in the "European Best Practice Guideline 14–16," and data on carnitine is also described (Table 14.11) [242]. According to the data, the proportion of patients receiving carnitine supplementation therapy for ESA-resistant anemia varies greatly from country to country, and it is relatively common in Greece and Germany, but not so much in other countries. This is probably due to differences in health care systems and policies, such as differences in insurance reimbursement for nutritional components such as vitamins and carnitine in each country, and drug pricing.

Table 14.11 Proportion of patients receiving carnitine supplementation therapy for more than 3 months for dialysis patients with ESA-resistant anemia in European countries

Country	% patients	Sample size (*n*)
Austria	1.5	682
Belgium, Luxembourg	0.7	1088
Denmark	0.9	221
Finland	0.3	321
France	3.7	3885
Germany	11.3	4365
Greece	15.6	1346
Italy	0.6	348
Netherlands	1.6	373
Norway	0	162
Spain	6.8	690
Sweden	0.2	556
Switzerland	5.2	327
Whole	**6.5**	**14,362**

Modified from Ref. [242]

In the United States, carnitine supplementation therapy for dialysis patients was approved in December 1999, but insurance reimbursement was not approved for a while. In November 2002, CMS, a medical insurance payer in the United States, approved insurance reimbursement for carnitine supplementation therapy for dialysis patients with intra-dialysis hypotension and ESA-resistant anemia from January 2003. (https://www.cms.gov/Regulations-and Guidance/Guidance/Transmittals/downloads/AB02165.pdf).

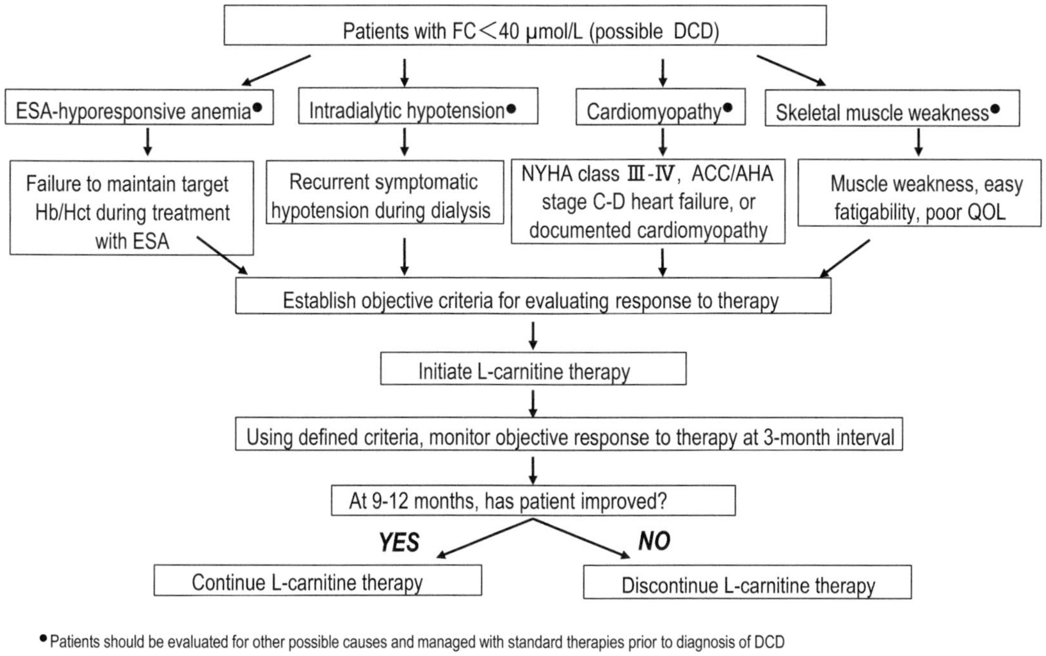

Fig. 14.66 Recommended approach to the evaluation and treatment of a patient with possible dialysis-related carnitine disorder (DCD) by "Carnitine Consensus Conference" panel in National Kidney Foundation. DCD: dialysis-related carnitine disorder, ESA: erythropoiesis stimulating agents, Hb: hemoglobin, Hct: hematocrit, NYHA: New York Heart Association, ACC: American College of Cardiology, AHA: American Heart Association, QOL: quality of life. (From Ref. [243])

However, insurance coverage is approved only when iron supplementation and fluid control cannot improve ESA-resistant anemia, and hypotension during dialysis cannot be controlled by normal methods during at least two sessions of dialysis within a month, and these records have to be documented. The policy also requests that a free carnitine concentration of 40 μmol/L or less in blood be recorded as a diagnosis of carnitine deficiency. Although this policy was issued to meet immediate insurance claims, the concept of what carnitine deficiency is and how it is diagnosed was unclear. Therefore, a panel in the National Kidney Foundation's Carnitine Consensus Conference discussed and published a recommendation for the use of carnitine supplementation therapy to dialysis-related carnitine disorders or deficiency (DCD) in clinical practice (Fig. 14.66) [243].

The outline of the recommendation is followings:

For dialysis patients with a blood free carnitine concentration of 40 μmol/L or less and ESA-resistant anemia, hypotension during dialysis, cardiomyopathy and heart failure, and skeletal muscle weakness, carnitine supplementation therapy should be started after establishing evaluation criteria, and the response should be observed every 3 months. The effect should be judged at 12 months and whether or not treatment can be continued should be determined (actual insurance reimbursement for CMS is only for ESA-resistant anemia and hypotension during dialysis).

14.6.3 Controversy Over the Pros and Cons of Carnitine Supplementation Therapy for Dialysis Patients in the United States

Thus, intravenous carnitine injectable for DCD was approved by the FDA, and some of it was reimbursed by insurance and recommended by a panel at the Carnitine Consensus Conference of the American Kidney Foundation, which increased the momentum for carnitine supplementation therapy. A series of special issues by nephrologists promoting carnitine replacement therapy were published in nephrology journals. One of them was a special supplement issue which featured on carnitine in the American Journal of Kidney Diseases, 2003. Eknoyan et al., who co-organized a panel at the Carnitine Consensus Conference in the American Kidney Foundation, wrote an overview of dialysis-associated carnitine deficiency [244], and Hoppel described the physiological role of carnitine and its position in energy metabolism [245]. Evans discusses the homeostasis of carnitine and the pharmacokinetics of carnitine in blood and tissues [246], Golper et al. investigated the potential of carnitine in the treatment of anemia in dialysis patients [162], Pauly et al. described the role of carnitine in cardiac dysfunction in dialysis patients [247], Miller et al. described muscular symptoms and mobility in dialysis patients, the role of carnitine in QOL was explained [2].

However, after a while, a controversy arose among nephrologists about the pros and cons of carnitine supplementation therapy for dialysis patients. This controversy is still ongoing. First, Steinman et al. criticized, the title is "L-carnitine use in dialysis patients: Is national coverage for supplementation justified? What were CMS regulators thinking—or were they." [248] They cited papers on the effects of carnitine and the effects of supplementation therapy in dialysis patients and concluded that the evidence was insufficient because some papers reported no effect or only some patients were effective, large-scale double-blind controlled trials had not been conducted, and the degree of deficiency could not be easily diagnosed by measuring blood levels alone, and the biggest problem was that the drug price of intravenous carnitine injection is extremely expensive in the United States, around $36–38.4 for 1 g/ampoule or vial, it is criticized that even if it is administered at 10 mg/kg, it will cost $5,994 per year, and if it is administered at 20 mg/kg according to the package insert, it will cost $11,980 per year. In 2005, Steinman published a similar critique to carni-

Table 14.12 Steinman's criticism for carnitine supplementation therapy in dialysis patients and Schreiber's rebuttal for it

Item	Steinman's criticism [248, 249]	Schreiber's rebuttal [250, 253]
1	There is a lack of large clinical trials or randomized controlled trials of carnitine supplementation therapy in dialysis patients	CMS selected and reviewed 36 high-quality studies from 186 publications since 1980 However, 16 of them were randomized controlled trials, with 1100 patients A meta-analysis also reviewed 21 randomized controlled trials and found that carnitine helped reduce EPO in six trials
2	The EPO-saving effect of carnitine was reported in an elderly study only, and iron and PTH levels were not explained	Caruso [163] and colleagues only in the elderly, but Kletzmayer [162] and Labonia [161] were in general dialysis patients, and most of the papers did not match the reported data
3	Clinical trials of carnitine for the treatment of anemia are small and heterogeneous, and no conclusions can be drawn	The meta-analysis of Hurot [1] showed an EPO reduction effect by using strict selection criteria and using only randomized controlled trials. There are heterogeneities, but the conclusions remain the same

(continued)

Table 14.12 (continued)

Item	Steinman's criticism [248, 249]	Schreiber's rebuttal [250, 253]
4	The cost of carnitine injections is prohibitively more than the EPO savings ($52/1 time for a 70 kg patient, $8086/year)	In fact, it's $2000/year, which is about 25% of Steinman's claim. EPO-resistant anemia is associated not only with hematology but also with various physiological abnormalities, and carnitine, which has therapeutic effects on them, should not be ignored
5	Measuring blood carnitine levels is unreliable and difficult	Fornasini et al. [257] report that there are several validated methods that make it easy to obtain measurement results by asking a dialysis service provider
6	CMS rejects large number of insurance claims	This is a misconception, and the CMS itself does not reject or approve insurance claims. Individual financial intermediaries approve or reject insurance claims. Some financial intermediaries set sloppy and fickle standards and require strict records documents, which can be accepted or rejected in a big way. This could set a bad precedent for the lack of access to valuable treatments
7	Treatment of EPO-resistant anemia and dialysis hypotension with carnitine are like to other treatments and should only be tried when they are ineffective	This widespread misconception in dialysis facilities stems from a lack of understanding of the role of carnitine in pathophysiology. EPO-resistant anemia is associated with the MIA symptomatic group, and it is not just a matter of increasing the Ht. Carnitine reduces the toxicity of accumulated acyl groups, has anti-inflammatory and antioxidant effects, and improves heart function, and is not only used to reduce EPO and hypotension during dialysis

Created using as Refs. [249–251, 254]
Schreiber's Conclusion:

Prevalent misconceptions concerning the use of L-carnitine in dialysis have limited the ability of nephrologists to fully understand the physiologic basis for its use and to appreciate the clinical data upon which the CMS NCD is based. Nephrologists are highly respected as medical scholars who favor decisions based on the best available data and practice observations rather than opinion, tradition, or prejudice. By being cognizant of these misconceptions related to L-carnitine and dialysis, nephrologists may hopefully refine their understanding of this therapeutic agent and employ it more effectively in patient care

tine supplementation therapy in the editorials of the journal Seminars in Dialysis [249]. On the other hand, Schreiber, who is a supporter of carnitine, specifically refuted the points listed by Steinman et al. as a correspondence, saying that the items pointed out by Steinman et al. are based on misunderstandings, and for example, the medical expenses for carnitine are actually about 25% (about $2000/year) of Steinman's claim. In addition to saving medical costs, it is also necessary to consider the effect of improving physiological abnormalities in various dialysis patients (Table 14.12) [250].

Hedayati also carefully explained the significance of carnitine supplementation therapy for dialysis patients [251]. In 2006, in the debate forum in a journal "Blood Purification," Handelman again argued that "there are no large-scale, double-blind studies of carnitine and there is insufficient evidence" [252]. In response to this, Schreiber asserted that carnitine supplementation therapy is a reasonable and justified treatment for at least some dialysis patients [253].

In 2013, Wasserstein echoed Steinman and Handelman, stating that "the use of carnitine for dialysis patients will disappear as dialysis care becomes more comprehensive in the face of limited medical costs" [254]. Mercadal et al. conducted a multicenter, randomized, double-blind, placebo-controlled trial in which 92 newly dialysis patients were divided into placebo and intravenous carnitine (1 g after dialysis three times a week for 1 year) (CARNIDIAL trial) and reported that carnitine administration had no effect on EPO dose or EPO resistance index [255]. Referring to the study in the journal's editorial, Yee described that "carnitine supplementation as a treatment for anemia in dialysis patients should only be considered as a last resort when other treatments are ineffective" [256]. However, in this study, patients with EPO resistance were only 0–6.6% in the placebo group and 0–2.2% in the carnitine group. This study examined the effects of carnitine supplementation in patients with almost no carnitine deficiency, with only 2.2% of patients and only about 11% lower blood carnitine levels in the placebo group, so the CARNIDIAL trial was the study of a kind of "putting the cart before the horse study." Fornasini et al. [37] and Evans [38] and others studied carnitine dynamics in dialysis patients, and considering the protocol of the CARDINAL trial, 3–5 years after the start of dialysis, there would have been a difference between the placebo group and the carnitine group, but it is unlikely that severe carnitine deficiency will appear within 1 year after the start of dialysis unless the patient is very malnourished.

There is a criticism of the lack of large-scale, placebo-controlled, double-blind trials of carnitine. Carnitine is approved as a drug, but unlike chemically synthesized drugs such as antihypertensive drugs and hyperlipidemia drugs, it is a substance that originally exists and functions in the body. As described in "Dobeneck's barrel," dialysis patients are not only deficient in carnitine but also deficient in various nutrients and minerals such as zinc, vitamin D_3, and folic acid, which cannot be easily proven by a large-scale placebo-controlled, double-blind study. Although it may be possible to draw conclusions by measuring such a large number of factors in large-scale trials and stratifying and analyzing them according to the degree of deficiency of various factors, the reality is that clinical trials cannot be conducted at such a high cost for carnitine preparations that have no patents. Large-scale, placebo-controlled, double-blind studies should be conducted to evaluate the efficacy and safety of chemically synthesized drugs, but it is rather difficult to evaluate nutrient substances such as carnitine. For example, the effect of early administration of vitamin B_1 on Wernicke's encephalopathy, which is common in alcoholics, has been established, but similar cases have been reported in undernourished elderly people [257]. There have been cases in which vitamin B_1, which was thought to be present in sufficient quantities and functions in the body, is deficient in undernourished elderly people, resulting in the development of Wernicke's encephalopathy. For pathologies caused by vitamin B_1 deficiency, vitamin B_1 administration is the only effective treatment. There is no effective treatment for carnitine deficiency other than carnitine administration. Dialysis patients are also extremely prone to malnutrition, inflammation, and atherosclerosis (called MIA syndrome due to malnutrition, inflammation, atherosclerosis) and PEW syndrome. Carnitine is positioned as an intermediate nutrient and a drug, and its supplementation therapy is often neglected or looked down, but those people who look down carnitine supplementation therapy should reflect the fact that the strict US FDA approved intravenous carnitine injectable for carnitine deficiency in dialysis patients without demanding some double-blinded pivotal studies.

14.7 Role of Carnitine Supplementation in Dialysis Treatment in Japan

As described above, the position of carnitine supplementation therapy for dialysis patients varies greatly from country to country, and there is a sense that the debate between carnitine supporters and carnitine deniers has developed among nephrologists and continues to date. This may be due to differences in drug prices and insurance reimbursement systems for carnitine preparations, as well as the comprehensive system for dialysis treatment medical cost. In the trend toward reducing medical costs, which are soaring with the aging of society in developed countries, there is a movement to encourage the use of supplements with co-payment and to restrict insurance reimbursement for medicines such as carnitine and vitamins that are positioned as nutrients.

According to statistics from The Japanese Society for Dialysis Therapy ("Current Status of Chronic Dialysis Therapy in Japan, December 31, 2020"), there were 347,671 dialysis patients in Japan as of the end of 2020. The majority of patients (47.5%) had been on dialysis for less than 5 years, but 27.5% had been on dialysis for more than 10 years, of which 8.5% had been on dialysis for 20 years or more, 2.3% had more than 30 years, and 0.4% had more than 40 years.

Table 14.13 Proportion of kidney transplantation in patients with ESRD in each country

Country	Number of patients with ESRD	Population ratio (vs. 1 million)	Number of dialysis patients	Dialysis patient ratio (%)	Number of kidney transplant patients	Kidney transplant patients (%)
USA[1]	527,283	1724	368,544	69.9	158,739	30.1
Germany[2]	82,305	998	60,992	74.1	21,313	25.9
United Kingdom[3]	37,848	638	20,854	55.1	16,994	44.9
Japan[4]	298,484	2341	297,000	99.5	1484	0.5

ESRD: end-stage renal disease

[1]Statistics in US are based on 2007 in the abstract of US Renal Data System 2009 Annual Data Report (Collins, AJ et al. Am J Kidney Dis 2010: 55 (1 Suppl 1); S1-420, A6-7. doi:10.1053/j.ajkd.2009.10.009)

[2]Statistics in Germany are based on Sumi Hidaka et al., Dialysis and Kidney Transplantation in Each Country (8) Germany, in 2004, The Japanese Journal of Clinical Dialysis 2007:23(2);215–221 (article in Japanese).

[3]Statistics in the United Kingdom are based on Hideaki Okabe, Dialysis and Kidney Transplantation in Each Country (6) United Kingdom, in 2004, The Japanese Journal of Clinical Dialysis 2007:23(2);201–207 (Article in Japanese)

[4]Renal Failure Treatment options and their practice (eds. Japanese Society of Nephrology/Japanese Society for Dialysis Therapy/Japan Society for Transplantation/Japanese Society for Clinical Renal Transplantation, 2012, booklet in Japanese), data of 2010, in p31

Percentage of dialysis patients (%) and percentage of kidney transplant patients shows each proportion of patients to end-stage renal disease

In this way, Japan is a "proud to the world dialysis powerhouse," and it is a rare country that can benefit from advanced dialysis treatment. On the other hand, kidney transplantation has been established as an excellent treatment for end-stage renal failure in developed countries, but the number of kidney transplantation cases in Japan is extremely small [258, 259]. As a result, the kidney transplantation rate of patients with end-stage renal disease is only 0.5%, which is extremely low compared to the United States, Germany, and the United Kingdom (Table 14.13).

The small number of kidney transplants that have been performed in Japan is mostly a living-donor kidney transplant, and the situation unique to Japan is that transplants from cadaveric-donner kidneys are extremely rare. It might be due to Japanese culture heavily influenced by Buddhism in which honoring the dead. The data in Table 14.13 is from the 2000s, but the number of kidney transplant patients in 2019 is 1,827 from living-donor, 176 from brain dead-donner, and 54 from cadaveric-donner, for a total of 2057 patients, which is only a slight increase (http://www.asas.or.jp/jst/general/number/). In other words, in Japan, kidney transplantation is rarely performed for end-stage renal failure patients, and majority of them are dependent on dialysis treatment for their entire lives. Therefore, it is possible that carnitine deficiency is more common among dialysis patients in Japan than in Europe and the United States, due in part to the long-term treatment with dialysis and the lower intake of meats and dairy products than in Europe and the United States.

Fatigue, muscle weakness, cardiomyopathy, heart failure, ESA-resistant anemia, chronic oxidative stress, hyperinflammatory, and malnutrition in dialysis patients are thought to be caused by multiple factors, but carnitine deficiency is considered to be a central component of the pathological condition. Only L-carnitine preparations are effective against carnitine deficiency. For carnitine replacement therapy in hemodialysis patients, it is important to be fully aware that carnitine becomes deficient by hemodialysis or peritoneal dialysis treatments and carefully observe each patient individually to see symptoms that are thought to be caused by carnitine deficiency, rather than seeking strict evidence from the perspective of general pharmaceutical drugs. We should then apply these findings to clinical practice. Isn't this true patient-oriented EBM? As introduced in this book, fortunately, many ambitious clinical studies on the usefulness and safety of carnitine supplementation therapy for various clinical symptoms of dialysis patients have been reported from Japan especially after expansion of the indication to "carnitine deficiency," and it is thought that clinical studies from dialysis facilities in Japan will lead carnitine supplementation therapy for dialysis patients in the future.

14.8 Summary

As described, it is evident that dialysis treatment causes carnitine deficiency and insufficiency. Carnitine deficiency and insufficiency are related to various symptoms in dialysis patients. Recently Takashima et al. reviewed many clinical studies which studied the effects of carnitine supplementation for anemia, cardiac function and hypotension, muscle symptoms and QOL, and lipid profiles and inflammation-related parameters [265]. Most clinical studies showed effectiveness of carnitine supplementation. Some studies denied the effect of carnitine supplementation, but it seems that the patients studied did not consider carnitine deficiency or insufficiency yet because these studies enrolled patients with very short dialysis history.

Acute kidney injury (AKI) is life-threatening disease and effective treatments are limited at present. Goto et al. reported the relationship between carnitine insufficiency and heat-related illness severity or AKI. They found that carnitine insufficiency (high AC/FC) is observed in sever heatstroke patients. Also, they studied same findings and effectiveness of carnitine pretreatment in animal model of heat-stress. Heatstroke is increasing according to global warming and it induces various critical illness such as AKI. They proposed that L-carnitine may be an effective preventive treatment for heat-related illness and AKI [266].

Diabetic kidney disease (DKD) is the leading cause of end-stage kidney disease. Ito et al. reported that impaired fatty acid oxidation in experimental and human diabetic kidney disease and carnitine supplementation might be potent therapeutic strategy for this disorder [267].

It is very important to study the pathogenic role of carnitine deficiency and insufficiency in renal tubular disorders such as AKI and DKD. Further studies are required to establish the therapeutic potential of carnitine supplementation.

References

1. Hurot JM, Cucherat M, Haugh M, et al. Effects of L-carnitine supplementation in maintenance hemodialysis patients: a systematic review. J Am Soc Nephrol. 2002;13(3):708–14. https://doi.org/10.1681/ASN.V133708.
2. Miller B, Ahmad S. A review of the impact of L-carnitine therapy on patient functionality in maintenance hemodialysis. Am J Kidney Dis. 2003;41(4 Suppl 4):S44–8. https://doi.org/10.1016/s0272-6386(03)00116-1.
3. Schreiber B. Levocarnitine and dialysis: a review. Nutr Clin Pract. 2005;20(2):218–43. https://doi.org/10.1177/0115426505020002218.
4. Guarnieri G, Biolo G, Vinci P, et al. Advances in carnitine in chronic uremia. J Ren Nutr. 2007;17(1):23–9. https://doi.org/10.1053/j.jrn.2006.10.006.
5. Lynch KE, Feldman HI, Berlin JA, et al. Effects of L-carnitine on dialysis-related hypotension and muscle cramps: a meta-analysis. Am J Kidney Dis. 2008;52:962–71.
6. Reuter SE, Faull RJ, Evans AM. L-carnitine supplementation in the dialysis population: are Australian patients missing out? Nephrology (Carlton). 2008;13(1):3–16. https://doi.org/10.1111/j.1440-1797.2007.00817.x.
7. Böhmer T, Rydning A, Solberg HE. Carnitine levels in human serum in health and disease. Clin Chim Acta. 1974;57(1):55–61. https://doi.org/10.1016/0009-8981(74)90177-6.
8. Chen SH, Lincoln SD. Increased serum carnitine concentration in renal insufficiency. Clin Chem. 1977;23(2 Pt 1):278–80.
9. Bartel LL, Hussey JL, Shrago E. Perturbation of serum carnitine levels in human adults by chronic renal disease and dialysis therapy. Am J Clin Nutr. 1981;34(7):1314–20. https://doi.org/10.1093/ajcn/34.7.1314.
10. Rodriguez-Segade S, de la Peña CA, Paz M, et al. Carnitine concentrations in dialysed and undialysed patients with chronic renal insufficiency. Ann Clin Biochem. 1986;23(Pt 6):671–5. https://doi.org/10.1177/000456328602300608.
11. Bertoli M, Battistella PA, Vergani L, et al. Carnitine deficiency induced during hemodialysis and hyperlipidemia: effect of replacement therapy. Am J Clin Nutr. 1981;34(8):1496–500. https://doi.org/10.1093/ajcn/34.8.1496.
12. Yano J, Ito S, Kodama G, et al. Kinetics of serum carnitine fractions in patients with chronic kidney disease not on dialysis. Kurume Med J. 2019;66(3):153–60. https://doi.org/10.2739/kurumemedj.MS663001.

13. Novoa D, Rodriguez-Segade S, Romero R, et al. Carnitine levels and hypertriglyceridemia in undialyzed patients. Nephron. 1987;47(3):231. https://doi.org/10.1159/000184499.
14. Wanner C, Förstner-Wanner S, Rössle C, et al. Carnitine metabolism in patients with chronic renal failure: effect of L-carnitine supplementation. Kidney Int Suppl. 1987;32:S132–5.
15. Aydogdu N, Atmaca G, Yalcin O, et al. Protective effects of L-carnitine on myoglobinuric acute renal failure in rats. Clin Exp Pharmacol Physiol. 2006;33(1–2):119–24. https://doi.org/10.1111/j.1440-1681.2006.04336.x.
16. Mohammadi M, Talasaz AH, Alidoosti M, et al. Nephroprotective effects of L-carnitine against contrast-induced nephropathy in patients undergoing percutaneous coronary intervention: a randomized open-labeled clinical trial. J Tehran Heart Cent. 2017;12(2):57–64.
17. Kunak CS, Ugan RA, Cadirci E, et al. Nephroprotective potential of carnitine against glycerol and contrast-induced kidney injury in rats through modulation of oxidative stress, proinflammatory cytokines, and apoptosis. Br J Radiol. 2016;89(1058):20140724. https://doi.org/10.1259/bjr.20140724.
18. Boyacioglu M, Turgut H, Akgullu C, et al. The effect of L-carnitine on oxidative stress responses of experimental contrast-induced nephropathy in rats. J Vet Med Sci. 2014;76:1–8.
19. Fan JP, Kim D, Kawachi H, et al. Ameliorating effects of L-carnitine on diabetic podocyte injury. J Med Food. 2010;13(6):1324–30. https://doi.org/10.1089/jmf.2010.1079.
20. Armaly Z, Abd El Qader A, Jabbour A, et al. Effects of carnitine on oxidative stress response to intravenous iron administration to patients with CKD: impact of haptoglobin phenotype. BMC Nephrol. 2015;16:135. https://doi.org/10.1186/s12882-015-0119-0.
21. Ye J, Li J, Yu Y, et al. L-carnitine attenuates oxidant injury in HK-2 cells via ROS-mitochondria pathway. Regul Pept. 2010;161(1-3):58–66. https://doi.org/10.1016/j.regpep.2009.12.024.
22. Bohmer T, Bergrem H, Eiklid K. Carnitine deficiency induced during intermittent haemodialysis for renal failure. Lancet. 1978;1(8056):126–8. https://doi.org/10.1016/s0140-6736(78)90422-1.
23. Battistella PA, Angelini C, Vergani L, et al. Carnitine deficiency induced during haemodialysis. Lancet. 1978;1(8070):939. https://doi.org/10.1016/s0140-6736(78)90714-6.
24. Moorthy AV, Rosenblum M, Rajaram R, et al. A comparison of plasma and muscle carnitine levels in patients on peritoneal or hemodialysis for chronic renal failure. Am J Nephrol. 1983;3(4):205–8. https://doi.org/10.1159/000166711.
25. Bizzi A, Mingardi G, Codegoni AM, et al. Accelerated recovery of post-dialysis plasma carnitine fall by oral carnitine. Biomedicine. 1978;29(6):183–4.
26. Bizzi A, Cini M, Garattini S, et al. L-carnitine addition to haemodialysis fluid prevents plasma-carnitine deficiency during dialysis. Lancet. 1979;1(8121):882. https://doi.org/10.1016/s0140-6736(79)91304-7.
27. Evans AM, Faull RJ, Nation RL, et al. Impact of hemodialysis on endogenous plasma and muscle carnitine levels in patients with end-stage renal disease. Kidney Int. 2004;66(4):1527–34. https://doi.org/10.1111/j.1523-1755.2004.00916.x.
28. Dębska-Ślizień A, Kawecka A, Wojnarowski K, et al. Correlation between plasma carnitine, muscle carnitine and glycogen levels in maintenance hemodialysis patients. Int J Artif Organs. 2000;23(2):90–6.
29. Dębska-Ślizień A, Kawecka A, Wojnarowski K, et al. Carnitine content in different muscles of patients receiving maintenance hemodialysis. J Ren Nutr. 2007;17(4):275–81. https://doi.org/10.1053/j.jrn.2007.05.002.
30. Golper TA, Wolfson M, Ahmad S, et al. Multicenter trial of L-carnitine in maintenance hemodialysis patients. I. Carnitine concentrations and lipid effects. Kidney Int. 1990;38(5):904–11. https://doi.org/10.1038/ki.1990.289.
31. Hiatt WR, Koziol BJ, Shapiro JI, et al. Carnitine metabolism during exercise in patients on chronic hemodialysis. Kidney Int. 1992;41(6):1613–9. https://doi.org/10.1038/ki.1992.233.
32. Bellinghieri G, Savica V, Mallamace A, et al. Correlation between increased serum and tissue L-carnitine levels and improved muscle symptoms in hemodialyzed patients. Am J Clin Nutr. 1983;38(4):523–31. https://doi.org/10.1093/ajcn/38.4.523.
33. Glöggler A, Bulla M, Puchstein C, et al. Plasma and muscle carnitine in healthy and hemodialyzed children. Child Nephrol Urol. 1988–1989;9(5):277–82.
34. Harper P, Wadström C, Cederblad G. Carnitine measurements in liver, muscle tissue, and blood in normal subjects. Clin Chem. 1993;39(4):592–9.
35. Savica V, Bellinghieri G, Di Stefano C, et al. Plasma and muscle carnitine levels in haemodialysis patients with morphological-ultrastructural examination of muscle samples. Nephron. 1983;35(4):232–6. https://doi.org/10.1159/000183087.
36. Siami G, Clinton ME, Mrak R, et al. Evaluation of the effect of intravenous L-carnitine therapy on function, structure and fatty acid metabolism of skeletal muscle in patients receiving chronic hemodialysis. Nephron. 1991;57(3):306–13. https://doi.org/10.1159/000186280.
37. Fornasini G, Upton RN, Evans AM. A pharmacokinetic model for L-carnitine in patients receiving haemodialysis. Br J Clin Pharmacol. 2007;64(3):335–45. https://doi.org/10.1111/j.1365-2125.2007.02926.x.
38. Evans AM, Faull R, Fornasini G, et al. Pharmacokinetics of L-carnitine in patients with end-stage renal disease undergoing long-term hemodialysis. Clin Pharmacol Ther. 2000;68(3):238–49. https://doi.org/10.1067/mcp.2000.108850.

39. Guarnieri G, Toigo G, Crapesi L, et al. Carnitine metabolism in chronic renal failure. Kidney Int Suppl. 1987;22:S116–27.
40. Constantin-Teodosiu D, Kirby DP, Short AH, et al. Free and esterified carnitine in continuous ambulatory peritoneal dialysis patients. Kidney Int. 1996;49(1):158–62. https://doi.org/10.1038/ki.1996.21.
41. Sakurabayashi T, Takaesu Y, Haginoshita S, et al. Dialysis patients and carnitine metabolism. J Jpn Assoc Dial Physicians. 2000;15(2):245–55. (Article in Japanese)
42. Winter SC, Zorn EM, Vance WH. Carnitine deficiency. Lancet. 1990;335(8695):981–2.
43. Tein I. Role of carnitine and fatty acid oxidation and its defects in infantile epilepsy. J Child Neurol. 2002;17(Suppl 3):3S57-3S83.
44. Carter AL, Abney TO, Lapp DF. Biosynthesis and metabolism of carnitine. J Child Neurol. 1995;10(Suppl 2):2S3-2S7.
45. Campos Y, Huertas R, Lorenzo G, et al. Plasma carnitine insufficiency and effectiveness of L-carnitine therapy in patients with mitochondrial myopathy. Muscle Nerve. 1993;16(2):150–3. https://doi.org/10.1002/mus.880160205.
46. Böhles H, Evangeliou A, Bervoets K, et al. Carnitine esters in metabolic disease. Eur J Pediatr. 1994;153(7 Suppl 1):S57–61. https://doi.org/10.1007/BF02138779.
47. Pons R, De Vivo DC. Primary and secondary carnitine deficiency syndromes. J Child Neurol. 1995;10(Suppl 2):2S8–2S24.
48. Hatanaka Y, Higuchi T, Akiya Y, et al. Prevalence of carnitine deficiency and decreased carnitine levels in patients on hemodialysis. Blood Purif. 2019;47(Suppl 2):38–44. https://doi.org/10.1159/000496720.
49. Sakurauchi Y, Matsumoto Y, Shinzato T, et al. Effects of L-carnitine supplementation on muscular symptoms in hemodialyzed patients. Am J Kidney Dis. 1998;32(2):258–64. https://doi.org/10.1053/ajkd.1998.v32.pm9708610.
50. Riley S, Rutherford S, Rutherford PA. Low carnitine levels in hemodialysis patients: relationship with functional activity status and intra-dialytic hypotension. Clin Nephrol. 1997;48(6):392–3.
51. Ito W, Uchiyama K, Mitsuno R, et al. Correlation between acylcarnitine/free carnitine ratio and cardiopulmonary exercise test parameters I patients with incident dialysis. Front Physiol. 14:1155281. https://doi.org/10.3389/fphys.2023.1155281.
52. Reuter SE, Faull RJ, Ranieri E, et al. Endogenous plasma carnitine pool composition and response to erythropoietin treatment in chronic haemodialysis patients. Nephrol Dial Transplant. 2009;24(3):990–6. https://doi.org/10.1093/ndt/gfn588.
53. Murphy WJ, Steiber A, Connery GC, et al. Altered carnitine metabolism in dialysis patients with reduced physical function may be due to dysfunctional fatty acid oxidation. Nephrol Dial Transplant. 2012;27(1):304–10. https://doi.org/10.1093/ndt/gfr334.
54. Kamei D, Tsuchiya K, Nitta K, et al. Association between resistance to erythropoiesis-stimulating agents and carnitine profile in patients on maintenance haemodialysis. Nephrology (Carlton). 2018;23(8):737–43. https://doi.org/10.1111/nep.13079.
55. Kamei Y, Kamei D, Tsuchiya K, et al. Association between 4-year all-cause mortality and carnitine profile in maintenance hemodialysis patients. PLoS One. 2018;13(8):e0201591. https://doi.org/10.1371/journal.pone.0201591.
56. Reuter SE, Evans AM, Faull RJ, et al. Impact of haemodialysis on individual endogenous plasma acylcarnitine concentrations in end-stage renal disease. Ann Clin Biochem. 2005;42(Pt 5):387–93. https://doi.org/10.1258/0004563054889954.
57. Csiky B, Bene J, Wittmann I, et al. Effect of hemodialysis session on the dynamics of carnitine ester profile changes in L-carnitine pretreated end-stage renal disease patients. Int Urol Nephrol. 2013;45(3):847–55. https://doi.org/10.1007/s11255-012-0209-x.
58. Yalkowsky SH, Zografi G. Some micellar properties of long-chain acylcarnitines. J Colloid Interface Sci. 1970;34(4):525–33. https://doi.org/10.1016/0021-9797(70)90215-8.
59. Bene J, Csiky B, Wittmann I, et al. Dramatic decrease of carnitine esters after interruption of exogenous carnitine supply in hemodialysis patients. Ren Fail. 2012;34(5):555–8. https://doi.org/10.3109/0886022X.2012.664509.
60. Marzo A, Arrigoni Martelli E, Mancinelli A, et al. Protein binding of L-carnitine family components. Eur J Drug Metab Pharmacokinet Spec No. 1991;3:364–8.
61. Albright RK, Kram BW, White RP. Carnitine status of continuous ambulatory peritoneal dialysis patients. Lancet. 1982;2(8291):218–9. https://doi.org/10.1016/s0140-6736(82)91067-4.
62. Leschke M, Rumpf KW, Eisenhauer T, et al. Quantitative assessment of carnitine loss during hemodialysis and hemofiltration. Kidney Int Suppl. 1983;16:S143–6.
63. Wanner C, Förstner-Wanner S, Schaeffer G, et al. Serum free carnitine, carnitine esters and lipids in patients on peritoneal dialysis and hemodialysis. Am J Nephrol. 1986;6(3):206–11. https://doi.org/10.1159/000167119.
64. Grzegorzewska AE, Mariak I, Dobrowolska-Zachwieja A. Continuous ambulatory peritoneal dialysis (CAPD) adequacy influences serum free carnitine level. Int Urol Nephrol. 1999;31(4):533–40. https://doi.org/10.1023/a:1007127614765.
65. Horie M, Hasegawa M, Nakamura Y. Carnitine in CAPD patients. Jpn J Clin Dial. 2000;16(2):183–90. (Article in Japanese)
66. Mochizuki T, Takahashi M, Abe R, et al. Carnitine metabolism and its relation to nutritional parameters

in patients with peritoneal dialysis. J Jpn Soc Dial Ther. 2002;35(3):165–70. (Article in Japanese)
67. Ohashi H, Oda H, Ohno M, et al. Characteristics of left ventricular hypertrophy estimated by MIBG and BMIPP cardiac scintigraphy in patients undergoing peritoneal dialysis. J Jpn Soc Dial Ther. 2002;35(13):1557–61. (Article in Japanese)
68. Shimizu S, Takashima H, Tei R, et al. Prevalence of carnitine deficiency and decreased carnitine levels in patients on peritoneal dialysis. Nutrients. 2019;11(11):2645. https://doi.org/10.3390/nu11112645.
69. Kaneko S, Hirai K, Morino J, et al. Association between carnitine deficiency and the erythropoietin resistance index in patients undergoing peritoneal dialysis: a cross-sectional observational study. Ren Fail. 2020;42(1):146–53. https://doi.org/10.1080/0886022X.2020.1719847.
70. Warady BA, Borum P, Stall C, et al. Carnitine status of pediatric patients on continuous ambulatory peritoneal dialysis. Am J Nephrol. 1990;10(2):109–14. https://doi.org/10.1159/000168064.
71. Koşan C, Sever L, Arısoy N, et al. Carnitine supplementation improves apolipoprotein B levels in pediatric peritoneal dialysis patients. Pediatr Nephrol. 2003;18(11):1184–8. https://doi.org/10.1007/s00467-003-1302-2.
72. Kawahara K, Sato T, Kawamura K, et al. Effect of carnitine administration on cardiomyopathy in children on peritoneal dialysis. J Jpn Soc Pediatr Ren Fail. 1995;15:224–7. (Article in Japanese)
73. Uemura O, Shimizu M, Iwasa M, et al. Effects of kidney transplantation on serum carnitine fractions in pediatric end-stage renal failure. J Jpn Pediatr Soc. 1999;103(8):869–70. (Article in Japanese)
74. Nakano M, Uemura O. Pediatric peritoneal dialysis and carnitine. In: Akiba T, Akizawa T, editors. Carnitine supplementation therapy for dialysis patients. Dialysis therapy Next XVII. Tokyo: Igakutosho Shuppan Co., Ltd; 2014. p. 99–107. (Book in Japanese).
75. Elisaf M, Bairaktari E, Katopodis K, et al. Effect of L-carnitine supplementation on lipid parameters in hemodialysis patients. Am J Nephrol. 1998;18(5):416–21. https://doi.org/10.1159/000013386.
76. Guarnieri GF, Ranieri F, Toigo G, et al. Lipid-lowering effect of carnitine in chronically uremic patients treated with maintenance hemodialysis. Am J Clin Nutr. 1980;33(7):1489–92. https://doi.org/10.1093/ajcn/33.7.1489.
77. Srivastava DK, Kumar S, Misra AP. Reversal of haemodialysis induced hypertriacylglycerolemia by L-carnitine. Indian J Clin Biochem. 1992;7:19–21.
78. Casciani CU, Caruso U, Cravotto E, et al. Beneficial effects of L-carnitine in post-dialysis syndrome. Curr Ther Res. 1982;32:116–27.
79. Caruso U, Cravotto E, Tisone G, et al. Long-term treatment with L-carnitine in uremic patients undergoing chronic hemodialysis: effects on the lipid pattern. Curr Ther Res Clin Exp. 1983;33:1098–104.
80. Galeone F, Giacomelli A, Rossi A, et al. L-carnitine activity on serum lipid patterns in chronic uraemia. A study of patients on dialysis. Clin Trials J. 1987;24:417–24.
81. Vacha GM, Giorcelli G, d'Iddio S, et al. L-carnitine addition to dialysis fluid. A therapeutic alternative for hemodialysis patients. Nephron. 1989;51(2):237–42. https://doi.org/10.1159/000185292.
82. Sohn HJ, Choi GB, Yoon KI. L-carnitine in maintenance hemodialysis. Clinical, lipid and biochemical effects. Korean J Nephrol. 1992;11(3):260–9. (Article in Korean)
83. Bellinghieri G, Savica V, Barbera CM, et al. L-carnitine and platelet aggregation in uremic patients subjected to hemodialysis. Nephron. 1990;55(1):28–32. https://doi.org/10.1159/000185914.
84. Rogerson ME, Rylance PB, Wilson R, et al. Carnitine and weakness in haemodialysis patients. Nephrol Dial Transplant. 1989;4(5):366–71. https://doi.org/10.1093/oxfordjournals.ndt.a091891.
85. Yderstraede KB, Pedersen FB, Dragsholt C, et al. The effect of L-carnitine on lipid metabolism in patients on chronic haemodialysis. Nephrol Dial Transplant. 1987;1(4):238–41.
86. Weschler A, Aviram M, Levin M, et al. High dose of L-carnitine increases platelet aggregation and plasma triglyceride levels in uremic patients on hemodialysis. Nephron. 1984;38(2):120–4. https://doi.org/10.1159/000183292.
87. Wanner C, Krane V, März W, et al. Atorvastatin in patients with type 2 diabetes mellitus undergoing hemodialysis. N Engl J Med. 2005;353(3):238–48. https://doi.org/10.1056/NEJMoa043545.
88. Vacha GM, Giorcelli G, Siliprandi N, et al. Favorable effects of L-carnitine treatment on hypertriglyceridemia in hemodialysis patients: decisive role of low levels of high-density lipoprotein-cholesterol. Am J Clin Nutr. 1983;38(4):532–40. https://doi.org/10.1093/ajcn/38.4.532.
89. Wanner C, Wieland H, Wäckerle B, et al. Ketogenic and antiketogenic effects of L-carnitine in hemodialysis patients. Kidney Int Suppl. 1989;27:S264–8.
90. Sloan RS, Kastan B, Rice SI, et al. Quality of life during and between hemodialysis treatments: role of L-carnitine supplementation. Am J Kidney Dis. 1998;32(2):265–72. https://doi.org/10.1053/ajkd.1998.v32.pm9708611.
91. Rathod R, Baig MS, Khandelwal PN, et al. Results of a single blind, randomized, placebo-controlled clinical trial to study the effect of intravenous L-carnitine supplementation on health-related quality of life in Indian patients on maintenance hemodialysis. Indian J Med Sci. 2006;60(4):143–53.
92. Fagher B, Cederblad G, Eriksson M, et al. L-carnitine and haemodialysis: double blind study on muscle function and metabolism and peripheral nerve function. Scand J Clin Lab Invest. 1985;45(2):169–78. https://doi.org/10.3109/00365518509160991.

93. Semeniuk J, Shalansky KF, Taylor N, et al. Evaluation of the effect of intravenous l-carnitine on quality of life in chronic hemodialysis patients. Clin Nephrol. 2000;54(6):470–7.
94. Spagnoli LG, Palmieri G, Mauriello A, et al. Morphometric evidence of the trophic effect of L-carnitine on human skeletal muscle. Nephron. 1990;55(1):16–23. https://doi.org/10.1159/000185912.
95. Giovenali P, Fenocchio D, Montanari G, et al. Selective trophic effect of L-carnitine in type I and IIa skeletal muscle fibers. Kidney Int. 1994;46(6):1616–9. https://doi.org/10.1038/ki.1994.460.
96. Ahmad S, Robertson HT, Golper TA, et al. Multicenter trial of L-carnitine in maintenance hemodialysis patients. II. Clinical and biochemical effects. Kidney Int. 1990;38(5):912–8. https://doi.org/10.1038/ki.1990.290.
97. Brass EP, Adler S, Sietsema KE, et al. Intravenous L-carnitine increases plasma carnitine, reduces fatigue, and may preserve exercise capacity in hemodialysis patients. Am J Kidney Dis. 2001;37(5):1018–28. https://doi.org/10.1016/s0272-6386(05)80019-8.
98. Rocchi L, Feola I, Calvani M, et al. Effects of carnitine administration in patients with chronic renal failure undergoing periodic dialysis, evaluated by computerized electromyography. Drugs Exp Clin Res. 1986;12(8):707–11.
99. Feinfeld DA, Briscoe AM, Nurse HM, et al. Myoglobinuria in chronic renal failure. Am J Kidney Dis. 1986;8(2):111–4. https://doi.org/10.1016/s0272-6386(86)80121-4.
100. Feinfeld DA, Kurian P, Cheng JT, et al. Effect of oral L-carnitine on serum myoglobin in hemodialysis patients. Ren Fail. 1996;18(1):91–6. https://doi.org/10.3109/08860229609052778.
101. Takeuchi M, Kiyohara M, Machida H, et al. Levocarnitine supplementation therapy for hemodialysis patients. J Jpn Soc Dial Ther. 2012;45(10):955–63. (Article in Japanese)
102. Takeuchi M. Muscular symptoms and carnitine supplementation therapy in dialysis patients. Dialysis therapy Next XVII. Tokyo: Igakutosho Shuppan Co., Ltd; 2014. p. 45–54. (Book in Japanese)
103. Muto M, Nodomi S, Manabe Y, et al. L-carnitine improved exercised capacity in hemodialysis patients during exercise training with a bicycle ergometer. J Jpn Soc Dial Ther. 2014;47(9):533–8. (Article in Japanese)
104. Kuwasawa-Iwasaki M, Io H, Muto M, et al. Effects of L-carnitine supplementation in patients receiving hemodialysis or peritoneal dialysis. Nutrients. 2020;12(11):3371. https://doi.org/10.3390/nu12113371.
105. Maruyama T, Maruyama N, Higuchi T, et al. Efficacy of L-carnitine supplementation for improving lean body mass and physical function in patients on hemodialysis: a randomized controlled trial. Eur J Clin Nutr. 2019;73(2):293–301. https://doi.org/10.1038/s41430-018-0348-y.
106. Takahashi A, Kikkawa Y, Okuno N, et al. The pathophysiology of hemodialysis-related muscle cramps. J Osaka Soc Dial Ther. 2002;20(1):61–4. (Article in Japanese)
107. Takahashi A. The pathophysiology of leg cramping during dialysis and the use of carnitine in its treatment. Physiol Rep. 2021;9(21):e15114. https://doi.org/10.14814/phy2.15114.
108. Kurata C, Wakabayashi Y, Shoda S, et al. Uremic heart and cardiac autonomic nervous abnormalities. Kidney and Dialysis. 1996;41(1):49–54. (Article in Japanese)
109. Ota M, Sato S, Kawaguchi Y. Does dialysis heart exist? Jpn J Clin Dial. 1996;12(8):1119–31. (Article in Japanese)
110. Ookawara S, Tabei K, Asano Y. X. Uremic cardiomyopathy. Including other kidney diseases. Jpn J Clin Med Suppl: By Area Syndrome Series No.17, Kidney Syndrome. 1997;2:98–101. (Article in Japanese)
111. Tamura T. Heart failure and dialysis heart. Jpn J Clin Dial. 2006;22(4):455–61. (Article in Japanese)
112. Ota M. Cardio-vascular complications in dialysis patients (1) Heart failure (dialysis heart). Jpn J Clin Dial. 2001;17:1297–306. (Article in Japanese)
113. Hinoshita F. Heart failure and dialysis heart. Kidney and Dialysis. 2010;69(5):557–61. (Article in Japanese)
114. Kudoh Y. The carnitine supplementation therapy and hypotension in dialysis patients. The role of L-carnitine in the pathogenesis of cardiomegaly in patients with chronic hemodialysis. Jpn Circ J. 1983;47(12):1391–7. https://doi.org/10.1253/jcj.47.1391.
115. Maruyama T, Abe M. Effects of levocarnitine treatment on the cardiac function in hemodialysis patients. Intern Med. 2018;57(24):3503–5. https://doi.org/10.2169/internalmedicine.1575-18.
116. Suzuki Y, Narita M, Yamazaki N. Effects of L-carnitine on arrhythmias during hemodialysis. Jpn Heart J. 1982;23(3):349–59. https://doi.org/10.1536/ihj.23.349.
117. Fujita Y, Shinzato T, Takai I, et al. Efficacy of L-carnitine administration for long-term dialysis patients with continuous hypotension. Jpn J Artif Organs. 1988;17(1):132–5. (Article in Japanese)
118. Sakurabayashi T, Takaesu Y, Haginoshita S, et al. Improvement of myocardial fatty acid metabolism through L-carnitine administration to chronic hemodialysis patients. Am J Nephrol. 1999;19(4):480–4. https://doi.org/10.1159/000013502.
119. Matsumoto Y, Sato M, Ohashi H, et al. Effects of L-carnitine supplementation on cardiac morbidity in hemodialyzed patients. Am J Nephrol. 2000;20(3):201–7. https://doi.org/10.1159/000013584.
120. Fagher B, Cederblad G, Monti M, et al. Carnitine and left ventricular function in haemodialysis

patients. Scand J Clin Lab Invest. 1985;45(3):193–8. https://doi.org/10.3109/00365518509160995.

121. Fricke L, Preuss R, Santjer G, et al. Effects of substitution treatment with L-carnitine on cardiac arrhythmias, echocardiographic parameters and clinical well-being in dialysis patients with cardiomyopathy. Nieren und Hochdruckkrankheiten Jahrgang. 1988;17(3):S114–9. (Article in German)
122. van Es A, Henny FC, Kooistra MP, et al. Amelioration of cardiac function by L-carnitine administration in patients on haemodialysis. In: Guarnieri G, Panzetta G, Toigo G, editors. Metabolic and nutritional abnormalities in kidney disease: scientific meeting of the European Study Group for the Conservative Management of Chronic Renal Failure, Trieste, September 1991, The book series of contributions to nephrology. Basel: S. Karger AG; 1992. https://doi.org/10.1159/000421598. ISBN (print): 978-3-8055-5582-1, ISBN (electronic): 978-3-318-03362-5.
123. Záruba J, Probst W, Binswanger U. L-carnitine improves vascular refilling in haemodialysis patients during ultrafiltration. Nephrol Dial Transplant. 1994;9(5):587–8. https://doi.org/10.1093/ndt/9.5.587.
124. Trovato GM, Iannetti E, Murgo AM, et al. Body composition and long-term levo-carnitine supplementation. Clin Ter. 1998;149(3):209–14.
125. Romagnoli GF, Naso A, Carraro G, et al. Beneficial effects of L-carnitine in dialysis patients with impaired left ventricular function: an observational study. Curr Med Res Opin. 2002;18(3):172–5. https://doi.org/10.1185/030079902125000606.
126. Kudoh Y, Aoyama S, Torii T, et al. Hemodynamic stabilizing effects of L-carnitine in chronic hemodialysis patients. Cardiorenal Med. 2013;3(3):200–7. https://doi.org/10.1159/000355016.
127. Kudoh Y. Hypotension in dialysis patients and carnitine supplementation therapy. Dialysis therapy Next XVII. Tokyo: Igakutosho Shuppan Co., Ltd; 2014. p. 74–80. (Book in Japanese)
128. Higuchi T, Abe M, Yamazaki T, et al. Levocarnitine improves cardiac function in hemodialysis patients with left ventricular hypertrophy: a randomized controlled trial. Am J Kidney Dis. 2016;67(2):260–70. https://doi.org/10.1053/j.ajkd.2015.09.010.
129. Higuchi T, Hotta S, Ishikawa Y, et al. A case of successful management of a hemodialysis patient with levocarnitine. J Jpn Soc Dial Ther. 2015;48(8):477–82. (Article in Japanese)
130. Kaneko M, Fukasawa H, Ishibuchi K, et al. L-carnitine improved the cardiac function via the effect on myocardial fatty acid metabolism in a hemodialysis patient. Intern Med. 2018;57(24):3593–6. https://doi.org/10.2169/internalmedicine.1055-18.
131. Naito J, Ohashi H, Ohno M, et al. Long-term levocarnitine ameliorates left ventricular diastolic as well as systolic dysfunction in hemodialysis patients – Multi-Center Study. Circ Rep. 2019;1(11):508–16. https://doi.org/10.1253/circrep.CR-19-0075.
132. Sada H, Kato A, Sumimoto R, et al. Effects of L-carnitine supplementation on nutritional, immunological, and cardiac parameters in hemodialysis patients: a pilot study. Ren Replacement Ther. 2015;1:3. https://doi.org/10.1186/s41100-015-0004-0.
133. Nishimura M, Tokoro T, Takatani T, et al. Effects of intravenous L-carnitine on myocardial fatty acid imaging in hemodialysis patients: responders or non-responders to L-carnitine. Springerplus. 2015;4:353. https://doi.org/10.1186/s40064-015-1119-z.
134. Sgambat K, Frank L, Ellini A, et al. Carnitine supplementation improves cardiac strain rate in children on chronic hemodialysis. Pediatr Nephrol. 2012;27(8):1381–7. https://doi.org/10.1007/s00467-012-2144-6.
135. Sgambat K, Moudgil A. Carnitine deficiency in children receiving continuous renal replacement therapy. Hemodial Int. 2016;20(1):63–7. https://doi.org/10.1111/hdi.12341.
136. Sgambat K, Clauss S, Moudgil A. Effect of levocarnitine supplementation on myocardial strain in children with acute kidney injury receiving continuous kidney replacement therapy: a pilot study. Pediatr Nephrol. 2021;36(6):1607–16. https://doi.org/10.1007/s00467-020-04862-3.
137. Kido J, Inoue H, Shimotsu H, et al. Effect of L-carnitine on amino acid metabolism in elderly patients undergoing regular hemodialysis. Blood Purif. 2020;49(5):614–21. https://doi.org/10.1159/000505609.
138. Stenvinkel P. Inflammation in end-stage renal failure: could it be treated? Nephrol Dial Transplant. 2002;17(Suppl 8):33–8. https://doi.org/10.1093/ndt/17.suppl_8.33.
139. Carrero JJ, Stenvinkel P. Persistent inflammation as a catalyst for other risk factors in chronic kidney disease: a hypothesis proposal. Clin J Am Soc Nephrol. 2009;4(Suppl 1):S49–55. https://doi.org/10.2215/CJN.02720409.
140. Zimmermann J, Herrlinger S, Pruy A, et al. Inflammation enhances cardiovascular risk and mortality in hemodialysis patients. Kidney Int. 1999;55(2):648–58. https://doi.org/10.1046/j.1523-1755.1999.00273.x.
141. Yeun JY, Levine RA, Mantadilok V, et al. C-reactive protein predicts all-cause and cardiovascular mortality in hemodialysis patients. Am J Kidney Dis. 2000;35(3):469–76. https://doi.org/10.1016/s0272-6386(00)70200-9.
142. Savica V, Santoro D, Mazzaglia G, et al. L-carnitine infusions may suppress serum C-reactive protein and improve nutritional status in maintenance hemodialysis patients. J Ren Nutr. 2005;15(2):225–30. https://doi.org/10.1053/j.jrn.2004.10.002.
143. Duranay M, Akay H, Yilmaz FM, et al. Effects of L-carnitine infusions on inflammatory and nutritional markers in haemodialysis patients. Nephrol Dial Transplant. 2006;21(11):3211–4. https://doi.org/10.1093/ndt/gfl356.
144. Suchitra MM, Ashalatha VL, Sailaja E, et al. The effect of L-carnitine supplementation on lipid parameters, inflammatory and nutritional markers in

maintenance hemodialysis patients. Saudi J Kidney Dis Transpl. 2011;22(6):1155–9.

145. Shakeri A, Tabibi H, Hedayati M. Effects of L-carnitine supplement on serum inflammatory cytokines, C-reactive protein, lipoprotein (a), and oxidative stress in hemodialysis patients with Lp (a) hyperlipoproteinemia. Hemodial Int. 2010;14(4):498–504. https://doi.org/10.1111/j.1542-4758.2010.00476.x.
146. Tabibi H, Hakeshzadeh F, Hedayati M, et al. Effects of L-carnitine supplement on serum amyloid A and vascular inflammation markers in hemodialysis patients: a randomized controlled trial. J Ren Nutr. 2011;21(6):485–91. https://doi.org/10.1053/j.jrn.2011.01.001.
147. Adachi T, Fukami K, Yamagishi S, et al. Decreased serum carnitine is independently correlated with increased tissue accumulation levels of advanced glycation end products in haemodialysis patients. Nephrology (Carlton). 2012;17(8):689–94. https://doi.org/10.1111/j.1440-1797.2012.01642.x.
148. Fukami K, Yamagishi S, Sakai K, et al. Potential inhibitory effects of L-carnitine supplementation on tissue advanced glycation end products in patients with hemodialysis. Rejuvenation Res. 2013;16(6):460–6. https://doi.org/10.1089/rej.2013.1459.
149. den Dekker MA, Zwiers M, van den Heuvel ER, et al. Skin autofluorescence, a non-invasive marker for AGE accumulation, is associated with the degree of atherosclerosis. PLoS One. 2013;8(12):e83084. https://doi.org/10.1371/journal.pone.0083084.
150. Meerwaldt R, Hartog JW, Graaff R, et al. Skin autofluorescence, a measure of cumulative metabolic stress and advanced glycation end products, predicts mortality in hemodialysis patients. J Am Soc Nephrol. 2005;16(12):3687–93. https://doi.org/10.1681/ASN.2005020144.
151. Nagano M, Fukami K, Yamagishi S, et al. Tissue level of advanced glycation end products is an independent determinant of high-sensitivity C-reactive protein levels in haemodialysis patients. Nephrology (Carlton). 2011;16(3):299–303. https://doi.org/10.1111/j.1440-1797.2010.01419.x.
152. Fukami K, Okuda S. Carnitine deficiency and oxidative stress in hemodialysis patients. Expectation its AGEs alleviation effects. Dialysis therapy Next XVII. Tokyo: Igakutosho Shuppan Co., Ltd; 2014. p. 13–20. (Book in Japanese)
153. Higuchi T, Abe M, Yamazaki T, et al. Effects of levocarnitine on brachial-ankle pulse wave velocity in hemodialysis patients: a randomized controlled trial. Nutrients. 2014;6(12):5992–6004. https://doi.org/10.3390/nu6125992.
154. Wolcott DL, Marsh JT, Rue AL, et al. Recombinant human erythropoietin treatment may improve quality of life and cognitive function in chronic hemodialysis patients. Am J Kidney Dis. 1989;14(6):478–85. https://doi.org/10.1016/s0272-6386(89)80148-9.
155. Lundin AP. Quality of life: subjective and objective improvements with recombinant human erythropoietin therapy. Semin Nephrol. 1989;9(1 Suppl 1):22–9.
156. Foley RN, Parfrey PS, Harnett JD, et al. The impact of anemia on cardiomyopathy, morbidity, and mortality in end-stage renal disease. Am J Kidney Dis. 1996;28(1):53–61. https://doi.org/10.1016/s0272-6386(96)90130-4.
157. Trovato GM, Ginardi V, Di Marco V, et al. Long-term L-carnitine treatment of chronic anaemia of patients with end-stage renal failure. Curr Ther Res. 1982;31(6):1042–9.
158. Albertazzi A, Capelli P, Di Paolo B, et al. Endocrine-metabolic effects of L-carnitine in patients on regular dialysis treatment. Proc Eur Dial Transplant Assoc. 1983;19:302–7.
159. Donatelli M, Terrizzi C, Zummo G, et al. Effects of L-carnitine on chronic anemia and erythrocyte adenosine triphosphate concentration in hemodialyzed patients. Curr Ther Res. 1987;41(5):620–4.
160. Kooistra MP, Struyvenberg A, van Es A. The response to recombinant human erythropoietin in patients with the anemia of end-stage renal disease is correlated with serum carnitine levels. Nephron. 1991;57(1):127–8. https://doi.org/10.1159/000186237.
161. Matsumura M, Hatakeyama S, Koni I, et al. Correlation between serum carnitine levels and erythrocyte osmotic fragility in hemodialysis patients. Nephron. 1996;72(4):574–8. https://doi.org/10.1159/000188942.
162. Golper TA, Goral S, Becker BN, et al. L-carnitine treatment of anemia. Am J Kidney Dis. 2003;41(4 Suppl 4):S27–34. https://doi.org/10.1016/s0272-6386(03)00114-8.
163. Labonia WD. L-carnitine effects on anemia in hemodialyzed patients treated with erythropoietin. Am J Kidney Dis. 1995;26(5):757–64. https://doi.org/10.1016/0272-6386(95)90439-5.
164. Kletzmayr J, Mayer G, Legenstein E, et al. Anemia and carnitine supplementation in hemodialyzed patients. Kidney Int Suppl. 1999;69:S93–S106.
165. Caruso U, Leone L, Cravotto E, et al. Effects of L-carnitine on anemia in aged hemodialysis patients treated with recombinant human erythropoietin: a pilot study. Dial Transplant. 1998;27(8):498–506.
166. Nilsson-Ehle P, Cederblad G, Fagher B, et al. Plasma lipoproteins, liver function and glucose metabolism in haemodialysis patients: lack of effect of L-carnitine supplementation. Scand J Clin Lab Invest. 1985;45(2):179–84. https://doi.org/10.3109/00365518509160992.
167. Matsumoto Y, Amano I, Hirose S, et al. Effects of L-carnitine supplementation on renal anemia in poor responders to erythropoietin. Blood Purif. 2001;19(1):24–32. https://doi.org/10.1159/000014474.
168. Higuchi T, Ishikawa Y, Yamazaki T, et al. Efficacy of levocarnitine on renal anemia in hemodialysis patients. J Jpn Soc Dial Ther. 2014;47(2):119–27. (Article in Japanese)
169. Kudoh Y, Aoyama S, Torii T, et al. Long-term effects of oral L-carnitine supplementation on anemia in chronic hemodialysis. Cardiorenal Med. 2014;4(1):53–9. https://doi.org/10.1159/000360865.

170. Murakami K, Suzuki T, Shirai K, et al. A study of the effects of L-carnitine for ESA resistant renal anemia. Kidney Dial. 2013;75(5):769–74. (Article in Japanese)
171. Maruyama T, Higuchi T, Yamazaki T, et al. Levocarnitine injections decrease the need for erythropoiesis-stimulating agents in hemodialysis patients with renal anemia. Cardiorenal Med. 2017;7(3):188–97. https://doi.org/10.1159/000462983.
172. Funakoshi S, Hayashida M, Teshima K, et al. Intravenous L-carnitine treatment for anemia in hemodialysis patients – evaluation for cost-effectiveness. J Jpn Soc Dial Ther. 2014;29(3):419–23. (Article in Japanese)
173. Aoki Y, Yamamoto T. Carnitine reduced erythropoietin dose required and improved cardiac function of patients on maintenance hemodialysis. Saudi J Kidney Dis Transpl. 2017;28(3):477–82. https://doi.org/10.4103/1319-2442.206445.
174. Suzuki A, Sakai Y, Hashimoto K, et al. Kinetics of carnitine concentration after switching from oral administration to intravenous injection in hemodialysis patients. Ren Fail. 2018;40(1):196–200. https://doi.org/10.1080/0886022X.2018.1455587.
175. Kanbay M, Perazella MA, Kasapoglu B, et al. Erythropoiesis stimulatory agent-resistant anemia in dialysis patients: review of causes and management. Blood Purif. 2010;29(1):1–12. https://doi.org/10.1159/000245041.
176. Fukushima T, Horike H, Fujiki S, et al. Zinc deficiency anemia and effects of zinc therapy in maintenance hemodialysis patients. Ther Apher Dial. 2009;13(3):213–9. https://doi.org/10.1111/j.1744-9987.2009.00656.x.
177. Kobayashi H, Abe M, Okada K, et al. Oral zinc supplementation reduces the erythropoietin responsiveness index in patients on hemodialysis. Nutrients. 2015;7(5):3783–95. https://doi.org/10.3390/nu7053783.
178. Shimizu S, Tei R, Okamura M, et al. Prevalence of zinc deficiency in Japanese patients on peritoneal dialysis: comparative study in patients on hemodialysis. Nutrients. 2020;12(3):764. https://doi.org/10.3390/nu12030764.
179. Fiedler R, Dorligjav O, Seibert E, et al. Vitamin D deficiency, mortality, and hospitalization in hemodialysis patients with or without protein-energy wasting. Nephron Clin Pract. 2011;119(3):c220–6. https://doi.org/10.1159/000328927.
180. Kiss Z, Ambrus C, Almasi C, et al. Serum 25(OH)-cholecalciferol concentration is associated with hemoglobin level and erythropoietin resistance in patients on maintenance hemodialysis. Nephron Clin Pract. 2011;117(4):c373–8. https://doi.org/10.1159/000321521.
181. Kumar VA, Kujubu DA, Sim JJ, et al. Vitamin D supplementation and recombinant human erythropoietin utilization in vitamin D-deficient hemodialysis patients. J Nephrol. 2011;24(1):98–105. https://doi.org/10.5301/jn.2010.1830.
182. Patel NM, Gutiérrez OM, Andress DL, et al. Vitamin D deficiency and anemia in early chronic kidney disease. Kidney Int. 2010;77(8):715–20. https://doi.org/10.1038/ki.2009.551.
183. Matias PJ, Ferreira C, Jorge C, et al. 25-hydroxyvitamin D3, arterial calcifications and cardiovascular risk markers in haemodialysis patients. Nephrol Dial Transplant. 2009;24(2):611–8. https://doi.org/10.1093/ndt/gfn502.
184. Argilés A, Lorho R, Mourad G, et al. High-dose alfacalcidol for anaemia in dialysis. Lancet. 1993;342(8867):378–9. https://doi.org/10.1016/0140-6736(93)91528-t.
185. Albitar S, Genin R, Fen-Chong M, et al. High-dose alfacalcidol improves anaemia in patients on haemodialysis. Nephrol Dial Transplant. 1997;12(3):514–8. https://doi.org/10.1093/ndt/12.3.514.
186. Goicoechea M, Vazquez MI, Ruiz MA, et al. Intravenous calcitriol improves anaemia and reduces the need for erythropoietin in haemodialysis patients. Nephron. 1998;78:23–7.
187. Aucella F, Scalzulli RP, Gatta G, et al. Calcitriol increases burst-forming unit-erythroid proliferation in chronic renal failure. A synergistic effect with r-HuEpo. Nephron Clin Pract. 2003;95(4):c121–7. https://doi.org/10.1159/000074837.
188. Date T, Shigematsu T, Hayashi K, et al. A clinical study of the effect of intravenous vitamin D_3 pulse therapy on renal anemia and its mechanism. Jpn J Clin Dial. 2011;27(6):743–8. (Article in Japanese)
189. Date T. Nutritional disorders and carnitine supplementation therapy in dialysis patients. A clinical experience of vitamin D_3, zinc, and carnitine supplementation. Dialysis Therapy Next XVII. Tokyo: Igakutosho Shuppan Co., Ltd; 2014. p. 33–44. (Book in Japanese)
190. Ida S, Iwasaki Y, Uchida K, et al. Guideline for diagnosis and treatment of carnitine deficiency 2018, https://www.jpeds.or.jp/uploads/files/20181207_shishin.pdf (Article in Japanese)
191. Deicher R, Hörl WH. Hormonal adjuvants for the treatment of renal anaemia. Eur J Clin Invest. 2005;35(Suppl 3):75–84. https://doi.org/10.1111/j.1365-2362.2005.01533.x.
192. Bonomini M, Zammit V, Pusey CD, et al. Pharmacological use of L-carnitine in uremic anemia: has its full potential been exploited? Pharmacol Res. 2011;63(3):157–64. https://doi.org/10.1016/j.phrs.2010.11.006.
193. Durak I, Akyol O, Başeşme E, et al. Reduced erythrocyte defense mechanisms against free radical toxicity in patients with chronic renal failure. Nephron. 1994;66(1):76–80. https://doi.org/10.1159/000187770.
194. McKenney J, Valeri CR, Mohandas N, et al. Decreased in vivo survival of hydrogen peroxide-damaged baboon red blood cells. Blood. 1990;76(1):206–11.
195. Wolfe LC, Byrne AM, Lux SE. Molecular defect in the membrane skeleton of blood bank-stored red cells. Abnormal spectrin-protein 4.1-actin complex

formation. J Clin Invest. 1986;78(6):1681–6. https://doi.org/10.1172/JCI112762.
196. Mohanty JG, Nagababu E, Rifkind JM. Red blood cell oxidative stress impairs oxygen delivery and induces red blood cell aging. Front Physiol. 2014;5:84. https://doi.org/10.3389/fphys.2014.00084.
197. Arduini A, Holme S, Sweeney JD, et al. Addition of L-carnitine to additive solution-suspended red cells stored at 4 degrees C reduces in vitro hemolysis and improves in vivo viability. Transfusion. 1997;37(2):166–74. https://doi.org/10.1046/j.1537-2995.1997.37297203519.x.
198. Arduini A, Mancinelli G, Radatti GL, et al. Role of carnitine and carnitine palmitoyltransferase as integral components of the pathway for membrane phospholipid fatty acid turnover in intact human erythrocytes. J Biol Chem. 1992;267(18):12673–81.
199. Arduini A, Tyurin V, Tyuruna Y, et al. Acyl-trafficking in membrane phospholipid fatty acid turnover: the transfer of fatty acid from the acyl-L-carnitine pool to membrane phospholipids in intact human erythrocytes. Biochem Biophys Res Commun. 1992;187(1):353–8. https://doi.org/10.1016/s0006-291x(05)81500-1.
200. Arduini A. Carnitine and its acyl esters as secondary antioxidants? Am Heart J. 1992;123(6):1726–7. https://doi.org/10.1016/0002-8703(92)90850-u.
201. Maeda N. Rheology and physiological function of blood. No. 3. Blood microcirculation and oxygen transport. J Physiol Soc Jpn. 2004;66(10):327–36. (Article in Japanese)
202. Shepherd JT, Vanhoutte PM. The human cardiovascular system: facts and concepts. New York: Raven Press; 1979. (Joint translation for Japanese edition: Imai S, Yamori Y, Kawai C, Takeuchi T, Nishimura Co., Ltd, Tokyo,1983 Books in Japanese)
203. Matsumura M, Hatakeyama S, Ito M, et al. Relationship between serum carnitine levels and erythrocyte osmotic fragility in chronic dialysis patients. J Jpn Soc Dial Ther. 1993;26(4):569–74. (Article in Japanese)
204. Nikolaos S, George A, Telemachos T, et al. Effect of L-carnitine supplementation on red blood cells deformability in hemodialysis patients. Ren Fail. 2000;22(1):73–80. https://doi.org/10.1081/jdi-100100853.
205. Ibrahim FF, Ghannam MM, Ali FM. Effect of dialysis on erythrocyte membrane of chronically hemodialyzed patients. Ren Fail. 2002;24(6):779–90. https://doi.org/10.1081/jdi-120015680.
206. Vlassopoulos DA, Hadjiyannakos DK, Anogiatis AG, et al. Carnitine action on red blood cell osmotic resistance in hemodialysis patients. J Nephrol. 2002;15(1):68–73.
207. Kobayashi A, Watanabe H, Fujisawa S, et al. Effects of L-carnitine and palmitoylcarnitine on membrane fluidity of human erythrocytes. Biochim Biophys Acta. 1989;986(1):83–8. https://doi.org/10.1016/0005-2736(89)90275-7.
208. Arduini A, Gorbunov N, Arrigoni-Martelli E, et al. Effects of L-carnitine and its acetate and propionate esters on the molecular dynamics of human erythrocyte membrane. Biochim Biophys Acta. 1993;1146(2):229–35. https://doi.org/10.1016/0005-2736(93)90360-c.
209. Hamasaki N. The role of erythrocyte membrane proteins. KAGAKU TO SEIBUTSU. 1995;33(9):566–74. (Article in Japanese)
210. Arduini A, Rossi M, Mancinelli G, et al. Effect of L-carnitine and acetyl-L-carnitine on the human erythrocyte membrane stability and deformability. Life Sci. 1990;47(26):2395–400. https://doi.org/10.1016/0024-3205(90)90483-8.
211. Butterfield DA, Rangachari A. Acetylcarnitine increases membrane cytoskeletal protein-protein interactions. Life Sci. 1993;52(3):297–303. https://doi.org/10.1016/0024-3205(93)90221-n.
212. Harigae H. Production and decay of red blood cells. J Jpn Soc Intern Med. 2006;95(10):1983–7. (Article in Japanese)
213. Testa U. Apoptotic mechanisms in the control of erythropoiesis. Leukemia. 2004;18(7):1176–99. https://doi.org/10.1038/sj.leu.2403383.
214. Kawabata H, Takaori A. Physiology and regulation of production of red blood cells. J Jpn Soc Intern Med. 2015;104(7):1367–74. (Article in Japanese)
215. Mutomba MC, Yuan H, Konyavko M, et al. Regulation of the activity of caspases by L-carnitine and palmitoylcarnitine. FEBS Lett. 2000;478(1-2):19–25. https://doi.org/10.1016/s0014-5793(00)01817-2.
216. Bonomini M, Sirolli V, Dottori S, et al. L-carnitine inhibits a subset of platelet activation responses in chronic uraemia. Nephrol Dial Transplant. 2007;22(9):2623–9. https://doi.org/10.1093/ndt/gfm212.
217. Matsumura M, Hatakeyama S, Koni I, et al. Effect of L-carnitine and palmitoyl-L-carnitine on erythroid colony formation in fetal mouse liver cell culture. Am J Nephrol. 1998;18(5):355–8. https://doi.org/10.1159/000013376.
218. Kitamura Y, Satoh K, Satoh T, et al. Effect of L-carnitine on erythroid colony formation in mouse bone marrow cells. Nephrol Dial Transplant. 2005;20(5):981–4. https://doi.org/10.1093/ndt/gfh758.
219. Calò LA, Pagnin E, Davis PA, et al. Antioxidant effect of L-carnitine and its short chain esters: relevance for the protection from oxidative stress related cardiovascular damage. Int J Cardiol. 2006;107(1):54–60. https://doi.org/10.1016/j.ijcard.2005.02.053.
220. Calò LA, Davis PA, Semplicini A, et al. L-carnitine and erythropoiesis: relationship with haeme oxygenase-1. Nephrol Dial Transplant. 2005;20(8):1769–70. https://doi.org/10.1093/ndt/gfh934.
221. Calò LA, Davis PA, Pagnin E, et al. Carnitine-mediated improved response to erythropoietin involves induction of haem oxygenase-1: studies in humans and in an animal model. Nephrol

Dial Transplant. 2008;23(3):890–5. https://doi.org/10.1093/ndt/gfm707.
222. Kanno H, Fujii H. Biochemical basis of senescence and premature death of red blood cells – phosphatidylserine on red cell phagocytosis as a marker of phagocytosis. Membrane. 2007;32(3):139–46. (Article in Japanese)
223. Lang F, Qadri SM. Mechanisms and significance of eryptosis, the suicidal death of erythrocytes. Blood Purif. 2012;33(1-3):125–30. https://doi.org/10.1159/000334163.
224. Lang E, Qadri SM, Lang F. Killing me softly – suicidal erythrocyte death. Int J Biochem Cell Biol. 2012;44(8):1236–43. https://doi.org/10.1016/j.biocel.2012.04.019.
225. Lang E, Lang F. Triggers, inhibitors, mechanisms, and significance of eryptosis: the suicidal erythrocyte death. Biomed Res Int. 2015;2015:513518. https://doi.org/10.1155/2015/513518.
226. Mizuguchi T. Red blood cell lifespan and oxidative stress. Kidney Dial. 2013;75(3):341–4. (Article in Japanese)
227. Sato Y, Mizuguchi T, Shigenaga S, et al. Shortened red blood cell lifespan is related to the dose of erythropoiesis-stimulating agents requirement in patients on hemodialysis. Ther Apher Dial. 2012;16(6):522–8. https://doi.org/10.1111/j.1744-9987.2012.01089.x.
228. Ly J, Marticorena R, Donnelly S. Red blood cell survival in chronic renal failure. Am J Kidney Dis. 2004;44(4):715–9.
229. Linde T, Sandhagen B, Wikström B, et al. The required dose of erythropoietin during renal anaemia treatment is related to the degree of impairment in erythrocyte deformability. Nephrol Dial Transplant. 1997;12(11):2375–9. https://doi.org/10.1093/ndt/12.11.2375.
230. Kalicki RM, Uehlinger DE. Red cell survival in relation to changes in the hematocrit: more important than you think. Blood Purif. 2008;26(4):355–60. https://doi.org/10.1159/000133838.
231. Nagata S. Apoptosis and clearance of apoptotic cells. Ann Rev Immunol. 2018;36:489–517. https://doi.org/10.1146/annurev-immunol-042617-053010.
232. Zwaal RF, Schroit AJ. Pathophysiologic implications of membrane phospholipid asymmetry in blood cells. Blood. 1997;89(4):1121–32.
233. Bonomini M, Sirolli V, Settefrati N, et al. Increased erythrocyte phosphatidylserine exposure in chronic renal failure. J Am Soc Nephrol. 1999;10(9):1982–90. https://doi.org/10.1681/ASN.V1091982.
234. Bonomini M, Sirolli V, Reale M, et al. Involvement of phosphatidylserine exposure in the recognition and phagocytosis of uremic erythrocytes. Am J Kidney Dis. 2001;37(4):807–14. https://doi.org/10.1016/s0272-6386(01)80130-x.
235. Andrieu-Abadie N, Jaffrézou JP, Hatem S, et al. L-carnitine prevents doxorubicin-induced apoptosis of cardiac myocytes: role of inhibition of ceramide generation. FASEB J. 1999;13(12):1501–10. https://doi.org/10.1096/fasebj.13.12.1501.
236. Arduini A, Bonomini M, Clutterbuck EJ, et al. Effect of L-carnitine administration on erythrocyte survival in haemodialysis patients. Nephrol Dial Transplant. 2006;21(9):2671–2. https://doi.org/10.1093/ndt/gfl155.
237. Mizuguchi T, Hojo C, Kanayama K, et al. Effects of substitution therapy using levocarnitine chloride on hemodialysis patients requiring high-capacity erythropoiesis-stimulating agents. J Jpn Soc Dial Ther. 2014;47(3):191–7. (Article in Japanese)
238. Mizuguchi T. Anemia and carnitine supplementation therapy in dialysis patients. Dialysis therapy Next XVII. Tokyo: Igakutosho Shuppan Co., Ltd; 2014. p. 21–32. (Book in Japanese)
239. Bevers EM, Williamson PL. Getting to the outer leaflet: physiology of phosphatidylserine exposure at the plasma membrane. Physiol Rev. 2026;96(2):605–45. https://doi.org/10.1152/physrev.00020.2015.
240. Kazmi WH, Obrador GT, Sternberg M, et al. Carnitine therapy is associated with decreased hospital utilization among hemodialysis patients. Am J Nephrol. 2005;25(2):106–15. https://doi.org/10.1159/000084853.
241. Weinhandl ED, Rao M, Gilbertson DT, et al. Protective effect of intravenous levocarnitine on subsequent-month hospitalization among prevalent hemodialysis patients, 1998 to 2003. Am J Kidney Dis. 2007;50(5):803–12. https://doi.org/10.1053/j.ajkd.2007.07.017.
242. Hörl WH, Jacobs C, Macdougall IC, et al. European best practice guidelines 14–16: inadequate response to epoetin. Nephrol Dial Transplant. 2000;15(Suppl 4):43–50.
243. Eknoyan G, Latos DL, Lindberg J. Practice recommendations for the use of L-carnitine in dialysis-related carnitine disorder. National Kidney Foundation Carnitine Consensus Conference. Am J Kidney Dis. 2003;41(4):868–76. https://doi.org/10.1016/s0272-6386(03)00110-0.
244. Eknoyan G, Lindberg JS. On the evolving nature of understanding dialysis-related disorders. Am J Kidney Dis. 2003;41(4 Suppl 4):S1–3. https://doi.org/10.1016/s0272-6386(03)00111-2.
245. Hoppel C. The role of carnitine in normal and altered fatty acid metabolism. Am J Kidney Dis. 2003;41(4 Suppl 4):S4–S12. https://doi.org/10.1016/s0272-6386(03)00112-4.
246. Evans A. Dialysis-related carnitine disorder and levocarnitine pharmacology. Am J Kidney Dis. 2003;41(4 Suppl 4):S13–26. https://doi.org/10.1016/s0272-6386(03)00113-6.
247. Pauly DF, Pepine CJ. The role of carnitine in myocardial dysfunction. Am J Kidney Dis. 2003;41(4 Suppl 4):S35–43. https://doi.org/10.1016/s0272-6386(03)00115-x.
248. Steinman TI, Nissenson AR, Glassock RJ, et al. L-carnitine use in dialysis patients: is national coverage for supplementation justified? What were CMS

regulators thinking – or were they? Nephrol News Issues. 2003;17(5):28–30, 32–4, 36.
249. Steinman TI. L-carnitine supplementation in dialysis patients: does the evidence justify its use? Semin Dial. 2005;18(1):1–2. https://doi.org/10.1111/j.1525-139X.2005.18104.x.
250. Schreiber B. Common misconceptions about levocarnitine and dialysis. Semin Dial. 2005;18(4):349–51. https://doi.org/10.1111/j.1525-139X.2005.18412.x.
251. Hedayati SS. Dialysis-related carnitine disorder. Semin Dial. 2006;19(4):323–8. https://doi.org/10.1111/j.1525-139X.2006.00180.x.
252. Handelman GJ. Debate forum: carnitine supplements have not been demonstrated as effective in patients on long-term dialysis therapy. Blood Purif. 2006;24(1):140–2. https://doi.org/10.1159/000089450.
253. Schreiber BD. Debate forum: levocarnitine therapy is rational and justified in selected dialysis patients. Blood Purif. 2006;24(1):128–39. https://doi.org/10.1159/000089449.
254. Wasserstein AG. L-carnitine supplementation in dialysis: treatment in quest of disease. Semin Dial. 2013;26:11–5.
255. Mercadal L, Coudert M, Vassault A, et al. L-carnitine treatment in incident hemodialysis patients: the multicenter, randomized, double-blinded, placebo-controlled CARNIDIAL trial. Clin J Am Soc Nephrol. 2012;7(11):1836–42. https://doi.org/10.2215/CJN.12431211.
256. Yee J. L-carnitine for anemia in hemodialysis patients: a last resort. Clin J Am Soc Nephrol. 2012;7(11):1746–8. https://doi.org/10.2215/CJN.09920912.
257. Yokote K, Yamanouchi H, Mizutani T, et al. Clinical characteristics of Wernicke's encephalopathy in the elderly. Jpn J Geriatr. 1992;29(1):35–40. (Article in Japanese)
258. Oi M, Maruhashi T, Asari Y. Carnitine deficiency in intensive care unit patients undergoing continuous renal replacement therapy: a single-center retrospective study. JMA J. 2024;7(1):70–6. https://doi.org/10.31662/jmzj.2023-0112. Epub 2023 Nov 16
259. Berger MM, Broman M, Forni L, et al. Nutrients and micronutrients at risk during renal replacement therapy: a scoping review. Curr Opin Crit Care. 2021;27(4):367–77. https://doi.org/10.1097/MCC.0000000000000851.
260. Saito K, Takahashi K. Kidney transplantation: a comparison of trends in Japan and the world. Jpn J Nephrol. 2004;46(1):2–11. (Article in Japanese)
261. Ishikawa I. Latest kidney transplantation statistics: contrast of treatment for end-stage renal failure in Western countries and Japan. Jpn J Nephrol. 2004;46(1):12–9. (Article in Japanese)
262. Van de Wyngaert C, Dewulf JP, Collienne C, et al. Carnitine deficiency after long-term continuous renal replacement therapy. Case Rep Crit Care. 2022;2022:4142539. https://doi.org/10.1155/2022/4142539.
263. Matsufuji S, Shoji T, Lee S, et al. Association between levocarnitine treatment and the change in knee extensor strength in patients undergoing hemodialysis: a post-hoc analysis of the Osaka Dialysis Complication Study (ODCS). Nutrients. 2022;14(2):343. https://doi.org/10.3390/nu14020343.
264. Chewcharat A, Chewcharat P, Liu W, et al. The effect of levocarnitine supplementation on dialysis-related hypotension: a systematic review, meta-analysis, and sequential analysis. PLoS One. 2022;17(7):e0271307. https://doi.org/10.1371/journal.pone.0271307.
265. Takashima H, Maruyama T, Abe M. Significance of levocarnitine treatment in dialysis patients. Nutrients. 2021;13(4):1219. https://doi.org/10.3390/nu13041219.
266. Goto H, Nakashima H, Mori K, et al. L-carnitine pretreatment ameliorates heat-stress acute kidney injury by restoring mitochondrial function of tubular cells. Am J Physiol Renal Physiol. 2024;326(3):F338–51. https://doi.org/10.1152/ajprenal.00196.2023.
267. Ito S, Taguchi K, Kamada G, et al. Improvement of impaired carnitine-induced fatty acid oxidation in experimental and human diabetic kidney disease. JCI Insight. 2025;10(13):e179362. https://doi.org/10.1172/jci.insight.179362.

15 Carnitine Deficiency in Patients with Liver Diseases

15.1 Introduction

Since carnitine is mainly biosynthesized in the liver and kidneys, carnitine deficiency is likely to occur when liver tissue is damaged by liver diseases such as cirrhosis and non-alcoholic fatty liver disease (NAFLD). The following is a review of previous studies on carnitine dynamics in blood and liver tissue in the field of liver diseases and carnitine supplementation therapy for clinical symptoms such as hepatic encephalopathy and cramps in cirrhotic patients.

15.2 Carnitine Kinetics in Liver Diseases

In 1974, Böhmer et al. studied serum carnitine concentrations in 30 healthy subjects and 109 patients with various diseases and reported that serum free carnitine concentrations in 7 cirrhotic patients (2 of whom had primary biliary cirrhosis: PBC) were not different from those in healthy subjects [1].

Chen et al. also examined serum free carnitine concentrations in 53 healthy patients and 76 patients with various diseases such as myocardial infarction, renal failure, cirrhosis, and cancer, and found that 3 of the 11 patients with cirrhosis were in the normal range, but 8 were higher than normal, especially in patients with cirrhosis who also had renal failure [2]. They considered that it may be due to a decrease in renal clearance of carnitine.

On the other hand, Rudman et al. examined serum carnitine concentrations in 16 healthy subjects and 247 patients hospitalized for 16 different diseases and found that only one group of patients with alcoholic cirrhosis (36 patients) had significantly lower total and free carnitine concentrations than healthy subjects [3]. In particular, 14 of these patients had subnormal serum carnitine concentrations, and 7 of them died during hospitalization. These patients were undernourished with symptoms such as jaundice, cachexia, anorexia, ascites, lethargy, and obtundation in the last month before death. Eight patients who died of acute diseases other than liver diseases and were in good nutritional condition were used as controls, and postmortems were performed to measure carnitine concentrations in myocardium, liver, and other tissues. The serum total carnitine and free carnitine concentration 10–42 days (mean 22 days) before death were 28.9 and 11.4 μmol/L, respectively, which are extremely low compared to 79 and 49 μmol/L in healthy subjects, and the total carnitine concentration in the tissues of the pectoral, myocardium, liver, kidney, and brain was extremely low at 15.6, 22.9, 34.5, 40.0, and 16.7%, respectively, of the concentration in the group of patients who died due to acute diseases and did not have problems with nutritional status.

N. Nakayama, *Carnitine Deficiency and Insufficiency in Medicine*,
https://doi.org/10.1007/978-981-95-7797-2_15

On the other hand, Fuller et al. compared the plasma carnitine dynamics of 20 patients with alcoholic cirrhosis with those of 30 healthy controls and reported that total carnitine, free carnitine, short-chain acylcarnitine, and long-chain acylcarnitine were all higher in cirrhotic patients, especially acylcarnitine was 2 to 3 times higher [4] (Fig. 15.1).

Subsequently, Fuller et al. examined the plasma carnitine fractions of 32 healthy subjects, 20 alcoholic patients without liver disease, and 28 alcoholic patients with liver disease, judging from liver hypertrophy, abnormal liver function, hypoalbuminemia, etc. Short-chain acylcarnitine and long-chain acylcarnitine levels were higher than in healthy subjects, but lower than in alcohol patients with liver disease [5]. This indicates that alcoholism itself affects carnitine metabolism independently of liver dysfunction. It is thought that ethanol suppresses the metabolism of long-chain fatty acids, and as a result, the acyl group that has not been degraded to the end binds to carnitine, resulting high long-chain acylcarnitines.

de Sousa et al. also reported plasma carnitine in 20 patients with chronic alcoholism and 12 healthy subjects with fatty liver [6]. According to the results, the concentration of total carnitine, free carnitine, and acylcarnitine in the plasma of alcoholics was 11–15% higher than healthy subjects, but there was no significant difference due to variation. Biopsies of the liver and skeletal muscle (healthy subjects and 7 alcoholic patients) showed that the free carnitine concentrations in the liver were 3.5 ± 2.1 versus 2.3 ± 0.7 nmol/mg of protein, respectively, which were slightly lower but not significant. On the other hand, the concentration of free carnitine in skeletal muscle (7 healthy subjects and 8 alcoholics) was 13.3 ± 5.2 versus 24.3 ± 9.3 nmol/mg of protein, respectively, significantly higher in alcoholics. They speculated that this may be due to the atrophy of type II muscle fibers and the higher content of type I muscle fibers with high carnitine content in the muscles of alcoholic patients.

Amodio et al. also measured plasma carnitine concentrations in 41 patients with cirrhosis (23 alcoholic cirrhosis, 4 positives for hepatitis B

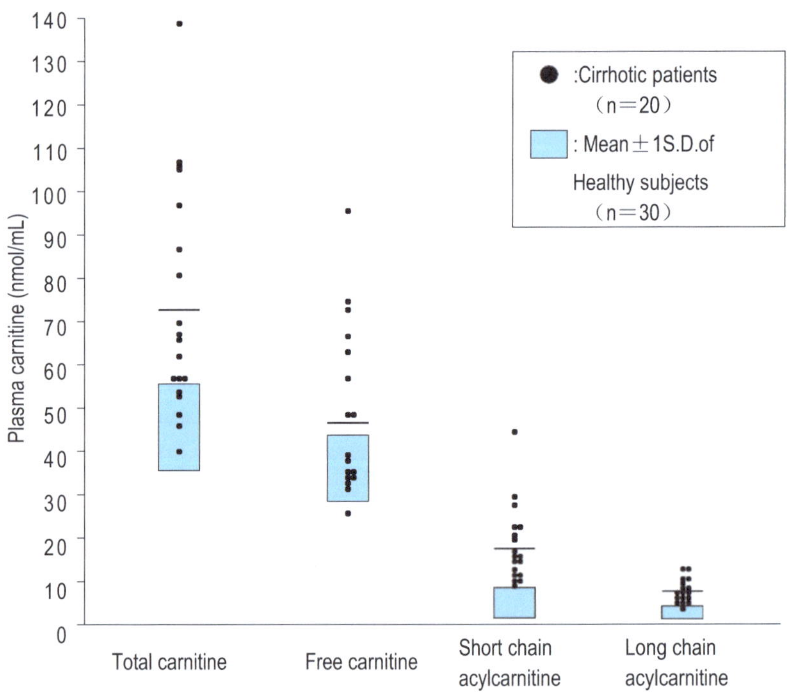

Fig. 15.1 Plasma carnitine and acylcarnitines in hepatic cirrhosis. The blue box represents the mean ± 1S.D. of values for carnitine and its metabolites for 30 healthy subjects. The solid dots are individual values for each of the 20 cirrhotic patients. The horizontal bar is the mean value for the cirrhotic patients. (From ref. [4])

virus (HBV), 13 non-B non-C cirrhosis, and 1 biliary cirrhosis) and 16 healthy subjects [7]. As a result, there was no difference in free carnitine, but the concentrations of total carnitine, short-chain acylcarnitine, and acylcarnitine were significantly higher in patients with cirrhosis. In addition, there was no difference between the type of cirrhosis and the severity of cirrhosis (Child-Pugh (CP) classification), and there was no particular association in nutritional indicators except that triceps subcutaneous fat thickness was weakly correlated with long-chain acylcarnitine concentration.

Alonso de la Peña et al. classified 167 patients with alcoholic cirrhosis in detail according to clinical and anatomic pathological criteria (fatty liver: Ia, alcoholic hepatitis: Ib, nodular cirrhosis: Ic by liver biopsy findings, mild liver damage: IIa by clinical criteria, chronic liver damage without portal hypertension: IIb, cirrhosis, e.g., IIc), and their serum carnitine concentrations were compared with those of 49 healthy subjects (Table 15.1) [8].

In this way, reports on the kinetics of carnitine in the blood of alcoholic cirrhosis patients have been reported to be significantly different from case to case compared to the healthy control group, except for Rudman's report, but there is often no significant difference, or the value is high, and the concentration of short-chain acylcarnitine and long-chain acylcarnitine is high. Rudman et al. reported a case of a patient with end-stage cirrhosis with cachexia. Similarly, Wennberg et al. reported that they examined the blood carnitine dynamics in 72 patients with protein-energy malnutrition (PEM), sepsis, renal failure, and long-term enteral nutrition with severe nutritional disorders [9]. In this study, plasma free carnitine and total carnitine levels in 10 patients with liver diseases such as hepatic cancer but not cirrhosis were more than twice as high as controls.

Regarding PBC (primary biliary cholangitis, formerly called primary biliary cirrhosis, the name of the disease was changed in 2016), 2 out of 7 patients in the aforementioned report by Böhmer et al. [1], but about 25% of the 28 cases reported by Palombo et al. on carnitine dynamics before and after liver transplantation were PBC patients, and pre-transplant carnitine levels including PBC patients were elevated compared to controls [10]. However, these reports do not provide separately the data of PBC patients and other patients, so we could not give assertive conclusion. Krähenbühl et al. presented data on carnitine concentrations in patients with chronic liver disease, categorized by liver diseases, and reported that blood free carnitine and acylcarnitine concentrations were no different from controls in 15 patients with PBC [11].

There are not too many reports of viral cirrhosis. In a response to the report of Fuller et al. on viral cirrhosis, D'Arienzo et al. examined serum carnitine in 15 patients with hepatitis B and found that it was normal in 5 patients with Child classification A, but elevated in 10 patients with Child classifications B to C [12]. Krähenbühl et al. also studied carnitine dynamics in 79 patients with liver disease, including 22 patients with cirrhosis caused by hepatitis B or hepatitis

Table 15.1 Serum carnitine levels in patients with alcohol-induced liver disease

Carnitine(μmol/L)				
Group	N	Total	Free	Acyl
Ia (Hepatic steatosis)	24	61.1 ± 12.0 *	45.1 ± 10.9	16.0 ± 3.9***
Ib (Alcoholic hepatitis)	15	64.4 ± 13.5 *	61.0 ± 13.8	3.8 ± 2.8
Ic (Micronodular cirrhosis)	16	135.0 ± 45***	96.0 ± 21***	39.0 ± 24***
IIa (Slightly hepatopathy)	20	53.8 ± 10.5	43.5 ± 10.2	10.4 ± 9.7
IIb (Chronic hepatopathy without portal hypertension)	26	56.3 ± 17.9	48.3 ± 15.5	8.0 ± 7.0
IIc (Hepatic cirrhosis)	31	104.0 ± 43***	81.0 ± 30***	23.0 ± 20***
Controls	49	53.3 ± 8.4	47.1 ± 7.7	6.2 ± 2.4

From ref. [8]
SD standard deviation
Mean ± SD, *$P < 0.01$, *** $P < 0.001$: significance of difference from controls

C virus (HCV), and found no significant difference except for significantly higher levels of long-chain acylcarnitines [11]. Eskandari et al. also reported significantly higher serum carnitine concentrations in hepatitis B (26 patients) and hepatitis C (15 patients) virus-infected patients without cirrhosis compared with healthy controls [13]. Aspartate aminotransferase (AST) and alanine aminotransferase (ALT) were also significantly higher in these patients than in healthy controls, suggesting that carnitine as well as AST and ALT may leak from hepatocytes as a result of hepatitis cell damage, leading to the higher levels.

Thus, there are various reports of carnitine concentrations in the blood of cirrhosis patients, but there are many reports that free carnitine concentrations and acylcarnitine concentrations are the same or increased as those of healthy subjects, except for the report by Rudman et al. In particular, there are many reports that short-chain acylcarnitine and long-chain acylcarnitine are increasing, suggesting that abnormalities in carnitine metabolism are occurring.

In order to search for good biomarkers to detect differences between chronic liver diseases such as hepatitis B and cirrhosis and hepatocellular carcinoma, Zhou et al. performed metabolomic analysis of free carnitine and acylcarnitine in the serum of these healthy individuals and patients with chronic liver diseases (hepatitis B and cirrhosis) and hepatocellular carcinoma using LC-MS [14] (Fig. 15.2).

They found that C16 and C18 acylcarnitines were markedly increased in patients with hepatitis B, cirrhosis, and hepatocellular carcinoma, while free carnitine and C2 to C12 short- and medium-chain acylcarnitines were decreased. Thus, in patients with chronic liver disease, not only free carnitine is decreased but also long-chain acylcarnitines are elevated, suggesting that so-called carnitine insufficiency occurs, and mitochondrial dysfunction occurs. This impaired energy production is thought to result in decreased albumin synthesis, hyperammonemia due to decreased ammonia processing by the urea cycle, and severe muscle cramps in patients with cirrhosis, and as described later, the efficacy of carnitine supplementation therapy for these symptoms is reported.

Shiraki et al. studied blood carnitine kinetics in 70 Japanese patients with cirrhosis (8 with hepatitis B, 37 with hepatitis C, 14 alcoholics, 11 other) and the relationship between liver tissue carnitine concentration and blood carnitine concentration in cases where partial biopsy was available (7 with cirrhosis, 5 without cirrhosis, 12 cases in total) [15]. The results showed that the

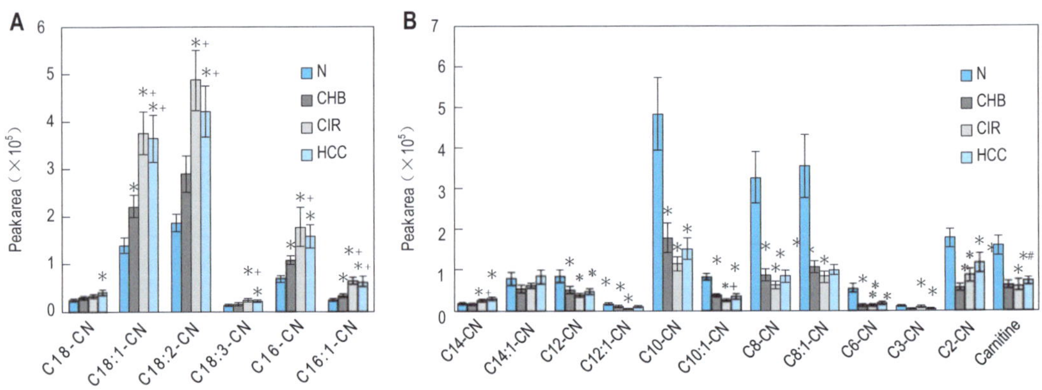

Fig. 15.2 Contents of acylcarnitines and free carnitine in CLD and HCC. (**a**): Acylcarnitines with long chain; (**b**): free carnitine and acylcarnitines with a medium or short chain. *CLD* chronic liver diseases, *CHB* chronic hepatitis B, *HCC* hepatocellular carcinoma, *CIR* hepatic cirrhosis, *N* healthy control. *: Peak area in corresponding liver disease group was significantly different from that in the control ($p < 0.05$), +: Peak area in CIR or HCC was significantly different from that in CHB, #: peak area in HCC was significantly different from that in CIR. (From ref. [14])

serum free carnitine concentration in Japanese patients with cirrhosis was 53.2 ± 2.6 μmol/L, within the normal range, and the total carnitine and acylcarnitine concentrations were also within the normal range. Suzuki et al. also reported that blood free carnitine concentration in cirrhotic patients tended to be in the normal range, while free carnitine and acylcarnitine tended to be higher with increasing severity of disease [16]. Therefore, it is difficult to diagnose the presence or absence of carnitine deficiency in patients with cirrhosis or liver failure based on blood levels alone, and the only way to diagnose carnitine deficiency is by clinical symptoms, clinical signs, and general laboratory findings, as stated in "Guidelines for Diagnosis and Treatment of Carnitine Deficiency 2018." [17]

Table 15.2 Proposed pathogenetic mechanisms of hepatic encephalopathy

Hyperammonemia Direct action of ammonia on the brain Glutamine-glutamate mediated action
Direct influx of neurotoxic substances such as ammonia into the bloodstream via Porto- systemic shunt (PS shunt)
Increase in pseudo-neurotransmitters due to decrease in BCAAs and increase in aromatic amino acids.
Abnormalities in glutamine, monoamine, and GABA metabolism in the brain
Disruption of neuroglia (astrocytes), which support neuronal homeostasis, due to growth, enlargement, and metabolic disorders
In addition, GABA/BZ/Cl^- channel complex theory, neuro-steroids $GABA_A$ receptor action theory, etc.

BCAA branched chain amino acid, *PS* porto-systemic, *GABA* γ-aminobutyric acid, *BZ* benzodiazepines

15.3 Hepatic Encephalopathy and Hyperammonemia in Patients with Cirrhosis

15.3.1 Causes of Hepatic Encephalopathy and Hyperammonemia in Patients with Cirrhosis

Decompensated cirrhotic patients with advanced cirrhosis often develop hyperammonemia and hepatic encephalopathy. The causes include hyperammonemia due to impaired ammonia detoxification in the liver, increased permeability of the blood–brain barrier to neurotoxic substances such as ammonia, abnormalities of metabolism of glutamine, monoamine, and γ-aminobutyric acid (GABA) due to elevated levels of aromatic amino acids, and abnormalities in brain neural transmission and collapse of defense mechanisms due to hyperplasia, swelling, and metabolic abnormalities of neuroglial cells (astrocytes) that maintain neuronal homeostasis, but there is no established theory yet (Table 15.2). However, ammonia concentration and the development of encephalopathy are not necessarily parallel. For example, some patients develop encephalopathy at a blood ammonia concentration of 80 μg/dL, while others do not develop encephalopathy even if their blood ammonia concentration rises to nearly 200 μg/dL. Thus, in addition to hyperammonemia, various other factors, such as the state of brain energy metabolism and vulnerability of astrocytes, may be involved in the development of hepatic encephalopathy, but elevated ammonia concentration is considered the most important factor [18, 19].

In patients with cirrhosis, various factors are thought to contribute to hyperammonemia, including decreased ammonia detoxification by the urea cycle due to the reduction and dysfunction of liver parenchymal cells caused by progressive liver fibrosis, ammonia generated in the intestinal tract flows directly into the bloodstream through a portosystemic shunt (PS shunt) without passing through the liver, and a decrease in muscle mass reduces the metabolic reaction that fixes ammonia to glutamine in skeletal muscle and temporarily suppresses the rise of ammonia, insufficient zinc intake due to protein-intake restriction therapy, and carnitine deficiency (Table 15.3).

In this book, I focus on the urea cycle localized in the liver which is the most important mechanism for processing ammonia in the body generated by amino acid metabolism and intestinal bacteria, and carnitine deficiency.

Cirrhosis is a chronic inflammatory process caused by hepatitis C or B virus, alcohol, and fatty liver, resulting in a decrease in liver paren-

Table 15.3 Causes of hyperammonemia in patients with cirrhosis

Reduction of hepatic parenchymal cells due to liver fibrosis
Decreased ammonia detoxification function by urea cycle in the liver
Decreased urea cycle activity due to chronic lack of energy (ATP)
Decreased acetyl-CoA production due to carnitine deficiency and decreased NAGA synthase activity due to accumulated long-chain acyl-CoA
⇒ Decrease in the positive cofactor NAGA ⇒reduced CPS1 activity
Zinc deficiency due to protein intake restriction ⇒ reduced OTC activity
Ammonia produced in the intestinal tract passes through the portosystemic shunt bypass the liver, so ammonia is not processed by the liver, but flows directly into the bloodstream and migrates to the brain
The following ammonia metabolism reactions in skeletal muscle are reduced due to loss of muscle mass
Ammonia + glutamic acid ⇒ glutamine

CPS1 Carbamoyl phosphate synthase 1, *NAGA* N-acetylglutamic acid, *OTC* Ornithine transcarbamylase

chymal cells and fibrosis over a period of 20–30 years, and a decrease in liver parenchymal cells that play a role in ammonia detoxification. The decrease in liver parenchymal cells is particularly prominent in decompensated cirrhosis, which is the last stage of cirrhosis.

In ammonia metabolism, the urea cycle in the liver converts ammonia to urea and excretes it in urine from the kidneys, which is the most important mechanism for ammonia detoxication. As described in detail in Fig. 15.6 in Chap. 5, the urea cycle is an energy-consuming metabolic system that requires a total of three molecules of ATP to fix one molecule of ammonia into one molecule of urea. On the other hand, in patients with decompensated cirrhosis, carbohydrate (glucose) metabolism is decreased, muscle mass is reduced, and glycogen stores in the muscle are also decreased. The metabolic system shifts to one that uses fatty acids to produce energy, and the patient is said to be in a state of chronic systemic energy deficiency, as if a normal person had fasted for 3 days [20]. Since albumin synthesis, an important function of the liver, also requires a large amount of energy (ATP), the low-energy state of cirrhotic patients is the basis for hypoalbuminemia due to decreased ammonia detoxification and decreased albumin synthesis. In addition, carnitine deficiency in cirrhotic patients results in decreased acetyl CoA production and accumulated long-chain acylcarnitines inhibit N-acetylglutamate synthase (NAGS) activity. These factors decrease N-acetylglutamate (NAGA), a positive co-factor for carbamoyl phosphate synthase 1 (CPS1), which is the most important primary enzyme in the urea cycle and fixes ammonia and synthesize carbamoyl phosphate, resulting in the decreased activity of the urea cycle. These findings suggest that urea cycle activity is greatly reduced in patients with cirrhosis. Patients with cirrhosis have been recommended to restrict protein intake to prevent hyperammonemia and are frequently deficient in zinc, which is abundant in meat and seafood, due to anorexia and other factors [21, 22]. Ornithine transcarbamylase (OTC) is a zinc enzyme, and zinc is essential for its activity. Therefore, patients with zinc deficiency are prone to hyperammonemia due to decreased urea cycle activity.

In addition, in patients with cirrhosis, PS shunts are formed due to portal hypertension, and ammonia produced in the intestinal tract does not pass through the liver but enters the general circulation and is transported into the brain, making them susceptible to encephalopathy [23]. Patients with shunts larger than 10 mm in diameter are more likely to develop hepatic encephalopathy and variceal hemorrhage [24]. In a retrospective study of approximately 2000 patients with cirrhosis, 28% of patients had shunts greater than 8 mm in diameter, 32% had shunts less than 8 mm, 60% of patients overall had PS shunts, and more patients retained larger shunts as the severity of cirrhosis increased. Furthermore, it has been reported that the prognosis for patients with large shunts is poor and that patients with large shunts often develop hepatic encephalopathy and ascites [25]. In addition, since cirrhotic patients often have multiple shunts rather than a single shunt, the same group recently measured the cross-sectional area of multiple shunts to examine the relationship between total shunt area and prognosis and found that patients with a total shunt area of 83 mm^2 or more had a worse prognosis than those with a total shunt area of 83 mm^2 or less [26]. Therefore, treatment of hepatic encephalopathy often requires shunt embolization as well as treatment of hyperammonemia [27].

In addition, branched chain amino acids (BCAA) such as valine, leucine, and isoleucine are used for energy production, ammonia detoxification, and albumin synthesis in skeletal muscle, and aromatic amino acids such as tyrosine and phenylalanine are increased, resulting in amino acid imbalance such as a decrease in c and BTR (blood BCAAs/tyrosine). This often causes clinical symptoms such as ascites due to hypoalbuminemia, muscle weakness due to skeletal muscle depletion, and general malaise and fatigue, and albumin and BCAA preparations are widely administered. These causes decrease skeletal muscle mass in patients with cirrhosis, resulting decrease the response of one of the ammonia disposal mechanisms in skeletal muscle to fix ammonia with glutamate as glutamine.

15.3.2 Protective Effect of Carnitine in Animal Models of Hepatic Encephalopathy

O'Connor et al. in Grisolía's group in Spain found that in a hyperammonemia model in which mice were administered a lethal dose of ammonium acetate (12 mmol/kg) intraperitoneally, administration of L-carnitine 30 min before administration of ammonium acetate reduced the toxicity of ammonia by 1 mmol/kg. It was shown that 100% of mice survived at 16 mmol/kg [28] (Table 15.4). It was suggested that the protective effect of carnitine against ammonia toxicity is due to the promotion of urea production and the reduction of ammonia concentrations in the blood and brain. Costell et al., in the same group, investigated the initial reaction of the urea cycle, in which bicarbonate ions labeled at ^{14}C are combined with ammonia and carbamoyl phosphate to be incorporated into urea by the CPS1 reaction, and showed that carnitine does not activate NAGS, but promotes the CPS1 response itself, increasing ^{14}C-urea production [29].

Table 15.4 Protective effect of L-carnitine on acute ammonia intoxication in mice

L-Carnitine (mmol/kg)	Mice injected (n)	Mice dead (*n*)
0	10	10
1	10	4
2	10	3
8	10	2
16	10	0
32	10	0

The animals received the indicated doses of L-carnitine in a single intraperitoneal injection, then 30 min later they were given intraperitoneally 12 mmol/kg body weight of ammonium acetate. (From ref. [28])

In addition, administration of ammonium acetate lowered glutamate in the brain and increased glutamine production, but carnitine administration reduced these changes [30]. When L-carnitine was administered intraperitoneally, intravenously, intramuscularly, and subcutaneously, the protective effect of ammonia toxicity was highest when administered intraperitoneally, in this case carnitine accumulated in the brain, muscles, and blood rather than in the liver, but in the case of intravenous administration with a low protective effect, the most carnitine accumulated in the liver [31]. From these studies, it was thought that carnitine exerts a protective effect against ammonia toxicity by promoting urea production in the liver and decreasing glutamine production in the brain. Costell et al. also examined the concentrations of free carnitine, acetylcarnitine, and long-chain acylcarnitine in the liver, muscles, and brain in a milder and more persistent hyperammonemia model in which urease was administered intraperitoneally. It was reported that free carnitine in the brain remained unchanged and acetylcarnitine increased [32]. They found that intraperitoneal administration of carnitine in this model suppressed the elevation of free carnitine and long-chain acylcarnitine in the liver, brain, and muscles. In other words, although the mechanism is unknown, it is thought that hyperammonemia itself affects carnitine metabolism such as acylation and distribution in various tissues. In addition, abnormalities in lipid metabolism are often reported in pathologies that cause hyperammonemia, such as Reye syndrome and urea cycle disorder, and hyperammonemia itself may cause abnormal carnitine metabolism and lipid accumulation. The protective effect of carnitine in a mouse model of hyperammonemia is thought to be mainly due to its ability to enhance urea production by promoting fatty acid oxidation and tricarboxylic acid (TCA) cycle, based on biochemical studies [33].

Kloiber et al. examined the protective effects of trimethylamine-N-oxide (TMAO), choline, and betaine, which also have quaternary amine structures and are known osmotic regulators, in addition to carnitine, in a hyperammonemia model of ammonium acetate loading (intraperitoneal administration, 12 mmol/kg, 0.8 M solution), because carnitine is a quaternary amine compound [34]. Similar to the report by O'Connor et al., ammonium acetate-only control mice died 100%, whereas 16 mmol/kg of intraperitoneal carnitine intraperitoneal administration of 16 mmol/kg carnitine resulted in 100% survival. In addition to this, 8 mmol/kg TMAO also resulted in 100% of mice survival, 1.6 mmol/kg choline in 87%, and 6 mmol/kg betaine in 67%. Surprisingly, 33% of mice survived by saline administration. These findings suggest that osmotic and other changes may be involved in the protective effects of these compounds against ammonia toxicity.

However, Deshmukh et al. reported that L-carnitine administration had no protective effect in ferret and mouse models of hyperammonemia treated with ammonium acetate, similar to O'Connor et al. [35] They also reported no effect in a model of urease administration [36].

Hearn et al. prepared a model of hepatic encephalopathy in which PS shunts were formed in rats by portal vein-vena cava apical anastomosis and examined the protective effect of L-carnitine on ammonia toxicity by O'Connor et al. [37] Intraperitoneal administration of L-carnitine (16 mmol/kg) 1 h before intraperitoneal administration of ammonium acetate (12 mmol/kg) exerted a protective effect, but experimental variation was remarkable. Namely, 5/5 (100%) survived in Experiment 1, 6/7 (86%) and 5/5 (100%) survived in Experiment 2, and 6/17 (35%) survived in Experiment 3. In Experiment 3, carnitine administration also extended the time to onset of clonic and tonic convulsions. Such protective effects were not observed with D-carnitine and deoxycarnitine, so that the effects of osmotic pressure by osmotic pressure regulators such as carnitine and betaine suggested by Kloiber et al. were denied. In addition, it was found that the protective effect of L-carnitine on ammonia toxicity was very short-lived, as administration of carnitine 2.5 h before administration of ammonium acetate prolonged the time to death by more than 2 times but eventually death, and when carnitine was administered 24 h before administration of ammonium acetate, the protective effect disappeared. In addition, hyperammonemia was observed when PS shunts rats were bred for 8, 12, and 16 weeks, but L-carnitine administration did not reduce this blood ammonia. Since ammonia toxicity is mainly a neurological symptom, the uptake of carnitine into the brain was examined using labeled carnitine, and its uptake was due to slow simple diffusion and was not parallel to the protective effect against ammonia toxicity. Therefore, it was inferred that the protective effect of carnitine was not exerted through its action on the brain, but rather by its peripheral effect. The concentrations of free carnitine, short-chain acylcarnitine, long-chain acylcarnitine, and total carnitine in the blood were found to be significantly lower in PS shunts rats than in normal rats.

Ohtsuka et al. also confirmed the protective effect of L-carnitine on ammonia toxicity in an ammonium acetate loading model in mice, and reported that D-carnitine and deoxycarnitine, which have quaternary amine structures, had no protective effect [38]. In addition, administration of hyperosmolar saline and sucrose solution has a protective effect, although not as protective as L-carnitine, and it is thought that the effect affects the absorption of ammonium acetate and reduces edema in the brain.

Matsuoka et al. used the freeze-clamp method to investigate the effects of saline, 10 mmol/kg and 20 mmol/kg L-carnitine on ammonia toxicity and the content of high-energy compounds in the brain 30 min before intraperitoneal administration of 15 mmol/kg ammonium acetate to mice [39]. The survival rates after 10 min of ammonium acetate administration were 80.0% in the saline group and 90.9 and 90.5% in the 10 mmol/kg and 20 mmol/kg carnitine groups, but the time to seizure and the number of convulsive seizures improved significantly in the carnitine group as shown in Fig. 15.3. Ammonia concentrations in the brain increased significantly by 12.4 times with ammonium acetate compared to controls but were significantly reduced by about 43 and 48% with carnitine. In addition, ammonium acetate administration significantly reduced phosphocreatine and ATP in the brain, as well as energy charge potential (ECP), which is an indicator of intracellular energy status, but carnitine administration prevented these decreases and improved energy metabolism in the brain. In addition, the concentration of ammonia in the blood was significantly increased by administration of ammonium acetate ($0.06 \pm 0.00 \Rightarrow 4.18 \pm 0.18$ µmol/mL), but the increase was significantly suppressed by administration of carnitine at 10 and 20 mmol/kg (2.48 ± 0.16 and 2.36 ± 0.16 µmol/mL).

After a while, Therrien et al. in Butterworth's group in Canada, well known for their long-standing research on the mechanism of ammonia toxicity in the field of hepatic encephalopathy, reported in an experimental system that L-carnitine (16 mmol/kg) was administered intraperitoneally, and ammonium acetate (8.5 mmol/kg) was administered subcutaneously to exclude the effects on osmotic pressure and absorption, and it was found that L-carnitine had a protective effect against acute ammonia toxicity [40] (Table 15.5). In this study, ammonia concentrations in the cerebrospinal fluid were significantly increased by administration of ammonium acetate to 0.93 ± 0.15 mmol/L in the precoma stage and 1.24 ± 0.15 mmol/L in the coma stage ($P < 0.01$) but were significantly reduced by pre-carnitine administration ($P < 0.01$), which was parallel to the protective effect on neurological symptoms

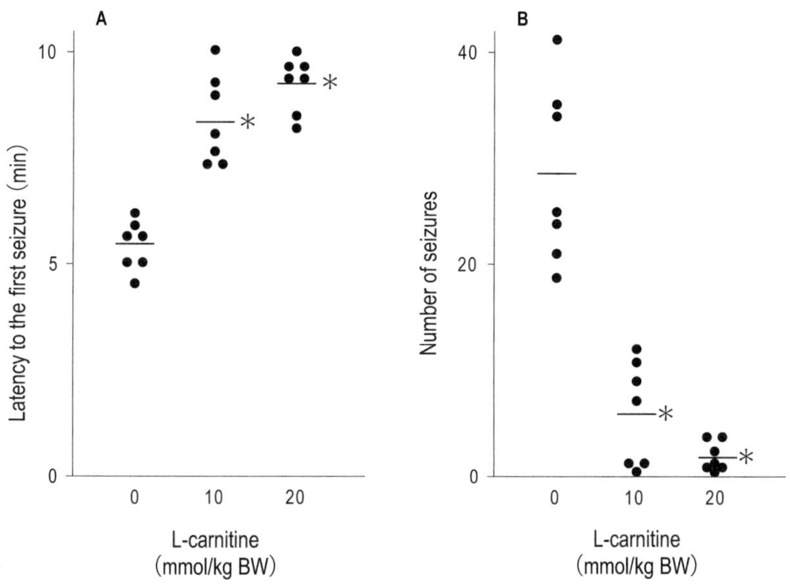

Fig. 15.3 Effects of L-carnitine on the latency to the first seizure (min, **a**), and number of the seizures during 10 min observation (**b**) Each mouse was injected with saline solution with or without L-carnitine and then with 15 mmol ammonium acetate/kg body weight. Bars indicate the mean values of each group ($n = 7$) and an asterisk significant difference ($P < 0.01$) from the untreated group (0 mmol/L-carnitine/kg body weight), (Wilcoxon test). (From ref. [39])

Table 15.5 Effect of L-carnitine pretreatment on the incidence of ammonia-induced encephalopathy and on mortality in PCS rats

Pretreatment	Loss of righting reflex (Precoma)	Loss of corneal reflex (Coma)	Mortality
Saline	36/36	36/36	21/22
L-carnitine	0/35[a]	0/35 [a]	0/20 [a]

From ref. [40]

L-carnitine (16 mmol/kg) or saline was administered intraperitoneally 60 min before ammonium acetate (8.5 mmol/kg, subcutaneously) to precipitate a reproducible sequence of neurological impairment progressing through loss of righting reflex (precoma) to loss of corneal reflex (coma). Mortality was assessed in a separate series of 22 animals. L-carnitine afforded complete protection and reduced mortality rate to zero

[a]: $P < 0.001$ vs. saline (Fisher's exact test), PCS rats: portacaval shunted rats

in encephalopathy. There was no change in plasma ammonia concentration (Fig. 15.4). In addition, carnitine suppressed the rise of lactic acid, alanine, and taurine in the cerebrospinal fluid during this coma, but did not affect the increase in glutamine. Glutamic acid and aspartic acid in cerebrospinal fluid did not change significantly. The concentration of carnitine in the cerebrospinal fluid was below the detection limit in the saline group, but it was significantly higher in the precoma and coma groups at 24.4 ± 12.9 and 31.4 ± 9.6 mmol/L, respectively ($P < 0.01$). These results showed that pre-administration of L-carnitine clearly had the effect of preventing ammonia-induced precoma and coma in PS shunts rat, and also had the effect of completely suppressing the fatality rate. The effect of carnitine was thought to be mainly to improve mitochondrial respiration in the centrally-mediated mechanism or by the action at the blood–brain barrier [40].

Thus, after O'Connor and Costell in Grisolía's group reported the protective effect of L-carnitine against ammonia toxicity in an in vivo animal model, many studies were reported, and as mentioned above, there was a period of confusion in the evaluation of the protective effect of carnitine against ammonia toxicity, but the study by Therrien et al. reaffirmed that L-carnitine has a protective effect against ammonia toxicity.

15.3.3 Ammonia Toxicity at the Cellular Level and Effects of Carnitine

In parallel with these studies in animal models of hyperammonemia, many studies have been published on ammonia toxicity at the cellular level and the protective effects of carnitine.

Maddaiah found that ammonia inhibits state 3 respiration (oxygen consumption when adenosine diphosphate (ADP) is added) of isolated liver mitochondria of rats with palmitic acid and octanoic acid as substrates, but not oxygen consumption with succinic acid or malic acid + glutamic acid as substrates [41]. Although ammonia does not inhibit the respiratory chain itself, it has been pointed out that it may inhibit some steps in the β oxidation of fatty acids, and it has been pointed out that it is related to ammonia toxicity in Reye syndrome which is common in children and urea cycle disorder.

Kosenko et al. reported that injection of ammonia into rats depleted brain ATP via activation of N-methyl-D-aspartate (NMDA) receptors [42]. Felipo et al. in the same group found in a series of studies using cultured neurons in the brain that ATP deprivation in brain neurons by ammonia administration is due to the activation of NMDA-type glutamate receptors, and that carnitine has the effect of preventing ATP deprivation, and that the mechanism of carnitine's protective action is the metabotropic glutamate receptor (mGluR, quisqualate receptor), which weakens the binding of glutamate to NMDA-type glutamate receptors [43–48].

Miñana et al. reported that carnitine reduces ammonia toxicity via activation of mGluR [45], and Koh et al. reported that activation of mGluR by trans-1-aminocyclopentyl-1,3-dicarboxylic acid (t-ACPD; a selective agonist of mGluR) reduces neurotoxicity of ammonia via NMDA receptors [49]. Albrecht introduced these studies by that time on the mechanism why glutamine receptor is overactivated by ammonia, and explained that when ammonia inhibits mGluR, the NMDA receptor becomes uncontrollable and ammonia further enhances NMDA receptormediated ammonia-induced neurotoxicity in his review [50].

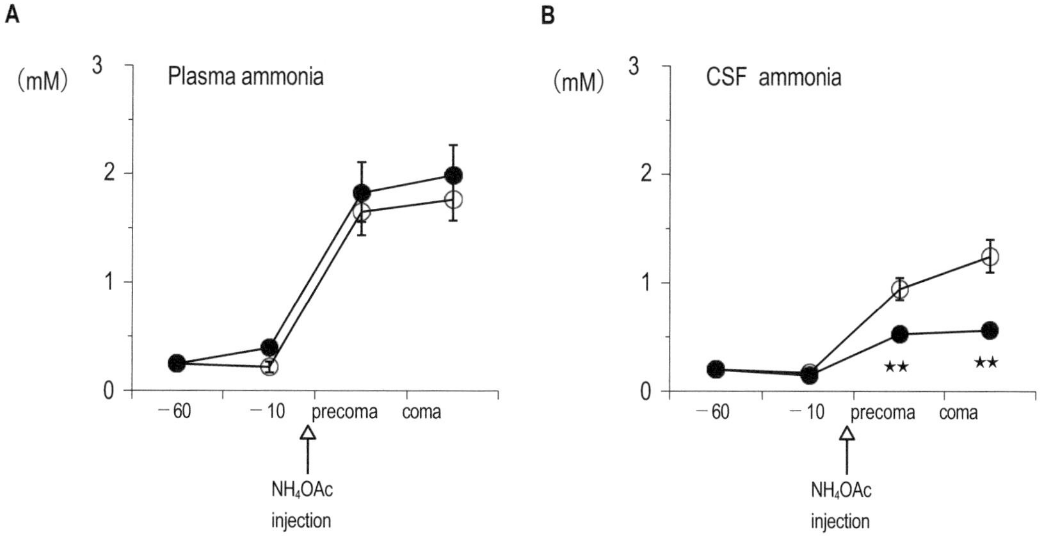

Fig. 15.4 Effect of L-carnitine (16 mmol/kg, i.p.) (●) and vehicle (○) on plasma (**a**) and CSF (**b**) concentration of ammonia in PCS rats administered ammonium acetate (NH_4OAc) to precipitate severe sign of encephalopathy. Precoma stage defined as stage at which animals lost their righting reflex; coma stage defined as loss of corneal reflex. Data points represent mean ± SD of duplicate determination from $n = 6$ per treatment group. Values significantly different from vehicle-injected controls indicated by ★★. $P < 0.01$ by Student's t-test with Bonferroni correction. CSF *cerebrospinal fluid*, *PCS rats* Portacaval shunted rats. (From ref. [40])

Norenberg et al. reported that the addition of glutamine, which is generated by the binding of ammonia to glutamate, induces the production of free radicals in astrocytes, leading to the mitochondrial permeability transition (MPT), a phenomenon in which cytochrome c is released through the opening of pores in the mitochondrial membrane whose permeability is originally controlled, resulting in mitochondrial swelling and cell apoptosis, causing the mitochondria in astrocytes to collapse and astrocytes undergo apoptosis and die, suggesting that glutamine plays the role as the "Trojan horse" in the Trojan War in Greek mythology [51]. Albrecht and Norenberg, who had been studying the role of glutamine and glutamate in the toxicity of ammonia in the brain, hit it off and jointly proposed the "Trojan horse" hypothesis, suggesting that ammonia combines with glutamate to form glutamine, which is temporarily detoxified [52, 53]. In other words, glutamine synthetase in the cytoplasm of astrocytes combines ammonia and glutamic acid to produce glutamine, which appears to reduce ammonia toxicity, but the glutamine is transported into the mitochondria by the mitochondrial glutamine transporter and is then converted to phosphate-activated glutaminase (PAG) that again produces glutamate and ammonia, which in turn generate free radicals (reactive oxygen species: ROS). When the amount of ammonia is high and this reaction occurs in excess, a large amount of ROS is produced, opening a permeation transition pore (PTP) in the mitochondrial membrane, from which cytochrome c is released, the mitochondria swell and collapse, and the astrocytes die by apoptosis. Namely, the Trojan horse (glutamine), which was thought to be safe, is transported into the castle of Elios (inside the mitochondria), and back to ammonia and glutamate again, which correspond to the Greek soldiers hiding inside the horse, are produced again, and ROS is produced to rampage around and kill the Troj (astrocytes) This mechanism is shown in Fig. 15.5.

Albrecht and Norenberg hypothesized that the conversion of ammonia to glutamine in brain cells such as astrocytes does not eliminate ammonia toxicity, but the same mechanism may apply to the idea that glutamine production in skeletal muscle reduces ammonia toxicity. In other words, it is unclear whether PAGs are present in the mito-

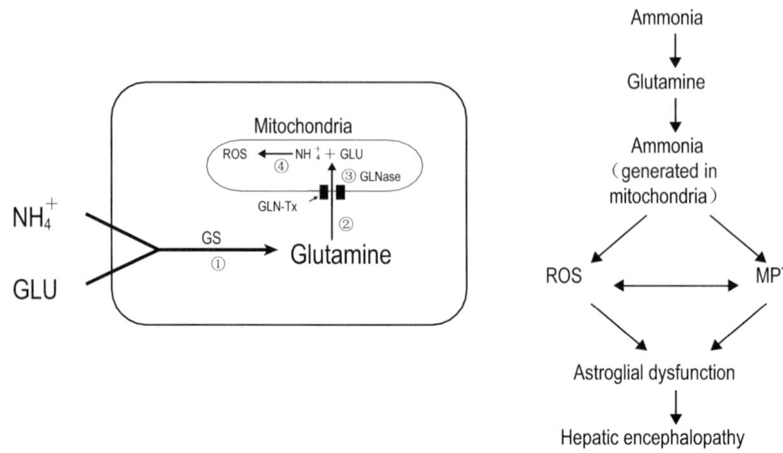

Fig. 15.5 The "Trojan horse hypothesis" in neurotoxicity of ammonia. Left: Diagrammatic sketch illustrating the synthesis of glutamine (①); its transport into mitochondria via the glutamine carrier (②); its hydrolysis by phosphate-activated glutaminase (PAG) resulting in glutamate (GLU) and ammonia (③); the generation of reactive oxygen species (ROS) from ammonia (④). Right: General scheme by which ammonia and glutamine lead to mitochondrial failure resulting in astrocyte dysfunction and clinical development of hepatic encephalopathy. *GS* glutamine synthetase, *GLU-Tx* glutamine transporter, *MPT* mitochondrial permeability transition, *ROS* reactive oxygen species. (From ref. [52])

chondria of skeletal muscle cells, but if they are present, the mitochondria of skeletal muscle cells are also damaged by oxidative stress, which leads to a decrease in energy production, leading to cramps and muscle weakness, and if skeletal muscle cells undergo apoptosis, sarcopenia, which is common in patients with cirrhosis, is thought to be promoted. As reported by Maddaiah, ammonia inhibits the β oxidation of long-chain fatty acids [41]. Then, medium-long chain acyl-CoA such as palmitoyl-CoA is thought to accumulate in the mitochondria. In fact, Zhou et al. reported high levels of C16 and C18 long-chain acylcarnitine in the blood in patients with chronic liver disease, including those with cirrhosis [14]. It is thought that free carnitine was used to eliminate palmitoyl-CoA accumulated in the mitochondria, and palmitoyl carnitine increased in the blood. Among MPT researchers, long-chain acyl-CoA, such as palmitoyl-CoA, is known to induce MPT in mitochondria, which in turn causes cell death [54]. Furuno et al. also reported that the addition of palmitoyl-CoA to rat liver mitochondria inhibited respiration and the release of cytochrome c, but carnitine was effective in blocking this effect [55]. In particular, MPT in the hepatic mitochondria is important for understanding the mechanism of hepatotoxicity of Reye syndrome and drugs [56]. Furthermore, it has been reported that carnitine prevents the hyperpermeability of the mitochondrial membrane due to palmitoyl-CoA and oleic acid, which has been reported in myocardial mitochondria [57, 58]. In other words, mitochondrial dysfunction caused by MPT and the resulting cell apoptosis and fibrosis are considered to be common mechanisms for organ fibrosis such as liver fibrosis and cardiomyopathy. Carnitine's action mechanism is thought to be mainly excretion of long-chain acyl-CoA from mitochondria, the suppression of apoptosis by direct inhibition of caspase, and the suppression of apoptosis by attenuation of oxidative stress and inflammatory cytokines via Nrf2. Further basic research and clinical application of carnitine to suppress MPT and cell death, which are the causes of various pathologies such as liver fibrosis and fatty liver, are expected in the future.

15.3.4 Effect of Carnitine on Hepatic Encephalopathy in Patients with Cirrhosis

15.3.4.1 Overseas Clinical Research

Malaguarnera et al. reported that serum carnitine concentrations (4.75 ± 1.5 mg/L) in 40 patients with hepatitis C were significantly lower ($P = 0.01$) than in 20 healthy controls (7.71 ± 0.92 mg/L), and that treatment with interferon α for 6 months was 5.9 ± 1.3 mg/L, a significant increase ($P = 0.0003$) compared to before treatment [59]. Patients' carnitine levels were negatively correlated with duration of disease ($r = -0.85$, $P = 0.00001$) and positively correlated with response to interferon α treatment ($r = 0.3$, $P = 0.04$). These results suggest that the concentration of free carnitine in the blood may be a new indicator of treatment response in patients with hepatitis C, and that it may be meaningful to supplement carnitine because hepatitis C may reduce the biosynthesis of carnitine in the liver.

In addition, Malaguarnera et al. examined the effects of carnitine administration on cirrhotic patients with mild and moderate hepatic encephalopathy, since basic studies using mice and rats suggest that carnitine may be effective in hyperammonemia in patients with cirrhosis as described above [60]. 120 randomely selected patients met the following inclusion criteria and were enrolled in this study: (1) cronic hepatitis with spontaneous manifest HE (mental state grade 1or 2according to West Haven criteria) and number connection test-A (NCT-A) performance time>30 s; (2) hyperammonemia (concentration of venous ammonia>50mmol/L,(3)cooperative,hospitalized adult patients with liver cirrhosis diagnosed clinical, histological and ultrasound findings. Patients were divided into two groups: carnitine (60 patients) and placebo (60 patients). Carnitine was administered orally at 2 g × 2 per day for 60 days. As a result, blood ammonia concentration and NCT-A time were significantly reduced in the carnitine group, but no decrease was observed in the placebo group (Table 15.6).

Similar to this clinical trial, Malaguarnera et al. conducted a randomized, double-blind, controlled trial in which 150 patients with cirrhosis with grade 0, grade 1, and grade 2 hepatic encephalopathy were randomized to receive carnitine and placebo in the West Haven criteria for minimal hepatic encephalopathy [61]. The etiology of cirrhosis patients in the study was alcoholic in 10 cases, hepatitis B in 41 cases, hepatitis C in 78 cases, and idiopathic disease in 21 cases. All patients took more than 30 s to complete NCT-A. The carnitine group received 2 g twice daily for 90 days. As a result, the concentration of ammonia in venous plasma on an empty stomach was significantly reduced compared to the predose value at 30, 60, and 90 days in the carnitine treatment group, and significantly decreased at 90 days compared to the placebo group (Fig. 15.6). In addition, in the trail making test-A, there was a significant decrease in the pre-dose value at 30, 60, and 90 days in the carnitine group, and a significant decrease at 90 days compared to the placebo group (Fig. 15.7). The same effect was observed in trail making test-B. In the symbol digit modalities test and block design test, there was a significant increase in scores in

Table 15.6 Effect of carnitine administration on cirrhotic patients with mild and moderate hepatic encephalopathy

		Carnitine group		Placebo group		
0 days		30 days	60 days	0 days	30 days	60 days
NH_4^+ fasting (mmol/L) HE1	80.2 ± 36.9	61.8 ± 29.2	41.8 ± 18.7	80.7 ± 34.7	76.4 ± 31.2	75.7 ± 32.8
HE2	88.7 ± 35.6	60.8 ± 31.4	45.6 ± 17.5	86.7 ± 37.2	81.4 ± 30.6	80.2 ± 31.4
NCT-A (sec) HE1	62.8 ± 18.1	50.1 ± 17.7	38.6 ± 17.4	60.7 ± 20.3	58.7 ± 19.4	56.7 ± 20.7
HE2	66.4 ± 20.2	51.5 ± 18.4	39.2 ± 18.5	65.8 ± 21.4	65.1 ± 19.7	63.2 ± 23.1

From ref. [60]
Carnitine group: HE1; 35 cases, HE2; 25 cases, Placebo group: HE1; 33 cases, HE2; 27 cases
NCT-A number connection test-A, *HE* hepatic encephalopathy, *HE1* grade 1 of West Haven criteria, *HE2* grade 2 of West Haven criteria

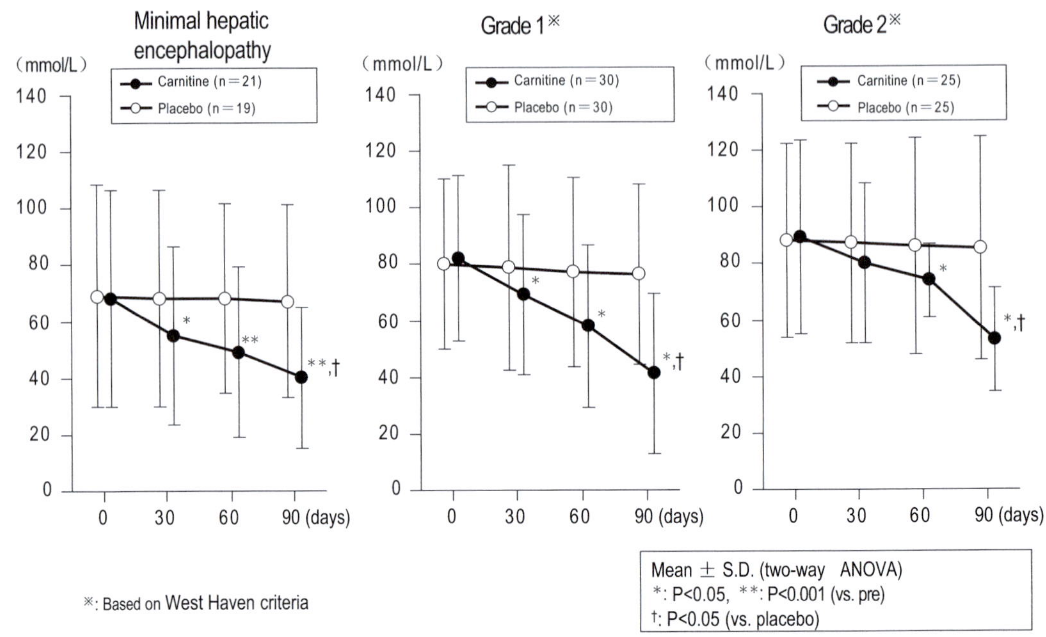

Fig. 15.6 Change in plasma ammonia level after carnitine or placebo administration in cirrhotic patients with hepatic encephalopathy. L-carnitine was administered orally (2 g, 2 times a day) for 90 days. (Created from Table 15.2. in ref. [61], with permission)

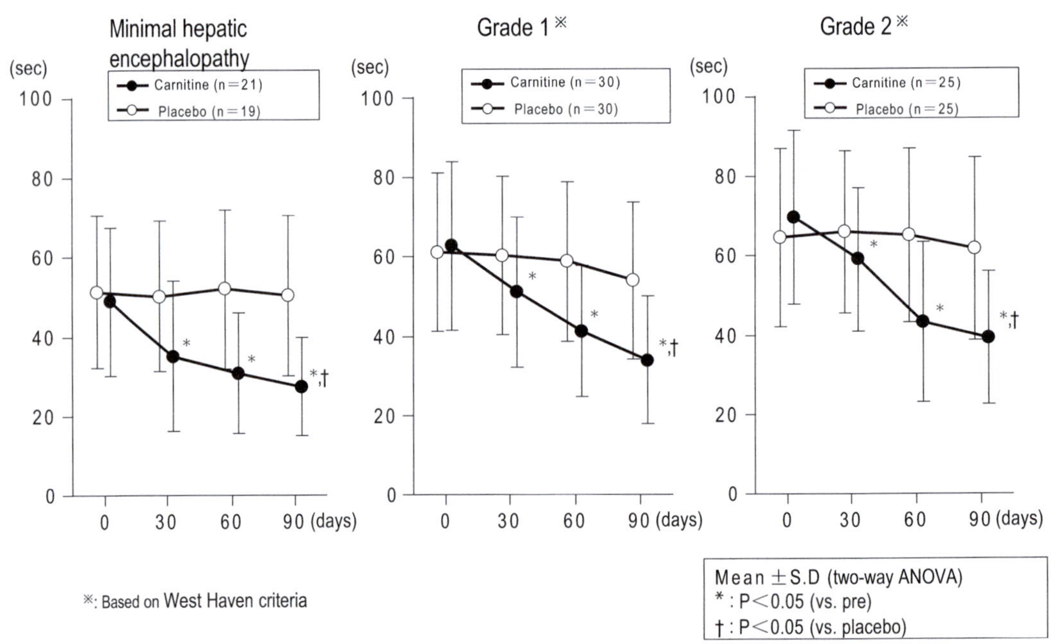

Fig. 15.7 Change in score of trail making test-A after carnitine or placebo administration in cirrhotic patients with hepatic encephalopathy. L-carnitine was administered orally (2 g, 2 times a day) for 90 days. (Created from Table 15.2. in ref. [61], with permission)

the carnitine group at 30, 60, and 90 days compared to the change in the placebo group, indicating an improvement in encephalopathy [61].

Subsequently, Malaguarnera et al. vigorously studied and reported the effects of acetyl-L-carnitine, which are a derivative of carnitine and is easily transferred to nerves and brain, on depressive symptoms, quality of life (QOL), and fatigue in patients with minimal and overt hepatic encephalopathy [62–65].

15.3.4.2 Clinical Studies of Carnitine in Patients with Liver Cirrhosis in Japan

After the aforementioned report of the effect of carnitine administration on patients with hepatic encephalopathy by Malaguarnera et al., clinical studies of carnitine administration to patients with liver cirrhosis were discontinued for a while. However, in March 2011, the indication for L-carnitine preparations L-Cartin® tablets was expanded in Japan to patients with carnitine deficiency, and several studies have been reported.

Kim reported in detail the course of two cases in which carnitine was administered to patients with refractory hyperammonemia in cirrhosis that could not be adequately controlled by existing treatments [66]. Both of the patients were men in their 70 s with type C compensated cirrhosis with ascites, jaundice, and liver cancer, and their blood ammonia levels were high at 435 and 192 μg/dL, and hepatic encephalopathy was repeated. When carnitine administration was discontinued, ammonia levels rose again, and when carnitine was re-administered, it decreased. In addition, since there are patients who do not show disorientation even if the ammonia level is extremely high, it is important not only to measure ammonia but also to observe symptoms in patients with hepatic encephalopathy.

Tamaki et al. reported the results of administering carnitine to seven patients with hepatic encephalopathy that could not be adequately controlled with lactulose or BCAA preparations [67]. Three patients were suspected of having minimal hepatic encephalopathy, four patients had poor control after the acute phase, and two of them were admitted to the hospital after recovery of consciousness after a visit to the emergency department due to grade III hepatic encephalopathy. After oral administration of carnitine at 900 mg/day (300 mg, 3 times daily), the mean blood ammonia level, which was initially 151.6 ± 18.9 μg/dL, significantly decreased to 97.6 ± 13.6 μg/dL after 30 days and to 91.4 ± 14.4 μg/dL after 60 days ($P < 0.015$), and a significant decrease was observed after BCAA infusion. The patient no longer had exacerbations requiring intravenous BCAA infusion and was able to receive outpatient treatment with weekly infusion. There was also a case in which encephalopathy improved, although ammonia levels did not decrease.

Oba administered carnitine (300–900 mg/day) to a total of seven patients (including six with hepatitis C and one with hepatitis B) who had undergone hepatic cancer resection (six patients) and splenectomy (one patient) due to splenic hyperfunction, etc., in liver surgery and investigated its effects on blood ammonia and muscle cramps [68]. The results showed that ammonia levels decreased in three of four patients with recurrent hyperammonemia and hepatic encephalopathy, one patient complained of lightheadedness and discontinued treatment, and one patient had recurrent hepatic encephalopathy triggered by urinary tract infection after rectal cancer surgery, but the frequency of such symptoms clearly decreased after carnitine administration. The frequency and severity of muscle cramps in three patients who had complained of muscle cramps became less frequent and less severe.

Murohisa investigated the effect of carnitine administration in eight patients with decompensated liver cirrhosis with repeated hepatic encephalopathy (5 with HCV, 1 with HBV, 1 with alcoholism, and 1 with other conditions, 3 of whom had hepatocellular carcinoma and 2 were on dialysis due to chronic renal failure) [69]. The median initial dose was 300 mg (200–900 mg), and the median maximum dose was 1200 mg (600–1800 mg). Concomitant medications included lactulose (100%), BCAA preparations (87.5%), kanamycin (50%), and zinc (polaprezinc, 37.5%). The results showed that after 2 weeks of carnitine administration, blood ammonia levels decreased significantly (from 118.1 to 63.6 μg/dL, author's note: the graph in the original report shows ammonia concentration in μg/mL, which may be an error) and the Child-Pugh score also decreased significantly (from 10.6 ⇒ 9.1). Patients who had repeated episodes of encephalopathy, which could not be controlled with existing medications, experienced almost no encephalopathy. All the patients experienced muscle cramps, but carnitine was effective in at least 50% of the patients judged by interview.

Mori et al. reported two cases of hepatocellular carcinoma complicated by decompensated cirrhosis in which hepatocellular carcinoma treatment was abandoned due to hepatic encephalopathy, but hepatocellular encephalopathy could be controlled by carnitine administration, and hepatocellular carcinoma treatment was possible [70]. Case 1 was a 70-year-old male patient with cirrhosis and hepatocellular carcinoma who underwent repeated hepatic artery chemoembolization but developed hepatic encephalopathy (degree II) and did not improve even with lactulose or 900 mg of carnitine. However, when the dose of carnitine was increased to 1800 mg, the blood ammonia level improved markedly, encephalopathy did not develop thereafter, and two hepatic artery chemoembolization procedures were performed (Fig. 15.8.). Case 2 was a 64-year-old man with type II diabetes mellitus, non-B non-C cirrhosis, and hepatocellular carcinoma who had undergone repeated hepatic artery chemoembolization, but an arterial portal vein shunt was formed near the tumor, making effective treatment difficult. On the third day of systemic chemotherapy (sorafenib 800 mg), he developed hepatic encephalopathy (degree III) and had high ammonia levels, improved with Aminoleban® infusion, discontinued sorafenib, and then performed one hepatic artery chemoembolization, but it was ineffective for hepatocellular carcinoma, and sorafenib was administered with carnitine 1800 mg. There was no development of hyperammonemia or encephalopathy, and treatment for hepatocellular carcinoma could be continued. These are cases where carnitine administration was utilized in cases where cancer cannot be treated due to refractory hepatic encephalopathy.

Tani et al. reported the results of carnitine administration to 10 cirrhotic patients with a history of hepatic encephalopathy (8 C-type, 1 alcoholic, 1 primary cholestatic, 3 of whom had hepatocellular carcinoma, 6 Child-Pugh B and 4 C) [71]. The blood ammonia level at the start was 137.8 ± 41.8 μg/dL, and in all cases, protein intake was restricted to 40 g/day or less, lactulose, BCAA preparations, oral nutrition for liver failure or intravenous drip, but the control of hepatic encephalopathy was poor. Therefore, eight patients were given 1800 mg/day of carnitine and two patients with renal insufficiency who were also on dialysis therapy, there was a significant decrease in ammonia levels (81.4 ± 24.8 and 85.3 ± 48.1 μg/dL) at 3 and 4 months after administration, and improvement in encephalopathy was observed in all patients. In addition, the frequency of painful muscle cramps was reduced in four cases.

Tsubata et al. treated five patients with BCAA infusions: four outpatients with cirrhosis who developed hepatic encephalopathy despite BCAA preparations, lactulose, kanamycin, and protein restriction and required treatment with BCAA infusions in the emergency room or emergency hospitalization, and one 62-year-old female patient who required daily BCAA administration and was hospitalized for a long time. Carnitine (900 mg/day) was administered [72]. As a result, in the four outpatients, ammonia levels decreased significantly ($P = 0.0317$) within 2 months of administration and did not increase thereafter, and the number of BCAA infusions was significantly reduced. Similarly, serum ammonia levels decreased in inpatient treatment patients, and the number of BCAA formulation infusions decreased significantly.

Adachi et al. reported the results of administering carnitine to 21 patients with cirrhosis of the liver with hyperammonemia [73]. The mean age of the 21 patients was 69.9 years (10 males, 11 females), and 1 was type B, 12 were C, 6 were alcoholic, 1 was autoimmune, and 1 was primary. The pre-administration value of ammonia was

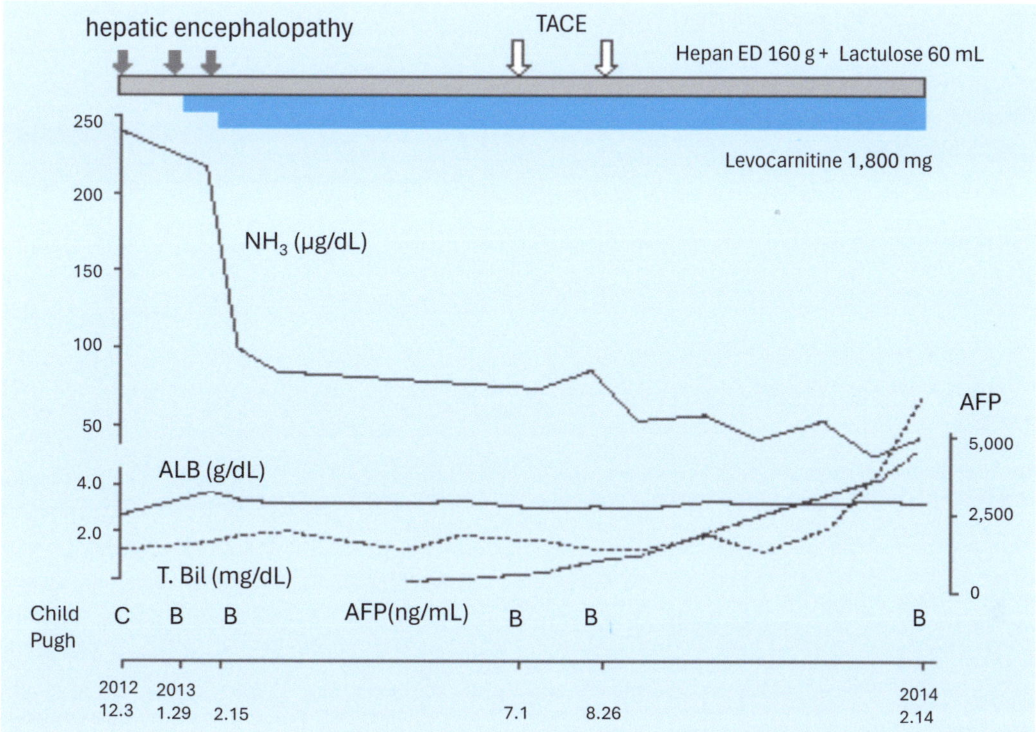

Fig. 15.8 Clinical course of case 1. (Male, 70 years old, cirrhosis type C, hepatocellular carcinoma). *ALB* albumin, *T. Bil* total bilirubin, *AFP* α-fetoprotein, *TACE* transcatheter arterial chemoembolization. (From ref. [70])

133.6 μg/dL, and 20 patients were BCAA preparations, 15 patients were laxatives, 10 patients were synthetic disaccharides, 13 patients were diuretics, 5 patients were zinc preparations, 12 patients were nutritional guidance, 14 patients were outpatients, and 7 patients were hospitalized. The dose of carnitine was 600–1800 mg/day, with an average of 840 mg/day. A decrease in ammonia was observed in 13 patients, especially in those with a Child-Pugh score of 10 or higher, and serum ammonia was significantly lower after 1 month of treatment ($P = 0.0038$). In addition, these patients tended to have low serum albumin levels and other nutritional indices, as well as low cross-sectional area of the psoas muscle, suggesting that carnitine may be effective in lowering ammonia in these patients.

Thus, many cases of oral administration of carnitine have begun to be reported for hepatic encephalopathy, which is difficult to treat with protein restriction, defecation control, synthetic disaccharide administration, BCAA preparations, and kanamycin administration, but there are some patients who do not have an effect, and the effective dosage varies.

Komada et al. reported an attempt at response-guided therapy, in which L-carnitine injections are first administered and subsequent oral administration is examined according to the reactivity [74]. Case 1 was an 85-year-old woman with cirrhosis C who had diabetes, hypertension, and chronic renal failure, and was hospitalized due to impaired consciousness and high levels of ammonia (hepatic encephalopathy III). Carnitine 1800 mg/day and lactulose enema for 3 days did not improve, and 1000 mg carnitine injection was intravenously injected once a day for 3 days to check the reactivity. As a result, the ammonia level decreased from 211 μg/dL on the fourth day to 129 μg/dL on the seventh day. As a result, it was considered that response to carnitine could be expected, and oral administration was continued. Due to the discontinuation of carnitine injections, a slight increase in ammonia levels, the

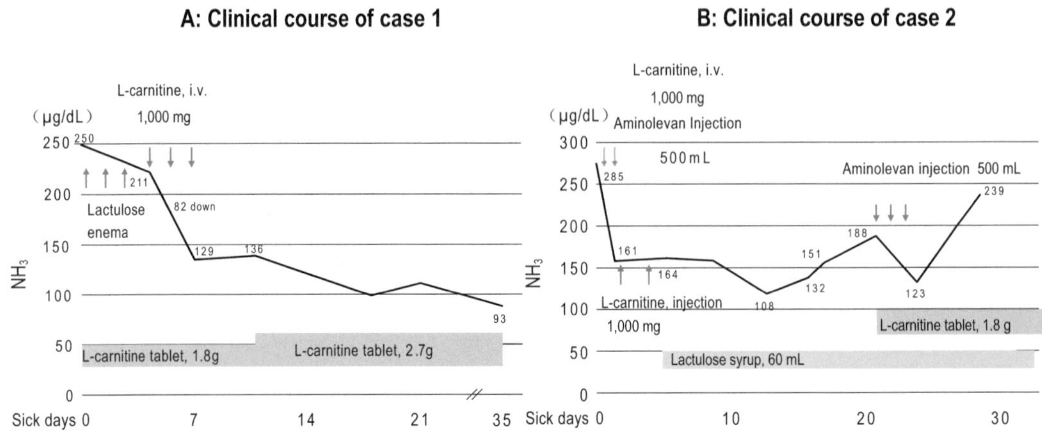

Fig. 15.9 Ammonia levels and clinical courses of case 1 (**a**) and case 2 (**b**) with severe hepatic encephalopathy. At the early stage of treatment, L-carnitine injection was administered, and judging the responsiveness of the treatment, dose of oral L-carnitine was determined. (From ref. [74])

dose of oral carnitine was increased to 2700 mg/day, and the ammonia was 99 μg/dL on the 18th day and 93 μg/dL on the 35th day, and there was no recurrence of encephalopathy, and the patient was under follow-up at a nearby physician (Fig. 15.9a). Case 2 was an 83-year-old female patient with cirrhosis C who was transported to an emergency hospital due to impaired consciousness (hepatic encephalopathy IV degree). After hospitalization, lactulose enemas, Aminoleban® infusions, and carnitine injections of 1000 mg were administered, and the ammonia level improved from 285 μg/dL at the time of admission to 161 μg/dL, and carnitine injection alone was used for the next 2 days. However, the ammonia level did not decrease further to 164 μg/dL, and the ammonia level decreased to 108 μg/dL on the 12th day after the start of lactulose syrup, but then gradually increased, and the ammonia level decreased with oral carnitine 1800 mg and Aminoleban® infusion, but the ammonia level increased again when carnitine was taken only orally. This is a case in which it was judged that oral carnitine alone was not effective (Fig. 15.9b). In this way, it was found that the dosage and effect of oral carnitine were possible to judge from the response of carnitine injections at an early stage.

The treatment method used here, in which carnitine injection is administered by intravenous drip in the case of emergency hospitalization due to severe hepatic encephalopathy, is similar to the recommendation of intravenous carnitine administration in the case of hyperammonemia due to valproic acid poisoning in the "Washington Manual," which is said to be the bible of the resident physicians (see Chap. 12, Sect. 12.11.3). Whatever the cause, it is extremely important to consider the administration of carnitine intravenous injectables that can increase blood and tissue carnitine concentrations early without going through the absorption process in the intestinal tract as one of the treatments to deal with hyperammonemia with impaired consciousness as soon as possible. In the future, further studies are needed on the efficacy of intravenous carnitine for severe hepatic encephalopathy. Carnitine deficiency is also often associated with deficiencies in other nutrients, and many patients with cirrhosis are deficient in zinc, which is essential for OTC activity in the urea cycle [21, 22]. Therefore, if a high dose of carnitine or intravenous carnitine injection does not have a sufficient effect, it is necessary to consider zinc supplementation at the same time.

Tajiri et al. retrospectively examined the effects of intravenous and oral carnitine administration in patients with overt hepatic encephalopathy [75]. Of the 53 patients with overt hepatic encephalopathy, 8 were excluded due to death. Ammonia levels and the Glasgow Coma Scale (GCS) were compared in 26 patients (Group A) who received conventional treatment such as intravenous infusion of BCAA preparations, lactulose, and non-absorbable antimicrobials and 19 patients who received intravenous L-carnitine in addition to conventional treatment (Group L, 1000 mg intravenous L-carnitine/500 mL BCAA). Patients in group L had worse ammonia levels (140 vs. 91 μg/dL, $P = 0.02$), median The Model for End-Stage Liver Disease (MELD) score, AST to platelet ratio index (APRI), and Fib-4 index at baseline than those in group A, but there were no significant differences in other patient characteristics. Thirty-four patients in groups A and L were discharged from the hospital, of which 28 were treated with conventional treatment and 6 were treated with orally carnitine. First, when comparing group A and group L during hospitalization, the blood ammonia level was higher in the L group than in group A, but it was significantly lower than baseline on both the third and seventh days of treatment, but only a significant decrease on the third day in group A. The rate of decrease in ammonia levels on day 7 was significantly lower in group L than in group A ($P < 0.01$). The coma grade of hepatic encephalopathy decreased in the same way on the third and seventh days of treatment in both groups. GCS was significantly worse in group L than in group A ($P = 0.03$) but was significantly higher in group L ($P = 0.02$) on the third day and was similar on the seventh day. There was no significant difference in the median survival of the overall survival curve between the two groups, and there was no significant difference in recurrence-free survival between the two groups. However, when patients were discharged from the hospital and were treated orally with carnitine (mean dose of 1833 ± 258 mg/day) compared with those receiving conventional treatment alone, the recurrence-free survival period was significantly longer in the oral carnitine group (735 vs. 497 days, $P = 0.03$). Thus, intravenous administration of L-carnitine preparations to hospitalized patients with reduced levels of consciousness and oral administration after discharge from the hospital were shown to be effective and may be a therapeutic option in patients with overt hepatic encephalopathy in extremely poor condition.

Cirrhotic patients with hepatic encephalopathy are discharged from the hospital after treatment, but are often re-hospitalized with repeated hepatic encephalopathy, and the suppression of these events is an important issue for hepatologists and emergency physicians. Tani et al. found that out of 68 patients with hepatic encephalopathy who received oral carnitine (300 mg × 3 times per day) between April 2013 and March 2016, 19 patients had received all standard treatments. Retrospective investigations were conducted on blood ammonia levels and the number of hospitalizations due to recurrence of hepatic encephalopathy over the medium to long term [76]. As a result, blood ammonia levels at 1 week, 12 weeks, 24 weeks, and 48 weeks at the start of treatment were significantly lower than at the start of treatment at 159, 119.5, 79, 75, and 82 μg/dL, respectively ($P < 0.0001$), and in the 3 years before the start of carnitine administration, 19 of these patients were hospitalized for a total of 29 times for hepatic encephalopathy. After the start of carnitine administration, the number of hospitalizations decreased significantly to 6 times over 3 years ($P < 0.001$, Fig. 15.10). In addition, the median survival time of these 19 patients in the Kaplan–Meier analysis was relatively long, 40.9 months, although the survival time of patients with decompensated cirrhosis is usually about 2 years.

Thus, it was suggested that the administration of carnitine to patients with repeated hepatic encephalopathy may significantly reduce subsequent hospitalization for hepatic encephalopathy.

15.3.4.3 Effect of Carnitine on Patients with Minimal Hepatic Encephalopathy

In the treatment of hepatic encephalopathy in cirrhosis, it is important to treat it as early as possible at the stage of minimal hepatic encephalopathy (MHE) before overt hepatic encephalopathy. Saito et al. examined the predictors of the effect of carnitine administration in patients with minimal hepatic encephalopathy, since some patients were effective, and others were

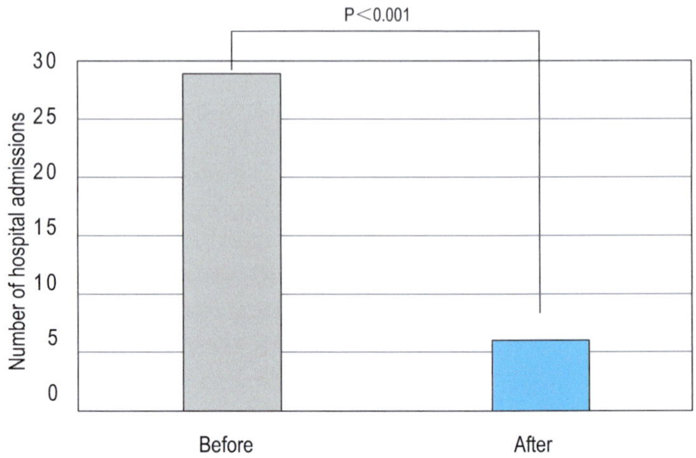

Fig. 15.10 Change in the number of hospital admissions for hepatic encephalopathy for 3 years before and after L-carnitine administration. This difference is statistically significant ($P < 0.001$). (From ref. [76])

not [77]. Neuropsychological tests (NPs) were performed on 64 outpatient patients with cirrhosis, and 24 MHE patients were administered carnitine (300 mg tablets×2 orally after meals 3 times a day for 3 months) to investigate the predictors associated with the effects. As NP tests, the number connection test (NCT)-A and B, figure position test, digit symbol test, block design test, reaction time test-A, B, and C were performed, and MHE was diagnosed if an abnormality was diagnosed in two or more tests. As a result, 11 (45.8%) of the 24 MHE patients were found to have improved after 3 months of carnitine administration. There was no improvement in 13 patients. Univariate analysis showed that the blood taurine concentration was significantly higher in the improvement group ($P = 0.006$). Backward stepwise multivariate logistic regression analysis showed that the concentration of taurine in the blood before carnitine administration was an independent factor in improving MHE ($P = 0.046$).

Nojiri et al. examined the effects of carnitine administration by collecting 76 patients with covert hepatic encephalopathy by NCT-B and dividing them into a carnitine group (38 patients, 1000 mg/day, 12 weeks) and a control group (38 patients) [78]. As a result, the blood ammonia level decreased significantly from 92.2 ± 31 ⇒ 72.5 ± 33 μg/dL in the carnitine group ($P < 0.01$) and remained unchanged in the control group from 91.8 ± 29 ⇒ 89.4 ± 39 μg/dL, indicating a significant difference between the two groups ($P < 0.05$). In addition, the number of NCT-B patients who improved was 18 out of 38 patients in the carnitine group and 9 cases in the control group, indicating that there was a significant improvement in the carnitine group ($P = 0.03$). These results indicate that carnitine administration is useful for the treatment of minimal hepatic encephalopathy.

Nakanishi et al. examined the effect of carnitine administration on brain function in patients with covert hepatic encephalopathy using near-infrared spectroscopy (NIRS) [79]. NIRS is a test method used in psychiatry to assist in the differential diagnosis of major depressive disorder, bipolar disorder, and schizophrenia, and is performed for 1 min on tasks such as a word fluency test, and the change in the increase in oxygenated hemoglobin (Hb) concentration in the brain over time, including before and after the test, is measured for 2 min and 30 s. It was found that in healthy subjects, an increase in the oxygenated Hb concentration in the frontal region was observed at the early stage of the task, but in patients with minimal hepatic encephalopathy, the increase in the oxygenated Hb concentration in the frontal region at the early stage of the onset of the task was observed. [80] Therefore, 80

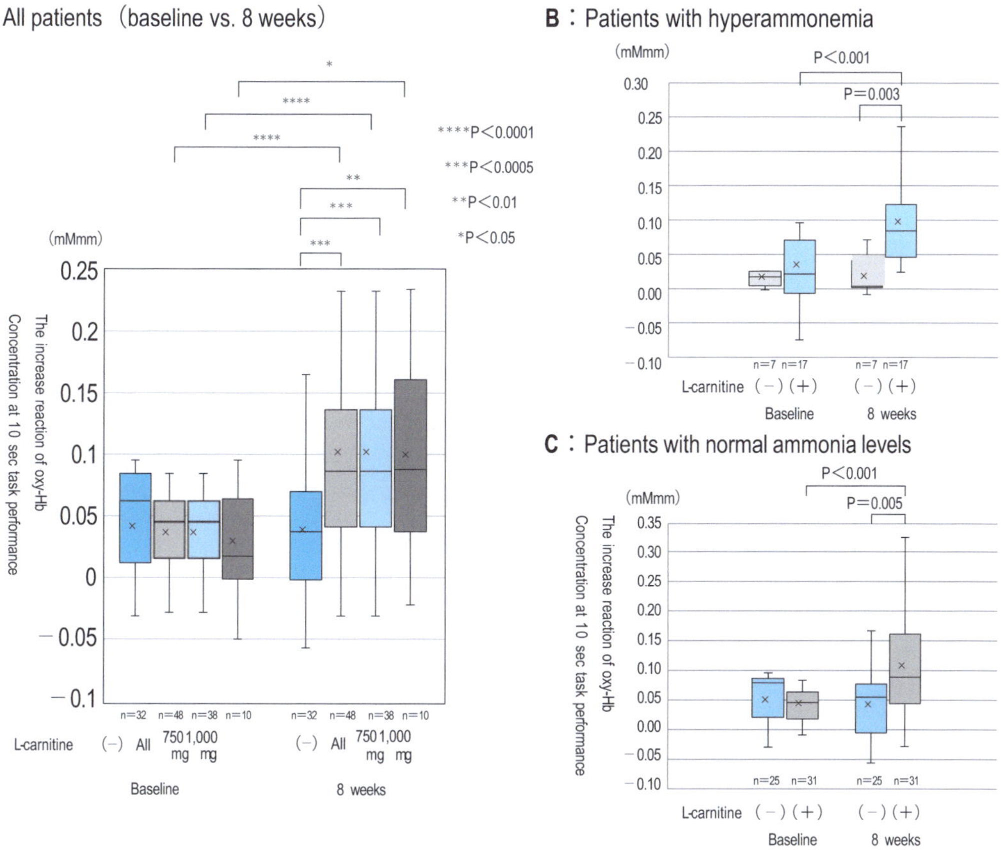

Fig. 15.11 Comparison of NIRS data between the L-carnitine and no treatment groups at baseline and 8 weeks post-treatment. (**a**) All patients. The box plot includes the stratified analysis according to the L-carnitine dose. (**b**) Patients with hyperammonemia. (**c**) Patients with normal ammonia levels at baseline. *NIRS* near-infrared spectroscopy. (From ref. [79])

patients with minimal hepatic encephalopathy were divided into a carnitine group (48 patients, oral administration, 38 patients receiving 750 mg/day, 10 patients receiving 1500 mg/day for 8 weeks) and a control group (32 patients), and the effect of carnitine administration on brain function in patients with occult hepatic encephalopathy was examined. There were no significant differences in patient characteristics between the two groups prior to treatment, except for low albumin, high bilirubin, and high Child-Pugh scores in the carnitine group. There was also no difference in baseline NIRS data. After 8 weeks, there was no change in the NIRS data in the control group, but the NIRS data in the carnitine group improved significantly both overall and by dose (Fig. 15.11a). Seven patients in the control group and 17 patients in the carnitine group had hyperammonemia. Therefore, the NIRS data were significantly improved in the carnitine treatment group regardless of the presence or absence of hyperammonemia (Fig. 15.11b, c). Thus, it was shown that administering carnitine to patients with minimal hepatic encephalopathy improved abnormal brain function [79].

15.4 Effect of Carnitine on Severe Muscle Cramps in Patients with Cirrhosis

15.4.1 Case Reports of Carnitine Administration

Cirrhotic patients often complain of various subjective symptoms such as general malaise, hepatic encephalopathy, and sleep disorders, and one of these symptoms is painful muscle cramps, which are said to significantly affect quality of life, including sleep disorders and decreased activity in daily life [81, 82]. The mechanism by which severe cramps occur in patients with cirrhosis is not well understood, and no particularly effective treatment has been developed. Nakanishi et al. have studied the mechanism of cramps in patients with cirrhosis, because carnitine administration is effective in treating muscle symptoms such as cramps in hemodialysis patients (see Chap. 14) [83]. Eleven cirrhotic patients (5 males, 6 females; mean age 72 years; type C: 9, type B: 1, non-B non-C: 1) with daily cramps despite Shakuyakukanzoto (Peony-Licorice Decoction) and BCAA preparations were treated with 600 mg (1 case), 900 mg (1 case), and 1200 mg (9 cases) of carnitine. The efficacy of carnitine was confirmed in 82% of the patients (2 cases with marked improvement and 7 cases with improvement), and ammonia levels decreased in 4 out of 6 patients with concurrent hyperammonemia (Table 15.7). The clinical course of a 72-year-old woman who complained of frequent and "painful" severe cramps, especially at night, is shown in Fig. 15.12.

Based on these results, Nakanishi et al. further investigated the effect of oral carnitine administration (300 mg 3 times daily, total 900 mg/day in 19 patients; 300 mg 4 times daily, total 1200 mg/day in 23 patients; administration period 8 weeks) in 42 cirrhotic patients with muscle cramps that repeatedly awakened them in bed due to pain, and the effects of carnitine on muscle cramps before and after carnitine administration were examined using the visual analog scale (VAS) [84]. The frequency of muscle cramps before and after carnitine administration was investigated using a questionnaire, and pain was evaluated using the VAS score. The frequency of muscle cramps decreased in 84.2% of the 900 mg group (19 patients), 91.3% of the 1200 mg group (23 patients), and 88.1% of all patients (42 patients) (Fig. 15.13a). Of these patients, 10.5% in the 900 mg group completely resolved muscle cramps, while 43.5% in the 1200 mg group ($P = 0.037$), a significant difference between the two groups, with the higher dose group being more effective (Fig. 15.13b). In addition, analysis of the 31 cases in which VAS evaluation of muscle cramp pain showed that carnitine administration significantly decreased the VAS score from 69.9 ± 22.5 to 26.2 ± 29.1 in 31 cases ($P = 0.0001$, Fig. 15.14a). By dose, the VAS score decreased significantly more in the 1200 mg group than in the 900 mg group ($P = 0.003$, Fig. 15.14b).

15.4.2 Multicenter Survey Study

Iwasa et al. reported a detailed multicenter survey study on the frequency and intensity of cramps, quality of life, and sleep disturbances in 1788 patients with chronic liver disease, including 671 patients with cirrhosis [85]. The questionnaire investigated (1) the frequency of cramps, (2) the relationship between cramps and QOL and sleep disturbances, (3) the characteristics of patients who require treatment for cramps, and (4) the characteristics of patients who often cause cramps. As a result, 51.8% of patients with cirrhosis often experienced cramps, and patients who have cramps every day or several times a week, and patients who have long cramps for several minutes to several hours, patients with cramps with a VAS intensity of ≥4 were significantly more likely to have a decrease in QOL and sleep disturbances ($P < 0.01$, Table 15.8). Older age, females, HCV positive, low albumin levels, and cirrhosis of the liver were independent factors associated with the severity of cramps. In addition, 12% of the patients had cramps every day and 32% of the patients had weekly cramps, and 11% had VAS intensity 5 and 32% had VAS intensity 4, and the calf was the overwhelming majority (81%), followed by

Table 15.7 Effect of L-carnitine on cramps and ammonia levels in patients with cirrhosis

Patient No.	Age	Gender	Duration (day)	LC dose (mg)	Cause of cirrhosis	Ascites control	Encephalopathy	CP score	Change in NH^3 after 8 weeks	Cramps
1	77	M	28	1200	Type C	–	–	5	Increase (19 → 70)	Improve
2	68	F	213	1200	Type C	–	–	5		Improve
3	81	F	14	600	Type C	Good	Grade I	5	No change (151 → 153)	Significantly improve
4	62	M	66	1200	Type B	–	Grade I	8	No change (87 → 84)	Improve
5	72	F	206	1200	Type C	Good	–	9	No change (41 → 40)	Significantly improve
6	71	M	234	1200	Type C	Good	Grade I	9	Decrease (124 → 71)	Improve
7	75	F	12	900	Non-B non-C	Refractory	–	9		
8	73	F	248	1200	Type C	Good	Grade I	10	Decrease (139 → 108)	Improve
9	77	F	123	1200	Type C	Good	–	10	Decrease (132 → 75)	No change
10	72	M	45	1200	Type C	Refractory	Grade I	11	Increase (43 → 54)	Improve
11	62	M	173	1200	Type C	Refractory	Grade I	12	Decrease (130 → 71)	Improve

From ref. [83]
Significantly improve; disappear the symptom, improve; decrease the appearance of the symptom in a week
LC L-carnitine, *CP* Child-Pugh

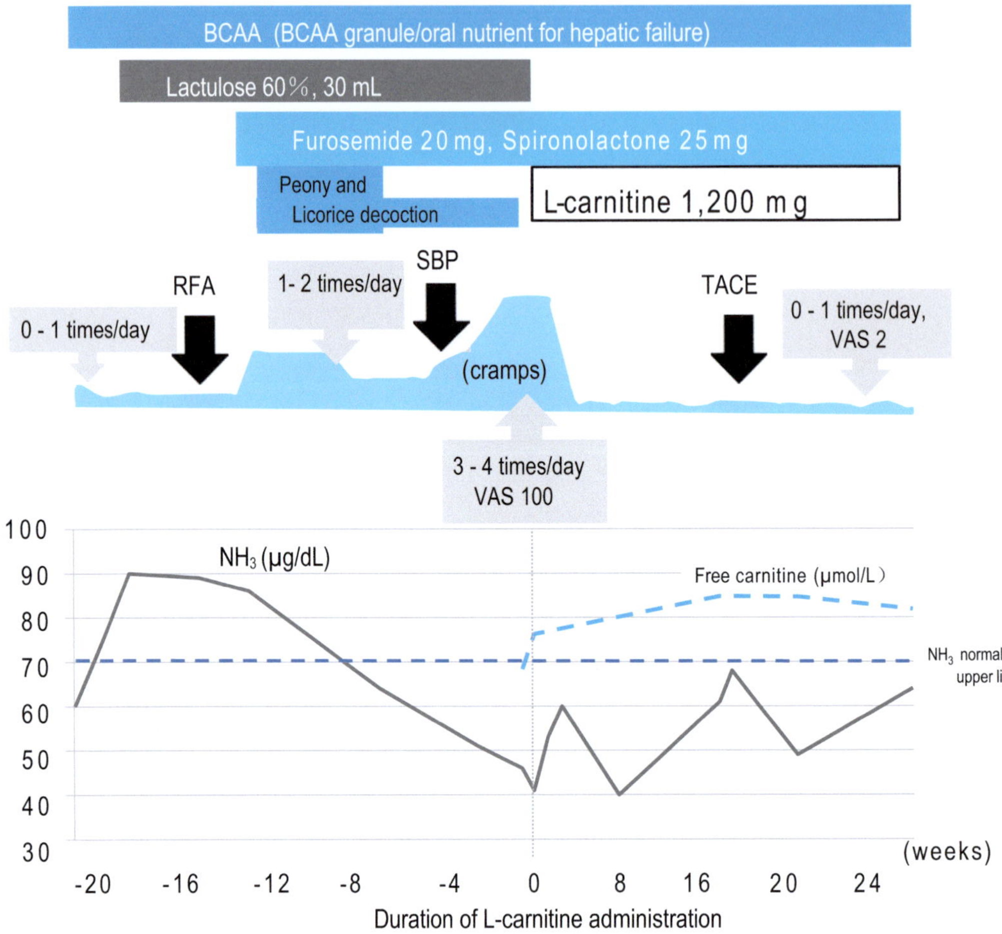

Fig. 15.12 Clinical course of a significant improvement case of severe muscle cramps of patient with hepatic cirrhosis (72 years-old, female, Child-Pugh B, cirrhosis of hepatitis virus type C) by L-carnitine administration. *VAS* visual analogue scale, *RFA* radiofrequency ablation, *SBP* spontaneous bacterial peritonitis, *TACE* transcatheter arterial chemoembolization. (From ref. [83])

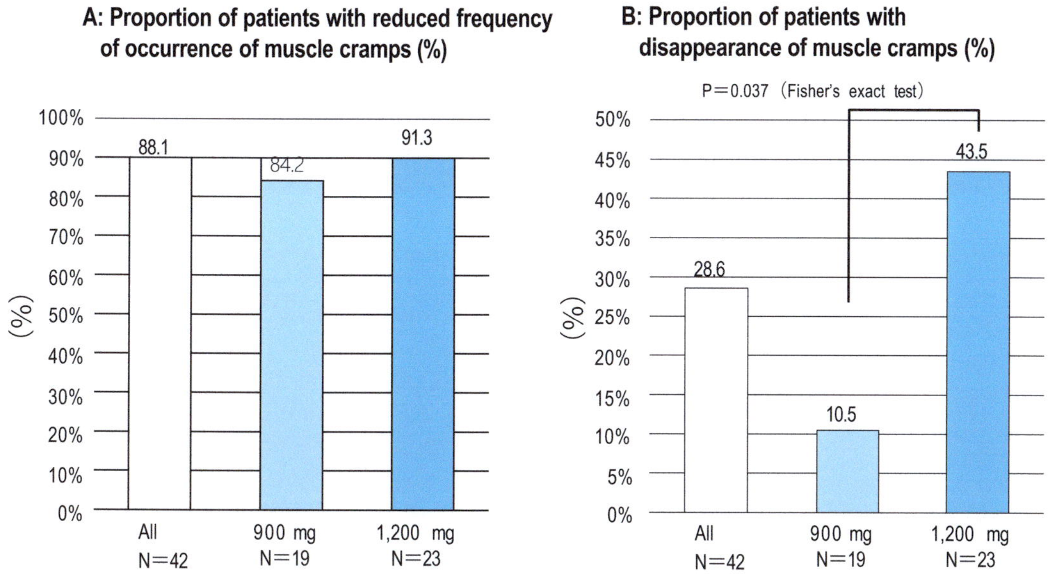

Fig. 15.13 Efficacy of L-carnitine treatment for the reduction or disappearance of muscle cramps. Percentages of (**a**) reduction and (**b**) disappearance of muscle cramps are shown. Muscle cramping was reduced in 88.1% and disappeared in 28.6% of patients after 8 weeks of therapy. The dose of L-carnitine was an important factor associated with disappearance of muscle cramps. L-carnitine was administered 900 mg/day (300 mg × 3 times/day) or 1200 mg/day (300 mg/day × 3 times/day). (From ref. [84])

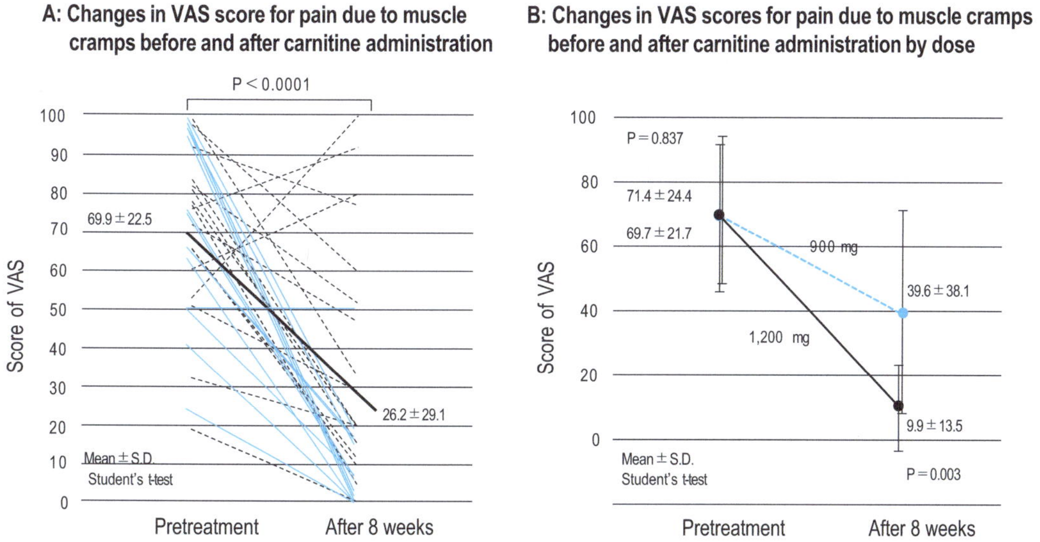

Fig. 15.14 Scores of VAS before and after 8 weeks of L-carnitine treatment. (**a**): Changes in VAS score for pain due to muscle cramps before and after L-carnitine administration (900 mg/day, black dotted lines, and 1200 mg/day, blue lines, n = 31). The VAS score of all cases significantly decreased from 69.9 ± 22.5 to 26.2 ± 29.1 (black bold line, $P < 0.0001$). (**b**): Changes in VAS scores for pain due to muscle cramps before and after L-carnitine administration by dose (900 mg/day, blue dotted line, 1200 mg/day, black line). The VAS score significantly decreased in 1200 mg group than in 900 mg group ($P = 0.003$). *VAS* visual analogue scale. (From ref. [84])

the toes (12%) and the thighs (6%). The time zones of cramps were 52% of patients experienced cramps at night, and 23% of patients experienced cramps during both the day and at night. The duration of the cramps was 58% for a few minutes and 40% for a few seconds, but 2% of the patients lasted several hours. The majority of patients (26.3%) felt a decrease in QOL due to cramps, but only 22.3% of patients received therapeutic drugs. The main therapeutic agents were BCAAs (132 cases), Shakuyakukanzoto (Peony-Licorice Decoction, 53 cases), carnitine (38 cases), and taurine (12 patients).

Table 15.8 Independent factors related to QOL and sleep disturbances

HR(95% CI)		*P* value
QOL disturbance		
Frequency, day or week	4.95(2.42–10.11)	<0.001
Duration, minutes or hours	4.65(2.00–10.82)	<0.001
Severity of pain, VAS > 4	3.75(1.83–7.68)	<0.001
T-Bil		0.060
CRP		0.079
Alb		0.091
Child-Pugh score		0.515
Prothrombin time (%)		0.670
Location, calf or thighs		0.929
Presence of cirrhosis		0.989
Sleep disturbances		
Severity of pain, VAS > 4	3.81(2.51–5.78)	<0.001
Duration, minutes or hours	2.40(1.53–3.76)	<0.001
Frequency, day or week	1.83(1.22–2.76)	<0.01
Liver diseases caused by HCV infection		0.064
CRP		0.140
Location, calf or thighs		0.339

From ref. [85]
Factors associated with QOL and sleep disturbances were analyzed using Cox proportional hazard modelling
Alb albumin, *CI* confidence interval, *CRP* C-reactive protein, *HCV* hepatitis C virus, *HR* hazard ratio, *QOL* quality of life, *T-Bil* total bilirubin, *VAS* visual analogue scale

Hiraoka et al. conducted a retrospective, multicenter study on the recurrence of 289 patients with cirrhosis (ages 70.6 ± 10.5 years, 149 males and 140 females) in Japan [86]. Questionnaire survey on the presence or absence of cramp symptoms in the past 3 months, their frequency, severity of pain (VAS), site of cramps, time of day, duration, sleep disturbance due to cramps, and whether they feel a decrease in daily living activities and QOL due to cramps was surveyed at baseline and 3–4 months later and summarized the actual conditions of patients with cirrhosis (Table 15.9). As a result, 160 patients (55.4%) had cramp symptoms, women were more frequent than men (53.8 vs. 41.9%), and patients with cramps had higher Child-Pugh scores and lower platelet counts. In addition, 48.8% of the patients with cramps had severe painful cramps with a VAS of 4 or higher, 37.5% of patients complained of sleep disturbances, 23.1% complained of decreased activities of daily living, and 31.9% of patients complained of decreased QOL. Of the 160 patients with cramps, 82 (51.3%) were treated with drugs, and carnitine was the most common (66 (80.5%), followed by Peony-Licorice Decoction (23, 28%), BCAA (19, 23.2%), and taurine (5, 6.1%). There were also 28% of cases of multi-drug treatment. When carnitine was administered as a single agent, it was significantly effective in all items of VAS,

Table 15.9 Therapeutic efficacy of monotherapy with L-carnitine for muscle cramps after excluding cases without data of second interviewing

	n = 45	Improvement/disappearance: Changeless: deterioration[a]	*P* value
Disappearance (%)	24 (53.3)	–	–
Improvement (%)	33 (73.3)	33:9:3	*P*<0.001
VAS	29 (64.4)	29:9:7	*P*<0.001
Frequency	31 (68.9)	31:9:5	*P*<0.001
Sleep disturbance	27 (60.0)	27:14:4	*P*<0.001
Daily activity	24 (53.3)	24:16:5	*P*<0.001
Duration of muscle cramps	30 (66.7)	30:12:3	*P*<0.001
QOL			

From ref. [86]
QOL quality of life, *VAS* visual analogue scale
[a]: Compared with baseline by Wilcoxon signed-rank test

frequency, sleep disturbance, daily activity, duration, and QOL ($P < 0.001$, Table 15.9). Peony-Licorice Decoction was significantly effective in sleep disturbance and quality of life ($P = 0.020$ and $P = 0.048$). BCAAs were used in combination with other drugs. In addition, taurine alone administration did not have a significant effect.

15.5 Abnormal Energy Metabolism in Patients with Cirrhosis and Effects of Carnitine

15.5.1 Abnormal Energy Metabolism in Cirrhosis Patients

Patients with cirrhosis have a high rate of PEM complication [87]. In patients with cirrhosis, resting energy expenditure is increased, while carbohydrate utilization is significantly reduced due to a decrease in glycogen stored in the liver and insulin resistance, and catabolism of body proteins is enhanced. Energy metabolism in such pathologies such as cirrhosis is studied by indirect calorimetry. An overview of the measurement principle is shown in Fig. 15.15. [88, 89] As the severity of cirrhosis progresses, among the three major energy substrates used for energy production, carbohydrates, lipids, and proteins, the use of carbohydrates, which are the most commonly used, decreases and the use of lipids increases (Fig. 15.16) [88]. This shift of energy metabolism pattern, in which more lipids are used for energy production than carbohydrates, is similar to the energy metabolism pattern of healthy people when they are fasted or starved for a long time, and it is considered to be a state in which intracellular energy (ATP) in organs such as muscles and liver is extremely low. Thus, in patients with cirrhosis, abnormal energy metabolism occurs, which is considered to be a chronic energy deficiency state similar to the starvation state of healthy people, and this is thought to be a predisposing factor to general malaise, decreased daily activity, decreased quality of life, muscle cramps, hyperammonemia, and hepatic encephalopathy.

Until now, the only way to deal with energy metabolism abnormalities in cirrhosis patients is "late evening snack" (LES), which consumes carbohydrate-rich snacks and BCAAs before going to bed at night, and LES is also recommended in the "Cirrhosis Treatment Guidelines 2020." [102] A study by Kato et al. has shown that BCAAs are useful for LES because BCAAs are 96% energy efficient when cirrhosis patients ingest the three major nutrients, which is the highest BCAA compared to glucose (41%) and fatty acids (27%) [88].

15.5.2 Effect of Carnitine on Abnormal Energy Metabolism in Cirrhosis Patients

Sakai et al. examined the effect of carnitine administration (600 mg × 3/day, orally for 4 weeks) on energy metabolism in 13 patients with cirrhosis using this indirect calorimetry [90]. The mean age of the patients was 67.2 ± 13.1 years, HBV 1, HCV 7, alcoholic 3, and 2 others. As a result, the non-protein respiratory quotient (npRQ) of all patients increased slightly with carnitine administration, but there was no significant difference. However, when they examined seven patients with a baseline npRQ of <0.85, which was considered to be abnormal in energy metabolism, the npRQ before and after administration was significantly increased from 0.76 ± 0.043 to 0.827 ± 0.030 ($P = 0.016$). When the changes in the burning ratios of carbohydrates, lipids, and proteins were 37.6% ⇒ 48.2%, 40.2% ⇒ 31.9%, and 22.2% ⇒ 19.9%, respectively, the lipid-first metabolism improved to prioritize carbohydrate combustion. In particular, when seven patients with cirrhosis with an npRQ < of 0.85 were examined, the carbohydrate burning ratio increased markedly from 15.3 ⇒ 34.2%, lipid burning decreased markedly from 59.9 ⇒ 45.1%, and protein burning decreased slightly from 24.8 ⇒ 20.6% (*Please see* Fig. 15.2 *of the ref.* [90]). It was also observed that the hsCRP

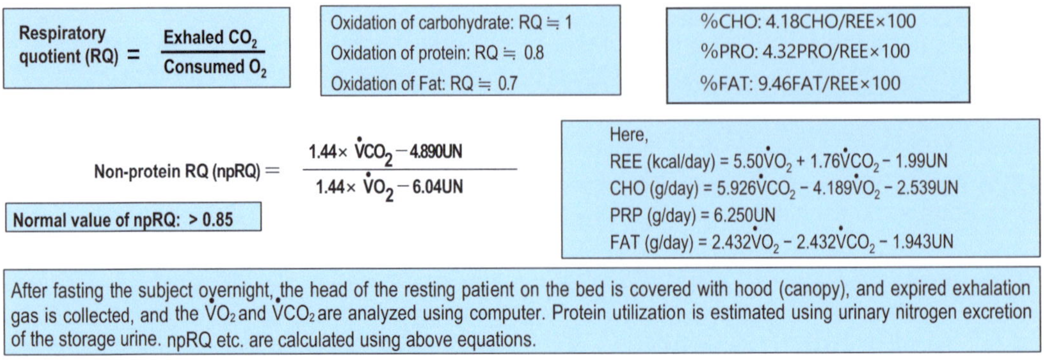

Fig. 15.15 Outline of measurement of energy metabolism using indirect calorimetry. *REE* resting energy expenditure, *RQ* respiratory quotient, VO_2 oxygen consumption per minute, VCO_2 CO_2 exhaled per minute, *UN* nitrogen excretion per 24 h. (Created as refs. [88, 89])

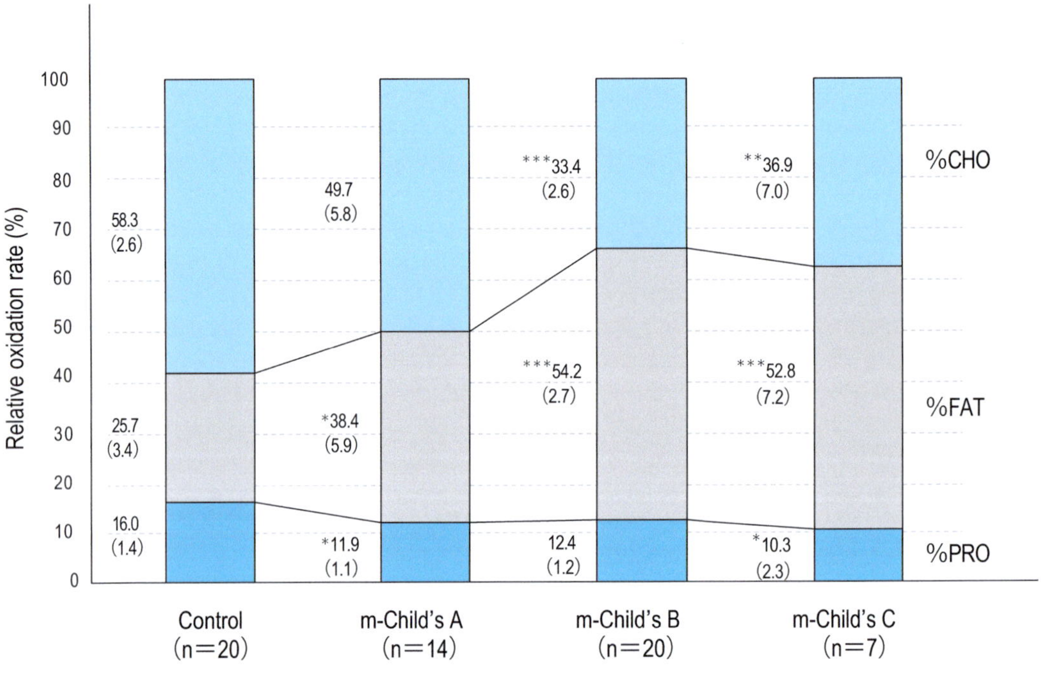

Fig. 15.16 Relative oxidation rates of carbohydrate (% CHO), fat (% FAT), and protein (% PRO) in controls and in patients with liver cirrhosis of modified Child's grade (m-Child) A, B, or C. Values are expressed as mean (SE). *: $P < 0.05$, **: $P < 0.01$, ***: $P < 0.001$ (Dunnett *t* test). (From ref. [88])

concentration significantly decreased after L-carnitine therapy for all cases. It was considered the improvement in energy metabolism by L-carnitine therapy may be associated with an improvement in systemic inflammation. These results indicate that carnitine administration to cirrhotic patients with advanced liver cirrhosis, reduced npRQ, and abnormal energy metabolism of the three major nutritional substrates may improve the abnormal energy metabolism.

15.6 Nutritional Status, QOL, and Effects of Carnitine on Sarcopenia in Patients with Cirrhosis

15.6.1 Nutritional Status, QOL Sarcopenia, and Life Prognosis of Patients with Cirrhosis

Patients with cirrhosis have poor nutrition status and are often restricted from protein intake. The liver is a central organ for the metabolism of sugars, lipids, and proteins, and as cirrhosis progresses and the number of hepatocytes decreases, the storage capacity of glycogen decreases, and glycogen and BCAAs in skeletal muscle are consumed for energy metabolism and albumin synthesis, so muscle mass tends to decrease. Furthermore, when cirrhosis progresses and decompensated cirrhosis occurs, hyperammonemia is likely to occur due to the formation of PS shunts, etc., and muscle BCAAs are consumed, protein synthesis decreases, and sarcopenia are likely to occur. Forty percent of patients with cirrhosis are said to have sarcopenia [91]. Cirrhosis patients with sarcopenia have been found to have a very poor prognosis [92–94], and although their treatment is extremely important, effective treatments are limited. In particular, cirrhotic patients with hepatocellular carcinoma are more likely to develop sarcopenia due to cachexia caused by cancer. In addition, invasive treatments such as transcatheter arterial chemoembolization (TACE) for the treatment of hepatocellular carcinoma often reduce liver function and further deteriorate nutritional status.

15.6.2 Clinical Studies of Carnitine Administration for Sarcopenia in Patients with Cirrhosis

According to a study of 149 patients with cirrhosis (55 non-sarcopenia and 94 sarcopenia) by Hanai et al., patients with a relative change in skeletal muscle area/year (ΔSMA/y) at the L3 level by computed tomography (CT) greater than −3.1% survival was significantly lower than in patients $\leq -3.1\%$ [94] ($P < 0.0001$).

Of the 53 patients who underwent TACE for hepatocellular carcinoma, Hassan et al. divided 50 patients who underwent TACE for hepatocellular cancer into a carnitine group (24 patients, mean age 71.6 ± 7.6 years, 600 mg/day, oral administration, 2 weeks before TACE and 12 weeks after TACE, for a total of 14 weeks) and a non-treatment group (26 patients, mean age 72.3 ± 6.8 years, control group). Changes in Child-Pugh (CP) scores, serum albumin, bilirubin, and prothrombin time before and after TACE at 1, 4, and 12 weeks after TACE were examined [95]. There were no significant differences in patient characteristics, type of anticancer drug, cancer stage, or baseline laboratory values at TACE between the two groups. A comparison of CP scores between the two groups showed that CP scores were maintained in the carnitine group at 1 week after TACE and significantly improved after 4 weeks ($P < 0.01$). After 12 weeks, the CP scores were unchanged from the previous level of TACE. On the other hand, in the control group, CP scores deteriorated significantly after 1 and 12 weeks ($P < 0.05$, Table 15.10). Serum albumin, prothrombin time, and bilirubin also deteriorated after TACE in the control group, but these deteriorations were lower in the carnitine group than control group (Table 15.10). The changes in CP score and serum albumin from baseline are shown in Fig. 15.17, and the changes in prothrombin time (PT) and bilirubin from baseline are shown in Fig. 15.18. It was shown that carnitine administration 2 weeks before TACE prevented deterioration of patients' CP scores, albumin, prothrombin time, and bilirubin. In addition, a comparison of the carnitine administration + BCAA group (16 patients) and the BCAA group (12 patients) showed that the hepatoprotective effect of carnitine was enhanced when BCAA was administered in addition to carnitine administration (Figs. 15.17b, d and 15.18b).

Iwasa et al. also administered carnitine (900 mg/day, from admission to discharge) to 69 cirrhotic patients with hepatocellular carcinoma undergoing transcatheter arterial infusion (TAI),

Table 15.10 Effects of L-carnitine in overall HCC patients who received TACE

Parameter(mean ± SD)		Pretreatment	After TACE		
			1 week	4 weeks	12 weeks
Control group (N = 26)	CP score	5.88 ± 0.99	6.24 ± 1.05†*	6.04 ± 1.07	6.39 ± 1.55†*
	S. albumin(g/dL)	3.35 ± 0.56	2.94 ± 0.57†***	3.29 ± 0.58‡***	3.21 ± 0.63‡**
	PT(%)	89.2 ± 17	85.9 ± 16.9†*	88.2 ± 17.7	87.1 ± 18.2
	T. bilirubin(mg/dL)	0.93 ± 0.46	1.13 ± 0.51†*	0.96 ± 0.55‡*	1.21 ± 0.7$^{†*,\ §*}$
	ALT(IU/L)	41.8 ± 28.8	56.2 ± 31†*	38.7 ± 21.1‡*	37.6 ± 18.8
	AST(IU/L)	57 ± 37.8	50.8 ± 26.5	55.3 ± 31.6	60.3 ± 44.3
	GGTP(IU/L)	66.3 ± 67.3	84.5 ± 100.3†*	91.8 ± 84.7†***	59.9 ± 74.7
	Ascites (no/moderate/massive)	21/5/0	20/5/1	19/5/2	17/5/4
	Encephalopathy (no/yes)	26/0	26/0	26/0	26/0
L-carnitine group (N = 24)	CP score	6.04 ± 1.04	6.17 ± 0.86	5.75 ± 0.73‡**	5.91 ± 0.92
	S. albumin(g/dL)	3.22 ± 0.57	3.05 ± 0.46†*	3.33 ± 0.5$^{†*,\ ‡**}$	3.26 ± 0.49‡*
	PT(%)	86.7 ± 19.4	91.5 ± 18.4	93.3 ± 16.8†*	90.7 ± 16.4
	T. bilirubin (mg/dL)	1.02 ± 0.57	0.97 ± 0.55	0.93 ± 0.46	1.05 ± 0.63
	ALT (IU/L)	32.6 ± 17.6	41.7 ± 23.1†*	34.7 ± 20.8	33.9 ± 11.6
	AST (IU/L)	51.5 ± 25.9	50.2 ± 32.1	54.5 ± 28	49.4 ± 15.6
	GGTP (IU/L)	70.7 ± 80.9	91.4 ± 124.4	97.1 ± 148.6	61.8 ± 43.4
	Ascites (no/moderate/massive)	21/3/0	22/2/0	24/0/0	23/0/1
	Encephalopathy (no/yes)	24/0	24/0	24/0	24/0

From ref. [95]

HCC hepatocellular carcinoma, *TACE* transcatheter chemoembolization, *SD* standard deviation, *CP* Child-Pugh, *S. albumin* serum albumin, *PT* prothrombin time, *T. bilirubin* total bilirubin, *ALT* alanine aminotransferase, *AST* aspartate aminotransferase, *GGTP* gamma-glutamyl transpeptidase

^{a}: Significant difference compared with baseline, ‡: significant difference compared with 1 week after TACE, §: significant difference compared with 4 weeks after TACE, *: $P<0.05$, **: $P<0.01$, ***: $P<0.001$

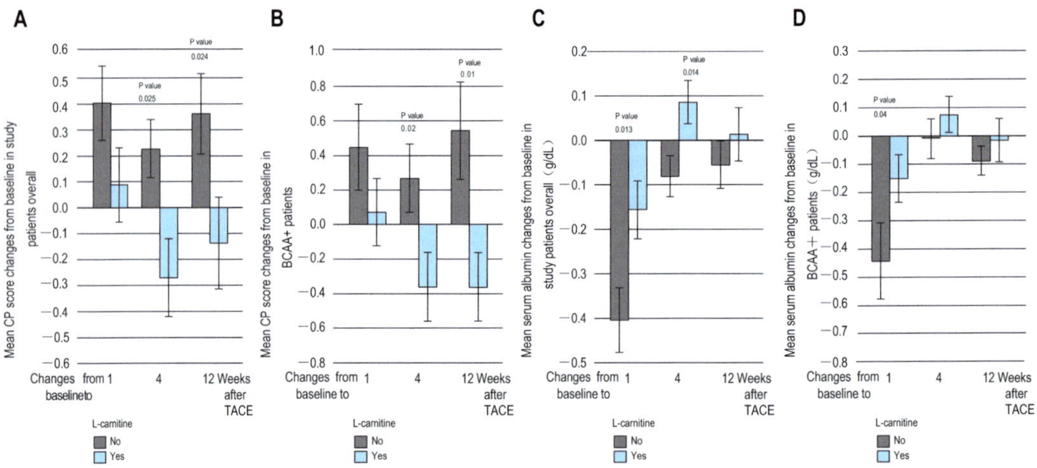

Fig. 15.17 L-carnitine effects on CP score and serum albumin. (**a**):L-carnitine effects on mean CP score changes from baseline in study patients overall. (**b**):L-carnitine effects on mean CP score changes from baseline in BCAA+ patients. (**c**):L-carnitine effects on mean serum albumin changes from baseline in study patients overall. (**d**):L-carnitine effects on mean serum albumin changes from baseline in BCAA+ patients. Error bars represent standard errors. *CP* Child-Pugh, *TACE* transarterial chemoembolization. (From ref. [95])

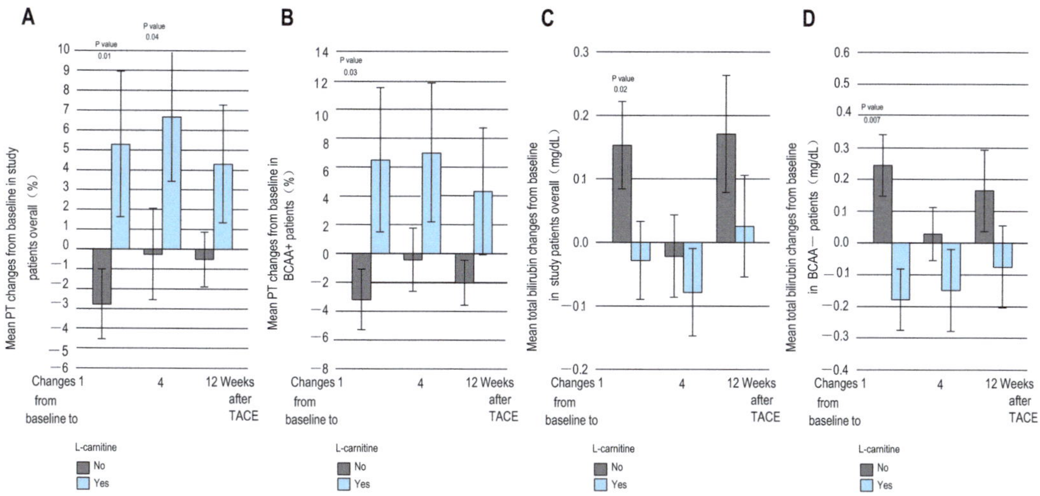

Fig. 15.18 L-carnitine effects on PT and total bilirubin. (**a**): L-carnitine effects on mean PT changes from baseline in study patients overall. (**b**): L-carnitine effects on mean PT changes from baseline in BCAA+ patients. (**c**): L-carnitine effects on mean total bilirubin changes from baseline in study patients overall. (**d**): L-carnitine effects on mean total bilirubin changes from baseline in BCAA-patients. Error bars represent standard errors. *PT* prothrombin time, *TACE* transarterial chemoembolization. (From ref. [95])

TACE, or percutaneous radiofrequency ablation (RFA), and compared the effects of carnitine on blood ammonia and serum albumin levels with those of 119 patients who were not treated with carnitine [96]. Since many of these patients were also receiving BCAA, the patients were divided into four groups: (A) the carnitine + BCAA group (31 patients), (B) the carnitine alone group (38 patients), (C) the BCAA alone group (36 patients), and (D) the no treatment group (83 patients) for analysis. The results showed that the decrease in ammonia from admission to discharge was greater in groups (A) and (C) than in group (D) ($P < 0.001$), and the decrease in ammonia was also significantly greater in group (B) than in group (D) ($P < 0.05$). Serum albumin was significantly lower in group (D), which did not receive any treatment, at discharge than at admission due to more invasive procedures such as TAI, TACE, and RFA ($P < 0.001$), while groups (A) and (C) had a significantly smaller decrease in albumin than group (D). A similar trend was observed only in patients with a CP score of A. These results indicate that carnitine and BCAAs may be useful in preventing a decrease in hepatic reserve and worsening of nutritional status during invasive treatments such as TACE and RFA.

Ishikawa et al. divided 77 cirrhotic patients (31 males and 46 females) into a sarcopenia group (56 patients) and a non-sarcopenia group (21 patients), administered carnitine (600 mg, 3 times daily) orally for 1 year, and investigated the effects on the skeletal muscle index (SMI) of the psoas major muscle by CT examination at the L3 level and the CONUT score (controlling nutritional status score) [97]. The CONUT score is based on four levels of serum albumin level, total cholesterol level, and peripheral lymphocyte count, and the nutritional level is evaluated by the summed score [139, 140] (Table 15.11). Baseline patient characteristics showed that there were significantly more patients with hepatocellular carcinoma ($P = 0.019$), higher Child-Pugh score ($P = 0.013$), and lower body mass index (BMI) ($P = 0.009$) in the sarcopenia group compared to the non-sarcopenia group. As a result of carnitine administration for 1 year, there was no significant change in SMI, but significant improvements in albumin, total cholesterol, and CONUT score were observed (Table 15.12).

Sato et al. administered carnitine to 30 patients with cirrhosis (1800 mg/day, orally, 6 months) and examined changes in blood ammonia, serum albumin, and oxidative stress markers, as well as the QOL of patients using a self-questionnaire called a chronic fatigue questionnaire [98]. Serum albumin levels significantly increased from 3.2 ± 0.5 μg/dL before treatment to 3.4 ± 0.3 and 3.6 ± 0.3 μg/dL at 3 and 6 months ($P < 0.05$), and ammonia levels significantly decreased from 101 ± 40 μg/dL before treatment to 56.5 ± 31 μg/dL at 6 months. The oxidative stress marker d-reactive oxygen metabolites (d-ROMs) also decreased significantly. In addition, when patients' QOL was investigated using a self-questionnaire, physical and mental scores related to fatigue and total score were significantly reduced at 3 and 6 months after treatment ($P < 0.05$). Thus, carnitine administration may improve the QOL of patients with liver cirrhosis.

Table 15.11 Calculation of CONUT (controlling nutritional status score) score

Nutritional status				
Variable	Normal	Light	Moderate	Severe
Albumin (g/dL)	≧3.5	3.0–3.49	2.5–2.99	<2.50
Score	0	2	4	6
Total lymphocyte count (/mm³)	≧ 1600	1200–1599	800–1199	<800
Score	0	1	2	3
Total cholesterol (mg/dL)	> 180	140–180	100–139	<100
Score	0	1	2	3
Screening total score	0–1	2–4	5–8	9–12

From refs. [139, 140]
CONUT controlling nutritional status
The CONUT score is a straightforward nutritional screening tool that relies on laboratory parameters, including total lymphocyte count, total cholesterol, and serum albumin. The calculation of CONUT score is the sum of serum albumin score, total lymphocyte count score, and total cholesterol score.

Sato et al. divided 41 patients with hepatitis C into group A (sofosbuvir + ribavirin, 22 patients) and group B group (sofosbuvir + ribavirin + carnitine, 19 patients) and examined Hb concentrations and sustained virological response (SVR) [99]. Sofosbuvir + ribavirin therapy is an effective drug treatment for virus eradication, but the treatment cannot be continued due to hemolytic anemia. In group A, 17 patients were able to complete treatment without changing the dose of ribavirin, but 4 patients had to change the dose due to hemolytic anemia. In group B, there was little decrease in Hb concentration, treatment was completed without changing the dose of ribavirin in all patients, and SVR24 was also high. When the quality of life of the patients was examined using the Short Form-8 (SF-8) questionnaire, a significant improvement in motor function was observed in group B compared to group A. Thus, it was suggested that ribavirin treatment for HCV patients could be continued due to the effect of reducing hemolytic anemia and improving QOL such as motor function when carnitine was administered in combination with ribavirin.

Ohara et al. screened 158 patients with cirrhosis and examined the effect of carnitine administration on muscle mass using a propensity score matching method [100]. The 35 patients in the carnitine group had been receiving carnitine for more than 6 months and had their muscle mass assessed by CT scans before and after carnitine administration (at least 6 months later). Seventy-nine patients who did not receive carnitine and underwent a paired CT

Table 15.12 Effects of L-carnitine treatment for cirrhotic sarcopenia patients on clinical variables

Variables	Baseline	After 1 year	*P* value
Albumin (g/dL)	3.63 ± 0.49	3.74 ± 0.46	*P* = 0.023 *
Total cholesterol (mg/dL)	163.74 ± 35.03	178.03 ± 31.08	*P* = 0.003 *
Total lymphocytes counts (/mL)	1363.31 ± 722.67	1446.05 ± 763.10	*P* = 0.261
Child-Pugh score	6.06 ± 1.67	5.91 ± 1.78	*P* = 0.474
CONUT score	3.41 ± 2.41	2.64 ± 1.88	*P* = 0.007 *

From ref. [97]
L-carnitine: 600 mg × 3 times/day, oral administration, for 1 year1, *: $P < 0.05$

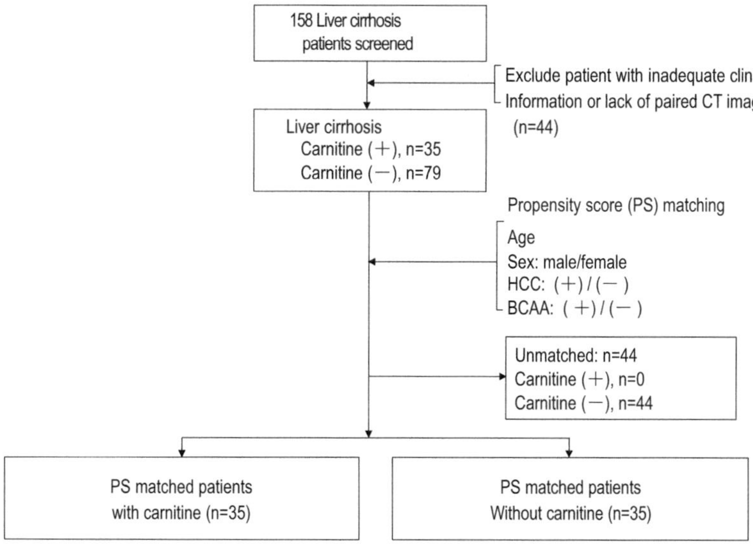

Fig. 15.19 Patient flow chart. Of 158 patients with liver cirrhosis who were screened, 44 were excluded due to inadequate clinical information or lack of paired CT imaging. Thus, 35 patients who were administered-carnitine and 79 control patients who were not administered L-carnitine were enrolled. After propensity score matching for age, sex, presence of HCC, and supplementation of BCAAs, 35 patients who received L-carnitine supplementation were selected as cases and 35 patients who did not receive carnitine supplementation were selected as matched controls. *HCC* hepatocellular carcinoma, *BCAA* branched chain amino acids. (From ref. [100])

scan during the study period and a second CT scan more than 6 months served as controls. Cases (patients with L-carnitine supplementation) and controls were matched for age, sex, branched-chain amino acids (BCAA) treatment, and the presence or absence of hepatocellular carcinoma, and retrospectively compared with matched control group (n=35, Fig. 15.19). On average, 1018 mg/day of carnitine was given orally in the carnitine group. Compared to the control group, the carnitine group had higher Child-Pugh scores and ammonia levels, lower platelet counts, white blood cell counts, prothrombin time, lower serum albumin, Hb and cholinesterase levels, and significantly worse liver function parameters. There was no significant difference in the psoas muscle mass index (PMI), which is an index for assessing the muscle mass of the psoas major muscle by the first and second CT, between men and women, in the carnitine administration group and the control group (non-carnitine group), but the rate of change in muscle mass per month (ΔPMI/m) by the first and second CT was higher in men. In women and overall, the carnitine group showed significantly less muscle mass loss than the control group (Fig. 15.20). Stratified analysis by age (under 65 years of age, 65 years of age or older), presence or absence of hepatocellular carcinoma, presence or absence of BCAA administration, presence or absence of hyperammonemia, etc., no effect of age was observed, and the presence of hepatocellular carcinoma significantly suppressed muscle mass loss but slightly weakened (no hepatocellular carcinoma: $P < 0.01$ versus with hepatocellular carcinoma: $P = 0.04$), there was no significant suppression of muscle mass loss without BCAA supplementation, but the effect of carnitine on inhibiting muscle loss was significant with BCAA supplementation (Fig. 15.21), and the effect of inhibiting muscle mass loss was slightly stronger in patients with ammonia reduction than in cases without ammonia reduction (Fig. 15.22). However, even in the group of patients who did not have a decrease in ammonia, carnitine administration significantly suppressed the loss of muscle mass compared to the non-treatment group. Thus, it was found that the administration of carnitine to cirrhosis patients with sarcopenia not only corrects hyperammonemia but also suppresses muscle mass loss.

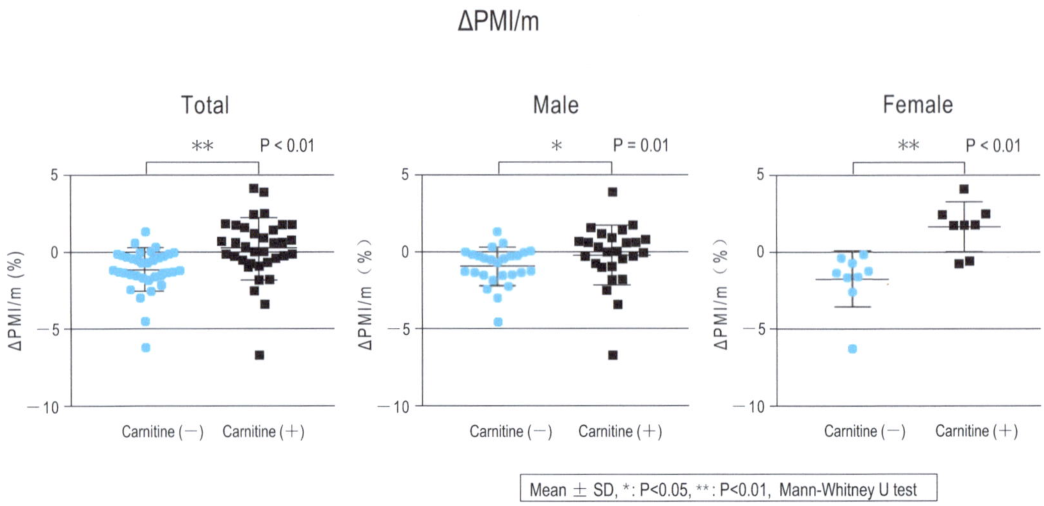

Fig. 15.20 Comparison of ΔPMI/month in patients with L-carnitine supplementation and controls. Left: overall cohort, middle: males, right: females. L-carnitine supplementation significantly suppressed loss of skeletal muscle mass. Data were analyzed with the Mann-Whitney U test, and shown as mean ± SD. Asterisk indicates a statistically significant difference (**P* < 0.05, ***P* < 0.01). *PMI* psoas muscle mass index (cross-sectional area of psoas muscle at L3 level of CT image). (From ref. [100])

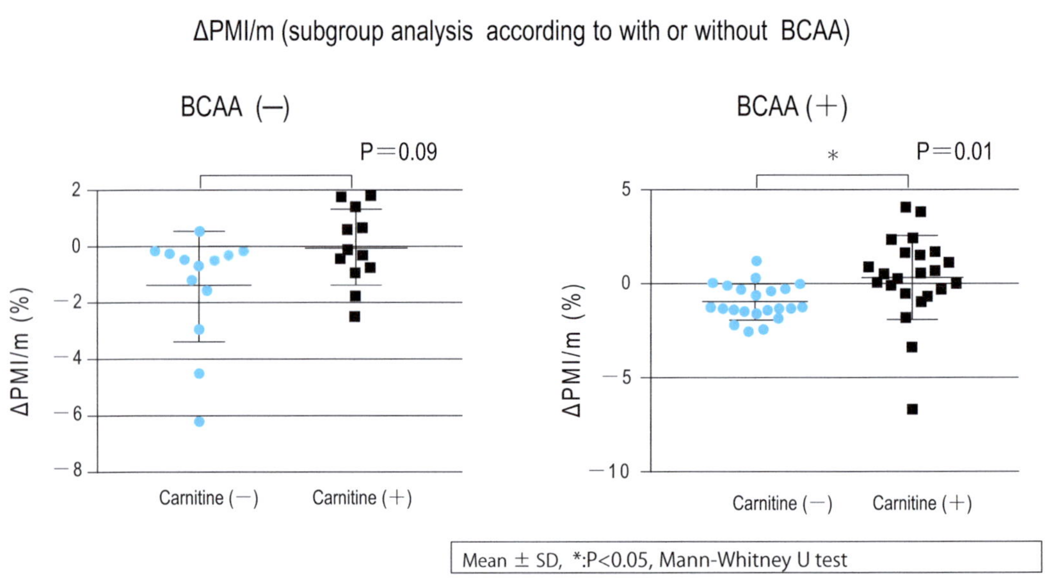

Fig. 15.21 Comparison of ΔPMI/month in patients with L-carnitine supplementation and controls in subgroup stratified according to BCAA supplementation. *PMI* psoas muscle mass index (cross-sectional area of psoas muscle at L3 level of CT image), *BCAA* branched chain amino acids. (From ref. [100])

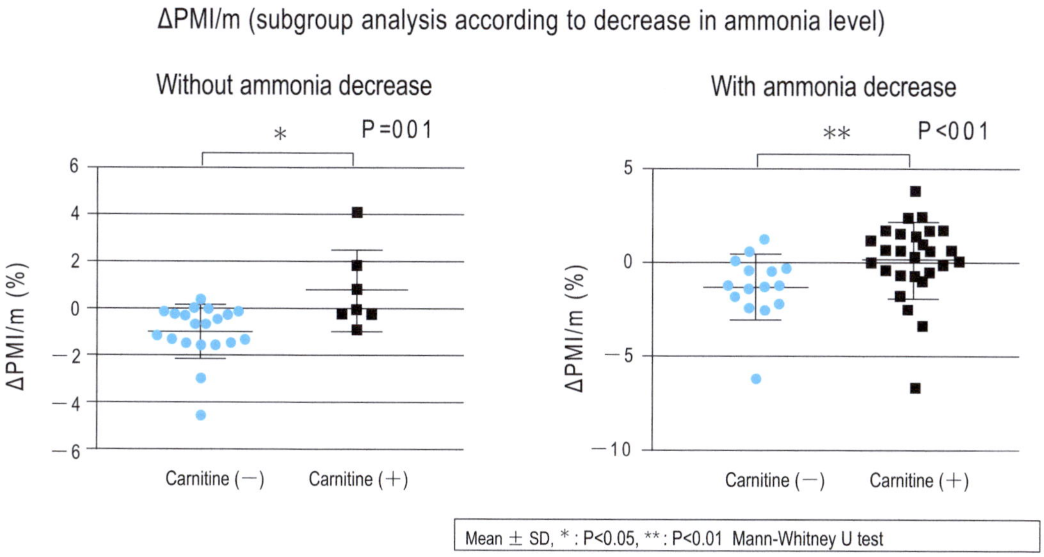

Fig. 15.22 Comparison of ΔPMI/month in patients with L-carnitine supplementation and controls in subgroup stratified according to patients with L-carnitine who did not experience an ammonia decrease (left), and patients with L-carnitine who experienced an ammonia decrease (right). Data were analyzed with the Mann-Whitney U test, and shown as mean ± SD. Asterisk indicates a statistically significant difference (*P < 0.05. **P < 0.01). PMI: psoas muscle mass index (cross-sectional area of psoas muscle at L3 level of CT image). (From ref. [100])

In the future, it is expected that the clinical application of carnitine supplementation therapy to sarcopenia in patients with cirrhosis will be further developed.

15.7 Positioning of Carnitine Supplementation Therapy in the Treatment of Patients with Cirrhosis

Thus, in patients with cirrhosis, the concentration of free carnitine in the blood is not necessarily reduced, and it may be high in some patients, and the concentration of acylcarnitine in the blood is often high. It is difficult to diagnose carnitine deficiency by measuring blood levels in patients with cirrhosis. In their review article, Hanai et al. discussed that carnitine deficiency in cirrhosis patients is not rare, and that carnitine deficiency should be suspected if there is hepatic encephalopathy, sarcopenia, cramps, decreased appetite, sleep disturbance, fatigue, decreased consciousness, or decreased cardiac function [101].

In this review, Hanai et al. described that carnitine deficiency in cirrhosis patients is formed mainly by a decrease in carnitine biosynthesis in the liver, with the addition of factors such as decreased food intake, PEM, skeletal muscle loss, and decreased renal function, and affects the function of the liver, which is the central organ such as gluconeogenesis, ammonia detoxification, and albumin biosynthesis. It is thought to be related to various clinical symptoms such as cramps (Fig. 15.23). Carnitine supplementation therapy for these symptoms in patients with cirrhosis is introduced by citing many papers in this review article, and its usefulness and potential are introduced, and it is said that it may be one of the important treatment methods for improving the QOL of patients with cirrhosis in the future.

In Clinical Question (CQ) 4–15 "Is carnitine useful for hepatic encephalopathy?" in the "Guidelines for the Treatment of Cirrhosis 2020 Revised 3rd Edition" edited by The Japanese Society of Gastroenterology and The Japan Hepatology Society, it is recommended that "Carnitine supplementation is proposed for

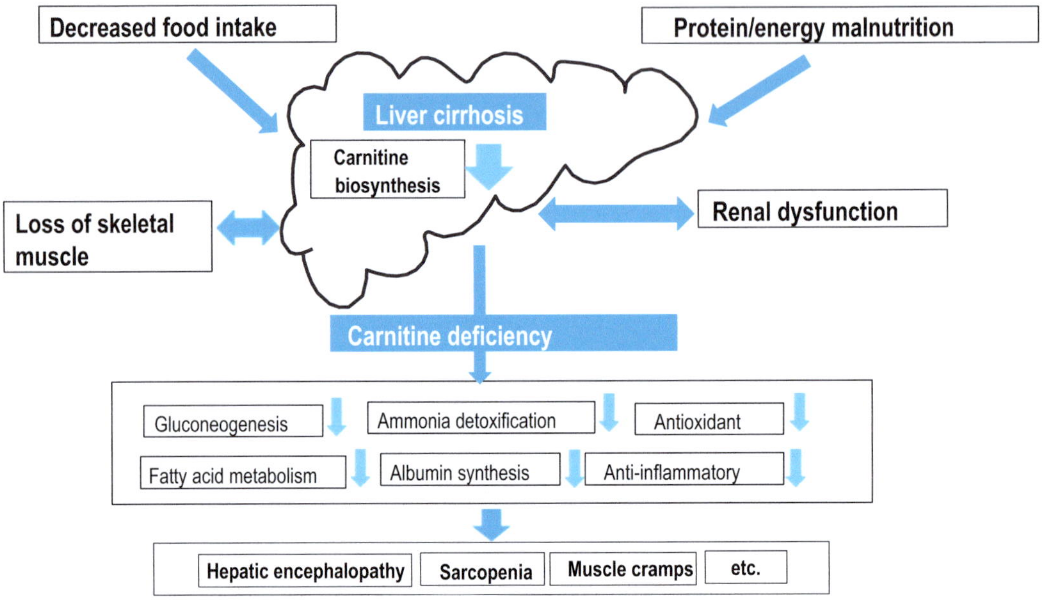

Fig. 15.23 Carnitine deficiency and liver cirrhosis. (From ref. [101])

patients with hepatic encephalopathy with carnitine deficiency because carnitine deficiency is often present in patients with cirrhosis. [Recommendation strength: weak (92% agreement rate, level of evidence: B)]." [102] CQ4–20 "Is there a useful treatment for muscle spasms complicated by cirrhosis of the liver?" also recommends that "Peony-Licorice Decoction, carnitine preparations, BCAA preparations, and zinc preparations be selected according to the pathology. [Strength of recommendation: Weak (80% agreement, level of evidence: C)]." [102]

15.8 NASH, NAFLD, and Carnitine

Non-alcoholic steatohepatitis (NASH) and non-alcoholic fatty liver disease (NAFLD) differ from cirrhosis derived from alcoholic hepatitis or viral hepatitis, which is derived from fat accumulation in the liver due to lifestyle-related diseases such as excessive fat intake, lack of exercise, obesity, and diabetes, is thought to be the underlying causes, with a high prevalence rate of approximately 30% in a large multicenter epidemiological survey of Japanese subjects from 2009 to 2010, 41% in men and 17.7% in women [103]. While progress in antiviral drugs has led to advances in the treatment of hepatitis C and other diseases and a decrease in the number of patients, the number of patients with NAFLD is on the increase, especially the high incidence of liver cancer from NASH with liver fibrosis, and the elucidation of its pathogenesis and the development of effective treatment methods are issues [104, 105].

15.8.1 Effect of Carnitine in Animal Models of NASH/NAFLD

15.8.1.1 Choline-Deficient Diet and Carnitine

Animal models of NASH and NAFLD include model animals induced by special diets such as the methionine-choline deficient diet (MCD), single-gene mutations and variants such as leptin-deficient mice (ob/ob mice), or STAM mice fed on a high-fat diet (HFD) after streptozotocin. In addition, a large number of animal models, such as those in which diabetic animal models are raised on a high-fat diet, have been developed and are being used to elucidate the pathophysiology of human NASH/NAFLD, which progresses over a very long period of time, and to search for treatments [106–108]. Most of these animal models of NASH/NAFLD show fatty liver, but it differs depending on the model, such as obesity, insulin resistance, steatohepatitis, the presence or absence of liver fibrosis, and the development of hepatocellular carcinoma, and it is not necessarily the same as the pathophysiology of human NASH/NAFLD. For example, NASH model fed MCD diet, which has been studied for a long time, does not show obesity or insulin resistance as seen in human NASH [109, 110]. In the MCD model animals, the uptake of free fatty acids into the liver is increased, and the deficiency of choline and methionine, which acts as a methyl group donor, stagnates the phospholipid phosphatidylcholine synthesis, which reduces VLDL and prevents the transport of triglycerides accumulated in the liver outside the liver in the form of VLDL, which is thought to lead to the formation of fatty liver [110].

Fritz et al. reported that choline administration to MCD rats suppressed fatty liver formation, but carnitine did not [111]. This may be because the main cause of fatty liver formation in the MCD animal model is a decrease in VLDL.

On the other hand, Corredor et al. reported that the β oxidation of long-chain fatty acids was reduced in the homogenates of the liver and heart of rats fed a choline-deficient diet, and that the addition of carnitine to this homogenate in vitro promoted fatty acid oxidation, but not when given choline [112]. In vivo, administration of carnitine or choline to choline-deficient rats increased long-chain fatty acid oxidation of tissue homogenates in the liver and heart. In addition, when choline was administered, the concentration of carnitine in these tissues increased. Therefore, they hypothesized that the decrease in the β oxidation of long-chain fatty acids in the liver and heart of choline-deficient rats may be at least partly due to a decrease in the concentration of carnitine in tissues.

Carter et al. found that a single intraperitoneal administration of choline (100 μmoles) to choline-deficient rats restored the carnitine concentration in the liver to normal values after 1.5 h, but the administration of methyl group donors such as betaine and methionine did not increase the concentration of hepatic carnitine. It was thought that it may be increasing transport of carnitine from other tissues and uptake into the liver [113]. Although it is not well understood why a choline-deficient diet reduces carnitine in tissues, it does not appear to be due to a decrease in carnitine biosynthesis due to a deficiency of choline, a simple methyl group donor.

Mollica et al. reported the effects of administering carnitine to a NAFLD model of MCD mice [114]. The progression of NAFLD was examined in three groups: the control group (control group), the MCD diet for 6 weeks (MCD group), and the carnitine (oral administration, 200 mg/kg) 3 weeks after the MCD diet (MCD + LC group). As a result, lipid accumulation and oxidative stress in the liver were reduced in the MCD + LC group, and fibrosis in the liver was suppressed by modulating the expression of α-smooth muscle actin (αSMA), peroxisome-activated receptor γ (PPARγ), and nuclear factor Kappa B (NF-κB). In addition, carnitine suppressed systemic inflammation, myocardial reactive oxygen production, and myocardial fibrosis.

Strilakou et al. also reported that myocardial inflammation and valvular heart membrane inflammation were pathologically observed in MCD animals, but these changes were suppressed by carnitine administration [115], and biochemical and histological studies of the effects of carnitine were reported [116, 117].

In the pathophysiology of NASH, mitochondrial dysfunction, especially β oxidation of long-chain fatty acids, has been reduced. Serviddio et al. reported that in MCD rats fed on a high-fat diet, the activity of CPT1, a rat-limiting enzyme of β oxidation, decreased, protein levels of CPT1 and mRNA levels were rather elevated, and post-translational oxidation of CPT1 molecules was the cause of the decrease in the activity [118].

15.8.1.2 Effects of Carnitine in Other NASH Model Animals

Ishikawa et al. investigated the effects of carnitine administration in STAM mice [119]. STAM mice sequentially developed NASH, liver fibrosis, and hepatocellular carcinoma in an animal

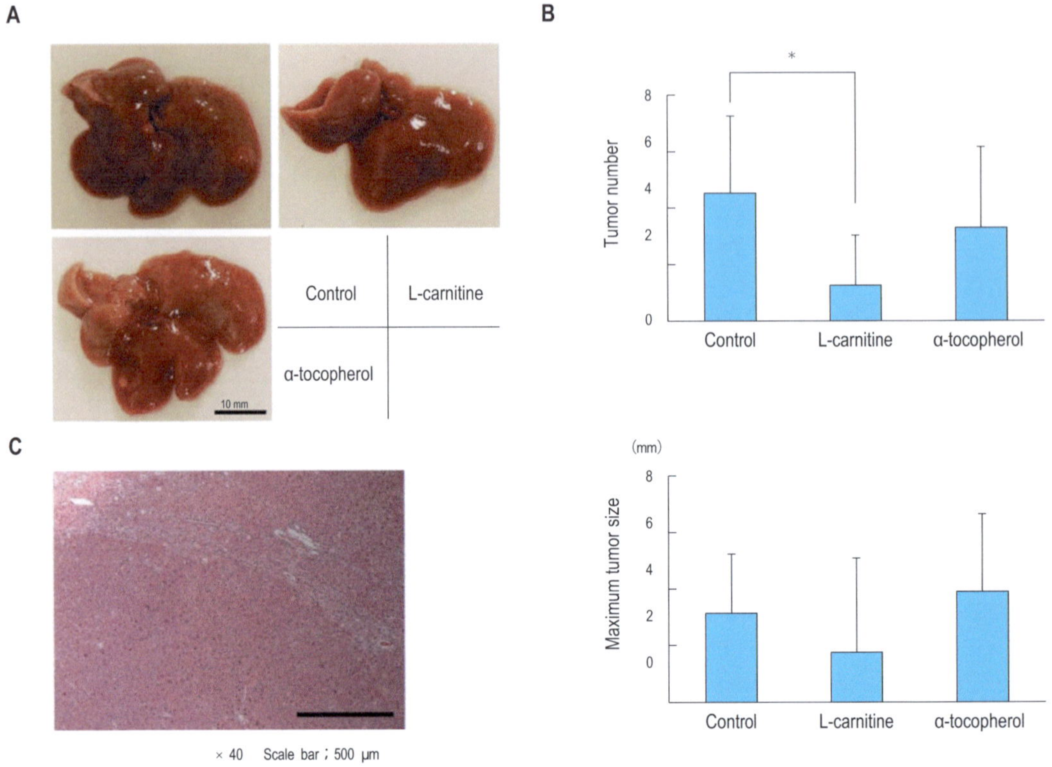

Fig. 15.24 Effects of L-carnitine and α-tocopherol on preventing hepatocarcinogenesis. (**a**) Control group developed numerous tumors on the liver surface, while groups receiving L-carnitine or α-tocopherol developed fewer tumors. (**b**) Number of tumors was significantly lower in the L-carnitine group. In the α-tocopherol group, the average tumor number and size were not reduced. (**c**) Histological findings showed that the tumors are hepatocellular carcinoma. Data are expressed as means ± SD. **P* < 0.05. (from ref. [119])

model of type I diabetes-inducing mice treated with streptozotocin on a high-fat diet. It is considered to be pathologically similar to human NASH [106](https://www.smccro-lab.com/jp/service/service_disease_area/stam.html). Eight-week-old STAM mice were divided into three groups: (1) a control group (high-fat diet), (2) a carnitine group (a high-fat diet +0.28% carnitine), and (3) a α-tocopherol group (a high-fat diet +0.01% α-tocopherols) and were kept for 4 and 8 weeks to perform biochemical and pathological studies and metabolome analysis, and the effects were compared with α-tocopherol. As a result, carnitine administration significantly suppressed fatty liver formation, and α-tocopherol administration significantly suppressed balloon-like degeneration of hepatocytes. In addition, carnitine administration enhanced the gene expression of transporters, enzymes, and antioxidant enzymes involved in long-chain fatty acid transport and mitochondrial β oxidation, thereby strongly suppressing oxidative stress and inflammatory cytokines in the liver. More interestingly, carnitine administration significantly reduced the number of hepatocellular carcinogenesis. α-Tocopherol was effective in pathological changes in NASH, but almost no carcinogenic suppression effect was observed (Fig. 15.24).

Since carnitine suppresses the development of hepatocellular carcinoma in NASH model animals, it is thought that carnitine, which is an agent that improves mitochondrial function, may be a simple new treatment strategy for NASH.

Diabetic animals fed a high-fat diet (HFD) are also often used as NASH models. Xia et al. examined the effects of carnitine administered to a mouse model of type II diabetes mellitus on a high-calorie diet/streptozotocin [120]. As a result, carnitine administration suppressed liver fatness, liver weight, and increase in triglycerides and free fatty acids, and histologically promoted clarification of cristae in hepatic mitochondria and decreased glycogen granules. Carnitine administration was thought to protect hepatic mitochondrial function.

Kon et al. investigated the effect of carnitine on fatty liver and fatty hepatitis caused by lipotoxicity due to free fatty acids in a high-fat diet-loaded KK-A^y mouse model [121]. KK-A^y mice were fed a high-fat diet for 4 weeks to produce fatty liver mice, and a group that continued the high-fat diet for another 4 weeks (HFD group). The fatty liver group was also treated with carnitine (1.25 mg/mL in drinking water) for 4 weeks, and the carnitine-treated group (LC group) was also fed with a high-fat diet for 4 weeks. The results of the study showed that the HFD group had higher body weight than the HFD group. The results showed that the HFD group showed increased liver weight per body weight, histologically advanced fatty liver, markedly increased serum AST and ALT, and progression of fatty hepatitis, while these were suppressed in the LC group (Fig. 15.25), and in addition, the LC group showed a strong suppression of fatty liver and hepatocyte apoptosis (M30-positive cells), as well as significantly reduced weight gain, hyperlipidemia, hyperglycemia, and hyperinsulinemia, which were observed in the HFD group. In lipid metabolism, serum free fatty acids and triglycerides were markedly increased in the HFD group, while free fatty acids and triglycerides were decreased in the LC group, and normalization of sterol regulatory element binding protein 1c (SREBP1c) expression in the LC group. This is possibly due to the removal of insulin resistance. These findings suggest that carnitine suppressed fatty hepatitis in high-fat diet-fed KK-Ay mice by enhancing fatty acid β-oxidation, normalizing insulin signaling, and suppressing lipogenesis in the liver. This suggests that carnitine administration may be a promising approach to prevent and treat NASH caused by metabolic syndrome.

15.8.1.3 Mechanism of Action of Carnitine in NASH/NAFLD Model

In this way, the protective effect of carnitine has been reported in the NASH/NAFLD model, and its action as a modulator of the increased permeability of mitochondrial membranes has attracted attention as a mechanism of action. In liver fibrosis seen in cirrhosis and NASH/NAFLD, there is a quality control system such as increased permeability of the mitochondrial membrane followed by mitochondrial swelling, division and fusion, and autophagy of impaired mitochondria, followed by necrosis. Cell death such as apoptosis or ferroptosis is thought to play a major role in the progression of disease conditions [122–128]. Carnitine has also been taken up as a modulator of increased permeability of mitochondrial membranes in various diseases, including liver diseases, and its mechanism of action is considered to be the suppression of caspase, antioxidant effect, and promotion of β oxidation of long-chain fatty acids [129]. As described in this chapter Sect. 15.3.3 and Chap. 7 carnitine excretes long-chain acyl-CoA such as palmitoyl-CoA accumulated in mitochondria as acylcarnitine from the mitochondria. In addition, it has been found that it suppresses apoptosis and maintains mitochondrial function by Nrf2-mediated suppression of inflammatory cytokine production, reducing oxidative stress, and suppressing caspase 3. In the future, it is expected that further research will be conducted to elucidate the pathophysiology and the significance of carnitine administration in this field.

15.8.2 Carnitine Kinetics and Effects of Carnitine Administration in NASH Patients

Pérez-Carreras et al. examined the carnitine content in liver tissue and the activity of respiratory chain enzymes in liver mitochondria by performing liver biopsies in 43 patients with NASH (mean age 44 ± 11 years) and healthy subjects (16 cases) [130]. As a result, long-chain acylcarnitine in liver

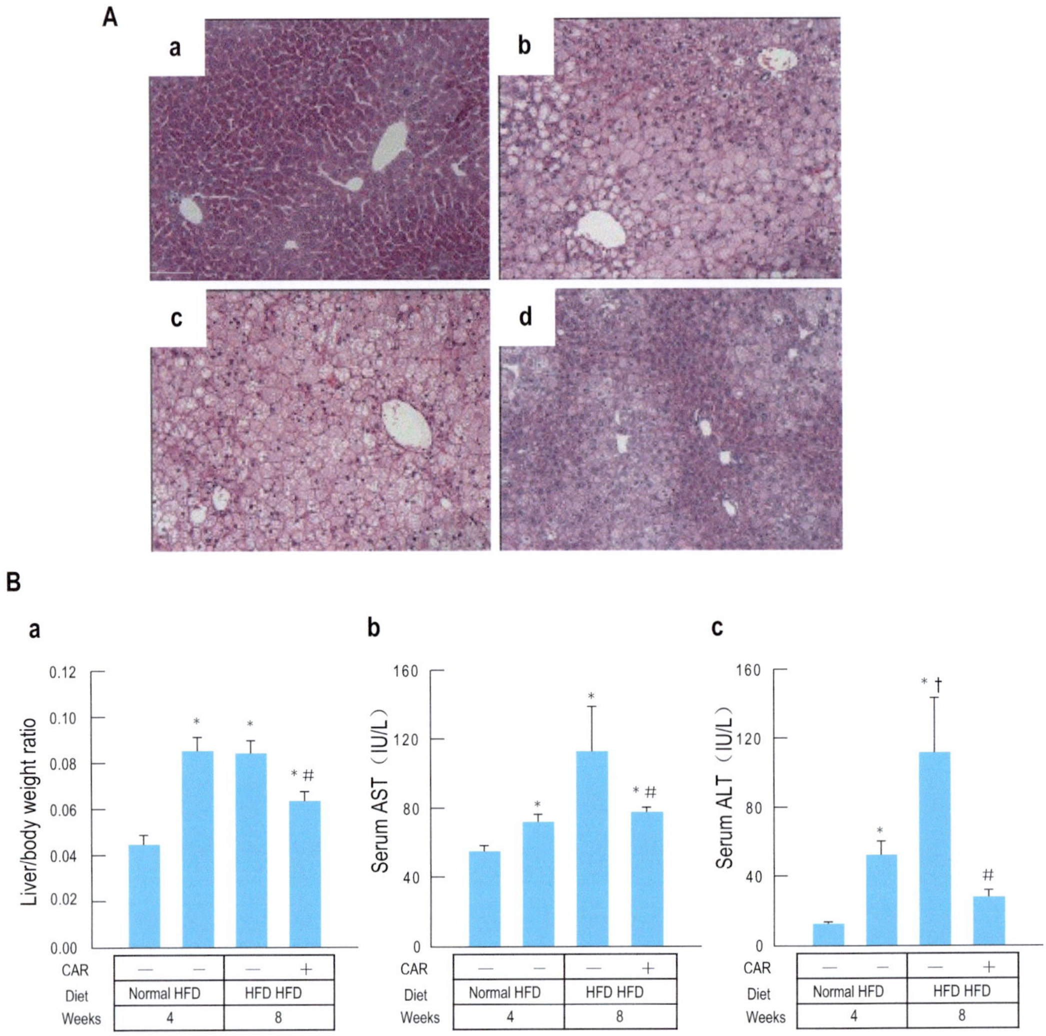

Fig. 15.25 Treatment with L-carnitine (CAR) prevents hepatic steatosis following high-fat diet (HFD) feeding in KK-Ay mice. Representative photomicrographs of liver histology in mice after 4 weeks of feeding with control chow (**a**), 4-week HFD (**b**), and 8-week HFD (**c**). (**d**) Eight-week HFD with L-carnitine for the latter 4 weeks (hematoxylin-eosin staining, original magnification × 100). Liver/body weight ratio (**e**), and the average levels of serum aspartate aminotransferase (AST) (**f**) and alanine aminotransferase (ALT) (**g**) are plotted (mean ± standard error of mean) ($n = 5$; $^*P < 0.05$ vs. controls; $^{\dagger}P < 0.05$ vs. 4 week HFD; $^{\#}P < 0.05$ vs. 8 week HFD alone, by analysis of variance (ANOVA) on ranks and Student-Newman-Keuls post-hoc test). (From ref. [121])

tissue was significantly increased ($P < 0.05$), short-chain acylcarnitine was significantly decreased ($P < 0.01$), and the long-chain acylcarnitine/free carnitine ratio was significantly increased in patients with NASH (Fig. 15.26). In addition, the respiratory chain enzyme activity of liver mitochondria in NASH patients was 63 ± 20% for complex I, 58.5 ± 16.7% for complex II, 70.6 ± 10.3% for complex III, 62.5 ± 13% for complex IV, and 42.4 ± 9.1% for complex V (ATP synthase); a significant decrease in these enzyme activities was observed in all cases (Fig. 15.27). In patients with NASH, serum tumor necrosis factor alpha (TNFα), Homeostatic metabolic assessment of insulin resistance (HOMA-IR), and BMI were significantly higher than controls, and serum TNFα and HOMA-IR were inversely correlated with mitochondrial respiratory chain enzyme activity. This suggests that long-chain acyl-CoA accumulation observed in animal NASH models also occurs in actual human hepatitis patients, causing mitochondrial dysfunction.

Although it is not in a study in patients with NASH/NAFLD, Romano et al. in Malaguarnera's group divided 70 patients with hepatitis C who were treated with interferon α + ribavirin (12 months) into two groups, and one group received carnitine (2000 mg/day, orally, 12 months) in addition to interferon α and ribavirin. The effects of carnitine administration were examined [131]. The results showed that the carnitine-treated group had a significantly greater reduction in ALT ($P < 0.05$) and a significantly greater reduction in total cholesterol and triglycerides ($P < 0.05$). The reductions in liver fibrosis and fatty liver scores were also significantly greater in the carnitine-treated group, and post-treatment response was also higher in the carnitine-treated group (48 vs. 56%). Based on these findings, Malaguarnera et al. hypothesized that carnitine treatment may reduce fatty liver formation during interferon alpha plus ribavirin treatment in hepatitis C patients.

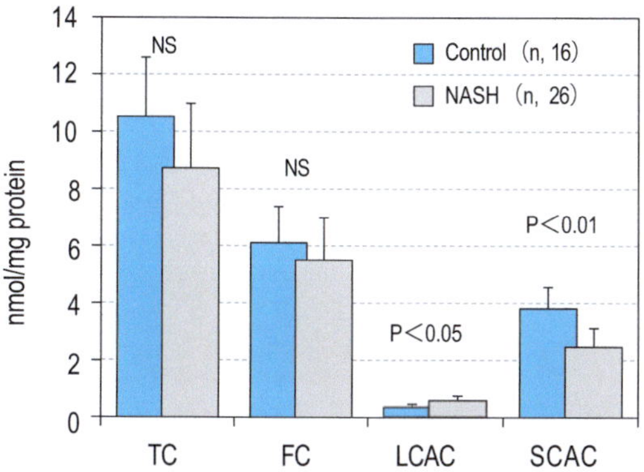

Fig. 15.26 Carnitine and carnitine ester hepatic contents. Total liver carnitine (TC), free carnitine (FC), long-chain acylcarnitine (LCAC), and short-chain acylcarnitine (SCAC) in liver specimens of 16 control subjects and 26 patients with NASH. *NS* not significant. (From ref. [130])

Therefore, Malaguarnera et al. divided NASH patients (74 patients) in two groups: a placebo + diet (1600 calories) + exercise (30 min total body stretch, 3 times a week) treatment only group (group A, 38 patients) and a similar diet + exercise + carnitine treatment (group B, 36 patients, 1 g morning and night, total 2 g/day, oral administration) and conducted a randomized, double-blind, placebo-controlled study to evaluate the effect of carnitine administration on regression of NASH by comparing liver function enzymes,

Fig. 15.27 Activity of the mitochondrial respiratory complexes (MRC) in liver specimens of 11 control subjects and 18 patients with NASH. Enzyme activities are expressed as nmol × min^{-1} × mg protein $^{-1}$ × 100/ nmol × mg protein^{-1} CS ([complex/CS] × 100). *CS* citrate synthase. (From ref. [130])

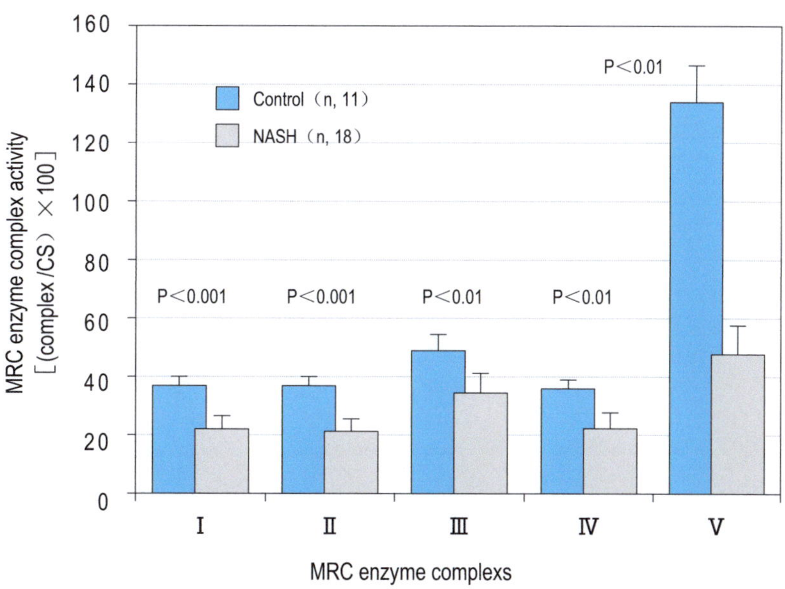

Table 15.13 Laboratory parameters in patients treated with placebo plus diet or L-carnitine plus diet before and after 24 weeks of treatment

	Group A (placebo + diet)	$n = 38$	Group B (L-carnitine + diet) $n = 36$	
	Before treatment	After 24 weeks	Before treatment	After 24 weeks
BMI (kg/m^2)	26.5 ± 3.8	25.4 ± 3.9	26.6 ± 3.7	25.3 ± 3.5
AST (IU/L)	124.2 ± 12.8	78.1 ± 21.4**	128.1 ± 13.9	56.4 ± 18.2$^{**\#\#}$
ALT (IU/L)	112.8 ± 13.1	75.4 ± 13.9**	110.2 ± 15.6	51.8 ± 16.4$^{**\#\#}$
γ-GT (IU/L)	98.2 ± 18.2	77.8 ± 19.4**	104.1 ± 17.2	66.5 ± 15.4$^{**\#}$
Albumin (g/L)	48 ± 5.1	48.7 ± 5.4	47.8 ± 5.4	48.1 ± 5.6
Total cholesterol (mmol/L)	6.04 ± 0.91	5.88 ± 0.98	6.09 ± 0.94	4.78 ± 0.82$^{**\#\#}$
HDL (mmol/L)	0.98 ± 0.12	1.10 ± 0.14**	0.96 ± 0.14	1.16 ± 0.12**
LDL (mmol/L)	4.45 ± 0.87	4.24 ± 0.88	4.52 ± 0.95	3.15 ± 0.66$^{**\#\#}$
Triglycerides (mmol/L)	3.04 ± 0.96	2.68 ± 0.92	3.07 ± 0.88	2.34 ± 0.66**
Glucose (mmol/L)	6.04 ± 0.66	5.97 ± 0.71	6.12 ± 0.69	5.32 ± 0.55$^{**\#\#}$
Insulin (mIU/L)	16.8 ± 5.4	15.2 ± 5.2	17.1 ± 5.6	13.8 ± 5.0*
C-peptide (mmol/L)	0.96 ± 0.4	0.94 ± 0.41	0.98 ± 0.41	0.93 ± 0.62
HOMA-IR	4.5 ± 0.67	4.03 ± 0.71*	4.65 ± 0.70	3.26 ± 0.56$^{**\#\#}$
CRP (mg/dL)	8.7 ± 3.4	7.4 ± 3.2	9.1 ± 3.2	5.2 ± 3.1$^{**\#}$
TNFα (pg/mL)	1.38 ± 0.22	1.30 ± 0.21	1.44 ± 0.28	1.08 ± 0.15$^{**\#\#}$

From ref. [132]
All values are x ± s.d.
Comparison within group A and within group B according to the values before the treatment; *: $P < 0.05$, **: $P < 0.001$
Comparison between groups A and B after 24 weeks of treatment; $^{\#}$: $P < 0.05$, $^{\#\#}$: $P < 0.001$

lipid profiles, fasting blood glucose, inflammatory markers, insulin resistance, and histological scores of the two groups after 24 weeks [132]. There were no significant differences in age, gender, BMI, or various laboratory tests between the two groups. Because both groups received diet and exercise as basic treatment, even the placebo-treated group A showed significant improvement in liver function enzymes, HDL, and HOMA-IR compared with the pretreatment. However, in the carnitine-treated group B, in addition to these parameters, various parameters after 24 weeks of carnitine treatment were significantly lower than before the treatment. In addition to these parameters, total cholesterol, LDL, triglycerides, blood glucose, insulin, CRP, and TNFα were also significantly improved in group B after 24 weeks of carnitine treatment compared with those before treatment (Table 15.13). Furthermore, a comparison of these parameters in groups A and B after 24 weeks of treatment showed that liver function enzymes, lipid parameters, blood glucose, HOMA-IR, CRP, and TNFα were significantly improved in group B (Table 15.13). In addition, histological parameters of liver tissue in both groups after 24 weeks of treatment were examined. Steatosis, hepatocellular injury, parenchymal inflammation, portal inflammation, fibrosis, Mallory bodies damage, and the NASH activity index, expressed as the sum of the scores for steatosis, parenchymal inflammation, and hepatocellular damage were all significantly improved in both groups before and after treatment, respectively in group A and group B. After 24 weeks of treatment, all parameters except the Mallory body damage score were significantly improved in the carnitine-treated group B (Table 15.14).

Table 15.14 Histological data in patients treated with placebo plus diet or L-carnitine plus diet before and after 24 weeks of treatment

	Group A	(Placebo + diet) $n = 38$	Group B (L-carnitine + diet) $n = 36$	
Parameter (0–4)	Before treatment	After 24 weeks	Before treatment	After 24 weeks
Steatosis	2.79 ± 0.47	1.68 ± 0.76**	3.22 ± 0.63	0.94 ± 0.88**##
Hepatocellular injury	2.66 ± 0.47	1.47 ± 0.64**	3.00 ± 0.33	1.05 ± 0.74**#
Parenchymal inflammation	2.76 ± 0.43	1.42 ± 0.63**	3.22 ± 0.48	1.94 ± 0.7**##
Portal inflammation	2.89 ± 0.38	1.82 ± 0.68**	2.77 ± 0.48	1.28 ± 0.80**#
Fibrosis	2.74 ± 0.44	1.89 ± 0.72**	2.86 ± 0.53	1.55 ± 0.72**#
Mallory bodies	2.74 ± 0.50	1.53 ± 0.72**	2.94 ± 0.47	1.42 ± 0.68**
NASH activity (0–12)[a]	8.21 ± 0.86	4.58 ± 1.31**	9.42 ± 0.76	3.19 ± 1.71**##

From ref. [132]
NASH non-alcoholic steatohepatitis
Comparison within group A and within group B according to the values before the treatment; *: $P < 0.05$, **: $P < 0.001$
Comparison between groups A and B after 24 weeks of treatment; #: $P<0.05$, ##: $P<0.001$
[a]: Sum of steatosis, hepatocellular injury, and parenchymal inflammation. All values are x ± s.d.

15.9 Pancreatic Cancer, NASH/NAFLD-Related Hepatocellular Carcinoma, and Carnitine Kinetics

Patients who are not with usual NASH/NAFLD but have undergone pancreaticoduodenectomy or total pancreatomy for malignant tumors of the pancreatic head region, lower bile ducts, duodenum, and duodenal papillae associated with liver adipose often develop fatty liver after surgery, which is a problem in the field of gastroenterological surgery. Although the pathogenesis mechanism is not yet understood, it has been pointed out that lipid uptake into hepatocytes, increase in lipid production and expression of PPARγ, disruption of VLDL secretion mechanism, and increase of oxidative stress occur [133].

Nakamura et al. administered carnitine (1800 mg/day, administration period 90.4 ± 44.6 days) to eight patients who had undergone pancreaticoduodenectomy or total pancreatic resection (two patients received carnitine at different times) and examined the effects by CT attenuation of CT images and blood tests [134]. As a result, CT value was significantly improved in seven out of eight patients ($P = 0.016$), and the mean CT value was also improved. In one case, CT levels in the liver improved after two different carnitine treatments but worsened again 3 months after discontinuation of carnitine. When carnitine was administered again, the CT value improved again.

In addition, Nakamura et al. retrospectively examined serum carnitine concentrations in patients who underwent pancreaticoduodenectomy (19 cases) and total pancreatic resection (2 cases) and found that many patients had low levels of free carnitine after removal surgery, and that the acylcarnitine/free carnitine ratio (AC/FC) was inversely correlated with hepatic attenuation (CT values, which were low as steatosis progress). Multivariate analyses also showed that AC/FC levels were the only risk factor for hepatic steatosis after pancreaticoduodenectomy or total pancreatectomy ($P = 0.020$) [135]. This indicates that carnitine insufficiency in which long-chain acyl-CoA accumulates in the mitochondria occurs in hepatic adipose after pancreatectomy.

Enooku et al. reported a profile analysis of 15 acylcarnitines (hereafter AC) in carbon chains 2–18 in 241 NAFLD patients, including 23 NAFLD-related hepatocellular carcinoma patients [136]. The results showed that AC14:1 and AC18:1 gradually increased with the progression of fibrosis and further increased in patients with hepatocellular carcinoma, while medium-chain acylcarnitines such as AC5:0 showed an opposite trend. In particular, AC18:1 was significantly higher in patients with hepatocellular carcinoma than in those without hepatocellular carcinoma. Long-chain acylcarnitine levels were positively correlated with inflammatory cytokine levels, and (AC16:0 + AC18:1)/AC2:0, an indicator of CPT2 deficiency, was independently associated with hepatocellular carcinoma after adjustment for age and liver fibrosis index. These findings suggest that accumulation of long-chain acylcarnitines and deficiency of free carnitine occur in patients with NASH/NAFLD and pancreatectomized patients and are closely related to pathological conditions such as decreased mitochondrial function and accelerated fibrosis and carcinogenesis, suggesting that carnitine supplementation therapy for NASH/NAFLD patients may be promising for suppressing the progression of the disease and inhibiting carcinogenesis [137].

15.10 Summary

As described above, there was some debate as to whether it could be said carnitine deficiency because the blood carnitine concentration of cirrhosis patients is often normal or high, except for patients with end-stage cirrhosis with cachexia. However, it has been found that the high concentration of carnitine in the blood is elevated due to the decay of hepatocytes and entering the blood and extracellular fluid pool, which is said to be about 0.5% of the total, and that the concentration of long-chain acylcarnitine is high not only in cirrhosis but also in NASH/NAFLD, pancreaticoduodenectomy, or fatty liver after pancreaticoduodenectomy or pancreatomy.

This high level of acylcarnitine in the blood (mainly long-chain acylcarnitine) is considered to be a pathological condition of carnitine insufficiency, which is a kind of carnitine deficiency in which long-chain acyl-CoA accumulates in the mitochondria and is relatively insufficient free carnitine. In this pathological condition, due to the suppression of mitochondrial β oxidation, long-chain acyl-CoA accumulates, and then suppresses mitochondrial energy metabolism and electron transport system, which cause mitochondrial dysfunction and energy metabolism abnormality.

Thus, not only cirrhosis but also animal models of NASH/NAFLD and clinical practice of patients with NASH/NAFLD, it was shown that carnitine administration promotes fatty acid oxidation in mitochondria, excretes long-chain acyl-CoA accumulated in mitochondria, and maintains mitochondrial function. Furthermore, it is thought that the anti-inflammatory, antioxidant, and anti-apoptotic actions of carnitine through Nrf2 suppress fatty liver, steatohepatitis, fibrosis, and their progression, thereby causing histological improvement. In addition, carnitine supplementation therapy in various liver disease pathophysiologists has been shown to improve the clinical symptoms and QOL of patients with liver disease by reducing ammonia, improving muscle symptoms such as cramps and fatigue, suppressing sarcopenia and improving nutritional indices such as serum albumin, and suppressing increased inflammation and oxidative stress. In addition, Miwa et al. report that carnitine administration improves the prognosis of life in patients with cirrhosis [138]. Carnitine is a substance that originally exists in the body, and it is considered that there is no problem even if it is administered intravenously at high concentrations, and its safety is excellent. In the future, clinical research on carnitine supplementation therapy in the field of liver diseases is expected to progress, and it is expected to contribute to treatment.

References

1. Böhmer T, Rydning A, Solberg HE. Carnitine levels in human serum in health and disease. Clin Chim Acta. 1974;57(1):55–61. https://doi.org/10.1016/0009-8981(74)90177-6.
2. Chen SH, Lincoln SD. Increased serum carnitine concentration in renal insufficiency. Clin Chem. 1977;23(2 Pt 1):278–80.
3. Rudman D, Sewell CW, Ansley JD. Deficiency of carnitine in cachectic cirrhotic patients. J Clin Invest. 1977;60(3):716–23. https://doi.org/10.1172/JCI108824.
4. Fuller RK, Hoppel CL. Elevated plasma carnitine in hepatic cirrhosis. Hepatology. 1983;3(4):554–8. https://doi.org/10.1002/hep.1840030413.
5. Fuller RK, Hoppel CL. Plasma carnitine in alcoholism. Alcohol Clin Exp Res. 1988;12(5):639–42.

https://doi.org/10.1111/j.1530-0277.1988.tb00256.x.

6. de Sousa C, Leung NW, Chalmers RA, et al. Free and total carnitine and acylcarnitine content of plasma, urine, liver and muscle of alcoholics. Dermatol Clin Sci (Lond). 1988;75(4):437–40. https://doi.org/10.1042/cs0750437.
7. Amodio P, Angeli P, Merkel C, et al. Plasma carnitine levels in liver cirrhosis: relationship with nutritional status and liver damage. J Clin Chem Clin Biochem. 1990;28(9):619–26. https://doi.org/10.1515/cclm.1990.28.9.619.
8. Alonso de la Peña C, Rozas I, Alvarez-Prechous A, et al. Free carnitine and acylcarnitine levels in sera of alcoholics. Biochem Med Metab Biol. 1990;44(1):77–83. https://doi.org/10.1016/0885-4505(90)90047-5.
9. Wennberg A, Hyltander A, Sjöberg Å, et al. Prevalence of carnitine depletion in critically ill patients with undernutrition. Metabolism. 1992;41(2):165–71. https://doi.org/10.1016/0026-0495(92)90146-2.
10. Palombo JD, Borum PR, Jenkins RL, et al. Blood carnitine status after orthotopic liver transplantation in patients with end-stage liver disease. Am J Clin Nutr. 1989;50(3):504–7. https://doi.org/10.1093/ajcn/50.3.504.
11. Krähenbühl S, Reichen J. Carnitine metabolism in patients with chronic liver disease. Hepatology. 1997;25(1):148–53. https://doi.org/10.1053/jhep.1997.v25.pm0008985281.
12. D'Arienzo A, Mattera D, Ambrogio G, et al. Hypercarnitinemia in cirrhosis. Hepatology. 1985;5(2):343. https://doi.org/10.1002/hep.1840050237.
13. Eskandari GH, Kandemir Ö, Polat G, et al. Serum L-carnitine levels and lipoprotein compositions in chronic viral hepatitis patients. Clin Biochem. 2001;34(5):431–3. https://doi.org/10.1016/s0009-9120(01)00234-x.
14. Zhou L, Wang Q, Yin P, et al. Serum metabolomics reveals the deregulation of fatty acids metabolism in hepatocellular carcinoma and chronic liver diseases. Anal Bioanal Chem. 2012;403(1):203–13. https://doi.org/10.1007/s00216-012-5782-4.
15. Shiraki M, Shimizu M, Moriwaki H, et al. Carnitine dynamics and their effects on hyperammonemia in cirrhotic Japanese patients. Hepatol Res. 2017;47(4):321–7. https://doi.org/10.1111/hepr.12750.
16. Suzuki K, Onodera M, Hidekatsu K, et al. Reevaluation of serum carnitine status in patients with liver cirrhosis. J Liver Res Disord Ther. 2016;2(2):58–64. https://doi.org/10.15406/jlrdt.2016.02.00025.
17. Ida S, Iwasaki Y, Uchida K, et al. Guideline for diagnosis and treatment of carnitine deficiency 2018. https://www.jpeds.or.jp/uploads/files/20181207_shishin.pdf (Article in Japanese).
18. Watanabe M. Hepatic encephalopathy. Acta Hepatol Jpn. 2002;43(2):77–83. (Article in Japanese)
19. Hadjihambi A, Arias N, Sheikh M, et al. Hepatic encephalopathy: a critical current review. Hepatol Int. 2018;12(Suppl 1):S135–47. https://doi.org/10.1007/s12072-017-9812-3.
20. Tajika M, Kato M, Miwa Y, et al. Prognostic value of energy metabolism in patients with liver cirrhosis. Gastroenterology. 1998;26(5):518–25. (Article in Japanese)
21. Katayama K. Abnormal nitrogen metabolism in cirrhosis and zinc. J Zn Nutr Therapy. 2010;1(1):26–34. (Article in Japanese)
22. Katayama K. Effects of zinc supplementation on hepatic encephalopathy and hyperammonemia: significance of zinc in chronic liver diseases. J Clin Exp Med. 2012;240(9):759–64. (Article in Japanese)
23. Shimizu Y, Kusano H. Porto-systemic shunt. Acta Hepatol Jpn. 2002;43:479–81. (Article in Japanese)
24. Nardelli S, Riggio O, Gioia S, et al. Spontaneous porto-systemic shunts in liver cirrhosis: clinical and therapeutical aspects. World J Gastroenterol. 2020;26(15):1726–32. https://doi.org/10.3748/wjg.v26.i15.1726.
25. Simón-Talero M, Roccarina D, Martínez J, et al. Association between portosystemic shunts and increased complications and mortality in patients with cirrhosis. Gastroenterology. 2018;154(6):1694–705. https://doi.org/10.1053/j.gastro.2018.01.028.
26. Praktiknjo M, Simón-Talero M, Römer J, et al. Total area of spontaneous portosystemic shunts independently predicts hepatic encephalopathy and mortality in liver cirrhosis. J Hepatol. 2020;72(6):1140–50. https://doi.org/10.1016/j.jhep.2019.12.021.
27. Philips CA, Rajesh S, Augustine P, et al. Portosystemic shunts and refractory hepatic encephalopathy: patient selection and current options. Hepat Med. 2019;11:23–34. https://doi.org/10.2147/HMER.S169024.
28. O'Connor JE, Costell M, Grisolía S. Protective effect of L-carnitine on hyperammonemia. FEBS Lett. 1984;166(2):331–4. https://doi.org/10.1016/0014-5793(84)80106-4.
29. Costell M, O'Connor JE, Míguez MP, et al. Effects of L-carnitine on urea synthesis following acute ammonia intoxication in mice. Biochem Biophys Res Commun. 1984;120(3):726–33. https://doi.org/10.1016/s0006-291x(84)80167-9.
30. O'Connor JE, Costell M, Grisolía S. Prevention of ammonia toxicity by L-carnitine: metabolic changes in brain. Neurochem Res. 1984;9(4):563–70. https://doi.org/10.1007/BF00964383.
31. O'Connor JE, Costell M, Míguez MP, et al. Influence of the route of administration on the protective effect of L-carnitine on acute hyperammonemia. Biochem Pharmacol. 1986;35(18):3173–6. https://doi.org/10.1016/0006-2952(86)90407-7.
32. Costell M, Míguez MP, O'Connor JE, et al. Effect of hyperammonemia on the levels of carnitine in

mice. Neurology. 1987;37(5):804–8. https://doi.org/10.1212/wnl.37.5.804.
33. O'Connor JE, Costell M, Míguez MP, et al. Effect of L-carnitine on ketone bodies, redox state and free amino acids in the liver of hyperammonemic mice. Biochem Pharmacol. 1987;36(19):3169–73. https://doi.org/10.1016/0006-2952(87)90628-9.
34. Kloiber O, Banjac B, Drewes LR. Protection against acute hyperammonemia: the role of quaternary amines. Toxicology. 1988;49(1):83–90. https://doi.org/10.1016/0300-483x(88)90178-3.
35. Deshmukh DR, Rusk CD. Failure of L-carnitine to protect mice against ammonia toxicity. Biochem Med Metab Biol. 1988;39(1):126–30. https://doi.org/10.1016/0885-4505(88)90068-0.
36. Deshmukh DR, Singh KR, Meert K, et al. Failure of L-carnitine to protect mice against hyperammonemia induced by ammonium acetate or urease injection. Pediatr Res. 1990;28(3):256–60. https://doi.org/10.1203/00006450-199009000-00021.
37. Hearn TJ, Coleman AE, Lai JC, et al. Effect of orally administered L-carnitine on blood ammonia and L-carnitine concentrations in portacaval-shunted rats. Hepatology. 1989;10(5):822–8. https://doi.org/10.1002/hep.1840100512.
38. Ohtsuka Y, Griffith OW. L-carnitine protection in ammonia intoxication: effect of aminocarnitine on carnitine-dependent metabolism and acute ammonia toxicity. Biochem Pharmacol. 1991;41(12):1957–61. https://doi.org/10.1016/0006-2952(91)90136-s.
39. Matsuoka M, Igisu H, Kohriyama K, et al. Suppression of neurotoxicity of ammonia by L-carnitine. Brain Res. 1991;567(2):328–31. https://doi.org/10.1016/0006-8993(91)90814-c.
40. Therrien G, Rose C, Butterworth J, et al. Protective effect of L-carnitine in ammonia-precipitated encephalopathy in the portacaval shunted rat. Hepatology. 1997;25(3):551–6. https://doi.org/10.1002/hep.510250310.
41. Maddaiah VT. Ammonium inhibition of fatty acid oxidation in rat liver mitochondria. A possible cause of fatty liver in Reye's syndrome and urea cycle defects. Biochem Biophys Res Commun. 1985;127(2):565–70. https://doi.org/10.1016/s0006-291x(85)80197-2.
42. Kosenko E, Kaminsky Y, Grau E, et al. Brain ATP depletion induced by acute ammonia intoxication in rats is mediated by activation of the NMDA receptor and Na+, K+-ATPase. J Neurochem. 1994;63(6):2172–8. https://doi.org/10.1046/j.1471-4159.1994.63062172.x.
43. Felipo V, Miñana MD, Cabedo H, et al. L-carnitine increases the affinity of glutamate for quisqualate receptors and prevents glutamate neurotoxicity. Neurochem Res. 1994;19(3):373–7. https://doi.org/10.1007/BF00971588.
44. Felipo V, Kosenko E, Miñana MD, et al. Molecular mechanism of acute ammonia toxicity and of its prevention by L-carnitine. Adv Exp Med Biol. 1994;368:65–77. https://doi.org/10.1007/978-1-4615-1989-8_7.
45. Miñana MD, Hermenegildo C, Llansola M, et al. Carnitine and choline derivatives containing a trimethylamine group prevent ammonia toxicity in mice and glutamate toxicity in primary cultures of neurons. J Pharmacol Exp Ther. 1996;279(1):194–9.
46. Llansola M, Felipo V. Carnitine inhibits hydrolysis of inositol phospholipids induced by activation of metabotropic receptors. Neurochem Res. 1998;23(12):1533–7. https://doi.org/10.1023/a:1020928020469.
47. Llansola M, Felipo V. Carnitine prevents NMDA receptor-mediated activation of MAP-kinase and phosphorylation of microtubule-associated protein 2 in cerebellar neurons in culture. Brain Res. 2002;947(1):50–6. https://doi.org/10.1016/s0006-8993(02)02905-0.
48. Llansola M, Erceg S, Hernández-Viadel M, et al. Prevention of ammonia and glutamate neurotoxicity by carnitine: molecular mechanisms. Metab Brain Dis. 2002;17(4):389–97. https://doi.org/10.1023/a:1021922305036.
49. Koh JY, Palmer E, Cotman CW. Activation of the metabotropic glutamate receptor attenuates N-methyl-D-aspartate neurotoxicity in cortical cultures. Proc Natl Acad Sci USA. 1991;88(21):9431–5. https://doi.org/10.1073/pnas.88.21.9431.
50. Albrecht J. Roles of neuroactive amino acids in ammonia neurotoxicity. J Neurosci Res. 1998;51(2):133–8. https://doi.org/10.1002/(SICI)1097-4547(19980115)51:2<133::AID-JNR1>3.0.CO;2-E.
51. Norenberg MD, Rama Rao KV, Jayakumar AR. Ammonia neurotoxicity and the mitochondrial permeability transition. J Bioenerg Biomembr. 2004;36(4):303–7. https://doi.org/10.1023/B:JOBB.0000041758.20071.19.
52. Albrecht J, Norenberg MD. Glutamine: a Trojan horse in ammonia neurotoxicity. Hepatology. 2006;44(4):788–94. https://doi.org/10.1002/hep.21357.
53. Albrecht J, Zielińska M, Norenberg MD. Glutamine as a mediator of ammonia neurotoxicity: a critical appraisal. Biochem Pharmacol. 2010;80(9):1303–8. https://doi.org/10.1016/j.bcp.2010.07.024.
54. Pastorino JG, Snyder JW, Serroni A, et al. Cyclosporin and carnitine prevent the anoxic death of cultured hepatocytes by inhibiting the mitochondrial permeability transition. J Biol Chem. 1993;268(19):13791–8.
55. Furuno T, Kanno T, Arita K, et al. Roles of long chain fatty acids and carnitine in mitochondrial membrane permeability transition. Biochem Pharmacol. 2001;62(8):1037–46. https://doi.org/10.1016/s0006-2952(01)00745-6.
56. Trost LC, Lemasters JJ. The mitochondrial permeability transition: a new pathophysiological mecha-

nism for Reye's syndrome and toxic liver injury. J Pharmacol Exp Ther. 1996;278(3):1000–5.
57. Oyanagi E, Yano H, Kato Y, et al. L-carnitine suppresses oleic acid-induced membrane permeability transition of mitochondria. Cell Biochem Funct. 2008;26(7):778–86. https://doi.org/10.1002/cbf.1506.
58. Oyanagi E, Yano H, Uchida M, et al. Protective action of L-carnitine on cardiac mitochondrial function and structure against fatty acid stress. Biochem Biophys Res Commun. 2011;412(1):61–7. https://doi.org/10.1016/j.bbrc.2011.07.039.
59. Malaguarnera M, Restuccia S, Di Fazio I, et al. Serum carnitine levels in chronic hepatitis C patients before and after lymphoblastoid interferon-α treatment. BioDrugs. 1999;12(1):65–9. https://doi.org/10.2165/00063030-199912010-00007.
60. Malaguarnera M, Pistone G, Astuto M, et al. L-carnitine in the treatment of mild or moderate hepatic encephalopathy. Dig Dis. 2003;21(3):271–5. https://doi.org/10.1159/000073347.
61. Malaguarnera M, Pistone G, Elvira R, et al. Effects of L-carnitine in patients with hepatic encephalopathy. World J Gastroenterol. 2005;11(45):7197–202. https://doi.org/10.3748/wjg.v11.i45.7197.
62. Malaguarnera M, Gargante MP, Cristaldi E, et al. Acetyl-L-carnitine treatment in minimal hepatic encephalopathy. Dig Dis Sci. 2008;53(11):3018–25. https://doi.org/10.1007/s10620-008-0238-6.
63. Malaguarnera M, Vacante M, Giordano M, et al. Oral acetyl-L-carnitine therapy reduces fatigue in overt hepatic encephalopathy: a randomized, double-blind, placebo-controlled study. Am J Clin Nutr. 2011;93(4):799–808. https://doi.org/10.3945/ajcn.110.007393.
64. Malaguarnera M, Bella R, Vacante M, et al. Acetyl-L-carnitine reduces depression and improves quality of life in patients with minimal hepatic encephalopathy. Scand J Gastroenterol. 2011;46(4):750–9. https://doi.org/10.3109/00365521.2011.565067.
65. Malaguarnera M, Vacante M, Bertino G, et al. The supplementation of acetyl-L-carnitine decreases fatigue and increases quality of life in patients with hepatitis C treated with pegylated interferon-α 2b plus ribavirin. J Interf Cytokine Res. 2011;31(9):653–9. https://doi.org/10.1089/jir.2011.0010.
66. Kim S. A new treatment option for refractory hyperammonemia. Front Gastroenterol. 2012;17(3):269–73. (Article in Japanese)
67. Tamaki K, Okubo A. Experience of using carnitine preparations for hepatic encephalopathy and refractory hyperammonemia. Front Gastroenterol. 2012;17(4):394–9. (Article in Japanese)
68. Oba N. Experience with the use of levocarnitine hydrochloride in liver surgery. Front Gastroenterol. 2013;18(3):267–71. (Article in Japanese)
69. Murohisa G. Experience with levocarnitine therapy for refractory hepatic encephalopathy. Front Gastroenterol. 2013;18(4):364–7. (Article in Japanese)
70. Mori A, Ohashi N, Hayashi S, et al. Two cases of hepatocellular carcinoma in which hepatic encephalopathy improved with levocarnitine and anticancer treatment became possible. Front Gastroenterol. 2014;19(2):154–6. (Article in Japanese)
71. Tani M, Sawai Y, Fukuda K, et al. Study of the usefulness of levocarnitine administration for patients with refractory hepatic encephalopathy. Front Gastroenterol. 2014;19(2):157–61. (Article in Japanese)
72. Tsubata S, Ogawa K, Hirano M, et al. Experience using levocarnitine chloride in treatment of hepatic encephalopathy. Acta Hepatol Jpn. 2014;55(1):76–8. (Article in Japanese)
73. Adachi T, Fujioka S, Shimomura Y, et al. Examination of the utility of the carnitine replacement therapy for hyperammonemia. Kanzo. 2014;55(8):459–67. (Article in Japanese)
74. Komada N, Sakoda S, Iguchi A, et al. Trial of response guide therapy with levocarnitine injection for hyperammonemia. Pharma Medica. 2015;33(7):77–80. (Article in Japanese)
75. Tajiri K, Futsukaichi Y, Kobayashi S, et al. L-carnitine for the treatment of overt hepatic encephalopathy in patients with advanced liver cirrhosis. J Nutr Sci Vitaminol (Tokyo). 2018;64(5):321–8. https://doi.org/10.3177/jnsv.64.321.
76. Tani J, Morishita A, Sakamoto T, et al. L-carnitine reduces hospital admissions in patients with hepatic encephalopathy. Eur J Gastroenterol Hepatol. 2021;32(2):288–93. https://doi.org/10.1097/MEG.0000000000001748.
77. Saito M, Hirano H, Yano Y, et al. Serum level of taurine would be associated with the amelioration of minimal hepatic encephalopathy in cirrhotic patients. Hepatol Res. 2016;46(2):215–24. https://doi.org/10.1111/hepr.12565.
78. Nojiri S, Fujiwara K, Matsuura K, et al. L-carnitine reduces ammonia levels and alleviates covert encephalopathy: a randomized trial. J Transl Sci. 2018;4(3):1–6. https://doi.org/10.15761/JTS.1000220.
79. Nakanishi H, Hayakawa Y, Kubota Y, et al. Impaired brain function improved by L-carnitine in patients with cirrhosis: evaluation using near-infrared spectroscopy. Sci Rep. 2020;10(1):13566. https://doi.org/10.1038/s41598-020-70585-y.
80. Nakanishi H, Kurosaki M, Nakanishi K, et al. Impaired brain activity in cirrhotic patients with minimal hepatic encephalopathy: evaluation by near-infrared spectroscopy. Hepatol Res. 2014;44(3):319–26. https://doi.org/10.1111/hepr.12127.
81. Angeli P, Albino G, Carraro P, et al. Cirrhosis and muscle cramps: evidence of a causal relation-

ship. Hepatology. 1996;23(2):264–73. https://doi.org/10.1002/hep.510230211.
82. Marotta PJ, Graziadei IW, Ghent CN. Muscle cramps: a 'complication' of cirrhosis. Can J Gastroenterol. 2000;14(Suppl D):21D–5D. https://doi.org/10.1155/2000/214916.
83. Nakanishi H, Kurosaki M, Izumi N. Treatment options for muscle symptoms in patients with cirrhosis: experience with the use of carnitine preparations for cramps. Front Gastroenterol. 2013;18(3):278–83. (Article in Japanese)
84. Nakanishi H, Kurosaki M, Tsuchiya K, et al. L-carnitine reduces muscle cramps in patients with cirrhosis. Clin Gastroenterol Hepatol. 2015;13(8):1540–3. https://doi.org/10.1016/j.cgh.2014.12.005.
85. Iwasa M, Karino Y, Kawaguchi T, et al. Relationship of muscle cramps to quality of life and sleep disturbance in patients with chronic liver diseases: a nationwide study. Liver Int. 2018;38(12):2309–16. https://doi.org/10.1111/liv.13745.
86. Hiraoka A, Yoshiji H, Iwasa M, et al. Clinical features of liver cirrhosis patients with muscle cramping: a multicenter study. Eur J Gastroenterol Hepatol. 2019;31(12):1557–62. https://doi.org/10.1097/MEG.0000000000001473.
87. Korenaga K, Sakaida I. Characteristics of abnormality in nutrition metabolic abnormality and treatment for the nutritional disorders. Jpn J Portal Hypertens. 2008;14(3):212–7. (Article in Japanese)
88. Kato M, Miwa Y, Tajika M, et al. Preferential use of branched-chain amino acids as an energy substrate in patients with liver cirrhosis. Intern Med. 1998;37(5):429–34. https://doi.org/10.2169/internalmedicine.37.429.
89. Moriwaki H, Kato M. Chapter 9. Energy metabolism in patients: dynamics of energy metabolism in cirrhosis patients. In: Hosoya N, editor. Why energy metabolism now: for the prevention of lifestyle-related diseases. Tokyo: Daiichi Shuppan Co., Ltd.; 2000. p. 205–26. (Book in Japanese, ISBN: 4-8041-0899-8).
90. Sakai Y, Nishikawa H, Enomoto H, et al. Effect of L-carnitine in patients with liver cirrhosis on energy metabolism using indirect calorimetry: a pilot study. J Clin Med Res. 2016;8(12):863–9. https://doi.org/10.14740/jocmr2734w.
91. Dasarathy S. Consilience in sarcopenia of cirrhosis. J Cachexia Sarcopenia Muscle. 2012;3(4):225–37. https://doi.org/10.1007/s13539-012-0069-3.
92. Iritani S, Imai K, Takai K, et al. Skeletal muscle depletion is an independent prognostic factor for hepatocellular carcinoma. J Gastroenterol. 2015;50(3):323–32. https://doi.org/10.1007/s00535-014-0964-9.
93. Masuda T, Shirabe K, Ikegami T, et al. Sarcopenia is a prognostic factor in living donor liver transplantation. Liver Transpl. 2014;20(4):401–7. https://doi.org/10.1002/lt.23811.
94. Hanai T, Shiraki M, Ohnishi S, et al. Rapid skeletal muscle wasting predicts worse survival in patients with liver cirrhosis. Hepatol Res. 2016;46(8):743–51. https://doi.org/10.1111/hepr.12616.
95. Hassan A, Tsuda Y, Asai A, et al. Effects of oral L-carnitine on liver functions after transarterial chemoembolization in intermediate-stage HCC patients. Mediat Inflamm. 2015;2015:608216. https://doi.org/10.1155/2015/608216.
96. Iwasa M, Sugimoto R, Ishihara T, et al. Usefulness of levocarnitine and/or branched-chain amino acids during invasive treatment for hepatocellular carcinoma. J Nutr Sci Vitaminol (Tokyo). 2015;61(6):433–40. https://doi.org/10.3177/jnsv.61.433.
97. Ishikawa T, Abe S, Watanabe T, et al. L-carnitine administration to cirrhotic patients with sarcopenia improves nutritional state including controlling nutritional status (CONUT) score. Int J Nutr Sci. 2016;1(1):1002.
98. Sato S, Namisaki T, Furukawa M, et al. Effect of L-carnitine on health-related quality of life in patients with liver cirrhosis. Biomed Rep. 2020;13(6):65. https://doi.org/10.3892/br.2020.1372.
99. Sato S, Moriya K, Furukawa M, et al. Efficacy of L-carnitine on ribavirin-induced hemolytic anemia in patients with hepatitis C virus infection. Clin Mol Hepatol. 2019;25(1):65–73. https://doi.org/10.3350/cmh.2018.0070.
100. Ohara M, Ogawa K, Suda G, et al. L-carnitine suppresses loss of skeletal muscle mass in patients with liver cirrhosis. Hepatol Commun. 2018;2(8):906–18. https://doi.org/10.1002/hep4.1207.
101. Hanai T, Shiraki M, Imai K, et al. Usefulness of carnitine supplementation for the complications of liver cirrhosis. Nutrients. 2020;12(7):1915. https://doi.org/10.3390/nu12071915.
102. Japan Society of Gastroenterology, Japan Society of Hepatology, eds.: Cirrhosis clinical practice guidelines 2020 Revised 3rd ed. CQ4-15 (p 109), CQ4-20 (p 124), Nankodo, Tokyo, 2020 (Book in Japanese, ISBN: 978-524-22545-3).
103. Eguchi Y, Hyogo H, Ono M, et al. Prevalence and associated metabolic factors of nonalcoholic fatty liver disease in the general population from 2009 to 2010 in Japan: a multicenter large retrospective study. J Gastroenterol. 2012;47(5):586–95. https://doi.org/10.1007/s00535-012-0533-z.
104. Tateishi R, Koike K. Hepatocellular carcinoma arising from NASH/NAFLD: present situation and future measures. Nihon Naika Gakkai Zasshi. 2020;109(1):42–6. (Article in Japanese)
105. Tateishi R, Okanoue T, Fujiwara N, et al. Clinical characteristics, treatment, and prognosis of non-B, non-C hepatocellular carcinoma: a large retrospective multicenter cohort study. J Gastroenterol. 2015;50(3):350–60. https://doi.org/10.1007/s00535-014-0973-8.
106. Tsuneyama K. Animal models with minimal intervention for the translational study of metabolic

syndrome-associated liver diseases. Acta Hepatol Jpn. 2018;59(2):92–101. (Article in Japanese)
107. Takahashi Y, Soejima Y, Fukusato T. Animal models of nonalcoholic fatty liver disease/non-alcoholic steatohepatitis. World J Gastroenterol. 2012;18(19):2300–8. https://doi.org/10.3748/wjg.v18.i19.2300.
108. Nakamura A, Terauchi Y. Lessons from mouse models of high-fat diet-induced NAFLD. Int J Mol Sci. 2013;14(11):21240–57. https://doi.org/10.3390/ijms141121240.
109. Rinella ME, Green RM. The methionine-choline deficient dietary model of steatohepatitis does not exhibit insulin resistance. J Hepatol. 2004;40(1):47–51. https://doi.org/10.1016/j.jhep.2003.09.020.
110. Rinella ME, Elias MS, Smolak RR, et al. Mechanisms of hepatic steatosis in mice fed a lipogenic methionine choline-deficient diet. J Lipid Res. 2008;49(5):1068–76. https://doi.org/10.1194/jlr.M800042-JLR200.
111. Fritz IB, Dupont P. Ineffectiveness of carnitine as a choline substitute in the prevention of fatty livers of rats maintained on a choline-deficient diet. Am J Phys. 1957;190(3):453–4. https://doi.org/10.1152/ajplegacy.1957.190.3.453.
112. Corredor C, Mansbach C, Bressler R. Carnitine depletion in the choline-deficient state. Biochim Biophys Acta. 1967;144(2):366–74. https://doi.org/10.1016/0005-2760(67)90165-8.
113. Carter AL, Frenkel R. The relationship of choline and carnitine in the choline deficient rat. J Nutr. 1978;108(11):1748–54. https://doi.org/10.1093/jn/108.11.1748.
114. Mollica G, Senesi P, Codella R, et al. L-carnitine supplementation attenuates NAFLD progression and cardiac dysfunction in a mouse model fed with methionine and choline-deficient diet. Dig Liver Dis. 2020;52(3):314–23. https://doi.org/10.1016/j.dld.2019.09.002.
115. Strilakou AA, Lazaris AC, Perelas AI, et al. Heart dysfunction induced by choline-deficiency in adult rats: the protective role of L-carnitine. Eur J Pharmacol. 2013;709:20–7.
116. Strilakou AA, Tsakiris ST, Kalafatakis KG, et al. Carnitine modulates crucial myocardial adenosine triphosphatases and acetylcholinesterase enzyme activities in choline-deprived rats. Can J Physiol Pharmacol. 2014;92(1):78–84. https://doi.org/10.1139/cjpp-2013-0165.
117. Strilakou A, Perelas A, Lazaris A, et al. Immunohistochemical determination of the extracellular matrix modulation in a rat model of choline-deprived myocardium: the effects of carnitine. Fundam Clin Pharmacol. 2016;30(1):47–57. https://doi.org/10.1111/fcp.12163.
118. Serviddio G, Giudetti AM, Bellanti F, et al. Oxidation of hepatic carnitine palmitoyl transferase-I(CPT-I)impairs fatty acid beta-oxidation in rats fed a methionine-choline deficient diet. PLoS One. 2011;6(9):e24084. https://doi.org/10.1371/journal.pone.0024084.
119. Ishikawa H, Takaki A, Tsuzaki R, et al. L-carnitine prevents progression of non-alcoholic steatohepatitis in a mouse model with upregulation of mitochondrial pathway. PLoS One. 2014;9(7):e100627. https://doi.org/10.1371/journal.pone.0100627.
120. Xia Y, Li Q, Zhong W, et al. L-carnitine ameliorated fatty liver in high-calorie diet/STZ-induced type 2 diabetic mice by improving mitochondrial function. Diabetol Metab Syndr. 2011;3:31. https://doi.org/10.1186/1758-5996-3-31.
121. Kon K, Ikejima K, Morinaga M, et al. L-carnitine prevents metabolic steatohepatitis in obese diabetic KK-Ay mice. Hepatol Res. 2017;47(3):E44–54. https://doi.org/10.1111/hepr.12720.
122. Rautou PE, Mansouri A, Lebrec D, et al. Autophagy in liver diseases. J Hepatol. 2010;53(6):1123–34. https://doi.org/10.1016/j.jhep.2010.07.006.
123. Amir M, Czaja MJ. Autophagy in nonalcoholic steatohepatitis. Expert Rev Gastroenterol Hepatol. 2011;5(2):159–66. https://doi.org/10.1586/egh.11.4.
124. Brenner C, Galluzzi L, Kepp O, et al. Decoding cell death signals in liver inflammation. J Hepatol. 2013;59(3):583–94. https://doi.org/10.1016/j.jhep.2013.03.033.
125. Tsurusaki S, Tsuchiya Y, Koumura T, et al. Hepatic ferroptosis plays an important role as the trigger for initiating inflammation in nonalcoholic steatohepatitis. Cell Death Dis. 2019;10(6):449. https://doi.org/10.1038/s41419-019-1678-y.
126. Hirschhorn T, Stockwell BR. The development of the concept of ferroptosis. Free Radic Biol Med. 2019;133:130–43. https://doi.org/10.1016/j.freeradbiomed.2018.09.043.
127. Joaquim M, Escobar-Henriques M. Role of mitofusins and mitophagy in life or death decisions. Front Cell Dev Biol. 2020;8:572182. https://doi.org/10.3389/fcell.2020.572182.
128. Li R, Toan S, Zhou H. Role of mitochondrial quality control in the pathogenesis of nonalcoholic fatty liver disease. Aging (Albany NY). 2020;12(7):6467–85. https://doi.org/10.18632/aging.102972.
129. Kroemer G, Galluzzi L, Brenner C. Mitochondrial membrane permeabilization in cell death. Physiol Rev. 2007;87(1):99–163. https://doi.org/10.1152/physrev.00013.2006.
130. Pérez-Carreras M, Del Hoyo P, Martín MA, et al. Defective hepatic mitochondrial respiratory chain in patients with nonalcoholic steatohepatitis. Hepatology. 2003;38(4):999–1007. https://doi.org/10.1053/jhep.2003.50398.
131. Romano M, Vacante M, Cristaldi E, et al. L-carnitine treatment reduces steatosis in patients with chronic hepatitis C treated with α-interferon and ribavirin. Dig Dis Sci. 2008;53(4):1114–21. https://doi.org/10.1007/s10620-007-9983-1.
132. Malaguarnera M, Gargante MP, Russo C, et al. L-carnitine supplementation to diet: a new tool

in treatment of nonalcoholic steatohepatitis – a randomized and controlled clinical trial. Am J Gastroenterol. 2010;105(6):1338–45. https://doi.org/10.1038/ajg.2009.719.
133. Nagaya T, Tanaka N, Kimura T, et al. Mechanism of the development of nonalcoholic steatohepatitis after pancreaticoduodenectomy. BBA Clin. 2015;3:168–74. https://doi.org/10.1016/j.bbacli.2015.02.001.
134. Nakamura M, Matsumoto H, Nakashima H, et al. L-carnitine supplementation improved hepatic steatosis after pancreatectomy. Pancreas. 2016;45(3):e7–9. https://doi.org/10.1097/MPA.0000000000000508.
135. Nakamura M, Nakata K, Matsumoto H, et al. Acyl/free carnitine ratio is a risk factor for hepatic steatosis after pancreatoduodenectomy and total pancreatectomy. Pancreatology. 2017;17(1):135–8. https://doi.org/10.1016/j.pan.2016.11.007.
136. Enooku K, Nakagawa H, Fujiwara N, et al. Altered serum acylcarnitine profile is associated with the status of nonalcoholic fatty liver disease(NAFLD) and NAFLD-related hepatocellular carcinoma. Sci Rep. 2019;9(1):10663. https://doi.org/10.1038/s41598-019-47216-2.
137. Savic D, Hodson L, Neubauer S, et al. The importance of the fatty acid transporter L-carnitine in non-alcoholic fatty liver disease (NAFLD). Nutrients. 2020;12(8):2178. https://doi.org/10.3390/nu12082178.
138. Miwa T, Hanai T, Nishimura K, et al. Survival benefit of l-carnitine supplementation in patients with cirrhosis. JPEN J Parenter Enteral Nutr. 2022;46(6):1326–34. https://doi.org/10.1002/jpen.2386.
139. de Ulibarri JI, Gonzalez-Madrono A, de Villar NGP, et al. CONUT: a tool for controlling nutritional status. First validation in a hospital population. Nutr Hosp. 2005;20(1):38–45.
140. Buglio AL, Bellanti F, Carmignano DFP, et al. Association between controlling nutritional status (CONUT) score and body composition, inflammation and frailty in hospitalized elderly patients. Nutrients. 2024;16(5):576. https://doi.org/10.3390/nu16050576.

16 Cardiovascular Diseases and Carnitine Deficiency

16.1 Introduction

The heart contracts about 100,000 times a day and pumps about 7000 liters of blood into the aorta per day or about 2.6 million liters per year for a person weighing about 60 kg. Therefore, a large amount of energy (ATP) is consumed, which amounts to about 6 kg. If one assumes that a lifetime is 80 years, the heart contracts about 3 billion times, and its total output is equivalent to that of 1000 of 200,000-ton class tankers. The energy metabolism of myocardium, which produces ATP, is 60–90% dependent on fatty acid oxidation and 10–40% on glucose oxidation [1, 2]. The efficiency of utilization of these substrates, such as fatty acids and glucose, and the supply of oxygen used for oxidation of these substrates are important factors regulating myocardial energy metabolism. Factors that influence metabolic pathways directly related to ATP production, such as the TCA cycle and the electron transfer system, are also very important.

Deficiency of carnitine, which is essential for the transport of long-chain fatty acids into the mitochondria, naturally reduces ATP production by fatty acid oxidation in the myocardium, and furthermore, the accumulated acyl-CoA cannot be exported out of the mitochondria as acylcarnitine, and various enzymes essential for energy metabolism such as pyruvate dehydrogenase (PDH), resulting in impaired ATP production. If this persists for a long time, cardiomyopathy and heart failure may develop. Thus, in primary carnitine deficiency, which is an abnormality of the carnitine transporter (OCTN2), heart failure and sudden death due to cardiomyopathy are characteristic clinical features, and sudden death and impaired cardiac function are also observed in fatty acid metabolism disorders (see "Chap. 10: Carnitine Deficiency in Inherited Metabolic Diseases"). What extent, then, are carnitine deficiency and carnitine insufficiency, in which acyl-CoA accumulates and free carnitine is relatively insufficient, and acyl-CoA toxicity cannot be removed or detoxified out of mitochondria as acylcarnitine, involved in heart failure and impaired cardiac function in adults in general?

In ischemic heart disease, such as angina pectoris and myocardial infarction, stagnation of coronary artery blood flow that perfuses the myocardium itself may reduce the oxygen supply necessary for the oxidation of fatty acids and glucose, and stagnation of beta-oxidation of long-chain fatty acids may result in accumulation of toxic long-chain acyl-CoA. The role of carnitine in ischemic myocardium is also described.

Arrhythmia is one of the most important cardiac diseases, but the development of therapeutic agents has been slow. Drugs that modulate the activity of ion channels in pacemaker cells have been developed for the treatment of arrhythmia, and the antiarrhythmic effects of carnitine are also discussed.

N. Nakayama, *Carnitine Deficiency and Insufficiency in Medicine*,
https://doi.org/10.1007/978-981-95-7797-2_16

Finally, the involvement of carnitine in chronic heart failure will be discussed, highlighting early and recent studies.

16.2 Ischemic Heart Disease and Carnitine

16.2.1 Exertional Angina Pectoris and Carnitine

The indication for L-carnitine preparations in Japan is "carnitine deficiency," but in the 1980s, Earth Pharmaceutical Co., Ltd. conducted several Phase II clinical trials [3–9]. and Phase III clinical trials [10] for the indication of "improvement of exercise tolerance in exertional angina pectoris," and confirmed a certain level of efficacy and high safety. These results were compiled into an application for approval as a metabolic improving for exertional angina pectoris. Dilazep hydrochloride hydrate (product name: Comelian® $_{\text{Kowa}}$ tablets) was used as a control drug. However, the application was withdrawn for various reasons. The main reasons were probably that further basic research was demanded, and that the clinical evaluation guidelines for angina pectoris were subsequently changed, requiring an additional clinical trial with a different control drug in the additional study.

In overseas, Cherchi et al. reported the results of a randomized, placebo-compared, double-blind study of carnitine in patients with chronic angina at rest (44 participants, crossover study of carnitine dose of 2 g/day, 4 weeks of carnitine and 4 weeks of placebo) [11]. As a result, a significant improvement in exercise tolerance and a decrease in ST depression on the electrocardiogram (ECG) were observed. Iyer et al. also reported that 47 patients with chronic resting angina pectoris were given 2 g/day of carnitine or placebo for 3 months, and that carnitine significantly improved exercise duration and time to ST change [12].

In this way, the administration of carnitine to patients with angina pectoris has been reported to have an effect on exercise tolerance and ST lowering of the electrocardiogram, but since then, more powerful coronary vasodilators such as calcium antagonists have been widely used, and there have been advances in catheter treatment for coronary blood vessels, and carnitine administration to patients with angina pectoris has been neglected.

16.2.2 Myocardial Infarction and Carnitine

16.2.2.1 Basic Research on Ischemic Myocardial Animal Models

In acute myocardial infarction, which is the most serious form of ischemic heart disease, oxygen supply and nutrient supply are interrupted by blockage of the coronary arteries that feed the heart muscle, so that the β oxidation of long-chain fatty acids, which is the main ATP production pathway, and the pyruvate produced by glycolysis cannot produce ATP through the TCA cycle and oxidative phosphorylation of the electron transport system of the inner mitochondrial membrane. Such a situation not only interferes with the pumping function of blood to the whole body but also affects the survival of cardiomyocytes themselves, which is clinically relevant to infarct size and remodeling after reperfusion.

Many studies using animal models of ischemic myocardium have shown that long-chain acyl-CoA and its carnitine ester, long-chain acylcarnitine, increase in ischemic myocardium as β-oxidation of long-chain fatty acids is inhibited, while short-chain acylcarnitines such as free carnitine and acetylcarnitine decrease [13–16]. The accumulated long-chain acyl-CoA inhibits adenine nucleotide translocase (ANT) in the mitochondrial inner membrane, thus preventing ATP and ADP from passing through the inner mitochondrial membrane, and myocardial energy metabolism is greatly impaired [17, 18]. In other words, in ischemic hearts, acute carnitine insufficiency develops, and the toxicity of accumulated acyl-CoA inhibits energy metabolism and other functions. Administration of carnitine prevented the decrease in ANT activity, decreased long-chain acyl-CoA and long-chain acylcarnitine, suppressed ROS production by long-chain acyl-CoA, and improved cardiac function [19–21]. When free fatty acids were added in excess to ischemia-perfused hearts, the concentration of long-chain acylcarnitines increased and cardiac function decreased, whereas the addition of carnitine improved cardiac function [22, 23]. Furthermore, it has been reported that administration of carnitine to ischemic hearts promotes glucose oxidation [24, 25] and that long-chain acyl-CoA is excreted from mitochondria as acylcarnitine, thereby removing the toxicity of long-chain acyl CoA on various metabolisms, including suppression of PDH [24–26]. Although ischemic myocardium is unable to produce large amounts of ATP by β-oxidation of long-chain fatty acids due to the cessation of oxygen supply, small amounts of ATP production from the glycolytic system are still extremely important for maintaining the viability of cardiomyocytes. However, accumulation of long-chain acyl CoA in ischemic myocardium inhibits PDH, which functions at the junction between the glycolytic system and the TCA cycle, as well as various enzymatic reactions and ion channels in the TCA cycle and mitochondria, ultimately resulting in mitochondrial morphological degeneration and collapse [27, 28].

The only way to release the toxicity of such long-chain acyl-CoA accumulated in mitochondria is to administer carnitine to increase free carnitine concentrations and to remove the long-chain acyl-CoA from the mitochondria, extracellularly, and ultimately excreted in the urine.

16.2.2.2 Effect of Carnitine Administration in Patients with Acute Myocardial Infarction

Davini et al. divided 160 patients with new myocardial infarction into a carnitine group (81 patients, 2 g orally every 12 h for a total of 4 g/day for 12 months) and a control group (79 patients) to examine cardiac function, angina attack frequency, rhythm disturbances, contractile disturbances, lipid mortality, and cause of death [29]. Both groups continued their usual treatment for acute myocardial infarction. There were no significant differences in medical history or laboratory values between the two groups, except for a slightly higher incidence of arterial hypertension in the control group. The results showed that the carnitine group was significantly superior in heart rate, systolic arterial pressure, number of angina attacks, improvement in lipid pattern, and mortality ($P < 0.005$). In particular, mortality was only one case (1.2%, thromboembolism: 1) in the carnitine group and 10 cases (12.5%, new infarction: 3, sudden death: 2, heart failure: 2, thromboembolism: 1, extracardiac causes: 2) in the control group. Based on these findings, Davini et al. concluded that carnitine administration to patients with acute myocardial infarction is useful in preventing subsequent cardiac events.

Iliceto et al. reported on a randomized, double-blind, placebo-controlled, multicenter study of the effect of carnitine administration on left ventricular remodeling in 472 patients with anterior wall acute myocardial infarction (The L-Carnitine Ecocardiografia Digitalizzata Infarto Miocardico: CEDIM-1 Study) [30, 31]. The placebo group (239 patients) and the carnitine group (233 patients) received conventional therapy for acute myocardial infarction, and the carnitine group received carnitine (9 g/day) by continuous intravenous infusion during hospitalization for 5 days and then switched to oral administration at a dose of 6 g/day (2 g three times) for 12 months. Two-dimensional echocardiography was used to evaluate left ventricular remodeling in terms of left ventricular end-diastolic volume and left ventricular end-systolic volume. Results showed that from 3 to 12 months after discharge from the hospital, left ventricular end-diastolic volume and left ventricular end-systolic volume were significantly lower in the carnitine group than in the placebo group (Fig. 16.1). Clinical events during hospitalization and at follow-up were not significantly different between the two groups, but during the 12-month follow-up, there were 10 and 13 deaths and 4 and 10 heart failures in the carnitine-treated and placebo-treated groups, respectively, with lower values in the carnitine group. These results indicate that intravenous carnitine administered during hospitalization and continued oral

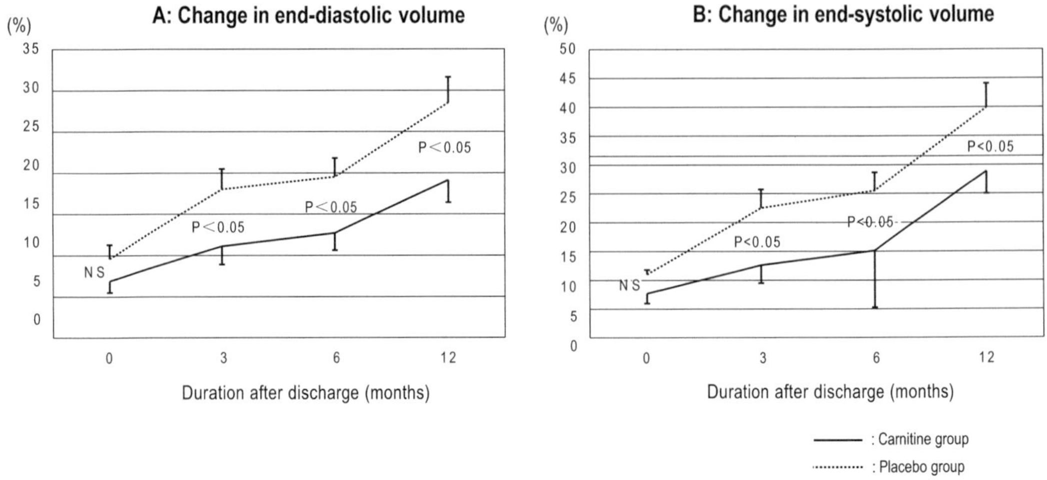

Fig. 16.1 Change in end-diastolic volume (left) and end-systolic volume (right) from baseline (hospital admission) to hospital discharge (3, 6, and 12 months) in the two treatment groups. Both groups received conventional treatment, and carnitine group was treated with intravenous carnitine infusion (9 g/day, for 5 days) and then oral carnitine (6 g/day, for 12 months). Mean ± 95% confidence interval. (From Refs. [30, 31], modified)

administration thereafter may prevent myocardial remodeling in patients with acute myocardial infarction.

Singh et al. also conducted a randomized, double-blind, placebo-controlled study to evaluate the effect of carnitine administration in 101 patients with suspected acute myocardial infarction, divided into two groups [32]. The carnitine group (51 patients) received oral carnitine (2 g/day for 28 days) in addition to conventional therapy, while the placebo group (50 patients) received conventional therapy alone. Results showed a significant decrease in QRS score (7.4 ± 1.2 vs. 10.7 ± 2.0) on ECG at day 10 and lipid peroxide in the carnitine group compared to the placebo group. Infarct size calculated from creatine kinase and the muscle-brain fraction of creatine kinase (CK-MB) were significantly lower in the carnitine group (95.5 ± 23.6 vs. 116.2 ± 26.2 g and 58.6 ± 16.6 vs. 73.3 ± 21.5 g, respectively, $P < 0.01$). In comparison of clinical events, angina was lower in 17.6 versus 36.0%, NYHA class III + IV in 7.8 versus 14.0%, arrhythmia in 13.7 versus 28.0%, and total cardiac events in 15.6 versus 26.0%, respectively, in the carnitine-treated group. Based on these results, the investigators concluded that carnitine in patients with suspected acute myocardial infarction may be protective against infarct size and clinical cardiac events.

Tarantini et al. of the Iliceto Group conducted the CEDIM-2 study, a randomized, double-blind, placebo-controlled, multicenter trial involving 153 centers to investigate the effect of carnitine administration in 4000 patients with anterior wall acute myocardial infarction. The study was terminated after 2330 patients, and the results were reported [33]. The objective of the study was to investigate the effect of carnitine administration (9 g/day continuous intravenous infusion during the initial hospitalization for 5 days and later 4 g/day orally for 6 months), with heart failure and mortality after 6 months as primary endpoints and mortality during the 5 days of hospitalization as secondary endpoints. The primary endpoint (mortality and heart failure) tended to be lower in the carnitine group (HR = 0.86, 95% CI: 0.66–1.11, $P = 0.25$), although there was no significant difference at any time point (HR = 0.86, 95% CI: 0.66–1.11, $P = 0.25$). In the CEDIM-1 study, carnitine was associated with a significantly lower incidence of death within 5 days of admission (27 deaths in the carnitine group and 44 deaths in the placebo group) ($P = 0.041$, HR = 0.61, 95% CI: 0.37–0.98). In the CEDIM-1 study, carnitine significantly prevented increases in left ventricular end-systolic volume and left ventricular end-diastolic volume at 3, 6, and 12 months after treatment. Left ventricular end-systolic volume is a strong predictor of mortality in patients with acute myocardial infarction, so that mortality and heart failure was set in the CEDIM-2 trial as primary endpoints, but carnitine treatment was not associated with left ventricular remodeling, perhaps because the treatment period was too short (6 months instead of 12 months). Perhaps because the duration of treatment was 6 months rather than 12 months, the CEDIM-1 trial did not replicate the results of the CEDIM-2 trial, which showed that carnitine treatment reduced mortality and heart failure by inhibiting left ventricular remodeling. Rather, in patients hospitalized for acute myocardial infarction, the early administration of carnitine reduced the risk of premature death. Although CEDIM-1 had a wide 95% confidence interval for deaths during first 5 days of hospitalization, it was not possible to conclude that carnitine administration reduced mortality with CEDIM-2 alone or in a pooled analysis of both trials, but it was concluded that early administration of carnitine reduced deaths during hospitalization [33].

Currently, the mainstream treatment for acute myocardial infarction is percutaneous coronary intervention (PCI) as early as possible after hospitalization due to chest pain attack, dilation of the stenosis site with a balloon catheter, and/or placement of a stent. Xue et al. reported a randomized, placebo-controlled trial (single-blind, open) investigating the effects of carnitine administration in combination with PCI treatment in 96 patients with non-ST-elevation acute myocardial infarction [34]. Both the carnitine group (48 patients, 5 g bolus +10 g/day intravenous infusion, 3 days) or placebo (48 patients) received PCI treatment within 48 h of admission and within 24 h of the onset of a chest pain attack. CK-MB and troponin I (TnI), which are thought to reflect infarct size, were compared between the two groups (Table 16.1). As a result, the 12-h and 24-h CK-MB values and the 8-h TnI values were significantly lower in the carnitine group than in the placebo group. Multivariate regression analysis also showed that carnitine administration was an independent predictor of CK-MB and TnI 24 h after PCI ($r = 0.596$, $P < 0.01$ and $r = 0.633$, $P < 0.001$, respectively). Therefore, carnitine administration in conjunction with PCI treatment may have the effect of reducing myocardial deviant enzymes by protecting the heart muscle. The composite endpoints of ischemic attack, re-infarction, and death during hospitalization or up to 30 days after discharge were 1 (2.1%) in the carnitine group and 3 (6.3%) in the control group, but there was no significant difference, probably due to the small sample size. It was thought that a larger trial would be necessary in the future.

Table 16.1 Changes in creatine kinase-MB (CK-MB) and troponin-I (TnI) after percutaneous coronary intervention

	Carnitine treatment group ($n = 48$)	Placebo group ($n = 48$)
CK-MB (ng/mL)		
Before	1.12 ± 1.66	1.03 ± 1.55
8 h	4.34 ± 9.01	5.58 ± 11.2
12 h	1.98 ± 4.15	3.21 ± 5.23[b]
24 h	0.66 ± 1.04	1.59 ± 2.72[a]
TnI (ng/mL)		
Before	0.21 ± 0.11	0.19 ± 0.33
8 h	0.44 ± 1.05	1.78 ± 5.11[b]
12 h	0.86 ± 2.45	1.35 ± 3.36
24 h	0.41 ± 1.34	0.89 ± 1.35

Compared to the carnitine group [a]$P < 0.05$, [b]$P < 0.01$, CK-MB creatine kinase MB, TnI troponin I
From Ref. [34]

16.3 Arrhythmias and Carnitine

16.3.1 Basic Research

In 1968, a correlation between free fatty acids in the blood and arrhythmia development in patients with acute myocardial infarction was reported [35], and it was found that free fatty acids and their acyl-CoA accumulate in the myocardium and exert a proarrhythmic effect, especially during reperfusion arrhythmia after revascularization [36, 37].

Suzuki et al. reported that carnitine administration (100 mg/kg, intravenous infusion, 5 min before coronary ligation) lowered the score of ventricular arrhythmias and prevented a decrease in free carnitine and ATP and the accumulation of long-chain acyl-CoA in the ischemic myocardium in an arrhythmia model in which heparin and intralipid were administered in addition to canine anterior descending branch ligation model [38].

Kobayashi et al. in the same group found that carnitine administration (100 mg/kg intravenously 5 min before ligation, 20 mg/kg/min from 5 min before reperfusion for a total of 20 min until the end of the experiment) was performed in a pre-descending ligation reperfusion model of dogs, and ventricular fibrillation after 15 min of reperfusion was significantly reduced from 29% in the control group to 5% in the carnitine group ($P < 0.05$ [39]. In this study, the concentration of free carnitine in the ischemic myocardium in the control group was significantly reduced compared to the non-ischemic area ($P < 0.01$), but no decrease in free carnitine in the ischemic site was observed in the carnitine group. In addition, the long-chain acyl-CoA of the ischemic myocardium in the control group was significantly elevated compared to the non-ischemic site ($P < 0.01$), but no such change was observed in the carnitine group. Therefore, it was thought that carnitine administration prevented the reduction of free carnitine and ATP and the accumulation of long-chain acyl-CoA in the ischemic myocardium, and that these metabolic effects prevented ventricular arrhythmias associated with ischemia-reperfusion. It has also been reported from the same group that the ATP content of ischemic myocardium is significantly inversely correlated with the long-chain acyl-CoA/free carnitine ratio and the long-chain acylcarnitine/free carnitine ratio [26, 40] ($r = -0.504$ and $r = -0.547$, both $P < 0.01$).

Imai et al. also evaluated the antiarrhythmic effect of L-carnitine chloride using a Harris coronary two-stage ligation arrhythmia model in conscious dogs [41]. As a result, it was considered that the free carnitine concentration decreases in ischemic myocardium, and that carnitine administration (300 mg/kg, intravenous injectionm) has antiarrhythmic effects and improves mitochondrial function (oxidative phosphorylating capacity, respiratory regulation function, etc.), and that carnitine improves mitochondrial function and exerts antiarrhythmic effects.

Subsequently, Kobayashi et al. showed that long-chain acyl-CoA actually accumulates in mitochondria in ischemic myocardium and that carnitine administration suppresses it [42].

In ischemic hearts, oxygen supply is insufficient, and long-chain fatty acid β-oxidation is inhibited, causing long-chain acyl-CoA to accumulate in the mitochondria, and its proarrhythmic effect makes arrhythmias more likely to occur. Carnitine administration is thought to reduce arrhythmias. Long-chain acyl-CoA also inhibits mitochondrial energy metabolism and PDH in the glycolytic system, and the resulting decrease in ATP may disrupt ion balance, making arrhythmias more likely to occur. If ATP production is maintained by carnitine administration, the occurrence of arrhythmias due to these causes may be reduced.

16.3.2 Clinical Research

Suzuki et al. administered carnitine (2 g orally 2 h before the start of dialysis) to 17 dialysis patients with frequent arrhythmias and examined its antiarrhythmic effects [43]. Of the 17 patients, 11 (65%) had ventricular premature contractions or supraventricular premature contractions from 20 to 30 min after the start of dialysis until the end of dialysis. The concentration of free carnitine in the blood of these dialysis patients before the start of dialysis was lower than that of healthy subjects (24.8 ± 7.9 vs. 46.1 ± 8.6 nmol/mL) and decreased to 8.2 ± 5.9 nmol/mL after the end of dialysis. In addition, blood free fatty acids increased after the start of dialysis and reached their maximum value in 20 min (0.23 ± 0.09 to 1.31 ± 0.64 mEq/L) and remained high until the end of dialysis (0.76 ± 0.29 mEq/L). A typical case is shown in Fig. 16.2. When carnitine was administered orally 2 h before the start of dialysis, the blood free carnitine concentration remained within the normal range during dialysis, and the increase in free fatty acids was suppressed. Therefore, when carnitine was

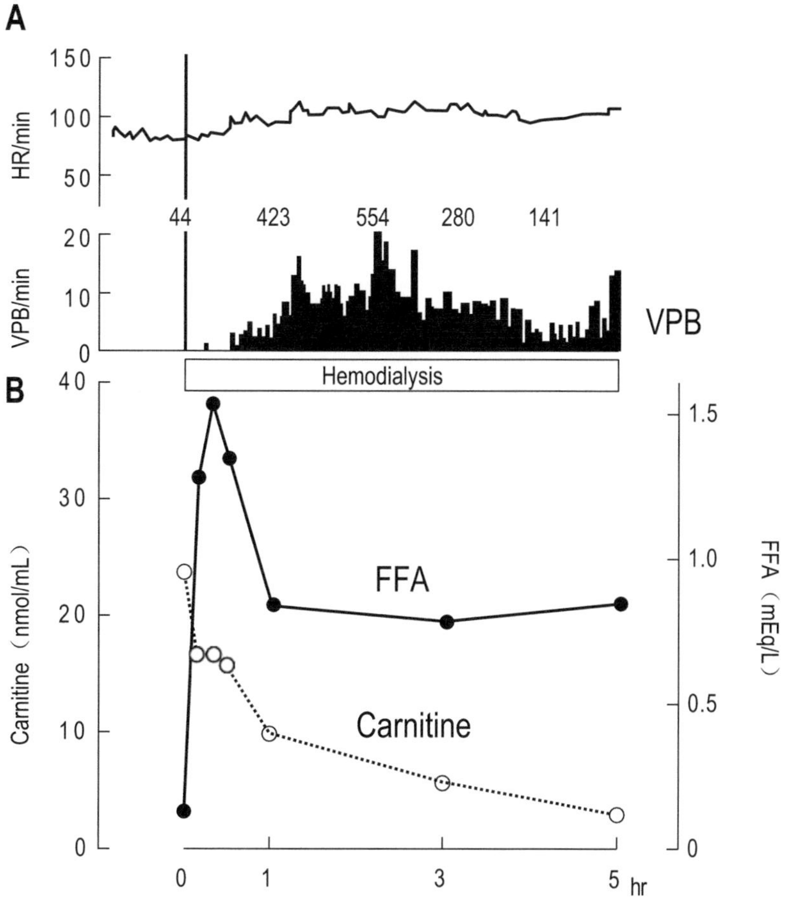

Fig. 16.2 A typical case of ventricular premature beats during hemodialysis in a 40-year-old female with 2.7-year history of hemodialysis (Case 4). Ventricular premature beats during hemodialysis were shown on a histogram as the number per minute (a part of the trend chart automatically analyzed by DYNA-GRAM Holter scanner Model 6000). HR heart rate, VPB ventricular premature beat, FFA free fatty acid. (From Ref. [43])

administered (2 g/day orally, 8 weeks) to eight patients who frequently experienced extrasystoles during dialysis, a significant reduction in the frequency and severity of extrasystoles was observed at weeks 4 and 8 (Fig. 16.3). These results suggest that carnitine administration is useful for extrasystole during dialysis.

Rizzon et al. reported a randomized, double-blind, placebo-controlled trial investigating the effect of carnitine administration on ventricular extrasystole and ventricular tachycardia development in 56 patients with acute myocardial infarction [44]. Fifty-six patients under the age of 75 with acute myocardial infarction within 3–12 h from chest pain attack to hospitalization were randomized into two groups of 28 patients each, and carnitine (100 mg/kg intravenous infu-

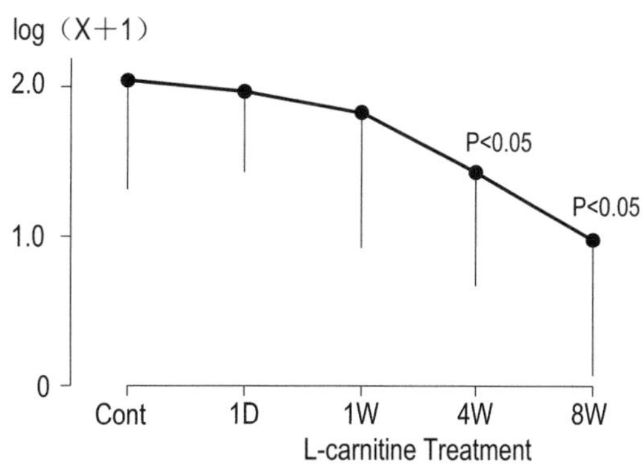

Fig. 16.3 Effects of L-carnitine on arrhythmias during hemodialysis. Values were represented as mean ± SD. A Student's *t*-test was used for the significance of changes, compared with the control. X: number of ventricular or supraventricular premature beats during hemodialysis. Cont control. (From Ref. [43])

Table 16.2 Premature ventricular beats and ventricular tachycardia in the placebo and L-carnitine groups

	1st day		2nd day	
	Placebo	L-carnitine	Placebo	L-carnitine
Recording time (h)	22.4 ± 3.4	22.5 ± 5	20.4 ± 7.2	20.4 ± 7.6
PVB h^{-1}	93 ± 272	60 ± 115	20 ± 36	4 ± 7[a]
Time with multiform PVBs (h)	11.1 ± 9	8.9 ± 8	7.3 ± 8	2.9 ± 4[a]
Time with paired PVBs (h)	6.6 ± 7.6	5.6 ± 6.8	2.8 ± 4.4	0.6 ± 1.2[b]
Number of paired PVBs (24 h)	23 ± 54	23 ± 46	7 ± 16	0.7 ± 1.5
Time with ventricular Tachycardia (h)	4.1 ± 5.9	3 ± 4	0.7 ± 1.1	0.2 ± 0.4[a]
Number of ventricular tachycardia episodes	17 ± 43	23 ± 97	1 ± 2	0.2 ± 0.5[a]

[a]$P < 0.05$, [b]$P < 0.02$, PVB: premature ventricular beats
From Ref. [44]

sion/12 h, 2 days, 4 doses) was given to the carnitine group, and placebo was administered similarly to the placebo group. At that time, conventional treatment (nifedipine, nitro compounds, diuretics) for patients with myocardial infarction was tolerated, but prophylactic antiarrhythmic drugs were not allowed in the protocol. However, in severe cases or for arrhythmias affecting cardiohemodynamics, antiarrhythmics administration was allowed. As a result, the incidence of ventricular extrasystole and ventricular tachycardia was significantly suppressed in the carnitine group compared to the placebo group, especially on the second day of hospitalization (Table 16.2). In addition, the number of patients with ventricular extrasystole after 10 h on the second day of hospitalization was significantly lower in the carnitine group than in the placebo group ($P < 0.01$, Table 16.3).

As for concomitant drug use, nifedipine was used in 14 patients in both groups on the first day, 25 in the carnitine group and 23 in the placebo group on the second day, nitro compound was used in 4 and 8 patients on the second day, the diuretic furosemide was used in 2 and 4 cases, respectively, and the use of antiarrhythmic drugs was 1 to 2 cases and was not high enough to affect the results of the analysis.

Martina et al. also reported a randomized, double-blind, placebo-controlled trial investigating the effects of carnitine in 20 patients with acute myocardial infarction [45]. Patients from chest pain attack between 4 and 12 h were assigned to receive carnitine (12 patients) and placebo (8 patients). In the carnitine group, carnitine (5 g, intravenous infusion, 2 h) was administered at 0, 12, 24, and 36 h, followed by 6 g (2 g × 3)/day, intravenous infusion, 2 h each administration, 3rd to 7th day. As a result, there was no difference in the overall incidence of ventricular premature contractions between the two groups, but the incidence of severe ventricular premature contractions on the second day was significantly lower in the carnitine group (4 out of 12 patients in the carnitine group, 7 out of 8 patients in the placebo group, $P = 0.028$).

Table 16.3 Number of patients with premature ventricular beats during the first and second day of treatment (placebo or L-carnitine)

	1st day			2nd day
	<10 h	>10 h	<10 h	>10 h
Placebo	10	18	15	13
L-carnitine	11	17	23	2

On the first and second days, data were collected and analyzed within 10 h and after 10 h, respectively
From Ref. [44]

16.4 Meta-Analysis of Carnitine Administration Trials in Ischemic Heart Disease

DiNicolantonio et al. conducted a meta-analysis of clinical trials investigating the effects of carnitine administration in ischemic heart disease such as acute coronary syndrome and acute myocardial infarction [46]. Electronic databases such as Ovid MEDLINE (1974–2012), PubMed (1973–2012), and Embase (1974–2012) were serched and extracted for the relevant clinical trials from the citations of relevant review articles, and also relevant trials and reviews were manually screened. They included placebo-controlled trials of carnitine in patients aged 18 years and older, clinical research papers assessing mortality, cardiovascular events, and the incidence of heart failure and arrhythmias, and excluded papers that did not describe all-cause mortality or morbidity outcomes. They collected 153 papers by literature search, extracted 18 papers, read them closely, and selected 13 papers. All-cause mortality was reported in 11 trials (n = 3579) with a significant 27% reduction compared to placebo or control (odds ratio: 0.73, 95% CI: 0.54–0.99, P = 0.05). Ventricular arrhythmias were reported in five studies (n = 229), and carnitine administration was associated with a highly significant 65% reduction in ventricular arrhythmias (relative risk: 0.35, 95% CI: 0.21–0.58, P < 0.0001). Reinfarction was reported in four studies (n = 829), and carnitine administration was not associated with a reduction in reinfarction (relative risk: 0.78, 95% CI: 0.41–1.48, P = 0.45). Heart failure was reported in six trials (3214 participants) and was not associated with a reduction in heart failure (relative risk: 0.85, 95% CI: 0.67–1.09, P = 0.21). Angina attacks were reported in two studies (n = 261), and carnitine reduced angina attacks by 40% (relative risk: 0.60, 95% CI: 0.50–0.72, P < 0.00001). It is also noted that there were no safety problems with early intravenous administration of large doses of carnitine to patients with acute myocardial infarction (see Figs. 9.4, 9.5, and 9.6 in Chap. 9, "Safety of Carnitine" for details). The study design, number of participants, details of the basic treatment and carnitine administration methods, and details of the excluded clinical trials are included in the URL of the journal in the Supplemental Online Material (Tables 16.1, 16.2, and 16.3 and Appendix 1, 2) (see https://www.mayoclinicproceedings.org/article/S0025-6196(13)00127-4/fulltext).

In addition, DiNicolantonio et al. provided an overview of the clinical trial reports discussed in this meta-analysis and showed that myocardial free carnitine decreases during myocardial infarction and long-chain acyl-CoA accumulates, inhibiting not only fatty acid oxidation but also energy metabolism such as glycolysis and mitochondrial function. It explains in an easy-to-understand manner that the toxicity of such long-chain acyl-CoA is canceled [47].

Shang et al. reported a meta-analysis of placebo-controlled trials investigating the effects of carnitine administration in patients with acute myocardial infarction, with particular focus on the dosage of oral carnitine during the maintenance phase [48]. Five trials (n = 3108) (Davini, 1992), Iliceto (1995), Singh (1996), Iyer (1999), and Tarantini (2006), which are also included in the meta-analysis by DiNicolantonio et al., were selected from among the 135 reports, and the all-cause mortality and cardiovascular disease incidence were analyzed. Three of the five trials administered 6–9 g/day of carnitine as a loading dose for 5–7 days during hospital stay, followed by an additional 2–6 g/day of oral carnitine. As a result of the interaction study, there was no significant difference in overall mortality between carnitine doses in the maintenance phase of 2, 3, 4, and 6 g/day (relative risk: 0.77, 95% CI: 0.57–1.03, P = 0.08). For heart failure, unstable angina pectoris, and re-infarction, it was judged that there was no significant marginal benefit at higher or lower doses than 3 g/day in the maintenance phase. However, this meta-analysis did not analyze intravenous carnitine administration during hospitalization immediately after the onset of disease.

16.5 Heart Failure and Carnitine

16.5.1 Carnitine Kinetics in Myocardial Tissue in Patients with Heart Failure

Suzuki et al. at the Department of Cardiology, Hamamatsu University School of Medicine, examined the carnitine profile of papillary muscle biopsied during valve replacement surgery in 16 patients with mitral valve disease and heart failure symptoms and compared with that of postmortem specimens from 5 controls without heart disease (Table 16.4) [49]. The results showed that free carnitine concentration in the myocardium of

Table 16.4 Levels of carnitine and acylcarnitine in human left ventricular papillary muscle

		Acylcarnitine		
	Free carnitine	Short chain	Long chain	Total carnitine
Control	868 ± 372	162 ± 116	231 ± 32	1261 ± 446
Heart failure	382 ± 168	373 ± 160	697 ± 355	1452 ± 363
P values	$P < 0.001$	$P < 0.05$	$P < 0.01$	NS

Mean ± SD, values are nmol/g wet tissue
From Ref. [49]

heart failure patients was significantly lower than in controls ($P < 0.001$), and acylcarnitine was significantly increased, especially long-chain acylcarnitine ($P < 0.01$). Total carnitine concentration, the sum of free carnitine, and acylcarnitine did not differ from that of controls.

Masumura et al. in the same group at first examined the postmortem changes in the concentrations of free carnitine and acylcarnitine in myocardial tissue in a study using seven dogs and confirmed that there was no significant change within 4 h [50]. Next, same as reported by Suzuki et al. [49], they biopsy the papillary muscles of eight patients with mitral valve disease with heart failure symptoms during valve replacement surgery, and measured carnitine and acylcarnitine in left ventricular papillary muscle samples within 4 h of the death of seven control patients without heart diseases who died from other causes [50]. As reported by Suzuki et al., the free carnitine in the myocardium was significantly reduced (412 ± 142 vs 769 ± 267 nmol/g, $P < 0.01$) and the long-chain acylcarnitine was significantly increased (532 ± 169 vs. 317 ± 72 nmol/g, $P < 0.01$) in patients with heart failure.

Regitz et al. also examined total carnitine levels in the myocardium of patients with dilated cardiomyopathy (DCM; 28 patients) and heart failure (8 patients) compared with controls (13 patients) [51, 52]. As a result, the total carnitine concentration of the control was 9.9 ± 0.8 nmol/mg noncollagen protein (NCP), which was significantly reduced to 6.1 ± 0.5 nmol/mg NCP in patients with dilated cardiomyopathy and 6.6 ± 1.1 nmol/mg NCP in patients with heart failure ($P < 0.0001$ and $P < 0.02$, respectively). Myocardial free carnitine concentrations were also significantly reduced to 4.6 ± 0.4 nmol/mg NCP in patients with dilated cardiomyopathy and 4.4 ± 0.5 nmol/mg NCP in patients with heart failure compared to the control level of 9.7 ± 0.7 nmol/mg NCP ($P < 0.0001$ and $P < 0.005$, respectively). On the other hand, the total and free carnitine concentrations in plasma were significantly higher in patients with heart failure than in controls.

On the other hand, Pierpont et al. also examined the myocardial carnitine content of 51 patients with end-stage congestive heart failure (idiopathic cardiomyopathy, coronary artery disease, myocarditis, and rheumatic heart disease, RHD) who underwent heart transplantation surgery compared with healthy controls (5 patients) [53]. First, when the total myocardial carnitine concentrations in the left ventricle and right ventricle were examined in healthy controls, there was no change, but the content of ventricular muscle was higher than that of atrial muscle ($P < 0.005$). In 30 patients with congestive heart failure, the total carnitine concentration of the left ventricular muscle was significantly higher than that of the right ventricular muscle ($P < 0.001$), and the total carnitine concentration of both ventricular muscles was significantly higher than that of the atrial muscle ($P < 0.005$). Of the 51 patients with congestive heart failure, only 7 had low levels of total carnitine in the myocardium, but the total concentration of carnitine in plasma was higher in patients with heart failure than in controls. In this study, free carnitine was not measured in the control because of postmortem deterioration.

Martín et al. also compared myocardial carnitine and CPT activity with healthy controls (21 patients, donor myocardium) in patients with end-stage congestive heart failure who underwent heart transplantation surgery (25 patients, including 5 patients with idiopathic dilated cardiomyopathy, 18 patients with coronary artery disease, and 2 patients with rheumatic heart disease) [54]. As a result, the total carnitine and free carnitine concentrations were significantly lower than the controls ($P < 0.001$) and the long-chain acylcarnitine/free carnitine ratio was significantly higher in myocardial patients with end-stage congestive heart failure ($P < 0.001$). In addition, the total CPT activity of myocardium was low, and the activity of malonyl-CoA insensitivity, which is considered to be CPT2 activity, or in the presence of the surfactant Triton X-100, was significantly reduced ($P < 0.05$, Table 16.5).

In addition, the left ventricular ejection fraction (LVEF) of these patients with congestive heart failure was positively correlated with myocardial free carnitine concentration ($r^2 = 0.676$, $P < 0.001$) and inversely correlated with the long-chain acylcarnitine/free carnitine ratio ($r^2 = 0.503$, $P < 0.001$).

El-Aroussy et al. reported plasma and urinary carnitine concentrations in 30 patients with congestive heart failure compared with those in controls (10 subjects) [55]. The primary diseases of patients with congestive heart failure were dilated cardiomyopathy (DCM) and patients with rheumatic heart disease (RHD). Plasma carnitine concentrations in DCM and RHD were significantly higher than controls, and urinary carnitine concentrations were also significantly higher ($P < 0.0001$). Plasma carnitine concentrations were significantly correlated with the degree of impairment of indicators of systolic function of the left ventricle, such as fractional shortening (FS) <25%, ejection fraction (EF) <55%, and left ventricular mass index (LVMI) >124 g/m2 (FS <25%:$r = -0.38$, $P = 0.038$, EF <55%:$r = -0.502$, $P = 0.005$, LVMI>124 g/m^2:$r = 0.436$, $P = 0.016$). Therefore, they considered that high levels of carnitine in plasma and urine may be indicators of myocardial damage and deterioration of left ventricular systolic function.

Table 16.5 Carnitine levels and CPT activities in normal controls and end-stage congestive heart failure patients

Controls (n)		End-stage congestive Heart failure patients (n)
TC	10.8 ± 1.8(10)	6.5 ± 2.2^a (19)
FC	10.3 ± 1.8(10)	5.4 ± 2.2^a (19)
SCAC	0.40 ± 0.12(10)	0.42 ± 0.24 (19)
LCAC	0.15 ± 0.09(10)	0.69 ± 0.27^a (19)
LCAC/FC	1.6 ± 1.1(10)	16.5 ± 10.1^a (19)
Total CPT	9.9 ± 1.5(11)	7.6 ± 2.3^b (19)
CPT1(mCoA Sens)	5.8 ± 0.8(11)	5.1 ± 1.9 (19)
CPT2(mCoA insens)	4.2 ± 0.6(11)	2.5 ± 1.3^b (19)
CPT2(triton X-100)	5.6 ± 1.0(4)	4.1 ± 1.3^b (6)

Mean ± S.D., TC, FC, SCAC and LCAC levels are nmol/mg noncollagen protein. CPT activity: nmol/min/mg noncollagen protein. $^aP < 0.001$ vs. controls, $^bP < 0.05$ vs. controls

TC total carnitine, *FC* free carnitine, *SCAC* short-chain acylcarnitine, *LCAC* long-chain acylcarnitine, *CPT* carnitine palmitoyltransferase, *mCoA sens* malonyl CoA sensitive, *mCoA insens* malonyl CoA insensitive

From Ref. [54]

Kobayashi et al. confirmed a decrease in myocardial free carnitine and an increase in long-chain acylcarnitine in cardiomyopathy hamsters (BIO 14.6) and found that intraperitoneal administration of carnitine improved myocardial damage. So, they studied effect of carnitine administration on exercise tolerance in 12 patients with angina pectoris (300 mg × 3 times/day, 12 weeks). As a result, there was no improvement in exercise tolerance during the first 4 weeks of placebo, but exercise tolerance was significantly improved after carnitine administration 4 and 12 weeks ($P < 0.05$ and $P < 0.01$). Then they studied the therapeutic efficacy of carnitine administration in chronic congestive heart failure patients ($n = 9$, 300 mg × 3 times/day, 8 weeks, for first 2 or 4 weeks they received placebo, then carnitine treatment). There were cases in which clinical indicators such as NYHA classification, cardiothoracic ratio (CTR), and EF of patients with heart failure improved, and overall improvement in the patient's condition was observed in seven out of nine patients [56].

Thus, free carnitine in the myocardium is decreased and long-chain acylcarnitine is increased in heart failure patients, which is associated with decreased cardiac function, and a small clinical study confirmed a certain effect of carnitine administration. In other words, myocardial carnitine deficiency and carnitine insufficiency occur in patients with heart failure, and this may have been a major factor in the decline in cardiac function. However, studies on abnormal carnitine dynamics and carnitine replacement therapy in heart failure patients have since faded away, and 20–30 years have passed before the dynamics of carnitine in heart failure patients has received renewed attention again.

16.5.2 Long-Chain Acylcarnitine in Blood and Prognosis in Patients with Heart Failure

After study by Kobayashi et al. [56], research on carnitine kinetics in patients with heart failure ceased for nearly 20 years, but since around 2013, long-chain acylcarnitine in patients with heart failure has been attracting attention as a prognostic marker.

Kalim et al. randomly selected 100 patients who died of heart disease within 1 year after dialysis introduction and 100 patients who lived for more than 1 year (controls) matched by age, sex, and race, and analyzed metabolites in plasma by LC-MS from two independent case-control study cohorts and reported an association with cardiovascular death [57]. As a result, four highly interrelated long-chain acylcarnitines reached a significant threshold adjusted by the multiple comparison method (adjusted P-value <0.0003). Among them, oleoyl carnitine (C18:1 carnitine) was most strongly correlated with cardiovascular death within 1 year in the unadjusted analysis and was associated with cardiovascular death within 1 year even after multivariate correction ($P = 0.001$). Oleoyl carnitine was also replicated as a factor associated with cardiovascular death in an independent cohort sample of 300 patients (100 patients who died cardiovascular death within 1 year of starting dialysis and 200 patients who survived at least 1 year after starting dialysis) ($P = 0.008$). Although weaker than oleoyl carnitine, linoleyl carnitine, palmitoyl carnitine, and stearoyl carnitine also reached significant thresholds adjusted by multiple comparisons (Fig. 16.4). Oleoyl carnitine is increased in patients with chronic renal failure compared to

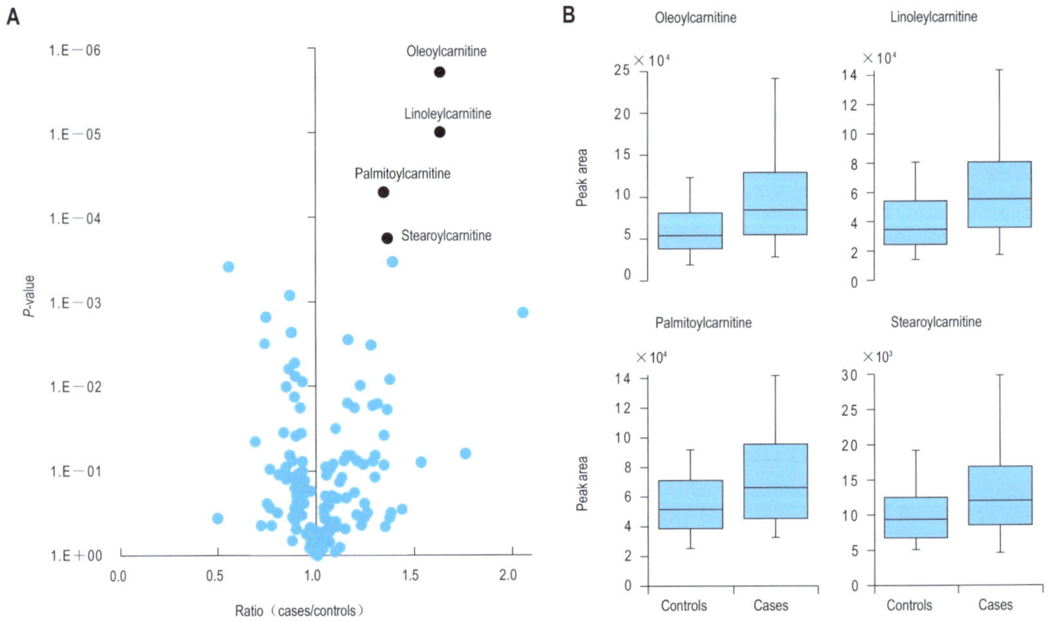

Fig. 16.4 Metabolite profiling identifies markers of cardiovascular death in end-stage renal disease (ESRD). (**a**) The mean ratio of each analyte for cases ($n = 100$) vs. controls ($n = 100$) in baseline plasma, with P values plotted on the y axis. The four metabolites that reached the Bonferroni significance threshold of $P < 0.0003$ are shown in black circles. (**b**) Median peak areas for oleoylcarnitine, linoleylcarnitine, palmitoylcarnitine, and stearoylcarnitine for cases and controls. Box plots show 75th and 25th percentiles; whiskers show 95th and 5th percentiles. (From Ref. [57])

patients with non-chronic renal failure ($P < 0.007$) and in patients with end-stage renal failure several times compared to patients with non-chronic renal failure ($P < 0.001$). It is known that carnitine deficiency occurs in dialysis patients with a longer dialysis period, but the plasma free carnitine concentration of dialysis patients in the early stages of dialysis was not different from that of healthy subjects. This study suggests that changes in the homeostasis of oleoyl carnitine and long-chain acylcarnitine in plasma may be predictors of early cardiovascular death in dialysis patients.

Ueland et al. examined the relationship between carnitine metabolism and prognosis such as progression and worsening of heart failure and all-cause mortality by examining plasma concentrations of carnitine and its derivatives in 183 patients with heart failure and controls (111 subjects) [58]. The mean age of patients with heart failure was 56 ± 13 years, the NYHA classification was 27/43/30% for II/III/IV, and the causative diseases of heart failure were dilated cardiomyopathy (94 patients), coronary artery disease (71 patients), and other 18 patients. As a result, the concentrations of free carnitine, acetylcarnitine, and palmitoyl carnitine were higher in patients with heart failurecompared to controls, and the concentrations of acetylcarnitine, octanoylcarnitine, and palmitoylcarnitine were significantly correlated with the severity of the NYHA classification ($P < 0.001$, $P = 0.003$, and $P < 0.001$, respectively, Fig. 16.5). In particular, the palmitoyl carnitine concentration was significantly correlated with "all-cause mortality + heart transplantation," which is a serious adverse event in patients with heart failure ($P < 0.001$), and the area under the curve was almost the same as that of NT-proBNP and CRP. In addition, the cumulative survival rate at 6 months was significantly reduced at a palmitoyl carnitine concentration of >0.09 μmol/L compared to a palmitoyl carnitine concentration of ≤0.09 μmol/L ($P < 0.0001$, Fig. 16.6).

These results suggest that abnormalities in carnitine metabolism are involved in the progression of heart failure, and that palmitoyl carnitine may be a prognostic marker for patients with heart failure.

Ahmad et al. reported that a subgroup of 2331 patients with systolic failure in the HF-ACTION (Heart Failure: A Controlled Trial Investigating the Outcomes of Exercise Training) trial who gave consent to provide a specimen (453 patients) had plasma metabolomics (45 acylcarnitines and 15 amino acids) analyzed by tandem flow injection mass spectrometry [59]. In addition, among this subgroup of patients (41 patients) with a left ventricular assist device (LVAD), changes in the metabolome were examined before and an average of 136 days (94–180 days) after the device was placed. Because of the large amount of data in such studies, the relevant metabolomes were analyzed in multiple clusters using a technique called principal component analysis (PCA) [60]. The results showed that Factor 5 (long-chain acylcarnitines) and Factor 7 (medium-chain acylcarnitines) were also associated with increased risk of total mortality or total hospitalization in a fully adjusted model. Factor 5 (C16, C18:1, and C18:2 long-chain acylcarnitines) was also associated with increased risk of total hospitalization and cardiovascular death/cardiovascular hospitalization. Factor 5 was significantly higher in patients with end-stage heart failure who had an LVAD before the LVAD was placed, but significantly lower after the LVAD was placed. These results suggest that blood levels of long-chain acylcarnitines are independently associated with serious clinical outcomes in heart failure patients and decrease with long-term LVAD placement, making these long-chain acylcarnitines not only useful prognostic markers but also potentially useful therapeutic targets.

16.5.3 Carnitine Insufficiency and Metabolic Inflexibility in Heart Failure

As mentioned above, there are many reports of decreased free carnitine and increased acylcarnitine (especially long-chain acylcarnitine) in myocardial tissue and high levels of free carnitine and acylcarnitine in the blood in heart failure patients. These are thought to be due to leakage of free carnitine into the blood from damaged myocytes associated with the progression of

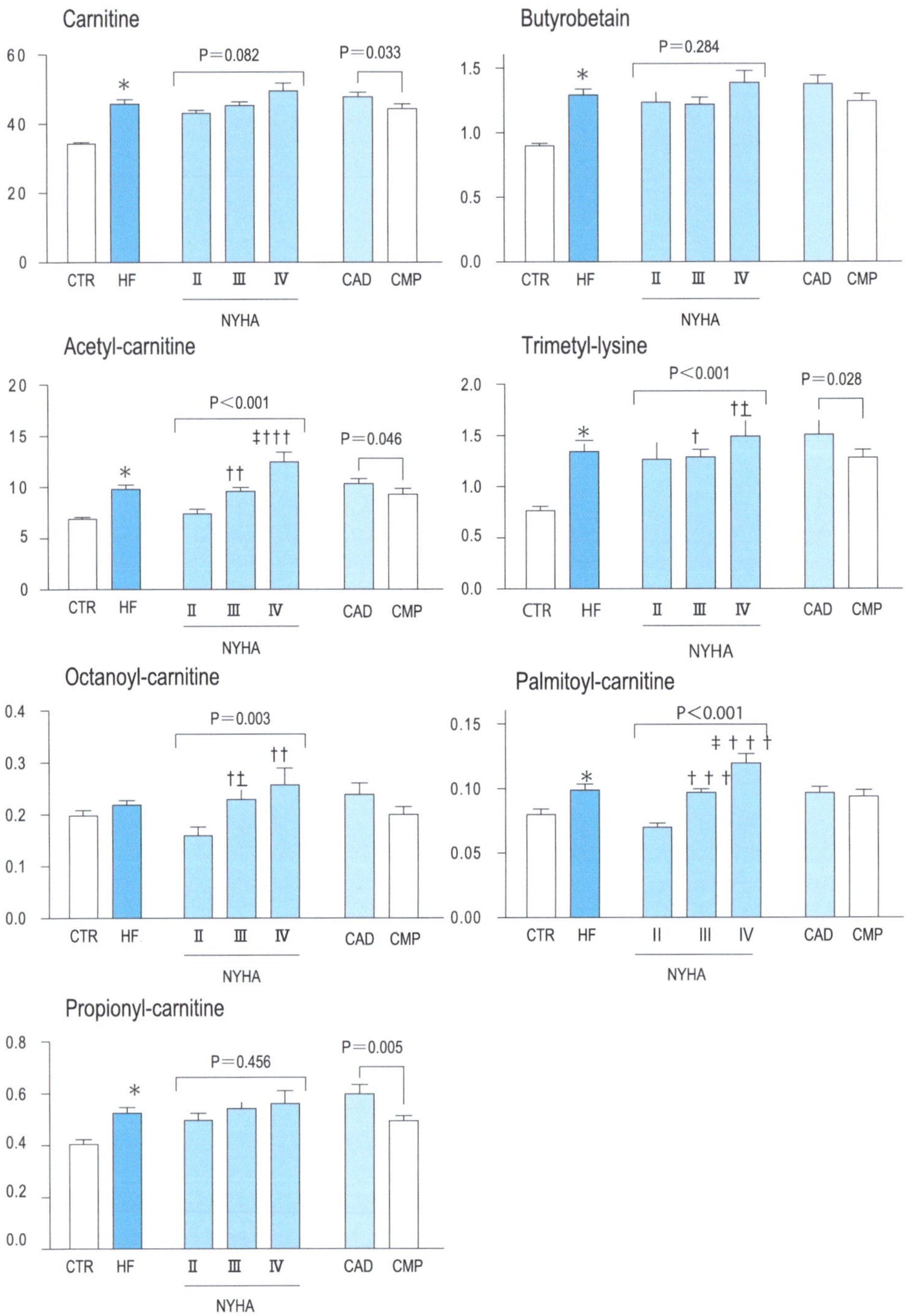

Fig. 16.5 Plasma levels of free L-carnitine, carnitine precursors, and derivatives (μmol/L) in HF patients ($n = 183$) and healthy controls (CTR, $n = 111$). In HF patients, the carnitine composition is also given in relation to clinical disease severity as assessed by NYHA classification and etiology (CAD vs. DCM). Data are given as mean ± SEM. $^{*}p < 0.001$ vs. controls and adjusted by age. BMI and gender distribution. $^{\dagger}p < 0.05$, $^{\dagger\dagger}p < 0.01$, and $^{\dagger\dagger\dagger}p < 0.001$ vs. NYHA II. $^{\ddagger}P < 0.05$ vs. NYHAIII. The p-value to the left denotes the result of the Kruskal–Wallis test comparing NYHA II/III/IV. (Note: original reference described "$^{\dagger}p < 0.05$, $^{\dagger\dagger}p < 0.01$, and $^{\dagger\dagger\dagger}p < 0.01$ vs. NYHA II" but the italic part supposed to be a mistake of "$^{\dagger\dagger\dagger}p < 0.001$ vs. NYHA II"). HF heart failure, CRT control, CAD coronary artery disease, CMP dilated cardiomyopathy (DCM), NYHA New York Heart Association classification for heart failure. (From Ref. [58], modified)

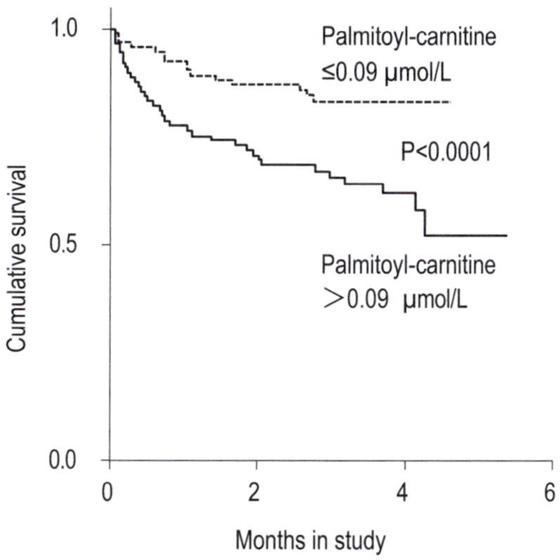

Fig. 16.6 Association between plasma levels of the carnitine derivative palmitoylcarnitine and serious adverse events (i.e., all-cause mortality and heart transplantation, $n = 49$) in patients with heart failure. Kaplan–Meier curves showing cumulative incidence of serious adverse events during the entire study (mean follow-up 34 ± 16[SD] months), according to dichotomized palmitoylcarnitine levels (cut-off median: 0.09 μmol/L). (From Ref. [58])

Table 16.6 Toxic properties of long-chain fatty acid esters

Long-chain acylcarnitine	Detergent action on cell membranes
	Promotes uncoupling of cells
	Inhibits membrane-bound enzymes Na^+-K^+-ATPase Na^+-Ca^{2+}exchanger
	Impairs electrical properties Delays repolarization Enhances afterpotentials Potentiates ectopic rhythms
	Inhibits EDRF in ischemia
Long-chain acyl CoA	Inhibits ANT resulting in sequestration of ATP Inhibits fatty acyl-CoA synthetase

EDRF endothelium dependent relaxing factor, *ANT* adenine nucleotide translocase
From Ref. [61]

heart failure and impaired β-oxidation of long-chain fatty acid, resulting in accumulation of long-chain acylcarnitines in mitochondria [61, 62]. Long-chain fatty acid esters such as long-chain acylcarnitines and long-chain acyl-CoA accumulated in cardiac myocytes destabilize cell membranes by surfactant action, disturb the coupling of between cells in conducting system, and induce arrhythmia by disrupting ion balance through inhibition of Na^+-K^+-ATPase and Na-Ca exchanger, etc., and inhibit ANT and fatty acyl-CoA synthetase causes serious disturbances in energy metabolism (Table 16.6) [61].

When acyl-CoA accumulates in the mitochondria, if there is enough free carnitine, it can carry toxic acyl-CoA out of the mitochondria as acylcarnitine as acylcarnitine, excrete it outside the cell, and finally excrete it in the urine. However, during ischemia when blood flow is blocked, long-chain acylcarnitine accumulates in the myocardium and exerts toxicity. ATP, which is consumed in large quantities by myocardium, is produced by β-oxidation of long-chain fatty acid oxidation and glycolysis. It is well known that glucose oxidation and fatty acid oxidation are reciprocal relationships which are called glucose/fatty acid cycle or Randle's cycle (Fig. 16.7). If blood flow and oxygen supply are sufficient, long-chain fatty acids are the main substrate, but when blood flow is reduced in ischemic heart, ATP production by glycolysis is important for the survival (viability) of cardiomyocytes (although it is not enough for the heart to function as a pump). In either case of glycolysis and fatty acid oxidation, a sufficient amount of free CoA is required. However, when long-chain acyl-CoA accumulates in ischemic heart or failing heart, the amount of CoA in the mitochondria is limited, and free CoA, which is important for energy metabolism, is insufficient. If acyl-CoA increases too much, the free CoA decreases relatively. In order to solve this condition, free carnitine receives the acyl group of acyl-CoA, and acyl carnitine increases, but if the amount of free carnitine is limited in the state of carnitine deficiency, or relative shortage compared to the acyl-CoA amount, it falls into the so-called "carnitine insufficiency," in which acylcarnitine is high and relatively free carnitine is low.

In addition, acetyl-CoA is the shortest acyl-CoA, and a large amount of acetyl-CoA is produced by fatty acid oxidation and glycolysis

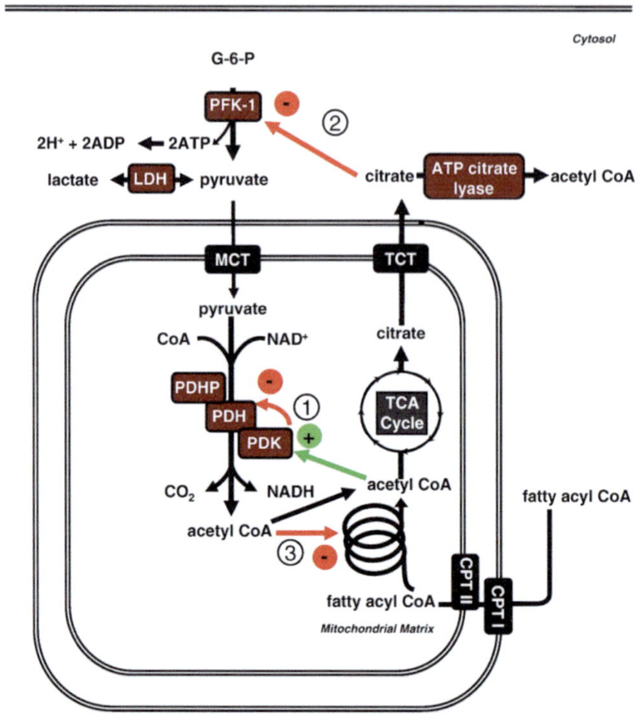

Fig. 16.7 The glucose/fatty acid cycle. Glucose and fatty acid are the major sources of cardiac energy production. The glucose/fatty acid cycle describes the reciprocal relationship between fatty acid and glucose metabolism. Acetyl-CoA and NADH produced from fatty acid β-oxidation can inhibit the pyruvate dehydrogenase (PDH) complex (1). Citrate derived from fatty acid β-oxidation-derived acetyl-CoA inhibits phosphofructokinase-1 (PFK-1), the rate determining step of glycolysis, which in turn can lead to an inhibition of hexokinase by glucose-6-phosphate (G-6-P) (2). Increasing the contribution of glucose oxidation to the generation of acetyl-CoA decreases fatty acid β-oxidation via feedback inhibition of 3-ketoacyl-CoA thiolase; in addition, NADH derived from glucose oxidation can decrease fatty acid β-oxidation via feedback inhibition of both acyl-CoA dehydrogenase and 3-hydroxyacyl-CoA dehydrogenase reaction (3). (From Ref. [62])

under good energy metabolism. When acetyl-CoA accumulates, malonyl-CoA is generated, and CPT1, which is a rate-limiting enzyme for fatty acid oxidation, is inhibited, fatty acid oxidation is suppressed, and metabolism moves in the direction of fat synthesis. In addition, the accumulation of acetyl-CoA, same as other long-chain acyl-CoA, inhibits PDH which is located at the junction site of glycolysis and TCA cycle. In the presence of a sufficient amount of carnitine, the accumulated acetyl-CoA is converted to acetylcarnitine (carnitine + acetyl-CoA ⇄ acetylcarnitine) by carnitine and carnitine acetyltransferase (CrAT) and is used for fat synthesis. However, in the state of carnitine deficiency, acetyl-CoA cannot be converted to acetylcarnitine. Then, the accumulated acetyl-CoA inhibits PDH, and the glycolysis system becomes stagnant. This is the concept of metabolic inflexibility. This concept has been proposed from the studies knocked out of CrAT, in skeletal muscle and has attracted attention and is actively studied as one of the causes of pathogenetic mechanisms of insulin resistance in diabetes and obesity, diabetic cardiomyopathy, liver diseases such as NASH and NAFLD, and heart failure [63–69]. (see Chap. 3, "What Is Carnitine?", Sect. 3.11.3. Types and roles of carnitine acyltransferases). Originally, normal cardiomyocytes and skeletal muscle cells switch between fatty acid oxidation and glucose glycolysis according to their metabolic state and prioritize energy metabolism in one of them (Randle's cycle). Since the myocardium cannot get supply oxygen and cannot oxidize fatty acids during ischemia, anaerobic ATP production by glycolysis is essential for maintaining the viability of cardiomyocytes, but if there is a relative lack of free carnitine due to carnitine deficiency or carnitine insufficiency, acetyl-CoA cannot be converted to acetylcarnitine. PDH inhibition by acetyl-CoA remains. In ischemic state, glucose metabolism is important but after reperfusion, there are several problems occurring such as acidosis and calcium overload (Fig. 16.8). As

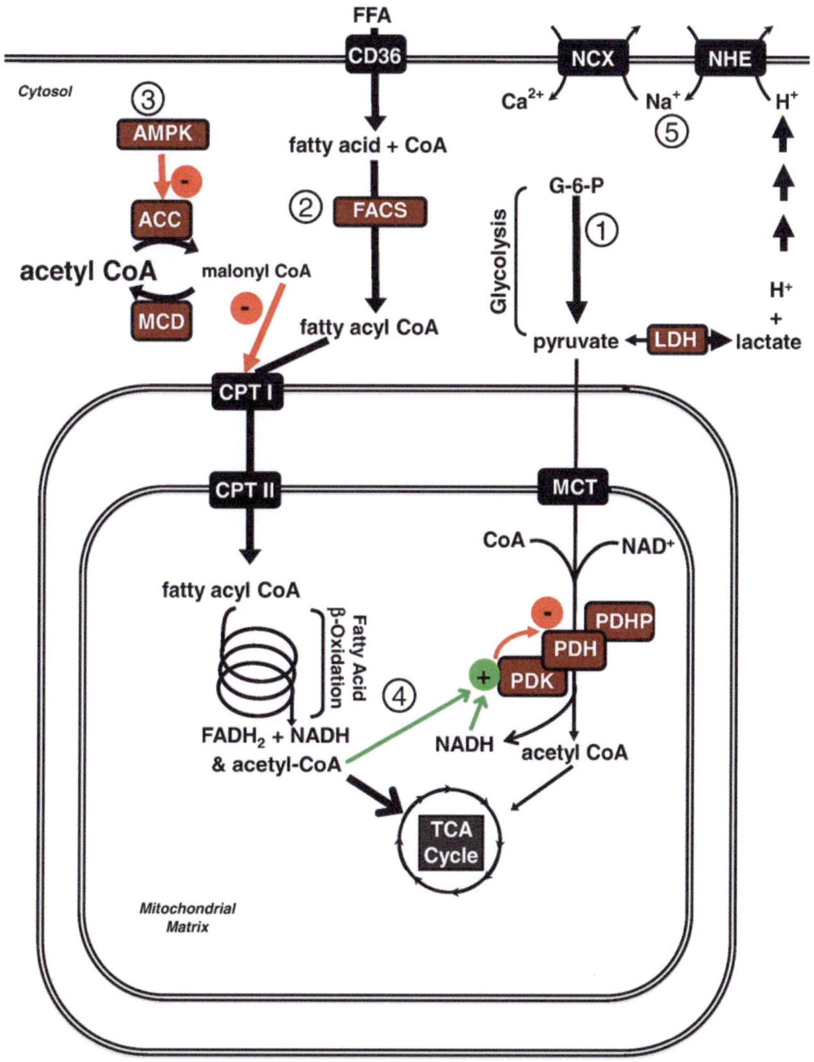

Fig. 16.8 Alterations in myocardial energy substrate metabolism in ischemia/reperfusion. During ischemia, glycolysis becomes an important source of ATP production in response to a decrease in supply of oxygen, or in the absence of oxygen (1). Fatty acid dominate as the substrate for residual oxidative metabolism due to increased plasma level of fatty acids (2) as well as the activation of 5′-adenosine monophosphate activated protein kinase (AMPK) which decreases the production of malonyl-CoA, the endogenous inhibitor of carnitine palmitoyltransferase-1 (CPT1) (3). During reperfusion, glycolysis rates remain high, while fatty acid oxidation dominates over glucose oxidation as the main source of oxidative metabolism. The dominance of fatty acid oxidation during reperfusion inhibits glucose oxidation (4). The uncoupling of glycolysis from glucose oxidation leads to increased proton production, which ultimately leads to myocardial acidosis and calcium overload (5). (From Ref. [62])

described previously (see Sect. 14.4.2 (2) "Mechanism of Onset of Muscle Cramps in Hemodialysis Patients"), the intracellular calcium ion concentration in muscle cells is kept very low at resting state, because increase of calcium ion concentration causes contraction of muscle cells and also activation of various phospholipases and proteases. Calcium overload causes increase in end-diastolic pressure and reduces filling volume resulting in diastolic dysfunction as shown in Fig. 16.1. Calcium overload also increases myocardial stiffness.

CrAT and free carnitine play an important role in regulating the amount of acetyl-CoA and making excess energy available to other metabolic systems. This breakdown in the regulation of acetyl-CoA levels by CrAT and free carnitine is a concept called metabolic inflexibility, which is similar to the state of carnitine insufficiency. The accumulated acyl-CoA inhibits various enzymes and functions such as the TCA cycle of the mitochondria, and further inhibits PDH and other functions that connect the glycolysis system and the TCA cycle, and the relative lack of free CoA causes impaired energy production. This is serious when this happens in the heart muscle, which acts as a pump that constantly delivers blood throughout the body. There is only one way to resolve carnitine insufficiency, namely, to carry the acyl group of acyl-CoA accumulated by carnitine out of the mitochondria as acylcarnitine and regenerate free CoA. Of course, it is also very important to resume blood flow as soon as possible not only for supplying oxygen but also to excrete these acylcarnitine outside of the mitochondria in myocardial infarction.

16.5.4 Effects of Carnitine Administration in Patients with Heart Failure

Jing et al. examined the effect of carnitine (1 g intravenously, twice daily, 7 days) on 265 patients with heart failure in a multicenter, placebo-compared, double-blind study [70]. The assessment was performed on day 7, and the primary endpoint was a decrease of at least one level in the NYHA class. Secondary endpoints included change in 6-min walking distance (alone or in combination with NYHA class reduction), LVEF, NT-proBNP, and adverse events. As a result, 60.9% of the patients in the carnitine group achieved the primary endpoint, but the achievement rate was significantly lower in the placebo group ($P = 0.012$, Table 16.7).

Among the secondary endpoints, the combination of 6-min walking distance alone and NYHA class reduction was significantly reduced in the carnitine group ($P = 0.0497$ and $P = 0.003$, respectively). NT-proBNP decreased significantly after 7 days in both groups, and there was no significant difference between the two groups. There were also no problems with the tolerability of carnitine. Patients with the lowest levels of free carnitine were most likely in NYHA classes III and IV, and they improved best with carnitine administration ($P = 0.002$). In addition, carnitine insufficiency was considered to be involved in the deterioration of the condition of patients with heart failure, as the effectiveness of carnitine administration in improving NYHA class and 6-min walking distance was significantly

Table 16.7 Efficacy according to change in NYHA classification in the per protocol population

Group	Excellent	Effective	Ineffective	Efficacy rate[a] (%)
L-carnitine group ($n = 128$)	6 (4.7)	72 (56.2)	50 (39.1)	60.9
Placebo group ($n = 114$)	3 (2.6)	48 (42.1)	63 (55.3)	44.7

[a]Significance of efficacy rate is patients rate of improvement in NYHA class (excellent + effective) between L-carnitine and placebo groups, $P = 0.012$, using χ^2 test. All values are shown as *n* (%)
Efficacy evaluation of NYHA classification: excellent (decrease 2 classes), effective (decrease 1 class), ineffective (no class change or aggravation)
L-carnitine group: L-carnitine, 1 g × 2 times/day, for 7 days, intravenously, placebo group: placebo, intravenously, for 7 days
From Ref. [70]

observed in patients with acylcarnitine/free carnitine (AC/FC) higher than 0.73 compared to patients with lower than 0.73 (Fig. 16.9). Thus, in Chinese heart failure patients, carnitine administration significantly improved the symptoms of patients with chronic heart failure, and the mechanism of action was probably due to the normalization of carnitine insufficiency.

Wang et al. reported a randomized controlled trial of 29 pediatric patients with dilated cardiomyopathy (17 boys and 12 girls, ages 1 month to 13 years) divided into a control group (10 patients) and a carnitine group (19 patients) to determine the effect of oral carnitine administration [71]. All patients received oral hydrochlorothiazide, enalapril, and spironolactone, plus digoxin for NYHA class IV patients, and intravenous dopamine and dobutamine for refractory heart failure. After cardiac function reached NYHA class II–III, oral metoprolol was administered. Patients in the carnitine group received carnitine (liquid, 50–100 mg/kg/day, oral) in addition to these basic treatments and were followed up for 1, 3, 6, and 12 months. The results showed that clinical symptoms improved, and arrhythmias disappeared in both groups. Echocardiographic evaluation showed that LVEF, fractional shortening (FS), left atrial and left ventricular diameters tended to improve in the control group, but there were no significant changes even at 1 year. In the carnitine group, all of these indices were significantly improved compared to

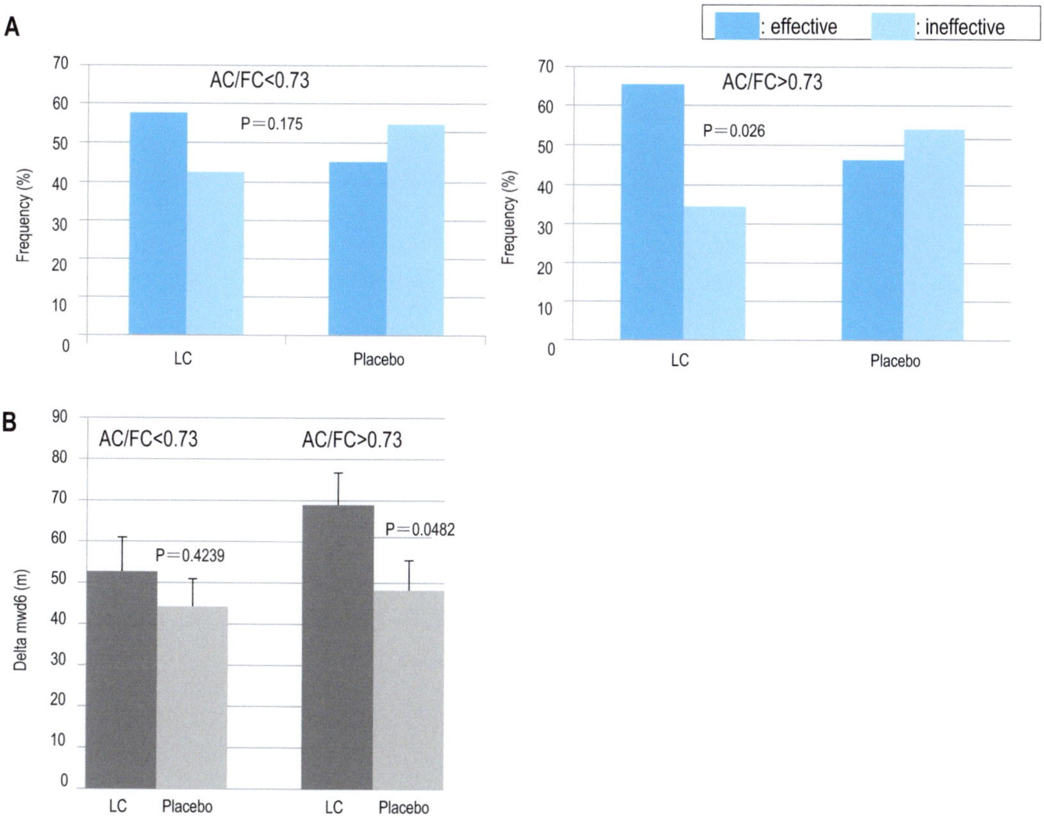

Fig. 16.9 (**a**) Efficacy evaluated as change in NYHA class in low AC/FC patients (left) and high AC/FC patients (right). Median values were measured at baseline for the whole study cohort. Data are shown as *n* (%). (**b**) Efficacy evaluated as 6-min walk distance (m) in low AC/FC patients and high AC/FC patients. 6-min walk distance changes are shown as mean ± SE. *P*-values calculated by logistic regression. LC L-carnitine, AC acylcarnitine, FC free carnitine. (From Ref. [70], modified)

the pre-treatment values (P = 0.003, P = 0.005, P = 0.021 and P = 0.046, respectively, Table 16.8).

Comparing echocardiographic measures at 1 year between the two groups, LVEF was significantly higher in the carnitine group (P = 0.035, Table 16.9).

Thus, improvement in cardiac function has been reported when carnitine is administered to patients with heart failure and dilated cardiomyopathy. Further studies are warranted.

16.5.5 Diastolic Heart Failure (HFpEF) and Carnitine

There are two types of heart failure: heart failure with reduced ejection fraction (HFrEF) and heart failure with preserved ejection fraction (HFpEF). HFpEF is thought to be caused by reduced diastolic function rather than contractility, and the pathogenesis of HFpEF has recently been the subject of intensive research. HFpEF occurs in a wide range of patients from young to elderly and is often caused by ischemic heart disease or cardiomyopathy, whereas HFpEF is more common in the elderly and women and is often caused by hypertensive heart disease [72].

Since the pathophysiology of HFpEF is unknown and there is no effective treatment, Omori et al. selected Dahl-salt-sensitive rats as an animal model of HFpEF and performed metabolomic analysis of their plasma and found that free carnitine levels were reduced [73]. In addition, the free carnitine content was also reduced in the left ventricular myocardium, and the expression of the carnitine transporter

Table 16.8 Changes of echocardiography indices following treatment in the control group and experimental group

Group	Control group (n = 10) Before treatment	after treatment	Z value	P value	Carnitine group (n = 19) Before treatment	After treatment	Z value	P value
EF (%)	32.5 (29.3–41.0)	47.0 (35.0–59.0)	−1.854	0.064	42.0(33.0–61.0)	61.0 (54.0–67.0)	−2.921	0.003
FS (%)	16.0 (12.0–18.5)	24.0 (15.8–31.5)	−1.831	0.067	20.0 (15.0–32.0)	32.5 (27.3–37.3)	−2.814	0.005
LA diameter increment values (mm)[a]	5.5 (3.0–8.5)	3.0 (0.0–7.5)	−1.109	0.267	7.0 (4.5 ~ 12.0)	3.0 (0.5–7.0)	−2.311	0.021
LV diameter increment values (mm)[a]	20.8 (11.7–23.4)	9.3 (3.9–22.6)	−1.589	0.112	7.5 (2.5 ~ 22.0)	4.5 (0.0–8.9)	−1.993	0.046

Data are presented as median (25th–75th percentile)
EF left ventricular ejection fraction, *FS* short axis shortening, *LA* left atrium, *LV* left ventricle
[a]LA diameter increment values and LV diameter increment values were the absolute increment values of LA diameter and LV diameter of patients compared with normal children
From Ref. [71]

Table 16.9 Echocardiography indices after treatment of the control group and carnitine group for 1 year in children with dilated cardiomyopathy

Group	Control group (n = 10)	Carnitine group (n = 19)	Z value	P value
EF (%)	47.0 (35.0–59.0)	61.0 (54.0–67.0)	−2.112	0.035
FS (%)	24.0 (15.8–31.5)	32.5 (27.3–37.3)	−1.836	0.066
LA diameter increment values (mm)[a]	3.0 (0.0–7.5)	3.0 (0.5–7.0)	−0.511	0.609
LV diameter increment values (mm)[a]	9.3 (3.9–22.6)	4.5 (0.0–8.9)	−1.881	0.060

Data presented as median (25th–75th percentile)
EF left ventricular ejection fraction, *FS* short axis shortening, *LA* left atrium, *LV* left ventricle
[a]LA diameter increment values and LV diameter increment values were the absolute increment values of LA diameter and LV diameter of patients compared with normal children
From Ref. [71]

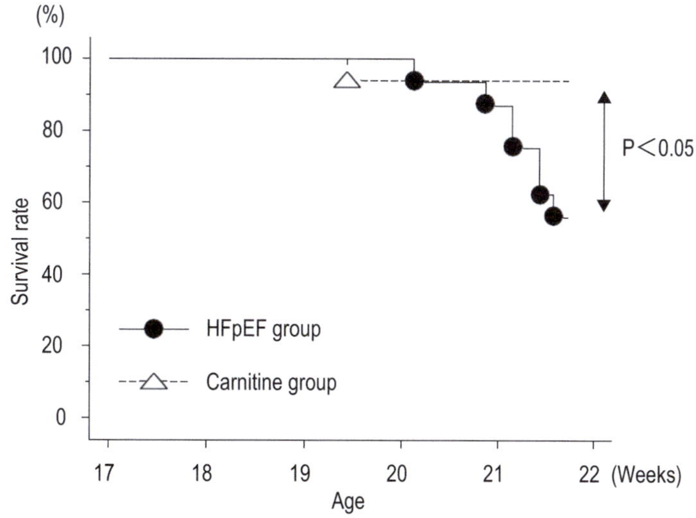

Fig. 16.10 Effect of L-carnitine on survival in the Dahl salt-sensitive rats. Kaplan–Meier survival curves in the HFpEF (closed circles, $n = 16$) and carnitine groups (open triangle, $n = 16$). HFpEF heart failure with preserved ejection fraction. (From Ref. [73])

(OCTN2) was decreased in the left ventricular myocardium and kidneys, and the urinary excretion of free carnitine was increased. Therefore, Dahl-salt-sensitive rats were divided into two groups starting at 17 weeks of age. The treatment group received carnitine (1% carnitine-containing diet, equivalent to approximately 390 mg/kg/day), and the other group received no treatment. The results showed that carnitine administration significantly improved survival ($P < 0.05$, Fig. 16.10). In addition, carnitine treatment improved fibrosis, stiffness, and pulmonary congestion in the left ventricular myocardium.

Experiments with cultured fibroblasts of myocardium have shown that carnitine increases the expression of fatty acid desaturase 1 and 2 (FADS1 and 2), which are rate-limiting enzymes in arachidonic acid biosynthesis. As a result, arachidonic acid was increased, and the antifibrotic effect of prostacyclin (prostaglandin I_2:PGI_2) prevented left ventricular stiffness and left ventricular diastolic insufficiency [73, 74]. I think this upregulation of FADS1 and FADS2 might be mediated by the action of carnitine on Nrf2 (Fig. 16.11). These studies have highlighted the prevention of left ventricular fibrosis in HFpEF as a therapeutic target, and investigations into the clinical efficacy of carnitine in patients with HFpEF have begun.

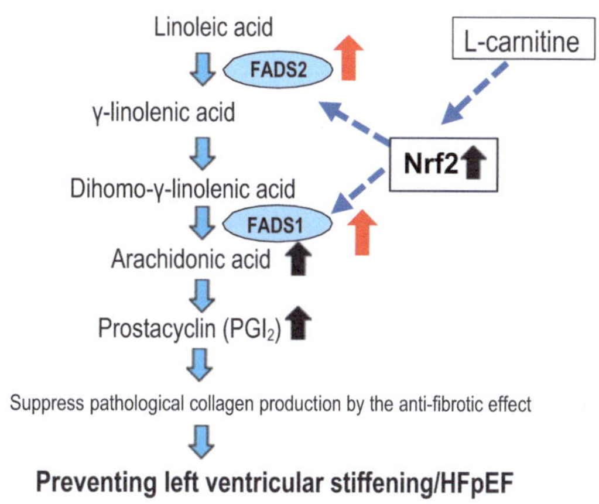

Fig. 16.11 Possible mechanism of the anti-fibrotic effect of carnitine. The beneficial effects of carnitine on left ventricular stiffening are likely provided through enhanced provision of arachidonic acid with upregulation of FADS1 and FADS2, which might be mediated through Nrf2 activation with L-carnitine (dotted arrows), and subsequent promotion of the production of prostacyclin, which suppresses pathological collagen production and prevents left ventricular stiffening. FADS fatty acid dismutase, Nrf2 nuclear factor erythroid 2-related factor 2 (see Chap. 7). (Created using as Refs. [73, 74])

Zordoky et al. examined serum samples from patients with diastolic heart failure (HFpEF, 24 patients), systolic heart failure patients with reduced ejection fraction (HFrEF, 20 patients), and non-heart failure controls (38 subjects) by LC/MS/MS and nuclear magnetic resonance (NMR) to identify the characteristic metabolites of each heart failure [75]. The results showed that in comparison between HFpEF and controls, higher levels of short, medium, and long-chain acylcarnitines, carnitine, creatinine, betaine, and arginine, and lower levels of phosphatidylcholine, lysophosphatidylcholine, and several types of sphingomyelins in HFpEF patients were observed. In comparison between HFrEF and controls, higher levels of short- and medium-chain acylcarnitines, carnitine, creatinine, betaine, and several amino acids were observed, while phosphatidylcholine, lysophosphatidylcholine, several types of sphingomyelins, acetic acid, and acetoacetic acid showed lower levels than controls. In a comparison of HFpEF and HFrEF, long-chain acylcarnitine, 2-OH-butyric acid, 3-OH-butyric acid and acetic acid showed high levels in HFpEF, and sphingomyelin (C24:1), and several types of phosphatidylcholines and lysophosphatidylcholines showed low levels in HFpEF. The combination of these metabolites with NT-proBNP or BNP results in an area under the curve (AUC) that is better than that of NT-proBNP or BNP alone, and the measurement of these metabolites may be useful for the diagnosis of HFpEF and HFrEF.

Assuming that metabolic disorders are closely related to the onset and progression of heart failure, Hunter et al. performed metabolome analysis of fasting plasma in patients with diastolic heart failure (HFpEF, 282 patients), systolic insufficiency heart failure patients with reduced ejection fraction (HFrEF, 279 patients), and non-heart failure controls (No-HF, 191 patients) [76]. As for the analysis method, the related metabolome was analyzed in multiple clusters using the method called principal component analysis (PCA) [59, 60] described above. Using this method, 63 metabolites were classified into 14 related clusters and analyzed. As a result, clusters consisting of six types of long-chain acylcarnitine were significantly higher in HFrEF ($P = 0.0004$), followed by HFpEF and No-HF ($P < 0.0001$ between groups). Table 16.10 shows the concentrations of these six individual long-chain acylcarnitines and a comparison between groups. Thus, it was found that the oxidative process of long-chain fatty acids was dysregulated in patients with heart failure, and that long-chain acylcarnitines were the highest in HFrEF and significantly higher in HFpEF than in the non-heart failure control group.

Quantifying individual acylcarnitine using tandem masses and investigating their association with cardiovascular events and all-cause mortality is a difficult task due to the enormous amount of data. As described in Chap. 5, "What Is Carnitine Deficiency?", AC/FC, which is an indicator of free carnitine deficiency and acylcarnitine increase, is often used as an indicator of carnitine deficiency and carnitine insufficiency. However, in patients with heart failure, myocardial free carnitine decreases and acylcarnitine (especially long-chain acylcarnitine) increases, but both free carnitine and acylcarnitine often increase in plasma, and the significance of AC/FC in plasma of patients with heart failure has not been examined much. Yoshihisa et al. measured free carnitine and total carnitine in plasma of 168 patients with heart failure using an enzyme cycling method, subtracted the free carnitine concentration from total carnitine to obtain acylcarnitine, and examined the association between AC/FC and cardiac death or worsening of heart failure [77]. Patients with heart failure were further divided into HFrEF and HFpEF and examined the relationship between the AC/FC tertile (1st tertile: 0.19 AC/<FC, 2nd tertile: 0.19≤AC/FC <0.27, 3rd tertile: 0.27 AC/≥FC, 56 cases each). As a result, the concentration of free carnitine in the third tertile was significantly lower than in the first tertile ($P < 0.05$), and the concentration of acylcarnitine and AC/FC was significantly higher than in the first and second tertile ($P < 0.01$). Cardiac death (23 cases) and worsening of heart failure (28 cases) occurred during the follow-up period (mean 1004 days). Kaplan–Meier plots of these cardiac events (cardiac death and worsening of heart failure) (Fig. 16.12) show that patients in the third tertile of AC/FC develop the event at a significantly higher rate among all patients with heart failure (Fig. 16.12a), especially in patients with HFpEF (Fig. 16.12b), and there is a tendency but not a significant tendency in HFrEF patients (Fig. 16.12c). These results

Table 16.10 Individual metabolites of long-chain acylcarnitine means and comparisons between HFpEF, HFrEF, and No-HF controls

Metabolites		ANCOVA		Pair-wise comparisons[a]			Mean concentration(μmol/L)[b]		
Structure	Trivial name	Basic[c]	Fully adjusted[d]	HFrEF vs. HFpEF	HFrEF vs. No-HF	HFpEF vs. No-HF	HFrEF (*N* = 273)	HFpEF (*N* = 263)	No-HF (*N* = 180)
C16	Palmitoyl-carnitine	<0.0001	<0.0001	0.001	<0.0001	0.0007	0.105 (0.04)	0.097(0.03)	0.084(0.03)
C18:2	Linoleyl-carnitine	<0.0001	<0.0001	<0.0001	<0.0001	0.04	0.099(0.05)	0.084(0.04)	0.072(0.03)
C18:1	Oleyl-carnitine	<0.0001	<0.0001	<0.0001	<0.0001	0.01	0.185(0.08)	0.161(0.07)	0.137(0.05)
C18	Stearoyl-carnitine	<0.0001	<0.0001	0.09	<0.0001	0.007	0.050(0.02)	0.047(0.02)	0.041(0.02)
C16:1- OH/ C14:1- DC	3-hydroxy-palmitoleoyl- carnitine or cis-5- tetradecenedioyl carnitine	<0.0001	<0.0001	0.002	<0.0001	0.21	0.011(0.01)	0.010(0.005)	0.009 (0.004)
C20:4	Arachidonoyl-carnitine	<0.0001	<0.0001	0.0007	<0.0001	0.49	0.010(0.01)	0.008(0.01)	0.007(0.004)

C indicates carbon chain length

HF heart failure, *HFpEF* heart failure with preserved ejection fraction, *HFrEF* heart failure with reduced ejection fraction

[a]*P* values reflect between-group pair-wise contrasts generated from the fully adjusted ANCOVA

[b]Values are unadjusted mean concentration. SD is provided beneath each value

[c]*P* values for basic model, adjusted for age, race, and sex

[d]*P* values for full model, adjusted for age, race, sex, body mass index, number of diseased coronary artery and history of diabetes mellitus, hypertension, dyslipidemia, smoking, glomerular filtration rate and batch

From Ref. [76]

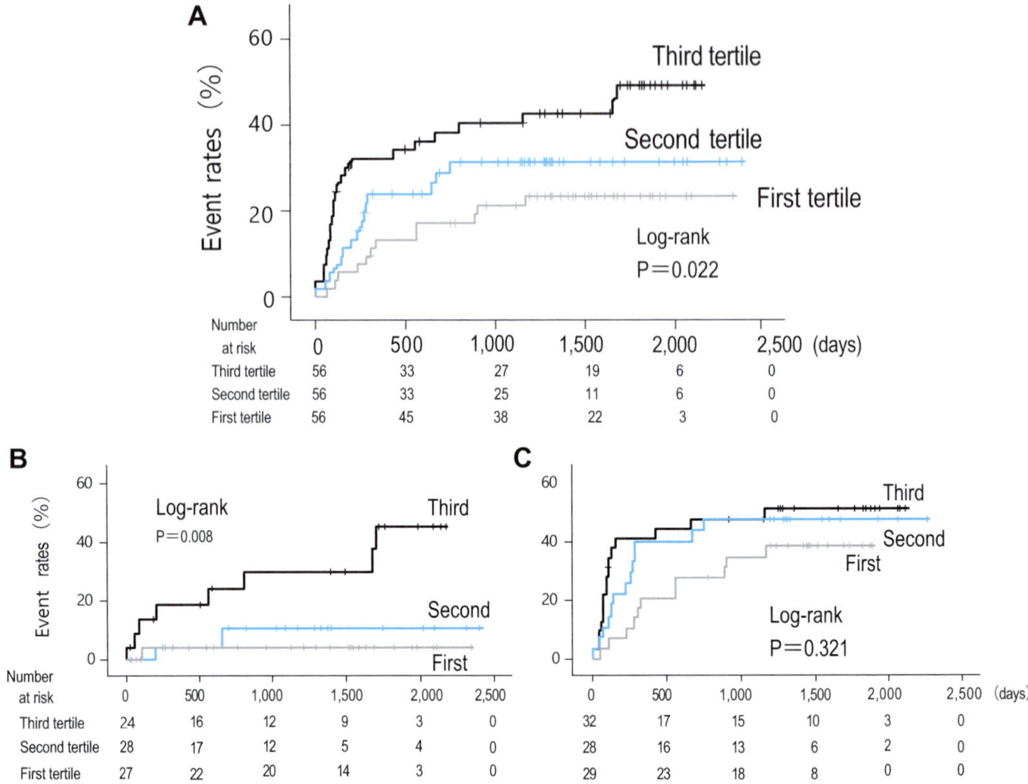

Fig. 16.12 Comparison of cardiac event rates according to the acylcarnitine to free carnitine ratio. Kaplan-Meier analysis for cardiac event rates. (**a**) All heart failure patients (n = 168); (**b**) heart failure with preserved ejection fraction patients (n = 79); and (**c**) heart failure with reduced ejection fraction patients (n = 89). (From Ref. [77])

suggest that AC/FC in plasma is particularly useful for predicting the prognosis of cardiac events in HFpEF patients.

Miyata et al. in the same group divided heart failure patients (131 patients) with sleep apnea syndrome into mild patients (group A, 45 patients), moderate patients (group B, 32 patients), and severe patients (group C, 54 patients) according to apnea–hypopnea index (AHI), and high-sensitive troponin T, which is an indicator of myocardial damage, and plasma free carnitine and acylcarnitine were measured [78]. As a result, the high-sensitive troponin T and total carnitine of group C were significantly higher than those of groups A and B, and the independent factors that determined log [high-sensitive troponin T] by multivariate analysis were high-sensitive CRP and AHI, and the independent factors that determined total carnitine were glomerular filtration rate (GFR) and AHI. It was suggested that potential myocardial damage and abnormalities in carnitine metabolism were occurring.

Kinugasa et al. conducted a prospective cohort study in 117 HFpEF patients because carnitine is said to be involved in HFrEF (e.g., reduction of free carnitine in the myocardium, accumulation of acylcarnitine), but the association of carnitine in diastolic heart failure is unknown [79]. Patients with HFpEF were hospitalized with LVEF ≥45% and heart failure symptoms, and serum free carnitine and total carnitine were measured by enzyme cycling method at the time of discharge. The geriatric nutritional risk index (GNRI), Barthel index (BI, a daily life index), and left ventricular function were also assessed by echocardiography. Hospitalization for heart failure during the follow-up period (472 ± 249 days) and heart-related deaths were used as composite endpoints. Carnitine insufficiency was defined as the lowest quartile of free carnitine (<56.3 μmol/L) or the highest quartile of AC/FC (≥0.35) (44.4% of patients). As a result, the subgroup with carnitine insufficiency had lower GNRI, lower BI, and higher BNP (all $P < 0.05$) and had lower early diastolic mitral annular velocity (E') in patients with sinus rhythm (73 patients, $P = 0.028$). During the follow-up period, 26.5% of patients developed a composite endpoint. In the Kaplan–Meier curve, the subgroup with carnitine insufficiency had a significantly lower event-free survival rate than the subgroup with preserved carnitine ($P = 0.022$, Fig. 16.13). Cox hazard analysis showed that carnitine insufficiency was an independent predictor of cardiac events ($P = 0.017$). Therefore, it was considered that carnitine insufficiency was strongly associated with the deterioration of the condition of HFpEF patients.

Kinugasa et al. found that carnitine insufficiency (carnitine deficiency or relative carnitine deficiency) is an independent predictor of diastolic insufficiency, physical frailty, and cardiac events in HFpEF patients, but the clinical significance of carnitine administration in HFpEF patients is unclear [72, 80]. So, they examined the effects of carnitine administration in patients with HFpEF in a small scale study [81]. Eighteen patients with HFpEF with 45% ≥ LVEF who were hospitalized with heart failure symptoms were treated in the control group (10 patients) and the carnitine group (L-carnitine (LC) group, 8 patients, 300 mg/day orally administered). If carnitine deficiency did not improve after 1 month, the dose was increased to 600 mg/day, to 12 months, and body composition and body function tests and cardiac function evaluation by echocardiography were performed. Seven patients in the control group and six in the LC group were able to complete the 12-month study. Serum free carnitine concentrations before administration did not differ between the LC and control groups (54.0 ± 22.3 vs. 43.8 ± 14.0 μmol/L, $P = 0.34$), but significantly higher in the LC group after 12 months (95.4 ± 26.3 vs. 57.4 ± 18.5 μmol/L, $P < 0.05$), and AC/FC was the same before treatment (0.45 ± 0.10 vs. 0.45 ± 0.13), and after 12 months, only the LC

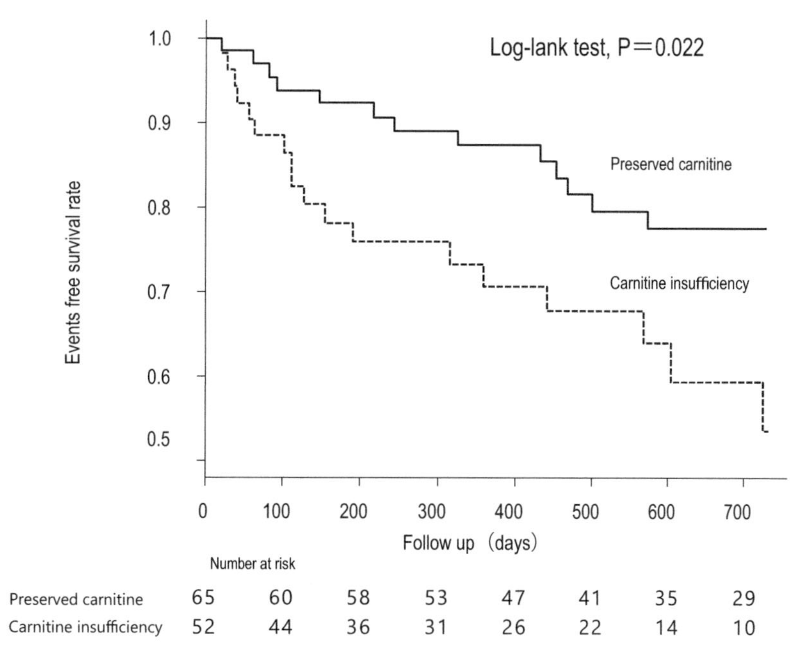

Fig. 16.13 Effect of carnitine insufficiency on clinical outcomes. Event-free survival rate in patients with or without carnitine insufficiency. Patients with carnitine insufficiency (44.4% of total patients) were defined as the lowest quantile of free carnitine level (<56.3 μmol/L) or the highest quantile of the acylcarnitine to free carnitine ratio (≥0.35). (From Ref. [79])

group decreased, but there was no significant difference (0.36 ± 0.14 vs. 0.45 ± 0.10, $P = 0.43$). In the LC group, the body fat percentage decreased after 12 months, but in the control group, it increased, which was a significant change between the two groups. In addition, the maximum walking speed of 10 m increased in the LC group and did not change in the control group, indicating a significant difference between the two groups. Each index of echocardiography did not change significantly. Thus, it was shown that carnitine administration to HFpEF patients may improve carnitine deficiency and carnitine insufficiency, suppress fat accumulation in skeletal muscle, improve muscle energy metabolism, and maintain and improve muscle strength and walking speed.

Kinugasa et al. further studied the relationship between carnitine insufficiency (FC <36 μmol/L or AC/FC ≥0.27) and muscle weakness in 124 patients with heart failure [82]. Skeletal muscle weakness was classified into two phenotypes: sarcopenia (low muscle strength with low muscle mass) and dynapenia (low muscle strength with normal muscle mass). Patients with carnitine insufficiency had a significantly higher prevalence of muscle weakness (Fig. 16.14), and a lower 6-min walk distance than without carnitine insufficiency ($P < 0.05$). However, there were only weak association between carnitine levels and dynapenia. They concluded that carnitine insufficiency is more closely associated with sarcopenia than with dynapenia in patients with heart failure, suggesting carnitine insufficiency as a potential therapeutic target for sarcopenia in these patients.

16.5.6 Micronutrient Deficiency in Patients with Heart Failure

There are various causes of heart failure, such as ischemic heart disease, heart valve disease, cardiomyopathy, and rheumatic heart disease. Among patients with heart failure, micronutrient deficiencies such as vitamin B_1 (thiamine), vitamin B_2 (riboflavin), vitamin D, D-ribose, taurine, creatine, and CoQ10, and mineral deficiencies such as iron, zinc, selenium, and magnesium may be the cause, and it has been reported that supplementation with these micronutrients and minerals may improve cardiac function and patient condition. Carnitine is one such micronutrient that has been discussed in many review articles [83–88]. These micronutrients are called "conditional nutrients" or "metabolic modulators," and although deficiency is usually rare, deficiency might occur due to malnutrition or anorexia, and energy metabolism such as carbohydrate metabolism and mitochondrial oxidative phosphorylation may be adversely affected. Studies using Syrian hamsters that develop cardiomyopathy have also reported improvements in cardiac function and cardiac microstructure in groups fed diets containing carnitine, taurine, CoQ10, selenium, vitamin E, vitamin C, creatine, thiamine,

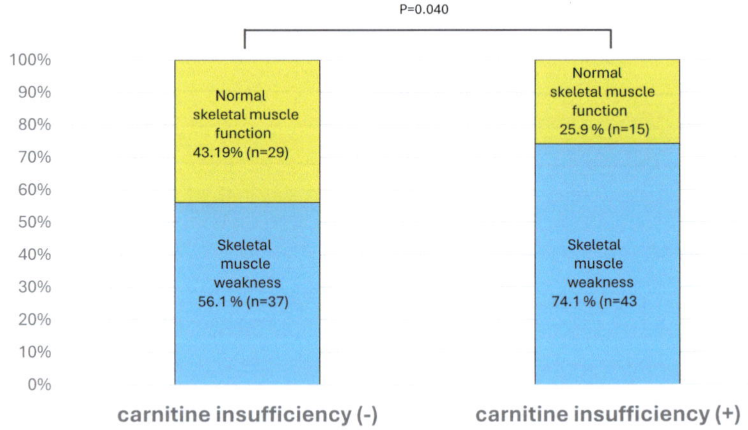

Fig. 16.14 Prevalence of skeletal muscle weakness in patients with and without carnitine insufficiency. (From Ref. [82])

and L-cysteine [89]. It has also been reported that administration of carnitine to Syrian hamsters with cardiomyopathy improves cardiac function and restores decreased amounts of high-energy phosphate compounds such as ATP and creatine phosphate [90].

In addition, selenium (selenium) and zinc are required for the active centers of various enzymes involved in oxidative stress reduction, such as glutathione peroxidase and SOD, and their deficiency leads to a decrease in the activity of these enzymes, and it is thought that oxidative stress and increased inflammation are involved in the progression of cardiomyopathy and heart failure. The relationship between these trace element deficiencies and the development of cardiomyopathy and heart failure has long been a problem in the field of postoperative enteral nutrition and central venous nutrition management in patients with severe mental and physical disabilities and children (see "Chap. 11: Tube Feeding, TPN, Special Milk, and Carnitine Deficiency"), but caution is required even in patients with general heart failure, especially in the elderly, who do not have adequate food intake.

It has been reported that inflammatory cytokines such as TNFα are elevated in patients with heart failure, and that these inflammatory cytokines and sphingosine cause atrophy of skeletal muscle and increase the concentration of carnitine in plasma in a rat heart failure model [91] and in human heart failure patients [92]. This is the basis of muscle atrophy and extreme fatigue in patients with heart failure, and it is considered necessary to treat patients other than cardiac function decline. Carnitine prevents apoptosis by inhibiting caspase 3 in skeletal muscle cells, reduces oxidative stress, and suppresses the decline in the expression of PGC-1α, a key regulator of mitochondria, which has been reported in many basic and clinical studies [93] (see Chap. 7, "Anti-inflammatory, Antioxidant, and Anti-apoptotic Effects of Carnitine"). Although rest was previously recommended for the treatment of heart failurepatients, it is now known that cardiac rehabilitation, such as active exercise, is associated with a better prognosis, and it is important to provide comprehensive rehabilitation, including not only exercises but also administration of conditionally nutrients including carnitine, or micronutrients called metabolic modulators [94].

Kumar et al. reported a randomized, double-blind, controlled trial in which 62 patients with heart failure were divided into two groups, one of which received carniQ-gel (a soft capsule containing 2250 mg/day of L-carnitine and 270 mg/day water-soluble ubiquinol) for 12 weeks, and the other received placebo [95]. Finally, they analyzed data from 29 patients in each of the carniQ-gel and placebo groups. Ischemic heart disease was the cause of heart failure in 34.5 and 37.9%, valve disease in 31 and 34.5%, cardiomyopathy in 34.5 and 27.6%, and ejection fraction on echocardiogram (38.8 ± 7.6 and 39.3 ± 6.7%). After 12 weeks, QOL scores such as 6-min walking distance, dyspnea, and fatigue improved significantly in the treatment group compared to before administration, and the proportion of patients with NYHA Class II to IV also decreased significantly from 100 to 62% ($P < 0.02$). These measures also improved significantly in the placebo group, but the proportion of patients in NYHA classes II to IV was small, ranging from 100 to 86.2% ($P < 0.05$). In addition, serum thiobarbituric acid reactive substances (TBARS) and malondialdehyde (MDA), which are indicators of oxidative stress, were significantly reduced in the treatment group ($P < 0.01$), but no such changes were observed in the placebo group.

Kinugasa et al. also discussed carnitine as a nutritional intervention in a review article on the treatment of frailty and sarcopenia in patients with HFpEF [96]. Because patients with carnitine insufficiency have low physical activity, left ventricular diastolic dysfunction, and a poor prognosis in previous studies at their own institution, they suggest that when studying the effect of carnitine administration in patients with HFpEF, patients should be selected from those with sarcopenia or pre-sarcopenia with carnitine deficiency or carnitine insufficiency, and effects of carnitine on physical activity and quality of life should be evaluated.

Thus, conditions of carnitine deficiency and carnitine insufficiency are often recognized in cardiac disease. As described in various parts of this book, carnitine supplementation therapy is the only way to correct carnitine deficiency and carnitine insufficiency. Further clinical studies will be needed to evaluate the effects of carnitine supplementation therapy on clinical symptoms, cardiac function and quality of life, cardiovascular events, life expectancy, and sarcopenia, not only in patients with HFrEF but also in patients with HFpEF.

16.6 Peripheral Arterial Disease and Carnitine

Peripheral arterial disease (PAD), also known as arteriosclerosis obliterans (obstructive arteriosclerosis (ASO), is a blockage of peripheral arteries in internal organs such as arteries in the extremities and renal arteries due to arteriosclerosis. It is a disease that causes movement disorders such as intermittent claudication (IC), numbness, and pain, and in severe cases, gangrene due to tissue necrosis and the need to amputate are required, which have a serious impact on QOL and prognosis. Although there are no exact statistics, it is believed that there are more than 3.5 million people in Japan, with an increasing trend due to the increase in lifestyle-related diseases and the westernization of dietary habits in addition to the aging of the population.

The basic treatment of PAD is endovascular treatment in which the occluded blood vessel is widened with a balloon catheter, or a stent is placed. However, carnitine deficiency and accumulation of acylcarnitine in skeletal muscles have been reported due to ischemia of the femoral artery, which feeds the muscles of the lower extremities, and carnitine supplementation therapy has also been attempted.

16.6.1 Effect of Carnitine Administration in Patients with Intermittent Claudication (IC)

As with the treatment of ischemic heart disease, endovascular treatment using a catheter is the main treatment method for the treatment of ASO. However, in ischemic heart disease, in addition to a decrease in blood flow, an abnormality in energy metabolism occurs due to a decrease in free carnitine in tissues and an increase in long-chain acylcarnitine, and carnitine supplementation therapy has been reported. Brevetti et al. investigated the effects of carnitine administration in patients with IC. First, in order to investigate the effect of carnitine administration on hemodynamics, 18 patients were divided into a carnitine group (3 g bolus intravenous infusion, followed by continuous administration at 2 mg/kg/min for 30 min) and a placebo group, and the blood flow rate of the diseased lower limb and the blood pressure ratio of the ankle/brachial artery were measured by the crossover method, but carnitine administration did not affect hemodynamics at all [97]. Next, the maximum walking distance before and after treatment was measured in a double-blind crossover study of carnitine administration (2 g divided into 2 doses per day, oral administration, 3 weeks) or placebo, and the degree of improvement in clinical symptoms was also scored by a medical interview. As a result, carnitine administration significantly extended the maximum walking distance (Fig. 16.15). In addition, improvement scores such as paresthesia, fatigue, pain while walking, and cold sensation improved slightly when placebo was administered, but improved significantly when carnitine was administered, and such symptom improvement disappeared when wash-out and placebo were administered.

By carnitine administration, the increase in lactate/pyruvate ratio in popliteal vein blood during the treadmill test, which was noticeable in the placebo group, was significantly suppressed in the carnitine group, and an increase in free carnitine and short-chain acylcarnitine in the muscles and a decrease in long-chain acylcarnitine were observed, indicating metabolic improvement. PDH is inhibited by acetyl-CoA and its activity is regulated by the acetyl-CoA/free CoA ratio. Acetyl-CoA, which had accumulated, became acetyl-carnitine by carnitine administration, and the inhibition of PDH by acetyl-CoA was canceled, and it was thought that muscle energy metabolism was improved. Thus, it was shown that carnitine had the effect of improving metabolism in IC patients even if the blood flow rate did not change.

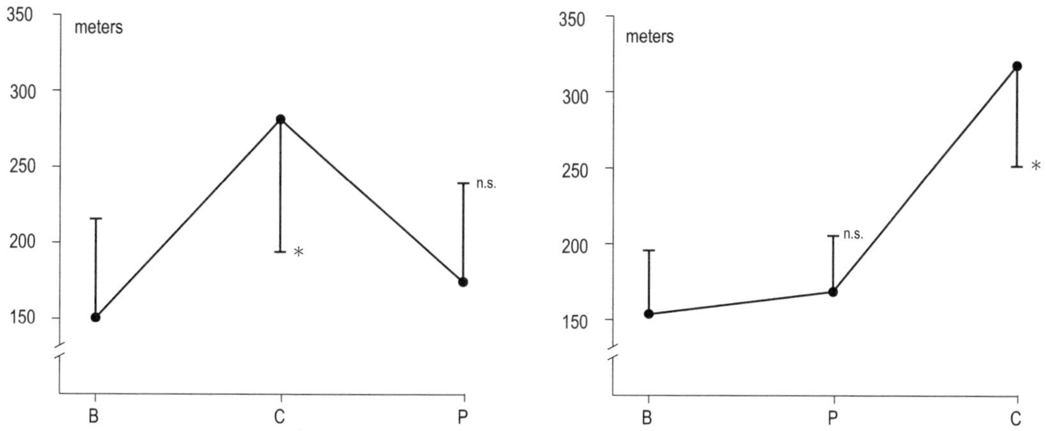

Fig. 16.15 Effect of the two treatments (placebo and L-carnitine) on absolute claudication distance in intermittent claudication patients during the entire study period. The effects of placebo or L-carnitine(oral administration, 2 g/day, divided 2 times, for 3 weeks) were studied in double-blind, cross-over manner. Note the sharp increase in walking capacity whenever L-carnitine is given. Data are expressed as mean ± SD. *Different from baseline value, $P < 0.01$, B: baseline value (after washout period), C: end of L-carnitine period, P: end of the placebo period. (From Ref. [96])

In addition, Brevetti et al. reported that total carnitine concentrations were significantly reduced in skeletal muscle in IC patients compared to controls (20.9 ± 5.2 vs. 11.6 ± 6.2 nmol/mg NCP, $P < 0.01$), and that CrAT activity was significantly reduced [98]. In other words, it was shown that carnitine deficiency and carnitine insufficiency occurred in the skeletal muscles of IC patients, and that PDH, a rate-limiting enzyme of glycolysis, was inhibited by acetyl-CoA, causing metabolic disorders. Brevetti et al. also reported the effect of administering propionyl-L-carnitine (PLC), which is said to be easily distributed in muscles, to restore the concentration of carnitine in muscles (see Chap. 3, "What Is Carnitine?" about PLC).

Brevetti et al. divided IC patients (22 patients) into those with low (IC1, 10 patients, acetylcarnitine concentration 3.7 ± 0.2 μmol/L) and those with high (IC2, 12 patients, acetylcarnitine concentration 7.9 ± 0.7 μmol/L) acetylcarnitine levels at rest. The IC2 patients showed pain earlier than IC1 patients when exercise loading, and the maximum walking distance was shorter ($P < 0.05$), but responded well to carnitine administration (500 mg, intravenous bolus), and the duration of pain onset and maximum walking time was prolonged. Stepwise multiple regression analysis showed that only changes in acetylcarnitine concentrations during exercise were associated with changes in pain onset time ($F = 6.700$, $r = 0.633$, $P = 0.027$) [99].

Hiatt et al. compared the carnitine fraction in muscle and exercise performance in 10 PAD patients with one leg artery occluded but the other leg normal and six age-matched healthy controls [100]. At rest, carnitine content and lactate levels in the gastrocnemius muscle of the normal legs of PAD patients did not differ from those of normal controls, but lactate and long-chain acylcarnitine were higher in the legs with ischemia, and short-chain acylcarnitine in the ischemic foot was inversely correlated with maximal oxygen uptake ($r = -0.75$, $P < 0.05$).

A treadmill stress test significantly increased the short-chain acylcarnitine content in the gastrocnemius muscle of the ischemic limb in patients with PAD (440 ± 130 vs. 900 ± 200 nmol/g, $P < 0.05$). In the non-ischemic limb of PAD patients, there was a positive correlation between the maximum blood lactate level and the change in the short-chain acylcarnitine content as in the healthy control ($r = 0.67$, $P < 0.05$), but in the ischemic limb, there was no correlation between lactate level and the change in the short-chain acylcarnitine content. The maximum exercise time and the change in short-chain acylcarnitine content were positively correlated ($r = 0.82$, $P < 0.01$). These results suggest that acylcarnitine accumulation occurs in the ischemic limbs of patients with PAD, and that this acylcarnitine amount is an indicator of motility, and that metabolic abnormalities occur in patients with PAD as well as blood flow.

Subsequently, the effects of administering carnitine to patients with PAD and IC on the maximum walking distance were actively studied, but most of the studies were conducted using PLC [101–104].

In addition, many studies have been reported on the mechanism of action of IC in PLC to improve walking distance, and it has been reported that it improves metabolism by eliminating acylcarnitine accumulation, as well as prevents capillary breakdown by reducing oxidative stress and enhancing endothelial function [105–110].

As described in Chap. 3, "What Is Carnitine?", carnitine and PLC have almost the same effect, and PLC is said to be more easily distributed in muscles and myocardium and is often studied in ischemic hearts [61]. In addition, as described in Chap. 7, "Anti-inflammatory, Antioxidant, and Anti-apoptotic Effects of Carnitine," both carnitine and PLC have the effect of protecting cells by reducing oxidative stress and inflammatory cytokines by Nrf2-mediated mechanisms. So, what is the difference between PLC and carnitine in improving walking distance in IC patients?

Brevetti et al. reported a direct comparison of carnitine and PLC on improving walking distance in IC patients [111]. First, they examined the dose of PLC to improve walking distance in 12 patients with PAD. PLC administration (300 mg, single intravenous administration) 30 min before the start of the treadmill test did not improve walking distance, but a single intravenous 600 mg dose significantly improved walking distance. Therefore, a comparative study was conducted in which the PLC dose was set at 600 mg and the carnitine dose was 500 mg, which was almost the same in molar ratio (Fig. 16.16).

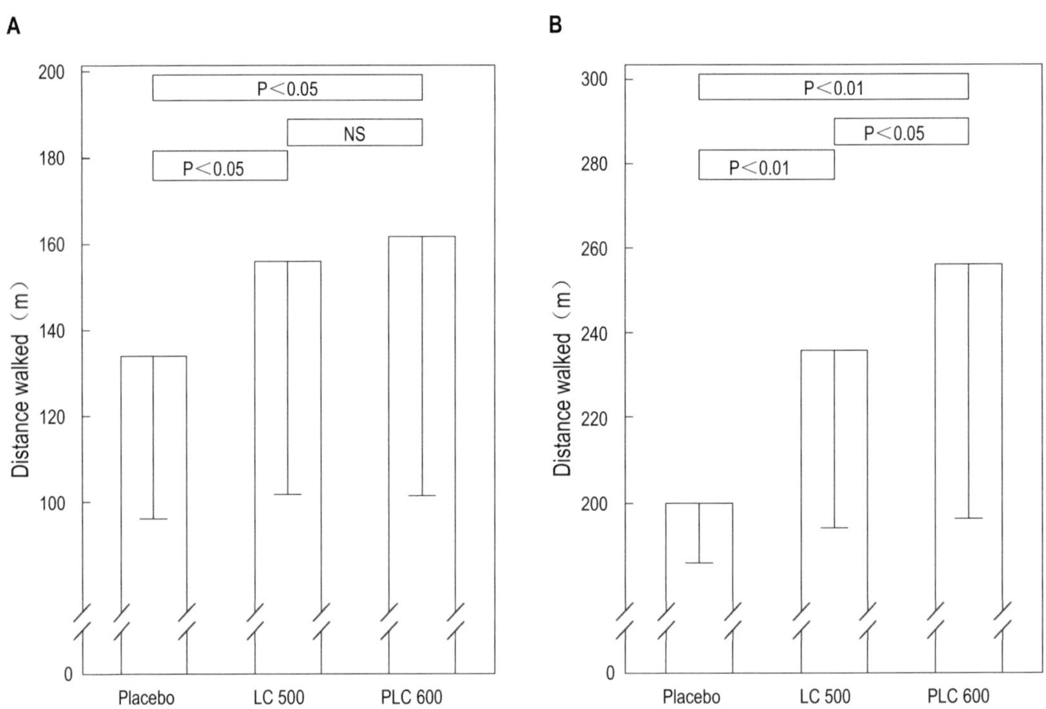

Fig. 16.16 Effect of placebo, L-carnitine 500 mg (LC 500) and L-propionylcarnitine 600 mg (LPC 600) on the claudication distance. (**a**) initial claudication distance. (**b**) absolute claudication distance. Note that the effect with LPC 600 is significantly greater than that with LC 500. Drugs were administered before 30 min of treadmill test (single, intravenous administration). (From Ref. [111])

PLC is slightly more effective, but it can be seen that carnitine also has the effect of improving walking distance. For this reason, there are more clinical studies using PLC than carnitine on the effects on IC, and Brass et al. have reported systematic reviews and meta-analyses [112]. In addition, PLCs seem to be approved in Italy, but not in other Western countries, and intravenous injectables are difficult to obtain. PLC has not been approved in Japan, but there are tablets, oral solutions, and injections as medicines for carnitine, and supplements can also be used.

16.6.2 Position of Carnitine Administration in the Peripheral Arterial Disease Guidelines

As described so far, carnitine and PLC have been reported to prolong walking distance in patients with PAD and IC and have been included in guidelines for peripheral arterial disease.

The 2007 The Trans-Atlantic Inter-Society Consensus for the Management of Peripheral Arterial Disease (TASC II) guidelines, carnitine and PLC are described as the drugs with evidence to improve claudication (Fig. 16.17) [113–115].

However, in the 2005 ACC/AHA guidelines for the management of patients with PAD, the recommendations for medical and pharmacological treatment of IC are treated as "Class II.b, Level of Evidence: B" because "the efficacy of improving walking distance in patients with intermittent claudication in PLC has not been sufficiently established" [116]. In addition, the "2022 Revised Guidelines for Peripheral Arterial Disease" of The Japanese Circulation Society/The Japanese Society for Vascular Surgery does not mention carnitine or PLC.

Cilostazol, an antiplatelet agent, has been reported to extend walking distance in patients with PAD [117, 118], and meta-analyses have shown its usefulness [119], and it is described in TASCII and Japanese guidelines. Goldenberg et al. compared the maximum walking distance of cilostazol alone (100 mg, twice daily) and cilostazol + carnitine (1000 mg, orally twice daily) in treadmill trials for patients with PAD [120]. As a result, the combination of cilostazol and carnitine significantly increased the maximum walking distance compared to the cilostazol alone group ($P < 0.048$ in the case of per protocol analysis).

16.6.3 Research and Future Prospects of Carnitine in PAD Patients in Japan

Although there are few studies on carnitine in patients with PAD in Japan, Murakami et al. administered carnitine to 33 ASO patients (900 mg/day, 3 months), and the effects on subjective symptoms and motor activity were examined [121]. As a result, in the Walking Impairment Questionnaire (WIQ) and Vascular Quality of Life (VascuQOL) scores, there was a slight improvement in walking speed, a slight decrease in pain scores, and an increase in scores on stairs,

Section C. Intermittent claudication

C4.2.2. Drugs with supporting evidence of clinical utility in claudication

Carnitine and propionyl-L-carnitine

Patients with peripheral arterial disease develop metabolic abnormalities in the skeletal muscles of the lower extremity. Thus, claudication is not simply the result of reduced blood flow, and alterations in skeletal muscle metabolism are part of the pathophysiology of the disease. L-carnitine and propionyl-L-carnitine interact with skeletal muscle oxidative metabolism, and these drugs are associated with improved treadmill performance. Propionyl-L-carnitine (an acyl form of carnitine) was more effective than L-carnitine in improving treadmill walking distance. In two multicenter trials of a total 730 patients, initial and maximal treadmill walking distance improved more with propionyl-L-carnitine than placebo. Additional trials in the broad population of patients with claudication will be necessary to establish the overall efficacy and clinical benefit of these drugs.

Fig. 16.17 Description in TASC II about L-carnitine and propionyl-L-carnitine for the treatment of claudication. TASC II: Inter-society consensus for the management of peripheral arterial disease (TASC II). (From Refs. [113–115])

although not significant in WIQ. VascuQOL showed a significant increase in activity score ($P = 0.0405$). The other scores were also improving, although there was no significant difference, and the walking distance for 6 min was also extended, although there was no significant difference. In addition, the pulse wave conduction velocity (baPWV) and augmentation index (AI) showed an improvement trend, suggesting that carnitine administration may improve arteriosclerosis itself. However, the reason why these indicators remained improving and there was no significant difference was probably due to the short treatment period of 3 months.

Komai et al. measured blood carnitine levels in 130 ASO patients [122]. They found that AC/FC was 0.26 ± 0.13 in the IC group, 0.42 ± 0.22 in the critical limb ischemia group, and 0.24 ± 0.10 and 0.31 ± 0.14 excluding dialysis patients, respectively, which were significantly higher in the critical limb ischemia group. Multivariate analysis showed that chronic maintenance dialysis ($P < 0.001$) and critical leg ischemia ($P < 0.01$) were significant factors affecting AC/FC, suggesting that AC/FC in blood carnitine test may be a marker for determining the severity of leg ischemia in ASO.

ASO is an arteriosclerosis of blood vessels throughout the body, including peripheral blood vessels, and is often neglected compared to coronary artery disease and cerebrovascular disease, but it is a serious disease that has a serious impact on QOL such as gait disorders, amputation due to ulcers, and necrosis, and has an extremely poor life prognosis [123]. In addition, the number of ASO patients is increasing due to various factors such as the increase in dialysis patients and diabetic patients and the westernization of diets in the aging society. Among them, there are many patients who cannot undergo endovascular treatment using a catheter.

Speaking of arteriosclerosis, in 2013, the Hazen group reported that the intake of carnitine and choline increases trimethylamine and trimethylamine-N-oxide by intestinal bacteria and promotes arteriosclerosis. Rather, carnitine administration has been reported to reduce cardiovascular events and improve baPWV in dialysis patients (see Chap. 9, "Safety of Carnitine" and Chap. 14, "Carnitine Deficiency in Patients with Renal Failure"). As described in the study by Komai et al. [122], patients with severe limb ischemia, which is often ulcerative and necrosis, have significant abnormalities in carnitine metabolism. Further studies are expected to be conducted to determine whether carnitine supplementation eliminates the accumulation of acylcarnitine and whether the anti-inflammatory, antioxidant, anti-apoptotic, and anti-arteriosclerotic effects of carnitine are clinically useful in patients with PAD such as ASO and severe limb ischemia.

References

1. Neubauer S. The failing heart — an engine out of fuel. N Engl J Med. 2007;356(11):1140–51. https://doi.org/10.1056/NEJMra063052.
2. Stanley WC, Recchia FA, Lopaschuk GD. Myocardial substrate metabolism in the normal and failing heart. Physiol Rev. 2005;85(3):1093–129. https://doi.org/10.1152/physrev.00006.2004.
3. Sotobata I, Noda S, Hayashi H, et al. Clinical evaluation of effects of levocarnitine chloride (LC-80) on exercise tolerance in angina pectoris by a serial multistage treadmill exercise testing: a multicenter, double blind study. Jpn J Clin Pharmacol Ther. 1989;20(3):607–18. (Article in Japanese).
4. Sato H, Matsuyama T, Ozaki H, et al. Long-term effects of L-carnitine on cardiac function and exercise capacity in patients with left ventricular dysfunction. Jpn J Clin Pharmacol Ther. 1990;21(2):439–47. (Article in Japanese).
5. Niizoe K. Clinical study of levocarnitine chloride (LC-80) for ischemic heart disease. Diagn Treat. 1989;77(6):1604–10. (Article in Japanese).
6. Yoshitoshi Y, Iimura O, Shintani H, et al. Clinical study of LC-80 for ischemic heart disease: a multicenter open trial. Jpn Pharmacol Ther. 1989;17(4):1689–711. (Article in Japanese).
7. Katsume H, Furukawa K, Kono Y, et al. Clinical effect of levocarnitine chloride (LC-80) on ischemic heart diseases. Jpn Pharmacol Ther. 1989;17(5):2203–11. (Article in Japanese).
8. Iimura O, Yoshida S, Shoji T, et al. Clinical effect of levocarnitine chloride (LC-80) on ischemic heart disease patients. Jpn Pharmacol Ther. 1989;17(5):2193–201. (Article in Japanese).
9. Watanabe T, Yamazaki N, Ogawa K, et al. Clinical evaluation of levocarnitine chloride (LC-80) for angina pectoris: focusing on cases that are difficult to treat with conventional antianginal drugs. Pharmacol Ther. 1989;17(5):2213–25. (Article in Japanese).

10. Yoshitoshi Y, Yamazaki N, Iimura O, et al. Efficacy of levocarnitine chloride (LC-80) in treatment of ischemic heart disease by the serial multistage treadmill exercise testing – a multicenter double-blind group-comparison study. Jpn J Clin Pharmacol Ther. 1989;20(4):687–707. (Article in Japanese).
11. Cherchi A, Lai C, Angelino F, et al. Effects of L-carnitine on exercise tolerance in chronic stable angina: a multicenter, double-blind, randomized, placebo controlled crossover study. Int J Clin Pharmacol Ther Toxicol. 1985;23(10):569–72.
12. Iyer RN, Khan AA, Gupta A, et al. L-carnitine moderately improves the exercise tolerance in chronic stable angina. J Assoc Physicians India. 2000;48(11):1050–2.
13. Schwartz A, Wood JM, Allen JC, et al. Biochemical and morphologic correlates of cardiac ischemia. I Membrane systems. Am J Cardiol. 1973;32(1):46–61. https://doi.org/10.1016/s0002-9149(73)80085-2.
14. Whitmer JT, Idell-Wenger JA, Rovetto MJ, et al. Control of fatty acid metabolism in ischemic and hypoxic hearts. J Biol Chem. 1978;253(12):4305–9.
15. Shug AL, Thomsen JH, Folts JD, et al. Changes in tissue levels of carnitine and other metabolites during myocardial ischemia and anoxia. Arch Biochem Biophys. 1978;187(1):25–33. https://doi.org/10.1016/0003-9861(78)90003-6.
16. Feuvray D, Plouët J. Relationship between structure and fatty acid metabolism in mitochondria isolated from ischemic rat hearts. Circ Res. 1981;48(5):740–7. https://doi.org/10.1161/01.res.48.5.740.
17. Shug AL, Shrago E, Bittar N, et al. Acyl-CoA inhibition of adenine nucleotide translocation in ischemic myocardium. Am J Phys. 1975;228(3):689–92. https://doi.org/10.1152/ajplegacy.1975.228.3.689.
18. Regitz V, Paulson DJ, Hodach RJ, et al. Mitochondrial damage during myocardial ischemia. Basic Res Cardiol. 1984;79(2):207–17. https://doi.org/10.1007/BF01908307.
19. Shug AL, Subramanian R. Modulation of adenine nucleotide translocase activity during myocardial ischemia. Z Kardiol. 1987;76(Suppl 5):26–33.
20. Folts JD, Shug AL, Koke JR, et al. Protection of the ischemic dog myocardium with carnitine. Am J Cardiol. 1978;41(7):1209–14. https://doi.org/10.1016/0002-9149(78)90877-9.
21. Silverman NA, Schmitt G, Vishwanath M, et al. Effect of carnitine on myocardial function and metabolism following global ischemia. Ann Thorac Surg. 1985;40(1):20–4. https://doi.org/10.1016/s0003-4975(10)61162-9.
22. Liedtke AJ, Nellis S, Neely JR. Effects of excess free fatty acids on mechanical and metabolic function in normal and ischemic myocardium in swine. Circ Res. 1978;43(4):652–61. https://doi.org/10.1161/01.res.43.4.652.
23. Liedtke AJ, Nellis SH. Effects of carnitine in ischemic and fatty acid supplemented swine hearts. J Clin Invest. 1979;64(2):440–7. https://doi.org/10.1172/JCI109481.
24. Broderick TL, Quinney HA, Lopaschuk GD. Carnitine stimulation of glucose oxidation in the fatty acid perfused isolated working rat heart. J Biol Chem. 1992;267(6):3758–63.
25. Broderick TL, Quinney HA, Barker CC, et al. Beneficial effect of carnitine on mechanical recovery of rat hearts reperfused after a transient period of global ischemia is accompanied by a stimulation of glucose oxidation. Circulation. 1993;87(3):972–81. https://doi.org/10.1161/01.cir.87.3.972.
26. Suzuki Y, Kamikawa T, Kobayashi A, et al. Effects of L-carnitine on tissue levels of acyl carnitine, acyl coenzyme A and high energy phosphate in ischemic dog hearts. Jpn Circ J. 1981;45(6):687–94. https://doi.org/10.1253/jcj.45.687.
27. Opie LH. Role of carnitine in fatty acid metabolism of normal and ischemic myocardium. Am Heart J. 1979;97(3):375–88. https://doi.org/10.1016/0002-8703(79)90440-x.
28. Neely JR, Feuvray D. Metabolic products and myocardial ischemia. Am J Pathol. 1981;102(2):282–91.
29. Davini P, Bigalli A, Lamanna F, et al. Controlled study on L-carnitine therapeutic efficacy in postinfarction. Drugs Exp Clin Res. 1992;18(8):355–65.
30. Iliceto S, Scrutinio D, Bruzzi P, et al. Effects of L-carnitine administration on left ventricular remodeling after acute anterior myocardial infarction: the L-carnitine Ecocardiografia Digitalizzata Infarto Miocardico (CEDIM) trial. J Am Coll Cardiol. 1995;26(2):380–7. https://doi.org/10.1016/0735-1097(95)80010-e.
31. Colonna P, Iliceto S. Myocardial infarction and left ventricular remodeling: results of the CEDIM trial. Am Heart J. 2000;139(2 Pt 3):S124–30. https://doi.org/10.1067/mhj.2000.103918.
32. Singh RB, Niaz MA, Agarwal P, et al. A randomized, double-blind, placebo-controlled trial of L-carnitine in suspected acute myocardial infarction. Postgrad Med J. 1996;72(843):45–50. https://doi.org/10.1136/pgmj.72.843.45.
33. Tarantini G, Scrutinio D, Bruzzi P, et al. Metabolic treatment with L-carnitine in acute anterior ST segment elevation myocardial infarction. A randomized controlled trial. Cardiology. 2006;106(4):215–23. https://doi.org/10.1159/000093131.
34. Xue YZ, Wang LX, Liu HZ, et al. L-carnitine as an adjunct therapy to percutaneous coronary intervention for non-ST elevation myocardial infarction. Cardiovasc Drugs Ther. 2007;21(6):445–8. https://doi.org/10.1007/s10557-007-6056-9.
35. Oliver MF, Kurien VA, Greenwood TW. Relation between serum-free-fatty acids and arrhythmias and death after acute myocardial infarction. Lancet. 1968;1(7545):710–4. https://doi.org/10.1016/s0140-6736(68)92163-6.

36. Sugiyama S, Ozawa T. Biochemical aspects of arrhythmia. J Jpn Biochem Soc. 1981;53(11):1221–4. (Article in Japanese).
37. Sugiyama S, Miyazaki Y, Kotaka K, et al. Mechanism of free fatty acid-induced arrhythmias. J Electrocardiol. 1982;15(3):227–32. https://doi.org/10.1016/s0022-0736(82)80023-x.
38. Suzuki Y, Kamikawa T, Yamazaki N. Effects of L-carnitine on ventricular arrhythmias in dogs with acute myocardial ischemia and a supplement of excess free fatty acids. Jpn Circ J. 1981;45(5):552–9. https://doi.org/10.1253/jcj.45.552.
39. Kobayashi A, Suzuki Y, Kamikawa T, et al. Effects of L-carnitine on ventricular arrhythmias after coronary reperfusion. Jpn Circ J. 1983;47(5):536–42. https://doi.org/10.1253/jcj.47.536.
40. Suzuki Y, Kamikawa T, Kobayashi A, et al. Effects of L-carnitine on tissue levels of free fatty acid, acyl CoA, and acylcarnitine in ischemic heart. Adv Myocardiol. 1983;4:549–57. https://doi.org/10.1007/978-1-4757-4441-5_53.
41. Imai S, Matsui K, Nakazawa M, et al. Anti-arrhythmic effects of (-)-carnitine chloride and its acetyl analogue on canine late ventricular arrhythmia induced by ligation of the coronary artery as related to improvement of mitochondrial function. Br J Pharmacol. 1984;82(2):533–42. https://doi.org/10.1111/j.1476-5381.1984.tb10790.x.
42. Kobayashi A, Fujisawa S. Effect of L-carnitine on mitochondrial acyl CoA esters in the ischemic dog heart. J Mol Cell Cardiol. 1994;26(4):499–508. https://doi.org/10.1006/jmcc.1994.1060.
43. Suzuki Y, Narita M, Yamazaki N. Effects of L-carnitine on arrhythmias during hemodialysis. Jpn Heart J. 1982;23(3):349–59. https://doi.org/10.1536/ihj.23.349.
44. Rizzon P, Biasco G, Di Biase M, et al. High doses of L-carnitine in acute myocardial infarction: metabolic and antiarrhythmic effects. Eur Heart J. 1989;10(6):502–8. https://doi.org/10.1093/oxfordjournals.eurheartj.a059519.
45. Martina B, Zuber M, Weiss P, et al. Anti-arrhythmia treatment using L-carnitine in acute myocardial infarct. Schweiz Med Wochenschr. 1992;122(37):1352–5. (Article in German).
46. DiNicolantonio JJ, Lavie CJ, Fares H, et al. L-carnitine in the secondary prevention of cardiovascular disease: systematic review and meta-analysis. Mayo Clin Proc. 2013;88(6):544–51. https://doi.org/10.1016/j.mayocp.2013.02.007.
47. DiNicolantonio JJ, Niazi AK, McCarty MF, et al. L-carnitine for the treatment of acute myocardial infarction. Rev Cardiovasc Med. 2014;15(1):52–62. https://doi.org/10.3909/ricm0710.
48. Shang R, Sun Z, Li H. Effective dosing of L-carnitine in the secondary prevention of cardiovascular disease: a systematic review and meta-analysis. BMC Cardiovasc Disord. 2014;14:88. https://doi.org/10.1186/1471-2261-14-88.
49. Suzuki Y, Masumura Y, Kobayashi A, et al. Myocardial carnitine deficiency in chronic heart failure. Lancet. 1982;1(8263):116. https://doi.org/10.1016/s0140-6736(82)90263-x.
50. Masumura Y, Kobayashi A, Yamazaki N. Myocardial free carnitine and fatty acylcarnitine levels in patients with chronic heart failure. Jpn Circ J. 1990;54(12):1471–6. https://doi.org/10.1253/jcj.54.12_1471.
51. Regitz V, Shug AL, Fleck E. Defective myocardial carnitine metabolism in congestive heart failure secondary to dilated cardiomyopathy and to coronary, hypertensive and valvular heart diseases. Am J Cardiol. 1990;65(11):755–60. https://doi.org/10.1016/0002-9149(90)91383-h.
52. Regitz V, Bossaller C, Strasser R, et al. Metabolic alterations in end-stage and less severe heart failure – myocardial carnitine decrease. J Clin Chem Clin Biochem. 1990;28(9):611–7. https://doi.org/10.1515/cclm.1990.28.9.611.
53. Pierpont ME, Judd D, Goldenberg IF, et al. Myocardial carnitine in end-stage congestive heart failure. Am J Cardiol. 1989;64(1):56–60. https://doi.org/10.1016/0002-9149(89)90653-x.
54. Martín MA, Gómez MA, Guillén F, et al. Myocardial carnitine and carnitine palmitoyltransferase deficiencies in patients with severe heart failure. Biochim Biophys Acta. 2000;1502(3):330–6. https://doi.org/10.1016/s0925-4439(00)00061-2.
55. El-Aroussy W, Rizk A, Mayhoub G, et al. Plasma carnitine levels as a marker of impaired left ventricular functions. Mol Cell Biochem. 2000;213(1–2):37–41. https://doi.org/10.1023/a:1007142919941.
56. Kobayashi A, Masumura Y, Yamazaki N. L-carnitine treatment for congestive heart failure – experimental and clinical study. Jpn Circ J. 1992;56(1):86–94. https://doi.org/10.1253/jcj.56.86.
57. Kalim S, Clish CB, Wenger J, et al. A plasma long-chain acylcarnitine predicts cardiovascular mortality in incident dialysis patients. J Am Heart Assoc. 2013;2(6):e000542. https://doi.org/10.1161/JAHA.113.000542.
58. Ueland T, Svardal A, Øie E, et al. Disturbed carnitine regulation in chronic heart failure – increased plasma levels of palmitoyl-carnitine are associated with poor prognosis. Int J Cardiol. 2013;167(5):1892–9. https://doi.org/10.1016/j.ijcard.2012.04.150.
59. Ahmad T, Kelly JP, McGarrah RW, et al. Prognostic implications of long-chain acylcarnitines in heart failure and reversibility with mechanical circulatory support. J Am Coll Cardiol. 2016;67(3):291–9. https://doi.org/10.1016/j.jacc.2015.10.079.
60. Shah SH, Kraus WE, Newgard CB. Metabolomic profiling for the identification of novel biomarkers and mechanisms related to common cardiovascular diseases: form and function. Circulation. 2012;126(9):1110–20. https://doi.org/10.1161/CIRCULATIONAHA.111.060368.

61. Arsenian MA. Carnitine and its derivatives in cardiovascular disease. Prog Cardiovasc Dis. 1997;40(3):265–86. https://doi.org/10.1016/s0033-0620(97)80037-0.
62. Jaswal JS, Keung W, Wand W, et al. Targeting fatty acid and carbohydrate oxidation — a novel therapeutic intervention in the ischemic and failing heart. Biochim Biophys Acta. 2011;1813(7):1333–50. https://doi.org/10.1016/j.bbamcr.2011.01.015.
63. Noland RC, Koves TR, Seiler SE, et al. Carnitine insufficiency caused by aging and overnutrition compromises mitochondrial performance and metabolic control. J Biol Chem. 2009;284(34):22840–52. https://doi.org/10.1074/jbc.M109.032888.
64. Muoio DM, Noland RC, Kovalik JP, et al. Muscle-specific deletion of carnitine acetyltransferase compromises glucose tolerance and metabolic flexibility. Cell Metab. 2012;15(5):764–77. https://doi.org/10.1016/j.cmet.2012.04.005.
65. Smith RL, Soeters MR, Wüst RCI, et al. Metabolic flexibility as an adaptation to energy resources and requirements in health and disease. Endocr Rev. 2018;39(4):489–517. https://doi.org/10.1210/er.2017-00211.
66. Athithan L, Gulsin GS, McCann GP, et al. Diabetic cardiomyopathy: pathophysiology, theories and evidence to date. World J Diabetes. 2019;10(10):490–510. https://doi.org/10.4239/wjd.v10.i10.490.
67. Chakravarthy MV, Neuschwander-Tetri BA. The metabolic basis of nonalcoholic steatohepatitis. Endocrinol Diabetes Metab. 2020;3(4):e00112. https://doi.org/10.1002/edm2.112.
68. da Silva Rosa SC, Nayak N, Caymo AM, et al. Mechanisms of muscle insulin resistance and the cross-talk with liver and adipose tissue. Physiol Rep. 2020;8(19):e14607. https://doi.org/10.14814/phy2.14607.
69. Gilbert M. Role of skeletal muscle lipids in the pathogenesis of insulin resistance of obesity and type 2 diabetes. J Diabetes Investig. 2021;12(11):1934–41. https://doi.org/10.1111/jdi.13614.
70. Jing ZC, Wu BX, Peng JJ, et al. Effect of intravenous L-carnitine in Chinese patients with chronic heart failure. Eur Heart J Suppl. 2016;18(Supplement A):A27–36. https://doi.org/10.1093/eurheartj/suw008.
71. Wang Y, Xu Y, Zou R, et al. Effect of levocarnitine on the therapeutic efficacy of conventional therapy in children with dilated cardiomyopathy: results of a randomized trial in 29 children. Paediatr Drugs. 2018;20(3):285–90. https://doi.org/10.1007/s40272-018-0284-2.
72. Yamamoto K. Heart failure in elder patients. Cardioangiology. 2016;80(5):347–51. (Article in Japanese).
73. Omori Y, Ohtani T, Sakata Y, et al. L-carnitine prevents the development of ventricular fibrosis and heart failure with preserved ejection fraction in hypertensive heart disease. J Hypertens. 2012;30(9):1834–44. https://doi.org/10.1097/HJH.0b013e3283569c5a.
74. Sakata Y, Ohtani T, Takeda Y, et al. Left ventricular stiffening as therapeutic target for heart failure with preserved ejection fraction. Circ J. 2013;77(4):886–92. https://doi.org/10.1253/circj.cj-13-0214.
75. Zordoky BN, Sung MM, Ezekowitz J, et al. Metabolomic fingerprint of heart failure with preserved ejection fraction. PLoS One. 2015;10(5):e0124844. https://doi.org/10.1371/journal.pone.0124844.
76. Hunter WG, Kelly JP, McGarrah RW 3rd, et al. Metabolomic profiling identifies novel circulating biomarkers of mitochondrial dysfunction differentially elevated in heart failure with preserved versus reduced ejection fraction: evidence for shared metabolic impairments in clinical heart failure. J Am Heart Assoc. 2016;5(8):e003190. https://doi.org/10.1161/JAHA.115.003190.
77. Yoshihisa A, Watanabe S, Yokokawa T, et al. Associations between acylcarnitine to free carnitine ratio and adverse prognosis in heart failure patients with reduced or preserved ejection fraction. ESC Heart Fail. 2017;4(3):360–4. https://doi.org/10.1002/ehf2.12176.
78. Miyata M, Yoshihisa A, Yamauchi H, et al. Impact of sleep-disordered breathing on myocardial damage and metabolism in patients with chronic heart failure. Heart Vessel. 2015;30(3):318–24. https://doi.org/10.1007/s00380-014-0479-6.
79. Kinugasa Y, Sugihara S, Yamada K, et al. Carnitine insufficiency is associated with adverse outcomes in patients with heart failure with preserved ejection fraction. J Aging Res Clin Pract. 2016;5(4):187–93. https://doi.org/10.14283/jarcp.2016.116.
80. Kinugasa Y, Yamamoto K. A new therapeutic agent for HFpEF. The target of its treatment. Medicina (B Aires). 2015;52(7):1119–21. (Article in Japanese).
81. Kinugasa Y, Sota T, Ishiga N, et al. L-carnitine supplementation in heart failure patients with preserved ejection fraction; a pilot study. Geriatr Gerontol Int. 2020;20(12):1244–5. https://doi.org/10.1111/ggi.14060.
82. Kinugasa Y, Sota T, Nakamura K, et al. Association of carnitine insufficiency with sarcopenia and dynapenia in patients with heart failure. Geriatr Gerontol Int. 2023;23(7):524–30. https://doi.org/10.1111/ggi.14621.
83. Sole MJ, Jeejeebhoy KN. Conditioned nutritional requirements and the pathogenesis and treatment of myocardial failure. Curr Opin Clin Nutr Metab Care. 2000;3:417–24.
84. Witte KK, Clark AL, Cleland JG. Chronic heart failure and micronutrients. J Am Coll Cardiol. 2001;37(7):1765–74. https://doi.org/10.1016/s0735-1097(01)01227-x.
85. Allard ML, Jeejeebhoy KN, Sole MJ. The management of conditioned nutritional requirements

in heart failure. Heart Fail Rev. 2006;11(1):75–82. https://doi.org/10.1007/s10741-006-9195-3.
86. Sinatra ST. Metabolic cardiology: an integrative strategy in the treatment of congestive heart failure. Altern Ther Health Med. 2009;15(3):44–52.
87. Soukoulis V, Dihu JB, Sole M, et al. Micronutrient deficiencies: an unmet need in heart failure. J Am Coll Cardiol. 2009;54(18):1660–73. https://doi.org/10.1016/j.jacc.2009.08.012.
88. Sciatti E, Lombardi C, Ravera A, et al. Nutritional deficiency in patients with heart failure. Nutrients. 2016;8(7):442. https://doi.org/10.3390/nu8070442.
89. Keith ME, Ball A, Jeejeebhoy KN, et al. Conditioned nutritional deficiencies in the cardiomyopathic hamster heart. Can J Cardiol. 2001;17(4):449–58.
90. Whitmer JT. L-carnitine treatment improves cardiac performance and restores high-energy phosphate pools in cardiomyopathic Syrian hamster. Circ Res. 1987;61(3):396–408. https://doi.org/10.1161/01.res.61.3.396.
91. Vescovo G, Ravara B, Gobbo V, et al. L-carnitine: a potential treatment for blocking apoptosis and preventing skeletal muscle myopathy in heart failure. Am J Physiol Cell Physiol. 2002;283(3):C802–10. https://doi.org/10.1152/ajpcell.00046.2002.
92. Vescovo G, Ravara B, Gobbo V, et al. Inflammation and perturbation on the L-carnitine system in heart failure. Eur J Heart Fail. 2005;7(6):997–1002. https://doi.org/10.1016/j.ejheart.2004.11.010.
93. Ringseis R, Keller J, Eder K. Mechanisms underlying the anti-wasting effect of L-carnitine supplementation under pathologic conditions: evidence from experimental and clinical studies. Eur J Nutr. 2013;52(5):1421–42. https://doi.org/10.1007/s00394-013-0511-0.
94. Gąsiorowski A, Dutkiewicz J. Comprehensive rehabilitation in chronic heart failure. Ann Agric Environ Med. 2013;20(3):606–12.
95. Kumar A, Singh RB, Saxena M, et al. Effect of carni Q-gel (ubiquinol and carnitine) on cytokines in patients with heart failure in the Tishcon study. Acta Cardiol. 2007;62(4):349–54. https://doi.org/10.2143/AC.62.4.2022278.
96. Kinugasa Y, Yamamoto K. The challenge of frailty and sarcopenia in heart failure with preserved ejection fraction. Heart. 2017;103(3):184–9. https://doi.org/10.1136/heartjnl-2016-309995.
97. Brevetti G, Chiariello M, Ferulano G, et al. Increases in walking distance in patients with peripheral vascular disease treated with L-carnitine: a double-blind, cross-over study. Circulation. 1988;77(4):767–73. https://doi.org/10.1161/01.cir.77.4.767.
98. Brevetti G, Angelini C, Rosa M, et al. Muscle carnitine deficiency in patients with severe peripheral vascular disease. Circulation. 1991;84(4):1490–5. https://doi.org/10.1161/01.cir.84.4.1490.
99. Brevetti G, di Lisa F, Perna S, et al. Carnitine-related alterations in patients with intermittent claudication. Indication for a focused carnitine therapy. Circulation. 1996;93(9):1685–9. https://doi.org/10.1161/01.cir.93.9.1685.
100. Hiatt WR, Wolfel EE, Regensteiner JG, et al. Skeletal muscle carnitine metabolism in patients with unilateral peripheral arterial disease. J Appl Physiol. 1992;73(1):346–53. https://doi.org/10.1152/jappl.1992.73.1.346.
101. Brevetti G, Diehm C, Lambert D. European multicenter study on propionyl-L-carnitine in intermittent claudication. J Am Coll Cardiol. 1999;34:1618–24.
102. Hiatt WR, Regensteiner JG, Creager MA, et al. Propionyl-L-carnitine improves exercise performance and functional status in patients with claudication. Am J Med. 2001;110(8):616–22. https://doi.org/10.1016/s0002-9343(01)00704-5.
103. Hiatt WR. Carnitine and peripheral arterial disease. Ann N Y Acad Sci. 2004;1033:92–8. https://doi.org/10.1196/annals.1320.008.
104. Luo T, Li J, Li L, et al. A study on the efficacy and safety assessment of propionyl-L-carnitine tablets in treatment of intermittent claudication. Thromb Res. 2013;132(4):427–32. https://doi.org/10.1016/j.thromres.2013.08.005.
105. Loffred L, Pignatelli P, Cangemi R, et al. Imbalance between nitric oxide generation and oxidative stress in patients with peripheral arterial disease: effect of an antioxidant treatment. J Vasc Surg. 2006;44(3):525–30. https://doi.org/10.1016/j.jvs.2006.05.023.
106. Stasi MA, Scioli MG, Arcuri G, et al. Propionyl-L-carnitine improves postischemic blood flow recovery and arteriogenetic revascularization and reduces endothelial NADPH-oxidase 4-mediated superoxide production. Arterioscler Thromb Vasc Biol. 2010;30(3):426–35. https://doi.org/10.1161/ATVBAHA.109.201533.
107. De Marchi S, Zecchetto S, Rigoni A, et al. Propionyl-L-carnitine improves endothelial function, microcirculation and pain management in critical limb ischemia. Cardiovasc Drugs Ther. 2012;26(5):401–8. https://doi.org/10.1007/s10557-012-6408-y.
108. Scioli MG, Giudice PL, Bielli A, et al. Propionyl-L-carnitine enhances wound healing and counteracts microvascular endothelial cell dysfunction. PLoS One. 2015;10:e0140697.
109. Silvestro A, Schiano V, Bucur R, et al. Effect of propionylcarnitine on changes in endothelial function and plasma levels of adhesion molecules induced by acute exercise in patients with intermittent claudication. Angiology. 2006;57(2):145–54. https://doi.org/10.1177/000331970605700203.
110. Milio G, Novo G, Genova C, et al. Pharmacological treatment of patients with chronic critical limb ischemia: L-propionyl-carnitine enhances the short-term effects of PGE-1. Cardiovasc Drugs Ther. 2009;23(4):301–6. https://doi.org/10.1007/s10557-009-6178-3.
111. Brevetti G, Perna S, Sabbà C, et al. Superiority of L-propionylcarnitine vs L-carnitine in improving

walking capacity in patients with peripheral vascular disease: an acute, intravenous, double-blind, cross-over study. Eur Heart J. 1992;13(2):251–5. https://doi.org/10.1093/oxfordjournals.eurheartj.a060155.
112. Brass EP, Koster D, Hiatt WR, et al. A systematic review and meta-analysis of propionyl-L-carnitine effects on exercise performance in patients with claudication. Vasc Med. 2013;18(1):3–12. https://doi.org/10.1177/1358863X12467491.
113. Norgren L, Hiatt WR, Dormandy JA, et al. Inter-society consensus for the management of peripheral arterial disease (TASC II). Eur J Vasc Endovasc Surg. 2007;33(Suppl 1):S1–S75. https://doi.org/10.1016/j.ejvs.2006.09.024.
114. Norgren L, Hiatt WR, Dormandy JA, et al. Inter-society consensus for the management of peripheral arterial disease (TASC II). J Vasc Surg. 2007;45(Suppl S):S5–S67. https://doi.org/10.1016/j.jvs.2006.12.037.
115. TASC 2 Working Group: Guidelines for diagnosis and treatment of arteriosclerosis obliterans II., Translated by Japanese College of Angiology, 1–109, Medical Tribune, Tokyo; 2007 (Book in Japanese).
116. Hirsch AT, Haskal ZJ, Hertzer NR, et al. ACC/AHA 2005 guidelines for the management of patients with peripheral arterial disease(lower extremity, renal, mesenteric, and abdominal aortic): executive summary. A collaborative report from the American Association for Vascular Surgery/Society for Vascular Surgery, Society for Cardiovascular Angiography and Interventions, Society for Vascular Medicine and Biology, Society of Interventional Radiology, and the ACC/AHA Task Force on Practice Guidelines (Writing Committee to Develop Guidelines for the Management of Patients With Peripheral Arterial Disease): endorsed by the American Association of Cardiovascular and Pulmonary Rehabilitation; National Heart, Lung, and Blood Institute; Society for Vascular Nursing; TransAtlantic Inter-Society Consensus; and Vascular Disease Foundation. J Am Coll Cardiol. 2006;47(6):1239–312. https://doi.org/10.1016/j.jacc.2005.10.009.
117. Dawson DL, Cutler BS, Meissner MH, et al. Cilostazol has beneficial effects in treatment of intermittent claudication. Results from a multicenter, randomized, prospective, double-blind trial. Circulation. 1998;98(7):678–86. https://doi.org/10.1161/01.cir.98.7.678.
118. Ishii H, Aoyama T, Takahashi H, et al. Treatment with cilostazol improves clinical outcome after endovascular therapy in hemodialysis patients with peripheral artery disease. J Cardiol. 2016;67(2):199–204. https://doi.org/10.1016/j.jjcc.2015.05.003.
119. Regensteiner JG, Ware JE Jr, McCarthy WJ, et al. Effect of cilostazol on treadmill walking, community-based walking ability, and health-related quality of life in patients with intermittent claudication due to peripheral arterial disease: meta-analysis of six randomized controlled trials. J Am Geriatr Soc. 2002;50(12):1939–46. https://doi.org/10.1046/j.1532-5415.2002.50604.x.
120. Goldenberg NA, Krantz MJ, Hiatt WR. L-carnitine plus cilostazol versus cilostazol alone for the treatment of claudication in patients with peripheral artery disease: a multicenter, randomized, double-blind, placebo-controlled trial. Vasc Med. 2012;17(3):145–54. https://doi.org/10.1177/1358863X12442264.
121. Murakami A, Haraguchi T, Kato M. Effects of L-carnitine on exercise performance and quality of life in patients with arteriosclerosis obliterans. J Jpn Coll Angiol. 2015;55(8):117–23. (Article in Japanese).
122. Komai H, Miyama N, Sakashita H, et al. Plasma carnitine level in peripheral arterial disease. J Jpn Coll Angiol. 2016;56(8):103–8. (Article in Japanese).
123. Miyata T. An approach to peripheral arterial disease from the viewpoint of systemic vascular disease management. Heart. 2018;50(6):669–72. (Article in Japanese).

Other Diseases and Carnitine

17

17.1 Introduction

Carnitine deficiency in the diseases discussed so far is also covered in the "Guidelines for the Diagnosis and Treatment of Carnitine Deficiency 2018," but the relationship between pathophysiology and fluctuations in carnitine metabolism has been reported in many other diseases, and carnitine supplementation therapy has been reported in some cases. In addition, there are a few effective treatments for these diseases. Therefore, for reference, the relationship between the main diseases and carnitine is introduced.

17.2 Influenza-Associated Encephalopathy and Carnitine

Influenza-associated encephalopathy is common in East Asia and Japan, and although the cause is not known, death and serious sequelae occur at high rates. Studies have shown that 5–15% of patients with influenza-associated encephalopathy are suspected to have organic acidemia or fatty acid metabolism abnormalities, and about 10% of patients taking valproic acid are also suspected to have underlying metabolic abnormalities. It has been reported that a polymorphism of the *CPT2* gene, a thermo-instable type, may cause mitochondrial energy metabolism disorders due to decreased activity upon exposure to high fever [1–4], and such polymorphisms of the *CPT2* gene have been reported to be associated with acute encephalopathy caused by various viral infections [5]. However, not all patients who developed influenza-associated encephalopathy were suspected to have inherited metabolic diseases, and the *CPT2* gene polymorphism alone cannot be the cause of all cases.

Kuroda et al. analyzed the pathophysiology of influenza-associated encephalopathy from the viewpoint of metabolic abnormalities and summarized the clinical characteristics of influenza-associated encephalopathy cases. According to the study, hypercytokinemia (so-called cytokine storm) and vascular endothelial damage progress rapidly, apoptosis progresses in the brain, and irreversible damage occurs due to vasculitis and cerebral edema [6]. Supportive treatments, specific treatments (antiviral drugs, methylprednisolone pulse therapy, γ-globulin high-dose therapy), and special treatments (cerebral hypothermia, plasmapheresis, cyclosporine therapy, antithrombin III high-dose therapy) described in the influenza-associated encephalopathy guidelines are not so effective.

First, organic acids, amino acids, acylcarnitine, and free carnitine were analyzed using urine, serum, and dried blood filter paper from 42 patients with influenza-associated encephalopathy in Japan. Of the 42 patients, six were suspected of having an inherited metabolic disorder (one case with CPT2 deficiency, one patient with

N. Nakayama, *Carnitine Deficiency and Insufficiency in Medicine*,
https://doi.org/10.1007/978-981-95-7797-2_17

glutaric acidemia type 2, and four patients with carnitine cycle disorder), and four patients with valproate metabolites were detected, for a total of 10 cases that were judged to have carnitine deficiency. In addition, blood-free carnitine levels were below normal in four patients, and totally 14 of the 42 patients may have developed carnitine deficiency. Based on these findings, they considered that carnitine deficiency may be an underlying risk factor for the development of influenza-associated encephalopathy. So that Kuroda et al. proposed a treatment method for the prevention and treatment of influenza-associated encephalopathy, aiming to energy supply from carbohydrates and energy supply from fatty acids correcting abnormality in fatty acid metabolism by carnitine administration. At that time, the indication of L-Cartin® tablets for carnitine deficiency had not yet been approved, and intravenous injectables had not been launched, so it was proposed to administer L-Cartin® tablets (100 mg/kg/day) orally or tube administration, but now that intravenous injectables are on the market, intravenous injection of 100 mg/kg/day is considered to be the best. Figure 17.1 shows the pathogenetic mechanism of influenza encephalopathy and a new proposal for the treatment with conventional treatment plus carnitine administration as soon as possible, proposed by Dr. Y. Kuroda et al. [6].

Purevsuren et al. also performed GC/MS analysis of urine samples of influenza-associated encephalopathy cases (70 cases) and screened them for inherited metabolic diseases such as organic acidemia and found six suspected inherited metabolic diseases (one case of methylmalo-

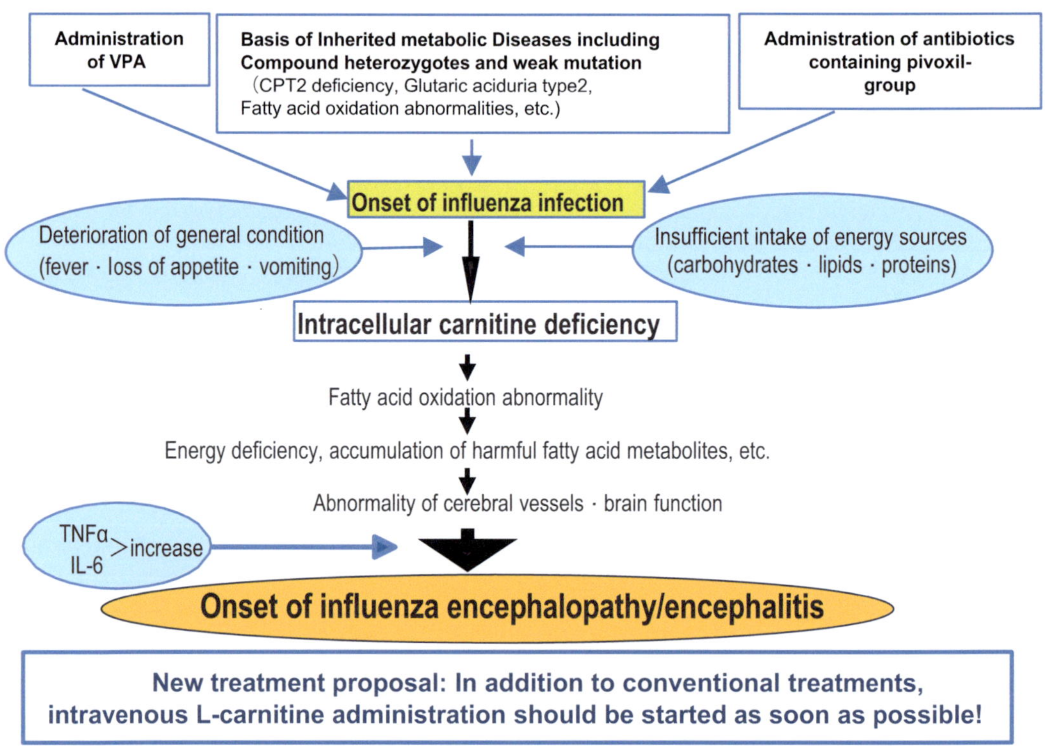

Fig. 17.1 Pathogenetic mechanisms of influenza encephalopathy/encephalitis and new treatment proposed by Dr. Y. Kuroda et al. Patients with influenza encephalopathy had a basis of carnitine deficiency, which was supposed to cause deterioration of energy metabolism in the brain. So, Dr. Kuroda proposed oral L-carnitine administration in addition to conventional treatments in his article because, at that time, only oral L-carnitine was available. Now, intravenous L-carnitine is available and suitable in the acute and comatose state of patients with influenza encephalopathy. CPT2 carnitine palmitoyltransferase 2, VPA valproic acid, TNF-α tumor necrosis factor-α, IL-6 interleukin 6. (Created using Ref. [6])

nic acidemia, one case of fructose-1, 6-diphosphatase deficiency, and four cases of suspected non-ketotic dicarboxylic aciduria). In addition, valproic acid metabolites were detected in ten cases [7].

Thus, in cases of influenza-associated encephalopathy, there is a high rate of suspected cases of carnitine deficiency that cannot be considered accidental. Carnitine is used for hepatic encephalopathy and hyperammonemic encephalopathy caused by valproic acid, and intravenous injections are used in particularly serious and urgent cases. It is also extremely safe (see Chaps. 12 and 15). In influenza-associated encephalopathy, ammonia elevation is rare, but if the essence of the pathology is the breakdown of vascular endothelial cells and brain cell apoptosis due to abnormal brain energy metabolism, inflammation, and increased oxidative stress, it is thought that carnitine's effects on improving energy metabolism and anti-inflammatory, antioxidant, and anti-apoptotic effects can be expected (Chap. 7). Propofol injection syndrome, which rarely develops due to propofol, an anesthetic and sedative drug commonly used in the field of pediatrics, is also a type of acute encephalopathy, which is considered to be an abnormality of energy metabolism and is said to be related to carnitine deficiency [8–11].

The number of patients with influenza-associated encephalopathy decreased to 134 in the 2019/2020 season (week 36 to week 52 of 2019), which was around 200 cases, but the prognosis is still severe due to the lack of effective treatment. I have heard that the Ministry of Health, Labour and Welfare's influenza-associated encephalopathy research group is calling for rigorous evidence on treatments. It is not feasible to seek evidence such as double-blind controlled trials for acute encephalopathy, which occurs very rarely, such as influenza-associated encephalopathy and propofol injection syndrome. True EBM is the best treatment for a patient based on the best evidence and science at the time. It is hoped that attempts will be made to administer carnitine to influenza-associated encephalopathy and propofol injection syndrome without any safety problems, and that more reliable evidence of efficacy and safety will be built up by accumulating cases.

17.3 Amyotrophic Lateral Sclerosis (ALS) and Carnitine

Amyotrophic lateral sclerosis (ALS) is the most severe neurological disease in which motor neurons degenerate, muscle atrophy and muscle weakness progress, and death occurs within 2–5 years of onset, and it is an intractable disease for which there is no fundamental cure. There are about 10,000 patients in Japan, and 1000–2000 people develop the disease every year. It is a neurodegenerative disease that causes damage to motor neurons and progresses to muscle atrophy of the upper and lower limbs, speech disorders, swallowing disorders, and respiratory muscle paralysis. Therefore, long-term nutritional management by tube feeding is required, and carnitine deficiency due to this has been reported [12–14].

Most ALS is sporadic ALS that develops suddenly after middle age, but about 10% have familial ALS, and some genes have been cited as causative genes. Among them, gene mutations in *SOD1* (Cu·Zn-superoxide dismutase) are the most common and are said to account for about 20% of familial ALS.

Kira et al. reported the effect of L-carnitine administration using mice overexpressing the mutated *SOD1* gene (G93A) (Tg mice) as an ALS model mice [15]. Although the mechanism of degeneration of motor neurons in ALS has not been elucidated, it is thought that mitochondrial function decline and degeneration are at the root of it, and the effect of L-carnitine, which has the effect of promoting mitochondrial function, was attempted. L-carnitine administration was given at 1 mg/mL of drinking water freely. As a result, the total carnitine concentration in the spinal cord decreased to about 50% of the control in Tg mice, but it was normalized by carnitine administration, and the onset of ALS was significantly delayed by carnitine administration ($P = 0.0008$, Fig. 17.2), and the decline in the motor function scale of the hind limbs of Tg mice at 33 weeks

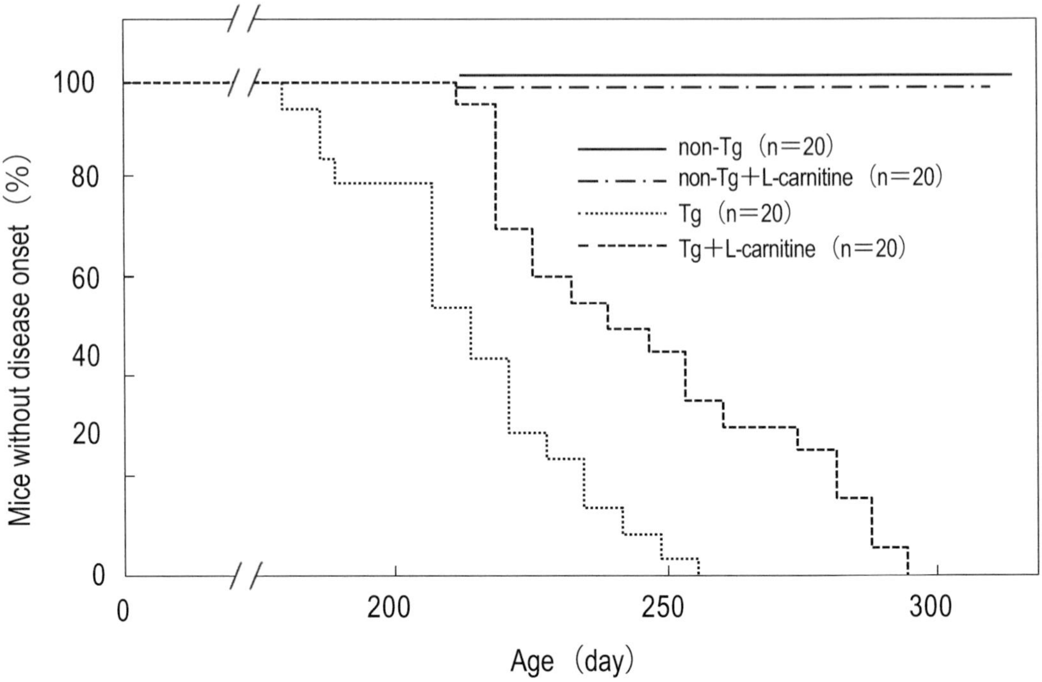

Fig. 17.2 Effects of L-carnitine on disease progression. Non-Tg and Tg littermates were allowed free access to either tap water or L-carnitine-containing water (1 mg/mL) from the age of 16 weeks. Proportions without disease onset were plotted with Kaplan-Meier curves, and differences between Tg and Tg groups treated with L-carnitine were tested by the log-rank test (P = 0.0008). Tg: transgenic mice carrying a human SOD1 gene with the G93A mutation. (From Ref. [15])

was normalized by carnitine administration (Fig. 17.3).

In Tg mice, protein carbonylation in the muscles and spinal cord increased by about 5–15 times compared to controls, but carnitine administration suppressed this. When apoptosis of hindlimb myocytes was examined by terminal deoxynucleotidyl transferase (TdT)-mediated deoxyuridine triphosphate (dUTP) nick end labeliing (TUNEL) staining, apoptotic cells increased markedly in Tg mice, but carnitine administration suppressed this (Fig. 17.4).

Carnitine administration significantly extended the lifespan of Tg mice (P = 0.0001, Fig. 17.5).

These results indicate that carnitine administration has a survival effect by delaying the onset of ALS in patients with familial ALS due to *SOD1* mutations. It may also be effective in disease progression due to muscle cell apoptosis and increased oxidative stress in patients with ALS not caused by *SOD1* mutations.

In their review of nutritional and exercise interventions in the treatment of ALS patients, Patel et al. cited the work of Kira et al. and described their expectation that the potential of carnitine will be clarified in clinical practice in the future, perhaps due to increased lipid peroxidation in ALS [16].

Although it is not carnitine, there are studies in which acetyl-L-carnitine (ALC) was administered. Beghi et al. conducted a randomized, double-blind, placebo-controlled study of 82 ALS patients 6–24 months post-onset, divided into two groups: riluzole (100 mg) or riluzole (100 mg) plus ALC (3000 mg), for 48 weeks [17]. The primary endpoint was the proportion of patients scoring two or less on at least one of the ALS functional rating scale-revised (ALSFRS-R) items of swallowing, use of eating utensils, or ambulation at 12 months (proportion of patients

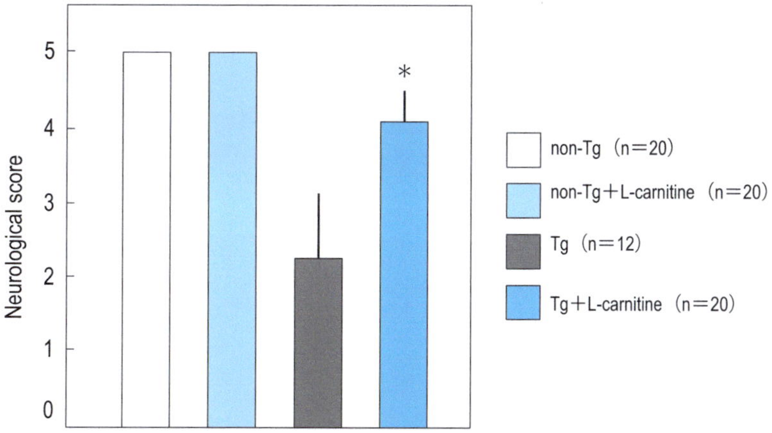

Fig. 17.3 Effects of L-carnitine on disease progression. Motor function in the hind limbs was scored at the age of 33 weeks using a rating scale. Values are means ± SD. *$P < 0.01$ versus Tg. (From Ref. [15])

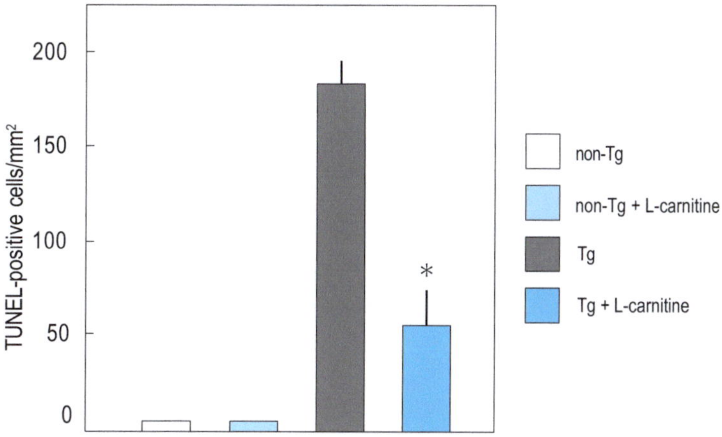

Fig. 17.4 Apoptotic cell death in hind limb muscles. Non-Tg and Tg littermates were treated as in Fig. 17.3. At the age of 33 weeks, gastrocnemius muscle samples were stained by the TUNEL method, and the number of TUNEL-positive cells was counted. Values are means ± SD ($n = 4$). *$P < 0.01$ versus Tg. (From Ref. [15])

not independent). Secondary endpoints included the Medical Research Council (MRC) scale, ALSFRS-R score, respiratory function score, forced vital capacity (FVC), and the proportion of patients who died or had a tracheotomy. As a result, 34 patients (80.9%) in the ALC group and 39 patients (97.5%) in the control group were unable to stand on their own ($P = 0.0296$) in the primary endpoint of the ITT analysis, and there was no significant difference between 84.4% and 100% in the PP analysis ($P = 0.0538$). At week 48, the mean ALSFRS-R scores were 33.6 ± 10.4 and 27.6 ± 9.9, which were significantly higher in the ALC group ($P = 0.0388$), and the FVC scores were 90.3 ± 32.6 and 58.6 ± 31.2, which were significantly higher in the ALC group ($P = 0.0158$). The mean survival was 45 and 22 months, with significantly longer survival in the ALC group ($P = 0.0176$). In addition, there were no safety problems.

In Japan, Yoshino et al. reported on a clinical trial in which carnitine was administered to 18 hospitalized ALS patients (1 g/day for 2 weeks, 4 g/day if there were no safety problems) in 2006 [18]. The main purpose of this study was to evaluate safety, and it is described that treatment was discontinued due to loose stools and diarrhea in three patients on ventilators (seven patients), and that there were no safety problems in patients with severity three or less.

In the absence of an effective treatment for ALS, clinical studies to confirm the efficacy and safety of carnitine are desired.

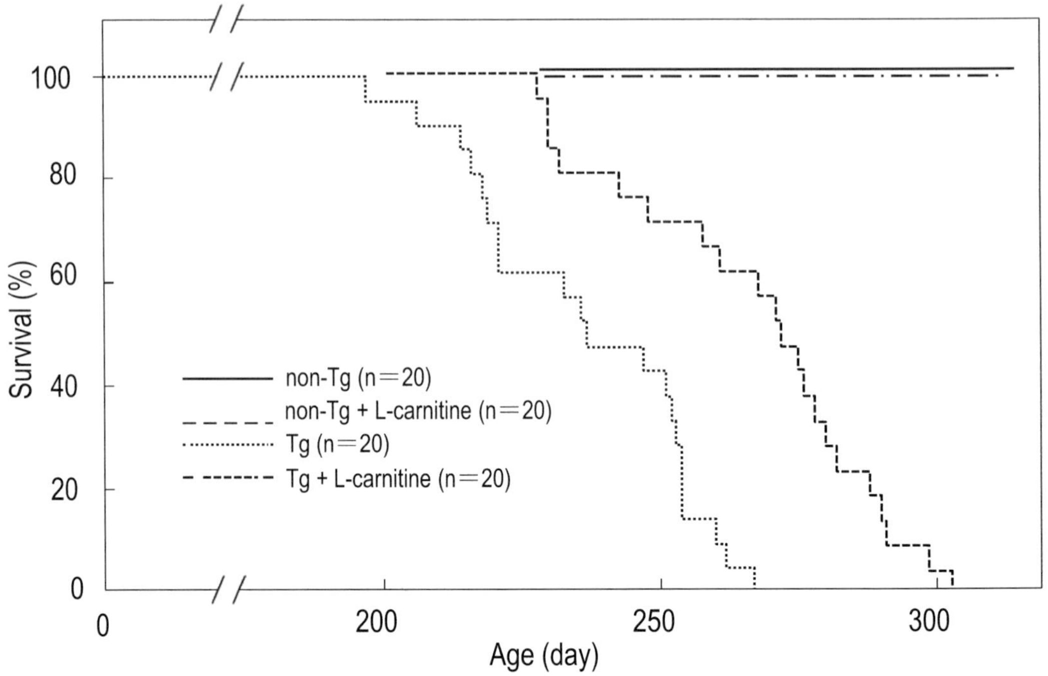

Fig. 17.5 Effects of L-carnitine on survival. Non-Tg and Tg littermates were treated as in Fig. 17.3. Proportions of survival were plotted with Kaplan-Meier curves, and the difference between Tg and Tg animals treated with L-carnitine was tested by the log-rank test ($P = 0.0001$). (From Ref. [15])

17.4 Chronic Fatigue Syndrome and Carnitine

Kuratsune et al. measured the serum carnitine fraction of 38 patients with chronic fatigue syndrome (CFS) and 308 healthy subjects using the enzyme cycling method [19]. As a result, it was reported that acylcarnitine was lower than the control in both men and women in CFS patients ($P < 0.001$). Kuratsune et al. reported in the next study that acylcarnitine concentrations were lower than controls in female and male patients, and that free carnitine concentrations remained unchanged (82 female patients, 9.4 ± 3.7 vs. 131 controls: 15.5 ± 4.5, 64 male patients: 9.7 ± 2.9 vs. 177 controls: 13.4 ± 4.6 μmol/L) [20].

Plioplys et al. reported that serum carnitine fractions of 35 CFS patients were not only low acylcarnitine but also low free carnitine and total carnitine levels [21]. Since the free carnitine concentration was inversely correlated with the CFS motor dysfunction index and the fatigue severity index, and the total carnitine concentration was inversely correlated with the fatigue severity index, they thought that carnitine deficiency may cause mitochondrial dysfunction and cause fatigue. Therefore, Plioplys et al. conducted a clinical trial in which 30 patients with CFS were given amantadine (100 mg/day for the first 4 weeks, 100 mg twice daily if well tolerated) or carnitine (1 g, three times a day) for 8 weeks, followed by a 2-week wash out, followed by a crossover method of carnitine or amantadine [22]. Amantadine was poorly tolerated, with only 15 patients completing 8 weeks of treatment, and none of the clinical parameters significantly improved. On the other hand, patients treated with carnitine showed significant improvement in 12 of the 18 clinical parameters after 8 weeks of treatment, and there were no cases of deterioration except for one patient who could not complete 8 weeks of treatment due to diarrhea. In addition, there were no problems with tolerability and safety.

Majeed et al. also measured serum carnitine and peripheral blood lymphocyte carnitine in 27 CFS patients [23]. As a result, there was no significant difference in serum, except that free carnitine tended to be lower than control in male CFS patients. However, the levels of free carnitine, short-chain acylcarnitine, and total carnitine in peripheral blood lymphocytes were significantly lower than those of the controls (about 1/3 of free carnitine and total carnitine, and about 1/4 of short-chain acylcarnitine).

On the other hand, Jones et al. examined serum carnitine in 31 CFS patients, 31 controls, 15 depressed patients, and 22 patients with rheumatoid arthritis, and reported that there was no significant difference in free carnitine, acylcarnitine, and total carnitine [24].

Reuter et al. measured the fatigue questionnaires and carnitine profiles of 44 CFS patients by LC/MS/MS (35 patients) and compared them with 49 age- and sex-matched controls [25]. The results showed that the concentrations of long-chain acylcarnitine were significantly decreased ($P < 0.0001$), such as oleoylcarnitine (C18:1) and linoleylcarnitine (C18:2) decreased by 30–40% of controls, and these long-chain acylcarnitine showed a significant negative correlation with fatigue severity scores. They hypothesized that this was probably due to decreased CPT1 activity and accumulation of ω-6 fatty acids and suggested that administration of ω-3 fatty acids and carnitine might be effective. However, clinical trials of the administration of ω-3 fatty acids and carnitine have not yet been conducted.

Watanabe et al. summarized the diagnosis and treatment of CFS patients and reported that microglial activation occurs in CFS patients due to neuroinflammation, which can be detected by PET scan using ^{11}C-PK11195 [26]. In addition, as medical treatment for patients with CFS, he described the use of Chinese herbal medicines such as Ho-Chu-Ki-To and Juzen-Dai-Ho-To for patients with impaired immunity, large doses of vitamin C, CoQ10, and imidapeptides for patients with impaired antioxidant capacity, and carnitine for patients with reduced carnitine.

It is not a CFS case, but Okubo et al. reported that fatigue during lenvatinib treatment in patients with hepatocellular carcinoma was associated with carnitine insufficiency (AC/FC), and carnitine supplementation improved the fatigue score [116].

Malaguarnera et al. reviewed the relationship between carnitine and fatigue and discussed the therapeutic effect [117].

17.5 Inflammatory Bowel Disease and Carnitine

Inflammatory bowel disease (IBD), such as Crohn's disease (CD) and ulcerative colitis (UC), is a chronic inflammation of the intestine that causes symptoms such as diarrhea, bloody stools, and abdominal pain. It is a disease that causes a decline in QOL, such as weight loss, due to the inability to consume enough food. It is thought to be an autoimmune disease, but the cause is unknown, and it is thought that there is a genetic component. It is an intractable disease that repeatedly recurs (flares up) even if the inflammation goes into remission. Among IBD patients, carnitine deficiency due to tube feeding and nutritional management by TPN, and the development of cardiomyopathy in short bowel syndrome, such as celiac disease, are described in Chap. 11, so IBD itself and carnitine are described here.

The mucosal cells of the intestine require a large amount of energy to absorb nutrients. The energy source is short-chain fatty acids (SCFAs) such as butyric acid and propionic acid produced by anaerobic bacteria in the intestine, and SCFAs are mainly used as an energy source, especially in the distal colon, but in UC patients, the oxidative capacity of butyric acid decreases, and the oxidation of glucose and glutamine is increased [27–29]. For this reason, many attempts have been made to treat patients with enemas containing SCFAs such as butyric acid, and they have been shown to be effective in some patients [30–34]. Attempts to add propionyl-L-carnitine (PLC) to enemas, although not carnitine, have also been reported [35, 36].

In addition, studies such as genome-wide association analysis of IBD have shown that the

IBD5 locus of human chromosome 5q31 is linked to IBD susceptibility, and many studies have been reported that *OCTN1* and *OCTN2* mutants located there have attracted attention [37–41]. In addition, it was reported that gene expression of *OCTN1* and *OCTN2* was significantly reduced in UC patients [42]. However, since the association between OCTN1 mutations and *OCTN2* single-nucleotide mutations and the risk of developing UC and CD has been reported in Caucasians, and these mutations are extremely rare in East Asians, such as Japan, there is a meta-analysis that *OCTN1* and *OCTN2* mutations and polymorphisms are not involved in CD sensitivity in East Asia [43].

On the other hand, the nine amino acid sequences of the OCTN1 (Leu503Phe) mutant are associated with the nine amino acid sequences of *Campylobacter jejuni*, which causes enterohemorrhagic infections, and *Mycobacterium paratuberculosis*, which is thought to be associated with the development of Crohn's diseas Crohn's disease, *Mycobacterium paratuberculosis* was reported to be equivalent to six amino acid sequences of OCTN1 [44]. It has been pointed out that antibodies are produced in the bodies of patients infected with these bacteria, and that the antibodies recognize and attack OCTN1, resulting in a decrease in the expression of OCTN1 and a decrease in carnitine.

Fortin et al. examined the effects of carnitine on antigen-presenting cells (APCs) and $CD4^+$ T cells in vitro and found that carnitine significantly suppressed APC and $CD4^+$ T cell function. Carnitine treatment not only suppressed the innate function of IL-1β and IL-6 production, but also suppressed the adaptive function of T cell proliferation in the inflow zone lymph node and histologically reduced the damage of TNBS-induced enteritis [45]. In other words, when intracellular carnitine is decreased in patients with IBD due to mutation or decreased expression of OCTN2, the hyper-immunosuppressive effect of carnitine is canceled, and $CD4^+$ T cells are activated, resulting in excess production of proinflammatory cytokines. Since carnitine has immunosuppressive, anti-inflammatory, antioxidant, and anti-apoptotic effects, they proposed that administering carnitine to patients with IBD may be a new therapeutic approach [46].

There are no clinical studies in which carnitine was subsequently administered to IBD patients, but there are studies in which PLC was administered. Mikhailova et al. reported a randomized, double-blind, controlled clinical trial of 121 patients with mild and moderate UC treated with usual oral medications (aminosalicylic acid or thiopurine) in three groups, placebo and PLC (1 g/day or 2 g/day, 4 weeks in addition to usual care) [47]. As a result, 57 out of 79 patients (72%) showed clinical and endoscopic improvement in PLC 1 and 2 g groups, which was a significant improvement compared to 20 out of 40 patients (50%) in the placebo group ($P = 0.02$). In particular, 30 out of 40 patients (75%) treated with PLC 1 g/day showed significant clinical and endoscopic improvement compared to the placebo group ($P = 0.02$). However, the PLC 2 g/day group had an improvement rate of 27 out of 39 patients (69%), but it was not significant compared to the placebo group ($P = 0.08$). There was no significant difference in the remission rate between 35% (14/40) in the placebo group, 55% (22/40) in the 1 g group, and 49% (19/39 patients) in the 2 g group, but only the 1 g group showed a significant remission rate compared to placebo in the mild cases. There were no safety issues.

Merra et al. also reported an open-label, proof-of-concept clinical trial in which patients with mild and moderate UC (10 patients) and CD (four patients) treated with aminosalicylic acid, mercaptopurine or azathioprine were switched to PLC administration (2 g/day, orally, twice daily, for 4 weeks), and the disease activity index (DAI) and histological index (HI) before and after administration were examined [48]. As a result, the DAI was 6.6 ± 2.4 in all cases, 6.6 ± 2.7 in UC patients, and 6.5 ± 1.7 in CD patients before administration, but after 4 weeks, the improvement was 1.9 ± 1.5, 2.0 ± 1.6, and 1.5 ± 1.3, respectively (the median change in UC patients was $P < 0.01$), and HI was also 0.5 ± 1.4 ± 0.5, 1.4 ± 0.5, and 1.5 ± 0.6 before administration, but it was improved to 0.5 ± 0.5, 0.5 ± 0.5 and 0.5 ± 0.6 ($P < 0.01$ in UC patients).

The results of these studies on PLC administration have problems such as the absence of dose-dependence and the possibility of placebo effects, and further examination is required.

Carnitine is not required for β oxidation of short-chain fatty acids, but in IBD patients, the pantothenic acid content is the same as that of the control, but the CoA activity is reduced to about 50% of the control [49]. Therefore, when free CoA is reduced due to carnitine deficiency or carnitine insufficiency, short-chain fatty acid metabolism may be greatly affected by CoA deficiency, and carnitine or PLC administration may be effective in these IBD patients. In addition, carnitine is highly safe and may be used in combination with existing drugs. In the future, it is expected that clinical research on carnitine in IBD patients will advance.

17.6 Alcoholism and Carnitine

Alcohol addiction and alcoholism are diseases that cannot be ignored by society, causing liver damage such as alcoholic hepatitis, alcoholic cardiomyopathy, and mental disorders such as depression, but research on the effects of alcohol consumption on the metabolism of living organisms has been surprisingly slow.

Fuller et al. examined the blood carnitine concentrations of alcoholics without hepatic impairment (Group A, 20 patients) and alcoholics with hepatic impairment (Group B, 28 patients) and compared them with healthy controls (Group C, 32 patients) [50]. As a result, free carnitine, total carnitine, short-chain acylcarnitine, and long-chain acylcarnitine were all significantly higher in group B than in group C, but only the long-chain acylcarnitine was significantly higher in group C than in group A. Six patients in group A abstained from alcohol for 3 weeks and measured the carnitine fraction, but none of them showed significant changes, although there was a slight decrease. Thus, it was shown that even in the absence of liver damage, elevations in long-chain acylcarnitine occurred in alcoholic patients. This may reflect the accumulation of long-chain acyl-CoA due to the inhibition of the metabolism of long-chain fatty acids. In alcoholics with hepatic impairment, both short-chain acylcarnitine and long-chain acylcarnitine were significantly elevated. The elevation of short-chain acylcarnitine is thought to reflect an increase in its intermediate metabolite, acyl-CoA, with an increase in branched-chain amino acids.

De Sousa et al. also measured plasma and urine carnitine in alcoholics with fatty liver (20 cases) and healthy controls (12 cases) [51]. In addition, carnitine in the livers of seven healthy controls and alcoholic patients was measured. Plasma and urine concentrations of free carnitine, acylcarnitine, and total carnitine varied widely in alcoholic patients but did not differ significantly. The concentrations of free carnitine, long and short chain acylcarnitine, and total carnitine in the liver were not significantly different from those in healthy controls, suggesting that carnitine deficiency was not the cause of fatty liver in alcohol-dependent patients. Carnitine concentrations in skeletal muscle were measured in alcoholic patients (eight patients), and healthy controls (seven patients), and the concentrations of free carnitine and total carnitine were significantly higher in alcohol-dependent patients. This was thought to be due to atrophy of type II muscle fibers, which was observed in half of the alcoholics, and a large number of carnitine-rich type 1 muscle fibers were collected in muscle biopsies. Thus, the kinetics of carnitine in alcoholics are still unclear.

On the other hand, there are interesting studies in basic research. Calabrese et al. reported that administration of 20% ethanol to rats (2 g/kg body weight, 1 week) significantly increased the amount of short-chain organic acids such as propionic acid, methylmalonic acid, and acetic acid in the urine, and that urinary excretion of propionic acid and other substances was proportional to the amount of ethanol administered [52]. At this time, the excretion of short-chain acylcarnitines such as acetylcarnitine and propionylcarnitine also increased. In addition, the increase in short-chain organic acids was not affected by the administration of pyrazole, an inhibitor of alcohol dehydrogenase, so the metabolism of alcohol was not involved, and the increase in short-chain organic acids was completely suppressed when the β blocker propranolol was administered. In a chronic model in which rats were given ethanol for 4 weeks, a large amount of short-chain organic acids, such as propionic acid and its carnitine esters, was similarly observed [53]. This 2 g/kg dose of ethanol is equivalent to about half a bottle of whiskey for a 70 kg human, so it is a sufficient amount for alcoholics. In patients with alcoholism, metabolic abnormalities such as branched-chain amino acids similar to those of propionic acidemia or organic acidemia are thought to occur.

Bykov et al. found that feeding rats ethanol for 10 weeks (gradually increasing the dose to eventually 5% w/v of the diet) increased liver steatosis several folds, accompanied by marked increases in liver inflammation, triglycerides, ALT, and TNFα, and pathological scores, but carnitine administration (2% of dry weight of the diet) suppressed these increases by about half, and the increase in TNFα was almost completely suppressed [54].

Dokmeci et al. reported that when rats were given ethanol (100% ethanol, 1 mL), gastric mucosal damage caused by oxidative stress and lipid peroxide, but such gastric mucosal damage was prevented by intragastric administration of 500 mg/kg carnitine 30 min before ethanol administration [55]. The protective effect of carnitine was remarkable from 50 mg/kg, and thiobarbituric acid reactive substances (TBARS), which are markers of lipid peroxide, were halved, and the glutathione concentration in gastric tissue, which was halved by ethanol, was maintained. In addition, it has been reported that carnitine administration reduces oxidative stress and lipid peroxidation caused by ethanol administration in serum, liver, central nervous system, and liver cell membranes [56–58].

Alcohol also has a significant effect on brain nerve cells, such as Wernicke's encephalopathy, and the mechanism is thought to be that ethanol inhibits the uptake of glucose by endothelial cells and astrocytes of the blood-brain barrier (BBB), and eventually kills neurons. In an animal model of alcohol poisoning, administration of ALC canceled the ethanol-induced inhibition of BBB glucose uptake [59, 60]. In addition, ALC protects endothelial cell function by stabilizing SOD in the BBB [61].

Alcohol cardiomyopathy in alcohol-dependent patients is also a major problem, and Jing et al. treated rats with 10% alcohol ad libitum + 60% alcohol intragastrically (5 mL/kg, once daily) for the first week, 20% alcohol ad libitum + 60% alcohol intragastrically (10 mL/kg, twice daily) for the second week, and 30% alcohol ad libitum + 60% alcohol intragastrically (15 mL/kg, three times daily) thereafter until 6 months [62]. Alcohol-loaded rats developed cardiomyopathy and showed a significant decrease in left ventricular ejection fraction, for example, from 64.3 ± 1.78% to 29.57 ± 1.5%, but there was no change in the group that received carnitine in addition to alcohol-loading, from 63.7 ± 1.72% to 57.71 ± 4.78%. In the alcohol-loaded group, free fatty acids increased, free carnitine and total carnitine decreased, myocardial ATP, ADP, and AMP decreased, and mRNA expression of PPARα, CPT1, MCAD, ANT1, etc. decreased, but carnitine administration prevented these decreases. The pathogenesis of alcohol cardiomyopathy is thought to be caused by apoptosis of cardiomyocytes due to mitochondrial oxidative stress [63, 64], and carnitine, which has anti-inflammatory, antioxidant, and anti-apoptotic effects, is thought to exert a protective effect.

In this way, the effect of carnitine administration on various clinical symptoms of alcoholics has been reported in basic studies, and clinical application is expected in the future.

17.7 Attention-Deficit/Hyperactivity Disorder (ADHD)/Autism and Carnitine

17.7.1 ADHD and Carnitine

Attention-deficit/hyperactivity disorder (ADHD) affects many adults as well as children. Various factors, such as environmental factors and genetic factors, are thought to be the cause, but there is no definitive theory yet.

Among them, phospholipids containing highly unsaturated fatty acids such as docosahexaenoic acid (22:6*n*-3: DHA) are considered to be essential for neuronal migration and morphogenesis in the developmental process of the central nervous system, and there is a theory that stagnation of biosynthesis of phospholipids containing such highly unsaturated fatty acids for some reason may be one of the pathologies of ADHD and autism [65]. DHA is biosynthesized (prolonged and unsaturated) by a carnitine-dependent system in mitochondria, but the biosynthesis of DHA in the body is originally small, and it is thought that carnitine deficiency or carnitine insufficiency impairs the biosynthesis of these phospholipids. Infante et al. reported that carnitine deficiency is a common feature of carnitine-dependent polyunsaturated fatty acid-containing phospholipids such as DHA, which is a common feature of Zellweger syndrome, which is a peroxisomal dysplasia, disordered oxidative phosphorylation, and disordered β oxidation, which may lead to the development of tissue disorders such as abnormal organogenesis [66, 67].

Van Oudheusden et al. found that in a preliminary study at their institute, four children with ADHD with low blood-free carnitine levels (<20 μmol/L) were administered carnitine (100 mg/kg/day, divided into two doses) and found that there was an improvement in impulsive behavior. A randomized, double-blind, placebo-controlled trial was conducted in which 26 boys with ADHD were divided into two groups [68]. Neurostimulants such as methylphenidate may be effective in treating children with ADHD, but concerns about side effects were one of the motivations for this study. One group (13 patients) received carnitine ⇒ placebo ⇒ and carnitine for 8 weeks each, and the other group received placebo ⇒ carnitine ⇒ placebo for 8 weeks each. The dose of carnitine was 100 mg/kg/day, divided into two doses per day, and a maximum of 4 g/day was used, and 6 g/20 mL of oral solution was used. Placebo was administered with an oral solution that tasted similar to the carnitine oral solution. The study lasted a total of 24 weeks, with 8 weeks for each step. As a result, 13 out of 24 children showed improvement in the child behavior checklist (CBCL, mainly at home behavior assessment), which is observed and scored by parents ($P < 0.02$), and 13 out of 24 patients showed improvement in the Conners teacher-rating score, which is graded by teachers ($P < 0.05$). The CBCL scores of these responders were significantly improved both in the total score and in each item compared to the previous value ($P < 0.0001$). The plasma free carnitine concentrations of these responders before carnitine administration averaged 42.3 ± 7.5 μmol/L and were 39.3 ± 8.1 μmol/L for non-responders, but significantly higher for responders than for non-responders after carnitine administration (81.0 ± 13.2 vs. 68.9 ± 11.9, $P < 0.03$). However, the children who participated in this study did not have particularly low levels of carnitine. There were no safety issues with carnitine.

Subsequently, attempts have been made to administer ALC to ADHD patients instead of carnitine. Arnold et al. studied ADHD patients (112 patients, 5–12 years old, 83 males, 29 females) with placebo or ALC (500–1500 mg/day, divided into two doses for 16 weeks) [69]. As a result, in the ITT analysis of all patients, there was no significant difference between the primary endpoint of the diagnostic interview schedule for children (DISC-IV), which was nine scores related to attention deficit symptoms scored by the teacher. However, when analyzed with attention deficit and mixed subtypes, ALC administration showed a significant improvement over placebo in the former ($P = 0.02$). There were no particular problems with safety.

Abbasi et al. conducted a clinical trial in which ADHD patients (40 cases, 28 males and 12 females, 7–13 years old) received methylphenidate (20–30 mg/day) plus ALC (500–1500 mg/day for 6 weeks) or placebo, but neither teacher scores nor parent scores were significantly different [70].

With regard to ADHD, there are concerns about the long-term administration of potent drugs such as methylphenidate to pediatric patients and the idea that nutrients and trace elements may be involved in the pathogenesis of the disease. Many clinical studies have been reported on nutrients and trace elements such as zinc, iron, and amino acids, as well as Chinese herbal medicines [71–73]. Richardson et al., as mentioned above, conducted clinical trials on ADHD patients and reported on the concept of highly unsaturated fatty acids [74–81]. However, the results of the study in which highly unsaturated fatty acids were effective were not reproduced in the next study, and certain conclusions have not been reached [79, 81]. As described in Chap. 14 in this book, there is a possibility that there are deficiencies in multiple nutrients and trace elements in the analogy of "Dobeneck's barrel." For example, if there are patients who are highly unsaturated fatty acid-deficient but not carnitine-deficient, patients who are carnitine deficient but not highly unsaturated fatty acid-deficient, and patients who are deficient in both, it may be necessary to try "highly unsaturated fatty acids + carnitine." Further research on the role of various nutrients, such as carnitine, in the prevention and treatment of ADHD is expected in the future.

17.7.2 Autism/Autism Spectrum Disorder and Carnitine

Autism and autism spectrum disorder (ASD) are also major problems for which there is no effective treatment. In 1979, the prevalence of autism and ASD was estimated to be five per 10,000 children (0–15 years old) and 21 with ASD, but new estimates later showed that autism was 41 per 10,000 (3–5 years old, 2008) and 157 per 10,000 (5–9 years old years, 2009) with ASD, which has increased more than eight times in about 30 years [82], and in the 2018 edition of the Journal of Child Psychology and Psychiatry, the number of cases was 0.4 per 1000 in the 1960s and 1970s. A study in South Korea reported an astonishing prevalence of 2.64% in 2011, and a study in South Carolina in 2017 (3.62%), and a 2018 study in New Jersey reported a surprising prevalence of 2.93% [83]. ASD is not only a pediatric patient, but also adolescent and adult patients, and in the absence of effective treatment, guidance, support, and rehabilitation have become a major social problem.

Recently, there have been many interesting reports on autism in relation to carnitine, and I will provide an overview of them.

In 2004, Filipek et al. retrospectively examined free carnitine and total carnitine levels in the serum laboratory data of 100 ASD patients and found that both levels were significantly lower ($P < 0.001$), and both were below the mean of the reference value of 83%, and the reference value of 1SD or less was 36% for total carnitine and 27% for free carnitine [84]. Pyruvate levels were also significantly lower ($P = 0.006$), and ammonia and alanine levels were significantly higher ($P < 0.001$). Based on these findings, they suspected that patients with ASD may have mild mitochondrial dysfunction. Mostafa et al. also examined plasma carnitine and DHA concentrations in children with ASD in Saudi Arabia and reported that 66% and 62% of patients, respectively, were lower than the fifth percentile of the reference value, and in a subsequent study in Egypt, that 90% and 76.6% of the patients were lower than the fifth percentile, respectively, too [85, 86]. In addition, since there are many cases of ASD symptoms in patients with mitochondrial disease, there is a belief that mitochondrial function abnormalities may be the basis of ASD, clinical studies, review articles, and meta-analyses from this perspective have been published [87–92]. In addition, studies of acylcarnitine and metabolites specific to acquired mitochondrial diseases in ASD patients have been reported [93, 94], but it seems that no significant results have been obtained.

In this context, a unique study has emerged that may lead to the pathogenesis of ASD. Celestino-Soper et al., in the Beaudet et al. research group, analyzed copy number variants (CNVs) in 3743 samples of families with ASD by comparative genomic hybridization (CGH) and found 55 pathogenic CNVs. They were validated and reported that there was a deletion of exon 2 of trimethyllysine hydroxylase epsilon (TMLHE) because it was a male proband [95, 96]. The TMLHE gene is located on the X chromosome, and the TMLHE encoded by it is an enzyme that is localized in the mitochondria and performs a reaction to convert 6-N-trimethyllysine to

3-OH-6-N-trimethyllysine in the first stage of the carnitine biosynthesis system (see Chap. 4, Fig. 4.3). The deletion of exon 2 is thought to cause a decrease in carnitine biosynthesis due to a significant decrease in activity.

Around the same time, Nava et al. found 33 new candidate genes by analyzing the whole exome of the X chromosome of 12 unrelated male ASD patients, including TMLHE [97]. And a nonsense mutation was found in two brothers with autism and intellectual disabilities. In addition, screening of the TMLHE coding sequence of 501 patients with ASD confirmed that two missense substitutions caused loss-of-function and increased trimethyllysine in the patients' plasma.

Geier et al. reported a double-blind, randomized trial of carnitine in autistic patients before these studies of autism and TMLHE mutations were reported [98]. Thirty patients with ASD were assigned to the carnitine group (19 patients) and the placebo group (11 patients), but the final group was the carnitine group (16 patients, 6.3 ± 2.4 years, 14 males, two women) and the placebo group (11 patients, 6.7 ± 1.6 years, nine males, two women). The breakdown of ASD was 10 cases of autism and six cases of PDD-NOS (unspecified pervasive developmental disorder)/Asperger's disorder in the carnitine group, and eight cases of autism and three cases of PDD-NOS/Asperger's disorder in the placebo group. The dose of carnitine was 50 mg/kg/day, and the duration of treatment was 3 months. They found significant improvements in the childhood autism rating scale (CARS) and modified clinical global impression (CGI) as scored by experts ($P = 0.02$ and $P = 0.03$), and the autism treatment evaluation as scored by the parents was significantly improved ($P = 0.02$ and $P = 0.03$). The checklist (ATEC) also improved significantly in speech ($P = 0.09$) and cognitive ($P = 0.009$). There were no safety issues.

Fahmy et al. conducted a clinical trial in which 30 autistic patients were randomly divided into 16 carnitine groups (14 men, two women, median age 69 months, liquid formula, 100 mg/kg/day divided into two morning and two evening) and 14 placebo groups (11 men, three women, median age 68.5 months) for 6 months. Significant improvement in CARS was observed in the carnitine group (carnitine group: 45.25 ± 6.191 ⇒ 37.06 ± 5.882, placebo group: 36.71 ± 5.594 ⇒ 34.71 ± 4.631, $P < 0.001$ vs. placebo, $P = 0.006$) compared to the previous value [99]. In the carnitine group, the number of severely ill patients with CARS decreased and the number of patients with mild or moderate disease decreased after 6 months (15 severe cases ⇒ 7 patients, 1 ⇒ 9 patients with mild or moderate disease), but no such change was observed in the placebo group (nine patients with mild or moderate disease, five patients with severe disease, no change after 6 months). There were no safety issues except for one case where a fishy odor was observed.

Ziats et al. of the Beaudet group attempted to administer carnitine (200 mg/kg/day) to a 4-year-old boy with a TMLHE gene mutation and developing ASD [100]. The child was diagnosed with pervasive developmental disorder when he was two and a half years old, but was otherwise healthy. Repeated regressions, such as behavioral disorders triggered by infection, etc., and parents made him take supplements such as vitamin B_{12}, B_6, folic acid, fish oil, and gluten and casein-free diets. After 4 days of carnitine treatment, the mother reported that she became "more docile," and after 2 weeks, there was an increase in verbal, non-verbal expression, and interaction with others, and there was an improvement, which lasted for more than 1 month. After 3 months, they were able to look directly at their parents, laugh, make eye contact, and show joint attention, awareness, and interest in others. After 4.5 months, eye contact and consciousness improved, but the hand fluttering behavior began. Attempts were made to increase or decrease the dose of carnitine, but 200 mg/kg/day was the most appropriate (with gastrointestinal discomfort at higher doses). A 13-month-old boy, who was a cousin of the patient, also had a TMLHE mutation, and his plasma free carnitine concentration was as low as 13 μmol/L, and trimethyllysine was high, so carnitine administration (50 mg/kg/day) was started to prevent the onset of the disease.

In addition, Goin-Kochel et al., in the same group, conducted an open-label pilot study of high-dose carnitine in young ASD patients (10 boys, mean age 59.8 months) and evaluated before and after administration [101]. The dosage of carnitine was started at 200 mg/kg/day in tablet or liquid form three times a day and increased to 400 mg/kg/day if there were no problems, with a maximum dosage of 6 g/day. If there were no problems, the dose was increased to 400 mg/kg/day, up to a maximum of 6 g/day, and reduced to 200 mg/kg/day if side effects such as fish odor or diarrhea occurred. The administration period was 8 weeks, and the results were evaluated at three points: before administration, 4, and 8 weeks after administration. Four cases of severe odor, four cases of diarrhea, and one case of spontaneous nausea were reported. Three cases could not increase the dose because of side effects. Three cases reported no side effects, but one of them was lost to follow-up after 4 weeks. Positive parental responses included calmer behavior (two cases), more activity (two cases), increased prosocial behavior (four cases), increased awareness (two cases), increased eye contact (two cases), and improved language skills. Six cases showed improvement on the CGI scale. The dose of 200–400 mg/kg/day, a dose used for primary carnitine deficiency, a carnitine transporter abnormality, was probably too high.

Beaudet, the leader of the TMLHE mutation research group in ASD, wrote a very interesting commentary on the relationship of carnitine deficiency and mitochondrial dysfunction with non-dysmorphic ASD (mild or essential ASD), in which there are no structural brain abnormalities or intellectual disability, and no abnormalities are seen for about 6 months after birth [102]. According to him, although common variants were the focus of genetic research on ASD in the 1990s, CNV and de novo point mutations have recently attracted more attention, and his group has found TMLHE mutations in CNV studies and TMLHE mutations in boys with ASD. Although less than 1% of boys with ASD have TMLHE mutations, carnitine deficiency in the brain and micronutrient deficiencies such as polyunsaturated fatty acids are still thought to be the underlying causes of ASD and ADHD. The idea is that carnitine deficiency and other micronutrient deficiencies may impair higher-order neurodevelopment of glial cells and neurons, as well as synaptic development. They hypothesized that carnitine depletion causes a decrease in mitochondrial copy number and that the reason that most ASD cases are symptomatic between the ages of 6 months and 3 years is because of the switch from carnitine-rich breast milk to a low-carnitine weaning diet of fruits, vegetables, cereals, and other low-carnitine foods. The carnitine concentration in the cerebrospinal fluid is said to be 1/10 to 1/25 of the blood level [103, 104], and mutations in OCTN2 and $ATB^{0,+}$, which control carnitine uptake at the blood-brain barrier, can cause a "brain carnitine deficiency." Brain carnitine levels are likely to be dissociated from blood carnitine levels. Therefore, while kidney and liver biosynthesis are important for carnitine homeostasis in the body as a whole, the brain carnitine biosynthetic system is extremely important during fetal and infantile life. Therefore, TMLHE mutation decreases carnitine biosynthesis in the brain, and *OCTN2* mutation decreases carnitine in the blood and its transport to the brain, even in heterozygous mutants. Mutations in *SLC6A14*, the $ATB^{0,+}$ gene, also reduce the passage of carnitine across the blood-brain barrier, possibly resulting in "brain carnitine deficiency." He advocated the need to promote research in animal models of ASD and to test this hypothesis in clinical trials.

Bankaitis et al., in their paper that won the American Society of Biochemistry and Molecular Biology Award, further developed Beaudet's "carnitine deficiency hypothesis" and proposed the "neural stem cell/carnitine malnutrition hypothesis." [105] The following is an overview of the hypothesis.

A risk factor that has not yet been evaluated in the pathogenesis of ASD may be the decrease in carnitine-dependent long-chain fatty acid β oxidation in developing embryonic neural stem cells. The premise is that the carnitine state of the brain during the fetal period is an extremely important element of neural stem cells, but it is extremely vulnerable to the disruption of the rapidly renewed nervous system development program, and if it is delayed, it is considered to be a risk of developing ASD. Fetal carnitine levels are affected by maternal genetic factors, environment, diet, and behavior. It is also well known that the concentration of carnitine in the blood of pregnant women decreases due to increased utilization. In addition, the concentration of carnitine in cerebrospinal fluid is less than 1/10 of the concentration in the blood. Under these circumstances, if there is a predisposition to TMLHE mutations in the carnitine biosynthesis system or inherited metabolic disorders that cause carnitine deficiency, even heterozygotes may have an effect (e.g., organic acidemia and fatty acid metabolism disorders). Although the genetic factors of ASD are diverse, effective prevention and treatment may be possible if a common underlying risk factor is found. Carnitine deficiency/malnutrition of neural stem cells is considered to be a risk, and as a public policy, carnitine levels should be evaluated for women planning pregnancy by 3 months of pregnancy, and carnitine supplementation should be given to those who need it, or a diet high in carnitine should be recommended. If the risk of ASD can be controlled, it will be possible to reduce the enormous cost of medical and social support for ASD patients, which is increasing rapidly. Based on basic and clinical research, it has been found that the neocortex of the frontal lobe, which controls perception, cognition, language, and behavior, has a laminate-shaped structure, and 70–80% are excitatory pyramidal neurons, which are made from neural stem cells in the dorsal frontal lobe during prenatal development. The homeostasis of neural stem cells during this period of brain development during the fetal period is extremely delicately regulated, and the exact number and type of neurons are formed sequentially. Carnitine deficiency in the fetal period and early birth is thought to cause hypoplasia due to abnormalities in energy metabolism in neural stem cells, neurons, and glial cells. In addition, the decrease in highly unsaturated fatty acids, such as DHA, which is required for phospholipids in the cell membrane, is also thought to be affected because carnitine is required. Maternal nutrition is extremely important for the development and formation of the neocortex of the frontal lobe of the fetus, and iron deficiency, which is common in pregnant women, increases the risk of ASD. This is thought to be due to the iron-dependent nature of TMLHE and γ-butyrobetaine hydroxylase 1, which are carnitine biosynthetic enzymes, and sufficient biosynthesis of carnitine is not possible if there is an iron deficiency. Just as recommend folic acid supplementation for pregnant women, we should also consider recommending carnitine.

Bankaitis not only describes the basis and background of the "neural stem cell/carnitine hypothesis" in this way, but also actually introduces TMLHE shRNA into mouse fetal neural stem cells by electroporation to inactivate TMLHE and investigates its effects. They demonstrated in animal models that a decrease in carnitine biosynthesis in the brain leads to a decrease in neural stem cells in the neocortex and impairs self-renewal programs.

In Chaps. 4 and 5, "Carnitine Biosynthesis", I state, "To date, carnitine deficiency due to genetic abnormalities in the carnitine biosynthetic system has not been reported." This is because carnitine deficiency due to abnormalities in biosynthetic enzymes has not been reported for the development of carnitine deficiency and carnitine insufficiency throughout the body, which have been discussed in this book so far. In addition, in the carnitine homeostasis of the entire body, the carnitine biosynthetic system of the brain was not considered important because it was quantitatively small. However, as in Beaudet's "carnitine deficiency hypothesis in the brain" and Bankaitis et al.'s " neural stem cell/carnitine malnutrition hypothesis," it was described that "carnitine deficiency in a region of the brain, especially during a specific period from 3 months of pregnancy to about 3 years of age, increases the risk of developing ADHD and ASD by affecting the development and structural formation of the neural network of the frontal neocortex of the brain." If it is true that carnitine deficiency occurs at a specific time from a month to about 3 years of age, carnitine deficiency and other nutrient deficiencies affect the development and structure formation of the neural network of the frontal lobe neocortex of the brain, and the risk of developing ADHD and ASD increases. It is necessary to study in more detail, such as deficiencies in certain stages of development in children. When considering carnitine deficiency, it might be important to recognize that such a local deficiency or a deficiency at a specific time may be the pathogenesis of serious diseases. For example, this perspective may be necessary not only for the developmental stage of children, but also for cognitive decline during the aging process in the elderly.

Kępka et al. overviewed the potential role of L-carnitine in ASD [118]. In the future, it is hoped that the importance of micronutrients such as carnitine, highly unsaturated fatty acids, iron, and zinc will be further studied and clinically applied in an interdisciplinary manner by not only pediatricians and psychiatrists but also nutrition researchers in cooperation in the prevention and treatment of ADHD and autism/ASD.

17.8 Infertility and Carnitine

17.8.1 Male Infertility

Carnitine is very abundant in the epididymis and sperm (2–100 mM, 40–2000 times or more of serum concentrations), it is thought to be deeply involved in sperm maturation, motility, respiration, etc., and has antioxidant and anti-apoptotic effects, so there are many basic studies and clinical studies [106]. Against this background, many attempts have been made to administer substances with antioxidant effects, such as carnitine and vitamin E, to male infertility caused by asthenospermia and oligospermia [107–109].

Khaw et al. conducted a systematic review and meta-analysis of clinical trials of carnitine or ALC for idiopathic male infertility [110]. In this meta-analysis, they excluded reviews and case reports from the papers reported between January 1, 2000, and April 30, 2020, and selected eight randomized controlled trials for qualitative analysis and seven trials for quantitative analysis. There were four studies in which carnitine 2 g/day + ALC 1 g/day (one was partially carnitine 3 g/day + ALC 3 g/day) and 1 g/day of carnitine was administered in 4 studies, and the duration of treatment was from 3 to 9 months. As a result, the administration of carnitine was effective in sperm motility and normal sperm production, but not on sperm count. In addition, pregnancy rates were reported in five trials, but one study had statistical errors, and only one of the remaining four studies reported a significant increase in pregnancy rate, but the other three reported no effect, and there was no definitive conclusion due to a too wide range of confidence limits.

Li et al. also conducted a meta-analysis of randomized controlled clinical trials assessing the parameters of sperm quality and pregnancy rates of antioxidants in patients with idiopathic male infertility [111]. Twenty-three studies (1917 patients in total) covered clinical trials of 10 antioxidants, including carnitine, carnitine + ALC, CoQ10, omega-3 fatty acids, and selenium. As a result, carnitine was most effective in terms of sperm motility and morphology, and omega-3 fatty acids were most effective in terms of sperm concentration (number). CoQ10 was moderately effective in sperm motility and concentration. As for pregnancy rates, the number of cases was still small, and none of the antioxidants were found to be effective (see the literature list for individual clinical trial reports, Khaw [110] and Li [111]).

The aforementioned clinical trials are using a single compound or two drugs, but in Italy, 24 types of products that combine 36 types of ingredients, such as carnitine, vitamins, minerals such as zinc and selenium, and unsaturated fatty acids such as DHA, are sold as supplements for male infertility. Garolla et al. categorized these individual components into A, B, C, and D based on reports of efficacy on parameters such as sperm count, motility, and morphology, and scored them using their own formulas based on the lowest effective dosage, etc., and examined the effectiveness of these nutritional supplements [112, 113]. Of the 36 ingredients, 18 were considered to be effective based on clinical study reports. Carnitine is included in these 18 ingredients, but the category was B/C, and the rank was low. The remaining 18 ingredients were considered ineffective. As a result of compiling the scores of 24 nutritional supplements, only three products (12.5%) were considered to be highly effective, nine products (37.5%) were considered to be low effective, and 12 products (50%) were not expected to be effective [113]. One of the three products that could be expected to have this effect contained carnitine. Szymański et al. studied L-carnitine fumarate and acetyl-L-carnitine therapy in patients with idiopathic infertility and found significant improvement in sperm concentration, total sperm count, sperm total and progressive motility [119].

In this way, micronutrients with several antioxidant effects, such as carnitine, omega-3 fatty acids, and CoQ10, are effective for sperm parameters, and further research is desired to see if they can really increase the pregnancy rate in the future.

17.8.2 Female Infertility

Vitagliano and colleagues from Garolla's group reviewed and compared the effects of 24 nutrient supplements for female infertility sold in Italy using a similar method [114]. Among them, the nutrients and components that were recognized as effective were melatonin, folic acid, myo-inositol, D-chiro-inositol, N-acetylcysteine, CoQ10, vitamin D, and carnitine. The following two studies were cited as clinical studies that provided the basis for evaluating the effectiveness of carnitine. Kitano et al. reported that 214 women who had not undergone successful in vitro fertilization (IVF) and embryo transfer were given carnitine (1 g/day) for an average of 82 days, and when IVF and embryo transfer were reperformed, the quality of embryos improved after 3 and 5 days of fertilization ($P < 0.05$), and several healthy newborns were born [115].

Pandey et al. conducted a literature review about the role of nutrient supplements in female infertility [120]. They concluded that available evidence is insufficient to recommend nutrient supplementation to improve female infertility.

Thus, reproductive medicine is a field in which the quality of sperm cells, oocytes, and fertilized eggs, which are single cells, is extremely important. It is quite predictable that oxidative stress, abnormal energy metabolism, and accumulation of fat and long-chain acyl CoA will cause deterioration in the quality of sperm, oocytes, and fertilized eggs. There is still much room for clinical application and development of carnitine, which has physiological effects such as improvement of energy metabolism, elimination of long-chain acyl groups, and antioxidant action, for infertility treatment and in vitro fertilization.

17.9 Summary

In this chapter, the relationship between carnitine deficiency or insufficiency and the pathogenesis of various disease which are not usually taken away.

Miwa et al. reported the effect of L-carnitine supplementation on muscle cramps induced by stroke [121]. A 79-year-old man with cerebral infarction suffered nocturnal leg cramps, which caused sleep disturbance. Supplementation of L-carnitine improved these symptoms. Also, Ueno et al. reported two cases of disorders of consciousness, convulsion, and muscle cramps, which were dissolved after L-carnitine supplementation [122]. The authors pointed out that carnitine deficiency has been overlooked in some patients in rehabilitation wards.

Thus, carnitine deficiency and insufficiency are related to severe symptoms which disturb QOL and clinical outcome. It is expected that hidden carnitine deficiency and insufficiency in various diseases will be discovered, and effects of carnitine supplementation will be studied in the future.

References

1. Chen Y, Mizuguchi H, Yao D, et al. Thermolabile phenotype of carnitine palmitoyltransferase II variations as a predisposing factor for influenza-associated encephalopathy. FEBS Lett. 2005;579(10):2040–4. https://doi.org/10.1016/j.febslet.2005.02.050.

2. Yao D, Mizuguchi H, Yamaguchi M, et al. Thermal instability of compound variants of carnitine palmitoyltransferase II and impaired mitochondrial fuel utilization in influenza-associated encephalopathy. Hum Mutat. 2008;29(5):718–27. https://doi.org/10.1002/humu.20717.
3. Kido H, Yao D, Chida J, et al. Pathogenesis of influenza-associated encephalopathy: disorder in fatty acid metabolism in mitochondria and increased membrane permeability in endothelial cells. Med Front. 2010;65(1):52–60. (Article in Japanese)
4. Kubota M, Chida J, Hoshino H, et al. Thermolabile CPT II variants and low blood ATP levels are closely related to severity of acute encephalopathy in Japanese children. Brain Dev. 2012;34(1):20–7. https://doi.org/10.1016/j.braindev.2010.12.012.
5. Shinohara M, Saitoh M, Takanashi J, et al. Carnitine palmitoyl transferase II polymorphism is associated with multiple syndromes of acute encephalopathy with various infectious diseases. Brain Dev. 2011;33(6):512–7. https://doi.org/10.1016/j.braindev.2010.09.002.
6. Kuroda Y, Matsuda J. Clinical characteristics and treatment for the influenza-associated encephalopathy and encephalitis in Japan. Neurol Med. 2004;60(2):128–33. (Article in Japanese)
7. Purevsuren J, Hasegawa Y, Kobayashi H, et al. Urinary organic metabolite screening of children with influenza-associated encephalopathy for inborn errors of metabolism using GC/MS. Brain Dev. 2008;30(8):520–6. https://doi.org/10.1016/j.braindev.2008.01.001.
8. Wolf A, Weir P, Segar P, et al. Impaired fatty acid oxidation in propofol infusion syndrome. Lancet. 2001;357(9256):606–7. https://doi.org/10.1016/S0140-6736(00)04064-2.
9. Uezono S, Hotta Y, Takakuwa Y, et al. Acquired carnitine deficiency: a clinical model for propofol infusion syndrome? Anesthesiology. 2005;103(4):909. https://doi.org/10.1097/00000542-200510000-00046.
10. Weinberg GL, Baughman V. Carnitine deficiency, mitochondrial metabolism, and abnormal response to anesthetics. Anesthesiology. 2006;104(6):1343. https://doi.org/10.1097/00000542-200606000-00036. (author reply 1343-1344 by Dr. Uezono S.).
11. Okazaki K. Propofol infusion syndrome. J Jpn Soc Clin Anesth. 2009;33(Supplement):329–48. (Article in Japanese)
12. Ueno K, Tanabe K, Toh N, et al. Current status of carnitine deficiency in patients with neuromuscular disease during long-term tube feeding. Jpn J Clin Nutr. 2014;125(5):635–9. (Article in Japanese)
13. Muguruma T, Omori H, Takeda I, et al. Nutritional evaluation in patients with intractable neurological diseases undergoing long-term tube feeding—in point of trace elements and carnitine. J Jpn Soc Clin Nutr. 2015;37(1):36–43. (Article in Japanese)
14. Isse N, Miura Y, Obata T, et al. Carnitine deficiency presenting with a decreased mental state in a patient with amyotrophic lateral sclerosis receiving long-term tube feeding: a case report. J Med Case Rep. 2013;7:286. https://doi.org/10.1186/1752-1947-7-286.
15. Kira Y, Nishikawa M, Ochi A, et al. L-carnitine suppresses the onset of neuromuscular degeneration and increases the life span of mice with familial amyotrophic lateral sclerosis. Brain Res. 2006;1070(1):206–14. https://doi.org/10.1016/j.brainres.2005.11.052.
16. Patel BP, Hamadeh MJ. Nutritional and exercise-based interventions in the treatment of amyotrophic lateral sclerosis. Clin Nutr. 2009;28(6):604–17. https://doi.org/10.1016/j.clnu.2009.06.002.
17. Beghi E, Pupillo E, Bonito V, et al. Randomized double-blind placebo-controlled trial of acetyl-L-carnitine for ALS. Amyotroph Lateral Scler Frontotemporal Degener. 2013;14(5–6):397–405. https://doi.org/10.3109/21678421.2013.764568.
18. Yoshino H, Kamei T. L-carnitine clinical trials in ALS. Clin Neurol. 2006;46(12):1090. (Meeting abstract, article in Japanese)
19. Kuratsune H, Yamaguti K, Takahashi M, et al. Acylcarnitine deficiency in chronic fatigue syndrome. Clin Infect Dis. 1994;18(Suppl 1):S62–7. https://doi.org/10.1093/clinids/18.supplement_1.s62.
20. Kuratsune H, Yamaguti K, Lindh G, et al. Low levels of serum acylcarnitine in chronic fatigue syndrome and chronic hepatitis type C, but not seen in other diseases. Int J Mol Med. 1998;2(1):51–6. https://doi.org/10.3892/ijmm.2.1.51.
21. Plioplys AV, Plioplys S. Serum levels of carnitine in chronic fatigue syndrome: clinical correlates. Neuropsychobiology. 1995;32(3):132–8. https://doi.org/10.1159/000119226.
22. Plioplys AV, Plioplys S. Amantadine and L-carnitine treatment of chronic fatigue syndrome. Neuropsychobiology. 1997;35(1):16–23. https://doi.org/10.1159/000119325.
23. Majeed T, De Simone C, Famularo G, et al. Abnormalities of carnitine metabolism in chronic fatigue syndrome. Eur J Neurol. 1995;2(5):425–8. https://doi.org/10.1111/j.1468-1331.1995.tb00151.x.
24. Jones MG, Goodwin CS, Amjad S, et al. Plasma and urinary carnitine and acylcarnitines in chronic fatigue syndrome. Clin Chim Acta. 2005;360(1–2):173–7. https://doi.org/10.1016/j.cccn.2005.04.029.
25. Reuter SE, Evans AM. Long-chain acylcarnitine deficiency in patients with chronic fatigue syndrome. Potential involvement of altered carnitine palmitoyltransferase-I activity. J Intern Med. 2011;270(1):76–84. https://doi.org/10.1111/j.1365-2796.2010.02341.x.
26. Watanabe Y, Kuratsune H. Pathophysiology and treatment of chronic fatigue syndrome/myalgic encephalomyelitis. Neurol Ther. 2016;33(1):40–5. (Article in Japanese)
27. Roediger WE. Role of anaerobic bacteria in the metabolic welfare of the colonic mucosa in man.

Gut. 1980;21(9):793–8. https://doi.org/10.1136/gut.21.9.793.
28. Roediger WE. The colonic epithelium in ulcerative colitis: an energy-deficiency disease? Lancet. 1980;2(8197):712–5. https://doi.org/10.1016/s0140-6736(80)91934-0.
29. Chapman MA, Grahn MF, Boyle MA, et al. Butyrate oxidation is impaired in the colonic mucosa of sufferers of quiescent ulcerative colitis. Gut. 1994;35(1):73–6. https://doi.org/10.1136/gut.35.1.73.
30. Breuer RI, Buto SK, Christ ML, et al. Rectal irrigation with short-chain fatty acids for distal ulcerative colitis. Preliminary report. Dig Dis Sci. 1991;36(2):185–7. https://doi.org/10.1007/BF01300754.
31. Breuer RI, Soergel KH, Lashner BA, et al. Short chain fatty acid rectal irrigation for left-sided ulcerative colitis: a randomised, placebo controlled trial. Gut. 1997;40(4):485–91. https://doi.org/10.1136/gut.40.4.485.
32. Scheppach W. German-Austrian SCFA study group: treatment of distal ulcerative colitis with short-chain fatty acid enemas. A placebo-controlled trial. Dig Dis Sci. 1996;41(11):2254–9. https://doi.org/10.1007/BF02071409.
33. Steinhart AH, Brzezinski A, Baker JP. Treatment of refractory ulcerative proctosigmoiditis with butyrate enemas. Am J Gastroenterol. 1994;89(2):179–83.
34. Vernia P, Marcheggiano A, Caprilli R, et al. Short-chain fatty acid topical treatment in distal ulcerative colitis. Aliment Pharmacol Ther. 1995;9(3):309–13. https://doi.org/10.1111/j.1365-2036.1995.tb00386.x.
35. Giancaterini A, Mingrone G, De Gaetano A, et al. Effects of propionil-L-carnitine topical irrigation in distal ulcerative colitis: a preliminary report. Am J Gastroenterol. 2001;96(7):2275–6. https://doi.org/10.1111/j.1572-0241.2001.03988.x.
36. Gasbarrini G, Mingrone G, Giancaterini A, et al. Effects of propionyl-L-carnitine topical irrigation in distal ulcerative colitis: a preliminary report. Hepato-Gastroenterology. 2003;50(53):1385–9.
37. Rioux JD, Silverberg MS, Daly MJ, et al. Genomewide search in Canadian families with inflammatory bowel disease reveals two novel susceptibility loci. Am J Hum Genet. 2000;66(6):1863–70. https://doi.org/10.1086/302913.
38. Vermeire S, Pierik M, Hlavaty T, et al. Association of organic cation transporter risk haplotype with perianal penetrating Crohn's disease but not with susceptibility to IBD. Gastroenterology. 2005;129(6):1845–53. https://doi.org/10.1053/j.gastro.2005.10.006.
39. Babusukumar U, Wang T, McGuire E, et al. Contribution of OCTN variants within the IBD5 locus to pediatric onset Crohn's disease. Am J Gastroenterol. 2006;101(6):1354–61. https://doi.org/10.1111/j.1572-0241.2006.00564.x.
40. Fisher SA, Hampe J, Onnie CM, et al. Direct of indirect association in a complex disease: the role of SLC22A4 and SLC22A5 functional variants in Crohn disease. Hum Mutat. 2006;27(8):778–85. https://doi.org/10.1002/humu.20358.
41. Silverberg MS, Duerr RH, Brant SR, et al. Refined genomic localization and ethnic differences observed for the IBD5 association with Crohn's disease. Eur J Hum Genet. 2007;15(3):328–35. https://doi.org/10.1038/sj.ejhg.5201756.
42. Yamamoto-Furusho JK, Mendivil-Rangel EJ, Villeda-Ramírez MA, et al. Gene expression of carnitine organic cation transporters 1 and 2 (OCTN) is downregulated in patients with ulcerative colitis. Inflamm Bowel Dis. 2011;17(10):2205–6. https://doi.org/10.1002/ibd.21621.
43. Xuan C, Zhang BB, Yang T, et al. Association between OCTN1/2 gene polymorphisms (1672C-T, 207G-C) and susceptibility of Crohn's disease: a meta-analysis. Int J Color Dis. 2012;27(1):11–9. https://doi.org/10.1007/s00384-011-1265-x.
44. Lamhonwah AM, Ackerley C, Onizuka R, et al. Epitope shared by functional variant of organic cation/carnitine transporter, OCTN1, campylobacter jejuni and mycobacterium paratuberculosis may underlie susceptibility to Crohn's disease at 5q31. Biochem Biophys Res Commun. 2005;337(4):1165–75. https://doi.org/10.1016/j.bbrc.2005.09.170.
45. Fortin G, Yurchenko K, Collette C, et al. L-carnitine, a diet component and organic cation transporter OCTN ligand, displays immunosuppressive properties and abrogates intestinal inflammation. Clin Exp Immunol. 2009;156(1):161–71. https://doi.org/10.1111/j.1365-2249.2009.03879.x.
46. Fortin G. L-carnitine and intestinal inflammation. Vitam Horm. 2011;86:353–66. https://doi.org/10.1016/B978-0-12-386960-9.00015-0.
47. Mikhailova TL, Sishkova E, Poniewierka E, et al. Randomised clinical trial: the efficacy and safety of propionyl-L-carnitine therapy in patients with ulcerative colitis receiving stable oral treatment. Aliment Pharmacol Ther. 2011;34(9):1088–97. https://doi.org/10.1111/j.1365-2036.2011.04844.x.
48. Merra G, Gasbarrini G, Laterza L, et al. Propionyl-L-carnitine hydrochloride for treatment of mild to moderate colonic inflammatory bowel diseases. World J Gastroenterol. 2012;18(36):5065–71. https://doi.org/10.3748/wjg.v18.i36.5065.
49. Ellestad-Sayed JJ, Nelson RA, Adson MA, et al. Pantothenic acid, coenzyme a, and human chronic ulcerative and granulomatous colitis. Am J Clin Nutr. 1976;29:1333–8.
50. Fuller RK, Hoppel CL. Plasma carnitine in alcoholism. Alcohol Clin Exp Res. 1988;12(5):639–42. https://doi.org/10.1111/j.1530-0277.1988.tb00256.x.
51. De Sousa C, Leung NW, Chalmers RA, et al. Free and total carnitine and acylcarnitine content of plasma, urine, liver and muscle of alcoholics.

Dermatol Clin Sci. 1988;75(4):437–40. https://doi.org/10.1042/cs0750437.
52. Calabrese V, Rizza V. Formation of propionate after short-term ethanol treatment and its interaction with the carnitine pool in rat. Alcohol. 1999;19(2):169–76. https://doi.org/10.1016/s0741-8329(99)00036-1.
53. Calabrese V, Calvani M, Butterfield DA. Increased formation of short-chain organic acids after chronic ethanol administration and its interaction with the carnitine pool in rat. Arch Biochem Biophys. 2004;431(2):271–8. https://doi.org/10.1016/j.abb.2004.08.020.
54. Bykov I, Jrveläinen H, Lindros K. L-carnitine alleviates alcohol-induced liver damage in rats: role of tumour necrosis factor-alpha. Alcohol. 2003;38(5):400–6. https://doi.org/10.1093/alcalc/agg109.
55. Dokmeci D, Akpolat M, Aydogdu N, et al. L-carnitine inhibits ethanol-induced gastric mucosal injury in rats. Pharmacol Rep. 2005;57(4):481–8.
56. Augustyniak A, Skrzydlewska E. L-carnitine in the lipid and protein protection against ethanol-induced oxidative stress. Alcohol. 2009;43(3):217–23. https://doi.org/10.1016/j.alcohol.2008.12.005.
57. Augustyniak A, Skrzydlewska E. The influence of L-carnitine supplementation on the antioxidative abilities of serum and the central nervous system of ethanol-induced rats. Metab Brain Dis. 2010;25(4):381–9. https://doi.org/10.1007/s11011-010-9217-7.
58. Dobrzyńska I, Szachowicz-Petelska B, Skrzydlewska E, et al. Effect of L-carnitine on liver cell membranes in ethanol-intoxicated rats. Chem Biol Interact. 2010;188(1):44–51. https://doi.org/10.1016/j.cbi.2010.07.004.
59. Abdul Muneer PM, Alikunju S, Szlachetka AM, et al. Ethanol impairs glucose uptake by human astrocytes and neurons: protective effects of acetyl-L-carnitine. Int J Physiol Pathophysiol Pharmacol. 2011;3(1):48–56.
60. Abdul Muneer PM, Alikunju S, Szlachetka AM, et al. Inhibitory effects of alcohol on glucose transport across the blood-brain barrier leads to neurodegeneration: preventive role of acetyl-L-carnitine. Psychopharmacology. 2011;214(3):707–18. https://doi.org/10.1007/s00213-010-2076-4.
61. Haorah J, Floreani NA, Knipe B, et al. Stabilization of superoxide dismutase by acetyl-L-carnitine in human brain endothelium during alcohol exposure: novel protective approach. Free Radic Biol Med. 2011;51(8):1601–9. https://doi.org/10.1016/j.freeradbiomed.2011.06.020.
62. Jing L, Zhou LJ, Li WM, et al. Carnitine regulates myocardial metabolism by peroxisome proliferator-activated receptor-α(PPAR α)in alcoholic cardiomyopathy. Med Sci Monit. 2011;17(1):BR1–9. https://doi.org/10.12659/msm.881311.
63. Jing L, Jin CM, Li SS, et al. Chronic alcohol intake-induced oxidative stress and apoptosis: role of CYP2E1 and calpain-1 in alcoholic cardiomyopathy. Mol Cell Biochem. 2012;359(1–2):283–92. https://doi.org/10.1007/s11010-011-1022-z.
64. Steiner JL, Lang CH. Etiology of alcoholic cardiomyopathy: mitochondria, oxidative stress and apoptosis. Int J Biochem Cell Biol. 2017;89:125–35. https://doi.org/10.1016/j.biocel.2017.06.009.
65. Richardson AJ, Ross MA. Fatty acid metabolism in neurodevelopmental disorder: a new perspective on associations between attention-deficit/hyperactivity disorder, dyslexia, dyspraxia and the autistic spectrum. Prostaglandins Leukot Essent Fatty Acids. 2000;63(1–2):1–9. https://doi.org/10.1054/plef.2000.0184.
66. Infante JP, Huszagh VA. On the molecular etiology of decreased arachidonic (20: 4n-6), docosapentaenoic (22, 5n-6) and docosahexaenoic (22, 6n-3) acids in Zellweger syndrome and other peroxisomal disorders. Mol Cell Biochem. 1997;168(1–2):101–15. https://doi.org/10.1023/a:1006895209833.
67. Infante JP, Huszagh VA. Secondary carnitine deficiency and impaired docosahexaenoic (22: 6n-3) acid synthesis: a common denominator in the pathophysiology of diseases of oxidative phosphorylation and β-oxidation. FEBS Lett. 2000;468(1):1–5. https://doi.org/10.1016/s0014-5793(00)01083-8.
68. Van Oudheusden LJ, Scholte HR. Efficacy of carnitine in the treatment of children with attention-deficit hyperactivity disorder. Prostaglandins Leukot Essent Fatty Acids. 2002;67(1):33–8. https://doi.org/10.1054/plef.2002.0378.
69. Arnold LE, Amato A, Bozzolo H, et al. Acetyl-L-carnitine (ALC) in attention-deficit/hyperactivity disorder: a multi-site, placebo-controlled pilot trial. J Child Adolesc Psychopharmacol. 2007;17(6):791–801. https://doi.org/10.1089/cap.2007.018.
70. Abbasi SH, Heidari S, Mohammadi MR, et al. Acetyl-L-carnitine as an adjunctive therapy in the treatment of attention-deficit/hyperactivity disorder in children and adolescents: a placebo-controlled trial. Child Psychiatry Hum Dev. 2011;42(3):367–75. https://doi.org/10.1007/s10578-011-0220-y.
71. Hurt EA, Arnold LE, Lofthouse N. Dietary and nutritional treatments for attention-deficit/hyperactivity disorder: current research support and recommendations for practitioners. Curr Psychiatry Rep. 2011;13(5):323–32. https://doi.org/10.1007/s11920-011-0217-z.
72. Sarris J, Kean J, Schweitzer I, et al. Complementary medicines (herbal and nutritional products) in the treatment of attention deficit hyperactivity disorder (ADHD): a systematic review of the evidence. Complement Ther Med. 2011;19(4):216–27. https://doi.org/10.1016/j.ctim.2011.06.007.
73. Rucklidge JJ, Johnstone J, Kaplan BJ. Nutrient supplementation approaches in the treatment of ADHD. Expert Rev Neurother. 2009;9(4):461–76. https://doi.org/10.1586/ern.09.7.

74. Richardson AJ, Puri BK. The potential role of fatty acids in attention-deficit/hyperactivity disorder. Prostaglandins Leukot Essent Fatty Acids. 2000;63(1–2):79–87. https://doi.org/10.1054/plef.2000.0196.
75. Richardson AJ, Puri BK. A randomized double-blind, placebo-controlled study of the effects of supplementation with highly unsaturated fatty acids on ADHD-related symptoms in children with specific learning difficulties. Prog Neuro-Psychopharmacol Biol Psychiatry. 2002;26(2):233–9. https://doi.org/10.1016/s0278-5846(01)00254-8.
76. Richardson AJ, Cyhlarova E, Ross MA. Omega-3 and omega-6 fatty acid concentrations in red blood cell membranes relate to schizotypal traits in healthy adults. Prostaglandins Leukot Essent Fatty Acids. 2003;69(6):461–6. https://doi.org/10.1016/j.plefa.2003.08.018.
77. Richardson AJ. Clinical trials of fatty acid treatment in ADHD, dyslexia, dyspraxia and the autistic spectrum. Prostaglandins Leukot Essent Fatty Acids. 2004;70(4):383–90. https://doi.org/10.1016/j.plefa.2003.12.020.
78. Richardson AJ. Omega-3 fatty acids in ADHD and related neurodevelopmental disorders. Int Rev Psychiatry. 2006;18(2):155–72. https://doi.org/10.1080/09540260600583031.
79. Richardson AJ, Burton JR, Sewell RP, et al. Docosahexaenoic acid for reading, cognition and behavior in children aged 7-9 years: a randomized, controlled trial (the DOLAB study). PLoS One. 2012;7(9):e43909. https://doi.org/10.1371/journal.pone.0043909.
80. Montgomery P, Burton JR, Sewell RP, et al. Low blood long chain omega-3 fatty acids in UK children are associated with poor cognitive performance and behavior: a cross-sectional analysis from the DOLAB study. PLoS One. 2013;8(6):e66697. https://doi.org/10.1371/journal.pone.0066697.
81. Montgomery P, Spreckelsen TF, Burton A, et al. Docosahexaenoic acid for reading, working memory and behavior in UK children aged 7-9: a randomized controlled trial for replication (the DOLAB II study). PLoS One. 2018;13(2):e0192909. https://doi.org/10.1371/journal.pone.0192909.
82. Tsuchiya K, Matsumoto K, Takei N. Recent progress of epidemiological research in understanding the causes of autism and autism spectrum disorder. Brain Sci Ment Disord. 2009;20(4):295–302. (Article in Japanese)
83. Fombonne E. Editorial: the rising prevalence of autism. J Child Psychol Psychiatry. 2018;59(7):717–20. https://doi.org/10.1111/jcpp.12941.
84. Filipek PA, Juranek J, Nguyen MT, et al. Relative carnitine deficiency in autism. J Autism Dev Disord. 2004;34(6):615–23. https://doi.org/10.1007/s10803-004-5283-1.
85. Mostafa GA, Al-Ayadhi LY. Reduced levels of plasma polyunsaturated fatty acids and serum carnitine in autistic children: relation to gastrointestinal manifestations. Behav Brain Funct. 2015;11:4. https://doi.org/10.1186/s12993-014-0048-2.
86. Mostafa GA, El-Gamal HA, El-Wakkad ASE, et al. Polyunsaturated fatty acids, carnitine and lactate as biological markers of brain energy in autistic children. Int J Child Neuropsychiatry. 2005;2(2):179–88.
87. Lombard J. Autism: a mitochondrial disorder? Med Hypotheses. 1998;50:497–500.
88. Palmieri L, Persico AM. Mitochondrial dysfunction in autism spectrum disorders: cause or effect? Biochim Biophys Acta. 2010;1797(6–7):1130–7. https://doi.org/10.1016/j.bbabio.2010.04.018.
89. Giulivi C, Zhang YF, Omanska-Klusek A, et al. Mitochondrial dysfunction in autism. JAMA. 2010;304(21):2389–96. https://doi.org/10.1001/jama.2010.1706.
90. Rossignol DA, Frye RE. Mitochondrial dysfunction in autism spectrum disorders: a systematic review and meta-analysis. Mol Psychiatry. 2012;17(3):290–314. https://doi.org/10.1038/mp.2010.136.
91. Legido A, Jethva R, Goldenthal MJ. Mitochondrial dysfunction in autism. Semin Pediatr Neurol. 2013;20(3):163–75. https://doi.org/10.1016/j.spen.2013.10.008.
92. Rose S, Niyazov DM, Rossignol DA, et al. Clinical and molecular characteristics of mitochondrial dysfunction in autism spectrum disorder. Mol Diagn Ther. 2018;22(5):571–93. https://doi.org/10.1007/s40291-018-0352-x.
93. Frye RE, Melnyk S, MacFabe DF. Unique acyl-carnitine profiles are potential biomarkers for acquired mitochondrial disease in autism spectrum disorder. Transl Psychiatry. 2013;3(1):e220. https://doi.org/10.1038/tp.2012.143.
94. Lv QQ, You C, Zou XB, et al. Acyl-carnitine, C5DC, and C26 as potential biomarkers for diagnosis of autism spectrum disorder in children. Psychiatry Res. 2018;267:277–80. https://doi.org/10.1016/j.psychres.2018.06.027.
95. Celestino-Soper PB, Shaw CA, Sanders SJ, et al. Use of array CGH to detect exonic copy number variants throughout the genome in autism families detects a novel deletion in TMLHE. Hum Mol Genet. 2011;20(22):4360–70. https://doi.org/10.1093/hmg/ddr363.
96. Celestino-Soper PB, Violante S, Crawford EL, et al. A common X-linked inborn error of carnitine biosynthesis may be a risk factor for non-dysmorphic autism. Proc Natl Acad Sci USA. 2012;109(21):7974–81. https://doi.org/10.1073/pnas.1120210109.
97. Nava C, Lamari F, Héron D, et al. Analysis of the chromosome X exome in patients with autism spectrum disorders identified novel candidate genes, including TMLHE. Transl Psychiatry. 2012;2(10):e179. https://doi.org/10.1038/tp.2012.102.
98. Geier DA, Kern JK, Davis G, et al. A prospective double-blind, randomized clinical trial of levocarnitine to treat autism spectrum disorders. Med Sci Monit. 2011;17(6):PL15–23. https://doi.org/10.12659/msm.881792.

99. Fahmy SF, El-hamamsy MH, Zaki OK, et al. L-carnitine supplementation improves the behavioral symptoms in autistic children. Res Autism Spectr Disord. 2013;7(1):159–66. https://doi.org/10.1016/j.rasd.2012.07.004.
100. Ziats MN, Comeaux MS, Yang Y, et al. Improvement of regressive autism symptoms in a child with TMLHE deficiency following carnitine supplementation. Am J Med Genet A. 2015;167A(9):2162–7. https://doi.org/10.1002/ajmg.a.37144. (Note: There is a corrigendum in the same journal, 167A(10), 2496).
101. Goin-Kochel RP, Scaglia F, Schaaf CP, et al. Side effects and behavioral outcomes following high-dose carnitine supplementation among young males with autism spectrum disorder: a pilot study. Glob Pediatr Health. 2019;6:2333794X19830696. https://doi.org/10.1177/2333794X19830696.
102. Beaudet AL. Brain carnitine deficiency causes nonsyndromic autism with an extreme male bias: a hypothesis. BioEssays. 2017;39(8):1700012. https://doi.org/10.1002/bies.201700012.
103. Rubio JC, de Bustos F, Molina JA, et al. Cerebrospinal fluid carnitine levels in patients with Alzheimer's disease. J Neurol Sci. 1998;155(2):192–5. https://doi.org/10.1016/s0022-510x(97)00314-6.
104. Shinawi M, Gruener N, Lerner A. CFS levels of carnitine in children with meningitis, neurologic disorders, acute gastroenteritis, and seizure. Neurology. 1998;50(6):1869–71. https://doi.org/10.1212/wnl.50.6.1869.
105. Bankaitis VA, Xie Z. The neural stem cell/carnitine malnutrition hypothesis: new prospects for effective reduction of autism risk? J Biol Chem. 2019;294(50):19424–35. https://doi.org/10.1074/jbc.AW119.008137.
106. Ng CM, Blackman MR, Wang C, et al. The role of carnitine in the male reproductive system. Ann N Y Acad Sci. 2004;1033(177–188):10.1196/annals.1320.017.
107. Cavallini G. Male idiopathic oligoasthenoteratozoospermia. Asian J Androl. 2006;8(2):143–57. https://doi.org/10.1111/j.1745-7262.2006.00123.x.
108. Walczak-Jedrzejowska R, Wolski JK, Slowikowska-Hilczer J. The role of oxidative stress and antioxidants in male fertility. Cent Eur J Urol. 2013;66(1):60–7. https://doi.org/10.5173/ceju.2013.01.art19.
109. Dutta S, Sengupta P, Roychoudhury S, et al. Antioxidant paradox in male infertility: 'a blind eye' on inflammation. Antioxidants. 2022;11(1):167. https://doi.org/10.3390/antiox11010167.
110. Khaw SC, Wong ZZ, Anderson RA, et al. L-carnitine and L-acetylcarnitine supplementation for idiopathic male infertility. Reprod Fertil. 2020;1(1):67–81. https://doi.org/10.1530/RAF-20-0037.
111. Li KP, Yang XS, Wu T. The effect of antioxidants on sperm quality parameters and pregnancy rates for idiopathic male infertility: a network meta-analysis of randomized controlled trials. Front Endocrinol (Lausanne). 2022;13:810242. https://doi.org/10.3389/fendo.2022.810242.
112. Garolla A, Petre GC, Francini-Pesenti F, et al. Dietary supplements for male infertility: a critical evaluation of their composition. Nutrients. 2020;12(5):1472. https://doi.org/10.3390/nu12051472.
113. Garolla A, Petre GC, Francini-Pesenti F, et al. Systematic review and critical analysis on dietary supplements for male infertility: from a blend of ingredients to a rationale strategy. Front Endocrinol (Lausanne). 2022;12:824078. https://doi.org/10.3389/fendo.2021.824078.
114. Vitagliano A, Petre GC, Francini-Pesenti F, et al. Dietary supplements for female infertility: a critical review of their composition. Nutrients. 2021;13(10):3552. https://doi.org/10.3390/nu13103552.
115. Kitano Y, Hashimoto S, Matsumoto H, et al. Oral administration of L-carnitine improves the clinical outcome of fertility in patients with IVF treatment. Gynecol Endocrinol. 2018;34(8):684–8. https://doi.org/10.1080/09513590.2018.1431769.
116. Okubo H, Ando H, Ishizuka K, et al. Carnitine insufficiency is associated with fatigue during lenvatinib treatment in patients with hepatocellular carcinoma. PLoS One. 2020;15(3):e0229772. https://doi.org/10.1371/journal.pone.0229772.
117. Malaguarnera M, Catania VE, Malaguarnera M. Carnitine derivatives beyond fatigue: an update. Curr Opin Gastroenterol. 2023;39(2):125–8. https://doi.org/10.1097/MOG.0000000000000906.
118. Kępka A, Ochocińska A, Chojnowska S, et al. Potential role of L-carnitine in autism spectrum disorder. J Clin Med. 2021;10(6):1202. https://doi.org/10.3390/jcm10061202.
119. Szymański M, Wandtke T, Wasilow K, et al. Comparison of 3- and 6-month outcomes of combined oral L-carnitine fumarate and acetyl-L-carnitine therapy, included in an antioxidant formulation, in patients with idiopathic infertility. Am J Mens Health. 2021;15(5):15579883211036790. https://doi.org/10.1177/15579883211036790.
120. Pandey C, Maunder A, Liu J, et al. The role of nutrient supplements in female infertility: an umbrella review and hierarchical evidence synthesis. Nutrients. 2025;17(1):57. https://doi.org/10.3390/nu17010057.
121. Miwa T, Hanai T, Morino K. Effect of L-carnitine supplementation on muscle cramps induced by stroke; a case report. Nutrition. 2020;71:110638. https://doi.org/10.1016/j.nut.2019.110638.
122. Ueno M, Miura S, Ohama R, et al. Disorders of consciousness after subacute stroke might partly be caused by carnitine deficiency: two case reports. Prog Reahabil Med. 2023;8:20230019. https://doi.org/10.2490/prm.20230019.

Guidelines for the Diagnosis and Treatment of Carnitine Deficiency

18

18.1 Introduction

Regarding the diagnosis and treatment of carnitine deficiency, the *Guideline for the Diagnosis and Treatment of Carnitine Deficiency 2016* was published, led by members of the Nutrition Committee of the Japan Pediatric Medical and Health Council (Four-party cooperation organization having regular meetings to exchange information and discuss various issues related to children's health. The four members are: Japan Pediatric Society, The Japan Council of Pediatric Surgery-related Societies, The Japan Pediatric Association, and The Japanese Society of Child Health). In November 2016, *Guideline for the Diagnosis and Treatment of Carnitine Deficiency 2016* was published on the website of the Japan Pediatric Society. In February 2018, the diagnostic tests using the enzyme cycling method was covered by insurance, with the push forward by this guideline. Then after adding new members from Japanese Society for Severe Mental and Physical Disorders, The Japan Society of Hepatology, and The Japanese Society for Dialysis Therapy, the guideline of 2016 was revised after soliciting public comments with a view to facilitate the diagnosis and treatment of carnitine deficiency in a wide range of age groups, including not only children but also the elderly, and in November 2018, the *Guideline for the Diagnosis and Treatment of Carnitine Deficiency 2018* was published. on the website of Japan Pediatric Society [1].

The *Guideline for the Diagnosis and Treatment of Carnitine Deficiency 2018* (hereinafter referred to as the *Guideline*) consists of two parts: "Summary" and "Main Text." The summary part is intended to help clinicians, nurses, pharmacists, clinical engineers, and nutritionists involved in daily clinical practice to understand about carnitine deficiency in a short time. In the "Summary," "I. Diagnosis Guideline" describes the clinical symptoms and clinical signs of carnitine deficiency and how to diagnose it using a blood carnitine test using an enzyme cycling method. In addition, "II. Treatment Guideline" summarizes in an easy-to-understand manner "Outline of treatment of carnitine deficiency," "Carnitine replacement therapy," "Matters to be considered when carnitine replacement therapy," and "Dosage and use of standard carnitine replacement therapy." Next, the beginning of the "Main Text" briefly summarizes the basics necessary to understand carnitine deficiency, such as "What Is carnitine?", "Physiological effects of carnitine," and "Homeostasis of carnitine," as described in the early parts of this book. In addition, there is a detailed explanation of the diagnosis of carnitine deficiency and the classification of carnitine deficiency, and a more detailed explanation of the pathogenesis mechanism and precautions for the treatment of carnitine deficiency in each cause and disease fields. In addition, case reports and clinical studies of carnitine deficiency in various causes and diseases that have actually developed in Japan are summarized

N. Nakayama, *Carnitine Deficiency and Insufficiency in Medicine*,
https://doi.org/10.1007/978-981-95-7797-2_18

in a table. Finally, the 375 important domestic and international research papers, review articles, and case reports cited in this *Guideline* are listed as "References." Therefore, in order to understand carnitine and carnitine deficiency, and to diagnose and treat carnitine deficiency, it is best to download and read this *Guideline* first, although it is described in Japanese. So I would like to introduce the outline of the *Guideline* and supplement the background of its formulation that is not described in the *Guideline* and explain points and questions that are difficult to understand in the description of the *Guideline*.

18.2 Necessity of Diagnostic Reagents for Carnitine Deficiency and History of Diagnostic Reagent Insurance Coverage in Japan

The background of why diagnostic reagents for carnitine deficiency were needed is supplemented: On March 10, 2011, the indication for L-Cartin® tablets (Otsuka Pharmaceutical Co., Ltd) in Japan was changed to "carnitine deficiency" by public-notice application. Until then, carnitine preparations could only be used for carnitine deficiency due to propionic acidemia and methylmalonic acidemia, but as a result of the expansion of the indication, now it can be used for carnitine deficiency due to primary carnitine deficiency (carnitine transporter disorder) and many other inherited metabolic diseases, and for other carnitine deficiency of dialysis patients, patients taking valproic acid, and various causes in Japan. However, the new package inserts for L-Cartin® tablets states that "At the time of administration, the deficiency status of carnitine will be ascertained. Patients receiving carnitine preparations should be tested for carnitine fractions on a regular basis." However, in 2011, the diagnosis of carnitine deficiency was not covered by insurance, and there were no diagnostic criteria led by the academic society on how to diagnose carnitine deficiency in clinical settings, and many problems remained in the administration of carnitine preparations in clinical settings. The Pharmaceuticals and Medical Devises Agency (PMDA) and the Ministry of Health, Labour and Welfare (MHLW) instructed Otsuka Pharmaceutical Co., Ltd. to lobby Kainos Co., Ltd., which sells reagents for research on blood carnitine 2 fractions testing using the enzyme cycling method, to be covered by insurance. In response to this, the two companies discussed and cooperated to conduct a clinical performance test, and an application was submitted by Kainos Co., Ltd., which was approved as an in vitro diagnostic reagent on February 26, 2014. In the case of in vitro diagnostics, unlike in the case of therapeutic drugs, an application for insurance coverage is made and reviewed after approval as an in vitro diagnostic drug. One of the reasons for the slow progress in terms of insurance coverage was that it would be difficult to approve insurance coverage without guideline on diagnosis and treatment of carnitine deficiency led by related academic societies. Under these circumstances, a member of the Nutrition Committee of the Japan Pediatric Medical and Health Council (Chairman Dr. Shinobu Ida) prepared diagnostic and treatment guideline, and in November 2016, *Guideline for Diagnosis and Treatment of Carnitine Deficiency 2016* was published on the website of the Japan Pediatric Society. This encouraged us to discuss with the relevant departments of the Ministry of Health, Labour and Welfare for insurance coverage, and although there were many twists and turns, diagnostic reagents were covered by insurance from February 2018. It took 7 years from the change in the indication for "carnitine deficiency" of L-Cartin® tablets, and 4 years after the blood carnitine 2 fractions test reagent was approved as an in vitro diagnostic agent.

18.3 Measurement of Blood Carnitine Two Fractions Suitable for the Diagnosis of Carnitine Deficiency and Insufficiency

Next, I will supplement why the blood carnitine 2 fractions test using enzyme cycling method is adopted in the *Guideline for Diagnosis and Treatment of Carnitine Deficiency 2016*. As described in detail in Chap. 5, carnitine deficiency is an absolute deficiency of free carnitine (FC), which is essential for the transport of long-chain fatty acids into the mitochondria, and a condition in which acyl-CoA accumulates in the mitochondria and relatively free CoA decreases. There is a condition in which there is a relative deficiency of sufficient free carnitine to carry and remove the acyl group of acyl-CoA from the mitochondria as acylcarnitine (AC) (this is called carnitine insufficiency). These two pathologies together are called "carnitine deficiency," but there are cases where one or the other is the main or a mixture of both. In the latter case, carnitine insufficiency, acyl-CoA increases due to the accumulation of harmful acyl groups in the mitochondria, free CoA decreases, which is important for energy metabolism, and the acyl-CoA/free CoA ratio is abnormally high. At this time, if there is enough free carnitine (or if carnitine is administered), the acyl group of acyl-CoA binds to carnitine to form acylcarnitine, and it is carried out of the mitochondria to remove the toxicity of acyl-CoA (detoxification). Originally, measurement of acyl-CoA and free CoA in mitochondria is desirable, but quantification of these is not easy. Therefore, considering that acyl-CoA/free CoA in the mitochondria has an almost equilibrium relationship with acylcarnitine/free carnitine (AC/FC) in the blood over a long period of time, one can judge the state of acyl-CoA/free CoA in the mitochondria by measuring acylcarnitine and free carnitine in the blood. As detailed in Chap. 8, there are various methods for the quantification of carnitine and acylcarnitine, each with its own advantages and disadvantages. Among them, the blood carnitine 2 fractions test using the enzyme cycling method and the tandem mass method are described in detail and compared.

18.3.1 Enzyme Cycling Method (Blood Carnitine 2 Fractions Test)

In the enzyme cycling method (blood carnitine 2 fractions test), the total carnitine concentration (TC) and the free carnitine concentration are measured with each kit. Since "total carnitine = free carnitine + acylcarnitine," the acylcarnitine concentration can be calculated by subtracting the free carnitine concentration from the total carnitine concentration. As described in Chap. 3, the acyl group refers to the R-CO- portion of R-COOH, an organic acid or fatty acid. Acylcarnitine, whose acyl group is ester-bonded to carnitine, includes short-chain acylcarnitine such as acetylcarnitine (C2) and octanoyl carnitine (there are many molecular species such as medium-chain acylcarnitine such as C8) and long-chain acylcarnitine such as palmitoylcarnitine (C16), as well as ultra-long-chain acylcarnitine. In this enzymatic cycling method, all the acyl groups attached to carnitine by ester bonds are cleaved with acylcarnitine esterase purified from soil bacteria, and all acylcarnitines of various carbon chain lengths are collectively calculated as "acylcarnitine." Therefore, there is a weakness that the contents of the acyl group (the length of the carbon chain) are unknown. Therefore, the tandem mass method is suitable for the initial diagnosis of inherited metabolic diseases, but the simple enzyme cycling method is suitable for daily clinical diagnosis such as whether a deficiency of free carnitine has occurred at follow-up after a definitive diagnosis. In the diagnosis of carnitine deficiency by causes other than inherited metabolic diseases, the enzyme cycling method is also suitable for measuring free carnitine concentrations and acylcarnitine concentrations. This is because the tandem mass method can only be used in newborn mass screening for inherited metabolic diseases in Japan. Therefore, if the advantages and disadvantages of enzyme cycling are understood and used, it can be said that it is a suitable quantitative method for diagnosing carnitine deficiency in daily clinical practice. This blood carnitine 2 fractions test measured total carnitine and free carnitine using carnitine dehydrogenase and acylcarnitine esterase isolated from soil bacteria of the genus *Alkaligenes* by Takahashi (Toyo Brewing, later Asahi Kasei Industries, Ltd.) and others [2] (see "Sect. 8.3.2 Principle of Determination of Free and Acylcarnitine by Enzyme Cycling Method" for details). These diagnostic kits were commercialized as a research reagent kit that is easy to use in daily clinical practice with the cooperation of clinicians who are familiar with carnitine deficiency, such as Dr. Naruji Sugiyama (at that time the Department of Pediatrics, Nagoya City University) as a collaborator. They were released in 1993 by Kainos Co., Ltd. (Total Carnitine Kainos and Free Carnitine Kainos). As a result, it has become possible to quickly and easily perform a 2 fractions test for carnitine in blood and urine using a general-purpose automatic analyzer in a clinical laboratory of a hospital. Furthermore, in 1994, major private contract testing institutions such as began to undertake quantitative carnitine testing using this method, and it has been used in clinical settings for diagnosis of carnitine deficiency and clinical research. The enzyme cycling method measures the free carnitine concentration and the total carnitine concentration with each kit, and acylcarnitine can be calculated by subtracting the free carnitine concentration from the total carnitine concentration, so it can be said to be the best quantitative method for diagnosing carnitine deficiency and carnitine insufficiency. However, only a small number of diagnostic kits for this enzyme cycling method have been exported overseas, and there are very few reports from overseas using this method. I deeply hope that this enzyme cycling method, which was developed in Japan, will be introduced worldwide because it is very useful to diagnose carnitine deficiency and insufficiency in clinical practice of various medical fields explained in this book.

18.3.2 Tandem Mass Method

In the diagnosis of carnitine deficiency due to inherited metabolic diseases such as organic acidemia and fatty acid metabolism disorders, free carnitine (C0) or acylcarnitine (such as C2–C18) is measured. In organic acidemia or fatty acid metabolism disorders, genetic mutations occur in the enzymes of amino acid metabolism or fatty acid metabolism processes, resulting in a decrease in their activity and the accumulation of metabolic intermediates upstream of the metabolic process. Since the metabolic intermediate binds to carnitine to form acylcarnitine, and the specific acylcarnitine is high for each disease, the tandem mass method, which can quantify acylcarnitine of individual molecular species, is an indispensable analytical method for diagnosing the disease name of inherited metabolic diseases. For example, in propionic acidemia, which is one of the disorders of organic acid metabolism, the activity of propionyl-CoA carboxylase decreases due to genetic mutations in the metabolic process of amino acids such as valine, isoleucine, methionine, and threonine, and propionyl-CoA and its upstream metabolites accumulate in the body, resulting in metabolic acidosis, hyperammonemia, and hypoglycemic attacks. Since propionyl-CoA is detoxified by carnitine conjugation, propionylcarnitine and upstream metabolites accumulate in the blood and are excreted in large quantities in the urine. Therefore, in order to diagnose disease names, it is necessary to quantify acylcarnitine and organic acid metabolites specific to these diseases. Since April 2012, blood carnitine analysis using tandem mass has been covered by insurance as a test for inherited metabolic diseases, and it is calculated when diagnosing and following up the disease name of "patients with strongly suspected inherited metabolic diseases" (in the "Revision of Medical Fees for 2020", the new diagnostic items were added as : "D010. Special analysis 8. Inherited Metabolic Diseases test B. Urinary organic acid analysis B. Blood polar long-chain fatty acids C. Tandem mass analysis D. Others", and these diagnosis are calculated at 1141 points, respectively for insurance coverage (in the "Revision of Medical Fees for 2020", the new diagnostic items were added as : "D010. Special analysis 8. Inherited Metabolic Diseases test B. Urinary organic acid analysis B. Blood polar long-chain fatty acids C. Tandem mass analysis D. Others", and these diagnosis are calculated at 1141 points, respectively for insurance coverage). A definitive diagnosis of inherited metabolic diseases is made in conjunction with the results of the related enzyme activity measurements and genetic tests of the enzymes suspected of genetic mutation. In addition, a neonatal mass screening project for the early detection of inherited metabolic diseases has been developed and has become popular through the efforts of Japanese Society for Inherited Metabolic Diseases, and the test fee is paid by the public at about 40 tandem mass screening facilities nationwide in Japan. The purpose of this mass screening of newborns is to detect inherited metabolic diseases as soon as possible and to reduce disability through early treatment and prevention of onset. In principle, carnitine measurement methods using tandem masses can also be used for dialysis patients, cirrhosis patients, severely mentally and physically handicapped, and general carnitine deficiency such as carnitine deficiency due to administration of valproic acid or pivoxil-containing antibiotics. Recent studies on blood carnitine analysis using tandem mass methods have yielded new findings that the accumulation of long-chain acylcarnitine occurs in dialysis patients, cirrhosis patients, and heart failure patients, and that it is related to disease progression and life prognosis, and future development is expected (see related chapters). Unfortunately, however, as far as I know, there are no facilities in Japan that are contracted to measure acylcarnitine using tandem masses of general carnitine deficiency patients other than "patients suspected of having inherited metabolic diseases" on a commercial basis. Therefore, the only way to measure carnitine in patients with carnitine deficiency other than inherited metabolic diseases by the tandem mass method is to find a private contract research institution that is contracted to conduct pharmacokinetic studies and has tandem mass equipment, or a laboratory that has a mass spectrometer at a university medical school or pharmaceutical school and outsource the blood carnitine analysis. In addition, although the tandem mass method can identify individual acylcarnitine molecular species, it is not easy to measure the amount of acylcarnitine as a sum, so in order to quantify the amount of acylcarnitine, it is probably necessary to quantify the amount of acylcarnitines and then sum them up. In the future, it is strongly hoped that a wide range of private contract analysis organizations will emerge in Japan that perform the quantification of acylcarnitine by tandem mass methods, and that research on the relationship between acylcarnitines and pathophysiology will be developed in fields other than inherited metabolic diseases.

18.3.3 Comparison of Both Measurement Methods

Let us compare the tandem mass method and blood carnitine 2 fractions method using enzyme cycling method. The tandem mass method has the advantage of being able to identify and quantify the molecular species of each acylcarnitine. On the other hand, as a disadvantage it requires very expensive and specialized LC/MS/MS (liquid chromatography-tandem mass spectrometry) equipment, specialized personnel with instrumental analysis skills that are familiar with the handling of LC/MS/MS, and stable isotope internal standards used to determine the molecular species of acylcarnitine and amino acids, but these internal standards are not in vitro diagnostic reagents. Therefore, there are still issues such as quality control and standardization as in vitro diagnostic reagents. On the other hand, the blood carnitine 2 fractions test using the enzyme cycling method has been used as a research reagent since 1993. If two types of commercially available kits are purchased, the free carnitine concentration and total carnitine concentration can be quickly measured using a general-purpose biochemical automatic analyzer that is widely used in hospital clinical laboratories, and the acylcarnitine concentration can also be calculated by subtracting the free carnitine concentration from the total carnitine concentration. In 1994, major contract clinical testing institutions also began to undertake testing, and it has been used for daily clinical practice and clinical research. Although the molecular species of acylcarnitines cannot be determined, the relative deficiency state can be diagnosed by determining the decrease in free carnitine and AC/FC. Against this background, the *Guideline* adopted the blood carnitine 2 fractions test using the enzyme cycling method.

Table 18.1 Comparison of enzyme cycling method and tandem mass method for carnitine measurement in Japan

Method		Enzyme cycling method	Tandem mass method
Insurance classification in Japan		D007 Blood chemistry test (23)	D010 Special analysis 8. Tandem mass analysis
Insurance points		Total carnitine and free carnitine, 95 points, respectively	1141 points (2022)
Objective	Use	Aids in the diagnosis of carnitine deficiency, and monitoring of blood carnitine levels	Diagnosis of inherited metabolic diseases by blood carnitine analysis using tandem mass
	Indications	Patients with suspected carnitine deficiency	Patients with a strong suspicion of inherited metabolic diseases
	Measurement items	Total carnitine (TC) and free carnitine (FC) are measured, and acylcarnitine (AC) is obtained by subtracting free carnitine from total carnitine to diagnose FC deficiency and relative deficiency (carnitine insufficiency) from AC/FC.	By quantifying free carnitine and acylcarnitine with individual carbon chain numbers, it is possible to estimate and diagnose the disease name of inherited metabolic diseases from the amount of free carnitine and abnormally accumulated acylcarnitine molecular species
Equipment used	Measuring instruments	Clinical laboratory general equipment of biochemical automatic analyzers	High performance liquid chromatography-tandem mass spectrometry (LC/MS/MS)
	Measuring facilities	Hospital clinical laboratory/contract clinical laboratory (about 7000 machines)	Facilities that meet the facility standards for inherited metabolic diseases screening (approximately 40 facilities)
	Technicians needed	Technicians who are proficient in the handling of medical laboratory techniques and general-purpose biochemical analyzers	Technicians proficient in instrumental analysis using tandem mass analyzers are required
Reagents	Required reagents	Total carnitine: "T-Carnitine reagent Kainos" Free carnitine: "F-Carnitine reagent Kainos" Standards required for calibration curves are included in the kit	Organic solvents and, also for derivatization, butyl esterification reagents are required, and a stable isotope acylcarnitine kit to be used as an internal standard is required[a]
	Sample	Serum or plasma (urine can also be measured)	Serum or plasma, filter paper blood[b], urine, etc.
	Quality control and standardization	They are in vitro diagnostic drugs, and quality control has been validated	It is necessary to standardize operation methods and quality control at each facility. Internal standard reagents are not in vitro diagnostic drug
Usage results		Kits have been sold as research reagents since 1993. Since 1994, it has been used as a carnitine fraction test in daily clinical practice and clinical research. They were approved as in vitro diagnostic drugs for diagnosis of carnitine deficiency in 2014 and covered by health insurance in 2018	It is used for follow-up on the diagnosis and treatment of inherited metabolic diseases at pediatric departments in university hospitals and children's hospitals that diagnose inherited metabolic diseases, and has been covered by insurance since March 2012

(continued)

Table 18.1 (continued)

Method	Enzyme cycling method	Tandem mass method
Past and future issues	Until the publication of the *Guideline for the Diagnosis and Treatment of Carnitine Deficiency 2016* in November 2016, there were no diagnostic and treatment guidelines led by academic societies and were not covered by insurance, but from February 2018, they were covered by insurance. Strictly, there are still issues such as the fact that patients with a high risk of carnitine deficiency, such as adults with severe physical and mental disabilities, are not covered by insurance	This measurement method has yielded interesting findings, such as the accumulation of long-chain acylcarnitine in fields other than inherited metabolic diseases, such as patients with cirrhosis, heart failure, and dialysis, but there are few contract laboratories in Japan that perform tandem mass carnitine analysis

[a]Internal standard reagents kit is available from Cambridge Isotope Laboratories and PerkinElmer

[b]Newborn mass screening, which uses the tandem mass method to analyze carnitine in filter paper blood, is widespread throughout Japan (the cost of the test is paid by the public)

Table 18.1 summarizes the comparison of carnitine measurement by enzyme cycling method and tandem mass method.

Regarding carnitine deficiency (or suspected cases) due to inherited metabolic diseases, as specified in the *Guideline*, the "Clinical Practice Guideline for Diseases Subject to Mass Screening for Newborns" was edited by Japanese Society for Inherited Metabolic Diseases in 2019 [3]. Carnitine deficiency caused by other than inherited metabolic diseases is diagnosed and treated in accordance with the *Guideline*. Although the tandem mass method is suitable for diagnosing the disease name of inherited metabolic diseases, the enzyme cycling method is also useful for quick and simple measurement when it is necessary to urgently diagnose whether carnitine supplementation therapy is necessary immediately when a child suspected of inherited metabolic disease is transported to the metabolic emergency department due to an acute attack. In addition, carnitine 2 fractions test by enzyme cycling is also useful for monitoring children with inherited metabolic diseases whose disease name has been definitively diagnosed to determine whether the dosage of the carnitine preparation administered is appropriate as they grow (see Sect. 18.6.7).

18.4 Actual Measurement of Blood Carnitine with 2 Fractions Tests

If carnitine deficiency is suspected, a blood carnitine 2 fractions test is performed by the enzyme cycling method described above. There are two types of reagent kits for blood carnitine 2 fractions test, "T-Carnitine Reagent Kainos" and "F-Carnitine Reagent Kainos," and are sold by Kainos Co., Ltd. (http://www.kainos.co.jp/). The total carnitine concentration and free carnitine concentration are measured with each reagent, and the acylcarnitine concentration is calculated by subtracting the free carnitine concentration from the total carnitine concentration. It is possible to purchase two kits and measure them at your own facility using a general-purpose automatic biochemistry analyzer in a clinical laboratory. Since the number of samples that can be measured with one kit of reagents varies depending on the model of the general-purpose automatic biochemistry analyzer, it is necessary to check with the person in charge of Kainos Co., Ltd., including various setting parameters. Although it is possible to measure only the free carnitine concentration in order to save costs, it is desirable to also measure the total carnitine concentration, calculate the acylcarnitine concentration, and evaluate the accumulation of acylcarnitine by calculating AC/FC, which is an index of carnitine insufficiency. In addition, since a major clinical laboratory contract organization is entrusted with the measurement of samples as a "blood carnitine fractionation test," it is possible to collect samples (about 0.5 mL of serum or plasma, refrigerated) and outsource the testing in Japan. Most major contract clinical laboratories are entrusted with a set of two items, total carnitine and free carnitine, as a "carnitine fraction-serum test" or "carnitine-serum test," and report the free carnitine concentration, total carnitine concentration, and acylcarnitine concentration as a test result (the number of days required varies depending on the clinical testing organization but is usually 1–14 days). AC/FC, which is an indicator of carnitine insufficiency, needs to be calculated from the test results.

This blood carnitine 2 fractions test is very useful to diagnose carnitine deficiency and insufficiency; however, this method is basically used only in Japan. I really hope that this method spreads in the world in near future.

18.5 Insurance Coverage of Carnitine 2 Fractions Test Using Enzyme Cycling Method in Japan

The carnitine 2 fractions test using the enzyme cycling method has been covered by insurance since February 1, 2018. The insurance score is 95 points as "D007. Blood Chemistry Test (19) Free Carnitine and Total Carnitine," respectively, and "D007 Blood Chemistry Test (19) Free Carnitine and Total Carnitine" is 95 points each, and "D007 Blood Chemistry Test Biochemical Test (1) Judgment Fee (144 points)" can be claimed only once a month together with other test items (see January 31, 30 "Hoi-Hatsu-0131 No. 3"). However, certain conditions are stipulated for insurance coverage. The *Guideline* also describes matters to note when applying for insurance, and the subsequent revision of medical fees on March 5, 2002, has been summarized in Table 18.2. The order of the items has been changed from the description in the *Guideline*, but the content is basically the same.

Table 18.2 Precautions for insurance coverage of blood carnitine 2 fractions test using enzyme cycling in Japan

(17) Total carnitine and free carnitine
A. Described in section "23," total carnitine and free carnitine shall be covered when measured by the enzyme cycling method in compliance with the guidelines for medical treatment established by the relevant academic society
B. If this test is performed to assist in the diagnosis of inherited metabolic diseases or to observe the progress, it is calculated up to once a month
C. When this test is performed to assist in the diagnosis or follow-up of carnitine deficiency in patients with muscular dystrophy and amyotrophic lateral sclerosis, children who are receiving venous or enteral nutrition for a long time, pediatric patients who are using artificial milk or special treatment formula, patients receiving sodium valproate, patients with Fanconi syndrome, or patients on chronic maintenance dialysis, the coverage is up to 1 time per 6 months
D. When this test and the "8" inherited metabolic diseases test of the special analysis of Category No. D010 are performed together for the same sample, only the main test is covered

Quoted from p. 296 of the revision of medical fees on March 4, 2022 (Annex 1 of 0304 No. 1 from Hoi)

This insurance coverage seems to cover most of the disease fields listed in the "Target Patients" Sect. 18.6.1 of the *Guideline*. However, if this points described in Table 18.2 are strictly interpreted, children with severe mental and physical disabilities who are fed by tube feeding are covered, but adult patients are not covered by insurance. Elderly people who are managed by tube feeding are also not covered by insurance. In addition, it is unclear whether insurance covers for patients with cirrhosis and liver failure, with heart failure or cardiomyopathy, patients taking antibiotics containing a pivoxil group, and patients receiving anticancer drugs. These disaeases are also prone to carnitine deficiency. In cirrhosis of the liver, blood free carnitine levels are often high, and blood carnitine 2 fractions test is not necessarily useful, but in patients with other diseases who are strongly suspected of carnitine deficiency, it is desirable to measure blood carnitine and diagnose whether carnitine supplementation should be performed.

18.6 Overview of Diagnosis and Treatment of Carnitine Deficiency and Insufficiency

For details on the diagnosis of carnitine deficiency, please refer to the *Guideline*, [1] but the outline is summarized and supplemented below. (Note: sentence described in italic type are quoted from the *Guideline*.)

18.6.1 Target Patients

The following patients are prone to carnitine deficiency:

1. Patients who are nutritionally controlled by tube feeding, central venous feeding, or some milk allergen removal formula (reduced intake of carnitine): Many of these nutritional products are carnitine-free, or the carnitine content in the product is relatively low, which may be insufficient for some patients.
2. Patients taking valproic acid (epilepsy patients, psychiatric patients, migraine patients, dementia patients with convulsions and agitation symptoms, patients receiving valproic acid to prevent epileptic seizures after brain surgery, etc.): Infants, young children, patients with multiple antiepileptic drugs, and the elderly who have reduced their dietary intake of carnitine-rich foods should be aware of carnitine deficiency.
3. Patients who have been receiving antimicrobials containing a pivoxil group for a long time (otitis media, pharyngitis, etc.): Caution should be exercised in patients who are unavoidably administered for a long time. There have been cases of hypoglycemic attacks developing in a short period of time.
4. Patients undergoing peritoneal dialysis or hemodialysis due to renal failure, patients with Fanconi syndrome, and patients undergoing continuous renal replacement therapy (increased excretion of carnitine, interruption or decrease in reabsorption): Patients undergoing long-term dialysis therapy, muscular symptoms such as cramps, Erythropoiesis Stimulating Agents (ESA)-resistant anemia, decreased cardiac function or heart failure, and loss of muscle mass are likely to be carnitine deficiency.
5. Patients receiving anticancer drugs (especially platinum preparations, anthracycline preparations, alkylating agents, etc.): Carnitine deficiency is more common in patients with extreme fatigue, decreased appetite, and cachexia.
6. Patients with cirrhosis and liver failure: Blood free carnitine levels are often normal or high, but acylcarnitine levels are often high. Clinical symptoms such as hyperammonemia, hepatic encephalopathy, severe muscle cramps, and fatigue suggest that carnitine deficiency is suspected, and therapeutic diagnosis with carnitine replacement therapy is the most effective.
7. Patients who are suspected of having carnitine deficiency based on underlying diseases, conditions, muscle mass, prescription drugs, nutritional forms, and dietary content (children with severe mental and physical disabilities, patients with neuromuscular diseases such as muscular dystrophy and amyotrophic lateral sclerosis, patients with anorexia nervosa, bedridden elderly people, patients with severe diseases such as chronic heart failure, undernourished patients, etc.): Refer to the relevant chapter of this book.

18.6.2 Clinical Symptoms and Signs of Suspected Carnitine Deficiency

If there are any of the following clinical symptoms or signs that are suspected of carnitine deficiency, or if there is a diagnosis of carnitine deficiency in the past, it is often necessary to diagnose carnitine deficiency.

1. Somnolence, emotional blunting, impaired consciousness, coma, etc.
2. Convulsions, tremor, hypoglycemic attack
3. Hypotonia, muscle weakness
4. Severe muscle cramps, severe malaise
5. Rhabdomyolysis
6. Encephalopathy, vomiting induced by hunger or infection, frequent vomiting
7. Delayed psychomotor development
8. Poor weight gain
9. Abnormal breathing (e.g., tachypnea)
10. Cardiac hypertrophy, cardiomyopathy, decreased cardiac function, and sudden death (or family history)
11. Repetitive Reye-like syndrome

18.6.3 General Clinical Laboratory Findings Suggesting Carnitine Deficiency

The following laboratory findings are suspected of carnitine deficiency.

1. Hypoketotic hypoglycemia
2. Metabolic acidosis
3. Hyperammonemia
4. Abnormal liver function
5. Fatty liver
6. Abnormal blood gas analysis (pH, HCO_3^-, BE)
7. Electrolyte abnormalities (Na, K, Ca, Cl)
8. Treatment-resistant anemia

Of course, since these clinical symptoms, clinical signs, and general clinical laboratory findings are often caused by causes other than carnitine deficiency, it is necessary to examine whether there are any other definite causes, and if they are denied, it is necessary to consider the diagnosis of carnitine deficiency.

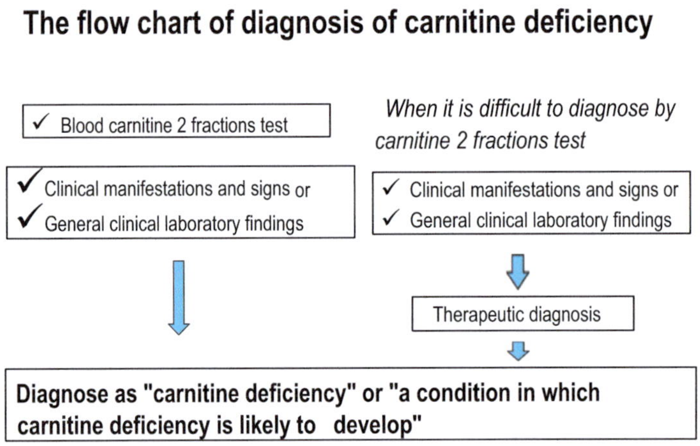

Fig. 18.1 The flow chart of diagnosis of carnitine deficiency. *Left*: In this case, it is possible to diagnose by blood carnitine 2 fractions test and clinical manifestation or signs. *Right*: When there are time constraints, or it is impossible to diagnose with blood carnitine 2 fractions test, administer carnitine preparations and judge by the responses (therapeutic diagnosis). (From Ref. [1])

18.6.4 Flow of Diagnosis of Carnitine Deficiency

Carnitine deficiency may be diagnosed based on the results of a blood carnitine 2 fractions test, clinical symptoms and clinical signs, and general clinical laboratory findings (left side of Fig. 18.1). In some cases, blood free carnitine concentration shows high value due to disease, or time constraints. Suspected carnitine deficiency may be diagnosed based on general clinical laboratory findings (Fig. 18.1, right side). This is the case of patients with liver cirrhosis or children with pediatric metabolic emergencies. In such cases, a "therapeutic diagnosis" must be made by administering a carnitine preparation and determining whether carnitine deficiency is present based on the response and improvement of clinical symptoms and other symptoms.

18.6.5 Reference Standard Value (Normal Value)

Next, I will describe how to diagnose carnitine deficiency from the free carnitine concentration and acylcarnitine concentration levels with the help of blood carnitine 2 fractions test.

There are several reports of blood carnitine levels in healthy individuals, but the values vary depending on age and gender [4–10]. In addition, the method of measuring the carnitine concentration used in these reports uses the radioenzyme method and the tandem mass method, which is different from the enzyme cycling method and requires careful consideration. In general, carnitine levels in the blood and tissues are higher in men than in women. This is thought to be due to the large amount of muscle mass in male, which is the storage site of carnitine. In addition, because the biosynthesis of carnitine is not sufficiently developed in infants, the value is low at a young age and reaches a plateau around the age of 20 years [8]. However, blood carnitine concentrations in healthy individuals are originally distributed over a wide range, and it is not practical to set a reference value that takes into account variations in blood carnitine concentrations due to differences in sex and age. Murakami et al. used the same enzymatic cycling kit and reported age-specific serum carnitine levels, but the number of cases in each age group was 10, which is too few to be adopted as a reference standard value for a diagnostic agent [11]. Under these circumstances, the enzyme cycling method (blood carnitine 2 fractions test) adopted in the *Guideline* has already been studied in healthy Japanese subjects (306 men and women in total), and the values have been used in many clinical studies as reference values for enzyme cycling method kits that were commercially available as research reagents [12] (Table 18.3). In 2014, when the kit was approved as an in vitro diagnostic reagent, it was considered appropriate to adopt this reference value by the Pharmaceuticals and Medical Devises Agency (PMDA). Therefore, this reference value is also adopted in the *Guideline*.

Table 18.3 Reference values for carnitine measurement by enzyme cycling method (blood carnitine 2 fractions test)

Concentration of carnitine in serum	
Total carnitine	45 ~ 91 μmol/L
Free carnitine	36 ~ 74 μmol/L
Acylcarnitine	6 ~ 23 μmol/L

Created using Ref. [12]

These reference values are set rather broadly and do not take into account gender and age differences. Furthermore, there are reports of diurnal variation in blood carnitine levels [13, 14], but this diurnal variation is also considered to be included in the wide range of reference values.

The normal value of AC/FC, which is an indicator of carnitine insufficiency, is said to be ≤0.2 or ≤0.25 based on clinical studies in various disease fields, and if it exceeds 0.4, it should be diagnosed as abnormal [15–21].

Although there is a wide range of normal reference values for blood carnitine concentrations, a significant decrease in free carnitine levels of less than 20 μmol/L or high levels of AC/FC above 0.4 should be diagnosed as likely to lead to the development of life-threatening carnitine deficiency. Conversely, fluctuations in blood carnitine concentrations of around

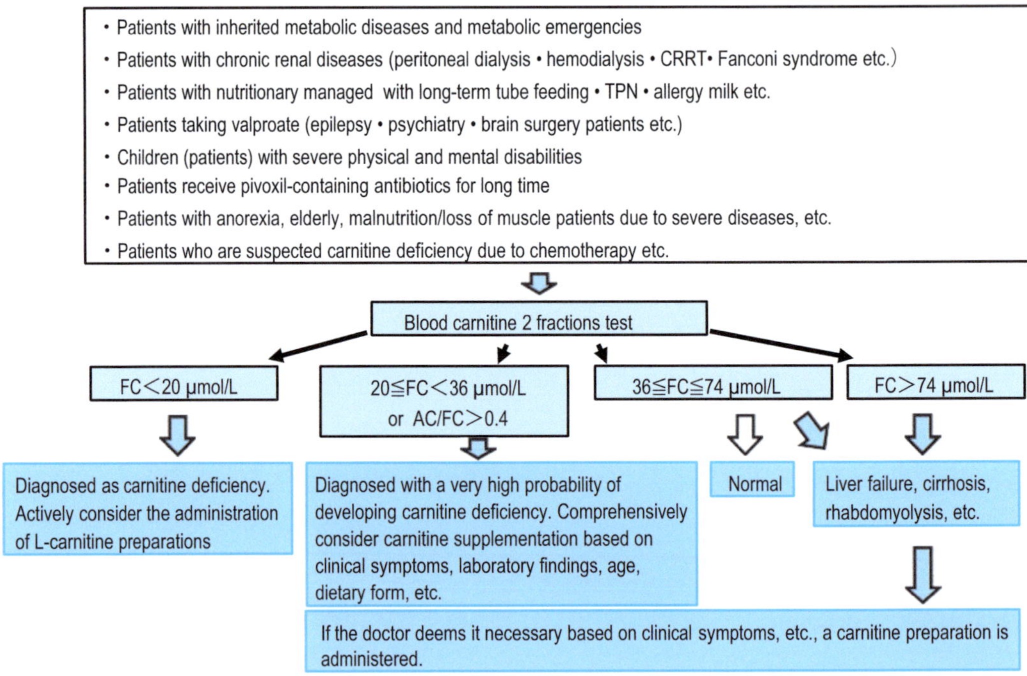

Fig. 18.2 Flow chart of diagnosis and treatment of carnitine deficiency by blood carnitine 2 fractions test. *AC* acylcarnitine, *CRRT* continuous renal replacement therapy, *FC* free carnitine, *TPN* total parenteral nutrition. (From Ref. [1])

10–20% are considered to be particularly unproblematic.

18.6.6 The Process from Diagnosis to Treatment of Carnitine Deficiency

For patients with suspected carnitine deficiency, free carnitine concentration (FC), total carnitine concentration (TC), and calculated acylcarnitine (AC) concentration are measured using the blood carnitine 2 fractions test, and the diagnosis is made as follows in conjunction with clinical symptoms, clinical signs, and general laboratory findings. If the diagnosis is confirmed, a carnitine preparation is administered (Fig. 18.2).

The diagnosis and treatment are described below.

1. When the free carnitine concentration is <20 μmol/L, it is judged that "a serious carnitine deficiency may develop at any time" based on many previous case reports of carnitine deficiency due to inherited metabolic diseases. The administration of the drug should be actively considered. This is because even if the patient is currently symptomatic or asymptomatic, there is a very high possibility that he or she will suddenly develop serious carnitine deficiency such as hypoglycemic attacks and hyperammonemia when the energy demand increases due to hunger, starvation, fever, infection, vomiting, and strenuous exercise.
2. If the free carnitine concentration is $20 \leq FC < 36$ μmol/L or $AC/FC > 0.4$ (normal is $0.25 \leq AC/FC$), the possibility of developing carnitine deficiency is extremely high, and supplementation with carnitine preparations should be considered in consideration of comprehensive clinical symptoms, clinical signs, general clinical laboratory findings, age, dietary form, etc. As in the case of a free carnitine concentration of <20 μmol/L, it is considered that even if the patient is currently symptomatic or asymptomatic, there is a possibility that serious carnitine deficiency may develop when energy demand increases such

as hunger, starvation, fever, infection, vomiting, and strenuous exercise. In the case of AC/FC > 0.4, acyl-CoA may accumulate in the mitochondria and carnitine insufficiency may occur, and hypoglycemic attacks and hyperammonemia may occur due to abnormalities in mitochondrial energy metabolism and urea cycles. Carnitine supplementation should be considered, especially in younger children and those who are being fed by tube feeding or intravenous feeding without carnitine.

3. When the free carnitine concentration is $36 \leq FC \leq 74$ μmol/L, the diagnosis is considered normal. However, in patients with cirrhosis, free carnitine leaks into the blood due to hepatocellular damage, so the free carnitine concentration may appear to be normal, but carnitine deficiency may cause hyperammonemia, extreme fatigue, and severe cramps (Fig. 18.2, diagonal arrow in the lower right).
4. When the free carnitine concentration is >74 μmol/L: Most of the carnitine is distributed in tissues such as muscles, and the pool in the blood is said to be about 0.6% (see Chap. 4). In patients with hepatic insufficiency, cirrhosis, rhabdomyolysis, and severe hepatic impairment such as anorexia nervosa, carnitine often leaks into the blood from tissue cells due to cell damage, resulting in high levels. It has been reported that chronic heart failure and cardiomyopathy also have high levels of free carnitine in the blood. Therefore, in the case of these diseases, it is not possible to diagnose from the free carnitine concentration alone, but if the AC/FC > is 0.4, acyl-CoA may accumulate in the mitochondria and carnitine insufficiency may occur. If the physician deems it necessary based on clinical symptoms, a carnitine preparation should be administered. In addition, carnitine palmitoyltransferase 1 (CPT1) abnormality, an inherited metabolic disease, also causes high levels of free carnitine in the blood (in this case, carnitine is not taken into the mitochondria even if it is administered, so carnitine preparations are not administered).
5. Carnitine replacement therapy should be used based on a comprehensive assessment according to the patient's condition and nutritional form, such as administration of levocarnitine preparations, switching to enteral nutrition containing carnitine, or a natural liquid diet (mixer diet). In addition, if it is judged that treatment is necessary, it is necessary to actively use carnitine preparations, and if the risk of developing carnitine deficiency is high, it is necessary to take care not to fall into a carnitine deficiency state by administering carnitine preparations or using a natural liquid diet (mixer diet) during the maintenance period.
6. If other causes cannot be ruled out based on clinical symptoms, clinical signs, and general clinical laboratory findings, and there is no time for blood carnitine 2 fractions test, a carnitine preparation may be administered without waiting for a blood carnitine 2 fractions test and a "therapeutic diagnosis" may be made to observe the response (for example, at the onset of acute encephalopathy or hepatic encephalopathy).

18.6.7 Monitoring After Carnitine Supplementation

The *Guideline* describes monitoring after carnitine supplementation as follows: "*Administration of carnitine preparations increases apparent blood carnitine levels, although there are differences depending on the disease. However, the administration of carnitine preparations is only a supplementation therapy, and for patients at risk of carnitine deficiency, continuous supplementation according to the deficiency state is necessary in consideration of clinical symptoms, laboratory findings, feeding status, age, physique, etc. In addition, it is desirable to discontinue the administration of carnitine preparations with caution based on a comprehensive assessment of the aforementioned monitoring items and the risk of recurrence of carnitine deficiency. At present, there are limited reports on monitoring of carnitine administration, and it is a subject for future consideration.*"

The implications of this statement and whether blood carnitine 2 fractions test can be used to monitor carnitine deficiency are described below.

The pool size of blood carnitine is only about 0.6% of the total, and even with oral administration with an absorption rate of about 20%, the concentration of carnitine in the blood increases significantly in a few hours after administration, and then it is absorbed into tissues or excreted in the urine by the kidneys and thus it slowly decreases (see Chaps. 4 and 8). Free carnitine is reabsorbed by more than 90% in the renal tubules, but when the blood concentration is considerably high, this reabsorption mechanism does not work and is excreted in the urine. This is because there is a threshold for the carnitine transporter (OCTN2) present in the renal tubules. Furthermore, since all the drugs enter the blood when carnitine is administered intravenously, the concentration of carnitine in the blood rises immediately after administration and decreases rapidly. This is also absorbed by tissues or excreted by the kidneys into the urine. Therefore, after administration of carnitine preparations, blood carnitine 2 fractions test cannot be used to monitor whether or not a person has carnitine deficiency because they are affected by the administered carnitine preparations.

18.6.7.1 Useful for Monitoring the Chronic Phase of Inherited Metabolic Diseases

In the case of inherited metabolic diseases, a blood carnitine 2 fractions test may be useful as a monitoring tool for adjusting the dose of carnitine preparations according to development. In the "Clinical Practice Guideline for Diseases Subject to Neonatal Mass Screening 2019" (edited by Japanese Society for Inherited Metabolic Diseases) published in 2019, (1) treatment of cases in which the disease has not developed after a definitive diagnosis in neonatal mass screening, (2) acute phase treatment, and (3) chronic stage management are summarized, and the management of the chronic stage is described as follows for each disease:

1. Control the concentration of free carnitine in serum or filter paper blood so that it does not fall below 20 μmol/L: Glutaric acidemia type 2, ultra-long-chain acyl-CoA dehydrogenase deficiency, trifunctional protein deficiency, medium-chain acyl-CoA dehydrogenase deficiency, OCTN2 disorders (systemic carnitine deficiency), CPT2 deficiency, CACT deficiency
2. Targeting a free carnitine concentration of 50 μmol/L in serum or filter paper blood: Urea cycle disorders
3. Keep the concentration of free carnitine in serum or filter paper blood at 50 μmol/L or more: Methylmalonic acidemia, propionic acidemia, isovaleric acidemia, methylcrotonylglycinuria, multiple carboxylase deficiency
4. Maintain the free carnitine concentration of serum or filter paper blood at 60–100 μmol/L: Glutaric acidemia type 1

The tandem mass method can also be used to measure the carnitine of serum or filter paper blood, but it takes time to obtain results due to outsourcing, unless it can be measured immediately at the facility. Blood carnitine 2 fractions test using the enzyme cycling method can be quickly diagnosed at the facility, and the results are available within a few days even if it is outsourced to an external clinical laboratory contract organization, and the measurement is covered by insurance at a frequency of once a month. In carnitine deficiency due to inherited metabolic diseases, the acyl group of acyl-CoA, an accumulated metabolic intermediate, is excreted in large quantities as acylcarnitine, or abnormalities in carnitine transporter (OCTN2) cause a large amount of free carnitine to be excreted in the urine, so even if carnitine preparations are administered, the excretion may not catch up and the blood concentration may fall below the normal limit. As described in the "Clinical Practice Guideline for Diseases Subject to Mass Screening of Newborns 2019," it is important to regularly monitor and adjust the dosage so that the concentration of free carnitine does not fall below the target value for each disease.

18.6.7.2 In the Case of Monitoring of General Carnitine Deficiency

Kumagai et al. studied effectiveness of monitoring FC in hemodialysis patients using the blood carnitine 2 fractions test [22]. Total 145 patients were divided into 3 groups. Group 1 consisted of patients ($n = 30$) who had been receiving supplementation of carnitine before this study and then discontinued at the beginning. Group 2 ($n = 13$) and Group 3 ($n = 102$) were categorized based on their baseline FC level, <20 or ≥20 μmol/L. Group 2 was started on supplementation, and Group 1 and Group 3 were observed without any intervention for the first 6 months. FC was measured every 6 months in all 3 groups up to 18 months. All patients in whom FC was less than 20 μmol/L at 6 and 12 months were supplemented carnitine. After 6 months, the FC level decreased from 262.5 ± 87.5 to 70.8 ± 33.6 μmol/L in Group 1 and increased in Group 2 from 17.4 ± 1.9 to 193.9 ± 43.3 μmol/L, with slight but significant decrease in Group 3. Thus, FC monitoring using the blood carnitine 2 fractions test is useful for appropriate carnitine supplementation in dialysis patients.

Unlike the case of monitoring blood carnitine levels in the management of chronic stages of inherited metabolic diseases, in dialysis patients, patients receiving valproate, and patients receiving nutritional management by tube feeding, a drastic decrease in carnitine concentration that exceeds the increase in blood carnitine concentration due to administered carnitine preparations does not usually occur, rather a gradual decrease in carnitine concentration takes place. Therefore, in most cases, the concentration of carnitine in the blood is apparently extremely high due to the administration of carnitine preparations, and it cannot be used to monitor carnitine deficiency. At present, as described in the *Guideline*, reports on monitoring at the time of carnitine administration are limited, and it must be said that it is a subject for future consideration. In the case of carnitine deficiency in children with severe mental and physical disabilities, the blood carnitine 2 fractions test is of course useful for diagnosing carnitine deficiency before administering carnitine preparations, but even if blood is collected and the blood concentration is measured while carnitine preparations are administered every day, the increase in blood concentration due to administration will have a significant effect. It is not possible to determine whether the tissue is deficient or not from the blood level.

18.6.7.3 Proposal for a New Monitoring Method

After administration of carnitine preparations, it is not possible to judge whether or not the patient is deficient due to the apparent increase in blood levels due to the effects of the supplementation of carnitine. In order to avoid this, the author believes that the best monitoring method is to stop carnitine supplementation for about 1 week and then measure the blood level. Considering the pharmacokinetics of carnitine, the half-life of orally administered carnitine is about 11 hours, and the effect of the final administration is negligible after 7 days after administration, during which time it is absorbed by the tissue or excreted by the kidneys, and after a few days, it is considered to have an equilibrium relationship between the tissue and the blood pool. Since the last carnitine preparation is in a stable state after it is absorbed by tissues or excreted in the urine, the author believes that it is possible to diagnose (i.e., monitor) whether the carnitine replacement therapy was effective or insufficient by measuring the carnitine concentration in the blood at that time. The problem is that the drug should be taken off for about 1 week. In diseases prone to severe carnitine deficiency, such as patients with inherited metabolic diseases, it may be dangerous to stop carnitine preparation for 1 week. However, in the case of carnitine deficiency such as children with severe physical and mental disabilities and chronic maintenance dialysis patients, it is extremely unlikely that the symptoms will worsen or become serious immediately after discontinuing carnitine supplementation for 1 week. For example, in a child with severe mental and physical disabilities who is tube fed, if a carnitine preparation is administered—but it is necessary to adjust the dosage or discontinue the carnitine supplementation—it may be possible to determine by stopping carnitine administration for 1 week and measuring the blood concentration.

In patients on chronic maintenance dialysis with end-stage renal disease, L-Cartin® tablets were administered orally after the indication was expanded in March 2011, but an intravenous injection was approved in December 2012, and carnitine preparations are now mostly administered intravenously in Japan (currently pre-filled intravenous injection syringe preparations). Immediately after its launch, the relationship between carnitine deficiency and clinical symptoms such as muscular symptoms (cramps, fatigue, etc.), anemia symptoms such as ESA-resistant anemia, cardiac function decline such as left ventricular ejection fraction, and cardiac index was studied by measuring the concentration of carnitine in the blood at many dialysis facilities, and many presentations were made at the annual meeting of Japan Society of Dialysis Therapy, etc. Summarizing the contents of these numerous conference reports, they were reported that the pre-dialysis blood carnitine concentration after 12 weeks was around 120 μmol/L with oral administration of 600 mg of carnitine preparation every day, and the pre-dialysis blood concentration was around 270 μmol/L after 12 weeks after intravenous administration of 1000 mg 3 times a week after dialysis was completed, which is considerably higher than the normal concentration, so most of the presentations concluded that carnitine supplementation should be discontinued or the dosage decreased. In other words, there were many discussions in the conference presentations of the Japan Society of Dialysis Therapy, such as at what point the medication should be discontinued or reduced with the apparently high concentration of carnitine immediately after administration, or whether carnitine replacement therapy should be continued if clinical symptoms do not improve. However, since there is no established monitoring method for patients who are actually taking carnitine supplementation except for carnitine deficiency due to the aforementioned inherited metabolic diseases, it is not possible to draw conclusions from the blood concentration measurements immediately after administration of carnitine presentations. The author believes that carnitine supplementation therapy can be stopped for about 1 week even in the case of intravenous administration, and blood carnitine 2 fractions test before dialysis can be performed to determine whether or not to resume carnitine preparation based on free carnitine concentration and AC/FC. In the case of dialysis patients, AC/FC is not significantly reduced by carnitine administration for various reasons, such as the difficulty of removing acylcarnitine from the dialysis membrane. However, since it is thought that the effects of the medication can be eliminated by stopping the drug for 1 week, it is considered that it is possible to judge whether the dose of carnitine preparations should be reduced, continued, or discontinued for a while by observing whether the blood concentration of free carnitine measured after 1 week of withdrawal is within the normal range of the reference value. As for dose reduction, I think that a simple method of dose reduction is to change the dosage number of times of carnitine from three times a week at the end of dialysis to twice a week or once a week. Of course, if medication is discontinued, as long as dialysis therapy, which is the cause of carnitine deficiency, continues, there is a high possibility that carnitine deficiency will reappear in 1–3 years, depending on the diet, and it will be necessary to consider resuming treatment (Fig. 18.3) [23].

Blood carnitine 2 fractions test can also be used to diagnose the risk of carnitine deficiency by using tube feeding products that do not contain carnitine for children (patients) with severe physical and mental disabilities. If you are using enteral nutritional preparations containing carnitine and still want to check the risk of carnitine deficiency, or if you want to confirm whether such enteral nutritional preparations are enough to prevent carnitine deficiency, there is currently no established monitoring method. However, as mentioned above, if the risk of developing hypoglycemic attack or hyperammonemia is low by discontinuation of carnitine administration for about 1 week, enteral nutritional preparations containing carnitine and mixer meals should be

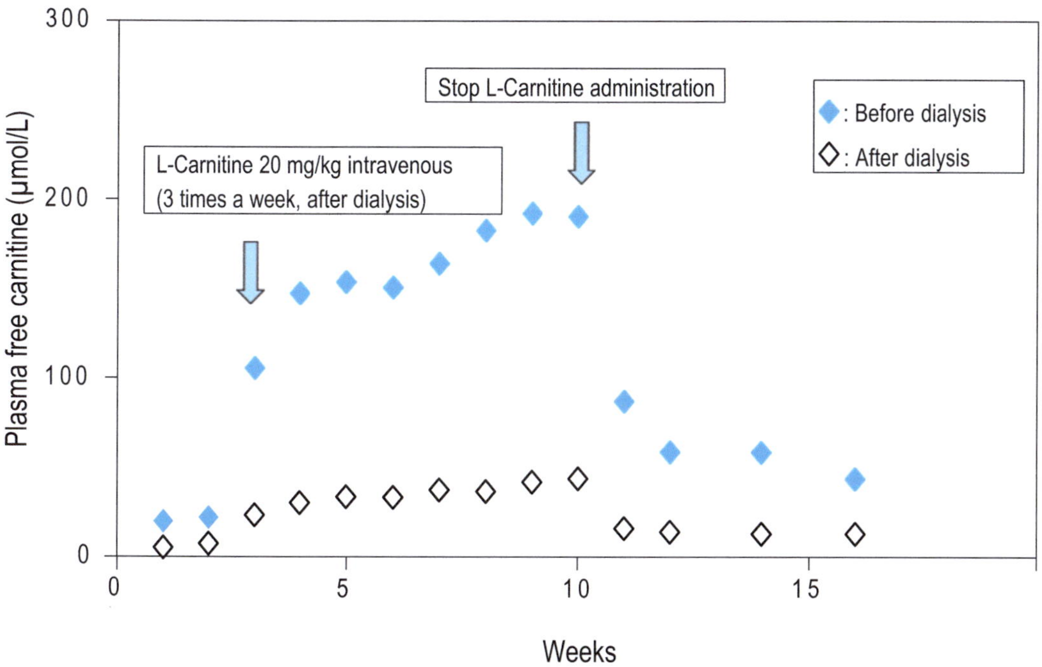

Fig. 18.3 Time-course of plasma free carnitine levels in chronic hemodialysis patients (6 female, 6 male, total 12 patients) when L-carnitine was administered intravenously (20 mg/kg, 3 times a week), after each dialysis session (weeks 3–10) and for 6 weeks after cessation of L-carnitine administration (in 6 of 12 patients only). (From Ref. [23])

discontinued for 1 week, and blood carnitine 2 fractions test should be performed to diagnose the carnitine status.

When hypoglycemic attacks occur caused by antibiotics containing a pivoxil group, the antibiotic containing the pivoxil group is absorbed and cleaved by intestinal esterase and pivalic acid is released, which is ester bound to free carnitine and excreted in the urine as pivaloylcarnitine. First of all, the administration of these antibiotics should be stopped immediately, and carnitine preparations should be administered. If antibiotic administration cannot be discontinued for various reason, it is important to switch to antibiotics that do not contain a pivoxil group. Switching from one pivoxil-containing antibiotic to another pivoxil-containing antibiotic has made no sense at all, and cases of recurrence of severe hypoglycemic attack have been reported (e.g., cefditoren ⇒ cefcapene pivoxil; see Chap. 13, Table 13.2). There is currently no established method for monitoring carnitine deficiency caused by antibiotics containing the pivoxil group. In principle, the number of moles of pivalic acid in the antibiotic is carnitine conjugated, and the concentration of free carnitine in the blood is greatly reduced even in healthy adults [24]. The half-life of a pivoxil-containing antibiotic is about 1 h, and if the administration of the pivoxil-containing antibiotic is discontinued and the pivalic acid remaining in the blood disappears, the cause of carnitine deficiency disappears. Carnitine should be continued to children who have suffered a hypoglycemic attack until the patient has recovered sufficiently, and there is no need for carnitine administration or monitoring. In addition, it is extremely dangerous to administer a pivoxil-containing antibiotic again to a child who has had a hypoglycemic attack after being administered a pivoxil-containing antibiotic, and even if antibiotic administration is required due to recurrence of an infectious disease, the patient should switch to an antibiotic that does not contain a pivoxil group.

Thus, at present, there are no sufficient reports on the dosage adjustment of carnitine preparations for patients with carnitine deficiency and the monitoring of carnitine preparations for the decision of dose reduction or discontinuation using blood carnitine 2 fractions test. Even in patients with general carnitine deficiency other than inherited metabolic diseases, it is possible to eliminate the apparent increase in carnitine concentration due to the medication itself by stopping the drug for about 1 week and measuring the blood level, and it may be used to judge whether carnitine supplementation therapy is still necessary, or whether the dose can be reduced or discontinued. I would like to see this study in various fields in the future. However, it goes without saying that this monitoring method should be limited to patients who are not considered to be at risk of developing serious carnitine deficiency during a 1-week withdrawal of carnitine preparations (or discontinuation of a mixer diet).

18.6.8 Standard Usage and Dosage in Carnitine Supplementation Therapy

The following is a summary of the usage and dosage of carnitine supplementation therapy described in the *Guideline*.

1. *Liver Cirrhosis*

   Although it is desirable to measure blood carnitine levels when initiating carnitine supplementation therapy for cirrhosis, it cannot be ruled out that patients with cirrhosis may be in a state of relative carnitine deficiency even when the carnitine levels are in the normal range or at high levels. So, administration should be initiated based on a comprehensive assessment of clinical symptoms and signs, ammonia levels, and complications of sarcopenia.

   - Hepatic encephalopathy: 750–1500 mg/day (L-Cartin® FF) orally, divided into 2 or 3 times (adult dose). In particular, in the case of hyperammonemia, 1500 mg/day should be administered over a period of 3 months. Although there are case reports of intravenous administration of carnitine for hepatic encephalopathy, the dosage and administration method have not been established, and future verification including efficacy is necessary.
   - Muscle cramps: 750–1000 mg/day, divided into 3 or 4 (added before bedtime) times. It was reported that the rate of symptom disappearance was higher in the 1000 mg group than in the 750 mg group, and that the addition of before bedtime administration improved the symptoms [25].
2. *Carnitine Deficiency in Patients with Chronic Renal Failure*
   - Hemodialysis: 10–20 mg/kg intravenously, 3 times a week, after completion of dialysis
   - Peritoneal dialysis: Start oral administration from 10 to 20 mg/kg/day and adjust the dose according to the data
   - Fanconi syndrome: 30–120 mg/kg/day, orally
3. *Supplementation to Carnitine-Free TPN (Total Parenteral Nutrition) or Enteral Nutrition*
   - Supplementation with TPN (neonates, nutritional management after surgery, etc.): 2–5 mg/kg/day (invariably use intravenous injectable formulation)
   - Supplementation to enteral nutrition (e.g., nutritional management by tube feeding for the elderly): 5 mg/kg/day (basically use oral solution)
4. *Carnitine Deficiency Associated with Neuromuscular Diseases and Psychiatric disorders*
   - 3–10 mg/kg/day as medication and 1–4 mg/kg/day as dietary intake [26]
   - In cases of severe mitochondrial dysfunction such as acute encephalopathy: 30–100 mg/kg/day, which is equivalent for metabolic diseases, is considered to be necessary
   - In addition, there are reports of efficacy of 100 mg/kg/day (up to 4 g) in Attention-Deficit/Hyperactivity Disorder (ADHD) [27]

   In any case, the dosage should be divided into two to three doses per day, and due to the aforementioned limitations of the absorption through the transporter, it is desirable that the maximum dose at one dose does not exceed 2 g.

5. *Drug-Induced Carnitine Deficiency Due to Sodium Valproate*
   - Prevention and treatment of side effects of valproic acid
   - Prophylactic administration (e.g., children without risk factors): 10–20 mg/kg/day or 100–300 mg/day; 250–750 mg/day for adults
   - Valproic acid-induced hepatotoxicity/valproic acid-induced hyperammonemic encephalopathy (treatment in the absence of impaired consciousness): 50–100 mg/kg/day orally (every 6 hours, up to a maximum of 3 g per day; increase or decrease appropriately depending on the severity)
   - In the case of impaired consciousness (when oral administration cannot be performed): Initial loading was 100 mg/kg intravenously (up to a maximum of 6 g), followed by an additional 15 mg/kg every 4–6 hours; when the ammonia level decreases, switch to oral (cases of valproic acid poisoning are also subject to this, and the dosage is increased or decreased as appropriate)
6. *Hypoglycemic Attacks Caused by Administration of Antibiotics Containing Pivoxil Group*

   There are no guidelines, but the following are proposed from the case reports so far.
   - If it can be administered orally: 40–60 mg/kg/day of carnitine preparations (duration of administration depends on the duration of antibiotic administration but should be administered at least until clinical symptoms and serum free carnitine levels become normal)
   - In the case of impaired consciousness, or when it cannot be administered orally, or when seriously ill and need emergency treatments: The initial dose is 100 mg/kg (up to a maximum of 6 g) intravenously, followed by an additional 15 mg/kg every 4 h; if it can be administered orally and the symptoms stabilize, it is the same as "when oral administration is possible"
   - There are no reports on prophylactic carnitine supplementation in the case of long-term administration of these antibiotics; perform a blood carnitine 2 fractions test and a blood glucose test regularly and pay attention to clinical symptoms and signs
7. *Carnitine Supplementation in Children with Severe Physical and Mental Disabilities*
   - As a nutritional supplement: 2–10 mg/kg/day when used for supplementation, and 20–50 mg/kg/day when used as a carnitine preparation for treatment (it is necessary to select the formulation according to the patient's situation)
   - However, it is desirable to make a comprehensive judgment on the dosage of carnitine preparations, taking into account the nutritional status of the affected child, the state of carnitine deficiency, and the use of antiepileptic drugs and antibiotics; in addition, blood carnitine 2 fractions tests should be performed regularly, and clinical symptoms and signs should be noted

18.6.9 Considerations for Carnitine Supplementation Therapy

In the "Summary" and "Main Text" of the *Guidelines*, it is described that when initiating carnitine supplementation therapy, the following should be considered in determining the method, dosage, route, and duration of administration of carnitine preparations. This is a summary of lessons learned and points of caution based on actual case reports in the diagnosis and treatment of carnitine deficiency. It is reproduced below and supplemented.

1. **Continuity of the Underlying Disease or Cause of Carnitine Deficiency**

   If the cause continues, such as inherited metabolic diseases or dialysis treatment, it is necessary to continue carnitine supplementa-

tion therapy in principle. However, in some cases, the dosage can be reduced after adequate supplementation. If carnitine deficiency is treated/prevented with drugs for a certain period of time, such as pivoxil-containing antibiotics, the administration of carnitine preparations is limited to the onset of symptoms or when the onset of symptoms is concerned.

As an example, when intravenous carnitine (20 mg/kg) is administered three times a week after dialysis to chronic maintenance dialysis patients (6 men and 6 women in total, 12 cases), the concentration of free carnitine in the blood before dialysis is high, around 160–200 μmol/L, as shown in Fig. 18.3, but when carnitine administration is discontinued after week 10, the blood free carnitine levels drop sharply during 11–16 weeks [23]. Therefore, if the treatment is discontinued as it is, it is thought that carnitine deficiency will eventually occur.

2. **Acute and Chronic Phases**

In the treatment of carnitine deficiency, the treatment for acute phase and chronic phase may differ (choice of dosage and dosage form).

In acute phase, time-sensitive treatments are required, and oral administration is often not possible, and intravenous injectables and nasogastric tube administration are performed. In addition, if the symptoms are severe and there is no time to spare, high-dose administration of carnitine preparations should be considered because of the high safety of the preparation. Of course, it is also essential to search for and treat pathologies other than carnitine deficiency, including symptoms of hypoglycemic attacks, hyperammonemia, and metabolic acidosis. There is an easy-to-understand description of these diagnoses, tests, and treatment methods in pediatric metabolic emergencies [28]. This book describes in detail the tests and treatments that must be performed first in metabolic emergencies, such as hypoglycemia, ketosis, hyperammonemia. For example, even in the case of a hypoglycemic attack, not only blood glucose levels but also insulin, ketone body, and free fatty acid levels must be tested to determine the cause of the hypoglycemic attack and how to deal with it. In addition, carnitine preparations are also described as the "Medicines to be kept in the pediatric department of the main hospital."

Carnitine deficiency often manifests (develops) when the demand for energy increases rapidly, such as during infection, starvation, and strenuous exercise, and patients with diseases or conditions that are prone to carnitine deficiency or those who are receiving medication should be careful on a daily basis. In the chronic phase when the condition is stable, oral administration of low doses is often sufficient, but since the absorption rate of carnitine from the gastrointestinal tract is low, it is necessary to consider administering it in several divided doses rather than administering it in large quantities at once. It is also necessary to adjust the dosage according to the growth of the child.

3. **Severity of Carnitine Deficiency**

Depending on the severity of carnitine deficiency, the dosage of levocarnitine preparations is different. Even with the same disease name, the dosage of carnitine preparations varies depending on the severity of the individual patient. In severe cases, a high dose is administered, and a dose reduction is considered after recovery of clinical symptoms.

Even if the cause of carnitine deficiency is the same, the severity of the disease varies depending on the duration of the disease, complications, and concomitant medications. For example, hyperammonemia, encephalopathy, and severe hepatotoxicity in patients taking valproate are more common in young people, in patients with multiple antiepileptic drugs, and in patients with a history of neurological disorders and metabolic diseases [29, 30].

4. **Composite Factors of Individual Patient**

Even if the underlying disease is the same, the dosage of levocarnitine differs depending on age, sex, duration of illness, amount of food, nutritional status, muscle mass, concomitant medications, complications, etc. *For example, in the case of primary carnitine deficiency due to a genetic mutation in the carnitine transporter (OCTN2), the dosage varies greatly depending on the severity of the consequence of the mutation.*

In primary carnitine deficiency, the most severe form of carnitine deficiency caused by inherited metabolic diseases, a genetic abnormality in the carnitine transporter gene (*SLC22A5*) causes extremely low levels of free carnitine in blood and tissues, resulting in hypoglycemic attacks, cardiomyopathy, and sudden death. As many as 138 genetic variants were reported in 2016 (among which 97 missense mutations, a total of 29 insertion of immature stop codons as a result of frameshifts (15) or direct mutations (14), 7 mutations predicted to cause abnormal splicing, and 2 single amino acid in-frame deletion, and 3 large gene deletion), and the clinical phenotype was thought to be determined by the effect of the mutation on transporter activity [31]. However, two homozygous patients with the same mutation (R399Q) in the same family have been reported to have completely different clinical phenotypes [32]. Thus, the actual clinical picture is not simple, as the phenotype (e.g., severity) may differ even for the same genetic mutation. Therefore, the dosage in the chronic phase of primary carnitine deficiency can vary as much as fourfold, from 100 to 400 mg/kg/day.

In the field of psychiatry, there have been reports of cases in which adult hospitalized patients with schizoaffective disorder and bipolar affective disorder who had been receiving valproic acid as a mood stabilizer became vegetarian after the death of their mother, and subsequently developed hyperammonemia, which is thought to be carnitine deficiency due to valproic acid [33]. It is important to note that serious carnitine deficiency often develops when there is an increased demand for energy due to dietary patterns, dietary content, concomitant medications, hunger, starvation, fever, infection, vomiting, or strenuous exercise.

5. **Urgency of Treatment**

   In emergency situations, higher doses or, in some cases, intravenous formulations should be administered. When the patient becomes stable, switch to a lower dose or to an oral formulation.

   Intravenous administration is often used in acute cases of inherited metabolic diseases, but intravenous administration or high-dose administration should be considered in cases of hyperammonemic encephalopathy or severe hepatotoxicity induced by valproic acid, valproic acid poisoning, and severe hepatic encephalopathy.

6. **Selection of Optimal Formulations**

   Carnitine preparations include tablets, oral solutions, and intravenous preparations, and the most suitable dosage form is selected according to age and pathological condition.

   The following is the case of L-Cartin® FF, a free form preparation of levocarnitine from Otsuka Pharmaceutical Co., Ltd.

   - Tablets—L-Cartin® FF Tablets 100 mg: Used for infants and children and frequently for dividing dosing
   - L-Cartin® FF Tablets 250 mg: Adolescents, adults
   - Internal liquid—L-Cartin® FF Oral Solution 10% (100 mL/bottle; 10 mL contains 1000 mg levocarnitine): Suitable for tube feeding or for elderly patients who have difficulty using tablets
   - L-Cartin® FF Oral Solution 10% in 5 mL packets (aluminum stick packets, 5 mL × 10 packets): For dividing dosing for children or the elderly who have difficulty taking tablets; no need for metered dosing; convenient for carrying
   - L-Cartin® FF Oral Solution 10% (10 mL aluminum stick packets, 10 mL × 10 packets): For divided administration in children or elderly people who have difficulty tak-

ing tablets; no need for measuring the volume; easy to carry

- Intravenous formulation—L-Cartin® FF 1000 mg syringe for intravenous infusion (pre-filled syringe containing 1000 mg levocarnitine in 5 mL): For acute cases of inherited metabolic diseases, metabolic emergencies, hepatic encephalopathy, valproic acid poisoning, other cases where patients are unconscious and cannot take oral formulations, hemodialysis patients, etc.

Currently, several generic levocarnitine are commercially available in Japan (the author will omit the detailed description in this book of English version).

Note: Levocarnitine Chloride and Levocarnitine in Free Form L-Cartin® tablets have been a levocarnitine chloride preparation since its launch from Earth Pharmaceutical Co., Ltd. in 1990 (it was succeeded by Otsuka Pharmaceutical Co., Ltd. in 1992). On March 10, 2011, the indication was changed to "carnitine deficiency" by public notice application. At this time, Japanese Society for Inherited Metabolic Diseases and others requested the development of intravenous formulations for use in the acute phase of inherited metabolic diseases and dialysis patients, as well as oral solutions for children who have difficulty taking them orally. In December 2012, free form of internal liquid (L-Cartin® FF oral solution 10%, 10 mL small bottle) and free form intravenous formulation (L-Cartin® FF intravenous solution 1000 mg, 1000 mg/5 mL, in glass ampoule) were approved. In order to supply intravenous preparations and oral solutions to the medical field as soon as possible in response to requests from academic societies, the author introduced a free form levocarnitine preparation from Sigma-Tau (then; now Alphasigma) in Italy, which is widely used for carnitine deficiency worldwide. For this reason, for a short period of time, the tablets were in a twisted state with levocarnitine chloride and oral solutions and intravenous preparations with levocarnitine free forms. Therefore, soon a free form tablet was developed and approved in June 2014, and all L-Cartin® preparations of Otsuka Pharmaceutical Co., Ltd. became free form levocarnitine. In addition, the release of the internal solution has made it possible to eliminate the time and effort required to dissolve film-coated tablets and administer them through a tube. However, the initial oral solution was a glass "brown small bottle" containing 10 mL, and it was troublesome and inconvenient to open and dispose of a large number of small vials when the dose was adjusted according to the patient at a facility for the severely mentally and physically handicapped and when administered to the tube feeding. Therefore, Otsuka have developed a large-capacity bottled oral solution (100 mL/bottle, containing 1000 mg of levocarnitine in 10 mL) that is easy to dispense at facilities for the severely mentally and physically disabled patients, and a 5 mL and 10 mL aluminum stick packaging formulation that is easy for children to take at home or school, does not require a measuring of the volume, and is easy to dispose of in containers, which was approved in September 2017. Subsequently, Otsuka Pharmaceutical Co., Ltd.'s levocarnitine chloride tablets were discontinued. Therefore, Otsuka Pharmaceutical Co., Ltd.'s levocarnitine preparations are all free form of tablets, oral solutions, and intravenous preparations. Subsequently, levocarnitine chloride tablets (both 100 mg and 300 mg tablets) were released from the various generic pharmaceutical companies, but now these generic companies have also launched free form carnitine preparations (the author will omit the detailed description in this book of English version). There is no difference in efficacy between levocarnitine chloride and levocarnitine free form, except for the difference in molecular weight. However, intravenous formulations of levocarnitine chloride have a risk of causing hyperchloremic acidosis, especially in cases of emergency acute phase with metabolic acidosis, so free form are considered to be suitable for intravenous preparations

7. **Multiple Nutrient Deficiencies**

   In addition to carnitine, other nutrients, trace elements, and vitamins are also likely to be *deficient in various pathologies that cause carnitine deficiency (selenium deficiency, zinc deficiency, iodine deficiency in severely mentally and physically handicapped children who are fed through tube feeding, iron deficiency, zinc deficiency and vitamin D_3 deficiency, in dialysis patients, and biotin deficiency in some milk allergen-eliminated formula patients). Although it is difficult to evaluate and establish evidence of supplementation therapy of these multiple deficient nutrients for improvement of clinical symptoms, it is necessary to actively diagnose and treat such multiple deficiencies, including those of other factors.*

In many cases, patients with severe mental and physical disabilities, patients who have undergone surgical procedures such as short bowel syndrome, and patients with severe illnesses such as sepsis, who are managed with tube feeding or TPN products that do not contain carnitine, may cause carnitine deficiency, resulting in decreasing cardiac dysfunction or hypoglycemic attacks. In such cases, not only carnitine deficiency but also selenium deficiency, zinc deficiency, iodine deficiency, and vitamin deficiencies such as thiamine may occur, and close examination is necessary [34–38]. In addition to carnitine deficiency, selenium, zinc, and copper deficiencies are also frequently observed in dialysis patients and continuous renal replacement therapy patients due to the removal of trace elements from the dialysis membrane, and may be associated with heart failure, impaired cardiac function, ESA-resistant anemia, and life prognosis [39–44]. Carnitine exerts antioxidant and anti-inflammatory effects by enhancing the expression of antioxidant enzymes such as Superoxide Dismutase (SOD), catalase, and glutathione peroxidase through the activation of Nrf2, but selenium and zinc are also essential trace elements for the activity of glutathione peroxidase, etc. and their deficiency might be associated with increased oxidative stress in dialysis patients and others. So, their deficiency may contribute to increased oxidative stress, inflammation, cardiovascular disease, decreased cardiac function, and cardiomyopathy in patients with chronic renal failure such as dialysis patients [45].

In patients with liver cirrhosis, not only carnitine deficiency but also zinc deficiency is very common, especially in patients with hepatic encephalopathy and hyperammonemia, and zinc supplementation therapy may be effective [46, 47]. Carnitine is essential for the synthesis of cofactor of carbamoyl phosphate synthase 1 (CPS1), the first enzyme in the urea cycle, but zinc is also required for the active center of the enzyme ornithine transcarbamylase (OTC), which is important in the urea cycle.

Thus, in the pathologies where carnitine deficiency occurs, other micronutrients and trace elements might be often deficient, and attention should be paid to deficiencies of other trace elements during carnitine supplementation therapy.

Patients with intractable neurological diseases are also prone to zinc, selenium, and carnitine deficiencies. Muguruma et al. reported that serum concentrations of zinc, selenium, and carnitine were examined in 24 patients with intractable neurological diseases who had been tube fed for a long period of time, and that they showed decreases in 70.8%, 58.3%, and 87.5%, respectively, and recommended regular measurement of trace elements and carnitine [48].

8. **Supplement Food with Carnitine or Natural Liquid Diet (Mixer Food)**

   Many enteral nutrition as medicines do not contain carnitine. If carnitine deficiency is caused by nutritional methods, supplementation with carnitine-containing foods or naturally liquid diets (mixer food) should be considered.

When carnitine supplementation therapy is given to children with severe mental and physical disabilities who are nutritionally managed by tube feeding, in addition to the administration of a pharmaceutical carnitine preparation as described here, (1) switch to an enteral nutritional preparation of a pharmaceutical containing carnitine, (2) switch to a nutritional supplement containing carnitine treated as a food instead of a medicine, and (3) administer a mixer food containing carnitine by mixing a meal containing carnitine (e.g., meats) as a mixer liquid meal by blender. Enteral nutritional preparations of pharmaceuticals containing carnitine include Enevo® (10.7 mg/100 kcal) and Enoras® (16.7 mg/100 kcal) as semi-digestible nutrients (carnitine content per 100 kcal is shown in parentheses). There are a large number of carnitine-containing nutrient foods, such as Isocal Junior® (20 mg/100 kcal) (see Table 11.2 in Chap. 11). In addition, mixer food is very useful because they can be used to administer meats containing a large amount of carnitine, but it requires a lot of effort to cook for each individual patient in a facility with many patients. Even if tube feeding preparations are used for regular nutritional management, it is desirable to use a mixer food even once a week.

9. **Age Group of Onsets**
*Carnitine deficiency is more likely to occur in children with low biosynthesis and low muscle mass, which is the site of storage, but it is by no means unique to children. In addition, if the cause of carnitine deficiency continues, it can develop in a wide range of age groups from children to adults (*e.g., *tube feeding in people with inherited metabolic diseases and severe physical and mental disabilities, peritoneal dialysis and hemodialysis have a lifelong risk of developing it, and valproic acid is also widely used in adult patients).*

Carnitine deficiency has been found to be more common in infants and children. However, as we become a super-aging society, it is thought that carnitine deficiency among the elderly will increase in the future. Tanaka et al. reported that a 77-year-old elderly female patient who was bedridden due to muscle weakness and rhabdomyolysis, developed severe hyperlipidemia when she was fed by tube feeding with Ensure® liquid, and her hyperlipidemia improved in a short period of time when she switched to a normal diet [49]. The total serum carnitine concentration at the onset of hyperlipidemia was 22 μmol/L (normal value 45–91 μmol/L), which was considered hyperlipidemia due to carnitine deficiency. Thus, carnitine deficiency may develop in elderly patients with reduced skeletal muscle mass, which is the storage site of carnitine, if factors that cause carnitine deficiency, such as tube feeding without carnitine, are added. In addition, attention should be paid to carnitine deficiency, such as the occurrence of fatty liver due to TPN in the nutritional management of pediatric patients after surgery [50]. Sarcopenia and locomotive syndrome in the elderly are increasing, and the relationship between life prognosis and maintenance of physical function is attracting attention, especially in patients with cirrhosis and dialysis [51–53]. Further research on the association between sarcopenia and carnitine deficiency in these diseases is expected.

Thus, carnitine deficiency can occur in any age group, and it is difficult to understand that the blood carnitine 2 fractions test is not covered by insurance in patients with severe physical and mental disabilities and the elderly who are nutritionally managed by tube feeding. Carnitine deficiency and carnitine insufficiency are also a problem in patients with heart failure. The author believes that the current restrictions on age and disease area should be abolished, and that diagnostic drugs should be covered by insurance for all patients who are considered to need a diagnosis of carnitine deficiency.

References

1. Ida S, Iwasaki Y, Uchida K, et al. Guideline for diagnosis and treatment of carnitine deficiency 2018, https://www.jpeds.or.jp/uploads/files/20181207_shishin.pdf. Article in Japanese.
2. Takahashi M, Ueda S, Misaki H, et al. Carnitine determination by an enzymatic cycling method with carnitine dehydrogenase. Clin Chem. 1994;40(5):817–21.
3. Japanese Society for Inherited Metabolic Diseases, editor. Clinical practice guideline for diseases subject to mass screening of newborns 2019. Tokyo:

SHINDAN TO CHIRYO SHA Inc; 2019. Book in Japanese.
4. Mitchell ME. Carnitine metabolism in human subjects II. Values of carnitine in biological fluids and tissues of "normal" subjects. Am J Clin Nutr. 1978;31(3):481–91. https://doi.org/10.1093/ajcn/31.3.481.
5. Lambert ME, Shipley K, Holbrook I, et al. Serum carnitine levels in normal individuals. JPEN J Parenter Enteral Nutr. 1988;12(2):143–6. https://doi.org/10.1177/0148607188012002143.
6. Schmidt-Sommerfeld E, Werner D, Penn D. Carnitine plasma concentrations in 353 metabolically healthy children. Eur J Pediatr. 1988;147(4):356–60. https://doi.org/10.1007/BF00496410.
7. Glöggler A, Bulla M, Puchstein C, et al. Plasma and muscle carnitine in healthy and hemodialyzed children. Child Nephrol Urol. 1988;9(5):277–82.
8. Takiyama N, Matsumoto K. Age- and sex-related differences of serum carnitine in a Japanese population. J Am Coll Nutr. 1998;17(1):71–4. https://doi.org/10.1080/07315724.1998.10720458.
9. Chace DH, Pons R, Chiriboga CA, et al. Neonatal blood carnitine concentrations: normative data by electrospray tandem mass spectrometry. Pediatr Res. 2003;53(5):823–9. https://doi.org/10.1203/01.PDR.0000059220.39578.3D.
10. Reuter SE, Evans AM, Chace DH, et al. Determination of the reference range of endogenous plasma carnitines in healthy adults. Ann Clin Biochem. 2008;45(Pt 6):585–92. https://doi.org/10.1258/acb.2008.008045.
11. Murakami K, Sugimoto T, Nishida N, et al. Examination of serum and urine carnitine levels using a carnitine assay kits. J Pediatr Pract. 1995;58(6):1083–6. Article in Japanese.
12. Tsushima K, Kase N. Methods for the determination of carnitine in human serum. Jpn J Clin Dial. 2000;16(2):167–73. Article in Japanese.
13. De Palo E, Gatti R, De Palo C, et al. Plasma carnitine and acetyl-carnitine levels at different times of the day. Clin Physiol Biochem. 1987;5(2):95–102.
14. Rebuzzi AG, Schiavoni G, Lanza GA, et al. Rhythm of carnitine levels in serum and urine of normal subjects. Drugs Exp Clin Res. 1989;15(1):43–6.
15. Winter SC, Zorn EM, Vance WH. Carnitine deficiency. Lancet. 1990;335(8695):981–2.
16. Campos Y, Huertas R, Lorenzo G, et al. Plasma carnitine insufficiency and effectiveness of L-carnitine therapy in patients with mitochondrial myopathy. Muscle Nerve. 1993;16(2):150–3. https://doi.org/10.1002/mus.880160205.
17. Böhles H, Evangeliou A, Bervoets K, et al. Carnitine esters in metabolic disease. Eur J Pediatr. 1994;153(7 Suppl 1):S57–61. https://doi.org/10.1007/BF02138779.
18. Carter AL, Abney TO, Lapp DF. Biosynthesis and metabolism of carnitine. J Child Neurol. 1995;10(Suppl 2):2S3–7.
19. Tein I. Role of carnitine and fatty acid oxidation and its defects in infantile epilepsy. J Child Neurol. 2002;17(Suppl 3):3S57–82, discussion 3S82–3.
20. Bellinghieri G, Santoro D, Calvani M, et al. Carnitine and hemodialysis. Am J Kidney Dis. 2003;41(3 Suppl 1):S116–22. https://doi.org/10.1053/ajkd.2003.50099.
21. Reuter SE, Evans AM. Carnitine and acylcarnitines. Clin Pharmacokinet. 2012;51(9):553–72. https://doi.org/10.1007/BF03261931.
22. Kumagai E, Hosohata K, Furumachi K, et al. Effectiveness of monitoring free carnitine levels for L-carnitine supplementation in hemodialysis patients to maintain carnitine sufficiency and nutritional factors. Ther Apher Dial. 2021;25(5):595–606. https://doi.org/10.1111/1744-9987.13621.
23. Evans AM, Faull R, Fornasini G, et al. Pharmacokinetics of L-carnitine in patients with end-stage renal disease undergoing long-term hemodialysis. Clin Pharmacol Ther. 2000;68(3):238–49. https://doi.org/10.1067/mcp.2000.108850.
24. Ito T, Sugiyama N, Kobayashi M, et al. Alteration of ammonia and carnitine levels in short-term treatment with pivalic acid-containing prodrug. Tohoku J Exp Med. 1995;175(1):43–53. https://doi.org/10.1620/tjem.175.43.
25. Nakanishi H, Kurosaki M, Tsuchiya K, et al. L-carnitine reduces muscle cramps in patients with cirrhosis. Clin Gastroenterol Hepatol. 2015;13(8):1540–3. https://doi.org/10.1016/j.cgh.2014.12.005.
26. Wakisaka A. Risk management for hypocarnitinemia in children with severe mental and physical disabilities: from the standpoint of pediatrics. J Severe Ment Phys Disabil. 2015;40(1):103–8. Article in Japanese.
27. Van Oudheusden LJ, Scholte HR. Efficacy of carnitine in the treatment of children with attention-deficit hyperactivity disorder. Prostaglandins Leukot Essent Fatty Acids. 2002;67(1):33–8. https://doi.org/10.1054/plef.2002.0378.
28. Igarashi T, Takayanagi M, editors. Pediatric clinical Pyxis 23: can't-missed inherited metabolic diseases. Tokyo: Nakayama Shoten; 2010. Book in Japanese.
29. Coulter DL. Carnitine deficiency in epilepsy: risk factors and treatment. J Child Neurol. 1995;10(Suppl 2):S32–9.
30. De Vivo DC, Bohan TP, Coulter DL, et al. L-carnitine supplementation in childhood epilepsy: current perspectives. Epilepsia. 1998;39(11):1216–25. https://doi.org/10.1111/j.1528-1157.1998.tb01315.x.
31. Longo N, Frigeni M, Pasquali M. Carnitine transport and fatty acid oxidation. Biochim Biophys Acta. 2016;1863(10):2422–35. https://doi.org/10.1016/j.bbamcr.2016.01.023.
32. Wang Y, Korman SH, Ye J, et al. Phenotype and genotype variation in primary carnitine deficiency. Genet Med. 2001;3(6):387–92. https://doi.org/10.1097/00125817-200111000-00002.

33. Carr RB, Shrewsbury K. Hyperammonemia due to valproic acid in the psychiatric setting. Am J Psychiatry. 2007;164(7):1020–7. https://doi.org/10.1176/ajp.2007.164.7.1020.
34. Terui K, Komatsu K, Shinozuka S, et al. Heart failure during total parenteral nutrition recovered by administration of carnitine and selenium. J Japanese Soc Pediatr Surg. 2011;47(7):1026–32. Article in Japanese.
35. Berger MM, Pantet O, Schneider A, et al. Micronutrient deficiencies in medical and surgical inpatients. J Clin Med. 2019;8(7):931. https://doi.org/10.3390/jcm8070931.
36. Kodama H, Asagiri K, Etani Y, et al. Treatment guideline for selenium deficiency 2018. Diagnosis and treatment of selenium deficiency. J Jpn Soc Clin Nutr. 2018;40(4):239–83. Article in Japanese.
37. Kodama H, Itakura H, Omori H; Japanese Society of Clinical Nutrition, et al., editors. Clinical guideline for zinc deficiency 2018. J Jpn Soc Clin Nutr. 2018;40(2):120–67. Article in Japanese.
38. Ozawa H. Nutritional management in children (and persons) with severe motor and intellectual disabilities: focusing on carnitine. J Severe Motor Intellect Disabil. 2012;37(1):101–6. Article in Japanese.
39. Tonelli M, Wiebe N, Hemmelgarn B, et al. Trace elements in hemodialysis patients: a systematic review and meta-analysis. BMC Med. 2009;7:25. https://doi.org/10.1186/1741-7015-7-25.
40. Berger MM, Shenkin A, Revelly JP, et al. Copper, selenium, zinc, and thiamine balances during continuous venovenous hemodiafiltration in critically ill patients. Am J Clin Nutr. 2004;80(2):410–6. https://doi.org/10.1093/ajcn/80.2.410.
41. Fujishima Y, Ohsawa M, Itai K, et al. Serum selenium levels are inversely associated with death risk among hemodialysis patients. Nephrol Dial Transplant. 2011;26(10):3331–8. https://doi.org/10.1093/ndt/gfq859.
42. Fukushima T, Horike H, Fujiki S, et al. Zinc deficiency anemia and effects of zinc therapy in maintenance hemodialysis patients. Ther Apher Dial. 2009;13(3):213–9. https://doi.org/10.1111/j.1744-9987.2009.00656.x.
43. Kalantar-Zadeh K, Kopple JD. Trace elements and vitamins in maintenance dialysis patients. Adv Ren Replace Ther. 2003;10(3):170–82. https://doi.org/10.1053/j.arrt.2003.09.002.
44. Kobayashi H, Abe M, Okada K, et al. Oral zinc supplementation reduces the erythropoietin responsiveness index in patients on hemodialysis. Nutrients. 2015;7(5):3783–95. https://doi.org/10.3390/nu7053783.
45. Podkowińska A, Formanowicz D. Chronic kidney disease as oxidative stress- and inflammatory-mediated cardiovascular disease. Antioxidants (Basel). 2020;9(8):752. https://doi.org/10.3390/antiox9080752.
46. Katayama K. Zinc replacement therapy for hepatic encephalopathy and hyperammonemia: significance of zinc in chronic liver disease. J Clin Exp Med. 2012;240(9):759–64. Article in Japanese.
47. Moriyama M, Matsumura H, Tanaka N, et al. Clinical significance of Zinc supplementation for patients with type C hepatitis related cirrhosis. Clin Gastroenterol. 2008;23(6):751–7. Article in Japanese.
48. Muguruma T, Omori H, Takeda I, et al. Nutritional evaluation in patients with intractable neurological diseases undergoing long-term tube feeding in point of trace elements and carnitine. J Jpn Soc Clin Nutr. 2015;37(1):36–43. Article in Japanese.
49. Tanaka S, Miki T, Hsieh ST, et al. A case of severe hyperlipidemia caused by long-term tube feedings. J Atheroscler Thromb. 2003;10(5):321–4. https://doi.org/10.5551/jat.10.321.
50. Tanaka Y, Yagi M. Pathogenesis of hepatobiliary dysfunction caused by TPN. J Clin Exp Med. 2006;218(5):409–12. Article in Japanese.
51. Hanai T, Shiraki M, Ohnishi S, et al. Rapid skeletal muscle wasting predicts worse survival in patients with liver cirrhosis. Hepatol Res. 2016;46(8):743–51. https://doi.org/10.1111/hepr.12616.
52. Ohara M, Ogawa K, Suda G, et al. L-carnitine suppresses loss of skeletal muscle mass in patients with liver cirrhosis. Hepatol Commun. 2018;2(8):906–18. https://doi.org/10.1002/hep4.1207.
53. Maruyama T, Maruyama N, Higuchi T, et al. Efficacy of L-carnitine supplementation for improving lean body mass and physical function in patients on hemodialysis: a randomized controlled trial. Eur J Clin Nutr. 2019;73(2):293–301. https://doi.org/10.1038/s41430-018-0348-y.

19 Clinical Pharmacological Issues Related to Carnitine

19.1 Introduction

From 2007 to 2017, I have been engaged in expanding the indications for L-Cartin® tablets, developing intravenous and oral internal liquid formulations, developing diagnostic agents for carnitine deficiency and their insurance coverage, and enlightening by lecturing about carnitine deficiency in various fields concerning clinicians and medical professionals. In the course of these activities, I have encountered and solved several clinical pharmacological issues. In addition, carnitine as a pharmaceutical product is often used in the medical treatment of inherited metabolic diseases, pediatric neurology, and severely mentally and physically handicapped patients, and I felt the difficulty of a drug development in these fields. I have experienced several issues that may be helpful for the future development of other drugs for other rare diseases. On the other hand, although carnitine is widely used in adult patients on dialysis, many doctors asked me why there is scarcely clear-cut data of double-blind comparative studies, and why the Food and Drug Administration (FDA) approved carnitine for dialysis patients, because they are skeptical about efficacy of carnitine. I have been struggling to convince many clinicians about importance of carnitine, a kind of mere simple nutrient, in the era of evidence-based medicine (EBM). So, I would like to summarize some of the points and some of the clinical pharmacological issues I have faced, learned, and noticed in my activities to date.

19.2 History of the Development of Carnitine as a Pharmaceutical Product

First, let us review briefly the history of the development of L-carnitine, one of the nutrients, as a pharmaceutical product (drug). Table 19.1 summarizes clinical studies, approvals status in the world and Japan, and remarks on carnitine preparations as a pharmaceutical product.

Overseas, in the 1950s and 1960s, DL-carnitine preparations were found to stimulate appetite and digestive enzyme secretion and were applied clinically. In Japan, there was no time lag in this, and many DL-carnitine chloride preparations were launched in the 1960s (see Chap. 6). Subsequently, the use of the D-body in DL-carnitine preparations as a drug or food, including importation, was banned in Europe and the United States because it was reported to cause myasthenia, and in Italy and other countries, the use of the DL-body was switched to the L-body. However, for unknown reasons, the use of carnitine preparations as an appetite enhancer was almost completely discontinued. Also in Japan, by around 2004, DL-carnitine products other than "Entomin® Injection" were discontinued

N. Nakayama, *Carnitine Deficiency and Insufficiency in Medicine*,
https://doi.org/10.1007/978-981-95-7797-2_19

Table 19.1 History of important clinical research on carnitine and its approvals in overseas (mainly in the United States) and in Japan

Year	Important clinical research and reporting	Overseas	Japan	Remarks
1950 ~ 1960s	Discovery of actions on appetite increase and digestive enzyme secretion promotion	Launched as an appetite stimulant and digestive enzyme secretion stimulant in Europe (DL-body)	Many products were released as appetite stimulants and digestive enzyme secretion stimulants (indications for anorexia and gastric secretion disorders, DL-body)	
1973	Reports of carnitine deficiency in muscle myopathy and CPT deficiency			
1975 ~ 1980	Reports of systemic carnitine deficiency			
1978	Report of carnitine deficiency in dialysis patients			
1980	Report of carnitine deficiency with valproic acid			
1983	Report on the concept of carnitine insufficiency in organic acidemia			
1984		Approval for the treatment of primary and secondary carnitine deficiency due to inherited metabolic diseases (FDA)		In 1981, FDA banned the use of D-bodies, then other countries by the mid-1980s. In Japan, DL-carnitine sales were gradually discontinued by around 2004, except "Entomin injection;" it is still available (DL-intravenous injection)
1989		Approval for the prevention and treatment of carnitine deficiency with valproic acid (FDA)		
1990			Approval for the treatment of carnitine deficiency due to methylmalonic acidemia and propionic acidemia	

(continued)

Table 19.1 (continued)

Year	Important clinical research and reporting	Overseas	Japan	Remarks
1993		Approval for the treatment of cardiomyopathy in children		
1997		Approval of indication for mitochondrial myopathy with zidovudine (FDA)		
1999		Approval for carnitine deficiency in dialysis patients (FDA)		
2002				Lifting the ban on use as food unless it claims to be effective (supplements)
2011/3/10			The indication for "L-Cartin" tablets was changed to carnitine deficiency (the dose was also changed) by a public knowledge application	It can now be administered to all patients with inherited metabolic diseases that cause carnitine deficiency, dialysis patients, valproate patients, tube feeding patients, etc.
2013			"L-Cartin FF" oral solution and intravenous agent released (free form levocarnitine preparation without chloride)	Conducting clinical trials because it was considered as new active ingredient drugs (only clinical pharmacological studies in healthy individuals were conducted)
2014			"L-Cartin FF" tablets released (free body)	L-carnitine chloride formulation was discontinued, and all formulations became free body
2016			*Guidelines for Diagnosis and Treatment of Carnitine Deficiency 2016* was published	
2017			L-Cartin FF Injection 1000 mg pre-filled syringe and oral internal liquid packets 5 and 10 mL were released	Changed to safe pre-filled syringe from glass ampoule; easy-carriable oral internal liquid aluminum sticks were released
2018			Blood carnitine 2 fractions test using enzyme cycling method was covered by insurance	The Diagnostic and Treatment Guidelines were revised in 2018

maybe due to declining demand and other factors.

In 1973, carnitine deficiency in muscle myopathy and carnitine palmitoyltransferase (CPT) deficiency were reported, and then carnitine deficiency and carnitine supplementation therapy for various inherited metabolic diseases were reported in the 1980s. In particular, in propionic acidemia and methylmalonic acidemia, the concept of carnitine deficiency and carnitine insufficiency was established, in which propionic acid, an intermediate metabolite, binds to free CoA and accumulates a large amount of propionyl-CoA, resulting in a relative decrease in free CoA required for energy metabolism and various metabolic processes. For these reasons, in 1984, the FDA approved the indication of L-carnitine for primary and secondary carnitine deficiencies due to genetic predisposition. In 1980, the usefulness of carnitine for hyperammonemia and hepatotoxicity caused by valproic acid was reported, and in 1989 the FDA approved the administration of L-carnitine for the prevention and treatment of carnitine deficiency due to valproic acid toxicity (hepatotoxicity and hyperammonemic encephalopathy). Carnitine deficiency due to dialysis was reported in 1978, and in 1999 the FDA approved carnitine supplementation therapy for carnitine deficiency in patients with end-stage renal disease requiring dialysis therapy (reimbursement by health insurance began in 2003). In addition, the FDA approved carnitine for cardiomyopathy in children in 1993 and for zidovudine-induced mitochondrial myopathy in 1997.

Thus, clinical studies on carnitine deficiency and the benefits of carnitine supplementation therapy for various causes have progressed in the world except Japan, and carnitine supplementation therapy has been approved for these conditions in turn.

In Japan, the clinical application of appetite stimulants and digestive enzyme secretion stimulants in the 1960s seems to have approved almost without time lag compared to overseas. This may be due to the relative loose and ease of obtaining drug approval at that time. However, in contrast, the subsequent approval for carnitine deficiency in inherited metabolic diseases in Japan was complicated. At that time, Earth Pharmaceutical Co., Ltd. was conducting clinical trials of L-Cartin® for exertional angina. However, at the strong request of the Japan Society for Inherited Metabolic Diseases, led by Dr. Naruji Sugiyama of the Department of Pediatrics at Nagoya City University (at that time), 19 institutions nationwide participated in clinical trials of 7 patients with propionic acidemia and 13 patients with methylmalonic acidemia in improving symptoms of carnitine deficiency in propionic acidemia and methylmalonic acidemia [1, 2]. At the time, there was no "public knowledge application" (or public knowledge-based application) system and clinical trials were required to obtain approval. Initially, development efforts were focused on carnitine deficiency due to propionic acidemia and methylmalonic acidemia, two rare inherited metabolic diseases for which there were relatively enough number of patients in Japan and the efficacy and safety of carnitine were almost well established. As a result of these efforts, the drug was approved in 1990 for the treatment of carnitine deficiency in methylmalonic acidemia and propionic acidemia. Dr. Sugiyama told me that it was a very difficult undertaking, taking nearly 10 years just to obtain indications of carnitine for these two diseases. Therefore, it was considered that focusing on two diseases of inherited metabolic diseases was the right development strategy considering the situation at the time. However, it was known that carnitine deficiency occurred in many other inherited metabolic diseases, dialysis patients, and patients receiving drugs such as valproic acid, etc. It was not possible for the company and clinicians to conduct several clinical trials for patients of these various diseases and medical conditions. In addition, Earth Pharmaceutical Co., Ltd. was aiming for clinical development for exertional angina pectoris, but the application was withdrawn after the authorities demanded to conduct clinical trials again using more appropriate control drug and additional basic research. Subsequently, there was a reorganization of the R&D system and field in Earth Pharmaceutical Co., Ltd., and L-Cartin® tablet was transferred and succeeded by Otsuka Pharmaceutical Co., Ltd, a same group company.

Otsuka Pharmaceutical had no one in the company who was familiar with carnitine or carnitine deficiency, or who could actively collect information on carnitine deficiency at that time. The only personnel involved were those in charge of safety to conduct post-marketing surveillance in preparation for reexamination, and production and sales for these two rare diseases. L-Cartin® tablet was not developed as an orphan drug, but the reexamination period was set at 10 years in accordance with orphan drugs.

As a result, the use of carnitine products for a wide range of carnitine deficiency disorders was left behind only in Japan compared to Europe and the United States, and L-carnitine products were frozen in a situation where carnitine deficiency disorders other than carnitine deficiency due to propionic acidemia and methylmalonic acidemia were not officially treated with L-carnitine products. The carnitine-related personnel of Earth Pharmaceutical Co., Ltd. remained in Earth Pharmaceutical Co., Ltd. to work in other fields or were transferred to the Ophthalmology and Dermatology Division of Otsuka Pharmaceutical Co. The expansion of the indication of L-Cartin® tablets for the treatment of carnitine deficiency in general was not taken into consideration by anyone, and L-Cartin® had a really "orphan" status in the company too.

19.3 Trends of Resolution of Drug-Lag and Expansion of the Indication for L-Carnitine in Japan

19.3.1 "2 Section Chief's Notification" and "Public Knowledge Application"

Not only carnitine, but also other drugs that are commonly used in the world are not approved or not indicated for use in Japan (drug-lag issue). In order to solve this issue, the Ministry of Health, Labour and Welfare (MHLW) issued "Handling of Ethical Drugs for Off-label Use" (Kenkyu No. 4, Pharmaceutical Affairs Council No. 104, so-called "2 section chief's notification") on February 1, 1999 [3, 4]. The main purpose of this notice is that if there is a request for development from the relevant academic societies, if it is recognized as medically necessary, and if the authorities request it, clinical trials should be conducted and an application for approval (or a partial change application) considered, and if there is sufficient evidence overseas, clinical trials should not be conducted by building on that information, or the application for approval should be submitted with the minimum necessary clinical trials. The so-called "public knowledge application" system was intended to eliminate the drug-lag by allowing approval without clinical trials or with the minimum number of clinical trials required if sufficient overseas evidence was available. If the drug has already been indicated for another disease, it is likely that the drug will be exempt from clinical trials unless racial differences are considered because the authorities know information on the manufacture, quality, safety, and pharmacokinetics of the drug. Clinical trials are essential for drugs being developed in Japan for the first time, but from the viewpoint of eliminating drug-lag at an early stage, it is necessary to conduct the minimum necessary clinical trials, and it is possible to fully discuss the necessary trials and protocols with the authorities in advance. It is necessary to construct a theory by collecting information on the approval status and efficacy and safety overseas and through descriptions in textbooks and guidelines, solid review articles, meta-analyses, clinical trial reports, case reports, etc., but there is a possibility that it is possible to avoid conducting multiple full clinical trials, and the hurdle has been greatly lowered.

After the issuance of the "2 section chief's notification," Otsuka Pharmaceutical Co., Ltd. was approached by the authorities regarding the expansion of the indication of L-Cartin® tablets. However, at that time, internal officials of Otsuka Pharmaceutical thought that it would be difficult to conduct clinical trials because only 1 year remained in the reexamination period, that it would be difficult to set up patients and evaluation indices for clinical trials in dialysis patients and valproate patients, that an increase in sales could not be expected due to the off-label use at the time, and that conducting trials in various fields would require considerable clinical development costs. Therefore, Otsuka answered that Otsuka did not plan to add any additional indications. Perhaps they thought that although the number of patients with inherited metabolic diseases was small and clinical trials would be exempted, clinical trials would surely be required for carnitine deficiency caused by dialysis or valproic acid. In addition, although the author had experienced some difficulties at that time, the "Good Clinical Practice" (GCP), which is a legally binding international standard, had just come into effect in April 1998, replacing the "Good Clinical Practice" (the Japanese version of the old GCP) that had been in effect since 1990. This was a major turning point in the state of clinical trials in Japan. Therefore, clinical development departments of pharmaceutical companies were struggling to establish the new GCP system and to submit applications for clinical trials conducted under the old GCP, and there was no time to work on drug-lag elimination. Since L-Cartin® tablets were already approved for propionic acidemia and methylmalonic acidemia, it should have been possible to establish efficacy and safety based on overseas approval status, case reports, review articles, textbooks, and guidelines without conducting clinical trials. However, as you can see from the Table of Contents of this book, carnitine deficiency disorders are spread over a very wide range of fields, and it is necessary to collect and read through overseas approval status and literature to compile evidence, even if the application is public knowledge without clinical trials. At that time, few clinicians were familiar with carnitine deficiency except for specialists of the Japanese Society for Inherited Metabolic Diseases and The Japanese Society of Child Neurology, and only a few clinicians were studying carnitine deficiency in the field of dialysis therapy after the approval of carnitine for dialysis in the United States. Given this situation, no one would have said, "Let us expand the indication!"

According to Saio et al., from 1999 to 2004, there were 30 drugs that were thought to have been approved in accordance with the "2 section chief's notification" [5]. In the field of pediatrics, I have heard that about ten products have been developed and approved after the "2 section chief's notification." Therefore, although it is considered that the "2 section chief's notification" has accomplished a certain achievement, but it could not fundamentally solve the issues of unapproved drugs, off-label drugs, and drug-lag.

19.3.2 Development of the "Priority List"

Many academic societies have also made efforts to eliminate the issues of unapproved drugs, off-label drugs, and drug-lags. It is natural for clinicians and others who are in daily contact with patients suffering from intractable diseases and difficult-to-treat diseases to wish to use therapeutic drugs that are as useful as possible. In the field of pediatrics, since 2004 in Japan, 21 subcommittees of pediatric societies, including the Japanese Society for Inherited Metabolic Diseases and the Japan Society of Child Neurology, have participated in the "Collection of Efficacy and Safety Information on Drugs and Medical Devices and Risk Assessment and Management Methods Based on the Information" program (Principal Investigator: Dr. Ichiro Matsuda) facilitated by the Grant-in-Aid for Pharmaceutical and Medical Device Regulatory Science Grants (https://mhlw-grants.niph.go.jp/project/11797) [6]. Dr. Toshihiro Ohura (Department of Pediatrics, Tohoku University, at the time) took the lead in developing the "Priority List" of drugs that are urgently needed or essential for the expansion of indications in the field of treatment of inherited metabolic diseases, including five priority list drugs from the Pediatric Pharmacotherapy Evidence Collection Project as Group I, six priority list drugs from the priority list of candidate drugs of the Study Council for the Use of Unapproved Drugs as Group II, and six priority list drugs from the Study Council for the Use of Unapproved Drugs as Group III. Group III consists of 12 drugs that are either converted from reagents or chemically synthesized at facilities. The number of patients with inherited metabolic diseases detected at an early stage is expected to increase with the spread of newborn mass screening, and Nelson's textbook of pediatrics describes carnitine supplementation therapy for inherited metabolic diseases, carnitine deficiency due to valproic acid administration, etc. L-carnitine preparations are already approved and used as standard drugs overseas. Therefore, L-carnitine was placed at the top of the priority list in Group I and should be given the highest priority. In addition, since L-Cartin® is used not only for the treatment of inherited metabolic diseases but also for the prevention and treatment of carnitine deficiency caused by valproic acid, it was jointly submitted by the Japanese Society for Inherited Metabolic Diseases and the Japanese Society of Child Neurology. Table 19.2 summarizes the subsequent approval status of the off-label and unapproved drugs featured in the "Priority List."

Table 19.2 Status of approvals of off-label drugs and unapproved drugs covered in "Formulation of Priority List of Off-Label Drugs and Unapproved Drugs in the Field of Inherited Metabolic Diseases" in Japan (as of the end of August 2022)

Drug name	Target diseases	Status
Group I. Items on the priority list for the Pediatric Pharmacotherapy Rationale Information Collection Project		
L-Cartin tablet (levocarnitine chloride)	Carnitine deficiency	Approved on March 10, 2011, by public knowledge application (Otsuka Pharmaceutical). The internal solution and intravenous injection for which development request was made were approved in 2013 (free body). In 2014, the tablets were also changed to free form, and all of Otsuka Pharmaceutical's L-Cartin preparations were unified into free form
Bioptene (Bioptene granules, Sapropterin hydrochloride)	Reduction of phenylalanine in BH4-reactive PAH deficiency	Approved in November 2013 (Daiichi Sankyo)
Biotin	Holocarboxylase synthase deficiency; biotinidase deficiency; mitochondrial metabolism disorders including mitochondrial encephalomyopathy	
Flagyl oral tablets (metronidazole)	Sterilization of intestinal anaerobes in methylmalonic acidemia and propionic acidemia	Approved in August 2012 (Shionogi)
Argi-U granules and intravenous infusion 20 g (arginine hydrochloride)	Treatment and prevention of acute-phase stroke attacks in Mitochondrial Encephalomyopathy, Lactic Acidosis, and Stroke-Like Episodes (MELAS)	Efficacy could not be confirmed in investigator-initiated clinical trials. The effect of prolonging life was confirmed by 9 years of follow-up. Aiming to apply for approval in the future (Professor Yasutoshi Koga, Department of Pediatrics, Kurume University)
Group II: Priority items of candidate drugs for the Review Committee on the use of unapproved drugs		
NTBC(2(- 2-nitro-4-trifluoromethyl benzoyl)-1,3-cyclohexanedione, nitisinone)	Tyrosinemia type 1	Approved in December 2014 (Astellas Pharmaceutical applied for development offering)
Cysteamine	Cystine storage disease (cystinosis)	Approved in July 2014 (Nephrogenic cystinosis, Mylan Pharmaceuticals)
Betaine	Homocystinuria	Approved in January 2014 (Reqmed). In January 2022, succeeded and transferred to Recordati Rare Diseases Japan
Idursulfase	Mucopolysaccharidosis type II	Approved in October 2007 (Eraprase, intravenous infusion 6 mg, Sanofi)
Alglucosidase alfa	Pompe disease	Approved in April 2007 (Myozyme, intravenous injection 50 mg, Sanofi)

(continued)

Table 19.2 (continued)

Drug name	Target diseases	Status
N-carbamoyl-L-glutamic acid	N-acetylglutamate synthase (NAGS) deficiency	Approved in September 2016 (Carbaglu dispersion tablets 200 mg, Pola Pharma), succeeded and transferred to Recordati Rare Deaths Japan (at that time) in October 2018
Group III: Chemically synthesized chemicals and reagents used in the hospital		
Sodium benzoate (oral, injection)	Hyperammonemia	Developing the sodium benzoate and sodium phenylacetate combination (Takeda Pharmaceutical)
Betaine	Homocystinuria	Approved in January 2014 (Reqmed). In January 2022, succeeded and transferred to Recordati Rare Diseases Japan
Cysteamine	Cystine storage disease	Approved in July 2014 (nephrogenic cystinosis, Mylan Pharmaceuticals)
Histidine copper	Menkes disease	
Biotin powder	Holocarboxylase synthase deficiency; biotinidase deficiency	
Sodium dichloroacetate	Hyperlactacidemia	
Neutral phosphorus preparations	Hypophosphatemic rickets	
5-Hydroxytryptophan	BH4 deficiency	
Citrulline	Urea cycle disorder; lysinuric protein intolerance	
Sodium phenylacetate (oral, injection)	Hyperammonemia	Developing the sodium benzoate and natrium phenylacetate combination (Takeda Pharmaceutical Company Limited)
Sodium succinate	Hyperlactacidemia	
Glycine	Isovaleric acidemia	

Created from Ref. [5] and the home page of "Evaluation Committee on Unapproved or Off-labeled Drugs with High Medical Needs" (https://www.mhlw.go.jp/stf/shingi2/0000198856_00024.html)

19.3.3 Study Council on the Use of Unapproved Drugs

MHLW also gathered experts in various fields and held a total of 22 "Study Councils on the Use of Unapproved Drugs" from 2005 to 2009 to survey the actual situation of what drugs are being used unapproved in what medical fields (so-called off-label use) and the demands of the medical field (https://www.mhlw.go.jp/stf/shingi/other-iyaku_128702.html). During this meeting, the medical necessity and overseas approval status of unapproved drugs and off-label use in various fields were also discussed. Dr. Makiko Osawa (Professor of Pediatrics at Tokyo Women's Medical University, at the time) attended the meeting as a member from the field of pediatrics and expressed her opinions on unapproved drugs and off-label drugs in the field of pediatrics, including L-Cartin®, and discussed them. This expert panel was dissolved and taken over by the "Evaluation Committee on Unapproved or Off-labeled Drugs with High Medical Needs" (hereinafter referred to as the "Evaluation Committee") in 2010.

19.3.4 New Framework for Eliminating Drug-Lag and Price Maintenance Premium

Even if the drug-lag issue in Japan was raised, only a small number of pharmaceuticals were developed using the "2 section chief's notification," etc., and it was difficult to reach a substantial solution. The main reason for this is that drug development, including clinical trials, requires a large amount of costs, and pharmaceutical companies are only willing to develop drugs for projects that can be expected to generate substantial profits. Therefore, MHLW also exchanged opinions with the Japan Pharmaceutical Manufacturers Association (JPMA) and decided to create a new framework in which pharmaceutical companies that have worked to solve drug-lag issues by adding benefits through the "price maintenance premium" will be given penalties for those who do not take action. The basic concept of the "price maintenance premium" is as follows. In Europe and the United States, drug prices of new drugs are maintained during the patent period, so pharmaceutical companies can secure sufficient profits, and a part of the profits can be allocated on drugs for rare diseases. However, in Japan, even new drugs with patent periods are devalued every 2 years, so profits decline and there is no room for drug development of rare diseases. Therefore, the new drug price is maintained during the patent period, and the drug price declines rapidly due to the launch of generic drugs after the patent is expired. The funds secured during the patent period can be used to accelerate the development of innovative new drugs, address unmet medical needs, and address unapproved and off-label drugs and drug-lag elimination. However, as a condition for granting the new drug creation surcharge, if there is a development request from the "Evaluation Committee," which is organized by experts in various fields as an expert committee member of the MHLW, the pharmaceutical company must submit the development schedule within a certain period of time and the development must be undertaken unless there is a reasonable reason. If not, the company will be penalized for not adding the "price maintenance premium" for all new drugs. In other words, the aim was to eliminate the drug-lag with "carrot and stick." For details on the "price maintenance premium," please refer to the book and papers [7–10]. In addition, the websites of JPMA, MHLW, Pharmaceuticals and Medical Devices Agency (PMDA), and Office of Pharmaceutical Industry Research have been updated with many related updates.

The framework for eliminating "price maintenance premium" is as follows (Fig. 19.1). Academic societies and patient groups submit requests for the development of unapproved drugs or drugs with off-label indications that are necessary for medical treatment to the "Evaluation Committee." The "Evaluation Committee" deliberates on whether it is reasonable to consider each drug to be medically necessary, and if the drug meets the criteria and is deemed necessary, the committee asks the pharmaceutical company for its corporate opinion on whether it is willing to develop the drug. The pharmaceutical company cannot refuse unless there is a particular rational reason and submits a development process chart and the necessary clinical trials or "public knowledge application" materials, if any. The "Evaluation Committee" evaluates the appropriateness of the clinical trial or public knowledge application, and the company initiates the clinical trial or "public knowledge application." The pharmaceutical company's seriousness in the development process is regularly checked. The company then submits a "public knowledge application" or the necessary studies, and a fast-track approval review is conducted. Once approved, the company's new drug is granted a "price maintenance premium." This framework was established with the approval of the Central Social Insurance Medical Council (Chuikyo). This framework for drug-lag elimination, which includes an additional "price maintenance premium," has forced pharmaceutical companies, which until now have been reluctant to eliminate drug-lags, to engage in the development of unapproved drugs or drugs for unapproved indications. In addition, a pharmaceutical company specializing in orphan drugs was newly established, undertaking new developments such as the approval of sodium phenylbutyrate (product name: Buphenyl tablet 500 mg, Buphenyl granules 94%; Orphan Pacific Co.). As of August 2022, 180 items have been approved, including 103 items in the first development request (May 2010), 72 items in the second development request (December 2010), and 5 items in the third request, with 1 item having been submitted for

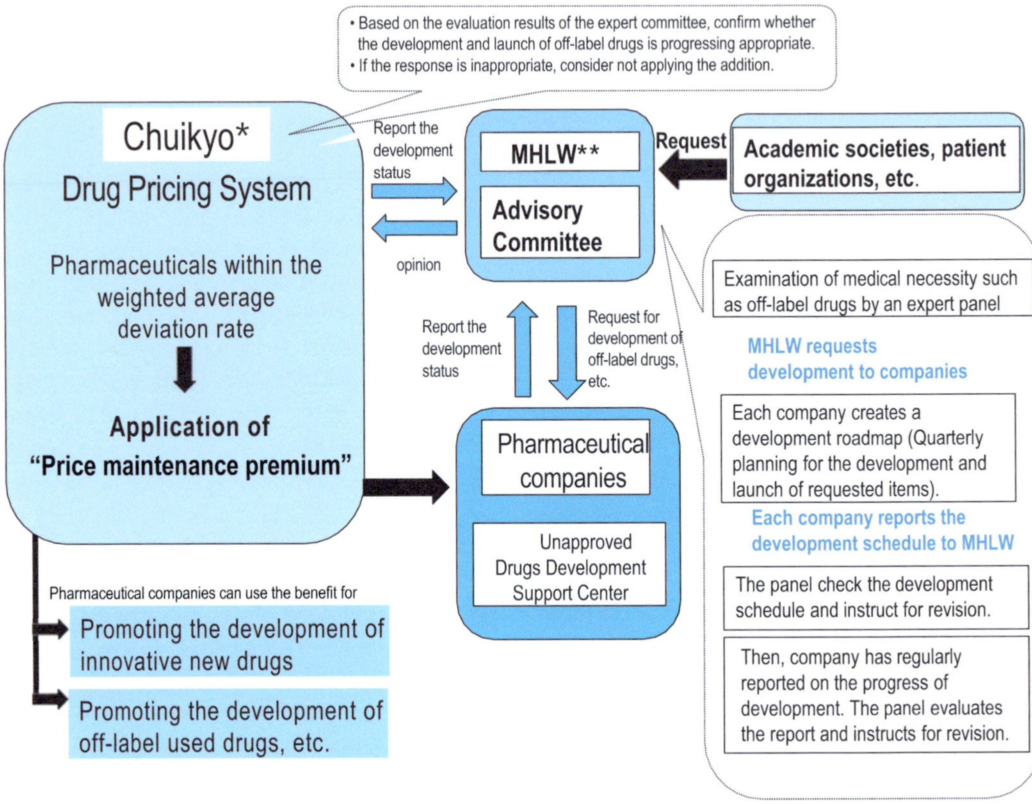

Fig. 19.1 Scheme to promote the development of off-label drugs and unapproved drugs through "Price maintenance premium" in Japan. *Chuikyo: Central Social Insurance Medial Council, **MHLW: Ministry of Health, Labour and Welfare. (From Medical Representatives' training materials, Distribution Optimization Committee, Japan Pharmaceutical Manufacturers Association)

approval and 2 items having had their clinical trial plans notified. This new framework continues to eliminate the drug-lag, and it can be said that this system has made great progress in eliminating the drug-lag in Japan.

19.3.5 Background of the Expansion of Indications for L-Cartin® Tablets Through "Public Knowledge Application"

Due to the expansion of the indication of L-Cartin® tablets for carnitine deficiency, it was listed as a product for the first development request in May 2010 by the "Evaluation Committee," and although the application for public knowledge was judged to be appropriate and materials were submitted and approved, the development had actually begun before that by myself, so the background is briefly summarized.

As mentioned above, MHLW approached Otsuka Pharmaceutical Co., Ltd. in December 2000 about expanding the indications for L-Cartin® tablets, but the company responded that there were no plans to develop them. However, in December 2006 and March 2007, the presidents of Japanese Society for Inherited Metabolic Diseases and Japan Pediatric Society submitted a request to Otsuka Pharmaceutical Co., Ltd. to expand the indication. Under such circumstances, although I will omit the details, it was decided that the Diagnostics Division, not the original Pharmaceuticals Division, would work on expanding the indication of L-Cartin® tablets (assuming that the "public knowledge application" would be approved). In addition, work had been started to collect literature on overseas approval status and domestic and international carnitine deficiency and carnitine supplementation therapy. At that time, I was transferred from the research institute, and as the head of the development department of the Diagnostics Division, I was involved in many projects as well as the expansion of the indications for L-Cartin® tablet. In August 2007, we compiled a draft of the "public knowledge application" form by summarizing the information we had collected so far, and consulted with the MHLW to see if it was possible to apply by the "public knowledge application." The authority said that it was possible to apply by "public knowledge application." The authorities instructed us to describe and construct theories in an easy-to-understand manner about what carnitine deficiency is, how to diagnose it, the basis for its use and dosage, and its safety, and to compile all case reports of carnitine deficiency in individual diseases to summarize its efficacy and safety. In addition, since there was a request from the academic society for the development of internal liquid and intravenous injection, it was also necessary to confirm whether the company was willing to develop these dosage forms within the company. We took it back to the company and discussed it with related departments such as the Pharmaceuticals Division, and all but the Diagnostics Division were negative about the development. Therefore, we continued to collect information such as case reports and review articles within the Diagnostics Division, but it was necessary to read, understand, and organize a large number of papers written in English, and when the members were busy with other projects, almost only the author was left to work on the L-Cartin® indication expansion project.

In order to hear the opinions of people in the real medical field, I participated in the annual meeting of Japanese Society for Inherited Metabolic Diseases in November 2007 and interviewed Dr. Fumio Endo (Kumamoto University, at that time), the president of the society, to ask for his opinions. Dr. Endo encouraged me, saying, "I would like you to work on this and I will introduce you to doctors who are familiar with carnitine deficiency, and the society will fully support it!". He introduced me to Dr. Seiji Yamaguchi (Shimane University), Dr. Yosuke Shigematsu (Fukui University), Dr. Toshihiro Oura (Tohoku University), and Dr. Masaki Takayanagi (Chiba Children's Hospital), who were experts in carnitine deficiency, and created a system where I could consult with them at any time if I had any questions. After that, I steadily collected and organized literature and information, and as the professors taught me on my questions, I came to understand what carnitine is, what kind of disease carnitine deficiency is, and why carnitine deficiency is a problem in various fields. I was also fortunate that I had been researching mitochondria when I was in graduate school, and that I had worked on energy metabolism research during research on heart failure drugs after joining the Otsuka. In addition, when I was enrolled in the New Drug Development Division, I worked on the development of an internal system for responding to the new GCP, and at that time, Dr. Eiji Uchida (then Department of Second Pharmacology, Showa University School of Medicine) and Dr. Kiichiro Tsutani (Institute of Intractable Diseases, Tokyo Medical and Dental University, then The University of Tokyo) guided me about the drug-lag issues in Japan, the orphan drug issues, and compassionate use program of drugs in the United States and the European Union. Unfortunately, some of the pharmaceutical compounds that I have worked on since joining the Otsuka have not been able to contribute to medical care after being released to the world as pharmaceuticals for various reasons. Therefore, I had a strong desire to do work that would contribute to the medical field as much as possible in the short time before retirement, so I spent a lot of time and made effort on expanding the indications of L-Cartin®, positioning it as my own project, even though it was a drug that was "treated as an orphan" within the company.

On July 31, 2009, Dr. Fumio Endo contacted me and said, "MHLW has allocated a supplementary budget of 75 billion yen to address unapproved drugs and off-label drug use and has decided to issue a public call to academic societies, patient groups, and pharmaceutical companies for unapproved drugs and off-label drugs to be resolved. Do you intend to apply as a company?" (personal communication). This was his question. I replied that I did not think it was necessary to apply because we were already preparing to submit an application for public knowledge under the guidance of the authorities, and that if that was the case, there would be no need for the society to apply as well. However, Dr. Fumio Endo commented, "The current call for candidate drugs seems to be a trend toward a national effort to resolve the long-standing drug-lag issues under a new system, and we should not miss this opportunity. The society would like to submit a request mainly for items on the priority list, including L-Cartin®" (personal communication). The Japanese Society for Inherited Metabolic Diseases thus requested the development of nine items to the "Evaluation Committee." Among them, the expansion of the indication for L-Cartin® tablets and the development of internal liquid and intravenous injection were also requested as the first priority list.

In the case of L-Cartin®, not only the indication expansion for tablets, but also the development of an internal liquid that is easy for children to swallow and an intravenous injection for use in acute phases are requested. In other words, the expansion of the indication of tablets alone is not sufficient to obtain "price maintenance premium," and it is necessary to develop internal liquid and intravenous injection. In facilities for the severely mentally and physically handicapped, tablets were crushed and dissolved in the facility, and an oral solution was prepared and administered through a tube, but carnitine is a film-coated tablet because it has high hygroscopicity as a physical property, and it takes a long time to dissolve. In addition, intravenous injections are not commercially available, and when used in patients in the acute phase, it is necessary for hospitals to import them and use them, or to dissolve the reagent-grade L-carnitine for research use as in-house preparations and sterilize and prepare them. The author had heard from Dr. Endo that injecting arginine hydrochloride, which is used for hyperammonemia in urea cycle abnormalities, may cause hyperchloremic acidosis, and that patients with inherited metabolic diseases often cause metabolic acidosis, so it is preferable to use a free body instead of hydrochloride salt as an intravenous injection if possible. However, there was no prospect of procuring free bodies immediately, and the problem of developing internal solutions and intravenous injection was shelved, thinking about concentrating on expanding the indications of tablets, at first.

Sigma Tau of Italy was the first company in the world to work on carnitine deficiency and is currently supplying tablets, internal liquid, and intravenous injection in countries around the world (now Alpha sigma). I wanted to obtain safety information on carnitine preparations from Sigma Tau, a pioneer in carnitine deficiency, for the purpose of applying for public knowledge, and tried to contact them by sending several emails, but did not receive a reply. However, a kind of serendipity, or rather an unexpected story came to me. At the beginning of September 2009, the vice president of Otsuka Pharmaceutical Co., Ltd. asked me to accompany him to Sigma Tau. Dr. Cavazza, the president of Sigma Tau, wanted to co-develop with Otsuka a carnitine-containing new nutritional product, but the vice president was not familiar with carnitine so he wanted someone who was familiar with carnitine to go to Italy with him and found me. Then, five people including the vice presidents and me flew to Rome for a meeting with Dr. Cavazza and other Sigma Tau executives. At the meeting, although it was not included in the original meeting agenda, I made presentation about my L-Cartin® project expanding the indication and developing internal liquid and intravenous injection to the Japanese medical field as soon as possible in response to requests from academic societies. I asked about the possibility to introduce internal liquid and intravenous injection products from Sigma Tau. Dr. Cavazza immediately replied, "After all, it does not matter if you have a disease with only 300 patients or 3 million patients, both the patients are waiting for medicine. Therefore, if Sigma Tau has the medicine, it will make an effort to deliver it to the disease where there are only 300 patients." President Cavazza agreed that if it needs an internal solution or intravenous injection, Sigma Tau will cooperate with Otsuka no matter how small the supply is. Because of this statement by the president of Sigma Tau, who founded Sigma Tau Co., Ltd. in 1957, we were able to smoothly conclude license negotiations in a short period of time after the decision was made to introduce the products of internal liquid and intravenous injection from Sigma Tau. I was deeply impressed by the brilliant judgment of President Cavazza, who has been working on carnitine deficiency since the 1970s and has a deep knowledge of rare diseases. In addition, President Cavazza taught me that "there are two pathologies of carnitine deficiency—carnitine deficiency and carnitine insufficiency—and it is important to understand them thoroughly before treatment." In the subsequent literature and information collection, I always paid attention to whether it was carnitine deficiency or insufficiency.

On October 9, 2009, MHLW instructed Otsuka Pharmaceutical Co., Ltd. to submit a corporate opinion on the product for which the academic society had requested development, and the New Drug Development Division submitted it using the materials collected and compiled by me. However, surprisingly, it seems that it was submitted with the statement that "there is no intention to develop an intravenous injection." In January 2010, MHLW instructed Otsuka to submit additional company opinions, stating that it was necessary to submit an application within 6 months in the case of a public knowledge application, otherwise the "price maintenance premium" would not be approved, and that there was a high possibility that the development of internal liquid and intravenous injection would also be requested. On February 8, 2010, the first meeting of the "Evaluation Committee" was held. A total of 374 items were put on the table. Around this time, the president of Otsuka ordered to "actively respond to requests as a pharmaceutical company," and Otsuka decided to work on the development of internal liquid and intravenous injection. The "price maintenance premium" is granted not only for the product requested for development, but also for all drugs whose patent has not expired and whose deviation rate in the market price does not exceed the weighted average deviation rate of all listed products. Otsuka Pharmaceutical Co., Ltd. has 11 products, including intravenous injections and oral solutions for L-Cartin®, and this "price maintenance premium" may be the reason behind the change in the company decision from "not develop" to "develop" internal liquid and intravenous injection. It is believed that other pharmaceutical companies also emphasized to the "price maintenance premium."

On April 27 in 2010, the 3rd "Evaluation Committee" was held, and L-Cartin® was officially selected as a requested product, and on May 21, the requested product was published on the website of the MHLW. Although it was necessary to prepare a corporate opinion and development schedule within 1 month, the collection and maintenance of literature by myself so far was very useful, and the corporate opinion was compiled according to the format specified by the authorities for carnitine deficiency due to various causes such as many inherited metabolic diseases, tube feeding, dialysis, and drug-induced one. In fact, it seems that this "format specified by the authorities" was the "format for public knowledge application" itself. This format requested to summarize cases of off-label use in Japan, and there was little need to include the large number of overseas case reports that were collected so far. In other words, emphasis was placed on cases where they are widely used in Japan as off-label use.

Since intravenous injection and internal liquid need to be developed and supplied to the market as soon as possible, it was decided internally to introduce them from Sigma Tau, and in June and July 2010, I visited Sigma Tau with other personnel in charge of regulatory affairs, production, formulation, metabolic analysis, toxicology research, licensing, interpretation, etc.; to introduce them we performed due-diligence and inspected the manufacturing plants of Sigma Tau.

On October 6, the 5th meeting of the "Evaluation Committee" was held, and it was concluded that "the applicability of the application for the expansion of the indication of L-Cartin® tablets is reasonable," and on October 29, the Pharmaceutical Affairs and Food Sanitation Council's Pharmaceutical Subcommittee 1 evaluated the expansion of the indication for L-Cartin® in advance, approved the applicability of the application for public knowledge, and issued a notice of insurance coverage (insurance coverage was issued from this point before official approval). Subsequently, on November 2, 2010, the company submitted a public knowledge application for the expansion of the indication for carnitine deficiency and a change in dosage (in almost the same format as the previous corporate opinion), and the package insert was discussed with the PMDA, and it was officially approved on March 10, 2011. Against this background, L-Cartin® tablets were expanded from the indication for carnitine deficiency in propionic acidemia and methylmalonic acidemia to carnitine deficiency due to many inherited metabolic diseases, tube feeding, valproic acid, dialysis, etc., but since there was no one in the Pharmaceutical Sales Division who was familiar with carnitine, the author was suddenly transferred to the Pharmaceutical Sales Division, and I was busy for product education and creation of sales promotion materials, thanking clinicians who taught me about carnitine deficiency. In the company, only a small number of people knew about the expansion of the indication of L-Cartin® tablets, and suddenly it came to the surface, and it was decided to conduct sales activities for carnitine deficiencies such as inherited metabolic diseases other than propionic acidemia and methylmalonic acidemia, patients taking valproate, and dialysis patients. There were a lot of issues to be addressed, such as education about carnitine deficiency to medical science liaisons and sales staff, and how to approach with related academic societies. On March 11, the day after the approval of the expansion of the indication of L-Cartin®, the Great East Japan Earthquake struck, and March 2011 was literally a violent tough-year.

19.3.6 Development of Internal Liquid and Intravenous Injection of L-Cartin®

The oral solution and intravenous injection to be in-licensed from Sigma Tau were free L-carnitine preparations. The L-Cartin® tablet that has been approved in Japan and has been expanded is L-carnitine chloride. Therefore, under the Pharmaceutical Affairs Law, free forms are considered to be a new active ingredient drug and are unapproved drug, so clinical trials are required in principle. In the consultation with the PMDA, when I said, "I would like to ask about the addition of dosage forms for internal liquid and intravenous injection," the PMDA officer said, "Please do not use the term addition of the dosage form, and this project is not about the addition of a dosage form of an already approved drug, but about the development of a new active ingredient drug." In that case, it would be a problem if a number of full-scale clinical trials such as normal phase I, phase II, and phase III were requested, but since this project was developed by the request from the "Evaluation Committee" based on the request of the academic society, it seems that PMDA also intended to request the minimum necessary clinical trials. As a result, it was instructed to conduct an equivalence study between tablets and oral solutions in healthy subjects, and a clinical pharmacological study to examine the pharmacokinetics of intravenous injections. Without going into details, the free internal solution and intravenous solution were launched on February 26, 2013. However, the internal solution is a "small brown bottle" containing 10 mL, and there is no longer a need to dissolve the tablet as before, but it takes time and effort to open many small bottles at facilities for the severely mentally and physically disabled, and there has been a demand for the development of large-capacity bottles and preparations that are easy to carry. The product manager in charge of L-Cartin® was also in charge of asthma drugs and was a kind of professional in the development of new dosage forms who had released products in various dosage forms and devices to the market. In December 2017, Otsuka launched large-capacity (100 mL) bottles and aluminum stick packets (5 mL and 10 mL) of L-Cartin® for internal use. In addition, the intravenous injection was originally a glass ampoule formulation, but it was changed to a pre-filled syringe in December 2017 and released in order to save from the trouble of transferring it to a syringe and to prevent injury when opening the ampoule.

When I interviewed Prof. Susumu Ito (then at Kagawa University) in 2012 to discuss carnitine deficiency caused by pivoxil-containing antibiotics, I also reported to him that we were developing an intravenous formulation and an internal liquid formulation. He said, "I am a member of the Pharmaceutical Affairs Committee of the Japan Pediatric Society, so I have been watching carefully to see if your company is really developing the internal liquid and intravenous injections requested by the society. If they were not, the academic societies would have reported to the "Evaluation Committee," and Otsuka Pharmaceutical Co., Ltd. would not have been granted the "price maintenance premium" because they had not completed the request."

For a while after February 2013, there were a twisted state: L-carnitine chloride for tablets and free form L-carnitine for internal liquid and intravenous injection, but in December 2014, free form tablets were launched and all formulations of L-carnitine products from Otsuka Pharmaceutical Co. were unified into free form of L-carnitine.

19.4 Development of Medicines for Pediatric Use

The 34th Annual Meeting of the Japanese Society of Clinical Pharmacology was held on December 4–6, 2013, and I attended the meeting after a long absence because Dr. Eiji Uchida, who served as a medical expert in conducting the aforementioned clinical pharmacological studies of free oral solution and intravenous drug, is the president of the society and I was interested in greeting him, as well as in the symposium entitled "Pediatric Drugs and the Future of Clinical Pharmacology and Clinical Pharmacy." At the symposium, there were presentations on interesting topics such as the fact that the dosage of drugs used for children is mostly empirical and that strict dosage setting studies are rarely conducted, the fact that the PMDA is working on these issues in the Pediatric Drug Working Group and strengthening cooperation with Europe and the United States, and the current status of the pediatric pharmacist system certified for pediatric pharmacotherapy. There were also presentations on interesting topics such as the current status of the pediatric pharmacist system. In a presentation on "Easy-to-Take Medicines Desired for Children (in terms of dosage form, taste, and ease of use)," the chairperson introduced the fact that "There are many complaints that medicines commonly used for children have a bad taste or are difficult to swallow." The chairperson pointed out, "I think that people from pharmaceutical manufacturers are participating in this meeting. I would like you to address these points as well." I involuntarily raised my hand and spoke up. "Among the pharmaceutical companies, we are also making efforts to develop dosage forms that are easier to take, taste better, and are easier to use, while fighting internally. However, the dosage forms we have developed are often not adopted due to restrictions imposed by hospital pharmacy departments, such as 'one increase, one decrease', and 'one drug, one dosage form only'. This is especially strict in the pharmacy departments of large hospitals such as university hospitals. It is understandable that it is difficult to manage a large number of drug items, but I suspect that the dosage forms that were finally developed at the request of academic societies are being squashed in a way that ignores the existence of patients." One of the two chairpersons was a Japanese Professor working in Canada, and he asked, "What means 'one increase, one decrease'? (Another chairman explained about it to him). There is no such strange regulation in Canada! It is not acceptable to deny the use of a dosage form that is easy for children to swallow and cut off other dosage forms." I strongly hope that these problems should not be uniformly dismissed as "one increase, one decrease" or "one drug, one dosage form," but the officials of hospital pharmacy have to always make a judgment based on an understanding of why such a dosage form was developed and think about patients, not only about drug management efficacy. Regarding these issues, the Pharmaceutical Affairs Committee of the Japan Pediatric Society (Chairperson: Prof. Susumu Ito, Kagawa University) has also made a recommendation to reframe the "one increase, one decrease" rule for drugs developed at the request of the academic society (http://www.jpeds.or.jp/modules/guidelines/index.php?content_id=61).

19.5 Insurance Coverage for Diagnostic Reagents for Carnitine Deficiency

From February to March 2011, due to a partial change in the approval of L-Cartin® tablets, the indication was changed to "carnitine deficiency," the dosage was also changed, and the contents of the package insert were discussed with the PMDA. At that time, officers in the authority were concerned about the reagent for measuring carnitine deficiency in blood that it is not covered by the insurance. He said, "I would like you to discuss with Kainos Co., Ltd., which sells blood carnitine measurement reagents, and work toward insurance coverage."

Blood carnitine 2 fractions test by enzyme cycling has been used in clinical practice as a research reagent since 1993 and is considered to be the most suitable for diagnosing general carnitine deficiency (see Chaps. 8 and 18). Therefore, we consulted with the relevant parties of Kainos Co., Ltd., and formulated a protocol for clinical performance testing with guidance on the development of diagnostic reagents through PMDA consultation, and the Diagnostic Division took the lead in conducting clinical performance tests on dialysis patients, patients with severe mental and physical disabilities taking valproate, and healthy subjects [11]. Kainos Co., Ltd. was in charge of the clinical performance test results, manufacturing method, stability, and quality, and the two companies collaborated to apply for manufacturing and marketing approval from Kainos Co., Ltd. on June 27, 2013, and was promptly approved as an in vitro diagnostic reagent on February 26, 2014. However, unlike pharmaceutical drugs, even if a diagnostic reagent is approved as an in vitro diagnostic reagent, it is not covered by insurance at the same time. We will not be able to move forward unless we separately discuss with the relevant departments within MHLW for insurance coverage. Ask the officer in charge, "What is carnitine?" and "What is carnitine deficiency?" I explained many times, "Why do we need to measure two types of free carnitine and acylcarnitine?," and when we thought we had made progress, the officer in charge was transferred in the next fiscal year, and the explanation had to be redone from the beginning, and it was difficult to move forward. In addition, we have worked on requests for insurance coverage from several related academic societies, but we have not been able to talk about insurance coverage for a long time. The development of diagnostic reagents began with the instruction of the authorities to "work on the problem that diagnostic reagents are not covered by insurance," and it was a big mistake to think that insurance coverage would proceed smoothly after approval as a diagnostic reagent. On the contrary, the indication of L-Cartin® has been expanded to a wide range of carnitine deficiency conditions, such as dialysis patients and patients taking valproate, and the development of an in vitro diagnostic reagent has been accelerated and quickly approved under the guidance of the authorities. Two years later, the officer in charge said, "Insurance coverage for inherited metabolic diseases is permitted, but it is difficult to cover insurance for dialysis patients, and insurance coverage for patients taking valproate needs to be justified," and after more than 2 years, I was told that "in order to be covered by insurance for diagnostic reagents in the future, it is necessary to have a basis for general clinicians to use them, such as guidelines led by academic societies and clinical practice guidelines." Should not such an important thing be communicated at the beginning? Against this background, at the request of the Japan Pediatric Society, the Nutrition Committee of the four-party association affiliated with the Japan Pediatric Society took the lead in publishing the *Guideline for the Diagnosis and Treatment of Carnitine Deficiency 2016* by Dr. Shinobu Ida, Deputy Director of the Osaka Maternal and Child Medical Center (at that time), and revised it in 2018.

In the first place, the Ministry of Health, Labour and Welfare (MHLW) and the PMDA considered that it was necessary to have insurance coverage for diagnostic reagents in order to prevent the unnecessary administration of relatively expensive carnitine preparations without diagnosis. Periodic carnitine fractionation testing is necessary, but there was an awareness of the problem that diagnostic reagents were not covered by insurance. Since diagnostic reagents are also part of medical expenses, it is understandable that allowing insurance coverage unnecessarily will lead to a rise in medical costs, but the insurance coverage of in vitro diagnostic reagents for blood carnitine 2 fractions test, which had been developed under the guidance of the MHLW's Examination and Management Division and PMDA, has now been partially restricted at the discretion of the department in charge of insurance within the same MHLW. Although I have done my best to cover the insurance this time, there are some precautions described in Table 18.2 of Chap. 18, and if interpreted strictly, it is assumed that even if you want to diagnose whether you are carnitine deficient, it will not be covered by insurance (adult patients with severe mental and physical disabilities, heart failure patients, etc.). Essentially, diagnostic reagents should be available when a doctor deems it necessary to diagnose. In the future, if it is considered necessary to test for blood carnitine 2 fractions for various diseases but it is not covered by insurance, I would like to request that the doctors of the relevant academic societies make efforts such as accumulating clinical data and submitting a request for insurance coverage.

19.6 Summary

In expanding the indications and changing the dosage of L-Cartin® tablets, I have faced and dealt with various clinical pharmacological issues, such as the drug-lag problem, the orphan drug problem, "price maintenance premium," the difference between the chloride and free form of the active ingredient, development of pediatric drugs, development of patient-friendly dosage forms and regulation by the pharmacy department, and therapeutic drugs and diagnostic drugs. These issues will not be solved unless many parties including relevant academic societies, pharmaceutical companies, MHLW, and PMDA cooperate beyond their positions.

However, as the demand for L-carnitine increased among dialysis patients and patients using valproic acid after the indication was expanded, several companies quickly introduced generic versions of L-carnitine. The pharmaceutical industry is in a sense an information industry, and in the case of carnitine, it has been only a short time since the indication of carnitine deficiency became publicly known. Therefore, companies that understand what carnitine deficiency is, how it should be used, and what kind of clinical reports are available should manufacture and market carnitine. It seems to me that companies with little information are launching generic drugs only for the sake of profit. While it is understandable that a system that encourages the use of generics with low drug prices in order to reduce medical costs is in place, the pursuit of profit alone may have the same root as the various problems that have arisen recently in the generic drug industries, such as fraudulent quality control. In addition, despite the fact that intravenous DL-chlorinated carnitine products containing the D-body have been banned overseas, these are still sold in Japan, although sales are low. It is hard to believe in this era of the matured International Council for Harmonization of Technical Requirements for Pharmaceuticals for Human Use (ICH), an important regulatory science issue in Japan.

It is often pointed out by doctors who emphasize EBM that they do not accept the efficacy of a drug without a large double-blinded comparative study. Are large-scale double-blind comparative studies the only true EBM? There are rare diseases for which large-scale studies are not possible. In some cases, a single drug may not have a clear effect on a disease state in which multiple nutrients are deficient. In the United States, Dr. Susan Winter et al. submitted a series of case reports on the efficacy and safety of carnitine supplementation therapy in various inherited metabolic diseases, which were accepted by the FDA. The FDA approved the application because it was considered that there was evidence established using the best available methods for these diseases.

The author hopes that this book will deepen your understanding of carnitine, help you avoid carnitine deficiency in your clinical practice, and further promote the clinical application of carnitine's unique actions in a variety of clinical fields without prejudice.

References

1. Sugiyama N, Matsuda I, Wada Y, et al. Effects of Levocarnitine chloride (L-carnitine) on propionic acidemia, and vitamin B_{12} unresponsive and responsive methylmalonic acidemia in Japanese nineteen institutes. (I) Clinical evaluation. J Jpn Pediatr Soc. 1989;93(8):1808–17. Article in Japanese.
2. Sugiyama N, Matsuda I, Wada Y, et al. Effects of levo-carnitine chloride (l-carnitine) on propionic acidemia, and vitamin B_{12} unresponsive and responsive methylmalonic acidemia nineteen Japanese institutes: (II) Biochemical evaluation. J Jpn Pediatr Soc. 1989;93(8):1818–27. Article in Japanese.
3. Ministry of Health, Labour and Welfare. Handling of ethical drugs for off-label use. Kenkyu No.4, Pharmaceutical Affairs Council No. 104. 1999. https://www.mhlw.go.jp/shingi/2005/01/dl/s0124-9h1.pdf. Article in Japanese.
4. Doi O. Effective operation of the 2 section chief's notification. Pharm Med Devices Regul Sci. 2011;42(5):442–3. Article in Japanese.
5. Saio T, Kurihara K. Supplemental approval of drug indication without carrying further clinical trial and the expanding role of the so-called "2 section chief's notification": the case of sodium valproate. Clin Eval. 2004;31(3):587–601. Article in Japanese.
6. Matsuda I. Grants-in-Aid for Scientific Research on Health, Labour and Welfare Comprehensive Research Project for Regulatory Science of Pharmaceuticals and Medical Devices. A study on the collection of efficacy and safety information on drugs for special patient groups such as children and risk assessment and management methods based on such information (2004-Pharmaceutical-001). Ohura T. Formulation of a priority list of off-label drugs and unapproved drugs in Japan in the field of inherited metabolic diseases. 2005 Research Report. 2006. p. 645–74. https://mhlw-grants.niph.go.jp/project/11797. Article in Japanese.
7. Idaka Y. Price maintenance premium and pharmaceutical industry. Elucidation of the system and influence—What is Soushutu KASAN? Tokyo: Iyaku Keizai Co., Ltd; 2010. ISBN: 978-4-902968-34-7. Book in Japanese.
8. Yoshino T. Development of unapproved drug or indication–current status in Japan. Folia Pharmacol Jpn. 2012;139(3):113–6. Article in Japanese.
9. Ito T, Urushihara H, Matsushima Y, et al. Mode of regulatory applications of drugs used off-label reviewed by the evaluation committee on unapproved or off-labeled drugs with high medical needs. Jpn J Clin Pharmacol Ther. 2015;46(5):233–41. Article in Japanese.
10. Yoshida M. Drug-lag: the status of unapproved drugs in Japan and their characteristics. JPMA News Letter. 2021;205. https://www.jpma.or.jp/news_room/newsletter/205/05pc-01.html. Article in Japanese.
11. Itami N, Aoyama S, Yasuo Kudoh Y, et al. Clinical performance of KDK-1201, the carnitine measurement reagent based on the principle of enzyme cycling method. J Clin Lab Med. 2014;58(5):651–8. Article in Japanese.

Afterword

When I started to write a specialized book on carnitine, carnitine deficiency, and insufficiency, I had picked up and sorted out the questions and misconceptions that medical professionals had about carnitine. Also, I had collected a wide range of literatures on carnitine from basic to clinical and organized them by field. Especially, I have carefully examined the literatures for points that I did not fully understand or could not fully explain. I told several close clinicians that I wanted to write a book on carnitine, and they said, "By all means, please write the book. I look forward to reading it!". They encouraged me to do so.

However, on a personal note, in the spring of 2013, when the intravenous formulation was launched and I was very busy traveling around Japan giving lectures and presentations, I was diagnosed with gastroesophageal junction cancer (with liver metastasis, stage IV, inoperable, life expectancy was 6 months) at a certain university hospital. The prognosis for this cancer was extremely poor. Fortunately, I miraculously achieved a complete remission after only one course of 3 weeks inpatient treatment (standard treatment: cisplatin infusion + TS-1 oral administration, and clinical research: radiation therapy). Then my doctor strongly recommended to take low-dose anticancer drugs, but I had refused it and instead had undergone regular examinations every month at first and now every 6 months, and it past almost 13 years without recurrence. I finally found a way to get insurance coverage for a diagnostic drug for carnitine deficiency, and I had enough time after retiring to write this book in earnest, but it took more than 5 years for Japanese version and 3 years for this English version.

In most medicines, such as antihypertensive drugs, there are various drugs which have various mechanisms of action and trends of the use come and go by a certain period of times. However, L-carnitine is the only one drug that is effective for carnitine deficiency/insufficiency, which causes serious conditions, and there are no alternative drugs. Therefore, if carnitine deficiency is present, carnitine preparations are indispensable drugs in clinical practice. Despite this, there is no full-fledged specialized textbook on carnitine and carnitine deficiency/insufficiency. I thought that no one other than myself would write a book that covered a wide range of fields beyond the medical field, from basic to clinical, and I decided to do my best because I had learned a lot about carnitine deficiency/insufficiency, and I am luckily alive from the end stage of cancer. As described in "Preface," there were two nice old books specialized on carnitine deficiency. However, it is difficult to obtain these books now and they are too old. My book is the first book on carnitine that covers a wide range of medical fields, including past and latest research in earnest. I hope that this book will deepen your understanding about carnitine and carnitine deficiency/insufficiency and contribute to preventing your own patients from becoming carnitine deficiency in clinical practice. Also, I hope this book will contribute to accelerate clinical application of the unique

N. Nakayama, *Carnitine Deficiency and Insufficiency in Medicine*,
https://doi.org/10.1007/978-981-95-7797-2

actions of carnitine in various clinical fields in future.

In compiling this book, I received guidance and instruction from a great many pysicians in various fields, including inherited metabolic diseases, severely mentally and physically disabled children, pediatric neurology, dialysis, liver disease, heart failure, clinical nutrition, and psychiatry. I would like to express my heartfelt gratitude to them. Especially I would like to thank Dr. Naruji Sugiyama (then a pediatrician at Nagoya City University and currently a professor at Aichi Medical College for Physical and Occupational Therapy), who conducted the first clinical study for carnitine deficiency in methylmalonic acidemia and propionic acidemia and worked hard to obtain the approval. Without this first indication in 1990, it would not have been possible to expand the indication of L-Cartin® tablets to carnitine deficiency by filing a public knowledge application. I would also like to express my gratitude to my colleagues and the relevant departments of Otsuka Pharmaceutical Co., Ltd. who worked together with the author to expand the indication of carnitine deficiency, develop new internal liquids and intravenous injections, establish a production system, develop diagnostic reagents, and promote the use in carnitine deficiency.

Finally, I would like to thank my wife Eriko and my two daughters for their devoted support during the difficult times when the author faced.

Spring 2026, Natsuki Nakayama, Ph.D.

Index

N. Nakayama, *Carnitine Deficiency and Insufficiency in Medicine*,
https://doi.org/10.1007/978-981-95-7797-2

N

O

MIX
Papier aus verantwortungsvollen Quellen
Paper from responsible sources
FSC® C105338

If you have any concerns about our products,
you can contact us on
ProductSafety@springernature.com

In case Publisher is established outside the EU,
the EU authorized representative is:
Springer Nature Customer Service Center GmbH
Europaplatz 3, 69115 Heidelberg, Germany

Printed by Libri Plureos GmbH
in Hamburg, Germany